HANDBOOK OF
Turfgrass
Management
and Physiology

HANDBOOK OF
Turfgrass Management
and Physiology

Edited by

Mohammad Pessarakli

University of Arizona
Tucson, Arizona, U.S.A.

CRC Press
Taylor & Francis Group
Boca Raton London New York

CRC Press is an imprint of the
Taylor & Francis Group, an **informa** business

CRC Press
Taylor & Francis Group
6000 Broken Sound Parkway NW, Suite 300
Boca Raton, FL 33487-2742

First issued in paperback 2019

ISBN-13: 978-0-8493-7069-4 (hbk)
ISBN-13: 978-0-367-38850-8 (pbk)

Library of Congress Cataloging-in-Publication Data

Handbook of turfgrass management and physiology / editor, Mohammad Pessarakli.
    p. cm. -- (Books in soils, plants, and the environment ; v. 122)
   ISBN 978-0-8493-7069-4 (hardback : alk. paper)
   1. Turf management--Handbooks, manuals, etc. 2. Turfgrasses--Physiology--Handbooks, manuals, etc. I. Pessarakli, Mohammad, 1948- II. Series.

SB433.H29 2007
635.9'642--dc22   2007029428

Visit the Taylor & Francis Web site at
http://www.taylorandfrancis.com

and the CRC Press Web site at
http://www.crcpress.com

Dedication

In memory of my beloved parents, Fatima and Vahab, who regrettably did not live to see this work, which in no small part resulted from their gift of many years of love to me.

Contents

SECTION I Turfgrass Management/Maintenance-Related Issues

SECTION II Turfgrass Growth, Management, and Cultural Practices

SECTION III Sports-Turf Management

SECTION IV Turfgrass Growth Regulators: Promoters and Inhibitors

SECTION V Turfgrass Breeding, Genetics, and Biotechnology

SECTION VI Turfgrass Growth Disorders, Pathology, Diseases, and Turf Disease Management and Control

SECTION VII   Turfgrass Integrated Pest Management

SECTION VIII   Turfgrass Management and Physiological Responses under Environmental Stress Conditions

SECTION IX Turfgrass Management and Physiological Issues in the Future

Preface

The importance of the dynamic, ever-expanding, multibillion-dollar turfgrass industry—which has tripled in the past ten years (this growth is more than any other agricultural industry)—to the economy and the public, especially golf and sports turf players, as well as demands for better-quality grasses, has resulted in the continuous desire for better knowledge of the management and physiology of turfgrasses. This includes finding new species and modern cultivars, particularly environmental stress–tolerant genotypes. This has resulted in the compilation of a large volume of information since the revolutionary period of the modernization of the turfgrass industry after World War II.

Since the early age of turfgrass evolution, turfgrass experts/scientists have observed that new cultural and management practices and careful attention to the physiological factors affecting turfgrass growth and quality must be constantly taken into consideration for the progress and growth of the turfgrass industry to continue for future generations. Reduction in grass growth and undesirable turfgrass quality were reported as a result of modification in the physiological process and environmental conditions that control turfgrass growth and development. Therefore, all the factors influencing turfgrass physiology and management must be constantly observed. Thus, a greater awareness of the proper management and cultural practices as well as a sound understanding of the physiological processes of turfgrass growth and development are essential to scientists, experts, and all those who are involved in the field of turfgrass management and physiology.

The above facts have necessitated the development of this unique, comprehensive source of information, and all the new findings in this field are included in this single volume entitled *Handbook of Turfgrass Management and Physiology*.

This handbook is a comprehensive wealth of information and an up-to-date reference book, effectively addressing issues and concerns related to turfgrass management and physiology. While many reference books and scientific articles are in circulation on this subject matter, they exist in relative isolation from each other, covering only one specific topic.

Efficiency and effectiveness in gaining proper knowledge on turfgrass management and physiology are dependent on the accountability and coordination of all the factors and the interrelationships involved in this knowledge and practice.

While other authors of the existing sources on this subject matter have indeed competently covered the many areas separately, the areas are, nonetheless, interrelated and should be covered comprehensively in a single text. Thus, the main purpose of this book is to fill this niche.

This book is a compilation of several contributions from among the most competent and knowledgeable scientists, specialists, and researchers in the area of turfgrass management and physiology. It is intended to serve as a resource for both lecture and independent purposes. Turfgrass scientists, researchers, practitioners, students, and all those who are involved with turfgrasses (i.e., home-lawn owners, keepers of parks and recreation grounds, landscapers, golf-course superintendents, sports turf managers, etc.) will benefit from this guide.

As with other fields, accessibility of knowledge is among the most critical factors in turfgrass management and physiology. For this reason, as many factors as possible are included in this handbook. To further facilitate the accessibility of the desired information, the volume has been divided into several sections: "Turfgrass Management/Maintenance-Related Issues"; "Turfgrasses Growth, Management, and Cultural Practices"; "Sports Turf Management"; "Turfgrass Growth Regulators"; "Turfgrass Breeding, Genetics, and Biotechnology"; "Turfgrass Growth Disorders, Pathology, Diseases, and Turf Disease Management and Control"; "Turfgrass Integrated Pest Management"; "Turfgrass Management and Physiological Responses under Environmental Stress

Conditions"; and "Turfgrass Management and Physiological Issues in the Future." Each of these sections consists of one or more chapters that discuss, independently, as many aspects of turfgrass management and physiology as possible.

"Turfgrass Management/Maintenance-Related Issues" contains one chapter, entitled "Preference of Golf Course Operators for Various Turf Varieties and Their Perceived Importance of Selected Problems in Turf Maintenance." This chapter introduces turf-related issues.

"Turfgrasses Growth, Management and Cultural Practices" consists of seven chapters. These chapters explain in detail various turfgrass-management subjects. These chapters are as follows: "Carbon Metabolism in Turfgrasses"; "Warm-Season Turfgrass Fertilization"; "Growth Responses of Bermudagrass to Various Levels of Nutrients in Culture Medium"; "Weed Management in Warm-Season Turfgrass"; "Weed Management Practices Using ALS Herbicides for a Successful Overseeding and Spring Transition"; "Management of Tropical Turfgrasses"; and "Turfgrass Culture for Sod Production."

"Sports Turf Management" explains this topic in a chapter entitled "Culture of Natural Turf Athletic Fields."

"Turfgrass Growth Regulators" has one chapter, entitled "Applied Physiology of Natural and Synthetic Plant Growth Regulators on Turfgrasses," which presents in-depth information on turfgrass growth regulators.

"Breeding Turfgrasses" is the only chapter in the section entitled "Turfgrass Breeding, Genetics, and Biotechnology," and it explains turfgrass breeding in detail.

There are four chapters included in the section "Turfgrass Growth Disorders, Pathology, Diseases, and Turf Disease Management and Control." These are as follows: "Nutritional Disorders of Turfgrass"; "Biological Control of Turfgrass Fungal Diseases"; "Current Understanding and Management of Rapid Blight Disease on Turfgrasses"; and "A Review of Dead Spot of Creeping Bentgrass and a Road Map for Future Research." These chapters explain in-depth issues related to these subjects.

The "Turfgrass Integrated Pest Management (IPM)" section contains seven chapters that give detailed explanations of the IPM practices for turfgrasses. These chapters are as follows: "Herbicide Formulation Effects on Efficacy and Weed Physiology"; "Techniques to Formulate and Deliver Biological Control Agents to Bentgrass Greens"; "Biological Control of Some Insect Pests of Turfgrasses"; "Microbial Control of Turfgrass Insects"; "Integrated Pest Management (IPM) of White Grubs"; "Biology and Management of the Annual Bluegrass Weevil"; and "Understanding and Managing Plant-Parasitic Nematodes on Turfgrasses."

Since turfgrasses, like any other living organisms (plants, animals, human), are not immune to environmental stresses, a section entitled "Turfgrass Management and Physiological Responses under Environmental Stress Conditions" is included in this volume. It explains the effects of the major environmental stresses on turfgrasses in 11 chapters. These chapters include "Acid Soil and Aluminum Tolerance in Turfgrasses"; "Relative Salinity Tolerance of Turfgrass Species and Cultivars"; "Physiological Adaptations of Turfgrasses to Salinity Stress"; "Salinity Issues Associated with Recycled Wastewater Irrigation of Turfgrasses Landscapes"; "Turfgrass Drought Physiology and Irrigation Management"; "Turfgrass Shade Stress Management and Physiology"; "Cold Stress Physiology and Management of Turfgrasses"; "Cold Stress of Cool-Season Turfgrass: Antioxidant Mechanism"; "Responses of Turfgrass to Low Oxygen Stress"; "Identification of Turfgrass Stress Utilizing Spectral Reflectance"; and "Enhancing Turfgrass Nitrogen Use under Stresses."

"Turfgrass Management and Physiological Issues in the Future" is the final section of the volume. This section consists of three chapters that explain future and potential turfgrasses and the related management and physiological issues. These chapters are "Saltgrass (*Distichlis spicata*), a Potential Future Turfgrass Species with Minimum Maintenance/Management Cultural Practices"; "Native Grasses as Drought-Tolerant Turfgrasses of the Future"; and "The History, Role, and Potential of Optical Sensing for Practical Turf Management."

Hundreds of figures and tables are exhibited in this technical guide to facilitate comprehension of the material presented. This volume was developed by utilizing thousands of references. Several hundreds of index words are also included to further increase accessibility of the information.

Mohammad Pessarakli, Ph.D.

Acknowledgments

I would like to express my appreciation for the assistance that I received from the secretarial and administrative staff of the Department of Plant Sciences, College of Agriculture and Life Sciences, The University of Arizona. The continuous encouragement and support of the department head Dr. Robert T. Leonard and the division chair Dr. Dennis T. Ray for my editorial work, especially the books, is greatly appreciated. Dr. Leonard and Dr. Ray, your encouraging words have certainly been a driving force for the successful completion of this project.

In addition, my sincere gratitude is extended to Mr. John Sulzycki (executive editor, Taylor & Francis Group, CRC Press) who supported this project from its initiation to its completion. Certainly, this job would not have been completed as smoothly and rapidly without his most valuable support and sincere efforts.

Also, I sincerely acknowledge Ms. Randy Brehm (editor, Taylor & Francis Group, CRC Press) who professionally and patiently exercised her efforts in the completion of this project. Randy's prompt actions and hard work certainly had a significant effect on the completion of this work.

I am also indebted to the project coordinator, Ms. Jill Jurgensen, for the professional and careful handling of the volume. Jill, many thanks to you for your extraordinary patience and carefulness in handling this huge volume. Also, the sincere efforts and the hard work of the copy editor, the acquisition editor, as well as production editor will never be forgotten.

The collective sincere efforts and invaluable contributions of several competent scientists, specialists, and experts in the field of turfgrass science made it possible to produce this unique source, which is presented to those seeking information on this subject. Each and every one of these contributors and their contributions are greatly appreciated.

Last, but not least, I thank my wife, Vinca, a high school science teacher, and my son, Mahdi, a medical college student at The University of Arizona, College of Medicine, who supported me during the course of the completion of this work.

Editor

Dr. Mohammad Pessarakli, the editor, is a research associate professor and teaching faculty in the Department of Plant Sciences, College of Agriculture and Life Sciences at the University of Arizona, Tucson. His work at the University of Arizona includes research and extension services as well as teaching courses on turfgrass science, management, and stress physiology. He is the editor of the *Handbook of Plant and Crop Stress,* the *Handbook of Plant and Crop Physiology,* and the *Handbook of Photosynthesis* (all three titles, except the second edition of the *Handbook of Photosynthesis* [CRC Press] published by, formerly Marcel Dekker, Inc., currently Taylor & Francis Group, CRC Press); has written about ten book chapters; is an editorial board member of *Journal of Plant Nutrition and Communications in Soil Science and Plant Analysis* as well as *Journal of Agricultural Technology*; a member of the Book Review Committee of Crop Science Society of America; and reviewer of the journals of the Crop Science Society of America, Agronomy Society of America, and Soil Science Society of America. He is an active member of these societies, among others. Dr. Pessarakli is the author or coauthor of about 70 journal articles. He is an executive board member of the American Association of the University Professors (AAUP), Arizona Chapter. Dr. Pessarakli is an esteemed member (*invited*) of Sterling Who's Who, Marques Who's Who, Strathmore Who's Who, Madison Who's Who, and Continental Who's Who as well as numerous honor societies (i.e., Phi Kappa Phi, Gamma Sigma Delta, Pi Lambda Theta, Alpha Alpha Chapter). He is a certified professional agronomist and certified professional soil scientist (CPAg/SS), designated by the American Registry of the Certified Professionals in Agronomy, Crop Science, and Soil Science. Dr. Pessarakli is a United Nations consultant in agriculture for underdeveloped countries. He received the B.S. degree (1977) in environmental resources in agriculture and the M.S. degree (1978) in soil management and crop production from Arizona State University, Tempe, and the Ph.D. degree (1981) in soil and water science from the University of Arizona, Tucson.

For more information about the editor, please visit http://ag.arizona.edu/pls/faculty/pessarakli.htm

Contributors

Christian M. Baldwin
Department of Horticulture
Clemson University
Clemson, South Carolina

Giorgio M. Balestra
Department of Plant Protection
Faculty of Agriculture
University of Tuscia Viterbo
Italy

Gregory E. Bell
Department of Horticulture
and Landscape Architecture
Oklahoma State University
Stillwater, Oklahoma

Prasanta C. Bhowmik
Department of Plant, Soil,
and Insect Sciences
University of Massachusetts
Amherst, Massachussetts

Stacy R. Blazier
Department of Entomology
and Plant Pathology
Oklahoma State University
Stillwater, Oklahoma

Barry J. Brecke
West Florida Research
and Education Center
University of Florida
Jay, Florida

James J. Camberato
Department of Agronomy
Purdue University
West Lafayette, Indiana

John L. Cisar
Environmental Horticulture
Fort Lauderdale Research
and Education Center
University of Florida
Fort Lauderdale, Florida

Stephen T. Cockerham
Agricultural Operations
University of California, Riverside
Riverside, California

Kenneth E. Conway
Department of Entomology
and Plant Pathology
Oklahoma State University
Stillwater, Oklahoma

William T. Crow
Entomology and Nematology Department
University of Florida
Gainesville, Florida

S. Gary Custis
PBI Gordon, Inc.
Kansas City, Missouri

Erik H. Ervin
Crop and Soil Environmental Sciences
Virginia Polytechnic Institute
and State University
Blacksburg, Virginia

Shui-zhang Fei
Department of Horticulture
Iowa State University
Ames, Iowa

Wojciech J. Florkowski
Department of Agricultural
and Applied Economics
University of Georgia
Griffin, Georgia

J. Howard Frank
Department of Entomology and Nematology
University of Florida
Gainesville, Florida

David S. Gardner
Ohio State University
Columbus, Ohio

Ali Harivandi
University of California
Alameda, California

Senhui He
Department of Agricultural
  and Applied Economics
University of Georgia
Griffin, Georgia

Asghar Heydari
Plant Diseases Research Department
Iranian Research Institute for Plant Protection,
  Agricultural Research, and Education
  Organization
Tehran, Iran

Bingru Huang
Department of Plant Biology and Pathology
Rutgers University
New Brunswick, New Jersey

Alberto A. Iglesias
Laboratory for Molecular Enzymology
Faculty of Biochemistry
  and Biological Sciences
UNL, Santa Fe
Argentina

Yiwei Jiang
Department of Agronomy
Purdue University
West Lafayette, Indiana

Paul G. Johnson
Department of Plants, Soils, and Climate
Utah State University
Logan, Utah

John E. Kaminski
Department of Plant Science
University of Connecticut
Storrs, Connecticut

Albrecht M. Koppenhöfer
Department of Entomology
  and the Center for Turfgrass Science
Rutgers University
New Brunswick, New Jersey

Haibo Liu
Department of Horticulture
Clemson University
Clemson, South Carolina

Hong Luo
Department of Genetics and Biochemistry
Clemson University
Clemson, South Carolina

Kenneth B. Marcum
Department of Applied Biological Sciences
Arizona State University
Mesa, Arizona

S. Bruce Martin
Department of Entomology, Soils,
  and Plant Sciences
Pee Dee Research and Education Center
Clemson University
Florence, South Carolina

Stephen E. McCann
Department of Plant Biology
  and Pathology
Rutgers University
New Brunswick, New Jersey

Benjamin A. McGraw
Department of Entomology
  and Center for Turfgrass Science
Rutgers University
New Brunswick, New Jersey

Sowmya Mitra
Department of Plant Sciences
California State Polytechnic University
Pomona, California

Dara M. Park
Horticulture Department
Pee Dee Research and Education Center
Clemson University
Florence, South Carolina

Mohammad Pessarakli
Department of Plant Sciences
The University of Arizona
Tucson, Arizona

Paul D. Peterson
Department of Entomology,
  Soils, and Plant Sciences
Pee Dee Research
  and Education Center
Clemson University
Florence, South Carolina

Tim D. Phillips
Department of Plant
  and Soil Sciences
University of Kentucky
Lexington, Kentucky

Florencio E. Podestá
Center for Photosynthetic
  and Biochemical Studies (CONICET)
Faculty of Biochemical
  and Pharmaceutical Sciences
Rosario, Argentina

Yaling Qian
Department of Horticulture
  and Landscape Architecture
Colorado State University
Fort Collins, Colorado

Dipayan Sarkar
Department of Plant, Soil,
  and Insect Sciences
University of Massachusetts
Amherst, Massachusetts

Kalidas Shetty
Department of Food Science
University of Massachusetts
Amherst, Massachusetts

George H. Snyder
Everglades Research and Education Center
University of Florida
Belle Glade, Florida

John Clinton Stier
Department of Horticulture
University of Wisconsin
Madison, Wisconsin

Kai Umeda
University of Arizona
Tucson, Arizona

Greg Wiecko
University of Guam
UOG Station/CNAS
Mangilao, Guam

Xi Xiong
Department of Agronomy
University of Florida
Gainesville, Florida

Xunzhong Zhang
Virginia Polytechnic Institute
  and State University
Blacksburg, Virginia

Section I

Turfgrass Management/
Maintenance-Related Issues

1 Preference of Golf-Course Operators for Various Turf Varieties and Their Perceived Importance of Selected Problems in Turf Maintenance

Wojciech J. Florkowski and Senhui He

CONTENTS

1.1   INTRODUCTION

Golf is a popular game with the public, and new golf courses are continually being constructed (Figure 1.1). However, in recent years the increase in the number of players has shown a mixed trend. The rate of new golf course construction outpaces the growth in the number of golfers,

3

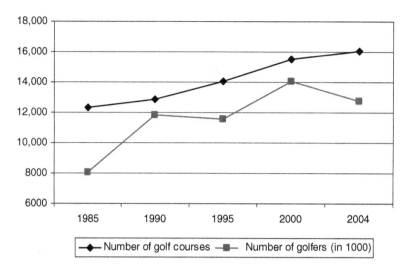

FIGURE 1.1   (Color Figure 1.1 follows p. 262.) The number of golf courses and players in the United States, 1985–2004. (From U.S. Department of Commerce, Statistical Abstract of the United States. 2004–2005. Bureau of the Census, Washington, D.C., 2005.)

suggesting that golfers have an increasingly wide course selection. New golf courses tend to be located in areas of high population density, i.e., where other golf courses likely exist. Therefore, golfers exercising their choice of a facility shift the revenue flow away from a facility they abandon in favor of a new one.

The quality of the playing surface is a major factor attracting players to a new course or encouraging the return of players to an existing facility. Although the day-to-day maintenance of a course is important, the essential decision associated with the construction of a new course or renovation of an existing course is the selection of appropriate turfgrass varieties for tees, fairways, and greens. Turfgrass selection is an investment decision because it affects the construction cost and the course characteristics, while the replacement of a turfgrass variety with another on any portion of a course results in renovation expenses and a loss of revenue due to suspended use of the course. In general, renovation is expected to provide golfers with a consistent playing surface [14]. The course condition reflected in, for example, turf density, weed presence, poor drainage, or green speed affects the demand for golf at a facility, assuming that other factors are constant [14,17]. Consequently, the choice of the turfgrass impacting the condition has direct implications for revenue. Older facilities can be under pressure to renovate, especially if a new facility has been constructed nearby [1]. In the survey results reported in this chapter, the average distance from the nearest competing golf course was 11.3 mi., but 51% of respondents indicated that the nearest competing golf course was no more than 5 mi. away.

The objective of this chapter is to summarize responses from a survey of golf courses in 14 southern states with regard to turfgrass-type use, the perceived importance of selected turfgrass issues such as various aspects of maintenance, turfgrass attribute importance from the standpoint of maintenance and surface condition, and opinions about the selected varieties. Turfgrass selection is a key investment decision made at the time of design and construction. The choice of a variety from the available portfolio is based on the bundle of attributes a specific variety represents, given the constraints related to the climate, soil, and the budgeted expenses for the turfgrass. Turfgrass varieties differ in their attributes, including their physiological requirements, appearance, and playability. Besides the commonly expected environmental risks, the growing number of golf courses suggests the ever-increasing presence of market risk demonstrated by golfers willing to try a game at different locations. The pleasure derived from the game is associated with performance, which is affected by the surface quality and condition that is directly related to the attributes of the turfgrass. Players blame poor turf and inadequate maintenance for their performance, and meeting course conditioning expectations places an ever-increasing burden on golf-course operators [11]. Failure to meet golfers' expectations affects the bottom line of a facility.

1.2 THE SURVEY OF SOUTHERN GOLF-COURSE OPERATORS

Information about the opinions of golf-course operators regarding the need for new turf varieties, preferred turfgrass attributes, and performance of varieties available on the market is fragmentary, although the contribution of breeding programs to advancements in the course surface conditions has been recognized [14]. The general tendencies in such opinions can be inferred from observations of variety selection decisions for planting on new courses, when undertaking renovation, or the interest may be gauged by attendance at seminars, trade shows, and field days. In addition, systematically collected information about turfgrass preferences is essential in guiding breeding programs, outreach, and marketing efforts of turfgrass growers. To fill the information gap, a survey among golf-course operators was undertaken in 2001.

The survey of golf courses was implemented to obtain information, among other things, about the turfgrass attribute importance, performance of selected varieties, and course maintenance issues. The survey included 14 southern states of the United States, i.e., the golf courses located in the warm grass zone. However, Florida was excluded from the survey because of specific climatic conditions, some of which were not shared by other southern states (e.g., subtropical climate of southern Florida). The research focus recognized the use of cool- and warm-zone turfgrasses (or C_3 and C_4 types, respectively) in the surveyed states, although Bermuda hybrids are predominantly used on golf courses in the southern United States [19]. The survey instrument inquired about the knowledge and use of *Paspalum* varieties, which were relatively new at the time of the survey implementation.

The list of 4892 golf courses located in the 14 chosen states was purchased from the National Golf Course Association (NGCA). Only members of the NGCA were surveyed. The survey was implemented in a single mailing, owing to cost considerations. A total of 720 completed questionnaires were returned, generating a return rate of 15%. This return rate was acceptable for a survey of business operators for whom responding to a survey is seldom a priority. Questionnaires were returned from all 14 states, but the number returned per state varied.

Among survey respondents, 76% were superintendents or assistant superintendents. Respondents had adequate educational background on which to base their opinions about the turfgrass attribute preferences and variety performance because 334 studied agronomy, crop, and soil sciences and 191 studied horticulture and landscaping after completing high school. It has been noted that having a well-qualified professional is important for maintaining the course in an excellent condition [15]. On average, a superintendent responding to the survey worked about eight years at the golf course where he received the questionnaire, while having about 17 years of experience working in the golf industry.

1.3 TURFGRASS TYPES ON FAIRWAYS, ROUGHS, AND GREENS

This section considers only turfgrass types and their use on fairways, roughs, and greens because teeing areas are relatively small and have less effect on the turfgrass selection decision or the game. Fairways cover the largest area of a course, while putting greens are managed differently because of their importance in the game. Their condition directly influences the ball's traveling speed and playability and, often, the outcome of the game.

1.3.1 FAIRWAYS

Given the geographical scope of the survey, the information was collected from states located in the range of bentgrasses, bermudagrasses, and the transition zone where both types are used [12]. In reality, the division is not strictly followed and superintendents use a variety that they are familiar with and which yields expected results. Not surprisingly, bermudagrasses and bentgrasses were the most often reported as used on fairways (Table 1.1). Bermuda was used most often on acreage ranging from 21 to 40 acres, although one-fifth of respondents to the survey indicated growing Bermuda on no more than 10 acres of fairways. Almost one-half of the respondents indicated bentgrass covering between 21 and 30 acres of fairways, while about 12% of respondents listed bentgrass as covering 20 acres or less and between 31 and 40 acres. However, a far larger number of respondents (547 individuals) reported the use of bermudagrass than bentgrass (47 persons). The dominance of Bermuda hybrids is consistent with the climate, adaptability to soil conditions, and ease of maintenance.

A small number of respondents (23 individuals) reported a strong preference for zoysiagrass (Table 1.1). Among those who used zoysiagrass, more than 39% used it on acreage ranging from 21 to 30 acres and more than a fifth on acreage ranging from 31 to 40 acres. About 17% reported its use on no more than a 10-acre area (Table 1.1). The use of fescue on fairways was minimal (11 individuals reporting), but the use of perennial ryegrass was more frequent (98 respondents) and, typically, on a larger area than fescue. For example, about 43% of those reporting using perennial ryegrass used it on an area ranging from 21 to 40 acres.

The predominant use of Bermuda hybrids on fairways in golf courses in the southern United States reflects climatic conditions, which guide the focus of breeding, research, and outreach programs at private and public (mostly land-grant universities) institutions. Workshops, golf clinics, and field days often include topics on the performance and maintenance of bermudagrass. For example, regularly held turf field days at the University of Georgia in Griffin devote a substantial portion of time to bermudagrass maintenance issues.

TABLE 1.1

Types of Grass Used on Golf Course Fairways According to Survey Respondents

Acres	Bermuda[a]	Bent[b]	Rye[c]	Zoysia[d]	Fescue[e]	Paspalum[f]	Other[g]
10 or less	2.6	12.8	10.2	17.4	45.5	33.3	10.0
11–20	8.7	12.7	17.4	8.7	18.1	16.7	25.0
21–30	30.2	49.0	24.4	39.1	0.0	33.3	15.0
31–40	28.9	12.7	18.4	21.8	18.2	16.7	22.5
41–50	12.2	4.3	12.3	4.3	0.0	0.0	12.5
51–60	7.0	6.4	5.1	4.4	0.0	0.0	7.5
61 or more	10.4	8.5	12.2	4.3	18.2	0.0	7.5

[a] $n = 547$, frequency missing 173.
[b] $n = 47$, frequency missing 673.
[c] $n = 98$, frequency missing 622.
[d] $n = 23$, frequency missing 697.
[e] $n = 11$, frequency missing 709.
[f] $n = 6$, frequency missing 714.
[g] $n = 40$, frequency missing 680.

Note: A golf course could use more than a single type of grass on fairways.

Source: Georgia Agricultural Experiment Station, University of Georgia.

TABLE 1.2
Types of Grass Used on Golf Course Roughs

Acres	Type of Grass						
	Bermuda[a]	Bent[b]	Rye[c]	Zoysia[d]	Fescue[e]	Paspalum[f]	Other[g]
10 or less	5.7	20.0	19.0	78.6	32.7	0.0	24.7
11–20	10.3	10.0	16.4	3.5	17.6	40.0	21.4
21–30	14.7	20.0	10.2	7.2	19.6	20.0	17.9
31–40	13.1	10.0	16.4	3.6	8.5	0.0	7.9
41–50	10.9	10.0	10.2	0.0	5.3	0.0	10.1
51–60	12.8	0.0	7.5	0.0	3.9	0.0	5.6
61 or more	32.5	20.0	20.2	7.1	12.4	40.0	12.4

[a] Based on 505 provided responses out of the total 720 returned questionnaires.
[b] Based on 10 provided responses out of the total 720 returned questionnaires.
[c] Based on 79 provided responses out of the total 720 returned questionnaires.
[d] Based on 30 provided responses out of the total 720 returned questionnaires.
[e] Based on 153 provided responses out of the total 720 returned questionnaires.
[f] Based on 6 provided responses out of the total 720 returned questionnaires.
[g] Based on 89 provided responses out of the total 720 returned questionnaires.
Note: A golf course could use more than a single type of grass on roughs.
Source: Georgia Agricultural Experiment Station, University of Georgia.

1.3.2 ROUGHS

Table 1.2 shows the use of various turfgrass types and the figures suggest a much different use of turfgrasses on roughs than fairways. Bermudagrass was used fairly evenly across the middle acreage categories, but about one third of 505 survey participants indicated that Bermuda was used on more than 60 acres at their facility. The use of cold-season grasses, i.e., perennial ryegrass and fescue, by 79 and 153 responding facilities, respectively, suggests that roughs are maintained to assure green color, enhancing the visual appeal of a course in months when warm-zone grasses are dormant. Bentgrass was used only on ten facilities, although its use varied widely in terms of acreage (Table 1.2), but zoysiagrass was most often used on areas of roughs not exceeding 10 acres, as reported by 30 facilities.

The frequent use of Bermuda on southern golf courses reflects its adaptability to soil and climate conditions and ability to maintain the course condition. Bermudagrass forms a dense turf and its ability to send stolons allows it to heal many torn spots. Results of national turfgrass trials suggest that several Bermuda varieties showed good healing 4–6 weeks after damage.

1.3.3 GREENS

The surface is expected to allow for speed. Mowing height is low and some varieties, for example, Penncross, Tifgreen, and Tifdwarf, suffer injury, becoming susceptible to disease [13]. However, Bermuda hybrids and bentgrass continue to be used on greens according to survey results, but bentgrass is more often used. The survey instrument probed respondents to indicate the area of greens in square foot because greens take a small portion of the course despite their key role in the game. Out of the 720 responding golf courses, 422 indicated the use of bentgrass on greens and 59% of those used bentgrass on area ranging from 100,000 to 200,000 sq. ft. About 8% used it on even larger area. Although the largest number of those growing bermudagrass on greens fell in categories similar to the case with bentgrass, a smaller percentage (42%) used bermudagrass on area ranging from 100,000 to 200,000 sq. ft., while 31% reported having bermudagrass on an area larger than

50,000 sq. ft. but smaller than 100,000 sq. ft. Other types of grasses were used sporadically on greens, according to the survey results.

1.4 IMPORTANCE OF SELECTED GENERAL TURFGRASS ISSUES

A golf-course operator faces multiple challenges in maintaining turf. Therefore, turf agronomics and the interest in new varieties are important, yet to be of any assistance to the breeders, turf growers, distributors, and outreach programs, the relative importance of the selected maintenance issues is particularly meaningful. Morris [13] noted that turfgrass buyers search for varieties having the desirable attributes for their location. Seven general issues were presented to respondents, who could assign a varying degree of importance to each issue (Table 1.3). The category "very important" was the highest offered choice. Disease control was viewed as a "very important" issue by over 60% of respondents, followed by the training in proper turf care (56%). Disease control is an obvious choice because it directly affects the visual appeal of the course and requires an application of fungicides before any visible damage occurs [9], or it will be ineffective in stopping the progress of the disease. The high placement of training in the proper care of turf is unexpected. Training in proper care requires allocation of time away from maintenance and away from the job, or on-the-job instruction, and implies a direct investment in the skills of the maintenance crew. On-the-job training and external training at workshops or field days are suitable [2]. A properly trained employee is expected to apply all cultural practices with efficiency and effectiveness that offsets the time and money spent on instruction. However, it appears that a high turnover discourages investment in training, at least of some members of maintenance crews. A general issue, turf agronomics, was "very important" to 53% survey participants.

A different set of issues was highly ranked in the category "slightly important," where the "need for new turfgrass varieties" was selected most frequently, i.e., by the 44% of respondents (Table 1.3). The golf industry continues to be interested in new varieties and in this respect it is similar to other sectors of the "green industry" that show a great interest in new plant varieties released

TABLE 1.3
The Importance of Selected Issues to Survey Respondents

Issue	Very Unimportant (%)	Slightly Unimportant (%)	Neither Important nor Unimportant (%)	Slightly Important (%)	Very Important (%)
Need for new turfgrass varieties	4.8	10.2	23.9	43.6	17.5
Turf agronomics	4.4	3.8	8.4	30.9	52.5
Disease control	5.8	1.9	5.4	25.2	61.7
Insect control	5.9	3.7	8.4	32.9	49.1
Need to renovate turf at the course	9.8	11.3	23.1	33.2	22.6
Need to limit typical maintenance expenditure	5.1	8.5	21.9	40.4	24.1
Training of maintenance personnel in proper turf care	6.2	2.3	6.0	29.6	55.9

Note: Based on 700 responses.

Source: Georgia Agricultural Experiment Station, University of Georgia.

by research and breeding programs. The interest in new turf varieties is explained by the second most often selected issue in the category "slightly important," namely, the "need to limit typical maintenance expenditures." More than 40% of respondents named this issue, which together with the interest in new turfgrass varieties outdistanced other issues among the "slightly important." Both environmental concerns [13] and the budgetary pressures [11] drive interest in lowering maintenance expenditures. The remaining issues were selected by a fairly even percentage of respondents.

Course renovation was at least "slightly important" to 56% of respondents, but almost 10%, the largest number across all other issues, thought renovation was "very unimportant" (Table 1.3). A renovation decision is affected by the number of rounds played, the quality of maintenance, course condition, and the type of turfgrass, among others. An important aspect is the anticipated change in the revenue flow of a facility due to the reduced use of a course during the renovation and the expected fall in revenue if the renovation is not undertaken in relation to the course potential to attract new players once the renovation is completed. More than one-fifth of respondents felt turf renovation was "very important" but a similar number perceived this issue as being "neither important nor unimportant" or assigned this issue into one of two categories, implying that the issue was unimportant. Because renovation of turf is also linked to the age of a course, sooner or later, it is needed and, therefore, one may expect that the distribution of responses across the five categories of importance in Table 1.3 will change over time. Moreover, the rapid growth in the construction of privately owned courses in the 1980s and 1990s suggests a booming renovation business in the current and the next decades.

Table 1.4 shows figures of those who reported a renovation in 3 years prior to the survey implementation. Among the specific renovation projects selected from a provided list, 98 respondents indicated an installation of an irrigation system. A golf course cannot be properly maintained and cared for without managing water availability and a system is installed at every golf course during construction. Over time, the system wears down and when repairs become too frequent or too costly, an overhaul of a system is necessary. Owing to technology improvements, old systems become obsolete, while new research and observations from thousands of golf courses lead to new designs, the use of new equipment, and construction of irrigation systems. For example, the single irrigation line on fairways was replaced by dual lines to reduce areas along the edges of the fairways, which were not receiving adequate water using the previous system. New irrigation equipment including moisture sensors, nozzles, computer-guided application systems, and other novel devices leads to improved water distribution, reduced water loss, and limited runoff. Consequently, a renovation project that involves updating or changing an irrigation system lowers the maintenance costs and water use, yet improves the course appearance and recovery; but an irrigation system is a costly capital investment [2].

Resodding or resprigging greens was a part of renovation projects reported by 86 participants out of 720 responding to the survey. Greens are a key area for golfers and about 50% of the game is played on greens [12, p. 215]. Also, because of the type of grass and shallow roots, any damage to greens may heal slower and remain visible for a longer period of time than in other parts of a

---

TABLE 1.4
Facility Renovation in 1999, 2000, or 2001

Nature of Renovation	Number of Respondents
Installation of the irrigation system	98
Resodding and resprigging of fairways	59
Resodding and resprigging of greens	86
Landscaping with ornamental plants	75
Other	198
Total number of respondents	518

Source: Georgia Agricultural Experiment Station, University of Georgia.

---

course. A problem arises because renovation of a green disrupts the game. Some operators build a temporary green, which allows for the continuing use of a course, but this is a costly solution and used only after a careful consideration of expenses and lost revenue.

Fairways, which take a far larger area than greens, were subject to renovation reported by 59 respondents during the 3-year period prior to the survey implementation. Depending on the type of turfgrass, resodding or resprigging may be preferred. Furthermore, resodding tends to be costlier, although it generally allows for a faster return of the game than after resprigging. Given the predominant use of Bermuda and bentgrass on fairways and roughs (Tables 1.1 and 1.2) both sod and sprigs are used and, in case of bentgrass, some courses use seeds. According to Waltz and Johnson [19], the average price per square foot of bermudagrass sod reported by Georgia producers fluctuated in the period 2001–2006. Prices declined in the first part of that period and since 2003 have been recovering, but the 2006 prices did not exceed the prices of 2001. The projected level of "adequate to excellent" supply of bermudagrass sod by Georgia producers declined from 78% in 2001 to 63% in 2006, helping prices to recover [19]. In spite of the slowly increasing bermudagrass sod prices, this turfgrass type remains the least expensive. Zoysiagrass and St. Augustinegrass sod were most expensive, with an average price per square foot of 30¢ and 32¢, respectively, in 2006.

Table 1.5 shows the approximate total amount spent on renovation in the 3 years preceding the survey, i.e., 1998–2000. Almost three out of five respondents (58.5%) reported spending in excess of $100,000. Renovations can be expensive, but the scope of a project depends on the costs of renovation, possible loss of revenue, and the expected rebound of revenue afterwards. The approximate total spending on resodding or resprigging was also sizable, but generally did not exceed $50,000. Given the cost per square foot or bushel of sprigs, although the renovated area covered thousands of square feet, it involved a small portion of a course in general. The small size of the renovated turf area is on the one hand a limitation, which may prevent operators from trying a new variety, but on the other hand they may be more inclined to try a new variety because it is planted on a small area. Renovation creates an opportunity to try a new variety.

The final decision as to what variety to use in what form (i.e., sprigs, sod, or seeds) results from the consideration of several issues. Although the cost of sod and sprigs is considered in the context of the speed of renovation, the attributes of a variety used to renovate a course are important and may override other factors. The next section describes in detail attribute preferences of golf-course operators responding to the survey and provides unique insights into what attributes influence the turfgrass selection decision.

TABLE 1.5

Approximate Expenditure for Renovation and Amount Spent on Only Resodding and Resprigging

Amount Spent ($)	Approximate Total Spent on Renovation[a]	Approximate Total Spent on Only Resodding or Resprigging[b]
5,000 and under	7.8	21.4
5,001–10,000	4.4	16.2
10,001–25,000	9.3	18.0
25,001–50,000	9.6	19.6
50,001–100,000	10.4	7.7
100,001 and more	58.5	17.1

[a] Based on 270 provided responses out of the total 720 returned questionnaires. Percent of respondents reporting renovation.

[b] Based on 117 provided responses out of the total 720 returned questionnaires. Percent of respondents spending on sod or sprigs.

Source: Georgia Agricultural Experiment Station, University of Georgia.

1.5 TURFGRASS ATTRIBUTE IMPORTANCE

The progress in turfgrass breeding and improvements in maintenance practices permit course conditions that could not be achieved in the past. To guide research efforts, information from the industry is needed about the importance of specific traits. A summary of responses allows a relative ranking of the golf course industry preference for various traits that would influence decisions on grass selection, and these insights are invaluable to breeders and turfgrass growers. A list of 15 turfgrass attributes was compiled and presented to respondents, asking them to select a level of importance associated with each attribute. A respondent attached a level of importance to each attribute independently and, therefore, more than one attribute could have been ranked "very important." To simplify the presentation of the results, the attributes were grouped into five broad categories after the summaries of responses were compiled.

1.5.1 APPEARANCE

Attributes describing appearance are vital because players expect turf to be green and often intensely green. These expectations were formed over a long period of time [3] and are slow to change. In the southern states, where the warm-zone grasses turn brown as they become dormant, attempts are being made to seed perennial ryegrass or fescue grass to provide some vivid green color in winter. Not surprisingly, 62% of respondents named "color" as an "important" attribute, more than any other attribute in this category (Table 1.6). The color of turf in the United States differs from the color of many European golf courses, where the color is often less intense and in a wide range of browns and greens. In contrast to the importance of color, the smooth transition from overseeding to spring greenup had the highest percentage of respondents, more than 9%, classifying this attribute as "not important at all."

The highest ranked attribute in this category was turf density (Table 1.6). Turf density is a major component determining the overall turf quality and is routinely taken into account by researchers and golf professionals when visually rating grasses. Only a negligible number of respondents rated turf density as unimportant. Leaf texture is relevant in the game of golf and the attribute's importance was reflected in the large number (50%; Table 1.6) of respondents listing it as "important," while more than a third of respondents thought it was a "very important" attribute. Leaf texture influences the evaluation of turf quality in the visual assessment of any turfgrass recommended for a golf course.

1.5.2 DISEASE-STRESS RESISTANCE

About 55% of respondents selected disease resistance as "very important," placing it among the top three issues in this category (Table 1.6). Only 1% of respondents thought disease resistance was not important and about 6% were ambivalent about it. The importance of disease resistance was already noted in responses to questions summarized in Table 1.3. Among many diseases are fungal diseases, caused by many indigenous strains of fungi, impossible to eradicate and constantly posing a threat to turf. The "wear and tear" and improper irrigation or fertilization may increase susceptibility of turf to fungi in a short period of time. Because of the nature of fungi growth, damage may occur before any visible signs become obvious, rendering treatment useless at that time. Therefore, an application of fungicides on a prescribed schedule is often practiced, especially if the course is located in the humid area of the southern U.S. However, there is an increasing public and regulatory pressure to limit the use of pesticides to reduce exposure and water pollution.

Drought resistance was "important" or "very important" to the largest number of respondents. Ability to survive prolonged periods of drought has proven essential during the period of below-normal precipitations between 2000 and 2003 in the southeastern United States. In 2006, several southern states also experienced drought conditions; the largest area included parts of Texas and Oklahoma. Carrow [7] provided an overview of earlier studies on drought resistance and

TABLE 1.6

Importance of Selected Turfgrass Attributes by Respondents to the Survey of Golf Courses Located in the Southern States, 2001

Category/Attribute	Not Important at All (%)	Rank in Category	Not Important (%)	Rank in Category	Neither Important nor Unimportant (%)	Important (%)	Rank in Category	Very Important (%)	Rank in Category	Percent Rating Important or Very Important	Rank[a]
Appearance											
Color	0.4	6	1.7	6	9.7	61.8	1	26.4	13	88.2	8
Smooth transition from overseeding to spring greenup	9.4	1	4.1	3	14.3	30.3	15	41.9	8	72.2	13
Leaf texture	0.3	8	1.3	9	13.8	50.3	3	34.3	11	84.6	10
Turf density	0.0	11	0.3	13	4.2	40.5	9	55.0	2	95.5	3
Plant Stress Resistance											
Drought resistance	0.0	11	0.4	12	2.7	43.0	7	53.9	4	96.9	1
Cold tolerance	1.8	4	3.4	5	15.9	38.9	11	40.0	9	78.9	12
Disease resistance	0.0	11	1.0	10	5.8	38.4	12	54.8	3	93.2	5
Wear resistance	0.1	9	0.0	14	3.6	43.1	6	53.2	5	96.3	2
Turf Management											
Mowing requirements	0.4	6	1.6	8	10.4	50.5	2	37.1	10	87.6	9
Fertilizer needs	0.1	9	1.6	8	18.4	49.2	4	30.7	12	79.9	11
Thatch management	0.0		1.7	6	9.6	40.0	10	48.7	7	88.7	7
Game Performance											
Playability—upright leaf-blade orientation	0.0	11	0.6	11	5.2	37.9	13	56.3	1	94.2	4
Playability—speed	0.6	5	3.6	5	8.0	40.6	8	49.3	6	89.9	6
Pricing and Certification											
Price	2.0	3	4.5	2	23.3	45.4	5	24.8	14	70.2	14
Licensed grower association	8.5	2	8.5	1	32.1	32.3	14	18.6	15	50.9	15

[a] Rank based on the percentage of respondents rating this attribute as "important" or "very important."

provided evidence of tolerance of selected varieties. According to that study, Tifway outperformed other bermudagrass varieties, tall fescue, and zoysiagrass. The results coincide with the use of bermudagrass on fairways and roughs (Tables 1.1 and 1.2).

1.5.3 TURF MANAGEMENT

Three attributes were included in this category. Fertilizer management was often selected less than other attributes as "very important," and slightly less than 50% of respondents perceived it as "important." Fertilization regimes for turfgrass types and varieties within each type have been well researched and the accumulated experience allows the control of nutritional needs, while assuring healthy condition and color. Mowing requirements were selected by slightly more than 50% of respondents as "important" but considerably fewer respondents perceived it as "very important" (37%). The importance of mowing results from its frequency and the nature of the job, which must be performed without interrupting the game. The frequency of mowing affects maintenance costs because it is labor-intensive and requires skills and the proper equipment. Poorly maintained equipment may actually injure turf, leading to infections and disease, especially the greens. Thatch management was the most important attribute in this category. It was ranked in the middle among the "very important" attributes and among the top 10 attributes selected as "important." Thatch buildup reduces water and nutrient penetration, while the accumulating organic matter may encourage the development of diseases.

1.5.4 GAME PERFORMANCE

Speed of the game is very important to players [14]. Although maintenance practices, for example, mowing height [13], can enhance the speed, the lasting achievements come from new releases from the breeding programs. Speed, which was determined by the turf type and variety, was "very important" to nearly one-half of survey participants and its overall importance reminds the reader about the purpose of the game. Although the final outcome of a played round depends on the player's skills, equipment, weather conditions (e.g., wind), and other factors, ultimately the turf quality may be blamed for the below-expectation performance. Not surprisingly, the top-ranked attribute in the "very important" category was the upright leaf blade orientation (56%). Fewer than 6% of all respondents perceived this attribute as unimportant or chose the neutral category with regard to this attribute. Obviously, golf-course operators are under pressure to reduce any imperfections of the game environment, and leaf blade orientation, which is to some extent genetically determined, must be taken into account in variety selection.

1.5.5 PRICING AND CERTIFICATION

The two characteristics included in this category received a ranking that placed them, in importance, below all the other turfgrass attributes presented to the surveyed golf-course operators (Table 1.6). However, the price of turfgrass was still viewed as "important" or "very important" by about 70% of respondents. The licensing of a variety and its production by a licensed grower association was selected as "important" or "very important" by little more than one-half of survey participants. This selection placed it far below any other of the 15 attributes included in the list.

A license attached to a variety by a grower association received little recognition in the context of all other attributes. It was expected that a license could perform as a quality guarantee and differentiate the variety from other varieties on the market. Although the differentiation may still be achieved, it was clearly not a characteristic that could offset other attributes essential to course performance and maintenance. It is possible that a licensed grower association certificate matters in case of new varieties being introduced which are unknown to golf-course operators. Buyers may be encouraged to try a variety if they are given some assurances that the produced turfgrass is supported by a structured organization broader than a single company, implying a reduced risk of failed performance on a course.

1.6 TURFGRASS VARIETIES, THEIR POPULARITY, AND ATTRIBUTE RANKING

To examine the familiarity of golf-course operators with specific varieties, a list of the commonly grown warm-zone turfgrass varieties was included in the questionnaire. Respondents selected among five categories indicating the degree of familiarity with each variety (Table 1.7) and, next, ranked them in terms of selected attributes, i.e., fertilizer needs, density, mowing requirements, water needs, and turf quality. The varieties included in the study were Crenshaw, Penncross, SR1020, Meyer Zoysia, five "Tif" Bermudas (Tifsport, Tifdwarf, Tifeagle, Tifgreen, and Tifway), and five varieties developed from *Paspalum* (Sea Isle, Salam, Seagreen, Seadwarf, and Seaway). Because the last five varieties were rather new at the time of the survey implementation, less than 5% of respondents reported any degree of familiarity with these varieties. This percentage of respondents was considered too small to allow for a meaningful assessment and comparison with other varieties and the five *Paspalum* varieties were omitted from Tables 1.8 through 1.14. The number of respondents indicating the degree of familiarity with a particular variety ranged from 687 in the case of Tifgreen to 706 in the case of Tifdwarf. Tables 1.8 through 1.14 summarize the responses organized according to the primary use of a variety on the course, i.e., a variety used mostly on fairways as opposed to greens.

1.6.1 FAMILIARITY WITH A VARIETY

Respondents were "very familiar" with Penncross (49%), a bentgrass used mostly on greens, followed by Tifway (42%). These two varieties, available on the market for about 50 years, outdistanced all other varieties. Tifdwarf was "very familiar" to little more than 35% of respondents. Four varieties, SR1020, Crenshaw, Meyer Zoysia, and Tifgreen, were "very familiar" to 21– 25% of survey participants (Table 1.7). Tifeagle and Tifsport were generally perceived as not familiar.

Once the two top levels of familiarity ("familiar" and "very familiar") were combined, varieties were ranked. Penncross (about 80%), Tifdwarf (about 78%), and Tifway (about 76%) were the most familiar to respondents. Not surprisingly, the top two places were occupied by varieties used mostly on greens, followed by a variety used mostly on fairways but released several decades ago. The next in terms of familiarity was Crenshaw (about 70%), a bentgrass used on greens, while Tifsport and Tifeagle were familiar to no more than one-half of respondents.

1.6.2 FERTILIZER NEEDS

Among the participants of the survey, a significant percentage was not familiar with the fertilizer needs of selected varieties (Table 1.8). However, it is difficult to draw solid conclusions about the knowledge of fertilizer requirements and the percentage of respondents indicating they were familiar with the variety (Table 1.7). In the case of six varieties, more than a third of respondents did not know their fertilizer needs, yet the vast majority of respondents were superintendents responsible for course maintenance. This observation can be linked to the degree of familiarity with a specific variety and the fact that some varieties were released relatively recently. Meyer Zoysia, used mostly on fairways, was ranked high in terms of fertilizer needs followed by SR1020. Penncross, Tifway, and Crenshaw were generally ranked low with regard to their fertilizer needs. The three bentgrass varieties used mostly on greens and the Bermuda hybrid used on fairways seem to have high fertilizer requirements.

Fertilizer needs were ranked among the bottom third of the important turfgrass attributes, but still almost 80% of respondents classified it as an "important" or "very important" attribute. Golf-course operators generally have adequate knowledge about how much and when to fertilize

TABLE 1.7

Familiarity with Selected Turfgrass Varieties by Respondents to the Survey of Golf Courses Located in Southern States, 2001

Variety	N[a]	Not at All Familiar (%)	Not Familiar (%)	Neither Familiar nor Unfamiliar (%)	Familiar (%)	Very Familiar (%)	Percent Rating Familiar or Very Familiar	Ranking[b]
Used Typically on Fairways								
Tifsport	691	16.4	19.8	15.3	39.1	9.4	48.5	9
Tifway	699	6.3	7.9	10.0	33.9	41.9	75.8	3
Meyer Zoysia	692	11.7	13.3	15.5	37.1	22.4	59.5	5
Used Typically on Greens								
Tifdwarf	706	5.9	6.7	9.8	42.3	35.3	77.6	2
Crenshaw	699	10.2	8.0	12.0	46.2	23.6	69.8	4
Penncross	696	7.9	4.6	7.8	30.7	49.0	79.7	1
Tifgreen	687	9.3	15.0	17.9	36.3	21.5	57.8	6
SR1020	688	16.1	14.7	12.8	31.0	25.4	56.4	7
Tifeagle	693	14.6	16.4	18.6	35.2	15.2	50.4	8

[a] Number of respondents providing an answer about a particular variety.

[b] Rank based on the sum of percentages of respondents that were "familiar" or "very familiar" with the specific variety.

Note: Tifgreen is currently considered a variety for use in other areas.

Source: Georgia Agricultural Experiment Station, University of Georgia.

TABLE 1.8

Ranking of Turfgrass Varieties in Terms of Their Fertilizer Needs by Respondents to the Survey of Golf Courses Located in Southern States, 2001

Variety	Ranking										
	1 Not Familiar at All	2	3	4	5	6	7	8	9	10 Very Familiar	Don't Know
	Percentage of Respondents										
Used Typically on Fairways											
Tifsport	0.7	0.8	4.9	5.5	17.5	4.5	3.4	1.9	0.6	0.2	60.0
Tifway	1.4	1.2	6.7	8.7	26.6	10.7	8.3	6.5	0.6	1.0	28.3
Meyer Zoysia	0.2	1.0	3.6	3.6	9.0	7.5	10.3	14.9	5.0	6.5	38.4
Used Typically on Greens											
Tifdwarf	1.6	1.7	8.6	9.2	28.8	6.7	7.5	4.8	0.7	0.8	29.6
Tifgreen	1.5	1.5	6.8	6.9	17.8	7.8	6.9	2.2	1.1	0.0	47.5
Tifeagle	0.9	1.9	6.1	3.8	11.9	6.5	8.9	6.9	1.3	0.4	51.4
Crenshaw	1.8	2.7	7.6	6.2	22.3	8.8	7.3	6.0	1.2	0.6	35.5
Penncross	1.4	4.3	9.8	9.5	27.4	8.6	7.4	5.6	0.8	1.2	24.0
SR1020	0.3	1.4	5.1	5.1	15.2	10.9	9.1	6.1	1.1	0.6	45.1

Note: Tifgreen is currently considered a variety for use in other areas.

their course and feel that they can control this aspect of course maintenance well. In terms of expenses, the cost of fertilizer is relatively small when compared with pesticide use. An indirect but one-time expense associated with fertilizer use, i.e., the construction of holding ponds, is part of the chemical input management and the appearance and design of a course and, therefore, not exclusively attributable to fertilizers.

1.6.3 Perceived Turf Density

Table 1.9 shows how respondents ranked the nine varieties with regard to enhanced density of turf. The two extreme rankings were bentgrass varieties: Penncross, ranked in the bottom half of the scale, and Crenshaw, ranked mostly in the upper half of the scale. Tifeagle was also ranked high, but a substantial number of survey participants (47%) selected the option "don't know," suggesting a lack of adequate knowledge of the variety (see Table 1.7). Tifdwarf was the next ranked variety among those used on greens. The ranking of Tifsport was relatively high, but similar to Tifeagle, more than 50% of respondents did not know the variety to evaluate it in terms of density. Tifgreen was not viewed as establishing a particularly dense turf and is currently considered a variety for other use areas. Among bentgrasses, highly ranked varieties in terms of enhanced turf density included two varieties used mostly on fairways: Meyer Zoysia and Tifway. Their rankings placed them in the top half of the evaluated varieties in terms of turf density.

1.6.4 Perceived Reduction of Mowing Frequency

Mowing frequency affects labor requirements and scheduling, and can interfere with access to the course by golfers. Also, because of increasing fuel costs and equipment maintenance linked directly to mowing frequency, reduced mowing frequency facilitates course management. Mowing was ranked higher than fertilizer management in the importance of turf attributes (Table 1.6). Respondents familiar with Meyer Zoysia, a variety used primarily on fairways, ranked it very high with regard to this attribute (Table 1.10). Two bentgrasses used mostly on greens, Penncross and Crenshaw, and Tifway, another variety used mostly on fairways, seemed to require frequent mowing. Among the remaining varieties used on greens, Tifdwarf was viewed more favorably in terms of mowing needs, but the evaluations of Tifeagle, scattered across many rank categories, suggested that this variety was not yet well assessed. Indeed, more than 48% of respondents selected the "don't know" category.

1.6.5 Purchase Price

Only one in four respondents thought price was a "very important" attribute (Table 1.6). Given that the turf is bought at the time of construction or later, when renovation of the course in undertaken, the relatively low ranking of price is not a surprise. Still, it was interesting to see how respondents ranked the nine varieties (Table 1.11) in terms of their price. Penncross, a variety used mostly on greens, seemed to have been ranked the highest followed by Tifway, a fairway-use variety. Among varieties used mostly on greens, Tifdwarf was also ranked high while Crenshaw was ranked in the middle. Meyer Zoysia was ranked in the bottom portion of the scale. The low ranking was expected due to its reputation as an outstanding turfgrass once it becomes established and, therefore, a highly desirable variety. Prices of Tifeagle and Tifsport, varieties of different intended use, were not well known to more than a half of respondents.

The relative unimportance of the turfgrass price reflects the ample supply. However, the supply and demand conditions for turfgrass vary, especially with regard to the form, i.e., seeds, sprigs, or sod. Among varieties considered in this study, bermudagrass is best distributed as sod or sprigs, while the cool-zone grasses, for example, Crenshaw, a bentgrass, are distributed as sod or seeds. Table 1.12 shows the average prices for sod, sprigs, and seeds of selected varieties. All varieties were sold as sod, but prices ranged from $0.24 per square foot for Tifsport to $1.00 per square

TABLE 1.9

Ranking of Turfgrass Varieties in Terms of Their Perceived Enhanced Density by Respondents to the Survey of Golf Courses Located in Southern States, 2001

| | Ranking | | | | | | | | | | |
Variety	1 Not Familiar at All	2	3	4	5	6	7	8	9	10 Very Familiar	Don't Know
					Percentage of Respondents						
Used Typically on Fairways											
Tifsport	0.0	0.2	0.4	1.6	9.7	9.8	8.6	12.6	3.0	2.7	51.4
Tifway	0.0	0.6	2.9	4.1	22.0	11.3	16.5	13.0	2.2	2.7	24.7
Meyer Zoysia	0.0	1.4	3.3	2.7	13.7	8.2	12.0	10.4	8.2	6.0	34.1
Used Typically on Greens											
Tifdwarf	0.0	0.9	2.8	5.9	17.7	14.7	16.7	12.0	2.9	2.2	24.2
Tifgreen	1.7	1.6	5.1	6.9	14.8	8.1	10.7	7.1	1.4	0.5	42.1
Tifeagle	0.2	0.5	1.3	0.9	4.4	3.2	6.4	16.2	10.3	10.0	46.6
Crenshaw	0.4	0.6	1.7	2.9	12.0	10.0	14.1	19.5	4.0	4.1	30.7
Penncross	2.0	6.3	11.4	13.2	20.2	10.5	6.6	5.4	1.6	1.4	21.4
SR1020	0.0	.05	1.3	1.8	9.5	11.7	16.4	11.5	4.0	2.4	40.9

Note: Tifgreen is currently considered a variety for use in other areas.

TABLE 1.10

Ranking of Turfgrass Varieties in Terms of Their Perceived Reduction of Mowing Frequency by Respondents to the Survey of Golf Courses Located in Southern States, 2001

Variety	Ranking										Don't Know
	1	2	3	4	5	6	7	8	9	10	
	Not Familiar at All									Very Familiar	
					Percentage of Respondents						
Used Typically on Fairways											
Tifsport	1.1	1.9	5.2	4.8	20.7	7.5	3.0	2.0	0.5	0.0	53.3
Tifway	2.6	4.3	9.3	11.1	29.5	8.1	4.7	3.1	1.0	0.6	25.7
Meyer Zoysia	0.3	1.9	2.3	3.7	12.6	11.2	12.8	12.9	3.9	4.0	34.4
Used Typically on Greens											
Tifdwarf	2.8	3.9	8.7	7.9	28.7	11.2	6.0	2.9	0.8	1.2	25.9
Tifgreen	3.2	4.0	7.9	10.4	17.2	6.7	4.6	2.1	0.5	0.0	43.4
Tifeagle	4.6	5.0	5.5	5.6	12.4	6.2	6.0	4.3	1.4	0.7	48.3
Crenshaw	2.9	5.4	10.5	7.3	25.4	5.6	5.6	3.6	0.8	0.2	32.7
Penncross	2.9	7.0	12.4	10.4	29.1	6.6	5.6	2.4	0.6	0.6	22.4
SR1020	1.5	5.3	7.6	7.3	20.2	5.5	5.8	3.5	0.7	0.2	42.4

Note: Tifgreen is currently considered a variety for use in other areas.

TABLE 1.11
Ranking of Turfgrass Varieties in Terms of Their Price by Respondents to the Survey of Golf Courses Located in Southern States, 2001

	1 Not Familiar at All	2	3	4	5	6	7	8	9	10 Very Familiar	Don't Know
Variety					Percentage of Respondents						
Used Typically on Fairways											
Tifsport	2.1	2.0	8.7	4.7	11.9	5.5	3.8	3.0	0.7	0.2	57.4
Tifway	0.5	0.2	0.8	3.5	28.9	7.9	13.1	9.0	3.3	2.7	30.1
Meyer Zoysia	2.6	4.0	6.5	9.5	17.4	6.0	5.3	4.9	1.1	1.1	41.6
Used Typically on Greens											
Tifdwarf	1.2	1.2	2.4	4.0	22.8	13.5	9.4	9.2	1.6	1.4	33.3
Tifgreen	0.7	0.5	2.1	3.3	17.7	8.5	9.2	5.2	1.4	1.2	50.2
Tifeagle	1.4	2.5	7.3	5.8	15.6	5.5	4.2	3.5	1.4	0.7	52.1
Crenshaw	0.9	1.3	5.3	6.5	26.8	7.9	4.6	7.9	0.6	1.5	36.7
Penncross	0.4	1.8	3.3	5.0	21.9	8.7	10.9	13.6	3.6	4.5	26.3
SR1020	0.2	3.4	2.2	3.6	19.7	9.6	5.8	5.8	2.1	1.3	46.3

Note: Tifgreen is currently considered a variety for use in other areas.

TABLE 1.12

Prices of Selected Turfgrass Varieties per Square Foot, Bushel, and Pound as Reported by Respondents of the Golf Course Survey, 2001

Variety	Price Paid by the Respondent in Dollars		
	Per Square Foot of Sod	Per Bushel of Sprigs	Per Pound of Seeds
Used Typically on Fairways			
Tifsport	0.24 (24)	6.49 (5)	–
Tifway	0.30 (66)	2.36 (12)	–
Meyer Zoysia	0.39 (32)	–	–
Used Typically on Greens			
Tifdwarf	0.62 (27)	6.41 (9)	–
Crenshaw	1.00 (14)	–	8.11 (34)
Penncross	0.73 (23)	–	6.77 (40)
Tifgreen	0.32 (21)	3.10 (6)	–
SR1020	0.84 (18)	–	7.66 (36)
Tifeagle	0.78 (22)	15.45 (10)	–

Note: Number of responses used to calculate the mean is given in parentheses. Respondents reported prices for all three categories, but the means were not reported if fewer than five answers were provided.

Tifgreen is currently considered a variety for use in other areas.

foot for Crenshaw. In general, varieties used on greens were priced higher than varieties used on fairways. Tifgreen, a variety used less often on greens and more in other areas, was among the lowest priced varieties. Meyer Zoysia was medium priced and given its desired attributes important to the game, perhaps only its high fertilizer needs and inconsistent tolerance of drought prevent it from wider use on fairways.

1.6.6 IRRIGATION NEEDS

Because of the increasing importance of water use and quality, the golf industry has developed a list of critical points on a golf course, which require attention to improve water management [8]. According to respondents, Penncross, used most often on greens, was a variety with relatively high water needs (Table 1.13). Two varieties used on fairways, Meyer Zoysia and Tifway, and Tifdwarf, grown on greens, were ranked high, implying that their water needs were limited. A similar tendency was observed in responses about Crenshaw and Tifeagle, but less so in case of Tifsport. The latter variety was unknown in terms of its water needs to 60% of respondents. SR1020 variety's evaluations were fairly scattered across available rankings. It appears that the use of Tifway and other varieties with relatively low water needs may increase in the future. Recent observations from Georgia suggest the increased use of zoysiagrass in renovation projects [19]. In some areas, *Paspalum* varieties become popular, among others, because of their high tolerance to poor-quality water used in irrigation.

Water needs are a characteristic of growing importance. Daily irrigation needs range from 250,000 to 1,000,000 gallons [12, p. 379] for an 18-hole golf course. Florkowski and Landry [10] reported the average use of water at 53.3 million gallons per course per year. Respondents to the survey ranked drought resistance high, implying the importance of this attribute (Table 1.6). Periodic dry spells are not uncommon in southern states. Moreover, periods of reduced precipitation seem to persist for years. For example, the southeastern United States experienced a period of prolonged drought between 2000 and 2002. In 2006, parts of Texas and Oklahoma also experienced a long drought. Various areas of California have been affected by drought in recent years as well. Insufficient subsoil moisture causes direct damage to turf and slows its recovery from normal wear

Handbook of Turfgrass Management and Physiology

TABLE 1.13

Ranking of Turfgrass Varieties in Terms of Their Water Needs by Respondents to the Survey of Golf Courses Located in Southern States, 2001

Variety	1 Not Familiar at All	2	3	4	5	6	7	8	9	10 Very Familiar	Don't Know
						Percentage of Respondents					
Used Typically on Fairways											
Tifsport	0.7	0.7	2.5	3.9	17.3	6.1	2.8	4.5	0.7	0.7	60.1
Tifway	1.2	0.4	3.9	5.3	28.4	8.7	11.9	7.1	2.9	1.2	29.0
Meyer Zoysia	2.0	2.6	4.4	3.7	15.6	7.9	11.1	8.6	4.2	2.6	37.3
Used Typically on Greens											
Tifdwarf	1.0	1.7	6.5	5.9	23.5	10.0	10.8	5.4	2.3	0.8	32.1
Tifgreen	0.7	2.1	4.4	5.1	18.8	8.8	8.1	4.1	1.2	0.2	46.5
Tifeagle	1.8	1.1	4.6	3.5	14.6	7.1	7.8	6.2	1.1	0.0	52.2
Crenshaw	1.7	1.4	5.2	8.2	23.8	9.7	5.8	6.8	2.5	0.6	34.3
Penncross	3.2	6.2	12.9	11.0	22.3	7.9	6.6	4.0	0.8	0.6	24.5
SR1020	0.7	1.3	4.4	7.9	16.6	7.7	8.3	6.4	1.8	1.1	43.8

Note: Tifgreen is currently considered a variety for use in other areas.

and tear. Irrigation schedules must be altered and the sources of irrigation water may become exhausted; for example, 67% of golf courses used their own ponds or lakes for irrigation in Georgia [10]. Many municipalities are unwilling to share water resources with golf courses in times of reduced water availability, and access to water becomes costly. States also regulate the drilling of new private wells and their capacity to meet the competing urban, agricultural, and recreation needs for water. Because the population in southern states continues to grow, the ability to access water and the cost of irrigation make turf water needs a very relevant attribute.

1.6.7 TURF QUALITY

Finally, Table 1.14 shows the rankings of the selected varieties in terms of turf quality. Because the question included varieties used either on fairways or greens, the evaluations may reflect the emphasis the respondents placed on a specific variety at the time of conducting the survey. The variety often ranked topmost was SR1020, used mostly on greens. Although typically ranked in the middle of the group in terms of attributes discussed earlier, this variety was able to produce high-quality turf, assuring its commercial success. This success is particularly important because one-third of respondents chose the category "unfamiliar" when evaluating their degree of SR1020 knowledge. Crenshaw, another variety grown on greens, was the second-highest ranked, followed by the three varieties Penncross, Tifdwarf, and Tifway, which received high marks in the top three rank categories (i.e., ranked 1, 2, or 3). Meyer Zoysia was, in general, ranked lower than the relatively little known Tifeagle and Tifsport. It appears that the quality of greens was very much on the minds of the respondents, reflecting the key role greens play in the operation of a golf course.

Turf quality is the synthetic attribute that reflects both the satisfaction derived from the game and an attribute of direct relevance to all maintenance efforts. As in case of many other products of durable use, it takes time for a variety to gain reputation. The more recently released varieties, for example, Tifeagle or Tifsport, received good evaluations, but a large percentage of respondents were insufficiently familiar with their ability to produce quality turf. The ability of a variety to penetrate the industry is limited because the turf selection decision is made infrequently, i.e., prior to construction or, on a much smaller scale, when undertaking a renovation project. This nature of the industry makes the introduction of new varieties particularly challenging.

When you recall the ranking of the familiarity with each variety (Table 1.7), Crenshaw was in the middle of the group and SR1020 was ranked second to last. Therefore, the summary of responses to this question (Table 1.14) has to be treated with caution. It is possible that the general nature of the question asking for opinions about turf quality after asking specific questions about the narrow aspects of turfgrass performance affected the way respondents made their selections.

1.7   IMPLICATIONS OF TURFGRASS SELECTION PREFERENCES

Turfgrass selection is driven by climate and player expectations [14]. In this chapter, the focus was on turfgrasses suitable to conditions in the 14 southern states. The desired variety would have superior quality and enhanced speed, especially on greens, despite heavy use [13]. Moreover, varieties with multiple stress tolerances will be preferred [5]. The varying individual evaluations, which were presented here only in the form of summaries, indicate that the same variety may perform differently at various locations because of site-specific conditions.

The interest in learning about new turfgrass varieties remains strong because 94% of survey participants responded positively to the question probing for interest in new varieties. The pressure on golf-course operators to supply continually improved playing surfaces is a result of the availability of alternative facilities and player expectations shaped by cumulative experience and newly gained knowledge. Golf course appearance influences players' choice [16]. In some instances, even lowering

TABLE 1.14

Ranking of Varieties According to Turf Quality by Respondents to the Survey of Golf Courses Located in Southern States, 2001

Variety	Ranking														
	1 Prefer Most	2	3	4	5	6	7	8	9	10	11	12	13	14 Prefer Least	Not Familiar
Used Typically on Fairways															
Tifsport	7.7	8.1	8.5	6.6	8.6	5.9	5.1	3.7	0.3	0.0	0.0	0.0	0.0	0.0	45.5
Tifway	10.3	11.1	12.0	12.6	10.0	9.2	7.4	4.1	0.5	0.2	0.0	0.0	0.0	0.0	22.6
Meyer Zoysia	5.7	7.0	8.9	7.8	10.2	7.6	5.0	10.9	1.6	0.0	0.0	0.0	0.0	0.0	35.3
Used Typically on Greens															
Tifdwarf	15.7	11.3	12.3	11.7	9.8	7.3	5.1	4.6	0.3	0.2	0.0	0.0	0.0	0.0	21.7
Tifgreen	4.9	5.8	8.3	8.4	9.8	8.9	8.6	8.8	0.6	0.0	0.0	0.0	0.0	0.0	35.9
Tifeagle	16.7	11.1	6.3	6.0	3.7	2.2	2.7	5.5	1.5	0.3	0.0	0.4	0.0	0.0	43.6
Crenshaw	18.1	15.6	10.3	7.0	6.7	7.1	4.1	6.3	0.1	0.7	0.0	0.0	0.0	0.0	24.0
Penncross	13.1	13.1	16.1	7.1	8.4	7.4	6.0	9.1	0.9	0.5	0.0	0.0	0.0	0.0	18.3
SR1020	19.7	12.9	9.9	5.2	4.9	4.2	3.3	5.2	0.7	0.2	0.0	0.0	0.0	0.0	33.8

Note: Tifgreen is currently considered a variety for use in other areas.

the green fees did not attract players if they could find another surface in the area in better condition [15]. At the same time, the regulatory environment at the local and higher levels forces changes in water management and the use of chemical inputs in agriculture. In some regions of the United States, golf courses are strongly encouraged to reduce pesticide use [11]. Because weeds and fungal diseases are the most common pests in the golf course, managers and superintendents desire new varieties with improved disease resistance (Table 1.3).

Drought-resistant varieties are highly valued. The survey question posed to respondents did not distinguish between drought avoidance and drought resistance [4], but focused on the issue at large. Great progress has been made in managing water by using improved equipment and evapotranspiration data to schedule irrigation [6], but these tools only remedy the conditions of water deficit, and the right approach is to breed for drought-resistant traits. New *Paspalum* varieties offer an alternative for courses where water quality is a problem.

Reaching the golf course industry with the news of new turfgrass varieties is essential, yet the actual purchase may be long in coming. First, existing golf courses undertake large renovation projects infrequently and, in general, such projects are limited in scope. Second, golf-course designers, builders, and incoming superintendents should work together from project conception to the first game at a new facility, so that those who will be responsible for the course condition and know the desired turfgrass variety can present their input. However, if a superintendent is hired when construction is under way and the turfgrass variety already chosen, he can only focus on the optimal maintenance of the surface. Reaching the key decision makers with information about newly released varieties is essential. According to the survey results, superintendents were most likely to make such decisions. Although the question did not deal with the timing of such decisions, it is most likely that superintendents were responsible for selections of varieties used to renovate the course. The primary source of knowledge about new varieties chosen by survey respondents were turf magazines named as "frequent" or "most often" source by more than 81% of respondents. Peers and United States Golf Association (USGA) research reports were virtually tied for the second most-often-named source about new varieties. Club members were in the second to last position among 15 choices offered in our survey, but the number of private club golf courses participating in the survey was limited because this category is in itself rather small. When asked where the respondents would like to learn about the new varieties, 74% selected the USGA and 53% indicated "other golf-course operators" (or peers). None of the remaining eight choices was chosen by more than 50% of survey participants.

Familiarity with *Paspalum* turfgrass varieties was very limited among survey respondents. However, this type of turfgrass was available through several varieties at the time of conducting the survey. Despite highly desirable attributes of this type of turfgrass, the golf industry was unaware of new varieties. *Paspalum* varieties are suitable for specific growing conditions, especially where seawater or alkaline soils prevented the use of Bermuda hybrids. Recently, *Paspalum* has become more popular and enjoys commercial sales.

On a broader scale, the need for new varieties with greater drought resistance and disease susceptibility remains high. Moreover, although the drought resistance path involves two established mechanisms (avoidance or tolerance), disease resistance will require continuous advancements because of the evolutionary nature of multiple organisms causing diseases. Moreover, the introduction of varieties based on turfgrasses from one ecosystem to another ecosystem creates unknown risks of disease susceptibility, while the changing environmental stress related to global climate instability may accelerate the spread of turfgrass pathogens. Monitoring variety performance and sharing the observations is a prerequisite for the development of varieties that will ensure the expected surface condition and tame the maintenance expenses. Research programs at private and public institutions continue to release new varieties that offer improved turf quality. The measure of turf quality is either lower maintenance costs or improved turf attributes that directly affect the game.

ACKNOWLEDGMENTS

This study was supported by financial assistance from the Georgia Seed Development Commission and Hatch funds. The author acknowledges and thanks Robert N. Carrow and Clint Waltz for comments on the earlier draft of this chapter. The author thanks Marilyn Slocum and Harvey Witt for their assistance in the preparation of the manuscript.

REFERENCES

1. Anonymous. Bootstrap Journey. Course Renovation Teaches Valuable Lessons. Jacobsen Golf and Turf, www.jacobsen.com. Accessed August 22, 2006.
2. Beard, J. B. *Turf Management for Golf Courses.* Ann Arbor Press, Chelsea, MI, 2002.
3. Borman, F. H., D. Balmori, and G. T. Geballe. *Redesigning the American Lawn.* Yale University Press, New Haven, CT, 1993.
4. Carrow, R. N. Drought Avoidance Characteristics of Diverse Tall Fescue Cultivars. *Crop Science,* 36: 371–377, 1996.
5. Carrow, R. N. Seashore Paspalum Ecotype Responses to Drought and Root Limiting Stress. *Turfgrass and Environmental Research Online,* 4(13): 1–9, 2005. Available online at http://usgatero.msu.edu/currentpastissues.htm. Accessed August 25, 2006.
6. Carrow, R. N. Drought Resistance Aspects of Turfgrasses in the Southeast: Evapotranspiration and Crop Coefficients. *Crop Science,* 35: 1685–1690, 1995.
7. Carrow, R. N. Drought Resistance Aspects of Turfgrasses in the Southeast: Root-Shoot Responses. *Crop Science,* 36: 687–694, 1996.
8. Fletcher, K. A. The Business Value of Environmental Stewardship. *Green Section Record,* September–October 2003, pp. 26–27. Available online at http://turf.lib.msu.edu/gsr/. Accessed September 29, 2006.
9. Florkowski, W. J. *Lawn-Care Treatments (Plant Pathogens). Encyclopedia of Pest Management.* Marcel Dekker, Inc., New York, 2002, pp. 439–441.
10. Florkowski, W. J. and G. Landry. *An Economic Profile of Golf Courses in Georgia: Course and Landscape Maintenance.* The University of Georgia, College of Agricultural and Environmental Sciences, The Georgia Agricultural Experiment Stations, Research Report 681, April 2002, p. 14.
11. Hopp, K., J. Skorulski, J. Baird, D. Bevard, B. Broune, D. Oatis, and S. Zantik. Survival 101: Dealing with Ever-Increasing Expectations. *Green Section Record,* May–June 2004, pp. 15–16. Available online at http://turf.lib.msuedu/gsr/. Accessed September 29, 2006.
12. McCarty, L. B. *Best Golf Course Management Practices.* Pearson-Prentice Hall, Upper Saddle River, NJ, 2nd edition, 2005.
13. Morris, K. Bentgrasses and Bermudagrasses for Today's Putting Greens. *Green Section Record,* January–February 2003, pp. 8–12. Available online at http://turf.lib.msu/gsr/. Accessed September 29, 2006.
14. Nelson, M., L. Gilhuly, B. Vavrek, and P. Vermulen. Promoting Reliable Turf. *Green Section Record,* May–June 2001, pp. 20–22. Available online at http://turf.lib.msu/gsr/. Accessed September 29, 2006.
15. O'Brien, P. M. Golf Course Superintendent: Expense or Investment? *Green Section Record,* September–October 2003. Available online at http://turf.lib.msu/gsr/. Accessed September 29, 2006.
16. Shamanske, S. *Golfonomics.* World Scientific, New York, 2004.
17. Trenholm, L. E., R. R. Duncan, and R. N. Carrow. Wear Tolerance, Shoot Performance, and Spectral Reflectance of Seashore Paspalum and Bermudagrass. *Crop Science,* 39: 1147–1152, 1999.
18. U.S. Department of Commerce. Statistical Abstract of the United States, 2004–2005. Bureau of the Census, Washington, D.C., 2005.
19. Waltz, C. and B. J. Johnson. Annual Georgia Sod Producers Inventory Survey. *Georgia Sod Producers Association News* 16(1): 1–5, 7, 2006.

Section II

Turfgrass Growth, Management, and Cultural Practices

2 Carbon Metabolism in Turfgrasses

Alberto A. Iglesias and Florencio E. Podestá

CONTENTS

2.1 CARBON ASSIMILATION IN HIGHER PLANTS

What is known as photosynthesis is the process by which inorganic carbon (i.e., atmospheric CO_2) is captured by certain living organisms and converted to organic forms, primarily carbohydrates [1–3]. The general equation to synthesize one molecule of glucose (the most abundant constitutive monosaccharide) can be written as follows:

$$6CO_2 + 6H_2O \rightarrow C_6(H_2O)_6 + 6O_2 \qquad (2.1)$$

The energy and reductive power necessary to make possible this highly endergonic reaction is taken by photosynthetic organisms from sunlight. A more detailed view allows division of photosynthesis into a light-dependent and a synthetic phase [1,3,4]. The former comprises the absorption of radiant energy from sunlight and its conversion into chemical intermediates containing high energy (ATP) and reductive power (NADPH). In fact, radiant energy is utilized to split the water molecule to produce oxygen and generate electrons. The transport of such electrons through specialized membranes is utilized to synthesize ATP and reduce $NADP^+$ to NADPH. The synthetic phase includes the utilization of ATP and NADPH to effectively convert inorganic CO_2 into biomolecules known as photoassimilates [4]. Formerly, the carbon assimilation process was denominated the dark step of photosynthesis [3]. However, the current name "synthetic phase" is more accurate since it is not completely "dark" (meaning absolute independence from light) because it needs photogenerated ATP and NADPH, and the whole process is regulated to be operative in the light, largely by light-dependent events [1,4].

Higher plants can be classified on the basis of the biochemistry exhibited to perform the synthetic phase of photosynthesis. The main metabolic route by which all plants ultimately fix atmospheric CO_2 into photoassimilates is the reductive pentose phosphate (RPP) pathway [1–4]. This common plant metabolism is also known as the Calvin–Benson cycle, after the researchers who elucidated it, or the C_3 route, considering the first metabolic product of carbon assimilation [1–3]. Most plants solely have the Calvin–Benson cycle for carbon fixation, and they are named C_3. Other plants possess an additional (not alternative) route for photosynthetic carbon assimilation that operates spatially (C_4 plants) or temporarily (CAM plants) separated from the RPP pathway [1–3]. Under this general classification, a few species are found within the borders, as they exhibit C_3–C_4 characteristics mainly related to nontypical C_4 plants [1,5].

All the variants exhibited by plants for carbon assimilation were generated by the evolutionary pressure toward a more efficient photosynthesis adapted to the diverse habitats in ever-changing environmental conditions, where carbon assimilation would be a lot less flexible if it were solely performed by the C_3 route [6,7]. The differences involve morphological, physiological, and biochemical changes that entail a coordinate operation and regulation for an optimal operation [6]. The knowledge of all the components, especially at a molecular level, is a key issue for the understanding of the performance and potentiality of a plant exhibiting one particular metabolic pathway for carbon fixation. It is worth pointing out that this understanding is critical for the design of strategies to optimize photosynthetic efficiency and thus plant productivity.

2.2 THE C_3 PATHWAY OF PHOTOSYNTHESIS

2.2.1 THE CALVIN–BENSON CYCLE, OR REDUCTIVE PENTOSE PHOSPHATE PATHWAY

The RPP pathway was elucidated in the late 1940s by Calvin, Benson, and coworkers, who utilized the recently discovered (by Ruben and Kamen in 1940) ^{14}C isotope as a stable radioactive marker [1]. The experiments included incubation of green algae cells in the light with $^{14}CO_2$ during a "pulse" time, and after further incubation in its absence during different time periods, metabolites were separated and identified to "chase" radioactivity [8]. Doing so, it was established that the metabolic route is cyclic, as shown in Figure 2.1. In Table 2.1, the 13 enzymatic reactions (with the corresponding 11 enzymes) involved in the RPP cycle are detailed. Table 2.1 also exemplifies that the cycle can be operatively divided into three stages [1–3]. The first is a carboxylation step, comprising the reaction of CO_2 fixation catalyzed by ribulose-1,5-bisphosphate carboxylase/oxygenase (Rubisco). In this way, inorganic carbon is incorporated into a pentose to give rise to two molecules of a three-carbon metabolite: 3-phospho-glycerate (3PGA). This is followed by a reductive phase, where NADPH and ATP are utilized to reduce 3PGA to glyceraldehyde-3-phosphate (Ga3P) in two reactions catalyzed consecutively by 3PGA kinase and Ga3P dehydrogenase (Ga3P-DHase) (Table 2.1). In the third stage, the initial acceptor of CO_2, Rub-1,5-bisP, is regenerated after the interconversion of triose-, tetrose-, pentose-, hexose-, and heptulose-phosphates involving ten reactions catalyzed by eight different enzymes (note in Table 2.1 that Reactions 2.6 and 2.9 are catalyzed by a same aldolase, as well as those of Reactions 2.8 and 2.11 that involve the same transketolase). In the last reaction of the regenerative stage, ATP from the light phase of photosynthesis is consumed to phosphorylate Rub5P (Table 2.1).

The whole functioning of the RPP pathway indicates that the net fixation of one CO_2 molecule occurs in every complete turn of the cycle. Thus, theoretically, in six turns a molecule of glucose is synthesized *de novo*. It is from intermediates of the RPP cycle that photoassimilates are derived to the production of the two major products of photosynthesis: starch and sucrose [3,4]. Starch buildup is initiated mainly from fructose-6-phosphate (Fru6P) within the chloroplast, whereas triose-phosphates are the metabolites predominantly exported to the cytosol for sucrose synthesis. In this way, an intracellular partitioning of photoassimilates is established between chloroplast and

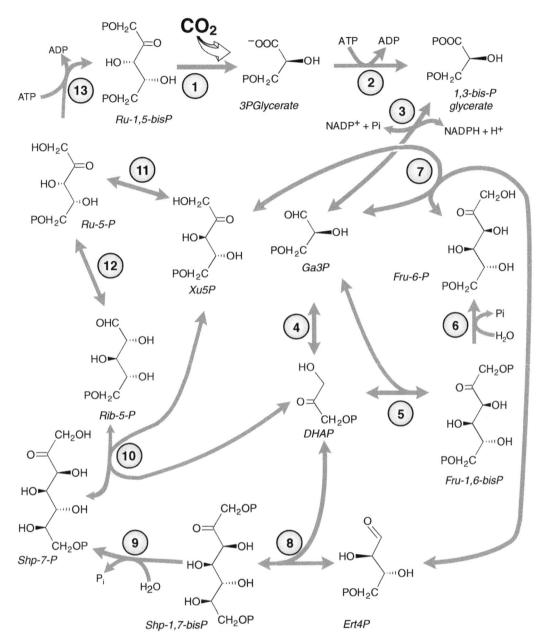

FIGURE 2.1 Schematic drawing of the reductive pentose phosphate pathway. Numbered reactions are catalyzed by (1) Rubisco; (2) 3PGA kinase; (3) Ga3P-DHase; (4) triose-P isomerase; (5) aldolase; (6) FBPase; (7) transketolase; (8) aldolase; (9) SBPase; (10) transketolase; (11) P-pentose 3-epimerase; (12) P-riboisomerase; (13) PRKase.

cytosol within the same source tissue. In contrast, sucrose is the major carbohydrate shuttled to nonphotosynthetic parts of a plant (sink tissues) to provide carbon and energy. Its transport constitutes a second level of photoassimilate partitioning, in this case at the intercellular level. The last is the partitioning between photosynthetic (source) and nonphotosynthetic (sink) tissues of the plant. Also, in certain heterotrophic tissues, the carbon transported via sucrose is the basis for synthesizing reserve products such as fats and starch.

TABLE 2.1

Reactions and Stages Involved in the RPP Cycle

Stage	Reaction	Equation	Enzyme
Carboxylative	Rub-1,5-bisP + CO_2 → 2 3PGA	2.2	Rubisco, EC 4.1.1.39
Reductive	3PGA + ATP ⇌ 1,3-bisPGA + ADP	2.3	3PGA kinase, EC 2.7.2.3
	1,3-bisPGA + NADPH + H^+ ⇌ Ga3P + $NADP^+$ + Pi	2.4	Ga3P-DHase, EC 1.2.1.13
Regenerative	Ga3P ⇌ DHAP	2.5	Triose-P isomerase, EC 5.3.1.1
	Ga3P + DHAP ⇌ Fru1,6bisP	2.6	Aldolase, EC 4.1.2.13
	Fru-,6-bis-P + H_2O → Fru-6-P	2.7	FBPase, EC 3.1.3.11
	Fru-6-P + Ga3P ⇌ Xyl5P + Ert-4-P	2.8	Transketolase, EC 2.2.1.1
	DHAP + Ert-4-P ⇌ Shp-1,7-bisP	2.9	Aldolase, EC 4.1.2.13
	Shp-1,7-bisP + H_2O → Shp-7-P	2.10	SBPase, EC 3.1.3.37
	Ga3P + Shp-7-P ⇌ Xyl-5-P + Rib-5-P	2.11	Transketolase, EC 2.2.1.1
	Xyl-5-P ⇌ Rub-5-P	2.12	P-pentose 3-epimerase, EC 5.1.3.1
	Rib-5-P ⇌ Rub-5-P	2.13	P-riboisomerase, EC 5.3.1.6
	Rub-5-P + ATP → Rub-1,5-bisP + ADP	2.14	PRKase, EC 2.7.1.19

Note: Rub-1,5-bisP, ribulose-1,5-bisphosphate; 3PGA, 3-phospho-glycerate; Rubisco, ribulose-1,5-bisphosphate carboxylase/oxygenase; 1,3-bisPGA, 1,3-bisphospho-glycerate; Ga3P, glyceraldehyde-3-phosphate; DHase, dehydrogenase; DHAP, dihydroxy-acetone-phosphate; Fru-1,6-bisP; fructose-1,6-bisphosphate; Fru-6-P, fructose-6-phosphate; FBPase, fructose-1,6-bisphosphate 6-phosphatase; Xyl-5-P, xylulose-5-phosphate; Ert-4-P, erytrose-4-phosphate; Shp-1,7-bisP, sedoheptulose-1,7-bisphosphate; Shp-7-P, sedoheptulose-7-phosphate; SBPase, sedoheptulose-1, 7-bisphosphate 1-phosphatase; Rib-5-P, ribose-5-phosphate; Rub-5-P, ribulose-5-phosphate; PRKase, ribulose-5-phosphate 1-kinase.

2.2.2 Regulation of the C₃ Pathway

Carbon assimilation is highly regulated in such a way as to coordinate its operation with the whole process of starch and sucrose synthesis [1–4,9–11]. A general way to modulate the action capacity and optimal functioning of the RPP pathway is determined by the fact that the enzymes are organized as a multienzyme complex [12–14]. Furthermore, the complex is loosely associated with thylakoid membranes near the sites of ATP and NADPH. Light plays a main role in regulation of the RPP cycle, as the pathway is fully operative during the day and essentially inactive at night. In full sunlight, production of ATP and NADPH by the light phase of photosynthesis is in excess of the needs of the RPP pathway, so that it becomes the limiting process of the rate of photoassimilation [2,3]. A general form of favoring the activity of the synthetic phase of photosynthesis in the light is determined by the increase in pH and Mg^{2+} concentration, together with changes in the level of different metabolites occurring in the chloroplast stroma and induced by photosynthetic electron transfer. Many enzymes of the cycle require Mg^{2+} as an essential cofactor, and they are more active at the more alkaline pH existing in the stroma in the presence of light [1,2].

A more complex regulation is achieved by mechanisms modulating the activity of certain enzymes by chemical modification. Regulation is exerted at the three stages of the RPP pathway, which are the main targets controlling the essentially irreversible steps of the cycle [1]. Thus, key points for regulation are the reactions catalyzed by Rubisco at the carboxylative stage, and by FBPase, SBPase, and PRKase at the regenerative stage (see Figure 2.1 and Table 2.1). Also, the step involving NADP–Ga3P-DHase (at the reductive stage; see Table 2.1) is regulated in a concerted manner. Although reversible *in vitro*, this reaction operates mainly in the reductive way owing to the high concentration of NADPH occurring in the stroma during the day [1–3].

Activation of Rubisco involves the ATP-dependent carbamylation of a lysyl residue mediated by a 200-kDa protein called Rubisco activase [15–19]. This enzyme exhibits ATPase activity, and during the carbamylation process the free energy of hydrolysis is utilized to induce conformational changes on Rubisco, thus altering the strength in the binding of substrates and catalytic intermediates. Rubisco activase exerts a chaperone-like action to promote and maintain the catalytic activity of its target enzyme [18]. As a main result, the correct formation of the catalytically competent complex Rubisco–CO_2–Mg^{2+}–Rub-1,5-bisP is facilitated [17,18,20]. The process is also affected by pH, phosphoesters, the carboxylase substrates (principally, Rub-1,5-bisP), and 2-carboxyarabinitol-1-P, a potent, naturally occurring inhibitor of Rubisco, which is derived from chloroplastic FBPase [20]. In most plants, further regulation is exerted via thioredoxin-mediated redox changes on Rubisco activase, which modify the action capacity of this protein (see below). Rubisco is also allosterically regulated by Pi, which activates the enzyme by reversibly stabilizing the binding of CO_2 and Mg^{2+} [2]. This activation is particularly important under conditions of limited demand of photoassimilates by sink tissues.

Concerning regulation of the reductive and regenerative stages of the RPP pathway, NADP–Ga3P-DHase, FBPase, SBPase, and PRKase are selectively reduced (and activated) by thioredoxins [9–11,21]. These low-molecular-mass (12 kDa) proteins are reduced by a specific reductase (ferredoxin thioredoxin reductase) at the expense of reduced ferredoxin, which is the final acceptor of the photosynthetic electron transport [9,10]. In this manner, the light signal generates reductive power that is ultimately utilized to convert definite disulfides into sulfhydryl groups in target regulatory enzymes of the RPP cycle, which become active at the reduced state. Four different forms of thioredoxin were identified in chloroplasts, two major forms (*m* and *f*) and two less abundant forms (*x* and *y*) [9,21]. Thioredoxin *f* is the one primarily involved in the regulation of enzymes from the RPP cycle *in vivo* [21]. A noncovalent manner of interaction between the target enzyme and thioredoxin appears to facilitate the exchange of reducing equivalents [22].

As pointed out above, regulation of carbon assimilation by the thioredoxin system is also exerted by redox modulation of Rubisco activase by thioredoxin *f*. However, ATP hydrolysis by Rubisco activase is not triggered by the presence of Rubisco [16,18]. To avoid the possibly wasteful ATP hydrolysis, most plants have two isoforms of activase, with a larger isoform more sensitive to ADP inhibition (meaning inhibition of the activity hydrolyzing ATP) than a smaller one. The two isoforms of activase differ only at their carboxyl termini [16–18]. It has been demonstrated that the sensitivity of the larger protein to ADP inhibition is diminished by the thioredoxin *f*–mediated reduction of a disulfide formed by two cysteine residues in the carboxyl terminus of this isoform. Also, the redox changes in the large isoform tightly regulate ATP hydrolysis of both isoforms through cooperative interactions in the activase complexes [16,18]. In this way, the function of Rubisco activase is synchronized with the rest of the synthetic phase and optimized in accordance with light conditions and levels of different metabolites.

2.3 PHOTORESPIRATION—THE C$_2$ OXIDATIVE CARBON PATHWAY OF PHOTOSYNTHESIS

Rubisco displays a dual catalytic ability, after which it has been defined as a "schizophrenic" enzyme [19]. It catalyzes two conflicting reactions on the Rub-1,5-bisP substrate as, in addition to the above described carboxylase activity (see Figure 2.1 and Table 2.1), it also exhibits an oxygenase activity:

$$\text{Rub-1,5-bisP} + O_2 \rightarrow 3\text{PGA} + 2\text{-phospho-glycolate} \tag{2.15}$$

The formation of 2-phospho-glycolate initiates an oxidative process (opposite to photosynthetic carbon assimilation) that constitutes a metabolic pathway that spreads between chloroplast, cytosol, and mitochondria; and is named photorespiration because it finally drives to the release of CO_2 during the day [23]. Since CO_2 and O_2 are alternative (competitive) substrates, the occurrence of one or another of the reactions is dependent on their relative levels in the place where Rubisco is found. In such a way, the enzyme exhibits characteristics of confused substrate specificity. Although this characteristic, together with Rubisco being a slow catalyst, could be considered as disadvantages and inefficiency in an enzyme, it has been pointed out that all Rubiscos may have already been nearly completely optimized during evolution [24].

In C_3 plants, photosynthesis takes place mostly in leaf mesophyll cells and, more specifically, within the chloroplasts of these cells, in plastids that contain all the machinery to carry out both the light and the synthetic phases of this metabolism. In these plants, two main alternatives can take place during the day: the occurrence of the productive carbon assimilation process that renders effective synthesis of carbohydrates, or the wasteful metabolism of photorespiration that oxidizes carbon [23]. When the concentration of CO_2 is relatively higher than that of O_2, carboxylation of Rub-1,5-bisP is favored and carbon assimilation is conducted through the reductive RPP cycle. Conversely, when the CO_2:O_2 ratio decreases, the oxygenase activity of Rubisco predominates, driving Rub-1,5-bisP to phostorespiration. This oxidative scenario becomes particularly important at temperatures above 30°C because the decrease in solubility of O_2 with temperature is relatively lower than that of CO_2. Consequently, the productivity of C_3 plants is critically challenged under certain environmental conditions that stress their ability to fix carbon via photosynthesis [7,25].

2.4 THE C_4 PATHWAY OF PHOTOSYNTHESIS

2.4.1 REACTIONS AND OPERATION OF C_4 METABOLISM

The evolutionary pressure for a higher photosynthetic efficiency in warmer and dryer habitats that constrain plant growth led to the diversification of photosynthesis into variants in which photorespiration was greatly reduced or absent. One of these variants is the C_4 photosynthetic pathway, which has existed for at least 12–13 mya but gained in importance after a global CO_2 decline 7 mya ago [6]. Plants able to perform the C_4 pathway of photosynthesis have two types of photosynthetic cells in their leaves: mesophyll and bundle-sheath cells [2,6]. This particular anatomy, named Kranz anatomy, is strictly linked to the functionality of the C_4 metabolism. In fact, the whole process involves not only a series of biochemical steps but also implies the coordinate operation of chemical reactions and cell physiology taking place in the different cells. In recent years, however, it has been confirmed that Kranz anatomy is not essential for C_4 photosynthesis [26,27], and many grasses departing from or lacking classical Kranz anatomy have been characterized [28,29]. Ultimately, the C_4 syndrome is an adaptive mechanism to pump CO_2 and thus increase the CO_2:O_2 ratio in the photosynthetic cell where Rubisco localizes, a feat that greatly decreases photorespiration. Figure 2.2 illustrates the operation of the pump mechanism. Atmospheric CO_2 is primarily fixed in mesophyll cells into phosphoenolpyruvate (PEP), producing a C_4 compound from which the metabolism receives its name. This is the molecular species that is transported to the bundle-sheath cells where it is decarboxylated. The CO_2 thus released is definitively assimilated via Rubisco and the RPP cycle, whereas the other product, a C_3 metabolite, is transported back to mesophyll cells to regenerate PEP. The C_4 metabolism was discovered by Hatch and Slack after performing experiments of "pulse" and "chase" using $^{14}CO_2$ (see above), similar to those made by Calvin and Benson. In this way, it was found that several grasses exhibited particular patterns in the distribution of the radioactive mark. Hatch and Slack went on to demonstrate that these grasses possessed an additional pathway for carbon fixation.

The operation of the C_4 metabolism is better understood if it is divided into three stages: a carboxylative, decarboxylative, and regenerative stage (Table 2.2). The pathway starts with the fixation of inorganic carbon to render a C_4 dicarboxylic acid (oxaloacetic acid, OAA) in a reaction catalyzed by PEP-Case. This Mg-dependent enzyme localizes in the cytosol of mesophyll cells, uses

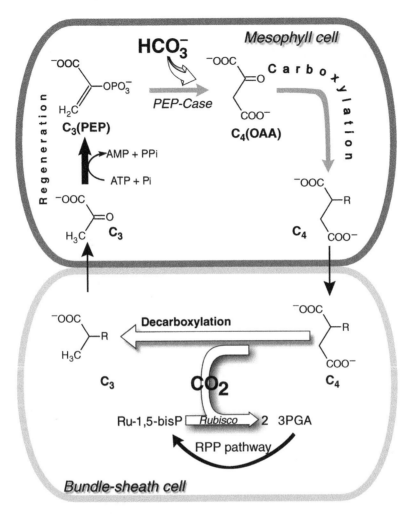

FIGURE 2.2 Diagram of C_4 metabolism, depicting the two cell layers arrangement. C_3 and C_4 refer to three- and four-carbon compounds involved in the pathway (see text).

HCO_3^- rather than CO_2 as a substrate and displays only a carboxylase activity that is not affected by O_2. The carboxylative stage is completed within the mesophyll cell through the conversion of OAA to a more stable compound by two different routes. One comprises its reduction to malate by a chloroplastic NADP-MDHase and the other is its transamination to aspartate by cytosolic Asp-ATase (Table 2.2). The corresponding products (malate or aspartate) are transported to the bundle sheath cells for the decarboxylative stage.

The variants that exist for the decarboxylation step in C_4 metabolism gives rise to three types of C_4 plant species, named after the enzyme catalyzing the decarboxylative reaction as NADP-ME, NAD-ME, and PEP-CKase types. In the first, the main pathway for OAA is the reduction to malate, which is decarboxylated in bundle sheath chloroplasts by NADP-ME (Table 2.2). In the NAD-ME type, OAA is transaminated to aspartate in the mesophyll cell and transported to bundle-sheath mitochondria, where it is converted back into OAA by an isoform of the transaminase (Equation 2.18 in Table 2.2) and then reduced to malate by a mitochondrial NAD-MDHase (Equation 2.19 in Table 2.2). Malate is the substrate of decarboxylating mitochondrial NAD-ME (Equation 2.22 in Table 2.2). In the PEP-CKase variant, OAA is mainly derived to aspartate, which is transaminated back to the keto acid (Equation 2.18 in Table 2.2, in both the cases) and finally decarboxylated by PEP-CKase (Equation 2.21 in Table 2.2) in the bundle-sheath cytosol.

TABLE 2.2

Reactions and Stages Involved in the C_4 Pathway for Carbon Assimilation

Stage	Reaction	Equation	Enzyme
Carboxylative	$PEP + HCO_3^- \rightarrow OAA + Pi$	2.16	PEP-Case, EC 4.1.1.31
	$OAA + NADPH + H^+ \rightleftharpoons \text{L-Mal} + MADP^+$	2.17	NADP-MDHase, EC 1.1.1.82
	$OAA + \text{L-Glu} \rightleftharpoons \text{L-Asp} + \alpha\text{-KGA}$	2.18	Asp-ATase, EC 2.6.1.1
	$OAA + NADH + H^+ \rightleftharpoons \text{L-Mal} + NAD^+$	2.19	NAD-MDHase, EC 1.1.1.37
Decarboxylative	$\text{L-Mal} + NADP^+ \rightleftharpoons Pyr + CO_2 + NADPH + H^+$	2.20	NADP-ME, EC 1.1.1.40
	$OAA + ATP \rightleftharpoons PEP + CO_2 + ADP$	2.21	PEP-CKase, EC 4.1.1.49
	$\text{L-Mal} + NAD^+ \rightleftharpoons Pyr + CO_2 + NADH + H^+$	2.22	NAD-ME, EC 1.1.1.39
Regenerative	$PEP + ADP \rightarrow Pyr + ATP$	2.23	Pyr-Kase, EC 2.7.1.40
	$Pyr + \text{L-Glu} \rightleftharpoons \text{L-Ala} + \alpha\text{-KGA}$	2.24	Ala-ATase, EC 2.6.1.2
	$Pyr + Pi + ATP \rightleftharpoons PEP + AMP + PPi$	2.25	Pyr-Pi diKase, EC 2.7.9.1
	$PPi + H_2O \rightarrow 2 Pi$	2.26	Inorganic pyrophosphatase, EC 3.6.1.1

Note: PEP, phospho-enolpyruvate; OAA, oxaloacetic acid; PEP-Case, phosphoenolpyruvate carboxylase; L-Mal, L-malic acid; NADP-MDHase, NADP-dependent malate dehydrogenase; L-Glu, L-glutamic acid; L-Asp, L-aspartic acid; Asp-ATase; aspartate aminotransferase; NAD-MDHase, NAD-dependent malate dehydrogenase; Pyr, pyruvic acid; NADP-ME, NADP-dependent malic enzyme; PEP-CKase, phosphoenolpyruvate carboxykinase; NAD-ME, NAD-dependent malic enzyme; Pyr-Kase, pyruvate kinase; Ala-ATase, alanine aminotransferase; Pyr-Pi diKase, pyruvate-orthophosphate dikinase.

In all the cases, the CO_2 released in the decarboxylative stage serves as the substrate of Rubisco, entering the RPP cycle operating in bundle-sheath chloroplasts. The C_3 compound produced in NADP-ME type C_4 plants and shuttled back to the mesophyll is pyruvate. In the other two types of C_4 plants, the principal metabolite transferred is alanine. In PEP-CKase type, the decarboxylation product, PEP, is converted into pyruvate and alanine by the sequential action of Pyr-Kase and Ala-ATase (see Table 2.2, Equations 2.23 and 2.24, respectively). Ala-ATase is also responsible for the reconversion of alanine into pyruvate in the cytosol of mesophyll cells. For all the variants of C_4 metabolism, regeneration of PEP from pyruvate occurs in mesophyll chloroplasts by Pyr-Pi diKase (Equation 2.25) in Table 2.2, in a reaction that is displaced from equilibrium by the hydrolysis of PPi by inorganic pyrophosphatase (Equation 2.26 in Table 2.2).

It must be noted at this point that C_4 photosynthesis is a phenomenon that is not taxonomically determined, as it has arisen many times during evolution in distantly related species [30].

2.4.2 REGULATION OF THE C_4 METABOLISM

Efficient operation to assimilate CO_2 by the C_4 pathway of photosynthesis requires an adequate activity of the enzymes, the congruent flux of metabolites within and between photosynthetic cells, and coordination with the operation of the Calvin–Benson cycle that ultimately incorporates inorganic carbon into organic metabolites [2,4,31]. The C_4 route is operative in light, and different mechanisms for its regulation have been identified. These mechanisms include *de novo* synthesis of proteins, and posttranslational modification as well as allosteric regulation by metabolites of enzymes catalyzing key steps of the pathway [1,2]: PEP-Case, NADP-MDHase, NADP-ME, PEP-CKase, and Pyr-Pi diKase.

The transport of metabolites is critical for the adequate operation of C_4 metabolism [1]. The intercellular diffusion of intermediates operates through plasmodesmata connecting the cells [31].

At the intracellular level, C_4 plants possess specific translocators in chloroplasts and mitochondria with unique or altered properties to allow the transport of intermediates of the C_4 metabolism. The general characteristics of these translocators favor a greater metabolite transport rate compared with C_3 species [31]. Key features in the transport system in C_4 grasses include (i) a specific protein carrying pyruvate in mesophyll chloroplasts by a light dependent process [1,2]; (ii) the transport of PEP mediated by the Pi-triose-phosphate translocator of the chloroplast envelope [31,32]; and (iii) a dicarboxylate carrier transporting malate and OAA, and a very active translocator specific for OAA located in the mesophyll chloroplast envelope [31].

The step of carboxylation catalyzed by PEP-Case is a point of regulation in the C_4 pathway of photosynthesis [33]. Allosteric regulation of the enzyme from C_4 plants is mainly exerted by glucose-6-phosphate (Glc6P) and triose-phosphates behaving as activators, and by malate and aspartate, which are inhibitors [34]. Regulation of the enzyme by these metabolites is pH dependent, and for malate the inhibitory effect is greater at pH 7.0 than at 8.0. A cross talk between the allosteric activators and inhibitors of the enzyme, as Glc-6-P decreases the inhibition caused by malate, has been identified. A main mechanism for regulating PEP-Case is its phosphorylation by a soluble and specific protein-serine kinase and dephosphorylation by protein phosphatase 2A [33,35,36]. The phosphorylation state of PEP-Case is largely controlled by the activity of the specific kinase, more active during the light period in C_4 plants. As a consequence of phosphorylation of a serine residue located at the N-terminal region, PEP-Case becomes less sensitive to inhibition by malate, more susceptible to being allosterically activated by Glc-6-P, and exhibits higher activity when assayed under conditions occurring physiologically [33,36].

NADP-MDHase is also an enzyme of C_4 metabolism under regulation [1,2,37]. The homotetrameric enzyme from mesophyll chloroplasts has an alkaline pH optimum, is strongly inhibited by $NADP^+$, and is light-activated by a mechanism involving reduction/oxidation of cysteine residues mediated by ferredoxin and thioredoxin m (see Section 2.2.2). The rates of reversible activation and inactivation of this enzyme are strongly influenced by $NADP^+$ and NADPH [37]. The overall picture shows NADP-MDHase active in the light and inactive in the dark. Activation occurs after reduction of two disulfides per enzyme monomer, one located at the N-terminus and the other at the C-terminus of the polypeptide. Complete activation of the enzyme requires reduction of both disulfides.

Different studies have shown the regulatory properties of the enzymes involved in the decarboxylative stage of C_4 metabolism [1,2,38–41]. Different isoforms of NADP-ME have been characterized, each one playing specific metabolic functions [39,40]. The activity for decarboxylation of the enzyme involved in C_4 metabolism is substantial at the pH and Mg^{2+} level found in the bundle-sheath chloroplast in the light period, and it is significantly higher than that of the enzyme from C_3 plants. NADP-ME involved in the C_4 route can adopt different oligomeric states, with Mg^{2+} favoring aggregation of the subunits to form a more active enzyme. The interconversion of the oligomers seems to be responsible for hysteretic behavior found for the purified enzyme [40]. It has been described that NADP-ME from leaves of the C_4 plant maize undergoes light activation, with marked changes in regulatory properties, through a mechanism that encompasses reduction of dithiols [38]. In contrast, PEP-CKase was found at a phosphorylated state in darkened leaves but dephosphorylated in illuminated leaves of C_4 plants. This mechanism of posttranslational modification was identified as having physiological significance, with the dephosphorylated enzyme (light form) being more active and exhibiting higher affinity toward substrates and altered sensitivity to regulation by adenylates [41,42].

The stage regenerating PEP is rate-limiting of the C_4 metabolism, with Pyr-Pi diKase being strictly regulated by a posttranslational reversible phosphorylation [1,43,44]. The enzyme undergoes dissociation induced by cold conditions or in the absence of Mg^{2+}, and is also affected by changes in pH and ATP levels in the chloroplast stroma. In the light, a threonine residue of the active site of the dikinase is phosphorylated by the so-called Pyr-Pi diKase regulatory protein, which displays a bifunctional kinase/phosphatase activity. This regulatory protein exhibits odd characteristics not only considering its dual kinase and phosphatase activities, but because each activity occurs with unique features. Thus, the regulatory polypeptide utilizes ADP instead of ATP to phosphorylate the threonine residue in the dikinase; and the mechanism for dephosphorylation comprises not a hydrolytic but a phosphorolytic mechanism, generating PPi [43,44]. The single gene coding for the Pyr-Pi diKase regulatory protein in maize leaves has been recently cloned [43], which allows the expression of the recombinant polypeptide that is critical for a full characterization of this protein with peculiar functional properties.

2.5   RESPIRATION OF ASSIMILATED CARBON

2.5.1   CHARACTERISTICS OF PLANT RESPIRATION

Plant respiration encompasses the processes by which the assimilated carbon is used with different purposes. These include the synthesis of precursors for a variety of biosynthetic pathways, the optimization of photosynthesis, and the more obvious ATP synthesis, among others. Thus, plant respiration is not simply a reversal of photosynthetic carbon fixation, but constitutes the nucleus of secondary metabolism and is essential for the coordinated flow of carbon and metabolites throughout the plant. A comprehensive study of plant carbohydrate metabolism must convey the notion that biomass generation by photosynthetic organisms is the balance between carbon assimilation and respiration. The importance of this is highlighted by the fact that about half the total carbon fixed by plants is released upon respiration, constituting a huge proportion of total CO_2 emissions (as much as 10 times the amount generated by human activities). As a consequence, small changes in the carbon balance by plants alone may have a vast impact at a global dimension. With respect to the relevance of grasses in this process, it must be taken into account that grasslands cover approximately 40% of the global land surface [45,46] and therefore constitute a key component of the C balance in the biosphere.

Respiration is achieved by the glycolitic breakdown of carbohydrates, the tricarboxylic acid (TCA) cycle, and the mitochondrial electron transport chain (miETC). Although the basic organization apparently emulates classical animal or yeast respiration, research within the past 20 years has produced a very different picture. That is, plant respiration differs in many important aspects from other organisms. To name a few, some glycolitic enzymes are unique to plants; plants have the ability to use PPi instead of ATP as phosphoryl donor; mitochondria can respire substrates other than pyruvate; and nonenergy conserving bypasses in some steps of glycolysis and the miETC have been evolved.

2.5.2   PLANT GLYCOLYSIS: DISTINCTIVE REACTIONS AND REGULATION

First of all, it must be brought to mind that plants contain two sets of glycolitic enzymes—one in the cytosol and the other in the plastids. The focus of this revision will be on the aspects of cytosolic glycolysis that clearly differentiate it as a plant specific pathway.

Plants carry out the degradation of sugars using the same set of 10 reactions as other organisms, with a few key variations in some reactions and enzymes. Owing to the particularity of photosynthetic metabolism, the substrates of glycolysis come from the hexose phosphates or triose phosphates pool instead of glucose (Figure 2.3).

The first step where plant glycolysis displays a striking divergence from other organisms is the conversion of Fru6P to Fru1,6P$_2$. This step is irreversibly catalyzed by the ATP-dependent phosphofructokinase (ATP-PFK) in all organisms and plants as well. But in addition, a pyrophosphate-dependent enzyme (PPi-PFK) can also catalyze this conversion, albeit in a reversible fashion (Figure 2.3). The use of PPi instead of ATP in this step represents a means of conserving energy, and becomes specially relevant under conditions of Pi starvation, when Pi levels in the cell can shrink up to 50-fold. Pi limitation leads to a rapid and drastic decline in adenylates, which thus becomes a limiting factor for metabolism. Instead, PPi levels are remarkably impervious to Pi nutrition status and tend to be constant regardless of the metabolic status of the cell. Since the discovery of

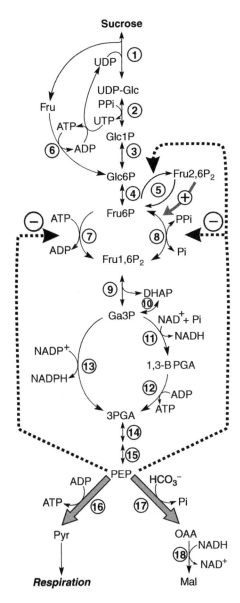

FIGURE 2.3 Reactions of glycolysis as it occurs in the cytosol of plant cells. Not all reactions are depicted. Numbered reactions are catalyzed by (1) sucrose synthase; (2) UDP-glucose pyrophosphorylase; (3) phosphoglucomutase; (4) phosphoglucose isomerase; (5) Fru6P 2-kinase/Fru2,6-bisphosphatase; (6) fructokinase; (7) ATP-PFK; (8) PPi-PFK; (9) aldolase; (10) triose-phosphate isomerase; (11) phosphorylating Ga3P-DHase; (12) phosphoglycerate kinase; (13) NP Ga3P-DHase; (14) phosphoglyceromutase; (15) enolase; (16) Pyr-Kase; (17) PEP-Case; (18) NAD-MDH.

PPi-PFK, the many studies carried out on the enzyme have yielded the view that it can be considered an adaptive enzyme, mainly involved in glycolysis rather than gluconeogenesis, capable of lending the plant the flexibility needed to adjust its metabolism to stress situations.

The second reaction that presents a "plant-type" signature in glycolysis is the conversion of Ga3P into 3PGA. This is an energy conservation site, which in classical glycolysis requires NAD-dependent Ga3P-DHase and 3PGA kinase and yields 3PGA, NADH, and ATP. In plants, this is still the main route to the lower part of glycolysis, but it can be circumvented by the presence of a non-phosphorylating, NADP-dependent Ga3P-DHase (GAPN) (Figure 2.3) [47,48]. Although activity of this enzyme is usually low, it can be increased under certain stress situations. It is believed that GAPN participates in a shuttle system for the transport of photosynthetically generated NADPH from the chloroplast to the cytosol, even though the synthesis of one ATP molecule is resigned [4]. Wheat heterotrophic GAPN has been demonstrated to be subject to a delicate mechanism of regulation involving reversible phosphorylation and binding to 14-3-3 proteins [48], which underscores the weight of the function of GAPN for central metabolism. The next two reactions yield phosphoenolpyruvate (PEP), which is an energy-rich compound used to generate ATP. PEP occupies a central role in cytosolic plant central metabolism, being situated at a metabolic branchpoint and behaving as a regulatory molecule on a number of important processes [49]. PEP partitioning is dependent on the activities of the enzymes that can use it. In the typical glycolytic fate, PEP is used by Pyr-Kase to yield pyruvate and ATP (Figure 2.3). This is the second energy-conserving reaction of glycolysis. Pyruvate can then be taken up and respired by mitochondria. In plants, PEP can also be transformed into OAA by PEP-Case. This enzyme exists in various plant- and tissue-specific isoforms and has multifaceted functions that encompass the housekeeping, anapleurotic replenishment of the TCA cycle to primary CO_2 fixation in C_4 and CAM species [50]. The very abundant C_4 isoform present in warm-season grasses has been dealt with before (see Section 2.4), so the housekeeping C_3 isoform will be discussed here. PEP-Case is induced in plant cells in response to a number of stress situations, such as Pi deprivation and cold stress, and in fact constitutes yet another bypass reaction for classical glycolysis. The action of PEP-Case produces OAA and releases Pi, which in a phosphate-stressed plant can be reused by other metabolic processes (Figure 2.3). Coupled to the very abundant cytosolic NAD-MDHase, PEP-Case behaves as a glycolytic enzyme that produces malate. It is also possible for PEP-Case to act as an ancillary fermentative enzyme in anoxic or cold-stressed tissue by switching glycolysis to the production of OAA and malate (with NAD-MDHase), thereby regenerating NAD^+ [51].

Two other enzymes can compete for cytosolic PEP–PEP phosphatase and PEP-CKase. PEP phosphatase is a vacuolar enzyme that hydrolyzes PEP to pyruvate and Pi. Its activity is increased more than 10-fold in response to Pi stress and since it is strongly inhibited by Pi, it seems conceivable that it acts especially under Pi-starvation conditions. PEP-CKase, although catalyzing a reversible reaction, is probably more linked with gluconeogenesis than it is with glycolysis.

Regulation of plant glycolysis also shows some atypical attributes. The main differences lie in the role of $Fru2,6P_2$, the role of PEP, and the site of primary control. ATP-PFK is insensitive to $Fru26P_2$, while it is a potent activator of PPi-PFK. Both enzymes' activities are downregulated by PEP by different means. The ATP-PFK is allosterically inhibited by this metabolite, whereas for PPi-PFK PEP lessens the activation afforded by, and inhibits the synthesis of, $Fru2,6P_2$ [49,52]. $Fru2,6P_2$ synthesis and degradation is effected by a single protein carrying both the kinase and phosphatase activity acting over the C_2 carbon of the fructose ring [52]. $Fru2,6P_2$ synthesis is inhibited by PEP, PPi, DHAP, and 3-PGA, while activation is afforded by Fru6P. Meanwhile, Pi and Fru6P inhibit the phosphatase activity [52]. Pyr-Kase, in turn, is not affected by $Fru2,6P_2$ or $Fru1,6P_2$. Many plant Pyr-Kases are inhibited by Glu and are activated by Asp [49], while other isozymes are virtually not responsive to single metabolites but rather are regulated in a concerted manner by pH and several metabolite effectors [53]. Since Pyr-Kase is regulated independent of the state of activation of the upper part of glycolysis, and given that PEP is an important inhibitor of the phosphorylation of Fru6P to $Fru1,6P_2$, the regulation of glycolysis proceeds "bottom up," in which the lower part of glycolysis exerts control over the upper part [49] (Figure 2.3). This is exactly the opposite of the regulation found in animals and yeast.

2.5.3 THE TCA CYCLE AND AEROBIC RESPIRATION

The TCA cycle links glycolysis to the aerobic degradation of its products to CO_2 and water. Pyruvate enters the mitochondria through a specific transporter of the inner mitochondrial membrane and is decarboxylated and converted into AcetylCoA by the pyruvate dehydrogenase complex. Additionally, plants can respire other substrates, such as malate and formate and even amino acids, especially under stress conditions [49]. PDC is subject to control by covalent modification. The plant enzyme complex contains a specific protein kinase that almost completely inactivates PDC and a type 2C protein phosphatase that dephosphorylates the complex and restores activity [54]. Inactivation is promoted by ATP and NH_4^+.

The miETC in plants, while structured roughly as in other organisms, exhibits several unique features, the most conspicuous being the existence of non-energy-conserving bypasses to the electron transport chain that drives electrons to O_2.

One prominent peculiarity is the existence of non-proton-pumping NAD(P)H dehydrogenases (npNAD(P)H-DHases), associated with the mitochondrial inner membrane and facing the inner or outer surface [54]. The two outer npNAD(P)H-DHases are rotenone and antimycin A–insensitive, Ca^{2+}-dependent enzymes of uncertain function. The inner npNAD(P)H-DHases are different; one is Ca^{2+}-independent and diphenyleneiodonium-insensitive, while the other is Ca^{2+}-dependent and very sensitive to diphenyleneiodonium. With affinities for NADH that are 10-fold lower than the NADH-DHase from complex I, the probable function of these enzymes is to control NAD(P)H concentration at high NADH levels that could thwart the operation of the TCA cycle. Some of the components of the TCA cycle, particularly the PDC, are very sensitive to inhibition by high NADH. Thus, the continued operation of the TCA cycle under high ATP levels would permit the generation of precursors for the biosynthesis of amino acids [54].

The second bypass involves the alternative oxidase (AOX), a protein of the matrix-side of the inner mitochondrial membrane that can transfer electrons from ubiquinol to O_2. AOX is encoded by a multigene family [55]. The three isoforms identified to date present membrane-embedded and matrix domains with the active site in the matrix-protruding part, and are formed by two subunits. Sulfhydryl residues, located at the N-terminals of each subunit, reversibly link the two subunits in a covalent manner and play a key role in the enzyme's regulation. Since AOX does not pump protons, the energy-conserving sites of complex III and IV of the miETC are bypassed, greatly decreasing the ATP yield. The activity and amount of AOX are tightly regulated in plant cells. Apart from its role in generating heat in some plants, AOX serves a variety of functions in respiratory metabolism. First, AOX can oxidize NAD(P)H during times of low ADP to maintain the TCA cycle active by avoiding NADH inhibition of its components. A very important function of AOX is to limit reactive oxygen species (ROS) formation by preventing overreduction of the miETC [56]. Also, it has been shown that AOX levels are sensitive to a variety of biotic and abiotic stresses [49,54]. The state of reduction of the sulfhydryl groups determines the activation state of AOX—inactive as disulfide and active AOX when reduced [57]. Reduction is performed by thioredoxin h. This mechanism provides a link between the general reduction state of the mitochondria and one of the proteins that have an effect on it. AOX is potently activated by pyruvate [56]. Although it is yet to be demonstrated that pyruvate is a significant effector of AOX *in vivo*, this would provide a means to synchronize the activities of glycolysis, of which pyruvate is a product, and the miETC.

A third component of the miETC that bypasses energy-conservation sites is the uncoupling protein (UCP), which diffuses the proton gradient generated by electron transport [58]. UCP expression levels are dependent on variables such as developmental stage, tissue, and stress [49,54]. Its activity helps keep superoxide levels low and enhances tolerance to H_2O_2 [59].

Through the very specific characteristics that plant respiration has evolved, this metabolism plays a vital role in the optimization of photosynthesis. One of the main roles of mitochondrial function in plants has to do with the oxidation of the photorespiratory product glycine. During active photosynthesis, the massive amounts of glycine produced by photorespiration are oxidized in the mitochondria, producing correspondingly large amounts of NADH. Much of this excess reductant is disposed off by the miETC, and the overflow mechanisms described above are essential for maintaining the respiratory apparatus free of damage by overreduction. The same mechanism also helps the chloroplast photosynthetic electron chain to dissipate the reductive power produced in excess of the needs of the RPP cycle [60]. Finally, it must be mentioned that the TCA cycle provides the carbon skeletons needed for N-fixation and amino acid synthesis, and is thus a major player in the C and N metabolism cross talk.

2.6   SOME CONSIDERATIONS REGARDING THE PHOTOSYNTHETIC CARBON ASSIMILATION PATHWAY AND THE ADAPTATION OF GRASSES TO THE ENVIRONMENT

By offering a way to assimilate CO_2 with a higher efficiency relative to the C_3 pathway, C_4 photosynthesis requires less stomatic aperture. As a consequence, less water will be lost by evapotranspiration and thus C_4 plants will be better suited to prosper in habitats with less water availability. Similarly, the higher photosynthetic efficiency at higher temperatures and illumination rates sets an important criterion for the ecogeographical distribution of grasses. The early recognition that different taxonomic groups are distributed according to environmental cues [61] can be explained with the discovery of the different photosynthetic types. Thus, C_4 grasses predominate in warmer climates and C_3 grasses are prevalent in more temperate regions [62]. However, many of the initial observations on grasses distribution could not be explained by the C_3 versus C_4 syndrome alone. Further research has provided unequivocal evidence that the different C_4 subtypes also have a say in determining the optimal habitat for the different grass species [63,64]. So far, it has become clear that there is a direct correlation between the C_4 subtype distribution and the annual precipitation. Results point to a positive correlation of NADP-ME and negative correlation for NAD-ME variants with annual mean precipitation [63,64]. However, other traits different from the C_4 variant type may affect the ecophysiology of C_4 grasses, because for PEP-CKase the response to the rain regime differs among location [63,64].

Another aspect that has to be taken into consideration in view of the global changes in temperature and CO_2 levels is whether C_4 plants will effectively respond positively. It is conceivable that a C_3 plant should fare better in such an environment due to a more favorable CO_2:O_2 ratio, but it is less clear how a C_4 will behave, given that it is already optimized in this aspect. However, recent research in the area shows signs that C_4 plants will also be affected advantageously by higher CO_2 levels, although this improvement will have the same limitations as in C_3 plants, such as an adequate N supply [65–67]. This is of no minor significance, since much of the land surface is covered by grasses, about half the known grasses are C_4, and C_4 plants contribute to about 20% of the total biomass production [67].

ACKNOWLEDGMENTS

FEP and AAI are members of the Researcher Career of the CONICET. This work was supported by grants PICTO'03 1-13241, PICT'03 1-14733, and PAV'03 137 from ANPCyT to AAI and PIP 6358 from CONICET to AAI and FEP.

REFERENCES

1. Iglesias, A.A., Podestá, F.E., and Andreo, C.S., Structural and regulatory properties of the enzymes involved in C_3, C_4, and CAM pathways for photosynthetic carbon assimilation, in *Handbook of Photosynthesis,* Pessarakli, M. Ed., Marcel Dekker Inc., New York, 1996, p. 481.
2. Bhagwat, A.S., Photosynthetic carbon assimilation of C_3, C_4, and CAM pathways, in *Handbook of Photosynthesis,* Pessarakli, M. Ed., Taylor & Francis Group, Boca Raton, FL, 2005, p. 367.
3. Edwards, G.E. and Walker, D.A., C_3, C_4: *Mechanisms of Cellular and Environmental Regulation of Photosynthesis,* University of California Press, Berkeley, 1983, p. 542.
4. Iglesias, A.A. and Podestá, F.E., Photosynthate formation and partitioning in crop plants, in *Handbook of Photosynthesis, 2nd Ed.,* Pessarakli, M. Ed., CRC Press, Taylor & Francis Group, Boca Raton, FL, 2005, p. 681.
5. Lara, M.V. and Andreo, C.S., Photosynthesis in nontypical C_4 species, in *Handbook of Photosynthesis,* Pessarakli, M. Ed., Taylor & Francis Group, Boca Raton, FL, 2005, p. 391.
6. Edwards, G.E. et al., What does it take to be C_4? Lessons from the evolution of C_4 photosynthesis, *Plant Physiol.,* 125, 46, 2001.

7. Sage, R.F. and McKown, A.D., Is C_4 photosynthesis less phenotypically plastic than C_3 photosynthesis? *J. Exp. Bot.*, 57, 303, 2006.
8. Bassham, J.A., Benson, A.A., and Calvin, M., The path of carbon in photosynthesis, *J. Biol. Chem.*, 185, 781, 1950.
9. Buchanan, B.B. and Luan, S., Redox regulation in the chloroplast thylakoid lumen: a new frontier in photosynthesis research, *J. Exp. Bot.*, 56, 1439, 2005.
10. Gelhaye, E. et al., The plant thioredoxin system, *Cell Mol. Life Sci.*, 62, 24, 2005.
11. Schurmann, P. and Jacquot, J.P., Plant thioredoxin systems revisited, *Annu. Rev. Plant Physiol. Plant Mol. Biol.*, 51, 371, 2000.
12. Sainis, J.K., Dani, D.N., and Dey, G.K., Involvement of thylakoid membranes in supramolecular organisation of Calvin cycle enzymes in *Anacystis nidulans*, *J. Plant Physiol.*, 160, 23, 2003.
13. Suss, K.H. et al., Calvin cycle multienzyme complexes are bound to chloroplast thylakoid membranes of higher plants in situ, *Proc. Natl. Acad. Sci. USA*, 90, 5514, 1993.
14. Winkel, B.S., Metabolic channeling in plants, *Annu. Rev. Plant Biol.*, 55, 85, 2004.
15. Zhang, N., Schurmann, P., and Portis, A.R., Jr., Characterization of the regulatory function of the 46-kDa isoform of Rubisco activase from Arabidopsis, *Photosynth. Res.*, 68, 29, 2001.
16. Wang, D. and Portis, A.R., Jr., Increased sensitivity of oxidized large isoform of Rubisco activase to ADP inhibition is due to an interaction between its carboxyl-extension and nucleotide-binding pocket, *J. Biol. Chem.*, 281, 25, 241, 2006.
17. Salvucci, M.E., Potential for interactions between the carboxy- and amino-termini of Rubisco activase subunits, *FEBS Lett.*, 560, 205, 2004.
18. Portis, A.R., Jr., Rubisco activase—Rubisco's catalytic chaperone, *Photosynth. Res.*, 75, 11, 2003.
19. Hartman, F.C. and Harpel, M.R., Structure, function, regulation, and assembly of D-ribulose-1,5-bisphosphate carboxylase/oxygenase, *Annu. Rev. Biochem.*, 63, 197, 1994.
20. Andralojc, P.J. et al., Elucidating the biosynthesis of 2-carboxyarabinitol 1-phosphate through reduced expression of chloroplastic fructose 1,6-bisphosphate phosphatase and radiotracer studies with $^{14}CO_2$, *Proc. Natl. Acad. Sci. USA*, 99, 4742, 2002.
21. Buchanan, B.B. and Balmer, Y., Redox regulation: a broadening horizon, *Annu. Rev. Plant Biol.*, 56, 187, 2005.
22. Balmer, Y. et al., Proteomics uncovers proteins interacting electrostatically with thioredoxin in chloroplasts, *Photosynth. Res.*, 79, 275, 2004.
23. Tolbert, N.E., The C_2 oxidative photosynthetic carbon cycle, *Annu. Rev. Plant Physiol. Plant Mol. Biol.*, 48, 1, 1997.
24. Tcherkez, G.G., Farquhar, G.D., and Andrews, T.J., Despite slow catalysis and confused substrate specificity, all ribulose bisphosphate carboxylases may be nearly perfectly optimized, *Proc. Natl. Acad. Sci. USA*, 103, 7246, 2006.
25. Brown, N.J., Parsley, K., and Hibberd, J.M., The future of C_4 research-maize, *Flaveria* or *Cleome*? *Trends Plant Sci.*, 10, 215, 2005.
26. Voznesenskaya, E.V. et al., Kranz anatomy is not essential for terrestrial C_4 plant photosynthesis, *Nature*, 414, 543, 2001.
27. Sage, R.F., C_4 photosynthesis in terrestrial plants does not require Kranz anatomy, *Trends Plant Sci.*, 7, 283, 2002.
28. Wakayama, M., Ohnishi, J., and Ueno, O., Structure and enzyme expression in photosynthetic organs of the atypical C_4 grass *Arundinella hirta*, *Planta*, 223, 1243, 2006.
29. Edwards, G.E., Franceschi, V.R., and Voznesenskaya, E.V., Single-cell C_4 photosynthesis versus the dual-cell (Kranz) paradigm, *Annu. Rev. Plant Biol.*, 55, 173, 2004.
30. Giussani, L.M. et al., A molecular phylogeny of the grass subfamily Panicoideae (Poaceae) shows multiple origins of C_4 photosynthesis, *Am. J. Bot.*, 88, 1993, 2001.
31. Leegood, R.C., C_4 photosynthesis: principles of CO_2 concentration and prospects for its introduction into C_3 plants, *J. Exp. Bot.*, 53, 581, 2002.
32. Iglesias, A.A., Plaxton, W.C., and Podestá, F.E., The role of inorganic phosphate in the regulation of C4 photosynthesis, *Photosynth. Res.*, 35, 205, 1993.
33. Chollet, R., Vidal, J., and O'Leary, M.H., Phosphoenolpyruvate carboxylase: an ubiquitous, highly regulated enzyme in plants, *Annu. Rev. Plant Physiol. Plant Mol. Biol.*, 47, 273, 1996.
34. Andreo, C.S., González, D.H., and Iglesias, A.A., Higher-plant phosphoenolpyruvate carboxylase—Structure and regulation, *FEBS Lett.*, 213, 1, 1987.

35. Bakrim, N. et al., Phosphoenolpyruvate carboxylase kinase is controlled by a similar signaling cascade in CAM and C_4 plants, *Biochem. Biophys. Res. Commun.*, 286, 1158, 2001.

36. Nimmo, H.G., Control of the phosphorylation of phosphoenolpyruvate carboxylase in higher plants, *Arch. Biochem. Biophys.*, 414, 189, 2003.

37. Carr, P.D. et al., Chloroplast NADP-malate dehydrogenase: structural basis of light-dependent regulation of activity by thiol oxidation and reduction, *Structure*, 7, 461, 1999.

38. Murmu, J., Chinthapalli, B., and Raghavendra, A.S., Light activation of NADP malic enzyme in leaves of maize: marginal increase in activity, but marked change in regulatory properties of enzyme, *J. Plant Physiol.*, 160, 51, 2003.

39. Drincovich, M.F., Casati, P., and Andreo, C.S., NADP-malic enzyme from plants: a ubiquitous enzyme involved in different metabolic pathways, *FEBS Lett.*, 490, 1, 2001.

40. Edwards, G.E. and Andreo, C.S., NADP-malic enzyme from plants, *Phytochemistry*, 31, 1845, 1992.

41. Leegood, R.C. and Walker, R.P., Regulation and roles of phosphoenolpyruvate carboxykinase in plants, *Arch. Biochem. Biophys.*, 414, 204, 2003.

42. Walker, R.P. et al., Effects of phosphorylation on phosphoenolpyruvate carboxykinase from the C_4 plant Guinea grass, *Plant Physiol.*, 128, 165, 2002.

43. Burnell, J.N. and Chastain, C.J., Cloning and expression of maize-leaf pyruvate, Pi dikinase regulatory protein gene, *Biochem. Biophys. Res. Commun.*, 345, 675, 2006.

44. Chastain, C.J. et al., Pyruvate, orthophosphate dikinase in leaves and chloroplasts of C_3 plants undergoes light-/dark-induced reversible phosphorylation, *Plant Physiol.*, 128, 1368, 2002.

45. White, R., Murray, S., and Rohweder, M., Pilot analysis of global ecosystems: Grassland ecosystems, in *Pilot Analysis of Global Ecosystems*, World Resources Institute, Washington, DC, 1, 2000.

46. Olson, J.S., Watts, J.A., and Allison, L.J., Carbon in live vegetation of major world ecosystems, in *Environmental Sciences Division Publication No. 1997*, Oak Ridge National Laboratory, Environmental Science Division, Oak Ridge, TN, 1983.

47. Bustos, D.M. and Iglesias, A.A., Non-phosphorylating glyceraldehyde-3-phosphate dehydrogenase is post-translationally phosphorylated in heterotrophic cells of wheat (*Triticum aestivum*), *FEBS Lett.*, 530, 1, 2002.

48. Bustos, D.M. and Iglesias, A.A., Phosphorylated non-phosphorylating glyceraldehyde-3-phosphate dehydrogenase from heterotrophic cells of wheat interacts with 14-3-3 proteins, *Plant Physiol.*, 133, 2081, 2003.

49. Plaxton, W.C. and Podestá, F.E., The functional organization and control of plant respiration, *Crit. Rev. Plant Sci.*, 25, 159, 2006.

50. Podestá, F.E., Plant glycolysis, in *Encyclopedia of Plant and Crop Science,* Goodman, T. Ed., Marcel Dekker, New York, 2004, p. 547.

51. Falcone Ferreyra, M.L. et al., Carbohydrate metabolism and fruit quality are affected in frost-exposed Valencia orange fruit, *Physiol. Plant.*, 128, 224–236, 2006.

52. Nielsen, T.H., Rung, J.H., and Villadsen, D., Fructose-2,6-bisphosphate: a traffic signal in plant metabolism, *Trends Plant Sci.*, 9, 556, 2004.

53. Podestá, F.E. and Plaxton, W.C., Kinetic and regulatory properties of cytosolic pyruvate kinase from germinating castor oil seeds., *Biochem. J.*, 279 (Pt 2), 495, 1991.

54. McDonald, A.E. and Vanlerberghe, G.C., The organization and control of plant mitochondrial metabolism, in *Control of Primary Metabolism in Plants,* Plaxton, W.C. and McManus, M.T. Eds., Blackwell Publishing, Oxford, 2006, p. 290.

55. Whelan, J., Millar, A.H., and Day, D.A., The alternative oxidase is encoded in a multigene family in soybean, *Planta*, 198, 197, 1996.

56. Maxwell, D.P., Wang, Y., and McIntosh, L., The alternative oxidase lowers mitochondrial reactive oxygen production in plant cells, *Proc. Natl. Acad. Sci. USA*, 96, 8271, 1999.

57. Vanlerberghe, G.C. and McIntosh, L., Alternative oxidase: from gene to function, *Annu. Rev. Plant Physiol. Plant Mol. Biol.*, 48, 703, 1997.

58. Laloi, M. et al., A plant cold-induced uncoupling protein, *Nature*, 389, 135, 1997.

59. Smith, A.M.O., Ratcliffe, R.G., and Sweetlove, L.J., Activation and function of mitochondrial uncoupling protein in plants, *J. Biol. Chem.*, 279, 51,944, 2004.

60. Raghavendra, A.S. and Padmasree, K., Beneficial interactions of mitochondrial metabolism with photosynthetic carbon assimilation, *Trends Plant Sci.*, 8, 546, 2003.

61. Hartley, W., Studies on the origin, evolution and distribution of the Graminae. 1. The tribe Andropogoneae, *Aust. J. Bot.* 6, 115, 1958.

62. Ehleringer, J.R., Implications of quantum yield differences on the distribution of C_3 and C_4 grasses, *Oecologia*, 31, 255, 1978.

63. Taub, D.R., Climate and the U.S. distribution of C_4 grass subfamilies and decarboxylation variants of C_4 photosynthesis, *Am. J. Bot.*, 87, 1211, 2000.

64. Hattersley, P.W., The distribution of C_3 and C_4 grasses in Australia in relation to climate, *Oecologia*, 57, 113, 1983.

65. LeCain, D.R. et al., Soil and plant water relations determine photosynthetic responses of C_3 and C_4 grasses in a semi-arid ecosystem under elevated CO_2, *Ann. Bot.*, 92, 41, 2003.

66. Anderson, L.J. et al., Gas exchange and photosynthetic acclimation over subambient to elevated CO_2 in a C_3-C_4 grassland, *Global Change Biol.*, 7, 693, 2001.

67. Wand, S.J.E. et al., Responses of wild C_4 and C_3 grass (Poaceae) species to elevated atmospheric CO_2 concentration: a meta-analytic test of current theories and perceptions, *Global Change Biol.*, 5, 723, 1999.

3 Warm-Season Turfgrass Fertilization

George H. Snyder, John L. Cisar, and Dara M. Park

CONTENTS

3.1 WARM-SEASON TURFGRASSES

As used in this chapter, warm-season turfgrasses are those that utilize the C_4 photosynthesis pathway, which allows them to be more efficient photosynthetically under high light intensity and high temperatures than their counterparts, cool-season C_3 plants, which are more efficient under cool and moist conductions and under lower light. Warm-season (C_4) turfgrasses generally have better water use efficiency with greater fixation of carbon dioxide per water molecule lost via transpiration. Moreover, C_4 turfgrasses, unlike C_3 turfgrasses, do not have apparent photorespiration, which is a wasteful process from a physiological perspective. Warm-season turfgrasses generally grow best at temperatures between 27°C and 35°C, whereas shoot growth ceases, leaf chlorophyll is lost, and dormancy occurs when soil temperatures are below 10–13°C [1]. Some of the more commonly used warm-season turfgrasses are bermudagrass (*Cynodon* sp. L.), St. Augustinegrass (*Stenotaphrum secundatum* [Walt] Kuntze), bahiagrass (*Paspalum notatum* Fluegge), zoysiagrass (*Zoysia* sp. L.), centipedegrass (*Eremochloa ophiuroides* [Munro] Hackel), and seashore paspalum (*Paspalum vaginatum* Swartz).

3.2 FERTILIZATION

3.2.1 NITROGEN

Fertilization of the 13 essential elements not supplied by air probably has a bigger impact on the growth of warm-season turfgrasses than any other factor when moisture and temperatures are favorable. Nitrogen (N) fertilization, in particular, can be used to influence green coloration, root and shoot growth, disease and abiotic stress tolerance, recuperative ability, the various developmental processes of plants, and is required in greatest amounts relative to plant-essential elements other than carbon, hydrogen, and oxygen [1]. Nitrogen is found in essential building-block amino acids and thus proteins (e.g., enzymes), which regulate metabolic reactions in the cytoplasm. Root-absorbed nitrate is assimilated into organic compounds in the plant [2]. Nitrate is rapidly reduced in the cytoplasm of both shoots and roots by nitrate reductase forming nitrite, which is highly toxic. Plants overcome this as nitrite is immediately transported to chloroplasts in shoots or plastids in roots and reduced to ammonium in a reaction catalyzed by nitrite reductase and the ammonium is assimilated into amino acids rapidly via sequential actions of glutamine synthetase and glutamate synthase to reduce toxic buildup.

Nitrogen is a constituent of chlorophyll, the light-trapping pigment, nucleic acids, and thus a part of the hereditary material of life. Nitrogen is mobile within the plant, so N deficiency is first noticed in older foliage as pale-green to yellow leaves lacking chlorophyll (chlorosis), accompanied by poor growth. Fertilization programs generally are built around this element. Bermudagrasses have the greatest responses to N without suffering deleterious effects, whereas N rates for bahia-grass should be much lower. St. Augustinegrass and zoysiagrass are intermediate in this regard. Seashore paspalum, which may be grown as an alternative to bermudagrass, generally is considered to require less N than bermudagrass. Excessive N fertilization may result in increased susceptibility to some diseases and insects, resulting in mower-induced scalping, and lead to N leaching.

Bermudagrass N fertilization rates vary with the grass usage. Nitrogen rates typically range between 5 and 10 g m^{-2} month^{-1} during the growing season for heavily trafficked areas such as golf greens and athletic fields. Lower N rates can be used on golf course fairways and roughs, and on home lawns, but seldom fall below 2.5 g of N m^{-2} month^{-1} during the growing season. St. Augustine-grass lawns generally require 2.5 g of N m^{-2} month^{-1} during the growing season, or less, and many receive only 5–10 g of N m^{-2} year^{-1}. Bahiagrass can be maintained with 5–10 g of N m^{-2} year^{-1}, or less. It appears that N-fixing bacteria such as *Azotobacter* sp. form associative relationships with bahiagrass roots, providing small amounts of N to the grass [3,4]. Consequently, bahiagrass often can be maintained with no N fertilization in low-maintenance situations such as roadsides. Centipedegrass also can be maintained with 5–10 g of N m^{-2} year^{-1}, or less, and zoysiagrass with 2.5 g of N m^{-2} month^{-1} during the growing season, or less. Seashore paspalum N fertilization rates generally are somewhat less than those for bermudagrass used for the same purpose. As with all grasses, excessive N fertilization during favorable growing conditions can lead to excessive growth, with resultant scalping during mowing. Seashore paspalum responds rapidly to N fertilization when growing conditions are favorable, and does not recover rapidly from scalping. Thus, N rates may be lower during warm weather than in the spring or fall. For all grasses, when clippings are not removed, N fertilization rates may be reduced 20–30% because N is recycled back into the soil.

Fertilization recommendations for various warm-season turfgrasses are primarily given as ranges because the fertilizer rates change on a seasonal basis to account for growth patterns. Regardless of the species, warm-season turfgrasses are actively grown during late spring and summer months. Active growth is desirable for developing a healthy turfgrass system, which includes the development of the grass inflorescence. The development of inflorescence (or seedheads) is generally a nuisance only for turfgrasses maintained for sports fields and golf courses and may be tolerated in residential lawns, roadsides, and parks. Fertilizing at the lower range of recommended rates during the months when the turfgrass species actively produces seedheads minimizes seedhead development. For example, zoysiagrass produces seedheads in a single pulse, whereas bermudagrass and bahiagrass tend to produce seedheads over a longer period of time [5].

In regions where warm-season grasses are dormant during part of the year and overseeding is not employed, the need for N fertilization will vary throughout the year. When the grass is breaking winter dormancy during the spring and utilizing the remaining stored carbohydrates, applying N too early or at too high rates to promote the growth of new foliage (spring green-up) should be avoided so as to not excessively deplete carbohydrate reserves needed for developing a new root system [6]. If this occurs, while the turfgrass may look established above ground, the root system may be very susceptible to other stresses [5].

Nitrogen fertilization also has traditionally been reduced or suspended in the late fall and winter to improve cold hardiness in regions where freeze damage may occur. However, contrary to traditional concerns over late-season fertilization of bermudagrass, recent research conducted in the transition zone of Virginia showed that late fall applications of N and Fe increased turf color without decreasing cold tolerance of bermudagrass [7]. Munshaw et al. [7] attributed cold tolerance differences in bermudagrasses to the amount of linolenic acid and proline concentrations, and asserted that neither N nor Fe impacted linolenic acid or proline in cells. Especially in regions where winterkill is unlikely, N fertilization can be used to maintain desirable color in warm-season grasses, even though little growth will occur.

Frequency of N application during favorable growing conditions can vary considerably, depending on the N source. The N in water-soluble sources, such as urea, ammonium nitrate, ammonium sulfate, and potassium nitrate, can be rapidly absorbed by the grass, and is easily leached in percolating waters. Consequently, water-soluble sources need to be applied at lower rates and more frequently than less soluble, "slow-release" or "controlled-release" sources, particularly during periods of potentially excess percolation soon after fertilization. Some "best management practices" (BMPs) documents being produced by regulatory agencies call for application of no more than 2.5 g of water-soluble N per square meter in any one application. Fertigation, the application of fertilizers through the irrigation system, often is used on golf courses to supply very low daily rates of N, which reduces growth surges and N leaching relative to infrequent, higher rates of N application [8].

A wide range of controlled-release or slow-release N sources is marketed for turfgrass. Some such as methylene ureas, ureaformaldehydes, natural organic materials, and biosolids (e.g., Milorganite®) release N by microbial decomposition, while others such as isobutylidene diurea (IBDU) are slowly soluble in water and require no microbial action. Soluble N sources, particularly urea, may be coated with sulfur, resins, and other polymers to slow the release of N to the turfgrass roots. When using these materials, the general goal is to apply the amount of N that will be released by the product to equate to the desired monthly rates of N availability. Thus, if a rate of 5 g of N m^{-2} $month^{-1}$ is required, a product releasing N over a 2-month period can be applied at the rate of 10 g of N m^{-2} per application.

Nutritional foliar sprays are commonly used to supplement fertilizations with N and certain other nutrients, particularly micronutrients. The foliar applications are particularly useful when root systems are very limited, such as occurs when nematodes are active or the turfgrass is mowed very closely. Ultradwarf bermudagrasses, which receive close mowing, respond well to foliar fertilization.

3.2.2 PHOSPHORUS

Phosphorus (P) is a major essential element important as a constituent of adenosine triphosphate (ATP), nucleic acids, and for carbohydrate transformation [1]. It affects establishment, rooting, maturation, and reproduction. It is an important component of phospholipids that make up cell membranes, and is required for the proper function of the membranes [9]. It is important for root growth. Nevertheless, warm-season turfgrasses generally do not have unusually high P requirements, although P deficiencies have been observed and are particularly critical during turfgrass establishment. For example, Wood and Duble [10] observed that the greatest effect of St. Augustinegrass P fertilization occurred during the first 8 weeks of establishment, with a much lesser effect thereafter. When produced on a low-soil-test organic soil, Cisar et al. [11] not only obtained better quality ratings for cut St. Augustinegrass sod pieces grown with P fertilization, but also better establishment when the sod was placed on a sandy soil, even though the soil had been P fertilized before sodding.

Phosphorus is mobile within the plant, and older leaves generally express deficiency symptoms before younger leaves. The leaves may take on a purplish coloration, which is accentuated by cool temperatures, in combination with poor growth. The foliage of adequately fertilized turfgrasses usually contains only 10% as much P as N (approximately 25% as much on a P_2O_5 basis), and turf-grass fertilizers have traditionally contained about 25% as much P as N, or more, on a P_2O_5 basis. However, because applied P rapidly converts into slowly soluble, slowly available forms in most soils, the initial P fertilization rate of low-P soils usually must be considerably greater than that of N. But after extensive periods of P fertilization, particularly when clippings are returned, soil-accumulated P may provide sufficient plant-available P to obviate the need for additional fertil-izations. Calibrated soil tests provide the best guide for assessing the P fertilization needs for turfgrasses. Unfortunately, most P soil tests have not been as extensively and thoroughly calibrated for turfgrasses as they have been for major crop species. Nevertheless, when soil tests indicate "adequate" or "medium" levels of P, routine P fertilization can generally be substantially reduced or eliminated with no adverse consequences. Avoiding excessive P fertilization of turfgrasses is particularly important in regions concerned with algal and aquatic plant growth in hydrologically connected surface waters. Periodic analyses of clippings also can be used as a guide for P fertiliza-tion. Tissue P should remain fairly constant over time. Phosphorus adequacy generally occurs in the range of 1.5–5.0 g of P kg^{-1} dry weight of clippings. Consistent increases or decreases over time suggest decreases or increases in P fertilization.

There are situations where P fertilization is especially important [6]. Unless soil P is high, when establishing warm-season grasses vegetatively or by seeding, incorporation of P into the soil at the rate of approximately 10 g of P_2O_5 m^{-2} should prove beneficial to developing roots and tillers. Whenever root growth lags top growth, such as in early spring green-up, or when the grass root system dies back, P fertilization at the rate of 1.0 g of P_2O_5 m^{-2} may be warranted.

3.2.3 POTASSIUM

Potassium (K) is essential in plant growth and developmental processes [1]. It affects carbohydrate synthesis and translocation, amino acid and protein synthesis, acts as an enzyme catalyst of numer-ous plant metabolic reactions, regulates stomatal opening and closing, affects cellular osmotic adjustment, controls the uptake rate of several nutrients, and regulates the respiration rate of plants. When deficient, the margins of older leaves become scorched and senesce prematurely. Turfgrass clippings generally contain approximately half as much K as N, with a normal range of 10–20 g kg^{-1} of dry clippings, and turfgrass fertilizers have traditionally contained less K than N. However, in many instances K fertilization rates have increased in recent years, especially for bermudagrass. It is widely recognized that K-deficient turfgrass is more susceptible to diseases, cold, heat, drought, and perhaps to other stress conditions, and K fertilization in the fall is commonly practiced since this is the time when carbohydrates and proteins are actively being synthesized to prepare the turf-grass for cold hardening. The evidence generated by scientific research does not, however, support the idea that very high rates of K fertilization will confer stress resistance to bermudagrass beyond that possessed by turf that does not display K deficiency [12,13]. In fact, excessive K fertilization can reduce soil and plant tissue calcium (Ca) and magnesium (Mg) [13,14]. Nevertheless, K fertil-ization rates used by turf managers frequently equal or exceed those of N. Some recommend these higher K rates for seashore paspalum, particularly when irrigated with saline water. As with N, K fertilization can be reduced when clippings are returned because K is readily recycled.

Soils containing appreciable clay retain K in plant-available exchangeable forms. On these soils, calibrated soil tests can be used as a basis for K fertilization. Sandy soils, excluding those containing appreciable mica, do not retain K. Consequently, soil testing does not provide useful information for K fertilization on these soils. On very sandy soils, K should be applied reasonably frequently, at rates approximately half that used for N. However, for bermudagrass, and likely for other rhizominous and stoloniferous grasses, K deficiencies arise much more slowly than N deficiencies. This likely occurs because of translocation of K from storage organs to leaves. However, correction of turfgrass K deficiencies occurs quite rapidly following K fertilization. Even on sandy soils, it appears that K can be easily managed without the use of controlled-release fertilizers.

3.2.4 CALCIUM

Calcium (Ca) is very important for plant structure. It is a major constituent of plant cell walls as calcium pectate in the middle lamellae of cell walls, making Ca relatively immobile in the plant [1]. It is also involved in cell division and Ca deficiencies result in reduced mitotic division. Calcium is a catalyst of enzymatic reactions, influences K and Mg uptake, and acts as a neutralizing factor of toxic substances. Calcium is essential for normal membrane function, probably as a binder of phospholipids to each other and or to other proteins [15]. For these reasons, severe Ca deficiency results in newer leaves being distorted, withering, and dying. Warm-season turfgrasses grown in moderate-to high-pH soils generally do not require Ca fertilization both because grasses have a relatively low Ca requirement, in contrast to dicots, and because limestone ($CaCO_3$) is used to correct soil acidity. Furthermore, Ca dominates the soil exchange complex of most moderate- to high-pH soils. In addition, Ca may be supplied in irrigation water. However, Ca fertilization, particularly as a foliar application, may be required for seashore paspalum irrigated with saline water even when there appears to be ample soil Ca.

Soil tests can be used to judge the Ca status of the soil and the need for Ca fertilization. However, caution must be exercised when evaluating Ca soil tests because some extractants, such as Mehlich-I, dissolve Ca compounds in the soil which are not readily plant available [16].

3.2.5 MAGNESIUM

Magnesium (Mg) is a relatively mobile element in plants and is a constituent of chlorophyll, affecting photosynthesis and plant color, functions in P translocation and utilization, and is a cofactor of enzyme systems such as dehydrogenases, carbozylases, and transphosphorylases [1]. The Mg status of most turfgrass soils can also be judged from soil tests. Magnesium deficiencies are rare, though not completely unknown, in turfgrasses. Deficiencies are most likely to occur on very acid soils, particularly if they are also sandy, and on soils that have only received calcitic limestone ($CaCO_3$) for pH correction. Magnesium deficiency symptoms include chlorosis of older leaves and cherry red firing along the margins of leaves. The sufficiency range for turfgrass clippings is 2.0–5.0 g kg^{-1}, dry weight. If deficiency is suspected, as a diagnostic test a response should be observed to an application of magnesium sulfate ($MgSO_4$) at approximately 0.5 g m^{-2}. Seashore paspalum appears to be quite efficient in absorbing most nutrients, but not Mg. When irrigated with saline water, Mg may be required to offset competition from Ca, K, and sodium (Na) ions.

3.2.6 SULFUR

Sulfur (S) is a constituent of several essential amino acids (methionine and cysteine) and is important for bonding in protein structure and in vitamins thiamine and biotin as well as coenzyme A, a compound essential for respiration and breakdown of fatty acids [15]. Deficiency symptoms look much like those of N, except that they occur first on older leaves, since S is not mobile within the plant. In addition to native soil S and atmospheric S deposition, the S needs of plants traditionally have been satisfied by S coincidentally contained in fertilizers. Examples for N, P, and K fertilizers, respectively, are ammonium sulfate ($NH_4)_2SO_4$, a calcium sulfate by-product of superphosphate production $Ca_3(PO_4)_2 + 2H_2SO_4 \rightarrow Ca(H_2PO_4)_2 + 2CaSO_4$, and potassium sulfate K_2SO_4. However, with the expanded use of more concentrated fertilizer sources, such as urea, triple superphosphate, and muriate of potash, less S is coincidentally contained in many fertilizers. Furthermore, vigorous attempts to reduce S emissions from power plants and automobile exhaust have reduced S deposition that provided plant-available S on cropland. Consequently, turf managers need to ensure that adequate S is available. Unfortunately, calibrated soil tests for gauging S availability for turf-grass are rare. A general guideline that should ensure adequacy of S is to provide approximately 10% as much S as N.

3.2.7 MICRONUTRIENTS

Plant availability of the micronutrients iron (Fe), manganese (Mn), zinc (Zn), and copper (Cu) is greatly affected by soil pH, with decreasing availability as pH increases. When high soil pH cannot be adjusted downward, foliar sprays are a good way of providing these micronutrients.

Iron is immobile in plants and is believed to precipitate in cells of older leaves [15]. It is essential for chlorophyll synthesis, and as a constituent of respiratory enzymes such as catalase, peroxidase, and cytochrome oxidase [1] and as an oxidation/reduction electron carrier in proteins [15]. Similar to Fe, Mn is involved in the synthesis of chlorophyll, and acts as a cofactor of enzyme systems including dehydrogenases and carboxylases [1]. It also acts as a structural component of chloroplasts, is important for splitting water during photosynthesis, and has an oxidation/reduction role in electron transport [15]. Iron and Mn deficiencies are frequently observed in warm-season grasses. Zinc and Cu deficiencies occur less often.

Iron and Mn deficiency symptoms can range from overall chlorotic leaves to highly demarcated interveinal chlorosis. In addition to high soil pH, susceptibility to Fe and Mn deficiencies varies both with turfgrass species and with variety. For example, centipedegrass and bahiagrass are particularly susceptible to Fe deficiency at elevated soil pH. The problem is exacerbated by excessive N fertilization, or other factors that accelerate growth. Even in more acid soils, bahiagrass may display Fe deficiency in spring when warm weather stimulates vegetative growth that exceeds the ability of the roots to extract Fe from relatively cool soil. As root growth increases in warmer soil, the Fe deficiency likely will be corrected. St. Augustinegrass, zoysiagrass, and bermudagrass are somewhat less susceptible to Fe deficiencies, but cultivar variations exist [17,18]. Iron can be foliarly applied to overseeded turfgrasses as a means of improving turfgrass color without applying N that might harm the dormant warm-season grass when dormancy breaks.

Manganese deficiency has been corrected in bermudagrass by either lowering soil pH or by fertilizing with Mn [19]. However, in high-pH soils, Fe and Mn fertilizers, even in chelated form, are unlikely to have lasting availability. For this reason, periodic foliar fertilization with these elements generally provides better deficiency correction. Diagnosing micronutrient deficiencies is fairly easy through the use of foliar sprays on relatively small test areas. Alleviation of chlorosis should be apparent in 1 week, or less. A solution containing 1 g of Fe or Mn L^{-1} (5 g of $FeSO_4 \cdot 7H_2O$ or 3 g of $MnSO_4 \cdot H_2O$ L^{-1}) sprayed on foliage, or even drenched into the soil, will provide a response in the presence of a deficiency. Although many soil tests provide information on Fe and Mn availability, few, if any, are calibrated for turfgrass. Accurate micronutrient soil tests generally require different extracting agents than those used for macronutrients, yet to save money and time micronutrient status often is reported on the basis of the same extractant that is used for macronutrients. These tests are not reliable. Tissue analyses can be used for indicating Mn deficiencies, with 25 mg kg^{-1} being a reasonable critical value, but tissue analyses are not reliable indicators of Fe deficiency or sufficiency. Supplemental application of Fe to bermudagrass has provided increased color during fall and improved turf recovery in the spring [7,20].

Zinc is an essential micronutrient involved as a constituent of plant enzymes such as carbonic anhydrase and is a cofactor in the synthesis of plant-growth substances and oxidation/reduction enzymes [1]. It also plays a role in chlorophyll synthesis and is tightly bound to many plant enzymes [15]. Copper (Cu) functions as a component of enzyme systems and plays a role in the synthesis of plant growth substances [1] and electron transport [15]. It is present in cytochrome oxidase, a respiratory enzyme in mitochondria, and plastocyanin, a chloroplast protein. Although molybdenum (Mo) is required in very small amounts, it is required as a cofactor for the enzyme system (nitrate reductase) for nitrate reduction, and it accumulates in leaf blades and actively growing points [1]. Boron (B) is immobile in plants and it accumulates in leaf tips. Boron is believed to play a role in nucleic acid synthesis, which is essential for cell division in apical meristems [15]. Deficiencies of Zn, Cu, Mo, and B rarely occur in warm-season turfgrasses. An exception could be Cu deficiency on virgin soils, especially if they are very sandy or very high in organic matter. Generally, one application of Cu at the rate of 15 kg ha^{-1} will correct the deficiency for many years. Excessive Cu fertilization is to be avoided because toxicity can occur, and the soils can remain toxic for long periods.

3.2.8 BENEFICIAL ELEMENTS

3.2.8.1 Silicon

Silicon (Si) appears to be essential for certain plants such as rice (*Oryza sativa* L.) and *Equisetum* sp. L. [2], but probably not for warm-season turfgrasses. Nevertheless, some warm-season turfgrasses have been shown to respond favorably to Si fertilization. Silicon is naturally absorbed by plants from soil as monosilicic acid, $Si(OH)_4$. Although Si is not considered to be a macronutrient, the content of Si in some plants can be in the range of certain macronutrients. However, the Si content of most turfgrass leaf tissue is much lower than that of rice, for example, generally ranging between 5 and 16 g kg^{-1}. Silicon is transported from roots via active transport primarily to leaves where it forms deposits that help stabilize cell wall and lignin polymers [21]. Because the deposits take time to develop, they are most prominent on older leaves. Cell walls of grass roots, stems, and inflorescence can also accumulate Si [15]. These Si deposits are commonly found where their presence acts as a surface-area defense system (on leaf and stem hairs, and on the outer epidermal walls). The location of the deposits aids in coping with water stress. Plants become more drought tolerant because the location of Si deposits reduces cuticle transpiration losses and strengthens cell walls to prevent the xylem from collapsing. In addition, location of Si deposits guards against the establishment of foliar fungal disease pathogens.

Silicon fertilization has been shown to be an effective means of controlling common diseases on warm-season turfgrasses. In both greenhouse and field studies, Si reduced the progression of gray leaf spot caused by *Magnaporthe grisea* on St. Augustinegrass [22,23]. Tifway bermudagrass has also been documented to show a greater resistance to bermudagrass leaf spot caused by *Bipolaris cynodontis* when fertilized with calcium silicate slag [24]. While warm-season turfgrass responses to Si fertilization have been demonstrated in greenhouse and field trials [25], research has not progressed to the point that routine Si fertilization practices have been established.

3.2.8.2 Sodium

Sodium (Na) is an essential element for bermudagrass (*Cynodon* sp.) and for certain other warm-season (C_4) plants [26]. It can substitute for K, and K-deficiency symptoms of many plants are reduced, not increased, by Na [15,27,28]. It is thought that sodium has a role in the C_4 plant pathway of photosynthesis, particularly in the transport of carbon dioxide to bundle-sheath cells, light harvesting and energy transfer systems, and transport of pyruvate across membranes in mesophyll chloroplasts [15] and the regeneration of phosphoenolpyruvate [2].

Sodium can be detrimental to the maintenance of soil structure, since soil colloids tend to disperse in the presence of excess Na [29]. Of course, very sandy soils that often are used for golf greens and athletic fields have no structure and are largely unaffected by Na. Nevertheless, excessive salinity is detrimental to the growth of warm-season turfgrasses, even those that are fairly salt tolerant, such as seashore paspalum, and deliberate Na fertilization is not recommended for turfgrass maintenance. However, it has been demonstrated that bermudagrass takes up little Na in the presence of K fertilization, and that high rates of K fertilization are not needed to offset moderate amounts of Na in soils [30].

ACKNOWLEDGMENTS

The authors express their appreciation to Dr. R. N. Carrow, University of Georgia, for providing fertilization information about seashore paspalum turfgrass.

REFERENCES

1. Beard, J. B. *Turfgrass Science and Culture.* Prentice-Hall, Inc., Englewood Cliffs, NJ, 1973.
2. Taiz, L. and Zeiger, E. *Plant Physiology,* 3rd Ed. Sinauer Associates, Inc. Sunderland, MA, 2002.
3. Dobereiner, J., Day, J. M., and Dart, P. J. Nitrogenase activity in the rhizosphere of sugarcane and some other tropical grasses. *Plant Soil* 37:191–196, 1972.
4. Baldani, J. L. and Baldani, V. L. D. History of the biological nitrogen fixation research in graminaceous plants: special emphasis on the Brazilian experience. *Ann. Brazilian Acad. Sci.* 77(3):549–579, 2005.
5. Dunn, J. and Diesburg, K. *Turf Management in the Transition Zone.* Wiley, Hoboken, NJ, 2004.
6. Carrow, R. N., Waddington, D. V., and Rieke, P. E. *Turfgrass Soil Fertility and Chemical Problems. Assessment and Management.* Ann Arbor Press, Chelsea MI, 2001. 400 pgs.
7. Munshaw, C. G., Ervin, E. H., Shang, C., Askew, S. D., Zhang, X., and Lemus, R. W. Influence of late-season iron, nitrogen, seaweed extract on fall color retention and cold tolerance of four bermudagrass cultivars. *Crop Sci.* 46:273–283, 2006.
8. Snyder, G. H., Augustin, B. J., and Cisar, J. L. Fertigation for stabilizing turfgrass nitrogen nutrition. *Int. Turfgrass Soc. J.* 6:217–219, 1989.
9. Christians, N. *Fundamentals of Turfgrass Management.* Ann Arbor Press, Chelsea, MI, 1998.
10. Wood, J. R. and Duble, R. L. Effect of nitrogen and phosphorus on establishment and maintenance of St. Augustinegrass. *Texas Agric. Exp. Stn.* PR-3368C, 1976.
11. Cisar, J. L., Snyder, G. H., and Swanson, G. S. Nitrogen, phosphorus, and potassium fertilization for Histosol-grown St. Augustinegrass sod. *Agron. J.* 84:475–479, 1992.
12. Sartain, J. B. Tifway bermudagrass response to potassium fertilization. *Crop Sci.* 42:507–512, 2002.
13. Snyder, G. H. and Cisar, J. L. Nitrogen/potassium fertilization ratios for bermudagrass turf. *Crop Sci.* 40:1719–1723, 2000.
14. Miller, G. K. Potassium application reduces calcium and magnesium levels in bermudagrass leaf tissue and soil. *Hort. Sci.* 34:265–268, 1999.
15. Salisbury, F. B. and Ross, C. W. *Plant Physiology,* 4th Ed. Wadsworth, Inc., Belmont, CA, 1992.
16. Sartain, J. B. Soil testing and interpretation for Florida turfgrasses. IFAS Extension Publication SL 181, University of Florida, Gainesville, 2001.
17. Kurtz, K. W. Use of 59 Fe in nutrient solution cultures for selecting and differentiating Fe-efficient and Fe-inefficient genotypes in zoysiagrass. *Int. Turfgrass Soc. Res. J.* 4:267–275, 1981.
18. McCaslin, B. D., Samson, R. F., and Baltensperger, A. A. Selection for turf-type bermudagrass genotypes with reduced iron chlorosis. *Commun. Soil Sci. Plant Anal.* 12:189–204, 1981.
19. Snyder, G. H., Burt, E. O., and Gascho, G. J. Correcting pH-induced manganese deficiency in bermudagrass turf. *Agron. J.* 71:603–608, 1979.
20. White, R. H. and Schmidt, R. E. Fall performance and post-dormancy growth of "Midiron" bermudagrass to nitrogen, iron, and benzyladenine. *J. Am. Soc. Hortic. Sci.* 115:57–61, 1990.
21. Marschner, H. *Mineral Nutrition of Higher Plants,* 2nd Ed. Academic Press, London, 1995.
22. Datnoff, L. E. and Nagata, R. T. Influence of silicon on gray leaf spot development in St. Augustinegrass. *Phytopathology* 89:S19, 1999.
23. Brecht, M. O., Datnoff, L. E., Kucharek, T. A., and Nagata, R. T. 2004. Influence of silicon and chlorothalonil on the suppression of gray leaf spot and increase plant growth in St. Augustinegrass. *Plant Dis.* 88:338–344, 2004.
24. Datnoff, L. E. and Rutherford, B. A. Accumulation of silicon by bermudagrass to enhance disease suppression of leaf spot and melting out. *USGA Turfgrass Environ. Res.* (online) 2(18):1–6, 2003.
25. Datnoff, L. E. Silicon in the life and performance of turfgrass. *Applied Turfgrass Science* (online) doi:10.1094/ATS-2005-0914-01-RV
26. Brownell, P. F. and Crossland, C. J. The requirement for sodium as a micronutrient by species having the C_4 dicarboxylic photosynthetic pathway. *Plant Physiol.* 49:794–797, 1972.

27. Hylton, L. O., Ulrich, A., and Cornelius, D. R. Potassium and sodium interrelations in growth and mineral content of Italian ryegrass. *Agron. J.* 59:311–315, 1967.
28. Amin, J. V. and Joham, H. E. The cations of the cotton plant in sodium substituted potassium deficiency. *Soil Sci.* 105:248–254, 1968.
29. Brady, N. C. and Weil, R. R. *Nature and Properties of Soils.* Prentice-Hall, Inc. Upper Saddle River, NJ, 1999.
30. Snyder, G. H. and Cisar, J. L. Potassium fertilization responses as affected by sodium. *Int. Turfgrass Soc. Res. J.* 10:428–435, 2005.

4 Growth Responses of Bermudagrass to Various Levels of Nutrients in the Culture Medium

Mohammad Pessarakli

CONTENTS

4.1 INTRODUCTION

Bermudagrass (*Cynodon dactylon* L.), a common warm-season turf species, grows worldwide in a wide range of climates, soils, and environmental conditions. The grass can be manipulated to produce dwarf varieties that require less mowing and can tolerate more stress- (i.e., salinity and drought) tolerant cultivars that require less maintenance (i.e., fertigation, fertilization, and irrigation) and can grow in poor soil conditions. These characteristics can be significantly beneficial for turfgrass growth in arid and semiarid regions where the soils are usually saline/sodic and water for irrigation and other agricultural uses is limited.

Golf-course operators, superintendents, sports-turf managers, and the general public always like and prefer greener grasses. Therefore, there are tremendous amounts of fertilizers, particularly nitrogen fertilizers, applied to turfgrasses. A large quantity of the applied fertilizers (i.e., nitrogen) is in excess of what can be used by the grass and is usually transported downward in the soil to the aquifer. This is on the one hand waste of the fertilizer and consequently an economic loss and on the other hand a major cause of groundwater contamination, resulting in health hazards and environmental pollution. To prevent or minimize these problems, an accurate assessment of the fertilizer (i.e., nitrogen) requirements of turfgrasses is needed.

One of the common methods of assessing fertilizer (i.e., nitrogen) requirements of the plants/grasses is via controlled-environment (greenhouse) hydroponics systems as model studies. On the basis of the results of these types of investigations, recommendations are made for the field applications of the fertilizers. During the course of the study of the hydroponics experiments, usually half-strength Hoagland nutrient solutions [1] are provided to the plants/grasses. The culture solutions are normally changed (renewed) every other week to ensure adequate nutrient elements for normal plant/grass growth, development, and physiological processes. Analysis of the nutrient solutions after two weeks of growth (of either agronomic or non-agronomic plants/crops) and before renewing the culture solutions has shown that there are substantial amounts of unused nutrient elements still remaining in the culture solutions [2–35,38]. Discarding the nutrient solutions during renewal of the culture medium process is a substantial waste of the nutrient elements and contaminates the environment/groundwater.

Therefore, an investigation was conducted to compare growth responses of bermudagrass, a model plant/turfgrass species, grown under different levels of nutrients in the culture medium in terms of shoot and root lengths and dry weights (shoot clippings weight). The objectives of this study were to find the necessary amounts of nutrients in the culture medium for optimum growth of this common turfgrass species.

4.2 MATERIALS AND METHODS

Bermudagrass, cv. Arizona common, was used in a greenhouse experiment to evaluate its growth in terms of shoot and root lengths and dry weights (shoot clippings weight) grown under various levels of nutrients in the culture medium, using a hydroponics technique.

The plants were vegetatively grown in cups of diameter and height, 9 and 7 cm, respectively. Silica sand was used as the plant anchor medium. Each cup was fitted into a 9-cm-diameter holes cut in a rectangular plywood sheet 46 cm × 37 cm × 2 cm. The plywood sheets served as lids for the hydroponics tubs and supported the cups above the solution to allow for root growth, and were placed on 42 cm × 34 cm × 12 cm Carb-X polyethelene tubs containing half-strength Hoagland solution No. 1 [1–35,38]. Eight replications of each treatment were used in a randomized complete block (RCB) design in this investigation.

Plants were allowed to grow in this nutrient solution for 42 days. During this period, the plant shoots (clippings) were harvested weekly to allow the grass to reach full maturity and develop uniform and equal-size plants. The harvested plant materials (clippings) were discarded. The culture solutions were changed biweekly to ensure adequate amounts of essential nutrient elements for normal plant growth and development. On day 42, roots were also cut to obtain plants with uniform roots and shoots prior to the initiation of the various nutrient treatments desired.

The various nutrient levels consisted of full nutrients (FN), half nutrients (1/2 N), quarter nutrients (1/4 N), one-eighth nutrients (1/8 N), and one-sixteenth nutrient (1/16 N). The culture solution levels in the tubs were marked and maintained at the 10-L level by adding the respective nutrient solutions as needed. Culture solutions were changed twice per week to maintain the desired nutrient concentration levels, especially for the low-level nutrient (1/8 and 1/16 N) solutions. The plant shoots (clippings) were harvested weekly for evaluation of the dry-matter production. At each weekly harvest, both shoot and root lengths were measured and recorded. The percent of the visual green cover was also estimated and recorded.

The harvested plant materials were oven-dried at 60°C and the dry weights measured and recorded. The recorded data were considered the weekly plant dry-matter production. At the termination of the experiment, the last harvest, plant roots were also harvested, oven-dried at 60°C, and the dry weights determined and recorded.

4.3 STATISTICAL ANALYSIS

The data were subjected to analysis of variance, using the SAS statistical package [37]. The means were separated, using the Duncan multiple range test.

4.4   RESULTS AND DISCUSSION

The results for both the cumulative and average weekly growth responses were essentially the same and are presented in Tables 4.1 through 4.3.

TABLE 4.1
Bermudagrass Growth Responses (Cumulative) to Various Nutrient Levels of the Culture Medium

Nutrient Levels	Length (cm)		Dry Weight (g)	
	Root	Shoot	Root	Shoot
FN	41.4b	39.1a	1.06a	3.09a
1/2 N	35.5b	42.2a	1.14a	3.35a
1/4 N	54.9a	38.8a	1.28a	3.07a
1/8 N	50.8a	32.3b	0.75b	2.50b
1/16 N	54.0a	25.6c	0.68b	2.07c

Note: The values are means of eight replications. All the values followed by the same letter in each column are not statistically different at the 0.05 probability level.

TABLE 4.2
Bermudagrass Growth Responses (Average Weekly Growth) to Various Nutrient Levels

Nutrient Levels	Length (cm)		Dry Weight (g)	
	Root	Shoot	Root	Shoot
FN	5.2b	4.9a	0.13a	0.39a
1/2 N	4.4b	5.3a	0.14a	0.42a
1/4 N	6.9a	4.8a	0.16a	0.38a
1/8 N	6.4a	4.0b	0.09b	0.31b
1/16 N	6.8a	3.2c	0.08b	0.26c

Note: The values are means of eight replications. All the values followed by the same letter in each column are not statistically different at the 0.05 probability level.

TABLE 4.3
Percent of Canopy Green Cover Responses of Bermudagrass to Various Nutrient Levels

Nutrient Levels	Harvests					
	3	4	5	6	7	8
FN	100	100	100	100	100	100
1/2 N	100	100	100	100	100	100
1/4 N	100	100	100	100	95	90
1/8 N	100	95	90	85	80	75
1/16 N	95	90	85	80	70	65

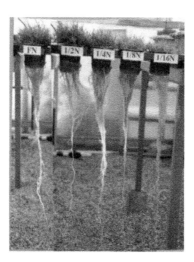

FIGURE 4.1   (Color Figure 4.1 follows p. 262.) Photo of bermudagrass shoot and root growth under various levels of nutrients in the culture medium.

4.4.1   ROOT LENGTH

Root length was stimulated at lower nutrient levels (1/4, 1/8, and 1/16 N) of the culture medium (Tables 4.1 and 4.2; Figure 4.1). This is in agreement with the Sagi et al. [36] and Pessarakli [7,10] reports as well as the common knowledge in plant physiology that under stress conditions (nutrient deficiency stress in the present study) roots grow more, in search of water and nutrients. No significant difference was detected among the root lengths at these levels of nutrients (Tables 4.1 and 4.2). Also, as shown in these tables, statistically, there was no difference in the root lengths at the FN and 1/2 N levels.

4.4.2   SHOOT LENGTH

In contrast to the root length, shoot length decreased significantly at lower levels (1/8 and 1/16 N) of nutrients in the culture medium (Tables 4.1 and 4.2; Figure 4.1). This reduction was more pronounced at the lowest level (1/16 N) of the nutrients. The values for the shoot lengths were statistically the same for the FN, 1/2 N, and 1/4 N levels in the culture medium.

4.4.3   ROOT DRY WEIGHT

Root dry weight responses followed the same pattern as shoot length, significantly decreased at lower nutrient (1/8 and 1/16 N) levels of the culture medium (Tables 4.1 and 4.2). However, statistically, there was no difference detected in the root dry weights at these two nutrient levels. Also, the differences in root dry weights among the FN, 1/2 N, and 1/4 N levels of the culture solutions were not significant.

4.4.4   Shoot Dry Weight

Shoot dry weight responses followed exactly the same pattern as shoot length. The values decreased significantly at lower nutrient (1/8 and 1/16 N) levels of the culture medium (Tables 4.1 and 4.2). The differences in shoot dry weights among the FN, 1/2 N, and 1/4 N levels of the culture medium were not significant. However, reduction in shoot dry weights was more pronounced at the lowest (1/16 N) level of nutrients in the culture medium.

4.4.5   Percent Canopy Green Cover

The percent of canopy green cover decreased only at the lower nutrient (1/8 and 1/16 N) levels (Table 4.3; Figure 4.1). This reduction was more pronounced as the growth period progressed.

4.5   SUMMARY AND CONCLUSIONS

Bermudagrass (cv. Arizona Common) was studied in a greenhouse to evaluate its growth responses in terms of shoot and root lengths and shoot and root dry weights under various levels of nutrients in the culture medium. The plants were grown hydroponically under five levels of nutrients in the growth medium (FN, 1/2 N, 1/4 N, 1/8 N, and 1/16 N), using Hoagland solution No. 1. Plant shoots (clippings) were harvested weekly, oven-dried at 60°C, and the dry weights recorded. At each harvest, both shoot and root lengths were measured and recorded. At the last harvest, plant roots were also harvested, oven-dried at 60°C, and the dry weights determined and recorded. The results show that shoot length, shoot and root dry weights, and the percent of canopy green cover significantly decreased at lower (1/8 and 1/16 N) nutrient levels. This reduction was more pronounced as the growth period progressed. Root length was stimulated at lower (1/4, 1/8, and 1/16 N) nutrient levels of the culture solutions. The differences in shoot lengths and shoot and root dry weights were not significant among the FN, 1/2 N, and 1/4 N levels of the culture solutions. The above results were observed for both the cumulative and weekly growth responses.

Since no significant differences were detected in the study parameters at the FN, 1/2 N, and 1/4 N levels in the culture solutions in this study, with culture solutions renewed biweekly, lower levels of nutrient solutions in the culture medium can be used for hydroponics studies. This will not only save a tremendous amount of nutrients/fertilizers, but also significantly prevent groundwater and environmental contamination via movement of the excess nutrient elements down the soil profile to the aquaphere.

REFERENCES

1. Hoagland, D.R. and D.I. Arnon, 1950. *The Water Culture Method for Growing Plants without Soil.* California Agricultural Experiment Station Circular 347 (Rev.).
2. Marcum, K.B., D.M. Kopec, and M. Pessarakli, 2001. Salinity tolerance of 17 turf-type saltgrass (*Distichlis spicata*) accessions. International Turfgrass Research Conference, July 15–21, Toronto, Canada.
3. Marcum, K.B. and M. Pessarakli, 2000. Tolerance to effluent salinity of 35 bermudagrass turf cultivars. ASA-CSSA-SSSA Annual Meetings, November 5–9, Minneapolis, Minnesota.
4. Marcum, K.B., M. Pessarakli, and D.M. Kopec, 2005. Relative salinity tolerance of 21 turf-type desert saltgrasses compared to bermudagrass. *Hort. Sci.* 40(3):827–829.
5. Pessarakli, M, 1991. Dry matter yield, nitrogen-15 absorption, and water uptake by green beans under sodium chloride stress. *Crop Sci. Soc. Amer. J.* 31(6):1633–1640.
6. Pessarakli, M, 1993. Response of green beans (*Phaseolus vulgaris* L.) to salt stress. In: *Handbook of Plant and Crop Stress* (M. Pessarakli, Ed.), Marcel Dekker, New York, pp. 416–432.

7. Pessarakli, M, 1994. Physiological responses of cotton (*Gossypium hirsutum* L.) to salt stress. In: *Handbook of Plant and Crop Physiology* (M. Pessarakli, Ed.), Marcel Dekker, New York, pp. 679–693.

8. Pessarakli, M, 1995. Selected plant species responses to salt stress. *J. Fac. Sci.*, U.A.E. Univ., 8(2):14–27.

9. Pessarakli, M, 1999. Response of green beans (*Phaseolus vulgaris* L.) to salt stress. In: *Handbook of Plant and Crop Stress*, 2nd Edition, Revised and Expanded (M. Pessarakli, Ed.), Marcel Dekker, New York, pp. 827–842.

10. Pessarakli, M, 2001. Physiological responses of cotton (*Gossypium hirsutum* L.) to salt stress. In: *Handbook of Plant and Crop Physiology*, 2nd Edition, Revised and Expanded (M. Pessarakli, Ed.), Marcel Dekker, Inc., New York, pp. 681–696.

11. Pessarakli, M. and J.T. Huber, 1991. Biomass production and protein synthesis by alfalfa under salt stress. *J. Plant Nutr.* 14(3):283–293.

12. Pessarakli, M., J.T. Huber, and T.C. Tucker, 1989. Dry matter yield, nitrogen uptake, and water absorption by sweet corn under salt stress. *J. Plant Nutr.* 12:279–290.

13. Pessarakli, M., J.T. Huber, and T.C. Tucker, 1989. Protein synthesis in green beans under salt stress conditions. *J. Plant Nutr.* 12:1105–1121.

14. Pessarakli, M., J.T. Huber, and T.C. Tucker, 1989. Protein synthesis in green beans under salt stress with two nitrogen sources. *J. Plant Nutr.* 12:1361–1377.

15. Pessarakli, M. and D.M. Kopec, 2003. Response of saltgrass to environmental stress. United States Golf Association (USGA), New or Native Grasses Annual Meeting, June 17–19, Omaha, NE.

16. Pessarakli, M. and D.M. Kopec, 2003. Survey of growth responses of various turfgrasses under environmental stress conditions. WRCC-11 Turfgrass Annual Meeting, States Turfgrass Research Reports, June 1–3, Logan, UT.

17. Pessarakli, M. and D.M. Kopec, 2004. Growth responses of bermudagrass and seashore paspalum to different levels of FerroGrow, multinutrient fertilizer. *J. Agr. Food Environ.* (*JAFE*) 2(3, 4):284–286.

18. Pessarakli, M., D.M. Kopec, and J.J. Gilbert, 2002. Growth responses of bermudagrass to different levels of nutrients in the culture medium. ASA-CSSA-SSSA Annual Meetings, November 10–14, Indianapolis, IN.

19. Pessarakli, M., D.M. Kopec, and J.J. Gilbert, 2004. Growth responses of bermudagrass and seashore paspalum influenced by FerroGrow, multi-nutrient fertilizer. ASA-CSSA-SSSA Annual Meetings, October 31–November 4, Seattle, WA.

20. Pessarakli, M., D.M. Kopec, and J.J. Gilbert, 2004. Growth responses of bermudagrass to different levels of nutrients in the culture medium. Turfgrass Landscape and Urban IPM Research Summary, Cooperative Extension, Agricultural Experiment Station, The University of Arizona, Tucson, U.S. Department of Agriculture.

21. Pessarakli, M., D.M. Kopec, and J.J. Gilbert, 2005. Competitive growth responses of three cool-season grasses to salinity and drought stresses. ASA-CSSA-SSSA Annual Meetings, November 6–10, Salt Lake City, UT.

22. Pessarakli, M., D.M. Kopec, J.J. Gilbert, A.J. Koski, Y.L. Qian, and D. Christensen, 2005. Growth responses of twelve inland saltgrass clones to salt stress. ASA-CSSA-SSSA Annual Meetings, November 6–10, Salt Lake City, UT.

23. Pessarakli, M., D.M. Kopec, and A.J. Koski, 2003. Establishment of warm-season grasses under salinity stress. ASA-CSSA-SSSA Annual Meetings, November 2–6, Denver, CO.

24. Pessarakli, M., D.M. Kopec, and A.J. Koski, 2004c. Competitive responses of three warm-season grasses to salinity stress. ASA-CSSA-SSSA Annual Meetings, October 31–November 4, Seattle, WA.

25. Pessarakli, M. and K.B. Marcum, 2000. Growth responses and nitrogen-15 absorption of Distichlis under sodium chloride stress. ASA-CSSA-SSSA Annual Meetings, November 5–9, Minneapolis, MN.

26. Pessarakli, M., K.B. Marcum, and D.M. Kopec, 2001. Growth responses of desert saltgrass under salt stress. Turfgrass Landscape and Urban IPM Research Summary, Cooperative Extension, Agricultural Experiment Station, The University of Arizona, Tucson, U.S. Department of Agriculture, AZ1246 Series P-126, pp. 70–73.

27. Pessarakli, M., K.B. Marcum, and D.M. Kopec, 2005. Growth responses and nitrogen-15 absorption of desert saltgrass (*Distichlis spicata*) to salinity stress. *J. Plant Nutr.* 28(10):1–18.

28. Pessarakli, M., K.B. Marcum, D.M. Kopec, and R.L. Ax, 2001. Growth response and nitrogen-15 uptake of bermudagrass under salt stress. ASA-CSSA-SSSA Annual Meetings, October 27–November 2, Charlotte, NC.

29. Pessarakli, M., K.B. Marcum, D.M. Kopec, and Y.L. Qian, 2002. Interactive effects of salinity and Primo on the growth of bluegrass. ASA-CSSA-SSSA Annual Meetings, November 10–14, Indianapolis, IN.

30. Pessarakli, M., K.B. Marcum, D.M. Kopec, and Y.L. Qian, 2004. Interactive effects of salinity and primo on the growth of Kentucky bluegrass. Turfgrass Landscape and Urban IPM Research Summary, Cooperative Extension, Agricultural Experiment Station, The University of Arizona, Tucson, U.S. Department of Agriculture.

31. Pessarakli, M. and T.C. Tucker, 1985. Uptake of nitrogen-15 by cotton under salt stress. *Soil Sci. Soc. Am. J.* 49:149–152.

32. Pessarakli, M. and T.C. Tucker, 1985. Ammonium (^{15}N) metabolism in cotton under salt stress. *J. Plant Nutr.* 8:1025–1045.

33. Pessarakli, M. and T.C. Tucker, 1988. Dry matter yield and nitrogen-15 uptake by tomatoes under sodium chloride stress. *Soil Sci. Soc. Am. J.* 52:698–700.

34. Pessarakli, M. and T.C. Tucker, 1988. Nitrogen-15 uptake by eggplant under sodium chloride stress. *Soil Sci. Soc. Am. J.* 52:1673–1676.

35. Pessarakli, M., T.C. Tucker, and K. Nakabayashi, 1991. Growth response of barley and wheat to salt stress. *J. Plant Nutr.* 14(4):331–340.

36. Sagi, M., N.A. Savidov, N.P. L'vov, and S.H. Lips, 1997. Nitrate reductase and molybdenum cofactor in annual ryegrass as affected by salinity and nitrogen source. *Physiol. Plant.* 99:546–553.

37. SAS Institute, Inc. 1991. SAS/STAT User's Guide. SAS Inst., Inc., Cary, NC.

38. Zhou, M., T.C. Tucker, M. Pessarakli, and J.A. Cepeda, 1992. Nitrogen fixation by alfalfa with two substrate nitrogen levels under sodium chloride stress. *Soil Sci. Soc. Am. J.* 56(5):1500–1504.

5 Weed Management in Warm-Season Turfgrass

Barry J. Brecke

CONTENTS

5.1 INTRODUCTION

A weed is any plant that is objectionable or interferes with the activities or welfare of man [1]. Approximately 30,000 of the 250,000 plant species in the world are considered to be weeds but only about 250 species are troublesome enough to be classified as weeds throughout the world (Table 5.1) [2,3]. Turfgrass quality is determined by uniformity, smoothness, texture, and color. The most undesirable characteristic of weeds in turf is the disruption of visual turf uniformity (texture and/or density) that occurs when weeds with a leaf width or shape, growth habit, or color that is different from the turf are present. Broadleaf weeds such as dandelion (*Taraxacum officinale* Weber in Wiggers), plantains (*Plantago* sp.), and pennywort (*Hydrocotyle* sp.) have leaves with different size and shape than the desirable turf species. Goosegrass (*Eleusine indica* (L.) Gaertn.) and annual bluegrass (*Poa annua* L.) grow in clumps or patches and disrupt turf uniformity. The lighter green color typically associated with certain weeds, such as annual bluegrass, in a golf green often detracts from the playing surface. In addition to being unsightly, weeds compete with turfgrass for light, nutrients, and soil moisture. Weeds can also be hosts for other pests such as diseases, insects, and nematodes.

5.2 WEED OCCURRENCE IN TURF

Weeds are often the result of a weakened turf, but are never the cause of the initial growth impairment. Weeds can readily invade low-quality, low-density turf. To reduce the potential for weed encroachment, turfgrass managers must follow good cultural practices, including proper irrigation and fertility; mow at the height and frequency recommended for the turfgrass species; avoid damage from other turfgrass pests such as diseases, insects, and nematodes; and minimize physical injury and compaction from excessive traffic. Unless proper cultural practices are followed, turfgrass quality can be expected to decline, hence increasing the potential for weeds to invade and become established.

TABLE 5.1
Turfgrass Weed Identification Guides

Title	Authors and Address
Weeds of Southern Turfgrass	Murphy, T.R., L.B. McCarty, D.L. Colvin, R. Dickens, J.W. Everest, and D.W. Hall. 1992. University of Florida Cooperative Extension Service, Gainesville, FL
Color Atlas of Turfgrass Weeds	McCarty, L.B., J.W. Everest, D.W. Hall, T.R. Murphy, and F. Yelverton. 2001. Sleeping Bear Press, Ann Arbor Press, Chelsea, MI
Weeds of the West	Whitsion, T.D., L.C. Burrill, S.A. Dewey, D.W. Cudney, B.E. Nelson, R.D. Lee, and R. Parker. 1991. Western Weed Science Society, University of Wyoming. Pioneer of Jackson Hole, Jackson, WY
Weed Identification Guide	Southern Weed Science Society, Champaign, IL

Many weed species can be described as opportunistic owing to their ability to rapidly invade turf with even the smallest voids. Characteristics of successful weeds include rapid germination under conditions favorable for weed growth along with rapid and aggressive growth habit and growth rate. Most weeds, especially annuals, produce large numbers of seeds. Hard-to-control perennial species often have vegetative reproductive structures (tubers, rhizomes, stolons, bulbs) in addition to producing seeds. These propagules can be disseminated by wind, water, or animals or by human activities, including movement with turfgrass seed, sod, vegetative plant material, soil, or turfgrass equipment.

5.3   PREDOMINANT WEED SPECIES

A diverse group of weeds infests warm-season turfgrass in the southeastern United States. Included in this group are both monocots and dicots and both annuals and perennials. A survey conducted by the Southern Weed Science Society in 2004 [4] listed the 10 most common and 10 most troublesome turfgrass weeds in 11 southeastern states. Of the total of 15 species that were listed as most common by at least two states, crabgrass (*Digitaria* sp.) was listed by all 11 states while annual bluegrass (*Poa annua* L.) and chickweed (*Stellaria* sp.) were listed by nine states (Table 5.2). The most frequent species identified as most troublesome were Virginia buttonweed (*Diodia virginiana* L.), listed by all 11 states, and dallisgrass (*Paspalum dilitatum* Poir) and nutsedge (*Cyperus* sp.), listed by 10 states (Table 5.3).

5.4   WEED IDENTIFICATION

With the variety of warm-season turfgrass weed problems, proper identification of the weed species infesting the turf is imperative for devising an effective weed management program. Since weeds are often indicators of fertilizer, drainage, traffic, irrigation, or pest problems, correct weed identification can aid turf managers in determining underlying causes of certain weed infestations. For example, goosegrass indicates compacted soil, nutsedge suggests drainage problems or

TABLE 5.2
Most Common Warm-Season Turfgrass Weeds

Common Name	Scientific Name	Number of States
Crabgrass	*Digitaria* sp.	11
Annual bluegrass	*Poa annua* L.	9
Chickweed	*Stellaria* sp.	9
Goosegrass	*Eleusine indica* (L.) Gaertn.	8
Henbit	*Lamium amplexicaule* L.	8
Dandelion	*Taraxacum officinale* Weber in Wiggers	8
Nutsedge	*Cyperus* sp.	7
Wild garlic	*Allium vineale* L.	6
Dallisgrass	*Paspalum dilitatum* Poir	6
White clover	*Trifolium repens* L.	5
Prostrate spurge	*Euphorbia humistrata* Engelm. Ex Gray	4
Virginia buttonweed	*Diodia virginiana* L.	4
Lawn burweed	*Soliva pterosperma* (Juss.) Less.	2
Bahiagrass	*Paspalum notatum* Fluegge	2
Broadleaf plantain	*Plantago major* L.	2

Source: Adapted from Webster, T.M., Weed survey—southern states, *Proc. South. Weed Sci. Soc.*, 57, 420, 2004.

TABLE 5.3
Most Troublesome Warm-Season Turfgrass Weeds

Common Name	Scientific Name	Number of States
Virginia buttonweed	*Diodia virginiana* L.	11
Dallisgrass	*Paspalum dilitatum* Poir	10
Nutsedge	*Cyperus* sp.	10
Goosegrass	*Eleusine indica* (L.) Gaertn.	8
Annual bluegrass	*Poa annua* L.	8
Bermudagrass	*Cynodon dactylon* (L.) Pers.	7
Violet	*Viola* sp.	6
Wild garlic	*Allium vineale* L.	5
Torpedograss	*Panicum repens* L.	4
Chamberbitter	*Phyllanthus niruri* L.	3
Nimblewill	*Muhlenbergia schreberi* J.J.Gmel.	3
Tufted lovegrass	*Eragrostis pectinacea* (Michx.) Nees	2
Florida betony	*Stachys floridana* Shuttlex.	2
Crabgrass	*Digitaria* sp.	2
Doveweed	*Murdannia nudiflora* (L.) Brenan	2

Source: Adapted from Webster, T.M., Weed survey—southern states, *Proc. South. Weed Sci. Soc.*, 57, 420, 2004.

TABLE 5.4
Weeds as Indicators of Specific Poor Soil Conditions

Soil Condition	Indicator Weed(s)
Low pH	Red sorrel
High pH	Plantains
Soil compaction	Goosegrass, path rush, knotweed, moss
Low nitrogen levels	Legumes (clover, lespedeza)
High nitrogen levels	Annual bluegrass, ryegrass, chickweed
Poor (sandy) soils	Poorjoe, sandbur
Wet sites	Sedges, algae, moss, alligatorweed
High nematodes	Spurges, Florida pusley, knotweed
Low mowing	Algae, annual bluegrass, chickweed

overwatering, spurge species (*Euphorbia* sp.) suggest nematode infestation, while red sorrel (*Rumex acetosella* L.) indicates low pH (Table 5.4). Continued weed problems can be expected until these underlying growth conditions are corrected.

When the weed management strategy selected includes the use of preemergence herbicides, weed problems need to be anticipated by using historical records that identify the species present and their location. Proper weed identification becomes even more important when postemergence-applied herbicides are a component of the weed management program. Accurate identification allows proper herbicide and rate selection, which leads to more economical and judicious chemical use. Several weed identification aids are available to assist with weed species identification and some are listed in Table 5.1.

Identification begins with classifying the weed type. Monocots include grasses, sedges, and rushes and have only one cotyledon or seed leaf present when seedlings emerge from the soil. Grasses have hollow, rounded, or flattened stems with noticeable nodes, parallel veins in their true leaves, and a two-ranked leaf arrangement. Examples include crabgrass, goosegrass, dallisgrass (*Paspalum dilatatum* Poir.), thin paspalum (*Paspalum setaceum* Michx.), and annual bluegrass. Sedge stems are usually three-sided and solid while stems of rushes are round and solid. Both have a three-ranked leaf arrangement. Broadleaves, or dicotyledonous plants, have two seed cotyledons at emergence and have netlike veins in their true leaves. Broadleaves often have colorful flowers. Examples include clover (*Trifolium* sp.), spurges, lespedeza (*Lespedeza striata* (Thunb. Ex Murray) Hook. & Arn.), plantain, henbit (*Lamium amplexicaule* L.), Florida pusley (*Richardia scabra* L.), and pennywort, among others.

Weeds can be further classified according to life cycle. Annual species complete their life cycles in one growing season. Summer annuals such as crabgrass and goosegrass germinate in spring, grow during summer, and die in fall. Winter annuals such as annual bluegrass germinate in fall, grow during winter, and die when temperatures rise in spring.

Weeds that complete their life cycle in two growing seasons, such as thistles (*Cirsium* sp.), are biennials. The first year's growth is vegetative typically as a rosette with a flower stalk and seed produced the second year. Perennials live for more than 2 years and often reproduce by both seed and vegetative propagules, making them more difficult to control than annual or biennial weeds. Examples of perennials include torpedograss (*Panicum repens* L.), dallisgrass, pennywort, and plantains.

5.5 WEED MANAGEMENT

Weed management employs a combination of practices to prevent weeds from growing with desired plants. This integrated approach utilizes all available techniques to minimize the negative impact of weeds. For turfgrass managers, the goal is to encourage ground cover of the desirable turfgrass while minimizing interference from undesirable plants (weeds). Practices used to restrict the growth and spread of weeds can be grouped into three categories: prevention, eradication, and control.

5.5.1 PREVENTION

Prevention includes all measures taken to limit the introduction, establishment, and spread of weeds into areas not currently infested. This concept is often overshadowed by the development of modern herbicides, which gives the false impression that effective chemical control can always be achieved. Prevention can be the least expensive component of a weed management program, especially when weed management is viewed as a long-term process. Local, state, and federal regulations have been enacted to prevent the spread of weeds. Seed and sod certification programs have been established to ensure that turfgrass propagation materials meet standards to minimize the potential for introduction and spread of weeds.

Methods to prevent weed infestation include using clean seed and vegetative planting material that is free of weed propagules and following good sanitation practices. Sanitation practices are designed to prevent weed movement by equipment or humans. Equipment such as mowers or core aerifiers can carry weed seed and vegetative propagules to weed-free areas. Equipment should be rinsed to prevent transport of weed propagules to adjacent areas. Other sanitation practices include using only weed-free soil when renovating and weed-free materials for topdressing, keeping adjacent fencerows, cart paths, roadways, and ditchbanks weed-free, and keeping the weeds already present from setting seed. Installing filter screens on irrigation inlet pipes when using ponds as a source of irrigation water will also prevent weed seed and vegetative propagules from being dispersed through the irrigation system.

5.5.2 ERADICATION

Some consider eradication to be the ideal method of weed control, but it is rarely achieved. Eradication involves complete removal of all plants, plant parts, and plant propagules from the infested area so that the weed cannot reappear without being reintroduced. Owing to the difficulty and high cost of eradication, this method of weed management is usually restricted to small areas. Eradication may be feasible for small, high-value situations—areas such as golf-course greens. The greatest difficulty is ridding the soil of all weed seeds and vegetative plant parts.

5.5.3 Control

Control encompasses those practices that reduce but do not necessarily eliminate weed infestations. Weed control is a matter of degree, ranging from poor to excellent. The level achieved depends on effectiveness of the weed control tactic and the abundance and tenacity of the weed species present [3]. The four general weed control methods or tactics are mechanical, cultural, biological, and chemical (herbicides).

5.5.3.1 Mechanical

Mechanical control involves the utilization of some type of device to kill or reduce the growth of weeds. Tillage prior to turfgrass establishment, hand removal, and mowing are the mechanical weed-control methods most often used in turfgrass. The tools used generally tear apart, uproot, or cut unwanted plants, leading to desiccation, exhaustion of food reserves, and plant death. Primary tillage (tillage prior to turf establishment) destroys any vegetation present by breaking, cutting, or tearing weeds from the soil, thus exposing the vegetation to desiccation, and by smothering the weeds with the soil. Multiple tillage operations may be required to control perennial weeds that have vegetative propagules. In addition, primary tillage prepares a seedbed and smooths the soil surface for turfgrass establishment. Disadvantages from tillage include increased soil erosion, spread of weed propagules, bringing weed seed to the surface, an increase in evaporative loss from soil surface, and loss of soil structure.

Mowing is part of normal production practices in turf and is often used for cosmetic purposes rather than for control. Mowing is effective on tall growing weeds and can be used to prevent weed seed production. It destroys apical dominance, altering stature from a single upright stem to a plant with several branches and a more prostrate growth habit. Branch development stimulated by frequent mowing can deplete plant carbohydrate reserves and achieve control if done often enough. Mowing is not effective for species such as annual bluegrass, which are able to produce seed below the height of mowing. If a weed exhibits this plastic response to mowing, it may be able to produce enough seed so that the soil seedbank may actually increase in turf with repeated mowing.

5.5.3.2 Cultural

The use of cultural practices as a component of a weed management program is one of the most effective but often overlooked strategies and is the foundation of effective weed control [5–7]. Cultural methods that can aid in weed suppression are often part of the normal turfgrass production process. Any practice that enables vigorous uniform turfgrass establishment and competitive turfgrass growth will usually reduce weed prevalence. It is important to select well-adapted turfgrass species/cultivars, use optimum planting date, maintain the optimum pH and fertility level, use proper mowing height and mowing frequency, irrigate properly, manage thatch, utilize core cultivation as needed, and control traffic.

Cultural practices that provide adequate soil fertility and water availability are necessary to ensure good, vigorous turfgrass growth. Poor irrigation or fertility creates turfgrass stress, which may favor weed infestation. Excess water or fertility may also reduce turf growth and allow water-tolerant or nitrophilous weeds to invade. Since light is required for the germination of many weed species such as crabgrass and goosegrass, cultural practices that increase turf density will prevent light from reaching the soil surface and thus reduce weed infestation. Utilizing the highest cutting height possible and maintaining adequate fertility levels will encourage a high shoot density and will also minimize light penetration to the soil surface.

Winter overseed in a warm-season turfgrass species can aid in winter-weed suppression. Annual bluegrass density was reduced when hybrid bermudagrass (*Cynodon dactylon* [L.] Pers. × *C. transvalensis* Burtt-Davy) was overseeded with perennial ryegrass (*Lolium perenne* L.) [8]. Overseed used in combination with herbicides provided better annual bluegrass control than herbicides alone, because of competition between the winter overseed and annual bluegrass. A negative aspect of overseeding with perennial ryegrass is that it also competes with the bermudagrass turf and may result in thinning of the bermudagrass in the summer following the winter overseeding.

5.5.3.3 Biocontrol

Biological control is the deliberate use of natural enemies to reduce the population of a target weed to below a threshold level [9]. The objective is not eradication but reduction of the population to an acceptable level. In the classical approach, a natural enemy is imported from the native range of the weed species that has been introduced and is to be controlled. Once introduced, the biocontrol agent is self-sustaining. The classical approach is used for control of perennial species in undisturbed areas such as aquatic systems and rangeland but not in managed turfgrass [10]. When the biocontrol agent is used as a bioherbicide, a natural enemy is used from the native range of a native weed, sometimes in a cropping or disturbed system. The bioherbicide is applied using equipment and application techniques similar to those used with a synthetic herbicide. Bioherbicides are targeted toward a wide range of weed problems, including those of turfgrass. For a bioherbicide to be successful, it must deliver acceptable levels of weed control within a relatively short time frame [11]; it must be patentable; it must have sufficient market potential for return on investment; and there must be a cost-effective manufacturing process of a stable, easy-to-use, and reliably effective formulated product. The principal problem with the development of bioherbicides has been reliable field efficacy [10].

Several biocontrol agents have been discovered that either have been commercialized or have the potential for being developed into bioherbicides for weeds that infest turfgrass. The bacterium *Xanthomonas campestris* pv. *poannua* controls annual bluegrass postemergence and has been commercialized for this purpose [12–14]. However, inconsistent levels of success have been reported because of the extreme sensitivity of the bacterium to environmental changes. A mixture of three fungal pathogens (*Drechslera gigantean*, *Exserohilum longirostratum*, and *E. rostratum*) is being developed for grass control in turfgrass. The fungal mixture controls several grass species, including crowfootgrass (*Dactyloctenium aegyptium* [L.] Willd.), guineagrass (*Panicum maximum* Jacq.), johnsongrass (*Sorghum halepense* [L.] Pers.), crabgrass, southern sandbur (*Cenchrus echinatus* L.), Texas panicum (*Panicum texanum* Buckl.), and yellow foxtail (*Setaris glauca* [L.] Beauv.), 83–100% within 2 weeks of inoculation [15]. This mixture is nonpathogenic to several turfgrass species tested. A mixture of pathogens with different modes or sites of action is more effective for providing broadspectrum weed control than a single pathogen, and the risk of weeds developing resistance is minimized.

Another fungal pathogen *Dactylaria higginsii* is being evaluated as a bioherbicide for purple nutsedge [16,17]. Three sequential applications of this potential bioherbicide provide greater than 90% control.

The term *bioherbicides* can be expanded to include other nonsynthetic pesticide products. An example of such a bioherbicide is corn gluten. Corn gluten provides preemergence control of annual grass weeds such as crabgrass (though somewhat less control than synthetic herbicides) without injuring perennial turfgrass species [14,18,19]. Application rates in excess of 1000 kg/ha and high cost limit the use of corn gluten to small areas such as home lawns where homeowners do not wish to use synthetic herbicides.

It is also possible to use salt for selective weed control in salt-tolerant turf species such as seashore paspalum (*Paspalum vaginatum* Sw) [20]. Both grass and broadleaf weeds can be selectively controlled in seashore paspalum with salt and it may be an alternative to traditional herbicide applications in this turf species.

5.5.3.4 Chemical

Even when preventative, mechanical, cultural, and biological weed management practices have been utilized, weeds may still invade turfgrass areas. When this occurs, appropriate chemical weed-control tactics should be employed. Herbicides are phytotoxic chemicals that are used for weed control [21] and are an integral part of modern-day weed management. The discovery of 2,4-D during World War II, a chemical that provided selective control of weeds when applied at relatively low rates, led to the rapid growth of chemical tools for weed management.

5.6 HERBICIDE CLASSIFICATION

To achieve maximum effectiveness from herbicides, it is necessary to select the appropriate herbicide for the weed and turfgrass species present. Some understanding of how herbicides are categorized is necessary to ensure that the proper herbicide is chosen to correct the existing problem. Herbicides can be classified according to selectivity, whether or not they are translocated, and by timing of application.

5.6.1 SELECTIVITY

A selective herbicide is more toxic to some plant species than to others and thus has the ability to control or suppresses certain weeds without seriously affecting the growth of the desirable turfgrass [21]. Selectivity may be due to differences in absorption, translocation, and metabolism between the weeds and the turfgrass. Many of the herbicides used in turfgrass are classified as selective [22]. For example 2,4-D selectively controls many broadleaf weeds without significant injury to the turfgrass. In contrast to a selective herbicide, a nonselective herbicide is phytotoxic to all plant species. Nonselective herbicides are generally used to kill all vegetation in an area prior to turfgrass renovation or establishment, for trimming along walkways, or to spot-treat an area infested with weeds for which no selective herbicide is available. Glyphosate, glufosinate, and diquat are examples of nonselective herbicides. The selectivity of some herbicides can vary depending on the turfgrass species involved. The relative selectivity (tolerance) of warm-season turfgrass species to selected herbicides is listed in Table 5.5 for preemergence herbicides and in Table 5.7 for postemergence herbicides.

5.6.2 TRANSLOCATION

Another way to classify herbicides is by whether or not they move within the plant once they are absorbed [21,22]. Systemic herbicides are translocated from the point of entry (leaves, stems, roots) to the site of action through a plant's vascular system along with nutrients, water, and organic materials necessary for normal growth and development. Translocation can sometimes be slow, so systemic herbicides may require from several days to several weeks to completely kill the target species. For translocation to occur (and for a translocated herbicide to be effective), the plant must be actively growing. Since systemic herbicides often move to all above- and belowground plant parts, many are effective for the control of perennial weed species that possess vegetative reproductive structures such as rhizomes, stolons, tubers, or rootstocks. Most herbicides used in turfgrass are systemic. Systemic herbicides can be either selective or nonselective. Glyphosate is an example of a nonselective, systemic herbicide while 2,4-D, dicamba, imazaquin, and sethoxydim are examples of selective, systemic herbicides.

TABLE 5.5
Warm-Season Turfgrass Tolerance to Preemergence Herbicides (Refer to Herbicide Label for Species-Specific Listings)

Herbicides	Bahiagrass	Bermudagrass[a]	Centipedegrass	St. Agustine-grass	Seashore Paspalum	Zoysiagrass	Overseed Perennial Ryegrass
Atrazine	NR	NR	S	S	NR	I-S	D
Benefin	S	S	S	S	NR	S	NR
Benefin + oryzalin	S	S	S	S	NR	S	NR
Benefin + trifluralin	S	S	S	S	NR	S	NR
Bensulide	S	S	S	S	NR	S	I-S
Bensulide + oxadiazon	NR	S	NR	NR	NR	S	NR
DCPA	S	S	S	S	NR	S	NR
Dithiopyr	S	S	S	S	S	S	I
Ethofumesate[b]	NR	S—dormant	NR	I	NR	NR	S
Isoxaben	S	S	S	S	NR	S	NR
Fenarimol	NR	S	NR	NR	NR	NR	NR
Metolachlor	S	S	S	S	NR	S	D
Napropamide	S	S	S	S	NR	NR	NR
Oryzalin	S	S	S	S	NR	S	NR
Oxadiazon	NR	S	NR	S	NR	S	I
Pendimethalin	S	S	S	S	S	S	I
Prodiamine	S	S	S	S	S	S	I
Pronamide	NR	S	NR	NR	NR	NR	D
Simazine	NR	I	S	S	NR	S	D

Note: S, safe at labeled rates on mature, healthy turf; I, intermediate safety—may cause slight damage to mature, healthy turf; D, damaging—do not use; NR, not registered for use on this turf species.

[a] Non-golf green only.

[b] Ethofumesate is labeled only for dormant bermudagrass overseeded with perennial ryegrass.

Source: Adapted from Unruh, J.B., *Pest Control Guide for Turfgrass Managers*, University of Florida, Gainesville, 2006.

Contact herbicides kill only the plant parts that come in contact with the chemical. They are absorbed into plant tissue but do not move throughout the plant. Therefore, vegetative propagules of perennial species such as rhizomes, stolons, or tubers are not killed. Complete coverage of the weed foliage is required for contact herbicides to be effective, and surfactants, which aid in wetting of the plant foliage, are often added to the spray solution to improve coverage. Contact herbicides kill plants rapidly, often within a few hours of application. Contact herbicides may be either selective or nonselective. Bromoxynil and bentazon are examples of selective contact herbicides while diquat is an example of a nonselective contact chemical.

5.6.3 HERBICIDE APPLICATION TIMING

Herbicides can be classified according to the time they are applied in relation to turfgrass and weed emergence [22]. Preplant herbicides are applied prior to turfgrass establishment. Often, the goal of a preplant application is to kill any vegetation present at the time of application but it can also serve to provide weed control after the turfgrass is propagated from seed, sprigs, or sod. Soil fumigants such as metham-sodium or methyl bromide and nonselective herbicides such as glyphosate or glufosinate are often used as nonselective preplant herbicides.

Preemergence herbicides are applied to turfgrass after establishment but prior to weed-seed germination. Many of the herbicides in this classification require rainfall or irrigation to move them into the soil where the weed seeds germinate. The herbicide then forms a barrier near the soil surface, which prevents weeds from emerging. This group of herbicides prevents cell division during weed-seed germination as the emerging seedling comes into contact with the herbicide layer. Preemergence herbicides generally have little effect on weeds once they have emerged.

Preemergence herbicides form the basis for a successful chemical weed-control program in turfgrass. These herbicides control annual grass and some broadleaf weed species. Preemergence herbicides usually have some residual activity that provides control for 60–120 days after application. The observed length of control depends on parameters such as the herbicide and rate of application used, soil texture, rainfall, and temperature. When a herbicide is applied to coarse-textured, low-organic-matter soils in an area of high rainfall and warm temperatures, it will remain active in the soil for a shorter time than if applied to finer-texture, higher-organic-matter soils in an area of lower rainfall and cooler temperatures. Most preemergence herbicides cannot be applied immediately before or immediately after turfgrass establishment because of potential damage to the turfgrass. The appropriate interval between turfgrass seeding, sprigging, or sodding and preemergence herbicide application is listed on the herbicide label.

Postemergence herbicides are applied to emerged, actively growing weeds. These herbicides are usually absorbed through the foliage, though some postemergence herbicides can also be root absorbed. Several postemergence-applied herbicides are available that provide selective control of broadleaf weeds in warm-season turfgrass species. Fewer herbicides are available that provide postemergence control of grass weed species and some warm-season turfgrass species are damaged by these graminicides. Depending upon the weed species present and the turfgrass species being grown, complete weed control can be achieved with postemergence herbicides, provided that multiple applications are used throughout the year. However, it is often more cost-effective to utilize a program of preemergence followed by postemergence herbicides as needed for season-long weed control. Postemergence herbicides are useful for control of perennial grasses and broadleaf weeds that are not controlled by preemergence herbicides. Some postemergence herbicides such as carfentrazone may also be used on newly established turfgrass.

5.7   SOIL FUMIGATION

Soil fumigation is used for preplant control of a wide range of soil-borne pests, including most weed species along with many nematodes, fungi, and insects. Owing to the expense of soil fumigation, it is more commonly used on intensively managed sites such as golf greens and tees but is also used in other situations where it is vital to ensure genetic purity of turfgrasses [23]. Important considerations before choosing a particular soil fumigant include expense of the treatment, soil moisture level, soil temperature, and time interval available before planting. Site preparation prior to fumigation usually includes minimum 20-cm-deep tillage to facilitate movement of the fumigant. Soil temperature should be above 15°C and the soil should be moist but not saturated to improve fumigant movement and biological activity [5]. Depending on the fumigant used, the interval between application and turfgrass planting can be as long as 3 weeks. Fumigants should be applied only by properly trained and licensed professionals using the appropriate personal protective equipment.

5.7.1 Methyl Bromide

Methyl bromide has been the predominant broad-spectrum soil fumigant used in the turf industry, including golf courses, sports fields, and sod farms [23]. Methyl bromide fumigation is used primarily to eliminate weeds and to ensure genetic purity of turfgrasses. This is especially important during reconstruction and regressing of existing turfed areas. Recently, a phaseout of methyl bromide was begun because of environmental concerns and this fumigant will eventually no longer be available. As the availability of methyl bromide declines, the cost of the fumigant increases and eventually it will no longer be an economic option for turfgrass uses.

Methyl bromide is a colorless, nearly odorless liquid under pressure that forms a vapor when released. It should be injected below the soil surface and immediately covered with gas-impermeable polyethylene tarp. The soil should be moist but not saturated when methyl bromide is applied. Up to a depth of 4 in., soil temperatures should be a minimum of 15°C for best results, and fumigation will not be effective if soil temperatures are below 10°C. Turfgrass can be planted 2–3 days after application. Methyl bromide is a toxic material and should only be applied by licensed professional applicators.

5.7.2 Metham-Sodium

Metham-sodium is a water-soluble, preplant soil fumigant that controls turfgrass pests including weeds. Upon contact with moist soils, it forms the biologically active and volatile compound methyl isothiocyanate (MITC). Because metham-sodium must first react with moist soil to become active, inconsistent weed control has sometimes been observed, especially under extremes of soil temperature and moisture [23]. Metham-sodium is most effective when covered with a polyethylene tarp. Turfgrass can be planted 2–3 weeks after application.

5.7.3 Dazomet

Dazomet is a microgranular material that reacts with soil moisture to produce the same biologically active compound (MITC) as metham-sodium [24]. Consequently, dazomet has the same environmental constraints as metham-sodium. The physical characteristics of dazomet (ultrafine powder) make it difficult to apply and limit its utility. It must be evenly distributed and incorporated into the soil. More consistent results are achieved if the treated area is covered with a polyethylene tarp. Turfgrass can be planted from 10 to 27 days after application, depending on soil temperature.

5.7.4 1,3-Dichloropropene

The area to be treated should be sprayed with nonselective herbicides at least two times (2-week interval) prior to fumigation. Soil should be tilled to a depth of 20 cm so that soil clods are less than 1.25 cm in diameter. Organic debris should be minimal and soil moisture should be at a level that allows the soil to maintain shape when squeezed together in one's hand. This fumigant is applied by injecting the liquid formulation below the soil surface and covering with a gas-impermeable polyethylene cover. The tarp should remain on the treated area for as long as possible and planting should occur no sooner than 3 weeks after treatment.

5.8 PREEMERGENCE WEED CONTROL

Preemergence herbicides are often the foundation for an effective herbicide program for controlling weeds in turfgrass [5,7,25,26]. Timing of application is critical for preemergence-applied herbicides. These herbicides must be in the soil prior to weed-seed germination. If applied after weed emergence, most preemergence herbicides will have little effect. Conversely, if the chemical is applied too early, it may dissipate before the peak germination period ends. Application 1–2 weeks prior to weed-seed germination ensures that the herbicide will have ample opportunity to move into the upper layer of soil prior to weed germination.

A wide array of herbicides is available for controlling weeds preemergence. The first step is to determine which herbicides can be used without causing injury to the turfgrass. Tolerance to herbicides varies among turfgrass species, and relative rankings of warm-season turfgrass susceptibility to preemergence herbicides are listed in Table 5.5. While most of the turf species listed have numerous herbicides that can be applied without causing damage, only a limited number of herbicides are listed as safe for seashore paspalum. Seashore paspalum is a relatively new turfgrass species and has only recently been developed for large-scale use. As a result, seashore paspalum is not listed on many of the common preemergence herbicide labels even though there may be no tolerance issues. As more tolerance information becomes available, seashore paspalum will be added to more of the preemergence herbicide labels.

After turfgrass tolerance to preemergence herbicides has been determined, the next step is to choose the most effective herbicide for the weeds that are expected to be present (remember that these herbicides must be applied prior to weed-seed germination to be effective). Maps from the previous year that delineate locations of various weed species can be useful in selecting the most appropriate preemergence herbicide. The relative effectiveness of herbicides for preemergence control of many weed species that are problems in warm-season turf is listed in Table 5.6.

5.8.1 BROADLEAF WEED CONTROL

Relatively few herbicides are available for broadleaf control in warm-season turfgrass [5,7,25,26]. Atrazine and isoxaben provide good preemergence control of several broadleaf weeds. These can be mixed with a dinitroaniline herbicide, dithiopyr, or oxadiazon to broaden the spectrum of control.

5.8.2 ANNUAL GRASS CONTROL

Most preemergence herbicides are applied in spring and are aimed at summer annuals. Crabgrass, goosegrass, and sandbur are examples of annual grasses that infest warm-season turf and are often controlled with preemergence herbicides. Herbicides in the dinitroaniline family (benefin, oryzalin, oxadiazon, pendimethalin, prodiamine, and trifluralin) are commonly used for preemergence annual grass control, as are herbicides in other families such as bensulide, DCPA, dithiopyr, and metolachlor [27–29].

There are differences in annual grass susceptibility to preemergence herbicides as listed in Table 5.6. For example, if crabgrass is the target species any of the dinitroaniline herbicides or dithiopyr will provide control; however, if goosegrass were the predominant species then oxadiazon would be the herbicide of choice. In addition, crabgrass germinates at cooler temperatures (55°F) than does goosegrass and will emerge 3–4 weeks earlier than goosegrass. Therefore, if the target species is goosegrass it would be best to delay herbicide application 2–3 weeks later than if crabgrass were the species to be controlled.

TABLE 5.6
Efficacy of Preemergence Herbicides

	Atrazine	Benefin	Benefin + Oryzalin	Benefin + Trifluralin	Bensulide	Bensulide + Oxadiazon	DCPA	Dithiopyr	Ethofumesate	Fenarimol	Isoxaben	Metolachlor	Napropamide	Oryzalin	Oxadiazon	Pendimethalin	Prodiamine	Pronamide	Simazine
Perennial Weeds																			
Bahiagrass	F	P	P	P	P	P	P	P	P	P	P	P	P	P	P	P	P	P	P
Bermudagrass	P	P	P	P	P	P	P	P	P	P	P	P	P	P	P	P	P	P	P
Dallisgrass	P	P	P	P	P	P	P	P	P	P	P	P	P	P	P	P	P	P	P
Tall fescue	F	P	P	P	P	P	P	P	P	P	P	P	P	P	P	P	P	G	F
Wild garlic/onion	P	P	P	P	P	P	P	P	P	P	P	P	P	P	P	P	P	P	P
Annual Grasses																			
Annual bluegrass	E	E	E	E	F	F-G	G	G-E	G-E	G	P-F	G	G	G	G	G	E	E	E
Crabgrass	F	E	E	E	E	E	G	G-E	G	P	P	G	G-E	E	E	E	E	F	F
Crowfootgrass	P-F	G	G	G	G	G		G		P	P	F	F	G	G	G	G		
Goosegrass	P	F	F-G	G	F	G	F	F-G		P	P	F	G	F-G	E	F-G	F-G	F	F
Sandbur	P	F	F-G	G	G	F-G	F	F		P	P	F	F	G	F	G	G		F
Broadleaf Weeds																			
Chamberbitter	G									P	G								
Common chickweed	E	G	G	G	P	P	E	G	G	P	E	F	E	G	P	G	G	E	E
Corn speedwell	E	E	E	E	P	P	G	G		P	G-E	G	E	P	G	E	E	G	G
Cudweed	E	P	P	P		P				P	G		G	P	P	G	P	P	
Dandelion	F	P	P	P	P	P				P	G			P	P	P	P	P	P
Dichondra	G	P	P	P	P	P				P				P	P	P	P	P	P
Docks	G	P	P	P	P	P			P	P				P	P	P	P	P	P
Doveweed	G	P	P	P	P	P				P				P	P	P	P	P	G
Florida betony	E	P	P	P	P	P				P				P	P	P	P	P	P

(Continued)

TABLE 5.6
(Continued)

	Atrazine	Benefin	Benefin + Oryzalin	Benefin + Trifluralin	Bensulide	Bensulide + Oxadiazon	DCPA	Dithiopyr	Ethofumesate	Fenarimol	Isoxaben	Metolachlor	Napropamide	Oryzalin	Oxadiazon	Pendimethalin	Prodiamine	Pronamide	Simazine
Ground ivy	E	P	P	P	P	P				P				P	P	P	P	P	P
Henbit	E	G	G	G	P	P	F			P	G		P	G	P	P	G	F	E
Hop clovers	E	P	G	G	P	F		G		P	G			F	G	G	P	F	E
Knotweed	E	P	G	G	G	G		G		P	G		G	F	G	G		G	G
Lespedeza	E	P	P	P	G			G		P				P		G	P		E
Mallow		P	P	P	P	P				P	G			P	P	P	P	F	P
Mock strawberry		P	P	P	P	P				P				P	P	P	P	P	P
Mouseear chickweed	E	E	E	E	P	P	G	G		P	G			P	P	G	G	G	G
Mugwort		P	P	P	P	P	P			P				P	P	P	P	G	P
Mustards	E				G	G-E		G		P	G				E			F	F
Parsley piert	E	P	P	P	E	G-E		G		P					G	P	P	P	G
Pennywort (dollarweed)	E	P	P	P	P	P				P	G			P	P	P	P	P	P
Plantains	G	P	P	P	P	P				P	G			P	P	P	P	P	P
Spurges	E	P	F	F	P	P	F	G		P	G	F	P	F	P	F	F	F	G
Spurweed (burweed)	E	P	P	P	P	P	P			P	E		E	P	P	P	G	P	E
VA buttonweed		P	P	P	P	P				P				P	P	P	P	P	P
Violets	E	P	P	P	P	P				P				P	P	P	P	P	P
White clover	E	P	P	P	P	P				P	G			P	P	P	P	P	G
Yellow woodsorrel (*Oxalis*)	E	P	F-G	F-G	P	F		G		P	G	P		F	G	F-G	F	P	P

Note: E, excellent control (90–100%); G, good control (80–89%); F, fair control (70–79%); P, poor control (<70%). A blank space indicates that weed response is not known.

Source: Adapted from Unruh, J.B., *Pest Control Guide for Turfgrass Managers*, University of Florida, Gainesville, 2006.

It is usually necessary to use sequential repeat applications of preemergence herbicides to achieve season-long control in warm-season turf. Most herbicides begin to dissipate quickly and by 8–12 weeks after application approach the threshold below which they will no longer be effective for annual grass control. Sequential applications must be made before the herbicide dissipates to below this threshold level. Labels of many of the preemergence herbicides will suggest retreatment 6–10 weeks after the initial application.

5.8.3 ANNUAL BLUEGRASS CONTROL

Annual bluegrass is a winter annual than can be very difficult to control in warm-season turf [7,25]. Because of its low growth habit and ability to produce seed close to the soil surface, it can flourish even in closely mowed turf. Annual bluegrass disrupts the texture of desirable turfgrass and negatively impacts the playing surface owing to the tendency of annual bluegrass to form clumps and the large number of seedheads that it produces.

5.8.3.1 Overseeded Bermudagrass

Warm-season turf is often overseeded in fall with cool-season species such as perennial ryegrass. Until recently, it was difficult to find preemergence herbicide treatments that could be applied soon before the overseed was planted. The interval from application to when perennial ryegrass or other overseed species can be safely planted is quite long for some herbicides. Pronamide is one herbicide that can be used for preemergence control of annual bluegrass in overseeded bermudagrass. It should be applied prior to annual bluegrass germination but has a minimum interval of 90 days between application and overseeding. Application should not be made where drainage flows onto overseeded areas or bermudagrass greens because of potential injury to the turfgrass.

Ethofumesate also provides preemergence and early postemergence annual bluegrass control in bermudagrass [30–32]. However, to prevent turfgrass injury, the bermudagrass turf must be completely dormant. If applied in fall before bermudagrass dormancy, an immediate cessation of bermudagrass growth occurs and a delay in spring transition from ryegrass to bermudagrass also occurs. In the more southern areas where warm-season turfgrass species are grown, bermudagrass may never be completely dormant and ethofumesate should not be used for annual bluegrass control.

Fenarimol is a systemic fungicide that reduces the infestation of annual bluegrass [33]. A treatment regimen of three applications spaced 10–14 days apart should be followed with the last application at least 2 weeks prior to overseeding. Fenarimol does not affect either perennial ryegrass overseed or bermudagrass. Annual bluegrass control with fenarimol can sometimes be inconsistent.

The dinitroaniline family of herbicides can provide preemergence annual bluegrass control but the interval between application and overseeding has a wide range—16 weeks for prodiamine, 12 weeks for pendimethalin and oryzalin, and 6 weeks for benefin. Depending on environmental conditions, annual bluegrass control can be inconsistent with these herbicides. In addition, dinitroaniline-resistant annual bluegrass biotypes have been identified in areas where this family of herbicides has been applied for several years [34]. Once a weed population has developed resistance to a herbicide, that herbicide is no longer effective for control of that weed species.

Dithiopyr also controls annual bluegrass when applied preemergence. The interval between application and overseeding is 6 weeks and a second application can be made 120 days after overseeding to provide season-long control. Several herbicides in the sulfonylurea family control annual bluegrass when applied prior to overseeding [8,35]. The intervals between application and overseeding are 7–10 days for sulfosulfuron, 10–14 days for rimsulfuron, 14 days for foramsulfuron, and 21 days for trifloxysulfuron. The intervals between application and when overseed can be safely planted is much shorter for the sulfonylurea herbicides than for the other annual bluegrass preemergence herbicides with a level of control equal to or better than with the other herbicides.

Bispyribac-sodium is the only herbicide that can be applied to both perennial ryegrass overseed and annual bluegrass postemergence. The application should be made no sooner than when the perennial ryegrass is well established (minimum 60 days after perennial ryegrass seeding) but not later than when the first seedheads form on the annual bluegrass. Repeat applications at 14- to 21-day intervals will improve annual bluegrass control [8].

5.8.3.2 Non-Overseeded Bermudagrass

Preemergence annual bluegrass control in non-overseeded areas can be achieved with the same herbicides used for preemergence crabgrass control without concern for the interval between application and seeding that is necessary in an overseeded situation. The only consideration is to make sure that the preemergence herbicides are applied 2 weeks prior to annual bluegrass germination. Pronamide or simazine will provide preemergence and early postemergence control of annual bluegrass in non-overseeded bermudagrass fairways. However, biotypes resistant to simazine have been identified on many golf courses where simazine was used repeatedly over a period of several years [36]. Trifloxysulfuron, rimsulfuron, foramsulfuron, and sulfosulfuron will control annual bluegrass when applied postemergence. These sulfonylurea herbicides will also control many winter broadleaf weeds.

5.9 POSTEMERGENCE HERBICIDES

Postemergence herbicides are utilized for selective control of broadleaf and grass weed species in turf [5,7]. These herbicides are generally more effective when applied to small weeds (two- to four-leaf stage) that are actively growing. Many postemergence herbicides are systemic in nature and are translocated throughout the plant. Most postemergence herbicides are relatively ineffective as preemergence herbicides.

As with preemergence herbicides, turfgrass tolerance to postemergence herbicides must be determined when selecting a herbicide. Turfgrass species tolerance to postemergence herbicides is listed in Table 5.7. Similar to the list of preemergence herbicides, a few postemergence herbicides are listed as safe for seashore paspalum. As more tolerance information becomes available, seashore paspalum will be added to the label of more postemergence herbicides. Tolerance varies widely with turfgrass species other than seashore paspalum. Of the 38 herbicides or herbicide combinations listed in Table 5.7, 27 are listed as safe for bermudagrass and 23 for zoysiagrass (*Zoysia* sp.) but only 9 are listed as safe for St. Augustinegrass (*Stenotaphrum secundatum* (Walter) Kuntze). As a result, it is much more difficult to control weeds postemergence in St. Augustinegrass than in bermudagrass or zoysiagrass.

5.9.1 BROADLEAF WEED CONTROL

There are many herbicides available for postemergence weed control in turfgrass. Table 5.8 lists the effectiveness of commonly used postemergence herbicides for broadleaf weed control in warm-season turfgrass.

Until relatively recently, growth-regulator-type herbicides in the phenoxy family of herbicides, including 2,4-D, mecoprop, dichloroprop, and MCPA, and the benzoic herbicide dicamba applied alone or in various combinations accounted for most of the herbicides applied for postemergence broadleaf control in turfgrass. Triclopyr, clopyralid, and fluroxypyr, which are growth-regulator herbicides in the pyridine family, have been introduced as alternatives to the older herbicides and are less likely to volatilize [37–40]. The growth-regulator herbicides are readily translocated throughout both the annual and perennial broadleaf weeds that they control. One disadvantage with these products is that all—except for clopyralid, which can be used for weed control in ornamentals—will damage nearby trees, shrubs, and flowers if they come in contact with these desirable plants.

Several herbicides in the sulfonylurea family including chlorsulfuron, trifloxysulfuron, and sulfosulfuron provide postemergence control of many broadleaf weeds. In addition, carfentrazone and sulfentrazone, PPO inhibitor herbicides, also provide control of many problem broadleaf weeds. With the more recently registered products added to the list of older herbicides, there are several herbicide options available for postemergence broadleaf weed control in turfgrass.

5.9.2 Grass Weed Control

Annual grasses are generally more easily controlled with preemergence herbicides while postemergence control is often the only option available for the control of perennial grass species. Perennial grasses are the most difficult-to-control weeds in turfgrass due to physiological similarities between most perennial grass weeds and the desirable turfgrass species. Depending on the turfgrass species and the species of perennial grass, it may not be possible to selectively remove the perennial grass from the turfgrass. In this situation, it will be necessary to use a nonselective herbicide and then either re-seed or re-sod the desirable turfgrass.

Organic arsenical herbicides such as MSMA have been used as single and repeat applications and in combination with other herbicides for many years to control both annual and perennial grasses in turfgrass species that are tolerant to this family of herbicides [27–29]. However, the availability of arsenical herbicides for use in turfgrass is in jeopardy because of a recent decision by the U.S. Environmental Protection Agency (EPA) to declare this family of herbicides ineligible for re-registration. This action will effectively remove the arsenical herbicides from the marketplace unless EPA reverses this decision. If EPA agrees to allow re-registration of arsenical herbicides, it may be limited to only MSMA with reduced application rates and fewer applications allowed per year.

Herbicides from families other than the arsenicals will also selectively control both annual and perennial grasses in turfgrass. Tolerance to postemergence grass herbicides (graminicides) varies with herbicide and turfgrass species and tolerance levels are listed in Table 5.7. The following discusses herbicides available for grass control in various turfgrass species.

5.9.2.1 Bermudagrass and Zoysiagrass

Annual grass control has traditionally been achieved with postemergence applications of MSMA. MSMA controls many annual grass species such as crabgrass and is sometimes mixed with metribuzin to improve control of goosegrass [27,28,41]. However, the potential for turfgrass injury is greater with the combination than with MSMA alone.

Alternatives for postemergence annual grass control include diclofop-methyl and fenoxaprop-methyl, members of the aryl-oxy-phenoxy herbicide family, and foramsulfuron, a member of the sulfonylurea family. Both diclofop and foramsulfuron provide good goosegrass control, though they act slowly over 2 weeks or longer [42]. In addition, they provide better control of smaller goosegrass and require multiple applications for larger-size goosegrass. Fenoxaprop-ethyl provides control of annual grass weeds, especially crabgrass. Both diclofop-methyl and foramsulfuron are tolerated well by bermudagrass, while zoysiagrass tolerates fenoxaprop and foramsulfuron.

Quinclorac provides excellent postemergence control of the perennial torpedograss and good control of the annual crabgrass [43]. Multiple applications improve long-term control of torpedograss. Both bermudagrass and zoysiagrass tolerate quinclorac.

5.9.2.2 Centipedegrass

Two herbicides from the cyclohexanedione family, clethodim and sethoxydim, can be safely applied to centipedegrass (*Eremochloa ophiuroides* [Munro] Hackel) but not the other warm-season turf species. Both herbicides control annual and perennial grasses including crabgrass, goosegrass, bermudagrass, and bahiagrass. Multiple applications may be necessary for control of perennial

TABLE 5.7

Warm-Season Turfgrass Tolerance to Postemergence Herbicides (Refer to Herbicide Label for Species-Specific Listing)

Herbicide	Bahiagrass	Bermudagrass	Centipedegrass	Seashore Paspalum	St. Augustinegrass	Zoysiagrass	Overseed Perennial Ryegrass
Atrazine	NR	NR	S-I	NR	S-I	I	D
Bentazon	S	S	S	S	S	S	S-I
Bispyribac-sodium	NR	S	NR	NR	NR	NR	S
Bromoxynil	S	S	S	NR	S	S	S
Carfentrazone	S	S	S	S	S	S	S
Chlorsulfuron	I	I	NR	NR	NR	NR	NR
Clethodim	NR	NR	NR	NR	NR	NR	NR
Clopyralid	S	S	S	S	S	S	S
2,4-D	S	S	I	NR	I	S	S-I
2,4-D + dicamba	S	S	I	NR	I	S	S-I
2,4-D + 2,4-DP	S	S	I	NR	I	S	I-D
2,4-D + MCPP	S	S	I	NR	I	S	I-D
2,4-D + MCPP + dicamba	S	S	I	NR	I	S	I-D
2,4-D + MCPP + 2,4-DP	S	S	I	NR	I	S	I-D
2,4-D + MCPP + dicamba + MSMA	D	S-I	D	NR	D	S-I	D
2,4-D + clopyralid + dicamba	S	S	I	NR	S-I	S	S
Dicamba	S	S	I	NR	I	S	I
Diclofop	NR	S	NR	NR	NR	NR	NR
MSMA	D	S	D	NR	D	I	D

Ethofumesate	NR	D	NR	NR	I	NR	D
Fenoxaprop	I-D	I-D	D	NR	D	I	I
Fluazifop	NR	NR	NR	NR	NR	S	NR
Fluroxypyr	S	I	S	NR	D	S	S
Foramsulfuron	NR	S	D	NR	NR	S	D
MCPA + MCPP + 2,4-D	S	S	I	NR	I	I	I-D
MCPP	S	S	I	NR	I	S	I
Metribuzin	D	S-I	D	NR	D	NR	D
Metsulfuron	D	S	S	NR	S-I	S	D
Pronamide	NR	S	NR	NR	NR	NR	D
Quinclorac	D	I-S	D	S	D	S	S
Rinsulfuron	NR	S	NR	NR	NR	NR	NR
Sethoxydim	D	D	S	NR	D	D	D
Simazine	NR	I	S	NR	S	S-I	D
Sulfentrazone	S	S	S	S	NR	S	S
Sulfometuron	I	I	NR	S	NR	NR	NR
Sulfosulfuron	NR	S	S	S	S	S	D
Triclopyr+clopyralid	I	I	I	NR	D	I	S
Trifloxysulfuron	NR	S	NR	NR	NR	S	D

Note: S, safe at labeled rates; I, intermediate safety, use at reduced rates; D, damaging, do not use; NR, not registered for use on this turfgrass. Safe when applied to overseeded bermudagrass—January 1 to April 15.

Source: Adapted from Unruh, J.B., *Pest Control Guide for Turfgrass Managers*, University of Florida, Gainesville, 2006.

TABLE 5.8

Efficacy of Postemergence Herbicides

	Lifecycle	Atrazine	Bentazon	Bispyribac	Bromoxynil	Carfentrazone	Clethodim	Clopyralid	2,4-D	2,4-D + Dicamba	Dicamba	Diclofop	Ethofumesate	Fenoxaprop
Perennial Weeds														
Bahiagrass	P	P	P		P	P	F	P	P	P	P	P	F	P
Bermudagrass	P	P	P		P	P	F-G	P	P	P	P	P		P
Dallisgrass	P	P	P		P	P	F	P	P	P	P	P		P
Torpedograss	P					P		P						
Wild garlic/onion	P	P	P		P	P	P	P	G	G	F	P		P
Annual Grasses														
Annual bluegrass	WA	E	P	G	P	P	F	P	P	P	P	P	G	P
Crabgrass	SA	F	P		P	P	E	P	P	P	P	P	F	G-E
Crowfootgrass	SA	P	P		P	P	E	P	P	P	P			G-E
Goosegrass	SA	P	P		P	P	F-G	P	P	P	P	E		G
Sandbur	SA	P	P		P	P	G	P	P	P	P			G-E
Broadleaf Weeds														
Bittercress, hairy	WA					P	F		E	E	P			P
Black medic	A				G	P		E	E	P	E	P		P
Buttercups	WA,B,P	F	P		P	G	P	F-P	E-F	E-F	E-F	P		P
Carpetweed	SA	E					P		E	E	E	P		P
Carrot, wild	A,B						P	F	F	E-F	E	P		P
Chamberbitter (niruri)	SA,P	G	P				P	P	P	P-F	P-F	P		P
Common chickweed	WA	E	G		P		P	E-F	P	G	E	P		P
Corn speedwell	WA	E	P		G		P	P	F	F	F	P		P
Cudweed		G			G	F	P	E	G-E	E	E	P		P
Dandelion	P	F	P		P		P	F	E	G	E	P		P
Dayflower, spreading	SA	G-E	G				P		F	F	F	P		P
Dichondra	P	E	P		P	F	P		G	G	G	P		P
Docks	P	G	P			G	P	E	F	G	E	P		P
Doveweed	SA	G-E	P		P		P		P	F	P	P		P
Florida betony	P	F-G	P		P	P	P		F	G	G	P		P
Geranium, carolina	WA	E				G	P	P	E	E	E	P		P
Ground ivy	P		P		P		P		P-F	F	G	P		P
Hawkweed	P					G	P		E-F	E-F	E-F	P		P
Henbit	WA	E	P		G	G	P		P	G	E	P		P
Hop clovers	WA	E			F	G	P	E	F	G	E	P		P
Knotweed	SA	E			F		P	E	P	G	E	P		P
Lespedeza	SA	E					P	P	P-F	G	E	P		P
Mallow	P	P	P				P		F	F-G	E	P		P
Mouseear chickweed	WA,P	G	P			G	P	F	P-F	G	E	P		P
Mugwort	P	P	P				P		F	F	G	P		P
Mustards	WA	E	G		G		P	P	E	G	E	P		P
Parsley piert	WA	E	G		G		P		P		E	P		P
Pennywort (dollarweed)	P	E	P		P	P	P	F	G	G	E	P		P
Pepperweed, VA	WA	E			F	G	P		E	E	E	P		P
Pigweed	WA	G	P		F-G	G	P		E	E	E	P		P
Plantains	P	F	P		P		P	F	E	G	F	P		P
Shepherdspurse	WA		G		G		P	F	E	E	E	P		P
Spurges	SA	E	P		F	G	P	F	F	G	G	P		P
Spurweed (burweed)	WA	E	E		G		P	E	G	E	P			P
Thistles	B,P	P					P	E	E-F	E-F	E-F	P		P
VA buttonweed	P	P	P		P	G	P	F-P	P	F	F	P		P
Violets	P	P	P			G	P		P	F	F	P		P
White clover	P	E	P			G	P	E	F	G	E	P	F	P
Yellow woodsorrel (*Oxalis*)	P	G	P		F	G	P	F-P	P	F	G	P		P

Note: A, annual; B, biennial; P, perennial; SA, summer annual; WA, winter annual; E, excellent control (90–100%); G, good control (80–89%); F, fair control (70–79%); P, poor control (<70%). A blank space indicates weed response is not known.

Source: Adapted from Unruh, J.B., *Pest Control Guide for Turfgrass Managers*, University of Florida, Gainesville, 2006.

Fluazifop	Fluroxypyr	Foramsulfuron	Glufosinate	Glyphosate	Imazaquin	Metribuzin	Metsulfuron	MSMA	Pronamide	Quinclorac	Rimsulfuron	Sethoxydim	Simazine	Sulfentrazone	Sulfometuron	Sulfosulfuron	Triclopyr + Clopyralid	Trifloxysulfuron
F	P	P	P	G	P	P	E	F	P	P		F	P			F	P	F
G	P	P	P	E	P	P	P	P	P	P		F	P			P	P	P
	P	P	P	E	P	P	P	G	P	F		P	P				P	F
P	P	P	F							E		P	P				P	G
P	P	P	P	G	E	P	E	P	P			P	P				P	
P	P	E	E	E	P-F	E	P	P	E	P	G	P	E		G	G	P	G
G	P	P	E	E	P	F	P	E	P	E		E	P				P	F
G	P	P	E	E	P	G	P	E	P			F-G	P				P	
G	P	G	E	E	P	G	P	F	P	P		G	P				P	
G	P	P	E	E	F	G	P	G	P			G	P				P	
P		P				G	E					P		G			E	
P	F	P								E		P		G			E	
P		P			G		E					P	F	G		G	E	
P		P					P					P	G	G				G
P		P					E					P						
P		P		E	P			P-F				P	F					
P	F	P	G	E	E	G	E	P	G		F	P	E	G	G		E	
P	P	P		E	P	E	E	P	G			P	E				P	
P	F	P		G	F		F-G					P	G	G			G-E	
P	F	P		E			E	P	P			P	P	G			E-F	G
P		P			G		P-F			P		P	G					
P		P		E				P	P			P	P				E	G
P		P	G	E			E	P	P			P	P	G			E	
P	P	P		G		F	P		P			P	F				P	
P	F	P		E				P	P			P	P				G	
P	P	P			G		F-G					P	G	G			F	
P	P	P		G				P	P			P	P	G			E-F	
P		P										P					E	
P	G	G		E	G	G	G	P	P		F	P	E	G		G	E	G
P	G	P	G	E		G	P-F	P				P	E				E	G
P		P		E		G		P				P	G	G			E	
P	F	P		E		E	E	P				P	G	G			E	
P		P						P	P			P	P	G			E	
P	F	P	G	E	G	E	E	P	P			P	P	G		G	E	
P		P		G				P	P			P	P					
P		P		E		F	F	P	P			P	G			G		
P	F	P		E	G	E		P	P			P	E	G			G	
P	F	P		E			G	P	P	E	F	P	P				E	
P	P	P						P	P			P	G					
P		P	G				E-G					P	F	G				
P		P		E			G	P	P			P	P	G			E	
P		P					G					P			G	G	E-F	
P	F	P		E		E	E	P			F	P	G	G			E-F	G
P	F	P		E	E	G	E	P	P			P	E	G			E	
P		P			G		F					P	P				E	
P	F	P		G				P	P			P	P				F	G
P	P	P						P	P			P	P	G			F-G	
P	E	P	G	F	F	F	E	P	P			P	P	G		G	E	G
P	F	P	G	E			F-G	G	P			P	P	G			F-G	G

grass species. Sethoxydim can be used on residential, golf-course, and commercial turfgrass while clethodim is restricted to use for sod production only.

5.9.2.3 St. Augustinegrass

This turfgrass species is sensitive to most postemergence grass herbicides and postemergence grass control/suppression in St. Augustinegrass is limited to asulam and ethofumesate. Asulam provides control of annual grasses but can only be used on St. Augustinegrass grown for sod. Ethofumesate provides only suppression of bermudagrass with repeat applications beginning when bermudagrass first breaks dormancy. Postemergence grass control in other than sod production requires the use of nonselective herbicides with subsequent reestablishment of the turfgrass.

5.9.2.4 Bahiagrass

Selective control of either annual or perennial grass in bahiagrass (*Paspalum notatum* Fluegge) is not possible with currently registered herbicides. The only option is to spot-treat with a nonselective herbicide and reestablish the bahiagrass.

5.9.2.5 Seashore Paspalum

Quinclorac is the only herbicide registered for use in seashore paspalum that provides postemergence grass control (see Section 5.9.2.1). Most other postemergence grass herbicides including MSMA, sethoxydim, clethodim, and diclofop cause significant injury to seashore paspalum. As more tolerance information becomes available, additional postemergence grass herbicides may be registered for use in seashore paspalum.

5.10 OVERSEED REMOVAL FOR SPRING TRANSITION

Removing perennial ryegrass that was used for winter overseeding in warm-season turfgrass is important for timely and rapid recovery of the warm-season turf. Delay in removal of the winter overseed can slow initial warm-season turfgrass growth in spring, resulting in thinned turfgrass and a potential for summer weed infestation. Pronamide has traditionally been used to remove perennial ryegrass but can require several weeks for complete removal [44]. Several herbicides in the sulfonylurea family have been registered for perennial ryegrass removal from warm-season turfgrass. These include foramsulfuron, trifloxysulfuron, rimsulfuron, and sulfosulfuron. Foramsulfuron, trifloxysulfuron, and rimsulfuron usually require only a single application while sulfosulfuron may require a repeat application to remove the perennial ryegrass. These herbicides generally remove the perennial ryegrass much faster than pronamide, usually within 14 days of application [45]. Environmental conditions can play a role in the speed with which the perennial ryegrass is controlled, with more rapid activity when temperatures are warmer and slower perennial ryegrass death when temperatures are cooler. Since death of the perennial ryegrass can be rapid, the warm-season species (bermudagrass) should be out of dormancy by the time one of the sulfonylurea herbicides is applied to aid in transition from cool-season overseed to warm-season turfgrass.

5.11 SEDGE CONTROL

The predominant sedge species that infest warm-season turfgrass are yellow nutsedge (*Cyperus esculentus* L.), purple nutsedge (both perennials), annual sedge (*Cyperus compressus* L.), and kyllinga species (both annual and perennial) (*Cyperus* sp.). Sedge species generally infest wet areas (over-irrigated or poorly drained). It is important to correct the underlying cause of the wet condition prior to employing herbicide control treatments. Sedge control and warm-season turfgrass tolerance rankings for sedge herbicides are listed in Table 5.9.

TABLE 5.9
Sedge Control and Turf Tolerance to Various Herbicides (Refer to Herbicide Label for Species-Specific Listing)

Herbicide	Sedge Control				Turfgrass Tolerance						
	Purple	Yellow	Annual	Kyllinga	Bermuda-grass	St. Augustinegrass	Bahiagrass	Centipedegrass	Zoysiagrass	Seashore Paspalum	Overseed Perennial Ryegrass
Bentazon	P	G	G	F-G	S	S	S	S	S	S	S-I
Sulfosulfuron	E	E	E	E	S	S	NR	S	S	S	D
Imazaquin	G	G	G	G	I	S	D	S	S	S	D
Halosulfuron	G	G	G	G	S	S	S	S	S	S	–
Sulfentrazone	G	G	–	G	S	NR	S	S	S	S	S
Trifloxysulfuron	E	E	E	E	S	NR	NR	NR	S	NR	D
MSMA/DSMA	F	F	F-G	F	S-I	D	D	D	I	NR	D
Imazaquin + MSMA	G	G	G	G	I	D	D	D	S	NR	D

Note: E, excellent (>89%) control; F, fair to good (70–89%); G, good control sometimes with high rates; however, a repeat treatment 1–3 weeks later each at the standard or reduced rate is usually more effective; P, poor (<70%) control in most cases; S, safe at labeled rates; I, intermediate safety; use at reduced rates; D, damaging, do not use; NR, not registered for use on this turfgrass; G, good; F = fair; P = poor.

Source: Adapted from Unruh, J.B., *Pest Control Guide for Turfgrass Managers*, University of Florida, Gainesville, 2006.

As with the annual grasses, the organic arsenicals were the key herbicides for sedge control (see Section 5.9.2) but can be used only in bermudagrass and zoysiagrass. Bentazon, another option for sedge control, is safe on all warm-season turf species and selectively controls yellow nutsedge and other sedge species but not purple nutsedge. Imazaquin provides control of sedge species and is safe on all warm-season turf except bahiagrass [46–48]. These herbicides often require multiple applications to achieve control. Mixing MSMA with either bentazon or imazaquin can improve control but these combinations can only be used on bermudagrass or zoysiagrass.

Several herbicides in the sulfonylurea family of herbicides control sedge species in warm-season turfgrass. Halosulfuron, sulfosulfuron, and trifloxysulfuron control most sedge species including yellow and purple nutsedge [49–51]. Halosulfuron and sulfosulfuron can be applied to bermudagrass, St. Augustinegrass, centipedegrass, zoysiagrass, and seashore paspalum while trifloxysulfuron can be applied to only bermudagrass and zoysiagrass. Repeat applications, especially for the more difficult-to-control purple nutsedge, may be necessary to provide season-long control. Sulfentrazone, a PPO inhibitor herbicide, also provides postemergence control of yellow and purple nutsedge and can be safely applied to all warm-season turfgrass except St. Augustinegrass [52–54].

TABLE 5.10

List of Common and Scientific Names of Plant Species Referenced in This Chapter

Common Name	Scientific Name
Annual bluegrass	*Poa annua* L.
Annual sedge	*Cyperus compressus* L.
Bahiagrass	*Paspalum notatum* Fluegge
Bermudagrass, common	*Cynodon dactylon* (L.) Pers.
Bermudagrass, hybrid	*Cynodon dactylon* (L.) Pers. × *C. transvalensis* Burtt-Davy
Bittercress, hairy	*Cardamine hirsute* L.
Black medic	*Medicago lupulina* L.
Buttercups	*Ranunculun repens* L.
Carpetweed	*Mollugo verticillata* L.
Carrot, wild	*Daucus carota* L.
Centipedegrass	*Eremochloa ophiuroides* (Munro) Hackel
Chamberbitter	*Phyllanthus niruri* L.
Common chickweed	*Stellaria media* (L.) Villars
Corn speedwell	*Veronica arvensis* L.
Crabgrass	*Digitaria* sp.
Crowfootgrass	*Dactyloctenium aegyptium* (L.) Willd.
Cudweed	*Gnaphalium* sp.
Dallisgrass	*Paspalum dilatatum* Poir.
Dandelion	*Taraxacum officinale* Weber
Dayflower, spreading	*Commelina diffusa* Burrm. f.
Dichondra	*Dichondra micrantha* Urban
Docks	*Rumex* sp.
Doveweed	*Murdannia nudiflora* (L.) Brenan
Florida betony	*Stachys floridana* Shuttlew
Geranium, carolina	*Geranium carolinianum* L.
Goosegrass	*Eleusine indica* (L.) Gaertner
Ground ivy	*Glechoma hederacea* L.
Hawkweed	*Hiedracium pilosella* L.
Henbit	*Lamium amplexicaule* L.
Hop clovers	*Trifolium aureum* Pollich
Knotweed	*Polygonum aviculare* L.

(Continued)

TABLE 5.10
(Continued)

Common Name	Scientific Name
Lespedeza	*Lespedeza striata* (Thumb. Ex Murray) Hook. & Arn.
Mallow	*Malva rotundifolia* L.
Mock strawberry	*Duchesnea indica* (Andreus) Focke
Mouseear chickweed	*Cerastium vulgatum* L.
Mugwort	*Artemisia vulgaris* L.
Mustards	*Brassica kaber* (DC) L. Wheeler
Parsley piert	*Alchemilla arvensis* (L.) Scop.
Pennywort (dollarweed)	*Hydrocotyle* sp.
Pepperweed, Virginia	*Lepidium virginicum* L.
Perennial ryegrass	*Lolium perenne* L.
Pigweed	*Amaranthus* sp.
Plantains	*Plantago* sp.
Purple nutsedge	*Cyperus rotundus* L.
Sandbur	*Cenchrus* sp.
Seashore paspalum	*Paspalum vaginatum* Sw
Shepherdspurse	*Capsella bursa-pastoris* (L.) Medikus
Spurges	*Euphorbia* sp.
Spurweed (burweed)	*Soliva pterosperma* (Jussieu) Lessing
St. Augustinegrass	*Stenotaphrum secundatum* (Walter) Kuntze
Tall fescue	*Festuca arundinacea* Schreber
Thistles	*Cirsium* sp.
Torpedograss	*Panicum repens* L.
VA buttonweed	*Dioda virginiana* L.
Violets	*Viola* sp.
White clover	*Trifolium repens* L.
Wild garlic	*Allium vineale* L.
Wild onion	*Allium canadense* L.
Yellow nutsedge	*Cyperus esculentus* L.
Yellow woodsorrel	*Oxalis stricta* L.

TABLE 5.11
Common and Trade Names of Turf Herbicides

Common Name	Trade Name(s)
Atrazine	Aatrex, Atrazine Plus, Purge II, and others
Benefin	Balan, Crabgrass Preventer, and others
Benefin and oryzalin	XL 2G
Benefin and oxadiazon	RegalStar
Benefin and trifluralin	Team 2G
Bensulide	Betasan, Pre-San, Bensumec 4, Weedgrass Preventer, and others
Bentazon	Basagran T&O, Lescogran 4L
Bentazon and atrazine	Prompt
Bispyrabac	Velocity
Bromoxynil	Buctril 2L

(Continued)

TABLE 5.11
(Continued)

Common Name	Trade Name(s)
Carfentrazone	Quicksilver/Aim
Carfentrazone + 2,4-D + MCPP + dicamba	SpeedZone, SpeedZone—Southern
Carfentrazone + MCPA + MCPP + dicamba	Power Zone
Clethodim	Envoy
Clopyralid	Lontrel T&O
2,4-D	Many
2,4-D + clopyralid + dicamba	Millennium Ultra 2
DCPA	Dacthal 75WP, Garden Weed Preventer, and others
Dicamba	Vanquish 4L, K-O-G Weed Control, and others
Diclofop	Illoxan 3EC
Dithiopyr	Dimension Ultra
Diquat	Reward LS
DSMA	Ansar 6.6
Ethofumesate	Progress 1.5L
Fenoxaprop	Acclaim
Fluazifop	Fusilade II
Fluroxypyr	Spotlight
Foramsulfuron	Revolver
Glufosinate	Finale 1L
Glyphosate	RoundUp, Touchdown PRO, Glyphomaxx
Halosulfuron	Sedgehammer 75DG/Sandea 75DG
Imazaquin	Image 70DG
Isoxaben	Gallery 75DF
MCPP	Mecomec 4
MCPP, 2,4-D + dicamba + MCPA and/or 2,4-D	Trimec Southern/3-Way Selective/Eliminate DG/33, and others
Metribuzin	Sencor 75DF
Metolachlor	Pennant Magnum
Metsulfuron	Manor
MSMA	Daconate 6, Bueno 6, others
Napropamide	Devrinol 50DF, Ornamental Herbicide 2G
Oryzalin	Surflan AS
Oxadiazon	Ronstar 2G, 50WP
Pendimethalin	Pre-M, Pendulum, Southern Weedgrass
Prodiamine	Barricade 65WG, 4FL/ProClipse
Pronamide	Kerb 50WP
Quinclorac	Drive
Rimsulfuron	TranXit GTA
Sethoxydim	Vantage
Simazine	Princep Liquid, others
Sulfentrazone	Dismiss
Sulfentrazone + 2,4-D + MCPP + dicamba	Surge Broadleaf Herbicide
Sulfometuron	Oust
Sulfosulfuron	Certainty
Triclopyr + clopyralid	Confront
Trifloxysulfuron	Monument

5.12 NEWLY SEEDED OR SPRIGGED AREAS

Weed control in newly seeded or sprigged warm-season turfgrass is difficult to achieve, especially if preplant fumigation was not utilized. A few herbicides are available that can be applied before the turfgrass is well established. Oxadiazon can be applied immediately prior to or immediately after sprigging of bermudagrass or zoysiagrass for annual grass control with minimal injury to the turfgrass. For seashore paspalum, oxadiazon can be applied 10–14 days after sprigging.

Quinclorac provides postemergence control of grass and broadleaf weeds and can be applied anytime prior to or immediately after seeding or sprigging of bermudagrass or zoysiagrass. For seashore paspalum, quinclorac cannot be applied until 14 days after emergence of the turfgrass. Carfentrazone controls broadleaf weeds postemergence and can be applied 7 days after emergence of newly seeded or sprigged bermudagrass and St. Augustinegrass. Carfentrazone should not be applied until at least 14 days after emergence of seeded or sprigged zoysiagrass.

The common and scientific names of plant species referenced in this chapter are listed in Table 5.10. All chemicals mentioned in this chapter are for reference only (Table 5.11). Some products may be restricted by some state, provinces, or federal agencies; thus, be sure to check the current status of the herbicide being considered for use. Always read and follow the manufacturer's label. Mention of a proprietary product does not constitute a guarantee or warranty of the product by the author or the publishers and does not imply approval to the exclusion of other products that also may be suitable.

REFERENCES

1. Ahrens, W.H., *Herbicide Handbook*, 7th Ed., Weed Science Society of America, Champaign, IL, 1994, 318.
2. McWhorter, C.G. and Chandler, J.M., Conventional weed control technology, in *Biological Control of Weeds with Plant Pathogens*, Charudattan, R. and Walker, H.L., Eds., Wiley, New York, 1982, 5.
3. Radosavich, S.R. and Holt, J.S., *Weed Ecology: Implications for Vegetation Management*, Wiley, New York, 1984, 3.
4. Webster, T.M., Weed survey—southern states, *Proc. South. Weed Sci. Soc.*, 57, 420, 2004.
5. Beard, J.B., *Turf Management for Golf Courses*, 2nd Ed., Wiley, Hoboken, NJ, 2002, 447.
6. Christians, N.E., *Fundamentals of Turfgrass Management*, 2nd Ed., Wiley, Hoboken, NJ, 2004, chap. 11.
7. McCarty, L.B. and Murphy, T.R., Control of turfgrass weeds, in *Turf Weeds and Their Control*, Turgeon, A.J., Ed., American Society of Agronomy, Madison, WI, 1994, chap. 8.
8. Brecke, B.J., Stephenson, D.O. IV, and Unruh, J.B., Annual bluegrass control in overseeded bermudagrass, *Weed Sci. Soc. Am. Abstr.*, 45, 101, 2005.
9. Watson, A.K., The classical approach with plant pathogens, in *Microbial Control of Weeds*, TeBeest, D.O., Ed., Chapman and Hall, New York, 1991, 3.
10. Hallett, S.G., Where are the bioherbicides? *Weed Sci.*, 53, 404, 2005.
11. Weston, V.C.M., The commercial realization of biological herbicides, in *Proceedings of the Brighton Crop Protection Conference—Weeds,* Vol. 1, BCPC, Farnham, U.K., 1999, 281.
12. Johnson, B.J., Biological control of annual bluegrass with *Xanthomonas campestris* pv. *poannua* in bermudagrass, *HortScience*, 29, 659, 1994.
13. Zhou, T. and Neal, J.C., Annual bluegrass (*Poa annua*) control with *Xanthomonas campestris* pv. *poannua* in New York State, *Weed Technol.*, 9, 173, 1995.
14. McCarty, L.B., and Tucker, B.J., Prospects for managing turf weeds without protective chemicals, *Int. Turfgrass Res. J.*, 10, 34, 2005.
15. Chandramohan, S. and Charudattan, R., Control of seven grasses with a mixture of three fungal pathogens with restricted host ranges, *Biol. Control Theor. Appl. Pest Manage.*, 22, 246, 2001.
16. Kadir, J.B., Charudattan, R., Stall, W.M., and Brecke, B.J., Field efficacy of *Dactylaria higginsii* as a bioherbicide for to control of purple nutsedge (*Cyperus rotundus*), *Weed Technol.*, 14, 1, 2000.
17. Kadir, J. and Charudattan, R., *Dactylaria higginsii*, a fungal agent for purple nutsedge (*Cyperus rotundus*), *Biol. Control Theor. Appl. Pest Manage.*, 17, 113, 2000.
18. Christians, N.E., The use of corn gluten meal as a natural preemergence weed control in turf, *Int. Turfgrass Soc. Res. J.*, 7, 284, 1993.

19. Bingamen, B.R. and Christians, N.E., Greenhouse screening of corn gluten meal as a natural control product for broadleaf and grass control, *HortScience*, 30, 1256, 1995.

20. Wiecko, G., Ocean water as a substitute for postemergence herbicides in tropical turf, *Weed Technol.*, 17, 788, 2003.

21. Anderson, W.P., *Weed Science Principles and Applications*, 3rd Ed., West Publishing Company, St. Paul, MN, 1996, chap. 3.

22. McCarty, L.B. and Murphy, T.R., Control of turfgrass weeds, in *Turf Weeds and Their Control*, Turgeon, A.J., Ed., American Society of Agronomy, Madison, WI, 1994, chap. 8.

23. Unruh, J.B., Brecke, B.J., Dusky, J.A., and Godbehere, J.S., Fumigant alternatives for methyl bromide prior to turfgrass establishment, *Weed Technol.*, 16, 379, 2002.

24. Brecke, B.J., Unruh, J.B., and Stephenson, D.O. IV, Dazomet blended with rootzone mix for fumigation, *Int. Turfgrass Res. J.*, 10, 1176, 2005.

25. Emmons, R.D., *Turfgrass Science and Management*, Delmar Publishers, Albany, NY, 1995, chap. 14.

26. Fermanian, T.W., Shurfleff, M.C., Randell, R., Wilkinson, H.T., and Nixon, P.L., *Controlling Turfgrass Pests*, Prentice Hall, Upper Saddle River, NJ, 1997, chap. 3.

27. Wiecko, G., Sequential herbicide treatments for goosegrass (*Eleusine indica*) control in bermudagrass (*Cynodon dactylon*) turf, *Weed Technol.*, 14, 686, 2000.

28. Johnson, B.J., Sequential herbicide treatments for large crabgrass (*Digitaria sanguinalis*) and goosegrass (*Eleusine indica*) control in bermudagrass (*Cynodon dactylon*) turf. *Weed Technol.*, 7, 674, 1993.

29. Johnson, B.J., Preemergence and postemergence herbicides for large crabgrass (*Digitaria sanguinalis*) control in centipedegrass (*Eremochloa ophiuroides*), *Weed Technol.*, 11, 144, 1997.

30. Coats, G.E. and Krans, J.V., Evaluation of ethofumesate for annual bluegrass (*Poa annua*) and turfgrass tolerance, *Weed Sci.*, 34, 930, 1986.

31. Dickens, R., Control of annual bluegrass (*Poa annua*) in overseeded bermudagrass (*Cynodon* pp.) golf greens, *Weed Sci.*, 27, 642, 1979.

32. Johnson, B.J., Response to ethofumesate of annual bluegrass (*Poa annua*) and overseeded bermudagrass (*Cynodon dactylon*), *Weed Sci.*, 31, 385, 1983.

33. McElroy, J.S., Walker, R.H., Wehtje, G.R., and van Santen, E., Annual bluegrass (*Poa annua*) populations exhibit variation in germination response to temperature, photoperiod, and fenarimol, *Weed Sci.*, 52, 47, 2004.

34. Isgrigg, J. III, Yelverton, F.H., Brownie, C., and Warren, L.S. Jr., Dintroaniline resistant annual bluegrass in North Carolina, *Weed Sci.*, 50, 86, 2002.

35. Wehtje, G. and Walker, R.H., Response of two annual bluegrass varieties to preemergence and postemergence rimsulfuron, *Weed Technol.*, 16, 612, 2002.

36. Hutto, K.C., Coats, G.E., and Taylor, J.M., Annual bluegrass (*Poa annua*) resistance to simazine in Mississippi, *Weed Technol.*, 18, 846, 2004.

37. Ni, H., Wehtje, G., Walker, R.H., Belcher, J.L., and Blythe, E.K., Turf tolerance and Virginia buttonweed (*Diodia virginiana*) control with fluroxypyr as influenced by the synergist diflufenzopyr, *Weed Technol.*, 20, 511, 2006.

38. Kelly, S.T. and Coats, G.E., Virginia buttonweed (*Diodia virginiana*) control with pyridine herbicides, *Weed Technol.*, 14, 591, 2000.

39. Lawson, R.N., Unruh, J.B., and Brecke, B.J., Lawn burweed (*Soliva pterosperma*) control in hybrid bermudagrass (*Cynodon dactylon* × *C. transvalensis*) and common centipedegrass (*Eremochloa ophiuroides*), *Weed Technol.*, 16, 84, 2002.

40. Main, C.L., Robinson, D.K., Teuton, T.C., and Mueller, T.C., Star-of-Bethlehem (*Ornithogalum umbellatum*) control with postemergence herbicides in dormant bermudagrass (*Cynodon dactylon*) turf, *Weed Technol.*, 18, 1117, 2004.

41. Johnson, B.J., Tank-mixed postemergence herbicides for large crabgrass (*Digitaria sanguinalis*) and goosegrass (*Eleusine indica*) control in bermudagrass (*Cynodon dactylon*) turf, *Weed Technol.*, 10, 716, 1996.

42. McCarty, L.B., Goosegrass (*Eleusine indica*) control in bermudagrass (*Cynodon* spp.) turf with diclofop, *Weed Sci.*, 39, 255, 1992.

43. Chism, W.J. and Bingham, S.W., Postemergence control of large crabgrass (*Digitaria sanguinalis*) with herbicides, *Weed Sci.*, 39, 62, 1991.

44. Johnson, B.J., Effects of pronamide on spring transition of a bermudagrass (*Cynodon dactylon*) green overseeded with perennial ryegrass (*Lolium perenne*), *Weed Technol.*, 4, 322, 1990.

45. Teuton, T.C., Sorochan, J.C., Main, C.L., Campbell, B.N., and Mueller, T.C., Selective control of overseeded perennial ryegrass in bermudagrass with ALS herbicides, *Proc. South. Weed Sci. Soc.*, 57, 348, 2004.
46. Belcher, J.L., Walker, R.H., van Santan, E., and Wehtje, G., Nontuberous sedge and kyllinga species' response to herbicides, *Weed Technol.*, 16, 575, 2002.
47. Coats, G.E., Munoz, R.F., Anderson, D.H., Herring, D.C., and Scruggs, J.W., Purple nutsedge (*Cyperus rotundus*) control with imazaquin in warm-season turfgrasses, *Weed Sci.*, 35, 691, 1987.
48. Lowe, D.B., Whitwell. T., Martin, B.S., and McCarty, L.B., Yellow nutsedge (*Cyperus esculentus*) management and tuber reduction in bermudagrass (*Cynodon dactylon* × *Cynodon transvaalensis*) turf with selected herbicide programs, *Weed Technol.*, 14, 72, 2000.
49. McElroy, J.S., Yelverton, F.H., Troxler, S.C., and Wilcut, J.W., Selective exposure of yellow (*Cyperus esculentus*) and purple nutsedge (*Cyperus rotundus*) to postemergence treatments of CGA-362622 and MSMA. *Weed Technol.*, 17, 554, 2003.
50. McElroy, J.S., Yelverton, F.H., Gannon, T.W., and Wilcut, J.W., Foliar vs. soil exposure of green kyllinga (*Kyllinga brevifolia*) and false-green kyllinga (*Kyllinga gracillima*) to postemergence treatments of CGA-362622, halosulfuron, imazaquin, and MSMA, *Weed Technol.*, 18, 145, 2004.
51. McElroy, J.S., Yelverton, F.H., and Warren, L.S. Jr., Control of green and false-green kyllinga (*Kyllinga brevifolia* and *Kyllinga gracillima*) in golf course fairways and roughs, *Weed Technol.*, 19, 824, 2005.
52. Brecke, B.J., Stephenson, D.O. IV, and Unruh, J.B., Control of purple nutsedge (*Cyperus rotundus*) with herbicides and mowing, *Weed Technol.*, 19, 809, 2005.
53. Blum, R.R., Isgrigg, J. III, and Yelverton, F.H., Purple (*Cyperus rotundus*) and yellow nutsedge (*C. esculentus*) control in bermudagrass (*Cynodon dactylon*) turf, *Weed Technol.*, 14, 357, 2000.
54. Bunnell, B.T., McCarty, L.B., Lowe, D.B., and Higingbottom, J.K., *Kyllinga squamulata* control in bermudagrass turf, *Weed Technol.*, 15, 310, 2001.

6 Weed Management Practices Using ALS-Inhibiting Herbicides for Successful Overseeding and Spring Transition

Sowmya Mitra, Prasanta C. Bhowmik, and Kai Umeda

CONTENTS

6.1 INTRODUCTION

In the desert southwest and other warm-season turfgrass-growing regions of the United States, cool-season turfgrasses like perennial ryegrass and *Poa trivialis* are overseeded into dormant warm-season bermudagrass during fall. It is extremely important to control weeds prior to the overseeding to have a good stand of cool-season turf. During spring, when the temperatures start rising, herbicides may have to be used to control the cool-season turfgrasses so that the transition of the warm-season turf is not affected. Several herbicides can be used before overseeding and during spring transition, but the sulfonylurea (SU) class of herbicides has proven to be extremely versatile and useful. The SUs and imidazolinone classes of herbicides inhibit the acetolactate synthase (ALS) enzyme in the branched-chain amino acid biosynthesis pathway. These herbicides are extremely selective and can be used very effectively for weed management strategies before overseeding and can be used to control the cool-season turf during spring transition.

6.2 SULFONYLUREA HERBICIDES

The herbicidal properties of SUs were first reported in 1966 [1]. The early SUs were derivatives of triazine herbicides. George Levitt of E.I. du Pont de Nemours and Company noted that the SUs having an aniline as the aryl group exhibited weak plant-growth-regulatory activity while the aminopyrimidine derivative displayed very high biological activity [2,3]. Levitt et al. [2] soon produced SUs having up to 100 times the activity of conventional herbicides, and thus began one of the most exciting breakthroughs in the field of herbicide research in several decades [4].

6.2.1 SU CHEMISTRY

The SUs are represented by a general structure that has three distinct parts: an aryl group, the sulfonylurea bridge, and a nitrogen-containing heterocycle (Figure 6.1). High herbicidal activity was reported with a phenyl group and a substituent at the ortho position [4]. Except for the ortho carboxyl and hydroxyl substituents, the other types of electron-donating and electron-withdrawing ortho substituents have potential for herbicidal activity. SUs containing aryl groups other than a phenyl, such as thiophene [5–7], furan [5,8], or naphthalene [9], were reported to have herbicidal activity.

SUs usually exhibit the greatest herbicidal activity when the heterocyclic portion of the molecule is a symmetrical pyrimidine or symmetrical triazine containing lower alkyl or lower alkoxy substituents [4]. When the nitrogen heterocycles are triazoles [10,11], asymmetrical triazines [5], fused-ring pyrimidines [12,13], and pyridines [14], the SUs were reported to have less activity. SUs with a modified bridging group, such as $SO_2NHC(S)NH$ [5] or $SO_2NHCON(CH_3)$ [14], are also active.

6.2.2 TOXICOLOGICAL PROPERTIES

Most of the SUs have low acute oral, dermal, and inhalation toxicity in mammals. The acute oral LD_{50} value of table salt for rats is 3000 mg kg^{-1} while most of the SUs have LD_{50} values greater than 4000 mg kg^{-1} of body weight [15]. Most SUs are not mutagenic or teratogenic, and they exhibit low toxicity to fish, wildlife, honeybees, and dogs [4]. Low toxicity, combined with very low application

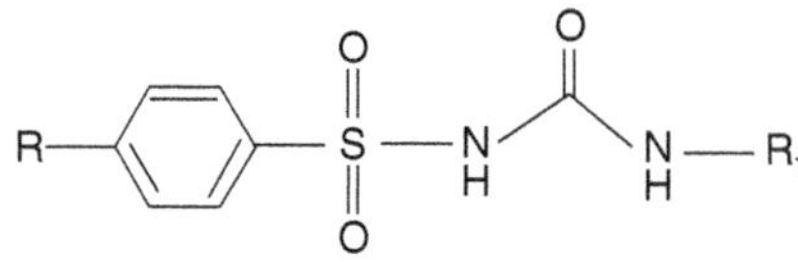

FIGURE 6.1 Chemical structure of SUs.

rates (2–75 g a.i. ha^{-1}) makes them especially attractive from an environmental and human-health standpoint. The potential for groundwater contamination through seepage, percolation, runoff, or infiltration is low.

6.2.3 BIOLOGY

SUs are potent inhibitors of plant growth—root and shoot growth in sensitive seedlings. Depending on the plant species, dose, and environmental conditions, various secondary plant responses often develop. Secondary responses such as enhanced anthocyanin formation, loss of leaf nyctinasty, abscission, vein discoloration, terminal bud death, chlorosis, and necrosis have been reported [4]. These secondary effects are often slow to develop and sometimes do not occur until 2 weeks or longer following treatment.

6.2.3.1 Uptake and Translocation

Herbicides can enter plant cells by three different mechanisms [16]: (1) passive, nonfacilitated diffusion into plant cells, as is the case for atrazine [17], norflurazon [18], monuron [19], and amitrole [20]; (2) carrier-mediated active uptake, as is the case for 2,4-D [19] and IAA [21]; and (3) ion trapping of weak acids, as is the case for chlorsulfuron [22] and imidazolinone herbicides [16].

The SUs are weak acids with acid dissociation constants (pK_a) ranging from 3.3 to 5.2. At lower pH, they are more permeable to cell membranes owing to a greater proportion of the molecules being in an undissociated or protonated state. They follow the weak acid theory of phloem mobility. Sulfonylurea molecules reach the acidic cell wall area (pH 5.5–6.0) in the neutral form, to which the phloem tube or companion cell plasma membrane is highly permeable. Upon entering the phloem (pH 8.0), the molecule associates and becomes trapped in a relatively anionic form. Once trapped, the herbicide moves systemically by "mass flow" mechanism with sucrose and other phloem solutes.

SUs are taken up by both roots and foliage. Reddy and Bendixen [23,24] reported the pattern and distribution of root and tuber absorption of ^{14}C-labeled chlorimuron, 2-[[[[(4-chloro-6-methoxy-2-pyrimidinyl)amino]carbonyl]amino]sulfonyl] benzoic acid. They stated that the ^{14}C absorbed by the tuber remained in the tuber while that absorbed by the roots was translocated to the shoots.

Nandihalli and Bhowmik [25] studied the absorption of chlorimuron by excised velvetleaf roots and the efflux of chlorimuron-ethyl by excised soybean root tissue [26]. They reported that uptake of chlorimuron-ethyl in soybean was similar in both tolerant and susceptible species. The uptake of the herbicide was directly proportional to the external herbicide concentration. The linear relationship indicated that the herbicide uptake is a nonfacilitated process. Since the uptake and efflux of chlorimuron-ethyl appear to be similar in the root tissues of both tolerant and susceptible species, crop selectivity does not seem to be due to differential uptake. Hall and Devine [27] reported that the herbicide chlorsulfuron, 2-chloro-N-[[(4-methoxy-6-methyl-1,3,5-triazin-2-yl)amino]carbonyl], was not translocated readily in plants because of an inhibitory effect on phloem translocation. They found that more chlorsulfuron was translocated in a chlorsulfuron-resistant (R) biotype of *Arabidopsis thaliana* than in a susceptible (S) biotype, indicating that the effect on translocation is secondary to the primary site of action of the herbicide. Chlorsulfuron pretreatment inhibited rapid sucrose uptake into leaf discs of *Arabidopsis* by 41% in the S biotype but only by 17% in the R biotype. The result suggested that chlorsulfuron inhibited phloem transport by restricting sucrose uptake into the phloem [27].

6.2.4 MODE OF ACTION

6.2.4.1 Growth and Metabolism Inhibition Studies

Study of the mode of action of SUs was first reported by Ray [28] and Hatzios and Howe [29]. Ray reported that chlorsulfuron did not have any inhibitory effect on photosynthesis or respiration.

Concentrations of chlorsulfuron as low as 2.8 nM (1 ppb) significantly inhibited root growth and higher concentrations effectively reduced shoot growth within 2–4 h of treatment. The results indicated an effect of chlorsulfuron on either cell division or cell enlargement [28]. Studies have revealed that cell enlargement was not affected by chlorsulfuron but cell division was highly sensitive to herbicide application. Ray [30] reported that chlorsulfuron blocked progression from second gap (G_2) to mitosis (M) in the cell-division cycle and, secondly, reduced movement from first gap (G_1) to S (DNA synthesis). No aberrant mitotic figures were observed and there was no change in the distribution pattern of the mitotic stages (prophase through telophase). These results suggested that chlorsulfuron was a potent and rapid inhibitor of plant-cell division but did not directly interfere with the mitotic apparatus. Ray [28] suggested that the inhibitory effect of chlorsulfuron on cell division was not due to a direct effect on DNA synthesis or the synthesis of nucleosides.

6.2.4.2 Plant Hormone Studies

According to the cell enlargement studies, chlorsulfuron did not directly block the growth-promoting action of auxins, cytokinins, and gibberellins, but there was a potential for the intervention of ethylene in chlorsulfuron action. Suttle and Schreiner [31] observed that treatment of soybean seedlings with chlorsulfuron inhibited growth and greatly stimulated anthocyanin formation. They also observed that phenyl alanine ammonia lyase activity and ethylene production were also stimulated. The presence, however, of a 3- to 4-day lag period between chlorsulfuron application and an increase in ethylene production led to ruling out ethylene as an important primary factor. Thus, it was demonstrated that ethylene was not involved in the primary action of the sulfonylurea herbicides. However, "stress-ethylene," formed in some plant species as a result of the phytotoxic action of the SUs, can contribute to the secondary symptoms that develop following herbicide treatment. Hageman and Behrens [32] reported that the defoliation of velvetleaf caused by chlorsulfuron treatment was due to enhanced ethylene production.

6.2.4.3 Studies with Bacteria

A major advance in identifying the target site of action of the SUs came from studies involving bacteria. Since bacteria and plants share many common biochemical pathways, studies with bacteria provided means for localizing the site of sulfonylurea action. On the basis of various experiments by La Rossa and Schloss [33], it was suggested that the SUs inhibited some step in the biosynthesis of branched-chain amino acids. In a series of experiments using mutants resistant to sulfometuron-methyl, N-[[[(4,6-dimethyl-2-pyrimidinyl amino]carbonyl]-2-methoxycarbonyl], the site of action of the latter was identified as the enzyme ALS [33].

ALS, also known as acetohydroxyacid synthase (AHAS), is a key enzyme in the branched-chain amino acid biosynthetic pathway of bacteria, fungi, and higher plants. Valine, isoleucine, and leucine are synthesized in plants and microbes by a common pathway (Figure 6.2).

The enzyme requires thiamine pyrophosphatase and Mg^{2+} as well as FAD, even though the reactions catalyzed by this enzyme involve no net oxidation or reduction. ALS catalyzes (a) the condensation of two molecules of pyruvate to form CO_2 and α-acetolactate, which leads to valine and leucine synthesis, and (b) the condensation of one molecule of pyruvate with α-ketobutyrate to form CO_2 and α-aceto-α-hydroxybutyrate, which leads to isoleucine formation. The pathways and enzymes are apparently restricted to the plastid [34].

All of the SUs display unusual "slow-binding" behavior with the enzyme and this behavior may help explain the efficacy of the herbicides. Resistance to these herbicides developed through a number of different procedures, and the mechanism of resistance is through changes in sensitivity of the enzyme to the herbicides [35]. The herbicides have been reported to be competitive with the amino acids for binding to the enzyme. ALS-inhibiting herbicides bind to the regulatory site on the enzyme [36]. Chaleff and Ray [37] selected some herbicide-resistant mutants of tobacco in culture.

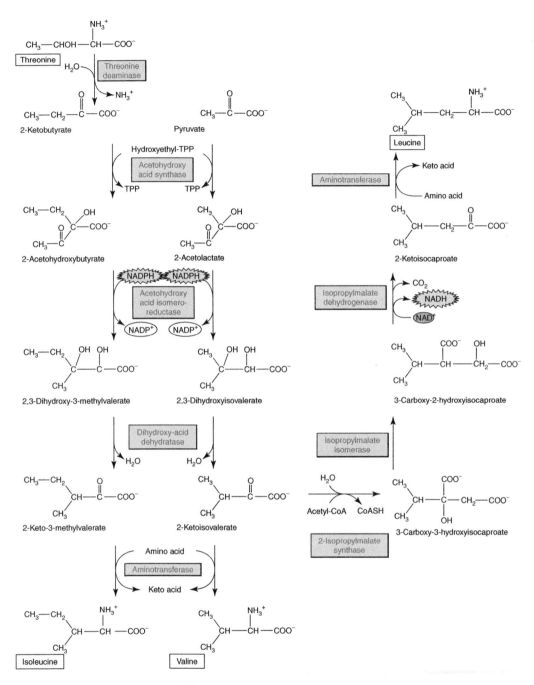

FIGURE 6.2   (Color Figure 6.2 follows p. 262.) Biosynthetic pathway of the branched-chain amino acids. (From Buchannan, B.B., Gruissem, W., and Jones, R.L., *Biochemistry and Molecular Biology of Plants*, American Society of Plant Biologists, Maryland, 2000. With permission.)

They reported that the herbicide resistance trait and herbicide-insensitive enzyme co-segregated during breeding of herbicide-resistant tobacco.

Various experiments have been designed to characterize the molecular and genetic details of ALS inhibition in bacteria. In *Salmonella typhimurium* and *Escherichia coli*, there are several isozymes of ALS, each encoded by a separate gene. In contrast, only a single gene encoding for ALS has been detected in yeast [38]. Isozyme II from *S. typhimurium* [33] and isozyme III from *E. coli* [39] have been shown to be sensitive to sulfometuron-methyl. Inhibition of ALS isozyme II from *S. typhimurium* exhibits slow, tight-binding kinetics with an initial K_i of 660 ± 60 nM and a final steady-state K_i of 65 ± 25 nM. It has been suggested that sulfometuron-methyl binds tightly but reversibly to the ALS-FAD-TPP-Mg^{2+}-decarboxylated pyruvate complex and competes for the second pyruvate binding site.

6.2.4.4 Studies with Plants

Crude preparations of the ALS enzyme from pea shoots revealed that it was strongly inhibited by chlorsulfuron; in fact, the plant enzyme was significantly more sensitive than the bacterial and yeast ALS enzymes [29]. Wheat is highly tolerant to chlorsulfuron but the herbicide is a very potent inhibitor of the ALS from this crop, with an I_{50} value of 19–22 nM. Similarly, none of the other selective SUs such as metsulfuron-methyl, 2-[[[[(4-methoxy-6-methyl-1,3,5-triazin-2-methyl)amino] carbonyl]amino]sulfonyl]benzoic acid; thifensulfuron-methyl, 3-(4-methoxy-6-methyl-1,3,5-triazin-2-carbamoylsulfamoyl) thiophene-2-carboxylic acid; bensulfuron-methyl, 2-[[[[[(4,6-dimethoxy-2-pyrimidinyl)amino]carbonyl]amino]sulfonyl]methyl]benzoic acid; and chlorimuron-ethyl owe their crop tolerance to differential ALS sensitivity. The tolerance to different SUs by various crops is due to the rapid inactivation or detoxification of the herbicide by the crop [29,40]. Through a series of genetic crosses, it was demonstrated that all normal, sensitive segregants possessed a chlorsulfuron-sensitive ALS, whereas all homozygous mutants contained a highly resistant form of the enzyme. Heterozygotes contained ALS with an intermediate degree of resistance to chlorsulfuron. Thus, resistance at the whole plant level paralleled resistance at the ALS enzyme level [40].

6.2.5 Biological Selectivity

In 1987, an SU-resistant prickly lettuce (*Lactuca serriola*) biotype was identified in a no-till, continuous-winter wheat field that had been treated with SUs for 5 years near Lewiston, Idaho [41]. SU resistance has also been reported in natural populations of kochia (*Kochia scoparia* [L.] Schrad.), Russian thistle (*Salsola iberica,* Sennen and Pau), common chickweed (*Stellaria media* [L.] Vill.) [42], perennial ryegrass (*Lolium perenne* L.) [43], and rigid ryegrass (*Lolium rigidum* Gaudin) [44]. Mallory-Smith et al. [41] reported that the SU resistance trait was controlled by a single nuclear gene with incomplete dominance (susceptible:intermediate:resistant = 1:2:1).

6.2.5.1 Differential Response of Crops and Weeds to SUs

There is ample evidence that corn cultivars differ in susceptibility to SUs [45–47]. The differential sensitivity of sweet corn cultivars to nicosulfuron, 2-[[[[(4,6-dimethoxy-2-pyrimidinyl)amino] carbonyl]amino] sulfonyl]-*N,N*-dimethyl-3-pyridine carboxamide [48,49], has been reported, particularly among cultivars containing the shrunken-2 (sh_2) endosperm mutant. Stall and Bewick [50] reported that the sweet corn cultivars intolerant to nicosulfuron contained sh_2 or sugar-enhanced (se) endosperm mutant. Cultivars that were most tolerant contained the sh_2 gene.

A soybean (*Glycine max* [L.] Merr.) mutation conferring resistance to a wide range of SUs has been reported [51]. The resistance was monogenic, semidominant, and not allelic to any of the recessive genes hs1, hs2, or hs3. Several mutants resistant to the herbicides chlorsulfuron, benzenesulfonamide, and sulfometuron-methyl have been isolated from cultured cells of *Nicotiana tabacum*. Chaleff and Ray [37] reported that resistance was inherited as a single dominant or semi-dominant mutation in all cases of resistance.

6.2.5.2 Basis for Selectivity

The differential response of SUs to different plants led to the study of selectivity of this class of herbicides. Differences in the sensitivity to chlorsulfuron of up to 4000-fold were observed between highly sensitive broadleaf plants, such as mustard, sugarbeet, soybean, and cotton [52]. These large differences in sensitivity could not be explained in terms of differences in penetration or translocation, nor could they be explained by differences in the sensitivities of the ALS enzymes from these plants to chlorsulfuron [30].

Metabolism studies, however, indicated a highly positive correlation between tolerance and the rate of [^{14}C] chlorsulfuron metabolism. Sweetser et al. [52] reported that chlorsulfuron is first metabolized in wheat and other tolerant grasses to the 5-hydroxyphenyl intermediate and then rapidly conjugated with glucose. There was no 5-hydroxychlorsulfuron detected in wheat extracts, indicating that the final conjugation step was very rapid. Synthesis and whole-plant testing of the conjugate demonstrated that it was herbicidally inactive. Hence, they concluded that rapid metabolism was the basis of chlorsulfuron selectivity. Oxygenase and glucosyltransferase are involved in this two-step detoxification process in wheat. With the exception of the resistant tobacco mutants, rapid metabolism was the principal factor responsible for the differential responsiveness of plants to SUs. The pathway of sulfonylurea detoxification can vary greatly depending on the herbicide and crop involved.

6.2.5.3   Resistance to SUs

SUs have been used in agricultural crops for over 25 years and numerous weed species have been reported to demonstrate resistance to these ALS-inhibiting herbicides (SUs and imidazolinones). Various weeds have been reported to be resistant to ALS-inhibiting herbicides, which might involve two mechanisms: increased metabolism of the herbicides or a herbicide-insensitive ALS [41,45]. Continual or overuse of SUs should be avoided to minimize the risk of developing resistance by weeds to the ALS-inhibiting herbicides, and herbicides with different modes of action should be used in rotation.

6.2.6   APPLICATIONS OF SUs

The discovery of SUs along with the imidazolinones was one of the most exciting breakthroughs in the field of herbicide research in several decades. With their unprecedented herbicidal activity, the application rates plummeted to grams rather than kilograms per hectare. The environmental and ecological profiles were highly favorable for the low-dose applications of these new chemistries.

The site of action of the SUs was pinpointed to be the enzyme ALS. Inhibition of this enzyme results in cessation of production of the three essential amino acids, valine, leucine, and isoleucine. This leads to retarded growth and eventually plant death. The absence of this enzyme in mammals explains the low toxicity of the SUs. Hence, they can be regarded as safe herbicides ($LD_{50} > 4000$ mg kg^{-1} of body weight) [4]. All plants contain the target enzyme, making them susceptible, but the ability of these herbicides to control grass weeds in monocotyledonous crops without affecting the latter has remained a challenge. Crop tolerance has been credited to the ability of the crop plants to rapidly convert the herbicide to inactive products. This inactivation occurs so rapidly that the active molecule never reaches the enzyme in sufficient quantities to effectively inhibit it.

SUs degrade under field conditions at rates similar to and often faster than conventional herbicides. Chemical hydrolysis and microbial breakdown are the main modes of degradation. The SUs are weak acids and under acidic soil conditions often undergo rapid degradation by chemical hydrolysis. Under alkaline soil conditions, microbial breakdown is the predominant degradation method. With the help of mutant forms of the ALS gene that code for insensitive forms of the enzyme, it has been possible to work on the genetic engineering of crops with high levels of SU resistance.

6.2.6.1   Applications in Turfgrasses

Several SUs and imidazolinone herbicides were introduced into the turf market for selective control of broadleaf weeds, difficult-to-control grasses like annual bluegrass (*Poa annua*), clumpy ryegrass (*Lolium perenne*), creeping bentgrass (*Agrotis palustris* Huds.), and sedges like yellow nutsedge (*Cyperus esculentus*), purple nutsedge (*Cyprus rotundus*), green kyllinga (*Kyllinga brevifolia*), and false green kyllinga (*Kyllinga gracillima*). The highly versatile SUs provide golf-course superintendents and professional turf managers with tools to control weeds in cool- and warm-season turfgrasses. They can be used on warm-season bermudagrass before overseeding with cool-season turfgrasses like perennial ryegrass or *Poa trivialis* in fall, or they can be used during spring transition to remove the cool-season turfgrass. Several SUs have been registered for use on golf courses and sports turf facilities such as metsulfuron (Manor or Blade), chlorsulfuron (Corsair), foramsulfuron (Revolver), halosulfuron (SedgeHammer, formerly Manage), rimsulfuron (TranXit GTA), trifloxysulfuron-sodium (Monument), flazasulfuron (proposed Katana), and sulfosulfuron (Certainty). Another new herbicide bispyribac-sodium (Velocity) for *P. annua* control works on the same ALS enzyme.

Some SUs like halosulfuron can be used to selectively control sedges in cool- and warm-season turfgrasses while some like trifloxysulfuron-sodium can control sedges safely in only warm-season turfgrasses [53]. Trifloxysulfuron-sodium controls various sedges and can be applied as a spring-transition aid in removing overseeded winter turf since it controls perennial ryegrass rapidly [54]. When perennial ryegrass escapes from an overseeded area and infests areas of dormant bermudagrass or survives through the excessive heat of desert summers, it becomes clumpy and is unsightly. It is more difficult to control than a dense stand of well-managed perennial ryegrass. Foramsulfuron has been reported to be very effective in controlling clumpy ryegrass [55].

SUs do not necessarily leach downward through the soil profile but tend to move laterally, so caution should be exercised for applications of SUs in saturated soils. To reduce lateral movement, a short irrigation (0.6 cm) should be applied after herbicide application [54]. Hydrolysis of SUs leads to degradation of the parent herbicide molecule and is favored under acidic soil-pH conditions compared with neutral and basic soil-pH conditions [56].

SUs are translocated via the phloem in the plants to storage organs like rhizomes, stolons, tubers, or bulbs and hence they are very effective in controlling hard-to-control weeds like quackgrass (*Elytrigia repens*) or different sedges. Herbicides like trifloxysulfuron-sodium have been reported to control yellow (*Cyperus esculentus*) and purple nutsedge (*C. rotundus*) as well as green kyllinga (*Kyllinga brevifolia*) and false green kyllinga (*K. gracillima*) [54]. Sulfosulfuron, bispyribac-sodium, foramsulfuron, and trifloxysulfuron-sodium have been reported to effectively control *P. annua* in non-overseeded bermudagrass turf [54,55]. The efficacy of SUs can be enhanced by adding an adjuvant like methylated seed oil or a nonionic surfactant like organo-silicone surfactants [57].

6.2.6.2 Benefits of Using SUs

The SUs have very low rates of application, low mammalian toxicity, and are very selective among several turfgrass species. They can be used for hard-to-control sedges, grasses, and broadleaf weeds with single or sequential applications. Since these herbicides affect the production of essential amino acids unique to plants, the herbicidal symptoms develop slowly and the speed at which they control cool- or warm-season weeds can vary based on the products and environmental conditions. Golf-course superintendents, sports-turf managers, and landscape-maintenance managers can use these herbicides as tools to achieve a successful overseeding program, spring transition, or control over a broad spectrum of weeds.

6.2.6.3 Precautions of Using SUs

Some of the SUs adsorb very strongly on clay and organic matter under low-pH soil conditions. In higher alkaline-pH conditions, these herbicides are desorbed into the soil solution and can become available to be taken up by the turfgrasses to cause injury. The herbicide present in the soil solution generally causes phytotoxicity since it is easily taken up by the weed or turfgrass roots. Lateral movement of these herbicides as runoff with excessive irrigation can cause injury to sensitive turf.

"Tracking" or off-site movement by golf-cart tires or on the soles of golf shoes may occur on wet turf from irrigation or dew. Turfgrasses grown on coarse-textured sandy soils and low-organic-matter soils are more prone to injury with SUs. Applications should be made to actively growing weeds for achieving optimum weed control since the herbicide has to be absorbed and translocated by plant tissues to work effectively. SUs or any other ALS-inhibiting herbicides should not be used continuously over a long period of time. These herbicides should be rotated with other herbicide chemistries that have different modes of action like pronamide ([3,5-dichloro-*N*-(1,1-dimethyl-2-propynyl)benzamide], Kerb), which inhibits mitotic cell division in sensitive plants.

6.2.6.4   *P. annua* Control before Overseeding

P. annua infestation in winter-overseeded perennial ryegrass is a major problem for golf-course superintendents and other professional turf managers. Introduction of various SUs provide new tools in managing *P. annua* [58]. One strategy to control *P. annua* is to apply SUs before over-seeding, but care should be taken not to apply the herbicides too close to the time of overseeding. Trifloxysulfuron-sodium, foramsulfuron, rimsulfuron, and sulfosulfuron applied preemergence prior to overseeding have shown to be effective in controlling *P. annua*. Since SUs are systemi-cally active in plants, the absorption, translocation, and inhibition of the ALS enzyme takes at least 14 days to observe optimum control of *P. annua*.

Trifloxysulfuron-sodium application at 23 g a.i. ha^{-1}, foramsulfuron at 30 g a.i. ha^{-1}, or sulfosul-furon at 75 g a.i. ha^{-1} can control over 90% of the *P. annua* population within 28 days of treatment (DAT) [59]. Optimum control is achieved between 30 and 60 DAT (Figures 6.3 through 6.5). Effi-cacy of a single application of SUs is sometimes reduced to less than acceptable levels after 90 DAT [59]. Sequential applications of SUs within 4–6 weeks of the first application are more effective in controlling *P. annua* compared with a single application [58,59].

Sulfosulfuron has also been shown to be very effective in controlling *P. annua* when applied at 75 g a.i. ha^{-1}. Optimum control is achieved between 21 and 45 DAT with a single application of sul-fosulfuron (Figure 6.4). The efficacy of a single application of sulfosulfuron declines after 75 DAT. Sequential applications would expand the window of controlling *P. annua* control. *P. annua*

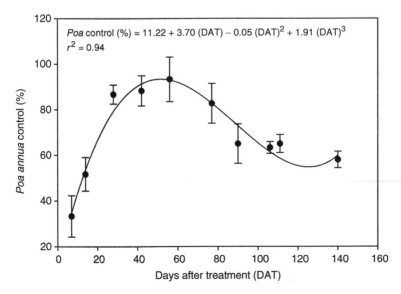

FIGURE 6.3   *P. annua* control with a single application of foramsulfuron at 30 g a.i. ha^{-1} 14 DBO. The first rating date was the day of overseeding. The data points were mean values of the visual rating of *P. annua* control while the curve was fitted to a quadratic regression model.

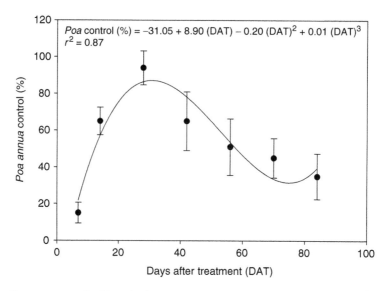

FIGURE 6.4   *P. annua* control with a single application of sulfosulfuron at 75 g a.i. ha^{-1} applied 10 DBO. The first rating date was the day of overseeding. The data points were mean values of the visual rating of *P. annua* control while the curve was fitted to a quadratic regression model.

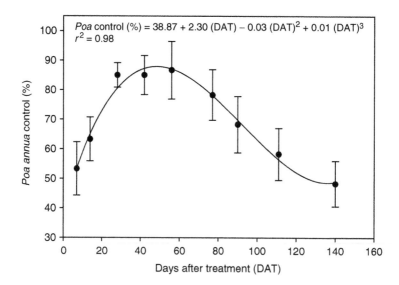

FIGURE 6.5   *P. annua* control with a single application of trifloxysulfuron-sodium at 23 g a.i. ha^{-1} applied 14 DBO. The first rating date was the day of overseeding. The data points were mean values of the visual rating of *P. annua* control while the curve was fitted to a quadratic regression model.

plants are very aggressive and are prolific seedhead producers, so they start producing new plants as soon as the efficacy of SUs decline. Hence sequential applications are needed to achieve season-long control of *P. annua* populations [59].

Trifloxysulfuron-sodium application at 23 g a.i. ha^{-1} has been reported to control *P. annua* effectively [54], and the optimum level of control was achieved between 30 and 70 DAT [59] (Figure 6.3). The efficacy of a single application of trifloxysulfuron-sodium declined after 80 DAT. In an experiment conducted on dormant bermudagrass to evaluate the efficacy of *P. annua* control

TABLE 6.1

Number of *Poa annua* Plants in a 100-m² Area after a Single and Sequential Application of Foramsulfuron (30 g a.i. ha⁻¹) and Trifloxysulfuron-Sodium (23 g a.i. ha⁻¹) on a Hybrid Bermudagrass Fairway

Treatments	111 DAT	140 DAT
Trifloxysulfuron-sodium; single application	280 b	348 b
Trifloxysulfuron-sodium; sequential application	148 c	280 c
Foramsulfuron; single application	148 c	228 c
Foramsulfuron; sequential application	120 d	165 d
Untreated control	908 a	1052 a

Note: Means were separated using Duncan's New Multiple Range Test (DNMRT) ($p = 0.05$). Means within columns followed by the same letter were not significantly different at $p = 0.05$ level. DAT indicates days after treatment.

with trifloxysulfuron-sodium and foramsulfuron at 23 and 30 g a.i. ha⁻¹, respectively, sequential applications of the herbicides were very effective in controlling *P. annua* up to 140 DAT [59] (Table 6.1).

6.2.6.5 Perennial Ryegrass Injury

SUs are effective tools for controlling weeds before overseeding bermudagrass tees and fairways. A potential drawback with these products is the likelihood of injury to overseeded ryegrasses. Hence, the application timing of the SUs is very critical. Minimum injury to perennial ryegrass was observed with 23 g a.i. ha⁻¹ of trifloxysulfuron-sodium when applied 21 days before overseeding (DBO) compared with the application made 10 DBO [59] (Table 6.2). The extent of injury was not very severe at only 12% [59]. The ryegrass was slightly stunted and showed some yellowing 8 weeks after overseeding (WAO). The reason for the delayed response could be due to slower desorption of the herbicide molecule into the soil solution over a period of time after application since the experiments were conducted on a clay loam soil. The herbicide adsorbed strongly on the fine-textured soil after application and was not readily available in the soil solution. Slowly over a period of time, the herbicide desorbed from the soil surface and dissolved in the soil solution. The ryegrass plants absorbed the herbicide and chlorosis was observed. Under colder temperatures, ryegrass plants have reduced metabolism and the plants might not show injury symptoms immediately. Once temperatures start increasing, the ryegrass plants grow actively and hence absorb the herbicide from the soil solution, and injury results.

Maximum ryegrass injury of 12% at 3 WAO with sulfosulfuron applied at 36 g a.i. ha⁻¹ was observed in our experiments when the herbicide was applied 7 DBO [59] (Table 6.2). No injury to ryegrass was observed when sulfosulfuron was applied at 27 g a.i. ha⁻¹ at 10 DBO. The perennial ryegrass recovered from the injury by 7 WAO. Rimsulfuron at 54g a.i. ha⁻¹ applied 7 DBO caused over 45% injury to perennial ryegrass at 3 WAO (Table 6.2). The ryegrass injury increased to 60% at 5 WAO and at 7 WAO, 50% of the perennial ryegrass stand was lost [59]. Rimsulfuron was very effective in controlling *P. annua* within 3 WAO. To minimize injury to ryegrass, rimsulfuron should be applied 14–21 DBO at lower rates.

6.2.6.6 Using ALS-Inhibiting Herbicides to Control Difficult-to-Control Weeds

6.2.6.6.1 Clumpy Ryegrass Control

When perennial ryegrass survives extreme high temperatures of summer or when it escapes the overseeded area, this turf species can become clumpy weed and is unsightly [54]. Clumpy ryegrass

TABLE 6.2

Perennial Ryegrass Injury with Different SUs

SU Herbicides (g a.i. ha^{-1})	Days before Overseeding (DBO)	Weeks after Overseeding (WAO)					
		3	5	7	9	12	15
		% Injury					
Sulfosulfuron (36)	10	3.3	3.0	2.5	1.0	0	0
	7	13.3	15.0	12.0	10.0	5.0	2.0
	3	1.7	1.7	1.0	0	0	0
Trifloxysulfuron-sodium (23)	21	0	0	5.0	6.5	3.5	0
	14	2.5	1.8	6.5	8.8	4.5	0
	10	5.5	7.5	8.0	12.5	10.5	5.0
Rimsulfuron (54)	10	25.0	35.0	25.0	20.5	10.5	10.0
	7	32.0	45.0	40.0	35.5	24.5	15.5
	3	45	61.5	50.0	45.5	35.0	27.5
LSD ($p = 0.05$)		1.25	2.20	5.50	6.50	3.55	4.55

Note: The SUs were applied either 21, 14, 10, 7, or 3 DBO and the ryegrass injury was visually rated on a 0–100 scale where
0 = no injury and 100 = complete loss of stand at 3, 5, 7, 9, 12, and 15 WAO. Means were separated using least
significant difference (LSD) ($p = 0.05$).

is more difficult to control than a well-managed overseeded perennial ryegrass turf. Pronamide is not effective in controlling clumpy ryegrass but some of the SUs have been reported to be very effective [54].

In our experiments, foramsulfuron applied at 45 g a.i. ha^{-1} was very effective in controlling clumpy ryegrass at 5 weeks after treatment (WAT) [58] (Figure 6.6). A sequential application 4–6 weeks after the initial application provided superior control of clumpy ryegrass compared with a single application. Foramsulfuron applied during late spring or early summer provided optimum control of clumpy ryegrass compared with early spring applications. Foramsulfuron uptake by plants is through the foliage; hence under higher temperatures, when the plant is actively growing, the herbicide is absorbed and translocated within the plant faster compared with minimal translocation at lower temperatures.

The early spring application of foramsulfuron (April–May) was not as effective as the later application during early summer (June–July) and late summer (August–September) in controlling clumpy ryegrass for all rates of application, as can be inferred from the slopes of the regression curves in Figure 6.7. For all the application times, the optimum rate was around 45 g a.i. ha^{-1}, as observed from the maxima of the regression curves for percent control of clumpy ryegrass 4 WAT. The highest rate of foramsulfuron at 90 g a.i. ha^{-1} controlled 100% of the clumpy ryegrass 8 WAT [59].

The sequential application of foramsulfuron at 45 g a.i. ha^{-1} during the early application timing (spring) did not result in significantly higher percentage control of clumpy ryegrass. At 4 WAT, the 45 g a.i. ha^{-1} rate of application controlled 68% of the clumpy ryegrass when applied during spring, while it controlled 88% when applied during early summer and 100% control when applied during late summer [59]. Hence, a higher level of control was achieved when foramsulfuron was applied later in summer. The higher summer temperatures exposed the clumpy ryegrass plants to heat stress and may have increased the activity of foramsulfuron owing to increased uptake and translocation of the herbicide when applied at the later application date. The role of sequential application became more critical during the later application timings, which resulted in higher levels of clumpy ryegrass

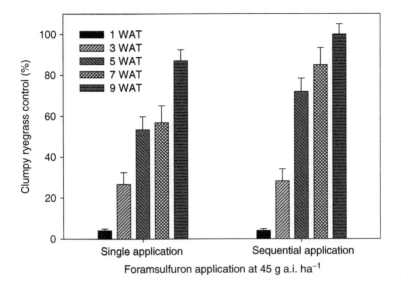

FIGURE 6.6 Clumpy ryegrass control 4 weeks after initial application of single and sequential applications of foramsulfuron at 45 g a.i. ha^{-1}.

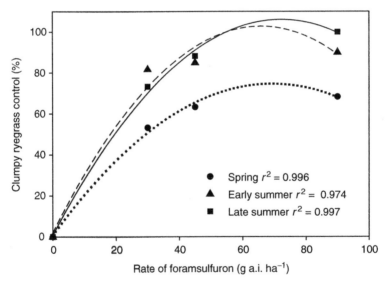

FIGURE 6.7 Effect of application timing and rate of foramsulfuron 4 weeks after each application on clumpy ryegrass control.

control. Following the late summer timing of application, the sequential applications resulted in significantly higher level of clumpy ryegrass control at all rating dates [59].

6.2.6.6.2 Nutsedge Control

Purple nutsedge (*Cyperus rotundus* L.) control strategies in the desert southwest and other warm-season turfgrass-growing regions of the United States include the use of SUs and imidazolinone herbicides along with monosodium acid methanearsonate (MSMA). Currently registered products that are effective against nutsedge in most warm-season turfgrasses include halosulfuron (Sedge-Hammer, formerly Manage), trifloxysulfuron-sodium (Monument), sulfosulfuron (Certainty), and imazaquin (Image). Flazasulfuron is in development stages and shows promise to effectively control nutsedge. Imazaquin, an imidazolinone herbicide, was the first of the ALS-inhibiting herbicides to be marketed for nutsedge control in turf in the mid-1980s. The effective use rate is 0.56 kg a.i. ha^{-1} and usually multiple applications during the active growing season are required to reduce nutsedge populations [60,61]. Imazaquin combined with MSMA is enhanced in activity and reduces nut-sedge more rapidly than when applied alone. The addition of a nonionic surfactant or oil adjuvant improves the efficacy of imazaquin, but turfgrass injury could occur. Imazaquin activity in nutsedge is uptake by both shoots and roots with rapid translocation to the growing points [62]. Plant growth stops immediately and chlorosis can be observed within 3–7 DAT. Addition of MSMA causes foliar burning. Complete death of treated nutsedge plants occurs within 2 WAT.

Shortly after the introduction of imazaquin, halosulfuron entered the marketplace as another effective tool for nutsedge control. The effective use rate for halosulfuron is 70 g a.i. ha^{-1}. Its activity is similar to imazaquin, rapidly entering and translocating to growing points in nutsedge; however, halosulfuron exhibits differential selectivity to perennial ryegrass. In the desert region, purple nut-sedge can begin to appear in overseeded winter ryegrass turf as early as March. Halosulfuron can be safely applied in the early spring to initiate nutsedge control in ryegrass before spring emergence of bermudagrass [63,64].

More recently, since 2000 new SUs have become available for turf managers to effectively control nutsedge. Trifloxysulfuron-sodium at an effective rate of 30 g a.i. ha^{-1} and sulfosulfuron at 70 g a.i. ha^{-1} have shown excellent selectivity in hybrid bermudagrasses while effectively reduc-ing nutsedge populations [65,66]. Unlike halosulfuron, trifloxysulfuron-sodium and sulfosulfuron are not selective to perennial ryegrass. Trifloxysulfuron-sodium is labeled a transition-aid product to remove overseeded winter ryegrass during spring transition to enable bermudagrass to emerge from winter dormancy. Sulfosulfuron is currently being developed for use during spring transi-tion. Sulfosulfuron has been reported to be very effective in controlling yellow nutsedge (*Cyperus esculentus* L.) [67–69].

The number of applications necessary to control purple nutsedge using ALS-inhibiting herbi-cides depends on the severity of the infestation in turf and the timing of initiating applications. An application of halosulfuron, trifloxysulfuron-sodium, or sulfosulfuron will control existing nutsedge plants, but if there is a high population, tubers will continue to sprout and new plants will emerge within 4–6 weeks. A repeat application will be necessary to control the newly emerged population. Generally, two sequential applications spaced 4–6 weeks apart in July and August appear to reduce nutsedge populations in desert turfs [63,66]. When nutsedge populations emerge earlier during spring, SUs can be applied while up to four applications may be required during spring and summer to reduce nutsedge populations, and may not be economically feasible. MSMA could be utilized for earlier spring applications with repeated applications every 2 weeks prior to initiating treatments using SUs [70]. The multiple applications will need to be repeated for multiple seasons as long as viable tubers remain in the soil profile and new plants emerge.

6.2.6.7 Using SUs for Spring Transition

The SUs are very versatile herbicides that have utility for several aspects of weed control and turf management during overseeding and transition. Transition during the spring is the greenup and regrowth of warm-season bermudagrass while the overseeded winter turf recedes and diminishes. Generally, cool-season perennial ryegrass, *Poa trivialis*, and creeping bentgrass cannot tolerate the extreme heat of the desert summers and will eventually die out to allow bermudagrass to be the dominant summer turfgrass. Ideally, bermudagrass should have about a 100-day growing season to adequately grow a solid root system for winter carbohydrate storage in the rhizomes to enable regrowth in the following spring season to emerge from dormancy [71]. The pressures to maintain lush green turfgrasses on golf courses year-round have led to the development of hardy perennial ryegrass cultivars that withstand higher temperatures at fall overseeding time and survive well into the summers. The season for growing and establishing a bermudagrass turf foundation is getting shorter as winter grass varieties are improved.

To provide an adequate bermudagrass-growing season, SUs can complement cultural management practices to eliminate the competitive overseeded winter turfgrasses. At the onset of higher summer temperatures, bermudagrass grows out of winter dormancy while competing for light, water, and nutrients with the winter turfgrass. Lowering the height of cut will thin the ryegrass canopy to allow light penetration for bermudagrass growth. A light vertical mowing or verticutting will also thin out and place additional stress on the ryegrass while stimulating bermudagrass growth. Under conditions where cool temperatures or shady conditions exist to delay competitive bermudagrass growth, transition-aid herbicides can hasten the removal of ryegrass. Pronamide (Kerb) is a herbicide used to effectively remove ryegrass, but a major disadvantage in its use is the lateral movement or runoff away from the target site of application. It is also very slow to remove ryegrass. The SUs were found to effectively and rapidly remove ryegrass to allow bermudagrass transition. The SUs marketed for transition include metsulfuron-methyl (Manor, Blade), chlorsulfuron (Corsair), foramsulfuron (Revolver), rimsulfuron (TranXit), and trifloxysulfuron-sodium (Monument). Newer compounds currently under development are sulfosulfuron (Certainty) and flazasulfuron. Trifloxysulfuron-sodium can be used at rates as low as 5–16 g a.i. ha^{-1} for eliminating ryegrass, compared with 30 g a.i. ha^{-1} that controls nutsedge. Foramsulfuron is most effective at 30 g a.i. ha^{-1} and has also shown efficacy to provide postemergence control of *P. annua*. The SUs often have commercial labels that suggest a range of rates for transition. Efficacy for removing ryegrass is dependent on temperature and ideal growing conditions that optimize the uptake, translocation, and metabolism of SUs. Spring conditions that are cold are not conducive for optimal plant growth and applications of low or high rates of SUs will exhibit minimal herbicidal activity and a prolonged period for ryegrass removal. Warmer weather conditions will encourage a greater degree of herbicide activity where both low and high rates can be equally active to remove ryegrass rapidly.

The SUs are subject to off-site movement and caution must be exercised when they are used near sensitive areas such as golf-course greens composed of desired cool-season grasses. Lateral movement on slopes or "tracking" on soles of shoes or vehicle tires when the treated turf is wet can be detrimental. Caution must be used to apply SUs with proper irrigation to place and set the herbicide so that movement is minimized.

The transition-aid herbicides are new tools that golf-course superintendents and professional turf managers can use to enhance turf conditions [72–74]. The SUs applied early in the spring can effectively remove winter-overseeded grasses to enable bermudagrass to grow out of winter dormancy without ryegrass competition. If bermudagrass is weak or removed, ryegrass presents significant-sized bare soil patches; such areas are readily identified for early renovation or repair by sodding, sprigging, or seeding. The newly planted bermudagrass will be offered a long season of growing to establish a solid root and rhizome foundation [75]. The SUs have shown utility for weed control while establishing bermudagrass by seeding. The SUs caused injury that was generally short-lived and did not significantly reduce the rate of bermudagrass coverage in most treatments [76,77].

The SUs can supplement ryegrass removal in golf-course roughs where mowing heights cannot be lowered to reduce ryegrass competition [78]. In shady areas, under the dense canopy of trees where ryegrass can survive well into the summer, it can be eliminated by spraying SUs; however, there must be a stand of bermudagrass that can effectively fill the void when ryegrass is eliminated. When used in concert with cultural management practices, the SUs offer effective removal of winter-overseeded ryegrass to allow bermudagrass emergence and growth into summer.

6.3   CONCLUSIONS

- Sequential application of SUs is more effective than a single application in controlling *P. annua* and other difficult-to-control perennial weeds like sedges.
- Single applications of SUs are effective when conditions are optimal for weed growth.

- Optimum control of weeds is observed when SUs are applied to actively growing weeds that can absorb, translocate, and metabolize the herbicide.
- Lateral movement and tracking of SUs can occur with high soil moisture conditions or when foliage is excessively wet.
- SUs adsorb on the soil more under acidic soil conditions and desorb to the soil solution under basic soil conditions. Hence, basic soils are subject to causing phytotoxicity or injury to plants more than acidic soils.
- Herbicide resistance of several weeds to SUs has been observed; so golf-course superintendents and turf managers should be cautious not to use SUs excessively and continuously. SUs should be used in rotation with other herbicides with a different mode of action.
- SUs should be applied at least 10–14 DBO to reduce injury to ryegrass.
- Addition of an adjuvant such as oils or a nonionic surfactant increases the efficacy of SU herbicides.
- Most of the SUs are toxic to cool-season turfgrasses and care should be taken to reduce off-site movement.
- SUs are effective transition-aid tools.

REFERENCES

1. Koog, H.J., Jr. Netherlands Patent 121,788. *Deutsche Gold and silver- Scheidean-stalt Vormals*, 1966.
2. Levitt, G. U.S. Patent 4,398,939. *DuPont.* 1983.
3. Sauers, R.F. and Levitt, G. Sulfonylurea synthesis, in *Pesticide Synthesis through Rational Approaches*, Magee, P.S., Kohn, G. K., and Mean, J.J. Eds., American Chemical Society, Washington D.C., 1984, 21.
4. Kearney, P.C. and Kaufman, D.D. *Herbicides Chemistry, Degradation and Mode of Action*. Marcel Dekker Inc., New York, 1988, 118–183.
5. Levitt, G. U.S. Patent 4,169,719. *DuPont.* 1979.
6. Levitt, G. Human welfare and the environment, in *Pesticide Chemistry*, Miyamato, J. and Kearney, P.C. Eds., Pergamon Press, New York, 1983, 1–243.
7. Levitt, G. U.S. Patent 4,481,029. *DuPont.* 1984.
8. Levitt, G. U.S. Patent 4,435,206. *DuPont.* 1984.
9. Levitt, G. U.S. Patent 4,370,479. *DuPont.* 1983.
10. Selby, T.P. and Wolf, A.D. U.S. Patent 4,421,550. *DuPont.* 1983.
11. Topfl, W., Kristinsson, H., and Meyer, W. Australian Patent Appl. 16890/83. *Ciba-Geigy.* 1983.
12. Levitt, G. U.S. Patent 4,339,267. *DuPont.* 1982.
13. Zimmerman, W.T. U.S. Patent 4,293,320. *DuPont.* 1981.
14. Levitt, G. U.S. Patent 4,293,330. *DuPont.* 1981.
15. Sax, N.I. *Dangerous Properties of Industrial Materials*, Van Norstrand Reinhold, New York, 1979, 5–978.
16. VanEllis, M.R. and Shaner, D.L. Mechanism of cellular absorption of imidazolinones in soybean (*Glycine max*) leaf discs, *Pestic. Sci.*, 23, 25–34, 1988.
17. Price, T.P. and Balke, N.E. Characterization of atrazine accumulation by excised velvetleaf (*Abutilon theophrasti*) roots, *Weed Sci.*, 31, 14–19, 1983.
18. Mersie, W. and Singh, M. Norflurazon absorption by excised velvetleaf (*Abutilon theophrasti*) roots, *Weed Sci.*, 30, 633–639, 1987.
19. Donaldson, T.W., Bayer, D.E., and Leonard, O.A. Absorption of 2,4-dichlorophenoxyacetic acid and 3-(p-chlorophenyl)-1,1-dimethyl urea (monuron) by barley roots, *Plant Physiol.*, 52, 638–645, 1973.
20. Lichtner, F.T. 1983. Amitrole absorption by bean (*Phaseolus vulgaris* L. cv "red kidney") roots, *Plant Physiol.*, 71, 307–312, 1973.
21. Minocha, S.C. and Nissen, P. Uptake of 2,4-dichlorophenoxyacetic acid and indole acetic acid in tuber slices of Jerusalem artichoke and potato, *J. Plant. Physiol.*, 120, 351–362, 1985.
22. Devine, M.D., Bestman, H.D., and VandenBorn, W.H. Uptake and accumulation of the herbicide chlorsulfuron and clopyralid in excised pea root tissue, *Plant Physiol.*, 85, 82–86, 1987.
23. Reddy, K.N. and Bendixen, L.E. Toxicity, absorption, translocation and metabolism of foliar-applied chlorimuron in yellow and purple nutsedge (*Cyperus esculentus* and *C. rotundus*), *Weed Sci.*, 36, 707–712, 1988.

24. Reddy, K.N. and Bendixen, L.E. Toxicity, absorption and translocation of soil-applied chlorimuron in yellow and purple nutsedge (*Cyperus esculentus* and *C. rotundus*), *Weed Sci.*, 37, 147–151, 1989.

25. Nandihalli, U.B. and Bhowmik, P.C. Chlorimuron absorption by excised velvetleaf (*Abutilon theophrasti*) roots, *Weed Sci.*, 37, 29–33, 1989.

26. Nandihalli, U.B. and Bhowmik, P.C. Uptake and efflux of chlorimuron ethyl by excised soybean (*Glycine max* L. Merr) root tissue, *Weed Res.*, 31, 295–300, 1991.

27. Hall, L.M. and Devine, M.D. Chlorsulfuron inhibition of phloem translocation in chlorsulfuron-resistant and susceptible *Arabidopsis thaliana, Pestic. Biochem. Physiol.*, 45, 81–90, 1993.

28. Ray, T.B. The mode of action of chlorsulfuron: A new herbicide for cereals, *Pestic. Biochem. Physiol.*, 17, 10–17, 1982.

29. Hatzios, K.K. and Howe, C.M. Influence of the herbicides hexazinone and chlorsulfuron on the metabolism of isolated soybean leaf cells, *Pestic. Biochem. Physiol.*, 17, 207–214, 1982.

30. Ray, T.B. Site of action of chlorsulfuron, *Plant Physiol.*, 75, 827–831, 1984.

31. Suttle, J.C. and Schreiner, D.R. Effects of DPX-4189 [2-chloro-N{(4-methyl-1,3,5-triazin-2-yl)amino-carbonyl}benzenesulfonamide] on anthocyanin synthesis, phenylalanine ammonia lyase activity and ethylene production in soybean hypocotyls, *Can. J. Bot.*, 60, 741–745, 1982.

32. Hageman, L.H. and Behrens, R. Basis for response differences of two broadleaf weeds to chlorsulfuron, *Weed Sci.*, 32, 162–167, 1984.

33. La Rossa, R.A. and Scloss, J.V. The sulfonylurea herbicide sulfometuron methyl is an extremely potent and selective inhibitor of acetolactate synthase is *Salmonella typhimurium*, *J. Biol. Chem.*, 259, 8753–8757, 1984.

34. Jones, A.V., Young, R.M., and Leto, K.J. Subcellular localization and properties of acetolactate synthase, target site of the sulfonylurea herbicides, *Plant. Physiol.*, 77, 293, 1985.

35. Stidham, M.A. Herbicides that inhibit acetohydroxyacid synthase, *Weed Sci.*, 39, 428–434, 1991.

36. Subramanian, M.V., Gallant, V.L., Dias, J.M., and Mireles, L.C. Acetolactate synthase inhibiting herbicides bind to the regulatory site, *Plant. Physiol.*, 96, 310–313, 1991.

37. Chaleff, R.S. and Ray, T.B. Herbicide-resistant mutants for tobacco cell cultures, *Crop Sci.*, 223, 1148–1151, 1984.

38. Falco, S.C., Dumas, K.S., and McDevitt, R.E. *Molecular Form and Function of the Plant Genome*, Plenum Press, New York, 1985.

39. Schloss, J.V. Interaction of the herbicide sulfometuron methyl with acetolactate synthase: a slow binding inhibition, in *Flavins and Flavoproteins*, Bray, R.C., Engel, P.C., and Mayhew, S.G. Eds., Walter De Gruyter Inc., New York, 1984, 737–740.

40. Chaleff, R.S. and Mauvais, C.V. Acetolactate synthase is the site of action to two sulfonylurea herbicides in higher plants, *Crop. Sci.*, 224, 1443–1445, 1984.

41. Mallory-Smith, C.A., Thill, D.C., and Dial, M.J. Identification of sulfonylurea herbicide resistant prickly lettuce (*Lactuca serriola*), *Weed Technol.*, 4, 163–168, 1990.

42. Thill, D.C., Mallory-Smith, C.A., Saari, L.L., Cotterman, J.C., and Primiani, M.M. Sulfonylurea resistance—mechanism of resistance and cross resistance, in *Proc. Weed Sci. Soc. Am.*, 9, 132, 1989.

43. Smith, W.F., Cotterman, J.C., and Saari, L.L. Effect of ALS-inhibitor herbicides on sulfonylurea-resistant weeds, in *Proc. West. Soc. Weed Sci.*, 43, 24–25, 1990.

44. Dekker, J.H., Lux, J.F., and Burmester, R.G. Mulch-till corn weed control with experimental sulfonylurea herbicides, in *North Central Weed Cont. Conf. Res. Rep.*, 43, 165–166, 1986.

45. Heap, I. and Knight, R. The occurrence of herbicide cross-resistance in a population of annual ryegrass, *Lolium rigidum,* resistant to diclofop-methyl, *Aust. J. Agric. Res.*, 37, 149–156, 1986.

46. Eberlein, C.V. and Miller, T.L. Corn tolerance and weed control with thiameturon, *Weed Technol.*, 3, 255–260, 1989.

47. Eberlein, C.V., Rosow, K.M., Geadelman, J.L., and Openshaw, S.J. Differential tolerance of corn genotypes to DPX-M6316, *Weed Sci.*, 37, 651–657, 1989.

48. Monks, D.W., Mullins, C.A., Johnson, K.E., and Onks, D.O. Effect of Accent (DPX-V9360) and Beacon (CGA-136872) on sweet corn and johnsongrass, in *Proc. Southern Weed Sci. Soc.*, 43, 179, 1990.

49. Stall, W.M. and Bewick, T.A. Tolerance variability among sweet corn cultivars to DPX-V9360, in *Proc. Southern Weed Sci. Soc.*, 43, 170, 1990.

50. Stall, W.M. and Bewick, T.A. Sweet corn cultivars respond differentially to the herbicide nicosulfuron, *Hort. Sci.*, 27(2), 131–133, 1992.

51. Sebastian, S.A., Fader, G.M., Ulrich, J.F., Forney, D.R., and Chaleff, R.S. Semidominant soybean mutation for resistance to sulfonylurea herbicides, *Crop. Sci.*, 29, 1403–1408, 1989.

52. Sweetser, P.B., Schow, G.S., and Hutchinson, J.M. Metabolism of chlorsulfuron by plants: Biological basis for selectivity of a new herbicide for cereals, *Pestic. Biochem. Physiol.*, 17, 18–23, 1982.

53. Murphy, T., Waltz, C., Ferrell, J., and Yelverton, F. Sulfonylurea herbicides: How do different turfgrasses tolerate them? *Turfgrass Trends, Golfdom*, March, 69–72, 2004.

54. Yelverton, F. New weed control in warm-season grasses, *Golf Course Management*, January, 203–206, 2004.

55. Yelverton, F. A new herbicide for weeds in bermudagrass and zoysiagrass, *Golf Course Management*, May, 119–122, 2003.

56. Sarmah, A.K., Kookana, R.S., Duffy, M.J., Alston, A.M., and Harch, B.D. Hydrolysis of triasulfuron, metsulfuron-methyl and chlorsulfuron in alkaline soil and aqueous solutions, *Pest Manage. Sci.*, 56(5), 463–471, 2000.

57. Mitra, S., Bhowmik, P.C., and Xing, B. Synergistic effects of an oil and a surfactant in influencing the activity of rimsulfuron in weed control, in *Proc. Fifth Int. Symp. Adjuvants Agrochem.*, Memphis, U.S.A., 5(1), 298–304, 1998.

58. Mitra, S. Sulfonylureas target ALS enzyme to control grasses, broadleaf weeds, *Turfgrass Trends, Golfdom*, March, 68–72, 2005.

59. Mitra, S. SU herbicides control weeds, *Poa* in advance of overseeding program, in *Turfgrass Trends, Golfdom*, October, 72–76, 2005.

60. Umeda, K. and Towers, G. Purple nutsedge control in turfgrass with ALS-inhibiting herbicides, in *Proc. Western Soc. Weed Sci.*, 58, 65, 2005.

61. Umeda, K. and Towers, G. Comparing and determining effective rates of herbicides for purple nutsedge control in turfgrass, in *Res. Prog. Rep. Western Soc. Weed Sci.*, 87, 2006.

62. Shaner, D.L. and O'Connor, S.L. *The Imidazolinone Herbicides*, CRC Press, Boca Raton, FL, 1991.

63. Umeda, K. and Towers, G. Purple nutsedge control in turfgrass with various timings and combinations of herbicides, in *Res. Prog. Rep. Western Soc. Weed Sci.*, 21, 2005.

64. Umeda, K. and Towers, G. Spring initiated application of herbicides for nutsedge control and effects on overseeded ryegrass, in *Res. Prog. Rep. Western Soc. Weed Sci.*, 91–92, 2006.

65. Umeda, K. and Towers, G. Efficacy of herbicides for nutsedge control in turf, in *The 2004 Turfgrass, Landscape and Urban IPM Research Summary*, Kopec, D. and Casler, R. Eds., College of Agriculture and Life Sciences, University of Arizona, vol. AZ1359, series P-141, 60–64, 2004.

66. Umeda, K. and Towers, G. Comparison of sulfonylurea herbicides for spring transition, in *The 2004 Turfgrass, Landscape and Urban IPM Research Summary*, Kopec, D. and Casler, R. Eds., College of Agriculture and Life Sciences, University of Arizona, vol. AZ1359, series P-141, 167–174, 2004.

67. Bhowmik, P.C., Sarkar, D., and Riego, D. Activity of sulfosulfuron in controlling yellow nutsedge and quackgrass, in *Proc. Northeast. Weed Sci. Soc.*, 60, 86, 2006.

68. Bhowmik, P.C. Yellow nutsedge control with MON 44951 in cool-season turfgrass. (low volume), in *Massachusetts Weed Sci. Res. Results*, 24, 21–23, 2005.

69. Bhowmik, P.C. Yellow nutsedge control with MON 44951 in cool-season turfgrass (high volume), in *Massachusetts Weed Sci. Res. Results*, 24, 25–27, 2005.

70. Brecke, B.J., Stephenson, D.O., and Unruh, J.B. Control of purple nutsedge (*Cyperus rotundus*) with herbicides and mowing, *Weed Technol.*, 19(4), 809–814, 2005.

71. Keese, R., Spak, D., and Sain, C. A practical user's guide to the sulfonylurea herbicides. New Tools for the Golf Course Superintendent, *USGA Green Section Record*, July–August, 16–18, 2005.

72. Howard, H.F. Proactive vs. passive transition. Part I. Pay now or pay later, *Golf Course Management*, 74(2), 93–96, 2006.

73. Howard, H.F. and Elwood, P. Proactive transition. Part II. Making it happen, *Golf Course Management*, 74(3), 111–114, 2006.

74. Kopec, D.M., Gilbert, J., Pesserakli, M., Evans, P., Ventura, B., and Umeda, K. Response of common bermudagrass sports turf to select herbicides used for spring transition enhancement, in *The 2004 Turfgrass, Landscape and Urban IPM Research Summary*, Kopec, D. and Casler, R. Eds., College of Agriculture and Life Sciences, The University of Arizona, vol. AZ1359, series P-141, 254–278, 2004.

75. Kopec, D.M. and Gilbert, J.J. Spring transition of Tifway 419 bermudagrass as influenced by herbicide treatments, in *The 2004 Turfgrass, Landscape and Urban IPM Research Summary*, Kopec, D. and Casler, R. Eds., College of Agriculture and Life Sciences, The University of Arizona, vol. AZ1359, series P-141, 319–329, 2004.

76. Richardson, M.D., Karcher, D.E., Boyd, J.W., McCalla, J.H., and Landreth, J.W. Tolerance of "Riviera" bermudagrass to MSMA tank-mixtures with postemergence herbicides during establishment from seed, *Plant Management Network, Appl. Turfgrass Sci.*, 2005. Available at http://www.plantmanagementnetwork. org/sub/ats/research/2005/riviera/

77. Murphree, T.A., Rodgers, C.A., Umeda, K., and Towers, G.W. Postemergence herbicide effects on establishment of seeded bermudagrass turf, in *Proc. Annu. Meet. Southern Weed Sci. Soc.*, 59, 79, 2006.

78. Umeda, K. Jumpstarting bermuda, *Turfgrass Trends, Golfdom*, September, 68–69, 2006.

7 Management of Tropical Turfgrasses

Greg Wiecko

CONTENTS

7.1   INTRODUCTION

Grasses are divided into two major groups of species, those suitable for warm climates and those suitable for cool climates [1]. The first, called warm-season grasses, are best adapted to temperatures between 27°C and 35°C (80–95°F) and the second, called cool-season grasses, are best adapted to temperatures between 18°C and 24°C (65–75°F). A unique characteristic of warm-season grasses is that they undergo dormancy when soil temperature drops below 10°C (50°F). Tropical grasses are essentially warm-season grasses that never become dormant and often manifest little tolerance for cold stress [11]. This chapter will mostly discuss warm-season species grown in the climates where temperature is relatively uniform and warm year round.

7.2   TURFGRASS SPECIES

All grasses belong to a single family of plants, the *Poaceae*. This family is divided into six sub-families, which incorporate 25 tribes, 600 genera, and 7500 species. Only less than 40 species are used as turfgrasses. All turfgrass species are within the three subfamilies: *Festucoideae, Panicoideae,* and *Eragrostoideae* [5]. Turfgrasses in the *Festucoideae* are usually adapted to cool climates. Those in the *Panicoideae* and *Eragrostoideae* are adapted to warm climates. Cool- and warm-season grasses require substantially different management practices, so the information provided in this chapter may not be appropriate for cool locations.

7.2.1 WARM-SEASON TURFGRASSES

7.2.1.1 The Genus *Cynodon* (Eragrostoideae)

Bermudagrass (*Cynodon dactylon*) is the most widely distributed turfgrass around the world and in the warm climate; it grows fast and continuously, usually staying dense and medium to dark green year round [7]. Where the soil temperature drops to 10°C (50°F) for 10–14 days, bermudagrass loses its chlorophyll and turns yellow to light brown and remains dormant until the soil temperature rises again and remains above 10°C (50°F). Bermudagrass is drought tolerant and has a good tolerance to wear. It is well adapted to sunny conditions and has a medium-coarse texture with a greyish-green color. Common bermudagrass establishes a deep root system and produces long and lively rhizomes and stolons. The only stress it does not tolerate well is low light intensity and therefore performs poorly under tree shade.

The genus *Cynodon* includes eight species besides *C. dactylon*. Several of them and their hybrids are used as turfgrasses and range in importance from major to marginal.

The *Cynodon dactylon* × *Cynodon transvaalensis* hybrid has become the most important bermudagrass hybrid around the world. It can be found mostly on golf courses and other recreational and sport turfs. *Cynodon transvaalensis* Burtt Davey, a very fine-textured turf species, has the softest leaf blades of all bermudagrass species. *Cynodon magennisii* Hurcombe, a natural hybrid of *Cynodon dactylon* and *Cynodon transvaalensis*, is very fine textured and is used for golf-course greens in many parts of the world, especially Africa and Central and South America. *Cynodon bradleyi* Stent, a fine-leaf turf species, is used mostly in eastern and central Africa for golf-course greens.

7.2.1.2 The Genus *Zoysia* (Eragrostoideae)

The genus *Zoysia* includes 10 species, but only three are considered important turfgrasses: *Zoysia japonica* Steud., *Zoysia matrella* (L.) Merr., and *Zoysia tenuifolia* Willd. ex Thiele. *Zoysia japonica* is a creeping turfgrass that produces both stolons and rhizomes. It grows well in a wide range of soils, in the full sun, and under moderate shade. *Zoysia japonica* tolerates heat and drought and is also the most cold tolerant of the zoysias. Overall, if properly managed, zoysiagrass is relatively resistant to diseases but often suffers from insect damage. *Zoysia japonica* produces seeds, but its germination rate is low. Therefore, turfs are usually established vegetatively with sod, sprigs, or plugs rather than by seeding [4]. *Zoysia matrella* is similar to *Zoysia japonica*, except that it has a finer leaf texture and is less cold tolerant. It does not produce many viable seeds and generally must be propagated vegetatively. *Zoysia tenuifolia* is the finest textured of all zoysias and also does not produce viable seeds.

7.2.1.3 The Genus *Stenotaphrum* (Panicoideae)

The single species in this genus used as a turfgrass is St. Augustinegrass (*Stenotaphrum secundatum* [Walter] Kuntze). This turfgrass is widely adapted to the warm, humid regions [8], has a very coarse leaf texture, and produces stolons but not rhizomes. It can be grown in a wide variety of soils, except very alkaline soils, where it may require supplemental iron fertilization. Under proper maintenance and a mowing height of 5–7 cm (2–3 in.), it produces a dark green, dense turf with exceptional shade tolerance. St. Augustinegrass has poor tolerance for heavy traffic. In the tropics, weed control in St. Augustinegrass is somewhat challenging because there are only a few herbicides that effectively control weeds without injuring the turf.

7.2.1.4 The Genus *Eremochloa* (Panicoideae)

Centipedegrass, *Eremochloa ophiuroides* (Munro) Hack., is very popular for use on residential lawns. It is a creeping grass of medium-coarse texture that produces aboveground stolons. It grows well in full sun and modest shade, but it does not survive heavy shade under trees. It is also not recommended for areas subject to traffic. It tolerates low soil fertility and ordinarily has a light-green color [6]. High fertilization, especially with nitrogen, darkens the color but results in numerous problems ranging from low stress tolerance to reduced resistance to weeds, insects, and especially fungal diseases. The most popular centipedegrass is a common type that can be established from seeds. It can be also established vegetatively from sprigs, plugs, or sod.

7.2.1.5 The Genus *Paspalum* (Panicoideae)

Seashore paspalum, *Paspalum vaginatum* Sw., spreads by rhizomes and stolons. It produces seeds, but in many varieties seeds are not viable. The stolons and leaves of seashore paspalum are slightly coarser than those of bermudagrass, but in fact these two turfs resemble one another [3]. When mowed regularly and often, paspalum produces a very dense, superb putting surface, sometimes better than those of hybrids of bermudagrass. Paspalum needs less fertilizer than bermudagrass and can grow under conditions of low nutrient availability and severe nutrient imbalances. A deep root system makes it relatively drought resistant. It also tolerates wide pH ranges and low oxygen conditions. One of the outstanding characteristics of paspalum is its tolerance to saline conditions. Chemical weed control in paspalum is somewhat challenging. Postemergence herbicides are often toxic to this species.

Bahiagrass, *Paspalum notatum* Flüggé, can be established from seeds or propagated vegetatively, mostly from sprigs or plugs [2]. Bahiagrass develops a sizeable root system, which makes it one of the most drought-tolerant turfgrasses. It does well in poor soils, especially sandy soils, and does not require much, if any, fertilizer. Despite low maintenance requirements, bahiagrass must be mowed quite often; otherwise it produces tall and unsightly seed heads. Bahiagrass does not perform well in high-pH soils and does not have good tolerance to shade, traffic, or salinity. It also has quite low tolerance for many herbicides, making chemical weed control difficult.

7.2.1.6 The Genus *Axonopus* (Panicoideae)

Axonopus affinis Chase, known as carpetgrass, is a coarse-textured, low-growing, light-green turf-grass that spreads by stolons. It can be established from seeds or vegetatively, and is used mainly in low-quality turfs, along roadsides, on slopes, etc. Carpetgrass tolerates low fertility and very acidic soils and needs little maintenance. Its importance is increasing especially in coastal areas with sandy soil.

7.2.1.7 The Genus *Distichlis* (Eragrostoideae)

Distichlis spicata (L.) Greene, known as saltgrass, alkali grass, or spike grass, is a low-growing species that looks similar to low-quality bermudagrass. It can be used on all types of lawns and is often used in the restoration of salt marshes. It can inhabit extremely salty and alkaline soils that are poorly drained and have a high water table.

7.2.1.8 The Genus *Pennisetum* (Panicoideae)

Kikuyugrass, *Pennisetum clandestinum* Hochst. ex Chiov., is a coarse-textured, light-green grass that spreads by both rhizomes and stolons. It requires high fertility and moist soil. Its importance is very minor.

7.3 SOIL

Soils of the tropical regions of the world are extremely diverse. They range from infertile arid and semiarid soils, which are low in organic matter, to relatively fertile, humid tropical soils. Because high temperatures persist year-round, as does high humidity in some areas, microbiological and chemical activities in tropical soils are much quicker than those of the temperate regions.

Rainfall is often seasonal, sometimes excessive, and in some areas abundant year-round. The information included in this chapter pertains to the general properties of soil that are typical for most tropical regions.

Soil consists of four major components: minerals, air, water, and humus. Soil components influence each other and form the environment in which plants can grow.

The basic functions of soil are to provide physical support for the turfgrass plants and to supply moisture, air, and nutrients to the roots.

7.3.1 Physical and Chemical Properties of Soil

7.3.1.1 Soil Texture

The term soil texture describes the size of the individual mineral particles and the proportion of each size in the soil. The largest soil particles are called sand; intermediate particles, silt; the smallest particles, clay. Sand consists of particles ranging from 2 to 0.05 mm in diameter. Silt consists of particles ranging from 0.05 to 0.002 mm. Clay consists of particles smaller than 0.002 mm in diameter.

Soil types are based on texture. For example, sandy soils are those that contain a high proportion of sand particles, and clayey soils are those containing a high proportion of clay. Soils containing a high proportion of silt, however, are usually called loamy rather than silty. On the basis of the proportions of sand, silt, and clay particles, distinct soil textural classes have been defined. The textural triangle (Figure 7.1) presents 12 textural soil classes commonly used around the world [9].

7.3.1.2 Soil Structure

Soil structure is the way in which individual soil particles are bound together into larger units called aggregates. Development of a stable soil structure often helps ease the challenges resulting from imperfect soil texture. Soils are usually classified as structural, moderately structural, or weakly structural. Aggregates are generally described as blocky, granular, platy, or prismatic (Figure 7.2).

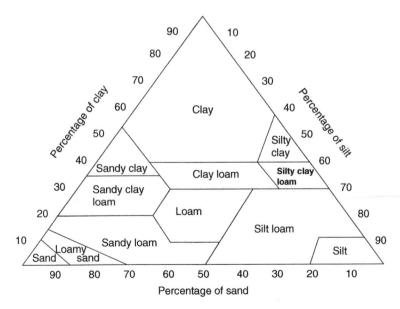

FIGURE 7.1 Widely accepted textural triangle showing the names commonly assigned to soil textural classes according to their relative percentages of sand, silt, and clay. (From Saxton, K.E., W.J. Rawls, J.S. Romberger, and R.I. Papendick, *Soil Sci. Soc. Amer. J.*, 50, 1031, 1986. With permission.)

Blocky

Granular

Platy

Prismatic

FIGURE 7.2 Four basic shapes of soil aggregates. (From Wiecko, G., *Fundamentals of Tropical Turf Management*, CAB International, Oxfordshire, UK, 2006. With permission.)

Organic matter is the most important factor contributing to stable soil aggregation. Organic matter in the soil undergoes complex biological and chemical transformations, binds soil particles together, and helps create small pores within each aggregate and larger pores between aggregates. Its adequate presence improves aeration, moisture retention, and air movement. Soil organic matter also supplies nutrients, decreases loss of soil to erosion, and improves the soil's cation exchange capacity.

7.3.1.3   Soil pH, Soil Cation Exchange Capacity, and Soil Buffering Capacity

Soil pH is an important chemical property that defines soil. It is a measure of acidity or alkalinity existing in the soil solution. Most warm-season turfgrasses require a soil pH between 5.5 and 7.5. Excessively low or high pH affects availability of nutrients and the biological composition of the soil. During physical and chemical weathering, soil minerals and organic matter break up and finally form colloids. Colloids are generally negatively charged and therefore attract and hold cations such as potassium (K^+), ammonium (NH_4^+), calcium (Ca^{2+}), or magnesium (Mg^{2+}). The ability of a soil to hold cations is known as cation exchange capacity (CEC) and is one of the most important chemical properties of soil. The cations held in exchangeable form can become available to plants.

Another important CEC-related property is the soil buffering capacity, which reveals how resistant the soil is to rapid change of pH. Soils high in clay usually have high CEC and possess high buffering capacity. Knowing a soil's CEC and buffering capacity is essential to determine the amount of liming material required to adjust soil pH. For example, raising the pH of a loamy soil (high CEC) into the acceptable range requires the addition of much more highly alkaline lime than raising the pH of a sandy soil (low CEC).

7.3.1.4   Effect of Soil pH on Nutrient Availability

The availability of some nutrients is greatly influenced by soil pH [10]. Correct pH is especially important for phosphorus availability, which is greatest in neutral soils (Figure 7.3). In acidic soils, phosphate ions bond with iron and aluminum to form insoluble iron and aluminum phosphates. At pH levels above 7.0, insoluble calcium and magnesium phosphates are formed. Soil pH affects

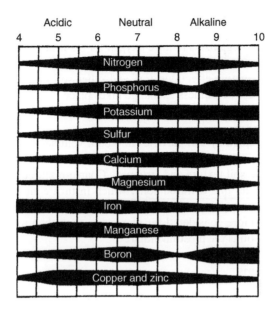

FIGURE 7.3   Availability of essential nutrients depends strongly on soil pH. (From Troug, E., *Soil Sci. Soc. Amer. Proc.*, 11, 305, 1946. With permission.)

availability of other nutrients as well. In general, the availability of nitrogen, potassium, calcium, and magnesium is high in neutral soil and decreases rapidly when pH drops below 6.0 or rises above 8.0. Iron, manganese, boron, and several others are most available when soil pH is in the acidic range. As the pH of soil approaches the otherwise optimal 7.0, their availability decreases, and deficiencies can become a problem. Turfgrasses in the tropics often suffer from iron deficiency, especially on alkaline, coral soils along the coast or on sodic soils in arid climates. Sometimes, just foliar application of iron and other micronutrients is effective.

7.3.1.5   The Soil Water

Most of the water in soil originates from natural precipitation, irrigation, or (occasionally) flooding and is held in the pore space between the solid particles of minerals and organic matter. After an intense rain or plentiful irrigation, the soil is saturated with all the pores filled up. With time, water is removed from the large pores by gravitational forces, and air takes its place. Gravitational forces remove water only until an equilibrium is reached at which the force of cohesion between water molecules and the force of adhesion between water molecules and soil particles are sufficient to balance gravitational forces and prevent further drainage. As the soil's water content continues to decrease, mostly as a result of evapotranspiration, soil capillary forces become strong enough to counteract the ability of roots to absorb water. Finally, the point is reached at which the turfgrass plant wilts because its roots are unable to absorb the remaining water, which is held too tightly by the small pores.

One of the main objectives in turfgrass irrigation is to balance soil capillary forces and root water-absorption forces. The amount of irrigation should be such that the soil always maintains enough suction to prevent water drainage but not enough to counteract root suction and subject plants to water stress. The ease of achieving this goal depends on soil texture, structure, and organic-matter content. Sandy soils having low organic-matter content are the most challenging, but heavy clays are also difficult. Clays may hold the largest quantity of water, but very strong capillary forces may prevent most of it from being taken up by roots; so availability of water to plants may be restrained. Loamy soils are usually the best. They hold large amounts of water readily available to plants.

7.3.1.6 The Soil Air

Turfgrass roots and most soil organisms require oxygen for respiration. During the respiration process, roots and microorganisms release mostly carbon dioxide and also some other gases, including toxic ones that may have negative effects on turf and other soil organisms. Carbon dioxide and toxic gases will build up in the soil unless exchange takes place between the air in the soil and the aboveground atmosphere. This process of gas exchange, called soil aeration, occurs primarily by diffusion through the soil pores.

The total space not occupied by soil particles in a bulk volume of soil is called the pore space and can range from 30% to 70% of the total soil volume. The size distribution of pores in the soil also varies depending on the soil texture and structure. Coarse-textured soils contain a higher percentage of large pores than do fine-textured soils, which in turn contain large numbers of small pores. Large pores help aeration, but small pores are needed for water retention. The ratio of large pores to small pores is therefore considered far more important than total pore space. Numerous studies indicate that a 1:1 ratio is the most favorable.

7.3.1.7 The Soil Organisms

A fertile soil may contain many organisms, such as microscopic bacteria, fungi, actinomycetes, algae, protozoa, and nematodes, as well as relatively large earthworms. The soil organisms perform various activities, mostly beneficial but sometimes harmful to turfgrass growth and development. They break up organic matter, fix nitrogen, transform essential elements from one form to another, help in soil aggregation, and sometimes improve soil aeration and drainage. Bacteria are the smallest and most abundant organisms in the soil. Together with fungi and actinomycetes, they contribute substantively to organic matter decomposition and are mostly beneficial. Particular fungi, nematodes, or insects are sometimes pathogenic and undesirable. Some of them infect or feed on roots, shoots, or leaves, causing damage to turfgrass plants.

7.4 TURFGRASS ESTABLISHMENT

After soil cultivation, turfgrass should be established as quickly as possible. Soil should be cultivated and in some instances modified before planting. Proper preparation of the seedbed is essential for the quick establishment of uniform turf. The goal is to create a fertile homogeneous root zone with acceptable infiltration, aeration, and drainage.

Establishment of turf area can be relatively simple or complex. A small residential lawn may require only general gardening skills and some common sense. Golf courses will likely require input from many people, detailed sets of drawings, lists of specifications, construction schedules, etc.

In general, turfs can be established from seeds or vegetatively from other living parts of the plant. Vegetative establishment refers to any of four basic methods: sodding, plugging, stolonizing, and sprigging. Choice of the most appropriate of these methods depends on the turfgrass species and the particular situation. Steps in producing the optimum root zone for turf establishment may include any or all of the following: control of existing weeds, removal of rocks and debris, rough grading, surface and subsurface drainage, soil modification, fertilization and liming, and final soil preparation.

Hard-to-control weeds should be chemically controlled with herbicides or soil fumigation. Nonselective herbicides such as glyphosate could be used to eliminate all growing vegetation. Because weed control with herbicides is most effective when chemical control is combined with mechanical control, cultivation such as plowing or disking should follow shortly afterward. Less common but more effective than herbicides is soil fumigation, which eliminates not only weeds but also insects, diseases, and nematodes. After weed control, rocks and other debris should be removed from the soil surface after cultivation. As a general rule, a few rocks smaller than a golf ball usually do not interfere with turf performance or maintenance and can be left in place. In some situations, especially on recreational or sport turfs, even small rocks are not acceptable since they may cause damage to specialized turf machinery.

Rough grading is needed when terrain chosen for turf establishment is not even. The objective of grading is to provide a relatively smooth, firm surface, which assures both pleasant appearance and adequate drainage of surface water. For establishment of a small area such as a residential lawn, rough grading will be needed occasionally. Large turf areas, such as golf courses, almost always need rough grading. Effective surface and sometimes subsurface drainage can eliminate many potential problems in turfgrass culture. Turf areas should slope by at least 1%. Athletic turf such as football fields should slope about 2% from the center. Sometimes, despite proper slope, surface drainage is not sufficient, especially on soils containing large amounts of clay. If the soil is poorly drained and impermeable, or the water table is shallow, the area is likely to flood periodically, so a drainage system may be needed. Properly designed drainage systems should ensure rapid removal of water from the turfgrass root zone. Factors such as soil texture, topography of the area, and desired removal rate should be taken to consideration. Drainage systems should be designed by professionals.

If the topsoil was removed before rough grading, it should be brought back and redistributed over the surface. If the topsoil is of poor quality, high-quality topsoil from another site should be brought in and spread over the surface in a layer at least 10–15 cm (4–6 in.) thick. After the topsoil is redistributed or modified, it should be allowed to settle undisturbed for about 1–2 weeks. Therefore, if an underground irrigation system is to be installed, the installation should take place before redistribution of topsoil.

If lime is needed, it should be applied at this stage and mixed with the soil before seeding or planting. Lime does not move down the soils with water, so later applications on established turfgrass are not effective. The same rule applies to phosphorus. Soil tests can determine how much of each nutrient should be applied, but application of phosphorus and potassium at 100 kg/ha (2 lbs/1000 ft^2) is generally sufficient. Timing of potassium application is less critical, but nitrogen should be applied only a day or two before planting. Nitrogen applied earlier can be lost to leaching and volatilization, and can also promote the growth of competitive weeds.

The final seedbed should be firm, moist but not wet, and free of clods, stones, and other rubble. If possible, it should be lightly tilled less than 24 h before planting or seeding. If weeds have germinated since the last tillage, a light application of glyphosate can be used.

7.4.1 SEEDING

Most of the warm-season turfgrasses are propagated vegetatively, but centipedegrass, common bermudagrass, some types of zoysiagrass, and several other less commonly used species can be seeded as well. In cool climates, seeding is an essential method of turf establishment, but in warm climates seeding is optional. Species differ substantially in seeding rates and germination rates.

The best time to seed is before the longest period of optimum temperature and moisture conditions, which in the tropics may mean any time of the year. The seeding rate depends upon turfgrass species, germination rate, germination conditions, and the desired rate of establishment. In the tropics, high seedling density during the establishment period is desired. It reduces weed pressures and produces more competitive turfgrass plants. However, seeding rates that are too high can extend the period of seedling immaturity. Delicate, juvenile seedlings are less able to compete with weeds and more easily infected by diseases. Most turfgrass seeds should be planted quite shallow, at about 0.6–0.8 cm (1/4 in.) depth. Applying the seeds to the surface and lightly raking afterward can produce the desired planting depth. Seeds planted too deep may not germinate, or their germination may be delayed.

The primary goal in seeding is to distribute seeds uniformly. Seeding by hand is popular on small areas. Push-type spreaders, both broadcast and drop types, are used on areas up to 1–2 ha (2–4 acres). Tractor-mounted spreaders can be used on larger areas. Overall, seeding warm-season grasses requires more effort than seeding cool-season grasses. The seeds are very small and often slippery, so filling the spreader with seeds and applying them directly can be very risky. The seeds should therefore be thoroughly mixed with dry, preferably white, sand and the spreader should be calibrated for the seed–sand mixture.

Areas that are difficult to reach, such as steep hillsides or roadsides, are sometimes hydro-seeded with special equipment designed to spread seeds, paper pulp, and fertilizer mixed with water. The paper pulp sticks to the soil, prevents seed runoff, retains moisture for days without rain, and lasts until the grass is well established.

Sometimes, pregermination is used where a very rapid establishment is desired or to reduce weed pressure. In the pregermination process, seeds are soaked in water, usually for 5–10 days. When the first seeds show signs of germination, the water-soaked seeds are mixed with dry sand and spread over the desired area.

7.4.1.1 Germination and Postgermination Care

Seeds need moisture to germinate, but the most critical time comes just after germination, when seedlings have begun to root but have not yet developed a root system. The seedbed surface must be kept moist in the absence of rain by light applications of water several times a day. After this critical period, the grass root system develops to the extent that water can be obtained from the underlying soil. From this time, irrigation should gradually increase in volume and decrease in frequency.

Mulching is often practiced, especially in dry tropics. Straw mulching is inexpensive and provides excellent results by retaining surface moisture. The straw can be raked away several weeks later or left to decompose naturally. After germination, proper irrigation is essential for quick establishment. Seedlings should also be protected from the effects of traffic. The first mowing should occur when seedlings reach a height one-third greater than the anticipated mowing height. From this point, mowing should continue at standard frequency and should be guided by the same "one-third rule." Mowing less frequently removes too much leaf area at one time and can set the plant back severely. Shortly after the first mowing, a light application of nitrogen may substantially speed up turf establishment. Not more than 50 kg/ha (0.5 lb N per 1000 ft^2) should be applied and watered into the soil. The soil surface should be moist as long as the root system remains poorly developed. At the time of turf establishment (especially in humid tropics), weeds may massively invade newly established turf. Fortunately, the appearance of the newly seeded turf greatly improves after the first mowing and keeps improving with time. Mowing at the correct height removes most broadleaf weeds without harming the crowns of the turfgrass plants. In many cases, competition from thriving turfgrasses weakens weeds so greatly that herbicides may not be needed.

7.4.2 Vegetative Establishment

7.4.2.1 Sodding

Sod is established turf that is harvested, with roots and soil attached to it, and transplanted from its place of origin to grow in another place. Sodding is the most expensive method of turf establishment but produces an established turf instantly. During the installation process, sod should be laid in a bricklike pattern without gaps between pieces to prevent encroachment of weeds. Sod should be well watered immediately after being installed. Lighter watering should continue usually for about 2 weeks until rooting is adequate. The first mowing should take place as soon as the turf can tolerate traffic without rupturing. In most cases, within 4–5 weeks, the sodded area can receive maintenance similar to that suitable for mature turf.

7.4.2.2 Stolonizing

Every bud on a stolon, just like every seed, can potentially become a new plant. Stolonizing is spreading pieces of stolons rather than spreading seeds. Stolons are commonly harvested from mature turf with a vertical mower. After they are harvested, stolons should be kept moist and broadcasted not later than the following day. After broadcasting, stolons should be covered by 3–6 mm (1/8–1/4 in.) of soil but not buried completely—approximately 15–25% of each stolon should extend above the soil surface. The stolonized area should be watered and maintained in a way similar to that for a seeded area.

7.4.2.3 Sprigging

Sprigging is a modification of stolonizing. It involves placing stolons in narrow furrows spaced 15–20 cm (6–8 in.) apart. A stolon should be placed every 5–10 cm (2–4 in.) in the furrow and covered by soil such that 15–25% extends above the surface. The major advantage of this method is minimal loss due to desiccation. The major disadvantage is higher cost of labor.

7.4.2.4 Plugging

Only vigorously stoloniferous or rhizomatous turfgrasses can be established by plugging. Plugging, sometimes called spot sodding, is the planting of small pieces of sod spaced apart over a large area. Establishment by plugs is chosen mostly for small areas and involves a considerable amount of labor. Plugs are usually about 5 × 5 cm (2 × 2 in.) and are placed in a grid pattern on 15- to 30-cm (6–12 in.) centers. Major advantages of plugging are low costs of planting material and minimal losses. The major disadvantage is high cost of labor.

7.4.2.4.1 Strip Sodding

Strip sodding is a modification of plugging. Instead of plugs, strips of sod are laid in long rows. This method of turfgrass establishment is somewhat faster, is frequently performed on larger areas, involves less manual labor, but uses more plant material.

7.5 TURF NUTRITION AND FERTILIZATION

7.5.1 ESSENTIAL NUTRIENTS

Sixteen chemical elements have been identified as necessary for the growth of turfgrasses: carbon (C), hydrogen (H), oxygen (O), nitrogen (N), phosphorus (P), potassium (K), calcium (Ca), magnesium (Mg), sulfur (S), iron (Fe), manganese (Mn), zinc (Zn), boron (B), copper (Cu), molybdenum (Mo), and chlorine (Cl). These are the essential nutrients required for growth and development of turf. Without all of them, the grass plant cannot complete its life cycle. Carbon, hydrogen, and oxygen are taken from air and water. The other 13 essential elements are categorized into two groups: macronutrients and micronutrients. Nitrogen, phosphorus, and potassium are frequently called primary macronutrients. Turf requires them in the largest quantities. The secondary macronutrients, calcium, magnesium, and sulfur, are just as important but are required in smaller quantities. The micronutrients, iron, manganese, zinc, boron, copper, molybdenum, and chlorine, are required by plants in only very small quantities.

7.5.1.1 Nitrogen

Unlike other essential plant nutrients, nitrogen is not found in significant amounts in minerals. In the soil, quite large amounts of nitrogen are present in organic form and unavailable to plants. Soil organic matter contains about 5% nitrogen and must break down to release it to the soil (Figure 7.4). In the tropics, organic nitrogen is continuously converted into its inorganic forms through the process of mineralization. A large part of the mineralized organic nitrogen undergoes a process called nitrification and is finally converted into the soluble nitrate form (NO_3^-). Because nitrates exist in soil solution as negatively charged ions, they are not held by negatively charged soil

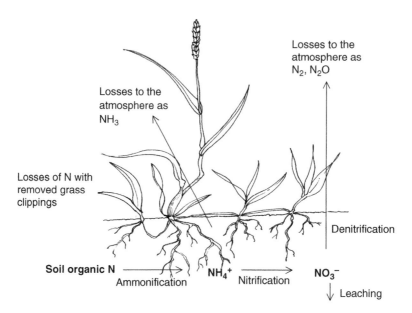

FIGURE 7.4   Fate of nitrogen in turf. (From Wiecko, G., *Fundamentals of Tropical Turf Management*, CAB International, Oxfordshire, UK, 2006. With permission.)

colloids and can easily be leached from the root zone. The rate of nitrogen mineralization depends on microbial activity; so in warm climates, substantial amounts of nitrogen are frequently lost before they can be taken up by roots. If the amounts of inorganic nitrates that originate from mineralization or direct application during fertilization are too large for the plants to absorb immediately, heavy rainfall can leach them from the soil into the groundwater. In addition to leaching, nitrogen can be lost from the soil in gaseous form. Organic matter and, to an even larger extent, urea and other fertilizers containing ammonia (NH_4^+) undergo the process called ammonification. Ammonium ions get converted into ammonia gas (NH_3), which lost to the atmosphere. In some cases, when the turf remains flooded for long periods, nitrate fertilizers may be lost through a process called denitrification. Under anaerobic conditions, nitrate ions can be biologically reduced to nitrous oxide (N_2O) or nitrogen gas (N_2) and lost to the atmosphere.

7.5.1.2   Phosphorus

Phosphorus forms insoluble compounds with aluminum and iron at low soil pH and with calcium at high soil pH. Even in neutral soil, phosphorus is almost always associated with other elements. Insoluble compounds slowly change into soluble orthophosphates (HPO_4^{2-} and $H_2PO_4^-$), which can then be taken up by plants. Unlike the compounds of nitrogen, amounts of soluble phosphorus are small, practically immobile in the soil, and very little is lost to leaching or other processes. Unlike many other agronomic crops, turfgrass does not need high quantities of phosphorus in the soil.

7.5.1.3   Potassium

Potassium is involved in growth processes and influences tolerance to environmental stresses. Potassium occurs as a positively charged cation (K^+) and usually constitutes a large portion of the soil's cation exchange capacity. Potassium is relatively mobile but not as mobile as nitrogen. Management of potassium is more critical than management of phosphorus but still far less essential than management of nitrogen.

7.5.1.4 Calcium

Calcium exists in the soil as positively charged cations (Ca^{2+}). Most soils contain enough calcium to support growth of turfgrasses. Calcium is applied in a process known as liming to increase low soil pH rather than to support plant growth. Calcium deficiencies in plants are extremely rare.

7.5.1.5 Magnesium

Magnesium, like calcium, exists in the soil as positively charged cations (Mg^{2+}). Deficiencies occur sometimes, especially on coarse-textured sandy soils. Magnesium deficiency is frequently associated with low soil pH.

7.5.1.6 Sulfur and Microelements

Sulfur deficiency is sporadic. It sometimes happens on sandy soils, especially if low in organic matter. Plants are rarely deficient in micronutrients such as copper, boron, manganese, zinc, and molybdenum. Iron is the only micronutrient in which plants are frequently deficient, especially in alkaline soils. Micronutrients are seldom applied in turfgrass management and may be required under very unusual circumstances. Applications of micronutrients should always be based on confirmed observations of deficiency symptoms in the field, soil tests, and tissue test results. Soil pH is an important tool for managing micronutrients. Except for molybdenum, levels of plant-utilizable forms of these elements decrease with the rise of pH.

7.5.2 FERTILIZATION

Different turfgrass species have different nutrient requirements, as different levels of maintenance require different nutrient management. A single fertility program cannot be designed to benefit all types of turfs and all management regimes.

Turfgrass can obtain nutrients through root absorption from the soil solution or by direct foliar absorption. Most nutrients are absorbed by roots from the soil. Foliar applications are used when nutrients are immobilized in the soil (as iron is in alkaline soils), or sometimes when a quick improvement in appearance is needed.

7.5.2.1 Turf Fertilizers

Most inorganic fertilizers dissolve quickly in the soil water and are instantly available to plants. Plants absorb the majority of nutrients from the soil solution in the form of simple inorganic ions (Table 7.1). However, organic fertilizers are more complex, require time to be broken down into forms usable by plants, release their nutrients more slowly, and progressively meet the demand of the growing plants.

Many fertilizers are being produced. Fertilizer manufacturers offer products with a wide variety of nutrients, various nutrient proportions, combinations of organic and inorganic fertilizers, various types of nutrient release, various formulations, and so on. Some fertilizers, for example, urea, ammonium nitrate, superphosphate, potash, and ammonium sulphate, often contain only one of the primary nutrients. Others, such as ammonium phosphate or potassium nitrate, may contain two. Still others are blends of several nutrient carriers.

7.5.2.2 Fertilizer Sources

Even though 16 different mineral elements are essential for the growth of turfgrass, nitrogen is by far the most important. It has a dramatic impact on turfgrass color, growth, density, tolerance to stress, and recuperative ability. A wide selection of nitrogen sources is available for use in turfgrass situations. An understanding of the various nitrogen sources is critical in determining the best nitrogen source for a particular turfgrass situation.

TABLE 7.1

Forms in Which Essential Mineral Elements Are Available to Turfgrasses

Element	Available Forms
Nitrogen	NH_4^+, NO_3^-
Phosphorus	HPO_4^{2-}, H_2PO^-
Potassium	K^+
Sulfur	SO_4^{2-}
Calcium	Ca^{2+}
Magnesium	Mg^{2+}
Iron	Fe^{2+}, Fe^{3+}
Manganese	Mn^{2+}
Boron	$H_2BO_3^-$
Copper	Cu^{2+}
Zinc	Zn^{2+}
Molybdenum	MoO_4^{2-}
Chlorine	Cl^-

Source: Wiecko, G., *Fundamentals of Tropical Turf Management*, CAB International, Oxfordshire, UK, 2006. With permission.

In general, nitrogen sources can be separated into two groups: inorganic and organic. Ammonium sulphate, ammonium nitrate, calcium nitrate, and potassium nitrate are common inorganic fertilizers. Once they are applied to the soil, they quickly dissolve. Ammonium ions (NH_4^+) are held in the soil by the negatively charged clay particles and by organic matter. They can be taken up by the turf either immediately or at a later time. Dissolved nitrate ions (NO_3^-) become available immediately for uptake but since they are not held, may be lost from soil by leaching.

Organic nitrogen carriers can be divided into two major categories: natural and synthetic. In natural organic fertilizers, nitrogen originates from plant or animal sources. Composts, sewage sludge, and similar types of products belong to this group. In synthetic organics, the nitrogen originates from atmospheric air, and fertilizers are mass produced in the process of organic synthesis. Urea is the most important fertilizer in this group.

7.5.2.2.1 Synthetic Organics

Urea is a low-cost nitrogen source for turfgrass and is the most widely used. Commonly used granulated urea is a fast-release, completely soluble product. Granulated urea looks similar to ammonium nitrate and contains even more nitrogen, but as an organic compound it has much less potential to burn turfgrass leaves. After application, it dissolves in the soil water but cannot be taken up by plants before the enzyme urease, which is naturally present in the soil, breaks urea's molecules and releases ammonia. Ammonia reacts with water, forming ammonium ions (NH_4^+). Ammonium ions are held by soil colloids and in this form are available for plants. Unfortunately, part of the ammonia can be lost from the soil in its gaseous form through the process of volatilization. Sometimes, leaching also occurs before dissolved urea is converted into ammonia ions. Volatilization losses occur especially in sandy soils and leaching under high-rainfall conditions. Even though urea's leaf-burning potential is lower than those of inorganic carriers, it is still a concern. To reduce the leaf-burning potential as well as volatilization and leaching losses, fertilizer researchers have developed numerous ways to slow down the release of nitrogen from the urea.

One major approach is to combine urea chemically with some other relatively inexpensive compound. Another is to encapsulate urea pellets within less permeable material.

Examples of products that have resulted from the first approach are urea formaldehyde (also called UF), triazone, and isobutylidine diurea (called IBDU). Examples of encapsulated products are sulfur-coated urea, resin-coated urea, and combinations of the two.

7.5.2.3 Phosphorus Carriers

Far fewer sources of phosphorus than of nitrogen are used in turfgrass situations. Phosphorus is usually applied to the soil as superphosphate (20% P_2O_5) or triple superphosphate (46% P_2O_5). Two other water-soluble ammonium phosphate sources, known as MAP (monoammonium phosphate 12-61-0) and DAP (diammonium phosphate 18-46-0), supply both nitrogen and phosphorus. In addition to easy soluble forms of phosphorus, sulfur- or polymer-coated fertilizers are available, but their importance is marginal.

7.5.2.4 Potassium Carriers

Two sources of potassium are especially popular, potassium chloride (60% K_2O) and potassium sulphate (50% K_2O). A third source is potassium nitrate (13-0-44), which supplies both potassium and nitrogen. Slow-release potassium is available in the form of sulfur- or polymer-coated products.

7.5.2.5 Iron Carriers

Several sources of iron are applied to the soil or foliage. On acidic soils (pH below 6.5), ferrous sulphate (20% Fe) or ferrous ammonia sulfate (14% Fe) is used most often. These forms are insoluble at higher pH, and under those conditions should be replaced with a so-called chelated form of iron. A chelator is a large organic molecule that holds iron bound to its carbon ring structure. Chemical binding forces are relatively weak but strong enough to prevent iron from rapidly changing into forms that are unavailable to plants. Fertilizers made up of iron chelate contain 7–10% Fe.

7.5.3 FERTILIZATION FREQUENCY

Theoretically, it would be most beneficial to apply a very small amount of fertilizer every day; however, this practice is rarely used because it is impractical and expensive. More feasible is to apply fertilizers frequently enough to maintain consistently adequate and balanced levels of nutrients in the soil. Overloading the soil with nutrients, then depleting it to the level of deficiency, and then overloading it again should be avoided. In the tropics or warm climates, when fast-release fertilizers are used under relatively stable weather conditions, high-maintenance turfs such as golf greens should usually be fertilized once a month; medium maintenance turfs such as fairways, bimonthly; home lawns, 3–4 times a year. In the arid or semiarid tropics, intervals will be somewhat longer. It is important to remember that if slow-release fertilizers are used, rapid changes of fertilization program are not possible. A heavy load of slow-release fertilizer cannot be washed out of the soil for months, so turf managers are often unwilling to use slow-release nitrogen carriers on high-value turfs and prefer instead to apply otherwise less-convenient fast-release fertilizers.

7.6 MOWING

Sound mowing is perhaps the single most important factor contributing to the attractiveness and longevity of any turfgrass area. Cutting leaves off at a uniform height produces smooth turf with an attractive appearance. Cutting through stems (stolons, rhizomes, and tillers) causes the turfgrass plants to produce more stems—which grow roots of their own to produce more individual plants—and therefore maintains or increases turf density. Turf managers ordinarily use mechanical mowers powered by internal combustion or electrical engines, but occasionally, especially on small areas, hand-pushed mowers are still used. Among the major benefits that result from frequent mowing are improvements of turfgrass appearance and preservation of the plants' health. Properly mowed turf is frequently denser, more resistant to invasion by weeds, and more resistant to traffic damage, diseases, pests, and numerous other stresses.

7.6.1   Turf Responses to Mowing

Cutting leaf tissue disrupts physiological processes and creates open wounds in the tissue through which unwanted pathogens or other organisms can enter the plant. Removal of leaf area reduces the plant's capacity to carry out photosynthesis and consequently lowers production of carbohydrates. Reduced production of carbohydrates results in decreased production of new roots, which in turn results in decreased ability to draw water and nutrients from the soil. All these negative factors would seem to indicate that grasses should be mowed as infrequently as possible. This is correct for over 10,000 species of wild and forage grasses, but the 40–50 species of turfgrasses are different. They have the unique ability to compensate for the loss of leaf tissue to mowing by increasing their density below the mowing height. For these species and varieties, mowing is still somewhat stressful but is not damaging if constant mowing height is maintained. The proper mowing height and frequency are those that provide the optimal balance between desired appearance of the turf and the physiological abilities of turfgrasses to withstand mowing stress.

7.6.2   Mowing Height

The height to which a given grass can be mowed is directly related to its ability to produce enough leaves and to keep up with the production of carbohydrates. Factors such as the length of internodes, the number of stolons or rhizomes, the height of the crown above the soil surface, and natural vigor all influence the amount of leaf mass produced by a particular grass species and determine its ability to withstand low mowing heights.

Mowing height is directly correlated with the growth of the root system, and higher mowing promotes both greater total root mass and greater rooting depth (Table 7.2). Larger numbers of deep roots increase the turfgrass's ability to draw water from deeper soil zones, and a larger root system overall promotes absorption of nutrients from the most fertile zone, the topsoil. Mowing height is sometimes imposed by the purpose for which the turf is maintained. Golf greens, tennis courts, and similar areas must provide surfaces of particular quality, on which a ball would roll or bounce at a specific speed or in a particular manner. These turfs must often be mowed lower than is desirable from a physiological standpoint. Maintenance of such areas can be extremely challenging.

7.6.3   Mowing Frequency

It has been determined that removal of more than one-third of total leaf area results in severe physiological shock to the plant, greatly restricting carbohydrate production. The plant must use all its carbohydrate reserves to repair the damage, as well as to build leaf tissue for restoring

---

TABLE 7.2
Recommended Mowing Heights for Tropical Turfgrass Species

Species	Mowing Height (mm; in.)
Bermudagrass	5–38 (3/16–1½)
Paspalum grass	7–50 (1/4–2)
Zoysiagrass	7–50 (1/4–2)
Carpetgrass	25–75 (1–3)
Centipedegrass	25–75 (1–3)
Bahiagrass	50–75 (2–3)
St. Augustinegrass	75–100 (3–4)

Source: Wiecko, G., *Fundamentals of Tropical Turf Management*, CAB International, Oxfordshire, UK, 2006. With permission.

---

photosynthesis as quickly as possible. All these resources are used at the expense of the roots. The greater the percentage of leaf removal, the longer the disruption of root growth.

Unlike mowing height, mowing frequency cannot be specified on the basis of turfgrass variety. Only the growth rate determines mowing frequency. Because no more than one-third of the leaf area should be removed at any one mowing, mowing frequency generally increases as mowing height is lowered.

7.7   IRRIGATION

Precipitation in the tropics ranges from just about 0 to over 500 cm (200 in.) per year. Warm-season turfgrasses are able to use from 2 to 5 mm of water per day, depending upon location, species, weather conditions, type of maintenance, and several other factors. Unfortunately, the highest usage of water occurs in arid regions, where rainfall is low, and the lowest occurs in the humid tropics, where rainfall is high. In most localities in the tropics, medium- and high-quality turfs require supplemental irrigation.

7.7.1   IRRIGATION NEEDS

The water required by growing turf may originate from rainfall, irrigation, or a combination of the two. By far, the most important factor that determines the plant's need for water is evapotranspiration. In turf, where the soil is usually completely covered by growing leaves and stems, evaporation is minor and most of the water loss is due to transpiration.

Water enters turfgrass through the roots and exits through the stomata. Stomata can open and close in response to changing environmental conditions and therefore can regulate water loss from the plant. Many factors influence evapotranspiration, but the most important are humidity, temperature, wind, and canopy resistance.

Water loss from the plant occurs because of the gradient that exists between the water-saturated cells and the moisture in the surrounding air. In drier air, the gradient is steeper and water is therefore lost faster.

Increasing temperature directly accelerates evaporation, but temperature's effect on transpiration is more complex. Initially, leaves transpire more to cool themselves down, but when the temperature is very high, partial closure of the stomata can be triggered to conserve water. Turfgrasses adapted to the tropics use this type of physiological defense less than cool-season turfgrasses, and unless the temperature is very high, they increase their evapotranspiration when the temperature rises.

Movement of air above the turf canopy is another important factor influencing evapotranspiration. In the absence of wind, the leaf surface is surrounded by a thick boundary layer of air molecules that block free movement of water molecules diffusing through leaf stomata and thus reduce water loss. Wind disrupts this boundary layer, increasing evapotranspiration.

Shoot density, leaf orientation, leaf area, and growth rate all affect water loss through the turf canopy. These factors, described as canopy resistance, may either increase or decrease evapotranspiration (Figure 7.5).

7.7.2   THE TIME TO IRRIGATE

Water stress occurs when the rate of water loss through evapotranspiration exceeds the rate of absorption through the root system. Therefore, from the physiological standpoint, the best time to irrigate is just before the turf experiences water stress, which is manifested by wilting. Because irrigation systems cannot be turned on anytime (e.g., during athletic events or when golfers are on the course), turf is usually irrigated only as near the ideal time as is practical. The simplest visual evidence that wilt is imminent is visibility of prints on the turf canopy after walking across the

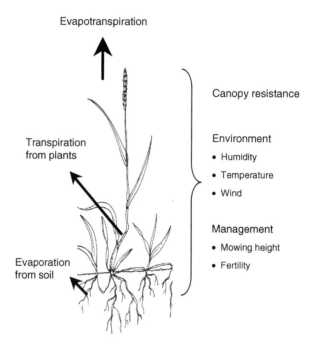

FIGURE 7.5 Factors influencing evapotranspiration of turf. (From Wiecko, G., *Fundamentals of Tropical Turf Management*, CAB International, Oxfordshire, UK, 2006. With permission.)

area and observing how long the turfgrass leaves take to return to their original, upright position. Depending on the turf species, turgid leaves return within one to several minutes, whereas wilted leaves take 15–20 min or more. An additional guide is turf color. Water-stressed patches of turf turn bluish green and can be easily distinguished from areas that have not yet undergone water stress. Another method is observation of turfgrass leaves. Water-stressed leaves roll or fold to conserve moisture. Examination of soil is also reliable. If soil sampled to a depth of 15 cm (6 in.) feels dry, the turf is probably experiencing water stress.

7.7.3 IRRIGATION FREQUENCY

The first principle of irrigation management is "deep and infrequent." At each irrigation event, soil capillaries in the entire root zone should be filled with water to soil-field capacity and then gradually depleted to the point at which turfgrass approaches slight water stress. Excessive irrigation results in water loss, whereas excessive irrigation frequency results in development of shallow root systems. Roots seek water where it can be found and therefore often elongate to depths of 50 cm (20 in.) or even more.

7.7.4 WATER SOURCES

The majority of home lawns, parks, landscapes around businesses, sport turfs, and other relatively small turf areas are irrigated with municipal water. Golf courses, resorts, large parks, and other similar areas generally strive for independent water sources. Connecting to municipal water is the easiest and does not require investments in pumping stations or water distribution systems, but the water is controlled by municipal government and may be restricted at the time of the highest demand. The most common independent sources of water for irrigation are wells, lakes, reservoirs, ponds, streams, and rivers. Artificial ponds should be constructed small and deep rather than large and shallow and should hold enough water for an extended period of drought. They should be located at the lowest point on the property and recharged from surface drainage or possibly some other source such as natural springs.

Wells provide an excellent water source but require large, high-quality pumps and ongoing maintenance to be dependable. Streams and rivers are good sources of water if access is unrestricted. Finally, municipal effluent discharged from sewage treatment plants is actually quite a dependable source of irrigation water. Usage of household water does not fluctuate greatly during the year, so it remains available during drought or other periods when potable water may be restricted. Municipal effluent used for irrigation is usually free from pathogenic bacteria, heavy metals, and other compounds posing a hazard to humans and animals, but it may contain elevated amounts of certain elements such as sodium and boron, which can have detrimental effects on turf. Irrigation with effluent is becoming a global issue.

7.7.5 Irrigation Methods

Turf is irrigated by three main methods: overhead, surface, and subsurface. Overhead irrigation is used on an overwhelming majority of turfs. Water is distributed through some type of irrigation system, either pipes or garden hoses, and sprayed by a sprinkler head. The purpose of a sprinkler head is to disperse water into fine droplets that fall uniformly on the turf surface, as would a light rain. A great variety of sprinkler heads is available on the market. Some are designed for use with high-pressure water lines for irrigating large areas, some for intermediate areas such as parks, some for small areas such as home lawns, some for athletic fields, and so on. They vary greatly in size, design, efficiency, methods of spraying water, and material used for their assembly. Except for a few unique types, most can be classified as either rotary or fixed.

7.7.5.1 Overhead Irrigation

Rotary sprinkler heads shoot water as one or more streams of spray. Water flowing through the sprinkler head makes it rotate to cover a circular area or a set portion of a circle (Figure 7.6).

When adjacent heads are properly spaced, their overlapping pattern can provide relatively uniform coverage along a line drawn between them. Rotating sprinklers produce the desired

FIGURE 7.6   Sprinklers on the edge overlap with sprinklers in the center. (From Wiecko, G., *Fundamentals of Tropical Turf Management*, CAB International, Oxfordshire, UK, 2006. With permission.)

wedge-shaped pattern of irrigation when they are operated within the proper range of water pressure. Excessively high pressure causes the water to form a fine spray instead of droplets. Insufficient water pressure results in streams of water that do not disperse sufficiently and irrigate only a narrow band at some distance from the sprinkler head. When water pressure fluctuates, irrigation can be excessive in some areas and insufficient in others. Wind strongly influences irrigation patterns and water efficiency, often more than water pressure, sprinkler head design, or sprinkler spacing. Its influence should be minimized as much as possible. The largest sprinkler heads should operate around sunrise, when wind velocity is usually lowest.

7.7.5.1.1 Fixed Sprinkler Heads

Fixed sprinkler heads have no moving parts and usually operate at low water pressure. Each produces a fine spray of water covering a relatively small area, commonly 3–6 m (10–20 ft) in diameter. They are often used to water home lawns, flower beds, and other landscape plants. Some sprinkler heads can be adjusted to control the amount of water delivered. They can also be adjusted to deliver water in a variety of spatial patterns. A full head delivers water in a full circular pattern; other typical heads irrigate half and quarter circles. Adjustable heads can water any part of a circle, from 0° to about 360°. Some fixed sprinkler heads are designed to water rectangular and square areas of turf, and end, center, and side strip patterns are available for grassy pathways, side yards, and other tight spaces. Fixed sprinkler heads are most frequently installed as pop-up systems; they are relatively trouble free and are the least affected by wind.

7.7.6 IRRIGATION SYSTEMS

7.7.6.1 Portable Irrigation Systems

Portable irrigation systems are those that can be moved from one turf area to another. They are used on many turfs, especially relatively small ones, but cannot be considered as efficient as installed systems. Installed irrigation systems are those in which the system that delivers water to the sprinkler heads is fixed in place, usually underground. Besides sprinkler heads, they include pipes, control systems, valves, and if the system is independent, water pumps. Piping is the basis of a turfgrass irrigation system. Pipes transport water from the water source to the sprinkler head, where it is dispersed onto the turf.

A control system coordinates operation of the entire irrigation system. In manual systems, someone must turn the control valves by hand. An automated system typically includes controllers, which integrate a clock, a timer, and a series of terminals called stations. Some automatic systems are equipped with soil-moisture sensors that can override the controller program and prevent the system from being turned on during or shortly after rain, when soil remains moist. Modern, fully computerized, automatic control systems analyze essential weather and soil moisture data, account for all programmed variations and modifications, and open selected valves at the best times and for the proper periods.

Larger irrigation systems usually include water pumps. Two types of pumps are used: booster and system supply. Booster pumps raise in-line water pressure without affecting flow rate. In large irrigation systems, such as on golf courses, booster pumps are used where water pressure drops considerably as a result of elevation changes. System-supply pumps are used to draw water from a water source, such as a well, pond, or river. A system usually includes several of them, and each supplies a specific flow rate at a specific pressure. Installation of an automated sprinkler system requires considerable investment and careful planning, and should be left to a professional who understands the capabilities and limitations of all the system's components.

7.8 TURF CULTIVATION, COMPACTION, AND THATCH

In agriculture, the term "cultivation" refers mainly to the tilling of soil, but in turfgrass management, it refers to the mechanical processes used to loosen the soil, break up undesired soil layers, remove thatch, and stimulate turf growth, as well as to modifying the soil surface. Cultivation is usually conducted on higher-quality turfs and involves techniques such as coring, slicing, spiking, forking, water injection, verticutting, top dressing, and rolling. Cultivation, although necessary, also causes a certain amount of stress, temporarily worsens turf appearance, and may create conditions favorable for invasion by pests.

7.8.1 SOIL COMPACTION

Compaction is defined as reduction in soil volume. Compressing forces from vehicular and foot traffic press individual soil particles closer together, resulting in a denser soil mass. Compaction does not directly reduce turfgrass activity but rather affects its growth by influencing soil properties. The primary effect of soil compaction is reduction in pore volume and redistribution of pore sizes. These changes influence many other physical properties of soil, such as aeration, water retention, drainage, and mechanical impedance to root growth. A reduced ratio of macropores to micropores may result in higher soil-moisture content, but the water may be held so tightly that its availability to plants is limited. Further, the reduction of pore diameters is often accompanied by loss of pore continuity, which reduces even further the diffusion of air and soil-water movement and results in further growth reduction. Compaction also increases soil strength, especially when the soil is dry. The mechanical resistance of compacted soil significantly limits the growth of deep roots and therefore reduces turfgrass drought resistance.

Compaction of turf soils typically occurs close to the soil surface, and can usually be corrected by cultivation devices that penetrate the compacted soil relatively shallowly, not more than 10–15 cm (4–6 in.). Several cultivation techniques can be used to alleviate surface compaction, but core aerification, sometimes simply called coring, is the most frequently used. Besides reducing compaction, coring also controls thatch accumulation; so one operation offers double benefit.

7.8.1.1 Coring

Coring is a cultivation method in which small holes are made in the soil, usually by removal of small cylinders or plugs of soil and turf. These plugs are pulled out with hollow tines, spoons, or screws, discharged on the surface and then either removed or be broken up and worked back into the turf. The density of cores depends upon the type of machine but generally ranges from 10 to 20 per sq. ft. The removal of the core permits nearby soil to slide into the opening and overall become less compacted. Within a week or two, the core holes are filled with roots growing vigorously in a loosened soil (Figure 7.7).

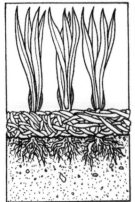 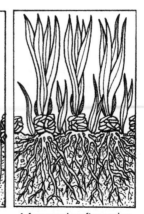

Before coring   A few days after coring   A few weeks after coring

FIGURE 7.7 Effect of coring on development of roots. (From Wiecko, G., *Fundamentals of Tropical Turf Management*, CAB International, Oxfordshire, UK, 2006. With permission.)

Solid tines are sometimes used in place of hollow tines and can be mounted on the same machines. The solid tines penetrate the soil and open holes, but no soil plugs are removed. The solid tines are less disruptive to the turf surface, but they are less effective in loosening the soil and relieving compaction.

7.8.1.2 Other Cultivation Methods

Deep-tine aerification helps alleviate the problem of subsurface compaction. Deep-tine machines with hollow or solid tines can penetrate to a depth of about 30 cm (12 in.). Slicing is a cultivation method in which rotating flat tines or stationary blades slice vertically through the turf and soil. Slicing is usually done on moist soil, penetrates to a depth of 10–15 cm (4–6 in.), does not relieve soil compaction as efficiently as hollow tine aerification, but causes much less surface disruption. Spiking is a cultivation method in which non-power-driven, solid, flat or pointed blades penetrate the turf and soil surface. The blades are similar to slicing blades but have smaller, knifelike tines and do not penetrate more than 5 cm (2 in.) into the turf. Subsoil aerification is a cultivation method in which bullet-shaped devices are forced through the deeper soil zones to break up compacted layers below the surface. Subsoil aerification is usually performed on moderately dry soil and requires a tractor generating a substantial amount of mechanical power. Forking is a cultivation method in which a fork or a similar, solid tine device is used to make holes in the turf. Forking is usually performed on golf greens and some sport fields. The holes are usually slender and deep, up to 20 cm (8 in.), and the surface is not disrupted. Water injection aerification is very effective but also quite expensive to carry out. A special machine injects "bullets" of water, under very high pressure, into the turf surface. These big "drops" or streams of water relieve soil compaction without any disruption to the surface.

7.8.2 Thatch Control

The second most important reason for turfgrass cultivation is thatch control. Thatch is a layer of living and dead grass stems, roots, and other organic matter that is found between the soil surface and the grass blades, and its buildup indicates an imbalance between the amount of organic matter produced by the turfgrass plants and the rate of its decomposition. Excessive thatch is disadvantageous. It harbors potentially destructive diseases and turf insects, limits root penetration into the soil, and in some cases, causes hydrophobicity. Heavily thatched turfs may also affect games in which the bouncing response of the ball is critical to the performance of the players.

Certain turfgrass species as well as management practices such as nitrogen fertilization, soil pH, mowing height, pesticide applications, or soil type may influence the rate of production or decomposition of organic matter therefore the buildup of thatch.

When the thickness of the thatch layer exceeds 1.5 cm (0.5 in.), it can become a problem and should be reduced. If its thickness significantly exceeds 1.5 cm, dethatching is recommended. One of the most common methods of thatch control is vertical mowing. Golf greens in the tropics are dethatched every several weeks but lightly; other turf areas, especially when highly fertilized, may need extensive detaching every 1–3 years. Dethatching should be performed at times of vigorous growth and should be avoided before anticipated periods of environmental stress, such as drought.

7.9 SUMMARY

This chapter presents basic turf-management topics and emphasizes the application of basic principles to recreational and residential turfgrasses. The chapter has been organized into eight sections that provide an overview of major warm-season turfgrass species, soils, turf establishment, turf nutrition and fertilization, mowing, irrigation, and cultivation. This text is addressed to horticulture students and practitioners working on golf courses, residential turfs, sport fields, parks, and recreational grounds.

REFERENCES

1. Beard, J.B. 1973. *Turfgrass: Science and Culture*, Prentice-Hall, Engelwood Cliffs, N.J., chap. 17.
2. Christians, N.E. and M.C. Engelke. 1994. Choosing the right grass to fit the environment, in *Handbook of Integrated Pest Management for Turfgrass and Ornamentals*. Lewis Publishers, Boca Raton, FL, pp. 99–112.
3. Duncan, R.R. and R.N. Carrow. 2000. *Seashore Paspalum: The Environmental Turfgrass*. Ann Arbor Press, Chelsea, MI.
4. Gary, J.E. 1967. The vegetative establishment of four major turfgrasses and the response of stolonized "Meyer" zoysiagrass (*Zoysia japonica* var. Meyer) to mowing height, nitrogen fertilization, and light intensity. MS Thesis. Mississippi State University, pp. 1–50.
5. Gould, F.W. and R.B. Shaw. 1969. *Grass Systematics*, 2nd ed. McGraw-Hill, New York, p. 397.
6. Horn, G.C. 1967. Turfgrass variety comparisons. *Proceedings of the Florida Turf-Grass Management Conference*, 15:91–99.
7. Huffine, W.W. 1966. Bermudagrass around the globe. *Turf-Grass Times*, 1:8–24.
8. Mathews, J.W. 1935. Lawn grasses on trial at Kirstenbasch. *Journal of the Botanical Society of South Africa*, 21:11–13.
9. Saxton, K.E., W.J. Rawls, J.S. Romberger, and R.I. Papendick. 1986. Estimating generalized soil–water characteristics from texture. *Soil Science Society of America Journal*, 50:1031–1036.
10. Truog, E. 1946. Soil reaction influence on availability of plant nutrients. *Soil Science Society of America Proceedings*, 11:305–308.
11. Wiecko, G. 2006. *Fundamentals of Tropical Turf Management*. CAB International, Oxfordshire, UK.

8 Turfgrass Culture for Sod Production

Stephen T. Cockerham

CONTENTS

8.1 INTRODUCTION

Turfgrass sod production is the commercial growing of turfgrasses for transplanting into lawns, sports fields, golf courses, and other swards. Preparing turf to become a sod product requires the application of fundamental turfgrass management plus cultural practices unique to the needs of the crop.

Sod is a term used here to describe strips of turfgrass plus the adhering soil, which are used for vegetative planting.

8.2 LAND PREPARATION

Land preparation is very important to a turfgrass sod grower. Proper preparation will not guarantee a high-quality product, but improper preparation will guarantee a quality problem and almost sure crop losses.

The soil types preferred for sod production are loams. Loam soil has good drainage, moisture-holding capacity, nutrient retention, and is compatible with a wide range of the soils on construction sites. Heavier soils or those with higher clay content are slow to drain so that field cultural activities after a rain are limited and scheduling field work around irrigation is more difficult. The impact upon sod production includes the fact there is a narrow range of compatible soils for installation and the product is much heavier to haul on the trucks. Lighter soils or those with higher sand content than loams are easier to work with than the heavier soils. Light soils have good drainage, but may require closer irrigation management due to a lower water-holding capacity.

Water does not move readily between soil layers if the textures of the layers are very different. Water will stand in a sand layer if the underlying soil is clay and it will resist moving into sand from a cover layer of clay. For this reason, sod produced on either a very sandy or very clayey soil will not

139

be compatible with as wide a range of installation site soils as a loam soil, which will be a problem for the consumer.

Stony soils are difficult to work with and become major impediments to harvest. The sod cutter blade must be sharp and true to harvest uniformly. Stones in the soil disrupt machine action and damage the blade. It is very difficult to produce a quality product on this type of soil.

Soils with pH between 6.0 and 8.0 are satisfactory, with pH below 7.0 preferred. Soil chemistry modification may be necessary if the pH is too low or high. Salinity is difficult to modify and will limit the type of grasses that can be grown. Good turf can be grown on a soil with salinity of less than 2.0 dS (decisiemens). Soil of 4.0 dS or above will be difficult to manage for some grasses.

Land preparation for sod includes the various tillage operations necessary to make a seedbed. The soil needs to be deeply tilled, particularly after the compaction of harvest. Plowing, disking, or even ripping is the first step. The clods are broken with a disk or harrow to a medium to fine tilth. The surface grade is made with a float, landplane, or leveler followed by the final seedbed preparations with a harrow. Growers often use a light, flat roller as the final operation before seeding or stolonizing. Using a laser leveler to true the surface every other crop or so can be very cost-effective in helping keep crop losses to a minimum. The final grade should allow surface runoff with no puddling or ponding. The sod cutter is unforgiving to uneven surfaces. The use of the big roll makes the losses from unevenness quite significant.

The heavier soils tend to form hard clods that resist breaking. Rotary tillers can be useful in breaking up the clods. Working the soil too dry or too wet increases the problems of getting a good seedbed.

Clean, weed-free soil is ideal. Fumigation for nonselective weed control may be possible in some areas. Although expensive, the cleanup can be amortized over several crops and the benefit to crop productivity and product quality is quite significant.

There may be environmental and legal issues with some fumigants which preclude their use. Although time-consuming, pregerminating particularly troublesome weeds and killing them to reduce the weed population potential can increase returns on the crop.

A preplant fertilizer application is beneficial prior to seeding or planting, to have nitrogen and phosphorus available to the emerging plants. The rule of thumb is to apply 440 lb ac^{-1} (500 kg ha^{-1}) 16-20-0 (ammonium-phosphate-sulfate) fertilizer or an equivalent. The uniformity of the stand and time to maturity are improved when the nitrogen and phosphorus combination is applied preplant.[1]

8.3 PLANTING

After land preparation, the next most important operation to establish the crop is planting. The ultimate objective is a uniform stand. Several factors can reduce the stand if the grass is not planted correctly. Fixing the consequences of poor planting in time for harvest is rarely successful.

The most widely used cool-season grass species—Kentucky bluegrass, perennial ryegrass, and tall fescue—are generally established from seed. There are a number of sources of information as to which specific varieties or cultivars (cultivated varieties) perform best in a given market. The local Cooperative Extension personnel from the state universities are usually an excellent source for current information and the National Turf Evaluation Program publishes an annual report that would include data for the major markets.

The seed industry makes *sod-quality* seed available to sod growers, at a premium price. Whatever the cost, it is a good value. Sod-quality seed has an exceptionally high germination percentage and is clean from contaminants such as weeds, weedy grasses, and other crop seed. Even using the premium seed, this is the least expensive input and cutting corners makes no sense.

Seed *blends* are combinations of two or more varieties or cultivars of the same species. Seed *mixtures* are combinations of two or more species such as Kentucky bluegrass and perennial ryegrass. Each grass will respond differently to climate and various stresses and the best-adapted grass will predominate. The seed distributor usually makes blends in seed-mixing equipment for uniformity. Seed mixtures may be made prior to bagging, but if there is a big seed size differential, separation in the bag or in the seeder may affect stand uniformity. Some growers will seed one grass and immediately overseed with the second grass to be sure of uniform distribution in the stand. The blends and mixtures are made in an effort to put the best sod product into the market place.

Seeding rates for cool-season grasses vary widely among the various climate regions. In the northern states, a sod crop may take over a year to mature enough to harvest. Most growers prefer fall seeding; fields are seeded either in spring or fall, and seeding rates are low, such as 20–40 lb ac^{-1} (23–45 kg ha^{-1}) for Kentucky bluegrass.

In the warm-climate regions, a crop of cool-season grasses may mature in as little as 3 months. Seeding may be in any time of the year, avoiding the very hot periods, and seeding rates for cool-season grasses are high, such as 80–120 lb ac^{-1} (91–136 kg ha^{-1}). Perennial ryegrass is seeded at 30–50 lb ac^{-1} (34–57 kg ha^{-1}) and tall fescue at 200–250 lb ac^{-1} (227–284 kg ha^{-1}).

The seeder should be calibrated and in good working condition. Gaps and skips usually reduce the stand uniformity and increase losses at harvest. Some growers prefer seeding in one pass over the field; others choose to seed at half rate twice over at right angles.

Reinforcing netting is widely used in some markets. The netting adds to the cost, but significantly reduces the growth time to harvest. Netting is installed either before or after seeding at the preference of the grower. It is usually not covered with soil and the grass plants germinate through it and begin to spread, locking in the netting.

Irrigation of the seeded field can be applied several hours to several days after seeding, but once it starts it must be carefully monitored. Initially light, frequent irrigations, even twice per day if practical, are required. After the plants emerge and the roots grow the irrigation, though still critical, can be less frequent.

Warm-season grasses that are established from seed are planted just as the cool-season species. Many warm-season grasses are established vegetatively with *stolons* (sprigs) or *plugs*. Stolons are the runners that creep across the ground from the plants. At each joint or node of the stolon is a bud that can grow a new plant, which will be genetically the same as the parent plant. Plugs are small pieces of turf, usually about 2-in. (5 cm) squares that are planted in the field. The plugs are planted a given distance apart, such as 12 in. on center (30 cm), and grow to fill in between. Of the two methods, plugging to establish sod fields is rare.

Stolons are harvested by shredding sod pieces. The planting rate is given in bushels (bu), not a well-defined unit of measure, being that contained in 2 ft^3 (33 cm^3) or approximately 8 lb (3.6 kg), or the stolons from shredding 1 yd^2 (0.8 m^2) of sod. The planting rate depends upon the growth establishment speed of the grass, but is generally considered to be about 200 bu ac^{-1} (227 bu ha^{-1}). The stolons are distributed uniformly by hand or machine and covered with soil or pressed into the soil with a disk, ring roller, or other method. A key to success is the irrigation, which must be very soon after planting as dry soil will quickly draw the moisture from the stolons, killing the buds. On a hot day, an hour may be too long to wait.

Warm-season grasses must be planted with the soil temperature consistently above 55°F (13°C). Planting too late in the fall increases the risk of stand failure. September 15 is a guideline last date for planting in most of the warm-season-grass areas. In spring, when the soil temperature gets consistently in the range, planting will be a success.

8.4 IRRIGATION

The uniform distribution of water is a key factor in producing quality turf. Growing turfgrass sod in arid climates makes irrigation absolutely the most important operation in the production cycle, where getting the maximum efficiency out of the irrigation system and the water application is a full-time job. In areas where annual rainfall is 100 cm or more, irrigation is not as important, but still can be a valuable management tool.[2]

Uniformity of stand is very important to the sod grower, and irrigation is often the key. Within a field and from field to field the land preparation, planting, irrigation system distribution, soil, topography, and macroenvironment will vary, often quite significantly. The science of irrigation may consider most of these variations, but it is left to the irrigator to recognize the crop needs and make adjustments.

Irrigation should be nonlimiting, adequate to avoid drought stress to the plants, and requires attention to avoid wasting water. Young sod can be completely lost due to desiccation in hot wind, drought, or even freezing winter winds. Wilting turf is the first symptom of drought stress. The grass has a gray-blue patina seen easily from some distance. Up close, the turf readily shows footprints, since the grass blades tend to not spring back up after they are compressed. In the early stages of stress, the turf will recover with immediate water application. As wilt progresses, the blades may die, but the turf will recover even though its growth is temporarily retarded. If drought continues, turf plants will die, and the density of the planting will decline. A severely desiccated area usually does not recover enough to be harvested with the rest of the field and it ultimately becomes waste. Seemingly minor drought stress of the turfgrass can cause heavy waste at harvest, weeks or even months later.

Drought stress seldom occurs uniformly over a field. Normally, there are distinct patterns of stress that conform to irrigation or soil variations. It is impossible to apply water perfectly, so some areas will always receive more or less water than others. Soils may vary a great deal throughout a field. Heavy clays that stay wet can be found beside sandy areas that do not hold moisture. Poorly designed or operated irrigation systems magnify these differences.

Drought patterns in the shapes of circles, rings, triangles, or strips show that something is wrong with the irrigation system. Too low or too high a pressure, plugged sprinklers, sprinklers cocked at an angle, and broken sprinklers could be causes. Laterals that are too far apart will leave strips of stressed turf.

Irrigation systems employed on sod farms are most often portable systems adapted from those developed for general agriculture. Permanent systems used on sod farms are prone to problems from mowers, sod cutters, and land preparation equipment working around and over the sprinklers and water lines.

The wind is an important influence on sprinkler irrigation. The patterns are significantly affected and system design, hardware selection, and irrigation schedules must consider the wind. High and exposed areas dry fast. Soil texture shows up in wind with the lighter soils drying faster. Stand uniformity is a result of how well the water is managed in the wind. During the crop growth cycle wind can burn the leaf tips, slow growth, and cause a thinning of stands.

It would be impractical to have a water supply, pumps, and mainline sufficient to run all the sprinklers at the same time. Therefore, the supply must be balanced between the various fields so that turf plantings in several stages of maturity receive the correct amount of water. Too little water will stress the grass. Too much water can injure the turf, leach nutrients, and increase costs unnecessarily by wasting water. All goes into scheduling irrigation.

Operations are scheduled to coordinate with irrigation. A harvest field must receive some water the night before harvest. Fertilizer, iron, and some pesticide applications must be watered in to prevent burn or to activate the material. Mowing is scheduled on dry fields. Irrigation scheduling impacts the entire production operation.

A useful tool in irrigation management is the soil probe. The observer can see to what depth the soil is dry or wet—there is no guessing. If the soil is dry after 4 days, irrigation may be needed every 3 days.

Evapotranspiration (ET) is used as an estimate of the water use of a crop. Reference evapotranspiration (ET_0) is calculated from various accumulated climatic data and reported in historic and real time terms. ET_0 is reported in millimeters per day and used to help schedule irrigation. Historic ET_0 is good for planning, but the irrigator still needs to make adjustments as needed for the crop. Real-time ET_0 is used as a part of a water budgeting program. Irrigation scheduling is a management responsibility. Knowledge of the plant needs, soil characteristics, irrigation system efficiency, and climate is used.

Water budgeting requires filling the soil water reservoir by irrigation, using real-time ET_o to estimate sod-crop water use, and applying irrigation to refill the reservoir based on crop use. In practice, to avoid stressing the turf the soil reservoir should not be allowed to deplete over about 30% of capacity. Distribution uniformity must be considered in calculating the water to be applied and the precipitation rate should not be greater than the soil percolation rate.

Soil moisture-sensing devices can be useful tools for sod farms. As with any technology being applied to crop production, there is a learning process in how to use them and in understanding the meaning of the readings in relation to the sod crop with the shallow rooting, immature turf, and soil variability. Sod irrigation usually puts water on as fast as the soil will take it, as frequently as necessary. Irrigation sets seldom last longer than 1 h.

The irrigation system must be capable of applying the ET_o of the highest-use period. Irrigation is scheduled to replace the amount of water lost or used, a relatively simple approach in practice requiring good management and efficient irrigation equipment. Wasting water is expensive and difficult to track.

Mismanagement of irrigation encourages pest invasion, which ultimately reduces the potential for meeting the production schedule for getting the sod crop mature and marketable.

Water quality can be a problem, and the impact increases when irrigating for the relatively shallow root zone of sod production. Irrigation water evaporates leaving the salts, allowing salinity in the root-zone soil to build up. Several problems occur to reduce plant growth and soil infiltration decreases. Grasses must be grown that are tolerant of the respective level of salinity. Removal of salts by leaching with excess water is temporary.[2]

8.5 MOWING

The most expensive and labor-intensive operation on a sod farm is mowing. Timely mowing at the right height helps the crop to mature when it is supposed to; therefore, frequency and height of cut are the management factors.

Mowing frequency is determined by the growth rate of the grass. The rule of thumb is when the grass has grown one-third of its height of cut it should be mowed, which may be 2 weeks or 3 days. For example, if the height of cut is to be 1.0 in. (25 mm) the turf should be mowed before the grass is 1.5 in. (38 mm) tall.

Staying on top of the turf growth is not easy. Growth spurts caused by weather changes or rain making fields too wet for mowing are common reasons that the growth can get away. If the turf grows too high, it may be less shock to the grass to take a couple of mowings to get down to the desired height. Shocking the plant by cutting off too much at one time will slow the growth of the entire plant including the roots. If it happens often, the crop maturity can be set back.

Each turfgrass has a best height of cut for performance such as sports traffic tolerance. The sod grower should maintain the turf at a given height of cut within the optimum range for steady growth (Table 8.1 shows the optimum height of cut for sod production). The performance characteristic most important to sod production is the plant growth rate, including root growth, to make

TABLE 8.1
Optimum Height of Cut for Sod Production

Species	Inches	Millimeters
Hybrid bermudagrass	0.5–0.75	13–19
Kentucky bluegrass	1.5–2.0	38–51
Perennial ryegrass	1.5–2.0	38–51
Tall fescue	1–2	25–51

the sod product. If growth needs to be pushed, it should not include changing the mowing height or frequency.

The mower height of cut is set on a bench in the shop. Old, very mature turf with significant thatch will cause the mower to "ride up" and the grass can be longer than the height of cut set on the unit. Turfgrasses grown as sod rarely get that old, so that the mower usually cuts a true height.

Mowers should be kept sharp and well adjusted. Reel mowers produce the best cut, but are more expensive to buy and maintain than rotary mowers. Shredded and torn grass leaf blades dry back and stop growth for a short time. Operating costs of dull and poorly maintained mowers are much higher.

Mowers used on warm-season grasses should either not be used on other grasses or be thoroughly cleaned between fields. These grasses easily propagate vegetatively and can be carried into a field, contaminating other turf. Casual washing is not good enough. A power washer or steam cleaner works best.

Turf wet with dew, rain, or irrigation is difficult to mow. The clippings build up on rollers and wheels and cause poor cutting. Turf under clumps of clippings that fall from the mower will yellow and thin after a few days.

Clipping removal is not always necessary, but high-quality sod usually had them removed in the latter stages of growth. Short clippings from grass mowed frequently can sift down into the canopy. In areas where the sod crop takes a year or more to mature, clippings have time to decompose. In the short-time crops, the turf appearance is usually better with clippings picked up.

Turf that is being prepared for harvest should be groomed with a very sharp, adjusted mower. The first view that the customer gets of the sod product is right after installation. A poorly mowed turf gives an immediate negative impression.[1]

8.6 NUTRITION

Fertilizer applications should be balanced with the sod needs, the potential response from the turf, and the risk. Turfgrasses are often forced to produce their maximum growth rate for sod production. Even though the sod is sold on the basis of the quality and appearance of the topgrowth, the grower must be sure that it is balanced with root and rhizome development; otherwise it will not provide an adequate turf for the end user.

The preplant fertilizer application (see Section 8.2) will carry the new seeding and young turf well into the stage I inventory (see Section 8.10). As the turf approaches or enters stage II inventory, nitrogen fertilizers are applied in soluble form at 50 lb N ac^{-1} (44 kg N ha^{-1} mo^{-1}) and care must be taken to avoid overwatering. Loss of nitrogen to leaching is not only an environmental problem but ultimately slows growth. In stage II inventory, the nitrogen fertilizer is applied to keep the color of the turf. Overapplication of nitrogen increases the topgrowth and reduces lateral strength.[3] Overwatering at this stage of maturity leaches the nitrogen, requiring more frequent applications to keep the turf color.

Temperature has an impact upon the sod response to fertilizers. The optimum temperature for cool-season turfgrass growth is 18–27°C for shoots and 13–18°C for roots; for warm-season turfgrass growth, 27–35°C for shoots and 21–27°C for roots. As the temperature in spring increases into the optimum range, the turf response to fertilizer will be fast and the growth can be pushed as long as the temperature is in the optimum range. As it moves above the optimum range, the turf growth decreases and the response to fertilizer application also decreases. When the temperature drops into the optimum range in late summer and fall, the turf will respond to fertilizer, but probably cannot be pushed.

The risk of volatility and leaching loss of fertilizer applied when the turf is not likely to respond or responds minimally is higher than when the turf is growing. New sod field plantings are quite bare of vegetation and as such are subject to nutrient leaching. As the turf matures, the leaching potential decreases. In good market climate, the turf on sod fields does not get mature enough to stop the risk of nutrient leaching. Water management is critical to the minimizing of the losses.

8.7 PEST MANAGEMENT

The sod grower has an obligation to the customer to deliver a clean, pest-free product. To do that, integrated pest management (IPM) has been an important management strategy in turfgrass sod production, incorporating a multifront approach that includes cultural practices to promote growth, reducing potential pest population buildup, biological controls, species selection, and pesticides.

In IPM, pesticides are used as one of many tools. The sod grower has an opportunity between sod crops for sanitation operations that can significantly help control pests. Avoid burying old sod pieces. Collect them and remove them from the field. Turn the soil by plowing or disking when the opportunity occurs. Weeds and insects are easily spread throughout a field by tilling operations and care should be taken to avoid dragging soil from the edges into the field. Equipment that must be used in weedy or diseased areas should be thoroughly cleaned before moving into clean fields. Field edges should be kept clean and weeds should not be allowed to go to seed in ditches and fence rows.

Diseases such as *Pythium* blight (*Pythium aphanidermatum*), necrotic ring spot (*Leptosphaeria korrae*), and bermudagrass decline (*Gueamannomyces graminis*) can be devastating to a sod field. Recovery from the diseases is usually very slow. One alternative is to destroy the field and start over, which is not very attractive to most growers. Another would be to scalp the turf and overseed with the same grass or another species. For example, Kentucky bluegrass damaged by necrotic ring spot can be scalped down and overseeded with perennial ryegrass. The resulting mixture will make a good turf and the disease will not be a problem in the Kentucky bluegrass/perennial ryegrass lawn for the consumer.

Grubs, cutworms, and armyworms are the primary insects pests that show up in sod fields, with sod webworms being a problem in some areas. Birds observed working in established turf are usually due to the build up of worms. If the population is getting to problem size, worms can be found under things lying in the field, such as irrigation pipe. The grower should apply an insecticide to control the worms. It is not a good thing for the consumer to see worms crawl onto the patio out of new sod.

Weeds are the most common pests for sod growers to manage. Although growing a vigorous, dense turf will help keep most weeds from becoming a major problem, keeping weeds under control is a full-time job. Even though the harvesting operation can remove about 0.5 in. (1.3 cm) or more of soil with the turf and mat, a weed seed bank will build up. Flooding, wind, equipment movement, and other means increase the risk of contamination.

Germination of the weeds and knocking them down by cultivating or spraying a nonselective herbicide before planting the crop will reduce the population. Some weeds, such as broadleaf weeds, can be controlled after the crop is up. Grassy weeds are much more difficult to control after they germinate, but some selective herbicides are available or in development.

Most herbicides that are recommended for use on turf can be used on grasses grown for sod with only minimal effect on growth. The grower should be aware, though, of the potential impact of a specific herbicide on turf roots, rhizomes, and stolons before making the decision to apply.

Although the application of soil fumigants is regulated, with the use of some materials being tightly controlled, some growers are able to use a fumigant for nonselective weed control between crops. Where possible, fumigation is an excellent tool for weed management.[4]

8.8 QUALITY

The most important factor in turfgrass sod quality is based upon how the product looks at or immediately after installation. The product quality values density, color, freedom from weeds and pests, and minimum thatch. The cut of the sod piece is important as variation in width and thickness makes installation more difficult and affects appearance. Lateral strength for handling is important for efficient harvest and factors into quality in that thin, torn sod does not have an acceptable appearance. Quick rooting is the final checkpoint for quality.

Sod-product quality standards vary among the markets. Weeds and crop contaminants, such as other grasses, are considered acceptable to some degree in specific markets, whereas in others there can be no contaminants. Quality is frequently used as a means of product segmentation between competitors or within the pricing structure of one firm. The higher quality generally commands a premium price.

8.9   PRODUCT FINISH

Growers of high-quality sod conduct specific cultural practices to apply a finish to the product. Grooming takes place the final 3–4 weeks before harvest and is more than just mowing the grass. The turf is mowed more frequently and the height of cut is a little lower than the regular mowing cycle. The operator gives more care to the pattern, the clippings are removed, and a well-sharpened, properly adjusted reel mower is used.

Rolling with a large-diameter roller, such as 1-ton 4-ft diameter by 8-ft wide (900-kg 1.2-m diameter by 2.4-m wide) roller just before harvest smooths the surface. Mowing quality is higher and the sod is cut more uniformly. Sod handling can also be helped on some soils by rolling just before harvest.

Use of a nitrogen fertilizer at 20 lb N ac^{-1} (23 kg N ha^{-1}) a month before harvest will help keep the color. Nitrogen fertilizers should not be used closer than 2 weeks to harvest due to the risk of burning the turf inside the sod roll or folded piece. Iron can be used to enhance turf color right up to harvest. An application of 20 lb Fe ac^{-1} or 100 lb ferrous sulfate ac^{-1} (23 kg Fe ha^{-1} or 115 kg Fe ha^{-1}) will deepen the turf color in a couple of days with no risk in the sod roll or fold. The use of a nonburning iron product, such as chelated iron, lowers the risk of injury to the turf.

8.10   INVENTORY

The sod inventory is managed in four distinct stages. (Table 8.2 shows the definition of each stage of inventory in relation to turf growth stage.) Beginning with planting through harvest, the turf has value, which gains with growth and maturity. As the grass becomes sod, the accumulated input becomes the investment in the product.[2]

Stage I is newly seeded or stolonized grass up to fully covering the soil, which is particularly vulnerable to the environment and needs close attention by the irrigator.

Stage II is very young grass having reached full cover through growth to mature turf that has harvest strength, which is less vulnerable to the environment and somewhat forgiving of mistakes by management.

TABLE 8.2
Sod Inventory

Stage	Definition
I	New planting
II	Work in process
III	Ready for harvest
IV	Held in inventory

Stage III is the sod being groomed for harvest, requires close attention by management, and needs nonlimiting irrigation.

Stage IV is the mature sod that is ready for harvest, continues to be groomed, and requires close management to maintain quality while minimizing investment of resources.

The grower has to manage the sod inventory to meet the market needs. If sales are brisk, the decision may be to try to push or accelerate the sod growth to speed maturity. If sales are slow, the decision may be to slow the turfgrass growth to reduce mowing and thatch development. If growth is to be slowed for inventory, it may take 3–6 weeks to get growth started and the field groomed for harvest.

8.11 SOD HARVEST

Turfgrass sod is commercially harvested in pads or rolls 1.5 ft (0.5 m) to 4 ft (1.25 m) wide and up to 40 ft (12 m) long. Sod is ready to harvest as soon as the rhizomes and roots have knitted sufficiently to permit handling the sod pieces without tearing. The time required to produce a marketable crop of sod varies from 3 months in more favorable climates up to 2 years under less favorable conditions. Sod harvested with the plastic netting installed at seeding may be immature turf, but is fully capable of rooting.

Sod is cut with 0.5–0.6 in. (12–16 mm) of soil attached. If the cut is much thicker, it will lead to slow rooting. If thinner, its moisture-holding capacity will be inadequate to keep the sod fresh until installation.

Sod growers supply thick-cut sod with 1.5–2 in. (3.8–5.0 cm) of soil for sports fields and events when there is no time window for the normal sod to establish roots. Weighing 110–165 lb yd^{-1} (60–90 kg m^{-1}), the heavy weight of the sod will hold it in place for even the most vigorous athletic contest. Thick-cut sod is usually considered to be a temporary surface to be removed after a few weeks or months. On occasion, it is left as the permanent surface, but it is very slow to root into the soil or root-zone media.

Irrigating for the harvest of thick-cut sod requires a deft touch to wet to the depth necessary while allowing the surface to be dry enough to support the equipment.

Sod harvest shows up the efficacy of the irrigation throughout crop life. If possible the harvest should start on the highest side of the field to prevent runoff from the field irrigation into the harvest area.

Most sod is lifted with about 0.5 in. (1.5 cm) of soil. The spots that had been dry in the field tend to fall apart and not lift. The wet spots may smell the grass, may be off color, and may often have ruts that become holes in the lifted sod.

Perhaps the most closely coordinated team in the sod company is the irrigators and the harvest crew. Irrigation management for harvest requires a great deal of flexibility from irrigators that are trained to keep to a schedule.

There is a relatively narrow range of acceptable soil moisture with which the harvester will work efficiently. If too dry, the sod cutter blade cuts erratically and the product may not have enough moisture to get through the time until it gets water after installation. If too wet, the sod will not cut cleanly and the sod harvester will have difficulty tracking straight, so the sod pads will not be uniform. The loaded forklifts and trucks tend to leave deep ruts and frequently get stuck. Wet sod causes problems for the stackers, automatic or human, because of the weight and accumulation of mud on conveyers, rollers, and platform. Extra-heavy sod reduces the amount of sod a truck can haul and the sod tends to heat in the roll.

Sod harvesters are practically inoperative when the soil is dry and hard. The cutter blade bounces out of the ground, the sod has little strength if it can be cut, and the turf loss in transit is high.

High temperatures and drying winds require that extra attention be paid to harvest irrigation. Under these conditions, keeping the soil soft enough to cut may necessitate interrupting the harvest to water the field. The exposed sod in rolls or on pallets should not be allowed to dry out and can be watered with sprinklers or a hose.

Fields with 2% slope or more should be harvested beginning on the low side working across the slope, moving bottom to top. Unless the field can be lifted in one day, some irrigation will be needed before the field is finished, and the new water should run downhill, away from the harvest and harvested area. The harvested area can then dry so that it can be worked with soil-preparation equipment before the entire field is harvested. Harvest has to run parallel to the irrigation laterals to allow for efficient harvest irrigation management. Where center pivots are used, the harvester follows the track of the towers and usually begins on the outside of the pattern.

8.12   SOD DELIVERY

The delivery of the sod is usually the only part of the entire operation that is visible to the customer. Clean equipment, good tarps covering the pallets, and neatly stacked sod are good for business. The most significant issue, though, is on-time delivery. Sod installers have employees standing around waiting for the truck, so that not being on time costs the installer money and can be a strong word-of-mouth disincentive for customers. Having a reputation for being on time is a huge marketing tool.

Once a high-quality product is grown and harvested, the customer is to receive it in the best of condition. Delivery should be made within 36 h of harvest. Turfgrass as sod sitting too long on the pallet will become discolored, dry out, and be slower rooting.

Turfgrass grown to be lifted as sod and transported to another site is a unique crop and a unique segment of the turfgrass industry. The turfgrass manager sees the operation as an investment with profit potential. Sod production has its own cultural practices to meet the demands.

REFERENCES

1. Cockerham, S.T., *Turfgrass Sod Production*, ANR Publications, University of California, Oakland, CA., 1988, 84 pp.
2. Cockerham, S.T. and Burger, D.W., Irrigation of turfgrass sod, container, and nursery stock, in *Irrigation of Agricultural Crops*, Lascano, R.J. and Sojka, R.E., Eds., American Society of Agronomy, Madison, WI, 2006, chap. 12.
3. Cockerham, S.T. and Minner, D.D., Turfgrass nutrition and fertilizers, in *International Turf Management Handbook*, Aldous, D.E., Ed., CRC Press, Boca Raton, 1999, chap. 8.
4. Cockerham, S.T., Integrating cultural and pest management practices for sod production, in *Handbook of Integrated Pest Management for Turf and Ornamentals*, Leslie, A.R., Ed., Lewis Publishers, Boca Raton, 1994, chap. 12.

Section III

Sports-Turf Management

9 Culture of Natural Turf Athletic Fields

Stephen T. Cockerham

CONTENTS

9.1   INTRODUCTION

Sports fields that are well designed and constructed, and also well maintained, are likely to be safe, highly playable, tough, and look good. The sports-field owner (may be an entity), user, and spectator each have expectations of field performance. The expectations of the sports-field owner—public or private—are subject to the demands made by the users and spectators. The decisions for the degree of management support and the scope of the investment to meet the expectations are made by the owner. Users are the player, and they expect a safe, playable surface. Spectators expect the field to look good and support the game or activity.

9.2   SPORTS-FIELD PERFORMANCE CHARACTERISTICS

The characteristics by which sports-field performance is judged are safety, playability, aesthetics, and durability.

9.2.1   SAFETY

Field hardness and traction are measured to estimate impact absorption (ability of the turf to take shock) and shear resistance (ability of the turf to resist the tearing of the shoe cleats sliding over the turf), which relates to footing.

9.2.2   PLAYABILITY

Field playability is both measurable and perceptual. For various sports, playability concerns uniformity, footing, and speed (actual and perceived), ball speed, hop, deflection, and surface elasticity.

9.2.3   AESTHETICS

Turf is supposed to look good, and the appearance of the field reflects the pride of the maintenance personnel. The perception of safety and playability hinges on the aesthetics and uniformity of the field.

9.2.4   DURABILITY

The number of activities and their time on the field composes the demands. Traffic tolerance is the measure of field durability and the capability of meeting the demands.

9.3   DEFINING PERFORMANCE LEVELS

The purpose of the field is support for the basic movement patterns that an athlete makes in sports: walking, running, cutting, veering, stopping, pivoting, dodging, lunging, jumping, and landing. The energy return of turf is the elasticity for the enhancement of athlete performance. The performance of the field is in how it supports the athlete.

9.3.1   GAME FIELDS

Sports fields can be segregated into four levels of quality and performance expectations: *premium*, *choice*, *standard*, and *play*. (Table 9.1 shows quality and performance expectation levels.) *Premium* fields have high visibility and as such are expected to be of very high quality. They typically have high traffic from sports and events. Management intensity and owner investment are very high to meet the expectations. These fields generally support major professional league and major college sports teams.

   Optimum care of a premium-level sports field reflects best management practices, including optimum fertilizer for the climate and the turfgrass species, nonlimiting irrigation, mowing suitable

TABLE 9.1
Sports-Field Quality and Performance Expectation Levels

Grade	Quality	Traffic	Management Intensity
Premium	Very high	High	Very high
Choice	High	Moderate to high	High
Standard	Moderate	Very high	Moderate
Play	Low	Very high	Low

Source: Cockerham, S.T., et al., *Establishing and Maintaining the Natural Turf Athletic Field: Publication 21617*, University of California Agriculture and Natural Resources, Oakland, CA, 2004.

for the sport, frequent aeration and topdressing, rolling, overseeding, and repair of traffic injury. Premium sports fields are managed by having a close attention to detail and by meeting the needs of the grass around the schedule of events.

Choice sports fields have high visibility in a community, and quality is expected to be high. They have moderate to high traffic from sports and events. Management intensity is high. These fields generally support minor league professional, college, and high school sports teams. Traffic pressure on local school stadiums can be high when they are faced with community demands for access to the field for junior leagues, bands, and other events.

Optimum care of a choice-level sports field includes enough fertilizer applied as needed to meet the performance expectation, timely uniform irrigation, mowing, aeration, topdressing, rolling, overseeding, and repair of traffic injury. A high-traffic-level expectation for a game field performance would be an actual use of approximately 18 game-time hours of soccer per week, 12 game-time hours of youth football per week, or 30 game-time hours of baseball per week.

Standard sports fields may have high community visibility with moderate quality expectations. These fields typically have very high traffic from a community college and several high school sports teams as well as use as practice fields at all levels, including professional, college, and high school. Resource input is restricted, investment is low, and management intensity is moderate. Practice fields receive less attention during design, construction, and care but are subjected to greater use than game fields and generally have lower maintenance budgets. The use intensity on practice fields is very high, and compaction and wear reduce the turf surface performance. Care of a standard sports field includes enough fertilizer applications to allow the grass to grow, timely uniform irrigation, mowing, and aeration as needed. For this minimum investment, the performance expectation of actual use would be approximately 10 game-time hours of soccer per week, 6 game-time hours of youth football per week, or 20 game-time hours of baseball per week, which can be met with the minimum input on a field built on at least a loam soil and with reasonable drainage. However, it is not unusual for traffic levels on standard fields to far exceed this.

Play fields are park and school fields with very high traffic. Quality is low due to restricted resource input, resulting in low management intensity.[1] Play fields are probably the most common sports-fields in any community and are very important in the role of providing space for sports activities not associated with organized leagues and organizations. The construction is simple and inexpensive, and the use is varied and high.

Care of play fields should include fertilizer at least once per year, in spring, and irrigation as needed to keep the turf growing. Mowing is the one cultural practice that cannot be ignored.

9.3.2 Practice Fields

Practices, rehearsals, and physical education classes create such high traffic over a short time period that they are every bit as damaging and often more so than game action. Any kind of practice tends to concentrate the traffic. Practice fields are expected to take great abuse to prevent damage to the game field. The use of a game field for practices should be closely controlled; for example, one band practicing 45 min twice per week on the field will not be a problem if the surface is dry.

9.3.3 EVENTS TRAFFIC

Sports-fields are often used for other events, such as commencements, concerts, and rallies. The turf is often covered in an effort to protect the playing surface.

Turf that is covered for more than 4 days will have long-term damage. Turf damage can be reduced significantly by syringing the turf with a hose or an irrigation system immediately after removing the cover. Syringing should be applied within 30 min of uncovering for maximum effect.

Commercial covers that provide maximum protection for the turf are available. The most common cover, though, is a geotextile on the turf alone or under plywood. Geotextile is nonwoven, needle-punched, uniform polyester fiber in the form of a big blanket. There will always be some damage, but the geotextile helps distribute the effects of the compaction and protects the turf from the scuffing and tearing caused by foot traffic, chairs, and other hardware associated with special events.

Events traffic that moves across the cover, particularly trucks and forklifts, increases the damage by rubbing and abrading the grass. Plywood, usually 3/4 in. or 1 in. thick, on top of the geotextile affords some protection from heavy traffic. Covering the plywood with additional cloth helps extend its life and provides a safer surface for walking.

If soil is to be piled on top of the cover, a nonpermeable polymer layer at 4–10 mil (101.6–254 microns) thickness is placed between the geotextile and the plywood. This keeps the transient soil from filtering through to the sports-turf rootzone. The turf is left on the surface and is stripped off after soil removal and resodded. At soil removal, the plywood provides a guide for the loader blade or bucket to keep it from tearing out the field and provides some protection for the irrigation system.

9.4 CONSTRUCTION BASICS

Sports fields can come into being by simply scraping off a flat spot and trying to grow grass on it, by creating a well-designed facility with state-of-the-art construction, by creating and everything in between. How the field starts out sets the bar for the upper limit of possible field performance. The maintenance determines how close it gets to the bar.

The native soil field that is very basic construction is very common. Design additions such as subsurface drainage, an irrigation system, addition of soil amendment, and surface grading increase the potential for better performance. It is important that the owners, builders, and users generally accept the limitations inherent in the performance of a basic native soil sports field.

Unfortunately, far too many sports fields are built with good intentions and significant resource input, but poor judgment and lack of accountability result in field failure and a tremendous misuse of resources. Poor construction usually originates with administrative decisions, which include the politics of city councils or school boards and the mistaken concept that the low bid saves money. The selection or nonselection of a designer and the selection of a contractor who lacks experience or shows little concern with sports fields is a sure way to waste funds.

The results of construction errors are increased maintenance costs, high repair costs, and ultimately field failure. During planning, design, and construction, compromises are necessary, and even though decisions are usually based on budgetary versus technical needs, these compromises should be approached with thought and knowledge.

Good construction is the result of the land preparation in drainage and irrigation with the use of available technology in modified rootzones, species selection, and proper planting.

9.4.1   DRAINAGE

Most high-traffic turfgrass failures are directly related to inadequate drainage.[5] Poorly drained sports fields wear excessively and quickly lose playability and quality. Rutting and soil displacement occur when the soil cannot support traffic.

Sports fields are crowned for surface drainage. The type of sport and the type of rootzone media in the construction impact the height of the crown. Use over time will beat down a crown and potentially create areas of ponding. When that happens, the crown must be reestablished.

Sideline drains are important for protection of the field. The surface drainage as a result of the crown or slope is picked up by the sideline drains and taken from the field. In stadiums, the stands may drain onto the field, so sideline drains are necessary to keep excess water off the playing surface.

Saturated soil on high-traffic facilities decreases the turf durability and the capacity of the surface to support activity. Water that has moved through rootzone media must have an outlet, or the soil remains saturated and then becomes a mud hole. Drain tiles are made of clay, concrete, or perforated flexible plastic and are the most effective outlet for internal drainage and carrying the water away from the site.

Soils with poor physical structure and little internal drainage can sometimes be helped with the use of sand-filled slits. A machine cuts a slit about an inch (2.5 cm) wide, and a foot (30 cm) deep and then fills the slit with sand in a single operation. Sand-filled slits are quite effective by improving drainage and infiltration and can be completed quickly with minimum disruption of the playing surface. French drains are trenches filled with coarse gravel or rock. French drains are quick to build for a relatively small area but are not considered a permanent solution, though they can last a long time. Just as with any drain, the bottom of the trench must have a fall to carry the water away from the site being drained.

9.4.2   SAND ROOTZONE

The inability of most native soils to provide adequate surface and internal drainage through normal traffic is the cause of most sports-field failure, which has lead to the development of sand-based rootzone media for construction. The sand used in sports-field construction has a rigidly specific particle-size distribution, the specifications for which have been developed through extensive research and field experience by several universities and professional groups. Table 9.2 shows the particle-size distribution of a sand specification for a sports-field construction. The sand has

TABLE 9.2
Rootzone Media Sand for Sports Field Construction

Designation	Particle Size (mm)	Sieve No.	% Retained
Fine gravel	>2	10	0
Very coarse sand	1–2	18	<10
Coarse sand	0.5–1	35	
Medium sand	0.25–0.5	60	80–100
Fine sand	0.15–0.25	140	
Silt	0.05–0.15	270	<8
Clay	<0.5		<2

Source: Cockerham, S.T. and Minner, D.D., *International Turf Management Handbook*, CRC Press, Boca Raton, FL, 1999.

drainage, has the ability to hold moisture in the rootzone, and resists compaction. The amount of sand used in sports fields varies from pure sand to mixtures of various percentages of sand and organic amendments.

A primarily sand rootzone of a new field can be unstable if the sand is not settled and can "move" with traffic. Stability improves relatively quickly with the development of the biomass. Polymer mesh and fibers have been mixed into the upper profile to stabilize sand rootzone media.

9.4.3 Amending Soil

Sports fields built of native soil can be improved by amending with topdressing after aerification, although it is a long and slow process. Organic amendments such as peat and weathered sawdust can be used. Sand as a mineral amendment increases porosity, while calcined clay increases the porosity when added to fine soils and holds water when added to sand.

When using sand as a soil amendment, it is important to use enough. Too little sand mixed into a fine soil can decrease the drainage capability by setting up a concrete-like combination. To have any chance at being effective, the rootzone mix should have over 50% sand.

9.4.4 Species Selection

For maximum sports-turf performance, the grass species must be matched to the demands of the specific field use. Plant characteristics that favor wear tolerance are typically found in Kentucky bluegrass (*Poa pratensis*), perennial ryegrass (*Lolium perenne* L.), bermudagrass (*Cynodon dactylon* L.), and zoysiagrass (*Zoysia* spp., Willd.). Perennial ryegrass is a very durable turf species for winter sports, particularly in moderate climates.

In contrast to wear tolerance, plant characteristics that favor traffic tolerance and injury recovery are the turfgrass growth habits that result from the production of rhizomes, stolons, and tillers. Rhizomes are underground lateral stems found in Kentucky bluegrass, zoysiagrass, and bermudagrass. Stolons are aboveground lateral stems or runners also found in zoysiagrass and bermudagrass. Rhizomes and stolons each produce new plants at the joints or nodes. Tillers are side shoots from the base of a grass plant, which form roots and are found on the turfgrasses used on sports fields. These are the plant structures that give turfgrasses the ability to increase turf density and to recover from traffic injury.

For summer sports, bermudagrass where adapted is the best warm-season grass. Bermudagrass has higher recuperative potential than zoysiagrass, and the hybrid bermudagrasses (*C. dactylon*, L. × *C. transvaalensis*, Burtt-Davy) is higher than common bermudagrass.

9.4.5 Seeding Sports Fields

The seedbed soil should be loose enough to push a shovel into easily but firm enough to walk on. A fertilizer application prior to planting assures that adequate nutrients, especially phosphorus, will be available to the emerging seedlings. A rule of thumb is to apply 6–8 lb per 1000 ft^2 (2.9–3.9 kg per 100 m^2) of 16–20–0 (ammonium–phosphate–sulfate) or an equivalent. The uniformity of the stand and time to maturity are vastly improved when the nitrogen and phosphorus combination is applied preplant. On sand-based sports fields, potassium should be applied at a rate approximately equal to the nitrogen, such as 6–8 lb per 1000 ft^2 (2.9–3.9 kg per 100 m^2) of 15–15–15 (N–P–K).

Preplant fertilizer is applied even if soil tests show adequate phosphorus. However, soil tests are not to be ignored or discounted—they are indispensable for determining pH and salinity levels and for setting up the maintenance nutrition program.

Seeders should be calibrated to apply the seed accurately and uniformly. In soil, if the planter is a seed drill that places seed in a shallow furrow or trench and covers the seed, a mulch generally would not be necessary. Seeders that broadcast the seed over the surface require the seed to be covered by raking or mulching for a dense stand. Sand sports fields may need to be mulched regardless of planting technique.

The final step in planting grass seed is irrigation. Once the seed has been wet, it must be kept moist. Perennial ryegrass seed can germinate in 2–3 days, while bermudagrass seed may take 3 weeks to germinate. In hot weather, the irrigation pattern, wind, and irrigation timing will result in success or failure of a turfgrass stand.

9.4.6 STOLONIZING

Hybrid bermudagrass and zoysiagrass are planted vegetatively. Preparing soil for vegetative planting is the same as preparation for seeding. The application of preplant fertilizer is also the same.

Two common methods for vegetative planting of turfgrasses are sprigging and stolonizing. Stolons are the runners that grow above the ground as the mechanism by which certain grasses spread. Harvested stolons, in different markets, are called either stolons or sprigs. Along the length of the stolon are several joints, or nodes. Each node contains a bud capable of growing a new plant exactly like the parent.

The objective in planting stolons is to distribute the stolons uniformly and cover them with soil. Mechanical stolon-planting machines are available, and hydromulching machines have been used quite successfully to plant stolons.

Turf managers planting their own stolons can spread them by hand and then press them into the seedbed with a light disk or a cultipacker. More than one pass over the area may be necessary. Disk blades should be set straight (closed) to prevent the normal soil-turning action. Stolons buried deeper than 2 in. (2.5 cm) probably will not survive.

Planting rates for stolons vary depending upon the grass and the speed of turf coverage desired. A practical and economical planting rate is 200 bushels per acre.*

Irrigation is the single most important factor in successful stolon planting. Since stolons are planted into dry soil, they desiccate very quickly. On a hot day, an hour may be too long to wait before getting water on the planting. It is critical to not plant more than can easily be irrigated. Hold the stolons in bags or bins and sprinkle frequently to keep them moist and cool, rather than plant them and then not get the water on fast enough. The most common cause of stolon-planting failure is improper watering after planting.

The optimum planting time for stolons of warm-season grasses is from late April or when soil temperatures reach 60°F (about 15°C) through the first week of September.

9.4.7 SOD

Turfgrass sod is a viable alternative for sports fields that are being resurfaced or newly constructed. A sodded sports field can support play within a few weeks, or under certain circumstances, a few hours. The effort (labor, cost, expertise, and time) of establishing turf from seed or stolons often more than offsets any differences in the cost of the sod over other establishment methods.

Thin-cut sod with 1/2–5/8-in. (1.3–1.6-cm) soil can be expected to root within a few days if soil temperatures are above 60°F (15°C) and be ready for play within 4–6 weeks.

A field that must be ready for immediate use can be sodded with turf cut as much as 2 in. (5 cm) thick. The weight can be up to 15 lb per ft^2 (68 kg per m^2) to hold the turf in place. Thick-cut sod roots slowly into the underlying soil and may not root at all.

The big-roll sod harvest has been a boon to sports-turf managers. Sod cut over 40 in. (>1 m) wide and 30 ft (>9 m) long goes down quickly with minimum seams. It requires specialized equipment for the installation. Thick-cut sod in the big roll may leave a rough surface that must be rolled, especially

---

* A bushel of stolons has been defined as a square yard or square meter of sod, less the soil, chopped.

for sports such as baseball where ball bounce is important. Because time is usually a problem and the reason for using thick-cut big-roll sod, topdressing takes too long to smooth the surface.

A successful sod installation can be assured by following some simple guidelines. Put the sod down the same day that it arrives. Arrange to receive only the amount that the crew is capable of installing. Normally, one man can lay 500 ft^2 (46 m^2) of thin-cut sod per hour for 1 day.

Sod should be laid within an area bounded by sprinklers controlled by one irrigation station so that it will be possible to get water on immediately. When sod is irrigated, the water should soak through the sod into the soil underneath. New sod should be kept very moist for the first week, if traffic permits.

Fertilizer can be placed under the sod at installation. This is one of the few opportunities a turf manager has to place fertilizer in the rootzone where the plant needs it.

Some cool-season grass sod is grown with netting as an integral part of the product. The netting is at the soil surface below the crown of the grass and becomes exposed when the grass wears. A disadvantage for sports fields is that on occasion, the turf will shear at the netting layer. The netting itself tears easily and is not a physical problem with the cleats; however, it may be a perceptual problem with the athletes. Given a choice, netting-free sod is preferred for a sports field.

9.5   SPORTS-FIELD CULTURAL PRACTICES

The minimum effort for the care of the sports field constitutes six cultural practices that are fundamental to sports-field management. Sports-field quality and performance are increased by the use of five additional cultural practices.

9.5.1   MINIMUM MAINTENANCE

The maintenance practices that are essential to sports fields that are managed to meet the minimum requirements for safety, playability, aesthetics, and durability are (1) mowing, (2) irrigation, (3) nutrition, (4) compaction relief, (5) rolling, and (6) repair.

9.5.1.1   Mowing

Turf density is affected by height of cut and mowing frequency. Height of cut affects a number of turf-growth factors. Lowering the height of cut reduces carbohydrate production and the depth of rooting. It causes the leaves to be narrower and decreases the rhizome and stolon number, weight, and internode length. Turf vigor gradually decreases with decreasing plant size. Even though thatch and puffiness are reduced, durability declines. Both the shoot density and speed of playability increase with lower mowing. As a result of lowering the height of cut, the texture is finer with an increase in turf density and speed at the expense of shorter roots and lower traffic tolerance. Raising the height of cut increases the traffic tolerance while reducing turf density and speed. Table 9.3 shows the recommended height of cut for selected turfgrasses for the optimum growth and for the minimum that will sustain sports traffic, the height of cut below which a turfgrass fails under traffic pressure.

Each grass has a minimum height of cut at which point traffic tolerance declines rapidly. Changing species on the sports field may be required to meet the demands of the activity if, for example, the desired mowing height is too low for the existing turfgrass.

Cool-season species maintained at a high mowing height and then cut to a lower mowing height are more resistant to wear than the same grasses maintained at a low mowing height and allowed to grow taller. Scalping the sports field reduces appearance and playability.

The mowing frequency is easily determined by the rule of thumb to remove no more than one-third of the blade length in a single clipping. For example, a turf mowed at 1.0 in. (25 mm) should be cut before the turf reaches 1.5 in. (38 mm). If the time needed to grow the one-third length is 2 weeks, 1 week, or 2 days, then that becomes the mowing frequency.

TABLE 9.3
Height of Cut for High-Traffic Grasses

	Growth Optimum		Traffic Minimum[a]	
	(in.)	(mm)	(in.)	(mm)
Kentucky bluegrass	1½–2.0	38–50	5/8	15
Perennial ryegrass	1½–2.0	38–50	1/2	13
Tall fescue	1.0–2.0	25–50	3/4	19
Common bermudagrass	3/4–1.0	19–25	5/8	15
Hybrid bermudagrass	1/4–3/4	6–19	1/2	13
Zoysiagrass	1/2–1.0	12–25	5/8	15
Kikuyugrass	3/4–1.0	19–25	5/8	15

[a] Height of cut below which a turfgrass fails under traffic pressure.

Source: Cockerham, S.T., et al., *Establishing and Maintaining the Natural Turf Athletic Field: Publication 21617*, University of California Agriculture and Natural Resources, Oakland, CA, 2004.

The aesthetics of a sports field can be greatly enhanced by creating ribbon or striped mowing patterns. Patterns are often done to give a unique detailing to the field and are a source of pride. Patterns are created when the grass is pressed down to lay in opposite directions. Grass that lies away from the observer is light in color, while the darker grass lies toward the observer. The rollers on reel mowers inherently press the grass leaves down. The sports-turf manager can increase the intensity of the pattern with brushes or a drag mat. A skillful operator is aware of the pattern and careful as to its appearance.[3]

Field repair and other problems can often be masked by an intense pattern. Cross-hatching squares is the easiest to apply. The first passes, at least, must be laid out with string to assure straight lines. Many managers will string out every other pass to be sure that the pattern is true.

9.5.1.2 Irrigation

The successful management of a quality sports field is largely dependent upon the attention paid to irrigation. Wear tolerance and recuperative ability of sports turf require optimum irrigation and good drainage.

The most efficient irrigation is that which is scheduled to replace the water used by the turfgrass plant. The water use is estimated by the measurement of evapotranspiration (ET), the evaporation of water from soil and plant surfaces, and transpiration, which is the moisture released through plant leaves.

ET is measured under conditions of nonlimiting water availability to the plant. Cool-season turfgrasses are acceptable when irrigated at an annual average of 80% ET, and warm-season grasses are acceptable at an annual average of 60% ET. A high-performing sports turf will require more water to generate the fast growth required for recovery from injury. The art of irrigation is critical for matching the water application to the needs of the turf.

The most effective water conservation measures on high-traffic turf include being sure that the distribution is uniform and that there is no waste. There are no savings in keeping the moisture level below optimum for turfgrass growth on a sports field, particularly when safety to athletes may be compromised.

The first visual expression of drought stress is a gray-blue patina across the surface and, up close, the grass appears dry and limp. Footprints show when the grass is walked upon, as the blades do not spring back. In the early stages of wilt, the turf recovers fairly quickly with irrigation or syringing. If allowed to continue, the turf will reach the permanent wilting point, and there will be a loss of turf density.

Moderate water stress may actually stimulate deeper rooting. Sports-turf managers can produce turf better able to withstand drought by gradually lengthening the intervals between irrigations to create gradual water stress for deeper rooting. The manager must understand that there will be a risk associated with this.

Bermudagrass and zoysiagrass are warm-season grasses used on sports fields; have well-developed mechanisms for drought avoidance, including extensive root systems and the ability to tolerate high concentrations of solutes in internal water; and use less water than cool-season grasses for the same performance. Tall fescue has a fairly high water-use rate, but good drought tolerance results from the rolling of the leaves and an extensive root system.

Irrigation without uniform distribution not only wastes water but also nullifies the benefit from other cultural practices. A water audit using collection containers placed in a uniform grid will quickly demonstrate where problems exist, even down to specific sprinkler heads. Cans or plastic cups are set on 20-ft (6-m) centers, and the irrigation system is run for several minutes. The turf manager can see the distribution uniformity by measuring water in each container or just by looking into the containers.

Irrigation distribution patterns indicating poor uniformity are usually seen as arcs or circles of wilting grass or, in the advanced stages, low turf density. Where there are irrigation patterns, traffic injury is often excessive, the turf stand is clumpy, and the ground has hard and soft spots. Patterns are caused by malfunctioning hardware, poor design, low pressure, incorrect installation, and poor maintenance. Irrigation system deficiencies must be diagnosed and corrected.

Soil compaction is inherent to sports-turf facilities, and it is a major cause in the reduction of water infiltration. The standard cultural approach to increasing infiltration is core cultivation. If the decrease in infiltration rate is due to the surface compaction from traffic or the formation of turfgrass thatch, core cultivation is beneficial. If the soil has inherently poor physical properties for traffic, there will be no long-term and little short-term benefit from core cultivation. A practical short-term aid to problems with infiltration is the use of surfactants—chemicals that reduce the surface tension of water enough to allow infiltration and movement through marginally permeable soil conditions. Surfactants are useful tools for the turf manager working with high-traffic turfs.

Moisture sensors are instruments used to detect the soil moisture present and available to plants. The turf manager can use moisture sensors for periodic readings as a guide to adjust the irrigation schedule. Of more use is the role the moisture sensor can play in the automatic operation of an irrigation system.

The most critical factor in the use of moisture sensors is the location and depth of the instruments within the turf site. On high-traffic turf, the traffic-affected areas have such dramatically different moisture requirements from nontraffic areas that moisture sensors should be used only as one of several tools for irrigation scheduling.

The scheduling of water applications is the most misunderstood and difficult factor in irrigating high-traffic turfs. Sports activities, entertainment events, weather, and politics may have priority over the irrigation needs of the turf. Irrigation controllers, moisture sensors, and ET data are important tools that help in scheduling around activities.

An important irrigation management tool is syringing, the application of a light spray of water either by hand or by a short irrigation cycle. Syringing cools the grass and the environment around the grass blades. It slows wilt and increases the turgidity of grass blades causing them to stand up. It perks up the grass. Syringing is useful in taking care of the "hot" spots.

When the turf has been covered for an event such as a concert, syringing immediately upon removal of the cover can mean the difference between the grass reviving and not reviving in a reasonable period of time.

Applying fertilizer through the irrigation system, fertigation, is an acceptable technique used throughout agriculture and the turfgrass industry. There is low labor input required and minimal equipment. Irrigation distribution flaws will show up dramatically as the grass response to the fertilizer will emphasize the patterns.

TABLE 9.4

Nutrition for Sports Performance (Month of Growing Season)

Level	N (lb per 1000 ft^2)	N (Kg per 100 m^2)
Premium	1.0	1.1
Choice	0.75	0.8
Standard	0.5	0.2
Play	0.25	0.1

9.5.1.3 Turf Nutrition for Sports Performance

Turfgrasses require nitrogen (N) in the largest amount of any of the fertilizer nutrients. Since turf-grass nutrition is most critical in the quality of a sports field and is frequently ignored, the nutrients are often deficient. Nitrogen is the nutrient that is most often required.

Nitrogen levels that are too low on soil or sand do not produce enough turf biomass to sustain wear. Too much nitrogen is also undesirable. On both sand and soil sports fields, high-nitrogen turf is less wear tolerant and deteriorates faster than turf with intermediate levels of applied nitrogen. Optimum wear tolerance of cool-season grasses is usually at about an application rate of 1.0 lb N per 1000 ft^2 per month (1.1 kg per 100 m^2 per month) of growing season. Table 9.4 shows the recommended rate of application for nitrogen fertilizer to a sports field per month of growing season for performance level.

The turf manager can increase the biomass and cushion on the sports field by increasing N applications, although the tradeoff is at the expense of a reduction in root mass, lower recovery potential, and weaker shear strength. Increasing the cushion can make the field safer if footing is not seriously reduced.

Since N tends to increase topgrowth, under rapid growth conditions, shoots take priority over roots and rhizomes. If this occurs in the spring, excess N will cause the plant to enter the summer stresses with reduced root development and increased succulence and disease susceptibility.

Phosphorus (P) is used in relatively high quantities by plants. Since, in most forms, it is slowly soluble, it resists leaching. It can become unavailable to the plant if the soil pH gets too high or too low. P-deficient turf is stunted and may show a red color beginning at the leaf tips. Root development is increased from applying P to new seedings and new sod installations (applied under the sod). On cool- and warm-season turfgrasses, phosphorus is applied at the rate of 0.5–1.0 lb P_2O_5 per 1000 ft^2 per year (0.25–0.5 kg P_2O_5 100 per m^2 per year).

Potassium (K) improves turfgrass wear tolerance, disease tolerance, and aesthetic quality. Keeping the K in balance with the N will produce a tougher sports turf, reducing the water stress in turf and increasing capability to recover from drought.

Iron (Fe) is a valuable tool for providing color. If N is not deficient, an application of soluble Fe such as ferrous sulfate at 0.25–1.0 Fe lb per 1000 ft^2 (0.12–0.5 kg Fe per 100 m^2) will almost always darken the color of the turf within 2–3 days. Because the highly soluble products tend to burn the turf, irrigation should follow application immediately. Application on a hot day or at a high rate will cause burn where the applicator tires scuff the turf. Effective iron formulations are available that produce quick color with much lower burn potential.

Occasional use of sulfur-containing nitrogen fertilizers (e.g., ammonium sulfate) usually takes care of the sulfur needs of sports turf. Many sports fields are built on marginal soils, including landfill, where sulfur applications may be beneficial. Sulfur fertilizers tend to lower pH due to their acid reaction in the soil and are usually preferred for alkaline soils.

Nitrogen applied in the fall before the last mowing of the cool-season grasses improves spring greenup. Nitrogen applied late in the season on warm-season grasses before they go dormant improves fall color retention. A late-season application of high N to prolong color may have a cost in the increase of spring dieback of bermudagrass roots.

As soil temperatures in fall measured at 2 in. (5 cm) depth move below 50°F (10°C), dormant warm-season grasses do not benefit from fertilizer applications. For cool-season grasses, the corresponding low soil temperature is about 40°F (5°C). As the soil temperatures increase in spring, the turf will begin to grow and benefit from fertilizer applications.

The unique demands of the sports field often call for forcing extraordinary growth. Frequent high N-rate applications will cause a rapid flush of primarily topgrowth, but the roots also respond. This response is useful in peaking for a certain activity or recovering from a particularly damaging event. The cost in forcing growth is a risk of long-term problems with the turf and reduction in injury recovery potential.[4]

9.5.1.4 Compaction Relief

Sports traffic, especially cleated shoe traffic, results in varying degrees of soil compaction and surface sealing primarily in the upper inch of soil. It is expressed in reduced rooting depth when the soil is moist and a reduction in total root growth when the soil is dry. Traffic also causes sealing on compaction-resistant sand rootzone media by compacting the mat, thatch, and organic matter.

Water moves very slowly into compacted soil. Runoff occurs quickly testing the surface drains. Once the soil is wet, the rootzone does not drain well and does not allow oxygen exchange. The soil begins to sour from anaerobic activity. The soil and sand get blue colored streaks and pockets, which results in a strong sulfurous odor.

Compaction and sealing that result from sports traffic does not occur uniformly over the entire field. The traffic patterns of each sport are indicators of areas that need attention. Field hardness increases with increased compacting and can contribute to injuries.

Sports fields are also subject to damage from other types of traffic such as events set up, performance, and tear down. Vehicles moving across the turf to erect stages, sound towers, and to install seating units usually are not well organized and are heavy and prone to making a lot of trips.

Safety, playability, aesthetics, and durability of the sports field depends upon controlling compaction to achieve surface uniformity. Keeping unnecessary traffic off the field, building biomass to cushion the soil, and cultivation of the sealed layers are the basics of controlling compaction.

Compaction relief by core cultivation is the tilling of the soil to provide aeration without destroying the turf. It is used to (1) relieve soil compaction, (2) relieve surface sealing, (3) aid in thatch control, (4) disrupt undesirable soil layers, (5) prepare for overseeding, (6) enhance fertilizer and pH applications, (7) stimulate turf density by severing stolons and rhizomes, and (8) aid in soil modification.

Soil and sand sports fields should be core cultivated at least in early spring, summer, and fall to reduce localized dry spots and promote turf growth. Severely compacted fields need frequent aeration. On compacted fields that are particularly hard, this may require repeated core cultivation treatments followed by irrigation to eventually get adequate penetration. Practice fields receive more use than game fields and are improved by frequent core cultivation.

Compaction relief equipment other than core cultivators is in the marketplace. As with the core cultivators, once the procedures' upsides and downsides are understood they are very effective.

Wet soil can be aerated to help dry the soil profile. Muddy soil must be allowed to dry enough to support a vehicle before aerating. Aerating wet soil will create a tine sole, which is the sealing of the walls in the core holes. It will be necessary to aerate again when the field dries further.

Dragging the field with a steel mat after core aeration to break up the cores and work soil back into the holes will help reduce thatch. Aeration may reduce turf quality for a short term due to turfgrass crown injury and loss. Also, it may increase turf susceptibility to cold injury, and may increase weed encroachment. Nevertheless, overall, there is a general improvement of turf quality with compaction relief.

Core cultivation can be performed with a drum aerator using open and hollow tines or by a vertically operated hollow-tine aerator. As the tine is pushed into the turf, the soil forces the previous core out of the tine, contributing to a cultivation sole or a thin compacted soil layer.

Solid tines are used to shatter the soil below the surface, and the process is known as shatter coring. This is most effective on dry soil; wet soil does not shatter but compacts. Properly used, there is essentially no turf performance difference between the use of the hollow-tine core and solid-tine core.[5]

Spiking and slicing are effective to increase infiltration and reduce surface crusting. Spikes are nail-like steel tines, and slicers are flat steel triangles. The penetration is generally shallow and the openings small, thereby not disturbing the turf surface. Spiking and slicing are quick temporary operations that can be performed just before a game. A tine harrow can be used to break up aeration cores with the back used as a drag. It is also used to relieve compaction, especially on wet fields. The harrow does disturb the turf surface.

9.5.1.5 Rolling

Use of a heavy, flat, steel roller is effective in smoothing the field and improving the real and perceived turf quality. Rolling can correct what athletes, such as soccer players, feel as a loose field and will increase the speed of a baseball infield or football field. Frost heaving and foot printing can be relieved.

Rolling is also intense traffic and has a cost in wear and compaction. Aeration is an important associated program. Moisture level is critical in the successful use of the roller. Soil that is too dry is hard and does not respond to rolling. Soil that is too wet compacts too readily and is very difficult to manage.

The roller should weigh at least a ton. Self-propelled construction rollers work quite well. When it is necessary to roll a field, the operation should be in two directions. Fields should not be rolled more than two times in a month and two months in a year to give the turf a chance to recover. Coring for compaction relief should be done after each operation.

9.5.1.6 Repair

Sports fields need attention for the repair of damage from normal traffic, unusual activities, vandalism, pests, and even weather-related events.

Traffic damage initially shows as discolored turf. In baseball, outfielders tend to stand in a small area, which maximizes cleat injury. Other areas commonly damaged include the overrun at first base, the front of the pitcher's mound, and the paths to and from the mound and home plate. The area around the pregame batting practice cage is another frequent wear area.

Football traffic injury is most significant in the field center, between the 40-yard lines and the hash marks. The areas where the coaches and players stand also wear significantly. Soccer injury is severe in the center of the field and around the goalmouth. Linesmen run over a narrow path along the sidelines wearing it down.

Sodding may be a quick, simple solution to repair of sports-field damage. Thick-cut sod is used if there is little time for rooting. If the soil temperatures are warm, thin-cut sod can be ready for play in 3–4 weeks. Apply 1 lb N per 1000 ft^2 (0.5 kg N per 100 m^2) of a fertilizer high in N and P, for example, 6.25 lb per 1000 ft^2 (3.0 kg per 100 m^2) of 16–20–0 or 15–15–15. The tough decision is often where to draw the line when turf has to be replaced. It is generally preferable to take out all of the damaged turf to match new turf to old turf without a worn zone in between.

Overseeding is less expensive, though it requires a longer window between activities. Perennial ryegrass works well for overseeding, germinating in a few days and growing fast. Overseed at the rate of 10–15 lb per 1000 ft^2 (4.9–7.3 kg 100 m^{-2}), and keep the seed moist.

Divoting and tearing usually requires replacement of the turf as spot repair with sod. The sod should be thick cut and be as close to the color and texture of the surrounding turf as possible. Another patching technique is to mix sand and perennial ryegrass seed, which can be poured into the divots. Unless the turf is the same species, it will be a patchwork appearance.

If the field is used for events other than sports activities, repair is critical during the season. Many events require the field to be covered for a day or several days. The turf recovery is significantly improved by syringing as the cover is removed. Usually, some damage occurs to the irrigation system, so hoses with spray nozzles should be readily available to syringe the turf. After syringing, if traffic has been severe and the field covered for over 3 days, spiking the field opens the surface. It is usually not practical to core aerate, although coring is very beneficial. Mow the grass the day after uncovering; otherwise, keep traffic off of the field. An application of N at 1.0 lb per 1000 ft^2 (0.5 kg N per 100 m^2) and Fe at 0.5 lb per 1000 ft^2 (0.3 kg N per 100 m^2) following mowing will increase the recovery rate.

Heavily trafficked sports fields will wear from the intense use. If the field condition has deteriorated beyond localized repair, renovation or even complete rehabilitation or replacement may be necessary.

Light renovation involves vertical mowing and aerating, followed by overseeding into the turf. If the grass is normally established by seed, overseeding with the same species will speed turf recovery. Hybrid bermudagrass stolons can be planted in the old turf surface with a planter that uses disks or coulters to cut through the old turf. To stimulate the recovery, apply 1 lb N per 1000 ft^2 (0.5 kg N per 100 m^2) of a fertilizer high in N and P, for example, 6.25 lb per 1000 ft^2 (3.0 kg per 100 m^2). Weeds should be treated with herbicides.

Heavy renovation involves the killing of the surface plant material, including weeds and the existing turfgrass, with a nonselective herbicide. Remove the old sod with a sod cutter, and rototill the soil. *Never lay sod over old turf even if it has been rototilled.*

9.5.2 Cultural Practices for Performance

The performance of a sports field can be increased with the application of additional cultural practices: (1) vertical mowing, (2) topdressing, (3) overseeding, (4) pest and weed management, (5) temperature modification, and (6) colorants.

9.5.2.1 Vertical Mowing

The vertical mower is used to comb out the thatch and slice through lateral growth. Vertical mowing is an important tool to keep thatch and puffiness under control.

Thatch is an accumulation of undecomposed organic matter forming between the soil surface and the turfgrass crown, which develops naturally with the vigorous growth of plant leaves and roots.

Thatch is an asset on sports turf because it increases wear resistance, increases footing, and acts as an impact-absorbing safety pad. The sports-turf manager commonly develops and maintains a thatch layer. Use helps keep the thatch from excess buildup.

Turfgrass species that have a vigorous lateral growth habit, such as bermudagrass, zoysiagrass, and Kentucky bluegrass, tend to be moderate to heavy thatch producers. Perennial ryegrass and tall fescue have lower thatching tendencies, unless fertilized heavily with high nitrogen applications, although some cultivars have a considerable amount of thatch.

Thatch buildup over 0.75 in. (1.9 cm) becomes excessive. Excess thatch inhibits water and fertilizer getting to the soil, and rootzone oxygen exchange can be reduced, which weakens and reduces roots and rhizomes. Turf that is spongy and scalps easily may have too much thatch. High thatch will increase traffic tolerance but slow the surface.

Fast growing, vigorous turf can develop a springy feel often referred to as being puffy. Puffiness is the result of the growth habit of the grass where the mostly stemmy topgrowth gets tall and lays over. Puffiness is the result of excessive lateral growth, mostly from stems. The undesirable effects are a lower plant density, a strong tendency to scalp, and the turf tears easily, which results in poor footing. Hybrid bermudagrasses have a tendency to get puffy with a high nitrogen program or if mowed too high. Vertical mowing slices the lateral stems for control of puffiness. Puffiness can easily be detected by brushing a hand or shoe across the turf in several directions. The grass will stand up in one direction and lay smooth in the opposite.

Vertical mowing cuts stolons and tillers and combs dead material from the turf leaving the debris on the surface. In the short term, it is a thinning process. In the long term, turf density can be increased with lateral growth stimulation as a result of the slicing. Vertical mowing is a harsh operation with the potential of causing considerable turfgrass injury. The grass should be actively growing so a full and rapid recovery will result. When the turf is under stress or temperatures are out of the optimum range—too high or too low—recovery is slow.

9.5.2.2   Topdressing

Turf is topdressed to produce and maintain a smooth playing surface; provide a firm, uniform profile; and control thatch. Topdressing is the application of a growth medium, usually specified rootzone sand, over an established turf surface and working it into the mat with brushes or drags. It is a method of detailing a sports field to get a true surface.

Topdressing a cored surface and using a drag to work the material into the holes is a technique that can be used to gradually modify an existing soil medium. A topdressing that is widely accepted for general turf is a medium sand described in Section 9.4.2 and applied at 1/8 to 1/4 in. (3.2–6.4 mm) depth four times per year.

Some benefits from topdressing are immediate, such as increasing speed, while others are normally quite slow, such as truing a rough surface. Topdressing is most useful on fields that are in reasonable playing condition where the surface drainage is adequate. Repeated heavy topdressing can be used over time to bring surface depressions to grade level.

9.5.2.3   Overseeding

Playing surfaces can be overseeded for winter color or to increase the turf density. The process is to vertical mow, apply nitrogen, overseed, and lightly irrigate until the seedlings are established. Stand density can often be improved by making several passes over the field with an aerator, followed by overseeding with a slit seeder.

For winter color, perennial ryegrass is the most durable species for overseeding bermudagrass. A caution is that spring transition is poor for bermudagrass overseeded with the perennial ryegrass and subjected to traffic. Perennial ryegrass as an overseed is not compatible with zoysiagrass. Zoysiagrass should be overseeded with tall fescue or possibly bluegrass.

Turfgrasses that are normally established on sports fields by seed can be overseeded to help repair traffic injury. Broadcasting seed before a game and allowing the players to work the seed into the mat is a technique that may help keep density, particularly if there is a window of no traffic for the seedlings to emerge.

Very high seeding rates are required for successful overseeding. The principle is to generate a cover of seedlings very quickly. Most of the seed is exposed on the surface regardless of the technique used, and through the wetting and drying that normally occurs, many of the weaker seedlings die. The ryegrasses germinate in a few days, and Kentucky bluegrass in 3–4 weeks.

Germination of overseed grasses can be speeded up by presoaking or pregerminating seed. One technique is to place approximately 150 lb (65 kg) of loose seed in a steel drum capable of holding about 50 gal (190 L) of water, with a drain in the bottom that can be screened and plugged. Ten to twenty drums are used to pregerminate enough seed for a football field. The seed is put in the drums, and then filled with water. The water is changed two to three times per day for 2 days. On the third day, moist seed that is to be planted after presoaking is spread on a surface and mixed with sand as a carrier (4:1 of sand to seed). A common substitute for sand is a commercial low-nitrogen organic turf fertilizer. The seed/carrier mixture is spread on the turf and kept lightly irrigated until the seedlings emerge. Germination time will be reduced to half the normal time for the species used. Seed that is to be pregerminated is treated by the filling and draining process until about 25% of the seed sprouts. The handling of pregerminated seed is similar to the presoaked seed, but the pregerminated seed must be gently mixed with the carrier. Pregerminated ryegrass seed can produce seedlings on the turf within hours. If the soil temperature is below that for normal germination, a vented tarp can be used to warm the surface and assist seedling growth.

Seed for overseeding is usually broadcast over the surface. After overseeding, the field is dragged with a drag mat to shake the seed down into the mat. Irrigation should be such that the surface will not dry out for the first week. This may require irrigation two or three times per day. The new seed lies uncovered on the surface and is very susceptible to drying out and dying. Even at that, only a small percentage of the seed will germinate, and this is the reason for the heavy seeding rate. Water management is the key to successful overseeding.

9.5.2.4 Pest and Weed Management

The options for pest management on sports fields have been reduced by public perceptions about the use of pesticides. Managing a vigorous turf helps prevent disease and insect infestation. Weeds, however, are invited by the damage caused by traffic.

Sports fields with weeds not only look bad; they have different play and wear characteristics than weed-free fields. Actively growing turfgrasses maintained at optimum cutting height, irrigated, and with balanced fertility are competitive against weed establishment and invasion.

A high weed population on a sports field usually indicates that use has exceeded the tolerance of the field. The field construction and maintenance determine the use capability of the surface and may not be in the same performance level as the expectations.

Hand weeding is a viable option for sports turf. In half an hour, a couple of teams with weeders in their hands can cut out a lot of weeds. When a herbicide is necessary, the label must be checked to be sure that the chemical can be used on athletic fields.

9.5.2.5 Temperature Modification

The temperature of the air and soil has a significant impact upon the growth of turfgrasses: there is little biological activity above 120°F (50°C) or below 32°F (0°C), with optimum air temperatures for shoot growth of warm-season grasses at 80–95°F (47–35°C) and cool-season grasses at 60–75°F (16–24°C). Optimum soil temperatures for root growth taken in the upper 2 in. (5 cm) for warm-season grasses are 75–85°F (24–29°C) and 50–65°F (10–18°C) for cool-season grasses.

Chilling injury on turfgrasses occurs at temperatures just above freezing. Bermudagrass roots continue to grow after the shoot has lost color as it goes into dormancy. Some root growth of all turfgrasses occurs as long as the soil is not frozen. In the summer months, sports fields are frequently subjected to traffic during periods of high temperatures resulting in a decrease in durability and recuperative potential.

The temperature of a sports turf, when outside the optimum range, can sometimes be modified to allow or even encourage growth. The turf can be heated or cooled as needed. Soil warming with buried electrical heating cables or tubes for circulation of heated fluids is very effective. Soil warming is particularly valuable for major league facilities that have an early-season turf performance requirement or late-season demand. The installation and use of a soil-warming process is not a simple alteration of management practices; the turfgrass "system" has changed so there is a learning process to maximize the benefits of the heating system.

For fields that do not have rootzone warming capability, the use of a vented tarp is a practical and inexpensive means of increasing the turf temperature. Acting much like a greenhouse, the vented tarps allow heat buildup during daylight hours and reduce loss through the night. The condensation under the tarp keeps the turf and microenvironment moist, which can be beneficial for germinating overseeded grass. The tarps are usually translucent polyethylene of 10-mil (254 microns) thickness with 0.5 in. (1.3 cm) diameter vent holes punched every 6–8 in. (15–20 cm). Venting is required for oxygen and carbon dioxide exchange as well as temperature management. Woven polyethylene and geotextiles can be used since they are good insulators, but heat does not build up as quickly under these materials as it does with the vented polyethylene. The disadvantages of any of the tarps are related to the labor input. The tarps must be removed for mowing and field use; there is high potential for diseases such as *Pythium* spp.

Heating systems and tarps can raise the soil and turf surface temperature too high. Rootzone systems should be thermostatically controlled, but the tarps are not, so frequent monitoring of soil temperatures is essential. Very high soil temperatures can be devastating by destroying roots, rhizomes, and stolons, thereby reducing playability and the ability of the turf to recover from injury. As soil temperatures approach 95°F (35°C), the turf is susceptible to high-temperature damage.

High-temperature stress on turfgrasses is usually associated with moisture stress. Syringing or light applications of water lowers the surface temperature of the turf and reduces hot-weather desiccation. Syringing with about 0.2 in. (5 mm) water will reduce the surface temperature 2–5°F (4–9°C) for up to 2 h. Drought resistance of various species correlates with the high-temperature killing time. Heat stress is moderated with minimum nitrogen, higher mowing height, and longer mowing frequency. Although frequently impractical in the field, heat-stress tolerance can be increased by hardening off a turfgrass with exposure to hot, dry conditions.

9.5.2.6 Turf Colorants

The use of turf colorants is a valuable management tool for sports-field care particularly where visibility is high. Turf colorants are permanent, colorfast materials, which are used for turf color on dormant or damaged turf. Colorants are applied as dilute solutions in water. Application techniques are critical to make the turf look as natural as possible. The color is built up gradually on the grass blades, and streaks are avoided by repeated spraying.

The investment of work and resources made to provide a safe place to play a game, which looks good and stands up to the use, is rewarded by the pride of the community. It is something that pays off for everyone.

REFERENCES

1. Cockerham, S.T., Gibeault, V.A. and Silva, D.B., *Establishing and Maintaining the Natural Turf Athletic Field: Publication 21617*, University of California Agriculture and Natural Resources, Oakland, CA, 2004, 56 pp.
2. Cockerham, S.T. and Minner, D.D., Turfgrass nutrition and fertilizers, in Aldous, D.E., Ed., *International Turf Management Handbook*, CRC Press, Boca Raton, 1999, chap. 8.
3. Davis, W.B. and Paul, J.L., *A Guide for Evaluating Sands for Use as a Growing Medium for High Traffic Turf*, University of California Cooperative Extension, University of California, Davis, CA, 1985, 4 pp.
4. Mellor, D.R., *Picture Perfect: Mowing Techniques for Lawns, Landscapes, and Sports*, Sleeping Bear Press, Chelsea, MI., 2001, 160 pp.
5. Puhalla, J., Kranz, J. and Goatley, M., *Sports Fields: A Manual for Design, Construction, and Maintenance*, Ann Arbor Press, Chelsea, MI, 1999, 464 pp.

Section IV

Turfgrass Growth Regulators:
Promoters and Inhibitors

10 Applied Physiology of Natural and Synthetic Plant Growth Regulators on Turfgrasses

Erik H. Ervin and Xunzhong Zhang

CONTENTS

10.1 INTRODUCTION

Altering the growth and development of turfgrass through exogenous applications of various chemicals began with maleic hydrazide in 1945 and has been a research endeavor of turfgrass scientists for more than 60 years.[1,2] The primary objective has been to restrict leaf growth and the subsequent need for frequent mowing. The earliest plant growth regulators (PGRs) labeled for use on turfgrasses, maleic hydrazide and mefluidide, provided adequate shoot growth suppression but often caused unacceptable phytotoxicity and stand-density losses because of their ability to rapidly stop cell division and differentiation in meristematic regions.[1] These cell-division inhibitors were

TABLE 10.1

Classification of Natural and Synthetic Plant Growth Regulators Used on Turfgrasses[a]

Class	Active or Base Ingredient	Brand Name (Manufacturer)
GA inhibitors, late		
A	Trinexapac-ethyl	Primo Maxx (Syngenta) Governor (Andersons)
A	Prohexadione-calcium	Not labeled for turf
GA inhibitors, early		
B	Flurprimidol	Cutless (SePro)
B	Paclobutrazol	Trimmit (Syngenta) TGR (Andersons)
Cell-division inhibitors		
C	Mefluidide	Embark (PBI/Gordon)
Herbicides		
D	Glyphosate	Roundup Pro (Monsanto); many others
D	Imazapic	Plateau (BASF)
D	Sethoxydim	Poast (Micro Flo)
D	Diquat	Reward (Syngenta)
D	Chlorsulfuron	Corsair (Riverdale)
Hormones		
E	Ethephon (ethylene)	Proxy (Bayer)
E	Indole butyric acid (auxin)	Kickstand (Helena)
E	Gibberellic acid	ProGibb T&O (Valent U.S.A.) GibGro (NuFarm Americas)
E	Benzyl-adenine (cytokinin) Kinetin (cytokinin)	None labeled for turf
E	Salicylic acid	None labeled for turf
E	Acibenzolar-S-methyl	Actigard (Syngenta); not labeled for turf
E	Jasmonic acid	None labeled for turf
Natural-source PGRs		
F	Humic substances: humic and fulvic acids	Many sources
F	Seaweed extracts	Many sources
F	Composted organic residuals	Many sources

[a] This is not intended to be an all-inclusive list of available Class D, E, or F PGRs for turf.

originally classified as type I PGRs but are now known as class C materials.[3] In fine turf, their use is now primarily limited to seedhead suppression. Our focus in this chapter will be on beneficial effects of PGRs on stress physiology and performance of fine turf, so no further mention of class C PGRs will be made.

By the late 1970s, type II PGRs were developed to reduce cell elongation by inhibiting gibberellic acid (GA) biosynthesis while not inhibiting cell division. These materials are much safer on fine turf and continue to be widely used for various purposes. Type II PGRs are now divided into two classes (see Table 10.1). Class A PGRs, represented by trinexapac-ethyl (TE) and prohexadione-calcium (PE), are foliar-absorbed and inhibit GA late in its biosynthetic pathway, while class B PGRs, represented by flurprimidol and paclobutrazol, are root-absorbed and inhibit GA early in its biosynthetic pathway. Considerable research has been conducted on the physiological effects of these compounds, especially TE, and these data will be reviewed here.

Class D PGRs consist of various herbicides used at low (i.e., sublethal) rates to suppress growth or seedhead development of turfgrasses. Growth suppression can be accomplished with herbicides that interfere with amino acid biosynthesis (e.g., glyphosate, chlorsulfuron, imazapic) or fatty acid biosynthesis (e.g., sethoxydim). However, turfgrass tolerance can be marginal and is highly rate dependent, so use of these herbicides as PGRs is limited to low-maintenance sites where turfgrass injury can be tolerated.[4] As such, class D PGRs are not used to improve the performance of fine turf and will not be discussed in this chapter.

Class E PGRs are compounds that are phytohormones or mimic the action of phytohormones. All the major hormone groups, except abscisic acid, are potentially represented on the turfgrass market in compounds such as ethephon (ethylene), benzyl-adenine (cytokinins), GA, and indole-3-butyric acid (auxin). Exogenous application effects of these primary phytohormones on stress tolerance of fine turfgrasses will be reviewed in this chapter along with reported effects of secondary phytohormone compounds such as salicylic acid (SA), acibenzolar-S-methyl (ASM), and jasmonic acid (JA).

We would like to suggest a final PGR type in this chapter, class F. We define class F materials broadly as organic compounds of animal, algal, or plant origin that have been mined or processed and been shown to have various PGR effects on turfgrasses. Examples of class F materials whose growth-regulating effects will be reviewed in this chapter are humic substances and seaweed extracts. Although these materials are often base ingredients in commercially formulated biostimulants or fertilizers, these terms are too broad for our purposes. We therefore suggest that class F materials be discussed under the heading of natural-source PGRs.

10.2 TURFGRASS GROWTH REGULATORS: APPLIED PHYSIOLOGY

10.2.1 CLASS A PGRs (LATE GA INHIBITORS)

10.2.1.1 Mode of Action

TE (4-[cyclopropyl-α-hydroxy-methylene]-3,5-dioxo-cyclohexane-carboxylic acid ethyl ester) and PC (calcium salt of 3,5-dioxo-4-propionylcyclohexane-carboxylic acid) are structural mimics of 2-oxoglutaric acid and act competitively with 2-oxoglutaric acid to inhibit the 3β-hydroxylase conversion of gibberellic acid-20 (GA_{20}) to GA_1 resulting in reduced shoot elongation.[5] These PGRs are foliarly absorbed, with radioactive tracing studies showing that uptake occurs in 1 h or less, with between 50% and 75% of labeled TE or PC remaining in the foliage and 10% recovered in the roots of Kentucky bluegrass (*Poa pratensis* L.) at 24 h after application.[6,7] TE has been labeled for use on cool- and warm-season turfgrasses since 1991, while PC is not currently registered for turfgrass use. Multiple reports are available documenting the rates required for an approximate 50% reduction in clipping yields on many turfgrass species due to TE,[8–12] while only one report is available for PC.[13] Readers interested in the details of reducing mowing requirements with class A PGRs should consult the references cited as our focus is on the morphological and physiological side effects pertinent to fine turf performance under stress.

10.2.1.2 Structural and Physiological Effects

Inhibition of cell elongation due to TE has been shown to increase mesophyll cell density and chlorophyll concentration,[14–16] resulting in dwarfed shoots that are darker green. Net photosynthesis may be increased by TE as photosynthetic rates are reportedly unchanged due to TE on cool-season[17] and warm-season[18] turfgrasses, while maintenance respiration may be decreased.[19] Photosynthate not used for leaf elongation must be stored or used in other organs for growth and maintenance. Hybrid bermudagrass (*Cynodon dactylon* x *C. transvaalensis*) treated with ^{15}N-labeled ammonium nitrate allocated approximately 50% more N to roots and rhizomes when treated with TE.[20] These data and reports of postinhibition shoot growth enhancement[21,22] led to the hypothesis that total nonstructural carbohydrates (TNC) may accumulate during the period of TE inhibition or be used to fuel other developmental events such as increased tillering, stem growth, or rooting.

TABLE 10.2

Total Nonstructural Carbohydrates in Creeping Bentgrass, Kentucky Bluegrass, and Hybrid Bermudagrass as Influenced by Labeled Rates of TE Applied Every 2 Weeks Over a 2-Month Period

		TNC (g kg⁻¹ DW)				
		0 WAIT[a]	2 WAIT	4 WAIT	6 WAIT	8 WAIT
Bentgrass	+TE[b]	34.8	35.1	50.2*[c]	42.8*	47.0*
Bentgrass	− TE	33.7	34.7	43.5	36.4	32.6
Bluegrass	+TE	44.9	43.0	44.2*	32.3	58.3
Bluegrass	− TE	49.8	48.0	32.0	29.9	54.1
Bermuda	+TE	83.5	86.4	85.3*	81.6*	88.9*
Bermuda	− TE	86.1	89.9	72.1	70.5	74.0

[a] WAIT: Weeks after initial treatment with TE.

[b] +TE: TE applied every 2 weeks beginning at 0 WAIT, with the fourth and last application at 6 WAIT; − TE: control; TE rates were 23, 108, and 45 g/ha for bentgrass, bluegrass, and bermudagrass, respectively.

[c] After a mean, "*" indicates significance relative to the control (within each species) at $p = 0.05$.

Han et al.[23,24] reported increased TNC content in the verdure of creeping bentgrass (*Agrostis stolonifera* L.) 2 weeks after the initial TE application, with decreased TNC levels from weeks 4 to 16. Richie et al.[25] reported no differences in tall fescue (*Festuca* arundinacea Schreb.) leaf, crown, or root TNC levels at 6–7 weeks posttreatment. More recently, Ervin and Zhang (2003, unpublished data) found that application of TE every 2 weeks for a total of four applications resulted in increased leaf TNC from 4 to 8 weeks after the initial application on L-93 creeping bentgrass and "Tifway" hybrid bermudagrass but were not as pronounced in "Midnight" Kentucky bluegrass (Table 10.2). Waltz and Whitwell[26] also reported increased TNC in Tifway hybrid bermudagrass root and shoot tissues during sequential TE applications. Variable TNC status findings between studies are not easily explained due to differences in TE rates studied, sampling timing as regards plant metabolism of TE, or environments. Reported increases in tillering and, less frequently, root mass, along with improved carbohydrate status of TE-treated turf in reduced-light environments, appear to support a hypothesis that leaf carbohydrates are conserved and partitioned for use in stem and root tissues.

Numerous reports are available documenting increased tiller density of cool- and warm-season turfgrasses due to sequential TE applications.[8,27–29] In developing perennial ryegrass (*Lolium perenne* L.), TE application at a 3-week interval resulted in a 67% tiller density increase.[8] A 22% tiller density increase was reported on Kentucky bluegrass that was mowed at 5.7 cm and treated with TE at 0.27 kg ha⁻¹ on a 6-week interval.[27] "Meyer" zoysiagrass (*Zoysia japonica* L.) mowed at 1.6 cm and treated with TE at 0.1 kg ha⁻¹ on a 4-week interval resulted in a 12% increase in tiller density.[28] Lastly, sequential TE treatment resulted in a 6% tiller density increase in hybrid bermudagrass.[29] The rooting data concerning effects of TE are less clear.

Root-length density of tall fescue grown in sand-filled greenhouse rooting tubes was reported to be decreased due to TE at a 10- to 20-cm depth only.[30] TE applications had no effect on perennial ryegrass root-length density when the field was maintained in a silt loam and allowed to undergo moderate wilting over two summer trial periods.[31] Similarly, Ervin and Koski[8,27] reported no effect of TE on root mass of perennial ryegrass seedlings grown in an 22/18°C growth chamber or on mature field-grown Kentucky bluegrass sampled to a depth of 60 cm over a 3-year period. The root mass of creeping bentgrass also was unaffected when maintained as a putting green and treated with TE monthly for 24 consecutive months.[32]

All these cool-season turfgrass data were taken on a plant area/soil volume basis, not accounting for TE effects on a per tiller basis. Beasley et al.[33] investigated the effects of one TE application over a 7-week period on hydroponically grown Kentucky bluegrass responses on a per tiller basis. Total root length and root surface area on a whole plant basis was increased due to TE from 1 to 4 weeks after treatment (WAT). However, because the tiller number was significantly greater for the TE-treated plants 4 WAT (21 versus 13.5), total root length and surface area per tiller was significantly decreased. These data appear to illustrate a usually unseen chain of developmental events caused by TE on cool-season grasses. As leaf elongation is inhibited, the crown is favored as a sink pulling extra photosynthate to the meristem, triggering the development of new tillers followed by their new roots. During inhibited leaf elongation, growth of existing roots may be slowed while these new organs develop. The question of whether these developmental changes confer an adaptive advantage or disadvantage will be indirectly addressed as the stress tolerance research is reviewed in Section 10.1.2.3.

Warm-season turfgrasses, especially hybrid bermudagrass, appear to respond in a slightly different manner morphologically as compared to cool-season grasses. Tillering appears to be stimulated to a lesser extent, whereas rooting has been more consistently reported to be increased. Application of TE at 0.0125 kg ha^{-1} every 10 days resulted in 23% and 27% greater root mass for Miniverde and Floradwarf hybrid bermudagrasses, respectively, while the root mass of four other cultivars was unchanged.[34] Similarly, when the hybrid bermudagrass Tifeagle was treated with TE at 0.05 kg ha^{-1} every 3 weeks, root mass increased by 43%.[35] Further field verification of increased rooting due to TE on bermudagrass and other warm-season turfgrasses is needed as both these studies were conducted in greenhouse environments.

It has been noted that enhanced assimilate partitioning to the crown due to TE may be responsible for accelerated development of new tillers and roots. Cell division must occur for these new tissues to develop; a process that is primarily directed by cytokinins. Rademacher[5] in a review of the literature pertaining to the side effects of GA inhibitors summarizes their effects on the levels of other hormones: "Typically, these compounds induce increased contents of cytokinins, whereas ethylene levels are lowered. ABA concentrations may be significantly increased under distinct conditions whereas the auxin status is not significantly affected. Resulting primarily from these effects, a delay in senescence and increases in resistance to environmental stresses are often found." Ervin and Zhang (2003, unpublished data) found that sequential TE applications at labeled rates to creeping bentgrass, Kentucky bluegrass, and hybrid bermudagrass resulted in increased leaf tissue levels of the cytokinin, zeatin riboside (Table 10.3). No data are available concerning the effects of TE on levels of ethylene, auxins, or ABA. Although the chain of metabolic events leading to increased cytokinin levels remains unclear, the fact that they can be increased helps explain some of the positive morphological and physiological effects that have been reported.

Cytokinins are known to strongly delay senescence via a number of mechanisms, including acting as antioxidants, inhibiting respiration, decreasing enzymes (lipase and lipoxygenase) involved in membrane breakdown, limiting stomatal closure, and preventing chlorophyll loss.[36] Increased cytokinins due to TE might then be expected to confer improvements in some physiological functions. Zhang and Schmidt[37] report that monthly TE application to Penncross creeping bentgrass resulted in leaf tissue that maintained consistently higher levels of the antioxidant superoxide dismutase, along with higher canopy photochemical efficiency as measured with a chlorophyll fluorometer. Such positive physiological responses have led to the hypothesis that TE may enhance resistance of turfgrasses to environmental stresses.

TABLE 10.3

Cytokinin (Zeatin Riboside) Content in Leaves of Creeping Bentgrass, Kentucky Bluegrass, and Hybrid Bermudagrass as Influenced by Labeled Rates of TE Applied Every 2 Weeks over a 2-Month Period

		Cytokinin, Zeatin Riboside (ng g^{-1} Fresh Weight)				
		0 WAIT[a]	2 WAIT	4 WAIT	6 WAIT	8 WAIT
Bentgrass	+TE[b]	7.7	8.3	13.8*[c]	19.6*	22.2*
Bentgrass	− TE	7.8	8.4	12.5	16.2	18.3
Bluegrass	+TE	13.7	14.2	25.0*	26.3	31.8*
Bluegrass	− TE	12.9	13.5	21.0	21.0	26.0
Bermuda	+TE	10.9	12.5	29.3*	35.4*	42.8*
Bermuda	− TE	14.4	13.7	21.1	27.0	22.2

[a] WAIT: Weeks after initial treatment with TE.

[b] +TE: TE applied every two weeks beginning at 0 WAIT, with the fourth and last application at 6 WAIT; − TE: control; TE rates were 23, 108, and 45 g ha^{-1} for bentgrass, bluegrass, and bermudagrass, respectively.

[c] After a mean, "*" indicates significance relative to the control (within each species) at $p = 0.05$.

10.2.1.3 Stress Resistance Effects

10.2.1.3.1 Evapotranspiration, Drought, and Salinity

Data are available regarding the effects of TE on evapotranspiration (ET) but are extremely limited in terms of effects on drought resistance. Marcum and Jiang[30] reported that TE reduced tall fescue ET by 11% over a 6-week period. King et al.[38] also indicated that a TE-treated Kentucky bluegrass/tall fescue mixed turf used approximately 20% less water than the control from 0 to 4 WAT. Ervin and Koski[39] found that TE reduced the ET of Kentucky bluegrass weekly in only 5 out of 34 weeks as determined over 3 years. In terms of drought resistance, Jiang and Fry[31] reported that TE-treated perennial ryegrass maintained higher quality than the controls during dry-down in the greenhouse and during moderate drought events in the field.

Kentucky bluegrass subjected hydroponically to salinity stress (5 dS m^{-1}) decreased in shoot dry weight and percent canopy green cover when treated with three rates of TE as compared to the salt-only control.[40] However, Baldwin et al.[41] reported that biweekly TE applications to salt-stressed (12–26 dS m^{-1}) Champion and Tifeagle ultradwarf bermudagrasses resulted in greater turf-quality maintenance and 25% greater root mass. The available research data on the effect of TE on salinity and drought stress effects provide no clear consensus for cool- or warm-season grasses, pointing to the need for further research.

10.2.1.3.2 Heat Tolerance

Cytokinins have been shown to function in a regulatory role for mitigating cool-season turfgrass responses to heat stress.[42] Reported increases in antioxidant activity[37] and cytokinin levels (Ervin and Zhang, 2003; unpublished data) coupled with decreased mitochondrial respiration[19] due to TE appear to strongly support the hypothesis that TE increases turfgrass heat tolerance. Accelerated soil and plant respiration leads to excessive heat accumulation of sod stacked on pallets, causing considerable turfgrass damage during summer transport and storage. Heckman et al.[43] reported that sequential TE applications at 6 and then 2 weeks before Kentucky bluegrass sod harvest reduced heating of stacked sod within 4 h of harvest, with temperatures remaining 3°C cooler than the untreated sod at 12 h of storage. In a similar study, these researchers reported that a single TE application 2 weeks before harvest resulted in Kentucky bluegrass sod that was 10°C cooler than the controls in the center of sod stacks after 48 h of storage.[44] Reduced sod temperatures led to 30% greater tensile strength and 17% better quality ratings in treated sod after 24 h of storage.

Follow-up research by this group investigating the direct effects of TE on Kentucky bluegrass heat tolerance, however, provided some interesting results. In a growth-chamber temperature gradient block experiment, the temperature needed to kill 50% of the Kentucky bluegrass population was 35.5°C (TE-treated) versus 36.1°C (control).[45] Additionally, TE-treated Kentucky bluegrass had less cell-membrane thermostability when prestressed at 35°C and then tested for electrolyte leakage of leaf tissues while exposed to 50°C.[46] Reasons behind the discrepancy between reduced sod heating while on the pallet and decreased heat tolerance due to TE are not immediately apparent, although some literature may provide an insight as to the differences.

Structural mimics of 2-oxoglutaric acid such as TE have been shown to inhibit the activity of aminocyclopropane carboxylic acid oxidase (ACC oxidase).[5] Ethylene is generated from ACC in a reaction catalyzed by ACC oxidase. Reduced ethylene generation by treated plants and microbes in the sod could play a large role in temperature reduction, while not playing a primary role in improving the direct heat tolerance of TE-treated plants. As Heckman et al.[45] point out, the structural similarity of TE to herbicides in the cyclohexandione chemical family that kill grasses by inhibiting acetyl coenzyme-A activity and blocking lipid synthesis could play a role in the reports of decreased heat tolerance. However, because the reductions in sod heating due to TE appear to be of a greater magnitude than its effects on tissue heat tolerance, the overall effect remains positive in the case of Kentucky bluegrass sod viability.

10.2.1.3.3   Freezing Tolerance

TE's effects on hormone and carbohydrate metabolism have been thought to potentially affect turfgrass adaptation to freezing temperatures. Limited research, however, has been conducted as concerns the effects of late-season TE applications on turfgrass freezing tolerance. Sequential low-rate TE applications to 80% shaded *Poa supina* increased winter survival in 1 of 2 years tested, with the authors suggesting that TE amelioration of nonstructural carbohydrate losses during the winter were associated with improved freezing tolerance.[47] Tifway bermudagrass that received either three summer TE applications or a single early fall application had increased stolon freezing tolerance when sampled in October but not in November.[48] Richardson[49] found that late-season TE application did not have a consistent effect on freeze tolerance of Tifway bermudagrass, improving rhizome survival to −4°C in one year but having no effect in the other. A lack of negative effects of TE on bermudagrass freezing tolerance, coupled with Richardson's observations of greater retention of fall color and faster spring greenup, make this a beneficial turfgrass management practice. As the research to date has provided scarce data as to how TE may affect cool- and warm-season physiological responses to freezing temperatures, more detailed investigations are warranted.

10.2.1.3.4   Reduced Irradiance Tolerance

Using shade cloth, Tan and Qian[50] reported that 87% reduced irradiance increased bioactive GA_1 levels by ~45%, while monthly TE applications reduced Kentucky bluegrass GA_1 levels by 47%, resulting in a clipping yield reduction of 50%. These data clearly indicate that increased GA_1 levels under reduced irradiance play a major role in promoting wasteful leaf elongation, leading to turfgrass canopy thinning. Thus, numerous researchers have investigated the potential of TE for improving turfgrass shade tolerance.

Qian et al.[51] showed that repeated monthly TE applications on Diamond zoysiagrass (*Zoysia matrella* [L.] Merr.), under 88% reduced irradiance, resulted in increased tiller density, rhizome mass, rhizome TNC content, and canopy photosynthetic rate. In a further study, Qian and Engelke[52] showed that 88% shaded Diamond zoysiagrass receiving monthly TE applications maintained acceptable quality for over 7 months, while the untreated fell below acceptable after 3 months. Improved quality persistence due to TE was closely associated with increased root mass, tiller density, canopy photosynthesis rate, and TNC content. Ervin et al.[28] showed similar results on Meyer zoysiagrass under 77% or 89% reduced irradiance. On this fairway-height zoysiagrass, monthly TE treatment significantly delayed loss of tiller density and quality. TE treatment of Tifeagle bermudagrass receiving only 4 h of direct sunlight a day resulted in ~25% more chlorophyll and the maintenance of acceptable turf quality when mowed at 4.7 mm.[53]

Similar positive results have been reported on cool-season turfgrasses. TE increased the chlorophyll content of supina bluegrass and Kentucky bluegrass under 85% reduced irradiance, providing 50% increased tiller density for supina but no tillering change for Kentucky bluegrass.[54] Similarly, Goss et al.[55] reported that TE applications to bentgrass under 80% reduced irradiance resulted in 50% greater tiller density, shoot plus stem fructose content increases of up to 40%, but no effect on root mass. Another study under 80% reduced irradiance reported that monthly or bimonthly TE applications improved the quality, density, and chlorophyll levels of supina bluegrass, Kentucky bluegrass, and creeping bentgrass.[56] Creeping bentgrass under 88% reduced irradiance and treated monthly with TE maintained 35–45% better quality, showing increased photochemical efficiency and antioxidant superoxide dismutase activities in August of each year.[57] A final positive result was 28% improved root strength at the end of each season.

In the only study conducted under natural (deciduous) shade, Gardner and Wherley[58] reported increased tall fescue color but no difference in visual quality due to TE application on a 6-week interval. Natural shade both reduces photosynthetic photon flux (PPF) and alters spectral composition, in particular red:far red (R:FR) ratio, while shade-cloth (or neutral-density) shade merely lowers PPF. Wherley et al.[59] reported that deciduous shade (8% of full sun) resulted in less tall fescue tillering and chlorophyll content, but equivalent reductions in root mass relative to neutral density shade (7% of full sun). These results may indicate that tillering in natural shade is primarily a phytochrome-mediated response rather than a GA-mediated one. However, both shade types result in typical shade-avoidance responses such as increased shoot elongation and reduced tillering, and the plant physiological literature is unclear as to whether phytochrome and GA act independently or multiple signaling pathways intersect in regulating responses to changes in the light environment.[60] Further research detailing GA and phytochrome responses in turfgrasses subjected to natural shade and treated with TE are warranted given such uncertainty. In summary, use of TE to block GA_1 overproduction in artificially shaded environments (next to buildings, partially enclosed stadia, etc.) clearly improves turfgrass tolerance to reduced PPF, but it is still unclear if TE will improve persistence in natural shade characterized by reductions in both the PPF and R:FR ratio.

10.2.1.3.5 Mechanical Damage

Turfgrasses, by definition, are plants that persist under regular mowing and traffic.[61] Then, restricting leaf elongation or the growth of other tissues due to the use of TE may be viewed as a negative following mechanical damages (divoting, crushing, or tearing) common to sports turf. Any practice, however, that increases the verdure and root density of a turf increases the probability of tolerance to wear stress.[62] Repeated TE use has been shown to consistently increase tiller density, while reported improvements in rooting have been less common. Ervin and Koski[27] reported a reduction in visual quality of TE-treated Kentucky bluegrass at the end of two consecutive summer seasons of heavy simulated traffic. While tiller density and root mass were equivalent under traffic, whether treated with TE or not, repeated mechanical damage to elongation-inhibited leaf blades resulted in reduced visual quality. TE-treated supina bluegrass under 90% reduced irradiance and subjected to weekly cleated traffic (human jogging) provided higher quality for the first 4 weeks following traffic initiation.[54] After 4 weeks, the combination of traffic and reduced irradiance caused all treatments to thin excessively.

As reported previously, hybrid bermudagrass treated with ^{15}N-labeled ammonium nitrate allocated approximately 50% more N to roots and rhizomes when treated with TE.[20] Enhanced allocation of N and photosynthate for tillering due to TE might also be expected to be available for other stem growth such as is needed for stolon or rhizome spread following divoting. Steinke and Stier[56] reported that divot recovery of creeping bentgrass, Kentucky bluegrass, and supina bluegrass was not influenced by TE.

10.2.1.3.6 Disease Tolerance

Disease has been reported to be more severe on class C regulated grasses when compared with non-treated grasses because reduced growth rate and recuperative potential prevents the masking of disease-blighted leaves.[1] Does TE have the same type of negative effects? Burpee et al.[63] demonstrated that TE is fungistatic, significantly suppressing mycelial growth of *Sclerotinia homoeocarpa* F.T. Bennett (dollar spot), *in vitro*. Severity of dollar spot was significantly less in creeping bentgrass plots from 6 to 14 days (year 1) and 13 to 17 days (year 2) after TE treatment. It was further shown that pretreatment of turf with TE in plots treated with chlorothalonil, iprodione, or propiconazole resulted in significantly lower values of area under the disease progress curve from 5 to 16 days after fungicide treatment compared with plots treated with fungicide alone. These results prompted the suggestion that TE acts in two ways to reduce dollar spot severity: as a fungistat and as a growth regulator, limiting removal of fungicides by mowing. Golembiewski and Danneberger[64] confirmed this benefit, reporting that monthly TE applications consistently decreased dollar spot incidence on creeping bentgrass by 5–10%. In contrast, TE had no effect on the severity of Rhizoctonia blight of tall fescue.[65] In summary, growth regulation due to TE does not appear to be so strong as to exacerbate recovery of diseased turf; instead, TE acts as a mild fungistat, suppressing dollar spot development and potentially enhancing fungicide longevity.

10.2.2 Class B PGRs (Early GA Inhibitors)

10.2.2.1 Mode of Action

Class B PGRs, represented by flurprimidol and paclobutrazol, are primarily root-absorbed, translocated in the xylem, and act by inhibiting GA biosynthesis early in its biosynthetic pathway.[5] Flurprimidol (α-[1-methylethyl]-α-[4-[trifluoro-methoxy]phenyl]5-pyrimidine-methanol) and paclobutrazol ([+/−]-(R*,R*)-β-[[4-chlorophenyl]methyl]-α-[1,1-dimethyl]-1H-1,2,4,-triazole-1-ethanol) both specifically act as inhibitors of cytochrome P-450−dependent monooxygenases catalyzing oxidative steps from *ent*-kaurene to *ent*-kaurenoic acid early in the GA biosynthesis pathway.[5] As inhibitors of cytochrome P-450−dependent monooxygenases, both compounds have also been shown to interfere with sterol biosynthesis.[66] Sterol interference has led, in some instances, to cell-division inhibition and phytotoxicity and is mostly likely the reason for their observed fungicidal activity.[5]

As GA inhibitors, neither compound inhibits seedhead development,[1] and they exhibit differential effectiveness in extent and duration of foliar growth suppression on various turfgrass species. Compared with foliar-absorbed PGRs, paclobutrazol and flurprimidol are slower (10–14 days) in suppressing foliar growth, but their duration of activity is usually longer, lasting from 4 to 8 weeks, depending on application rate.[4]

10.2.2.2 Structural and Physiological Effects

Although not demonstrated directly on a turfgrass species, class B PGRs have been shown to increase palisade mesophyll cell density by up to 40%[67] and increase specific leaf weight.[68] Increased chlorophyll content has been reported in cool- and warm-season turfgrasses.[68,69] After a period of initial discoloration that is rate- and species-dependent and likely due to slight sterol inhibition, darker-green leaves are often observed.[1] Increased chlorophyll density did not correspond to greater carbon dioxide exchange rates (CER) on flurprimidol-treated creeping bentgrass. In fact, increasing flurprimidol rates caused a linear decline in CER.[68] No other reports are available regarding class B effects on CER of turfgrasses, clearly warranting further research to begin to understand the implications of this supposed negative effect.

Paclobutrazol and flurprimidol have been shown to cause a significant decline in photosynthate partitioning to the roots of Kentucky bluegrass 4 WAT, with transient increases in partitioning to the crown and axillary shoots.[70] Flurprimidol and paclobutrazol significantly increased the TNC content of creeping bentgrass verdure 2 weeks after their initial application, but TNC levels decreased relative to the control by week 6.[23] Given these results, the authors speculated that the reduction in TNC content of treated plants may be attributed to continuous use of photosynthate for supporting the production and growth of tillers. Thus, it is not surprising that increased tiller density has been reported due to class B PGRs.[1,71,72]

Hanson and Branham[70] found that the proportion of photosynthate transported to the roots (9–10%) in paclobutrazol- and flurprimidol-treated Kentucky bluegrass dropped significantly below that of control plants (27%) 4 WAT. No differences in photosynthate partitioning to the roots was measured 1 or 2 weeks after treatment. The PGRs in this radioactive labeling study were applied only once prior to analysis. In a study where flurprimidol was applied twice annually for 4 consecutive years to a mixed stand of Kentucky bluegrass and red fescue *(Festuca rubra* ssp. *rubra* L.), significantly higher root weights were found.[72] Greater root weight in this case was positively correlated with increased tiller density. Repeated autumn, then spring treatments of creeping bentgrass with paclobutrazol resulted in 11–35% more root mass at a 0–15-cm depth.[73] In summarizing these results, Koski, in agreement with Dernoeden, stated that the greater plant density that is produced and sustained by the more frequent PGR applications may result in increased root density. Fagerness and Yelverton[74] found no root-mass differences on creeping bentgrass due to repeated paclobutrazol treatment at labeled rates and frequencies. However, when applied at twice the labeled rate and with greater application frequency than is recommended, root mass was decreased.

Very limited data are available as regards class B PGR effects on rooting of warm-season turfgrasses. McCullough et al., however, conducted three studies of their effects on ultradwarf hybrid bermudagrasses. When applied at the labeled rate (0.42 kg ha^{-1} per 3 weeks) for Tifgreen hybrid bermudagrass, flurprimidol reduced Tifeagle root mass by 39%, while paclobutrazol was not different from the control.[69] Repeated application of both class B PGRs resulted in unacceptable quality. In a follow-up rate response study, it was found that severe Tifeagle phytotoxicity occurred when paclobutrazol was applied at more than 0.14 kg ha^{-1} per 6 weeks.[75] Root mass, as opposed to their previous finding,[70] was decreased as the paclobutrazol rate was increased from 0.14 (−28%) to 0.28 (−45%) to 0.42 kg ha^{-1} (−61%). In a final study comparing the response of two dwarf cultivars to paclobutrazol and flurprimidol, it was found that Champion had poor tolerance to both PGRs, exhibiting unacceptable turf injury and reductions in rooting regardless of rate.[76] Tifeagle rooting in this study was not affected by either paclobutrazol (0.14 kg ha^{-1}) or flurprimidol (0.14–0.42 kg ha^{-1}). Given these potential negative effects, TE is clearly the better choice for safe and effective growth regulation of ultradwarf bermudagrass putting greens. Further research is required to begin to understand the differential phytotoxic sensitivity of turfgrasses such as *Poa annua*[77] and ultradwarf bermudagrasses to these class B PGRs. Detailed investigations of class B PGR effects on sterol inhibition of tolerant and intolerant species might be revealing as to the mechanism of selectivity.

Similar to class A PGRs, it has been suggested that the PGR properties of triazoles and pyrimidines are mediated by interfering with the isoprenoid pathway and thus modulating the balance of primary plant hormones, including GA, ABA, and cytokinins. Inhibiting the activity of cytochrome P-450 monooxygenases not only interferes with the formation of bioactive GA but also inhibits the catabolism of ABA to phaseic acid.[78] The stress resistance implications of a transient increase in ABA levels will be discussed in the next section. In triazole-treated rice *(Oryza sativa* L.), the increased cytokinins were identified as *trans*-zeatin, dihydrozeatin, and its ribosides.[79] It is not currently understood how increased cytokinins occur due to treatment with these GA inhibitors as there are no obvious metabolic links.[5] Additionally, no reports are available in the turfgrass literature as regards class B PGR effects on ABA or cytokinin levels. Evidence of possible enhanced cytokinin activity due to triazoles was reported by Goatley and Schmidt[80] who showed a delay in senescence and improved CER of excised Kentucky bluegrass leaves due to treatment with triadimefon or propiconazole.

Increased activities of important antioxidants such as superoxide dismutase, ascorbate peroxidase, glutathione reductase, peroxidase, and catalase due to triazole treatment have been shown to protect wheat (*Glycine max* L.) seedlings from high-temperature damage.[78] Monthly application of propiconazole has been shown to result in greater superoxide dismutase activity of creeping bentgrass.[37] Higher antioxidant activity most likely functioned to reduce oxidative stress with the result being greater maintenance of photochemical efficiency of the propiconazole-treated bentgrass. Whether the same physiological benefits are present in paclobutrazol- or flurprimidol-treated turfgrasses is unknown at this time. A few studies, however, have reported on the stress-protective effects of class B or similar compounds.

10.2.2.3 Stress Resistance Effects

10.2.2.3.1 Evapotranspiration, Drought, and Salinity

One application of flurprimidol (0.84 kg ha^{-1}) significantly reduced the ET rate of St. Augustinegrass (*Stenotaphrum secundatum* [Walt.] Kuntze) 3–8 WAT when maintained at an optimal soil moisture level (–0.01 MPa) but decreased ET for an even longer period (weeks 2–17) when maintained at a suboptimal soil moisture level (–0.8 MPa).[81] These water savings were accompanied by decreased visual quality for up to 15 weeks, which the authors attributed to severe reductions (80%) in leaf elongation rates. Reduced ET under both limited and nonlimited soil moisture conditions in this study was most likely strongly influenced by the large reduction in water required for leaf elongation, but increased ABA levels signaling greater stomatal closure may have also played a role. Quality recovered once the inhibitory effects of flurprimidol subsided. In contrast, Koski[73] observed long-term visual quality and density benefits on creeping bentgrass treated in autumn and spring with paclobutrazol but noted no differences from the control in resistance to wilting or rate of recovery from drought over two summers. Marcum and Jiang[30] reported that paclobutrazol had no effect on tall fescue ET rate over a 6-week postapplication period. Paclobutrazol had no effect on perennial ryegrass leaf relative water content, leaf water potential, osmotic potential, or soil moisture content when irrigation was withheld in greenhouse and field trials.[31] Additionally, visual quality of the ryegrass during drought periods was not different from the control. These results do not support the hypothesis that triazole-induced increases in ABA result in improved plant water status under conditions of limited soil moisture availability. Alternatively, Nabati et al.[82] reported improved salinity stress tolerance of Kentucky bluegrass due to propiconazole. At the highest level of soil salinity, propiconazole-treated plants possessed much stronger root development that appeared to be associated with greater maintenance of leaf water content and sustained shoot growth. Although increased drought resistance due to class B compounds has been reported in many crop and tree species,[78] this effect in turfgrass species is inconsistent at best. Differences in turfgrass species responses to the sterol-inhibiting effects of these materials and the interacting effects of other environmental stresses (heat, wear, etc.) may often function to negate any positive drought stress effects.

10.2.2.3.2 Heat Tolerance

As opposed to TE, almost no heat-tolerance-specific research has been conducted for class B–treated turfgrasses. Paclobutrazol-treated wheat seedlings have been shown to have increased heat tolerance attributed, in part, to increased transpiration and activities of various antioxidant enzymes.[78] Treatment of tall fescue with propiconazole resulted in significantly greater canopy photochemical efficiency one day prior to sod harvest.[83] Treated sod that was then stored at 40°C for 72 h resulted in 35% greater posttransplant root strength and less visual shoot injury.

10.2.2.3.3 Freezing Tolerance

Class B PGR use in fall on cool-season turfgrasses is not usually primarily intended for shoot growth reduction but directed toward selective suppression of *Poa annua*. For reasons unspecified, application for this purpose past the period of active shoot growth in fall is not generally recommended.[4] Maintenance of membrane fluidity is thought to be one of the prerequisites for unimpaired survival at low temperature.[84] Application of class B compounds during cold acclimation that may negatively affect membrane components (sterols) presents the potential of reducing turfgrass freezing tolerance. While no published research is available as regards class B effects on cool-season turfgrass freezing tolerance, one study is available as concerns warm-season grasses. Shepard[85] reported that neither paclobutrazol nor flurprimidol altered carbohydrate partitioning in such a way as to have any effect on the cold tolerance of common bermudagrass, St. Augustinegrass, or bahiagrass (*Paspalum notatum* Flugge).

10.2.2.3.4   Reduced Irradiance Tolerance
Applications of sublabel rates (0.56 kg ha⁻¹) of flurprimidol repeated every 30–40 days under 80–95% reduced irradiance was shown to extend the period of acceptable Kentucky bluegrass quality by 30 days than if not treated.[86] Prolonged quality maintenance appeared to be related to increased tillering, while shear resistance and rooting were unaffected by flurprimidol.

10.2.2.3.5   Mechanical Damage
Under reduced irradiance, flurprimidol (0.56 kg ha⁻¹) was shown to maintain acceptable Kentucky bluegrass quality for up to 6 weeks longer than the untreated.[86] As stated previously, this effect was most likely due to flurprimidol-induced increases in tiller density. Calhoun[87] reported that repeated paclobutrazol or flurprimidol treatments had no significant effect on the rate of creeping bentgrass divot closure. He did, however, note a slight decrease in wear tolerance due to flurprimidol under a low nitrogen regime. Paclobutrazol delayed the rate of divot closure of Tifway bermudagrass when sequentially applied.[88] Delayed divot recovery was related to reduced emergence of shoots from rhizome nodes, suggesting persistent soil activity of paclobutrazol. These results imply that increased density due to repeated use of class B PGRs may provide short-term increases in wear tolerance. However, severely restricted leaf regrowth and reduced stem elongation on continuously regulated areas that receive intense wear or divoting may quickly override any tiller density advantages that these PGRs may provide.

10.2.2.3.6   Disease Tolerance
Given that flurprimidol and paclobutrazol are chemically related to several fungicides, it is not surprising that Burpee et al.[63] reported both compounds to be fungistatic to *S. homoeocarpa*. Additionally, both compounds were fungistatic at approximately 30× lower concentrations than TE. Accordingly, rates of dollar spot epidemics were significantly lower in creeping bentgrass treated with either class B compound compared with nontreated or TE-treated areas. Class B–regulated bentgrass also resulted in enhanced efficacy of chlorothalonil, iprodione, and propiconazole. Flurprimidol and paclobutrazol failed to decrease the severity of Rhizoctonia blight of tall fescue in 1 year, whereas in year 2, paclobutrazol significantly suppressed the disease 15 and 22 days after treatment.[65] Turfgrass literature regarding class B PGR effects on other diseases is lacking.

As stated in the introduction, class C and D PGRs will not be discussed in this chapter as their use is primarily for vegetative or seedhead suppression, which is often phytotoxic in nature. As our focus is on growth-regulating compounds that may improve the quality or persistence of fine turf, we next turn to a discussion of class E or hormone-type PGRs.

10.2.3   Class E PGRs (Hormonal Compounds)

Class E PGRs are compounds that are phytohormones or mimic the action of phytohormones. All the major hormone groups, except abscisic acid, are potentially represented on the turfgrass market in compounds such as ethephon (ethylene), benzyl-adenine (cytokinins), GA, and indole-3-butyric acid (auxin). Within turfgrass science, we are also aware of studies reporting on the effects of compounds classified as secondary (or stress-induced) hormones such as SA and JA, or compounds (ASM) that have been shown to induce higher levels of secondary hormones.

10.2.3.1 Ethephon

Ethephon (2-chloroethyl phosphonic acid) is a PGR registered for use on turfgrass in 1998 under the trade name Proxy (Table 10.1). Research on its potential use as a growth retardant for turfgrasses began in the early 1970s.[89] It is a synthetic form of ethylene, which is stable when formulated at an acidic pH but decomposes to release ethylene at physiological pH over an extended period of time, estimated at 4–10 days.[90] Ethylene, a hormone, is a simple unsaturated hydrocarbon (C_2H_4), which has a role in regulating many physiological processes, including fruit ripening, dormancy, abscission, flowering, senescence, induction of defense- or pathogenesis-related (PR) proteins, shoot epinasty, and various rooting responses.[60] Because ethephon is active within turfgrass foliage, growth responses begin quickly. A change in growth habit is usually not detected until the second week after application.[91]

The major structural effect of ethylene is to alter the direction of cell enlargement in stems and roots by changing the orientation of cellulose microfibrils from transverse to longitudinal, causing cells to swell up rather than elongate.[92] As a result, stems and roots, in general, become shorter and thicker. These generalizations, however, have not always held true for turfgrasses. Enhanced internode elongation of Kentucky bluegrass stems[89,92] has been reported to result in a raised canopy (witches' broom effect) and increased scalping potential. More often, however, mowing reductions occur due to reduced leaf elongation and shoot epinasty, resulting in a more decumbent growth habit that orients the canopy below that of the mower bench setting. Ethephon has also been reported to increase Kentucky bluegrass density by nearly doubling the number of leaves per shoot.[91]

Various studies have shown ethephon as improving bermudagrass performance,[93] and having no effect[94] or increasing[95,96] Kentucky bluegrass rooting. On creeping bentgrass at putting green height, ethephon (3.8 or 7.6 kg ha^{-1}) reduced quality at every weekly observation and reduced root length by 30% and root mass by 35%.[97] Marcum and Jiang[30] found that although ethephon treatment reduced ET rate of tall fescue, it also resulted in shorter maximum rooting depth, less root length density, total root length, and root water extraction from soil, thereby increasing vulnerability to water stress. Similarly, ethephon (3.8 or 7.6 kg ha^{-1}) caused chlorosis and severe thinning of Tifeagle bermudagrass, with root mass reduced by 20–33%.[98] More apparent leaf yellowing of closely cut turf canopies due to the senescence-promoting effects of ethylene is not surprising, but reductions in rooting indicate plant damage in excess of what might be expected from the senescence of older leaves. Investigation of ethephon effects on photosynthesis, respiration, and photoassimilate are required to gain a better understanding of these negative effects. Addition of N, Fe, or TE have been shown to mask or reduce ethephon-induced chlorosis on creeping bentgrass and bermudagrass putting greens, but their safening effects are variable and often insufficient for maintaining aesthetic acceptability.[91]

Efforts at chlorosis reduction in putting greens have been made because of ethephon's efficacy in reducing *Poa annua* seedhead development. Proper spring application timings of the class C cell-division inhibitor, mefluidide, has been shown to provide up to 90% *Poa* seedhead suppression, but unacceptable phytotoxicity often occurs.[99] Ethephon has been shown to be slightly less effective, providing approximately 75% seedhead suppression, but with less phytotoxicity than mefluidide. Research aimed at refining application timings, rates, and mixtures for most effective spring *Poa* seedhead suppression is ongoing.[100,101] The reason why higher ethylene concentrations suppress seedhead development is not understood. One hypothesis, however, is that a greater ratio of ethylene to cytokinins in plant tissue functions to suppress cytokinin activation of cell division required for inflorescence development.[102–104]

Although ethylene as a "stress-response" hormone has been shown to upregulate plant defense against various abiotic and biotic stresses,[60] the negative adaptive responses reviewed previously call into question the utility of ethephon for improving turfgrass stress tolerance. Nevertheless, Larkindale and Huang[105] have shown that exogenous treatment of creeping bentgrass with the ethylene precursor ACC prior to exposure to heat stress (35°C) increased thermotolerance by enhancing antioxidant enzyme activities, resulting in less lipid peroxidation and greater sustained photosynthesis. Acceptable visual quality was maintained for 14 days longer for the ACC-treated plants relative to the controls. Whether increased thermotolerance due to exogenous ACC or ethephon can be measured in a field situation remains to be investigated. As previous field experience suggests, the stress of frequent, close mowing; a dynamic edaphic and atmospheric environment; and variable species sensitivity to ethylene has often led to decreases in stress tolerance.

In a study where ethephon was applied in autumn to promote early senescence of cold-acclimating bermudagrass, it was found that freezing tolerance was unaffected.[106] While late-season ethephon promoted bermudagrass color loss and delayed spring greenup, it did not affect the ratio of saturated to unsaturated fatty acids (i.e., membrane permeability) or proline content in stolons. This research appears to be incongruous with reports showing that ethephon applications prior to chilling temperatures improve cold hardiness by increasing membrane permeability.[107]

As a turfgrass PGR, ethephon has been shown to have two effective roles: reducing mowing requirements and suppressing seedhead development. These advantages, depending on species and climate, may also be diminished due to negative issues of phytotoxicity, scalping, and loss of roots. Much more research into the applied stress physiology of turfgrass responses to ethephon is needed before any solid use recommendations can be made.

10.2.3.2   Indole Butyric Acid (Auxin)

Indole butyric acid (IBA) is a synthetic auxin primarily used for promoting root initiation. Auxin is a generic term representing a class of compounds that are characterized by their capacity to induce elongation in shoot cells in the subapical region, resembling indole-3-acetic acid (IAA) in physiological action. Auxins are involved in many physiological processes in plants, including cellular elongation, phototropism, geotropism, apical dominance, rooting, ethylene production, fruit development, abscission, and sex expression.[108] The IBA product Kickstand, from Helena, is currently labeled for use in turfgrass (Table 10.1).

Auxins profoundly influence root morphology, inhibiting root elongation but increasing lateral root production and inducing adventitious roots.[109] IBA is more effective than IAA at lateral root induction, perhaps because unlike IAA, IBA efficiently induces lateral roots at concentrations that only minimally inhibit root elongation (Ervin and Zhang, 2006; unpublished data).[110] Although small auxin-induced increases in shoot elongation would not generally be desirable on fine turf surfaces, auxin-induced increases in rooting density and subsequent efficiency of soil water extraction would.

In an early study, Hoveland (1963)[110] reported that dipping sprigs in IBA (3000 mg kg^{-1}) increased transplant rooting of Coastal and Suwanee bermudagrass. In a glasshouse experiment, IBA (0.34 g mL^{-1}) was applied as a soil drench to tall fescue maintained under either nonlimiting (−0.01 MPa [soil water potential]) or limiting (−0.5 MPa) soil moisture conditions. It was found that IBA did not impact growth or metabolism under nonlimiting soil moisture (Table 10.4). Under water stress, however, IBA increased root density, leaf proline content, and activity of antioxidant enzymes (superoxide dismutase and ascorbate peroxidase) when compared with the untreated control (Ervin and Zhang, 2006; unpublished data). Increased rooting and improved abiotic stress resistance due to exogenous IBA has been reported in other cultivated plants.[111]

10.2.3.3   Gibberellic Acid

The gibberellins are a large group of related compounds, all of which have some biological activity and share the presence of a gibbane ring structure. The number of known gibberellins exceeds 100 and they are classified as GA_1, GA_2, GA_3, and so on. GA_1 and GA_3 are the most biologically active in terms of promoting elongation growth in stems and grass leaves at the intercalary meristem.[92] GA_3 was the first commercially available gibberellin. GA in pure form is not readily soluble in water, so water-soluble formulations generally contain a water-miscible solvent or the acid is converted into a water-soluble salt form. Movement of GAs over short distances is by diffusion, while over longer

TABLE 10.4

Tall Fescue Responses to Exogenous IBA[a] without Moisture Stress[b] or Moderate Moisture Stress[c]

Treatment	Clipping Yield (g)		Root Density (mg cm^{-3})		Leaf Proline (mg g^{-1} FW)		APX (activity unit per mg protein)		SOD (activity unit per mg protein)	
	Wet	Dry	Wet	Dry	Wet	Dry	Wet	Dry	Wet	Dry
Control	18.7a[d]	12.1a	0.31a	0.37b	0.37a	1.45b	232b	115b	3.2a	2.9b
IBA	15.9b	11.3a	0.35a	0.69a	0.48a	2.08a	435a	418a	4.4a	4.5a

Note: IBA, indolebutyric acid; APX, ascorbate peroxidase; SOD, superoxide dismutase.

[a] Soil drench: 0.34 g mL^{-1}.
[b] Wet: −0.01 MPa.
[c] Dry: −0.5 MPa.
[d] Means followed by the same letter in the same column are not significantly different at $p = 0.05$.

distances it occurs in the phloem.[92] Several GA products are labeled for use in turf, such as RyzUp, ProGibb T&O, and GibGro (Table 10.1).

Turfgrasses treated with GA generally have an etiolation response where leaves quickly elongate and become chlorotic. Use of GA products is currently recommended to reverse the effects of an overapplication of anti-GA (class A and B) PGRs and to initiate or maintain growth and prevent color changes (e.g., purpling) during periods of cold stress and light frosts on bermudagrasses.[4,93] As early as 1958, Juska[112] reported that treatment with GA during bermudagrass spring greenup promoted a dark greening within 10 days. Thus, on warm-season grasses, GA products are recommended to retain fall leaf color and promote the rate of leaf greening in spring.

Other uses that have been investigated are to increase lateral spread of stolons during vegetative establishment. Juska[112] also reported that stolons of bermudagrass, bentgrass, and zoysiagrass soaked 24 h in only water or in solutions containing 5, 10, 50, 100, and 500 mg L^{-1} GA$_3$ increased bentgrass spread with each concentration of GA up to 50 mg L^{-1}, after which inhibition of stolon spread was observed. Bermudagrass, with exception of the 5 mg L^{-1} treatment, showed a continuous increase in stolon growth through the entire range of treatments, while Meyer zoysiagrass showed little response to GA. In a related field experiment, application at 10 mg L^{-1} increased bentgrass growth compared with the control, whereas considerable etiolation and yellowing began to appear at 50 mg L^{-1}. Concentrations over 50 mg L^{-1} affected growth adversely, causing thin, anemic growth. In particular, when treated with higher concentrations of GA, zoysiagrass grew taller and yellowed extensively. A similar response was observed in bermudagrass. In Kentucky bluegrass, increased growth, yellowing, and etiolation were roughly proportional to the concentrations of GA applied.

In a study with Tifeagle bermudagrass, repeated application of GA every 15 days at 0.03 and 0.06 kg ha^{-1} did not alleviate turf quality decline due to fenarimol use. Repeated GA applications promoted shoot growth, while decreasing chlorophyll, visual quality, and root mass.[113] In a simulated-shade experiment, GA$_3$ (0.06 kg ha^{-1} per 14 days) reduced Tifeagle visual quality and chlorophyll levels when exposed to 12 or 8 h of full sunlight, but did not affect lateral regrowth of stolons or TNC content of roots and rhizomes.[53] Investigations regarding responses of GA-treated turfgrasses to other stresses are lacking, but the shoot-growth-promoting effects of exogenous GA appear to be adaptive in only a few unique cases.

10.2.3.4 Cytokinins

Natural cytokinins are a series of adenine molecules modified by the addition of five-carbon side chains off the C-6 position, with the two main bioactive groups being zeatin and dihydrozeatin. Kinetin and benzyl adenine (BA) are the two most common bioactive synthetic cytokinins, although none are currently labeled for turf use (Table 10.1). Cytokinins as plant hormones are mostly known for their ability to promote cell division, but they have a wide spectrum of biological activities. As was reviewed with class A PGRs, enhanced endogenous cytokinin promotes development of lateral buds (i.e., tillers) and delays senescence.[92,108] Cytokinins have also been reported to enhance nitrate reduction by increasing nitrate reductase activity.[104]

White and Schmidt[114] showed that BA treatment increased gross CO_2 exchange rate in Tifgreen bermudagrass grown under chilling stress. Goatley and Schmidt[80] found that BA increased gross CO_2 exchange rate of Kentucky bluegrass 10 days after treatment, indicating antisenescence activity. Kane and Smiley[115] indicated that BA-treated Kentucky bluegrass exhibited higher chlorophyll and nonstructural carbohydrate contents than untreated leaves.

Application of BA (6 mg m^{-2}) significantly increased shoot and root growth of Kentucky bluegrass seedlings 4 WAT in only one out of three experiments.[116] In another study, BA (6 mg m^{-2}) did not enhance sod strength or rooting of Kentucky bluegrass.[117] Their results suggest that BA was an effective antisenescence agent but may not be a reliably active growth-promoting compound. Positive, or sometimes neutral, physiological effects of exogenous synthetic cytokinins indicate excellent potential for improving stress performance of fine turfgrasses. Increased density, nitrate reductase activity, photosynthesis, and carbohydrate contents appear to provide fine turfgrasses with an adaptive advantage when faced with temperature and wear stress. Research regarding the interactions of rates, timings, environmental factors, and species is required to determine if effects of biological and practical significance can be consistently achieved.

10.2.3.5 Salicylic Acid

SA (2-hydroxy-benzoic acid), named after plants in the Salix genus, was first discovered as a major component in extracts from willow tree bark that had been used as a natural pain reliever and anti-inflammatory drug from ancient times to the eighteenth century.[118] Acetylsalicylic acid, widely known as aspirin, was the first synthetic drug introduced by the Bayer Company in 1897.[118] In 1992, SA was first defined as a plant hormone functioning as an endogenous signaling molecule active in controlling defense against biotic and abiotic stresses.[119] Exogenous SA action in plant systems has received increasing attention since the 1970s. Exogenous SA rapidly enters cells and initiates plant responses as a signal molecule inducing gene expression to produce pathogenesis-related (PR) proteins, resulting in systemic acquired resistance (SAR) to pathogen infection.[118,119] However, no reports are available in the turfgrass literature concerning exogenous SA effects on disease incidence.

Abiotic stress tolerance effects due to exogenous SA have also been reported to improve monocot and dicot tolerances to heat, chilling, and salt stress.[120–122] Heat-tolerance research conducted on cool-season turfgrass species has shown a clear association between SA application and protection against oxidative damage by promoting the activities of antioxidant enzymes.[105,123,124]

Pretreatment of creeping bentgrass with a series of root drenches of SA (0.01 mM) prior to exposure to heat stress (35/30°C) for 1 month resulted in maintenance of more green leaves, decreased membrane leakage, and reduced oxidative damage compared with control plants.[123] SA appeared to play a signaling role in improved thermotolerance as its endogenous concentration rose dramatically an hour after heat exposure. In another study with creeping bentgrass, Larkindale and Huang[105] reported that SA pretreatment (0.01 mM) maintained ascorbate peroxidase activity at a significantly higher level than the controls after 24 h of heating, reducing oxidative damage and maintaining higher leaf photosynthetic rates and visual quality. He et al.[124] reported similar increases in antioxidant activity and subsequent thermotolerance of Kentucky bluegrass that were SA-rate dependent, with the 0.25 mM rate providing better results relative to lower (0.1 mM) or higher (0.5–1.5 mM) rates.

Similar increases in antioxidant responses due to pretreatment with SA prior to exposure to oxidative stresses (UV-B, heat, and drought) have been reported in a series of greenhouse and field trials. One foliar application of SA (30 mM) alleviated injury and loss of photochemical efficiency of Kentucky bluegrass and creeping bentgrass under UV-B stress.[125] SA enhanced photochemical efficiency by 86% when measured 10 or 12 days after UV-B initiation relative to the control. In addition, SA increased α-tocopherol concentration and superoxide dismutase and catalase activities at 10 days after UV-B initiation.[126] In a field study,[127] SA was applied at 30 mM to tall fescue and Kentucky bluegrass foliage 10 days before sod harvest. Rolled sod was subjected to high-temperature stress (38–40°C for 72 or 96 h), transplanted into the field, and had its injury and root strength determined. Application of SA enhanced preharvest photochemical efficiency, reduced posttransplant visual injury, and enhanced posttransplant root strength in both years. Schmidt and Zhang[125] noted that SA-treated Kentucky bluegrass exhibited greater photochemical efficiency and chlorophyll content than the control under drought stress. These reports strongly suggest that exogenous SA applications at small concentrations prior to significant occurrences of summer oxidative stresses could improve cool-season turfgrass stress tolerance.

10.2.3.6 Acibenzolar-S-Methyl

ASM (Actigard, Syngenta Crop Protection) is a plant-defense activator that induces SAR, and has been shown to have antifungal, antibacterial, and antiviral activity.[128] ASM has been shown to be a successful SA mimic, competing successfully with SA for binding to a SA-binding protein, activating PR gene expression, and signaling SAR. ASM is not currently labeled for turfgrass use, but its disease-control potential has been studied in creeping bentgrass.

Lee and Fry[129] examined the effects of ASM on controlling dollar spot and brown patch in a creeping bentgrass putting green. The results indicated that ASM (0.035 kg ha^{-1}) reduced the number of dollar spot infection centers by up to 38% but did not impact brown patch in a 50:50 blend of Crenshaw and Cato when applied on a 14-day interval. Although ASM provided a moderate reduction in dollar spot, it did not suppress the disease to a level that was commercially acceptable. In another study,[130] ASM (0.035 kg ha^{-1}) reduced dollar spot by 27% in Cobra and 26% in L-93 creeping bentgrass when applied at a 14-day interval. Combining chlorothalonil with ASM resulted in 44% fewer dollar spot infection centers and improved the visual quality of Cobra compared with turfgrass receiving the chlorothalonil alone. Positive dollar spot suppression results with ASM and abiotic stress tolerance enhancement with SA indicate that further research into the physiological effects of each under abiotic and biotic stresses is warranted.

10.2.3.7 Jasmonic Acid

Jasmonates are ubiquitously occurring hormones that are assumed to function as a "master switch" in biotic and abiotic stress responses.[131] They are lipid-derived cyclopentanone compounds that have been shown to occur in all plants tested. The principal naturally occurring jasmonates in plants are *trans*-jasmonic acid (JA) and methyl jasmonate (MJ).[132] Although JA seems to be more prevalent in plant tissues, MJ is the more active compound when applied exogenously because it is volatile and lipophilic.[60] Initially, jasmonates were recognized for their shoot- and root-growth-inhibiting activity, but more recently, they have been shown to elicit a wide variety of physiological responses such as activation of defense genes in response to wounding or microbial attack and promoting senescence.[108] It has been further proposed that UV protection might be enhanced via a jasmonate-mediated synthesis of anthocyanins. When exogenously applied, MJ has been shown to promote beta-carotene synthesis and improve chilling tolerance.[133] Zhang and Ervin[126] reported that improved Kentucky bluegrass UV-B tolerance due to foliar MJ treatment (at 6.5 mM) was associated with higher antioxidant activities and photochemical efficiency when measured at 10 days after UV-B initiation. Late September and early October MJ foliar applications (at 15 mM) to Princess-77 and Riviera bermudagrasses in Blacksburg, Virginia, appeared to improve cold hardiness as stolons sampled during dormancy (February) had significantly greater proline and water-soluble carbohydrate contents that were associated with a faster rate of spring greenup.[134] Similar results were measured when an NaCl solution (20 dS m^{-1}) was applied to the bermuda-grasses, providing evidence that the application of a "mild stressor" during cold acclimation may signal greater accumulation of defense chemicals. Much more research is required as the data to date on the effects of exogenous MJ on turfgrass responses to stress are minimal and inconclusive.

10.2.4 CLASS F PGRs (NATURAL SOURCE MATERIALS)

10.2.4.1 Humic Substances: Humic and Fulvic Acids

Humic substances are naturally occurring, biogenic, heterogeneous organic substances that can generally be characterized as being yellow to black in color, of high molecular weight (2000–300,000 g mol^{-1}), and refractory.[135] They are characterized by aromatic, ring-type structures that include polyphenols and polyquinones that are highly resistant to microbial degradation. In general, humic substances are classified into three chemical groupings based on solubility:[136] (1) fulvic acid, lowest in molecular weight and lightest in color, soluble in both acid and alkali, and most susceptible to microbial degradation; (2) humic acid, medium in molecular weight and color, soluble in alkali but insoluble in acid, and intermediate to degradation; and (3) humin, highest in molecular weight, darkest in color, insoluble in both acid and alkali, and most resistant to microbial degradation. Soil organic matter is comprised of 60–80% humic substances that exert beneficial effects on plant-nutrient availability, soil structure, microbial populations, cation-exchange capacity, pH buffering of the soil, and water-holding capacity.[137] Apart from these effects, there is considerable literature showing that an exogenous supply of humic substances can play a biologically active role, influencing plant growth and metabolism. Positive growth responses are generally observed only when the application of humic substances is directed to soils low in organic matter, such as sand-based turf-soil systems, or plants grown in a nutrient solution.[137] Reported biochemical and physiological effects of humic substances include increased membrane permeability resulting in a more rapid and selective entry of essential elements into the root; activation of respiration and the Krebs cycle with a concomitant increase in ATP production; an increase in chlorophyll content and photosynthesis; and stimulation or inhibition of enzyme activity.[138] At the whole plant scale, stimulation of root growth and enhancement of root initiation have commonly been found, often to a greater extent than that of shoots.[137]

Two main theories as to the mechanisms involved in growth enhancement due to humic substances continue to be debated. The first hypothesis is that growth enhancement is due to increased micronutrient availability, Fe and Zn in particular, either via complexing or through binding of colloidal conformations of the metal hydroxide to the humic substance.[139] This effect has been shown to be dependent on pH and soil levels of dissolved organic matter.[140] Increased Fe availability due to humic substances is not required in acidic soils (pH < 5) where Fe solubility is high or in soils with high levels of dissolved organic matter (>100 mg L^{-1}) that provides adequate micronutrient chelation. The second hypothesis is that humic substances serve as precursors or substrates for the synthesis of hormones by microorganisms, increasing, in particular, plant bioavailablity of auxin or auxin-like compounds in the growing medium.[141–143] Readers who are interested in this issue should consult the cited references as no attempt will be made here to resolve this debate.

A number of turfgrass-specific studies have reported on the potentially beneficial effects of exogenously supplied humic substances. Humic acid (HA) may increase plant uptake of certain mineral nutrients and improve rooting and various turfgrass physiological functions. As early as 1964, Gaur[144] reported increased N, P, and K uptake in perennial ryegrass grown in sand amended with HA extracted from compost. Varshovi[145] found no increase in bermudagrass N uptake following application of a commercial humate material at 263 or 803 kg ha^{-1}. Cooper et al.[146] reported that soil-, peat-, and leonardite-based foliar HA applications (400 mg L^{-1}) significantly increased creeping bentgrass leaf P concentration when grown in sand culture (pH not reported), but not in solution culture (pH = 6), and did not impact N, Ca, Mg, and Fe uptake. None of the foliar HA treatments affected root growth. In another solution culture experiment, HA (400 mg L^{-1}) placed in quarter-strength Hoagland's nutrient solution (pH = 6) significantly enhanced the net photosynthesis, root dehydrogenase activity, and root mass of creeping bentgrass.[147] HA did not consistently impact chlorophyll content or tissue mineral concentrations. Liu and Cooper[148] used the same system to examine the effects of HA on creeping bentgrass salt tolerance. Quarter-strength Hoagland's solution was dosed with HA (400 mg L^{-1}) and 0, 8, or 16 dS m^{-1} salinity levels. HA did not affect tissue water content, net photosynthesis, root growth, or mineral nutrient uptake of salt-stressed bentgrass.

Alternatively, our research group has reported increased drought resistance and summer performance of cool-season turfgrasses following foliar applications of a leonardite-based HA. Creeping bentgrass, Kentucky bluegrass, and tall fescue subjected to a constant soil moisture potential of −0.5 MPa had increased shoot and root growth when treated with HA relative to the control.[149] Improved growth under limited soil moisture availability was associated with greater antioxidant (α-tocopherol, ascorbic acid, and β-carotene) contents in all three species. Monthly HA (150 mg m^{-2}) applications throughout summer to field plots of creeping bentgrass resulted in greater root mass, antioxidant enzyme (superoxide dismutase) activity, and photochemical efficiency on a number of measurement dates. These results were measured for both the low N (25 kg N ha^{-1} per month) and the high N (50 kg N ha^{-1} per month) treatments. The soil used in these experiments was a silt loam with a pH of 6.2 and no apparent nutrient deficiencies.[150] From these data, there is no clear way to determine whether positive HA responses were from improved plant nutritional status, enhanced hormone status, or neither.

A more recent study by our group, however, did indicate a potential link between cytokinins identified in leonardite HA and improved creeping bentgrass metabolic responses to drought. Enzyme-linked immunosorbent assay (ELISA) estimated that the HA contained 57 μg g^{-1} of the cytokinin-zeatin riboside.[151] This HA source, when foliarly applied prior to drought imposition, did not affect shoot growth, but delayed wilting and loss of leaf color, a response that was associated with increases in leaf zeatin riboside, α-tocopherol, and photochemical efficiency. The influence of HA on improved nutritional status due to increased nutrient availability, either from the HA itself or from its possible chelating effects, was controlled by combusting the HA and applying the resulting inorganic fraction. This "ashed HA control" failed to have any of the positive physiological effects of the nonashed HA. In this study, root mass of the drought-stressed, HA-treated bentgrass was not affected, and the question of whether this HA source contained auxin or displayed auxin-like activity was not addressed. In summary, although the research information to date shows excellent potential of foliar applications of HA to improve turfgrass growth and persistence, the conditions under which positive responses may be seen and how these effects occur remain unclear.

10.2.4.2   Seaweed Extracts (SWE)

Extracts of marine brown algae have been used to improve the growth and persistence of agricultural and horticultural plants for centuries.[152,153] Since the early 1950s, the alga most commonly used for the production of commercial extracts in Europe and North America is *Ascophyllum nodosum*, commonly referred to as rockweed. In the southern hemisphere, the brown algae *Eklonia maxima* and *Durvillea potatorum* are used in making extracts.[154] In the preparation of liquid extracts, the seaweed is usually fragmented in a heated (<100°C), alkaline (KOH) solution and then filtered. Resulting commercial liquid extracts contain from 5% to 15% dissolved solids. Soluble powders are often prepared by dehydrating these liquid extracts. Analyses of the chemical composition of KOH-extracted seaweed indicate that in general, they contain 1–2% N and P, 15–20% K, 1–7% Na, and trace amounts of micronutrients. Additionally, they have been shown to contain a suite of amino acids, carbohydrates, betaines, and cytokinin.[154] Reported beneficial plant effects are not thought to be nutritional in origin as the amounts applied as foliar sprays (0.5–4 kg ha^{-1}) form an insignificant proportion of a crop's annual mineral requirements.[154–156] Enhanced metabolic activity, rather, has been attributed to the levels of betaines and cytokinins found in the SWEs.[155,158]

Improved root and shoot growth has been reported due to foliar SWE applications when grown under conditions of adequate soil moisture availability. Zhang[149] noted that foliar application of SWE (soluble powder at 33 mg m^{-2}) increased shoot and root growth of Kentucky bluegrass, creeping bentgrass, and tall fescue. In a 2-year field study, monthly application of SWE at 50 mg m^{-2} increased root mass of creeping bentgrass in both years.[150] When SWE was applied to creeping bentgrass at a lower rate (16 mg m^{-2}), visual quality was improved in 1 out of 2 years.[159] Foliar application of SWE (2.5 g L^{-1}) increased spinach (*Spinacia oleracea* L.) total fresh-leaf weight by 12–15% after 8 weeks of growth[160] and increased maize (*Zea mays* L.) seedling root and stem mass by 15–25% over the controls.[161]

Enhanced chlorophyll content due to SWE treatment has been found in other crops such as tomato (*Solanum lycopersicum* L.), dwarf French bean (*Phaseolus vulgaris* L.), wheat (*Triticum aestivum* L.), barley (*Hordeum vulgare* L.), and maize.[162] Increased chlorophyll content may be partially related to the suite of betaines that have been found in SWE.[157,162] This antisenescence, or reduction of oxidative-stress effect is also commonly observed in plants with enhanced cytokinin levels.[36,132] It is not our intention, herein, to argue for an exclusive role of either suite of compounds as regards the positive growth-regulating effects of SWEs; rather, our focus will be to review the physiological effects of SWE with specific reference to turfgrass stress resistance.

Foliar application of SWE has been found to increase photochemical efficiency, CO_2 exchange rate, enzymatic antioxidant activities, and nonenzymatic antioxidant (ascorbic acid, α-tocopherol, and β-carotene) contents in creeping bentgrass, Kentucky bluegrass, and tall fescue.[80,125,149–151,159,163–168] SWE promoted enzymatic antioxidants (peroxidase and catalase) in rice.[169] Application of SWE was also found to enhance contents of photosynthetic pigments, free amino acids, leaf soluble protein, and starch in maize.[170]

Application of SWE has been shown to increase drought tolerance of turfgrasses.[125,149,151,163–165] Zhang[149] found that foliar application of SWE (33 mg m^{-2}) increased leaf moisture content and shoot and root growth of Kentucky bluegrass seedlings grown under a low soil moisture (−0.5 MPa) regime for 6 weeks. Improved drought tolerance due to SWE was found to be related to increases in leaf tissue levels of α-tocopherol. Price and Hendry[171] have noted a close correlation between drought tolerance and α-tocopherol in several grass species. In a glasshouse study, SWE was applied to creeping bentgrass and irrigation was withheld for 28 days to permit progressive development of drought stress. The results showed that SWE-treated bentgrass had greater concentrations of zeatin riboside and α-tocopherol and exhibited greater wilting resistance and visual quality when compared with the control.[151] These data support the contention that SWE improves drought tolerance by increasing endogenous cytokinins and upregulating antioxidant defense systems, which can delay the degradative effects of drought-induced oxidative stress.

Liu and Huang[42] have demonstrated that exogenous application of 10-μM zeatin riboside (ZR) was most effective in slowing creeping bentgrass leaf senescence and alleviating heat-induced lipid peroxidation when exposed to constant heat stress (35°C). It was further posited that alleviation of heat injury by exogenous ZR was related to the maintenance of the scavenging ability of antioxidants (superoxide dismutase and catalase) at high temperatures. Recently, our lab[172] tested whether foliar application of a similar ZR concentration (10 μM) from two SWE sources might show the same improvement in creeping bentgrass heat tolerance. The first SWE tested contained 14.4% solids and a ZR content of 867 μg g^{-1} (Acadian Seaplants, Dartmouth, Nova Scotia), while the second contained 8% solids and 1554 μg g^{-1} ZR (Ocean Organics, Waldoboro, Maine). Each treatment was applied to the foliage in a solution to provide ZR at 10 μM and compared with a water control, a pure ZR control (10 μM), and an ashed SWE control. Treatment applications occurred at day 1 and day 28 of two 56-day trials. Plants were grown in a hydroponic nutrient solution culture system under constant heat stress (35°C) for 49 days. On day 14 of heat stress, leaf-tissue ZR contents were greater than the control for all three cytokinin-containing compounds. Greater ZR continued to be measured due to the Acadian and Ocean Organics extracts after 42 days of heat stress. Less leaf senescence due to all three cytokinin-containing compounds was seen visually and confirmed physiologically as greater photochemical efficiency was measured at 42 days of heat stress. Greater stability of the photosynthetic apparatus due to the SWE and ZR treatments appeared to be associated with higher superoxide dismutase activities and less lipid peroxidation at 35 and 49 days of heat stress. Root fresh weight was greater for the Acadian SWE at 35 and 49 days of heat stress, while greater root viability was measured on these sample dates for the Ocean Organics SWE. Although the Acadian and Ocean Organics SWE sources differed in ZR concentration, when applied at equivalent ZR concentration (10 µM), they improved creeping bentgrass heat tolerance similarly. The responses measured appeared to be a cytokinin effect as the synthetic ZR treatment gave similar results to the two SWEs, while the ashed SWE treatment responded similar to the control.

Several researchers have reported that exogenous application of SWE suppressed nematode damage to roots.[173–176] Sun[177] applied SWE to nematode-infected creeping bentgrass as a soil drench or foliar treatment and found that the SWE (25% SWE solution at 4 L ha^{-1}) drenched twice a month reduced yellowing, thinning, and wilting and improved leaf water status and clipping yields of infected creeping bentgrass. Rooting improvement of more than 40% was associated with the SWE treatment. Healthier growth of the SWE-treated plants occurred even though it was shown that nematode populations did not differ between the control and SWE-treated pots.[177,178] Wu et al.[179] noted that SWE resulted in a significant reduction in the number of second-stage juveniles of both *Meloidogyne javanica* and *M. incognita* invading the roots of tomato plants, compared with those of plants treated with water. Egg recovery from SWE-treated plants was also significantly lower. Exogenous application of the three major betaines found in this SWE (γ-aminobutyric acid betaine, δ-aminovaleric acid betaine, and glycinebetaine) also led to significant reductions in both the nematode invasion profile and egg recovery when applied at concentrations equivalent to those present in the SWE.

As is apparent from the many studies reviewed in the preceding text, plants treated with SWE often show improved adaptive responses to abiotic and biotic challenges that appear to be related to constitutive compounds (cytokinins and betaines) that have PGR rather than nutritional effects.

10.2.4.3 SWE + HA

Many commercially available "plant-growth promoters" or "biostimulants" contain SWE and HA as two main ingredients, most likely because research and experience have led to the contention that the two applied together have greater beneficial effects than either alone.[180] In this section, we review a number of studies, conducted primarily by our research group over the past 15 years, which serve to support this contention.

UV-B radiation coupled with high temperature is considered a major factor in quality decline and death of turfgrass sod during summer transplanting. Schmidt and Zhang[125] noted that SWE + HA improved leaf chlorophyll content of creeping bentgrass under UV-B stress. In field studies, SWE + HA alleviated injury and loss of photochemical efficiency and enhanced rooting and recovery of Kentucky bluegrass and tall fescue sod from heat injury.[83,181] Increased UV-B stress tolerance due to SWE + HA was found to be associated with increased enzymatic antioxidants (superoxide dismutase, catalase, ascorbate peroxidase) and nonenzymatic antioxidants (α-tocopherol, carotenoids) that served to delay pigment degradation.[83,181,182]

Yan[183] reported that application of SWE + HA to perennial ryegrass increased unsaturation of fatty acids and total free sterol content, resulting in increased salinity tolerance. Kentucky bluegrass treated with SWE + HA was better able to maintain leaf water content, leaf color, and root development when subjected to salinity stress.[82] Salinity-stressed creeping bentgrass treated with SWE + HA showed increased proline content, root mass, and photochemical efficiency while Na$^+$ uptake was reduced.[184] Foliar SWE + HA application to drought-stressed creeping bentgrass resulted in greater antioxidant content (α-tocopherol), photochemical efficiency, and subsequent wilting resistance (turf quality) when compared with application of either compound alone.[151]

In summary, it is our contention that these natural compounds (together or alone) function as PGRs, not fertilizers, because small exogenously applied concentrations have been shown to affect plant growth and development and improve metabolic responses to oxidative stress. Effects measured following applications at parts-per-million concentrations, along with evidence of bioactive concentrations of phytohormones (cytokinins and auxins) and osmoregulants (betaines) in these compounds, support our PGR hypothesis.

10.3 SUMMARY AND CONCLUSIONS

Class A PGRs (TE and PC) are known to inhibit late-stage GA biosynthesis and safely retard shoot elongation for reduced mowing requirements of fine turfgrasses. Class A PGRs and particularly TE was shown to consistently increase tiller density and leaf color. These positive effects were discussed as possibly being associated with measured increases in leaf cytokinins and antioxidants that may confer improved thermo- and osmotolerances. Restriction of GA-induced etiolation by TE in reduced light conditions also was reviewed as a potential positive use. Finally, TE was shown to act as a mild fungistat and extender of fungicide efficacy for control of dollar spot. TE's safety, consistent effectiveness, and physiological benefits imply that it will remain a preferred PGR for repeated use on fine turfgrasses across the stresses of summer.

Class B PGRs (flurprimidol and paclobutrazol) are known to inhibit early-stage GA biosynthesis and reduce mowing requirements. Because they act on monoxygenases that also interfere with sterol biosynthesis, considerable phytotoxicity potential that is rate-, species-, and temperature-dependent has been noted. *Poa annua* is injured to a greater extent than more desirable cool-season turfgrasses, making selective *Poa annua* suppression one of the major targeted uses of class B PGRs. While these compounds have been found to confer similar increases in tillering as those seen with class A PGRs, reports of improved responses to oxidative stresses are less common. Enhanced cytokinins and antioxidants due to related triazoles have been reported in turfgrasses and other crops, but caution as concerns application rates and timings should be employed when considering class B PGR use during summer stress periods on fine turfgrasses.

Class C and D PGRs were not discussed in this chapter because their phytotoxicity potential implies few reasonable uses for improving physiological performance of fine turfgrasses. Class E PGRs were discussed as enhancing endogenous levels of a number of plant hormones and potentially improving fine turfgrass stress tolerance. Release of ethylene due to ethephon application was shown to reduce mowing requirements, but a number of negative side effects such as leaf yellowing, root mass reductions, and increased scalping potential speak against its use for improving stress tolerance. Reports of increased cold hardiness in other crops due to ethephon application prior to chilling temperatures were not measured in bermudagrasses. Ethephon has been shown to be an effective seedhead suppressor of *Poa annua* and this effect, along with reductions in mowing requirements, appears to be its safe arena of use.

Sparse turfgrass-specific research is available as concerns exogenous IBA, but it has shown some potential to increase rooting and metabolic adjustment during drought stress. Such reports indicate that much more turfgrass-specific research should be conducted as concerns the potential of boosted auxin availability for improving stress tolerance.

Exogenous GA could be used to negate the growth-retarding effects of an overapplication of a class A or B compound, retain fall color and promote leaf greening of bermudagrass in spring, or enhance stolon spread. However, its etiolation and yellowing effects on actively growing turfgrasses are primarily negative in terms of increased turfgrass stress resistance.

Synthetic cytokinin use is not currently a common practice on fine turfgrasses, but the few reports reviewed here indicate that it has potential to improve quality and stress resistance by increasing density, photosynthesis, and antioxidant responses. Salicylic acid is also not currently in common use on fine turfgrasses. However, our review of the literature suggests its adoption as a prestress conditioning material to upregulate defense responses prior to periods of high oxidative stress. Similar statements can be made as regards the utility of the SA mimic, acibenzolarASM. Suppression of dollar spot and improved creeping bentgrass summer quality due to ASM application suggest that both compounds (ASM and SA) be investigated further for their potential to improve summer performance of fine turfgrasses.

The final class E PGR reviewed was JA. Jasmonates have been shown to be hormones that activate defense genes in response to wounding or microbial attack and promote senescence. In turfgrasses, only two reports of its exogenous use were available. In one, MJ improved antioxidant defense responses to UV-B stress. In the other, improved osmoregulant and soluble carbohydrate status of MJ-treated bermudagrass was associated with improved cold hardiness. These reports justify further research into the potential of jasmonates for improved stress tolerance of fine turfgrasses.

In this chapter, we introduced a new PGR group, class F, or natural-source PGRs. Humic substances, particularly foliar applications of humic and fulvic acids, have been shown to have growth-promoting effects that have been tied to their ability to chelate micronutrients and improve hormone (auxin and cytokinin) availability. Enhanced photosynthesis and root mass in nutrient solution culture due to humic acid, coupled with improved rooting, antioxidant status, and wilting resistance in drought-stressed cool-season turfgrasses indicate that these materials are functioning as PGRs. SWEs are another natural-source material that research has shown to have growth-regulating effects on turfgrasses. In particular, it has been shown that improved shoot and root growth is most likely associated with bioactive concentrations of cytokinins and betaines contained within the SWE, rather than being due to the small amounts of nutrients supplied. Improved thermo- and osmotolerances reported due to SWE foliar applications have been most closely tied to improved antioxidant capacity to scavenge free radicals. While the research is far from conclusive, it is apparent that combining these two natural-source PGRs, as is done in many commercial "biostimulants," can result in additive gains in turfgrass stress tolerance.

In conclusion, turfgrass practitioners have a number of synthetic and natural PGRs at their disposal to not only reduce mowing, but also, in some instances, to improve aesthetics and stress resistance of fine turfgrasses. In the case of the class A and B PGRs, much research and practical information is available to guide their safe and effective use. In the case of the class E and F PGRs, much less is known. A primary reason is that these hormone-type PGRs act on multiple genes, while the anti-GA PGRs are much more specific in their action. As a result, discrete effects can be predicted and tracked in a much more efficient manner for class A and B PGRs. Regardless, progress made in uncovering the applied physiological effects of class E and F PGRs will slowly improve our use recommendations and eventually allow us to better understand how these compounds can further improve our efforts at improving the aesthetics and stress resistance of fine turf surfaces.

REFERENCES

1. Watschke, T.L., M.G. Prinster, and J.M. Breuninger. 1992. Plant growth regulators and turfgrass management. In D.V. Waddington, R.N. Carrow, and R.C. Shearman (Eds.) *Turfgrass*. American Society of Agronomy, Madison, WI, chap. 16.
2. Beard, J.B. *Turfgrass Science and Culture*. Prentice-Hall, Englewood Cliffs, NJ. 1973.
3. Watschke, T.L. and J.M. DiPaola. 1995. Plant growth regulators. *Golf Course Manage*. 63(3):59–62.
4. Murphy, T.M., B. McCarty, and F.H. Yelverton. 2005. Turfgrass plant growth regulators. In L.B. McCarty (Ed.) *Best Golf Course Management Practices* (2nd ed.). Pearson and Prentice Hall, Upper Saddle River, NJ, pp. 705–714.
5. Rademacher, W. 2000. Growth retardants: Effect of gibberellin biosynthesis and other metabolic pathways. *Annu. Rev. Plant Physiol. Plant Mol. Biol.* 51:501–531.
6. Fagerness, M.J. and D. Penner. 1998. ^{14}C-Trinexapac-ethyl absorption and translocation in Kentucky bluegrass. *Crop Sci.* 38:1023–1027.
7. Beam, J.B. 2004. Prohexadione calcium for turfgrass management and *Poa annua* control and molecular assessment of the acetolactate synthase gene in *Poa annua*. PhD dissertation, ETD-05112004-101527. Virginia Polytechnic Institute and State University, Blacksburg, VA.

8. Ervin, E.H. and A.J. Koski. 1998. Growth responses of *Lolium perenne* L. to trinexapac-ethyl. *Hort-Science* 33:1200–1202.

9. Fagerness, M.J. and D. Penner. 1998. Evaluation of V-10029 and trinexapac-ethyl for annual blue-grass seedhead suppression and growth regulation of five cool-season turfgrass species. *Weed Tech.* 12:436–440.

10. Johnson, B.J. 1993. Response of tall fescue to plant growth regulators and mowing frequencies. *J. Environ. Hortic.* 11(4):163–167.

11. Johnson, B.J. 1994. Influence of plant growth regulators and mowing on two bermudagrasses. *Agron. J.* 86:805–810.

12. Ervin, E.H. and C. Ok. 2001. Influence of plant growth regulators on suppression and quality of 'Meyer' zoysiagrass. *J. Environ. Hortic.* 19(2):57–60.

13. Beam, J.B. and S.D. Askew. 2004. Prohexadione calcium effects on bermudagrass, Kentucky bluegrass, perennial ryegrass, and zoysiagrass. *Int. Turfgrass Soc. Res. J.* 10:286–295.

14. Ervin, E.H. and A.J. Koski. 2001. Trinexapac-ethyl increases Kentucky bluegrass leaf cell density and chlorophyll concentration. *HortScience* 36:787–789.

15. Heckman, N.L., R.E. Caussoin, G.L. Horst, and C.G. Elowsky. 2005. Growth regulator effects on cellular characteristics of two turfgrass species. *Int. Turfgrass Soc. Res. J.* 10:857–861.

16. McCullough, P.E., H. Liu, L.B. McCarty, T. Whitwell, and J.E. Toler. 2006. Nutrient allocation of "Tifeagle" bermudagrass as influenced by trinexapac-ethyl. *J. Plant Nutr.* 29(2):273–282.

17. Stier, J.C., J.N. Rogers III, and J.A. Flore. 1997. Nitrogen and trinexapac-ethyl effects on photosynthesis of supina bluegrass and Kentucky bluegrass in reduced light conditions. *Agron. Abstr.* 89:126.

18. Qian, Y.L., M.C. Engelke, M.J.V. Foster, and S. Reynolds. 1998. Trinexapac-ethyl restricts shoot growth and improves quality of "Diamond" zoysiagrass under shade. *HortScience* 33:1019–1022.

19. Heckman, N.L., T.E. Elthon, G.L. Horst, and R.E. Gaussoin. 2001. Influence of trinexapac-ethyl on respiration of isolated mitochondria. In *Annual Meetings Abstracts* [CD-ROM]. ASA, CSSA, SSSA. Madison, WI.

20. Fagerness, M.J., D.C. Bowman, F.H. Yelverton, and T.W. Rufty Jr. 2004. Nitrogen use in "Tifway" bermudagrass as affected by trinexapac-ethyl. *Crop Sci.* 44:595–599.

21. Fagerness, M.J. and F.H. Yelverton. 2000. Tissue production and quality of "Tifway" bermudagrass as affected by seasonal application patterns of trinexapac-ethyl. *Crop Sci.* 40: 493–497.

22. Lickfeldt, D.W., D.S. Gardner, B.E. Branham, and T.B. Voigt. 2001. Implications of repeated trinexa-pac-ethyl applications on Kentucky bluegrass. *Agron. J.* 93:1164–1168.

23. Han, S.W., T.W. Fermanian, J.A. Juvik, and L.A. Spomer. 1998. Growth retardant effects on visual quality and nonstructural carbohydrates of creeping bentgrass. *HortScience* 33:1197–1199.

24. Han, S.W., T.W. Fermanian, J.A. Juvik, and L.A. Spomer. 2004. Total nonstructural carbohydrate storage in creeping bentgrass treated with trinexapac-ethyl. *HortScience* 39:1461–1464.

25. Richie, W.E., R.L. Green, and F. Merino. 2001. Trinexapac-ethyl does not increase total nonstructural carbohydrate content in leaves, crowns, and roots of tall fescue. *HortScience* 36:772–775.

26. Waltz, Jr., F.C. and T. Whitwell. 2005. Trinexapac-ethyl effects on total nonstructural carbohydrates of field grown hybrid bermudagrass. *Int. Turfgrass Soc. Res. J.* 10:899–903.

27. Ervin, E.H. and A.J. Koski. 2001. Kentucky bluegrass growth responses to trinexapac-ethyl, traffic, and nitrogen. *Crop Sci.* 41:1981–1877.

28. Ervin, E.H., C.H. Ok, B.S. Fresenburg, and J.H. Dunn. 2002. Trinexapac-ethyl restricts shoot growth and prolongs stand density of "Meyer" zoysiagrass fairway under shade. *HortScience* 37: 502–505.

29. Fagerness, M.J., Trinexapac-ethyl affects canopy architecture but not thatch development of Tifway bermudagrass. 2001. *Int. Turfgrass Soc. Res. J.* 9:860–864.

30. Marcum, K.B. and H. Jiang. 1997. Effects of plant growth regulators on tall fescue rooting and water use. *J. Turfgrass Manage.* 2(2):13–27.

31. Jiang, H. and J. Fry. 1998. Drought response of perennial ryegrass treated with plant growth regulators. *HortScience* 33:270–273.

32. Fagerness, M.J. and F.H. Yelverton. 2001. Plant growth regulator and mowing height effects on seasonal root growth of "Penncross" creeping bentgrass. *Crop Sci.* 41:1901–1905.

33. Beasley, J.S., B.E. Branham, and L.M. Ortiz-Ribbing. 2005. Trinexapac-ethyl affects Kentucky bluegrass root architecture. *HortScience* 40:1539–1542.

34. McCullough, P.E., H. Liu, and L.B. McCarty. 2005. Response of six dwarf-type bermudagrasses to trinexapac-ethyl. *HortScience* 40:460–462.

35. McCullough, P.E. H. Liu, L.B. McCarty, T. Whitwell, and J.E. Toler. 2006. Growth and nutrient partitioning of "Tifeagle" bermudagrass as influenced by nitrogen and trinexapac-ethyl. *HortScience* 41:453–458.
36. Mok, D.W.S. and M.C. Mok. 1994. *Cytokinins: Chemistry, Activity, and Function*. CRC Press, London.
37. Zhang, X. and R.E. Schmidt. 2000. Application of trinexapac-ethyl and propiconazole enhances superoxide dismutase and photochemical activity in creeping bentgrass (*Agrostis stoloniferous* var. palustris). *J. Am. Soc. Hortic. Sci.* 125:47–51.
38. King, R.W., C. Blundell, L.T. Evans, L.N. Mander, and J.T. Wood. 1997. Modified gibberellins retard growth of cool-season turfgrasses. *Crop Sci.* 37:1878–1883.
39. Ervin, E.H. and A.J. Koski. 2001. Trinexapac-ethyl effects on Kentucky bluegrass evapotranspiration. *Crop Sci.* 41:247–250.
40. Pessarakli, M., K.B. Marcum, and D.M. Kopec. 2006. Interactive effects of salinity and Primo on the growth of Kentucky bluegrass. *J. Food Agr. Environ.* 4(1):325–327.
41. Baldwin, C.M., H. Liu, L.B. McCarty, W.L. Bauerle, and J.E. Toler. 2006. Effects of trinexapac-ethyl on the salinity tolerance of two ultradwarf bermudagrass cultivars. *HortScience* 41(3):808–814.
42. Liu, X. and B. Huang. 2002. Cytokinin effects on creeping bentgrass response to heat stress: II. Leaf senescence and antioxidant metabolism. *Crop Sci.* 42(2):466–472.
43. Heckman, N.L., R.E. Gaussoin, and G.L. Horst. 2001. Multiple trinexapac-ethyl applications reduce Kentucky bluegrass sod storage temperatures. *HortTechnology* 11(4):595–598.
44. Heckman, N.L., G.L. Horst, R.E. Gaussoin, and K.W. Frank. 2001. Storage and handling characteristics of trinexapac-ethyl treated Kentucky bluegrass sod. *HortScience* 36:1127–1130.
45. Heckman, N.L., G.L. Horst, R.E. Gaussoin, and L.J. Young. 2001. Heat tolerance of Kentucky bluegrass as affected by trinexapac-ethyl. *HortScience* 36:365–367.
46. Heckman, N.L., G.L. Horst, R.E. Gaussoin, and B.T. Tavener. 2002. Trinexapac-ethyl influence on cell membrane thermostability of Kentucky bluegrass leaf tissue. *Scientia Horticulturae* 92(2):183–186.
47. Einke, K., and J.C. Stier. 2004. Influence of trinexapac-ethyl on cold tolerance and nonstructural carbohydrates of shaded supine bluegrass. *Acta Horticulturae* 661:207–215.
48. Gerness, M.J., F.H. Yelverton, D.P. Livingston, and T.W. Rufty. 2002. Temperature and trinexapac-ethyl effects on bermudagrass growth, dormancy, and freezing tolerance. *Crop Sci.* 42:853–858.
49. Richardson, M.D. 2002. Turf quality and freezing tolerance of "Tifway" bermudagrass as affected by late-season nitrogen and trinexapac-ethyl. *Crop Sci.* 42:1621–1626.
50. Tan, Z.G. and Y.L. Qian. 2003. Light intensity affects gibberellic acid content in Kentucky bluegrass. *HortScience* 38:113–116.
51. Qian, Y.L., M.C. Engelke, M.J.V. Foster, and S. Reynolds. 1998. Trinexapac-ethyl restricts shoot growth and improves quality of "Diamond" zoysiagrass under shade. *HortScience* 33:1019–1022.
52. Qian, Y.L. and M.C. Engelke. 1999. Influence of trinexapac-ethyl on "Diamond" zoysiagrass in a shade environment. *Crop Sci.* 39:202–208.
53. Bunnell, B.T., L.B. McCarty, and W.C. Bridges Jr. 2005. "Tifeagle" bermudagrass response to growth factors and mowing height when grown at various hours of sunlight. *Crop Sci.* 45:575–581.
54. Stier, J.C. and J.N. Rogers III. 2001. Trinexapac-ethyl and iron effects on supina and Kentucky bluegrass under low irradiance. *Crop Sci.* 41:457–465.
55. Goss, R.M., J.H. Baird, S.L. Kelm, and R.N. Calhoun. 2002. Trinexapac-ethyl and nitrogen effects on creeping bentgrass grown under reduced light conditions. *Crop Sci.* 42:472–479.
56. Steinke, K. and J.C. Stier. 2003. Nitrogen selection and growth regulator applications for improving shaded turf performance. *Crop Sci.* 43:1399–1406.
57. Ervin, E.H., X. Zhang, S.D. Askew, and J.M. Goatley Jr. 2004. Trinexapac-ethyl, propiconazole, iron, and biostimulant effects on shaded creeping bentgrass. *HortTechnology* 14:500–506.
58. Gardner, D.S. and B.G. Wherley. 2005. Growth response of three turfgrass species to nitrogen and trinexapac-ethyl in shade. *HortScience* 40:1911–1915.
59. Wherley, B.G., D.S. Gardner, and J.D. Metzger. 2005. Tall fescue photomorphogenesis as influenced by changes in spectral composition and light intensity. *Crop Sci.* 45:562–568.
60. Srivastava, L.M. 2002. *Plant Growth and Development: Hormones and Environment*, Academic Press, New York, chap. 26.
61. Turgeon, A.J. 2002. *Turfgrass Management* (6th ed.). Prentice Hall, Upper Saddle River, NJ.
62. Carrow, R.N. and G. Wieko. 1989. Soil compaction and wear stresses on turfgrasses: Future research directions. *Int. Turfgrass Soc. Res. J.* 6:37–42.

63. Burpee, L.L., D.E. Green, and S.L. Stephens. 1996. Interactive effects of plant growth regulators and fungicides on epidemics of dollar spot in creeping bentgrass. *Plant Dis.* 80:1245–1250.

64. Golembiewski, R.C. and T.K. Danneberger. 1998. Dollar spot severity as influenced by trinexapac-ethyl, creeping bentgrass cultivar, and nitrogen fertility. *Agron. J.* 90(4):466–470.

65. Burpee, L.L. 1998. Effects of plant growth regulators and fungicides on Rhizoctonia blight of tall fescue. *Crop Prot.* 17(6):503–507.

66. Grossmann, K. 1990. Plant growth retardants as tools in physiological research. *Physiol. Plantarum.* 78:640–648.

67. Benton, J.M. and A.H. Cobb. 1995. The plant growth regulator activity of the fungicide, epoxiconazole, on *Gallium aparine* L. *Plant Growth Regul.* 17:149–155.

68. Gaussoin, R.E., B.E. Branham, and J.A. Flore. 1997. Carbon dioxide exchange rate and chlorophyll content of turfgrasses treated with flurprimidol or mefluidide. *J. Plant Growth Regul.* 16:73–78.

69. McCullough, P.E., H. Liu, L.B. McCarty, and T. Whitwell. 2004. Response of "Tifeagle" bermudagrass to seven plant growth regulators. *HortScience* 39:1759–1762.

70. Hanson, K.V. and B.E. Branham. 1987. Effects of four growth regulators on photosynthate partitioning in "Majestic" Kentucky bluegrass. *Crop Sci.* 27:1257–1260.

71. Dernoeden, P.H. 1984. Four-year response of a Kentucky bluegrass-red fescue turf to plant growth retardants. *Agron. J.* 76(5):807–813.

72. Stier, J.C., J.N. Rogers, J.R. Crum, and P.E. Rieke. 1999. Flurprimidol effects on Kentucky bluegrass under reduced irradiance. *Crop Sci.* 39:1423–1430.

73. Koski, A.J. 1997. Influence of paclobutrazol on creeping bentgrass root production and drought resistance. *Int. Turfgrass Soc. Res. J.* 8:699–709.

74. Fagerness, M.J. and F.H. Yelverton. 2001. Plant growth regulator and mowing height effects on seasonal root growth of "Penncross" creeping bentgrass. *Crop Sci.* 41:1901–1905.

75. McCullough, P.E., H. Liu, L.B. McCarty, and T. Whitwell. 2005. Physiological response of "Tifeagle" bermudagrass to paclobutrazol. *HortScience* 40:224–226.

76. McCullough, P.E., H. Liu, and L.B. McCarty. 2005. Dwarf bermudagrass responses to flurprimidol and paclobutrazol. *HortScience* 40:1549–1551.

77. McCullough, P.E., S.E. Hart, and D.W. Lycan. 2005. Plant growth regulator regimens reduce *Poa annua* populations in creeping bentgrass. *Appl. Turfgrass Sci.* 4:1–6.

78. Fletcher, R.A., A. Gilley, N. Sankhla, and T.D. Davis. 1999. Triazoles as plant growth regulators and stress protectants. *Hortic. Rev.* 23:55–138.

79. Izumi, K., S. Nakagawa, M, Kobayashi, H. Oshio, A. Sakurai, and N. Takahashi. 1988. Levels of IAA, cytokinins, ABA, and ethylene in rice plants as affected by a gibberellin biosynthesis inhibitor, uniconazole-P. *Plant Cell Physiol.* 29:97–104.

80. Goatley, Jr., J.M. and R.E. Schmidt. 1990. Anti-senescence activity of chemicals applied to Kentucky bluegrass. *J. Am. Soc. Hortic. Sci.* 115:654–656.

81. Green, R.L., K.S. Kim, and J.B. Beard. 1990. Effects of flurprimidol, mefluidide, and soil moisture on St. Augustinegrass evapotranspiration rate. *HortScience* 25:439–441.

82. Nabati, D.A., R.E. Schmidt, and D.J. Parrish. 1994. Alleviation of salinity stress in Kentucky bluegrass by plant growth regulators and iron. *Crop Sci.* 34:198–202.

83. Zhang, X., E.H. Ervin, and R.E. Schmidt. 2003. Seaweed extract, humic acid, and propiconazole improve tall fescue sod heat tolerance and posttransplant quality. *HortScience* 38:440–443.

84. Samala, S., J. Yan, and W.V. Baird. 1998. Changes in polar lipid fatty acid composition during cold acclimation in "Midiron" and "U3" bermudagrass. *Crop Sci.* 38:188–195.

85. Shepard, D.P. 1992. Effects of four plant growth regulators on carbohydrate reserves and cold tolerance of warm-season turf. PhD dissertation, North Carolina State University.

86. Stier, J.C., J.N. Rogers, J.R. Crum, and P.E. Rieke. 1999. Flurprimidol effects on Kentucky bluegrass under reduced irradiance. *Crop Sci.* 39:1423–1430.

87. Calhoun, R.N. 1996. Effect of three plant growth regulators and two nitrogen regimes on growth and performance of creeping bentgrass. M.S. Thesis, Michigan State University.

88. Fagerness, M.J. and F.H. Yelverton. 1999. Divot recovery of "Tifway" bermudagrass treated with trinexapac-ethyl and paclobutrazol. *Proc. NE Weed Sci. Soc.* 53:58.

89. Poovaiah, B.W. and A.C. Leopold. 1973. Effects of ethephon on growth of grasses. *Crop Sci.* 13:755–758.

90. Biddle, E., D.G.S. Kerfoot, Y.H. Koo, and K.E. Russell. 1976. Kinetic studies of the thermal decomposition of 2-chloroethyl phosphonic acid in aqueous solution. *Plant Physiol.* 58:700–702.

91. Diesburg, K. 1999. A new growth regulator for golf course turfgrass. *Golf Course Manage.* 66(4):1–2.
92. Cleland, R.E. 1999. Introduction: nature, occurrence and functioning of plant hormones. In P.J.J. Hooykaas, M.A. Hall, and K.R. Libbenga (Eds.) *Biochemistry and Molecular Biology of Plant Hormones.* Elsevier Sciences, Amsterdam, pp. 3–23.
93. Dudeck, A.E. and C.H. Peacock. 1985. "Tifdwarf" bermudagrass growth response to carboxin and GA₃ during suboptimum temperatures. *HortScience* 20:936–938.
94. Diesburg, K.L. and N.E. Christians. 1989. Seasonal application of ethephon, flurprimidol, mefluidide, paclobutrazol, and amidochlor as they affect Kentucky bluegrass shoot morphogenesis. *Crop Sci.* 29:841–847.
95. Christians, N.E. 1985. Response of Kentucky bluegrass to four growth retardants. *J. Am. Soc. Hortic. Sci.* 110:765–769.
96. Christians, N.E. and J. Nau. 1984. Growth retardant effects on three turfgrass species. *J. Am. Soc. Hortic. Sci.* 109:45–47.
97. McCullough, P.E., H. Liu, and L.B. McCarty. 2005. Response of creeping bentgrass to nitrogen and ethephon. *HortScience* 40:836–838.
98. McCullough, P.E., L.B. McCarty, H. Liu, and T. Whitwell. 2005. Response of "Tifeagle" bermudagrass to ethephon and trinexapac-ethyl. *Weed Technol.* 19(2):251–254.
99. Kane, R. and L. Miller. 2003. Field testing plant growth regulators and wetting agents for seedhead suppression of annual bluegrass. *USGA Turfgrass Environ. Res.* (online) 2(7):1–9.
100. Askew, S.D., W.L. Barker, D.R. Spak, J.B. Willis, and D.B. Ricker. 2006. Degree days for predicting annual bluegrass seedhead emergence. *Proc. NE Weed Sci. Soc.* 60:114.
101. Calhoun, R. and A. Hathaway. 2005. Using growing degree day models to predict application timing of mefluidide or ethephon for suppression of annual bluegrass seedheads. Annual Meeting Abstracts [ASA/CSSA/SSSA].
102. Sanyal, D. and F. Bangerth. 1998. Stress induced ethylene evolution and its possible relationship to auxin-transport, cytokinin levels, and flower bud induction on shoots of apple seedlings and bearing apple trees. *Plant Growth Regul.* 24:127–134.
103. Rogers, H.J. 2006. Programmed cell death in floral organs: how and why do flowers die? *Ann. Bot.* 97:309–315.
104. Suty, L.T., M.T. Moureaux, B.T. Leydecker, and De la Serve. 1993. Cytokinin affects nitrate reductase expression through the modulation of polyadeylation of the nitrate reductase mRNA transcript. *Plant Sci.* 90:11–19.
105. Larkindale, J. and B. Huang. 2004. Thermotolerance and antioxidant systems in *Agrostis stolonifera*: Involvement of salicylic acid, abscisic acid, calcium, hydrogen peroxide, and ethylene. *J. Plant Physiol.* 161:405–413.
106. Munshaw, G.C. 2004. Nutritional and PGR effects on lipid unsaturation, osmoregulant content and relation to bermudagrass cold hardiness. PhD dissertation, Virginia Polytechnic Institute and State University, Blacksburg, VA.
107. Harber, R.M. and L.H. Fuchigami. 1989. Ethylene-induced stress resistance. In P.L. Li (Ed.) *Low Temperature Stress Physiology in Crops.* CRC Press, Boca Raton, FL, pp. 81–89.
108. Arteca, R.N. 1996. *Plant Growth Substances: Principles and Applications.* Chapman & Hall. New York, NY.
109. Woodward, A.W. and B. Bartel. 2005. Auxin: regulation, action, and interaction. *Ann. Bot.* 95:707–735.
110. Hoveland, C.S. 1963. Effect of 3-indole butyric acid on shoot development and rooting of several bermudagrasses. *Agron. J.* 55:49–50.
111. Yokoyama, M., S. Yamaguchi, and K. Kusakari. 2003. Stress enhances indole-3-butyric acid-induced rooting in *Bupleurum falcatum* L. (Saoko) root culture. *Plant Biotechnol.* 20:331–334.
112. Juska, F.V. 1958. Some effects of gibberellic acid on turfgrasses. *USGA J. Turf Manage.* July:25–28.
113. McCullough, P.E., T. Whitwell, L.B. McCarty, and H. Liu. 2005. Response of 'Tifeagle' bermudagrass to fenarimol and gibberellic acid. *Int. Turfgrass Soc. Res. J.* 10:1245–1250.
114. White, R.H. and R.E. Schmidt. 1988. Carbon dioxide exchange of "Tifgreen" bermudagrass exposed to chilling temperatures as influenced by iron and BA. *J. Am. Soc. Hortic. Sci.* 113:423–427.
115. Kane, R.T. and R.W. Smiley. 1983. Plant growth regulating effects of systemic fungicides applied to Kentucky bluegrass. *Agron. J.* 75:469–473.
116. Goatley, Jr., J.M. and R.E. Schmidt. 1990. Seedling Kentucky bluegrass growth responses to chelated iron and biostimulator materials. *Agron. J.* 82:901–905.

117. Goatley, Jr., J.M. and R.E. Schmidt. 1991. Biostimulant enhancement of Kentucky bluegrass sod. *HortScience* 26:254–255.
118. Kawano, T., T. Furuichi, and S. Muto. 2004. Controlled salicylic acid levels and corresponding signaling mechanisms in plants. *Plant Biotech.* 21:319–336.
119. Delaney, T.P. 2005. Salicylic acid. In P.J. Davies (Ed.) *Plant Hormones: Biosynthesis, Signal Transduction, Action.* Kluwer Academic Publishers, Dordrecht, pp. 635–653.
120. Dat, J.F., C.H. Foyer, and I.M. Scott. 1998. Changes in salicylic acid and antioxidants during induced thermotolerance in mustard seedlings. *Plant Physiol.* 118:1455–1461.
121. Janda, T., G. Szalai, I. Tari, and E. Paldi. 1999. Hydroponic treatment with salicylic acid decreases the effects of chilling injury in maize (*Zea mays* L.) plants. *Planta* 208:175–180.
122. Borsani, O., V. Valpuesta, and M.A. Botella. 2001. Evidence for a role of salicylic acid in the oxidative damage generated by NaCl and osmotic stress in *Arabidopsis* seedlings. *Plant Physiol.* 126: 1024–1030.
123. Larkindale, J. and B. Huang. 2005. Effects of abscisic acid, salicylic acid, ethylene and hydrogen peroxide in thermotolerance and recovery for creeping bentgrass. *Plant Growth Regul.* 47:17–28.
124. He, Y., Y. Liu, W. Cao, M. Huai, B. Xu, and B. Huang. 2005. Effects of salicylic acid on heat tolerance associated with antioxidant metabolism in Kentucky bluegrass. *Crop Sci.* 45:988–995.
125. Schmidt, R.E. and X. Zhang. 2001. Alleviation of photochemical activity decline of turfgrasses exposed to soil moisture or UV-B radiation. *Int. Turfgrass Soc. Res. J.* 9:340–346.
126. Zhang, X. and E.H. Ervin. 2005. Effects of methyl jasmonate and salicylic acid on UV-B tolerance associated with free radical scavenging capacity in Poa pratensis. *Int. Turfgrass Soc. Res. J.* 10:910–915.
127. Ervin, E.H., X. Zhang, and R.E. Schmidt. 2005. Exogenous salicylic acid enhances post-transplant success of heated Kentucky bluegrass and tall fescue sod. *Crop Sci.* 45:240–244.
128. Cole, D.L. 1999. The efficacy of acibenzolar-S-methyl, an inducer of systemic acquired resistance, against bacterial and fungal diseases of tobacco. *Crop Prot.* 18:267–273.
129. Lee, J. and J. Fry. 2003. Dollar spot and brown patch incidence in creeping bentgrass as affected by acibenzolar-S-methyl and biostimulants. *HortScience* 38:1223–1226.
130. Zhang, Q., J. Fry, and N. Tisserat. 2005. Evaluation of plant defense activators for dollar spot and brown patch control on creeping bentgrass putting green. *Int. Turfgrass Soc. Res. J.* 10:180–185.
131. Wasternack, C. and B. Parthier. 1997. Jasmonate-signaled plant gene expression. *Trends Plant Sci.* 2:302–307.
132. Srivastava, L.M. 2002. *Plant Growth and Development: Hormones and Environment.* Academic Press, New York, chap. 12.
133. Gonzalez-Aguilar, G., M. Tiznado-Hernandez, and C.Y. Wang. 2006. Physiological and biochemical responses of horticultural products to methyl jasmonate. *Stewart Postharvest Rev.* 2:1–9.
134. Zhang, X., G.C. Munshaw, and E.H. Ervin. 2004. Influence of late-season jasmonic acid and sodium chloride treatments on seeded bermudagrass cold hardiness. Annual Meetings Abstracts [CD-ROM]. ASA, CSSA, SSSA. Madison, WI.
135. Aiken, G.R., D.M. McKnight, R.L. Wershaw, and P. MacCarthy. 1985. Humic substances in soil, sediment, and water: geochemistry, isolation, and characterization. In G.R. Aiken, P. MacCarthy, R.L. Malcolm, and R.S. Swift (Eds.) *Humic Substances in Soil, Sediment, and Water.* Wiley, New York.
136. Brady, N.C. and R.R. Weil. 1999. *The Nature and Properties of Soils* (12th ed.). Prentice Hall, Upper Saddle River, NJ.
137. Chen, Y. and T. Aviad. 1990. Effects of humic substances on plant growth. In Y. Chen, T. Aviad, and P. MacCarthy (Eds.) *Humic Substances in Soil and Crop Sciences: Selected Readings. Proceedings of a Symposium Co-sponsored by the International Humic Substances Society.* Chicago, IL, pp. 161–186.
138. Vaughn, D. and R.E. Malcolm. 1985. Effects of humic acid on invertase synthesis in roots of higher plants. *Soil Biol. Biochem.* 11:247–252.
139. Chen, Y., H. Magen, J. Riov, N. Senesi, and T.M. Miano. 1992. Humic substances originating from rapidly decomposing organic matter: properties and effects on plant growth. In Y. Chen, H. Magen, J. Riov, N. Senesi, and T.M. Miano (Eds.) *Humic Substances in the Global Environment and Implications on Human Health: Proceedings of the 6th International Meeting of the International Humic Substances Society.* Monopli, Italy, September 20–25, pp. 427–443.
140. Chen, Y., C.E. Clapp, H. Magen, and V.W. Cline. 1999. Stimulation of plant growth by humic substances: Effects of iron availability. In E.A. Ghabbour and G. Davies (Eds.) *Understanding Humic Substances: Advanced Methods, Properties, and Applications.* Royal Society of Chemistry, Cambridge, pp. 255–264.

141. Nardi, S., G. Concheri, D. Pizzeghello, A. Sturaro, R. Rella, and G. Parvoli. 2000. Soil organic matter mobilization by root exudates. *Chemosphere* 41:653-658.

142. Pizzeghello, D., G. Nicolini, and S. Nardi. 2001. Hormone-like activity of humic substances in *Fagus sylvaticae* forests. *New Phytologist* 151:647–657.

143. Muscolo, A., F. Bovalo, F. Giofriddo, and S. Nardi. 1999. Earthworm humic matter produces auxin-like effects on *Daucus carota* cell growth and nitrate metabolism. *Soil Biol. Biochem.* 31:1303–1311.

144. Gaur, A.C. 1964. Influence of humic acid on growth and mineral nutrition in plants. *Bull. Assoc. Fr. Itude Sol.* 35:207–219.

145. Varshovi, A.A. 1996. Humates and their turfgrass applications. *Golf Course Manage.* 64(8):53–56.

146. Cooper, R.J., C.H. Liu, and D.S. Fisher. 1998. Influence of humic substances on rooting and nutrient content of creeping bentgrass. *Crop Sci.* 38:1639–1644.

147. Liu, C.H., R.J. Cooper, and D.C. Bowman. 1998. Humic acid application affects photosynthesis, root development, and nutrient content of creeping bentgrass. *HortScience* 33:1023–1025.

148. Liu, C.H. and R.J. Cooper. 2002. Humic acid application does not improve salt tolerance of hydroponically-grown creeping bentgrass. *J. Am. Soc. Hortic. Sci.* 127:219-223.

149. Zhang, X. 1997. Influences of plant growth regulators on turfgrass growth, antioxidant status, and drought tolerance. PhD dissertation, Virginia Polytechnic Institute and State University, Blacksburg, VA.

150. Zhang, X., R.E. Schmidt, E.H. Ervin, and S. Doak. 2002. Creeping bentgrass physiological responses to natural plant growth regulators and iron under two regimes. *HortScience* 37:898–902.

151. Zhang, X. and E.H. Ervin. 2004. Cytokinin-containing seaweed and humic acid extracts associated with creeping bentgrass leaf cytokinins and drought resistance. *Crop Sci.* 44:1737–1745.

152. Aiken, J.B. and T.L. Senn. 1965. Seaweed products as a fertilizer and soil conditioner. *Botanica Marina* 8:144–148.

153. Crouch, I.J. 1990. The effect of seaweed concentrate on plant growth. PhD Dissertation, Department of Botany, University of Natal, Pietermaritzburg.

154. Verkleij, F.N. 1992. Seaweed extracts in agriculture and horticulture: a review. *Biol. Agr. Hortic.* 8:309–324.

155. Blunden, G. 1977. Cytokinin activity of seaweed extracts. In D.L. Faulkner and W.H. Fenical (Eds.) *Marine Natural Products Chemistry.* Plenum Publishing, NY.

156. Crouch, I.J. and J. Van Staden. 1993. Evidence for the presence of plant growth regulators in commercial seaweed products. *Plant Growth Regul.* 13:21–29.

157. Blunden, G. and S.M. Gordon. 1986. Betaines and their sulphino analogues in marine algae. *Prog. Phycol. Res.* 4:39–79.

158. Tay, S.A.B., J.K. Macleod, L.M.S. Palni, and D.S. Letham. 1985. Detection of cytokinins in a seaweed extract. *Phytochemistry* 24:2611–2614.

159. Zhang, X., R.E. Schmidt, and E.H. Ervin. 2003. Physiological effects of liquid applications of a seaweed extract and a humic acid on creeping bentgrass. *J. Am. Soc. Hortic. Sci.* 128:492–496.

160. Cassan, L., I. Jeannin, T. Lamaze, and J.F. Morot-Gaudry. 1992. The effect of the *Ascophyllum nodosum* extract Goemar GA 14 on the growth of spinach. *Botanica Marina* 35:437–439.

161. Jeannin, I., J.C. Lescure, and J.F. Morot-Gaudry. 1991. The effects of aqueous seaweed sprays on the growth of maize. *Botanica Marina* 34:469–473.

162. Blunden, G., T. Jenkins, and Y.W. Liu. 1997. Enhanced leaf chlorophyll levels in plants treated with seaweed extract. *J. Appl. Phycol.* 8:535–543.

163. Zhang, X. and R.E. Schmidt. 1997. The impact of growth regulators on alpha-tocopherol status of water-stressed *Poa pratensis* L. *Int. Turfgrass Soc. Res. J.* 8:1364–1373.

164. Zhang, X. and R.E. Schmidt. 1999. Antioxidant responses to hormone-containing products in Kentucky bluegrass subjected to drought. *Crop Sci.* 39:545–551.

165. Zhang, X. and R.E. Schmidt. 2000. Hormone-containing products' impact on antioxidant status of tall fescue and creeping bentgrass subjected to drought. *Crop Sci.* 40:1344–1349.

166. Ayad, J.Y., J.E. Mahan, V.G. Allen, and C.P. Brown. 1997. Effect of seaweed extract and the endophyte in tall fescue on superoxide dismutase, glutathione reductase and ascorbate peroxidase under varying levels of moisture stress. *Proc. Amer. Forage Grassl. Council.* Fort Worth, TX, pp. 173–177.

167. Coelho, R.W., J.H. Fike, R.E. Schmidt, X. Zhang, and V.G. Allen. 1997. Influence of seaweed extract on growth, chemical composition and superoxide dismutase activity in tall fescue. *Proc. Amer. Forage Grassl. Council.* Fort Worth, TX, pp. 163–167.

168. Fike, J.H., V.G. Allen, J.P. Fontenot, and R.E. Schmidt. 1997. Influence of seaweed extract on wether lambs grazing endophyte-infected tall fescue. *Proc. Amer. Forage Grassl. Council.* Fort Worth, TX, pp. 153–157.

169. Kumar, R.A.S. and S.S. Babu. 2004. Effect of seaweed extract on oxidizing enzymes during the senescence of *Oryza sativa* var. Ambai-16. *Seaweed Res. Util.* 26:177–180.

170. Lingakumar, K., R. Jeyaprakash, C. Manimuthu, and A. Haribaskar. 2004. Influence of *Sargassum* sp. crude extract on vegetative growth and biochemical characteristics in *Zea mays* and *Phaseolus mungo*. *Seaweed Res. Util.* 26:155–160.

171. Price, A.H. and G.A.F. Hendry. 1989. Stress and the role of activated oxygen scavengers and protective enzymes on plants subjected to drought. *Biochem. Soc. Trans.* 17:493–497.

172. Ervin, E.H., X. Zhang, and G. Seaver. 2006. Impact of seaweed extract-based cytokinins and zeatin riboside on creeping bentgrass heat tolerance. Plant Growth Regulation Society of America, 33rd Annual Meeting, Abstract 20.

173. Stephenson, W.A. 1968. *Seaweed in Agriculture and Horticulture.* Faber and Faber, London, UK.

174. Tarjan, A.C. 1977. Kelp derivatives for nematode-infected citrus trees. *J. Nematol.* 9:287.

175. Trajian, A.C. and J.J. Frederich. 1983. Comparative effects of kelp preparations and ethoprop on nematode-infected bermudagrass (*Cynodon dactylon*). *Nematropica* 13:55–62.

176. Featonby-Smith, B.C. and J.J. Van Staden. 1983. The effects of seaweed concentrate and fertilizer on the growth of *Beta vulgaris*. *Pflanzenphysiologie* 112:155–162.

177. Sun, H. 1994. The effect of seaweed concentrate on turfgrass growth, nematode tolerance and protein synthesis under moisture stress conditions. PhD dissertation, Virginia Polytechnic Institute and State University, Blacksburg, VA.

178. Sun, H., R.E. Schmidt, and J.D. Eisenback. 1997. The effect of seaweed concentrate on the growth of nematode-infected bentgrass grown under low soil moisture. *Int. Turfgrass Soc. Res. J.* 8:1336–1342.

179. Wu, Y., T. Jenkins, G. Blunden, C. Whapham, and S.D. Hankins. 1997. The role of betaines in alkaline extracts of *Ascophyllum nodosum* in the reduction of *Meloidogyne javanica* and *M. incognita* infestations of tomato plants. *Fund. Appl. Nematol.* 20(2): 99–102.

180. Schmidt, R.E., E.H. Ervin, and X. Zhang. 2003. Questions and answers about biostimulants. *Golf Course Manage.* 71(6):91–94.

181. Zhang, X., E.H. Ervin, and R.E. Schmidt. 2003. Plant growth regulators can enhance recovery of Kentucky bluegrass sod from heat injury. *Crop Sci.* 43(3):952–956.

182. Ervin, E.H., X. Zhang, and J. Fike. 2004. Reducing ultraviolet radiation damage on *Poa pratensis* II: hormone and hormone-containing substance. *HortScience* 39:1471–1474.

183. Yan, J. 1993. Influence of plant growth regulators on turfgrass polar lipid composition, tolerance to drought and saline stresses, and nutrient use efficiency. PhD dissertation, Virginia Polytechnic Institute and State University, Blacksburg, VA.

184. Doak, S.O., R.E. Schmidt, and E.H. Ervin. 2005. Metabolic enhancers impact on creeping bentgrass leaf sodium and physiology under salinity. *Int. Turfgrass Soc. Res. J.* 10:845–849.

Section V

Turfgrass Breeding, Genetics, and Biotechnology

11 Breeding Turfgrasses

Tim D. Phillips

CONTENTS

11.1 INTRODUCTION

Conventional plant-breeding methods have been used successfully to transform several species into markedly improved turfgrasses. The choice of which species to use in a breeding program can be dictated by the region in which the cultivar will be grown, the specific breeding objectives, and market forces. The basic requirements for a possible turfgrass species for improvement are adaptation to the area of use, genetic potential (variability), and germplasm availability. Usually, a turfgrass breeder will specialize and work with one or a few related species, but some breeders work with numerous turf species over their careers. Other grass breeders have bred grasses for both turf and forage uses.

11.2 BREEDING OBJECTIVES

The National Turfgrass Evaluation Program (NTEP) conducts replicated turfgrass trials across the United States and Canada. One of its websites (http://www.ntep.org/reports/ratings.htm) gives explanations of the traits that are evaluated. Some of the traits are subjective and visual, like turfgrass quality, genetic color, leaf texture, density, and percent living ground cover. Other important traits include spring green-up, seedling vigor, winter hardiness, traffic tolerance, and pest resistance. Any of these traits can be targets for improvement in a plant-breeding program, and usually selection is for several traits simultaneously. Uniformity in height, color, and texture is also a goal of turfgrass breeders. Other traits not directly related to turfgrass performance can be very important. Seed yield generally is not of concern to most end users, but without adequate seed yield, a new cultivar will not be commercially successful. Similarly, turfgrass breeders may need to select for resistance to pests in the seed production environment as well as in the area of use as a turf cultivar.

11.3 BREEDING METHODS

Improved cultivars of turfgrasses have been developed using different breeding techniques, depending on the mode of propagation (seeded or vegetative) and breeding behavior of the grass species (self-pollinating, cross-pollinating, or apomictic). In this chapter, some of the plant-breeding methods utilized to develop improved turfgrass cultivars will be discussed.

Early turfgrass cultivars were ecotype selections (mass selection) of naturalized populations (e.g., "Kentucky 31" tall fescue [1], "Kenblue" Kentucky bluegrass [2], and "Highland" bentgrass [2]) with little or no selection for improved turf characteristics. Concerted breeding efforts over the past 50 years have resulted in vastly improved turf characteristics in several turfgrass species, while some recently introduced turfgrass species are similar to wild populations, with room for improvement. Early turf-breeding programs utilized germplasm collected from old turf (home lawns, golf courses, cemeteries, etc.) as the genetic foundation of new turf varieties [3]. Dr. C. Reed Funk at the New Jersey Agricultural Experiment Station (NJAES) at Rutgers University was a pioneer in the collection of turf germplasm from old turf areas across the eastern United States, providing a genetic base for several turfgrass species.

11.3.1 PEDIGREE AND OTHER METHODS FOR SELF-POLLINATING SPECIES

Annual bluegrass (*Poa annua* L.), a relatively recent subject of interest of turfgrass breeders, is the only known self-pollinating turfgrass species. True-breeding (homozygous) lines of annual bluegrass with turf characteristics have been identified [3]. Methods of breeding other self-pollinating plants (wheat, rice, soybean, etc.) would be applicable for annual bluegrass. The pedigree breeding method is often used for these species, and entails controlled mating of two selected plants, then selfing for several generations before selection (identification) of desirable lines [4]. Back-cross breeding also would be appropriate, if lines with outstanding performance are used as the recurrent parent, crossed to some genotype possessing one or a few desirable genes but otherwise unacceptable as a turf cultivar. Use of F_1 hybrids may be worthwhile if the cultivar is vegetatively propagated. Annual bluegrass possesses some limitations for its success as a commercial turfgrass species. One involves limited seed set and indeterminate flowering, and another is the difficulty of keeping undesirable types of annual bluegrass out of the seed production fields. Vegetative propagation of superior perennial turf-type genotypes may help overcome these challenges.

Vegetative propagation of a turfgrass species, regardless of its breeding habit, allows the use of any genotype as a potential cultivar. These genotypes can be "found" plants (ecotype cultivars) of unknown background, or they can be the result of designed, controlled crosses. Bermudagrass can be seeded or propagated vegetatively as sod or sprigs. Some turf-type bermudagrass cultivars are single genotypes of sterile hybrids from wide crosses ("Tifdwarf," "Tiffine," "Tifgreen," "Tiflawn," and "Tifway") [2]. As such, turf uniformity should be better than in seeded, heterogeneous cultivars.

11.3.2 RECURRENT SELECTION

Many turfgrass species are cross-pollinated (Table 11.1). For outcrossing grasses with self-incompatibility (pollen will not germinate on the plant that produces it, ensuring high levels of cross-pollination), synthetic cultivars can be produced. A synthetic cultivar is produced by blending equal amounts of seed from each parent in a crossing block [4]. The parents can be superior genotypes collected from old turf areas, or genotypes produced from an ongoing recurrent selection program [5]. Two common variations on recurrent selection are phenotypic recurrent selection and genotypic recurrent selection. With phenotypic selection, parents are chosen based on the performance of the plant being selected. Phenotypic selection works well with traits that have high heritability, such as heading date. A successful modification of phenotypic recurrent selection was devised by G.W. Burton at Tifton, Georgia, with bahiagrass [6]. He imposed a grid over the selection nursery, and selected the best 20% of plants (the best five plants out of each group of 25 plants). This procedure helps minimize the effects of soil heterogeneity, and the moderate selection intensity ensures that genetic variability will be maintained.

Selected plants from each cycle of recurrent selection must be identified prior to pollination. Plants can be dug from the nursery and moved to an isolated area to intermate, or the nonselected plants can be mown prior to pollen shed, leaving selected parents to intermate in the nursery. Selected plants need to have similar flowering dates for optimal polycrossing. Isolation techniques include using

TABLE 11.1
Some Cross-Pollinating Turfgrass Species

Common Name	Botanical Name	Cool/Warm Season
Velvet bentgrass	*Agrostis canina* L.	Cool
Colonial bentgrass	*Agrostis capillaries* L.	Cool
Creeping bentgrass	*Agrostis stolonifera* L.	Cool
Bahiagrass[a]	*Paspalum notatum* Flugge	Warm
Bermudagrass	*Cynodon dactylon* (L.) Pers.	Warm
Kentucky bluegrass[a]	*Poa pratensis* L.	Cool
Rough bluegrass	*Poa trivialis* L.	Cool
Texas bluegrass[b]	*Poa arachnifera* Torr.	Cool
Buffalograss[b]	*Bouteloua dactyloides* (Nutt.) J.T. Columbus	Warm
Centipedegrass	*Eremochloa ophiuroides* (Munro) Hack.	Warm
Tall fescue	*Schedonorus phoenix* (Scop.) Holub (=*Festuca arundinacea* Schreb.)	Cool
Fine-leaved fescues	*Festuca* spp.	Cool
Perennial ryegrass	*Lolium perenne* L.	Cool
St. Augustinegrass	*Stenotaphrum secundatum* (Walt.) Kuntze	Warm
Seashore paspalum	*Paspalum vaginatum* Swartz	Warm
Zoysiagrass	*Zoysia japonica* Steudel	Warm

[a] Apomixis is also important in this species.
[b] This species is dioecious.

small grains (cereal rye or triticale), mowing surrounding areas, or moving the selected parents to a greenhouse or other areas protected from unintentional pollination.

Most recent turfgrass cultivars were developed after two or more cycles of phenotypic recurrent selection [7], depending on the source material being used. Some turfgrass cultivars are named in ways that indicate ongoing recurrent selection (e.g., "Rebel," "Rebel Jr.," "Rebel II," "Rebel III," "Rebel IV" turf-type tall fescue cultivars) [8]. Recurrent selection allows the breeder to make improvements to an existing cultivar by continuing with additional cycles of selection within the cultivar and adding new germplasm from outside the genetic base of the cultivar. Early generations of selection often involve identifying relatively few genotypes and moving them to intermate in isolation. In later stages of recurrent selection cycles, a final cycle of selection can be made by eliminating the off-type plants or undesirable genotypes, then harvesting the remaining plants in place.

In genotypic recurrent selection, progeny from intermated parents are evaluated to determine which parents produce the best offspring. Small turf plots can be evaluated for several years, and then the best-performing half-sib families can indicate which parents are best. These superior parents (maintained in a nursery) then can be crossed in isolation from the mediocre or inferior parents. Alternatively, half-sib families can be evaluated as spaced plants in nurseries, and selection can be within or among half-sib families. Genotypic selection requires more time for each cycle, but generally is more successful for traits with lower heritability.

Texas bluegrass (*Poa arachnifera* Torr.) and buffalograss (*Bouteloua dactyloides* [Nutt.] J.P. Columbus) are the two North American-native turfgrass species, both of which are dioecious (separate male and female plants). Improved turf cultivars of buffalograss are promoted for low-maintenance use in drier areas of the United States (mainly the central great plains) [3]. Some genotypes of female buffalograss have been released and are vegetatively propagated ("Legacy," "Prairie," "Scout," and "609"), and in California, recommended as low-allergen plants [9]. Seeded cultivars of buffalograss are synthetics, produced by intermating selected male and female clones, then bulking seed from each female parent. Female plants of Texas bluegrass can be pollinated with Kentucky bluegrass plants to produce hybrids. This method was used to develop the cultivar "Reveille." Reveille is a facultative apomictic (a trait from the Kentucky bluegrass parent), so it produces seed identical to the parental hybrid [10].

To illustrate how a new cultivar is developed, one can review cultivar releases published in the journal *Crop Science*. For example, "Manhattan 4," released in 2004 [11], was developed by selecting 110 turf-type ryegrass plants from three ryegrass breeding nurseries in Oregon. These plants were crossed in isolation, and then seed of the best 83 plants was harvested. A nursery of 4000 plants was established from this seed, of which 836 plants remained after the second cycle of selection. The seed from the intermated 836 plants formed the Breeder class of seed for this new cultivar. The breeders note that maternal parentage traces to previous "Manhattan" cultivars (23% from "Manhattan 3," 18% from "Manhattan II," and 7% from "Manhattan"), 16 from germplasm collected in Missouri, Virginia, and Illinois, and 36% from a breeding line from NJAES at Rutgers University. Most of the sources of this cultivar were the product of several cycles of recurrent selection themselves.

"Rebel" tall fescue was released in 1980 after five cycles of selection and 19 years of testing in turf trials [2,12]. Since then, several cultivars have been released, using Rebel as a genetic base. For example, "Rebel, Jr.," released in 1990, contains progeny from plants collected from old turfs in 12 states and plants from or related to Rebel [2]. Rebel, Jr. is an advanced generation synthetic from 91 clones developed through recurrent selection. "Rebel II," released in 1986 [2] resulted from a modified backcross breeding program, using Rebel, and included several cycles of both restricted recurrent phenotypic and genotypic selection.

11.3.3 APOMIXIS

Kentucky bluegrass is often regarded as the premier turfgrass species [13]. Its breeding is complicated by the apomictic nature of its reproductive system [13]. Apomixis can be thought of as "vegetative reproduction via seed." Many genotypes of Kentucky bluegrass produce seed that are exact clones of the maternal parent (i.e., with no sexual reproduction). The type of apomixis in Kentucky bluegrass is pseudogamous apospory, in which pollination is generally required to initiate seed development, but the ovule is replaced by an unreduced embryo sac cell, so sexual fertilization does not occur. The seed develops with the embryo being genetically identical to its mother. Some techniques have been developed to aid in producing sexual hybrids [14], after which superior genotypes would be identified and released as a new cultivar if they were apomictic. Current commercial cultivars of Kentucky bluegrass are expected to maintain a level of apomixis of 95% or higher [13].

Early Kentucky bluegrass cultivars were essentially mixtures of seed harvested from pastures or other naturalized stands in the Midwest (especially in Kentucky). Kenblue Kentucky bluegrass was a composite mixture of seed from Kentucky bluegrass seed production fields in central Kentucky [2]. Some ecotype breeding, in which single, highly apomictic genotypes were identified and propagated as new cultivars, was common until the 1970s [14]. "Merion" Kentucky bluegrass was a breakthrough cultivar, released in 1947, for its combination of low growth habit and improved resistance to leafspot [2]. It originated as a single plant selection from Merion Golf Club in Ardmore, Pennsylvania, in 1936. After techniques were developed to aid production of sexual hybrids in facultative apomictic genotypes, hybridization and sexual recombination became more common [13,15]. Progeny from designed crosses are evaluated for turf performance and level of apomixis. Aberrant genotypes in an otherwise uniform progeny group provide an indication that there may be sexual fertilization involved [13]. Large numbers of progeny need to be evaluated to identify the off-type or aberrant plants, then progeny from these plants must be evaluated to determine their level of apomixis. Many intraspecific hybrids in Kentucky bluegrass are the result of fertilization of unreduced eggs, resulting in increases in the chromosome number [13]. There probably is an upper limit for chromosome number for optimal cultivar performance. Flow cytometry can be helpful in monitoring ploidy levels in Kentucky bluegrass, as well as in other turfgrass species (especially the fine fescue complex).

REFERENCES

1. Buckner, R.C. and Bush, L.P. 1979. *Tall Fescue*. ASA monograph No. 20 (ASA, CSSA, SSA, Madison, WI).
2. Alderson, J.A. and Sharp, W.C. 1994. *Grass Varieties in the United States*. Agriculture Handbook 170. Soil Conservation Service, U.S. Department of Agriculture.
3. Casler, M.D. and Duncan, R., Eds. 2003. *Turfgrass Biology, Genetics, and Breeding*. Wiley, New York.
4. Fehr, W.R., Ed. 1987. *Principles of Cultivar Development*. McGraw-Hill, New York.
5. Pedersen, J.F. and Vogel, K.P. 1993. Breeding systems for cross-pollinated perennial grasses. *Plant Breed. Rev.* 11: 251–274.
6. Burton, G.W. 1982. Improved recurrent restricted phenotypic selection increases bahiagrass forage yield. *Crop Sci.* 22: 1058–1061.
7. National Grass Variety Review Board. 2006. www.aosca.org/VarietyReviewBoards/2006GrassNVRB Report_Final.pdf
8. Oregon Seed Certification Office. 2007. Available at http://www.oscs.orst.edu/publications/varieties/ varietiesandclassescurrent.pdf
9. Ogren, T.L. 2003. Safe sex in the garden: and other propositions for an allergy-free world, pp. 59–60. Ten Speed Press, Berkeley, CA.
10. Read, J.C., Reinert, J.A., Colbaugh, P.F., and Knoop, W.E. 1999. Registration of "Reveille" hybrid bluegrass. *Crop Sci.* 39: 590.
11. Fraser, M.L., Rose-Fricker, C.A., Meyer, W.A., and Funk, C.R. 2004. Registration of "Manhattan 4" perennial ryegrass. *Crop Sci.* 44: 346–347.
12. Funk, C.R., Dickson, W.R., and Hurley, R.H. 1981. Registration of "Rebel" tall fescue. *Crop Sci.* 21: 632.
13. Huff, D.R. 2003. Kentucky Bluegrass. In *Turfgrass Biology, Genetics, and Breeding*. M.D. Casler and R. Duncan, Eds., Wiley, New York, pp. 27–38, Chapter 2.
14. Bashaw, E.C. and Funk, C.R. 1987. Apomictic grasses. In *Principles of Cultivar Development*. Vol. 2 Crop Species. W.E. Fehr, Ed., Macmillan Publishing Company, New York.
15. Riordan, T.P., Shearman, R.C., Watkins, J.E., and Behling, J.P. 1988. Kentucky bluegrass automatic hybridization apparatus. *Crop Sci.* 28: 183–185.
16. Funk, C.R., Engel, R.E., Pepin, G.W., Radko, A.M., and Peterson, R.J. 1973. Registration of Bonnieblue Kentucky bluegrass. *Crop Sci.* 14: 906.

Section VI

Turfgrass Growth Disorders, Pathology, Diseases, and Turf Disease Management and Control

12 Nutritional Disorders of Turfgrasses

Asghar Heydari and Giorgio M. Balestra

CONTENTS

12.1 INTRODUCTION

Nutritional disorders are common problems in all crops, and this can greatly influence plant health and productivity [1–15]. A balanced and good nutritional system can lead to healthy plant growth and result in higher yield and productivity [1–15]. In contrast, unbalanced and poor nutrition may create unhealthy and weak conditions that adversely affect plant health and growth and significantly reduce yield and productivity. Poor nutrition can also make the plant more susceptible to infectious and noninfectious diseases and disorders [5–14].

Like many other plants, turfgrass can also suffer from nutritional problems that may seriously alter its growth and vigor [1–43]. More evident and common symptoms of nutrient deficiency in turfgrass usually appear on the root, shoot, rhizome, or stolon which results in reduction in growth and production [1,7,12,14]. These symptoms may be similar to those caused by environmental stress or pest problems [1,2,5,8,9,11,14]. In turfgrass, unlike other crops, nutritional problems usually do not appear on leaves alone; therefore, it is important to recognize some typical and diagnostic symptoms induced by nutritional problems on turfgrass growth and quality [1–4,8,9,11,12,14].

In the turfgrass-cropping system, yield and production are usually determined qualitatively and not quantitatively; therefore, the impact of any nutritional disorder on the plant should be evaluated on the basis of turfgrass quality and functional characteristics [1,2,12,14,16,22]. Nutritional disorders and problems on turfgrass may be affected by several factors and parameters such as soil nutrients, soil texture, soil organic matter contents, turfgrass species and cultivar, duration of the growing season, amount of rainfall, pests and disease populations, and crop management strategies [1–25]. These factors in combination or alone may affect and complicate disorders and problems caused by deficiency of different nutrients.

Excessive levels of nutrients may also cause toxicity in turfgrass and result in nutritional problems, inducing symptoms on different parts of the plant [3,4,5,16,21]. To have a healthy turf, it is very important to apply suitable fertilizers at the optimum level and right time. This will provide the plant with a good nutritional environment, improve turfgrass growth and vigor, and enhance tolerance to abiotic and biotic stress factors [1,2,5,8,9,11,14]. The problems and disorders caused by unbalanced levels of different nutrients (minerals) in turfgrass are discussed below.

12.2 NUTRIENTS AND MINERALS REQUIRED FOR TURFGRASS GROWTH AND DEVELOPMENT

Turfgrasses require several chemical elements for growth and development [1–30]. These elements can be divided into two main groups based on where they are obtained by turf plants. The first group—carbon (C), hydrogen (H), and oxygen (O)—is obtained from atmospheric carbon dioxide and water and comprises most of the turfgrass body [17]. The second group is minerals absorbed from soil or fertilizers. This group can also be divided into three sub-groups based on the quantities in which the minerals are used by turf plants. The macronutrients or essential nutrients, which include nitrogen (N), phosphorus (P), and potassium (K), are very essential for plant growth and are used in relatively large quantities by turfgrasses [1–35]. The secondary nutrients, sulfur (S), calcium (Ca), and magnesium (Mg), are used in smaller amounts by turf plants [1,17,36]. Finally, the micronutrients, iron (Fe), manganese (Mn), zinc (Zn), copper (Cu), boron (B), and molybdenum (Mo), are used in the smallest amounts by the plants [1,17,30,37,38,40–42]. All these nutrient elements (primary and secondary macronutrients as well as micronutrients) are important for turfgrass growth and development and called plant/grass essential nutrient elements.

12.3 MINERAL NUTRIENTS AND THEIR IMPACT ON TURFGRASS GROWTH AND DEVELOPMENT

12.3.1 MACRONUTRIENTS

12.3.1.1 Nitrogen

Nitrogen deficiency is the most common nutritional problem in almost all turfgrass species [5,7–9,12–16,19–21,24,25,27,28,31]. This is due to high nitrogen requirement in most turfgrasses, low levels of nitrogen in some soils, and the critical impact of nitrogen on qualitative characteristics of turfgrass [7,12,15,20,21]. General yellowing or chlorosis of leaves, especially in the older ones, is the most evident symptom of nitrogen deficiency in turfgrass, which will then be followed by tip dieback, reduction in shoot density, and tillering [7–9,15,20,21,27,28,31]. If nitrogen deficiency is continued, older leaves will turn yellow, becoming darker yellow-brown until they die [15,20]. Nitrogen-deficient turf usually becomes less dense, encouraging weed infestation [20] and several diseases (e.g., dollar spot or red thread commonly occur in turf that is nitrogen deficient [8,9,11,13]). Finally, nitrogen-deficient turf grows slowly, producing fewer leaves and tillers [15].

The amount of nitrogen required depends on the turfgrass species and cultivar [7,9,12,15]. Application of nitrogen fertilizer to turfgrass first results in an increase in root and shoot growth and rhizome production. Excessive application of nitrogen may cause reduction of root growth due to the excess growth of shoot and reduction in carbohydrate levels, which in turn can influence the qualitative characteristics of turfgrass [12,16,20]. Optimum application of nitrogen enhances wear tolerance of turfgrass, but excess may result in reduced wear tolerance [15]. Almost all species of turfgrass can tolerate stressful conditions caused by scalping, drought, and herbicides by optimum application of nitrogen fertilizers [5,14,18,22]. However, it has been reported that excessive levels of nitrogen can reduce turf tolerance to heat, drought, and cold [18,22].

Many biotic problems caused by plant pests have also been reported to be related to nitrogen levels in turfgrass [8,9,11,13,19]. Application at required levels can increase turfgrass stand and density, which reduces weed infestation [3], but excessive levels of nitrogen increases weed populations [20]. Nitrogen deficiency can also increase turfgrass susceptibility to plant pathogenic agents. Several fungal diseases including dollar spot, red thread, take-all patch, and rusts have been reported to be more prevalent in nitrogen-deficient turfgrasses [6,19]. Excessive application of nitrogen fertilizers on turfgrass may also increase the incidence of some diseases including *Rhizoctonia* brown patch, *Drechslera* leaf spot, and anthracnose [8,19].

12.3.1.2 Phosphorus

Phosphorus (P) is another mineral element that is considered one of the essential and most important nutrients for every plant including turfgrass [1,4,6,16,27]. The first indication of low phosphorus concentration in turfgrass usually appears on older leaves as a dark-green color, which then turns purplish to reddish purple [27]. Phosphorus deficiency also causes wilted stand in turfgrass, which is very similar to that of drought stress [16]. The quality and performance of turfgrass can be seriously affected by phosphorus limitation, especially during the establishment phase. This is particularly important for some warm-season turfgrass varieties [27]. Phosphorus deficiency can also affect cool-season grasses during establishment and reduce significantly the growth and stand of seedlings [27]. Reduction in root growth and chlorophyll content is another symptom of phosphorus deficiency in turfgrass.

Physiologically, phosphorus is involved in holding and transferring the energy required by turfgrass plants for metabolic processes [27]. However, it is comprised in only a small portion of dried turf tissues, and the greatest growth response to phosphorus is usually observed with new turfgrass seedlings. Phosphorus deficiency is rarely observed in established turf, unless the phosphorus level in the soil is extremely low or an unfavorable soil pH exists [16,27].

Research has indicated that the phosphorus requirement varies among different species and varieties [16]. For example, among Kentucky bluegrass, perennial ryegrass, and Chewing fescue, it was found that Kentucky bluegrass was more sensitive to seedbed application of phosphorus [16,27]. Established turfgrasses are less sensitive to phosphorus deficiency compared with those in the establishing stage. However, in very sandy soil such as the putting green, high phosphorus concentration may affect and enhance root or top growth of established turfgrass [16,27]. Application of phosphorus fertilizers in soils with low concentrations of phosphorus results in a light-green appearance of both cool- and warm-season turfgrasses, including Kentucky bluegrass, perennial ryegrass, and creeping red fescue [16,27]. The qualitative characteristics of these grasses were not, however, significantly affected.

It has been reported that the environmental-stress tolerance of turfgrass, such as drought, heat, and cold tolerance, is not greatly affected by low concentrations of phosphorus [16]. However, research has shown that cold tolerance of bermudagrass and drought tolerance of Kentucky bluegrass can be increased by application of phosphorus fertilizers [27]. Incidence of some turfgrass diseases has been shown to be affected by essential nutrients, including phosphorus [10]. According to the results of research studies, fungal diseases such as take-all patch, Fusarium patch, red thread, and stripe smut have decreased on application of phosphorus fertilizers to the soil with adequate amounts of nitrogen and potassium nutrients [10].

Application of phosphorus fertilizers may also have a significant impact on the relative proportion of different species in a turfgrass mixture [27]. The proportion of annual bluegrass in creeping bentgrass increased with higher concentrations of phosphorus nutrients in the soil. These findings suggest that annual bluegrass requires more phosphorus fertilizers compared with other species [27].

12.3.1.3 Potassium

Like nitrogen and phosphorus, potassium is another mineral that is an essential nutrient element for turfgrass growth and health [1,32–35]. Potassium plays a vital role in healthy turfgrass growth and development and is second to nitrogen in the amounts required for turf growth. Physiologically, potassium is involved in cellular metabolism, environmental stress resistance, disease resistance, internal water management, and wear tolerance [32–35]. As with phosphorus, potassium applications should be based on soil tests [32]. The principal factors affecting the potassium requirement for turf are clipping removal, irrigation, and soil texture [33,34]. If the clippings are removed, larger and more frequent applications of potassium are generally required to maintain satisfactory growth. The specific requirement is usually about half the rate at which nitrogen is applied [34,35].

Symptoms of low potassium concentration in turfgrass usually appear first on leaves, as yellowing of older ones, necrosis along leaf margin, and tip dieback [33,34]. A good indicator of potassium deficiency in turfgrass is early spring chlorosis [33,34]. Burning of turf due to the creation and production of high levels of soluble salts may occur as a result of excessive applications of potassium [34]. Research has also shown that under high concentrations of nitrogen, excessive application of potassium may cause growth reduction in some species of turfgrasses [33,34]. This could be due to manganese deficiency induced by the excessive use of nitrogen and potassium nutrients [33].

Establishment of turfgrass is not very dependent on potassium fertilizers, although it has been reported that top growth, stolon growth, and spread of some species of turfgrasses have been enhanced by the addition of potassium [34]. However, research has shown that different turfgrass species vary in their response to potassium application during the establishment stage [34]. Initiation of new rhizomes and elongation of existing rhizomes of some species of turfgrasses may also be affected by potassium deficiency [33,34].

Tolerance to environmental stresses, including drought, heat, and cold, has been shown to be significantly correlated to potassium [33–35]. Several research experiments have been conducted in this regard and results have indicated that winter injury of turfgrass decreases after application of potassium fertilizers [33–35]. It has also been found that hardiness of centipedegrass and bermudagrass is enhanced by application of potassium [35]. Application of potassium fertilizers has also resulted in an enhanced tolerance of Kentucky bluegrass and bentgrass to drought and heat. Potassium may also play an important role in the tolerance of turfgrass to diseases by enhancing plant tissues growth [34,35]. Pathogenic agents causing wilt diseases may be less active in soils with high concentrations of potassium [35].

12.3.2 SECONDARY MINERALS

12.3.2.1 Calcium

Calcium (Ca) is an important element that is very beneficial for turfgrass health and growth [1,2,23]. Leaves, particularly younger ones, are the first part of turfgrass to be affected by calcium deficiency [1]. A reddish brown color appears along with leaf margins in calcium-deficient leaves. Research studies have shown that different species and cultivars respond variously to calcium limitation [2]. It has been reported that some diseases of turfgrass may be influenced by low calcium levels [43–48]. For example, there was higher incidence of pythium blight diseases in bentgrass low in calcium [48]. Low concentrations of calcium in soil can cause several problems including higher availability and toxicity of nonessential elements and increased or decreased concentration of essential nutrients, which result in lower populations of soil microorganisms and increased susceptibility of the turfgrass to environmental stresses and disease [48]. To solve these problems, especially in acidic soil, it is helpful to add a liming agent to the soil, which increases the calcium availability to turfgrass with little impact on soil microbial populations [1].

12.3.2.2 Magnesium

Magnesium deficiency in turfgrass is not very common. Results of a few studies that have been conducted on the effect of magnesium on turfgrass growth and health indicate that low magnesium concentration in the soil affects older leaves and causes a red color along their margins [1]. Research has shown that different species and varieties of turfgrass respond variously to magnesium deficiency. It was found that soil application of magnesium fertilizers had no significant beneficial effects on bermudagrass and perennial ryegrass [17]. However, lack of magnesium in the soil may have significant impacts on warm-season turfgrass species such as zoysiagrass [1].

12.3.2.3 Sulfur

Older leaves are the first part of turfgrass to be affected by sulfur deficiency [1,36]. Chlorosis and yellowing are typical symptoms induced by sulfur deficiency. These symptoms are very similar to those caused by nitrogen deficiency, except that the midvein may remain green in sulfurdeficient leaves [36]. Beneficial effects of sulfur application on turfgrass growth and quality have been studied [36]. It has been found that bentgrass growth increased significantly after application of sulfur to the soil that contained a high concentration of nitrogen fertilizers [36]. Results of studies have also shown that application of sulfur may reduce the incidence of some turfgrass fungal diseases [36,43,46,48–51].

12.3.3 MICRONUTRIENTS

12.3.3.1 Iron

Several conditions and factors may result in iron deficiencies in turfgrass soil and affect the growth and health of the plant [37,38,40,41,42]. For example, high concentrations of phosphorus and nitrogen in the soil and high soil pH can cause iron deficiency in turfgrass [38]. Iron-deficiency symptoms may also occur in sandy, cold, and wet soils. The most prevalent and important symptoms of iron deficiency in turfgrass include chlorosis of the interveinal area in younger leaves, which in severe cases may result in an almost white appearance of the leaves [40,41]. Upright growth in iron-deficient turfgrass may also occur due to reduced plant vigor. Color enhancement in iron-deficient turfgrass may happen by fertilizer application [40]. Foliar application of iron-containing compounds was found to significantly improve the green color of leaves in Kentucky bluegrass within a few hours [40]. In high-pH soils, reduction of pH has been found to significantly affect and enhance leaf color in some turfgrass species [42]. However, environmental factors and the species and cultivar may play important roles in the response of turfgrass to iron fertilizers [42]. In addition to leaf chlorosis, low concentration of iron in the soil may affect other growth characteristics of turfgrass, including root growth and clipping weight [40,41]. Results of research studies have indicated that application of iron fertilizers with nitrogen led to a significant increase in root weight and early spring growth of creeping bentgrass [38]. In addition, iron application enabled recovery from cold injury in creeping bentgrass.

Excessive application of iron fertilizers has been shown to cause some disorders in turfgrass, especially under stressful environmental conditions [38,40]. Symptoms of these include blackening of leaf tissues and discoloration, burning, and inhibition of rhizome formation [40]. Foliar dieback has also been observed under excessive iron application for some species of turfgrasses. The role of environmental factors such as air temperature in turfgrass response to iron-containing compounds, particularly in the case of foliar application, has also been emphasized [40].

12.3.3.2   Zinc

Low concentrations of zinc nutrients in soil may cause some typical symptoms on turfgrass, including stunted leaves, puckered leaf margins, and chlorosis [1,37]. Different species and varieties of turfgrass respond variously to zinc deficiency and excess. It has been reported that in creeping bentgrass and Kentucky bluegrass, excess zinc results in stimulation of root growth but inhibition of rhizome growth [1,37]. However, most varieties of turfgrass are not very sensitive to low or high concentrations of zinc in the soil.

12.3.3.3   Manganese

Manganese deficiency is not common in turfgrass, and manganese-deficient turfgrasses have rarely been observed or reported. However, if it occurs, it usually results in chlorosis of younger leaves, particularly in the interveinal area [37]. In the affected leaves, tissue necrosis may also develop later. Low absorption of manganese sometimes occurs under excessive application of nitrogen and potassium fertilizers, which may result in decreased growth in some turfgrasses, including bentgrass [37]. Environmental factors such as soil pH have been found to affect manganese application in turfgrass.

12.3.3.4   Copper

Like many other nutrients, copper deficiency symptoms also appear in younger turfgrass leaves and include tip dieback, on which the leaves become white [37]. Although very little research has been conducted on the impacts of copper fertilizers on turfgrass, it has been shown that these affect the growth of some turfgrass species [37]. It has also been shown that excessive application of copper fertilizers may cause root injury in some turfgrass species. The results of the above-mentioned studies indicate that different species and cultivars of turfgrass respond variously to copper deficiency and toxicity [37].

12.3.3.5   Boron

Reduction in overall growth and stunting are some common symptoms associated with boron deficiency in turfgrass [37]. Research studies have indicated that in soils with low fertility, application of boron fertilizers may result in rapid greening and enhancement of root growth in some turfgrass species, including Kentucky bluegrass [37]. High levels of boron may result in some nutritional disorders such as pale greening and leaf-tip death. Since boron will accumulate in leaf tips upon excessive application, clipping will remove excess boron from the turfgrass plant [37].

12.3.3.6   Molybdenum

Low concentrations of molybdenum (Mo) in turfgrass usually result in symptoms very similar to those of nitrogen deficiency. Older leaves of the plant show chlorosis, followed by tip dieback [1]. Interveinal chlorosis is a characteristic symptom of molybdenum deficiency in most turfgrasses.

12.4   FERTILIZATION PROGRAMS IN TURFGRASS

A successful turfgrass fertilization program depends highly on nitrogen utilization [5,9,12–15, 19,52–57]. The quantity and application schedule are very important factors in nitrogen application [9,12,13,54,55]. Nitrogen is used by turfgrass in large quantities, and due to its mobility in the soil, it should be applied to most turfs several times per year [55–57]. For cool-season turfgrasses such as creeping bentgrass, perennial ryegrass, and Kentucky bluegrass, the active growing time is from mid-spring to early summer and from late summer through mid-fall [55].

By following the schedule of three or four applications per year, turf of moderate to high quality can be maintained. For low- to medium-quality lawns, mineral nutrients should be applied one or two times per year [55]. For cool-season turfgrasses, late summer or early fall is the best period for nitrogen application. This helps turf recover from summer stresses and prepare for winter [55].

Water-soluble nitrogen sources can be used in moderate amounts during the mid-spring period after the early flush of growth and during the late summer or early fall [14,15,19,55]. This early summer application might consist of a combination of nitrogen sources: half as soluble nitrogen to provide a readily available source of nitrogen and half as slowly soluble nitrogen to provide a nitrogen carryover through the summer months [55]. Late-season fertilization can supply adequate fertility for early spring growth. Fertilizer applied at this time can enhance turf root growth, provide early spring green-up without a large flush of growth, and supply enhanced winter color [55]. Quick-release nitrogen fertilizers such as urea, ammonium nitrate, ammonium sulfate, ammonium phosphates, or slow-release nitrogen sources should be used in an adequate fertilization program [55].

Research studies have indicated that high application rates of nitrogen in early spring may increase the incidence of leaf spot diseases [57]. To sustain healthy and vigorous turf, two applications of fertilizer per year will usually be enough. In most cases, turf fertilization should not consist solely of nitrogen application [55,57]. Soil tests are required to determine the need for other mineral fertilizers, especially potassium and phosphorus [57].

12.5   ENVIRONMENTAL CONDITIONS AND FERTILIZATION MANAGEMENT

It is obvious that in the turf-cropping system, each growing season is different from the other. Fertilization practices should therefore be selected according to environmental conditions [55,58]. Certain conditions, such as extremely cool and wet or hot and dry weather, require different fertilization practices than more normal conditions. In cool and wet conditions that are favorable for cool-season turfgrass growth and nitrogen leaching, the plant may require additional fertilizer application [55]. In contrast, in hot and dry weather that is unfavorable for cool-season turf growth, usually less fertilizer should be applied [55].

In addition, variation in soil or in the amount of light and shade can also affect fertilization programs. For example, in sandy soils, nitrogen applications are usually greater than in more finely textured or highly organic soils because nitrogen leaching usually occurs in sandy soils [55]. Under shady conditions, lower amounts of fertilizers should be used since the nitrogen requirement of the grasses in shade is lower compared with those in full sunlight [55].

The mowing and irrigation system should also be considered for developing a fertilization program [55]. Less mineral nutrients are required where clippings are returned to the turf than where clippings are not returned, because clipping breakdown can supply the additional minerals to the turf. Heavily irrigated turfgrasses will generally require additional fertilizer due to leaching and increased turf-growth rate [55]. Other management practices may also affect the fertilization programs in turfgrass and should be taken into consideration.

12.6   SUMMARY

The health and productivity of plants depend on a number of biotic and abiotic factors, including pest presence and population, genetic characteristics, and agronomic practices. A balanced nutrition and fertilization program plays a very important role in a sustainable agricultural system for any given crop, including turfgrass. Several groups of mineral nutrients are required for turfgrass growth and development. Some of them, like oxygen, carbon, and hydrogen, can be taken from atmospheric carbon dioxide [1]. Other nutrients should be applied to the soil or foliage to be available and utilized by turfgrass.

Essential minerals or macronutrients are quantitatively the most important and primary elements necessary for turf growth and development, and include nitrogen, phosphorus, and potassium [1–35]. Availability of these minerals to the turfgrass can result in balanced growth and development and will increase and enhance plant vigor and health. However, deficiency in macronutrients may cause critical disorders and lead to poor and weak growth [5–20]. Insufficient application of macronutrients may also reduce the tolerance of the turfgrass to environmental stresses and increase its susceptibility to pests and pathogens [21–35].

Nitrogen (N) is the most important nutrient mineral required for turfgrass growth and development [1–28]. Lack of sufficient amounts of nitrogen is the most common nutrient deficiency problem in almost all species and varieties of turfgrasses. This is because of the shortage of nitrogen in most turf-growing soils [21–28]. Insufficient application of nitrogen can adversely affect both quantitative and qualitative characteristics of turfgrass and cause significant reduction in plant vigor and health factors [21–28]. Root and shoot growth, leaf chlorosis, and tip dieback are some common symptoms of nitrogen deficiency in most species of turfgrass [21–28]. Weed infestation and susceptibility to fungal diseases, including red thread, rust, and take-all patch, may also increase in nitrogen-deficient turfgrasses [21–28]. Decrease in the tolerance of nitrogen-deficient turfgrasses to environmental stresses such as heat, cold, and drought has also been reported [25–28]. Like nitrogen deficiency, excessive application of nitrogen fertilizers may also cause some disorders in turfgrass, including increase of weed infestation, increase in susceptibility to some diseases, and reduction in turfgrass tolerance to unfavorable environmental conditions [25–28].

Phosphorus (P) is the second essential nutrient element required for turfgrass growth and development [1,4,6,16,22,27]. It is involved in the energy transfer in the plant system and plays a very important role in metabolism [22,27]. Phosphorus deficiency first appears in older leaves as a dark-green color, which then turns purple and, finally, the leaf dies [16,22,27]. Stand wilting is the most important disorder caused by phosphorus deficiency and resembles drought-induced wilting [16]. Susceptibility to some fungal diseases, including Fusarium patch, take-all patch, red thread, and strip smut, has also been reported to be increased in phosphorus deficiency [27]. The response of different species and varieties of turfgrasses to phosphorus deficiency is not the same [16,22,27]. Some turfgrasses, such as Kentucky bluegrass, are more sensitive to phosphorus. Research has also shown that seedlings and turfgrasses in the establishing stage are more sensitive to phosphorus than established turf [27]. However, it has also been found that excessive application of phosphorus fertilizers may decrease the tolerance of turfgrasses to environmental stress, including drought stress [22,27].

The third essential element, potassium (K), is another macronutrient required for turfgrass growth and performance [1,32–35]. Potassium plays a critical role in water transport, metabolic activity, environmental-stress tolerance, and disease tolerance in almost all species of turfgrasses. The first and primary symptom of potassium deficiency is chlorosis of older leaves, which then become necrotic, and ends with tip dieback [33–35]. Early spring chlorosis is the most evident indicator of potassium deficiency in most turfgrass species [34]. Potassium fertilization is very important in some growth characteristics of turfgrass, including new rhizome initiation and elongation of existing rhizomes. Results of research studies have shown that potassium has significant impacts on the tolerance of turfgrass to environmental stresses such as cold, drought, and heat [32–35]. Application of potassium fertilizer has also been shown to increase turfgrass tolerance to several fungal diseases. However, excessive application of potassium, especially when nitrogen concentration is high, may cause a reduction in growth and development of some species of turfgrasses [32–35].

Although quantitatively not as important as essential nutrients, secondary elements, including calcium, magnesium, and sulfur, are also required for turfgrass vigor and performance [1,2,17,23,36]. Calcium is the most important element among secondary nutrients and can significantly affect turfgrass growth and health. A low concentration of calcium can cause redness of younger leaves and reduce their growth and vigor [1]. Calcium deficiency results in higher availability of nonessential elements and causes toxicity due to their excessive uptake by the plant. Low concentration of calcium fertilizers may also increase or decrease the availability of essential nutrients, which may affect turfgrass growth and decrease its tolerance to environmental stress and disease [1].

In addition to calcium, magnesium is another secondary nutrient mineral required for healthy growth and development of some turfgrass species [1]. Very few studies have been conducted on the effect of magnesium on different turfgrasses and the results of these have indicated that some species, such as perennial ryegrass and bermudagrass, are not very sensitive to magnesium application. In contrast, some other species, including zoysiagrass, were shown to be affected by magnesium, and their growth and development increased after application of magnesium fertilizers [1]. Sulfur is also important for turfgrass health [1,36]. Deficiency of sulfur compounds in the soil may cause yellowing and chlorosis of the older leaves [36]. Research has shown that application of sulfur to soil containing a high concentration of nitrogen carriers can significantly enhance the quantitative and qualitative growth of some turfgrass species including bentgrass [36]. It has also been found that sulfur application may increase the susceptibility of turfgrass to some fungal diseases, including Fusarium patch and take-all patch [36,43,46,48–51].

Micronutrients (iron, manganese, copper, zinc, boron, and molybdenum) are some other mineral nutrients that are important in very small quantities to turfgrass growth and development [1,37,38,40–42]. Among these, iron is perhaps the most required element, which can significantly affect growth characteristics of turfgrass. Iron deficiency may cause leaf discoloration, reduction in root growth, and reduction in clipping weight [37,38,40–42]. The application of iron fertilizers has been found to increase the growth of most turfgrass species and enhance the tolerance of the plant to environmental stress [37,38,40]. However, excessive application of iron fertilizers may also cause some disorders in turfgrass, including burning and inhibition of rhizome formation and foliar dieback [41,42]. Other micronutrients (zinc, copper, manganese, boron, and molybdenum) do not play very important roles in turfgrass growth and development [1,37]. However, their low concentration, especially in soils with unbalanced concentrations of essential nutrients, may cause some disorders, including leaf discoloration, growth reduction, and root injury [1,37]. Their limitation may also decrease turfgrass tolerance to environmental stress and increase its susceptibility to some fungal diseases [1,37].

Application of fertilizers to turfgrass requires a proper knowledge of plant nutrition and agronomic practices. Essential nutrients, especially nitrogen, play a critical role in the turfgrass fertilization program. In nitrogen application, timing and the amount of fertilizer are the most important factors and should be carefully considered [5,9,12–15,19,52–57]. For most turfgrasses, nitrogen should be applied several times per year [55]. Species and varieties of turfgrasses determine the type and time of application [52–55]. There is a great difference in the amount of nitrogen fertilizers required and their application time among warm- and cool-season turfgrasses [55]. The type of nitrogen fertilizer should be chosen carefully. Both quick- and slow-release nitrogen carriers should be used when necessary [55]. In addition to nitrogen, other mineral nutrients, especially potassium and phosphorus, should be applied to the soil to maintain a healthy and vigorous turfgrasses [55–57].

Environmental factors such as air temperature, soil type and temperature, and soil organic matter content are very important and can greatly influence the efficacy of the turfgrass fertilization program [55–58]. Studies have indicated that the amount and type of nutrients, especially nitrogen, are highly dependent on environmental conditions [55–58]. Therefore, for the establishment of a healthy and vigorous turfgrass, it is necessary to arrange a good fertilization program based on adequate soil tests and consideration of environmental and agronomic factors.

REFERENCES

1. Fry, J.O. and Dernoeden, P.H., Growth of zoysiagrass from vegetative plugs in response to fertilizers, *J. Am. Soc. Hortic. Sci.*, 112, 286, 1987.
2. Gilbert, W.B. and Davis, D.L., Influence of fertility ratios on winter-hardiness of bermudagrass, *Agron. J.*, 63, 591, 1971.
3. Johnson, B.J. and Bowyer, T.H., Management of herbicides and fertility levels on weeds and Kentucky bluegrass turf, *Agron. J.*, 74, 845, 1982.

4. Mehall, B.J., Hull, R.J., and Skogley, C.R., Cultivar variation in Kentucky bluegrass: P and K nutritional factors, *Agron. J.*, 75, 767, 1983.

5. Carroll, J.C. and Welton, F.A., Effects of heavy and late application of nitrogenous fertilizers on the cold resistance of Kentucky bluegrass, *Plant Physiol.*, 14, 297, 1939.

6. Cahill, J.V., Murray, J.J., O'Neill, N.R., and Dernoeden, P.H., Interrelationship between fertility and red thread fungal disease of turfgrass, *Plant Dis.*, 67, 1080, 1983.

7. Carrow, R.N., Johnson, B.J., and Landry, Jr., G.W., Centipedegrass response to foliar application of iron and nitrogen, *Agron. J.*, 80, 746, 1988.

8. Danneberger, T.K.,Vargas, J.M., Rieke, P.E., and Street, J.R., Anthracnose development on annual bluegrass in response to nitrogen carriers and fungicide application, *Agron. J.*, 75, 35, 1983.

9. Dernoeden, P.H., Management of take-all patch of creeping bentgrass with nitrogen, sulfur and phenyl mercury acetate, *Plant Dis.*, 71, 226, 1987.

10. Dernoeden, P.H. and Jackson, N., Managing yellow turf disease, *J. Sports Turf Res. Inst.*, 56, 9, 1980.

11. Goss, R.L. and Gould, C.J., Some interrelationship between fertility levels and Fusarium patch disease in turfgrasses, *J. Sports Turf Res. Inst.*, 44, 19, 1968.

12. Horst, G.L., Baltensperger, A.A., and Firkner, M.D., Effects of N and growing season on root-rhizome characteristics of turf-type bermudagrasses, *Agron. J.*, 77, 327, 1985.

13. Hull, R.J., Jackson, N., and Skogley, C.R., Influence of nitrogen on strip smut severity in Kentucky bluegrass turf, *Agron. J.*, 71, 553, 1979.

14. Johnson, B.J., Influence of nitrogen on recovery of bermudagrass (*Cynodon dactylon*) treated with herbicides, *Weed Sci.*, 32, 819, 1984.

15. Kohlmeier, G.P. and Eggens, J.L., The influence of wear and nitrogen on creeping bentgrass growth, *Can. J. Plant Sci.*, 63, 189, 1980.

16. Christians, N.E., Martin, D.P., and Wilkinson, J.F., Nitrogen, phosphorous and potassium effects on quality and growth of Kentucky bluegrass and creeping bentgrass, *Agron. J.*, 71, 564, 1979.

17. Butler, J.D. and Hodges, T.K., Mineral compositions of turfgrasses, *HortScience*, 2, 26, 1967.

18. Alexander, P.M. and Gilbert, W.B., Winter damage to bermuda greens, *Golf Course Rep.*, 31, 50, 1963.

19. Cock, R.N., Engel, R.E., and Bachelder, S., A study of the effect of nitrogen carriers on turfgrass disease, *Plant Dis. Rep.*, 48, 254, 1964.

20. Murray, J.J., Klingman, D.L., Nash, R.G., and Woolson, E.A., Eight years of herbicide and nitrogen fertilizer treatment on Kentucky bluegrass (*Poa pratensis*) turf, *Weed Sci.*, 31, 825, 1983.

21. Watschke, T.L. and Waddington, D.V., Effect of nitrogen source, rate and timing on growth and carbohydrates of Merion Kentucky bluegrass, *Agron. J.*, 66, 691, 1974.

22. Pellet, R.M. and Roberts, E.C., Effects of mineral nutrition on high temperature induced growth retardation of Kentucky bluegrass, *Agron. J.*, 55, 474, 1963.

23. Adams, W.E. and Twersky, M., Effect of soil fertility on winter killing of coastal bermudagrass, *Agron. J.*, 52, 325, 1960.

24. Schrader, L.E., Function and transformation of nitrogen in higher plants. In: *Nitrogen in Crop Production*, 1st ed., American Society of Agronomy, Madison, WI, 1984, 125.

25. Blue, W. G., Forage production and N nitrogen contents and soil changes during 25 years of continuous white clover-Pensacola bahiagrass Paspalum notatum growth on a Florida Spodosol Limed treatments, *Agron. J.*, 71, 795, 1979.

26. Busey, P., Bermudagrass germplasm adaptation to natural pest infestation and suboptimal nitrogen fertilization, *J. Am. Soc. Hortic. Sci.*, 111, 630, 1986.

27. Cisar, J.L., Snyder, G.H., and Swanson, G.S., Nitrogen, phosphorus, and potassium fertilization for Histosol-grown St. Augustinegrass sod, *Agron. J.*, 84, 474, 1992.

28. Pate, F.M. and Snyder, G.H., Effect of water table and nitrogen fertilization on tropical grasses grown on organic soil, *Trop. Grasslands*, 18, 74, 1984.

29. Peacock, C.H. and Dudeck, A.E., A comparison of sod type and fertilization during turf establishment, *HortScience*, 20, 189, 1985.

30. Peacock, C.H., Dudeck, A.E., and Wildmon, J.C., Growth and mineral content of St. Augustinegrass cultivars in response to salinity, *J. Am. Soc. Hortic. Sci.*, 118, 464, 1993.

31. Snyder, G.H. and Burt, E.O., Nitrogen fertilization of bermudagrass turf through an irrigation system, *J. Am. Soc. Hortic. Sci.*, 101, 145, 1976.

32. Keisling, T.C., Bermudagrass rhizome initiation and longevity under differing potassium nutritional levels, *Commun. Soil Sci. Plant Anal.*, 11, 629, 1980.

33. Miller, G.L. and Dickens, R., Bermudagrass carbohydrate levels as influenced by potassium fertilization and cultivar, *Crop Sci.*, 36, 1283, 1996.

34. Miller, G.L. and Dickens, R., Potassium fertilization related to freezing resistance in bermudagrass, *Crop Sci.*, 36, 1290, 1996.

35. Snyder, G.H. and Cisar, J.L., Controlled-release potassium fertilizers for turfgrass, *J. Am. Soc. Hortic. Sci.*, 117, 411, 1992.

36. Goss, R.L., Brauen, S.E., and Orton, S.P., Uptake of sulfur by bentgrass putting green turf, *Agron. J.*, 71, 909, 1979.

37. Broyer, T.C., Carlton, A.B., Johnson, C.M., and Stout, P.R., Micronutrient elements for higher plants, *Plant Physiol.*, 29, 526, 1954.

38. Brown, J.C., Summary of symposium on iron nutrition and interactions in plants, *J. Plant Nutr.*, 5, 987, 1982.

39. Sachs, P.D., *Handbook of Successful Ecological Lawn Care*, The Edaphic Press, Newbury, Vermont, 1996, 45.

40. Harivandi, M.A. and Butler, J.D., Iron chlorosis of Kentucky bluegrass cultivars, *HortScience*, 15, 496, 1980.

41. Horst, G.L., Iron nutrition for warm season grasses, *Ground Maintenance*, 44, 126, 1984.

42. McCaslin, B.D., Samson, R.F., and Baltensperger, A.A., Selection for turf-type bermudagrass genotypes with reduced iron chlorosis, Commun. *Soil Sci. Plant Anal.*, 12, 189, 1981.

43. Smiley, R.W., Dernoeden, P.H., and Clarke, B.B., *Compendium of Turfgrass Diseases*, 2nd ed., American Phytopathological Society, St. Paul, MN, 1992, 15.

44. Lo, C.T., Nelson, E.B., and Harman, G.E., Biological control of turfgrass diseases with a rhizosphere component strain of Trichoderma harzianum, *Plant Dis.*, 80, 736, 1996.

45. Lo, C.T., Nelson, E.B., and Harman, G.E., Improved biocontrol efficacy of Trichoderma harzianum for foliar phases of turf diseases by use of spray applications, *Plant Dis.*, 81, 1132, 1997.

46. Clarke, B.B. and Gould, A.B., *Turfgrass Patch Diseases Caused by Ectotrophic Root-Infecting Fungi*, APS Press, St. Paul, MN, 1993, 20.

47. Couch, H.B., *The Turfgrass Disease Handbook*, Krieger Publishing Co., Melbourne, FL, 2000, 30.

48. Allen, T.W., Martinez, A., and Burpee, L.L., Pythium blight of turfgrass, *Plant Health Instructor*, 10, 1094, 2004.

49. Allen, T.W., Martinez, A., and Burpee, L.L., Dollar spot of turfgrass, *Plant Health Instructor*, 10, 1094, 2005.

50. Bockus, W.W. and Tisserat, N.A., Take-all root rot, *Plant Health Instructor*, 10, 1094, 2001.

51. Tredway, L.P. and Burpee, L.L., Rhizoctonia diseases of turfgrass, *Plant Health Instructor*, 10, 1094, 2001.

52. Beard, J.B. and Green, R.L., The role of turfgrass in environmental protection and its benefits to humans, *J. Environ. Qual.*, 23, 452, 1994.

53. Milesi, C., Running, S.W., Elvidge, C.D., Dietz, J.B., Tuttle, B.T., and Nemani, R.R., Mapping and modeling the biogeochemical cycling of turfgrasses in the United States, *Environ. Manage.*, 36, 426, 2005.

54. Baltensperger, A.A., Schank, S.C., Smith, R.L., Littell, R.C., Bouton, J.H., and Dudeck, A.E., Effect of inoculation with Azospirillum and Azotobacter nitrogen fixing bacteria on turf-type bermuda genotypes, *Crop Sci.*, 18, 1043, 1978.

55. Busey, P., Seedling growth, fertilization timing, and establishment of bahiagrass, *Crop Sci.*, 32, 1099, 1992.

56. Busey, P. and Snyder, G.H., Population outbreak of the southern chinch bug is regulated by fertilization. *Int. Turfgrass Soc. Res. J.*, 7, 357, 1993.

57. McCoy, R.E., Relation of fertility level and fungicide application to incidence of Cercospora fusimaculans on St Augustinegrass, *Plant Dis. Rep.*, 57, 33, 1973.

58. Heydari, A. and Misaghi, I.J., Herbicide-mediated changes in the populations and activity of root associated microorganisms: a potential cause of plant stress. In: *HandBook of Plant and Crop Stress*, 2nd ed., Marcel Dekker, New York, 1999.

13 Biological Control of Turfgrass Fungal Diseases

Asghar Heydari

CONTENTS

13.1 INTRODUCTION

Plant diseases are mostly controlled by chemical pesticides and in some cases by cultural practices. However, the widespread use of chemicals in agriculture has been a subject of public concern and scrutiny due to the potential harmful effects on the environment, their undesirable effects on non-target organisms, and possible carcinogenicity of some chemicals [1]. Other problems include development of resistant races of pathogens, a gradual elimination and phasing out of some available pesticides, and the reluctance of some chemical companies to develop and test new chemicals due to the problems with registration process and cost [1]. The need for the development of nonchemical alternative methods to control plant diseases is therefore clear [1].

Biological control of plant diseases has been considered a viable alternative method to manage plant diseases [1–45]. Biological control is the inhibition of growth, infection, or reproduction of one organism using another organism [1]. Biocontrol is environmentally safe and in some cases is the only option available to protect plants against pathogens [1–10]. Biological control of disease employs natural enemies of pests or pathogens to eradicate or control their population. This can involve the introduction of exotic species, or it can be a matter of harnessing whatever form of biological control exists naturally in the ecosystem [1]. The induction of plant resistance using non-pathogenic or incompatible microorganisms is also a form of biological control [1]. Some diseases that can be successfully controlled using biological agents are pathogens of pruning wounds and other cut surfaces, crown gall, diseases of leaves and flowers such as powdery mildew, diseases of fruits and vegetables such as *Botrytis*, and fungal pathogens in the soil [1]. In this chapter, first disease biocontrol agents (antagonistic bacteria and fungi), their application strategies, their mode of action (mechanisms), and commercialization of biocontrol is reviewed; then biological control of turfgrass fungal diseases is discussed.

13.2   MODE OF ACTION OF BIOCONTROL AGENTS

The most common modes of actions for microbial antagonism of plant pathogens are competition, parasitism, cross-protection, and induced resistance, which are discussed below [1–6].

13.2.1   COMPETITION

Competition exists between organisms that require the same resource for growth and survival [1]. Use of the resource by one organism reduces its availability for the other organism. Competition for site or nutrients usually takes place between closely related species; therefore, it can be effective to treat plants or seeds with a nonpathogenic strain of a related species that can compete with the pathogenic organism [1–4]. In some cases, the treating species need not be closely related to the pathogen, as long as it uses the same resources. For example, bacteria and yeast can reduce fungal spore germination by competing with the spores for nutrients on the surface of leaves [1,13]. While microorganisms can produce secondary metabolites that have antimicrobial properties when grown in culture, these chemicals are rarely detected in natural environments [2–4,17,28–30]. Therefore, antibiotics would need to be produced in culture and then applied. However, antibiotics are easily lost to the atmosphere and are commonly broken down by organisms that are insensitive to them, and so they are not ideal biological agents against plant pathogens [1–3].

13.2.2   PARASITISM

Parasitism of one fungus by another (hyperparasitism or mycoparasitism) is well documented and is affected by environmental factors, including nutrient availability [1,11,15]. Formulations of some parasitic species of fungi are available commercially for the control of fungal plant pathogens in the soil and on the plant surface [1,15]. Parasitic fungi hyphae penetrate their victims, sometimes by secretion and production of wall-degrading enzymes [11,15]. Bacteria on the plant surface and in the soil are also known to parasitize plant pathogens, such as other bacteria and fungal spores [1,2,6,8,12].

13.2.3   CROSS-PROTECTION

An organism present on the plant can protect it from a pathogen that comes into contact with the plant later. This is known as cross-protection. For example, nonpathogenic strains of tobacco mosaic virus (TMV) can protect tomatoes from pathogenic strains of the same virus, which is very similar to immunization in animals [1].

13.2.4 INDUCED RESISTANCE

Induced resistance is a form of cross-protection where the plant is inoculated with inactive pathogens, low doses of pathogens, pathogen-derived chemicals, or nonpathogen species to induce resistance in plants [1–3,11,15,41]. This prepares the plant for an attack by pathogens, and its defense mechanisms are already activated when infection occurs.

13.3 APPLICATION OF BIOCONTROL AGENTS

13.3.1 GENERAL APPLICATION

Successful application of biological control strategies requires more knowledge-intensive management [1,3,6,8,32]. Understanding when and where biological control of plant pathogens can be profitable requires an appreciation of its place within integrated pest management (IPM) systems [1–3,10,12,20–22,24–26,32,37–39].

In general, the foundation of a sound pest and disease management program in an annual cropping system begins with cultural practices that alter the farm landscape to promote crop health [1,25,26,39,40,42]. These include crop rotations that limit the availability of host material used by plant pathogens [1]. Proper use of tillage can disrupt pest and pathogen life cycles, bury weeds, and prepare seedbeds of optimal moisture and bulk density [1,25]. Careful management of soil fertility and moisture can also limit plant diseases by minimizing plant stress [1,42]. In nurseries and greenhouses, environmental control can be more tightly regulated in terms of temperature, light, moisture, and soil composition, but the design of such systems cannot wholly eliminate pest problems [1,5,6,25,26,40].

The second layer of defense against pests consists of the quality of crop germplasm [1,15,25]. Breeding for pathogen resistance contributes substantially to crop success in most regions. Newer technologies that directly incorporate genes into crop genomes, commonly referred to as genetic modification or genetic engineering, are bringing new traits into crop germplasm [18]. The most widely distributed of these plant-incorporated protectants are the different insecticidal proteins derived from *Bacillus thuringiensis* [18]. Other technologies, such as seed washing, testing for pathogens, and treatment, are also used to keep germplasm pathogen-free [14,21]. In perennial cropping systems, such as orchards and forests, germplasm quality may be more important than cultural practices, because rotation and tillage cannot be used as regularly. Upon these two layers, growers can further reduce pest and pathogen pressure by considering both biological and chemical inputs [7,8,10]. Biologically based inputs such as pheromone traps and microbial pesticides can be used to interfere with pest activities. Registered biopesticides are generally labeled with short reentry intervals and preharvest intervals, giving greater flexibility to growers who need to balance their operational requirements and pest management goals [1,14,16,24–26]. When living organisms are introduced, they may also augment natural beneficial populations to further reduce the damage caused by targeted pathogens or pests [1,5,10].

Finally, synthetic chemicals may be added to limit crop damage by pests and pathogens. Because of growing concerns about health and environmental safety, the use of toxic, carcinogenic, and environmentally damaging chemicals is currently being discouraged [1–10]. Within the context of IPM programs, the use of such chemicals should be considered only after other management options have been fully implemented [1,5,25].

13.3.2 CLASSIFICATION OF BIOLOGICAL CONTROL BY APPLICATION STRATEGIES

13.3.2.1 Application Directly to the Infection Court

Application directly to the infection court at a high population level to "swamp" the pathogen (inundate application) includes application of *Agrobacterium radiobacter* K84 to transplant roots [1]; seed application (e.g., *Trichoderma harzianum* [1,11,15]); antagonists applied to fruit for protection in storage (e.g., *Pseudomonas fluorescens* [21]); and application to soil at the site of seed placement [11,34].

13.3.2.2 Application at One Place

Application at one place (each crop year) involves lower populations, which then multiply and spread to other plant parts and give protection (augmentative application) against pathogens. An example of this method is plant-growth-promoting rhizobacteria (PGPR) and atoxigenic *Aspergillus flavus* on wheat seed scattered on the soil to spread to cotton flowers where they displace aflatoxin-producing strains of *A. flavus* and fungal antagonists added to soil [1,19].

13.3.2.3 One-Time or Occasional Application

One-time or occasional application maintains pathogen populations below threshold levels. In theory, parasites of the pathogen, or hypovirulent (disease carrying) strains of the pathogen, might be used and not require yearly repetition (e.g., hypovirulent strains of the chestnut blight pathogen) [1].

13.4 COMMERCIAL USE OF BIOLOGICAL CONTROL

Commercial use and application of biological control has been slow mainly due to variable performance under different environmental conditions in the field [1,5,7,8,11,14]. Many biocontrol agents perform well in the laboratory and greenhouse conditions but fail to do so in the field [5,11]. This problem can only be solved by better understanding the environmental parameters that affect biocontrol agents [1,5,11]. In addition to this problem, there has been relatively little investment in the development and production of commercial formulation of biocontrol-active microorganisms, probably due to the cost of developing, testing, registering, and marketing of these products [1,11,12,27]. Biological control agents are generally formulated as wetable powders, dusts, granules, and aqueous or oil-based liquid products using different mineral and organic carriers [27].

Currently in the market, a number of biologically based products are being sold for the control of plant diseases [27]. A growing number of companies are also developing new products that are in the process of registration. Many of these companies are small, privately owned firms with a limited productline [27]. Others are publicly traded and have substantial capitalization values. In addition, larger companies with more diverse product lines that include a variety of agricultural chemicals and biotechnologies have played a significant role in the development and marketing of products for the control of plant pathogens [27].

Biocontrol products are either marketed as stand-alone products or formulated as mixtures with other microbials [27]. Some products with biocontrol properties may not be registered, but are sold instead as plant strengtheners or growth promoters without any specific claims regarding disease control [27]. To help improve the global market perception of biopesticides as effective products, the Biopesticide Industry Alliance is establishing a certification process to ensure industry standards for efficacy, quality, and consistency [27].

13.5 MAJOR FUNGAL DISEASES OF TURFGRASS

For turfgrass diseases to actively infect the turf, all the following factors need to be present: the disease organism, a host plant, favorable environmental conditions, and a means of distribution of the spores [45,46]. Other factors that trigger most turf diseases are high relative humidity and the movement of surface water over the grass for a long period of time (about 24 h), which allows spores to germinate and infect the grasses; nonoptimal range of temperature and nutrient levels; mowing in general, which opens the grass to pathogens, and mowing too short for the grass species; insufficient aeration; excessive thatch buildup; and frequent applications of insecticides, herbicides, and even fungicides (thatch decomposition is reduced due to the absence or low level of earthworms and decomposing microorganisms) [46].

Most turf-disease fungi live in the soil and thatch layer and feed off the living or dead plant or animal matter [46–56]. Most of these fungi produce spores, sclerotia, or stromata that are spread mechanically, by wind or by water, and then grow and multiply into mycelia [46–50]. The diseases break down plant tissues to the point that the tissues are unable to function normally. More spores or other reproductive parts are produced and the cycle continues [46–50]. The critical role of healthy turf and soil in minimizing the occurrence and severity of turf diseases is increasingly becoming better understood, and practices that create overall soil health are being more commonly applied for this reason, especially in regard to golf-course turf management [46]. Some of the major fungal diseases of turfgrass and their descriptions are discussed below.

13.5.1 Brown Patch

This disease is caused by *Rhizoctonia solani*, which is a plant-pathogenic fungus [46,49,56]. Ideal conditions for brown patch are when temperatures are above 29°C during the day and 16°C at night in combination with 6–8-h periods of humidity [46,56]. Symptoms typically appear as irregularly brown patches that range from several centimeters to a few meters in diameter [46]. When humidity is high, a ring of fungal mycelium that disappears when the foliage dries out may surround the patch. The centers of the patches sometimes recover, forming a halo of diseased grass [46]. Infected leaves appear dark and water-soaked initially, turn brown, and eventually dry out [46]. The disease is exacerbated with very low mowing heights and unbalanced soil fertility. Bentgrass, bluegrass, ryegrass, bermudagrass, and zoysiagrass are all susceptible to brown patch [46].

13.5.2 Dollar Spot

The causal agent of dollar spot disease is fungus *Sclerotinia homeocarpa* [46,52,54]. Dollar spot can occur at temperatures ranging from 15°C to 25°C. Disease spots are bleached to straw-colored, range in size from that of a quarter to a dollar, and commonly coalesce to form large diseased areas [46]. Infection takes place only on the foliage. Leaf spots occur over the entire width of the blade and are bleached to light tan with reddish-brown margins. The disease overwinters in thatch. For turfgrass with nutrient deficiency, warm days and cool nights are the factors that favor development [52,54]. Infected grass clippings, mowers, and foot traffic aid in spreading the disease. Dollar spot is one of the most destructive diseases of closely mown turf, especially creeping bentgrass. Most turfgrasses, especially bentgrasses, are susceptible to dollar spot disease [46,52].

13.5.3 Fusarium Blight

Two species of Fusarium including *F. culmorum* and *F. tricinctum* are the causal agents of this disease [46,52]. Fusarium blight is a warm-season disease that occurs at temperatures ranging from 20°C to 40°C. All parts of the host plant can show the symptoms [46]. Infected turf begins as dark blue to purple that eventually turns into light tan. These spots of dead and dying grass are initially 5 cm in diameter but can increase to as large as 60 cm [46]. A pinkish mycelium can be seen on the surface of the crown when soil moisture is high. Fusarium blight is very similar to summer patch and it is often hard to distinguish between these two diseases. Bentgrass, bluegrass, and fine fescue are very susceptible to this disease [52].

13.5.4 Gray Snow Mold

Fungus *Pyricularia grisea* is the causal agent of gray snow mold disease [46–52]. This is a cold-season disease and can occur at temperatures between 0°C and 4°C when the soil is unfrozen and moisture is high [52]. The fungus is inactive during summer but can survive on the infected leaves as small, dark-colored sclerotia. Circular, grayish to brown spots ranging from 7.5 cm to 0.6 m in diameter develop as the snow melts [46]. Immediately after the snow melts fuzzy gray-white fungal mycelium may be observed, especially along the margins of the spots. Most cool-season grasses, including fescues, ryegrass, and bluegrass, are highly susceptible to this disease [46,52].

13.5.5 RED THREAD

Red-thread disease caused by fungus *Laetisaria fuciformis* occurs during cool wet weather in spring and fall [46,52]. Temperatures ranging between 20°C and 28°C, high humidity, drizzle, fog, and prolonged periods of rain are the favorable conditions for the disease [46]. Leaves and leaf sheaths are the only parts of the plant affected and at points of infection small water-soaked spots can be seen. Infected grass has rusty, red, or pink threads of fungal mycelium that extend from the tips of the blades during final stages of the disease [46]. Spots take the form of circular to irregular patches of scorched leaf tips 5 cm to 1 m across. Perennial ryegrass, Kentucky bluegrass, colonial bentgrass, bentgrass, annual bluegrass, and bermudagrass are most susceptible to red-thread disease [52].

13.5.6 TAKE-ALL PATCH

Take-all patch or Bermuda decline is an important fungal disease of turfgrass caused by *Gaeumannomyces graminis* var. *graminis* [46,52,55]. Infection takes place in late spring and resembles other patch diseases until it has spread. During stressful summer conditions, circular or ring-shaped dead spots occur. The most important identification characteristic of this disease is red leaf blades in stressed grass at the perimeter of the patch along with bronze to reddish-brown grass that fades to a dull brown to gray [52,55]. Unlike brown patch, take-all disease patches recur in the same location year after year, enlarging with each successive year. The disease is particularly prevalent on wet, poorly drained, and unbalanced pH soils [55]. Bentgrass, bluegrasses, ryegrasses, and red and tall fescues are the most susceptible varieties to this disease [52,55].

13.5.7 ANTHRACNOSE

The causal agent of anthracnose disease is *Colletotrichum graminicola* [46,52]. The fungus attacks leaves and sometimes crowns and roots during wet and humid conditions [46]. Disease symptoms include stem lesions in cool, wet weather and reddish-brown lesions on the leaves in warm and wet conditions [46]. Infected plants become blighted, turning yellow to brown in irregular patches ranging from several centimeters to several meters in width [46]. Ruptured lesions of dead leaves have tiny black, hair-like fruiting bodies. The disease is most severe on annual bluegrass. Various fungal strains are very host-specific and infect only one of the susceptible grass species [52].

13.5.8 PYTHIUM BLIGHT

Pythium blight, also called cottony blight, is caused by *Pythium aphanidermatum* [46,52,53]. This fungal disease is favored during rainy weather and in low-lying areas where air circulation is poor [46]. Round, dark, greasy to slimy patches of matted grass, from 5 to 30 cm in diameter, appear suddenly [46]. When the fungal pathogen is very active, mycelium grows profusely over affected plants so that diseased areas have a cotton-like appearance. Pythium may cause seedling blight and poor stand development in perennial ryegrass overseedings [52]. It also can cause crown and root rots, which generally occur in early spring or late fall when soils are cool and excessively wet or saturated [46]. Symptoms of pythium root rot mimic melting out and anthracnose and there is no foliar mycelium. Most grass varieties are susceptible to this disease [53].

13.5.9 DOWNY MILDEW

The causal agent of this disease is *Sclerophthora macrospora* [46,52]. Leaf blades in affected hosts show white or yellow-green linear streaks [46]. Leaves turn yellow and there may be some browning of leaf tips. The disease is usually more severe in moist shaded areas because the causal fungal agent requires moisture for its reproduction and activity [46]. Most turfgrass varieties are susceptible to this disease when favorable conditions are present [52].

13.5.10 POWDERY MILDEW

Powdery mildew disease is caused by fungus *Erysiphe* spp. [46,52]. Powdery mildew is especially common on annual grasses used for overseeding [52]. It spreads rapidly in shaded areas [46]. Powdery mildew appears as a grayish-white fungal growth on the upper surface of leaves and leaf sheaths [46]. Infected leaves turn yellow and gradually die. Repeated infestations result in eventual death of plants. Surviving plants are weakened [52].

13.6 BIOCONTROL STRATEGIES FOR CONTROLLING TURFGRASS FUNGAL DISEASES

Microorganisms naturally present in the turfgrass ecosystem will help reduce disease potential or disease damage, but only if they are allowed to grow vigorously [46–50]. They accomplish these tasks by competing with the pathogens for food sources, producing metabolites that inhibit the growth of the pathogens and physically eliminating the pathogens from the plant by occupying the space and sites first [1–3,13,46–48]. Therefore, it is just as critical to keep the soil microbial population healthy as it is the turfgrass. Reduction of pesticide use is one way in which this may be accomplished.

Microorganisms not naturally present in the turfgrass environment can be introduced in an attempt to control diseases [46–49]. This can be done by application of organic materials that contain natural microbial populations such as composts or natural microbial populations added to them including natural organic fertilizers with microbial supplements [44,45,47,49]. In both cases, the products must be applied prior to disease development as they are preventive and not curative [44,49]. Natural organic fertilizers should be used for their nutrient value (nitrogen and potassium) and not for any possible secondary effects.

According to the results of several research projects, some species of fungal biocontrol agents, including *T. harzianum*, may be used as soil application to reduce the incidence of turfgrass diseases, especially those caused by soilborne pathogens [47–49]. Turfgrasses contain a high-density canopy that is mowed daily on putting greens [49]. Since many turfgrass pathogens can spread readily in turfgrass foliage, control of these diseases requires both suppression of initial plant infection and reduction of the infection rate [49]. Granular applications of strain 1295-22 of *T. harzianum* has been shown to significantly inhibit disease severity of dollar spot during the initial stage of disease development, most likely by reducing levels of the pathogen inoculum in soil and thatch [49]. However, dollar spot severity increased over the course of the season. It is apparent, therefore, that soil applications alone cannot effectively control the foliar phases of this disease [49]. Conversely, in greenhouse and field experiments, it was found that *T. harzianum* significantly reduced the foliar phases of pythium root rot, brown patch, and dollar spot when spray applications of conidial suspensions containing Triton X-100 were used [49]. Weekly spray applications were as effective as the standard (monthly) fungicide applications. These results indicate that the efficacy of *T. harzianum* against turf diseases, especially those involving secondary infections, is very strongly affected by the method of application [49].

Additives have been commonly used with fungicides to improve efficacy and they may also enhance the ability of biocontrol agents to reduce plant disease [47–49]. For example, it was reported that seed treatment using 10% Pelgel with solid matrix priming markedly enhanced the efficacy of Trichoderma strains to control *Pythium* spp. on various crops [49]. Research has indicated that for the three diseases of creeping bentgrass, greater control was obtained when Triton X-100 was included than when no additives, Pelgel, or Tween 20 were used [49]. The use of specific surfactants with Trichoderma strains seems essential to obtain levels of control equivalent to those achieved with chemical fungicides [49]. Detergents such as Triton X-100 may have several functions in biocontrol systems. They may slow the growth of pathogens more than that of the biocontrol agents, or they may enhance wetting and adhesion of spores to infection courts [49]. In preliminary experiments, both Tween 20 and Triton X-100 slowed the growth of both *T. harzianum* and the pathogens, but the ratio of the growth rates of *T. harzianum* and pathogens was greater with Triton X-100 than with Tween 20 [47,48].

Living organisms, in addition to yielding a large quantity of biomass of the bioprotectant fungus, must perform effectively in each application [1]. To examine this, different spore formulations of *T. harzianum* were compared in a study for controlling turfgrass diseases [49]. It was found that all formulations provided equivalent levels of control, indicating that the method of spore production may not be a key factor in the efficacy of this fungal biocontrol agent in controlling the diseases of creeping bentgrass [49]. To predictably and successfully use biological control agents for turfgrass disease control, it is critical that their biology and ecology be more completely understood. Therefore, effective antagonists must become established in turfgrass ecosystems and remain active against target pathogens during periods favorable for plant infection [47–49].

Broadcast application of granules of Trichoderma to control turfgrass diseases has resulted in the establishment of stable and effective populations in turfgrass soils [49]. Similarly, it was shown that the populations of *T. harzianum* in soils or thatch treated with spray applications were as high as those in soils or thatch treated with granular formulations. Population levels of strain 1295-22 in about 5×10^5 CFU/g of soil significantly reduced dollar spot, Pythium blight and root rot, and brown patch diseases of turfgrass [49]. Top-dressing putting greens with a granular formulation of strain 1295-22 apparently enhances the overall quality of creeping bentgrass relative to untreated plots or those sprayed with either fungicides or the biocontrol agent, as evidenced by enhanced greenness [49]. Spray applications, even though resulting in numerically similar levels of root colonization, did not provide the same benefit [49]. This may reflect the differences in the inoculum potential of granules versus spray applications. Granules are applied as a several-millimeter-diameter particle that is completely colonized by the fungus. Conidial inoculum, in contrast, is much smaller and would therefore be expected to possess lower inoculum potential than the granular formulation [49].

The ability to survive on the phylloplane is also a desirable trait for strains of Trichoderma used as biocontrol agents against foliar diseases [1,47,49]. Spray applications of strain 1295-22 of *T. harzianum* has resulted in disease-suppressive population levels on leaf [47]. These populations were sufficient to suppress Pythium root rot, brown patch, and dollar spot over the entire season [49]. The fungus survived on turf leaves at least 4 weeks in growth chamber trials (data not shown). Thus, *T. harzianum* 1295-22 may possess a measure of phylloplane competence on creeping bentgrass. The ideal biocontrol strategy attempts to introduce or promote the activity of biocontrol agents only when and where they are needed or are most effective, and minimizes wasteful application of inoculum to nontarget habitats [49]. Thus, for effective delivery, it is necessary to consider plant pathogen–antagonist interactions in terms of time and space. On the basis of the collective results of studies, a possible strategy for effective control of turfgrass diseases begins with granule application in spring to create disease-suppressive soils and possibly enhance plant vigor [47–49].

Pythium, Rhizoctonia, and Sclerotinia are important soilborne pathogens of turfgrass, and their survival structures in soil serve as primary inoculum. Consequently, suppression of the initial inoculum is the first step in managing these diseases of creeping bentgrass [49]. This granular application should be followed by monthly spray applications to suppress foliar phases of these diseases. Inhibition of the secondary infection and dissemination of these pathogens is also important for disease management [49]. Because of the nature of turfgrass ecosystems, these pathogens may escape the granule treatment, infect foliage, and spread from blade to blade. Monthly spray applications of *T. harzianum* can provide a second step in the protection of turfgrass foliage from attack by preventing these pathogens from initially infecting leaf blades and by reducing the spread of disease through mowing or other methods of inoculum dissemination. Finally, results have indicated that it is necessary to apply weekly sprays for highly effective control of these diseases under severe disease situations [49].

In addition to the diseases discussed above, research has indicated that soil inoculation of Trichoderma may slow or prevent the development of Fusarium patch and gray snow mold on turfgrass [51,52]. As mentioned in previous sections of this review, microbial biocontrol agents including *Trichoderma* spp. use different modes of action to overcome plant pathogens and have therefore the capability of inhibiting many fungal pathogens on a variety of crops including turfgrass [1–4,11,13,20,21,45]. For example, take-all disease caused by the fungus *Gaeumannomyces graminis*, which is an important turfgrass disease, has successfully been controlled on wheat by bacterial antagonists [31,33,35,36]. This means that it may be possible to apply fungal and bacterial antagonists against other soilborne fungal diseases of turfgrass including take-all. Although most research on the biological control of plant diseases has concentrated on soilborne pathogens [1–10], foliar diseases including leaf spots and mildews can also be the subject of biocontrol studies. Fungal and bacterial antagonists may be optimistically applied against non-soilborne diseases of different plants including turfgrass, to come up with an environmentally sound solution to protect these crops from serious damages and significant losses caused by plant-pathogenic agents.

13.7   APPLICATION OF COMPOSTS ON TURFGRASS

Research data and observations in nurseries have shown that the addition of composted organic matter to potting mixes results in the suppression of soilborne diseases [22–24,40,42]. The concentration of suppressive microorganisms in compost-amended substrates is very high but greatly reduced in "worn-out" soils or potting mixes 2 years after the amendment [22,23]. As a result, predictive disease suppression models have been developed based on the composition and concentration of microbial biomass [24,40,42]. Data are currently being developed to correlate β-glucosidase activity with suppressiveness. The concentrations of lignin and protected cellulose in the substrate predict suppressiveness.

The effectiveness of composts in the suppression of soilborne diseases is dependent on heat kill, organic-matter decomposition, recolonization of compost by suppressive microorganisms following heat kill, and physical and chemical factors [24,40,42]. Although previous work has focused on turf soilborne diseases, current research indicates that potting mixes containing composted organic materials that also have been inoculated with *T. hamatum* can be effective as a biocontrol alternative to foliar fungicides; however, the mechanism of this systemic type of induced resistance is not yet understood [42]. Although the turf industry has traditionally relied on aged pine bark and composted biosolids to provide the potential for disease suppression, research indicates that composted animal manure has the potential to replace some of these components, but a consistent quantity and quality of these materials will need to be incorporated [44]. The maturity (stability) of the composted manure and its salinity largely determine its ability to induce suppression.

13.8   SUMMARY

Owing to the serious environmental and health problems that widespread use of chemical pesticides has created in the world, search for alternative safe methods is unavoidable [1,43]. Biological control of plant diseases has been the subject of numerous research projects in recent years [1–25]. There is a growing demand for biologically based pest management practices. Recent surveys of both conventional and organic growers indicate an interest in using biocontrol products, suggesting that the market potential of biocontrol products will increase in the future [1,2,5,7,9,10,12–14,20]. Application of different biological control strategies has been successful in the greenhouse industry and continues to increase [1,2,5,7,9,10,12–14,20,40,42]. An upswing in commercial interests has also developed in the past few years, and prospects for increased growth are positive [27,32].

The Biopesticide Industry Alliance has been formed and is now actively promoting the value and efficacy of biopesticides (including those that control plant pathogens). Clearly, the future success of the biological control industry will depend on innovative business management, product marketing, extension education, and research [27,32].

Increased demand for organic products in home gardening activities by using nonchemical methods has enlarged the market for biocontrol products [27,32]. The field of plant pathology will contribute substantially to making the twenty-first century the age of biotechnology by the development of innovative biocontrol strategies. A variety of research questions about the nature of biological control and the means to most effectively manage it under production conditions has yet to be fully answered. Advanced molecular techniques are now being used to characterize the diversity, abundance, and activities of microbes that live in and around plants, including those that significantly impact plant health. Still, much remains to be learned about the microbial ecology of both plant pathogens and their microbial antagonists in different agricultural systems. Fundamental work remains to be done on characterizing the different mechanisms by which organic amendments reduce plant disease [1,3,4,7,8,10,12–14,16,20]. More studies on the practical aspects of mass production and formulation need to be undertaken to make new biocontrol products stable, effective, safer, and more cost-effective.

Plant pests, especially fungal pathogens, are among the most important factors that cause serious damage and loss to turfgrass [46,49,50,52–56]. Turf industries have been greatly dependent on chemical pesticides in the past and recent years [46,47,52]. Harmful impacts of the chemical pesticides on the environment and nontarget organisms have clearly been documented [1,4,5, 7–9,12,15,46,47,49]. The need for the development of nonchemical alternative strategies to protect turfgrasses against plant diseases, including fungal pathogens, is therefore clear. Biological control using fungal and bacterial antagonists to manage turfgrass diseases seems to be a promising alternative and has successfully been applied to control some diseases on turfgrass, including dollar spot, pythium root rot, and brown patch [47–49]. Biocontrol strategies may also be used to manage other turf diseases, including foliar ones. Some of the important factors that affect the efficacy of microbial biocontrol agents in controlling turfgrass diseases and should be carefully considered include the method of application, formulation of biocontrol microorganisms, and timing of application [46,47,49]. Various composts and organic amendments as other means of biological control have also been tested on turfgrass and have produced promising results.

There are many products composed of living organisms being sold, primarily bacteria and fungi, that claim that they will increase turfgrass health. However, for any material to be considered a biological fungicide or microbial biopesticide, the environmental protection agencies and organizations must register it. This registration indicates that the safety of the product to humans, nonhumans (e.g., fish), and the environment has been determined. Materials that have not been approved should be used with caution.

Complete elimination of chemical pesticides for controlling plant pests in modern agriculture may be impossible, but a logical reduction in their application is absolutely reachable. To have a sustainable agricultural system with minimum contamination and risks to the environment, a combination of all available methods should be applied to manage pest problems and this can be achieved by IPM. Implementation of IPM strategies may be the safest solution for the management of pest and disease problems in every cropping system including turfgrass, and biological control is no doubt one of the most important components of IPM.

REFERENCES

1. Cook, R.J. and Baker, K.F., *The Nature and Practice of Biological Control of Plant Pathogens*, American Phytopathological Society, St. Paul, MN, 1983.
2. Howell, C.R. and Stipanovic, R.D., Control of *Rhizoctonia solani* in cotton seedlings with *Pseudomonas fluorescens* and with an antibiotic produced by the bacterium, *Phytopathology* 69, 480, 1979.
3. Howell, C.R. and Stipanovic, R.D., Suppression of *Pythium ultimum* induced damping-off of cotton seedlings by *Pseudomonas fluorescens* and its antibiotic pyoluteorin, *Phytopathology* 70, 712, 1980.
4. Laville, J., Voisard, C., Keel, C., Maurhofer, M., Defago, G., and Haas, D., Global control in *Pseudomonas fluorescens* mediating antibiotic synthesis and suppression of black root rot of tobacco, *Proc. Natl. Acad. Sci.* 89, 1562, 1992.
5. Heydari, A. and Misaghi, I.J., Herbicide-mediated changes in the populations and activity of root associated microorganisms: a potential cause of plant stress, in *Handbook of Plant and Crop Stress*, 2nd ed., Marcel Dekker Press, New York, 1999.
6. Heydari, A., Fattahi, H., Zamanizadeh, H.R., Hassanzadeh, N., and Naraghi, L., Investigation on the possibility of using bacterial antagonists for biological control of cotton seedling damping-off in green house, *Appl. Entomol. Phytopathol.* 72, 51, 2004.
7. Heydari, A. and Misaghi, I.J., The role of rhizosphere bacteria in herbicide-mediated increase in *Rhizoctonia solani*-induced cotton seedling damping-off, *Plant Soil* 257, 2003.
8. Heydari, A. and Misaghi, I.J., Biocontrol activity of *Burkholderia cepacia* against *Rhizoctonia solani* in herbicide-treated soils, *Plant Soil* 202, 109, 1998.
9. Heydari, A. and Misaghi, I.J., Bacterial community structures in cotton rhizosphere in herbicide-treated soils, *Phytopathology* 88, 37, 1998.
10. Heydari, A., Misaghi, I.J., and McClouskey, W.B., Effects of three soil applied herbicides on populations of plant disease suppressing bacteria in the cotton rhizosphere, *Plant Soil* 195, 75, 1997.
11. Naraghi, L., Heydari, A., Karimi-Roozbehani, A., and Ershad, D., Isolation of *Talaromyces flavus* from Golestan cotton fields and its antagonistic effects on *Verticillium dahliae* the causal agent of cotton verticillium wilt, *Iranian J. Plant Pathol.* 39, 109, 2004.
12. Zaki, K., Misaghi, I.J., Heydari, A., and Shatla, M.N., Control of cotton seedling damping-off in the field by *Burkholderia cepacia*, *Plant Dis.* 82, 291, 1998.
13. Shahriari, F., Khodakaramian, G., and Heydari, A., Assessment of antagonistic activity of *Pseudomonas fluorescens* biovars toward *Pectobacterium carotovorum* subsp. *Atrosepticum*, *J. Sci. Technol. Agric. Nat. Resour.* 8, 201, 2005.
14. Batson, W., Caceres, J., Benson, M., Cubeta, M., Brannen, P., Kenny, D., Elliott, M., Huber, D., Hickman, M., Keinath, A., Dubose, V., Ownley, B., Newman, M., Rothrock, C., Schneider, R., and Summer, D., Evaluation of biological seed treatments for control of the see Kerry, B.R., Rhizosphere interactions and the exploitation of microbial agents for the biological control of plant-pathogenic fungi, *Ann. Rev. Phytopahtol.* 38, 423, 2000.
15. Larkin, R.P. and Fravel, D.R., Efficacy of various fungal and bacterial biocontrol organisms for control of Fusarium wilt of tomato, *Plant Dis.* 82, 1022, 1998.
16. Laville, J., Voisard, C., Keel, C., Maurhofer, M., Defago, G., and Haas, D., Global control in *Pseudomonas fluorescens* mediating antibiotic synthesis and suppression of black root rot of tobacco, *Proc. Natl. Acad. Sci.* 89, 1562, 1992.
17. Loper, J.E., Role of fluorescent siderophore production in biological control of *Pythium ultimum* by a *Pseudomonas fluorescens* strain, *Phytopathology* 78, 166, 1992.
18. Nowak-Thompson, B., Chaney, N., Wing, J.S., Gould, S.J., and Loper, J.E., Characterization of the pyoluteorin biosynthetic gene cluster of *Pseudomonas fluorescens* Pf-5, *J. Bacteriol.* 181, 2166, 1999.
19. Rodriguez, F. and Pfender, W.F., Antibiosis and antagonism of *Sclerotinia homoeocarpa* and *Drechslera poae* by *Pseudomonas fluorescens* Pf-5 *in vitro* and *in planta*, *Phytopathology* 87, 614, 1997.
20. Stutz, E.W., Defago, G., and Kern, H., Naturally occurring fluorescent pseudomonads involved in suppression of black root rot of tobacco, *Phytopathology* 76, 181, 1986.
21. Xu, G.W. and Gross, D.C., Selection of fluorescent pseudomonads antagonistic to *Erwinia carotovora* and suppressive of potato seed piece decay, *Phytopathology* 414, 1986.
22. Abbasi, P.A., Al-Dahmani, J., Sahin, F., Hoitink, H.A.J., and Miller, S.A., Effect of compost amendments on disease severity and yield of tomato in conventional and organic production systems, *Plant Dis.* 86, 156, 2002.

23. Bulluck III, L.R. and Ristaino, J.B., Effect of synthetic and organic fertility amendments on southern blight, soil microbial communities, and yield of processing tomatoes, *Phytopathology* 92, 181, 2002.

24. Fravel, D.R., Rhodes, D.J., and Larkin, R.P., Production and commercialization of biocontrol products, in *Integrated Pest and Disease Management in Greenhouse Crops*, R. Albajes, M.L. Gullino, J.C. van Lenteren, and Y. Elad, eds., Kluwer, Dordrecht, 1999, pp. 365–376.

25. Jacobsen, B., Role of plant pathology in integrated pest management, *Annu. Rev. Phytopathol.* 35, 373, 1997.

26. Ligon, J.M., Hill, D.S., Hammer, P.E., Torkewitz, N.R., Hofmann, D., Kempf, H.J., and Van Pee, K.H., Natural products with antifungal activity from *Pseudomonas* biocontrol bacteria, *Pest Man. Sci.* 56, 688, 2000.

27. Mathre, D.E., Cook, R.J., and Callan, N.W., From discovery to use: Traversing the world of commercializing biocontrol agents for plant disease control, *Plant Dis.* 83, 972, 1999.

28. Kloepper, J.W., Leong, J., Teintz, M., and Schroth, M.N., Pseudomonas siderophores: A mechanism explaining disease-suppressive soils, *Curr. Microbiol.* 4, 317, 1980.

29. Misaghi, I.J., Grogan, R.G., Spearman, L.C., and Stowell, L.J., Fungistasis activity of water-soluble fluorescent pigments of fluorescent pseudomonas, *Phytopathology* 72, 33, 1982.

30. Misaghi, I.J., Grogan, R.G., Spearman, L.C., and Stowell, L.J., Antifungal activity of fluorescent pigments produced by fluorescent pseudomonads, *Phytopathology* 71, 106, 1981.

31. Pierson, E.A. and Weller, D.M., Use of mixture of fluorescent pseudomonads to suppress take-all and improve the growth of wheat, *Phytopathology* 84, 940, 1994.

32. Shah-Smith, D.A. and Burns, R.G., Shelf-life of a biocontrol *Pseudomonas putida* applied to the sugar beet seeds using commercial coatings, *Biocontrol Sci. Technol.* 7, 65, 1997.

33. Thomashow, L.S. and Weller, D.M., Role of antibiotics and siderophores in biocontrol of take-all diseases of wheat, *Plant Soil* 129, 93, 1990.

34. Weller, D.M., Biological control of soil-born plant pathogens in the rhizosphere with bacteria, *Ann. Rev. Phytopathol.* 26, 379, 1988.

35. Weller, D.M. and Cook, R.J., Suppression of take-all of wheat by seed treatment with fluorescent pseudomonads, *Phytopathology* 73, 463, 1983.

36. Zaspei, I., Studies on the influence of antagonistic rhizosphere bacteria on winter wheat attacked by *Gaummannomyces graminis* var. *tritici, Bulletin* 15, 142, 1992.

37. Cook, R.J., Advances in plant health management in the 20th century, *Ann. Rev. Phytopathol.* 38, 95, 2002.

38. Hoitink, H.A.J. and Boehm, M.J., Biocontrol within the context of soil microbial communities: A substrate dependent phenomenon, *Annu. Rev. Phytopathol.* 37, 427, 1999.

39. Hutchinson, S.W., Current concepts of active defense in plants, *Annu. Rev. Phytopathol.* 36, 59, 1999.

40. Paulitz, T.C. and Belanger, R.B., Biological control in greenhouse systems, *Annu. Rev. Phytopathol.* 39, 103, 2001.

41. Van Loon, L.C., Bakker, P.A.H.M., and Pietrse, C.M.J., Systemic resistance induced by rhizosphere bacteria, *Annu. Rev. Phytopathol.* 36, 453, 1998.

42. Zhang, W., Han, D.Y., Dick, W.A., Davis, K.R., and Hoitink, H.A.J., Compost and compost water extract-induced systemic acquired resistance in cucumber and Arabidopsis, *Phytopathology* 88, 450, 1998.

43. Cook, R.J., Making greater use of introduced microorganisms for biological control of plant pathogens, *Annu. Rev. Phytopathol.* 31, 53, 1993.

44. Nelson, E.B. and Craft, C.M., Suppression of Pythium root rot with top dressings amended with composts and organic fertilizers, *Biol. Cult. Tests Control Plant Dis.* 7, 104, 1992.

45. Harman, G.E., Taylor, A.G., and Stasz, T.E., Combining effective strains of *Trichoderma harzianum* and solid matrix priming to improve biological seed treatments, *Plant Dis.* 73, 631, 1989.

46. Smiley, R.W., Dernoeden, P.H., and Clarke, B.B., *Compendium of Turfgrass Diseases*, 2nd ed., American Phytopathological Society, St. Paul, MN, 1992, p. 12.

47. Lo, C.T., Nelson, E.B., and Harman, G.E., Biological control of pythium, rhizoctonia and sclerotinia infected diseases of turfgrass with *Trichoderma harzianum, Phytopathology* 84, 1372, 1994.

48. Lo, C.T., Nelson, E.B., and Harman, G.E., Biological control of turfgrass diseases with a rhizosphere component strain of *Trichoderma harzianum, Plant Dis.* 80, 736, 1996.

49. Lo, C.T., Nelson, E.B., and Harman, G.E., Improved biocontrol efficacy of *Trichoderma harzianum* for foliar phases of turf diseases by use of spray applications, *Plant Dis.* 81, 1132, 1997.

50. Sachs, P.D., *Handbook of Successful Ecological Lawn Care*, The Edaphic Press, Newbury, VT, 1996, p. 45.
51. Clarke, B.B. and Gould, A.B., *Turfgrass Patch Diseases Caused by Ectotrophic Root-Infecting Fungi*, APS Press, St. Paul, MN, 1993, p. 20.
52. Couch, H.B., *The Turfgrass Disease Handbook*, Krieger Publishing Co., Melbourne, FL, 2000, p. 30.
53. Allen, T.W., Martinez, A., and Burpee, L.L., Pythium blight of turfgrass, *Plant Health Instruct.* 10, 1094, 2004.
54. Allen, T.W., Martinez, A., and Burpee, L.L., Dollar spot of turfgrass, *Plant Health Instruct.* 10, 1094, 2005.
55. Bockus, W.W. and Tisserat, N.A., Take-all root rot, *Plant Health Instruct.* 10, 1094, 2001.
56. Tredway, L.P. and Burpee, L.L., *Rhizoctonia* diseases of turfgrass, *Plant Health Instruct.* 10, 1094, 2001.

14 Current Understanding and Management of Rapid Blight Disease on Turfgrasses

Paul D. Peterson, S. Bruce Martin, and James J. Camberato

CONTENTS

14.1   INTRODUCTION

Since it was first diagnosed in 1995, rapid blight disease has been a frequent problem for the golf course industry. The first documented case was on *Poa annua* (annual bluegrass) putting greens in southern California. From 1995 through 1999, the disease was recognized sporadically in different parts of the western United States: on *P. annua* from locations throughout California and from Colorado, on *Poa trivialis* (rough bluegrass) used for overseeded bermudagrass putting greens from Nevada, and on *Lolium perenne* (perennial ryegrass) from Arizona [1,2]. In 2000, rapid blight was positively identified for the first time on *P. trivialis* from the eastern United States. Rough bluegrass on two golf courses in South Carolina was decimated that year, followed by a severe epidemic in 2001 in which over 14 golf courses were affected in South and North Carolina [3]. Martin and Stowell also identified the disease on salinity-stressed *Agrostis palustris* (creeping bentgrass) in 2001 and 2002 [1]. Since it was first diagnosed, rapid blight disease has been confirmed on *P. annua*, *P. trivialis*, *L. perenne*, and *A. palustris* from more than 100 golf courses in 11 states over the southern United States [L.J. Stowell, S.B. Martin, and P.D. Peterson, unpublished]. In 2004, the disease was confirmed on *Agrostis capillaries* (colonial bentgrass) and *P. annua* from a golf course in the United Kingdom [4].

14.2   THE CAUSAL ORGANISM

14.2.1   DESCRIPTION

On the basis of morphological [5] and molecular [6] data, the causal agent of rapid blight is *Labyrinthula terrestris*, a eukaryotic protist species recently described in the genus *Labyrinthula*.

First described in 1867 [7], *Labyrinthula* have been classified in a variety of ways over the past 139 years, as plants, fungi, slime molds, and even amoebae [8,9,10]. Currently, morphological, ultra-structural, and molecular data place the genus in the kingdom Stramenopila (Chromista) together with organisms such as diatoms, brown algae, and the water molds or Oomycetes [10]. There are major questions of taxonomic criteria with the *Labyrinthula*, however, and the classification of these organisms undoubtedly will be revised on the basis of continued morphological and molecular studies.

The most distinguishing characteristic of the *Labyrinthula* is an unusual communal cell organization with motile fusiform-shaped cells surrounded by a colorless wall-less network of interconnecting filaments (called the ectoplasmic network). Specialized cell-surface organelles called bothrosomes (also called sagenogens) form the netlike ectoplasmic network [11], enabling the cells to move freely with a gliding motion. Bothrosomes are not known to occur in any other organism. The ectoplasmic network consists of branching and anastomosing filaments capable of absorbing nutrients as well as attaching the organism to substrates. These networks explain why the *Labyrinthula* have often been inaccurately referred to and classified as slime molds, although they are not phylogenetically related to the true fungi. Other distinguishing morphological characteristics include mitochondria containing tubular cristae and biflagellate heterokont zoospores [8,10,11].

Ten species of *Labyrinthula* have been described [12], excluding the turfgrass pathogen. These are heterotrophic and generally occur as saprobes, although some are pathogens [8]. They can be found in marine and estuarine environments throughout the world, typically associated with benthic algae, detrital sediments, and particularly marine vascular plants, such as *Spartina*, *Zostera*, and *Thalassia* [8]. Some species are pathogenic on seagrasses. One species, *L. zosterae*, has been shown to cause a devastating wasting disease of eelgrass, an ecologically important marine vascular plant that serves as a nursery bed for larval shrimp, oysters, and scallops [13]. Catastrophic declines in eelgrass in North America and Europe during the 1930s and 1940s were associated with *L. zosterae*. It was not until 1988, however, that Muehlstein et al. [13] demonstrated that *L. zosterae* was the causal agent of the eelgrass wasting disease. Studies have shown that elevated salinities in eelgrass habitats were linked with outbreaks of the wasting disease [14]. In addition to the eelgrass pathogen, a different species of *Labyrinthula* has been associated with devastating die-offs of another sea-grass, *Thalassia testudinum* (turtle grass), in Florida Bay during the 1980s [9]. Prior to the discovery of rapid blight disease, *Labyrinthula* spp. were not known to parasitize terrestrial grasses, although these organisms have been isolated from inland saline soils [8].

The life cycle of *Labyrinthula terrestris* as presently known consists of a single vegetative stage. *L. terrestris* trophic cells are commonly fusiform, hyaline to dark yellow in culture, and vary in size averaging 6×16 μm, with division transverse and mostly longitudinal. Cells are uninuculeate with a large nucleolus, various-sized vacuoles, and conspicuous granules of lipid composition (Figure 14.1). The vegetative cells have a flexible cell membrane and exhibit a gliding type of motility within the ectoplasmic network produced by the specialized bothrosomes [5]. As the network forms, vegetative cells migrate along the filaments or "slimeways." In parallel, the entire colony undergoes constant change—cells increase by division, new filaments are formed, and old filaments are widened. The outer, limiting membrane of the ectoplasmic network is the boundary with the external environment and effectively encloses all the cells of the colony.

On nutrient medium (1% serum seawater agar) in petri plates [11], colonies of *L. terrestris* will appear from infected turfgrass leaves within 2–4 days at room temperature. Colonies grow as cells multiply readily by mitotic division and the ectoplasmic network expands for a period of approximately 14–21 days before expansion ceases [P.D. Peterson, unpublished]. Cells also form spherical aggregates within the network in senescent cultures or in cultures exposed to unfavorable environmental conditions [5, P.D. Peterson, unpublished]. In other *Labyrinthula* species, these aggregates have been shown to develop into reproductive sori as precursors to the formation of zoospores [15,16]. In *L. vitellina*, a partial sexual life cycle has been observed and characterized by the formation of zoospores produced by meiosis [17], although the fusion of gametes has not

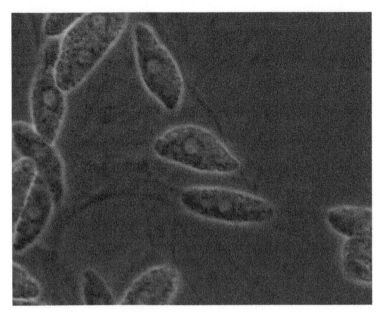

FIGURE 14.1　**(Color Figure 14.1 follows p. 262.)** The rapid blight organism growing in culture. Vegetative cells of *Labyrinthula terrestris* at 100× magnification.

been observed [9]. Reproductive sori and zoosporulation have not been observed in *L. terrestris*. Indeed, the complete life cycle of any species of *Labyrinthula* remains to be revealed.

14.3　THE DISEASE ON TURFGRASS

14.3.1　SYMPTOMS

When it emerges on golf courses, rapid blight symptoms initially appear as irregularly shaped patches of chlorotic or necrotic turf commonly ranging from 15 to 20 cm or larger. Sometimes, these patches have a darkened edge where the foliage appears water-soaked (Figure 14.2). Subsequent symptoms include a partial collapse of the turf and reduction in plant size [18,19]. With some turfgrass varieties, such as *P. trivialis*, the irregularly shaped patches may turn a yellowish brown or bronze color. As the disease progresses, these patches enlarge and coalesce to form large areas of infected turf (Figure 14.3) [18–21].

The disease is difficult to diagnose; there are no obvious signs of pathogen invasion, such as fungal mycelia or spores. Accurate diagnosis requires the use of light microscopy. Upon microscopic examination of infected plant tissue, *L. terrestris* vegetative spindle-shaped cells are located abundantly in leaf epidermal and mesophyll cells [P.D. Peterson and D. Porter, unpublished] (Figure 14.4). *L. terrestris* trophic cells also have been detected in the root tissue of infected plants.

Rapid blight can affect juvenile as well as mature turf. In the southern United States, winter-dormant bermudagrass is overseeded often with a temporary stand of cool-season grasses. Overseeded grasses in the seedling stage at first mowing are particularly vulnerable to infection. Extensive damage can result and stands may be eliminated within a week if disease occurs at establishment. However, rapid blight can be severe on putting greens of mature annual bluegrass and creeping bentgrass as well as established stands of overseeded perennial ryegrass [18,22]. Mowing and foot traffic appear to increase disease severity. Rapid blight is most severe on golf course greens where mowing and foot traffic patterns have been observed repeatedly [23, S.B. Martin and P.D. Peterson, unpublished].

FIGURE 14.2 **(Color Figure 14.2 follows p. 262.)** Rough bluegrass overseeding infected with rapid blight disease.

FIGURE 14.3 **(Color Figure 14.3 follows p. 262.)** Symptoms of rough bluegrass overseeding on putting greens infected with rapid blight disease.

14.3.2 FACTORS AFFECTING DISEASE OCCURRENCE

Rapid blight outbreaks have been associated with saline irrigation water, drought, and a buildup of salt in the soil. A salt or Na requirement for *L. terrestris* would not be unusual, since all other known *Labyrinthula* spp. are marine organisms. Rapid blight has been diagnosed primarily in the fall and spring months under moderate temperatures on both the east and west coasts of the United States, suggesting that cooler temperatures may promote disease development. Peterson conducted a series of controlled environment studies both *in vitro* and *in planta* to determine the relative growth of 14 different *L. terrestris* isolates collected across the United States under conditions of varying

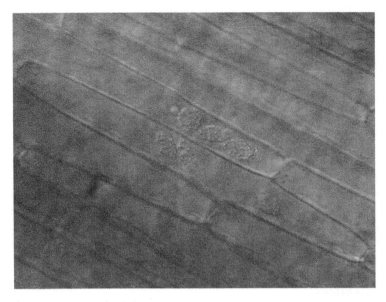

FIGURE 14.4   (Color Figure 14.4 follows p. 262.) Vegetative cells of *Labyrinthula terrestris* in leaf epidermal cells at 40× magnification.

temperatures and levels of salinity. The results indicated that *L. terrestris* grows best at salinity levels from 3.5 to 10.5 dS m^{-1} and at temperatures between 22°C and 26°C [24].

High salinity levels may also stress the host plant so that disease develops. Saline irrigation water and drought allow soluble salts to accumulate and predispose susceptible turfgrasses to the disease. All cool-season grasses examined under conditions of high salinity examined to date are susceptible to rapid blight but disease severity varies greatly, from <1% to >90% [22]. The species most tolerant to rapid blight were the slender creeping red fescues, creeping bentgrasses, and some alkaligrasses—grasses that are generally considered to be among the most salt-tolerant cool-season grasses. Rapid blight–sensitive species, such as annual and rough bluegrasses and colonial and velvet bentgrass were the most susceptible.

Subsequent greenhouse trials showed that increasing salinity increased disease in susceptible perennial ryegrass and Kentucky bluegrass cultivars [25]. Substantial disease occurred with increased salinity from 2.5 to 4.8 dS m^{-1} but little disease occurred with irrigation water ≤1.3 dS m^{-1}. Slender creeping red fescues showed significant disease only at 4.8 dS m^{-1}.

Peterson, Martin, and Camberato collected irrigation water from the western and eastern regions of the United States, where rapid blight occurred, and found it to differ in chemical composition. Western irrigation water from these sites was typically higher in Ca and lower in Na and HCO$_3$ than eastern irrigation waters from rapid blight–affected sites. Perennial ryegrass grown with irrigation water simulating eastern and western water sources had substantially more disease with the eastern water [26]. Shoot-tissue Na was higher and Ca and K were lower with eastern water. Disease correlated best with tissue Na across water compositions, salinity levels, and inoculation treatments [27]. Tissue K was negatively correlated with disease. Managing irrigation water to reduce salt and Na accumulation in soils is likely an effective way to reduce the occurrence of rapid blight.

14.3.3   GEOGRAPHIC DISTRIBUTION AND HOST RANGE

Rapid blight disease damages golf course greens, tees, fairways, and roughs cultured to rough bluegrass, perennial ryegrass, annual bluegrass, and sometimes creeping bentgrass in the southeastern and western United States. Although the disease has been diagnosed on creeping bentgrass, it rarely occurs, correlating with greenhouse experiments showing creeping bentgrass to be affected only at high salinities [22]. Rapid blight has been reported also on colonial bentgrass and annual bluegrass from one location in the United Kingdom as well [4]. All 49 cool-season turfgrass species across eight genera examined in greenhouse studies were found susceptible to rapid blight [22].

The host range of *L. terrestris* may not be limited to cool-season turfgrasses. In a survey of two golf courses where rapid blight had occurred in cool-season grasses used for overseeding, Olsen isolated the rapid blight pathogen from nonsymptomatic bermudagrass roots and stolons [28]. In a preliminary controlled study, Peterson isolated *L. terrestris* from bermudagrass roots and crown regions as well as from other warm-season grasses including St. Augustine and centipede [P.D. Peterson, unpublished]. All these grasses from which the rapid blight organism was isolated were nonsymptomatic, suggesting a potential oversummering mechanism for the pathogen.

The artificial host range of *L. terrestris* may not even be limited to turfgrass. Non-salt-tolerant cultivars of rice, lettuce, and radish as well as salt-tolerant varieties of alfalfa, barley, and wheat were screened in the greenhouse and laboratory to determine if *L. terrestris* could infect plants other than turfgrasses. Wheat, barley, and rice plants were infected, became symptomatic, and died. Radish and lettuce were infected but nonsymptomatic. Alfalfa was not infected and exhibited no symptoms [D.M. Bigelow and M.W. Olsen, unpublished].

14.4 MANAGEMENT

14.4.1 CHEMICAL CONTROL

During epidemics on the west coast of the United States in the late 1990s, there was no information on control of rapid blight on golf course turf. The causal organism had not been identified, although the fusiform vegetative cells were associated consistently with affected plants [2]. A suggestion by the late Houston Couch, turf pathologist at Virginia Polytechnic Institute, was that mancozeb as a broad-spectrum fungicide might have activity; applications seemed to lessen disease severity. However, controlled experiments were not conducted until 2001, when Martin and Camberato tested several fungicides for control of the disease in replicated trials on a putting green in South Carolina. Epidemics of the disease were concurrent with testing, and these replicated field experiments confirmed activity from trifloxystrobin and mancozeb. Later, an additional experiment at the same site confirmed excellent activity from pyraclostrobin, at that time tested as BAS 500, an experimental fungicide [3]. In 2004, Martin et al. [29] completed a series of fungicide control trials in the greenhouse for control of rapid blight on perennial ryegrass. Most of the currently labeled fungicides for control of turf diseases were tested in replicated and repeated experiments. Still, pyraclostrobin (Insignia 20WG), trifloxystrobin (Compass 50WG), and mancozeb (Fore 80WP) were most efficacious, with high labeled rates of pyraclostrobin, or low labeled rates of pyraclostobin tank-mixed with the low labeled rate of mancozeb being most effective. These results have been confirmed from Olsen et al. [20]. They also confirmed good activity from the copper-containing fungicide Kocide, which contains copper hydroxide as the active ingredient. Recommendations have been made to users to be cautious with copper-containing fungicides, which have a high potential for phytotoxicity. Currently, Insignia remains the most effective fungicide for control of rapid blight with fungicides.

14.4.2 CULTURAL CONTROL

Cultural control of rapid blight includes all practices that improve the vigor of susceptible grasses on golf courses [19]. These include thatch management, core aerification to provide channels for removal of salinity from root zones, and avoiding sand topdressing and aggressive cultivation when infections are "active." It has been observed commonly in east-coast environments that periodic rainfall reduces the potential of rapid blight epidemics. Presumably, this is due to the leaching of high levels of soluble salts from the rootzones of cultivated turfgrasses. Therefore, management of salinity, including periodic leaching with low-salinity irrigation water (if available), as well as acid buffering of irrigation waters high in bicarbonate are suggested. If irrigation waters are high in sodium, which has been associated with rapid blight on the east coast, then applications of gypsum, followed by leaching of the rootzone, are suggested to displace sodium from the affected soil profile. Additionally, overseeding with less susceptible turfgrass species may provide less risk of devastating epidemics when conditions are at high risk of disease outbreaks. These might include creeping bentgrass, certain cultivars of alkaligrass, or slender creeping red fescues. Blending of tolerant and susceptible species may lessen risk of severe damage as well.

14.5 CONCLUSION

Rapid blight is a unique disease of cool-season turfgrasses and perhaps other terrestrial plants. The nature of the organism itself, association of the disease with salt-affected sites, and difficulty of control using cultural as well as chemical approaches make management challenging and difficult. It is expected that rapid blight will spread or be documented in other sites in the United States and around the world as low-quality irrigation waters increase in usage.

REFERENCES

1. Martin, S.B., Stowell, L.J., Gelernter, W.D., and Alderman, S.C. Rapid blight: A new disease of cool-season turfgrasses. *Phytopathology*, 92, S52, 2002.
2. Stowell, L.J. and Gelernter, W.D. A new disease of annual bluegrass, ryegrass and *Poa trivialis. PACE Insights*, 7, 1, 2001.
3. Martin, S.B, Stowell, L.J., and Gelernter, W.D. Rough bluegrass, annual bluegrass and perennial ryegrass hit by new disease. *Golf Course Management*, April 2002, p. 61.
4. Entwistle, C.A., Olsen, M.W., and Bigelow, D.M. First report of a *Labyrinthula* spp. causing rapid blight of *Agrostis capillaris* and *Poa annua* on amenity turfgrass in the UK. *New Disease Reports*, 11, 1, 2005. www.bspp.org.uk/NDR
5. Bigelow, D.M., Olsen, M.W., and Gilbertson, R.L. *Labyrinthula terrestris* sp. nov. a new pathogen of turfgrass. *Mycologia*, 97, 185, 2005.
6. Craven, K.D., Peterson, P.D., Windham, D.E., Mitchell, T.K., and Martin, S.B. Molecular identification of the turf grass rapid blight pathogen. *Mycologia*, 97, 160, 2005.
7. Cienkowski, L. Ueber den Bau und die Entwicklung der Labyrinthuleen. *Archiv für mikroscopische Anatomie*, 3, 274, 1867.
8. Porter, D. Phylum Labyrinthulomycota. In *Handbook of Protoctista*, Margulis, L., Corliss, J.O., Melkonian, M., and Chapman, D.J., Eds., Jones and Bartlett Publishers, Boston, Massachusetts, 1990, p. 388.
9. Mozley, S.E., Porter, D., and Cubeta, M.A. Slime molds and zoosporic fungi. In *Plant Pathology: Concepts and Laboratory Exercises*, Trigiano, R.N., Windham, M.T., and Windham, A.S., Eds., CRC Press, Boca Raton, 2004, p. 91.
10. Leander, C.A. and Porter, D. The Labyrinthulomycota is comprised of three distinct lineages. *Mycologia*, 93, 459, 2001.
11. Porter, D. Labyrinthulomycetes. In *Zoosporic Fungi in Teaching and Research*, Fuller, M.S. and Jaworski, A., Eds., Southeastern Publishing Corp, Athens, Georgia, 1987, p. 110.
12. Dick, M.W. *Straminipilous Fungi*. Kluwer, Dordrecht, 2001.
13. Muehlstein, L.K., Porter, D., and Short, F.T. *Labyrinthula zosterae* sp. nov., the causative agent of wasting disease of eelgrass, *Zostera marina. Mycologia*, 83, 180, 1991.
14. Muehlstein, L.K, Porter, D., and Short, F.T. *Labyrinthula* sp., a marine slime mold producing the symptoms of wasting disease in eelgrass, *Zostera marina. Marine Biology*, 99, 465, 1988.
15. Perkins, F.O. and Amon, J.P. Zoosporulation in *Labyrinthula* sp.; an electron microscope study. *Journal of Protozoology*, 16, 235, 1969.
16. Amon, J.P. and Perkins, F.O. Structure of *Labyrinthula* sp. zoospores. *Journal of Protozoology*, 15, 543, 1968.
17. Margulis, L., Corliss, J.O., Melkonian, M., and Chapman, D.J. *Handbook of Protoctista*, Jones and Bartlett, Boston, 1990.
18. Peterson, P.D., Martin, S.B., and Camberato, J.J. Rapid blight disease of cool-season grasses. *USGA Green Section Record*, 42, 20, 2004.

19. Smiley, R.W, Dernoeden, P.H., and Clarke, B.B., Eds., *Compendium of Turfgrass Diseases*, 3rd ed. American Phytopathological Society Press, St. Paul, Minnesota, 2005.

20. Olsen, M.W., Bigelow, D.M., Kohout, M.J, Gilbert, J., and Kopec, D. Rapid blight: A new disease of cool-season turf. *Golf Course Management*, August 2004, p. 87.

21. Kopec, D., Olsen, M.W., Gilbert, J., Bigelow, D.M., and Kohout, M.J. Cool-season grass response to rapid blight disease. *Golf Course Management*, December 2004, p. 78.

22. Peterson, P.D., Martin, S.B., and Camberato, J.J. Tolerance of cool-season turfgrasses to rapid blight disease. *Applied Turfgrass Science*, doi:10.1094/ATS-2005-0328-01-RS. Available at: http://www. plantmanagementnetwork.org/pub/ats/research/2005/blight/

23. Stowell, L.J. and Gelernter, W.D. Progress in understanding rapid blight of cool season turf. *Pace Insights*, 9(3), 1, 2003. Available at: www.paceturf.org/PTRI/Documents/Diseases/0303.PDF

24. Peterson, P.D., Martin, S.B., and Fraser, D.E. The effect of temperature and salinity on growth of the turfgrass pathogen, *Labyrinthula terrestris*. *Phytopathology*, 95, S83, 2005.

25. Camberato, J.J., Peterson, P.D., and Martin, S.B. Salinity and salinity tolerance alter rapid blight in Kentucky bluegrass, perennial ryegrass, and slender creeping red fescue. *Applied Turfgrass Science*, doi:10.1094/ATS-2006-0213-01-RS. http://www.plantmanagementnetwork.org/pub/ats/research/2006/salinity/

26. Camberato, J.J., Peterson, P.D., and Martin, S.B. Irrigation water composition affects rapid blight of cool-season turfgrass. *Phytopathology*, 96, S18, 2006.

27. Camberato, J.J., Peterson, P.D., and Martin, S.B. Irrigation water composition alters plant mineral composition and rapid blight on perennial ryegrass. *Agronomy Abstracts*. Available at: http://a-c-s.confex. com/crops/2006am/techprogram/P24368.HTM

28. Stowell, L.J., Martin, S.B., Olsen, M.W., Bigelow, D.M., Kohout, M.J., Peterson, P.D., and Camberato, J.J. Rapid Blight: A new plant disease. 2005, http://www.apsnet.org/online/feature/rapid/

29. Martin, S.B., Olsen, M.W., Peterson, P.D., and Camberato, J.J. Evaluation of fungicides for control of rapid blight on cool season grasses. *Phytopathology*, 94, S66, 2004.

15 A Review of Dead Spot on Creeping Bentgrass and a Road Map for Future Research

John E. Kaminski

CONTENTS

15.1 INTRODUCTION

In 1998, Dernoeden et al.[1] discovered a new disease of creeping bentgrass (*Agrostis stolonifera* L.) incited by an unidentified species of *Ophiosphaerella*. Through morphological and molecular study, it was shown that the pathogen constituted a new species, *Ophiosphaerella agrostis*.[2] The disease is commonly referred to as dead spot. Research on various biological and genetic aspects of *O. agrostis* and epidemiology and management of dead spot was conducted at the University of Maryland between 1998 and 2004. A synthesis of these results and suggestions for future dead-spot research are presented herein.

15.2 BIOLOGY AND EPIDEMIOLOGY

Dead spot has only been observed on creeping bentgrass and bermudagrass (*Cynodon dactylon* [L.] Pers. x *C. transvaalensis* Burtt-Davy) grown on sand-based root zones. There are no reports of the disease in turf grown on native soil. Active dead-spot infection centers generally appear in areas with full sun and good air circulation. In particular, *O. agrostis* infection centers often appear initially along ridges and on mounds and south-facing slopes of putting greens. These areas are particularly prone to higher soil temperatures and often are the first to exhibit drought symptoms. Dead spot typically develops between 1 and 2 years following bentgrass establishment; however, outbreaks of the disease have been observed on creeping bentgrass that was less than 1 year old and as old as 6 years of age.[3]

Dead spot is most severe during the first or second year of symptom expression and the disease rarely recurs in areas previously infected by the pathogen.[3] The decline phase typically lasts from 1 to 3 years after the first year of disease expression, with the number of infection centers per green normally decreasing in subsequent years. Turf recovers very slowly, as stolons growing into dead patches often become infected and die. Dead spot is most prevalent on newly constructed putting greens, but the disease may also appear on putting greens established following fumigation with methyl bromide. In a pilot field study, the disease did not recur following fumigation with dazomet.[4]

Evidently, methyl bromide more effectively reduces populations of microbes that in some way antagonize or compete with *O. agrostis*. A slower decline occurs with take-all (*Gauemannomyces graminis* [Sacc.] Arx and D. Olivier var. *avenae* [E. M. Turner] Dennis) in *Agrostis* turf in response to a buildup of bacterial antagonists.[5,6] The speculated antagonists responsible for dead-spot decline are unknown. *O. agrostis* antagonists may reproduce more rapidly and be more competitive than those responsible for the decline of *G. graminis* var. *avenae*. For this reason, *O. agrostis* may serve as a model pathogen for examining the impact of soil microbial antagonists on natural disease suppression. Identification and enumeration of specific microbial populations (e.g., fluorescent *Pseudomonas* spp.) following initial construction of sand-based putting greens as well as after methyl bromide fumigation may provide some insight into their influence on dead-spot decline. Emerging techniques that aide in the identification and quantification of microbial communities (e.g., Biolog®, Inc., Hayward, CA) may provide a useful tool for studying decline phenomena.

In addition to an increase in microbial populations, a reduction in inoculum also may lead to a decrease in disease severity. In a 3-year cultivar study, dead spot was severe in the first and third year and only limited disease was observed in the second year.[3] Although monthly air and soil temperatures during all 3 years of the field study generally were similar (i.e., 1–4°C difference for each month), inoculum levels varied each year. Ascospores were not quantified in year 1, however, damage from *O. agrostis* was considered severe and numerous pseudothecia were observed within infection centers. In the second year following establishment, a reduction in the number of new infection centers as well as the number of pseudothecia produced resulted in low levels of ascospores. For the third and final year of the study, an adjacent portion of the research green was fumigated with methyl bromide and disease levels again were extremely and moderately severe in both the fumigated and reinoculated cultivar study, respectively. The number of infection centers and pseudothecia developing in year 3 resulted in a greater number of ascospores released, when compared with year 2. Environmental conditions appear to play an important role in the development of disease symptoms, but the pathogen likely is most damaging where soil microbial populations are low and ascospore levels are high. The maturity of the turf also likely influences the incidence and severity of dead spot. Growth chamber studies, in which bentgrass plants of varying maturity are inoculated with varying concentrations of ascospores, may reveal differences in the ability of the pathogen to infect plants and elicit dead-spot symptoms. An additional aspect of such a study should include an assessment of the influence of fumigated versus nonfumigated soil on dead-spot severity.

An important aspect of the biology of *O. agrostis* is the release and alighting of ascospores on bentgrass tissues. Ascospores can be forcefully released from pseudothecia and disseminated by wind or water. Dead spots generally only coalesce under severe pressure, suggesting that wind rather than water is the primary mechanism of dispersal. It is likely, however, that a majority of the ascospores released into wind currents are blown off-site and would not come into contact with a susceptible host. Ascospores that ooze from pseudothecia in the presence of water likely have the ability to infect bentgrass stolons, roots, and leaves within or along the periphery of dead-spot infection centers. Visual observations of the distribution of dead-spot infection centers revealed that new infection centers most often develop within close proximity (1–30 cm or greater) of older infection centers.[7] On the basis of the aforementioned observation, however, it appears possible that splashing rain or irrigation water may carry large numbers of ascospores to other plants. The rain-splash mechanism of ascospore dispersal also suggests that a critical mass of ascospores may be necessary to cause infection and subsequent symptom expression.

On creeping bentgrass grown in the mid-Atlantic region, dead spot is most severe between mid-June and late August. During this period, patch diameter as well as the number of pseudothecia produced per patch increase at a linear rate.[8] In growth-chamber studies, pseudothecia developed when incubated under constant light and were not produced when incubated in darkness.[9] In field studies, pseudothecia began to develop when day length increased to 14 h.[8] Although light is an important factor for the development of pseudothecia, the influence of photoperiod on their development remains unclear. Examination of diseased hybrid bermudagrass turf from putting greens in Texas and Florida, however, revealed the presence of numerous pseudothecia as early as March. Therefore, in addition to light, the accumulation of heat may play an important role in the development of pseudothecia. In a Maryland study, pseudothecia production generally began to increase at a linear rate in mid-to-late June.[8] Air and soil temperatures increase earlier in the year in the southern United States, when compared to the mid-Atlantic region. If the accumulation of heat were important to their development, pseudothecia likely would begin to develop later (i.e., June) in the mid-Atlantic region, when compared with Florida and Texas (i.e., March). Controlled experiments designed to examine the influence of temperature and photoperiod on *O. agrostis*-infested tall fescue/wheat bran mix would elucidate their role and possible interaction on pseudothecia development.

Dead-spot symptoms can appear as early as May in inoculated sites, however, the development of natural disease symptoms likely occurs following prolonged periods when soil temperatures are ≥20°C. Indeed, laboratory studies revealed that a 12–28-day incubation period at a constant temperature (20–30°C) was required for dead-spot reactivation in naturally infected, winter-dormant plants.[9] A model that utilized soil temperatures ≥20°C to predict the appearance of new *O. agrostis* infection centers (incidence) generally made false predictions during late spring.[10] In the autumn months, however, soil temperature generally declined dramatically (21°C in September to 14°C in October) and few new infection centers were observed after early October. In a 3-year field study, dead spot was severe in the initial year following establishment and following fumigation.[8,10] In those years when disease pressure was severe (2000 and 2002), dead-spot symptoms first appeared in either early June (2000) or early May (2002). In the 2000 study site, the area was inoculated the previous October, while inoculation of the study area used to monitor dead spot in 2002 occurred in March 2002. Despite nearly a 1-month difference in the appearance of new infection centers in the spring of 2000 and 2002, a linear increase in the expansion of new spots did not begin in either year until 218 and 213 total degree days (DDs) (biofix date = May 1; base temperature = 15°C).[7] Additionally, when inoculation occurred in autumn, *O. agrostis* infection centers immediately began to increase in size when they first appeared in late spring (June 8). When bentgrass was inoculated in the spring, however, new infection centers appeared, but remained relatively small in size for approximately 1 month prior to the linear increase in patch diameter. These observations indicate that there is an influence of inoculation timing on the appearance of dead-spot symptoms, but they also support the view that naturally developing symptoms appear later in the year (mid-to-late June) following the accumulation of DDs. Owing to the differences in inoculation timing, a cumulative DD model was not accurate in predicting the development of initial infection centers.[10] The linear increase phase for patch diameter and pseudothecia development, therefore, may be representative of the environmental conditions influencing natural symptom expression and peak dead-spot activity. As previously noted, natural dead-spot symptoms develop between 213 and 218 cumulative DDs. Hence, while a DD model for predicting the appearance of the initial infection centers in the spring was highly variable, the DD model predicting the start of the linear growth phase (i.e., peak period of disease) of the disease was more precise. Validation of the DD model for predicting the linear growth phase of dead spot, however, would require the collection of environmental data from various geographic regions and should be conducted under conditions in which inoculation occurs in autumn so that the pathogen overwinters and the disease is allowed to develop more naturally.

Dead spot is a polycyclic disease and the importance of *O. agrostis* ascospores in the spread and development of the disease was examined by Kaminski et al.[9] Ascospores of *O. agrostis* were observed to be forcefully ejected through ostioles of pseudothecia or exuded *en masse* in the presence of water. In field and growth-chamber studies, ascospores were rarely released when pseudothecia were dry. In the field, ascospores were released in large numbers at dawn and dusk, and also during precipitation events. Ascospore release events occurred when periods of low relative humidity coincided with periods of leaf wetness. As indicated by a growth-chamber study, ascospores were captured only when pseudothecia contained some level of moisture, but generally were not released from dry fruiting bodies.[11] During the morning hours in a field study, relative humidity was observed to sharply decrease while the canopy remained wet. Conversely, as dew and guttation fluid began to form in the evening hours, relative humidity remained low (~60%). In both instances, a moisture gradient would have been created in which dry atmospheric air presumably pulled water from saturated pseudothecia. The loss of moisture from saturated pseudothecia presumably caused a disruption of the two-layered asci, which resulted in the forceful discharge of ascospores. This likely occurs from contraction of bitunicate asci, which have a rigid outer layer (ectoascus) and an elastic inner layer (endoascus). As the moisture available within a pseudothecium decreases, the ectoascus layer ruptures, allowing the elastic endoascus to extend toward the ostiole, rupture, and forcefully release its ascospores.[12] The importance of declining atmospheric moisture levels on ascospore release was further supported by a growth-chamber study where it was shown that a sharp decrease (100% to ~50%) in relative humidity resulted in the immediate discharge of ascospores.[11] While it is clear that some level of moisture is needed for ascospore release, it appears that lower levels of relative humidity in conjunction with moist or saturated pseudothecia are essential for forceful ascospore release. It is possible, however, that high levels of relative humidity that saturate pseudothecia with water may result in an oozing rather than ejection of ascospores into the air. This may explain the lack of ascospores found in air samples collected from pseudothecia in a growth chamber where relative humidity was raised rapidly from ~50 to 100%. The rapid increase in relative humidity and resulting saturation of pseudothecia may have prevented the forceful release of Ascospores; however, pseudothecia were not monitored for *en masse* ascospore release at this time. Future growth-chamber studies designed to elucidate the impact of precipitation events during periods of varying levels of relative humidity would further clarify the mechanism of *O. agrostis* ascospore release. The study would include the application of water from simulated rainfall events to mature pseudothecia being maintained at a constant level of relative humidity ranging in increments from high (100%) to low (50%) relative humidity. To assess the impact of high levels of constant relative humidity on the *en masse* release of ascospores, the concentration of ascospores within the free water accumulating around mature pseudothecia could be determined through pipette and hemocytometer quantification techniques. Additionally, the impact of rain splash on the dissemination of ascospores could be assessed by quantifying the concentration of ascospores collected within petri dishes placed at varying distances from the inoculum source.

Regardless of whether ascospores are forcefully discharged or ooze from pseudothecia, the initiation of their release is rapid (≤1 h) and may continue as long as free moisture is present and relative humidity is below some unknown critical level. In the laboratory, ascospores were observed to germinate in as little as 2 h.[9] During the early hours of incubation, ascospores generally germinated in larger numbers in the presence of light and bentgrass leaves or roots. Using an ascospore suspension, ascospore germination and infection were observed to occur within 24 h, but the infection process may occur more rapidly.[9] Up to four germ tubes may develop from an individual ascospore. Each germ tube can either produce an appressorium or continue to grow as hyphae. Appressoria are capable of directly penetrating leaves and roots, while hyphae or germ tubes may enter open stomates on leaves. Tissue penetration from multiple appressoria from the same ascospore was observed in the lab.[9] Leaf-surface exudates appear to be an important factor in rapid germination and subsequent infection of leaves by *O. agrostis*. Although ascospores can germinate quickly (i.e., ≤2 h), a majority of the ascospores germinated within 8–12 h in water in the lab (25°C). On the basis of lab results, it appears likely that ascospores released in nature and just prior to an extended period of leaf wetness are more likely to complete the infection process. In the aforementioned study, percent ascospore germination was ≤36% after 4 h of incubation at 25°C. In a field study, large numbers of ascospores were collected between 0700 and 1000 h, and 1900 and 2300 h.[11] The bentgrass canopy, however, generally was dry by 1000 or 1100 h, leaving only a short period for germination and infection. Therefore, only small numbers of ascospores released in the morning or during short irrigation cycles (15–30 min) are likely to cause infection due to the rapid drying (1–2 h) of the bentgrass canopy. A majority of successful *O. agrostis* infections likely occur following the release of ascospores at dusk, prior to extended periods of leaf wetness during the evening hours.

Although unknown, the field environmental conditions necessary for ascospore germination and penetration of bentgrass leaves likely include prolonged periods of leaf wetness and average daily air temperatures $\geq 22°C$. The importance of leaf wetness duration and temperature on the incidence and severity of several plant pathogens has been widely studied.[13–17] The requirement for some minimum temperature for infection by *O. agrostis* to occur is supported by the observation that the large numbers of ascospores released in May, and again in September, did not coincide with the appearance of similarly large numbers of new infection centers.[10] While average air temperatures in May and September (2000–2002) ranged between 17°C and 21°C, air temperatures between June and August ranged between 22°C and 26°C. The aforementioned temperatures, as well as extended periods of leaf wetness, are common at night on creeping bentgrass putting greens in Maryland throughout the summer. Elucidation of the environmental conditions necessary for *O. agrostis* ascospores to infect tissue would provide a better understanding of the disease cycle. Growth-chamber studies in which ascospore-inoculated creeping bentgrass was exposed to varying hours of leaf wetness and temperatures and then incubated at varying postinoculation temperatures would better define both the conditions necessary for infection to occur as well as the conditions necessary for disease symptoms to appear. Additionally, inoculation using varying ascospore concentrations and direct placement of inoculum on various tissues (i.e., leaves, collar region, roots/stolons, etc.) would help determine the relative susceptibility of the various tissue types to infection. Results from the aforementioned studies would provide insight into the infection process. Furthermore, this new information would allow for the reexamination of data from the 3-year Maryland field study in which the environmental conditions surrounding the development and spread of dead spot were recorded.[8,10,11]

Although ascospores are often released in abundance, leaf spots rarely are observed on bentgrass plants. Dead-spot symptoms generally include reddish-brown or discolored leaves and darkened leaf tissue at the collar region of older leaves. On the basis of these observations, it appears likely that most ascospores alighting on bentgrass leaves are washed down into the collar (i.e., intersection of leaf and ligule) and potentially between the leaf sheaths, where they germinate and infect the leaf, sheath, or both. Following infection of bentgrass leaves, the pathogen apparently moves downward, colonizing tissue as it progresses toward stems and roots. It appears that *O. agrostis* enters a latent phase of several days between the time when infection is initiated and when dead spot symptoms first appear. It is likely that during this period, the pathogen continues to spread along the surfaces of the plant as hyphae and hyphal mats may be seen on stem bases and more commonly on nodal regions of bentgrass stolons of infected plants. Infection of these tissues does occur since simple hyphopodia and direct penetration have been observed. While traditional histological techniques may be useful in the examination of the infection process, these techniques are often time consuming and require familiarity with the pathosystem. For instance, examination of the infection process in potted plants often results in the necessity to differentiate fungal mycelia from both the target pathogen and other fungi that commonly contaminate the growing medium. Even when growing media are sterilized, these opportunistic fungi quickly reestablish themselves. If infection is successful with the target pathogen, the pathogen can be stained for visual examination. This process usually kills the organism, resulting in a snapshot of one particular phase of the infection process. Emerging molecular techniques (e.g., green fluorescence protein [GFP] transformation), however, allow for the visual observation of the infection process as well as the movement of the pathogen throughout the plant in real time and without disturbance to either the plant or the pathogen. Following insertion of a fluorescence gene, the pathogen will emit a fluorescent "glow" when exposed to various excitation wavelengths such as ultraviolet light. *O. agrostis* isolates transformed with the GFP protein allow for the monitoring of a single ascospore as it germinates, infects the host and moves throughout the plant without death of either the plant or pathogen.

During environmental conditions favorable for dead spot, plants survive for short periods following infection during what may be termed the "latent phase" of the disease. Symptoms appear when the pathogen colonizes large areas of the plant or the plant is weakened by some other environmental stress (e.g., heat or drought). Small dead spots (approximately 1–2 cm in diameter) generally appear in the field between 4 and 10 days following a large release of ascospores.[10] The length of this latent period (i.e., after infection, but before the appearance of symptoms) is likely dependent upon environmental conditions on days following infection. While extended periods of leaf wetness are likely necessary for infection to occur, dead-spot symptoms appear to be influenced by environmental conditions that promote plant stress (i.e., heat and drought). The environmental conditions defined below that influence epidemics and pseudothecia production help to better characterize the latent period between infection and symptom expression.

Conditions that influence dead-spot epidemics were defined by a multiparameter model including elevated air (ATMax $\geq$ 27°C) and soil (STMean $\geq$ 18°C) temperatures; low relative humidity (RHMean $\leq$ 80%); shortened periods of leaf wetness (LWD $\leq$ 14 h); and high levels of solar radiation (SOLMean $\geq$ 230 W m^{-2}).[10] The model accurately predicted 37 out of 40 dead-spot epidemics. It should be noted, however, that the models for predicting dead-spot incidence and severity were developed in Maryland and validation in other geographic regions has not been performed. In field situations, pseudothecia can be found within necrotic tissues when infection centers first appear. In growth-chamber studies, pseudothecia developed within 4 days following incubation in constant light on media consisting of tall fescue seed and wheat bran.[9] Combining the information obtained from the growth-chamber study and field observations, it is likely that pseudothecia formed by the time-field symptoms first appear had begun to develop at least 4 days prior to the appearance of new infection centers. Hence, the latent period between infection and symptom expression can occur in as little as 3 days under ideal environmental conditions that favor the pathogen.

Mature ascospores develop in pseudothecia within 7 days of incubation at 25°C in a growth chamber.[9] Pseudothecia development within necrotic bentgrass tissues occurred between late June and early September in Maryland and corresponded with mean soil temperatures between 24°C and 26°C in all 3 years.[8] This process of ascospore release, infection, symptom expression, and the subsequent production of pseudothecia, and maturity and release of new ascospores occurred in a cyclic pattern, which was repeated about every 12 days. In the field, major ascospore release events occurred as few as 9 days apart. Accounting for the release of ascospores and subsequent infection (possibly $\leq$ 24 h), the appearance of dead-spot symptoms (3–10 days following ascospore release), and the development of mature ascospores (3–4 days following appearance of mature pseudothecia), the entire cycle from ascospore release to the development of new ascospores may occur within 6–14 days. This timeline is consistent with field observations in which large numbers of ascospores were released on average every 11 (2001) to 12 (2002) days.[10] While it is apparent that the release of *O. agrostis* ascospores and the development of dead-spot symptoms occur in a cyclic fashion, it is unlikely that this cycle is dictated by anything more than the necessary time-line required between ascospore release and the development of new ascospores. It is likely, however, that the time required to accomplish this reproductive process is influenced by the previously described environmental conditions associated with infection, symptom expression, and pseudothecia development. Controlled growth-chamber experiments designed to assess the influence of temperature, relative humidity, soil moisture, and other factors on the development of dead-spot symptoms would more accurately define the disease cycle.

15.3   MOLECULAR ASPECTS OF *O. AGROSTIS*

Dead spot often is confused with other common turfgrass diseases or maladies. While the presence of pseudothecia often serves as a quick and accurate diagnostic aid, these sexual fruiting bodies are not always present in infected samples. To aid in the identification of the pathogen and diagnosis of the disease, species-specific primers were developed.[18] Each primer in the set was designed from previously sequenced internal transcribed spacer (ITS) regions of *O. agrostis*.[2] Both primers (OaITS1 and OaITS2) consist of 22 nucleotides and amplify a 445- or 446-bp DNA fragment specific to

O. agrostis and do not amplify the DNA of the three other known *Ophiosphaerella* species found in turf. Additionally, the primers did not amplify the DNA from several other pathogens commonly associated with various turfgrass diseases. The primers were capable of detecting the pathogen from both pure cultures as well as from field-infected creeping bentgrass. The entire process including DNA isolation, amplification, and visualization can be completed in approximately 4 h. Therefore, these primers would be beneficial to diagnostic labs that are equipped to perform polymerase chain reaction (PCR), and especially valuable to those having little or no experience in diagnosing dead spot.

In general, plant pathogens that reproduce solely by asexual means exhibit very little genomic diversity. In contrast, higher levels of genetic diversity would be expected from fungi that undergo sexual recombination. Additionally, many plant pathogens with a polycyclic disease cycle initiate disease by sexual means (e.g., ascospores) and then produce asexual structures (e.g., conidia), which serve as the sole source of secondary inoculum. An unusual characteristic of *O. agrostis*, however, is its lack of a known anamorph combined with its ability to produce numerous pseudothecia throughout the summer months. Ascospores within pseudothecia serve as the secondary source of inoculum. The exclusivity of the sexual stage of *O. agrostis* in nature likely adds to the genetic diversity of the species. Initial DNA fingerprinting of *O. agrostis* isolates collected in 1998 revealed an 87% or greater similarity among isolates from five different states.[2] On the basis of the DNA fingerprinting of 77 isolates collected between 1998 and 2003 from 11 states, the species was separated into three distinct clades with ≥69% similarity.[19] Variation in the results of the two studies was likely due to differences in the amplified fragment length polymorphism (AFLP) techniques utilized (fluorescent versus polyacrylamide gel) and the larger number of isolates examined (12 versus 77).

15.4   DEAD-SPOT CONTROL AND MANAGEMENT

Little is known about cultural and chemical strategies for managing dead spot. A few studies, however, have identified differences in disease severity among bentgrass cultivars, fungicides, and various nitrogen sources.[3,20–22] Kaminski and Dernoeden[3] reported that *O. agrostis* was capable of infecting the three common *Agrostis* species grown on golf courses, including creeping, velvet (*Agrostis canina* L.), and colonial (*Agrostis capillaris* L.) bentgrasses. Variation in cultivar susceptibility was reported among bentgrass cultivars. Newer bentgrass cultivars (L-93 and the Penn A and G series), generally exhibited the most susceptibility and the older cultivars (i.e., Penncross and Pennlinks) the least.

When applied after peak disease activity, various nitrogen (N) sources (ammonium sulfate; isobutylidene diurea, IBDU; sulfur coated urea, SCU; urea; Ringer Greens Super; and methylene urea) aided in the recovery of bentgrass from dead spot.[23] In the aforementioned study, the disease was allowed to naturally develop throughout summer and no control measures were implemented prior to application of the fungicide iprodione and the various fertilizer treatments. It was observed that large infection centers (6–10 cm in diameter) were not capable of completely recovering, regardless of treatment. Hence, none of the fertilizers provided for complete turf recovery prior to winter. Plots treated with ammonium sulfate, however, resulted in a more rapid turf recovery, when compared with the slow-release IBDU fertilizer. These results indicated that quick-release N sources may be more beneficial in aiding bentgrass recovery, but that larger infection centers will likely remain present for extended periods, regardless of N source.

In a second field study, the impact of five water-soluble N sources ($Ca(NO_3)_2$, KNO_3, $(NH_4)_2SO_4$, urea, and 20–20–20 on dead-spot severity and recovery were examined.[23] Anecdotal information obtained from Dr. R. White at Texas A&M indicated that the application of $(NH_4)_2SO_4$ may prevent and reduce the occurrence of dead spot in bermudagrass. When applied approximately 1 month (June 28, 2002) after infection centers first were observed, however, the aforementioned N-sources failed to prevent new infection centers from occurring. In fact, the number of infection centers among all treatments increased from an average of 21 to 93 infection centers per plot between 27 June and 10 August 2002. All the N-sources reduced disease severity prior to winter, when compared with unfertilized plots. In year 2 of this second field study, dead spot did not recur in plots receiving $(NH_4)_2SO_4$, but new infection centers were observed in plots treated with nitrate or urea-based fertilizers. The number of dead-spot infection centers was positively correlated with pH, and disease incidence appeared to be favored by a pH greater than 6.0 and 6.6 in the mat (0–2.5 cm) and underlying soil (2.6–5.0 cm), respectively. Ammonium-based N fertilizers have been shown to reduce the severity of several turfgrass diseases, presumably due to their ability to acidify soil. These results suggest that the acidifying $(NH_4)_2SO_4$ may reduce dead spot when applications are initiated during establishment and several months in advance of the time when initial symptoms would appear. Conversely, increases in the severity of some turfgrass diseases have been attributed to the application of alkaline-reacting nitrate forms of fertilizer such as $Ca(NO_3)_2$ and $NaNO_3$. Although not statistically different from the other N sources evaluated, applications of KNO_3 resulted in a general reduction in dead spot in the first year, but dead-spot symptoms recurred in those plots in the second year. The full impact of these N sources on dead-spot incidence and severity may be realized only following their repeated application during establishment (i.e., prior to symptom expression). Monitoring increases or decreases in soil pH accorded to the different N sources and relating them to dead-spot incidence and severity would be most informative.

In the second study,[23] the initial soil K_2O levels were low, and it remains unclear if the addition of K from sources other than KNO_3 would result in an overall reduction of the disease. Further research on the impact of K in reducing dead spot would provide a better understanding of cultural dead-spot-suppression methods. Various K sources and rates should be examined for their impact on dead-spot severity. It would be informative to apply K_2SO_4 (0–0–50) or KCl (0–0–60) with various N sources to determine the impact of N + K versus the impact of N or K alone on dead-spot severity. These studies may reveal an additive or synergistic effect on the reduction of dead spot from the application of N (e.g., $(NH_4)_2SO_4$ versus $NaNO_3$) in conjunction with K. A potential reduction in disease simply from the addition of K in K-deficient soils also should be assessed.

Additionally, growth-chamber and field experiments assessing the impact of soil type and varying pH levels would help elucidate their role in dead-spot incidence and severity. Greenhouse studies should examine the effect of sand-based soils (calcium versus silicate sands) modified to varying levels of pH on dead-spot severity. While it is clear that various nitrate or ammonium-based N sources change soil and mat pH after repeated applications, the direct impact of pH is unknown. Owing to the elevated pH levels of water used to irrigate many golf courses, acidifying agents are occasionally used to reduce irrigation water pH. Therefore, studies should be conducted to also address the impact of irrigation water pH on the disease. The repeated application of acidifying fertilizers (i.e., $(NH_4)_2SO_4$) may reduce the pH to levels unfavorable for bentgrass growth. In these instances, turfgrass managers may apply lime to ameliorate a low soil pH condition. The effect of lime or other alkaline-reacting materials applied alone or in combination with ammonium sulfate is unknown and requires further investigation.

Owing to the limited occurrence of dead spot and its rapid decline, few fungicide efficacy studies have been conducted. The ability of the fungicide iprodione to control dead spot has provided varying results. In North Carolina, iprodione was shown to reduce dead-spot severity when applied both preventively and curatively.[21] In a New Jersey study,[20] however, preventive applications of iprodione provided only fair control of dead spot. Iprodione applied after the peak period for dead-spot activity in Maryland had passed (i.e., September) resulted in little or no reduction in dead spot severity.[23] After this peak period in the Maryland study, however, few new infection centers appeared and increases in patch diameter generally did not occur. Therefore, the Maryland study indicated that iprodione, and perhaps other fungicides, are likely to be effective only when applied preventively or just after the initial symptoms are observed. Field fungicide evaluations reported by Towers et al.[20] and Wetzel and Butler[21,22] showed that boscalid, propiconazole, chlorothalonil, thiophanate methyl, fludioxonil, and pyraclostrobin effectively controlled dead spot, but little or no control was provided by triadimefon, trifloxystrobin, mefenoxam, or a formulation of *Bacillus subtilis*. Curative chemical management with the aforementioned fungicides is less efficacious.[22]

In the mid-Atlantic region, a preventive fungicide program for new putting-green constructions would begin in May, prior to the time when symptoms are most likely to appear. Applying fungicides during the period of increasing patch diameter and pseudothecia development (i.e., June through August), would likely result in reduced or poor dead-spot control. According to Wetzel and Butler,[21] weekly applications of urea in conjunction with early curative fungicides reduced the number and diameter of dead-spot infection centers. When applied weekly, however, urea alone did not significantly reduce dead-spot severity, when compared with the untreated control.[21] Hence, in an early curative fungicide program, chemicals need to be applied early enough to prevent new infections and nitrogen should be tank-mixed with fungicides to aid in the recovery of existing dead spots. The interaction of various fungicides and fertilizers needs to be investigated further to properly develop control programs directed toward the most effective management of dead spot.

ACKNOWLEDGMENT

The author thanks Dr. Peter H. Dernoeden for his knowledge, patience, and encouragement during the investigation of dead spot.

REFERENCES

1. Dernoeden, P.H., O'Neill, N.R., Câmara, M.P.S., and Feng, Y. A new disease of *Agrostis palustris* incited by an undescribed species of *Ophiosphaerella*, *Plant Dis.*, 83, 397, 1999.
2. Câmara, M.P.S., O'Neill, N.R., van Berkum, P., Dernoeden, P.H., and Palm, M.E. *Ophiosphaerella agrostis* sp. nov. and its relationship to other species of *Ophiosphaerella*, *Mycologia*, 92, 317, 2000.
3. Kaminski, J.E. and Dernoeden, P.H. Geographic distribution, cultivar susceptibility, and field observations on bentgrass dead spot, *Plant Dis.*, 86, 1253, 2002.
4. Kaminski, J.E., unpublished data, 2002.
5. Smiley, R.W., Dernoeden, P.H., and Clarke, B.B. *Compendium of Turfgrass Diseases*, 3rd ed., The American Phytopathological Society Press, St. Paul, MN, 2005, p. 103.
6. Smith, J.D., Jackson, N., and Woolhouse, A.R. *Fungal Diseases of Amenity Turf Grasses*, 3rd ed., E. & F.N. Spon, New York, 1989, chap. 11.
7. Kaminski, J.E. Biology of *Ophiosphaerella agrostis*, epidemiology of dead spot and a molecular description of the pathogen, Ph.D. diss., University of Maryland, College Park, 2004.
8. Kaminski, J.E. and Dernoeden, P.H. Severity of dead spot, pseudothecia development and overwintering of *Ophiosphaerella agrostis* in creeping bentgrass, *Phytopathology*, 95, 248, 2006.
9. Kaminski, J.E., Dernoeden, P.H., O'Neill, N.R., and Momen, B. Reactivation of bentgrass dead spot and growth, pseudothecia production and ascospore germination of *Ophiosphaerella agrostis*, *Plant Dis.*, 86, 1290, 2002.
10. Kaminski, J.E., Dernoeden, P.H., and Fidanza, M.A. Environmental monitoring and exploratory development of a predictive model for dead spot of creeping bentgrass, *Plant Dis.*, 91, 565, 2007.
11. Kaminski, J.E., Dernoeden, P.H., and O'Neill, N.R. Environmental influences on the release of *Ophiosphaerella agrostis* ascospores under controlled and field conditions, *Phytopathology*, 95, 1356, 2005.
12. Alexopoulos, C.J., Mims, C.W., and Blackwell, M. *Introductory Mycology*, 4th ed., Wiley, New York, 1996, chap. 7.
13. Carisse, O. and Kushalappa, A.C. Influence of interrupted wet periods, relative humidity and temperature on infection of carrots by *Cercospora carotae*, *Phytopathology*, 82, 602, 1992.
14. Evans, K.J., Nyquist, W.E., and Latin, R.X. A model based on temperature and leaf wetness duration for establishment of Alternaria leaf blight of muskmelon, *Phytopathology*, 82, 890, 1992.
15. Schuh, W. and Adamowicz, A. Influence of assessment time and modeling approach on the relationship between temperature-leaf wetness periods and disease parameters of *Septoria glycines* on soybeans, *Phytopathology*, 83, 941, 1993.
16. Sullivan, M.J., Damicone, J.P., and Payton, M.E. The effects of temperature and wetness period on the development of spinach white rust, *Plant Dis.*, 86, 753, 2002.
17. Wu, L., Damicone, J.P., Duthie, J.A., and Melouk, H.A. Effects of temperature and wetness duration on infection of peanut cultivars by *Cercospora arachidicola*, *Phytopathology*, 89, 653, 1999.

18. Kaminski, J.E., Dernoeden, P.H., O'Neill, N.R., and Wetzel III, H.C. A PCR-based method for the detection of *Ophiosphaerella agrostis* in creeping bentgrass, *Plant Dis.*, 89, 980, 2005.

19. Kaminski, J.E., Dernoeden, P.H., Mischke, S., and O'Neill, N.R. Genetic diversity among *Ophiosphaerella agrostis* strains causing dead spot in creeping bentgrass, *Plant Dis.*, 90, 146, 2006.

20. Towers, G.W., Majumdar, P.R., Weibel, E.N., Frasier, C.L., Vaiciunas, J.N., Peacos, M., and Clarke, B. Evaluation of chemical and biological fungicides for the control of bentgrass dead spot in creeping bentgrass, *Rutgers Turfgrass Proc.*, 32, 211, 2000.

21. Wetzel III, H.C. and Butler, E.L. Evaluation of fungicides and urea for the control of bentgrass dead spot in an 'L-93' putting green in Raleigh, NC, 1999, *Fungic. Nematic. Tests*, 55, 510, 2000.

22. Wetzel III, H.C. and Butler, E.L. Preventive versus curative control of bentgrass dead spot with fungicides and urea, 2000, *Fungic. Nematic. Tests*, 56, T22, 2001.

23. Kaminski, J.E. and Dernoeden, P.H. Nitrogen source impact on dead spot (*Ophiosphaerella agrostis*) incidence and severity in creeping bentgrass, *Int. Turfgrass Res. Soc. J.*, 10, 214, 2005.

Section VII

Turfgrass Integrated Pest Management

16 Herbicide Formulation Effects on Efficacy and Weed Physiology

David S. Gardner and S. Gary Custis

CONTENTS

16.1 WEEDS AS UNDESIRED COMPONENTS OF A TURF SYSTEM

A common definition of a weed is a plant grown out of place. In highly maintained turfgrass, the crop-weed dynamic is altered by frequent mowing. Only a select few plants can adapt to continuous mowing pressure to become weeds in managed turfgrass areas. However, the species that can survive are in many cases quite capable of competing with the turfgrass for available resources.

When attempting to control weeds in turfgrass, the first step is to properly identify the weedy species. This will aid in determining the most appropriate herbicide for control. Almost as important as identification of the weed is recognition of the life cycle of the weedy species. Weeds in turfgrass can have either an annual life cycle or persist as perennials. Annuals are those in which the weed germinates, matures, sets seed, and then dies within the course of 1 year. Annual species are subclassified into summer annuals, those that germinate in spring and die following frost in fall, and winter annuals, which germinate in fall, persist vegetatively over winter, then set seed and die before summer. Perennials are those weeds that persist for more than 3 years. Recognition of the life cycle of the weed governs the most appropriate time of year for control. Annuals should be controlled before the set seed, meaning a spring application for cool-season annuals, early summer for warm-season annuals, or early winter for winter annuals. Perennials are best controlled in mid- to late fall. Spring time applications will, in many cases, result in regrowth of the weed. The time of year in which the herbicide is applied will, in many cases, affect the choice of formulation of the herbicide.

16.2 OVERVIEW OF HERBICIDE FORMULATIONS

Herbicides may be classified as either nonselective or selective. Nonselective herbicides, such as glyphosate, are efficacious against all plant species. In contrast, selective herbicides will result in the death of one type or species of plant but leave others intact. Examples in turfgrass are materials that remove broadleaf species from the grass, such as 2,4-D (2,4-dichlorophenoxy acetic acid) or triclopyr (3,5,6-trichloro-2-pyridyloxyacetic acid). It is also possible to develop herbicide chemistry based on the metabolism of the plant. As such, selective herbicides are available to remove warm-season grasses from cool-season grasses and vice versa. However, the more closely related the weedy species, the more difficult it is to develop selective herbicides. Thus, annual grasses are more difficult to control than broadleaf weeds and perennial grasses, in many cases, can not be selectively controlled in turf.

Herbicides can also be classified according to their application timing relative to the life stage of the target weed. Preemergence herbicides are intended to form a barrier at the soil surface, affecting the weed following germination of the seed. These are effective against annual weeds that come from seed each year. They will not have activity on existing weeds. Postemergence herbicides are applied to the existing plant, preferably when still young. Postemergence herbicides are available for control of both grassy and broadleaf weeds.

Finally, herbicides may be classified according to whether they are contact or systemic. Contact materials will kill only those plant parts directly contacted by the herbicide. They tend to cause more rapid expression of herbicide-injury symptoms. Systemic materials are absorbed by the plant and translocated throughout the tissues. These herbicides are more effective on perennial weeds and those that persist from belowground structures.

16.2.1 HERBICIDE FORMULATIONS

Pure herbicidal compounds are rarely, if ever, available to the end user. In most cases, the pure herbicide is either too concentrated or in a form that is not easily applied. Instead, they are typically formulated with solvents, surfactants, and other additives. The same active ingredient may be available in several different formulations. Formulation chemistry has advanced considerably and it is now possible to formulate a herbicidal compound in a way that enhances its speed of activity and overall efficacy or to enhance storage and make mixing more easy and convenient.

Herbicides are available in either liquid or granular formulations (Table 16.1). The type of formulation will affect mixing and application methods. There will also be differences in volatility, phytotoxicity, uniformity and ease of application, and cost. Most important, perhaps, is that the type of herbicide formulation chosen can affect the performance of the herbicide either by altering its uptake from the plant surface or by effecting efficiency of root uptake. In many cases, variation in plant response after treatment with various formulations of the same herbicide is caused by differences in foliar absorption among formulations.

16.2.1.1 Liquid Formulations

Liquid formulations are made by mixing the pesticide concentrate, either in liquid or solid form, with water or another solvent to form either a suspension or a solution. Those materials that form suspensions must be agitated to prevent the active ingredient from settling to the bottom of the tank. Liquid formulations are used to deliver both pre- and postemergence herbicides. Preemergence

TABLE 16.1

Formulations of Herbicides Available in the Turfgrass Industry

Formulation	Abbreviations	Units	Notes
Liquid, aqueous suspension	S, SL	Pounds a.i./gallon	Stable suspensions of liquid active ingredients (a.i.) in an aqueous carrier
Granular	G	% a.i. by weight	Ready-to-use dry materials with a.i. formulated on clay or other carrier
Water-dispersible granules	WDG, DG	% a.i. by weight	Does not dissolve in water. Agitation required to keep product in suspension
Wettable powder	WSP, WP, W	% a.i. by weight	Does not dissolve in water. Agitation required to keep product in suspension
Soluble powder	SP	% a.i. by weight	Dry materials that form a true solution and do not require further agitation after mixing
Dry flowable	DF	% a.i. by weight	Materials sprayed on inert carrier that is finely ground. Does not dissolve in water
Liquid flowable	F, L, FL, FLO	Pounds a.i./gallon	Same as DF except material is suspended in small amount of liquid. Shake before use
Emulsifiable concentrate	EC, E	Pounds a.i./gallon	a.i. + petroleum solvents + emulsifier. Separates if solution is not agitated
Water-soluble concentrate	WS	Pounds a.i./gallon	Active ingredients readily dissolve in water. No agitation required
Dust	D	% a.i. by weight	Ready to use. Easily drifts into nontarget areas
Fumigant			Gaseous material or material that reacts to form gas. A tarp is usually required for use

herbicides, if in liquid form, are generally applied as a water-insoluble formulation, such as a liquid flowable (F), dry flowable (DF), wettable powder (WP), or water-dispersible granule (WDG).

WPs are finely ground formulations of a carrier, often clay, with herbicide and other materials designed to enhance the ability of the powder to suspend in water. They are easy to store and handle, but are abrasive to application equipment, require agitation, and may leave visible residues after application. DFs and WDGs are similar except that the herbicide is formulated onto a larger particle size. The main advantage of these formulations are ease of measuring and lower inhalation hazard when mixing. Like WPs, these formulations may leave visible residues. Soluble powder (SP) formulations form true solutions and do not require further agitation after mixing. However, these formulations are rare because few of the herbicide active ingredients used commonly in turf are very water soluble.

Postemergence herbicides are usually formulated as a liquid concentrate, such as a solution (S), emulsifiable concentrate (EC or E), liquid flowable (FL or L), or microemulsions (ME). Solutions are formulations of herbicides that readily dissolve in water and do not settle out of solution after tank mixing. Emulsifiable concentrates contain the herbicide active ingredient, other additives, and an emulsifying agent that allows for mixing with water. This is a very common formulation because it offers several advantages, such as ease of transport and storage, lower cost, and lower abrasiveness to application equipment. However, the solvents in emulsifiable concentrates result in higher potential phytotoxicity compared with other formulations. Liquid flowables are similar to DFs, except that the active ingredient and finely ground carrier are suspended in liquid. Liquid flowables are easy to handle and store, but like DFs, these formulations may leave visible residues on the plant surface.

16.2.1.2 Granular Formulations

There are two main formulations of pesticides applied dry: granular (G) and dust (D). Dusts are almost never used when applying herbicides, due to concerns with drift. The main advantages of granular formulations are ease and speed of application. Granular formulations are reserved almost exclusively for soil-applied preemergence herbicides that are taken up by plant roots, but can be used foliarly if applied to wet foliage. When using preemergence herbicides, application as a granular or as a liquid suspension (WDG, DF, or WP) aids in forming a barrier to weed emergence at the soil surface. Granular formulations are also used to deliver postemergence herbicides, particularly for broadleaf weed control in the homeowner market. Some advances have been made to make granular formulations more likely to stick to foliage, particularly dew-moistened foliage. However, granular formulations are generally not as effective as their liquid counterparts.

16.2.2 FUNCTIONAL GROUPS

Many of the herbicides commonly used for broadleaf weed control are members of two chemical classes, the phenoxies and the pyridinoxies. The phenoxies include the herbicides 2,4-D, 2,4-DP (2,4-dichlorophenoxy propionic acid), MCPA (2-methyl-4-chlorophenoxyacetic acid), and MCPP (2-[2-methyl-4-chlorophenoxy]propionic acid). The pyridinoxies include triclopyr, clopyralid (3,6-dichloropyridine-2-carboxylic acid), and fluroxypyr (4-amino-3,5-dichloro-6-fluoro-2-pyridyl-oxyacetic acid). These herbicides are available formulated as either granular materials or as liquids. The acid forms of these herbicides are not usually available due to their low solubility in either water or oil [1]. It was discovered soon after 2,4-D and related compounds were developed that formulating the herbicide by substituting the hydrogen on the carboxyl group with either an amine or an ester affected herbicidal activity [2]. The formulation can affect plant uptake and efficacy of the herbicide as well as volatilization potential and safety to the applicator.

16.2.2.1 Amine Formulations of Herbicides

The addition of an amine group to a herbicide acid molecule forms a relatively nonvolatile salt. For example, adding a dimethylamine group converts 2,4-D acid into an amine. The primary advantage lies in the reduction in volatility and thus reduced potential for nontarget drift injury to nearby ornamental plants. Interestingly, amine formulations are generally associated with a stronger odor, while ester formulations have lower odor. However, Brown et al. [3] concluded that trace impurities and additives contribute to the odor of the herbicide and not the active ingredient, which is odorless. The odor in most amine formulations comes from the use of dimethylamine in the formulation. Common forms of amines include the monoethanolamine, dimethylamine, and tiethanolamine salts.

16.2.2.2 Ester Formulations of Herbicides

The addition of an ester group to a herbicide acid molecule allows for easier penetration of the herbicide through the waxy cuticle of the leaf, thus increasing herbicide uptake and control efficacy. Esters are insoluble in water, but soluble in oil, and thus often formulated with emulsifying agents. Especially in warm weather, these compounds can volatilize and drift resulting in injury to non-target plants. Therefore, in turfgrass, esters are used primarily in fall, winter, or early spring.

There are several different types of ester formulations, all of which are more volatile than amine or sodium salts [4]. The volatility of esters is inversely proportional to the molecular weight and chain length of the alcohol branch. Methyl (one-carbon) and ethyl (two-carbon) esters are very volatile as is the isopropyl (three-carbon) ester. Butyl (four-carbon) and amyl (five-carbon) esters are not as volatile. Lower-weight esters, such as the methyl, ethyl, and isopropyl esters are no longer commonly commercially available [2].

The most common ester herbicides in turfgrass are the isooctyl (2-ethylhexyl), butoxyethyl, and octanoic acid ester formulations, though others are also used, such as the 1-decyl ester of clopyralid and the 1-methylheptyl ester of fluroxypyr. Low volatile esters of herbicides such as 2,4-D have been used for weed control since the 1940s [5]. The longer-chain ester formulations reduce, but do not eliminate, volatility [1]. Research continues to develop esters with lower volatility. The use of new solvents and emulsifiers has allowed ester formulations to be developed that have much lower volatilities that are similar to amine formulations.

The weight of the ester has a significant impact on the efficacy of the herbicide. Lighter esters of 2,4-D penetrate the leaf very rapidly. However, they also result in rapid injury and subsequent failure to translocate throughout the plant. In contrast, heavy esters such as glycol diesters or tetrahydrofufuryl esters will penetrate the cuticle too slowly to be effective [6].

Studies with ester formulations of 2,4-D found that upon uptake, the ester is hydrolyzed and the free acid form of the herbicide is moved through the plant [6,7]. This process occurs rapidly, with up to 75% of the 2-ethylhexyl ester formulation of 2,4-D found to be hydrolyzed within 30 min [8]. Complete metabolism (no detectible residues) of the herbicide within a forage plant was found to occur within 4 weeks [5].

16.2.3 ISOMERS OF HERBICIDES

Isomers have the same chemical formula but different structural arrangement of the atoms within the molecule. The manufacturing process can often result in a mixture of isomers of the same active ingredient. One isomer of a herbicidal compound will be effective as a herbicide and the others will have much reduced efficacy. Fenoxaprop-ethyl ([RS]-2-[4-[6-chloro-1,3-benzoxazol-2-yloxy]phenoxy]propionic acid), commonly used to control crabgrass postemergence in turfgrass, is an example of a herbicide in which the manufacturer developed a method of producing a formulation with a greater percentage of the active isomer. As a result, the recommended application rates were decreased on the product label.

16.2.4 SURFACTANTS

Effective control of weeds with postemergence herbicides is, in many cases, governed by the ability to effectively introduce sufficient quantity of the herbicide into the target plant tissue. Several factors associated with the herbicide spray affect this, including the volume of the spray, the size of the spray droplets, and the surface tension of the leaves [9]. An insufficient spray volume can result in decreased control by decreasing coverage of the target foliage. Conversely, if spray volumes are too large, control can be decreased because the target plant surface is not able to hold sufficient quantities of the spray.

Surfactants, or surface acting agents, are usually included in a herbicide formulation to improve penetration of the herbicide into the plant tissue. The mechanism by which this occurs reduces the surface tension on the leaf as well as the contact angle with the leaf surface [10]. As a result, the spray droplets spread over a larger portion of the leaf surface. This leads to increases in both the rate and total amount of herbicide absorption. Surfactants may also reduce or prevent crystallization of certain forms of the herbicide on the leaf surface, which will lead to improved uptake [11].

Depending on the herbicide, the addition of a surfactant may increase effectiveness by up to 50%. Changes in the concentration or formulation of a surfactant will affect herbicide uptake. For example, Kloppenburg and Hall [12] noted that lower surfactant concentration affected absorption of clopyralid. Bovey et al. [13] found similar amounts of 2,4,5-T in plants treated with amine or ester formulations when application rates were similar and a surfactant had been added.

16.3 FORMULATION EFFECTS ON ENVIRONMENTAL FATE

The ideal scenario is for 100% of the applied herbicidal compound to be absorbed by the target weed. However, herbicides can and do move from their intended target and this may cause either potential for damage to the health of the turf user or environmental contamination [14–16]. The fate and environmental impact of pesticides has received considerable attention since 1990.

After a pesticide has been applied to a turfgrass system, it is subject to numerous biological, chemical, and physical processes that will determine its fate. Many of these processes will interact and this complicates our ability to quantify or to predict pesticide fate. The study of pesticide fate is also complicated by the wide array of chemicals used on turfgrass. Each pesticide is unique. The dinitroanalines are an example of a family of herbicides with widely varying sorption and volatility potentials [17]. Differences in soil type and climate will also affect pesticide fate.

Processes that affect the environmental fate of herbicides include sorption of the pesticide to soil particles or organic matter, degradation of the compound by soil-borne microbes, leaching of the material through the soil profile, runoff, volatilization, uptake of the material by turfgrass or target weeds, photolysis, and chemical degradation of the pesticide [18]. A thorough review of fate processes was conducted by Branham [19] and the reader is referred to that source for more information. The fate processes most affected by alterations in herbicide formulation are volatilization, photodegradation, and plant uptake (herbicide efficacy).

The formulation of the herbicide will impact its fate immediately upon exiting the sprayer, due to differences in potential to volatilize. All herbicides may potentially drift as spray droplets. Physical drift is the result of mechanical effects on the droplet. High pressure and nozzle size can directly affect the potential of any material to drift.

Most volatilization losses are observed within the first few days of application. Volatilization losses tend to increase with increasing temperatures or solar radiation intensity [16]. The study of pesticide volatilization is time consuming and expensive and thus not practical when attempting to study the full range of herbicides used in turf as well as different site and weather conditions. However, Haith et al. [16] implicate direct volatilization from turfgrass foliage to be the most likely source of volatility losses. Therefore, the formulation of an applied herbicide will have a significant impact on the volatilization potential of that herbicide. For example, application of a granular herbicide would result in significantly less volatilization, particularly if irrigation or rainfall did not follow application.

Formulation differences will also affect other avenues of the environmental fate of a herbicide. For example, the ester formulation of MCPA is stable when the pH of the solution is acidic. However, it will hydrolyze under basic conditions [20]. With 2,4-D, however, the rate of breakdown of the ester and amine formulation is similar because either form is rapidly converted into the same acid form [1]. The butoxyethyl ester formulation of triclopyr may be lost to volatilization, photodegredation, or uptake by the plant [21].

Another possible route of herbicide loss, which has environmental ramifications, is for the herbicide to dislodge from the plant tissue. When equipment, people, etc. come into contact with recently treated turf, the herbicide may move from the treated plant. The application rate has a large impact on this, with increasing herbicide rates causing increases in dislodgeable residues. In fact, doubling or quadrupling the rate was reported by McLeod et al. [22] to cause a greater than 2 times or 4 times increase in dislodgeable residues. The formulation of the herbicide also has an impact. McLeod et al. [22] applied either the dimethylamine salt or isooctyl ester of 2,4-D and found less of the ester formulation dislodged, independent of application rate. This was likely due to either the faster rate of absorption into the plant cuticle or the greater volatility of esters.

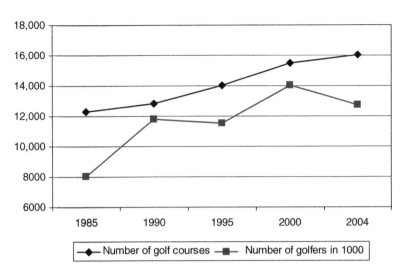

COLOR FIGURE 1.1 The number of golf courses and players in the United States, 1985–2004. (From U.S. Department of Commerce, Statistical Abstract of the United States. 2004–2005. Bureau of the Census, Washington, D.C., 2005.)

COLOR FIGURE 4.1 Photo of bermudagrass shoot and root growth under various levels of nutrients in the culture medium.

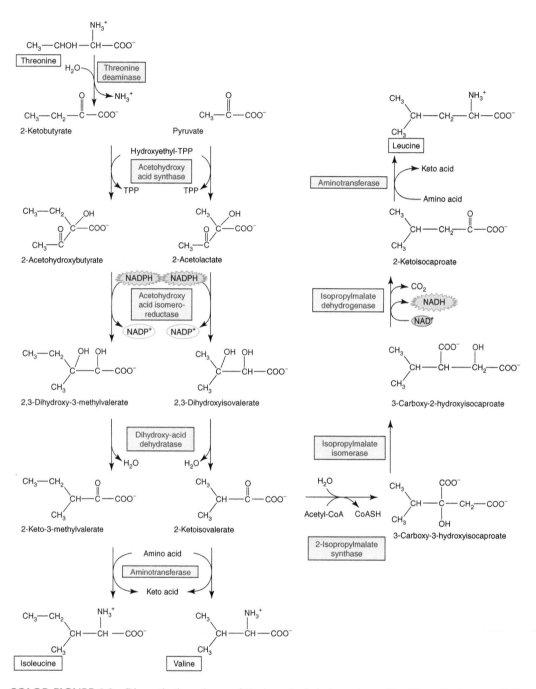

COLOR FIGURE 6.2 Biosynthetic pathway of the branched-chain amino acids. (From Buchannan, B.B., Gruissem, W., and Jones, R.L., *Biochemistry and Molecular Biology of Plants*, American Society of Plant Biologists, Maryland, 2000. With permission.)

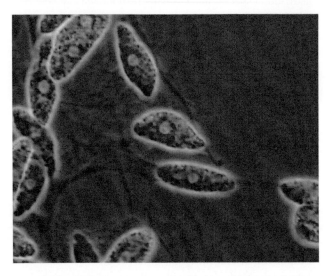

COLOR FIGURE 14.1 The rapid blight organism growing in culture. Vegetative cells of *Labyrinthula terrestris* at 100× magnification.

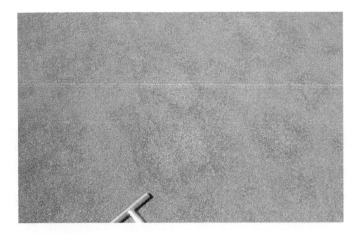

COLOR FIGURE 14.2 Rough bluegrass overseeding infected with rapid blight disease.

COLOR FIGURE 14.3 Symptoms of rough bluegrass overseeding on putting greens infected with rapid blight disease.

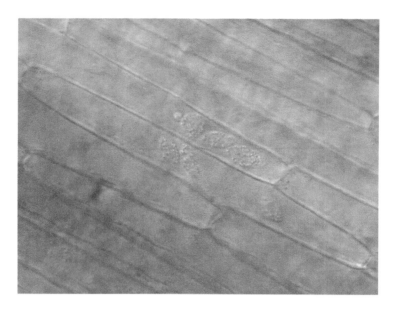

COLOR FIGURE 14.4 Vegetative cells of *Labyrinthula terrestris* in leaf epidermal cells at 40× magnification.

COLOR FIGURE 21.2 Damage to a golf-course fairway caused by high densities of first-generation ABW larvae.

COLOR FIGURE 21.4   Adult ABW.

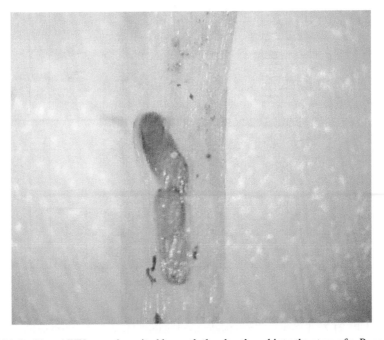

COLOR FIGURE 21.5   Two ABW eggs deposited beneath the sheath and into the stem of a *P. annua* plant.

COLOR FIGURE 21.6 ABW third, fourth, and fifth instars and pupa next to a 10 cent coin for comparison.

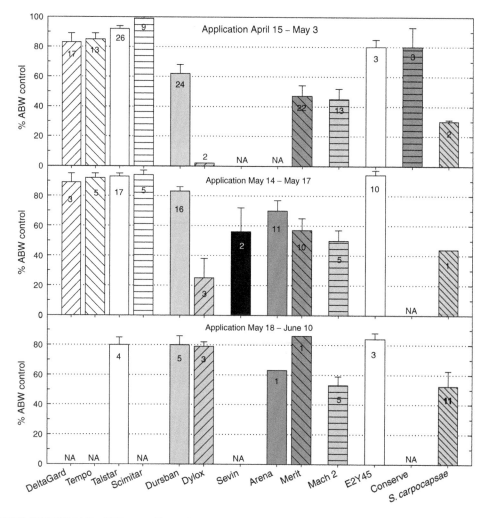

COLOR FIGURE 21.7 Efficacy of insecticides used against ABWs in the northeastern United States. Shown are means ±SE. Numbers in bars are number of treatments included in summary. (*Summary from Arthropod Management Tests*, vols. 18–28; McGraw, unpublished data, 2005.)

COLOR FIGURE 22.1 Bermudagrass roots damaged by sting nematode (*Belonolaimus longicaudatus*) (left) are abbreviated and "stubby" in appearance compared with healthy roots (right) (University of Florida).

COLOR FIGURE 22.2 Bermudagrass putting green infested with damaging numbers of sting (*Belonolaimus longicaudatus*) and lance (*Hoplolaimus galeatus*) nematodes (University of Florida).

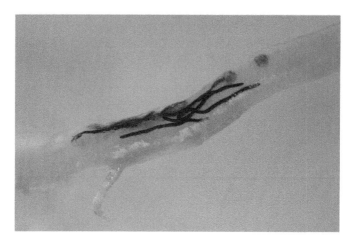

COLOR FIGURE 22.3 These lance nematodes (*Hoplolaimus galeatus*) (stained red) are migratory endo-parasites of bermudagrass (University of Florida).

COLOR FIGURE 22.4 Lance nematode–infected St. Augustinegrass roots (right) are dark, rotten appearing, and lack feeder roots compared with healthy roots (left).

COLOR FIGURE 22.5 St. Augustinegrass root infected by root-knot nematode (*Meloidogyne graminis*); the epidermis of the root has been peeled away to reveal the swollen female nematode within (University of Florida).

COLOR FIGURE 22.6 Necrotic patches on an annual bluegrass putting green caused by *Anguina pacificae* (Courtesy P. Gross, USGA).

COLOR FIGURE 22.7 Bermudagrass thinning and declining due to high infestation of sting nematodes (*Belonolaimus longicaudatus*) allows weeds to proliferate (University of Florida).

COLOR FIGURE 22.8   Irregular patches of wilting and declining centipedegrass on a sod farm caused by high numbers of ring nematodes (*Mesocriconema ornata*) (University of Florida).

COLOR FIGURE 22.9   Square plot treated with an experimental nematicide on a sting nematode (*Belono-laimus longicaudatus*)–infested putting green (University of Florida).

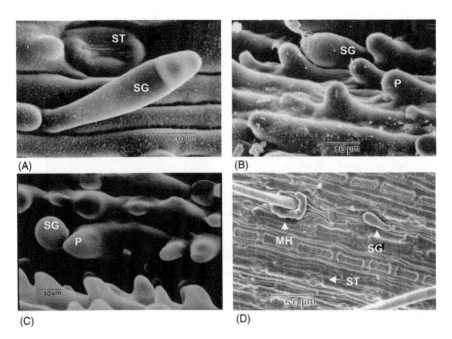

COLOR FIGURE 25.3 Scanning electron micrographs of adaxial leaf surfaces. (A) Salt gland of *Bouteloua* spp.SG, salt gland; ST, stomate. (B) Salt gland of *Cynodon* spp. SG, salt gland; P, papillae. (C) Salt gland of *Distichlis spicata* var. *stricta*. Only cap cell is visible—basal cell is imbedded in epidermis. SG, salt gland; P, papillae. (D) Overview of *Buchloë dactyloides* leaf surface, showing location of salt gland relative to other structures. SG, salt gland; MH, macrohair; ST, stomate.

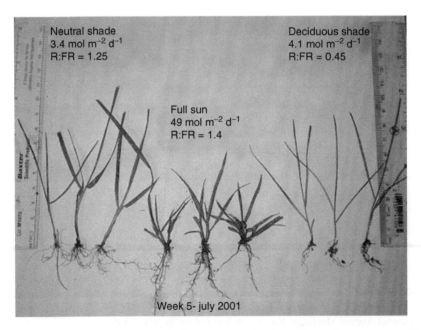

COLOR FIGURE 28.3 Cultivar Plantation tall fescue grown under different light treatments 5 weeks following seeding in Columbus, Ohio. Light intensity was the average of measurements at 15-min intervals. The R:FR ratio was measured using a spectroradiometer. (From Wherley, B.G., Turfgrass photomorphogenesis as influenced by changes in spectral composition and intensity of shadelight, Ph.D. thesis, The Ohio State University, Columbus, 2003. With permission.)

COLOR FIGURE 28.4 Cultivar Plantation tall fescue grown under different light treatments 14 months following seeding in Columbus, Ohio. Light intensity was the average of measurements at 15-min intervals. The R:FR ratio was measured using a spectroradiometer. (From Wherley, B.G., Turfgrass photomorphogenesis as influenced by changes in spectral composition and intensity of shadelight, Ph.D. thesis, The Ohio State University, Columbus, 2003. With permission.)

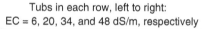

Tubs in each row, left to right:
EC = 6, 20, 34, and 48 dS/m, respectively

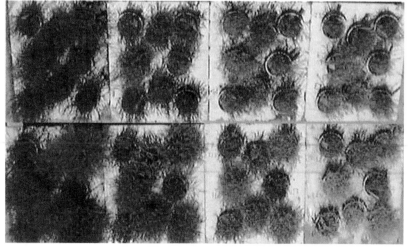

COLOR FIGURE 34.1 Saltgrass growth under various salinity levels of the culture medium. (From Pessarakli, M. et al., *Growth Responses of Twelve Inland Saltgrass Clones to Salt Stress.* ASA-CSSA-SSSA International Annual Meetings, Nov. 6–10, Salt Lake City, UT, 2005.)

16.4   FORMULATION EFFECTS ON HERBICIDE TOXICOLOGY

The formulation of a herbicide can affect not only its efficacy on target weeds but also nontarget organisms or humans and domestic animals. For example, most amine formulations of phenoxy herbicides carry a "Danger" signal word on the label. This is not due to a true toxicological effect, but due to the potential of the dimethylamine to cause injury to the eyes of humans. Ester formulations usually carry a "Caution" signal word. During application, there is potential for exposure to spray drift. Brown et al. [3] concluded that exposures of toxicological significance due to spray drift are unlikely. However, unintentional contact with the herbicide due to dislodgeable residues or spray drift is possible and much research has been conducted to determine the effect of various herbicide formulations on mammalian toxicology.

While formulation may have an effect on acute toxicity of the compound, van Ravenzwaay et al. [20] found that sublethal doses of either the dimethlyamine salt or the 2-ethylhexyl ester formulation of MCPA were rapidly metabolized in rats and that there were no differences attributed to formulation. Similarly, Pelletier et al. [23] and Knopp and Schiller [24] noted no differences in the metabolism in rats whether 2,4-D was administered as the dimethylamine salt or as the sodium salt. Klingman et al. [8] found that residues of 2,4-D esters on treated forage declined rapidly and that residues in milk from cows grazing the forage fell below detection limits within 4 days. Residues of the butyl ester disappeared more rapidly than did the 2-ethylhexyl ester residues on treated forage [8].

16.5   FORMULATION EFFECTS ON HERBICIDE UPTAKE AND EFFICACY

For optimal control of the weed, the ideal herbicide formulation should be absorbed slowly by the plant so that the tolerance level of the mesophyll and phloem to the herbicide is not exceeded. This is to allow for sufficient uptake of the herbicide and for translocation throughout the plant to the roots to occur over a period of several days [6]. The effects of formulation on the foliar penetration and efficacy of herbicides have received considerable attention [21,25]. Also, it is well known that ester formulations are more effective than amines at controlling weeds in turfgrass. Several researchers have found that nonpolar, e.g., oil-based or lipophilic, forms of a herbicide are more efficient than polar, e.g., water-based or hydrophilic, forms in penetrating leaf tissue [26–28].

Many research studies have investigated the dynamics of herbicide uptake. Many of these have found that the longer a pesticide remains in solution on the leaf surface, the greater the uptake of the herbicide. After the spray residue dries, differences have been observed among plant species as to the pattern of recrystallized amine or sodium salts of MCPA [9]. On bermudagrass, the deposits were much smaller than those observed on sugar beet (*Beta vulgaris* L.). However, the uniformity of distribution was the same. The authors speculated that amine salt formulations of herbicides could perhaps be improved if the herbicide could be formulated to improve the uniformity of distribution of the herbicide salts on the leaf surface. However, Hess et al. [29] noted that the addition of a surfactant alone would not result in uniform coverage of the leaf.

It has been hypothesized that glycerol or other humectants could enhance herbicide efficacy by delaying spray droplet drying, which would keep the herbicide in solution and therefore more available for uptake. Humectants were used successfully during the 1960s and 1970s. However, ethlylene oxide content of surfactants became a primary focus of formulation technology in the mid 1980s. This is controversial, however, and humectants are still hypothesized to be an important potential addition to the herbicide formulation [30].

In contrast to the patterns observed on sprayed leaves using the amine or sodium salt formulations, Hess et al. [9] observed that residues left when using the isooctyl ester formulation were small and discontinuous with granular deposits visible at high magnification. They speculated that as the spray carrier evaporated, the herbicide remained on the surface in congealed deposits, which could potentially affect uptake efficiency since only a small percentage of the total area of the deposit would be in direct contact with the leaf. Differences among manufacturers of the same formulation were also observed in this study with respect to the distribution of herbicide deposits. The authors concluded that keeping the active ingredient in solution as the spray evaporates by other ingredients in the formulation results in more uniform deposition of herbicide on the leaf surface [9].

The spray mix used can affect efficacy. A greater percentage of oil to water in the tank mix can enhance uptake by disrupting the cuticle on the leaf surface [21]. Increasing the rate of herbicide application or inclusion of surfactants in the formulation results in increased absorption of the herbicide [31].

16.5.1 FOLIAR ABSORPTION

The efficiency of penetration of the plant cuticle and uptake of a foliar applied herbicide is influenced both by the formulation of the herbicide and by characteristics of the plant surface to which the herbicide is applied [32]. Studies by Hull et al. [33] demonstrated that formulation differences in the performance of a herbicide could be explained by differences in the efficiency of foliar uptake. However, whereas the characteristics of the herbicide spray solution can be easily controlled, the characteristics of the plant surface cannot [9].

The cuticle is the first barrier to the uptake of herbicide or other substances into the plant [34]. The cuticle lies on the surface of higher plant leaves and is thought to have evolved as a mechanism for reducing water loss as plants evolved from the sea onto land [30]. The cuticle is heterogeneous and consists of a polyester matrix, cutin, cuticular lipids, carbohydrates, phenolic compounds, and polypeptides. The structure and composition of the cuticle varies among plant species [30]. Numerous studies have investigated the sorption and permeability of the cuticle to different organic and inorganic materials [32].

Research conducted during the last half of the twentieth century found that there are two primary avenues of herbicide entry into plants, depending on whether the herbicide is hydrophilic or lipophilic [30]. Lipophilic compounds are thought to move into the plant by a sorption/diffusion mechanism that can be modeled using Fick's first law in a manner similar to the way in which movement of herbicides through thatch and soil can be modeled. Hydrophilic compounds, in contrast, are thought to enter the plant through water-soluble paths through the plant cuticle. Acceptance of this notion has increased in recent years, due to studies that showed that certain surfactants increased the water permeability of the cuticle and the notion that if water can penetrate the cuticle, then so to can water-soluble compounds [30].

Kloppenburg and Hall [12] observed differences in the absorption of clopyalid depending on whether the free acid, monoethanolamine, or potassium salt formulation was used when studying Canada thistle (*Cirsium arvense* L.) and wild buckwheat (*Polygonum convolvulus* L.). Once absorbed, each of these formulations translocated similarly throughout the plant. In contrast, the translocation of either the 1-decyl or 2-ethylhexyl ester formulations was less than that of the acid or salt forms. This may have been due to selective retention of the ester formulations in the cuticle.

Further experiments by Kloppenburg and Hall [34] using isolated cuticular membranes suggest that foliar absorption of esters is different and that while their solubility in oil makes them more capable of absorption into the cuticle, this property may serve to limit further uptake by the apoplast and translocation throughout the plant. Kloppenburg and Hall [34] also noted that while absorption of the butoxyethyl ester of triclopyr, 1-decyl ester of clopyralid, and 1-methylheptyl ester of fluroxypyr into the cuticle was similar, that differential retention of these formulations by the cuticle may impact further translocation into the plant. Larger amounts of 2,4,5-T ([2,4,5-trichlorophenoxy]acetic acid) were found in treated leaves when using the butoxyethyl ester formulation, compared with the ammonium salt formulation. However, similar amounts of herbicide were found to translocate to the stems of the plant [35]. Curiously, Klingman et al. [8] found that almost all of the butyl ester formulation of 2,4-D and 75% of the 2-ethylhexyl ester formulation of 2,4-D found in or on foliage had hydrolyzed to the acid form within half an hour of application.

After absorption of a lipophilic herbicide into the epicuticular wax, diffusion occurs within the cuticle. The herbicide then partitions between the cuticle and apoplast and subsequently enters the plant cell [36]. Kloppenburg and Hall [12] found that after absorption, the free acid, monoethanolamine and potassium salt formulations of triclopyr were similar in their ability to diffuse through the cuticle and into the underlying cellular tissue. King and Radosevich [37] investigated the factors affecting uptake of triclopyr acid. They found that foliar penetration was greater in young leaves compared with mature ones. It was also noted that the amount of epicuticular wax was a factor, with increasing thickness decreasing penetration. Those parts of the plant that lack the cuticle or where it is thinner, such as stomata, trichomes, leaf veins, or petioles, have increased herbicide penetration rates [38]. Additionally, it was noted that the side of the leaf in which the herbicide was sprayed impacted efficacy, with herbicide delivery to the abaxial side resulting in greater uptake than if the adaxial side was sprayed.

It has also been noted that uptake of herbicide deposits is greater with deciduous broadleaf plants when compared with evergreen broadleaf plants [38]. Leaf size also impacts herbicide uptake, because with larger leaves the volume of herbicide deposits increases, as does the area available for absorption [21]. There is also evidence that in certain plants, there is no difference between the amine or ester formulation of a herbicide when measuring absorption or translocation [13].

Bentson and Norris [21] also observed that changes in leaf weight due to hydration can impact herbicide uptake. They reasoned that decreased hydration and subsequent shrinkage of the leaf tissue could result in increased cracking of epicuticular wax on the leaf surface, which would allow more herbicide to penetrate. It is also possible that as leaf hydration decreases, the hydrated tissue phase would decrease relative to the organic tissue phase, which could increase the proportion of the plant that the herbicide could partition into [21]. The time of the year also directly affects the thickness of the epicuticular wax on the leaf surface. Winter annual weeds develop a thicker wax surface as protection over the winter. When growth resumes in spring, the epicuticular wax cracks and thins with leaf expansion. Certain summer weeds also contain thicker wax coatings as a means of conserving moisture in hot weather.

After absorption by the plant, the isopropyl ester formulation of 2,4-D was found to hydrolyze to the acid and the acid is the form in which the herbicide translocates throughout the plant [6]. Ketchersid et al. [5] also observed this when using the isooctyl ester of 2,4-D. Because of this, it is thought that the form in which 2,4-D is translocated is similar, regardless of the type of ester formulation is applied [8].

16.5.2 ROOT ABSORPTION

Absorption of soil-applied herbicides can occur in root or shoot tissue that is in contact with the soil–water solution. Root absorption is the typical avenue of entry for granular applications of herbicides and for preemergence materials applied as a WP, WDG, or DF via a sprayer. Root absorption is significant for the activity of preemergence herbicides designed to prevent crabgrass and other annual weeds as well as postemergence applications of formulations intended to control broadleaf weeds.

Herbicides in direct contact with roots are thought to be absorbed by simple diffusion because unlike leaves, roots do not have structural components that offer a significant barrier to uptake [2].

16.5.3 FORMULATION/WEATHER INTERACTIONS

After deposition of the herbicide on the leaf surface, many environmental factors interact with the characteristics of the target plant and the chemical characteristics of the herbicide, thus determining the efficiency of plant uptake.

Environmental conditions at the time of application affect herbicide efficacy, especially humidity, which affects not only the rate of spray droplet drying, but also hydration of the cuticle [30]. In addition, Price [39] noted that plants grown under water stress have thickened and dehydrated cuticles. This results in reductions in absorption and translocation of several herbicides, including the butyl ester, dimethylamine salt, and potassium salt formulations of 2,4-D, dicamba (3,6-dichloro-2-methoxybenzoic acid), and picloram (4-amino-3,5,6-trichloropyridine-2-carboxylic acid) [40]. Not all plants are equally affected by reductions in humidity. Kloppenburg and Hall [12] found no reduction in absorption of the acid and ester formulations of clopyralid into water-stressed wild buckwheat plants. Similarly, Merkle and Davis [41] found that extreme moisture stress did not effect absorption of 2,4,5-T or picloram by bean (*Phaseolus vulgaris* L.). However, subsequent movement of the herbicide throughout the plant was reduced.

Many studies suggest that water-soluble herbicide formulations may be more sensitive to environmental conditions at the time of application than are the oil-soluble herbicides [30]. Absorption and translocation of some herbicides can be affected by changes in humidity [12]. This is not the case with all formulations, and absorption of acid formulations of weak acid herbicides is not generally thought to be affected by changes in humidity. O'Sullivan and Kossatz [42] found that the absorption and translocation into canada thistle (*Cirsium arvense*) of the monoethanolamine salt of clopyralid was doubled under conditions of higher relative humidity. Kloppenburg and Hall [12] observed similar results and noted that while the free acid was absorbed and translocated rapidly regardless of humidity, the monoethanolamine and potassium salt formulations showed marked decreases in absorption under decreased humidity.

Research suggests that highly water-soluble, ionic herbicides are more sensitive to lower humidity and rapid drying of spray droplets on the leaf surface compared with lipophilic formulations [30]. Another possible explanation for differences in efficacy among formulations is that different formulations result in spray deposition patterns that may result in more or less uniform distribution on the leaf surface. This was investigated by Hess et al. [9] who found that when either the sodium or amine salt formulation of MCPA was applied, the salts recrystallized on the leaf in patterns thought to represent the size of the original spray droplet. The authors used electron microscopy to determine that the concentration of herbicide was much higher in the depressions of the cell walls, forming an overall netlike pattern within the spray droplet area. When formulations result in suspensions in the spray tank, such as DF and WPS, the herbicide also tends to accumulate near the edges of the original spray drops [9].

There is some confusion in the literature with regard to the impact of humidity on efficacy of lipophilic herbicides. However, in another study, increased humidity or formation of dew resulted in increased loss of other pesticides, including the butoxyethyl ester formulation of triclopyr [21]. Amine formulations can rewet (redissolve) with dew on following mornings, but ester formulations will not redissolve after drying on the leaf surface.

The timing of application during the life cycle of the weed also impacts efficacy. Davis et al. [43] found that the highest concentration of herbicides in the leaf tissue corresponded to the seasonal period that resulted in best control. Auxin-mimicking herbicides like 2,4-D or triclopyr have the maximum effect when the weed is young and actively growing. When weeds flower and start producing seed, growth slows as the plant is in a reproduction cycle.

The differences in uptake efficiency among formulations have ramifications not only for effectiveness of the herbicide but also for whether rainfall soon after application will affect control [44]. Other factors that interact with formulation include the sensitivity of plant species, the rate of the herbicide applied and the addition of any surfactants to the spray. Herbicide efficacy is affected not only by the susceptibility of the target species, but also the timing of rainfall after the herbicide application and formulation of the herbicide [45]. Bovey and Diaz-Colon [46] found that water-soluble herbicides were generally more affected by rainfall than were oil-soluble herbicides.

Research suggests that herbicide selectivity may be affected by uptake efficiency; thus species that are more sensitive to a particular herbicide may be so due to increased efficiency of uptake.

In this case, rainfall would have less impact on overall control [44]. In their study, Behrens and Elakkad [44] observed that control of common lambsquarets (*Chenopodium album* L.) equivalent to 80% of that observed in the absence of rainfall occurred using a butoxyethyl ester formulation of 2,4-D even when rainfall occurred after 1 min. In contrast, 24 h were required to achieve 80% control using an alkanolamine formulation of 2,4-D applied to velvetleaf (*Abutilon theophrasti*). The amount of rainfall also impacted efficacy, with less than 1 mm required to reduce efficacy on velvetleaf compared with more than 15 mm required to reduce efficacy on lambsquarters [44].

16.6 CONCLUSION

As important as it is to use a herbicide that is efficacious against the target weed, it is also important to formulate the herbicide in a manner that will optimize its uptake and translocation into the weed. Our understanding of target weed plant physiology has advanced considerably. So too our understanding continues to advance, not only of how to synthesize compounds with selective herbicide activity, but also how to effectively move them into the weed to optimize control.

REFERENCES

1. Wilson, R.D., Geronimo, J., and Armbruster, J.A., 2,4-D dissipation in field soils after application of 2,4-D dimethylamine salt and 2,4-D 2-ethylhexyl ester, *Environ. Toxicol. Chem.*, 16, 1239, 1997.
2. Zimdahl, R.L., *Fundamentals of Weed Science*, Academic Press, Inc, San Diego, CA, 1993.
3. Brown, J.N., Gooneratne, S.R., and Chapman, R.B., Herbicide spray drift odor: Measurement and toxicological significance, *Arch. Environ. Contam. Toxicol.*, 38, 390, 2000.
4. Mullison, W.R. and Hummer, R.W., Some effects of the vapor of 2,4-dichlorophenoxyacetic acid derivatives on various field crops and vegetable seeds, *Botan. Gaz.*, 11, 77, 1949.
5. Ketchersid, M.L., Fletchall, O.H., Santelmann, P.W., and Merkle, M.G., Residues in sorghum treated with the isooctyl ester of 2,4-D, *Pestic. Monit. J.*, 4, 111, 1970.
6. Crafts, A.S., Evidence for hydrolysis of esters of 2,4-D during absorption by plants, *Weeds*, 8, 19, 1960.
7. Yip, G. and Ney, R.G., Jr., Analysis of 2,4-D residues in milk and forage, *Weeds*, 14, 167, 1966.
8. Klingman, D.L., Gordon, C.H., Yip, G., and Burchfield, H.P., Residues in the forage and in milk from cows grazing forage treated with esters of 2,4-D, *Weeds*, 14, 164, 1966.
9. Hess, F.D., Bayer, D.E., and Falk, R.H., Herbicide dispersal patterns: III. As a function of formulation, *Weed Sci.*, 29, 224, 1981.
10. Anderson, W.P., *Weed Science: Principles and Applications*, 3rd Ed., West Publishing, St. Paul, MN, 1992.
11. Sharma, M.P., Chang, F.Y., and Vanden Born, W.H., Penetration and translocation of picloram in Canada thistle, *Weed Sci.*, 19, 349, 1971.
12. Kloppenburg, D.J. and Hall, J.C., Effects of formulation and environment on absorption and translocation of clopyralid in *Cirsium arvense* (L.) Scop. and *Polygonum convolvulus* L., *Weed Res.*, 30, 9, 1990a.
13. Bovey, R.W., Hein H., Jr., and Meyer, R.E., Absorption and translocation of triclopyr in honey mesquite (*Prosopis juliflora* var. *glandulosa*), *Weed Sci.*, 31, 807, 1983.
14. Balogh, J.C. and Anderson, J.L., Environmental impacts of turfgrass pesticides, in *Golf Course Management and Construction: Environmental Issues*. Lewis Publishers, Boca Raton, FL., 1992, pp. 221–353.
15. Clark, J.M., Roy, G.R., Doherty, J.J., Curtis, A.S., and Cooper, R.J., Hazard evaluation and management of volatile and dislodgeable foliar pesticide residues following application to turfgrass, In: J.M. Clark and M.P. Kenna (eds.) *Fate and Management of Turfgrass Chemicals*, ACS Symp. Ser. 743, American Chemical Society, Washington, DC, 2000, pp. 294–312.
16. Haith, D.A., Lee, P.C., Clark, J.M., Roy, G.R., Imboden, M.J., and Walden, R., Modeling pesticide volatilization from turf, *J. Environ. Qual.*, 31, 724, 2002.
17. Weber, J.B., Behavior of dinitroanaline herbicides in soils, *Weed Technol.*, 4, 394, 1990.
18. Kenna, M.P., What happens to pesticides applied to golf courses? *USGA Green Section Record*, 33(1), 1, 1995.
19. Branham, B.E., Herbicide fate in turf, In: A.J. Turgeon (ed.) *Turf Weeds and Their Control*, American Society of Agronomy and Crop Science Society of America, Madison, WI, 1994.

20. Van Razenzwaay, B., Pigott, G., and Leibold, E., Absorbtion, distribution, metabolism, and excretion of 4-chloro-2-methylphenoxyacetic acid (MCPA) in rats, *Food Chem. Toxicol.*, 42, 115, 2004.

21. Bentson, K.P. and Norris, L.A., Foliar penetration and dissipation of triclopyr butoxyethyl ester herbicide on leaves and glass slides in the light and dark, *J. Agric. Food Chem.*, 39, 622, 1991.

22. McLeod, H.L., Bowhey, C.S., Swanson, N.M., and Stephenson, G.R., Dislodgeable residues of 2,4-D amine and ester formulations on turfgrass, *Int. Soc. Res. J.*, 7, 14, 1994.

23. Pelletier, O., Ritter, L., Caron, J., and Somers, D., Disposition of 2,4-dichlorophenoxyacetic acid dimethylamine salt by Fischer 344 rats dosed orally and dermally, *J. Toxicol. Environ. Health*, 28, 221, 1989.

24. Knopp, D. and Schiller, F., Oral and dermal application of 2,4-dichlorophenoxyacetic acid sodium and dimethylamine salts to male rats: Investigations on absorption and excretion as well as induction of mixed-function oxidaze activities, *Arch. Toxicol.*, 66, 170, 1992.

25. Bentson, K.P., Fate of xenobiotics in foliar pesticide deposits, *Rev. Environ. Contam. Toxicol.*, 114, 125, 1990.

26. Chamel, A., Foliar absorption of herbicides: study of cuticular penetration using isolated cuticles, *Physiol. Veg.*, 24, 491, 1986.

27. Richardson, R.G., A review of foliar absorption and translocation of 2,4-D and 2,4,5-T, *Weed Res.*, 17, 259, 1977.

28. Bovey, R.W., Meyer, R.E., and Bauer, J.R., Potential herbicides for brush control, *J. Range Manage.*, 34, 144, 1981.

29. Hess, F.D., Bayer, D.E., and Falk, R.H., Herbicide dispersal patterns: I. As a function of leaf surface, *Weed Sci.*, 22, 394, 1974.

30. Ramsey, R.J.L., Stephenson, G.R., and Hall, J.C., A review of the effects of humidity, humectants, and surfactant composition on the absorption and efficacy of highly water-soluble herbicides, *Pestic. Biochem. Physiol.*, 82, 162, 2005.

31. Stand, O.E. and Behrens, R., Effects of adjuvants on the phytotoxicity of foliarly applied atrazine. *Abstr. Weed Sci. Soc. Am.*, 67, 1970.

32. Schafer, W.E. and Bukovac, M.J., Effect of acid treatment of plant cuticles on sorption of selected auxins, *Plant Physiol.*, 83, 652, 1987.

33. Hull, H.M., Morton, H.L., and Wharrie, J.R., Environmental influence on cuticle development and resultant foliar penetration, *Bot. Rev.*, 41, 421, 1975.

34. Kloppenburg, D.J. and Hall, J.C., Penetration of clopyralid and related weak acid herbicides into and through isolated cuticular membranes of *Euonymous fortunei*, *Weed Res.*, 30, 431, 1990.

35. Morton, H.L., Davis, F.S., and Merkle, M.G., Radioisotope and gas chromatographic methods for measuring absorption and translocation of 2,4,5-T by mesquite, *Weed Sci.*, 16, 88, 1968.

36. Schafer, W.E., Morse, R.D., and Bukovac, M.J., Effect of pH and temperature on sorption of auxin by isolated tomato fruit cuticles, *HortScience*, 23, 204, 1988.

37. King, M.G. and Radosevich, S.R., Tanoak (*Lithocarpus densiflorus*) leaf surface characteristics and absorption of triclopyr, *Weed Sci.*, 27, 599, 1979.

38. Riederer, M. and Schoinherr, J., Accumulation and transport of (2,4-dichlorophenoxy)acetic acid in plant cuticle, *Exotoxicol. Environ. Saf.*, 9, 196, 1985.

39. Price, C.E., The effect of environment on foliage uptake and translocation of herbicides, In: *Aspects of Applied Biology 4, Influence of Environmental Factors on Herbicide Performance and Crop and Weed Biology*, The Association of Applied Biologists, Wellesbourne, Warwick, 1983.

40. Friesen, H.A. and Dew, D.A., The influence of temperature and soil moisture on the phytotoxicity of dicamba, picloram, bromoxynil, and 2,4-D ester, *Can. J. Plant Sci.*, 46, 653, 1966.

41. Merkle, M.G. and Davis, F.S., Effect of moisture stress on absorption and movement of picloram and 2,4,5-T in bean, *Weeds*, 15, 10, 1967.

42. O'Sullivan P.A. and Kossatz, V.C., Absorption and translocation of ^{14}C-3,6 dichloropicolinic acid in *Cirsium arvense* (L.) Scop., *Weed Res.*, 24, 17, 1984.

43. Davis, F.S., Meyer, R.E., Bauer, J.R., and Bovey, R.W., Herbicide concentrations in honey mesquite phloem, *Weed Sci.*, 20, 264, 1972.

44. Behrens, R. and Elakkad, M.A., Influence of rainfall on the phytotoxicity of foliarly applied 2,4-D, *Weed Sci.*, 29, 349, 1981.

45. Bovey, R.W. and Davis, F.S., Factors affecting the phytotoxicity of paraquat, *Weed. Res.*, 7, 281, 1967.

46. Bovey, R.W. and Diaz-Colon, J.D., Effect of simulated rainfall on herbicide performance, *Weed Sci.*, 17, 154, 1969.

17 Techniques to Formulate and Deliver Biological Control Agents to Bentgrass Greens

Kenneth E. Conway and Stacy R. Blazier

CONTENTS

17.1 INTRODUCTION

Connick et al.,[12] state that product formulation is key to successful biological control of plant diseases and that a successful biological control formulation is one that is:

- Safe
- Stable in the environment
- Easily applied using conventional equipment
- Can be produced economically

Little research work on methods to achieve biological control as part of a comprehensive turf management program has been done.[17] In order for biological control to be incorporated into a turf management system, it must readily fit with existing operating procedures. The target pathogen and site of interaction are other considerations when planning a biological control program.

Dollar spot caused by *Sclerotinia homeocarpa* F. T. Bennett (Figure 17.1) and brownpatch caused by *Rhizoctonia solani* Kuhn (Figure 17.2A,B) have been among the most common and destructive turfgrass pathogens since their identification in the early 1920s and 1913, respectively.[15] Most attempts at biological control on turf have focused on one or both of these pathogens. *R. solani* is a diverse group of isolates and represent several interrelated anastomosing groups of isolates[3] that makes relationships within the groups very complex and confusing. These pathogen cause patch disease on a number of turfgrasses and each disease may have different fungicide sensitivities.

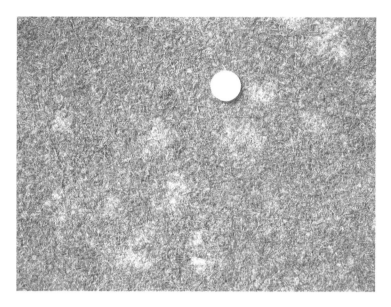

FIGURE 17.1 Dollar-spot symptoms on bentgrass. The pathogen, *Sclerotinia homeocarpa* produces a rounded spot, approximately the size of a U.S. quarter on greens.

Therefore, to successfully control the disease it is important not only to identify the pathogen, but also know which anastomosis group it represents.[3]

17.2 ATTEMPTS AT BIOLOGICAL CONTROL ON TURF

There have been several reports concerning biological control of turfgrass pathogens.[3,7,9,14,19,38,39] Research efforts with turf pathogens have indicated that some biocontrol organisms either did not control the pathogen or did not perform well as selected fungicides did.[48] However, our research[3,15] and others[9,19,22,31,32] have shown that biological control agents can be applied to turf or soil and can maintain populations larger than the natural population. Further, when our techniques and organisms were applied to turf,[14] we were able to follow population spread from points of inoculum to thatch and root systems. Although this research was conducted in growth chambers, other researchers had similar successes in the field.[7,9,19,22,38,39] As techniques and formulations are developed and refined, biological control of turfgrass pathogens will become more acceptable and practical.

17.3 MECHANISMS

In general, mechanisms for biological control agents are attributed to nutrient competition,[1,5,9,31] host-induced resistance,[7] hyperparasitism,[4,10,11,25,35,46,51] and antibiotic production.[6] Burpee et al.[8,9] were able to suppress gray snow mold on turfgrass which is caused by two species of *Typhula*: *Typhula ishikariensis* and *Typhula incarnata*. They used a nonpathogen of turfgrasses, *Typhula phacorrhiza* to suppress symptoms of gray snow mold. Grain infested with *Typhula phacorrhiza* was applied as a top dressing to turf before snow cover at a rate of 200 g m^{-2} and this suppressed *Typhula* blight as effectively as the recommended fungicide pentachloronitrobenzene. Later, they formulated *Typhula phacorrhiza* in alginate pellets[9] and also achieved suppression. They speculated that the mechanism used to suppress the pathogens was competition for nutrients or space. Mechanisms for *Trichoderma harzianum* and *Laetisaria arvalis* elucidated on other crops is often shown to be hyperparasitism[10,44] antibiotic production,[6] and, more recently, competition.[5] Possession of multiple mechanisms by biological control organisms may enhance their efficacy as control organisms since they can attack the pathogen on more than one level.

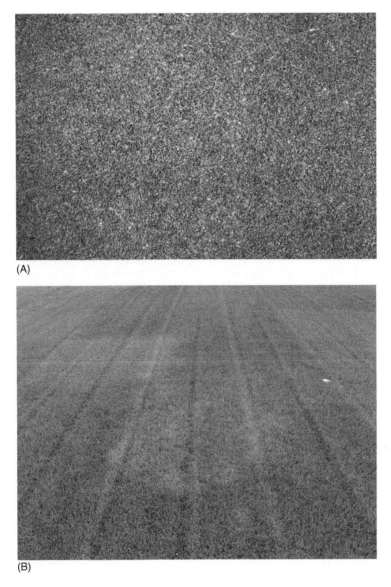

(A)

(B)

FIGURE 17.2 (A) Early symptoms of brown patch on bentgrass showing a diffuse brown coloration in the grass. (Courtesy of K. E. Conway, Department of Entomology and Plant Pathology, Oklahoma State University, Stillwater, OK.) (B) Older more developed symptoms of brown patch showing a large discrete circular patch infection on bentgrass. (Photo courtesy of Dr. D. L. Martin, Department of Horticulture and Landscape Architecture, Oklahoma State University, Stillwater, OK.)

17.4 FORMULATIONS

Biological control is an attractive nonchemical alternative to fungicides for control of turf diseases because it may pose fewer environmental risks.[34] This strategy involves the manipulation and utilization of antagonistic modes of action of biotic agents that provide beneficial reductions in pathogen inoculum and disease severity. Several reports[3,20,29] have indicated that certain individual microbial organisms are capable of reducing severity of turfgrass diseases. Many novel systems have been developed to deliver biological control agents to turf and soil including use of alginate pellets[9,12,18–20,30,45] and other carriers such as clay,[9,20] ground seeds,[9,23] diatomaceous earth granules,[47] and a mixture of cornmeal and sand[3,20] also have been used to deliver biocontrol agents. Blazier[3] formulated both *Trichoderma harzianum* and *L. arvalis* to control brown patch on bentgrass in cornmeal–vermiculite–sand topdressing. To control soilborne pathogens, biological control organisms must be delivered to the thatch and to the soil. The best time to apply these beneficial organisms would be after turf renovation when soil plugs are removed from the greens to allow for aeration and root growth. During the renovation, plugs of thatch and sand approximately three inches in depth are removed. Sand is normally spread over the greens to fill these holes. Therefore, one way to formulate biological control organisms is a sand base that could be included in the renovation operation. To insure the integrity of the putting surface, care must be taken to use sand of the same quality and type as that used in renovation.

A good formulation should provide a food base for the biological agent to provide energy for growth and colonization around the plant or plant part to be protected. Papavizas et al.[45] found that the type of bulking agent used to produce alginate pellets affected the total colony forming units formed in the soil. Care must be taken to ensure that the food base is completely colonized by the biological agent or the food base may act to stimulate the pathogen that actually increases the severity of disease. Biological control organisms whose mode of operation is competition should be formulated with a food base to provide a source of exogenous energy to initiate growth to colonize around the plant and occupy niches to exclude the pathogen. However, even though Punja et al.[48] used diatomaceous earth granules impregnated with a 10% molassses solution to formulate *Trichoderma* spp. to control *Sclerotium rolfsii* which significantly increased populations of *Trichoderma* spp. in treated plots, disease severity was not reduced.

17.5 POPULATION DYNAMICS

Good formulations should increase the population levels of the biological agents to an extent that can control the pathogens. However, Fravel and Lewis[18] contend that the efficacy of the propagule and the timing and placement are more important than the population size. Lewis and Papavizas[28,29] have shown that chlamydospores of *Trichoderma* spp. and *Gliocladium virens* are more effective as biological propagules compared to conidia. Knudsen et al.[26] demonstrated that soaking alginate pellets in polyethylene glycol 8000 compared to pellets soaked in water enhanced both the growth and sporulation of pelletized fungi when used as biological control agents. Blazier[3] quantified populations in both soil cores and in thatch. Populations of *Trichoderma* spp. in soil core samples containing roots and soil increased by approximately one order of magnitude compared to untreated soil. *Trichoderma* spp. survived and persisted at high populations (10^5 propagules/g soil) and remained stable or increased slightly during the course of the study. Population levels in the upper 3 cm soil sections were slightly greater than those in lower soil sections. Bentgrass, *Agrostis stolonifer* L. has rather shallow roots, so it seems possible that the topdressing application method was sufficient in establishing *Trichoderma* spp. in the rhizosphere since populations were higher in the soil portions where most of the roots and minor pathogens were likely to be located. *L. arvalis* populations were enumerated using selective media[46] and a soil pelleter.[24] *L. arvalis* populations in soil associated with creeping bentgrass did not reach the levels attained by *Trichoderma* spp. Throughout the course of this study, no fungal propagules of *L. arvalis* were present in either upper or lower 3 cm soil core sections of untreated plots. At day 0, prior to biocontrol applications, no *L. arvalis* propagules were detected in any soil core samples from any of the plots for any of the treatments. However, at day 30, following topdressing application, populations in soil core samples containing roots and soil increased 1000-fold and persisted at these levels during the course of the study. Populations in upper 3 cm soil sections were higher than those in lower soil sections. In the field, neither *Trichoderma harzianum* nor *L. arvalis* provided reduction in brown patch severity compared to the fungicide azoxystrobin. However, in laboratory investigations, *L. arvalis* was the most effective treatment compared to either the fungicide or *Trichoderma* treatments.

Turfgrass diseases like brownpatch and dollar spot can cause large economic losses due to unsightliness and overall reduction in turf quality and surface uniformity if control measures are not instituted in a timely and effective manner. Typically, control of these diseases is achieved by repeated applications of expensive fungicides, but problems with soil and water contamination, fungicide resistance, and other harmful effects of fungicides have prompted the need for alternative control strategies. Biological control is one alternative to the use of environmentally problematic fungicides, and the use of biological control agents for turfgrass diseases has been well documented.[3,7,9,32] However, reports of unreliability in biological control have limited the widespread application of this disease management strategy on turf. Walsh et al.[54] stated that turfgrass disease biocontrol is an "emerging technology." Basically, biological control as a disease management strategy is still in its infancy because many of the "kinks" are still being worked out and explored. There are ongoing efforts to determine the most effective methods of introduction, strategies for maintaining biological control population levels when and where they are needed, and the activities of biological control organisms in association with turfgrass. Biological control is unlikely to become the sole management strategy in any disease situation until these problems are resolved and the knowledge base is increased. Only two biological control organisms have been commercialized and registered with the United States Environmental Protection Agency for control of turfgrass diseases including brownpatch and dollar spot.[37] These products are *Trichoderma harzianum* strain 1295-22 that is marketed as Bio-Trek (Bioworks, Inc. Geneva, NY; Wilbur-Ellis, Fresno, CA) which has demonstrated a high degree of rhizosphere competence when introduced to turf, and Eo Soil Systems' TX-1, a strain of *Pseudomonas aureofaciens.*

17.6 INTEGRATED CONTROL

Although several researchers have documented that individual biological control organisms are capable of reducing severity of disease, Spurr[50] used the term "silver-bullet" to describe the predilection of plant pathologists to select one antagonist to control one pathogen. Such a silver-bullet approach to disease control may not always be successful, and combining these biological organisms with other organisms or fungicides may be necessary. However, there are some[22] who believe that a silver-bullet approach is possible. Several researchers have documented successful integration of chemical and biological control for plant disease management.[15,27,31] Lo et al.[33] suggested that integration of chemical and biological control for brownpatch and dollar-spot management is an attractive control strategy. Brown patch and dollar spot are both soilborne, monocyclic diseases. Zadoks and Schein[55] define "monocyclic" as only one infection cycle but high-density, high-maintenance turfgrass ecosystems create an environment in which these diseases may become polycyclic with several secondary infection cycles. Suppression of both initial and secondary infections of these diseases is, therefore, necessary.[32] Lo et al.[33] using granular or peat-based formulations of *Trichoderma harzianum* strain 1295-22 were able to suppress the initial infections of brownpatch and dollar spot on underground turf plant parts but were unable to control secondary, foliar infections of these diseases. They proposed that while soil applications of biological control organisms should be used to suppress initial turf infection, compatible fungicides should be incorporated with biological control strategies for reductions in secondary foliar phases of turfgrass diseases. Blazier[3] explored the possibility of using biological control topdresses consisting of vermiculite–cornmeal–sand to control the initial infections of brown patch and dollar spot in conjunction with a microbially derived fungicide, azoxystrobin, for control of secondary, foliar infections of brown patch. The timing of the fungicide application was also explored. Isolates of *Trichoderma harzianum* and *L. arvalis* were selected for tolerance to azoxystrobin. Disease control between the addition of the biologicals both before and after fungicide application was compared. On the basis of the research of others,[16,31,35] it was hypothesized that application of the fungicide to turf before application ("niche-clearing") of the biological might reduce competition from other soil fungi and allow the biological control to colonize the thatch and soil and therefore provide for greater control of the pathogens. Niche-clearing was effective for *Trichoderma harzianum* but was not evident for *L. arvalis*.

The addition of fungicide to soil to establish a previously rhizosphere-incompetent micro-organism on roots was suggested by Mendez-Castro and Alexander[35] and later verified by others.[30] It was envisioned that the fungicide would adversely affect the resident soil colonist and allow the fungicide-tolerant agent a competitive advantage.

We chose to use a bulk amendment solid substrate biocontrol topdressing formulation consisting of vermiculite-cornmeal-sand because it was inexpensive, relatively easy to prepare and deliver to the turfgrass, and because other researchers[20] indicated success using a similar type of formulation for control of various soilborne pathogens.

17.7   COMPOSTS TO MANAGE TURF DISEASES

Turfgrass managers have added composts and organic fertilizers to putting greens to control diseases. Milorganite® is a natural organic fertilizer that is commonly used by the turf industry as a topdressing. This commercial product is applied to greens in the fall. It is a dried municipal sludge that contains a small amount of a slow release fertilizer, 6–2–0 (NPK). Since it is a dried product, most microbes have been eliminated and the greening effect is probably due more to the small amount of N in the product. Microbial activity is generally necessary for disease suppressive activities and some biological control organisms can be added to compost to increase microbial activity.[37] Other composts have suppressed a number of pathogens, such as, *Typhula* blight,[42] dollar spot,[38,41] brownpatch,[39] and red thread.[40] However, there are barriers to the commercialization of compost for disease control,[13] which include governmental regulations that might require a particular product to be sterilized, thus removing all microbial activity.

17.8   PEST MANAGEMENT IN THE FUTURE

The development of remote sensing technology is and will usher in new and exciting tools for management of diseases on turf. Green et al.[21] used a hand-held multispectral radiometer Model MSR 87, manufactured by Cropscan, Inc., Fargo, ND to assess disease on tall fescue. The unit was mounted on a pole and the unit had 16 photodiodes that measured incidence and reflected light in eight 25–33-nm wide spectral bands with midpoints near 460, 510, 560, 610, 660, 710, 760, and 810 nm. The radiometer was held on a metal pole about 2 m above the plots and measured canopy reflectance from approximately a 1-m^2 circular area within each plot. The sensor head was linked to small Radio Shack computer located at the base of the pole. They found that reflectance within the 810-nm band exhibited the strongest relationship with visual severity estimates of *Rhizoctonia* blight and grey leaf spot. Trenholm et al.[52] using a Cropscan multispectral model MSR16 radiometer found that reflectance and ratios of different ranges were highly correlated to turf qualities such as photosynthetic active radiation, leaf area index, and plant response to stress. Green et al.[21] demonstrated that spectral reflectance could be used to detect severity of *Rhizoctonia* disease on tall fescue. Earlier, Nutter et al.[43] also using a Cropscan system showed that disease severity could be assessed on turfgrass and that multiple personnel could reliably use this system and accurately repeat reflectance readings. Miles[36] used a Cropscan system (Figures 17.3A,B,C) to assess loss of Scotch pine (*Pinus sylvestris*) seedlings due to the root pathogen, *Macrophomina phaseolina* at the State of Oklahoma Forest Nursery (Figure 17.3A). He found that wind gusts often interfere with radiometer readings and that lowering the radiometer head reduced the area of reflectance. Pinter[47] has demonstrated that dew and temperature can interfere in spectral measurements. Recently, researchers at Oklahoma State University, Department of Biosystems and Agricultural Engineering, have been developing sensor systems based on spectral reflectance (Figure 17.4) that can be used

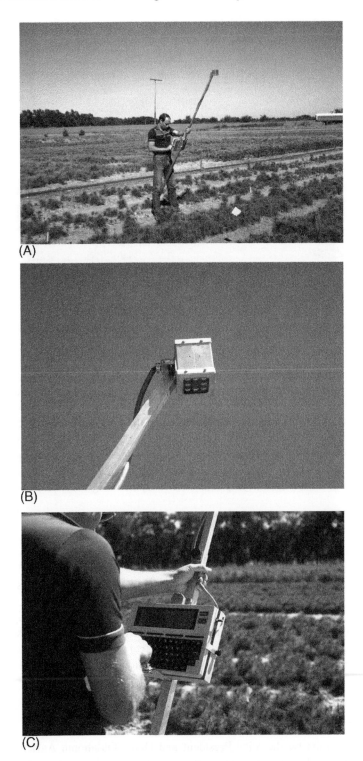

FIGURE 17.3 (A) Mark Miles using a Cropscan multispectral unit to study the loss of Scotch pine seedlings (*Pinus sylvestris*) due to the root pathogen *Macrophomina phaseolina*. (B) Multispectral head showing nine photodiodes. (C) Radio Shack computer mounted on the base of the pole. (All photos by Kenneth E. Conway, Department of Entomology and Plant Pathology, Oklahoma State University, Stillwater, OK.)

FIGURE 17.4 Dr. Marvin Stone, Department of Biosystems and Agricultural Engineering, Oklahoma State University, calibrates a prototype of a hand-held sensor system used to identify dollar spot on bentgrass greens. (Photo by Kenneth E. Conway, Department of Entomology and Plant Pathology, Oklahoma State University, Stillwater, OK.)

to identify diseases of turf and direct fungicide applications to the exact areas needing treatment. These systems can locate and identify diseased grass as well as determine turf quality and identify nutritional needs of the turf.[2] The final system could be mounted on a mowing machine and identify diseased areas on the green as the machine was mowing, thus, taking care of two operations at one time. Sensors should be sensitive enough to identify difference in reflectance caused by pathogen activity. We felt that recognition of discrete disease symptoms such as dollar spot would be easier that pathogens that produce diffuse symptoms such as brown patch but sensors were sensitive enough to also recognize the initial symptoms of brown patch (Figure 17.2A).

17.9   CONCLUSIONS

As formulations and strategies for deployment are developed, the use of biological control agents on turf will become acceptable and desirable for management of turfgrass diseases. The biggest barrier will be the reluctance of industry to develop the biologicals as a product. The success of a small company like BioWorks, Inc. to develop and deliver quality products for the turf industry should encourage others to join in this effort. Within the next 10–20 years, turfgrass managers could have a radiometric system linked to satellites that could map each green and with the sensors linked to a computer, diseased areas could be identified and the correct rate and concentration of the fungicide or biological control agent could be directed to specific areas on the greens in need of treatment before the disease becomes apparent to the manager.

ACKNOWLEDGMENTS

Approved for publication by the Vice President and Dean, Oklahoma Agricultural Experiment Station. Mention of a trademark, proprietary product, or vendor does not constitute a guarantee or warranty of the product by Oklahoma State University nor imply their approval to the exclusion of other products or vendors that may also be suitable.

REFERENCES

1. Ahmad, J.S. and Baker, R., Competitive saprophytic ability and cellulolytic activity of rhizosphere competent mutants of *Trichoderma harzianum*, *Phytopathology*, 77, 358, 1987.
2. Bell, G.E., Martin, D.L, Wiese, S.G., Dobson, D.D., Smith, M.W., Stone, M.L., and Solie, J.B., Vehicle-mounted optical sensing: an objective means for evaluation of turf quality, *Crop Sci.*, 42, 197, 2002.
3. Blazier, S.R., Characteristics of *Rhizoctonia solani* turf isolates and evaluation of an integrated control strategy for brown patch and dollar spot on creeping bentgrass, M.S. thesis, Oklahoma State University, Stillwater, 1999.
4. Blazier, S.R. and Conway, K.E., Characterization of *Rhizoctonia solani* isolates associated with patch diseases on turfgrass, *Proc. Okla. Acad. Sci.*, 84, 41, 2004.
5. Bobba, V. and Conway, K.E., Competitive saprophytic ability of *Laetisaria arvalis* compared with *Sclerotium rolfsii*, *Proc. Okla., Acad. Sci.*, 83, 17, 2003.
6. Bowers, W.S., Hoch, H.H., Evans, P.H., and Katayama, M., Thallophytic allelopathy: isolation and identification of laetisaric acid, *Science*, 1232, 105, 1986.
7. Burpee, L.L. and Goulty, L.G., Suppression of brown patch disease of creeping bentgrass by isolates of nonpathogenic *Rhizoctonia* spp., *Phytopathology*, 74, 692, 1984.
8. Burpee, L.L., Kaye, L.M., Goulty, L.G., and Lawton, M.B., Suppression of gray snow mold by an isolate of *Typhula placorrhiza*, *Plant Dis.*, 71, 97, 1987.
9. Burpee, L.L. and Lawton, M., Biological control of *Typhula* blight, *Golf Course Management*, 58, 76, 1990.
10. Chet, I., Mycoparasitism—recognition, physiology and ecology, in *New Directions in Biological Control*, Baker, R.R. and Dunn, P.E., Eds., Alan R. Liss, Inc., New York, pp. 725–733, 1990.
11. Chet, I., Harman, G.E., and Baker, R., *Trichoderma harzianum*: its hyphal interactions with *Rhizoctonia solani* and *Pythium spp.*, *Microb. Ecol.*, 7, 29, 1981.
12. Connick Jr., W.J., Lewis, J.A., and Quimby Jr., P.C., Formulation of biocontrol agents for use in plant pathology, in *New Directions in Biological Control*, Baker, R.R. and Dunn, P.E., Eds., Alan R. Liss, Inc., New York, pp. 345–372, 1990.
13. Conway, K.E., Barriers to implementation of biological control of plant pathogens, in *Regulations and Guidelines: Critical Issues in Biological Control*, Charudattan, R. and Browning, H.W., Eds., University of Florida, IFAS, p. 54, 1992.
14. Conway, K.E., Gerken, D.A., Sandburg, M.A., and Martin, D.L., Population dynamics of *Laetisaria arvalis* and *Burkholderia cepacia*, potential biocontrol agents in soil cores and thatch of creeping bentgrass (*Agrostis palustris*), *Proc. Okla. Acad. Sci.*, 80, 39, 2000.
15. Conway, K.E., Maness, N.E., and Motes, J.E., Integration of biological and chemical controls for *Rhizoctonia* aerial blight and root rot of Rosemary, *Plant Dis.*, 81, 795, 1997.
16. Couch, H.B., Diseases of turfgrasses caused by fungi, in *Diseases of Turfgrass*, 3rd ed., Krieger Publishing Co., Malabar, FL, pp. 21–199, 1995.
17. Doyle, J.M., Soil amendments and biological control, *Golf Course Manage.*, March, 86–97, 1991.
18. Fravel, D.R. and Lewis, J.J., Production, formulation and delivery of beneficial microbes for biocontrol of plant pathogens, in *Pesticide Formulation and Application Systems*, 11th vol., STM STP 1112, Casin, D.G. and Bode, L.E., Eds., American Society for Testing Materials, Philadelphia, PA, 1990.
19. Fravel, D.R., Marois, J.J., Lumsden, R.D., and Connick Jr., W.J., Encapsulation of potential biocontrol agents in an alginate–clay matrix, *Phytopathology*, 75, 774, 1985.
20. Goodman, D.M. and Burpee, L.L. Biological control of dollar spot disease of creeping bentgrass, *Phytopathology*, 81, 1438, 1991.
21. Green II, D.E., Burpee, L.L., and Stephenson, K., Canopy reflectance as a measure of disease in tall fescue, *Crop Sci.*, 38, 1603, 1998.
22. Harman, G.E., Myths and dogma of biocontrol, *Plant Dis.*, 84, 377, 2000.
23. Haygood, R.A. and Mazur, A.R., Evaluation of *Gliocladium virens* as a biological agent of dollar spot of bermudagrass, *Phytopathology*, 80, 435(Abstr.), 1980.
24. Henis, Y., Ghaffar, A., Baker, R., and Gillespie, S.L., A new pelleter soil sampler and its use for the study of population dynamics of *Rhizoctonia solani* in soil, *Phytopathology*, 68, 371, 1978.
25. Kitamoto, Y., Kono, R., Tokimoto, M.N., and Ichikawa, Y., Production of lytic enzymes against cell walls of basidiomycetes from *Trichoderma harzianum*, *Trans. Mycol. Soc. Japan*, 26, 69, 1984.

26. Knudsen, G.R., Eschen, D.J., Dandurand, L.M., and Wang, Z.G., Method to enhance growth and sporulation of pelletized biocontrol fungi, *Appl. Environ. Microbiol.*, 57, 2864, 1991.

27. Lewis, J.A. and Papavizas, G.C., Integrated control of *Rhizoctonia* fruit rot of cucumber, *Phytopathology*, 70, 85, 1980.

28. Lewis, J.A. and Papavizas, G.C., Production of chlamydospores and conidia by *Trichoderma* spp. In liquid and solid growth media, *Soil Biol. Biochem.*, 15, 351, 1983.

29. Lewis, J.A. and Papavizas, G.A., A new approach to stimulate population proliferation of *Trichoderma* species and other potential biocontrol fungi introduced into natural soils, *Phytopathology*, 74, 1240, 1984.

30. Lewis, J.A. and Papavizas, G.C., Characteristics of alginate pellets formulated with *Trichoderma* and *Gliocladium* and their effect on the proliferation of fungi in soil, *Plant Pathol.*, 34, 571, 1985.

31. Lifshitz, R., Lifshitz, S., and Baker, R., Decrease in the incidence of *Rhizoctonia* damping-off by use of integrated chemical and biological controls, *Plant Dis.*, 69, 431, 1985.

32. Lo, C.-T., Nelson, E.B., and Harman, G.E., Biological control of turfgrass diseases with a rhizosphere competent strain of *Trichoderma harzianum*, *Plant Dis.*, 80, 736, 1996.

33. Lo, C., Nelson, E.B., and Harman, G.E., Improved biocontrol efficacy of *Trichoderma harzianum* 1295-22 for foliar phases of turf diseases by use of spray applications, *Plant Dis.*, 81, 1132, 1997.

34. Lumsden, R.D. and Lewis, J.A., Selection, production, formulation, and commercial use of plant disease biocontrol fungi: problems and progress, in *Biotechnology for Improving Plant Growth*, Whipps, J.M. and Lumsden, R.D., Eds., Cambridge University Press, Cambridge, pp. 171–190, 1989.

35. Mendez-Castro, F.A. and Alexander, M., Method for establishing a bacterial inoculum on corn roots, *Appl. Environ. Microbiol.*, 45, 254, 1983.

36. Miles, M.N., Evaluation of soil solarization to control soilborne fungi and distribution studies of selected soilborne fungi in Oklahoma, M. S. thesis, Oklahoma State University, Stillwater, 1984.

37. Nelson, E.B., Using biological control strategies for turf, Part 2: Diseases, http://grounds.com/mag/grounds_maintenance_using_biological_control/index.html, 7/28/2006.

38. Nelson, E.B. and Craft, C.M., Introduction and establishment of strains of *Enterobacter cloaceae* in golf course turf for biological control of dollar spot, *Plant Dis.*, 75, 510, 1991.

39. Nelson, E.B. and Craft, C.M., Suppression of brown patch with top dressings amended with composts and organic fertilizers, *Biol. Cult. Tests Control Plant Dis.*, 6, 90, 1991.

40. Nelson, E.B. and Craft, C.M., Suppression of red thread with top dressings amended with composts and fertilizers, 1989, *Biol. Cult. Tests Control Plant Dis.*, 6, 101, 1991.

41. Nelson, E.B. and Craft, C.M., Suppression of Dollar spot on creeping bentgrass and annual bluestem turf with compost amended top dressings, *Plant Dis.*, 76, 954, 1992.

42. Nelson, E.B. and Craft, C.M., Suppression of *Typhula* blight with top dressings amended with composts and organic fertilizers, 1990–1991, *Biol. Cult. Tests Control Plant Dis.*, 7, 101, 1992.

43. Nutter, F.W., Gleason, M.L., Jenco, J.H., and Christians, N.C., Assessing the accuracy, inra-rater repeatability and inter-rater reliability of disease assessment systems, *Phytopathology*, 83, 806, 1993.

44. Odvody, G. H., Boosalis, M.G., and Kerr, E.D., Biological control of *Rhizoctonia solani* with a soil-inhabiting Basidiomycete, *Phytopathology*, 70, 655, 1980.

45. Papavizas, G.C., Fravel, D.R., and Lewis, J.A., Proliferation of *Talaromyces flavus* in soil and survival in alginate pellets, *Phytopathology*, 77, 131, 1987.

46. Papavizas, G.C., Morris, B.B., and Marois, J.J., Selective isolation and enumeration of *Laetisaria arvalis* from soil, *Phytopathology*, 73, 220, 1981.

47. Pinter, P.J., Effect of dew on canopy reflectance and temperature, *Remote Sens. Environ.*, 19, 187, 1986.

48. Punja, Z.K., Grogan, R.G., and Unruh, T., Comparative control of *Sclerotium rolfsii* on golf greens in northern California with fungicides, inorganic salts and *Trichoderma* spp., *Plant Dis.*, 66, 1125, 1982.

49. Sivan, A. and Chet, I., The possible role of competition between *Trichoderma harzianum* and *Fusarium oxysporum* on rhizosphere colonization, *Phytopathology*, 79, 198, 1989.

50. Spurr Jr., H.W. and Knudsen, G.R., Biological control of leaf diseases with bacteria, in *Biological Control on the Phylloplane*, Windels, C.E. and Lindow, S.E., Eds., The American Phytopathological Society, St. Paul, MN, 1985.

51. Sumner, D.R. and Bell, D.K., Antagonism of binucleate *Rhizoctonia*-like fungi and other Basidiomycetes to *Rhizoctonia solani* AG-4 and AG-2 type 2, *Phytopathology*, 78, 629, 1988.

52. Trenholm, L.E., Carrow, R.N., and Duncan, R.R., Relationship of multispectral radiometry data to qualitative data in turfgrass research, *Crop Sci.*, 39, 763, 1999.
53. Tu, J.C. and Vaartaja, O., The effect of the hyperparasite (*Gliocladium virens*) on *Rhizoctonia solani* and on *Rhizoctonia* root rot of white beans, *Can. J. Bot.*, 59, 22, 1981.
54. Walsh, B., Ikeda, S.S., and Boland, G.J., Biology and management of dollar spot (*Sclerotinia homeocarpa*), an important disease of turfgrass, *Hort. Sci.*, 34, 13, 1999.
55. Zadoks, J.C. and Schein, R.D., *Epidemiology and Plant Disease Management*, Oxford University Press, New York, 1979.

18 Biological Control of Some Insect Pests of Turfgrasses

J. Howard Frank

CONTENTS

18.1 INTRODUCTION

Biological control is the use of living organisms to control pests. When thought of this way, it is comparable, as an action that people can take, to chemical, physical, and cultural control. Its inception was the observation that living organisms regulate many potential pests in nature, so that no action by people is needed. The naturally occurring process is best called natural population regulation (although a few people call it "natural biological control," this is a misnomer if we define biological control as an action by people).

Healthy turf provides a habitat for many kinds of organisms. They range from soil bacteria through earthworms, mites and insects, to vertebrate animals (birds and mammals). All these organisms are the natural wildlife using turf as habitat. Insects and other invertebrates are regulated as wildlife by U.S. Federal law. This wildlife does and should include innocuous and beneficial insects. It may even include some pest insects. Provided that populations of those pest insects remain low, there is no need for action.

Several kinds of actions qualify as being biological control. Consider first (because it is most familiar to turf managers) those that require a turfgrass manager to buy a pesticide to control a pest. Industry produces and distributes the product. The purchaser applies the product to the area with the pest problem. When the product is a living organism, it is a biopesticide and its use is a form of biological control. The economics of research, production, and distribution are comparable to those of other forms of pesticides. The producer who makes an adequate profit can stay in business. The advantage and disadvantage of biopesticides is their specificity. High specificity of pesticides is an advantage to marketers when it comes to safety issues and lack of nontarget effects. High specificity of pesticides is a disadvantage to marketers when it comes to restricted markets; pesticides that compete well in many markets will be far more rewarding to their producers. Typically, biopesticides are more target-specific than chemical pesticides (Table 18.1).

Names additional to chemical pesticide and biopesticide have been coined. Manufacturers who developed pesticides from natural materials, or as synthetic analogs of natural materials, have called these biorational pesticides. For example, azadirachtin extracted from the neem tree (*Azadiractha indica* A. Juss., Meliaceae) is marketed as an insecticide, but this is not biological control. For another example, the insect-killing bacterium *Bacillus thuringiensis* Berliner (*Bt*) is sometimes cultured to express its toxin, and then killed and marketed; but because the marketed product is not a living organism, its use is not considered (at least here) to be biological control. Genes from *Bt* have been genetically engineered into crop plants, and now those plants produce the bacterium's toxin and kill some insects that feed on them. Such a process is considered here to be genetic engineering, not biological control. Selection for host-plant resistance is not considered to be biological control (Table 18.2).

Biopesticides as defined here are living organisms commercially produced in large numbers and marketed. They are antagonists to the pest, and are also called biocontrol agents. They may already be present in the turf that is to be protected, but in numbers too small to control the pest. Correction of that situation is called augmentative biological control. Obviously, it depends upon measured numbers present before application, with attempts to increase those numbers to the point where they are capable of adequate suppression of numbers of the pest. Alternatively, they may be released or applied in very large numbers, regardless of whether they are already present, to achieve local eradication in a situation called inundative biological control. Inundative biological control approaches the use of chemical pesticides. The biopesticide used may be a virus, bacterium, fungus, nematode, or various kinds of living insects released in large numbers for immediate effect.

TABLE 18.1
The Pesticide Producer's Dilemma

Typical of Chemicals	Typical of Biologicals
Broad spectrum of targets	Narrow spectrum of targets
Many markets	Niche markets
Commercially advantageous but environmentally risky	Environmentally advantageous but commercially risky
Greater nontarget effects	Few if any nontarget effects
Large-scale production may allow discounted price and increase competitive edge	Small-scale production may force high price and make product uncompetitive
Tend to store well	Tend to have short shelf life

TABLE 18.2
What Biological Control Is and Is Not

Biological Control Is	Biological Control Is Not
Classical biological control	Genetic engineering
Manipulative biological control ("conservation biocontrol")	Selection for host-plant resistance
Use of biopesticides, whether *augmentative*, or *inundative*	Use of chemical pesticides including "biorational" pesticides and bacterial toxins
Uses living organisms (antagonists or biocontrol agents) to control pests	Does not use living organisms to control pests

TABLE 18.3
Three Kinds of Biological Control Useful to Turf Managers

Kind	How Used
Biopesticides	Must be purchased and applied by turf managers
Manipulative biological control (conservation biocontrol)	Carried out by turf managers to conserve populations of beneficial organisms that are already present, or increase their effectiveness. There is no product to buy
Classical biological control (introduction biocontrol)	Carried out by university or state or federal scientists to bring additional beneficial organisms permanently to the area to suppress pest populations area-wide. There is no product to buy

With inundative biological control, there is no attempt to establish a continuing population of the beneficial organism, and there is no expectation of sustainable control, just as with most modern chemicals. A second's thought shows that the producer (seller) may benefit if the organism (the biopesticide) does *not* establish a continuing population, because thereby the producer may be able to sell the same product to the same buyer year after year, just as with a chemical pesticide.

Classical biological control (sometimes called introduction biocontrol or inoculative biocontrol) is the introduction from another country or region of a beneficial organism not previously present that establishes a permanent, reproducing population and to a greater or lesser extent suppresses populations of a target pest throughout the area where it becomes established. For example, a classical biological control organism might be introduced into the southern United States from Australia, establish a permanent population that spreads to many states, and control a pest (that had arrived years before from Australia as a result of human carelessness). Such programs typically are conducted by university, state, or federal scientists for public benefit; these programs do not result in a product for sale.

Manipulative biological control (often called conservation biocontrol) is the attempt to increase numbers, activity, or fitness of living biological control organisms at a locality by making conditions more favorable for them. This may be done by providing food, refuges, or various other actions. Development is typically conducted by university, state, or federal scientists and research results are often provided freely by statewide extension services (Table 18.3).

Use of biopesticides and manipulative biological control is site-specific. These actions are carried out by a turfgrass manager to enhance control of a pest or pests at a particular site (a playing field or a golf course, for example). Classical biological control has an area-wide objective, where the area may be an entire state or several states and the program uses beneficial organisms that were not previously present. To be really effective, the beneficial organism(s) must spread well of their own accord. If the turf that you manage gets invaded year after year from surrounding areas by a particular pest insect, classical biological control helps by suppressing populations of that pest in surrounding areas as well as on the turf that you manage.

Integrated pest management is the turf manager's program to use the best methods to control all the pests in the managed area in an integrated way. It begins with a knowledge of what the potential pests are and why some of them are seldom if ever abundant enough to cause a problem. Mismanagement might cause outbreaks of some of those minor pests. Then comes mapping and monitoring through time of the major pests. Third is a plan about how to manage all pests in a compatible way. For example, use of a biopesticide might be the best (most effective and lowest cost) way to control one of them, but a chemical pesticide might be the best way to control another. If both pests are too abundant and need control, there may or may not be a conflict: will use of the chemical pesticide interfere with use of the biopesticide?

18.2   THE MEANING OF PERCENT PARASITISM (MEASUREMENT OF SUCCESS)

Percent parasitism seems to represent the effectiveness of a parasitoid. "Parasitism rate" is an expression often used in the literature of economic entomology, apparently as a fuzzy rendition of percent parasitism (a "rate" in physics is something per unit of time). A third expression seen in the literature is "parasitization rate," which has nothing to recommend it because it just uses a long word "parasitization" when a shorter one, "parasitism," is adequate.

But, let us take a hard look at what percent parasitism means. It is usually applied to some sample taken in the field, such as the percentage of pests (hosts) infected by a parasitoid and doomed to die or (in some examples) to be castrated. What we really need to know is the percent of pests (hosts) that will be destroyed by the parasitoid in each host generation. When we think of the pest Japanese beetle and its parasitoid *Istocheta aldrichi* (see details below), the percent parasitism in the field has, in some instance in New England, been measured as about 20% (see below). This 20% is a reasonable approximation of the generational mortality caused to the host by the parasitoid because each has only one generation each year. But see also problems in estimation pointed out by Cherry and Klein,[9] explained below, and Van Driesche.[46]

When we read about 20% parasitism of mole cricket hosts by *Steinernema scapterisci* measured at a single time in turf plots, some 12 years after application of the nematode (see details below), we are looking at a radically different situation. This is because the host mole crickets have only one generation each year, whereas the nematodes release infective-stage juveniles into the soil within (usually) 10 days of their grandparents having infected the host (at ≈25°C air temperatures). This means that the nematode has many generations each year to the mole crickets' one, albeit that the "window of opportunity" for infection varies during the year. It represents a vastly higher level of generational mortality of the mole crickets.

To illustrate, assume (a) a constant climate during the year, (b) that a host has one generation each year, (c) that its parasitoid has four generations each year, (d) that all stages of the host are vulnerable to the parasitoid, and (e) that percent parasitism is 20%. Then, the first generation of the parasitoid kills 20% of the hosts, the second kills 20% of the survivors, etc., until by the end of the year only 41% of the hosts survive. Of course, assumptions (a) and (d) are unlikely to be valid; their complexities, especially because climate is likely to regulate the life cycle of both species, and the activity level of individuals of both are reasons why such calculations have been avoided by biologists. A further complication is that generations of the parasitoid may overlap.

Writers will continue to use the expression percent parasitism to report levels measured by field samples. But then they should indicate the result of a second calculation, "generational mortality of the host." This latter is the complicated calculation. On preliminary evidence and a rough calculation, I estimate that while percent parasitism of pest mole crickets caused by *Larra bicolor* in northern Florida was measured at about 25% throughout the autumn months of 2 years (see below), the generational mortality of the mole crickets was >60%. This was due to there being at least three generations of the wasp to the mole crickets' one generation.

18.3   THE WORST PEST INSECTS OF TURF

In 2000, a ranking of pest insects in turf by experts in the United States revealed that in the Pacific northwest, leather jackets (*Tipula* larvae) were rated worst; in the southeast, mole crickets; in the rest of the country, white grubs (larvae of scarab beetles, led by Japanese beetle wherever it occurs).[31] Although mole crickets were rated worst in 2000 in both Florida and South Carolina and had been so for decades, the situation has recently changed in Florida due to successful classical biological control. Now, white grubs, billbugs, and chinch bugs have all become more prominent in Florida as mole cricket populations have declined. It is not that several resident species of scarab beetles suddenly increased their population densities in Florida and thus became important (a very unlikely scenario), or that additional pest species arrived and became important. It is not that new chemical pesticides have sealed the fate of mole crickets in Florida (else why would they not have done that in South Carolina, too, where they are equally available?). A chemical explanation of decline on treated turf cannot explain why mole cricket populations have declined in pastures that have not been treated with chemicals. Rather, it is that scarab beetles always caused damage in Florida turf, but that damage had been secondary and was masked by the overwhelming problem with mole crickets. Now that mole cricket populations are decreasing in Florida, the damage caused by scarab larvae is being revealed. This chapter reviews biological control of *Tipula* larvae, mole crickets, various white grubs, Rhodesgrass mealybug, and some caterpillars.

18.4   CLASSICAL BIOLOGICAL CONTROL

Classical biological control has been attempted against a few turf pests of foreign origin. In most instances, it was undertaken because the pests attacked other commodities, but the projects also benefited and continue to benefit turf growers and managers.

18.4.1   Rhodesgrass Mealybug, *Antonina graminis* Maskell (Homoptera: Pseudococcidae)

Rhodesgrass mealybug is native to Asia. Its presence was detected in Hawaii before 1920, in Texas in 1942, and in Florida in 1945. Its effects on various pasture grasses were severe, and for that reason biological control agents were sought. However, it also attacks bermudagrass and St. Augustinegrass. A parasitoid wasp *Anagyrus antoninae* Timberlake (Hymenoptera: Encyrtidae) was detected in Hawaii attacking the mealybug; it had not been deliberately introduced and so is here considered an immigrant. It controlled the mealybug in Hawaii without need for further action. Of course, this wasp was the first biological control agent to be imported into Texas against the pest in 1949 and then into Florida in 1954. It adapted quite readily to Florida's climate and suppressed populations of the pest. It did not establish in the more arid areas of Texas, and so additional biological control agents were sought. A second parasitoid, *Neodusmetia sangwani* (Subba Rao) (Hymenoptera: Encyrtidae), was obtained in India and released in Florida in 1957 and Texas in 1959. Great effort was put into propagating this second wasp in Texas and distributing it over thousands of acres of pastureland by air because it is wingless, so unable to fly. This second wasp was so successful in Texas that it largely displaced the first and did so in Florida, too.

Rhodesgrass mealybug's control in Texas is an outstanding example of classical biological control. Dean et al.,[13] who summarized the project, estimated that annual savings due to effective biological control were almost $17 million for turf, and for pasture grasses more than $177 million, in Texas alone. Nobody collected and evaluated data for Florida or anywhere else. Meanwhile, *Neodusmetia sangwani* was released in Mexico, Bermuda, and Brazil and has been detected (presumably as an immigrant) in Bolivia, Costa Rica, Honduras, Jamaica, Puerto Rico, and Venezuela,[5] presumably saving untold tens of millions of dollars annually.

18.4.2 *Gryllotalpa* Mole Crickets (Orthoptera: Gryllotalpidae)

Gryllotalpa is a genus containing some 56 described species with representatives in all continents except Antarctica. Some of the species are rare, even endangered. The two native *Gryllotalpa* mole crickets in the United States are not turf pests. However, there are various poorly documented reports of the damage caused by *Gryllotalpa* to turf in Asia, most of them failing to identify the pests to the species level. In fairness to those reporting the damage, identification of Asian mole crickets is a difficult task because the taxonomy has not been completed and there are no adequate keys for most parts of Asia. In South Africa, *Gryllotalpa africana* Palisot de Beauvois is a pest of turf.[14] When *G. africana* was reported as a newly arrived pest in Hawaii, the Hawaiian Sugar Planters' Association began a classical biological control project to combat it. Eventually, a sphecid wasp *Larra luzonensis* Rohwer was imported from the Philippines and established in Hawaii.[47] This biological control attempt was judged successful on the grounds that mole crickets have not for many decades been important pests in Hawaii, although the level of control imparted by the wasp was not measured. Later, taxonomic studies revealed that the mole cricket in Hawaii is not *G. africana*, and it may be *G. orientalis* Burmeister, and the correct name for the wasp is *Larra polita* (Smith),[18] with additional evidence by de Graaf et al. that the mole cricket was not *G. africana*.[15] It may be that prompt importation of *Larra polita* by the Hawaiian Sugar Planters' Association forestalled *G. orientalis* from becoming an important turf pest in Hawaii.

18.4.3 *Scapteriscus* Mole Crickets (Orthoptera: Gryllotalpidae)

Scapteriscus is a genus containing some 23 described species native to South and Central America. Hundreds of years ago a South American winged species, *Scapteriscus didactylus* (Latreille), began a gradual northward invasion of West Indian islands by flight, island by island, eventually reaching Puerto Rico and Hispaniola. It became a devastating agricultural pest in Puerto Rico.[50] A short-winged flightless species, *Scapteriscus abbreviatus* Scudder, later began a gradual trek from southern South America northward along the subtropical and tropical eastern coast of Brazil, carried from port to port as a stowaway in ship ballast. In the late nineteenth and early twentieth centuries, due to increased commerce, it extended its range still farther north to some West Indian islands, including Cuba, Hispaniola, Jamaica, New Providence (Bahamas), Puerto Rico, and St. Croix (U.S. Virgin Islands) by the same method. It used the same method to reach Florida and extreme southeastern Georgia (United States). Two winged species, *Scapteriscus borellii* Giglio-Tos and *S. vicinus* Scudder, arrived in the southeastern United States from southern South America by stowing away in ship ballast and then spread widely north to North Carolina and west to Texas. Most recently, *S. didactylus* arrived in eastern Australia by ship or by air as a stowaway in some cargo. The first effects of these pest mole crickets (except *S. didactylus* in Australia[41]) were felt by growers and farmers. Only later, as turf management became important, was substantial damage to turf reported. Agricultural interests sponsored classical biological control projects, and turf managers benefited.

In the late 1930s, Puerto Rican researchers imported a wasp that they called *Larra americana* Saussure into their island from Amazonian Brazil, targeted against a mole cricket that they called *Scapteriscus vicinus*.[48,49] In reality, the wasp was *Larra bicolor* F., and the mole cricket was *S. didactylus*, the major pest mole cricket in Puerto Rico. The wasp became established and spread in Puerto Rico, but no measurement was made of its effects on the target mole cricket. Much later, without explanation, Cruz and Segarra[12] assessed the level of success as "partial." It is evident to this writer that population levels of *S. didactylus* in Puerto Rico in the early twenty-first century (visits to that island in 2001, 2002, 2003, 2004 to assess problems with mole crickets) are vastly lower than those reported in the early twentieth century, e.g., by Zwaluwenburg,[50] but was it correct to attribute that reduction to importation of *L. bicolor* in the late 1930s? Could there have been some other reason, such as importation of the infamous generalist predator toad, *Bufo marinus* L. (Amphibia: Bufonidae), in 1920 from Barbados (but originally from South America), intended to control white grubs (scarab larvae) in sugarcane? Certainly, *B. marinus* eats mole crickets. It is difficult to assess the evidence now, so many years later, when *S. didactylus* is a far less common pest of agricultural crops but still damages the irrigated turf of golf courses and sod farms. Another complication was that in 1917, the short-winged mole cricket *S. abbreviatus* was detected as a new invader in Puerto Rico, and much later it was shown to be a better host for *L. bicolor* than *S. didactylus*.[7] Was it *S. abbreviatus* rather than *S. didactylus* that acted as the main host for *L. bicolor* in Puerto Rico?

With banning of the insecticide chlordane for agricultural use in the United States in the 1970s, Florida cattle ranchers were left without an affordable and persistent method of controlling pest mole crickets in pastures. Political pressure from them caused the Florida legislature to order the University of Florida to begin a mole cricket research program to solve the problem. That program was begun in 1978, and by the following year it had turned its attention to the possibility of classical biological control. A quarter century later, mole cricket densities were much lower in parts of northern Florida, and biological control agents were spreading to other parts of the state.[20]

The first biological control agent to be imported and established in Florida was the wasp *L. bicolor*, from Puerto Rico, in 1981. It became established near Ft. Lauderdale, but failed to spread widely even with intervention. It had trivial effect on a local population of *S. abbreviatus* and no measurable effect on the other two pest species *S. borellii* and *S. vicinus*.[8] Additional wasps nominally of the same species[34] were imported from Bolivia in 1988 and 1989 and released and established in northern Florida. By late 2005, this Bolivian stock had spread to at least 25 counties. It caused an estimated generational loss to *S. vicinus* of at least 70% at the only two localities where measurement was made in one county. The Puerto Rican and Bolivian stocks of the wasp differ in microscopic characters that were recognized by Menke,[34] who did not deem them of sufficient importance for species-level separation. It remains to be determined whether their behavioral characteristics are of consistent difference that these would suffice for species-level separation. Wasps of this Bolivian stock were released in southern Georgia and became established there, and apparently spread naturally to coastal Mississippi. The wasps undergo several generations each year. Although adult wasps are killed by freezing temperatures, the immature stages dwell underground, insulated from low temperatures. Pupae overwintering underground enter diapause, a prolonged inactive "hibernating" stage, from which adult wasps emerge in late spring and early summer.

Another biological control agent imported and established in Florida was the parasitoid fly *Ormia depleta* (Wiedemann) (Diptera: Tachinidae) from Brazil.[21] The first releases were late in 1988 and more followed over the next 4 years as 28 Florida golf courses and a sod farm funded mass-rearing and distribution under the auspices of the Florida Turfgrass Research Foundation. By 1994, populations of the fly seemed to be established in all Florida counties from Orlando southward, and were present late each year in additional counties as far north as Gainesville. The problem seemed to be that the stock of the fly, from 23° S in Brazil, did not have a diapausing stage and could not survive winters in northern Florida. Perhaps the reason was that the winters were so cold that freezing temperatures killed aboveground parts of flowering plants that at other times provided nectar to the adult flies. Search in 1998–1999 produced a stock of the same fly from 30° S in Brazil (it was not detected farther south), and this new stock seemed to be only very slightly better adapted to cooler temperatures, and unable to diapause.

A third biological control agent to be imported and established in Florida was the entomogenous nematode *Steinernema scapterisci* Nguyen and Smart (Rhabditida: Steinernematidae).[36] Imported from Uruguay, it was first released in small plots in pastures in 1985. The need for very large numbers of mole crickets in laboratory production of the nematode limited the numbers available for field trials. After application for a use patent to control *Scapteriscus* mole crickets, a starter stock of the nematode was provided to industry, which developed a rearing method using artificial diet. By the late 1980s, two companies were producing sufficient numbers for field trials, and trials were initiated in pastures in Florida, and in turf in Florida and a few other states. The trials in most states were head-to-head with chemicals, as if the nematode were a biopesticide with brief persistence. Most of the trials in Florida were intended to evaluate establishment and long-term action of the nematode as a classical biological control agent. Both companies ceased production of this nematode a few years later, but a third company began production and marketing in the United States in 2002. Data showed that nematodes applied on Florida golf courses have established populations that persisted 12 years (and counting) despite use of chemical insecticides on those courses[4]; so essentially their populations are permanent and this species is indeed a classical biological control agent. Although the nematode's ability to spread by itself is extremely limited, it is spread by newly infected mole crickets. Winged adult mole crickets can spread it considerable distances by flight. The nematode overwinters successfully in Florida and southern Georgia, but whether it can do so farther north is unknown.

In northern Florida where *L. bicolor* and *S. scapterisci* are well established, a 95% reduction of pest mole cricket (*S. borellii* and *S. vicinus*) populations was recorded as compared with population levels in the 1980s.[20] In time, this level of decrease should be achieved everywhere in Florida, even without further action, as these two agents spread naturally. However, University of Florida entomologists and Extension agents in 2006 began an organized project to spread the wasp to all Florida counties. Distribution of the nematode was widened by demonstration projects and is being widened by commercial sales.

18.4.4   Japanese Beetle, *Popillia japonica* Newman (Coleoptera: Scarabaeidae)

Native to eastern Asia, this insect was detected for the first time in the United States in 1916, in a nursery in New Jersey. Its current distribution in America north of Mexico is in southern Ontario and Quebec, and all the New England states westward to Wisconsin and south to Alabama and Georgia. Localized infestations in western states of the United States have been eradicated. The adults, by feeding on foliage, flowers, and fruits, damage a long list of plants. The larvae are pests especially of turf, and this insect is consequently the most important pest of turf throughout its range in North America.

A parasitoid wasp, *Tiphia popilliavora* Rohwer (Hymenoptera: Tiphiidae), was detected in Japan by USDA explorers. Females oviposit on beetle larvae in the soil and the resultant larvae feed externally on the beetle larvae, eventually pupating in a cocoon in the soil. Stock of the wasp was imported from Japan, Korea, and China and was released first in New Jersey beginning in 1921 and then in another nine states. There is a single annual generation and adults feed on plant nectars. Female wasps preferentially attack third-instar beetle grubs and it was the stock from Korea that proved best synchronized with host populations, at least in New Jersey, with female wasps abundant at the same time as the third-instar hosts. No estimate of percent parasitism achieved by this wasp was given by Clausen.[10] Godfrey[24] cited a personal communication from Klein to say that the current status of *T. popilliavora* in the United States is unknown. Does it still persist?

A second species, *Tiphia vernalis* Rohwer (Hymenoptera: Tiphiidae), was detected in Korea by USDA explorers. The life cycle and behavior are similar to those of *T. popilliavora* but adults were reported to feed on aphid honeydew rather than plant nectars[17] (but see conflicting evidence by Rogers and Potter[43] below). Stock was first released in New Jersey in 1925 and then distributed to another 15 states as far south as North Carolina by 1953. Levels of parasitism by this species in Pennsylvania and New Jersey ranged up to 61% during 1935–1949.[10] It is now present at least as far south as Kentucky (see below).

The parasitoid fly *Istocheta aldrichi* (Mesnil) (Diptera: Tachinidae), elsewhere called *Hyperecteina aldrichi*, was brought from northern Japan by USDA explorers and released in 13 states from New Jersey northward, beginning in 1922. A population became established. Female flies oviposit on adult Japanese beetles, especially female beetles. The resultant fly larvae penetrate the beetles and feed at first on the gonads, and kill the host within 5 days. The established population was not well synchronized with that of the host beetles, becoming active too soon to encounter the bulk of the beetle population. Later releases of stock from central Japan in New England by USDA researchers in the 1930s, and then in North Carolina in the 1980s by researchers of the North Carolina Department of Agriculture, may have increased the genetic variability of the established stock and the geographical area occupied by the fly. In the 1970s, levels of parasitism of 20% were reported in New England. Still, this level is much less than the 50% reported from central Japan by USDA researchers in early explorations.[17] The big questions are whether the imported stock has enough genetic variability that it will in time adapt to conditions everywhere in North America where the beetle is established, and whether it will achieve as high a level of parasitism as was observed in central Japan in the 1920s.

A second fly, *Dexilla ventralis* (Aldrich) (Diptera: Tachinidae), was imported from Korea and released first in New Jersey in 1926, and then in Pennsylvania, Maryland, Illinois, New York, and Connecticut. It seems to be established in New Jersey but only with a small population.[17] A third fly, *Prosena siberita* (F.) (Diptera: Tachinidae) was imported from Japan, released in New Jersey in 1923, and established through 1929, but subsequently disappeared.

Was the big, long-running, biological control program inaugurated against Japanese beetle by 1920 by the USDA a success? Explorers collected biological control agents in Japan, Korea, China, and India from 1920 to 1933, importing well over a million specimens to the United States. The progeny of 14 species was released in up to 15 states (depending upon the species). Collaborating departments of agriculture in additional states participated and helped distribute them. Five insect species became established at least temporarily and now either two or three are still established. Milky spore disease (two species) was discovered in New Jersey and distributed widely in other states (see below); Clausen[10] questions whether the two milky spore species really were native to New Jersey, or whether they too originated in the Orient. Clausen[10] concluded that "the contribution of the two species of *Tiphia* to the decline of the pest populations in the older infested areas has been substantial; however, it is obscured by the presence of the milky spore diseases of the grubs" … "populations in the 270 sq. mile [area] occupied by the beetle in southern New Jersey and southeastern Pennsylvania in 1921, were estimated at more than 500 million per sq. mile but through the influence of the imported parasites and the disease they had declined by 1945 to the point where the pest was still present only at isolated sites and only in small numbers. The same cycle is being repeated in the more recently infested areas." This message was rosy, but the USDA project was discontinued in 1953. And now only two (perhaps three) insect biological control agents persist, milky spores may (see below) no longer be as effective as they once were reported to be, and turf managers are raising their standards for acceptable levels of damage. In the 1980s, as the beetle began to affect North Carolina, that state's department of agriculture (NCDA) began a program to import new genetic stock of some of the biocontrol agents from Japan. Notable success was not achieved, and the NCDA program ended. States to the west have imposed quarantine restrictions on commodities from the east to try to prevent arrival of the beetle, and have eradicated detected beetles. The emphasis has changed from classical biological control to the other forms of biological control. However, further distribution of classical biological control agents, especially *Istocheta aldrichi*, should reduce damage to agricultural crops in the relatively new range (mid-west) of Japanese beetle where its populations are very high. A side benefit of such distribution would be lower populations on turf.

18.4.5 OTHER ADVENTIVE SCARAB BEETLES (COLEOPTERA: SCARABAEIDAE)

Other adventive scarab beetles have been targeted as pests of turf and other plants in the northeastern United States. They are the Asiatic garden beetle, *Maladera castanea* (Arrow), first detected in New Jersey in 1921 and now distributed from Vermont to Georgia; oriental beetle, *Anomala orientalis* (Waterhouse), first detected in the northeastern United States in 1920 and subsequently spread south along the Atlantic coast and west to Ohio and Kentucky; and European chafer, *Amphimallon majale* (Razumovskii), (elsewhere called *Rhizotrogus majalis*), first detected in New York state in 1940 and now known from Ontario, Quebec, Connecticut, New York, and West Virginia. Of the various biological control agents imported and released against them, only a wasp, *Tiphia asericae* Allen and Jaynes (Hymenoptera: Tiphiidae) (against *Maladera castanea* in Pennsylvania) became established, but it was ineffective in providing control.[10] However, *Tiphia vernalis*, which was imported and established against Japanese beetle, also attacks oriental beetle. The classification of *Anomala* used here follows Jameson et al.[28] In southern Australia, the major pest is African black beetle, *Heteronychus arator* (F.).

The outcome differed in Hawaii, where oriental beetle was detected in 1912 and became a pest of sugarcane. Employees of the Hawaiian Sugar Planters' Association successfully imported and established two wasps from the Philippines in 1915–1917. The early promise of *Campsomeris marginella modesta* (Smith) (whose synonym is *Scolia manilae* Ashmead), was brought into question when it did not prevent heavy infestations of the beetle in the early 1930s. *Tiphia segregata* Crawford was the other established species. Then, in 1932, unfortunately, *Bufo marinus* was imported from Puerto Rico and was reported to feed voraciously on the beetle. Populations of the beetle then declined.[10] The precise roles of the two wasps and the toad in controlling the beetle seem not to have been measured.

18.4.6 *Tipula* Crane Flies or Leather Jackets (Diptera: Tipulidae)

Tipula paludosa Meigen (European crane fly) and *T. oleracea* L. (no "common" name yet approved by the Entomological Society of America) are European species adventive to the Pacific Northwest of North America. *Tipula paludosa* was detected in the 1950s in Canadian maritime provinces, then in 1965 in British Columbia, and *T. oleracea* was detected in 1999 in Washington. Both have spread. The states affected are Alaska, Oregon, Washington, and extreme northern California as well as the Canadian provinces of British Columbia, Newfoundland, and Nova Scotia. Both have recently been detected in New York.[37] The larvae of these flies, called leather jackets because of their tough skins, damage the roots of crop plants, forest trees in nurseries, and turfgrasses. *Siphona geniculata* DeGeer (Diptera: Tachinidae), parasitoid of *T. paludosa* and native to Europe, was introduced from Germany into British Columbia in 1968 and became established, although with no reported effect on populations of either crane fly.

18.4.7 Conclusion

The turfgrass industry, at least in America north of Mexico, depends on pesticides to control its pest problems. The classical biological control that has been achieved has been on the coattails of other agricultural industries.

The "marine toad" or "cane toad" *Bufo marinus* has been vilified as a generalist predator, very harmful to native faunas in places where it has been introduced. It is perhaps the ultimate example of biological control agent gone wrong. It was in almost all instances introduced by or for sugarcane growers to protect that crop from damage by larvae of scarab beetles. Places where it was introduced successfully (its origin was South America) include Barbados, 1820; Jamaica and Martinique, 1844; Puerto Rico, 1920; Hawaii, 1932; Queensland (Australia), 1935; New Guinea and Fiji, 1936; Guam, 1937; and Mauritius, 1938. Perhaps its effects are not all bad; perhaps it contributes substantially to the control of West Indian mole cricket, *Scapteriscus didactylus*, in Puerto Rico and oriental beetle, *Anomala orientalis*, and billbugs, *Sphenophorus venatus* (Say) (Coleoptera: Dryophthoridae), in Hawaii.

18.5 BIOPESTICIDAL CONTROL

The concept of a biopesticide as a living organism, applied much like a chemical pesticide and having the same function, is greatly stretched in some of the examples below. This is because some of the organisms establish long-lasting populations when they are applied, reproducing in the target host and releasing infective progeny into the soil for generation after generation. In this way, they resemble classical biological control agents, more so when they originate in a distant geographical area. They are almost always produced and marketed by commercial organizations, which must make a profit to stay in business.

18.5.1 *SCAPTERISCUS* MOLE CRICKETS (ORTHOPTERA: GRYLLOTALPIDAE)

The entomogenous nematode *Steinernema scapterisci* introduced to Florida as a classical biological control agent was developed by industry as a biopesticide for use against *Scapteriscus* mole crickets. It does not perform as well as the most effective chemicals in terms of speed of kill or percentage kill, but it has three considerable advantages. First, it can and often does establish essentially permanent populations, at least under Florida conditions of soils and climate. This is because it reproduces in pest mole crickets and releases progeny into the soil. Of course, its success at reproducing depends upon the continued existence of some mole crickets in the area treated. By having a brief generation time, the nematode can continue infecting and killing mole crickets month after month and, during an entire year, can kill a very high percentage of the pests (most of which have a single generation each year). Second, its population spreads. Thus, every square meter of a golf course or playing field does not have to be treated with it; a few treated swaths across a large area can be enough to get a population established and, in time, spread to every square meter. This also means that it will spread to surrounding properties, to provide area-wide pest control, thus reducing the number of winged adult mole crickets available to invade a property by flight. Third, it is harmless to vertebrate animals (and people), and so in contrast to the use of chemicals there is no period during which re-entry to the treated area is prohibited. It is remarkably unaffected by at least most chemical insecticides.[3] It should be applied when mole crickets are large, in spring or fall, not in summer when small nymphs prevail, because it is not very infective to small nymphs. If applied to the soil surface (rather than by injection into the soil), its application should be preceded and followed by irrigation to wash it into the soil where it is safe from excessive heat and damaging ultraviolet solar radiation. It should be kept at refrigerator temperature until it is applied. Only one company is licensed by the University of Florida to produce it. Annual applications on high-value turf may be worthwhile because they can increase the soil inoculum of the nematode, thus exposing more mole crickets to infection.[4]

18.5.2 JAPANESE BEETLE, *POPILLIA JAPONICA* NEWMAN (COLEOPTERA: SCARABAEIDAE)

Among the various pathogens found in Japanese beetle larvae, a bacterium has been used most successfully as a biopesticide. It was found in larvae of the beetle in New Jersey in 1933 and was named *Bacillus popilliae* Dutky (Eubacteriales: Bacillaceae). It is best known as "milky spore disease" or "milky disease" because of the opaque white appearance of larvae having many spores in the hemolymph. Spores of this species were the first microbial agent registered as a pesticide active ingredient. In 1995, the active ingredient was reassessed to ensure that it met the registration requirements. As of October 2004, there was one end-use product.[16] It is sold as a powder to be spread on soil sometime between spring and fall (when the ground is not frozen). Once the bacteria and their spores become established in a geographic area, they decrease the numbers of Japanese beetle larvae and consequently of adults, thereby reducing damage to plants.

The characteristics of *Bacillus lentimorbus* Dutky, which was discovered under the same conditions as was *B. popilliae*, and also called milky disease, have been questioned. Initially it was thought to be a distinct species, but others have thought it to be another strain of the same species. It is infective to Japanese beetle and was likewise produced *in vivo* and distributed in areas where the beetle had become a pest. Later, RNA gene sequences of both species were studied, confirming that they are two distinct species but belonging to the genus *Paenibacillus*.[38] The two species names are thus *Paenibacillus popilliae* (Dutky) and *P. lentimorbus* (Dutky). Neither of the two seems restricted to Japanese beetle; both seem to have activity against some other pest scarabs.

From 1939 to 1953, more than 100 tons of *P. popilliae* spores (and more than 3 tons of *P. lentimorbus* spores) were applied on turf in the eastern United States, in a government-sponsored program encompassing more than 160,000 sites in Connecticut, Maine, New Hampshire, Rhode Island, Vermont, New York, New Jersey, Maryland, Delaware, District of Columbia, Virginia, West Virginia, North Carolina, Ohio, and Pennsylvania. This was reported to cause a reduction of 90–95% in Japanese beetle populations at the application sites; the bacteria became established and suppression persisted for years.[17]

In contrast to the early reports of high efficacy of *P. popilliae*, Redmond and Potter[40] found a much lower level of efficacy of two commercial formulations. They questioned especially the efficacy of spores produced *in vitro*, and suggested that efficacy may be highest at very high densities of beetle larvae (as perhaps it had been in targeted early outbreaks). Or, perhaps *P. popilliae* is not very useful as a biopesticide, but if it establishes an inoculum in the soil it may suppress the pest for months or years. Cherry and Klein,[9] in a study of infection of *Cyclocephala parallela* Casey by *P. popilliae*, showed that it is easy to underestimate mortality inflicted by the bacterium. More than 10 years later, the efficacy of milky spore disease is still not clear.

Entomogenous nematodes, too, have been used as biopesticides against *P. japonica*. This is complicated because of the existence of many species and strains with varying virulence to the beetle larvae, but few of them are available commercially. *Heterorhabditis zealandica* Poinar proved to have the highest virulence among five species and 12 strains of *H. bacteriophora* tested by Grewal et al.[25] although a strain of *H. bacteriophora* Poinar caused substantial mortality. *Steinernema glaseri* (Steiner) and *S. kushidai* Mamiya have also been effective.[39]

18.5.3   OTHER ADVENTIVE SCARAB BEETLES (COLEOPTERA: SCARABAEIDAE)

These include Asiatic garden beetle, *Maladera castanea* (Arrow), European chafer, *Amphimallon majale* (Razumovskii), oriental beetle, *Anomala orientalis* Waterhouse, and African black beetle, *Heteronychus arator* (F.). *Heterorhabditis zealandica* proved to be highly virulent against oriental beetle but only feebly so against European chafer.[25]

18.5.4   NATIVE SCARAB BEETLES (COLEOPTERA: SCARABAEIDAE)

In western Europe, the main pest is the garden chafer, *Phyllopertha horticola* L., but Welsh chafer, *Hoplia philanthus* Fuessly, June beetle, *Amphimallon solstitiale* (L.), and cockchafer, *Melolontha melolontha* (L.), also can be important. In New Zealand, the overwhelming pest of pastures and turf is the grass grub, *Costelytra zealandica* (White). In America north of Mexico, native scarab beetles that damage turf belong to the genera *Anomala, Aphodius, Ataenius, Cotinis, Cyclocephala, Hybosorus, Tomarus,* and *Phyllophaga*.

Serratia entomophila (Bacteria: Enterobacteriaceae) is a bacterium native to New Zealand, which has been developed as a biopesticide for use there against the grass grub.[26] It is perhaps the most promising of all the biopesticides under development against white grubs because it was selected for its performance against just one pest species against which it is effective. All the other pathogens developed and under investigation in other countries have encountered a recurrent problem: they may provide the most effective biopesticide against pest species A (at least until another one is detected) but their performance against pest species B, C, and D is inferior.

Heterorhabditis zealandica proved to be highly virulent against the northern masked chafer, *Cyclocephala borealis* Arrow, as did a U.K. strain of *H. megidis* Poinar, Jackson and Klein.[25] Many other test results have been published, some of them conflicting. It is not clear whether researchers are looking for a nematode that is highly virulent to all scarab species; if they are, the objective is perhaps not a good idea, because the consequence is the antithesis of the idea of target specificity in favor of a broad range of targets (as for chemical pesticides), which would threaten nonpest scarabs including beneficial dung-burying species.

Bacillus thuringiensis var. *japonensis* Buibui strain is specific to scarabs. It is effective against several species of white grubs, including Japanese beetle, but its future is in doubt due to formulation and marketing issues.[39]

Fungal pathogens such as *Beauveria bassiana* (Balsamo) Vuillemin and *Metarhizium anisopliae* (Metchnikoff) Sorokin are variously pathogenic in laboratory tests to various white grubs and other pests. The major difficulty with their practical use has been in delivering enough of the spores at low cost to achieve a high level of control. The same also applies to their use against other pest insects of turf.

18.5.5 TIPULA CRANE FLIES OR LEATHER JACKETS (DIPTERA: TIPULIDAE)

Tipula paludosa Meigen and *T. oleracea* L., whose larvae are called leather jackets in England, have always been pests of turf grasses in northwestern Europe. Their presence in North America is a recent event. The entomogenous nematode *Steinernema feltiae* (Filipjev) is produced and marketed as a biopesticide for use against these pests in Europe. Among all the known species of entomogenous nematodes, this is the species best known for its activity against fly larvae.

18.5.6 CATERPILLARS—LEPIDOPTERA

Caterpillars harmful to turf include cutworms and armyworms (Noctuidae) (e.g., black cutworm, *Agrotis ipsilon* (Hufnagel); armyworm, *Pseudaletia unipuncta* (Haworth); fall armyworm, *Spodoptera frugiperda* (J.E. Smith); grass loopers, *Mocis* spp.; and lawn armyworm, *Spodoptera mauritia* (Boisduval) in Hawaii) as well as webworms (Pyralidae) (e.g., tropical sod webworm, *Herpetogramma phaeopteralis* Guenée; grass webworm, *H. licarsicalis* (Walker); and the various genera and species that are called sod webworms) and burrowing sod webworms (Acrolophidae) (*Acrolophus* spp.). Early instars are susceptible to *Bacillus thuringiensis* (Bt) var. *kurstaki*, which is the caterpillar-adapted variety of this pathogen species. The bacterium must be ingested to be effective and it acts as a true biopesticide because it is unlikely to establish a population. For the most part, these caterpillars feed on grass blades, which can readily be sprayed with *Bt* var. *kurstaki*, thus allowing contact of the biopesticide with the pest; so this biopesticide is the logical choice. Entomogenous nematodes, due to their exposure (and susceptibility) to UV radiation when they are sprayed on grass blades, seem a less logical choice. However, some of these caterpillars, such as the cutworms, are more subterranean in habit and because of this they make adequate targets for entomogenous nematodes that have been applied and washed by irrigation into the soil. Thus, the nematode *Steinernema carpocapsae* (Weiser) has been used effectively to control black cutworm.

18.5.7 CONCLUSION

Competition from chemical pesticides limits the use of entomogenous nematodes as biopesticides in the North American market, although the future appears bright.[23] Purchase and use of biopesticides depends to a large extent upon whether a would-be purchaser can find evaluative information about products. Producers and products may change rapidly, and yet there is no up-to-date online catalog of them. A database of pathogens (and nematodes) used as biopesticides was prepared by Shah and Goettel[44] and was published online by the Society for Invertebrate Pathology. Another, drawing heavily from the first but introducing transcription errors along with some expansion and updates, was prepared in 2002 and published online by the University of Oregon's Integrated Plant Protection Center.[27] When examined, it had not been modified since 2002 and it needed correction and expansion. So the potential purchaser, lacking access to an up-to-date and evaluative database and wanting to buy biopesticides, must rely on textbooks (soon outdated even if up-to-date when written), advertising in trade journals, and advice from university extension personnel. A big question is whether extension personnel have adequate information about biopesticides.

Another problem is that there are many pest species and many species, subspecies, and strains of potential "biopesticides," and the companies that produce and market them have few funds to spend on research. The consequence is that few species and strains are being marketed, and the limited research that has been done is constantly showing that more effective species and strains exist to combat this, that, or another pest. The companies cannot readily produce and market all the discoveries of research, and the consequence of this is that the products marketed are not necessarily the best for the purpose.

The definition of best is an uncertainty. Very often, a biopesticide is measured in terms of virulence to its target pest. Virulence is a measure of rapid kill and is appropriate for a head-to-head comparison with a chemical pesticide. It is much like percent parasitism mentioned above for classical biological control agents. Quite another measurement is generational mortality that a "biopesticide" can inflict upon its target if it establishes a population and can continue killing the target generation after generation. Such a measurement is not given by vendors. Lack of one suggests a role by mathematical modelers, although mathematical modeling of insect populations is no longer the fundable research topic that it was in the 1970s.

18.6   MANIPULATIVE BIOLOGICAL CONTROL (CONSERVATION BIOCONTROL)

This aspect of biological control can be considered to be any action that promotes the survival or abundance of resident natural enemies, or that promotes their contact with their target. Instead, here it is restricted to the attempt to enhance the presence and survival of parasitoid wasps by providing nectar sources. Other aspects are addressed under the heading of integrated pest management.

18.6.1   *Scapteriscus* Mole Crickets

Adult *Larra bicolor* wasps are especially attracted to flowers of *Spermacoce verticillata* (L.) (Rubiaceae) in their native range.[47] In Puerto Rico, where the wasp was introduced in the 1930s, its range seemed to depend on the abundance of this same wildflower.[49] In Florida, where the wasp was also introduced, the same wildflower attracts the wasps and provides them with nectar in the same way that other kinds of plants attract butterflies.[1] It still remains to be proved that deliberate plantings of this plant adjacent to turf will increase the number of wasps in the area and the percentage of mole crickets parasitized by the wasp; testing is now in progress. In southern Florida, the plant flowers all year and so is an ideal nectar source. In northern Florida, it freezes to the ground at the first hard frost but reemerges and flowers again by late April or May; its absence in the winter months hardly matters because wasp pupae are in diapause underground. Additional nectar sources are being sought.

18.6.2   Japanese Beetle, *Popillia japonica*

Tiphia vernalis wasps feed on honeydew secreted by Homoptera,[22,17] and on a Kentucky golf course having fairways bordered by mature oaks, maples, and other woody plants infested with aphids and soft scales, their parasitism of Japanese beetle grubs was as high as 50%.[43] The wasps also feed on nectar secreted by extrafloral nectars of peony, *Paeonia lactiflora* Pallas (Paeoniaceae).[32,43] Experimental planting of groups of four peony plants in a field of turf, and measuring parasitism by this wasp at 1, 10, 30, and 50 m distance, showed that the closest beetle grubs were far more heavily parasitized than were those at any other distance.[43] There is good experimental evidence but not yet a blueprint for turf managers.

18.7 INTEGRATED PEST MANAGEMENT (IPM)

All pest insects have natural enemies. Not all pest species existing in a turf habitat build up in numbers to cause substantial damage all the time. Instead, their populations are held in check (regulated) by natural enemies much of the time. IPM is the attempt to use this naturally occurring regulation, and only to take control action when the natural enemies do not achieve adequate regulation. Then, when control action is taken against a given pest, this action should be compatible with continued reliance on existing natural enemies to control other pests.

Studies such as that by Lopez and Potter[33] have documented that resident generalist natural enemies contribute to the control of pests. Others have shown that turfgrass management practices, especially use of chemicals, harm beneficial organisms.[2,6,11,30,45] Rogers and Potter[42] took this a step further in suggesting that the timing of application of imidacloprid can be adjusted to minimize harm to *Tiphia vernalis*, a specialist parasitoid of Japanese beetle, while Oliver et al.[35] found that imidacloprid and halofenozide are less harmful to it than are other chemicals. Other strategies are to rely on the survival of insect biological control agents in chemically untreated areas such as the roughs of many golf courses[19] or to apply spot treatments of chemicals instead of broadcast applications. But not all natural enemies are susceptible to chemical insecticides. For example, entomogenous nematodes may be tolerant to such chemicals,[3] or the chemicals and the nematodes may even have a synergistic effect.[29] Although the Green Section of the US Golf Association is sponsoring environmental research, there is a long way to go before IPM becomes a reality in highly managed turf.

ACKNOWLEDGMENTS

I thank my colleague Eileen Buss for critical review of a manuscript draft.

REFERENCES

1. Arévalo, H.A. and Frank, J.H., Nectar sources for *Larra bicolor* (Hymenoptera: Specidae), a parasitoid of *Scapteriscus* mole crickets (Orthoptera: Gryllotalpidae), in northern Florida, *Florida Entomol.*, 88, 146, 2005.
2. Arnold, T.B. and Potter, D.A., Impact of a high maintenance lawn-care program on nontarget invertebrates in Kentucky bluegrass turf, *Environ. Entomol.*, 16, 100, 1987.
3. Barbara, K.A. and Buss, E.A., Integration of insect parasitic nematodes (Nematoda: Steinernematidae) with insecticides for control of pest mole crickets (Orthoptera: Gryllotalpidae: *Scapteriscus* spp.), *J. Econ. Entomol.*, 98, 689, 2005.
4. Barbara, K.A. and Buss, E.A., Augmentative applications of *Steinernema scapterisci* (Nematoda: Steinernematidae) for mole cricket (Orthoptera: Gryllotalpidae) control on golf courses, *Florida Entomol.*, 89, 257, 2006.
5. Bennett, F.D., Biological control of miscellaneous pests, in *Pest Management in the Subtropics. Biological Control – A Florida Perspective*, Rosen, D., Bennett, F.D., and Capinera, J.L., Eds., Intercept, Andover, U.K., 1994, chap. 8.
6. Braman, S.K. and Pendley, A.F., Relative and seasonal abundances of beneficial arthropods in centipedegrass as influenced by management practices, *J. Econ. Entomol.*, 86, 494, 1993.
7. Castner, J.L., Suitability of *Scapteriscus* spp. mole crickets (Orthoptera; Gryllotalpidae) as hosts of *Larra bicolor* (Hymenoptera: Sphecidae), *Entomophaga*, 29, 323, 1984.
8. Castner, J.L., Evaluation of *Larra bicolor* as a biological control agent of mole crickets, Ph.D. dissertation, University of Florida, 1988.
9. Cherry, R.H. and Klein, M.G., Mortality induced by *Bacillus popilliae* in *Cyclocephala parallela* (Coleoptera: Scarabaeidae) held under simulated field temperature, *Fla. Entomol.*, 80, 261, 1997.
10. Clausen, C.P., Scarabaeidae, in *Introduced Parasites and Predators of Arthropod Pests and Weeds: A World Review*, Clausen, C.P., Ed., *USDA Agriculture Handbook*, 480, 277, 1978.
11. Cockfield, S.D. and Potter, D.A., Predatory arthropods in high- and low-maintenance turfgrass, *Can. Entomol.*, 117, 423, 1985.

12. Cruz, C. and Segarra, A., Potential for biological control of crop pests in the Caribbean, *Florida Entomol.*, 75, 400, 1992.

13. Dean, H.A. et al., Complete biological control of *Antonina graminis* in Texas with *Neodusmetia sangwani* (a classic example), *Bull. Entomol. Soc. Am.*, 12, 262, 1979.

14. de Graaf, J., Schoeman, A.S., and Brandenburg, R.L., Seasonal development of *Gryllotalpa africana* (Orthoptera: Gryllotalpidae) on turfgrass in South Africa, *Florida Entomol.*, 87, 130, 2004.

15. de Graaf, J., Schoeman, A.S., and Brandenburg, R.L., Stridulation of *Gryllotalpa africana* (Orthoptera: Gryllotalpidae) on turfgrass in South Africa, *Florida Entomol.*, 88, 292, 2005.

16. EPA, U.S. Environmental Protection Agency, *Bacillus popilliae* spores (054502) fact sheet, 2004, <http://www.epa.gov/oppbppd1/biopesticides/ingredients/factsheets/factsheet_054502.htm>, accessed April 2006.

17. Fleming, W.E., Biological control of the Japanese beetle, *USDA Tech. Bull.*, 1383, 1968.

18. Frank, J.H., Inoculative biological control of mole crickets, in *Handbook of Integrated Pest Management for Turf and Ornamentals*, Leslie, A.R., Ed., Lewis Publishers, Boca Raton, FL,1994, chap 42.

19. Frank, J.H. and Parkman, J.P., Integrated pest management of pest mole crickets with emphasis on the southeastern USA, *Integr. Pest Manage. Rev.*, 4, 39, 1999.

20. Frank, J.H. and Walker, T.J., Permanent control of pest mole crickets (Orthoptera: Gryllotalpidae: *Scapteriscus*) in Florida, *Am. Entomol.*, 52, 138, 2006.

21. Frank, J.H., Walker, T.J., and Parkman, J.P., The introduction, establishment and spread of *Ormia depleta* in Florida, *Biol. Contr.*, 6, 368, 1996.

22. Gardner, T.R., Influence of feeding habits of *Tiphia vernalis* on the parasitization of the Japanese beetle, *J. Econ. Entomol.*, 31, 204, 1938.

23. Georgis, R. et al., Successes and failures in the use of parasitic nematodes for pest control, *Biol. Contr.*, 38, 103, 2006.

24. Godfrey, K.E., *Tiphia* wasps, in *Japanese beetle white grubs and other beetle pests of turf*, in *Classical Biological Control in the Southern United States*, Habeck, D.H., Bennett, F.D., and Frank, J.H., Eds., Southern Coop. Ser. Bull., 355, 143, 1990.

25. Grewal, P.S. et al., Differences in susceptibility of introduced and native white grub species to entomopathogenic nematodes from various geographic locations, *Biol. Contr.*, 24, 230, 2002.

26. Grimont, P.A.D. et al., *Serratia entomophila* sp. nov. associated with amber disease in the New Zealand grass grub, *Costelytra zealandica*, *Int. J. Syst. Bacteriol.*, 38, 1, 1988.

27. IPPC Database of microbial biopesticides (DMB), University of Oregon, Integrated Plant Protection Center, 2002, <http://ippc.orst.edu/biocontrol/biopesticides>, accessed June 2006.

28. Jameson, M.L., Paucar-Cabrera, A., and Solis, A., Synopsis of the New World genera of *Anomalini* (Coleoptera: Scarabaeidae: Rutelinae) and description of a new genus from Costa Rica and Nicaragua, *Ann. Entomol. Soc. Am.*, 96, 415, 2003.

29. Koppenhöfer, A.M. et al., Comparison of neonicotinoid insecticides as synergists for entomopathogenic nematodes, *Biol. Contr.*, 24, 90, 2002.

30. Kunkel, B.A., Held, D.W., and Potter, D.A., Impact of halofenozide, imidacloprid, and bendiocarb on beneficial invertebrates and predatory activity in turfgrass, *J. Econ. Entomol.*, 92, 922, 1999.

31. Liskey, E., Turf's most (un)wanted pests, *Grounds Maintenance*, 35(4), 18, 2000.

32. Long, C., Grow peonies to defeat Japanese beetles, *Organic Gardening*, 47, 10, 2000.

33. Lopez, R. and Potter, D.A., Ant predation on eggs and larvae of the black cutworm (Lepidoptera: Noctuidae) and Japanese beetle (Coleoptera: Scarabaeidae) in turfgrass, *Environ. Entomol.*, 29, 116, 2000.

34. Menke, A.S., Mole cricket hunters of the genus *Larra* in the New World (Hymenoptera: Sphecidae; Larrinae), *J. Hymenoptera Res.*, 1, 175, 1992.

35. Oliver, J.B. et al., Effect of insecticides on *Tiphia vernalis* (Hymenoptera: Tiphiidae) oviposition and survival of progeny to cocoon stage when parasitizing *Popillia japonica* (Coleoptera: Scarabaeidae) larvae, *J. Econ. Entomol.*, 98, 694, 2005.

36. Parkman, J.P. et al., Classical biological control of *Scapteriscus* spp. (Orthoptera: Gryllotalpidae) in Florida, *Environ. Entomol.*, 25, 1415, 1996.

37. Peck, D.C., Hoebeke, E.R., and Klass, C., Detection and establishment of the European crane flies *Tipula paludosa* Meigen and *Tipula oleracea* L. (Diptera: Tipulidae) in New York: a review of their distribution, invasion history, biology, and recognition, *Proc. Entomol. Soc. Wash.*, 108, 985, 2006.

38. Pettersson, B. et al., Transfer of *Bacillus lentimorbus* and *Bacillus popilliae* to the genus *Paenibacillus* with emended descriptions of *Paenibacillus lentimorbus* comb. nov. and *Paenibacillus popilliae* comb. nov., *Int. J. System. Bacteriol.*, 49, 531, 1999.

39. Potter, D.A. and Held, D.W., Biology and management of the Japanese beetle, *Annu. Rev. Entomol.*, 47, 175, 2002.
40. Redmond, C.T. and Potter, D.A., Lack of efficacy of *in-vivo* and putatively *in-vitro* produced *Bacillus popilliae* against field populations of Japanese beetle (Coleoptera: Scarabaeidae) grubs in Kentucky, *J. Econ. Entomol.*, 88, 846, 1995.
41. Rentz, D.C.F., The changa mole cricket, *Scapteriscus didactylus* (Latreille), a New World pest established in Australia, *J. Austr. Entomol. Soc.*, 34, 303, 1996.
42. Rogers, M.E. and Potter, D.A., Effects of spring imidacloprid application for white grub control on parasitism of Japanese beetle (Coleoptera: Scarabaeidae) by *Tiphia vernalis* (Hymenoptera: Tiphiidae), *J. Econ. Entomol.*, 96, 1412, 2003.
43. Rogers, M.E. and Potter, D.A., Potential for sugar sprays and flowering plants to increase parasitism of white grubs (Coleoptera: Scarabaeidae) by tiphiid wasps (Hymenoptera: Tiphiidae), *Environ. Entomol.*, 33, 619, 2004.
44. Shah, P.A. and Goettel, M.S., *Directory of microbial products and services*, Society for Insect Pathology, 1999, <http://www.sipweb.org/directorymcp/directory.htm>, accessed June 2006.
45. Terry, L.A., Potter, D.A., and Spicer, P.G., Insecticides affect predatory arthropods and predation on Japanese beetle (Coleoptera: Scarabaeidae) eggs and fall armyworm (Lepidoptera: Noctuidae) pupae in turfgrass, *J. Econ. Entomol.*, 86, 871, 1993.
46. Van Driesche, R.G., The meaning of percent parasitism in studies of insect parasitoids, *Environ. Entomol.*, 12, 1611, 1983.
47. Williams, F-X., Studies in tropical wasps – their hosts and associates (with descriptions of new species), *Hawaiian Sug. Plrs. Assoc. Exp. Sta., Entomol. Ser. Bull.*, 19, 1, 1928.
48. Wolcott, G.N., The introduction into Puerto Rico of *Larra americana* Saussure, a specific parasite of the "changa," or Puerto Rican mole cricket, *Scapteriscus vicinus* Scudder, *J. Agric. Univ. Puerto Rico*, 22, 193, 1938.
49. Wolcott, G.N., The establishment in Puerto Rico of *Larra americana* Saussure, *J. Econ. Entomol.*, 34, 53, 1941.
50. Zwaluwenburg, R.H. van, The changa or West Indian mole cricket, *Porto Rico Agr. Exp. Sta. Bull.*, 23, 1, 1918.

19 Microbial Control of Turfgrass Insects

Albrecht M. Koppenhöfer

CONTENTS

19.1 INTRODUCTION

Numerous insect and mite species inhabit turfgrass areas, and the vast majority of these does not cause any damage or contribute to nutrient recycling, soil building, suppressing pests, and various other beneficial functions [1–3]. However, due to the often low damage thresholds, numerous insects are considered pests. Insect pests of turfgrass vary in their behavior and on which parts of the turfgrass plant they feed, and management practices have to take these differences into account. Larvae of scarab beetles (Coleoptera: Scarabaeidae), often referred to as white grubs, usually feed on the grass roots. Nymphs and adults of mole crickets (Orthoptera: Gryllotalpidae) burrow extensively near the soil surface and feed on roots and other parts of grass plants. Larvae of most caterpillar pests (Lepidoptera: Noctuidae and Pyralidae) initially skeletonize foliage; but as they get larger they create burrows in the soil from which they emerge at night to feed on the grass shoots (webworms, hepialids, cutworms) or they live on the surface feeding on grass foliage and stems (armyworms). Larvae of some weevils (Coleoptera: Curculionidae) (billbugs, annual bluegrass weevil) and flies (Diptera) (craneflies, sod fly, March fly) feed on roots and bore into stems and crowns, killing tillers or entire plants. As groups white grubs, weevils, and caterpillar pests are found worldwide, but other groups have a more limited distribution (Table 19.1).

Various stages of many turfgrass insect pests are naturally infected by a variety of insect pathogens including viruses, bacteria, fungi, protozoans, and nematodes. The close association between soil-dwelling pests and microbes seems to have led to a high number of diseases associated with those insects [4]. As result of their coevolution with these organisms, soil-dwelling pests appear to have developed varying degrees of resistance to generalist pathogens, but some specific species and strains of pathogens have overcome host defenses and proven effective as microbial control agents of soil-dwelling pests [5]. Several of these pathogens have already been developed into microbial control products for turfgrass insect pests or are being investigated to these ends. In fact, the milky

TABLE 19.1

Major Turf Pests, Parts of the Plant They Attack, Geographical Problem Areas, and Potential Microbial Controls

Part Attacked	Damage	Pests	Pest Life Stage	Location	Potential Microbial Controls
Root	Chewing	White grubs	Larva	Worldwide	Bacteria, nematodes, fungi
		Mole crickets	Nymph, adult	SE USA	Nematodes, fungi
Stem/crown	Chewing	Annual bluegrass weevil	Larva	NE USA	Nematodes, fungi
		Billbugs	Larva	USA, Japan, New Zealand, Australia	Nematodes, fungi
		Crane fly	Larva	NW/NE USA, SW Canada, Europe	Bacteria, nematodes
Leaf/stem	Chewing	Armyworms	Larva	Worldwide	Nematodes, bacteria
		Cutworms	Larva	Worldwide	Nematodes
		Sod webworms	Larva	USA	Nematodes, bacteria
Leaf/stem	Sucking	Chinch bugs	Nymph, adult	C/E USA, SE Canada, Japan	Fungi
		Greenbug aphid	Nymph, adult	USA	Fungi
		Mealybugs	Nymph, adult	S USA, New Zealand	—
		Mites	Nymph, adult	USA	—
		Scales	Nymph, adult	S USA, Japan	—
		Spittlebugs	Nymph, adult	E USA, Brazil	Fungi

Note: C, central; E, eastern; NE, northeastern; NW, northwestern; S, southern; SE, southeastern; SW, southwestern.
Source: After Klein et al. *Field Manual of Techniques in Invertebrate Pathology*, Kluwer, Dordrecht, 2007.

disease bacteria, *Paenibacillus* (=*Bacillus*) *popilliae* and *P. lentimorbus* were the first commercial microbial control products in the United States, and were used in augmentative releases to suppress Japanese beetle, *Popillia japonica*, populations in turfgrass starting in 1948.

Microbial control agents have a number of advantages over conventional synthetic chemical insecticides [6]. Due to their generally narrow host range, they have only a limited impact on beneficial insects and do not have the potential to cause secondary pest outbreaks. Because they are generally nontoxic they are safer to apply, nonhazardous to the environment and compatible with other biocontrol agents. Many microbial agents also have the potential to suppress insect pests for more than one pest generation due to their reproduction in the hosts. Finally, they are adaptable to genetic modification through biotechnology. Resistance has developed only in some target insect to *Bacillus thuringiensis* (*Bt*) and *B. sphaericus*. However, compared with chemical insecticides microbial control agents, to varying degrees, also have the disadvantages of relatively high cost, shorter shelf life, long period of lethal infection, and higher sensitivity to environmental factors (UV light, desiccation, temperature extremes, etc.).

Several microbial organisms are available for control of turf insect pests (Table 19.2). These products can be divided into two broad categories, pathogenic microbes and microbial derivatives. Pathogenic microbes, including entomopathogenic nematodes, viruses, fungi, and bacteria cause infection in a target host and have the potential to recycle and spread in host populations. Microbial derivatives are toxins, such as spinosyns and the delta endotoxin of *Bt*, recovered from cultures of bacteria and fungi. Products based on microbial derivatives often do not contain live microorganisms, and therefore cannot spread in pest populations.

TABLE 19.2

Some Microbial Products for Turfgrass Pest Management

Pathogenic Microbes	Target Insect	Product Name
Bacteria		
Bacillus thuringiensis (*Bt*)		
Bt subsp. *aizawai*	Sod webworm, armyworms	XenTari
Bt subsp. *israelensis*	*Tipula* spp.	Teknar, Bactimos
Bt subsp. *japonensis Buibui*	White grubs	MYX
Bt subsp. *kurstaki*	Sod webworm, armyworms	Deliver, Dipel, Javelin, Lepinox, Crymax
Paenibacillus popilliae	*Popillia japonica*	Milky Spore
Serratia entomophila	*Costelytra zealandica*	Invade
Fungi		
Beauveria bassiana	Chinch bugs	Naturalis H&G, Botanigard
B. brongniartii	*Melolontha melolontha,* *Hoplochelus marginatus*	Engerlingspilz, Betel™, Melocont®-Pilzgerste
Metarhizium anisopliae	*Adoryphorus couloni,* *Dermolepida albohirtum*	Biogreen, Bio-Cane™
Nematodes		
Heterorhabditis bacteriophora	Billbugs, White grubs	Symbion-Temperate, Heteromask, Nemasys G, nema-green, Terranem H&G
H. marelatus	White grubs	Symbion-North
Steinernema carpocapsae	Annual bluegrass weevil, armyworms, billbugs, cutworms, fleas, European crane fly, sod webworms	Biosafe, Carponem, Millenium, Nemastar, No Flea™, Ecomask, NEMAgräs
S. glaseri	White grubs	Biotopia
S. riobrave	Mole crickets	BioVector
S. scapterisci	Mole crickets	Nematac™ S

Source: After Klein et al. *Field Manual of Techniques in Invertebrate Pathology*, Kluwer, Dordrecht, 2007.

Below a brief description will be given on the pathogen groups within which species are commercialized as microbial insecticides or that have potential for commercialization in the future.

19.2 MICROBIAL CONTROL AGENT GROUPS

19.2.1 VIRUSES

Viruses in at least 14 families are known to affect arthropods [6–8]. Most viruses that have been used successfully in microbial control have occlusion bodies, which consist of a proteinaceous matrix in which the virus particle (virion) is occluded. Among viruses with occlusion bodies, the baculoviruses (BVs) are by far the most extensively used in insect management. They are double-stranded DNA viruses with rod-shaped virions that replicate in the nuclei of infected host cells and produce well-defined occlusion bodies in the nucleus, which eventually causes the infected cell to disintegrate. The BVs are natural control agents of many leaf-feeding caterpillars [8]. Their host range is generally narrow, often encompassing only 1–3 species with the widest known range in the *Autographa californica* MNPV (56 species in 12 families of Lepidoptera) [7]. The family Baculoviridae is divided into two genera: the nucleopolyhedroviruses (NPV) that form polyhedral occlusion bodies and the Granuloviruses that form granular occlusion bodies.

Infection by BVs occurs by ingestion of virions. After the occlusion body is dissolved in the alkaline midgut, the virus enters midgut epithelial cells. Replication of BVs can take place only in midgut cells or, as in all Lepidoptera hosts, as a systemic infection in various tissue types such as hemocytes, fat bodies, muscles, and hypodermis. Host death does not occur until several days to over 1 week after infection, allowing the host to continue to feed and grow for some time while the virus is replicating. Toward the end of the infection, the host cuticle becomes fragile and eventually ruptures releasing the virions in the liquefied body contents of the host cadaver. The infection often changes the host's behavior to seek out higher or more exposed positions in the vegetation (*wipfelkrankheit*—tree top disease), which increases the chances of the virions being spread to lower vegetation to reach new hosts through direct leaking or rainwater. In addition, the exposed infected hosts are more likely to be consumed by predators such as birds that can spread the virions with their feces over longer distances without significant loss of virulence [9,10].

The occlusion body provides the BV virions with some degree of protection from environmental extremes. However, they are very vulnerable to UV radiation and in extreme cases can be inactivated after only a few hours of exposure to sunlight. Survival will therefore be greater in shaded areas, even on the underside of foliage to which they are applied. The addition of UV protectants to the spray can provide some protection. Research in several systems has shown that stilbene-derived optical brighteners do not only provide UV protection but also increase host susceptibility to some NPVs by decreasing the normal sloughing of infected midgut epithelium cells [11]. Rainfall also decreases the amount of virion on the foliar surface to which the spray was applied, but at the same time spreads the virus and ultimately carries it into the soil where it is protected from UV radiation and can serve as a virus reservoir for many years [12]. Viruses can also be inactivated by temperatures above 40°C, but high temperature is not an important factor in the attrition of virus inoculum under most common field conditions.

BVs can be applied with most spray equipment; however, optimal coverage and droplet-size distribution are very important as the inoculum has to be ingested to infect the host. While there has been some progress in *in vitro* production techniques, production in living hosts is presently the only effective means for commercial production. The problem of continued feeding activity of infected hosts for several days after infection may be overcome by genetically engineering of BVs to express insect-specific toxins (e.g., arthropod venoms, *Bt* toxins) or insect hormones that result in quicker host death. The downside of quicker death is reduced production of secondary inoculum from the infected hosts.

To date, no virus-based insecticides are labeled for any turfgrass usage. However, BVs have been isolated from several insect species that can be major turfgrass pests. In Brazil, a BV of the fall armyworm, *Spodoptera frugiperda*, the *Sf* NPV, is used on approximately 20,000 ha annually to control this pest in corn [8]. On golf courses in Kentucky, a BV of the black cutworm, *Agrotis ipsilon*, the *Agi*MNPV, was recently found causing epizootics of this pest [13]. *Agi*MNPV was shown to provide 75–93% control of third to fourth instar black cutworms in field trials, and also persisted on mowed and irrigated bentgrass field plots for at least 4 weeks, and thus holds promise as a preventive bioinsecticide with some long-term effect against black cutworm [13].

19.2.2 FUNGI

The majority of the over 700 recognized species of entomopathogenic fungi (EPF) are Eumycota and are found in all four divisions (Chytridiomycota, Zygomycota, Ascomycota, Basidiomycota) [6,14]. Due to the need for effective and economic *in vitro* production systems, all but one of the about nine presently commercialized EPF species [15] belong to the artificial class Hyphomycetes in the formed division Deuteromycota (i.e., not a monophyletic unit). This classification is based on the absence of a known sexual state in most of these fungi. All these asexual Hyphomycetes produce spores termed "conidia"; some species also produce chlamydospores.

Host infection by Hyphomycetes most commonly occurs by attachment of spores to the external cuticle, followed by spore germination and penetration into the host's hemocoel. The fungus proliferates in the host as cell-walled hyphal bodies or wall-less amoeboid protoplasts, and ultimately kills the host through nutrient depletion, physical obstruction or invasion of organs, and toxin production. Under favorable conditions, fungal hyphae emerge from the host soon after its death and produce spores that are passively dispersed by wind and water.

The host range of many Hyphomycetes (e.g., *Beauveria bassiana*, *Metarhizium anisopliae*) is quite wide, often ranging several insect orders. However, these species are comprised of "species complexes," and individual isolates or pathotypes have much more restricted host ranges, particularly under field conditions.

Fungal spores are highly susceptible to solar radiation, particularly UV light, and persist longer in shaded areas. Most EPF can tolerate temperatures in the range approximately 0–40°C with optima for infection, growth, and sporulation around 20–30°C. Spores are also freeze tolerant. While high moisture conditions are required for fungal sporulation and spore germination, fungi may work under fairly dry conditions after inundative application of formulated spores, presumably due to the presence of moisture in microhabitats such as abaxial leaf surfaces or membranous folds of insect cuticle. Spore persistence is generally better under cool, dry conditions, but some EPF including *M. anisopliae* survive better at moderate temperatures with higher humidity.

Hydraulic spray systems are used to apply water-based formulations (including oil emulsions and wettable powders) onto a variety of crops including rangeland at moderate to high spray volumes (generally >150 L/ha). Commercial-scale technologies for efficient application of large granules are not well developed. Field tests have employed mechanical spreaders, modified blower equipment, incorporation into soil by tilling, or modified conventional seed drills for subsurface applications of granules and liquid suspension. EPF are compatible with many agrochemicals including most insecticides, but are negatively affected by fungicides. However, much of the compatibilities has been determined under laboratory conditions, and in the field compatibility may be better as long as EPF and fungicides are applied asynchronously [16].

B. bassiana and *M. anisopliae* naturally infect white grubs, mole crickets, chinch bugs, and other turfgrass insects. *B. bassiana* has been associated with large-scale natural mortality of chinch bugs, especially under hot and humid conditions. In the United States, *B. bassiana* is presently commercialized as Naturalis H&G and until recently as Botanigard, and is labeled for use against white grubs, mole crickets, sod webworm, and chinch bugs. However, there is little evidence that any of these pests is effectively suppressed. Particularly for soil insects, use of mycoinsecticides is rather questionable because it is difficult to get the fungal spores into the upper soil layers to their targets. Subsurface applications would alleviate this limitation but require highly specialized equipment. In addition, at least mole crickets can actively avoid soil contaminated with EPF [17,18].

In pasture turf where subsurface applications are more feasible, particularly at establishment of new pastures, EPF have shown good suppression of some pests. *Beauveria brongniartii* has shown good suppression of the European cockchafer grub, *Melolontha melolontha*, in pasture turf in Europe and is commercially available as Engerlingspilz™, Betel™, and Melocont®-Pilzgerste. *M. anisopliae* is commercially available for the control of froghoppers and spittlebugs on pasture turf in Brazil, and as Biogreen™ for control of *Adoyphorus couloni* grubs in Australian pastures.

19.2.3 BACTERIA

Entomopathogenic bacteria are found in the families Bacilliaceae, Pseudomonadaceae, Enterobacteriaceae, Streptococcaceae, and Micrococcaceae [6]. However, only bacteria from the genera *Bacillus*, *Paenibacillus*, and *Serratia* have been developed into commercial products.

19.2.3.1 *Bacillus thuringiensis*

B. thuringiensis is an aerobic, motile, endospore-forming bacterium that commonly occurs in soil and sediment. During sporulation, it forms a parasporal body that contains insecticidal crystal proteins (ICP) also called delta endotoxins [6,19]. Different strains of *Bt* contain varying combinations of ICPs that are specifically toxic to insects in the orders Lepidoptera, Diptera, or Coleoptera, or to nematodes. No evidence of mammalian toxicity or infectivity has been found to date.

Bt-based products may contain viable *Bt* cells that could propagate and persist in the environment, but many products contain the ICP alone or in nonviable cells of *Bt* or other bacteria and thus cannot recycle.

Bt ICPs must be ingested by a susceptible host insect to be effective. In the midgut, the ICPs are activated by midgut proteases and bind to specific receptors in the surface membrane of midgut epithelial cells. This results in the formation of pores that destroy the cell's ability to control exchange of molecules and causes lysis of the cell. Damage to the midgut epithelium is associated with cessation of feeding and paralysis in many insect species, and the midgut lesions allow mixing of hemolymph and gut contents, resulting in a septicemia that may contribute to the insect's death. Host death may not occur until several days after ingestion of ICPs, but feeding will stop much earlier. After death, *Bt* spores and ICPs are released from the disintegrating insect cadaver into the environment. Spores may also be released with the feces by infected hosts before death.

Bt ICPs and spores are rapidly inactivated by solar radiation (UV) with 8–80% reduction in viability 1 day after foliar applications. However, use of UV protectants can extend the residual activity half-life to 2 weeks. In the soil, ICPs have been observed to have half-lives of 3–6 days or longer while spores can persist for weeks or even months. Once in the soil spores and crystals are relatively immobile, and even excessive rainfall may not move them below 6 cm soil depth.

Bt products can be applied with most spray-application equipments. In the tank, it should not be allowed to stand for too long as the ICPs can deteriorate or settle in the bottom of the tank. For foliar applications, good coverage, and droplet size are critical. Optimal droplet size is about 40–100 μm diameter.

Products containing *Bt* subsp. *kurstaki* (Deliver, Dipel, Javelin, Lepinox, Crymax) or *Bt* subsp. *aizawai* or their ICPs are mass-produced through fermentation and are found in products registered for the control of turf pests such as armyworms and sod webworms (Table 19.2). While generally active against Lepidoptera, they are not very effective against this group of pests in turf because of the rapid photodegradation of the toxins, slow activity of the ICP, and lack of contact activity. The weak performance of these products against late instar larvae makes successful use of these products dependent on detection of early instars of the pests. If egg hatch occurs over an extended period, multiple treatments of *Bt* may be needed. Efficacy of *Bt* products against caterpillars can be improved by treating late in the day so that the night-active caterpillars acquire the toxins before they have been degraded by sunlight. Some turf Lepidoptera like black cutworm larvae are also generally resistant to the commercial *Bt* strains.

The Buibui strain of *Bt* subsp. *japonensis* has shown good activity against larvae of various scarab species including cupreous chafer, *Anomala cuprea* [20]; Japanese beetle, *Popillia japonica*; oriental beetle, *A. orientalis*; and green June beetle, *Cotinis nitida* grubs [21]. To be effective, *Bt* subsp. *japonensis* has to be applied against early instars in a preventive mode. Its commercialization in the United States has been hindered primarily by competition from highly effective reduced-risk synthetic insecticides such as imidacloprid and halofenozide that are also applied preventively. Commercial development of this strain continues primarily in Japan. Other scarab-active *Bt* strains (e.g., *Bt* subsp. *galleriae* SDS-502) [22] have also shown high toxicity to various scarab larvae but may face the same issues from competition.

Bt subsp. *israelensis* has activity against the European crane fly, *Tipula paludosa* [23], but to date has not been developed commercially for the U.S. turf market.

The use of *Bt* products in turfgrass may increase in the future if biotechnology can provide more versatile and persistent products. For example, the creation of recombinant ICPs could enhance the toxicity or the target spectrum of *Bt* toxins [24], as has already been shown with hybrid ICPs against the otherwise *Bt*-resistant black cutworm [25]. Field persistence may be improved by incorporating the ICP genes into other bacteria and stabilizing those agents to encapsulate the ICP within a nonviable cell [26].

19.2.3.2   *Paenibacillus*

Paenibacillus (=*Bacillus*) *popilliae* is an obligate pathogen that causes "milky disease" in scarab larvae in the subfamilies Melolonthinae, Rutelinae, Aphodinae, and Dynastinae [6,19,27–29]. *P. popilliae* strains are highly specific, infecting only one host species. Only the strain infecting *P. japonica* has been commercialized.

Infection is initiated when the host insect ingests bacterial spores. Spores germinate in the gut and release a parasporal crystal protein that may facilitate entry of the vegetative rods into the hemolymph by damaging the midgut wall [30]. The bacteria transverse the midgut epithelium, go through one multiplication cycle on the luminal side of the basal lamina in immature replacement cells and regenerative cells, and finally move into the host's hemolymph where they multiply through several cycles, and eventually sporulate and form parasporal bodies. The high concentration of refractile spores and parasporal bodies during the final stages of infection gives the hemolymph a milky-white color, thus the name "milky disease." Milky disease is inevitably fatal but the exact cause of host death is not fully understood; it seems to involve the depletion of fat body reserves [31]. When the infected grub dies, typically after several weeks or even months, up to 5×10^9 spores may be released into the soil from the disintegrating grub cadaver, leading to a buildup of the disease.

P. popilliae spores can persist in the soil for several years, making it an excellent pathogen for permanent establishment in turf. It has been used for more than 60 years for suppression of *P. japonica* populations in the United States. A massive colonization program conducted from 1939 to 1953 is credited with disseminating the pathogen in the northeastern United States [28]. However, due to the lack of an effective *in vitro* production system or an economically feasible *in vivo* production system, *P. popilliae*–based products have been expensive, with only small quantities available. Formulations are made by mixing macerated diseased larvae with talc (10^8 spores/g) and applied by placing 2 g formulation every 1.3 m in a grid pattern (11 kg/ha). This, along with a history of inconsistent performance and generally poor establishment against *P. japonica* [32,33], has limited the use of *P. popilliae* to a small number of acres, and to one-time, inoculative applications. Under optimal conditions, e.g., soil temperatures $\geq 21°C$ for several months per year and high larval densities ($\geq 300/m^2$), *P. popilliae* can establish and provide lasting control in a release site over a period of several years [28,34].

Current commercial production relies on the collection of grubs from nature to mass-produce the bacterium. Only one company currently produces *P. popilliae* (Table 19.2), and the product "milky spore" is sold through garden centers and mail-order catalogs mostly for the home-lawn-care market. Strains that infect grub species other than *P. japonica* have not been commercially developed. Until effective *in vitro* production and more virulent strains can be developed, *P. popilliae* use will remain very limited.

19.2.3.3   *Serratia*

Amber disease is caused by certain strains of *Serratia* spp., a genus of bacteria commonly found in turf soils. The pathogenic bacteria are ingested by the host, cause a cessation of feeding, and grow in the gut of susceptible insects, eventually leading to starvation and death of the host [35]. In New Zealand, *S. entomophila* infects larvae of the grass grub *Costelytra zealandica*. Infected *C. zealandica* stop feeding within several days of infection, and starve to death over a period of 1–3 months, during which time the larvae take on a characteristic amber appearance. During the final stages of infection, the bacteria also penetrate into the grub's hemocoel. *S. entomophila* is mass-produced *in vitro* and is used commercially as the product Invade™ against *C. zealandica* in New Zealand pastures [36]. Because *S. entomophila* does not produce spores, the product has to be kept refrigerated and has a 3-month shelf life. Similar bacteria have been isolated from *P. japonica* and *Cyclocephala* spp. grubs in the United States, and amber disease may be a significant natural mortality factor in grub populations in turfgrass soils. However, no commercial development has taken place in the United States.

19.2.3.4 Other

Spinosad is a microbial derivative produced by the soil bacterium *Sacharopolyspora spinosa* during fermentation [37]. It contains spinosyns, which are insecticidal macrocyclic lactones that target a novel site in the nicotinic acetocholine receptor of insect nerves [38]. Spinosad has some contact activity but primarily works after ingestion. It works at very low dosages but has a relatively short residual period. It is specifically toxic to caterpillars and has shown excellent activity against turf pests such as black cutworm, armyworms, and sod webworms.

19.2.4 Nematodes

Nematodes with parasitic associations with insects are found in 23 nematode families, and seven of these families contain species that have potential for biological control of insects: Mermithidae and Tetradonematidae (Order: Stichosomida); Allantonematidae, Phaenopsitylenchidae, and Sphaerulariidae (Order: Tylenchida); and Heterorhabditidae and Steinernematidae (Order: Rhabditida). Because of problems with culture and limited virulence of other nematodes, presently only the Heterorhabditidae and Steinernematidae are used as microbial insecticides.

Steinernematidae and Heterorhabditidae are obligate pathogens in nature and more than 55 species have been discovered worldwide from soils throughout the world [39,40]. All known entomopathogenic nematode (EPN) species have a similar biology. The only stage that survives outside of a host is the nonfeeding, nondeveloping third-stage infective juvenile (IJ) or dauer juvenile. The IJs carry cells of a species-specific bacterial symbiont (*Xenorhabdus* and *Photorhabdus* bacteria for Steinernematidae and Heterorhabditidae, respectively) in their intestines. After locating a suitable host, the IJs invade it through natural openings (mouth, spiracles, anus) or thin areas of the host's cuticle and penetrate into the host hemocoel. The IJs release their symbiotic bacteria that propagate, kill the host by septicemia within 1–4 days, and metabolize its tissues. The nematodes start developing and feed on the bacteria and metabolized host tissues. *Heterorhabditis* IJs develop into hermaphroditic adults but are male or female in following generations, whereas *Steinernema* adults are always male or female. The nematodes go through one to three generations until a new generation of IJs emerges from the depleted host cadaver.

EPN infections can be recognized by various signs. Soon after death, the cadavers turn flaccid and start changing color. In the case of wax moth, *Galleria mellonella*, larvae and depending on nematode species, steinernematid-killed insects turn various shades of brown, from ochre to almost black, whereas heterorhabditid-killed insects turn red, brick-red, purple, orange, yellow, and green. The color of the cadaver is attributed to the associated bacterium, especially for *Photorhabdus* that have pigments. The steinernematid-killed insects stay flaccid throughout nematode development, whereas heterorhabditid-killed insects become less flaccid. If the cuticle is transparent, nematodes are visible inside the cadaver. The cadavers do not putrify.

The range of insects infected by many of the well-studied EPN species (e.g., *S. carpocapsae*, *S. feltiae*, and *Heterorhabditis bacteriophora*) is broad under laboratory conditions but more restricted due to the ecology of the nematodes and their potential hosts as well as environmental factors. The effect of inundative applications of EPN on nontarget organisms is negligible [41,42]. Some known species have a restricted host range. For example, *S. scapterisci* is adapted to mole crickets [43], and *S. kushidai* [44] and *S. scarabaei* [45] are adapted to scarab larvae.

IJs of different species use different foraging strategies to find hosts. Sit-and-wait strategists or ambushers tend to stay near the soil surface where their specialized foraging behaviors (nictation, jumping) allow them to efficiently infect mobile host species (e.g., *S. carpocapsae*, *S. scapterisci*) [46,47]. Widely searching foragers or cruisers distribute themselves actively throughout the soil profile and are well adapted to infecting sessile or less mobile hosts (e.g., *S. glaseri*, *H. bacteriophora*). Most known species appear to be situated somewhere along a continuum between these two extremes (e.g., *S. riobrave*, *S. feltiae*) [46,48].

The performance of EPN applied for insect control is dependent on the motility and persistence of the applied IJs. Active IJ dispersal, although rather limited with up to 90 cm in both horizontal and vertical dispersal within 30 days, gives EPN the ability to actively seek out hosts. When IJs are applied onto the soil surface, losses can reach 50% within hours of application due primarily to UV radiation and desiccation [49]. Even those IJs that settle in the soil suffer 5–10% losses per day until usually only around 1% of the original inoculum survives after 1–6 weeks. Because of this limited persistence, EPN should only be applied when susceptible stages of the target pest are present. Because UV light can inactivate and kill nematodes within minutes, nematodes should be applied early in the morning or in the evening, using enough spray volume and postapplication irrigation to rinse the IJs into the thatch or soil.

Moderate soil moisture is essential for good nematode performance [50,51]. IJs need a water film for effective propulsion. In soil, IJs move through the water film that coats the interstitial spaces. In laboratory studies, best performance has been observed at soil water potential of approximately −10 to −100 kPa [52]. In water-saturated as well as dry soil, nematode movement is restricted. IJs can survive desiccation to relatively low moisture levels if water removal is gradual, giving them time to adapt to an inactive stage [53]. This is generally the case in natural soils except near the surface in sandy soils low in organic matter.

The effect of temperatures on nematode performance varies with nematode species and strains [54]. Generally, IJs become sluggish at low temperatures (<10–15°C) and will be inactivated at higher temperatures (>30–40°C). Good performance for most commercially available nematode species can be expected between 20°C and 30°C. Exceptions are *S. feltiae* with an efficacious performance between ca. 12°C and 25°C, and *S. riobrave* with exceptional performances between 25°C and 35°C. Extended exposure to temperatures below 0°C and above 40°C is lethal to most EPN species.

It has been assumed that dispersal and survival tend to be lower in fine-textured soils with the lowest survival in clay soils [55]. The pH value of the soil does not have a strong effect on IJ survival as pH values between 4 and 8 do not vary in their effect on IJs [55]. However, additional research has shown that different nematode species may be differently affected by various soil parameters (e.g., Ref. 56), which may in part be due to differences in behavioral and physiological adaptations among nematode species. Due to the conglomeration of many parameters within each soil type, it is difficult to make generalizations on the effect of specific soil parameters on nematode performance.

Nematodes are sold in many different formulations. Water-based formulations (e.g., sponge, vermiculite, aqueous suspensions) require continuous refrigeration to maintain nematode quality for extended periods of time. Formulations that reduce IJ metabolism by immobilization or partial desiccation (alginate, clays, activated charcoals, and polyacrylamide) improve IJ shelf life and resistance to temperature extremes [57,58]. Presently, the most promising formulation consists of water-dispersible granules that combine long nematode shelf life without refrigeration (9–12 and 4–5 months at 2–10°C and 22–25°C, respectively, for *S. carpocapsae*; less for other species) with ease of handling [57,59]. The partially desiccated IJs rehydrate after application to a moist environment.

EPN can be applied with most spray equipment including hand or ground sprayers (pressurized or electrostatic), mist blowers, and aerial equipment on helicopters [57,60,61]. Filters and sieves (preferably removed) should be at least 300 μm wide, nozzle apertures >500 μm, and operating pressure should not exceed 2000 kPa (295 PSI) for *S. carpocapsae* and 1380 kPa (204 PSI) for *H. bacteriophora* and *H. megidis* [60]. Nematodes can also be delivered via irrigation systems including drip, microjet, sprinkler, and furrow irrigation [60]. Care should be taken not to expose the IJs to high temperatures in the tank mix or in the application equipment.

Spray volumes of 750–1890 L/ha are usually required depending on soil cover [57]. To quickly remove the IJs from exposure to UV light and desiccating conditions on the grass foliage and soil surface, postapplication and, in the case of dry soil, also preapplication irrigation are recommended. The ideal amount of irrigation will depend on the soil type, temperature, soil cover, the soil moisture before application, and the depth of the target insect in the soil. For white grubs and other targets in the soil, a minimum of 6 mm irrigation is recommended. Maintaining moderate soil moisture for at least 1 week after application is essential for good performance. When water is limited, subsurface injection of nematodes appears to be an efficient delivery method but does not improve nematode efficacy compared to application with sufficient amounts of water [62].

EPN appear to be compatible with many herbicides, fungicides, acaricides, insecticides, nematicides, azadirachtin, *Bt* products, and pesticidal soap. Other pesticides have limited to strong toxic effect on IJs, e.g., oxamyl, fenamiphos, carbaryl, bendiocarb, diazinon, dodine, paraquat, or methomyl. However, synergistic interaction between EPN and other control agents has been observed for various insecticides such as imidacloprid or tefluthrin and pathogens such as *P. popilliae* or *Bt*. An extensive summary of compatibilities and interactions is provided by Koppenhöfer and Grewal [63].

Several EPN-based products have been developed for the control of turfgrass pests (Table 19.2). Two EPN species are effective against mole crickets in turf and pastures in the southern United States. *S. riobrave* can only provide short-term curative control because it does not reproduce in mole crickets. However, the mole cricket–specific *Steinernema scapterisci* can provide long-term suppression of mole cricket populations after inoculative releases and may become a major means of mole cricket management in pastures in Florida [43]. *Steinernema scapterisci* is more effective against *Scapteriscus borellii* than *S. vicinus* and particularly *S. abbreviatus*, and is substantially less effective against mole cricket nymphs than adults.

EPNs are the most extensively studied parasites of white grubs [62,64,65]. Four species, *H. bacteriophora*, *H. zealandica*, *H. marelatus*, and *S. glaseri*, are currently available commercially for grub control in the world. Among white grub species that are important pest of turfgrass in the United States, *P. japonica* appears to be the most EPN-susceptible species, whereas larvae of other white grub species including *cyclocephala* spp.; *A. orientalis*; European chafer, *Rhizotrogus majalis*; Asiatic garden beetle, *Maladera castanea*; and May/June beetles, *Phyllophaga* spp. appear to be less susceptible to the commonly used EPN, i.e., *H. bacteriophora* and *S. glaseri* [64,66,67]. Nematodes that have provided good field control of *P. japonica* include *S. scarabaei* (100%), *H. bacteriophora* (GPS11 strain) (34–97%), *H. bacteriophora* (TF strain) (65–92%), and *H. zealandica* (X1 strain) (73–98%). *S. scarabaei* is the only nematode species that has provided high field control of *A. orientalis* (87–100%), *M. castanea* (71–86%), and *R. majalis* (89%). Against *C. borealis*, *H. zealandica* (X1 strain) (72–96%), *S. scarabaei* (84%), and *H. bacteriophora* (GPS11 strain) (47–83%) appear to be the most promising nematodes. Unfortunately, *S. scarabaei* has proven to be difficult to mass-produce and is thus far not commercially available.

Larval stages of white grubs also may also differ in their susceptibility but the effect varies with white grub and EPN species [68]. For example, efficacy against *P. japonica* and *A. orientalis* is higher against second instars than third instars for *H. bacteriophora* but there is no significant effect for *S. scarabaei*. Application timing should consider the presence of the most-susceptible white grub stages but also environmental conditions, particularly soil temperatures. For example, *H. bacteriophora* applications against *P. japonica* or *A. orientalis* in the northeastern United States will tend to be more effective if applied by early September due to the presence of the more susceptible younger larvae and a longer period after application before soil temperatures become too cool for good nematode activity.

Billbug larvae and adults can be effectively controlled with *S. carpocapsae* and *H. bacteriophora* [69]. Field tests in Ohio indicated that *S. parvulus* can be controlled with *S. carpocapsae* (average 78%) or *H. bacteriophora* (average 74%) [70,71]. In Japan, *S. carpocapsae* has been more effective for control of the hunting billbug, *S. venatus vestitus*, than standard insecticides (average 84% versus 69% control) [71,72]. It was the primary means of controlling this important golf-turf pest until the recent registration of imidacloprid for turfgrass uses in Japan. *S. carpocapsae* has given variable control (average 51%) as a rescue treatment for annual bluegrass weevil, *Listronotus maculicollis*, larvae on golf-course fairways, but has given up to 98% in some field trials and thus deserves further investigation (see Chapter 21).

S. carpocapsae can be used effectively to manage various turf caterpillar pests. Against *A. ipsilon, S. carpocapsae* is highly effective (average 95%), whereas *H. bacteriophora* has not provided satisfactory control (average 62%) [70]. Both *S. carpocapsae* and *H. bacteriophora* are effective against sod webworms in turfgrass. Against the common armyworm, *Pseudaletia unipunctata, S. carpocapsae* is more effective than *H. bacteriophora* [73].

The larvae of two crane fly species, *T. paludosa*—also called the European crane fly—and *T. oleracea*, are susceptible to heterorhabditid nematodes and particularly to *S. feltiae* [74]. In both species, susceptibility to *S. feltiae* decreases with larval development [75].

19.3 CONCLUSIONS

The use of microbial insecticides in turfgrass is extremely limited at this time with sales constituting <0.1% of the U.S. turfgrass insecticide market, which is estimated at about $500 million total in 1998 [76]. Few professional turfgrass managers use microbials [37]. A large proportion of microbials used in turfgrass are used by homeowners, who often value reduced exposure to synthetic insecticides higher than product efficacy, ease of use, and price.

The major impediment to wider use of microbials is the competition from synthetic chemical insecticides that are generally cheaper, easier to use, and usually also more effective than presently available microbials. Newer chemistry such as neonicotinoids, anthranilic diamides, and particularly insect growth regulators have considerably lower vertebrate toxicity and are often also less detrimental to invertebrate natural enemies of turfgrass insect pests [77]. However, most of these chemicals are used in a preventive approach, being applied over much larger areas than usually necessary to avoid turfgrass damage by insect pests. These wide-area applications combined with the high efficacy of these chemicals against a wide range of insect pests are likely to have indirect negative effects on natural enemies by depriving them of hosts and prey. If prey and particularly host populations drop below a critical level, these natural enemies will become locally extinct. As a result, our dependency on these chemicals increases.

The use of microbials could be increased by improving their effectiveness, decreasing their cost, and improving their ease of use. As discussed for the different microbials above the efficacy of microbials could be improved by isolating and developing more virulent strains from field populations, increasing virulence through biotechnology, increasing our understanding on how to use them most effectively, and improving application technologies. The cost could further be reduced by improving production technologies. Ease of use could be improved with better formulations, particularly by increasing shelf life and tolerance to high temperatures. Many of these improvements would likely be only gradual and relatively small. However, they may be sufficient to significantly increase the use of microbials at least in situations where pesticide regulations, local ordinances, and public opinion already impinge on the use of synthetic insecticides. Only radical improvements in competitiveness with synthetic insecticides could lead to a significant increase in the use of microbials where insecticide use is not restricted.

Finally, the use of microbials could also increase in many situations if we could more systematically exploit their ability to recycle in hosts and provide long-term pest suppression. Most of the research on microbial control agents has concentrated on inundative application with the emphasis on quick and effective pest control. However, many of the agents developed for ease of production and commercialization may have lost virulence and fitness during laboratory culture [78] and with that be less likely to persist in the environment. In addition, species that have a potential for long-term suppression generally appear to be rather host-specific [5] such as *B. brongniartii* against European cockchafer grubs in Europe [79], *S. entomophila* against grass grubs in New Zealand [36], or *S. scapterisci* against mole crickets in the southeastern United States [43]. Two other species that may lend themselves for long-term pest suppression should they become commercially available could be the *Agi*MNPV for black cutworm suppression in surrounds of golf-course greens [13] and *S. scarabaei* that can give at least 2 years of white grub suppression in turf plots

(Koppenhöfer, A.M., unpublished data, 2006). However, since these specific pathogens depend on the presence of hosts for recycling and may become patchy in distribution over time, their use may be feasible only in areas with some tolerance for occasional pest damage.

REFERENCES

1. Potter, D.A., *Destructive Turfgrass Insects: Biology, Diagnosis, and Control*, Ann Arbor Press, Chelsea, MI, 1998.
2. Vittum, P.J., Villani, M.G., and Tashiro, H., *Turfgrass Insects of the United States and Canada*, Cornell University Press, Ithaca, NY, 1999.
3. Niemczyk, H.D. and Shetlar, D.J., *Destructive Turf Insects*, HDN Books, Wooster, OH, 2000.
4. Jackson, T.A. and Glare, T.R., *Use of Pathogens in Scarab Pest Management*, Intercept, Andover, UK, 1992.
5. Jackson, T.A., Soil dwelling pests—is specificity the key to successful microbial control? in *Proceedings of the 3rd International Workshop on Microbial Control of Soil Dwelling Pests*, Jackson, T.A. and Glare, T.R., Eds., AgResearch, Lincoln, NZ, 1996, pp. 1–6.
6. Tanada, Y. and Kaya, H.K., *Insect Pathology*, Academic Press, San Diego, 1993.
7. Evans, H.G., Viruses, in *Field Manual of Techniques in Invertebrate Pathology*, Lacey, L.A. and Kaya, H.K., Eds., 2000, pp. 179–208.
8. Moscardi, F., Assessment of the application of baculoviruses for control of Lepidoptera, *Annu. Rev. Entomol.*, 44, 257–289, 1999.
9. Andreadis, T.G., Transmission, in *Epizootiology of Insect Diseases*, Fuxa, J.R. and Tanada, Y., Eds., Wiley, New York, 1987, pp. 159–176.
10. Entwistle, P.F., Adams, P.H.W., Evans, H.F., and Rivers, C.F., Epizootiology of a nuclear polyhedrosis virus (Baculoviridae) in European spruce sawfly (*Gilpinia hercyniae*): spread of disease from small epicenters in comparison with spread of baculovirus disease in other hosts, *J. Appl. Ecol.*, 20, 473–487, 1983.
11. Washburn, J.O., Kirkpatrick, B.A., Haas-Stapleton, E., and Volkman, L.E., Evidence that the stilbene-derived optical brightener M2R enhances *Autographa californica* nucleopolyhedrovirus infection of *Trichoplusia ni* and *Heliothis virescens* by preventing sloughing of infected midgut epithelial cells, *Biol. Control*, 11, 58–69, 1998.
12. Thompson, C.G., Scott, D.W., and Wickman, B.E., Long-term persistence of the nuclear polyhedrosis virus of the Douglas-fir tussock moth, *Orgyia pseudotsugata* (Lepidoptera: Lymantriidae), in forest soil, *Environ. Entomol.*, 10, 254–255, 1981.
13. Prater, C.A., Redmond, C.T., Barney, W., Bonning, B.C., and Potter, D.A., Microbial control of black cutworm (Lepidoptera: Noctuidae) in turfgrass by using *Agrotis ipsilon* multiple nucleopolyhedrovirus, *J. Econ. Entomol.*, 99, 1129–1137, 2006.
14. Goettel, M.S., Inglis, G.D., and Wraight, S.P., Fungi, in *Field Manual of Techniques in Invertebrate Pathology*, Lacey, L.A. and Kaya, H.K., Eds., Kluwer Academic Publisher, Dordrecht, 2000, pp. 255–282.
15. Shah, P.A. and Goettel, M.S., *Directory of Microbial Control Products and Services*, Microbial Contr. Div., Soc. Invertebr. Pathol, Gainesville, FL, 1999.
16. Jaros-Su, J., Groeden, E., and Zhang, J., Effects of selected fungicides and the timing of fungicide application in *Beauveria bassiana*-induced mortality of the Colorado potato beetle (Coleoptera: Chrysomelidae), *Biol. Control*, 15, 259–269, 1999.
17. Villani, M.G., Krueger, S.R., Schroeder, P.C., Consolie, F., Consolie, N.H., Preston-Wilsey, L.M., and Roberts, D.W., Soil application effects of *Metarhizium anisopliae* on Japanese beetle (Coleoptera: Scarabaeidae) behavior and survival in turfgrass microcosms, *Environ. Entomol.*, 23, 502–513, 1994.
18. Villani, M.G., Allee, L.L., Preston-Wilsey, L., Consolie, N., Xia, Y., and Brandenburg, R.L., Use of radiography and tunnel castings for observing mole cricket (Orthoptera: Gryllotalpidae) behavior in soil, *Am. Entomol.*, 48, 42–50, 2002.
19. Siegel, J.P., Bacteria, in *Field Manual of Techniques in Invertebrate Pathology*, Lacey, L.A. and Kaya, H.K., Eds., Kluwer Academic Publisher, Dordrecht, 2000, pp. 209–230.
20. Suzuki, N., Hori, H., Tachibana, M., and Asano, S., *Bacillus thuringiensis* strain Buibui for control of cupreous chafer, *Anomala cuprea* (Coleoptera: Scarabaeidae), in turfgrass and sweet potato, *Biol. Control*, 4, 361–365, 1994.

21. Alm, S.R., Villani, M.G., Yeh, T., and Shutter, R., *Bacillus thuringiensis japonensis* strain Buibui for control of Japanese and oriental beetle larvae (Coleoptera: Scarabaeidae), *Appl. Entomol. Zool.*, 32, 477–484, 1997.

22. Asano, S., Yamashita, C., Iizuka, T., Takeuchi, K., Yamanaka, S., Cerf, D., and Yamamoto, T., A strain of *Bacillus thuringiensis* subsp. *galleriae* containing a novel *cry8* gene highly toxic to *Anomala cuprea* (Coleoptera: Scarabaeidae), *Biol. Control*, 28, 191–196, 2003.

23. Smits, P.H., Vlug, H.J., and Wiegers, G.L., Biological control of leatherjackets with *Bacillus thuringiensis*, *Proc. Exp. Appl. Entomol.*, N.E.V. Amsterdam, 4, 187–192, 1993.

24. Bosch, D., Schipper, D., van der Kleij, H., de Maagd, R.A., and Stiekema, W.J., Recombinant *Bacillus thuringiensis* crystal proteins with new properties: possibilities for resistance management, *Biotechnology*, 12, 915–918, 1994.

25. de Maagd, R.A., Weemen-Henriks, M., Molthoff, J.W., and Naimov, S., Activity of wild type and hybrid *Bacillus thuringiensis* δ-endotoxins against *Agrotis ipsilon*, *Arch. Microbiol.*, 179, 363–367, 2003.

26. Feitelson, J.S., Quick, T.C., and Gaertner, F., Alternative hosts for *Bacillus thuringiensis* delta-endotoxins genes, in *New Directions in Biological Control: Alternatives for Suppressing Agricultural Pests and Diseases*, Baker, R.R. and Dunn, P.E., Eds., Wiley-Liss, New York, 1990, pp. 561–571.

27. Dutky, S.R., *The Milky Diseases, in Insect Pathology: An Advanced Treatise*, Vol. 2, Steinhaus, E.A., Ed., Academic Press, New York, 1963, pp. 75–115.

28. Fleming, W.E., *Biological Control of the Japanese Beetle*, USDA Tech. Bull. 1383, U.S. Print. Office, Washington, DC, 1968.

29. Klein, M.G., Microbial control of turfgrass insects, in *Handbook of Turfgrass Insect Pests*, Brandenburg, R.L. and Villani, M.G., Eds., Entomol. Soc. Am., Lanham, MD, 1995, pp. 95–100.

30. Zhang, J., Hodgman, T.C., Krieger, L., Schnetter, W., and Schairer, H.U., Cloning and analysis of the first cry gene from *Bacillus popilliae*, *J. Bacteriol.*, 179, 4336–4341, 1997.

31. Sharpe, E.S. and Detroy, R.W., Fat body depletion, a debilitating result of milky disease in Japanese beetle larvae, *J. Invertebr. Pathol.*, 34, 92–94, 1979.

32. Klein, M.G., Use of *Bacillus popilliae* in Japanese beetle control, in *Use of Pathogens in Scarab Pest Management*, Glare, T.R. and Jackson T.A., Eds., Intercept, Andover, UK, 1992, pp. 179–189.

33. Redmond, C.T. and Potter, D.A., Lack of efficacy of in vivo- and putatively in vitro-produced *Bacillus popilliae* against field populations of Japanese beetle (Coleoptera: Scarabaeidae) grubs in Kentucky, *J. Econ. Entomol.*, 88, 846–854, 1995.

34. Fleming, W.E. and T.L. Ladd, *Milky Disease for Control of Japanese Beetle*, U.S. Dept. Agric. Leafl. 500, U.S. Department of Agriculture, Washington, DC, 1979.

35. Jackson, T.A., Huger, A.M., and Glare T.R., Pathology of amber disease in the New Zealand grass grub *Costelytra zealandica* (Coleoptera: Scarabaeidae), *J. Invertebr. Pathol.*, 61, 123–130, 1993.

36. Jackson, T.A., Pearson, J.F., O'Callaghan, M.O., Mahanty, H.K., and Willocks, M.J., Pathogen to product-development of *Serratia entomophila* (Enterobacteriaceae) as a commercial biological control agent for the New Zealand grass grub (*Costelytra zealandica*), in *Use of Pathogens in Scarab Pest Management*, Jackson, T.A. and Glare, T.R., Eds., Intercept Ltd., Andover, UK, 1992, pp. 191–198.

37. Racke, K.D., Pesticides for turfgrass pest management: uses and environmental issues, in *Fate and Management of Turfgrass Chemicals*, Clark, J.M. and Kenna, M.P., Eds., ACS Symp. Ser. 522, Am. Chem. Soc., Washington, DC, 2000, pp. 45–64.

38. Salgado, V.L., Studies on the mode of action of spinosad: insect symptoms and physiological correlates, *Pestic. Biochem. Physiol.*, 60, 91–102, 1998.

39. Adams, B.J., Fodor, A., Koppenhöfer, H.S., Stackebrandt, E., Stock, S.P., and Klein, M.G., Biodiversity and systematics of nematode-bacterium entomopathogens, *Biol. Control*, 37, 32–49, 2006.

40. Hominick, W.R., Biogeography, in *Entomopathogenic Nematology*, Gaugler, R., Ed., CABI Publishing, Wallingford, UK, 2002, pp. 115–144.

41. Akhurst, R.J., Safety to non-target invertebrates of nematodes of economically important pests, in *Safety of Microbial Insecticides*, Laird, M., Lacey, L.A., and Davidson, E.W., Eds., CRC Press, Boca Raton, FL, 1990, pp. 233–240.

42. Bathon, H., Impact of entomopathogenic nematodes on non-target hosts, *Biocontrol Sci. Technol.*, 6, 421–434, 1996.

43. Parkman, J.P. and Smart Jr., G.C., Entomopathogenic nematodes, a case study: introduction of *Steinernema scapterisci* in Florida, *Biocontrol Sci. Technol.*, 6, 413–419, 1996.

44. Mamiya, Y., Comparison of infectivity of *Steinernema kushidai* (Nematoda: Steinernematidae) and other steinernematid and heterorhabditid nematodes for three different insects, *Appl. Ent. Zool.*, 24, 302–308, 1989.

45. Koppenhöfer, A.M. and Fuzy, E.M., Ecological characterization of *Steinernema scarabaei*: a natural pathogen of scarab larvae, *J. Invertebr. Pathol.*, 83, 139–148, 2003.

46. Lewis, E.E., Behavioural ecology, in *Entomopathogenic Nematology*, Gaugler, R., Ed., CABI Publishing, Wallingford, UK, 2002, pp. 205–224.

47. Lewis, E.E., Campbell, J.C., Griffin, C., Kaya, H.K., and Peters, A., Behavioral ecology of entomopathogenic nematodes, *Biol. Control*, 37, 66–79, 2006.

48. Campbell, J.F. and Gaugler, R., Inter-specific variation in entomopathogenic nematode foraging strategy: dichotomy or variation along a continuum? *Fundam. Appl. Nematol.*, 20, 393–398, 1997.

49. Smits, P.H., Vlug, H.J., and Wiegers, G.L., Biological control of leatherjackets with *Bacillus thuringiensis*, *Proc. Exper. Appl. Entomol. N.E.V.-Amsterdam*, 4, 187–192, 1993.

50. Georgis, R. and Gaugler, R., Predictability in biological control using entomopathogenic nematodes, *J. Econ. Entomol.*, 84, 713–720, 1991.

51. Grewal, P.S., Power, K.T., Grewal, S.K., Suggars, A., and Haupricht, S., Enhanced consistency in biological control of white grubs (Coleoptera: Scarabaeidae) with new strains of entomopathogenic nematodes, *Biol. Control*, 30, 73–82, 2004.

52. Koppenhöfer, A.M. and Fuzy, E.M., Soil moisture effects on infectivity and persistence of the entomopathogenic nematodes *Steinernema scarabaei*, *Steinernema glaseri*, *Heterorhabditis zealandica*, and *Heterorhabditis bacteriophora*, *Appl. Soil. Ecol.*, 35, 128–139, 2006.

53. Womersley, C.Z., Dehydration survival and anhydrobiotic potential, in *Entomopathogenic Nematodes in Biological Control*, Gaugler, R. and Kaya, H.K., Eds., CRC Press, Boca Raton, FL, 1990, pp. 117–137.

54. Grewal, P.S., Selvan, S., and Gaugler, R., Thermal adaptation of entomopathogenic nematodes: niche breadth for infection, establishment, and reproduction, *J. Thermal. Biol.*, 19, 245–253, 1994.

55. Kaya, H.K., Soil ecology, in *Entomopathogenic Nematodes in Biological Control*, Gaugler, R. and Kaya, H.K., Eds., CRC Press, Boca Raton, FL, 1990, pp. 93–116.

56. Koppenhöfer, A.M. and Fuzy, E.M., Effect of soil type on infectivity and persistence of the EPNs *Steinernema scarabaei*, *Steinernema glaseri*, *Heterorhabditis zealandica*, and *Heterorhabditis bacteriophora*, *J. Invertebr. Pathol.*, 92, 11–22, 2006.

57. Grewal, P.S., Formulation and application technology, in *Entomopathogenic Nematology*, Gaugler, R., Ed., CABI Publishing, Wallingford, UK, 2002, pp. 265–287.

58. Grewal, P.S. and Peters, A., Formulation and quality, in *Nematodes as Biocontrol Agents*, Grewal, P.S., Ehlers, R.-U., and Shapiro-Ilan, D.I., Eds., CABI Publishing, Wallingford, UK, 2005, pp. 79–90.

59. Georgis, R. and Kaya, H.K., Formulation of entomopathogenic nematodes, in *Formulation of Microbial Biopesticides: Beneficial Microorganisms, Nematodes and Seed Treatments*, Burges, H.D., Ed., Kluwer, Dordrecht, The Netherlands, 1998, pp. 289–308.

60. Wright, D.J., Peters, A., Schroer, S., and Fife, J.P., Application technology, in *Nematodes as Biocontrol Agents*, Grewal, P.S., Ehlers, R.-U., and Shapiro-Ilan, D.I., Eds., CABI Publishing, Wallingford, UK, 2005, pp. 91–106.

61. Shapiro-Ilan, D.I., Gouge, D.H., Piggott, S.J., and Fife, J.P., Application technology and environmental considerations for use of entomopathogenic nematodes in biological control, *Biol. Control*, 37, 124–133, 2006.

62. Klein, M.G., Biological control of scarabs with entomopathogenic nematodes, in *Nematodes and the Biological Control of Insect Pests*, Bedding, R., Akhurst, R., and Kaya, H., Eds., CSIRO Press, East Melbourne, Australia, 1993, pp. 49–58.

63. Koppenhöfer, A.M. and Grewal, P.S., Compatibility and interactions with agrochemicals and other biocontrol agents, in *Nematodes as Biocontrol Agents*, Grewal, P.S., Ehlers, R.-U., and Shapiro-Ilan, D.I., Eds., CABI Publishing, Wallingford, UK, 2005, pp. 363–381.

64. Grewal., P.S., Koppenhöfer, A.M., and Choo, H.Y., Lawn, turfgrass, and pasture applications, in *Nematodes as Biocontrol Agents*, Grewal, P.S., Ehlers, R.-U., and Shapiro-Ilan, D.I., Eds., CABI Publishing, Wallingford, UK, 2005, pp. 115–146.

65. Georgis, R., Koppenhöfer, A.M., Lacey, L.A., Bélair, G., Duncan, L.W., Grewal, P.S., Samish, M., Tan, L., Torr, P., and van Tol, R.W.H.M., Successes and failures in the use of parasitic nematodes for pest control, *Biol. Control*, 37, 103–123, 2006.

66. Koppenhöfer, A.M., Fuzy, E.M., Crocker, R., Gelernter, W., and Polavarapu, S., Pathogenicity of *Steinernema scarabaei, Heterorhabditis bacteriophora* and *S. glaseri* to twelve white grub species, *Biocontrol Sci. Technol.*, 14, 87–92, 2004.

67. Koppenhöfer, A.M., Grewal, P.S., and Fuzy, E.M., Virulence of the entomopathogenic nematodes *Heterorhabditis bacteriophora, H. zealandica*, and *Steinernema scarabaei* against five white grub species (Coleoptera: Scarabaeidae) of economic importance in turfgrass in North America, *Biol. Control.*, 38, 397–404, 2006.

68. Koppenhöfer, A.M. and Fuzy, E.M., Effect of white grub developmental stage on susceptibility to entomopathogenic nematodes, *J. Econ. Entomol.*, 97, 1842–1849, 2004.

69. Klein, M.G., Efficacy against soil-inhabiting insect pests, in *Entomopathogenic Nematodes in Biological Control*, Gaugler, R. and Kaya, H.K., Eds., CRC Press, Boca Raton, FL, 1990, pp. 195–214.

70. Georgis, R. and Poinar Jr., G.O., Nematodes as bioinsecticides in turf and ornamentals, in *Integrated Pest Management for Turf and Ornamentals*, Leslie, A., Ed., CRC Press, Boca Raton, FL, 1994, pp. 477–489.

71. Smith, K.A., Control of weevils with entomopathogenic nematodes, in *Control of Insect Pests with Entomopathogenic Nematodes*, Smith, K.A. and Hatsukade, M., Eds., Food and Fertilizer Technology Center, Republic of China in Taiwan, 1994, pp. 1–13.

72. Kinoshita, M. and Yamanaka, S., Development and prevalence of entomopathogenic nematodes in Japan, *Jpn. J. Nematol.*, 28, 42–45, 1998.

73. Rosa, J.S. and Simões, N., Evaluation of twenty-eight strains of *Heterorhabditis bacteriophora* isolated in Azores for biological control of the armyworm, *Pseudaletia unipunctata* (Lepidoptera: Noctuidae), *Biol. Control*, 29, 409–417, 2004.

74. Ehlers, R.-U. and Gerwien, A., Selection of entomopathogenic nematodes (Steinernematidae and Heterorhabditidae, Nematoda) for the biological control of cranefly larvae *Tipula paludosa* (Tipulidae, Diptera), *Zeitschr. Pflanzenkrank. Pflanzensch.*, 100, 343–353, 1993.

75. Peters, A. and Ehlers, R.-U., Susceptibility of leatherjackets (*Tipula paludosa* and *T. oleracea*; Tipulidae: Nematocera) to the entomopathogenic nematode *Steinernema feltiae*, *J. Invertebr. Pathol.*, 63, 163–171, 1994.

76. Grewal, P.S., Factors in the success and failure of microbial control in turfgrass, *Integr. Pest Manag. Rev.*, 4, 287–294, 1999.

77. Kunkel, B.A., Held, D.W., and Potter, D.A., Impact of halofenozide, imidacloprid, and bendiocarb on beneficial invertebrates and predatory activity in turfgrass, *J. Econ. Entomol.*, 92, 922–930, 1999.

78. Wang, X. and Grewal, P.S., Rapid deterioration of environmental tolerance and reproductive potential of an entomopathogenic nematode during laboratory maintenance, *Biol. Control*, 23, 71–78, 2002.

79. Keller, S., Use of *Beauveria brongniartii* and its acceptance by farmers, *IOBC/wprs Bull.* 23, 67–71, 2000.

80. Klein, M.G., Grewal, P.S., Jackson, T.A., and Koppenhöfer, A.M., Lawn, turf, and grassland pests, in *Field Manual of Techniques in Invertebrate Pathology*, 2nd edition, Lacey, L.A. and Kaya, H.K., Eds., Kluwer, Dordrecht, 2007, pp. 655–675.

20 Integrated Pest Management of White Grubs

Albrecht M. Koppenhöfer

CONTENTS

20.1 WHITE GRUB BIOLOGY AND ECOLOGY

Root-feeding larvae of scarabaeid beetles are among the most damaging turfgrass and pastures pests in different parts of the world [1]. Important endemic scarab pests include *Cyclocephala* spp. and *Phyllophaga* spp. in many parts of the Americas; *Holotrichia* and *Heteronychus* spp. throughout Asia and Africa; *Melolontha*, *Amphimallon*, and *Phyllopertha* species in Europe; and *Anomala* species in Japan and Korea. Important exotic species include the Japanese beetle, *Popillia japonica*; the oriental beetle, *Anomala (=Exomala) orientalis*; the European chafer, *Rhizotrogus majalis*; and the Asiatic garden beetle, *Maladera castanea*, in North America; and the South African beetle, *Heteronychus arator*, in New Zealand and Australian pastures.

Most important grub species have annual life cycles with adults emerging in summer [2–4]. The adults of some species may cause extensive damage by feeding on foliage or flowers of ornamentals and fruit trees. However, many species cause only insignificant damage or do not feed at all during the adult stage. After mating, the females lay eggs among the roots of host plants such as grasses. The grubs feed on the roots, which at high larval densities and under warm, dry conditions can lead to wilting of plants, gradual thinning of the turf, and death of large turf areas. Vertebrate predators such as skunks, raccoons, wild pigs, moles, crows, and other mammals and birds often cause further damage by tearing up the sod while foraging for grubs; often before the grubs themselves cause significant damage. Both types of damage tend to occur more commonly when the larger larvae, especially the third instar, are present. For most North American annual white grub species, the majority of grubs reach the third instar by mid-September, but they may continue feeding well into October. Larvae move downward into the soil for overwintering before the soil surface freezes. As the soil warms up in spring, the grubs resume feeding before they pupate and emerge as adults in the summer. Some species, such as *Melolontha* spp., *Amphimallon* spp., and some *Phyllophaga* spp. have 2- or 3-year life cycles, and damage is dependent on the larval stage and species present.

20.2 INTEGRATED PEST MANAGEMENT

Integrated pest management (IPM) is the considered and coordinated use of pest-control tactics in turf management with the goal of maintaining healthy functional turf in an economically viable and environmentally sound manner. The concept of IPM was first developed in traditional agriculture, where the success of a crop can be measured in economic yield, i.e., the quantity and quality of produce. The establishment of reliable economic thresholds is the key to such IPM programs. These thresholds are pest population levels above which the cost of expected crop loss exceeds the cost of implementing control measures. Basic to the IPM concept is the shift from expectations of pest eradication to management of pest populations.

In turfgrass, it is much more difficult to determine thresholds because the economic value of turf areas cannot be measured as for a crop. Therefore, it is more appropriate to refer to "tolerance levels" or "action thresholds." These thresholds depend on numerous factors such as turfgrass type, location and use of the turf area, and the individual expectations or preferences of the turf manager and the clientele. As a result, any published "action thresholds" are just guidelines to be adapted to the specific conditions and expectations of a given turf area. Nevertheless, the concepts of IPM are the same in turf as in traditional agriculture. IPM is a decision-making and management system that works with the surrounding environment and takes advantage of the natural conditions that help keep the turf healthy [5].

There are still many misconceptions about IPM among turfgrass managers. IPM is not an organic or biological pest control program. It does not preclude the use of synthetic insecticides but emphasizes selective intervention rather than routine calendar applications. Adoption of a more IPM-oriented turf management program will initially incur some costs. However, if conducted properly the time and money invested in monitoring, sampling, and employee training should be offset through reduced pesticide costs, better long-term health of the turf, reduced liability, increased professionalism, and environmental stewardship [2].

The basic components of an IPM approach include site assessment, monitoring and predicting pest activity (sampling), setting thresholds (decision making), stress management, identifying and optimizing management options (appropriate intervention), and evaluating the results [6]. In all this, it is essential to keep records as these can be considered to be the memory of an IPM program. Records allow knowing when and where to expect certain pest problems and help optimize timing for monitoring and management activities.

20.3 MONITORING, THRESHOLDS, AND PREDICTING PEST ACTIVITY

Accurate identification of pest insects and their population densities and development of appropriate action thresholds are the cornerstone of a successful IPM program. Monitoring should also include an assessment of the agronomic conditions of a site as these can affect the likelihood and extent of damage by a given pest population. To enhance the turf manager's ability to make informed decisions in subsequent years, good records have to be kept of all the relevant observations (symptoms, location, time, recent weather patterns, soil texture and condition, etc.).

20.3.1 ADULT MONITORING

20.3.1.1 Adult Aggregations

Adults in some scarab species may form conspicuous feeding and mating aggregations on ornamental plants and trees surrounding turfgrass areas [2–4]. After mating and feeding, the female beetles may not venture far from these aggregation sites to lay eggs. However, the correlation between adult aggregations or generally high adult densities and ensuing larval densities in nearby turf areas is generally weak. Nevertheless, high adult densities increase the likelihood of high larval densities. Therefore, they may be used to predict potential problem areas and allow concentrating monitoring and management activities around such areas.

Adult *P. japonica* aggregate on their favorite food plants during warm, sunny days to feed and mate. *P. japonica* adults have been observed to feed on more than 300 woody ornamental plants but have clear preferences for certain plants including grape, roses, cherries, peaches, plum, mountain ash, linden, and sassafras [2]. Turf areas with perennially high *P. japonica* larval densities tend to be near plants that attract adult feeding aggregations [7]. Adults of the green June beetle, *Cotinis nitida*, prefer to feed on ripening thin-skinned fruit (e.g., figs, peaches, grapes), and heavy damage to these fruits and high adult densities observed on the fruit can indicate high larval densities in nearby turf. Adults of *Phyllophaga* spp. like to feed on leaves of oaks, persimmon, hickory, walnut, elm, and birch, and defoliation of these trees may indicate larval damage in nearby turf areas. However, due to the multiyear life cycle of most *Phyllophaga* spp., larval damage may not occur in the same year. Adult *R. majalis* congregate around trees for mating within half an hour after sunset, and large mating aggregations may sound like swarms of honey bees. Such high densities can result in oviposition in nearby turf areas high enough to lead to turf damage by larvae in fall and the following spring.

20.3.1.2 Adult Traps

The adult scarab beetle can be monitored with traps that contain the female's sex pheromone, attracting males only, and feeding attractants, attracting both sexes. Even though scarab attractants and sex pheromones have been widely studied, lures are commercially available for only a few species. Commercial lures for *A. orientalis* contain the female's sex pheromone and attract only the males [8]. *A. orientalis* may be a good candidate species for using traps to predict larval populations because the adults do not seem to fly very long distances and do not cause any significant feeding damage [9]. However, correlations between trap captures of male *A. orientalis* and ensuing larval populations still have to be developed. Because there is a correlation between pheromone load and the number of male *A. orientalis* captured, and with that of distance from which males are drawn in, lures used for predicting larval populations should probably use lower pheromone loads than the presently available commercial lures [8].

The Japanese beetle trap is probably the most widely used scarab trap. Since the adult beetles are voracious feeders, the trap uses both the female's sex pheromone and a floral scent, the latter attracting both males and females. While these traps can attract thousands of *P. japonica* adults per day, only a fraction of these beetles end up in the traps, so that susceptible plants along the flight path of the attracted beetles and in the vicinity of the traps typically sustain more damage by *P. japonica* than if the traps were not used [10]. In addition, females that are attracted but not trapped may lay eggs in nearby turf areas. Pheromone traps are typically useful as indicators of pest presence in the area, but the presence of abundant *P. japonica* populations around a turfgrass area should be obvious enough to most turfgrass managers without the use of traps. Thus, Japanese beetle traps may be most useful for detecting and monitoring new infestations, as is done on a large scale in several western states [11].

20.3.2 LARVAE MONITORING

If short-residual insecticides are to be used for curative white grub control, timing of the application and with that of sampling activities is very important. Most chemical and biological insecticides work better when white grubs are still small because the younger grubs are more susceptible than the more mature larvae. In the case of biological control agents, warmer soil temperatures for a longer period of time before cooler weather sets in will also allow the pathogens to be more effective for a longer period of time, particularly if they recycle in infected white grubs.

Obviously, sampling should be done before damage can occur. However, if sampling is done too early it will be more difficult to find the larvae as they will be even smaller or perhaps still in the egg stage. For white grubs with an annual life cycle, sampling is therefore preferable when most of the larvae are second instars. In the northeastern United States at the latitude of New Jersey, this would typically be in mid-August.

At present, there is no effective sampling technique for larval white grub populations that does not involve some kind of labor-intensive and turf-disrupting soil/sod sampling. Therefore, sampling activities are typically concentrated to high-risk areas, i.e., areas with very little tolerance for damage or areas more likely to be infested with white grubs, including open sunny locations, south-facing slopes, sites that were irrigated during beetle flight, tee and green banks on golf courses, and areas with a history of grub infestations. Adults of night-flying attracted to outdoor lights and *P. japonica* adults attracted to favorite food plants will often lay eggs in nearby turf [2,3,7]. And adult black turfgrass ataenius, *Ataenius spretulus*, and *C. nitida* like to oviposit in areas high in organic matter or where manure-based fertilizers were applied [12,13].

Because white grub populations are typically patchy in distribution, sampling should always involve taking at least a few samples from each to-be-examined area. Ideally, this should be done in a grid pattern, the density of which will depend on the size of the area, the size of the individual sample, the accuracy of the estimate to be obtained, and the amount of work the samplers are willing to invest. Typical tools used for sampling include flat blade spades (about 15 cm = 6 in. blade width) or a standard golf cup cutters (diameter: 10.8 cm = 4.25 in. $\rightarrow$ sample size: 916 cm^2 = 0.99 ft^2). Using a spade, the grass can be cut on three sides of a square, the flap peeled back, and the roots and soil examined to a depth of 7.5 cm (3 in.). A cup cutter allows taking and examining samples more quickly, and the smaller size allows taking more samples in a smaller grid in the same time, resulting in a more accurate estimate throughout the area. To minimize disturbance of the turf surface, the samples should be placed back carefully and the area irrigated if the soil is not moist [2–4].

Surveys of larger areas can be conducted to detect and track over time grub hot spots. On a general map of the site, the grid pattern should be marked on the map and on the turf, the samples checked for grubs, and grub densities and species recorded on the map. For lawns, sampling spots should be about 2–3 m (~6–10 ft) apart and on sports fields, 3–6 m (~10–20 ft). Fairways can be sampled in a zigzag pattern starting about 1.5 m (~5 ft) from the edge with samples every 9–13.5 m (~30–45 ft) throughout the length of the fairway, or taking samples along three or four parallel lines down the fairway with samples taken every 3–6 m throughout the length of the fairway [4]. Using these procedures, an 18-hole golf course can be surveyed in 2–4 man days. Using any of the above procedures, the number of grubs per sample should be transformed into grub density, and if several adjacent samples contain grub densities at or above the action thresholds, a hot spot can be assumed, and the area in question be treated as needed.

Using a sampling plan similar to the one described above for fairways using parallel lines, Dalthorp et al. found that directional kriging, a geostatistical algorithm for estimating local means as weighted averages of samples, was more accurate and consistent in estimating local *P. japonica* larval densities on golf-course fairways than the use of moving averages or inverse distances [14]. Because distribution of *P. japonica* larvae was best fitted by the negative binomial distribution, sequential sampling plans based on common *K*s from the binomial distribution have been proposed [15,16]. However, extensive sampling will generally be too time-consuming and destructive to be adopted by most turf managers. Acoustic estimation of soil insect populations using acoustic detection systems may offer an alternative to the destructive soil sampling methods [17]. A single-sensor acoustic method was more successful in detecting mixed infestation of *Phyllophaga crinita* and southern masked chafer, *Cyclocephala lurida*, than the cup-cutter method but did not allow estimating local grub densities [18]. A four-sensor array allowed estimating local densities but was much slower than either the single-sensor or cup-cutter method [18]. Future improvements in the design and software of this system may lead to more efficient and nondestructive sampling methods for white grubs.

20.3.3 THRESHOLDS

The ability of turfgrass to tolerate feeding by white grubs can be affected by so many variables that it is impossible to develop good general action thresholds for white grubs. An often-used action threshold for white grubs in turf is a density of 10 grubs/ft^2 (=107.5/m^2). However, an experienced turfgrass manager should adapt this threshold to site-specific conditions.

The feeding activity of white grubs reduces the amount of roots available to the plants to absorb nutrients and water from the soil. Obviously, the depth and extent of the root system as well as the plant's ability to replace damaged roots will affect the plant's tolerance to white grub feeding. The ability to regrow roots is particularly important when warm and dry conditions coincide with the voracious feeding of the maturing grubs, e.g., typically in late summer/early fall in much of the transition zone and northeastern United States. The ability of turf to tolerate grub feeding activity can be affected by the grass species and any management practice and environmental condition that affects the turf's root system and its ability to regrow roots, including mowing height, soil type, soil moisture and irrigation, soil compaction, and fertilization (see Sections 20.4.1 and 20.4.2).

The action threshold can also vary to some extent with white grub species because the larvae of different species differ in size, food preferences, and behavior [2–4]. Damage by *P. japonica* larvae does not normally occur with <10 larvae per 0.093 m^2 (=1 ft^2) but unhealthy, poorly maintained turf may show damage at 4–5 mature larvae per 0.093 m^2. In laboratory pot experiments, 20–60 larvae per 0.093 m^2 did not significantly affect the foliar yield of common turfgrass species if not stressed [19,20]. No damage thresholds have been developed for *A. orientalis* larvae but due to the similarity in size and behavior, thresholds should be similar to those in *P. japonica*. Because *M. castanea* larvae are smaller than *P. japonica* larvae, turf may normally tolerate 15 or more *M. castanea* larvae per 0.093 m^2 [2]. Another reason given for the higher tolerance for this species is that its larvae feed somewhat lower than other white grub species, leaving some root tissue attached to the grass plants [2]. However, in my personal observations over many years, *M. castanea* larvae tended to feed at the same depth as larvae of *A. orientalis*, *P. japonica*, and northern masked chafer, *Cyclocephala borealis*.

Tolerance for *R. majalis* larvae is lower (5–10 per 0.093 m^2) [2] than for *P. japonica* larvae because they feed longer in fall and earlier in spring when the grass grows more slowly [3]. Larvae of many *Phyllophaga* spp. are considerably larger than *P. japonica* larvae and feed accordingly more, resulting in lower action thresholds [3]. *A. spretulus* larvae, in contrast, are much smaller than *P. japonica* larvae and irrigated fairways should be able to tolerate 30–50 larvae per 0.093 m^2 unless the grass is further stressed by drought, soil compaction, or disease [2]. However, depending on the weather and environmental conditions, the *A. spretulus* action threshold can vary from 20 (Coachella Valley in southern CA in July) to 100 (August in western NY) per 0.093 m^2 [3].

Cyclocephala spp. larvae are larger than *P. japonica* larvae but are fond of decaying organic matter and feed accordingly less on living roots. Vigorous turf can tolerate >15 grubs per 0.093 m^2, but stressed turf may show damage with half as many larvae [2,3]. *C. nitida* larvae are considerably larger than *P. japonica* larvae but prefer feeding on decomposing organic matter, including compost, thatch, and grass clippings, over feeding on living roots [2,3]. However, their tunneling and burrowing disturbs the root system, dislodges grass, and accelerates moisture loss from surface soil by loosening it [2]; an average of 6–8 per 0.093 m^2 may warrant control measures [21].

20.3.4 Predicting Pest Activity

Phenology calendars, degree-day models, and monitoring traps can be used to predict the activity patterns of a pest over time, when which stage is present, when the best time for monitoring is, and when to apply treatments to control the susceptible stages. This is particularly important if the development of a pest is difficult to observe directly, and this is certainly the case with the soil-dwelling white grubs. At this time, such methods have been worked out only for a few white grub species and are rarely used by practitioners. One reason for this is that they are relatively cumbersome. The second reason is that timing of monitoring and treatment events in most annual white grub species is not as critical as for many other pest types. Temperatures in the soil where white grubs spend most of their life cycle fluctuate less than air temperatures. And monitoring or management activities occur later in the season when the effect of unusually cool or warm spring temperatures has often been balanced by warmer or cooler temperatures, respectively, during summer.

20.3.4.1 Phenology Calendars

Phenology-related seasonally recurring biological events relate to climate and other events that are happening at the same time. Because the seasonal development of plants and insects are affected in a similar way by weather patterns, particularly temperatures, the occurrence of conspicuous events in indicator or signal plants can be used to predict the occurrence of certain stages and activities of a pest. Flowering trees and shrubs with conspicuous, easily recognizable stages like bud break, early, full, and late bloom, petal drop, or leafing make the best indicators.

The development of *A. spretulus* is fairly well correlated with the phenology of several indicator plants [22]. Females returning from overwintering sites begin laying eggs about when Vanhoutte spirea, *Spirea vanhouttei*, and horse chestnut, *Aesculus hippocastanum*, come into full bloom, and black locust, *Robinia pseudoacacia*, is showing first bloom. Larvae of the first generation begin hatching when multiflora rose, *Rosa multiflora*, is in full bloom. The summer-generation adults begin emerging when summer phlox, *Phlox paniculata*, is in full bloom and start laying eggs when rose of Sharon, *Hibiscus syriacus*, is in full bloom.

R. majalis pupation coincides with full bloom of Vanhoutte spirea, adult flight begins at first bloom of hybrid tea and floribunda roses, and peak flight occurs around full bloom of common catalpa, *Catalpa bignonioides* [23].

20.3.4.2 Degree-Day Models

The developmental speed of insect is correlated to temperature. Below a lower threshold temperature (for most insects, around 10°C) no development occurs, and above the upper threshold temperature development begins to deteriorate. A method called degree-day accumulation calculates the average temperature of each day, subtracts the lower threshold (base) temperature from it, and accumulates the values for each day starting with a predetermined date (usually January 1). Degree-day models correlate the cumulative degree-day units with pest development and activities and give degree-day unit ranges when certain stages or activities occur. Such models have been developed for *Cyclocephala* spp. where first adult flight and 90% adult flight occur at 898–905 degree days and 1,377–1579 degree days, respectively, for *C. borealis*, and at 1,000–1,109 degree days and 1,526–1,679 degree days, respectively, for *C. lurida* [24].

20.3.4.3 Trapping

Egg-laying activity and with that, egg hatch can be predicted for many white grub species by observing adult flight using traps. *P. japonica* and *A. orientalis* flight can be observed using traps lured with their respective sex pheromone (see Section 20.3.1.2.). The activity of night-active scarabs such as *M. castanea*, *Cyclocephala* spp., *Phyllophaga* spp., *R. majalis*, but also of the evening-active *A. orientalis*, can be observed using light traps. Light traps, however, attract many insect species indiscriminately, and working one's way through large amounts of these insects is usually too cumbersome for the typical turfgrass manager. The benefit of trapping is that recording of peak adult flight allows predicting hatch of most of the eggs about 3–4 weeks later and with that, proper timing of insecticide applications.

20.4 MANAGEMENT TECHNIQUES/OPTIONS

20.4.1 Cultural Control

The most basic and effective cultural control for white grubs is good turf management. Healthy, vigorous turf can generally tolerate much higher densities of pests including white grubs than a weak, starved, and moisture-stressed turf. In addition, the various components of turf management including fertility, irrigation, mowing, sanitation, and thatch management may be slightly modified to increase the turf's tolerance to the activity of key pests or improve recuperation from damage.

Irrigation can have a strong effect on white grub densities and the turf's tolerance to white grub feeding. White grub eggs must absorb water from the soil to develop and hatch, and the young grubs are also susceptible to desiccation. If conditions during the egg-laying period are hot and dry, females will seek out areas with moist soil and lush turf cover [25]. Judicious irrigation during peak flight activity can therefore reduce egg-laying and survival of young grubs. This is particularly true where species are common that can fly longer distances such as *P. japonica*. Once the grubs start feeding, particularly when the voraciously feeding third instars start appearing in late summer, rainfall and irrigation increase the turf's tolerance to white grub feeding by alleviating the effects of root loss and promoting root regrowth [26–28].

Sound fertilization practices to promote a healthy and deep root system increase the turf's ability to tolerate and recover from white grub feeding [28]. In cool-season grasses, moderate fertilization in fall will also help the turf to recover from white grub damage [28]. Mowing height is inversely related to the turf's root depth and ability to regrow damaged roots [29]. Accordingly, mowing height will also be correlated to the turf's tolerance to white grub feeding. Some white grub species including *P. japonica*, *A. orientalis*, and *A. spretulus* also tend to reach higher densities in turf maintained at lower mowing heights, possibly because many natural enemies are more abundant in higher-cut grass [9,25,30,31].

Adults of some scarab species will be attracted to food plants over some distance. Turf areas near favorite food plants will often receive more eggs and be more prone to grub damage [7,32]. Avoiding or removing such food plants along sensitive turf areas will alleviate such problems. For example, *P. japonica* favor food plants such as linden, Norway maple, purple-leaf plum, and sassafras, and May/June beetle are fond of oaks. *M. castanea* are more common in or near weedy turf areas that contain food plants for the adults or orange hawkweed, *Hieracium aurantiacum*, a favorite hiding place for adults during the day.

Larvae of *A. spretulus* and *C. nitida* are fond of compost and organic fertilizers as a food source, and female *C. nitida* and probably also *S. spretulus* are attracted to sites treated with these to lay eggs [12,13,25]. Accordingly, these materials should be used carefully where these scarab species are problematic. Lime applications to elevate soil pH had no effect on *P. japonica* egg laying and survival of eggs and larvae or larval abundance [25,33–35]. And use of a heavy roller did not reduce subsequent larval populations or crush larvae in fall [25]. Nitrogen fertilization did not affect density and weight of *P. japonica* larvae in tall fescue [25,36]. However, *Cyclocephala* spp. larvae were consistently smaller and less abundant in turf that had been treated with aluminum sulfate to reduce soil pH and in high mown turf [25]. And computer simulations suggested that more than 40% of larvae may be killed by the use of turfgrass aerators with appropriate hole patterns, but the effect has not been field tested [37].

20.4.2 RESISTANT TURFGRASSES

White grubs feed on roots of all turfgrasses and no resistant cultivars have been found to date [2,38]. However, grasses with extensive root systems, higher degrees of drought tolerance, and a greater ability to regrow roots under heat/water stress conditions are generally also more tolerant to white grub feeding. This includes many warm-season grasses and among the cool-season grasses, tall fescue [39,40]. Grasses with a creeping, spreading growth habit through rhizomes or stolons tend to fill in dead patches and recover more quickly from grub damage than bunch-type grasses.

Turfgrasses that are infected with endophytic fungi have increased resistance to many of the aboveground-feeding insect pests because the endophytes produce various types of alkaloids that act as feeding deterrents or are toxic to insects that feed on stems, leaves, and leaf sheaths [41]. However, observations on the effect of endophytes on white grubs have varied from lower to higher white grub populations in endophyte-infected grasses [38,42,43]. Greenhouse and field observations suggest that first instar development and survival are negatively affected by endophytes in tall fescue, whereas third instars are not affected [38,43]. While the effect of endophytes on white grubs may vary with white grub species, turfgrass species, and management practices that affect the grass and endophytes, it is not surprising that any negative effects on root-feeding insects are limited because only a fraction of the endophyte-produced alkaloids appears to be transferred into the root system [44].

Grewal et al. observed a small increase in speed of kill and establishment by the entomopathogenic nematode *Heterorhabditis bacteriophora* into *P. japonica* third instars if they had fed on endophyte-infected tall fescue and chewing fescue under laboratory conditions, but did not observe a significant increase in mortality[45]. Similarly, Koppenhöfer and Fuzy observed a weak and variable enhancing effect of endophyte infection in tall fescue on *H. bacteriophora* susceptibility of second and third instar *A. orientalis* but not of third instar *P. japonica* and *C. borealis* under greenhouse conditions, and no effect under field conditions for any of the three species [46]. Walston et al. did not find any effect of endophyte infection in perennial ryegrass on the susceptibility of *P. japonica* larvae to the milky disease bacterium *Paenibacillus popilliae* [47].

20.4.3 TRAPPING

Trapping for management has only been attempted with *P. japonica*. As discussed in Section 20.3.1.2, Japanese beetle traps can attract thousands of adults per day using a combination of the female's sex pheromone and a floral scent that attracts both sexes, but only a fraction of the beetles end up in the traps, resulting in more foliar damage and egg laying in the vicinity of the traps than if the traps were not used [10]. Placing the traps in the ground with only the funnel portion remaining above ground (funnel rim height about 13 cm) rather than the typical 90 cm can increase trap capture in sunny turfgrass by increasing the percentage of attracted males being trapped [48,49]. Emptying the traps daily also increases trap captures by around 50% [49]. Whether these modifications increase trap captures of both sexes enough to decrease foliar damage and egg laying in the trap vicinity still needs to be determined. Using only the sex pheromone in these traps will minimize the number of females attracted and with that, egg laying, but males attracted and not trapped will still cause foliar damage to the surrounding host plants, and females already in the vicinity will not be kept from feeding and egg laying.

Mass trapping with hundreds of traps on golf courses, in suburban neighborhoods, and in parks may reduce *P. japonica* populations [50]. But the to-be-expected limited success is unlikely to justify the cost and labor involved in such intensive trap use. An exception may be the use of traps for *P. japonica* management or eradication in recently invaded areas [51,52]. This could be particularly effective when used in combination with other control measures.

Traps may also have use for autodissemination of pathogen by adult beetles, particularly in areas where direct application of the pathogens is not feasible. Modified traps were used to contaminate and infect adult *P. japonica* with the entomopathogenic nematode *S. glaseri* and the entomopathogenic fungus *Metarhizium anisopliae* [53,54]. Whether the infected adults can actually establish the diseases in the larval habitats still needs to be confirmed.

20.4.4 BIOLOGICAL CONTROL

Biological control is the suppression of pest populations through the activity of living organisms or their by-products. Such natural enemies may include vertebrate and invertebrate predators, parasitoids, and pathogens. Within the framework of IPM, biological control may involve different methods or approaches: conservation, classical, and augmentation.

20.4.4.1 Conservation Biological Control

The most practical form of biological control in turfgrass is to conserve natural enemies by applying insecticides only when and where necessary and using insecticides with a reduced impact on natural enemies (biological and biorational control agents). Healthy lawns that are not regularly broadcast-treated with broad-spectrum insecticides contain a plethora of organisms that parasitize or prey on turfgrass insect pests and thereby reduce the frequency and severity of pest outbreaks. There are many predatory species of ground beetles, rove beetles, and, particularly, ants that prey on eggs and young larvae of white grubs [2,3,11,55–57]. There are also numerous wasps and flies that parasitize white grub larvae or adults [2,3,11]. In addition, there are numerous organisms that cause weakening or fatal diseases in turfgrass insect pests including entomopathogenic nematodes, fungi, bacteria, and protozoans [11,58,59].

All these invertebrate natural enemies may be significantly reduced after application of broad-spectrum insecticides such as carbamates, organophosphates, and pyrethroids [56,57,60,61]. Use of these compounds should be restricted to areas with potentially damaging white grub densities as determined by sampling. Neonicotinoids and insect growth regulators have a much smaller and more transient effect on these predators [57,62,63], but their use should also be restricted to high-risk areas as their indiscriminate broadcast use is likely to reduce invertebrate natural enemies by depriving them of prey or hosts.

Proper timing of insecticide applications can help reduce negative effects on natural enemies during critical periods. Applications of imidacloprid during May can reduce parasitism of *P. japonica* larvae by the spring-active *Tiphia vernalis*, a parasitic wasp, through intoxication of the females. If applications are withheld until June or July, the wasps can be conserved in nearby untreated areas [64].

Habitat modifications can increase the densities of some natural enemies. Raising mowing heights where possible may provide refuges for predators as observed for *A. spretulus* natural enemies [30]. Incorporating wildflower beds in the turf landscape can provide supplemental food for predators and parasitoids [65,66].

20.4.4.2 Classical Biological Control

Classical biological control involves the importation and release of natural enemies to suppress typi-cally introduced or exotic pests. *P. japonica* has received the most attention of any white grub spe-cies with respect to classical biological control. Between 1920 and 1933, a total 49 species of natural enemies of *P. japonica* and related white grub species from Asia and Australia were introduced into the United States [11]. But only a few became established, among the most widely distributed ones being the parasitic wasps *Tiphia popilliavora* and *T. vernalis* that parasitize *P. japonica* larvae in late summer and spring, respectively, and the tachinid fly, *Istocheta aldrichi*, that parasitizes adult *P. japonica*. These parasitoids may locally have significant effects on *P. japonica* populations [64,67] but only occur sporadically within their established ranges and cannot be relied upon for site-specific suppression of their host.

20.4.4.3 Augmentation

Augmentation biological control aims at increasing the number of already present natural enemies. It can involve the release of additional natural enemies in (1) small numbers (inoculative release) expecting establishment and slow but long-term pest suppression and (2) in large quantities (inundative release) expecting quick but not necessarily long-term pest suppression. Augmentation biological control of white grubs has mainly focused on inundative releases of pathogens that can be mass-produced in sufficient quantities, i.e., entomopathogenic bacteria, fungi, and nematodes. Observations on long-term establishment after inoculative releases are limited.

More detailed information on the biology, ecology, and application technology of the below dis-cussed pathogen groups can be found in Chapter 19 of this book and the references therein [68–71].

20.4.4.3.1 Bacteria

Paenibacillus (=*Bacillus*) *popilliae* is the causative agent of "milky disease" and has been observed in white grubs in the subfamilies Melolonthinae, Rutelinae, Aphodinae, and Dynastinae [68,69,72,73]. Its many strains are host species–specific. Only the strain infecting *P. japonica* has been commercialized, with milky spore being the only product currently available. Formulations made by mixing macerated diseased larvae with talc (10^8 spores/g) are applied by placing 2-g for-mulation every 1.3 m in a grid pattern (11 kg/ha). However, when infected grubs die, typically after several weeks or even months, up to 5×10^9 spores may be released into the soil from the disinte-grating grub cadaver, leading to a buildup of the disease. Spores can persist in the soil for several years. Under optimal conditions, e.g., soil temperatures $\geq 21°C$ for several months per year and high larval densities ($\geq 300/m^2$), *P. popilliae* can establish and provide lasting control in a release site over a period of several years [73,74]. The limited availability combined with inconsistent perfor-mance and establishment has limited the use of *P. popilliae* [75,76].

The Buibui strain of *Bacillus thuringiensis* ssp. *japonensis* (*Btj*) has shown good activity against larvae of the cupreous chafer, *Anomala cuprea*, *P. japonica*, *A. orientalis*, and *C. nitida* [77,78]. Because *Btj* has to be applied against early instars, its commercialization in the United States has been hindered by competition from highly effective reduced-risk synthetic insecticides such as imidacloprid and halofenozide, which are also applied preventively. Commercial development of this strain continues primarily in Japan. Other highly scarab-active *Bt* strains may face the same issues from competition (e.g., *Bt* ssp. *galleriae* SDS-502) [79].

Amber disease is caused by certain strains of *Serratia* spp., a genus of bacteria commonly found in turf soils. Infected larvae stop feeding within several days of infection, and starve to death over a period of 1–3 months, during which time they take on a characteristic amber appearance [80]. During the final stages of infection, the bacteria also penetrate into the grub's hemocoel. In New Zealand, *Serratia entomophila* is mass-produced *in vitro* and is used commercially as the product Invade™ against the grass grub *Costelytra zealandica* in pastures [81]. Despite isolation of similar bacteria from *P. japonica* and *Cyclocephala* spp. grubs in the United States, no commercial development has taken place in the United States.

20.4.4.3.2 Fungi

The entomopathogenic fungi *Beauveria bassiana* and *Metarhizium anisopliae* naturally infect white grubs. However, use of mycoinsecticides in turfgrass, particularly against soil insects, is rather questionable because it is difficult to get the fungal spore into the upper soil layers to their targets. Subsurface applications that could alleviate this limitation require highly specialized equipment. In pasture turf where subsurface applications are more feasible, *Beauveria brongniartii* has shown good suppression of European cockchafer, *Melolontha melolontha*, grubs, in Europe [82,83], and *M. anisopliae* is used for control of *Adoyphorus couloni* grubs in Australia [84].

Entomopathogenic fungi are compatible with many agrochemicals including most insecticides, but are negatively affected by fungicides. However, much of the compatibilities have been determined under laboratory conditions, and in the field compatibility may be better as long as entomopathogenic fungi and fungicides are applied asynchronously [85].

20.4.4.3.3 Nematodes

Entomopathogenic nematodes are the most extensively studied parasites of white grubs [86–88]. Currently, the species *Heterorhabditis bacteriophora*, *H. zealandica*, *H. marelatus*, and *Steinernema glaseri* are commercially available for grub control. Most of the work with entomopathogenic nematodes for white grub management has been conducted with *P. japonica*. Nematodes that have provided the best field control of *P. japonica* include *S. scarabaei* (100%), *H. bacteriophora* (GPS11 strain) (34–97%), *H. bacteriophora* (TF strain) (65–92%), and *H. zealandica* (X1 strain) (73–98%) [87].

Recent research has shown that many common white grub species such as *Cyclocephala* spp., *A. orientalis*, *R. majalis*, *M. castanea*, and *Phyllophaga* spp. appear to be less susceptible than *P. japonica* to the commonly used nematodes, i.e., *H. bacteriophora* and *S. glaseri* [87,89,90]. *S. scarabaei* is the only nematode species that has provided high field control of *A. orientalis* (87–100%), *M. castanea* (71–86%), and *R. majalis* (89%). Against *C. borealis*, *H. zealandica* (X1 strain) (72–96%), *S. scarabaei* (84%), and *H. bacteriophora* (GPS11 strain) (47–83%) appear to be the most promising nematodes. Unfortunately, *S. scarabaei* has proven to be difficult to mass-produce and is thus far not commercially available.

Nematode susceptibility can vary significantly with larval stage, but the effect may vary with white grub and nematode species [91]. For example, efficacy against *P. japonica* and *A. orientalis* is higher against second instars than third instars for *H. bacteriophora* but there is no significant effect for *S. scarabaei*. Thus, nematode application timing should consider not only environmental conditions, particularly soil temperatures and moisture, but also presence of the most susceptible white grub stages. For example, *H. bacteriophora* applications against *P. japonica* or *A. orientalis* in the northeastern United States will tend to be more effective if applied by early September due to the presence of the more susceptible younger larvae and a longer period after application before soil temperatures become too cool for good nematode activity. Applications of *S. scarabaei*, should it become commercially available, against these species may be more effective when applied after late August as the species is more cold-tolerant, similarly pathogenic to second and third instars, and due to a more profuse reproduction in the larger larvae, will provide better long-term suppression (Koppenhöfer, unpublished data) [92].

The persistence of entomopathogenic nematodes beyond a season following their application against third instar white grubs has been reported, thus suggesting the potential impact of entomopathogenic nematodes on multiple generations of white grubs [93–95]. The white grub–specific *S. scarabaei* has shown very promising long-term effects in small plot trials against populations of *A. orientalis* [92]. *S. scarabaei* apparently reproduces very well in infected white grubs in the field and its progeny persists for long periods in the soil in the absence of hosts. Applications rates of 0.4×10^9 to 2.5×10^9 *S. scarabaei* per ha provide 86–100% *A. orientalis* control within 1 month of application, and rates of 0.1×10^9 to 2.5×10^9 *S. scarabaei* per ha provide 86–100%, 62–95%, 69–95%, and 0–94% control at 8, 13, 20, and 25 months after application, respectively (Koppenhöfer, unpublished data).

Entomopathogenic nematodes should generally be compatible with endemic natural enemies in turfgrass. While larvae and pupae of several species of carabids, cicindelids, staphylinids, and hover fly larvae entering the ground for pupation were found to be moderately susceptible to several species of nematodes, the effect of nematode application on natural populations of carabids, staphylinids, histerids, and formicids was generally negligible, or was temporally as well as spatially restricted [96,97]. No studies have been conducted on nematode interactions with white grub parasitoids. On the basis of observations with other parasitic wasps, it could be expected that there would be some negative effect of nematode infection of parasitoid hosts on parasitoid larval survival depending on larval age [98,99]. However, after cocoon formation, parasitoid pupae should be generally resistant to nematode infection [98,100].

EPN appear to be compatible with many herbicides, fungicides, acaricides, insecticides, nematicides, azadirachtin, *Bacillus thuringiensis* products, and pesticidal soap. Other pesticides have limited to strong toxic effect on nematodes, e.g., oxamyl, fenamiphos, carbaryl, bendiocarb, diazinon, dodine, paraquat, or methomyl. In case of negative interactions, it is usually recommended to wait for 1 and 2 weeks after the application of chemical insecticides and chemical nematicides, respectively, before nematode application. An extensive summary of compatibilities is provided by Koppenhöfer and Grewal [101].

20.4.5 MATING DISRUPTION

Flooding a pest's environment with large quantities of its sex pheromone may result in disruption of the normal mating behavior, particularly the ability of males to find females, ideally leading to a reduction in the pest's reproduction. Mating disruption is widely used for the management of several moth species [102]. Mechanisms postulated to explain mating disruption include camouflage (female pheromone plumes can no longer be separated from pheromone background), imbalance of sensory input in males, false-plume-following by males (reducing time and energy available to males for finding a mate), and adaptation or habituation of the males' receptor system [103].

Even though the sex pheromones of scarab beetles have been studied intensively [104] and are used for monitoring purposes, only recently has mating disruption technology been considered a possibility for the management of white grubs. Thus far, the *A. orientalis* has been the only white grub pests for which mating disruption has been tested as a management option. The feasibility of *A. orientalis* using a microencapsulated sprayable pheromone formulation has been demonstrated in large-scale field tests in ornamental nurseries and blueberries and turfgrass [105,106]. In turfgrass, *A. orientalis* larval populations could be reduced by up to 74%, but the limited persistence of the formulation made two applications during the beetle's flight period necessary [106]. In addition, clothing articles such as shoes coming into contact with treated grass were contaminated sufficiently to attract large numbers of male beetles outside of treated areas [106].

Retrievable, point-source pheromone dispensers have been shown to be more persistent in disrupting *A. orientalis* mate finding in blueberry and cranberry trials [107,108]. However, deployment of such dispensers is not feasible in turfgrass areas that have to be mown regularly and are used for recreational activities. However, dispersible pheromone formulations may offer an effective, persistent, and less contamination-prone alternative to sprays and dispensers as suggested by ongoing studies in turfgrass (Koppenhöfer, unpublished data).

20.4.6 CHEMICAL CONTROL

While IPM recognizes that there are situations when insecticides may be the only practical solution to a pest problem, insecticide should be used only when and where necessary as determined by monitoring of key pests. Where practical, less toxic and environmentally less hazardous products should be selected. There are two approaches to chemical white grub control: the preventive and the curative or corrective approach. In the preventive approach, an insecticide is applied as insurance before potential grub damage can even show up. In the curative approach, an insecticide is applied when larvae are present, typically in late summer, after monitoring and sampling has indicated that white grub densities are present that may cause damage or after signs of white grub activity have been detected.

Curative control has long been the standard approach for white grub control, in part due to the fairly short residual of most of the organophosphates and carbamates that were labeled for white grub management in turfgrass. Due to the implementation of the Food Quality Protection Act of 1996 (FQPA), many of these products have lost registration in the United States [109]. Presently, only the carbamate carbaryl (Sevin) and the organophosphate trichlorfon (Dylox) are still labeled. Other countries face similar or even more restricting situations. Most curatively used insecticides work best when applied against young larvae, second and, better, first instars. In the northeastern United States, this would typically be around mid- to late August. After this time, these products become less and less reliable for white grub control. Curative control can fit well into an IPM program as it can be restricted to areas that need treatment as determined by monitoring and sampling. However, the products available for curative control at this time are broad-spectrum insecticides that can negatively affect nontarget invertebrates in turfgrass including white grub predators [56].

Preventive control has become widely used for white grub management since the registration of the neonicotinoid imidacloprid and the insect growth regulator halofenozide for turfgrass in the mid-1990s. More recently, another neonicotinoid, clothianidin, was registered for turfgrass use. Because of their long residual, these products can be applied as much as 2 months before grubs are present without significant loss of efficacy. Imidacloprid and, to a lesser extent, halofenozide have been marketed for preventive white grub control, to be applied no later than mid-August but generally recommended for application around the time of peak egg-laying activity, i.e., around July in the northeastern United States. Typically applied in the early and mid-part of the labeled application window before sampling or even monitoring is possible, this preventive approach often leads to unnecessary treatment due to the sporadic and localized nature of white grub outbreaks. However, neonicotinoids and halofenozide have a much lower vertebrate toxicity and also more limited negative effects on nontarget and predatory invertebrates including white grub predators than carbamates and organophosphates [57,62,63]. Nevertheless, if applied over large areas for consecutive years, their high efficacy against many turfgrass insect pests is likely to reduce predators, parasitoids and pathogens of white grubs and other insect pests by depriving them of prey/hosts.

While the common use of preventive applications does not fit into an IPM program, there are ways of using neonicotinoids and halofenozide in a more IPM-compatible way. First, the application should be done more selectively only to high-risk areas, i.e., areas with high beetle activity and with a history of white grub problems. Second, there is an increasing body of data that shows that these products are at least as effective as trichlorfon when applied as late as early September (Koppenhöfer, unpublished data). This expands the application window of these products at least into late August, which allows monitoring and sampling before application.

20.5 INTERACTIONS BETWEEN CONTROL AGENTS

A number of studies have examined interactions of various biological control agents with other biological control agents or synthetic insecticide with the goal of improving white grub control by synergistic interactions between the control agents. *P. popilliae* infection of third-instar masked chafer, *Cyclocephala hirta*, enhanced the efficacy of *H. bacteriophora* and *S. glaseri* by facilitating penetration of the nematode into the midgut [110,111]. Neither *H. bacteriophora* nor *P. popilliae* reproduction in the host was affected [110]. Synergistic combinations of *S. glaseri* or *H. bacteriophora* with *B. thuringiensis* ssp. *japonensis* (*Btj*) may be feasible when applied against a *Btj*-sensitive scarab species [112]. The nematodes *Heterorhabditis megidis* and *S. glaseri* and the fungus *M. anisopliae* showed a strong synergistic interaction in third-instar *Hoplia philanthus* when the fungus was applied at least 3–4 weeks before the nematode; combination with *M. anisopliae* had no effect on *S. glaseri* reproduction but negatively affected *H. megidis* reproduction at higher *M. anisopliae* application rates [113].

The synergism between *Heterorhabditis* spp. and *S. glaseri* and the neonicotinoid imidacloprid in third instars of *P. japonica*, *A. orientalis*, *C. hirta*, and *C. borealis* [114,115] is well documented. Imidacloprid reduces the grubs' defensive behaviors, resulting in increased nematode attachment and penetration [116]. However, *S. kushidai* and *S. scarabaei*, two rather scarab-specific and highly virulent species, generally do not interact with imidacloprid [117,118]. Imidacloprid does not compromise nematode recycling in grubs [119]. However, to make the combination of these rather expensive control agents feasible, high control rates have to be achieved with application rates of both control agents reduced by at least 50% compared with the regular field rate. Ongoing studies indicate that this may be achieved by applying the combinations when the more nematode-susceptible second instars prevail (Koppenhöfer, personal observation).

20.6 CONCLUSIONS

At the present time, white grub management in turfgrass still relies heavily on the use of synthetic insecticides. Whether the replacement of organophosphates and carbamates with the less hazardous and less environmentally disruptive neonicotinoids and insect growth regulators can be considered as an improvement is questionable due to the often indiscriminate preventive large-area use of the latter insecticides (see Section 20.4.6). There is a dire need for effective and competitive curative control options that could be used for curative spot treatment against white grubs. Ideally, this void would be filled with nonchemical control options such as microbials. But due to issues with high cost, limited shelf-life, and reliability of presently available microbial white grub control options (see Chapter 19), it is more likely that yet to-be-developed fast-acting reduced-risk insecticides will fill the void, at least for the nearer future. However, the major stumbling block for replacement of preventive applications is the present lack of effective, minimally disruptive sampling methods or other methods to predict white grub outbreaks (see Section 20.3.2).

The ultimate goal of IPM should be to reduce control measure to a minimum, be they chemical or microbial in nature. Development of resistant or at least more tolerant turfgrass species/cultivars would be a cornerstone to achieve this increase in sustainability. Transgenic expression of white grub repellent or toxic substance such as bacterial toxins or endophyte-associated alkaloids could be an avenue to this end provided that it is possible to express these substances in the grass roots. Conservation of endemic or applied natural enemies would be another more sustainable approach. The buffering effect of natural enemies on white grub populations could be achieved through the use of less insecticide and less disruptive compounds, a better understanding of the interactions of natural enemies with other components of turfgrass management, and through habitat manipulation (see Sections 20.4.4.1 and 20.5).

Last but not the least, more effective education of turfgrass managers and, particularly, their clientele as to the benefits of reducing pesticide use and the need for relaxing standards of acceptable turfgrass to more reasonable levels could facilitate the wider adoption of IPM in turfgrass management [2,120].

REFERENCES

1. Jackson, T.A. Scarabs—pests of the past or future? In *Use of Pathogens in Scarab Pest Management*, Jackson, T.A. and Glare, T.R., Eds., AgResearch, Lincoln, NZ, 1992, 1–6.
2. Potter, D.A., *Destructive Turfgrass Insects: Biology, Diagnosis, and Control*, Ann Arbor Press, Chelsea, MI, 1998.

3. Vittum, P.J., Villani, M.G., and Tashiro, H., *Turfgrass Insects of the United States and Canada*, Cornell University Press, Ithaca, NY, 1999.
4. Niemczyk, H.D. and Shetlar, D.J., *Destructive Turf Insects*, HDN Books, Wooster, OH, 2000.
5. Schuman, G.L., Vittum, P.J., Elliott, M.L., and Cobb, P.P., *IPM Handbook for Golf Courses*, Ann Arbor Press, Chelsea, MI, 1997.
6. Grant, J., IPM of insects, *TurfGrass Trends*, August 1995, 3–7.
7. Dalthorp, D., Nyrop, J., and Villani, M.G., Spatial ecology of the Japanese beetle, *Popillia japonica*, *Entomol. Exp. Appl.*, 96, 129–139, 2000.
8. Alm, S.R., Vilani, M.G., and Roelofs, W., Oriental beetle (Coleoptera: Scarabaeidae): current distribution in the United States and optimization of monitoring traps, *J. Econ. Entomol.*, 92, 931–935, 1999.
9. Facundo, H.T., Villani, M.G., Linn, C.E., Jr., and Roelofs, W.L., Temporal and spatial distribution of the oriental beetle (Coleoptera: Scarabaeidae) in a golf course environment, *Environ. Entomol.*, 28, 14–21, 1999.
10. Gordon, F.C. and Potter, D.A., Efficiency of Japanese beetle (Coleoptera: Scarabaeidae) traps in reducing defoliation of plants in the urban landscape and effect on larval density in turf, *J. Econ. Entomol.*, 78, 774–778, 1985.
11. Potter, D.A. and Held, D.W., Biology and management of the Japanese beetle, *Annu. Rev. Entomol.*, 47, 175–205, 2002.
12. Brandhorst-Hubbard, J.L., Flanders, K.L., and Appel, A.G., Oviposition site and food preference of the green June beetle (Coleoptera: Scarabaeidae), *J. Econ. Entomol.*, 94, 628–633, 2001.
13. Potter, D.A., Held, D.W., and Rogers, M.G., Natural organic fertilizers as a risk factor for *Ataenius spretulus* (Coleoptera; Scarabaeidae) outbreaks on golf courses, *Int. Turfgrass Soc. Res. J.*, 10, 753–760, 2005.
14. Dalthorp, D., Nyrop, J., and Villani, M.G., Estimation of local mean population densities of Japanese beetle grubs (Scarabaeidae: Coleoptera), *Environ. Entomol.*, 28, 255–265, 1999.
15. Ng, Y.-S., Trout, J.R., and Ahmad, S., Spatial distribution of the larval populations of the Japanese beetle in turfgrass, *J. Econ. Entomol.*, 76, 26–30, 1983.
16. Ng, Y.-S., Trout, J.R., and Ahmad, S., Sequential sampling plans for larval populations of the Japanese beetle (Coleoptera: Scarabaeidae) in turfgrass, *J. Econ. Entomol.*, 76, 251–253, 1983.
17. Brandhorst-Hubbard, J.L., Flanders, K.L., Mankin, R.W., Guertal, E.A., and Crocker, R.L, Mapping of soil insect infestations sampled by excavation and acoustic methods, *J. Econ. Entomol.*, 94, 1452–1458, 2001.
18. Zhang, M., Crocker, R.L., Mankin, R.W., Flanders, K.L., and Brandhorst-Hubbard, J.L., Acoustic estimation of infestations and population densities of white grubs (Coleoptera: Scarabaeidae) in turfgrass, *J. Econ. Entomol.*, 96, 1770–1779, 2003.
19. Crutchfield, B.A. and Potter, D.A., Damage relationships of Japanese beetle and southern masked chafer (Coleoptera: Scarabaeidae) grubs in cool-season turfgrasses, *J. Econ. Entomol.*, 88, 1049–1056, 1995.
20. Crutchfield, B.A. and Potter, D.A., Tolerance of cool-season grasses to feeding by Japanese beetle and southern masked chafer (Coleoptera: Scarabaeidae) grubs, *J. Econ. Entomol.*, 88, 1380–1387, 1995.
21. Baker, J.R., *Insects and Other Pests Associated with Turf: Some Important, Common, and Potential Pests in the Southeastern United States*, North Carolina Agricultural Extension Service, AG-268, 1982.
22. Wegner, G.S. and Niemczyk, H.D., Bionomics and phenology of *Ataenius spretulus. Ann. Entomol. Soc. Am.*, 74, 374–384, 1981.
23. Tashiro, H. and Gambrell, F.L., Correlation of European chafer development with the flowering period of common plants, *Ann. Entomol. Soc. Am.*, 56, 239–243, 1963.
24. Potter, D.A., Seasonal emergence and flight of northern and southern masked chafers in relation to air and soil temperature and rainfall patterns, *Environ. Entomol.*, 10, 793–797, 1981
25. Potter, D.A., Powell, A.J., Spicer, P.G., and Williams, D.W., Cultural practices affect root-feeding white grubs (Coleoptera: Scarabaeidae) in turfgrass, *J. Econ. Entomol.*, 89, 156–164, 1996.
26. Ladd, T.L., Jr. and Buriff, C.R., Japanese beetle: influence of larval feeding on bluegrass yields at two levels of soil moisture, *J. Econ. Entomol.*, 72, 311–314, 1979.
27. Potter, D.A., Influence of feeding by grubs of the southern masked chafer on quality and yield of Kentucky bluegrass, *J. Econ. Entomol.*, 75, 21–24, 1982.
28. Crutchfield, B.A., Potter, D.A., and Powell, A.J., Irrigation and fertilization effects on white grub feeding injury to tall fescue turf, *Crop Sci.*, 35, 1122–1126, 1995.
29. Christians, N., *Fundamentals of Turfgrass Management*, Ann Arbor Press, Chelsea, MI, 1998.
30. Rothwell, N.L. and Smitley, D.R., Impact of golf course mowing practices on *Ataenius spretulus* (Coleoptera: Scarabaeidae) and its natural enemies, *Environ. Entomol.*, 28, 359–366, 1999.

31. López, R. and Potter, D.A., Biodiversity of ants (Hymenoptera: Formicidae) in golf course and lawn turf habitats in Kentucky, *Sociobiology*, 42, 701–714, 2003.

32. Choo, H.Y., Lee, D.W., Park, J.W., Kaya, H.K., Smitley, D.R., Lee, S.M., and Choo, Y.M., Life history and spatial distribution of oriental beetle (Coleoptera: Scarabaeidae) on golf courses in Korea, *J. Econ. Entomol.*, 95, 72–80, 2002.

33. Vittum, P.J., Effect of lime applications on Japanese beetle (Coleoptera: Scarabaeidae) grub populations in Massachusetts soils, *J. Econ. Entomol.*, 77, 687–690, 1984.

34. Vittum, P.J. and Tashiro, H., Effect of soil pH on survival of Japanese beetle and European chafer larvae, *J. Econ. Entomol.*, 73, 577–579, 1980.

35. Vittum, P.J. and Morzuch, B.J., Effect of soil pH on Japanese beetle (Coleoptera: Scarabaeidae) oviposition in potted turfgrass, *J. Econ. Entomol.*, 83, 2036–2039, 1989.

36. Davidson, A.W. and Potter, D.A., Response of plant-feeding, predatory, and soil-inhabiting invertebrates to *Acremonium* endophyte and nitrogen fertilization in tall fescue turf, *J. Econ. Entomol.*, 88, 367–379, 1995.

37. Blanco-Montero, C.A. and Hernandez, G., Mechanical control of white grubs (Coleoptera: Scarabaeidae) in turfgrass using aerators, *Environ. Entomol.*, 24, 243–245, 1995.

38. Potter, D.A., Patterson, C.G., Redmond, C.T., Influence of turfgrass species and tall fescue endophyte on feeding ecology of Japanese beetle and southern masked chafer grubs (Coleoptera: Scarabaeidae), *J. Econ. Entomol.*, 85, 900–909, 1992.

39. Braman, S.K. and Pendley, A.F., Growth, survival, and damage relationships of white grubs in Bermudagrass vs. tall fescue, *Intern. Turfgrass Soc. Res. J.*, 7, 370–374, 1993.

40. Bughrara, S.S., Smitley, D.R., Cappaert, D., and Kravchenko, A.N., Comparison of tall fescue (Cyperales: Gramineae) to other cool-season turfgrasses for tolerance to European chafer (Coleoptera: Scarabaeidae), *J. Econ. Entomol.*, 96, 1898–1904, 2003.

41. Breen, J.P., *Acremonium* endophyte interactions with enhanced plant resistance to insects, *Annu. Rev. Entomol.*, 39, 401–423, 1994.

42. Murphy, J.A., Sun, S., and Betts, L.L., Endophyte-enhanced resistance to billbug (Coleoptera: Curculionidae), sod webworm (Lepidoptera: Pyralidae), and white grub (Coleoptera: Scarabaeidae) in tall fescue, *Environ. Entomol.*, 22, 699–703, 1993.

43. Koppenhöfer, A.M., Cowles, R.S., and Fuzy, E.M., Effects of turfgrass endophytes (Clavicipitaceae: Ascomycetes) on white grub (Coleoptera: Scarabaeidae) larval development and field populations, *Environ. Entomol.*, 32, 895–906, 2003.

44. Siegel, M.R., Latch, G.C., and Johnson, M.C., Fungal endophytes of grasses, *Annu. Rev. Phytopathol.*, 24, 293–315, 1987.

45. Grewal, S.K., Grewal, P.S., and Gaugler, R., Endophytes of fescue grasses enhance susceptibility of *Popillia japonica* larvae to an entomopathogenic nematode, *Entomol. Exp. Appl.*, 74, 219–224, 1995.

46. Koppenhöfer, A.M., and Fuzy, E.M., Effects of turfgrass endophytes (Clavicipitaceae: Ascomycetes) on white grub (Coleoptera: Scarabaeidae) control by the entomopathogenic nematode *Heterorhabditis bacteriophora* (Rhabditida: Heterorhabditidae), *Environ. Entomol.*, 32, 392–396, 2003.

47. Walston, A.T., Held, D.W., Mason, N.R., and Potter, D.A., Absence of interaction between endophytic perennial ryegrass and susceptibility of Japanese beetle (Coleoptera: Scarabaeidae) grubs to *Paenibacillus popilliae*, *J. Entomol. Sci.*, 36, 105–108, 2001.

48. Alm, S.R., Yeh, T., Campo, M., Dawson, C.G., Jenkins, E., and Simeoni, A., Modified trap designs and heights for increase captures of Japanese beetle adults (Coleoptera: Scarabaeidae), *J. Econ. Entomol.*, 87, 775–780, 1994.

49. Alm, S.R., Yeh, T., Dawson, C.G., and Klein, M.G., Evaluation of trapped beetle repellency, trap height, and string pheromone dispensers on Japanese beetle captures (Coleoptera: Scarabaeidae), *Environ. Entomol.*, 25, 1274–1278, 1996.

50. Langford, G.S., Crostwait, S.L., and Whittington, F.B., The value of traps in Japanese beetle control, *J. Econ. Entomol.*, 33, 317–320, 1940.

51. Hamilton, D.W., Schwartz, P.H., Townsend, B.G., and Jester, C.W., Traps reduce an isolated infestation of Japanese beetle, *J. Econ. Entomol.*, 64, 150–153, 1971.

52. Wawrzynki, R.P. and Ascerno, M.E., Mass trapping for Japanese beetle (Coleoptera: Scarabaeidae) suppression in isolated areas, *J. Arboric.*, 24, 303–307, 1998.

53. Lacey, L.A., Kaya, H.K., and Bettencourt, R., Dispersal of *Steinernema glaseri* (Nematoda: Steinernematidae) in adult Japanese beetles, *Popillia japonica* (Coleoptera: Scarabaeidae), *Biocontrol Sci. Technol.*, 5, 121–130, 1995.

54. Klein, M.G. and Lacey, L.A., An attractant trap for autodissemination of entomopathogenic fungi into populations of the Japanese beetle *Popillia japonica* (Coleoptera: Scarabaeidae), *Biocontrol Sci. Technol.*, 9, 151–158, 1999.

55. López, R. and Potter, D.A., Ant predation on eggs and larvae of the black cutworm (Lepidoptera: Noctuidae) and Japanese beetle (Coleoptera: Scarabaeidae) in turfgrass, *Environ. Entomol.*, 29, 116–125, 2000.

56. Terry, L.A., Potter, D.A., and Spicer, P.G., Insecticides affect predatory arthropods and predation on Japanese beetle (Coleoptera: Scarabaeidae) eggs and fall armyworm (Lepidoptera: Noctuidae) pupae in turfgrass, *J. Econ. Entomol.*, 86, 871–878, 1993.

57. Zenger, J.T. and Gibbs, T.J., Identification and impact of egg predators of *Cyclocephala lurida* and *Popillia japonica* (Coleoptera: Scarabaeidae) in turfgrass, *Environ. Entomol.*, 30, 425–430, 2001.

58. Hays, J., Clopton, R.E., Cappaert, D.L., and Smitley, D.R., Revision of the genus *Stictospora* and description of *Stictospora villani*, n. sp. (Apicomplexa: Eugregarina: Actinocephalidae) from larvae of the Japanese beetle, *Popillia japonica* (Coleoptera: Scarabaeidae), in Michigan, *J. Parasitol.*, 90, 1450–1456.

59. Hanula, J.L. and Andreadis, T.G., Parasitic microorganisms of Japanese beetle (Coleoptera: Scarabeidae) and associated scarab larvae in Connecticut soils, *Environ. Entomol.*, 17, 709–714, 1998.

60. Cockfield, S.D. and Potter, D.A., Predation on sod webworm (Lepidoptera: Pyralidae) eggs as affected by chlorpyrifos application to Kentucky bluegrass turf, *J. Econ. Entomol.*, 77, 1542–1544, 1984.

61. Vavrek, R.C. and Niemczyk, H.D., Effect of isofenphos on nontarget invertebrates in turfgrass, *Environ. Entomol.*, 19, 1572–1577, 1990.

62. Kunkel, B.A., Held, D.W., and Potter, D.A., Impact of halofenozide, imidacloprid, and bendiocarb on beneficial invertebrates and predatory activity in turfgrass, *J. Econ. Entomol.*, 92, 922–930, 1999.

63. Kunkel, B.A., Held, D.W., and Potter, D.A., Lethal and sublethal effects of bendiocarb, halofenozide, and imidacloprid on *Harpalus pennsylvanicus* (Coleoptera: Carabidae) following different modes of exposure in turfgrass, *J. Econ. Entomol.*, 94, 60–67, 2001.

64. Rogers, M.E. and Potter, D.A., Effects of spring imidacloprid application for white grub control on parasitism of Japanese beetle (Coleoptera: Scarabaeidae) by *Tiphia vernalis* (Hymenoptera: Tiphiidae), *J. Econ. Entomol.*, 96, 1412–1419, 2003.

65. Braman, S.K., Pendley, A.F., and Corley, W., Influences of commercially available wildflower mixes on beneficial arthropod abundance and predation in turfgrass, *Environ. Entomol.*, 31, 564–572, 2002.

66. Rogers, M.E. and Potter, D.A., Potential for sugar sprays and flowering plants to increase parasitism of white grubs (Coleoptera: Scarabaeidae) by typhiid wasps (Hymenoptera: Tiphiidae), *Environ. Entomol.*, 33, 619–626, 2004.

67. Rogers, M.E. and Potter, D.A., Biology of *Tiphia pygidialis* (Hymenoptera: Tiphiidae), a parasitoid of masked chafer (Coleoptera: Scarabaeidae) grubs, with notes on the seasonal occurrence of *Tiphia vernalis* in Kentucky, *Environ. Entomol.*, 33, 523–527, 2004.

68. Tanada, Y. and Kaya, H.K., *Insect Pathology*, Academic Press, San Diego, 1993.

69. Siegel, J.P., Bacteria, In *Field Manual of Techniques in Invertebrate Pathology*, Lacey, L.A. and Kaya, H.K., Eds., 2000, 209–230.

70. Goettel, M.S., Inglis, G.D., and Wraight, S.P., Fungi, In *Field Manual of Techniques in Invertebrate Pathology*, Lacey, L.A. and Kaya, H.K., Eds., 2000, 255–282.

71. Koppenhöfer, A.M., Nematodes, In *Field Manual of Techniques for the Application and Evaluation of Entomopathogens*, Lacey, L.A. and Kaya, H.K., Eds., Kluwer, Dordrecht, The Netherlands, 2000, 283–301.

72. Dutky, S.R., The milky diseases, In *Insect Pathology: An Advanced Treatise, Vol. 2*, Steinhaus, E.A., Ed., Academic Press, New York, 1963, 75–115.

73. Fleming, W.E., *Biological Control of the Japanese Beetle*, USDA Technical Bulletin 1383, U.S. Printing Office, Washington, DC, 1968.

74. Fleming, W.E. and T.L. Ladd, *Milky Disease for Control of Japanese Beetle*, U.S. Department of Agriculture Leaflet 500, U.S. Department of Agriculture, Washington, DC, 1979.

75. Klein, M.G., Use of *Bacillus popilliae* in Japanese beetle control, In *Use of Pathogens in Scarab Pest Management*, Glare, T.R. and Jackson T.A., Eds., Intercept, Andover, UK, 1992, 179–189.

76. Redmond, C.T. and Potter, D.A., Lack of efficacy of *in vivo-* and putatively *in vitro*-produced *Bacillus popilliae* against field populations of Japanese beetle (Coleoptera: Scarabaeidae) grubs in Kentucky, *J. Econ. Entomol.*, 88, 846–854, 1995.

77. Suzuki, N., Hori, H., Tachibana, M., and Asano, S., *Bacillus thuringiensis* strain Buibui for control of cupreous chafer, *Anomala cuprea* (Coleoptera: Scarabaeidae), in turfgrass and sweet potato, *Biol. Control*, 4, 361–365, 1994.

78. Alm, S.R., Villani, M.G., Yeh, T., and Shutter, R., *Bacillus thuringiensis japonensis* strain Buibui for control of Japanese and oriental beetle larvae (Coleoptera: Scarabaeidae), *Appl. Entomol. Zool.*, 32, 477–484, 1997.

79. Asano, S., Yamashita, C., Iizuka, T., Takeuchi, K., Yamanaka, S., Cerf, D., and Yamamoto, T., A strain of *Bacillus thuringiensis* subsp. *galleriae* containing a novel *cry8* gene highly toxic to *Anomala cuprea* (Coleoptera: Scarabaeidae), *Biol. Control*, 28, 191–196, 2003.

80. Jackson, T.A., Huger, A.M., and Glare T.R., Pathology of amber disease in the New Zealand grass grub *Costelytra zealandica* (Coleoptera: Scarabaeidae), *J. Invertebr. Pathol.*, 61, 123–130, 1993.

81. Jackson, T.A., Pearson, J.F., O'Callaghan, M.O., Mahanty, H.K., and Willocks, M.J., Pathogen to product-development of *Serratia entomophila* (Enterobacteriaceae) as a commercial biological control agent for the New Zealand Grass grub (*Costelytra zealandica*), In *Use of Pathogens in Scarab Pest Management*, Jackson, T.A. and Glare, T.R., Eds., Intercept, Andover, UK, 1992, 191–198.

82. Keller, S., Use of *Beauveria brongniartii* and its acceptance by farmers, *IOBC/wprs Bull.*, 23, 67–71, 2000.

83. Enkerli, J., Widmer, F., and Keller, S., Long-term field persistence of *Beauveria brongniartii* strains applied as biocontrol agents against European cockchafer larvae in Switzerland, *Biol. Control*, 29, 115–123, 2004.

84. Rath, A.C., Worledge, D., Koen, T.B., and Rowe, B.A., Long-term field efficacy of the entomogenous fungus *Metarhizium anisopliae* against the subterranean scarab, *Adoryphorus couloni*, *Biocontrol Sci. Technol.*, 5, 439–451, 1995.

85. Jaros-Su, J., Groeden, E., and Zhang, J., Effects of selected fungicides and the timing of fungicide application in *Beauveria bassiana*-induced mortality of the Colorado potato beetle (Coleoptera: Chrysomelidae), *Biol. Control*, 15, 259–269, 1999.

86. Klein, M.G., Biological control of scarabs with entomopathogenic nematodes, In *Nematodes and the Biological Control of Insect Pests*, Bedding, R., Akhurst, R., and Kaya, H.K., Eds., CSIRO Press, East Melbourne, Australia, 1993, 49–58.

87. Grewal., P.S., Koppenhöfer, A.M., and Choo, H.Y., Lawn, turfgrass, and pasture applications, In *Nematodes as Biocontrol Agents*, Grewal, P.S., Ehlers, R.-U., and Shapiro-Ilan, D.I., Eds., CABI Publishing, Wallingford, UK, 2005, 115–146.

88. Georgis, R., Koppenhöfer, A.M., Lacey, L.A., Bélair, G., Duncan, L.W., Grewal, P.S., Samish, M., Tan, L., Torr, P., and van Tol, R.W.H.M., Successes and failures in the use of parasitic nematodes for pest control, *Biol. Control*, 37, 103–123, 2006.

89. Koppenhöfer, A.M., Fuzy, E.M., Crocker, R., Gelernter, W., and Polavarapu, S., Pathogenicity of *Steinernema scarabaei*, *Heterorhabditis bacteriophora* and *S. glaseri* to twelve white grub species, *Biocontrol Sci. Technol.*, 14, 87–92, 2004.

90. Koppenhöfer, A.M., Grewal, P.S., and Fuzy, E.M., Virulence of the entomopathogenic nematodes *Heterorhabditis bacteriophora*, *H. zealandica*, and *Steinernema scarabaei* against five white grub species (Coleoptera: Scarabaeidae) of economic importance in turfgrass in North America, *Biol. Control.*, 38, 397–404, 2006.

91. Koppenhöfer, A.M. and Fuzy, E.M., Effect of white grub developmental stage on susceptibility to entomopathogenic nematodes, *J. Econ. Entomol.*, 97, 1842–1849, 2004.

92. Koppenhöfer, A.M. and Fuzy, E.M., Ecological characterization of *Steinernema scarabaei*: a natural pathogen of scarab larvae, *J. Invertebr. Pathol.*, 83, 139–148, 2003.

93. Sexton, S. and Williams, P., A natural occurrence of parasitism of *Graphognathus leucoloma* (Boheman) by the nematode *Heterorhabditis* sp., *J. Aust. Entomol. Soc.*, 20, 253–255, 1981.

94. Poinar, G.O., Jr., Jackson, T.A., and Klein, M.G., *Heterorhabditis megidis* sp. n. (Heterorhabditidae: Rhabditida), parasitic in the Japanese beetle, *Popillia japonica* (Scarabaeidae: Coleoptera), in Ohio, *Proc. Helminthol. Soc. Washington*, 54, 53–58, 1987.

95. Klein, M.G. and Georgis, R., Persistence of control of Japanese beetle (Coleoptera: Scarabaeidae) larvae with steinernematid and heterorhabditid nematodes, *J. Econ. Entomol.*, 85, 727–730, 1992.

96. Georgis, R., Kaya, H.K., and Gaugler, R., Effect of steinernematid and heterorhabditid nematodes (Rhabditida: Steinernematidae and Heterorhabditidae) on nontarget arthropods, *Environ. Entomol.*, 20, 815–822, 1991.

97. Bathon, H., Impact of entomopathogenic nematodes on non-target hosts, *Biocontrol Sci. Technol.*, 6, 421–434, 1996.

98. Kaya, H.K. and Hotchkin, P.G., The nematode *Neoaplectana carpocapsae* Weiser and its effect on selected ichneumonid and braconid parasites, *Environ. Entomol.*, 10, 474–478, 1981.

99. Zaki, F.N., Awadallah, K.T., and Gesraha, M.A., Competitive interaction between the braconid parasitoid *Meteorus rubens* Nees and the entomogenous nematode, *Steinernema carpocapsae* (Weiser) on larvae of *Agrotis ipsilon* Hufn. (Lepidoptera, Noctuidae), *J. Appl. Entomol.*, 121, 151–153, 1997.

100. Lacey, L.A., Unruh, T.R., and Headrick, H.L., Interactions of two idiobiont parasitoids (Hymenoptera: Ichneumonidae) of codling moth (Lepidoptera: Tortricidae) with the entomopathogenic nematode *Steinernema carpocapsae* (Rhabditida: Steinernematidae), *J. Invertebr. Pathol.*, 83, 230–239, 2003.

101. Koppenhöfer, A.M. and Grewal, P.S., Compatibility and interactions with agrochemicals and other biocontrol agents, In *Nematodes as Biocontrol Agents*, Grewal, P.S., Ehlers, R.-U., and Shapiro-Ilan, D.I., Eds., CABI Publishing, Wallingford, U.K., 2005, 363–381.

102. Cardé, R.T. and Minks, A.K., Control of moth pests by mating disruption: successes and constraints, *Annu. Rev. Entomol.*, 40, 559–585, 1990.

103. Cardé, R.T., Principles of mating disruption, In *Behavior-Modifying Chemicals for Pest Management: Applications of Pheromones and Other Attractants*, Ridgway, R.L. and Silverstein, R.M., Eds., Marcel Dekker, New York, 1990, 47–71.

104. Leal, W.S., Chemical ecology of phytophagous scarab beetles, *Ann. Rev. Entomol.*, 43, 39–61, 1998.

105. Polavarapu S., Wicki, M., Vogel, K., Lonergan, G., and Nielsen, K., Disruption of sexual communication for oriental beetles (Coleoptera: Scarabaeidae) with a microencapsulated formulation of sex pheromone components in blueberries and ornamental nurseries, *Environ. Entomol.*, 31, 1268–1275, 2002.

106. Koppenhöfer, A.M., Polavarapu, S., Fuzy, E.M., Zhang, A., Ketner, K., and Larsen, T., Mating disruption of oriental beetle (Coleoptera: Scarabaeidae) in turfgrass using microencapsulated formulations of sex pheromone components, *Environ. Entomol.*, 34, 1408–1417, 1995.

107. Sciarappa, W.J., Polavarapu, S., Holdcraft, R.J., and Barry, J.D., Disruption of sexual communication of oriental beetle (Coleoptera: Scarabaeidae) in highbush blueberries with retrievable pheromone sources, *Environ. Entomol.*, 34, 54–58, 2005.

108. Wenninger, E.J. and Averill, A.L., Mating disruption of oriental beetle (Coleoptera: Scarabaeidae) in cranberry using retrievable, point-source dispensers of sex pheromone, *Environ. Entomol.*, 35, 458–464, 2005.

109. The Food Quality Protection Act (FQPA) of 1996, United States Environmental Protection Agency, Office of Pesticide Research, http://www.epa.gov/pesticides/regulating/laws/fqpa/, 1996.

110. Thurston, G.S., Kaya, H.K., Burlando, T.M., and Harrison, R.E., Milky disease bacterium as a stressor to increase susceptibility of scarabaeid larvae to an entomopathogenic nematode, *J. Invertebr. Pathol.*, 61, 167–172, 1993.

111. Thurston, G.S., Kaya, H.K., and Gaugler, R., Characterizing the enhanced susceptibility of milky disease-infected scarabaeid grubs to entomopathogenic nematodes, *Biol. Control*, 4, 67–73, 1994.

112. Koppenhöfer, A.M., Choo, H.Y., Kaya, H.K., Lee, D.W., and Gelernter, W.D., Increased field and greenhouse efficacy against scarab grubs with a combination of an entomopathogenic nematode and *Bacillus thuringiensis*, *Biol. Control*, 14, 37–44, 1999.

113. Ansari, M.A., Tirry, L., and Moens, M., Interaction between *Metarhizium anisopliae* CLO 53 and entomopathogenic nematodes for the control of *Hoplia philanthus*, *Biol. Control*, 31, 172–180, 2004.

114. Koppenhöfer, A.M., Brown, I.M., Gaugler, R., Grewal, P.S., Kaya, H.K., and Klein, M.G., Synergism of entomopathogenic nematodes and imidacloprid against white grubs: greenhouse and field evaluation, *Biol. Control*, 19, 245–251, 2000.

115. Koppenhöfer, A.M., Cowles, R.S., Cowles, E.A., Fuzy, E.M., and Baumgartner, L., Comparison of neonicotinoid insecticides as synergists for entomopathogenic nematodes, *Biol. Control*, 24, 90–97, 2002.

116. Koppenhöfer, A.M., Grewal, P.S., and Kaya, H.K., Synergism of imidacloprid and entomopathogenic nematodes against white grubs: the mechanism, *Entomol. Exp. Appl.*, 94, 283–293, 2000.

117. Koppenhöfer, A.M. and Fuzy, E.M., *Steinernema scarabaei* for the control of white grubs, *Biol. Control*, 28, 47–59, 2003.

118. Koppenhöfer, A.M. and Fuzy, E.M., Biological and chemical control of the Asiatic garden beetle, *Maladera castanea* (Coleoptera: Scarabaeidae), *J. Econ. Entomol.*, 96, 1076–1082, 2003.

119. Koppenhöfer, A.M., Cowles, R.S., Cowles, E.A., Fuzy, E.M., and Baumgartner, L., Effect of neonicotinoid insecticide synergists on entomopathogenic nematode fitness, *Entomol. Exp. Appl.*, 106, 7–18, 2003.

120. Potter, D.A., Prospects for managing destructive turfgrass insects without protective chemicals, *Int. Turfgrass Soc. Res. J.*, 10, 42–54, 2005.

21 Biology and Management of the Annual Bluegrass Weevil, *Listronotus maculicollis* (Coleoptera: Curculionidae)

Benjamin A. McGraw and Albrecht M. Koppenhöfer

CONTENTS

21.1 INTRODUCTION

The annual bluegrass weevil (*Listronotus maculicollis* Dietz, formerly *Hyperodes* sp. near *anthracina—anthracinus*) (ABW) is the single most destructive insect pest on golf-course turfgrass in many states in the northeastern United States. Damage from the weevil is especially evident in short-mowed areas on tees, fairways, collars, and greens with a high percentage of annual bluegrass (*Poa annua*). ABW injury to turfgrass was first reported in Connecticut in 1931 and until the past 15 years was concentrated around the metropolitan area of New York including northeastern New Jersey and southwestern Connecticut.[1] Severe infestations are now being reported from

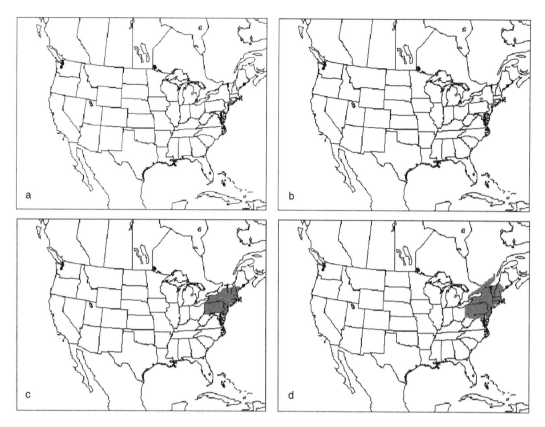

FIGURE 21.1   Known ABW distribution as of (a) 1971, (b) 1980, (c) 1999, and (d) 2005. (Adapted from Cameron, R.S. and Johnson, N.E., *J. Econ. Entomol.*, 64, 689, 1971; Vittum, P.J., Ph.D. thesis, Cornell University, Ithaca, NY, 1980; Vittum, P.J., et al., *Turfgrass Insects of the United States and Canada*, 2nd ed., Cornell University Press, Ithaca, NY, 1999, p. 230.)

all other states of the Northeast (Delaware, Massachusetts, Maine, New Hampshire, New Jersey, New York, Pennsylvania, Rhode Island, and Vermont), west into Ontario, and north into Quebec. In 2004, problems were reported for the first time in Maryland, representing the southern front of the weevil's expanding range of impact in the mid-Atlantic (Figures 21.1a through 21.1d).

Turfgrass managers around the epicenter of the weevil's distribution (the New York metropolitan area, northern New Jersey, and southern Connecticut) go to great extents to maintain pest densities below damaging levels. On some golf courses, as many as six applications of chemical pesticides a year may be made to reduce ABW populations.[2] Most management programs seek to control adults as they emerge from overwintering sites in spring in rough or leaf litter adjacent to fairways, greens, and tees.

Adult ABW cause only limited damage to the grass by chewing small holes or notches along the edge of the leaf blade when feeding and by chewing small holes into the stem to deposit eggs. Some yellowing of the plant will occur from adult feeding, however, severe damage arises from larval feeding and tunneling through the plant which will ultimately destroy the crown and cause the plant to die. Large populations of larvae cause serious damage to turf stands with high proportions of annual bluegrass, often leaving large patches of bare turf for extended periods during the season (Figure 21.2). Occasionally, there are reports of ABW feeding on other cool-season grasses, especially creeping bentgrass cultivars, but ABW has a clear preference for *P. annua*.[3]

Currently, the most efficacious means of reducing the density of first-generation progeny and damage from their feeding is by applying synthetic insecticides, in particular pyrethroids,

FIGURE 21.2 **(Color Figure 21.2 follows p. 262.)** Damage to a golf-course fairway caused by high densities of first-generation ABW larvae.

preventatively against adults in spring. Over the past 40 years, golf-course superintendents managing ABW have relied heavily on the use of chemical pesticides to meet the high aesthetic expectations of their clientele. The increase in the number of chemical applications and the expanded area of impact of ABW suggest that chemical management is not a sustainable approach to controlling populations. The changes in management practices in the past 40 years, such as drastic reductions in mowing height and fertilizer, have pushed turf to physiological extremes to achieve high standards. It is quite possible that the same changes in turf management have expanded the weevil's impact by increasing the amount of preferred hosts in the environment and decreasing the tolerance of turf stands to weevil feeding.

Current research seeks to address questions concerning basic biology and ecology of ABW to manage populations in a sustainable manner. As the weevil gains significance in new regions and environments, further research is needed to address how the insect's biology and seasonal phenology are affected. Furthermore, the expansion of ABW's area of impact suggests that resistance to chemical pesticides may be occurring in some areas, placing an increased need for precision monitoring for timing of management practices, as well as emphasizing a greater need for integrated management tactics.

21.2 HISTORY AND CLASSIFICATION

The history of the annual bluegrass weevil on golf-course turf is unclear at best. It is not known exactly how long the weevil has been damaging turfgrass in the northeastern United States or whether it is a native or introduced species, although the former is believed to be true. It is also not clear whether *L. maculicollis* as we know it today or related weevils were found damaging golf-course turfgrass beginning in the 1930s. Some of the ambiguity in the history of the ABW is related to the classification and reorganization of the genera *Hyperodes* and *Listronotus* within the Curculionidae family.

The genus *Listronotus* is classified in the order Coleoptera, superfamily Curculionoidea, family Curculionidae, subfamily Cyclominae. There has been much revision of the genus over the years. O'Brien states that the two genera are synonymous and transfers 35 species of *Hyperodes* into *Listronotus* for a total of 62 species in the United States and Canada.[4,5] Morrone provides an extensive analysis of the New World Listroderina and distinctively separates *Listronotus* and

Hyperodes based on 53 morphological characters.[6] Most recently, Arnett et al. maintain the synonymous grouping and place 81 species from the United States and Canada in the genus *Listronotus*.[7]

In the mid-1960s it became apparent that small weevils were damaging turf on several golf courses in Nassau and Westchester counties in New York. Cameron and Johnson suggest that the insect could have been causing the "spring die-out" that had been observed on Long Island golf courses for nearly 40 years.[8] Britton was the first to report a species of "*Hyperodes*" damaging golf-course turf in Farmington, CT, but due to the number of different species classified as *Hyperodes* found in the golf-course environment it is unclear whether ABW was responsible.[1,8]

Warner was the first to identify ABW in samples taken in 1957 and 1961 from damaged turf from Long Island, NY, and Pennsylvania.[9] She classified the insects as *Hyperodes anthracinus* (Dietz). H. Dietrich examined the same specimens and identified the specimens from Long Island to be *H. maculicollis*, a species that is present in all 50 states.[8] Around the same time, Schread reported damage to turfgrass in Connecticut by two separate species, *H. maculicollis* and *H. anthrosionus* and became the first to refer to the weevils as "annual bluegrass weevil" for its apparent preference for *P. annua*.[10] Warner later revised her classification of Long Island specimens in the Cornell University collection to "represent a species between *H. anthracinus* and *H. maculicollis*," and thus designated the species to be *Hyperodes* sp. near *anthracinus*.[8] In 1985, O'Brien placed ABW in its current genus *Listronotus*, and classified the weevil as *L. maculicollis*.[11] Although this classification has remained for over 20 years, the alternative or common name "*Hyperodes* weevil" is still more commonly used by superintendents than "annual bluegrass weevil."

21.3   SEASONAL HISTORY AND DESCRIPTION OF STAGES

The generalized seasonal phenology of ABW life stages (with the exception of the egg stage) is depicted in Figure 21.3. The figure illustrates approximate times of the year and relative densities of ABW by life stage that can be found during the different periods of the season based on data generated from the center of the insect's current distribution. In the most southern or northern areas of distribution, the weevil's phenology may diverge substantially. Though ABW geographical distribution is not extensive, the lifecycle can differ significantly within its range and between years,

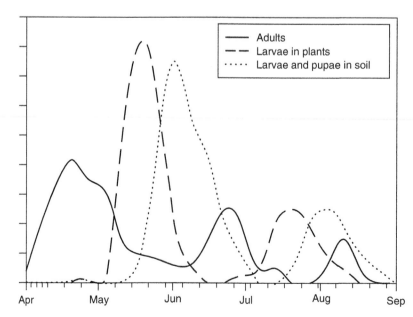

FIGURE 21.3   Generalized seasonal phenology of ABW stages in New Jersey. The first through third instars are typically found feeding within the plant, whereas the fourth and fifth instars and pupae are found in the soil.

largely based on the weather. The northern populations (central Connecticut to southern Vermont) most often will undergo two generations a year, whereas southern populations (central New Jersey to Maryland) may typically pass through three generations. In the epicenter of the distribution, three generations per year will occur in summers with above-average temperatures and three generations in colder summers.[11,12]

21.3.1   ADULT EMERGENCE—SPRING

The adult weevil, like many Curculionids, has a distinct, broad snout. The adults are approximately 3.5–5 mm in length and appear mottled, with tan scales against the underlying black elytra (Figure 21.4). The antennae of weevils in the genus *Listronotus* are attached to the apex of the snout and fold back toward the eyes in small grooves on the sides of the beak. ABW will play dead if disturbed, at which time the antennae will fold back into the grooves on the snout and the legs will fold under its body.

ABW spring emergence from overwintering sites occurs in mid to late April. Plant phenology is commonly used as an environmental indicator for estimating peak densities and timing chemical applications. Forsythia (*Forsythia* spp.) and flowering dogwood (*Cornus* spp.) do not have any direct effect on ABW development but are reliable indicators of adult presence on fairways.[13] Full bloom of forsythia indicates the initiation of the migration of adults, whereas once the full bloom of flowering dogwood has occurred, all adults will have moved out of their overwintering sites to food sources.

Rothwell determined that adults begin to move from overwintering sites in leaf litter off fairways by walking, rather than flying in search for hosts.[3] This was determined by significantly greater

FIGURE 21.4   (Color Figure 21.4 follows p. 262.) Adult ABW.

trap captures in pit-fall traps than on sticky traps designed to catch flying insects. Adults will move considerable distances from overwintering sites to food sources. ABW are infrequently detected in turf cut higher than 2.54 cm (1 in.). There can be a substantial distance between the preferred feeding sites of short-mowed fairway grasses and the overwintering sites in the leaf litter of trees on any given golf course. In our sampling of overwintering sites, we have found a majority of adults along the border of the rough and woods, 10–20 m away from the fairway, with considerable densities 30–50 m away from the nearest fairway. The effort to lay eggs in shorter turfgrasses appears to have consequences for larval fitness as larvae seem to develop better in shorter turfgrass.[3]

21.3.2 EGG LAYING

Once adults have reached the fairway, they spend their time looking for hosts, locating mates while occasionally feeding on the blades of grasses. The feeding damage to plants by adult weevils is not extensive, though yellowing of the leaves and notches can be seen in areas with high population densities. After mating, females chew small holes into the stem of the turfgrass plant, through one or two sheaths and deposit cylindrical eggs approximately 0.8 mm in length and 0.25 mm wide (Figure 21.5). The eggs first appear to be yellow and transparent, turning olive green to black before hatching. Eggs may be laid singly or end to end, in groups of up to six. Female weevils are capable of laying eggs for several weeks, resting for several days between oviposition events. Cameron and Johnson found that female ABW lay an average of 11.4 eggs in the laboratory but found upward of 62 oocytes in field collected females.[8] The eggs hatch in 4.6 days at 80°F in laboratory growth chambers.

21.3.3 LARVAL DEVELOPMENT AND PLANT DAMAGE

The larval stage consists of five instars, and time of development varies depending on the environmental conditions, particularly temperature and quality of food sources. Recently hatched neonate larvae are legless and creamy white. They burrow into the stem of the grass plant, protected from the outside environment while feeding internally. Larvae may feed in both directions along the stem and may exit the plant and reenter at another location along the stem. The larva will grow from

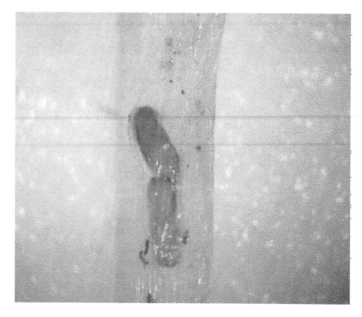

FIGURE 21.5 **(Color Figure 21.5 follows p. 262.)** Two ABW eggs deposited beneath the sheath and into the stem of a *P. annua* plant.

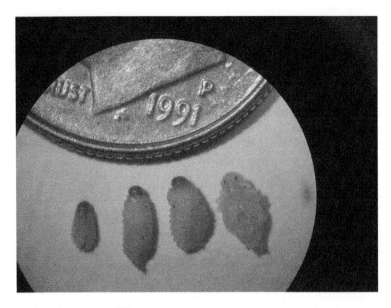

FIGURE 21.6 **(Color Figure 21.6 follows p. 262.)** ABW third, fourth, and fifth instars and pupa next to a 10 cent coin for comparison.

approximately 1 to 4.5 mm in length until fully mature, taking on an overall crescent shape and more rounded in the middle than the ends (Figure 21.6). Before pupation, the last instar larva turns into the prepupa, which is nearly identical to the fifth instar with the exception of losing several folds defining the thoracic segments.

The tunneling of the grass stem by early instar larvae causes yellowing of the plant. Late-instar larvae emerge from the plant into the soil and feed externally upon the crown and root system. Their feeding on the crown eventually kills the plant. Throughout its development, a typical larva can destroy between 12 and 20 stems.[8]

The spatial distribution of damage by the first-generation larvae is often localized around the perimeters of short-mowed turf (edges of fairways, collars, edges of tees, etc). If populations go untreated, high densities can damage a greater area of turf and the patches coalesce into large irregular patches several meters into the fairway from the rough border.

21.3.4 PUPAE AND TENERAL ADULTS

The prepupa readies the way for the initiation of the pupal stage by burrowing into the ground to 2–3 cm beneath the roots of the plant. To form an earthen cell around itself, the pupa performs a gyrating motion of the abdomen, while keeping the upper portion of body still, an activity that is referred to by many superintendents as the "*Hyperodes* dance." The pupa undergoes internal reconstruction, reforming body parts to bridge the development from the legless wormlike larva to the adult beetle. The pupa is initially creamy white, and then gradually darkens to reddish brown prior to molting to the adult stage.

The adult weevil emerges from within the pupal cuticle. Its cuticle is initially soft and reddish brown. The cuticle of this "teneral" or "callow" adult gradually turns black as it hardens. Cameron and Johnson determined that the amount of time spent as a nonfeeding pupa and teneral adult lasted approximately 10 days at 26.5°C in the laboratory.[8] After the cuticle hardens, the presence of mottled yellow to tan scales is apparent on the dorsal side of the insect, from the snout to the base of the abdomen. The adult then moves from its earthen cell to the soil surface to feed and mate. Over time, the scales are slowly worn off (most likely due to entering and reentering the thatch layer between periods of feeding), giving older adults a coal black appearance.

21.3.5 SUMMER GENERATIONS

The staggered emergence of the overwintering adults allows for oviposition to occur over several weeks. The developmental stages in the first generation are still somewhat synchronized. The first-generation adults spread out further and oviposit over a greater area. The developmental stages of the second generation are therefore even less synchronized and spread out over a larger area than in the first generation. Due to the second generation's greater spatial and temporal distribution, damage during summer is usually less severe and less localized than in spring. However, damage to *P. annua* can be extensive by the end of summer due to the combination of ABW feeding, disease pressure, and heat stress.

21.3.6 ADULTS—OVERWINTERING

The commencement of the adult weevil migration from the fairways to overwintering sites usually occurs in late August to early October in most areas. The journey is staggered within the population and it is not well understood what triggers the movement. It appears as though the weevils move to overwintering sites long prior to the onset of cold weather. Optimal host growth (*P. annua* trimodal seasonal root growth, cooler temperatures, and adequate moisture) coupled with warm days make it unclear as to why migration starts so early. Vittum found weevils in white pine leaf litter as early as mid-July in Westchester County, NY.[12] All of the adults that will successfully overwinter will be in the top 2 cm of soil and within the leaf litter and tall rough bordering infested sites by the end of October to early November.

Over the past 40 years several researchers have investigated the overwintering preferences of ABW. Prior to the late 1970s, it was thought that the primary overwintering sites were in tall rough and tufts of fescues. Vittum found extremely high densities of overwintering adults in white pine (*Pinus strobus*) leaf litter adjacent to infested fairways in Westchester, NY, and Fairfield, CT, counties.[12] Densities of weevils reached up to 450 adults per square meter under white pines and were significantly higher than in roughs and deciduous leaf litter in a 5-year study. More recent research suggests that this is not the case for all golf courses.[14,15] The preference for overwintering site may be influenced by the type of leaf litter but also by many site-specific variables including soil type and proximity to hosts.

Several studies have concentrated on the reproductive seasonality of ABWs in regard to overwintering.[3,8,12] Reproductive diapause has been shown to be variable in other *Listronotus* species, in particular, the Argentine stem weevil (ASW), *L. bonariensis*, a destructive pest of pasture ryegrasses in New Zealand and the carrot weevil (CW), *L. oregonesis*.[16,17] ASW, which was introduced to New Zealand from South America, requires increasing day lengths to become reproductively active. Conversely, CW experiences greater extremes in temperature over the course of the year and requires a minimum threshold of temperature units and increasing photoperiod before becoming reproductively active. Cameron and Johnson measured the size of ABW female reproductive structures and grouped them based on relative size.[8] They found two peaks in the year where females possessed large (well-developed) reproductive systems. Nearly 100% of females studied from Long Island, NY collections had large reproductive systems in mid-April to mid-June, with a smaller peak of approximately 50% of the females in mid-July. The reproductive system's size was also tightly correlated with sperm presence. The size of male prostrate glands showed similar trends. Vittum sought to use the seasonal peak in reproductive structures as a way of determining the number of ABW generations and was able to correlate the trends to the peaks in larval activity.[12]

21.4 HOST PLANT PREFERENCES

It was long believed that ABWs fed solely upon *P. annua*. *P. annua* is a highly invasive, bunch-type grass weed that is extremely prevalent in temperate regions such as the northeastern United States. It is considered a "winter annual" since it completes its life cycle in 1 year, producing seed from which it germinates in fall, and dying in the following summer. However, constant irrigation coupled with low mowing heights typically found on golf-course greens, tees, and fairways have turned this winter annual weed into a year-round problem. The invasion of *P. annua* can cause visible clumps of light green grass among darker grasses such as bentgrasses (*Agrostis* spp.) or Kentucky bluegrass (*Poa pratensis*). Damage resulting from ABW feeding in spring will often be in the same irregular clumping patterns along the edges of fairways where *P. annua* is densest, leaving patches of undamaged bentgrasses. Apart from undesirable aesthetic qualities, *P. annua* performs poorly in heat and drought periods during summer and requires intensive labor inputs (i.e., syringing and fungicides) to maintain healthy turf.

Only recently has ABW host preference for turfgrasses other than *P. annua* been examined. Rothwell monitored ABW in fairway turfgrasses and coupled her findings with small plot trials and laboratory choice and no-choice trials to determine host preference.[3] She found significantly more larvae in *P. annua* than in two creeping bentgrasses (*Agrostis stolonifera* c.v. L93 and c.v. 3way), Kentucky bluegrass (*Poa pratensis*), or a Kentucky bluegrass and perennial ryegrass (*Lolium perenne*) mix in choice and no-choice small plot trials. There was considerable variation in the density of ABW larvae between bentgrass cultivars, but Penncross, Penntrio, and Southshore contained fewer larvae than 3way and L93. Larval weight, a measure of performance, was significantly greater in *P. annua* plots than in the creeping bentgrass cultivars. Turf ratings in plots containing weevils showed that *P. annua* plots were also the lowest quality by the end of a 6-week study in choice and no-choice small plot trials.

Rothwell also determined that fertilizer and mowing height affect larval abundance in different cultivars.[3] She found more larvae in short-mowed grass than at higher cuts, yet no differences were found in larval weight. Damage by first-generation weevils in spring usually occurs on the borders between short- and tall-mowed turfgrasses such as between fairways and the rough or between collars and the rough. It is unclear why height of cut affects ABW, but it may affect the nutritional value of the grass. *P. annua* differed significantly from other grasses in the study as it had the lowest nitrogen content. Similarly, fertilizer was consistently negatively correlated with ABW abundance in all the tested cultivars. It appears that taken together, fertilizer, mowing height, and cultivar affect ABW host selection, but more work needs to be conducted on ABW ecology and behavior to accurately assess to what degree each contributes.

21.5 MANAGEMENT

21.5.1 Integrated Pest Management

Integrated pest management (IPM) is a philosophy that has been practiced for decades in production agriculture and more recently in golf-course management. Researchers and practitioners have had many definitions for IPM, which has led to confusion about what it actually is. Prokopy and Kogan define IPM as "a decision-based process involving the coordination of multiple tactics for optimizing the control of all classes of pests in an ecologically and economically sound manner."[18] Although many disagree on what the definition should contain, most agree that IPM is the concept that crops can be grown in an ecologically sustainable manner through the combination of cultural, biological and, when necessary, chemical pest-control methods.[19] IPM strives to reduce chemical pesticide inputs, but is not synonymous with organic agriculture, which is based on the elimination of chemical pesticides.

In the relatively brief history of ABW management, little progress has been made in regard to IPM. The intolerance to visible damage to high-valued turf has led superintendents to be dependent on preventative management through chemical pesticides. This dependence has steadily increased over the past 30 years. Where once a single springtime application was sufficient, now up to six applications are used in some operations.[3,13,20] Newer chemistries developed in the past 20 years have made significant increases in environmental safety without sacrificing efficacy, giving superintendents less urgency to use alternative solutions to ABW management. The pipeline of chemical products has provided superintendents "cheap insurance" when combating adults preventatively; however, it has not been able to reduce ABW populations to a long-term, sustainable, and manageable density in any region. In fact, the increase in applications and the pest's geographic area of impact in the past 10 years, suggest that chemical management alone is not a sustainable approach. The future holds promise for the integration of many different tools in combination with chemicals. As of late biological controls, cultural controls, and herbicides to reduce *P. annua* density have been given more attention as an ecological approach to suppress ABW populations.

21.5.2 CHEMICAL MANAGEMENT

Historically, ABW management has revolved around the use of broad-spectrum chemical pesticides targeting adults emerging in spring. In the past 10 years, this preventative approach has largely relied on the pyrethroid class of pesticides, although newer chemistries from other classes (i.e., neonicotinoids and anthranilic diamides) have recently shown promising efficacy against ABW.

Early control efforts with chlorinated hydrocarbons were largely unsuccessful since biological information and seasonal history of the weevil was unknown.[10,20] Most common failures resulted from applying broad-spectrum pesticides too late in the season, after the insect had caused damage to the turf and had pupated in the soil. After information on the biology of ABWs was gained, large-scale laboratory toxicity trials were conducted against the adults. Cameron and Johnson assayed 25 different compounds belonging to the organophosphate (OP), carbamate, and chlorinated hydrocarbon classes of pesticides.[20] OPs resulted in the lowest LD_{50}s in the laboratory and were also effective when tested in small plot trials on golf courses and grass tennis courts. The use of OPs for ABW management lasted until pyrethroids replaced them in the late 1980s.

The currently available chemical pesticides span several insecticide classes and differ in their ability to work preventatively or curatively. Rotations of chemicals by insecticide class are rarely employed even though several classes of pesticides are available to use at different times throughout the season. Most turf managers opt to apply pyrethroids to achieve adequate suppression of the springtime adults, several times if necessary, to achieve high levels of control, even though many of the modern ABW insecticides offer long residual control.

We have summarized data from insecticide efficacy tests published between 1993 and 2003 (Arthropod Management Tests Vols. 18–28) conducted by university researchers in the Northeast combined with our own unpublished data for 2004–2006 (Figure 21.7). The summary shows that pyrethroids were the most effective insecticides with no significant difference among the different compounds with an overall average of 90% control. The average control rates for the pyrethroids were 93% for bifenthrin (Talstar), 87% for cyfluthrin (Tempo), 84% for deltamethrin (DeltaGard), and 97% for lambda-cyhalothrin (Scimitar).

The timing of pesticide application is believed to be critical to achieve the highest levels of control and limit the number of applications needed. It is presently recommended to apply pyrethroids against the emerging adults between full bloom of forsythia and full bloom of flowering dogwood. However, our summary revealed no difference between pyrethroid applications in late April (April 15–May 3; 89%) and early May (May 4–May 15; 93%). In addition, recent research has shown no significant difference in bifenthrin efficacy between applications in late April (87%), early May (93%), or late May (80%), indicating that at least this pyrethroid has also good efficacy in curative applications against the larvae.

The OP chlorpyrifos (Dursban), the old standard for ABW control, appears to be more effective when applied in early May (83%) or late May (83%) than in late April (62%), suggesting a combined effect on adults and larvae. Another OP, trichlorfon (Dylox), was ineffective when applied in late April (0%) and early May (25%) but provided 79% control as a curative in late May.

Unfortunately, at least under high ABW pressure, most of the newer less-hazardous chemistries appear to lack the efficacy and consistency to replace pyrethroid applications, i.e., the neonicotinoids

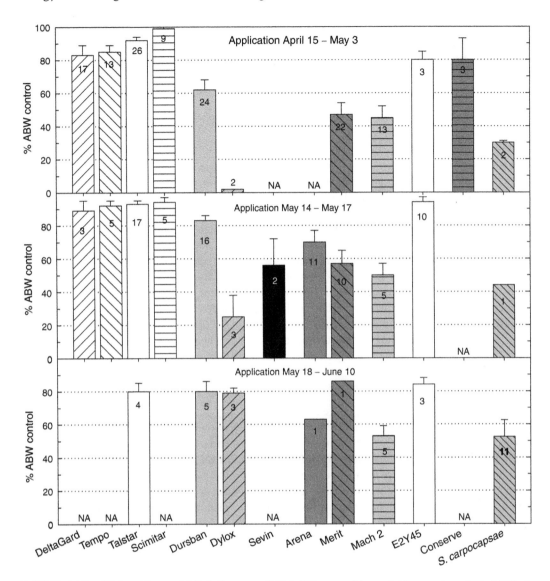

FIGURE 21.7 (Color Figure 21.7 follows p. 262.) Efficacy of insecticides used against ABWs in the northeastern United States. Shown are means ±SE. Numbers in bars are number of treatments included in summary. (*Summary from Arthropod Management Tests*, vols. 18–28; Koppenhöfer, unpublished data, 2005.)

imidacloprid (Merit) (52% control) and clothianidin (Arena) (70%, but variable and limited data thus far) and the insect growth regulator halofenozide (Mach 2) (48%). However, chlorantraniliprole, a new compound from a new insecticide class, the anthranilic diamides, has shown great promise in recent trials with 80%, 94%, and 84% control when applied in late April, early May, and late May, respectively.

In some areas, more applications are needed against later-generation adults and larvae. Multiple sprays against each weevil generation strongly suggest the development of insecticide resistance, particularly to the predominantly used pyrethroids. It is recommended that superintendents with high ABW populations regularly rotate insecticides from different insecticide classes, adhere to recommended label rates, and avoid "wall-to-wall" applications. It is quite common for two if not all three recommendations to not be followed. Early lab data from Connecticut suggest that the selection of pyrethroid-resistant ABW is a reality.[21] Preventative treatments can be conducted in an environmentally sensitive manner and in an IPM framework if data are collected on where infestations have occurred in the past and only treating areas of the highest value (greens and high-visibility turf). Insecticide resistance, the exposure to the public, and environmental fate are concerns that have researchers looking for alternative methods to control ABW stages, including biological and microbial controls.

21.5.3 BIOLOGICAL CONTROL

Biological control is the use of predators, parasites, or microbes to suppress pest populations below damaging thresholds. It has gained more attention as a component of turf IPM after successes have been observed when controlling pests such as billbugs, mole crickets, and black cutworms.[22–24] Biological control seeks to employ natural enemies against susceptible life stages to provide long-term suppression within an area. Researchers seek to use natural enemies that are specific to the pest biology and ecology, usually determined by culturing agents from naturally infected hosts. In the case of ABW, there are no known specific natural enemies of any life stage and therefore the use of biological controls for the suppression of populations would require the use generalist predators and parasites either inundatively (similar to conventional pesticide applications) or through a one-time inoculation, if the control organisms are able to recycle within the environment. The expansion of the distribution and the lack of effective natural enemies seem to challenge the belief of the primitive origin of ABW in the northeastern United States. It may, however, also be that changes in golf-turf management have increasingly reduced the impact of any existing natural enemies on ABW populations.

21.5.3.1 Entomopathogenic Nematodes

Recent ABW surveys in New Jersey have been able to identify entomopathogenic nematodes (EPNs) infecting large numbers of ABW larvae and pupae.[15] EPNs are microscopic, nonsegmented roundworms that are obligate parasites of insects. They are common in soil environments around the world and can infect and kill a wide array of insects. Part of their lifecycle is spent as a free-living juvenile that does not require feeding. These "dauer" or infective juveniles (IJs) seek out and move toward chemical cues given off by the insect. Once in contact with the insect, the IJs enter through natural openings (mouth, anus, or spiracles) or in some instances penetrate through the cuticle. After penetrating into the host's body cavity, the IJs release a symbiotic bacterium, and bacteria and nematode cooperate to kill the insect in 1–3 days. The nematodes feed on the multiplying bacteria and host tissues metabolized by the bacteria and develop through 1–3 generations. Depletion of food resources in the host cadaver triggers the development of a new cohort of IJs that emerge from the host cadaver in search of new hosts.

EPNs are attractive to pest managers for their potential to recycle within the environment, compatibility with many chemical pesticides and ability to be used in standard spray equipment. Commercially available EPNs have been used against ABW in the past, but limited efficacy was observed against the adult stage, which is at least in part attributable to low soil temperatures in spring. Adequate to excellent control was achieved when EPNs were used curatively against first-generation larvae (Figure 21.4). The success of EPN applications relies on optimal matching of nematode species to the weevil life stages and the environmental conditions (particularly soil temperature) present in the field at the time of application. The range of conditions is likely to differ drastically depending on the generation being targeted and the region where the control efforts are taking place. In addition, EPNs will perform better if moderate soil moisture and high relative humidity in the turf canopy can be maintained for at least 1 week after application. For further details, see Chapter 19.

Our ongoing efforts with EPNs aim at understanding the role of natural EPN populations in ABW population dynamics, finding the best nematodes species for the control of ABW at different times of the years, and ultimately provide sustainable suppression of ABW populations on golf-course fairways. A limited number of tests against ABW indicate that *Steinernema carpocapsae* is more effective when applied as a curative against the larvae in late May (51%) than against the adult in late April or early May (30% and 44%, respectively) (Figure 21.7). Since *S. carpocapsae* has achieved up to 98% in late-May applications, we are planning to further investigate this species, as we believe that we may be able to optimize its use. In addition, we are planning to test various other nematode species that are already commercially available, i.e., *Heterorhabditis bacteriophora*, *H. megidis*, *S. feltiae*, and *S. kraussei*. The latter three species are active at lower temperatures than *S. carpocapsae* and may therefore also be tested against adult ABWs in late April/early May. We have initiated laboratory studies to determine the virulence of these EPN species against the various ABW life stages and to test the effect of temperature on their virulence. Ultimately, we are planning to field-test the most promising EPNs under field conditions.

A statewide survey of New Jersey golf courses was initiated in June of 2005 to find EPNs that naturally infect ABWs. Nematode field-isolates may prove to be better adapted to ABWs as a host and the golf-course environment and will therefore be included in our above-described virulence studies. Soil samples were collected from historically ABW infested sites on 11 golf courses in five counties in central and northern New Jersey. Seven of the sites had ABWs present in the samples and two of the sites contained larvae infected by EPNs. EPN-infected stages were late-instar larvae or pupae. Ninety-eight percent were infected by *Heterorhabditis* sp. (probably *bacteriophora*) and 2% by *Steinernema* sp. (probably *carpocapsae*). EPNs were detected in the soil of 29% of all samples with 34% of isolates being *Heterorhabditis* sp. (probably *bacteriophora*), and 66% *Steinernema* sp. (probably *carpocapsae*).

21.5.3.2 Parasitoids

Parasitoids are insects whose larvae develop by feeding on the bodies of a single arthropods (usually insects) host, either internally or externally. Published surveys of ABW seasonal dynamics have not identified host-specific parasitoids to date, most likely due to the difficulties in finding parasitized eggs or larvae within cryptic habitats.

Parasitic wasps have been used to successfully control the pasture pest Argentine stem weevil (ASW) in New Zealand. This species is very similar to the ABW in biology and host utilization. McNeil et al. examined the potential for the braconid wasp, *Microtonus hyperodae*, a natural enemy of ASW and agent used in a classical biological control program, to control ABW larvae.[2] The braconid parasitized ABW larvae but parasitism rates were relatively low and the wasps were encapsulated and killed by the host.

21.5.4 Cultural Management

The most sustainable approach to managing ABW in the golf-course environment in an IPM context should begin with the management of *P. annua* rather than preventatively treating the symptoms from feeding damage. Reducing the amount of the *P. annua* in historically problematic areas should lead to reduced management inputs needed to maintain *P. annua* (i.e., increased fertilization, irrigation) as well as reducing the area and number of pesticide applications against ABW. Removing *P. annua* from stands where it comprises a high percentage of the total grasses is a difficult task and should be designed to reduce the weed over several seasons.

Newly developed herbicides and plant growth regulators (PGRs) allow superintendents to, in some cases, selectively remove *P. annua* from preferred grass stands. Vargas and Turgeon provide a complete list of the chemicals that can be used to suppress *P. annua* while trying to promote various turfgrasses.[25] PGRs are chemicals that alter the growth and development of a plant; however, the timing and strategy for targeting *P. annua* vary greatly (i.e., decreasing vegetative growth and competitiveness versus seedhead suppression). Occasionally, phytotoxicity to the selected grasses occurs with the use of herbicides and PGRs, and therefore the threshold for damage to turf stands must be determined prior to application. Biological controls for *P. annua* have been investigated, but have provided inconsistent control. Researchers have looked to host-specific plant pathogens to suppress *P. annua* in mixed stands. *Xanthomonas campestris* pv. *poannua* has been marketed as a "bioherbicide" inducing bacterial wilt on the plant and causing it to die.[26] Results have proven to be quite variable between isolates and field trials have not shown the same levels of control experienced in growth chambers.[27]

Cultivation practices such as core removal and "drill and fill" coupled with overseeding desired turfgrasses can help reduce *P. annua* densities in greens and fairways. The cultural practices have the added benefits of decreasing compaction, increasing water infiltration, and supplying oxygen to the root zone. As with chemical options, the timing of cultivations practices is critical. Coring should be done prior to the production of seedheads to reduce the seed bank over time and to allow bentgrass to have a competitive advantage.

21.6   MONITORING

Scouting for ABW is essential to the proper timing of management decisions, but often is not practiced in favor of less intensive management practices such as the use of calendar-based and long residual insecticide applications. Many factors complicate obtaining ABW population estimates, such as nocturnal movement, cryptic habitats, and small size of early instar larvae. The collection of data can be time consuming but is critical to employing proper strategies at the right time, reducing harm to nontarget and beneficial organisms, lessening exposure to humans, and decreasing costs associated with chemical treatments.

21.6.1   ADULTS

Many turfgrass managers rely on the use of environmental indicators, namely, plant phenology, to time the first pesticide application targeting emerging overwintering adults. The initiation of migration of the adults onto fairways coincides with *Forsythia* spp. full bloom and terminates with the full bloom of flowering dogwood (*Cornus spp.*) and redbud (*Cercis canadensis*).[11,28] Research has been conducted on developing degree-day models for timing overwintering adult emergence and egg laying as well as summer generation stage development.[3,12] The threshold temperatures for egg and larval development as well as adult movement was determined in the laboratory to be 13.3°C; however, temperature models have not been validated in the field.[3]

After adults have found their way into short turfgrass areas, flushes with household dishwashing detergents are effective in irritating adults hiding in the soil and thatch during the day to move to the surface. Soap flushes (1–2% soap in water) should be allowed to leach through the soil for 10–15 min to reach the adults and cause irritation. After the soap flush has been performed, the grass should be rinsed thoroughly with water to avoid scalding from the sun.

Adult movement onto and within fairways can be determined by pitfall trapping and vacuuming. A modified leaf blower can be used to vacuum adults from fairways but does not effectively pick them up from higher-cut turf.[3] Rothwell found no correlation between numbers of adults and larval densities, probably due to the adults' ability to move great distances.

Adult populations on greens and collars can be monitored by examining the collection of clippings in mower baskets. Although examination of basket collections can be quite reliable in determining the presence of adults on the surface of the green, many more adults may be hiding in the thatch during the day. Golf-course fairways and most tee boxes are not mowed to a height that would promote large captures or accurate estimates of weevil adults.

Vittum explored the use of blacklights in detecting adult populations in spring and summer.[12] Although few captures were recorded in spring and fall generations, many adults from the summer generation were captured across several locations. Captures, when tabulated in half-month increments, showed that the adult peak density occurred in early to mid-July in Westchester County, NY. The peak captures were consistent with soil and surface vacuuming data in various studies from the region and therefore might prove to be an alternative method for timing summer applications against adults.[3,12] Rothwell examined early season adult movement from overwintering sites.[3] Her study demonstrated that significantly more adults move on to fairways by walking rather than flying. It is possible that energy reserves are depleted or that wing muscles are not fully developed when adults emerge, and therefore they must walk to hosts, making blacklight trapping less efficient in spring.

Adults can be sampled for in overwintering sites in late fall and early spring; however, outside of research purposes this is not a common nor practical approach to estimate weevil abundance. Samples consisting of the top few centimeters of soil and leaf litter from overwintering sites adjacent to fairways can be submerged in water in 5-gal buckets for upward of 2 h. A piece of paper towel is placed on the surface of the water. As weevils emerge from the soil, they cling to the sides of the bucket or walk across the towel.

21.6.2   IMMATURE STAGES

Monitoring for immature stages is not a widespread practice but can be useful in assessing the efficacy of preventative practices and new chemical products or gauging densities in historically infested areas. Several chemical and biological control options for the management of larvae and pupae are available, in development, or being investigated, which would require monitoring of immature populations. Most management practices attempt to avoid larval damage by reducing adult populations, but some golf courses choose to allow the first generation larvae to feed in lower-value turf (i.e., fairways and tees) to accurately target the insects with curative controls. This approach will not be accepted by many golf courses with high aesthetic thresholds; however, it can be an effective method of assessing population size and the need for control tactics in lower-value turf. Population densities of immature stages can be gauged by using a standard cup cutter. Turf cores are placed on a tray and teased apart by hand. Late instar larvae and pupae can be readily seen with the naked eye. Young larvae that are still within the plant can be extracted by placing the turf core fragments in a beaker containing a lukewarm saturated saltwater solution. Larvae become irritated by the salt and float to the top of the container as they try to escape. Early instar larvae are very small (approximately 1 mm when they hatch) and detection might require the aid of a hand lens or dissecting microscope.

21.7   CONCLUSIONS

Over the past 40 years, the ABW has emerged as a serious threat to the aesthetics and playability of golf courses in the northeastern United States. In fact, ABW are considered the most severe insect pest on many courses within this region. The expansion of the insect's area of impact and overreliance on preventative chemical pesticides are of utmost concern for turfgrass entomologists. With the increasing pressure from government agencies and the general public to reduce pesticide use on golf courses and the absence of any alternatives at this time, there is a dire need to develop effective ABW control options with reduced environmental and health hazards and that are more IPM-compatible.

The reliance on chemical pesticides to preventatively control damage is a tactic that is likely to lead to resistant populations. Sadly, even with alternatives, the adoption of new management practices will not be embraced by all turf managers who have long favored the ease of use and cheap insurance that synthetic pesticides offer. The future of ABW management holds great potential for the development of novel management strategies as we gain more insight into the insect's biology and ecology. Our current understanding of the pest is based on information collected in relatively few research projects focusing on the epicenter of the insect's distribution. The dispersal of the weevil into new regions calls for new ecological studies in these regions and demonstrates the need to discover new and more sustainable approaches to manage ABW.

REFERENCES

1. Britton, W.E., Weevil grubs injure lawns, *Bull. Conn. Agric. Exp. Stn.*, 338, 593, 1932.
2. McNeil, M.R., Vittum, P.J., and Baird, D.B., Suitability of *Listronotus maculicollis* (Coleoptera: Curculionidae) as a host for *Microtonus hyperodae* (Hymenoptera: Braconidae), *J. Econ. Entomol.*, 92, 1292, 1999.
3. Rothwell, N.L., Investigation into *Listronotus maculicollis* (Coleoptera: Curculionidae), a pest of highly maintained turfgrass, Ph.D. thesis, University of Massachusetts, Amherst, MA, 2003.
4. O'Brien, C.W., *Hyperodes*, a new synonym of Listronotus, with a checklist of Latin American species (Cylindrorhininae, Curculionidae, Coleoptera), *Southwest. Entomol.*, 4, 265, 1979.
5. O'Brien, C.W., The larger (4.5mm+) *Listronotus* of America north of Mexico (Cylindrorhininae, Curculionidae, Coleoptera), *Trans. Am. Entomol. Soc.*, 107, 69, 1981.
6. Morrone, J.J., Cladistics of the New World genera of *Listroderina* (Coleoptera: Curculionidae: Rhytirrhinini), *Cladistics*, 13, 247, 1997.
7. Arnett, R.H., Thomas, M.C., Skelley, P.E., and Frank, J.H., *American Beetles*, vol. 2, Polyphaga: Scarabeoidea to Curculionoidea, CRC Press, New York, 2002, p. 766.
8. Cameron, R.S. and Johnson, N.E., Biology of a species of *Hyperodes* (Coleoptera: Curculionidae), a pest of turfgrass, *Search Agric.*, Geneva, New York, 1, 1, 1971.
9. Warner, R.E., *Hyperodes anthracinus* (Dietz) damaging golf greens (Coleoptera: Curculionidae), *Coleopts. Bull.*, 19, 32, 1965.
10. Schread, J.C., The annual bluegrass weevil, *Conn. Agric. Exp. Sta. Circ. no. 238*, New Haven, CT, 1970, p. 1.
11. Vittum, P.J., Villani, M.G., and Tashiro, H., *Turfgrass Insects of the United States and Canada*, 2nd ed., Cornell University Press, Ithaca, NY, 1999, p. 230.
12. Vittum, P.J., The biology and ecology of the annual bluegrass weevil, *Hyperodes* sp. near *anthracinus* (Dietz) (Coleoptera: Curculionidae), Ph.D. thesis, Cornell University, Ithaca, NY, 1980.
13. Tashiro, H., Murdoch, C.L., Straub, R.W., and Vittum, P.J., Evaluation of insecticides on *Hyperodes* sp., a pest of annual bluegrass turf, *J. Econ. Entomol.*, 70, 729, 1978.
14. Diaz, M., personal communication, 2005.
15. Koppenhöfer, unpublished data, 2005.
16. Goldson, S.L., *Hyperodes bonariensis* biology and grass resistance (Coleoptera: Curculionidae), in *Proc. 2nd Australasian Conf. Grassl. Invert. Ecol.*, Crosby, T.K. and Pottinger, R.P., Eds., Government Printer Wellington, New Zealand, 1979, p. 262.
17. Ryser, B.W., Investigations regarding the biology and control of the carrot weevil, *Listronotus oregonensis* (LeConte), in New Jersey, M.S. thesis, Rutgers University, New Brunswick, NJ, 1975.
18. Prokopy, R. and Kogan, M., Integrated pest management, in *Encyclopedia of Insects*, Resh, V.W. and Cardé, R.T., Eds., Academic Press, San Diego, CA, 2003, p. 589.
19. Kogan, M., Integrated pest management: historical perspectives and contemporary developments, *Annu. Rev. Entomol.*, 43, 243, 1998.
20. Cameron, R.S. and Johnson, N.E., Chemical control of the "annual bluegrass weevil," *Hyperodes* sp. nr. *anthracinus*, *J. Econ. Entomol.*, 64, 689, 1971.
21. Cowles, R.S., personal communication, 2005.
22. Smith, K.A., Control of weevils with entomopathogenic nematodes, in *Control of Insect Pests with Entomopathogenic Nematodes*, Smith, K.A. and Hatsukade, M., Eds., Food and Fertilizer Technology Center, Taiwan, 1994, p. 1.
23. Parkman, J.P. and Smart Jr., G.C., Entomopathogenic Nematodes, a case study: introduction of *Steinernema scapterisci* in Florida, *Biocontrol Sci. Techn.*, 6, 413, 1994.
24. Georgis, R. and Poinar Jr., G.O., Nematodes as bioinsecticides in turf and ornamentals, in *Integrated Pest Management for Turfgrass and Ornamentals*, Leslie, A.R., Ed., Lewis Publishers, Boca Raton, FL, 1994, p. 477.
25. Vargas, J.M. and Turgeon, A.J. *Poa annua: Physiology, Culture and Control of Annual Bluegrass*, John Wiley and Sons, Hoboken, NJ, 2004, p. 56.
26. Johnson, B.J., Biological control of annual bluegrass with *Xanthomonas campestris* pv. *Poa annua* in Bermudagrass, *HortScience*, 29, 659, 1994.
27. Zhou, T. and Neal, J.C., Annual bluegrass (*Poa annua*) control with *Xanthomas campestris* pv. *Poa annua* in New York state, *Weed Tech.*, 9, 173, 1995.
28. Tashiro, H and Straub, R.W., Progress in the control of turfgrass weevil, a species of *Hyperodes*, *Down Earth*, 29, 8, 1973.

22 Understanding and Managing Plant-Parasitic Nematodes on Turfgrasses

William T. Crow

CONTENTS

22.1 INTRODUCTION

Nematodes comprise the phylum Nemata, the unsegmented roundworms. These are the most abundant multicellular life on earth, occurring in almost every habitat, including all turfgrass environments. The vast majority of nematodes are beneficial, helping to cycle carbon and nitrogen in the soil. Some nematodes are even used by turfgrass managers to protect their turf from insect pests. However, plant-parasitic nematodes feed on plants and some can cause considerable damage to turfgrasses. These nematodes are among the least understood and most difficult to manage of the turfgrass pests. In addition to the damage they cause directly, plant-parasitic nematodes can increase the need for water and fertilizer inputs [1–3], and increase potential problems from weeds [4].

22.2 BIOLOGY OF PLANT-PARASITIC NEMATODES

All nematodes are aquatic animals; those that attack turf live in microscopic films of water surrounding soil particles or in the fluids inside the plant. The plant-parasitic nematodes that parasitize turfgrasses are all microscopic in size, most ranging from 0.25 to 3 mm in length. For identification and diagnosis, plant-parasitic nematodes are extracted from soil or plant tissue into water and generally observed at 40× magnification or greater.

All plant-parasitic nematodes have a stylet or mouth spear that is used for feeding on plant cells. This is a hollow tube that is functionally similar to a hypodermic needle. While all plant-parasitic nematodes have stylets, there are also nonplant parasites that have stylets. Fungal feeding nematodes specifically have stylets very similar in form and function to those of plant-parasitic nematodes. Therefore, the simple presence of a stylet does not always mean that a particular nematode is a plant parasite. The stylet is connected to esophageal glands that produce digestive secretions. The plant-parasitic nematode inserts its stylet into a plant cell and then injects these secretions through it into the cell. Different species of nematodes produce different digestive secretions, which cause a variety of reactions within the plant. Next, the nematode will withdraw cell contents through the stylet and esophageal lumen, and into the intestine for digestion.

Nematodes develop inside eggs as first-stage juveniles (J1), go through one molt inside the egg, and then hatch as second-stage juveniles (J2). After hatching, the developing nematode goes through three more molts into third (J3) and fourth (J4) juvenile stages, and finally into an adult nematode. Some species of nematodes parasitic on turfgrasses reproduce sexually, with copulation between male and female nematodes required to produce offspring. For these species, usually both male and female nematodes are abundant in a given population. Other turfgrass nematodes reproduce parthenogenically, with no mating required to produce viable eggs. With these species, males are rare or nonexistent. The time it takes to develop from a newly laid egg to an egg-laying adult generally varies from a couple of weeks to several months depending on the species and environmental conditions.

When nematodes hatch from their eggs, they locate host roots using chemosensory organs called amphids that are located on the nematode's head. They are attracted to host-plant roots by exudates produced by the root as it grows through the soil. Root tips are the primary initial point of attack for most plant-parasitic nematodes.

Some species of plant-parasitic nematodes exhibit a great deal of vertical mobility. They can move several feet deep into the soil to avoid adverse environmental conditions or the absence of a host. Then they move up when conditions change, or if they detect a host plant. Plant-parasitic nematodes exhibit little lateral movement, usually less than a foot during the life of the nematode.

Plant-parasitic nematodes are obligate parasites, but may feed on plants in different ways. These nematodes can be categorized as ectoparasites or endoparasites based on their feeding habit on plants. These groupings will be used to categorize the types of nematodes that feed on turf.

22.3 NEMATODE PARASITES OF TURFGRASSES

22.3.1 ECTOPARASITES

These are nematodes that remain in the soil and feed on roots from the outside by inserting their stylet into root cells. Ectoparasites typically feed near the root tips. If enough of these nematodes feed on a given root tip, it may cease growing and appear abbreviated or "stubby" (Figure 22.1).

FIGURE 22.1   (Color Figure 22.1 follows p. 262.) Bermudagrass roots damaged by sting nematode (*Belonolaimus longicaudatus*) (left) are abbreviated and "stubby" in appearance compared with healthy roots (right) (University of Florida).

Certain species of ectoparasitic nematodes migrate through soil from root tip to root tip, while others typically remain on a given root throughout their life. Because these nematodes remain exposed in the soil, they generally are more responsive to contact nematicides than are endoparasitic species.

22.3.1.1   Sting Nematodes

Sting nematodes comprise the family Belonolaimidae. These are large ectoparasitic nematodes capable of causing severe damage to turfgrasses. While there are several species, *Belonolaimus longicaudatus* is the most common, as well as the most studied in this group. Other sting nematodes known to attack turf are *B. lolii* and *Morulaimus gigas*, both found in Australia [5,6]. *B. longicaudatus* is native to the sandy coastal plains of the southeast United States, requiring sand contents of ≥85% to thrive [7]. Optimum soil temperature for activity of *B. longicaudatus* is around 30°C [7,8]. Because sand-based putting greens provide an ideal habitat for *B. longicaudatus*, it can do well in these greens even where the native soil is not conducive for it. In sandy regions such as Florida, *B. longicaudatus* is a common problem on golf courses, lawns, and athletic fields. A recent survey in Florida found this nematode present in damaging numbers on 60% of the golf courses in that state [9]. In the United States, *B. longicaudatus* has been reported from the Florida Keys to Delaware on the east coast and as far west as California [10–12]. Internationally, it has been spread to golf courses in Bermuda, Puerto Rico, and the Bahamas through introduction on infested planting material [10]. The host range for *B. longicaudatus* is extensive and includes most grasses. Known turfgrass hosts for *B. longicaudatus* include Italian (*Lolium multiflorum*) and perennial (*L. perenne*) ryegrasses, Bahiagrass (*Paspalum notatum*), seashore paspalum (*P. vaginatum*), creeping bentgrass (*Agrostis stolonifera*), bermudagrass (*Cynodon dactylon*), centipedegrass (*Eremochloa ophiuroides*), St. Augustinegrass (*Stenotaphrum secundatum*), and zoysia (*Zoysia japonica, Z. matrella*) [13].

FIGURE 22.2   (Color Figure 22.2 follows p. 262.) Bermudagrass putting green infested with damaging numbers of sting (*Belonolaimus longicaudatus*) and lance (*Hoplolaimus galeatus*) nematodes (University of Florida).

For plant-parasitic nematodes, sting nematodes are quite large, often achieving lengths of 3 mm or more. These are amphimictic species; so both males and females are abundant in most populations and are required to produce offspring. Eggs are laid in the soil in pairs, and under favorable conditions can mature into egg-laying adults in as little as 18 days [14]. Sting nematodes typically feed on growing root tips. Abundant populations of sting nematode can decimate turfgrass roots, causing severe decline and even death of the grass (Figure 22.2). Sting nematodes are among the most responsive nematodes to nematicide treatments and can be managed with either contact or systemic nematicides [15–22].

22.3.1.2   Stubby-Root Nematodes

Stubby-root nematodes are in the family Trichodoridae, which includes two genera with species known to damage turfgrasses, *Trichodorus* and *Paratrichodorus*. Nematodes in this group are unique in that they are the only nematodes with an onchiostyle, a curved, solid stylet that is used to puncture holes in root cells. Stubby-root nematodes secrete a hollow feeding tube into the cell, which is used to inject secretions and ingest cell contents. Numerous species of stubby-root nematodes have been associated with turfgrasses. However, only two species, *P. minor* (syn. *P. christiei*, *T. christiei*) and *T. obtusus* (syn. *T. proximus*) have been demonstrated to be turfgrass pathogens [23,24].

Paratrichodorus minor is of worldwide distribution and is found in both tropical and temperate environments [25]. This species reproduces parthenogenically, so males are rare and are not needed for reproduction. Known turfgrass hosts of *P. minor* are annual bluegrass (*Poa annua*), perennial and Italian ryegrasses, Bahiagrass, bermudagrass, centipedegrass, creeping bentgrass, St. Augustinegrass, seashore paspalum, and zoysia [25].

Trichodorus obtusus is known to occur only in North America. This species reproduces sexually, so both males and females are common. *Trichodorus obtusus* is a known pathogen of bermudagrass and St. Augustinegrass [24], and also has been associated with damage on centipedegrass, seashore paspalum, and Kentucky bluegrass (*P. pratensis*) [26–28].

While both these nematodes are damaging, *T. obtusus* has been shown to be more damaging than *P. minor* to bermudagrass and St. Augustinegrass [24]. Since these two species can be distinguished morphologically, separate risk thresholds can be used for diagnosing these nematodes [23]. Stubby-root nematodes typically respond well to organophosphate nematicides. However, *P. minor* has been shown to resurge to higher numbers following applications of fumigant nematicides than if no fumigant were used [29,30].

22.3.1.3 Ring and Pin Nematodes, and Their Relatives

Several nematodes in the families Criconematidae and Tylenchulidae are known to feed on and damage turfgrasses. These families are characterized by having an amalgamated procorpus (esophageal region) that is unique among nematodes. Nematodes in this group include ring (*Criconemoides* spp., *Mesocriconema* spp., etc.), sheath (*Hemicycliophora* spp., *Caloosia* spp.), sheathoid (*Hemicriconemoides* spp.), and pin (*Paratylenchus* spp. and *Gracilacus* spp.), all of which have species that are turfgrass parasites. While these are some of the most common plant-parasitic nematodes found associated with turfgrasses in North America [3,9,31–39], for most of them the amount of damage they cause has not been clearly defined.

Ring nematodes are known to cause damage on bermudagrass, bentgrass, St. Augustinegrass, and centipedegrass [26,40–42]. Centipedegrass is highly susceptible to damage from ring nematodes, and ring nematode is considered the most important nematode pest of centipede lawns and sod farms in Florida [43]. High numbers of ring nematodes occasionally are associated with the decline of putting greens throughout the United States.

22.3.1.4 Spiral Nematodes

The genera *Helicotylenchus*, *Peltamigratus*, and *Scutellonema* contain turfgrass ectoparasites. While similar in appearance, these three genera may differ in the amount of damage they cause and can be readily distinguished with low-power microscopy [44]. They are commonly called spiral nematodes because when dead their body curls into a spiral. *Helicotylenchus* spp. is the most common genus of spiral nematodes on turfgrasses, but high numbers are required to cause measurable damage [28,33,36,45,46]. *Helicotylenchus* spp. occasionally occur at damaging numbers in most locales, but commonly build up to very high numbers in western states [28,33,36]. In Florida, numbers of *Helicotylenchus* spp. that exceed risk thresholds are frequent on seashore paspalum but rare on other turf species [26]. It is theorized that *Peltamigratus christiei* may be more damaging to turfgrasses than *Helicotylenchus* spp. [44], but virulence comparison experiments have not yet been undertaken.

22.3.1.5 Stunt Nematodes

Nematodes in the genus *Tylenchoryhnchus* and related genera are commonly called stunt nematodes. Several species are ectoparasites of turfgrasses including *T. claytoni*, *T. dubius*, *T. nudus*, *T. maximus*, *T. annulatus*, and possibly others. Stunt nematodes are among the most common nematode problems on putting greens in temperate regions [3,28,34,37,47]. A survey in New England found them above threshold levels on 35% of samples collected [37]. While frequently encountered in southern states, surveys in Florida and Alabama did not find any populations exceeding threshold levels [9,31].

22.3.1.6 Other Ectoparasites

Other plant-parasitic ectoparasites that are occasionally encountered on turfgrasses and may cause damage are awl (*Dolichodorus* spp.), dagger (*Xiphinema* spp.), and needle (*Longidorus* spp.) nematodes.

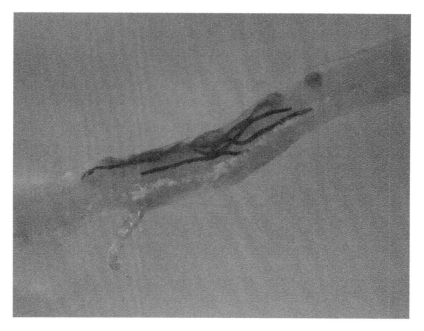

FIGURE 22.3 (**Color Figure 22.3 follows p. 262.**) These lance nematodes (*Hoplolaimus galeatus*) (stained red) are migratory endoparasites of bermudagrass (University of Florida).

22.3.2 ENDOPARASITES

Migratory endoparasites enter plant tissues and feed as they tunnel through them moving from cell to cell (Figure 22.3). They can also move out of the plant into the soil and then enter the plant again, or move to another plant. Typically, eggs are laid inside plant tissues. Because they migrate between the plant and the soil, various life stages may be detected from soil nematode assays and also from roots and plant tissue.

Sedentary endoparasites enter roots and establish a permanent feeding site. The feeding site is formed of specialized plant cells that develop in response to hormones injected into them by the nematodes. Once its feeding site develops, the nematode loses its worm shape, becomes swollen, and is no longer capable of moving. Some species lay their eggs in egg masses at the root surface or within root tissue, others retain eggs inside their body and form cysts. Males are required for mating with some species, but not others, so they may or may not be present. Juveniles may be extracted from soil samples before they enter a root, but adult females will only be found within plant tissue.

22.3.2.1 Lance Nematodes

Lance nematodes (*Hoplolaimus* spp.) are found throughout the world and are among the most common nematode problems on turfgrasses in the United States. These large migratory endoparasites are commonly associated with root discoloration and rotting, lack of feeder root development (Figure 22.4), decreased susceptibility to environmental stresses, and poorly growing grass [43,48,49]. Several species have been shown to parasitize turfgrasses in the United States, including *H. galeatus*, *H. columbus*, and *H. concaudajuvencus* [49–51]. However, *H. galeatus* is considered the most common lance nematode in turfgrass systems. *Hoplolaimus galeatus* is reported on turfgrasses throughout the continental United States and Canada and is a known parasite of most turfgrasses including bermudagrass, St. Augustinegrass, centipedegrass, seashore paspalum, creeping bentgrass, zoysia, Kentucky bluegrass, tall fescue (*Festuca arundinacea*), buffalograss (*Buchloë dactyloides*), and perennial ryegrass [26,49,52–54]. In sandy areas of the south, damage

FIGURE 22.4 (Color Figure 22.4 follows p. 262.) Lance nematode–infected St. Augustinegrass roots (right) are dark, rotten appearing, and lack feeder roots compared with healthy roots (left).

from lance nematodes can occur in lawns and all venues of sports turf. In other areas, damage occurs predominantly in putting greens. Lance nematodes were found above threshold levels on 50% of the golf courses in Florida [9], and are also the predominant nematode problem on St. Augustinegrass lawns in that state [26,43]. In southern New England, a survey found lance nematodes above threshold levels in about 10% of samples collected [37].

Hoplolaimus galeatus reproduces sexually, so both males and females are abundant in most populations. Adult and juvenile stages are usually found together in soil and root samples and look identical other than their size, and the development of sexual organs. Giblin-Davis et al. [55] found that 85% of *H. galeatus* vermiform stages and eggs occurred within roots of St. Augustinegrass and only 15% were recovered from soil. The majority of nematode diagnoses from turf are made based on nematode counts from soil, so these counts may grossly underestimate the total number of nematodes present.

Because of their endoparasitic habit, lance nematodes are easily spread in infested sprigs and sod, explaining their widespread distribution. They are also among the most difficult nematodes to manage on turfgrasses. Generally, systemic nematicides that are absorbed by plant roots are more effective than contact nematicides against lance nematode [56].

22.3.2.2 Lesion Nematodes

Species of lesion nematodes (*Pratylenchus* spp.) are adapted to a wide range of climactic habitats and can be found on turfgrasses throughout North America [3,9,34,36,39]. These are migratory endoparasites. Most species are not very damaging, but their pathogenicity on turfgrasses has not been well studied. *Pratylenchus zeae* is most common on turf in southern states [26], while other species adapted to cool climates are found in turf samples as far north as Canada.

22.3.2.3 Root-Knot Nematodes

Root-knot nematodes (*Meloidogyne* spp.) are the most well-known plant-parasitic nematodes worldwide, primarily because of the damage they cause to vegetable and agronomic crops. However, several species parasitize turfgrasses and can cause considerable damage. The root-knot nematodes that attack turfgrasses in the field are typically different species than those that are problems on

FIGURE 22.5   (Color Figure 22.5 follows p. 262.) St. Augustinegrass root infected by root-knot nematode (*Meloidogyne graminis*); the epidermis of the root has been peeled away to reveal the swollen female nematode within (University of Florida).

most other crops. *Meloidogyne graminis* is probably the most common root-knot nematode on turf-grasses throughout the United States [57,58]. Other species reported damaging turfgrasses in the field are *M. marylandi*, *M. hapla*, and *M. naasi* [58–61]. In recent years a new species, *M. minor*, has become a major problem on bentgrass greens in the United Kingdom [62].

Root-knot nematodes infect roots as J2 and penetrate the root with their entire bodies. Inside the root, the nematode injects hormones into cortical cells that cause a specialized feeding site to develop. After this time, the nematode is no longer capable of moving and must feed on the specialized cells of the feeding site. As the nematode matures and molts, its body swells. On most plants, the adult female root-knot nematode will protrude her posterior out of the root and lay her eggs at the root surface. The eggs are laid in an egg mass within a gelatinous matrix that contains antibiotics that protect the eggs from parasites. A single egg mass will contain several hundred eggs. However, on turfgrasses the females may remain completely inside the root and lay their eggs within the root tissue. Also, the gall or knot symptoms characteristic of root-knot nematodes on most hosts are often not visible with root-knot infection on turfgrasses [63]. A turfgrass root may look normal, but when the root is sectioned longitudinally, root-knot females (Figure 22.5) and eggs become apparent. Root-knot nematodes may also cause roots to be dark or rotten in appearance.

In addition to damaging roots, root-knot nematodes continually inject hormones into the plant that cause systemic effects. Photosynthates produced in the leaves that normally would be used for plant growth are translocated to the nematode feeding sites [64]. A frequent symptom of root-knot nematode infection on bentgrass is yellow blotches that do not respond to fertilizer or fungicide applications. This yellowing may be due to these systemic reactions to the nematodes.

22.3.2.4   Cyst Nematodes

The life cycle of cyst nematodes is similar to that of root-knot nematodes, except that at maturity the female nematode body will protrude out of the root. While a few eggs may be laid in an egg mass, the majority are retained inside the female nematode body. The female dies and her body hardens and becomes the cyst. Eggs inside of a cyst can remain viable for years and will only hatch in the presence of a host plant and proper environmental conditions. Cyst nematodes are very difficult to manage because the cysts are almost impervious to nematicides. Cysts can be observed as small (0.5 to 1.0 mm in length), pale yellow to brown colored, lemon-shaped spheres protruding from roots. Live female nematodes also may be observed protruding from roots, but will be white. The cysts can be crushed in water between a microscope slide and cover slip revealing the eggs within.

Two species of cyst nematodes are known to parasitize turfgrasses in the United States. The St. Augustinegrass cyst nematode (*Heterodera leuceilyma*) is only known to parasitize St. Augustinegrass and can cause considerable damage to this common lawn grass in the southeast [65]. The other species, *Heterodera iri*, is found primarily on creeping bentgrass putting greens in the northeast [66].

22.3.2.5   Seed, Stem, and Root-Gall Nematodes

Several nematodes in the family Anguinidae are parasites of turfgrasses. These nematodes are typically associated with gall formation. The adult nematodes form galls and lay their eggs inside of them. Each gall can contain hundreds of eggs and juvenile nematodes. Juveniles can exit the gall and move to other sites to infect, but the adults cannot leave the gall. While other species may infect turf, the species described below are the only ones known to cause economic damage to turfgrasses in the United States.

The bentgrass nematode, *Anguina agrostis*, causes galls of bentgrass seed. While not typically a turf maintenance problem, *A. agrostis* can reduce viable seed formation and be a problem for bentgrass seed production [67].

Anguina pacificae is a closely related species that causes galling of stem tissue and can be a major turf-maintenance problem. This nematode has only been found parasitizing annual bluegrass putting greens on coastal golf courses in northern California [36,68]. *Anguina pacificae* may be confined to this area due to the unique environmental conditions that occur there. Affected bentgrass develops characteristic galls or swelling at the base of the stem. Small patches of chlorotic turf develop, become necrotic (Figure 22.6), enlarge and coalesce, and eventually can decimate the entire green. Because this nematode primarily affects aboveground parts, it is especially difficult to control with anything other than systemic nematicides [36].

FIGURE 22.6   (Color Figure 22.6 follows p. 262.) Necrotic patches on an annual bluegrass putting green caused by *Anguina pacificae* (Courtesy of P. Gross, USGA).

Subanguina *radicicola*, the root-gall nematode, causes galls to develop on the roots of annual bluegrass putting greens in the northern United States and Canada [69,70]. Galls caused by the root-gall nematode may be mistaken for those caused by root-knot nematodes. When present, root-knot nematode galls on turf are typically rounded whereas those caused by root-gall nematodes tend to be elongated and curved in appearance [69,70].

22.4 ENVIRONMENTAL FACTORS AFFECTING NEMATODES ON TURF

Soil type can have a great impact on plant-parasitic nematodes. Sand-based putting greens make ideal habitats for most plant-parasitic nematodes. Certain species, such as sting nematodes, occur only in sandy soil [7,10]. Additionally, damage caused by a given number of nematodes will tend to be greater in sand than in heavy soil. Sand soils typically have decreased ability to retain water and nutrients compared with heavier soils. This gives nematode-damaged roots less opportunity to extract the water and nutrients and makes the turf less tolerant of the nematodes.

Temperature plays a major role in nematode population dynamics. Each species has its own optimum temperature; so some species are better adapted to warm climates and others to cooler climates. However, while the specific temperatures involved vary, generalizations can be made regarding seasonal temperature effects on nematodes. Each nematode species has a basal temperature, a temperature below which the nematode is inactive. As temperatures increase above the basal temperature, nematode activity (feeding, reproduction, etc.) increases until the optimum temperature for that species is reached. At temperatures above the optimum, nematode activity falls off precipitously. In a state like Florida, soil temperatures may remain such that nematodes can be active almost year-round. In northern states, nematode activity may be limited to only part of the year. For this reason, nematode problems tend to be more severe in southern states.

22.5 DIAGNOSING NEMATODE PROBLEMS ON TURFGRASSES

22.5.1 ROOT SYMPTOMS

Different types of root-feeding nematodes cause different types of symptoms to occur in turfgrasses. As mentioned previously, certain species can cause galls or swellings on roots. Endoparasites such as lance or root-knot nematodes may cause roots to appear dark or rotten and lack fine feeder roots. Ectoparasites like sting and stubby-root nematodes cause root tips to become abbreviated or stubby. Some nematodes may cause lesions or dark sunken areas on roots, or a general lack of root growth. On putting greens, high numbers of plant-parasitic nematodes may cause roots to appear "cropped off" just below the thatch layer.

Care should be taken to avoid over-reliance on root symptoms for diagnosis. Most of the visual symptoms on roots are general in nature and could be caused by other factors in addition to, or other than, nematodes. If any of the root symptoms described above are present, a nematode assay should still be conducted to verify the diagnosis.

22.5.2 ABOVEGROUND SYMPTOMS

Nematodes generally occur in an aggregated or patchy distribution. Nematode numbers can be extremely high in one spot and low just a few feet away [71]. Therefore, nematode damage typically occurs in irregularly shaped patches. Most plant-parasitic nematode species build up to higher numbers in sandy soil, so often damage will become evident in sandy areas first while areas with heavier soil may not appear damaged at all. Often, visual nematode symptoms will become evident on humps and then slowly move "downhill." Also, because plant-parasitic nematodes reduce turf tolerance to stresses, turf in stressed areas, i.e., areas that get less water, more traffic, lower mowing

FIGURE 22.7   (Color Figure 22.7 follows p. 262.) Bermudagrass thinning and declining due to high infestation of sting nematodes (*Belonolaimus longicaudatus*) allows weeds to proliferate (University of Florida).

height, etc., will show symptoms before less stressed areas. On putting greens, nematode damage may become evident on the edges of the green first and then move toward the center. This may be due to nematode migration into the green, or because the edges of the greens tend to be more stressed than the centers.

Because root-feeding nematodes reduce the ability of the turf to extract water from soil through the damaged roots, wilting is a common symptom (Figure 22.7). Often, when turf dies from nematode damage it is because of drought stress induced by the nematodes. This wilting is not because there is not enough moisture present, but rather because the nematode-damaged root system is not able to take up enough of the soil moisture to adequately supply the turf.

Discoloration, particularly yellowing, may result from nematode infestation on turf. This is usually due to nutrient deprivation resulting from nematodes damaging the root system. These damaged roots are not able to adequately extract nutrients from the soil to support turf health. Another type of yellowing can occur in response to physiological reactions within the plant to the hormones injected into it by certain types of nematodes.

Often, one of the first observable results of nematode problems on turf is an increase in weeds (Figure 22.8). Each species of plant-parasitic nematode has its own host range, and certain plants it will feed on. Often, the nematodes that feed on turfgrasses will not feed on some of the common weeds, or if they do, the weeds are more tolerant of them. This gives the weeds a competitive advantage, as the turf declines from nematodes weeds begin to predominate. In the southeastern United States, spotted spurge is often used as an indicator of nematode problems on golf-course turf. While the presence of spurge may indicate a nematode problem, this is not a substitute for conducting a nematode assay. The weeds are simply indicative of poorly growing grass, which could be caused by factors other than nematodes.

22.5.3   NEMATODE SAMPLING AND LABORATORY ANALYSIS

A nematode assay is the process by which nematode samples are collected, the nematodes extracted from soil or plant tissues, and then identified and counted. Typically, the samples are collected by

FIGURE 22.8   (Color Figure 22.8 follows p. 262.) Irregular patches of wilting and declining centipe-degrass on a sod farm caused by high numbers of ring nematodes (*Mesocriconema ornata*) (University of Florida).

the turfgrass manager or a professional consultant, and the extraction and diagnosis is conducted by professional laboratory staff. For an accurate nematode diagnosis, sample collection, submission, extraction, and identification must all be conducted in the proper manner.

Nematode assays may be for either pre- or postplant situations. A preplant assay is conducted before turf is planted. The goal of a preplant sample is to determine if nematodes are a potential threat, so that problem areas might be avoided, a nonsusceptible grass planted, or a preplant nema-tode treatment used. The purpose for a postplant nematode assay is typically to determine if nema-todes are contributing to a turfgrass problem, so that proper management tactics can be employed. Postplant samples include soil, but also may include roots and other plant tissues.

As stated earlier, nematodes typically occur in a clumped distribution. Therefore, a proper nematode sample will consist of multiple cores combined into a single sample rather than a large single core from one location. For statistical reasons, it is recommended that at least 12 cores be collected from the sampled area (lawn, field, green, fairway, etc.). However, the volume of the cores may necessitate an increased number of cores to provide an adequate amount of soil for analysis. For example, the most common type of sampling device is a 0.5-in.-diameter soil probe. Most labs request that nematode samples contain at least 250 cm^3 of soil. Therefore, to get enough soil with a 0.5-in.-diameter probe, about 20 cores should be taken. If a 1-in.-diameter probe is used, fewer cores are required.

Preplant or postplant samples collected from turf that is not showing visible symptoms should be collected in a pattern chosen such that cores are evenly collected across the sample area. Typi-cally, a zigzag pattern is recommended, but other patterns may be used. The critical factor is that the sampling pattern ensures that the entire area is represented equally.

Diagnostic postplant samples from areas where aboveground symptoms are evident should be taken from affected areas. Dead areas should be avoided. Plant-parasitic nematodes feed on live roots, so populations will be low in those areas. Declining areas where the turf is sick, but not dead, are best. Sample from the margins of dead areas where the turf is still alive. Regarding core depth, plant-parasitic nematode populations tend to be highest in the root zone of affected plants. There-fore, when taking diagnostic samples from turf, a sample depth of 3–4 in. is usually ideal.

After cores from a given area are collected, they should be placed in a single plastic bag to form a composite sample and sealed. Immediately remove the sample from direct sunlight and heat as this can kill any nematodes in the sample in a short period of time [72]. It is important that the samples go into plastic bags to prevent desiccation, since drying of the sample kills the nematodes. Samples should be stored in an air-conditioned area until they can be shipped. Try to send the samples as soon as possible; the longer the time interval between sample collection and processing the poorer the quality of the diagnosis is likely to be [72]. Unless instructed otherwise, include the plugs of turf and roots along with the soil as they can aid the diagnostic process later.

In the laboratory, one of several procedures will be used to separate nematodes from the soil into water for analysis [73]. The plant-parasitic nematodes in the sample will be identified and each kind counted. These nematodes will typically be identified to genus level, although some labs may be equipped to conduct species identification if required. Additionally, for some nematodes, i.e., cyst, root-knot, and gall-forming nematodes, direct observation of nematodes inside plant tissues may be conducted.

Unlike diagnosing fungal pathogens, where only the presence or absence of the pathogen is noted, nematode diagnosis is usually quantitative. Because nematodes are animals, the number of each kind can be determined. These numbers are then compared with some sort of "threshold" number, and if the quantity of nematodes exceeds this number, the turf is considered at risk of damage. Therefore, both the types of nematodes present and the abundance of each type are important for diagnosis.

Nematode thresholds are usually based on a combination of research data and field experience in a particular region. However, these should be considered guidelines rather than absolute values. In actuality the tolerance limit, the number of a particular nematode that the turf can tolerate before it is negatively impacted, is a variable and fluctuating quantity. For example, turf on a public golf course that gets lots of traffic will generally tolerate fewer nematodes than a private course that gets little wear. In a given lawn, sandier areas will tolerate fewer nematodes than areas with heavier soil. The same grass will be become less tolerant of nematodes during periods of temperature extremes. To account for this variability, some labs have threshold ranges that designate if the turf is at moderate or high risk of damage [43]. If the sample falls into the moderate risk range, it means that nematode damage may or may not become evident depending on variable factors. Nematode thresholds are generally developed for a particular region and may differ in other localities. Also, different labs will report in different units, such as nematodes per 100 cm^3 of soil, nematodes per 500 cm^3 of soil, nematodes/pint of soil, etc.

22.6 NEMATODE MANAGEMENT ON TURFGRASSES

22.6.1 CULTURAL PRACTICES

Nematodes are generally one of many stress factors affecting the overall health of turf. Often, nematode damage can be reduced by alleviating other stress factors to make the turf more tolerant of nematodes. While these strategies can help, even with the best overall turfgrass management practices, nematodes can still become a problem.

Increasing mowing height has been shown to increase the tolerance of turf to nematodes, especially on intensively managed turf [74]. While this might be a hard-sell to certain golfers, it may be a choice between having slightly slower putting greens and having large dead patches on them. Care should be taken to avoid scalping and going too long between mowing, as these practices add stress to the turf and decrease tolerance to nematodes.

As a preventative measure, applying irrigation only as needed and then applying enough water to wet the soil deeply can help avoid the development of a shallow root system. Deeply rooted turf is more tolerant of nematodes. However, once nematode damage occurs, turf may require frequent irrigation to survive because of its damaged root system. In Florida, putting greens with high numbers of sting nematode may require multiple irrigations per day to avoid desiccation. This is of great concern as water resources are becoming limited and many water management districts and municipalities are imposing mandatory water restrictions.

Proper nutrition is important for turfgrass health, and healthy turf is more tolerant of nematodes. Sometimes, the chlorosis resulting from nematode damage on turf can be masked with nitrogen or iron applications. With severe nematode infestation, where the root system is almost nonexistent, foliar fertilizer applications may be the only way nitrogen can be used by the turf. However, exceeding recommended nutrition levels has not been shown to enable turf to overcome effects of plant-parasitic nematodes. If the turf is unable to extract nutrients through damaged roots, there is concern that this might increase the potential for nitrate leaching, endangering water quality. In lysimeters, sting nematode–infested turf leached more nitrogen than noninfested turf [75].

Addition of certain organic amendments has been shown to reduce plant-parasitic nematode populations or to reduce nematode damage on turfgrasses. These beneficial effects could be caused by toxic materials released by the amendment, stimulation of nematode antagonists, or by the amendment improving plant health. Organic amendments that have reduced certain nematode populations in at least some research trials are municipal sludge, chitin amendments, and earthworm castings. However, the same or similar materials have not affected nematodes in other experiments.

Nematodes do not typically move from one location to another without assistance. Most commonly, they are moved from location to location on contaminated planting material, soil, or water. On turf, sprigs and sod are the most common means of nematode dispersal. Endoparasites especially are moved readily within the roots of planting material, but ectoparasites can be moved as well. To avoid a nematode problem, it is recommended that sod be assayed for nematodes and only be planted if the more damaging nematode species are absent. Equipment such as aerifiers or sod harvesters that may have soil adhering should be well washed after use to avoid moving nematodes to new areas. Similarly, areas with known nematode infestations should be aerified last to prevent nematode spread.

22.6.2 RESISTANT AND TOLERANT GRASSES

Different grass species vary in their susceptibility to damage from different plant-parasitic nematodes. Sometimes, it is possible to plant a different type of grass that is not susceptible to the major plant-parasitic nematodes in a given location. For example, the St. Augustinegrass cyst nematode is only known to parasitize St. Augustinegrass. So, if a St. Augustinegrass lawn is infested with this nematode it can be replaced with any other type of turf and that nematode will no longer be a problem. Similarly, Bahiagrass is generally tolerant of nematodes and can be planted in sites infested with most plant-parasitic nematodes without too much concern.

Cultivars of a given turfgrass species may vary in their susceptibility to nematodes. Evaluations of bermudagrass cultivars found that variations in relative tolerance and resistance to sting nematode occurred [76,77]. Diploid St. Augustinegrass cultivars were found to be more susceptible to sting nematode than were polyploid cultivars [78,79]. Similarly, St. Augustinegrass cultivars differed in susceptibility to lance nematode [52,55]. Root architecture and vigor is important. Cultivars with a vigorous-growing root system tend to be more tolerant to nematodes than poor-rooting cultivars.

Currently, some turfgrass breeders are seeking to develop cultivars with improved tolerance or resistance [80,81]. Additionally, efforts are underway to move nematode resistance genes from other plants into bermudagrass [82]. Hopefully, these efforts will soon pay off with commercially acceptable grasses that have fewer nematode problems.

22.6.3 BIOLOGICAL CONTROL

Plant-parasitic nematodes have many antagonists that in nature tend to hold their populations in check. These antagonists include other microscopic animals, fungi, and bacteria. Over the years, many attempts have been made to find ways to build up numbers of naturally occurring antagonists to reduce nematode populations. Unfortunately, to date none of these methods has proven consistent enough to be relied on for nematode control in most turfgrass situations. Possibly, with increased knowledge of the biology and interactions of these organisms in the soil, these methods may become more reliable in the future.

A more practical approach is to inundate an area with a biological control organism to either add it to an area where it is not present, or to supplement a naturally occurring population. To use this method, the organism needs to be easily culturable, and formulated for application.

Pasteuria spp. is a group of endospore-forming bacteria, many of which are obligate parasites of nematodes. These bacteria tend to be very selective, with different strains of *Pasteuria* attacking only certain species of nematodes [83]. The *Pasteuria* endospores attach to the cuticle of the nematode as it moves through the soil. The *Pasteuria* then penetrate into the nematode and proliferate inside the nematode's body. Infected nematodes have reduced movement, reduced fecundity, and eventually die. Naturally occurring *Pasteuria* are common, and are reported parasitizing many of the plant-parasitic nematode genera that attack turfgrasses, including root-knot, sting, lance, spiral, cyst, stunt, and ring nematodes [83]. *Candidatus Pasteuria usgae*, the species that parasitizes sting nematodes [84], was introduced into a putting green where it successfully reduced sting nematode populations and improved turf health [85]. Natural populations of other *Pasteuria* spp. have been observed parasitizing root-knot, stunt, spiral, and lance nematodes on turfgrasses [86–88]. Unfortunately, because they are obligate parasites, *Pasteuria* spp. are very difficult to culture [83]. However, there have been recent breakthroughs regarding culturing these bacteria that might lead to commercially available biological nematicides in the near future.

There are numerous species of entomopathogenic nematodes currently on the market that are effective biological controls of certain turfgrass insect pests such as mole-crickets and grubs. Some research has shown reduction of plant-parasitic nematodes on turf following application of these entomopathogenic nematodes [89,90]. However, a large body of research conducted in Florida found that these nematodes were not effective in controlling plant-parasitic nematodes on turf in that state [91].

Other organisms that have shown some potential as nematode biological controls are fungal parasites of nematode eggs [92]. Most of the research with this type of biological control has been done with nematode parasites of vegetables and agronomic crops. Their usefulness still needs to be established in turfgrass systems. Another area where more research is needed is with endophytic fungi. Research with some of these fungi on forage grasses has shown that they can impart resistance to certain nematodes, but their potential use on turfgrasses still needs to be established [93,94].

Many types of common bacteria produce nematicidal metabolites. Research with several of these bacteria, including *Pseudomonas fluorescens*, *Bacillus firmus*, and several bacterial symbionts of entomopathogenic nematodes, has revealed that high numbers of these bacteria or their metabolites can reduce numbers of plant-parasitic nematodes in soil [56,92,95]. However, the numbers required to affect nematodes in the field are much higher than can typically be achieved naturally. Current research is attempting to find ways to use these bacteria or their metabolites in a practical and effective manner in turfgrass systems.

22.6.4 NEMATICIDES

Nematicides were first used to control nematodes on turfgrasses in the 1950s. Most of these early turfgrass nematicides were fumigants such as 1,2-dibromo-3-chloropropane (DBCP) and 1,2-dichloropropane-1,3-dichloropropene (DD). These fumigants were very effective and served to raise the awareness of nematodes by the turfgrass community by showing the dramatic results achieved by controlling them. During the late 1960s and thereafter, organophosphate nematicides like fenamiphos and ethoprop were used to control nematodes on turf. Today, most of these older nematicides are no longer available due to environmental and health concerns. Nematicides can be used to reduce nematode populations before planting (preplant) or to manage infestations on existing turf (postplant).

22.6.4.1 Preplant Nematicides

Most preplant nematicide use is on sod farms, golf courses, and athletic fields. These are typically not true nematicides in that instead of specifically targeting nematodes, they are broad-spectrum in activity, killing a variety of organisms including nematodes. Methyl bromide is often used as a preplant fumigant to kill existing grasses and weeds before planting. In addition to being an herbicide, methyl bromide also kills nematodes, fungi, and insects in the soil. While methyl bromide has been phased out of general use according to the Clean Air Act of 1990, critical-use exemptions may allow it for certain turfgrass uses on a year-by-year renewable basis.

Other materials have been evaluated as methyl bromide replacements for use on turfgrasses, primarily for weed control, with varying degrees of success [96]. While most of these are not as effective as methyl bromide, several currently on the market, such as 1,3-dichloropropane (1,3-D), metam sodium and Dazomet, are nematicidal and may be used to reduce nematode populations prior to planting.

22.6.4.2 Postplant Nematicides

Unlike preplant nematicides, which are used to prevent a problem, postplant nematicides are used to treat an existing problem. An additional difference is that post-plant nematicides are usually targeted for nematodes instead of being broad-spectrum in activity. In recent decades, the organophosphate nematicides fenamiphos and ethoprop have been the primary nematicides used for this purpose on turfgrasses. However, turfgrass use was removed from the ethoprop label in the United States in 2000, and fenamiphos will no longer be manufactured after May 2007. The loss of fenamiphos (the active ingredient in the Nemacur® products) has increased the need for new turfgrass nematicides (Figure 22.9). The ideal turfgrass nematicide would have the following characteristics: (1) be consistently effective, i.e., the turfgrass manager should feel confident that when he uses it, it will work; (2) have a low potential for negative environmental effects; (3) be relatively safe for humans and grass; (4) be easy to work with and apply; (5) have contact and systemic activity to kill nematodes in soil and inside the turf plant; and (6) be economical.

FIGURE 22.9 (Color Figure 22.9 follows p. 262.) Square plot treated with an experimental nematicide on a sting nematode (*Belonolaimus longicaudatus*)–infested putting green (University of Florida).

1,3-D is the active ingredient in Telone® II and Curfew®, which are labeled for postplant turf-grass uses only in certain states in the southeastern United States. Telone II is the formulation registered for use on sod farms, and Curfew is the formulation registered for postplant application on golf courses and athletic fields. 1,3-D is a fumigant nematicide that is injected into the soil using tractor-mounted slit-injection rigs. The fumigant is a liquid when injected into the soil, but disperses through the soil as a gas. It is a contact nematicide and gives excellent control of most ectoparasitic nematodes, particularly sting nematode [15,16,20,21]. While effective, use of 1,3-D requires special equipment to apply. Because of this and its toxicity to humans, Curfew can be applied only by approved custom applicators. 1,3-D is also subject to buffer restrictions to buildings, reentry intervals, and may not be applied over certain geological formations.

Ditera® nematicides are derived from the fungus *Myrothecium* spp. This nematicide is labeled for turf and other crops in the United States. Results on other crops vary, but in general this material appears to give more consistent results in the western states. Various formulations have been evaluated against several genera of root-parasitic nematodes on turf, mostly with negative results [36,97–100]. However, in a trial in California against the stem gall nematode, Ditera was associated with some visual improvement and reduction in nematode eggs [36].

With the lack of synthetic nematicides registered for turfgrass uses, numerous botanically derived nematode management products have come on the market in recent years. These include extracts and essential oils from various plants, and dry-formulated plant materials. Many of these have little to no field efficacy data to evaluate their performance in reducing nematode problems on turfgrasses. Most of those that have been evaluated have not looked promising [36,97–104]. However, there are several that showed benefit in at least some tests.

Certain types of mustard plants and their relatives produce the nematicidal fumigant allyl iso-thiocyanate. Mustard-bran material reduced numbers of sting nematode in some experiments, but not in others [56,97,105–107]. Research is ongoing with this mustard material to improve its consistency. A commercial thyme-oil formulation was tested against ring and lance nematodes on turf. While thyme oil did not reduce nematode populations, it did slow down their reproduction [90]. Furfural is a byproduct of sugarcane processing and is used for nematode control on turf in South Africa. In the United States, it has reduced population densities of plant-parasitic nematode species in certain turfgrass research trials but not in others [56,108]. Extracts of poinsettia and spotted spurge suppressed sting nematode in laboratory and greenhouse trials [107].

22.7 SUMMARY

Plant-parasitic nematodes can play a key role in turfgrass health. Managing plant-parasitic nematodes on turfgrasses can promote healthier turf, reduce irrigation and fertilizer applications, and reduce problems with weeds, pests, and pathogens. Proper diagnosis of nematode problems on turf is critical as nematode damage is often mistaken for other turfgrass problems. When improper diagnosis occurs, the problem goes untreated, and effort and expense can be wasted treating the wrong problem.

When a nematode problem is identified, management strategies may be implemented. These include alleviating other stresses on the grass, replanting with resistant or tolerant grasses, using a biological control tactic or nematicide, or a combination of these. In the future new nematicides, biological controls, and resistant cultivars may also become options for nematode control on turf.

REFERENCES

1. Trenholm, L.E., Lickfeldt, D.W., and Crow W.T., Use of 1,3-dichloropropene to reduce irrigation requirements of sting nematode infested bermudagrass, *HortScience*, 40(5), 1543, 2005.
2. Luc, J.E. and Crow, W.T., Nematode and nitrogen management, *Golf Course Manage.*, 72(8), 97, 2004.

3. Walker, N.R. et al., Factors associated with populations of plant-parasitic nematodes in bentgrass putting greens in Oklahoma, *Plant Dis.*, 86, 764, 2002.
4. Busey, P., Cultural management of weeds in turfgrass: A review, *Crop Sci.*, 43, 1899, 2003.
5. Siviour, T.R., Biology and control of *Belonolaimus lolii* n. sp. in turf, *Aust. Plant Path Soc. News*, 7(1), 37, 1978.
6. Nobbs, J.M. and Eyres, M., *Morulaimus gigas* sp. n. from western Australia, *Afro Asian J. Nematol.*, 2, 115, 1992.
7. Robbins, R.T. and Barker, K.R., The effects of soil type, particle size, temperature, and moisture on reproduction of *Belonolaimus longicaudatus*, *J. Nematol.*, 6, 1, 1974.
8. Boyd, F.T. and Perry, V.G., Effects of seasonal temperatures and certain cultural treatments on sting nematodes in forage grass, *Soil Crop Sci. Soc. Fl. Proc.*, 30, 360, 1971.
9. Crow, W.T., How bad are nematode problems on Florida's golf courses? *Fl. Turf Digest*, 22(1), 10, 2005.
10. Perry, V.G. and Rhoades, H., The genus *Belonolaimus*, in *Nematology in the Southern Region of the United States*, Riggs, R.D., Ed., University of Arkansas Agricultural Publications, Fayetteville, AR, 1982.
11. Mulrooney, B., Sting nematode identified in Delaware, *Univ. Del. Coop. Ext. Weekly Crop Update*, 13(26), 3, 2005.
12. Mundo-Ocampo, M., Becker, J.O., and Baldwin J.G., Occurrence of *Belonolaimus longicaudatus* on bermudagrass in the Coachella Valley, *Plant Dis.*, 78, 529, 1994.
13. Crow, W.T. and Han, H.R., Sting nematode, *Plant Health Inst.*, doi:10.1094/PHI-I-2005-1208-01, 2005.
14. Han, H.R., Dickson, D.W., and Weingartner, D.P., Biological characterization of five isolates of *Belonolaimus longicaudatus*, *Nematropica*, 36, 25, 2006.
15. Crow, W.T., Giblin-Davis, R.M., and Lickfeldt, D.W., Slit injection of 1,3-dichloropropene for management of *Belonolaimus longicaudatus* on established bermudagrass, *J. Nematol.*, 35, 302, 2003.
16. Crow, W.T., Lickfeldt, D.W., and Unruh, J.B., Management of sting nematode (*Belonolaimus longicaudatus*) on bermudagrass putting greens with 1,3-dichloropropene, *Int. Turf. Soc. Res. J.*, 10, 734, 2005.
17. Giblin-Davis, R.M., Cisar, J.L., and Bilz, F.G., Evaluation of fosthiazate for the suppression of phytoparasitic nematodes in turfgrass, *Nematropica*, 23, 167, 1993.
18. Giblin-Davis, R.M., Cisar, J.L., and Bilz, F.G., Responses of nematode populations and growth of fairway managed bermudagrass to application of fertilizer and fenamiphos, *Nematropica*, 18, 117, 1988.
19. Lucas, L.T., Population dynamics of *Belonolaimus longicaudatus* and *Criconemella ornata* and growth response of bermudagrass and overseeded grasses on golf greens following treatment with nematicides, *J. Nematol.*, 14, 358, 1982.
20. Walker, N.R., Zhang, H., and Martin, D.L., Potential management approaches for the sting nematode in bermudagrass sod production, *Int. Turf. Soc. Res. J.*, 10, 793, 2005.
21. Waltz, F.C., Jr. et al., Effects of nematode management on harvest intervals in bermudagrass sod production, *Ann. Meet. Abstr.*, ASA, CSSA, and SSSA, Madison, WI. 2004.
22. Winchester, J.A. and Burt, E.O., The effect and control of sting nematodes on Ormond Bermuda grass, *Plant Dis.*, 48, 625, 1964.
23. Crow, W.T., Diagnosis of *Trichodorus obtusus* and *Paratrichodorus minor* on turfgrasses in the southeastern United States, *Plant Health Progress*, doi:10.1094/PHP-2005-0121-01-DG, 2005.
24. Crow, W.T. and Welch, J.K., Root reductions of St. Augustinegrass (*Stenotaphrum secundatum*) and hybrid bermudagrass (*Cynodon dactylon* × *C. transvaalensis*) induced by *Trichodorus obtusus* and *Paratrichodorus minor*, *Nematropica*, 34, 31, 2004.
25. Decreamer, W., Stubby root and virus vector nematodes, in *Manual of Agricultural Nematology*, Nickle, W.R., Ed., Marcel Dekker, New York, 1991.
26. Crow, W.T., *Florida Nematode Assay Lab Database*, unpublished data, 2006.
27. Hixson, A.C. and Crow, W.T., First report of plant-parasitic nematodes on seashore paspalum, *Plant Dis.*, 88, 680, 2004.
28. Smolik, J.D. and Malek, R.B., *Tylenchorhynchus nudus* and other nematodes associated with Kentucky bluegrass turf in South Dakota, *Plant Dis.*, 56, 898, 1972.
29. Rhoades, H.L., Efficacy of soil fumigants and nonfumigants for controlling plant nematodes and increasing yield of snap bean, *Nematropica*, 13, 239, 1983.
30. Weingartner, D.P. and Shumaker J.R., Effects of soil fumigants and aldicarb on corky ringspot disease and trichodorid nematodes in potato, *J. Nematol.*, 22, 665, 1990.

31. Sikora, E.J., Guertal, E.A., and Brown, K.L., Plant-parasitic nematodes associated with hybrid bermudagrass and creeping bentgrass putting greens in Alabama, *Nematropica*, 31, 301, 2001.
32. Cole, H., Jr. et al., Stylet nematode genera and *Fusarium* species isolated from Pennsylvania turfgrass sod production fields, *Plant Dis.*, 57, 891, 1973.
33. Chastagner, G.A. and McElroy, F.D., Distribution of plant-parasitic nematodes in putting green turfgrass in Washington, *Plant Dis.*, 68, 151, 1984.
34. Simard, L., Bélair, G., and Dionne, J., Survey of plant-parasitic nematodes on golf courses in Québec, *J. Nematol.*, 35, 364, 2003.
35. Lucas, L.T., Blake, C.T., and Barker, K.R., Nematodes associated with bentgrass and bermudagrass golf greens in North Carolina, *Plant Dis.*, 58, 822, 1974.
36. Westerdahl, B.B., Harivandi, M.A., and Costello, L.R., Biology and management of nematodes on turfgrass in California, *Green Sect. Rec.*, 43(5), 7, 2005.
37. Jordan, K.S. and Mitkowski, N.A., Population dynamics of plant-parasitic nematodes in golf greens turf in southern New England, *Plant Dis.*, 90, 501, 2006.
38. Fushtey, S.G. and McElroy F.D., Plant parasitic nematodes in turfgrass in southern British Columbia, *Can. Plant Dis. Survey*, 57(1), 54, 1977.
39. Yu, Q., Potter, J.W., and Gilby, G., Plant-parasitic nematodes associated with turfgrass in golf courses in southern Ontario, *Can. J. Plant Path.*, 20, 304, 1998.
40. Mullins, D.E., Ring nematode injury to centipedegrass lawns and a possible control, *Proc. Fl. St. Hort. Soc.*, 86, 438, 1973.
41. Newnam, M.R. and Lucas, L.T., The effects of ring nematode, *Criconemella ornata*, on the nutrient content of centipedegrass, *Phytopathology*, 77, 642, 1987.
42. Hagan, A.K., McLean, K., and Rivas-Davila, M.E., Occurrence of ring nematode on creeping bentgrass putting greens in Alabama, *Phytopathology*, 92, S33, 2002.
43. Crow, W.T. et al., *Florida Nematode Management Guide*, University of Florida Press, Gainesville, 2003.
44. Crow, W.T. and Walker, N.R., Diagnosis of *Peltamigratus christiei*, a plant-parasitic nematode associated with warm-season turfgrasses in the southern United States, *Plant Health Progress*, doi:10.1094/PHP-2003-0513-01-DG, 2003.
45. Perry, V.G., Darling, H.M., and Thorne, G., *Anatomy, Taxonomy, and Control of Certain Spiral Nematodes Attacking Blue Grass in Wisconsin*, University of Wisconsin, Madison, 1959.
46. Walmsley, W.H., A winter active 'persistent' disease of South Island golf courses associated with spiral nematode *Helicotylenchus*, *New Zeal. Turf Manage. J.*, 2(2), 19, 1988.
47. Snover, K.L. and Nelson, E.B., Evaluation of turfgrass phytonematode population distributions on a New York state putting green, *Phytopathology*, 91(6), S83, 2001.
48. Lukens, R.J. and Miller, P.M., Injury to turf grasses by *Tylenchorhynchus dubius* and *Hoplolaimus* spp., *Phytopathology*, 63(2), 204, 1973.
49. Settle, D.M. et al., Influence of soil temperature on host suitability of eight turfgrass species to the lance nematode (*Hoplolaimus galeatus*), *Int. Turf. Soc. Res. J.*, 10, 265, 2005.
50. Fassuliotis, G., Host range of the Columbia lance nematode, *Hoplolaimus columbus*, *Plant Dis.*, 58, 1000, 1974.
51. Golden, A.M. and Minton, N.A., Description and larval heteromorphism of *Hoplolaimus concaudajuvencus*, n. sp. (Nematoda: Hoplolaimidae), *J. Nematol.* 2, 161, 1970.
52. Henn, R.A. and Dunn, R.A., Reproduction of *Hoplolaimus galeatus* and growth of seven St. Augustinegrass (*Stenotaphrum secundatum*) cultivars, *Nematropica*, 19, 81, 1989.
53. Hixson, A.C., Crow, W.T., and Trenholm, L.E., Host status of 'SeaIsle 1' seashore paspalum (*Paspalum vaginatum*) to *Belonolaimus longicaudatus* and *Hoplolaimus galeatus*, *J. Nematol.*, 36, 493, 2004.
54. Settle, D.M. et al., Population dynamics of the lance nematode (*Hoplolaimus galeatus*) in creeping bentgrass, *Plant Dis.*, 90, 44, 2006.
55. Giblin-Davis, R.M., Busey, P., and Center, B.J., Parasitism of *Hoplolaimus galeatus* on diploid and polyploid St. Augustinegrasses, *J. Nematol.*, 27, 472, 1995.
56. Crow, W.T., unpublished data, 2006.
57. MacGowan, J.B., *Meloidogyne graminis, a Root-Knot Nematode of Grass*, Florida Department of Agriculture and Consumer Services, Gainesville, 1984.
58. Mitkowski, N.A., Root-knot nematodes on turf in the northeastern United States, *Turfgrass Trends*, 10(12), 1, 2001.

59. Jepson, S.B. and Golden, A.M., *Meloidogyne marylandi* n. sp. (Nematoda: Meloidogynidae), a root-knot nematode parasitizing grasses, in *Identification of Root-Knot Nematodes (Meloidogyne Species)*, Jepson, S.B., Ed., C.A.B. International, Wallingford, U.K., 1987.

60. Michell, R.E. et al., Races of the barley root-knot nematode, *Meloidogyne naasi*. III Reproduction and pathogenicity on creeping bentgrass, *J. Nematol.*, 5, 47, 1973.

61. Bélair, G., Simard, L., and Eisenback, J.D., First report of the barley root-knot nematode *Meloidogyne naasi* infecting annual bluegrass on a golf course in Quebec, Canada, *Plant Dis.*, 90, 1109, 2006.

62. Karssen, G. et al., Description of *Meloidogyne minor* n. sp. (Nematoda: Meloidogynidae), a root-knot nematode associated with yellow patch disease in golf courses, *Nematology*, 6, 59, 2004.

63. Heald, C.M., Pathogenicity and histopathology of *Meloidogyne graminis* infecting 'Tifdwarf' bermudagrass roots, *J. Nematol.*, 1, 31, 1969.

64. McClure, M.A., *Meloidogyne incognita*: a metabolic sink, *J. Nematol.*, 9, 88, 1997.

65. Perry, V.G. and Di Edwardo, A.A., *Heterodera leuceilyma* n. sp. (Nemata: Heteroderidae) a Severe Pathogen of St. Augustinegrass in Florida, University of Florida Agriculture Experiment Station, Gainesville, 1964.

66. La Mondia, J.A. and Wick, R.L., Occurrence of *Heterodera iri* in putting greens in the northeastern United States, *Plant Dis.*, 76, 643, 1992.

67. Pinkerton, J.N. and Alderman, S.C., Epidemiology of *Anguina agrostis* on 'Highland Colonial' bentgrass, *J. Nematol.*, 26, 315, 1994.

68. Cid del Prado Vera, I. and Maggenti, A.R., A new gall-forming species of *Anguina* Scopoli, 1777 (Nemata: Anguinidae) on bluegrass, *Poa annua* L., from the coast of California, *J. Nematol.*, 16, 386, 1984.

69. Vovlas, N., Histopathology of galls induced by *Subanguina radicicola* on *Poa annua* roots, *Plant Dis.*, 67, 143, 1983.

70. Mitkowski, N.A. and Jackson, N., *Subanguina radicicola*, the root-gall nematode infecting *Poa annua* in New Brunswick, Canada, *Plant Dis.*, 87, 1263, 2003.

71. Goodell, P. and Ferris, H., Plant-parasitic nematode distributions in an alfalfa field, *J. Nematol.*, 2, 136, 1980.

72. Barker, K.R., Nusbaum, C.J., and Nelson, L.A., Effects of storage temperature and extraction procedure on recovery of plant-parasitic nematodes from field soils, *J. Nematol.*, 1, 240, 1969.

73. Shurtleff, M.C. and Averre, C.W. III., *Diagnosing Plant Disorders Caused by Nematodes*, APS Press, St. Paul, MN, 2000.

74. Giblin-Davis, R.M. et al., Management practices affecting phytoparasitic nematodes in 'Tifgreen' bermudagrass, *Nematropica*, 21, 59, 1991.

75. Luc, J.E. and Crow, W.T., Sting nematode: Not a steward of the environment, *Golf Course Manage.*, 72(9), 86, 2004.

76. Giblin-Davis, R.M. et al., Host status of different bermudagrasses (*Cynodon* spp.) for the sting nematode, *Belonolaimus longicaudatus*, *J. Nematol.*, 24, 749, 1992.

77. Tarjan, A.C. and Busey, P., Genotypic variability in bermudagrass to damage by ectoparasitic nematodes, *HortScience*, 20(4), 675, 1985.

78. Busey, P., Giblin-Davis, R.M., and Center, B.J., Resistance in *Stenotaphrum* to the sting nematode, *Crop Sci.*, 33, 1066, 1991.

79. Giblin-Davis, R.M., Busey, P., and Center, B.J., Dynamics of *Belonolaimus longicaudatus* parasitism on a susceptible St. Augustinegrass host, *J. Nematol.*, 24, 432, 1992.

80. Genovesi, A.D., Engelke, M.C., and Elias, S., Interspecific hybridization between Pembagrass (*Stenotaphrum dimidiatum*) and St. Augustinegrass (*S. secundatum*) using embryo rescue, *Ann. Meet. Abstr.*, ASA, CSSA, and SSSA, Madison, WI. 2005.

81. Kenworthy, K.E., Turfrass breeding at the University of Florida, *Ann. Meet. Abstr.*, ASA, CSSA, and SSSA, Madison, WI. 2005.

82. Qu, R., Genetic transformation of bermudagrass for sting nematode resistance, *N. C. Turfgrass*, 18(3), 34, 2000.

83. Chen, Z.X. and Dickson, D.W., Review of *Pasteuria penetrans*: Biology, ecology, and biological control potential, *J. Nematol.*, 30, 313, 1998.

84. Giblin-Davis, R.M. et al., *Candidatus Pasteuria usgae* sp. nov., an obligate endoparasite of the phytoparasitic nematode *Belonolaimus longicaudatus*, *Int. J. Sys. Evol. Micro.*, 53, 197, 2003.

85. Giblin-Davis, R.M., *Pasteuria* sp. for biological control of the sting nematode, *Belonolaimus longicaudatus*, in turfgrass, in *Fate and Management of Turfgrass Chemicals*, Clark, J.M. and Kenna, M.P. Eds., Oxford University Press, Washington, DC, 2000.

86. Rungrassamee, W. and Wick, R.L., Relationship of the nematode hyperparasite *Pasteuria* spp. to *Meloidogyne graminis* and *Tylenchorhynchus* spp. in golf greens, *J. Nematol.*, 35, 361, 2003.

87. Ciancio, A., Mankau, R., and Mundo-Ocampo, M., Parasitism of *Helicotylenchus lobus* by *Pasteuria penetrans* in naturally infested soil, *J. Nematol.*, 24, 29, 1992.

88. Hewlett, T.E. et al., Occurrence of *Pasteuria* spp. in Florida, *J. Nematol.*, 26, 616, 1994.

89. Grewal, P.S. et al., Suppression of plant-parasitic nematode populations in turfgrass by application of entomopathogenic nematodes, *Biocontrol Sci. Tech.*, 7, 393, 1997.

90. Perez, E.E. and Lewes, E.E., Suppression of plant-parasitic nematodes in turfgrass after application of biologically-based treatments, *J. Nematol.*, 36, 340, 204.

91. Crow, W.T. et al., Entomopathogenic nematodes are not an alternative to fenamiphos for management of plant-parasitic nematodes on golf courses in Florida, *J. Nematol.*, 38, 52, 2006.

92. Kerry, B., Rhizosphere interactions and the exploitation of microbial agents for the biological control of plant-parasitic nematodes, *Ann. Rev. Phytopath.*, 38, 423, 2000.

93. Bacetty, A.A., Snook, M.E., and Bacon, C.W., Nematotoxicity of alkaloids from endophyte-infected tall fescue, *Phytopathology*, 96(6), S7, 2006.

94. Eerens, J.P.J. et al., Influence of the ryegrass endophyte on phyto-nematodes, in *Neotyphodium/Grass Interactions*, Bacon, C.W. and Hill, N.S. Eds., Plenum, New York, 1997.

95. Siddiqui, Z.A., Role of bacteria in the management of plant parasitic nematodes: A review, *Biosci. Tech.*, 69, 167, 1999.

96. Unruh, J.B. et al., Fumigant alternatives for methyl bromide prior to turfgrass establishment, *Weed Tech.*, 16, 379, 2002.

97. Crow, W.T., Biologically derived alternatives to Nemacur, *Golf Course Manage.*, 73(1), 147, 2005.

98. Crow, W.T., Alternatives to fenamiphos for management of plant-parasitic nematodes on bermudagrass, *J. Nematol.*, 37, 477, 2005.

99. Davis, R.F., Sting nematode management in a bermudagrass putting green with a biological nematicide, *Bio. Cult. Tests Cont. Plant Dis.*, 14, 137, 1999.

100. Giblin-Davis, R.M., Year-long test fails to find effective nematicide, *Fl. Turf Dig.*, 18(2), 24, 2001.

101. Wick, R.L. and Nissenbaum, T., Evaluation of commercially-available fenamiphos-alternatives for reducing plant pathogenic nematodes in putting greens, *Phytopathology*, 88(9), S137, 1998.

102. Wick, R.L. and Nissenbaum, T., Evaluation of two consecutive seasons of Neo-Trol® for managing nematodes in turf, *Bio. Cult. Tests Cont. Plant Dis.*, 14, 136, 1999.

103. Wick, R.L. and Massoni, S.C., Evaluation of Neo-Tec® for controlling nematodes in golf greens, *Bio. Cult. Tests Cont. Plant Dis.*, 20, 2, 2005.

104. Wick, R.L. and Abbot, E., Evaluation of Dragonfire® and Rutopia® for controlling nematodes in golf greens, *Bio. Cult. Tests Cont. Plant Dis.*, 19, 4, 2004.

105. Giblin-Davis, R.M. and Crow, W.T., Nematicidal potential of mustard-based products excites researchers, *Fl. Turf Dig.*, 20(1), 18, 2003.

106. Crow, W.T., Yu, Q., and Chiba, M., Mustard bran as a biochemical nematicide for turfgrasses, *J. Nematol.*, 36, 313, 2004.

107. Cox, C.J. et al., Suppressing sting nematodes with *Brassica* sp., poinsettia, and spotted spurge, *Agron. J.*, 98, 962, 2006.

108. Hensley, J. and Burger, G., Nematicidal properties of furfural and the development for nematode control in various crops for the Unites States markets, *J. Nematol.*, in press.

Section VIII

Turfgrass Management and
Physiological Responses under
Environmental Stress Conditions

23 Acid Soil and Aluminum Tolerance in Turfgrasses

Haibo Liu, Christian M. Baldwin, and Hong Luo

CONTENTS

23.1 INTRODUCTION

The entire turfgrass industry in United States is conservatively estimated as being a $60 billion per year industry with a total area of more than 50 million acres, in which home lawns comprise about 65% and the remaining 35% include utility turf areas, golf courses, sports fields, and turf production (turfgrass seed and sod) areas [96]. Some of these managed turfgrass areas have unfavorable soil conditions such as high salinity and lower soil pH, particularly in the southeastern states and the northern boarder states between United States and Canada. In the world, approximately 50% of the arable soils is acidic with a pH of 5.0 or lower, at which the soil-soluble aluminum (Al) becomes a significant limiting factor to plants [28,47,49].

Soil acidity is one of the major problems in the establishment and maintenance of turfgrass in many areas of the world. The earth's crust is comprised of about 8% Al by weight [38,78]. About one half of the world's cultivated land area has acid soil problems and [16,28,47] aluminum toxicity is a major growth-limiting factor in many of those acid soils [28,29]. Depending on species, an Al concentration can be toxic to a turfgrass when soil pH is below 5.0 [58]. Aluminum toxicity to crops was first reported more than 80 years ago [35]. However, the mechanisms of Al toxicity and tolerance still remain complicated with inconsistent results. Research has proved that Al toxicity levels to plants depend on the interaction among species, cultivars, environmental conditions, and soil conditions [23,47]. The mechanisms of Al tolerance among plants are genetically controlled, while plants respond to acid soil and Al toxicity in complicated ways [20,47,49].

A newly released acid soil–resistant cultivar of tall fescue (*Festuca arundinacea* Shreb.), "Southeast," [20] can tolerate acid soils down to pH 4.0 (Al saturation = 75%) and is more persistent compared with traditional cultivars. The new breakthroughs of transgenic Al-resistant plants or organisms [15,26,27] and diversities of turfgrass species [2,7,8,11,43,81,95,103,104,112] provide a promising future for turfgrass enhancement to acid soil stress. This chapter attempts to focus on the acid soil complex, Al toxicity resistance, and enhancement of turfgrasses for acid soil resistance.

23.2 ALUMINUM SPECIES IN SOILS

Aluminum is the third most abundant element, next to oxygen and silicon, in soils and the hydrolysis of Al species contributes to soil acidity by releasing H^+ into soil solutions [77]. Depending on soil pH, the main soil Al molecular forms exist in five major species of hydroxyaluminum: $Al(H_2O)_6^{3+}$, $AlOH^{2+}$, $Al(OH)_2^+$, $Al(OH)_3$, and $Al(OH)_4^-$. When soil pH is lower than 5.0 $Al(H_2O)_6^{3+}$— octahedral hexahydrate, known simply as Al^{3+}—is the dominant species. On the basis of several studies [33,47], Al^{3+} is the most toxic species to plants. However, two other species $AlOH^{2+}$ and $Al(OH)_2^+$ will become dominant when soil pH ranges from 5 to 6. In general, these two species are not as toxic as Al^{3+} to plants [44–47]. When soil pH is neutral, $Al(OH)_3$ (gibbsite) is formed, which is not toxic to plants and relatively insoluble. Another hydroxy Al ion called aluminate, $Al(OH)_4^-$, is also toxic to plants and is a dominant Al species in alkaline conditions (pH > 8.0) [44].

Under soil or hydroponic conditions, Al species combine to form potentially toxic polynuclear polymers including triskaidekaaluminum ($AlO_4Al_{12}(OH)_{24}(H_2O)_{12}^{7+}$), or simply Al_{13}. Kinraide [46] re-ranked the toxicity of Al species in wheat (*Triticum aestivum* L.) with the following order: $Al_{13} > Al^{3+} > AlF^{2+} > AlF_2^+$, with the two additional Al species of mononuclear fluroaluminum found toxic to wheat. Plants may vary in Al sensitivity to different Al species [13]. However, information on Al toxicity of Al species to a wide range of plants and interaction with other cations and anions is still lacking.

The concentration of soluble Al (toxic Al) in an acid soil can range from several micromoles (μM) to several millimoles (mM), depending on the soil condition. For turfgrass screening at the cultivar level using a hydroponic culture with a pH 4.0, an Al concentration of 160–320 μM has been suggested for sensitive species such as Kentucky bluegrass (*Poa pratensis* L.), 320–640 μM for resistant species such as fine fescues (*Festuca* spp.), and 640–960 μM for most warm-season turfgrasses [3,63,64].

23.3 ALUMINUM UPTAKE BY PLANTS

Aluminum uptake by root systems is a nonmetabolic exchangeable process at the Ca-binding site on the root cell surface [94]. The uptake process starts as soon as the root is exposed to Al. After only a very short period of time (in minutes), Al is found in root epidermal cells. The root apex is the major Al uptake site and removal of the root cap mucilage prior to Al exposure enhances Al uptake [94].

Cell walls are the major binding sites of Al before it enters the cytoplasm due to the negative charges of pectic substances inside cell walls. However, the plasma membrane functions as an additional barrier to passive Al movement into cells before it is saturated or destroyed. In root cells of Al-resistant plants, the production of Al-binding compounds such as inorganic phosphate, organic acids (citrate, malate, oxalate), and some proteins (phospholipase, glucanase, NAD kinase) can be induced to bind free Al to form nontoxic compounds [48,85,108].

The transport of Al through cortical cells is both symplasmic and apoplasmic. Aluminum enters the central cylinder of the root and moves upward to leaves. Accumulation of Al in roots follows a linear relationship with time of Al exposure to the root [118]. However, Al uptake is significantly reduced at a lower temperature (0°C) compared with a higher temperature (23°C) [118]. High cation NH_4^+ and H^+ concentrations in the rhizosphere will facilitate Al uptake with a soil pH < 5.0 [47,108].

23.4   ALUMINUM TOXICITY AND FACTORS AFFECTING Al TOXICITY

23.4.1   ACID SOIL CONDITIONS AND AFFECTING FACTORS

Major growth-limiting factors associated with acid soil stress conditions have been defined in Table 23.1 [31,47,49,50,77,78,105]. The relative importance and interactions between these factors depend on soil type and texture, parent material, soil pH, soil structure, soil organic matter, soil compaction and aeration, soil microbial activity, soil nutrient leaching characteristics, fertility management, soil amendment, plant species, cultivar, and climatic conditions.

Several limiting factors, listed in Table 23.1, impact turfgrass management under low soil pH (<5.0) conditions. Al or Mn toxicities are most likely the primary limiting nutrient factors on most acid soils. A turfgrass model relating Al stress and other abiotic stress conditions has been well documented by Duncan and Carrow [22]. Liming has traditionally alleviated low soil pH problems for turfgrasses; however, it does not efficiently reduce the acidity of subsoils, especially in perennial turfgrass swards. Excess application of acid-forming fertilizers, low-pH irrigation water, acid rainfall, poor soil aeration (black layer formation, anaerobic conditions), low soil organic matter, and drought conditions will enhance acid soil stress on turfgrasses.

The most significant factor affecting Al toxicity to plants is soil pH. A soil pH above 5.5 reduces Al solubility significantly and Al toxicity is normally not a problem. Soluble Al concentration in a soil is obviously critical for the level of Al toxicity and can be affected by the soil chemical and physical properties and availability of other minerals in ionic forms. An acid soil in most cases has a poor nutrient balance of Ca, P, and Mg, which play positive roles in reducing Al toxicity [38,78].

TABLE 23.1
The Major Growth-Liming Factors Associated with Acid Soil Conditions

Growth-Limiting Factors	Plant Symptoms and References
Al toxicity	Short and abnormal roots and eventually poor shoot growth [31,44,50]
H^+ toxicity	Excessive H^+ in the rhizosphere competes with Ca^{2+} and Mg^{2+} uptake to cause Ca and Mg deficiency [31,49,77]
Mn toxicity	Interveinal chlorosis and necrosis [31,50]
Mg, Ca, and K deficiency	Discoloration, chlorosis, poor efficiency in water use [31,49,105]
P, Mo, B, and Fe deficiency	Yellow, or purple coloration [106]
Reduced nitrogen uptake	Reduced uptake of nitrate: nitrogen deficiency with reduced microbial activities, chlorosis [106]
Reduced root growth	Short roots and reduced root hairs, brittle, stubbly, black in color [47,50]
Increased nutrient leaching potential	Lower nutrient concentrations in plant tissues and further nutrient deficiency due to poor nutrient removal by plants [77]
Poor water and nutrient uptake by roots	Plasma membrane damage and greater vulnerability to drought: poor retention of water in root cells [49,77]

TABLE 23.2

Possible Targets of Al Toxicity in Higher Plants

Categories	Evidences	References
Tissue Targets		
Root apex	Al inhibits root cell division and elongation at the root transition zone	[49,50,99,110,116]
Root cap	Progressive root cap vacuolation	[4]
Root hair	Al reduces root hairs	[10]
Cellular Targets		
Cell wall	Al increases pectin, hemicellulose, and cellulose contents in cell walls. It targets elongating cells by reducing cell wall extensibility in wheat roots	[55,67]
Plasma membrane	Al inhibits cation transport	[17,49]
Mitochondria	Al inhibits respiratory activities	[42]
Cytoskeleton and microtubules	Al causes reorganization and stabilization of cytoskeletons	[6,49,100,110]
Vacuole	Al increases vacuolation	[76]
Symplast	Al accumulation in nuclei	[94]
Apoplast	Al accumulation between cells	[110]
Molecular Targets		
DNA	Disturbance of DNA replication	[1]
ATP	Reduced ATP production: shortage of inorganic phosphate	[106]
GTP	Reduced GTP functions	[36]
Calmodulin	Reduced calmodulin functions	[1,36]
Proteins/enzymes	Impacts to membrane-bound proteins, ATPase, and NAD-kinase	[101,106,107]
Interacting with Other Inorganic Cations or Anions		
Ca^{2+}	Disruption of Ca signal pathways	[71]
H^+	Alternate cytosolic pH	[49]
Toxic free radicals of oxygen	Damage to membrane lipids	[41,115]

More severe Al toxicity to plants is found in clay subsoil and is often associated with Mn toxicity as well [23,28,31].

Nitrogen forms have impacts on soil pH; e.g., NH_4^+–N lowers soil pH, therefore increasing Al toxicity [112]. Excessive P in soil solutions will detoxify Al [3], while silicon has been reported to be a beneficial nutrient reducing Al toxicity [25,40].

23.4.2 Aluminum Toxicity

The most puzzling question of Al toxicity is where exactly Al damages plants. Possible Al toxicity targets in higher plants are categorized in Table 23.2. Excess Al interferes with cell division in roots; inhibits the respiratory activity in mitochondria; inhibits cation transport across plasma membranes; increases pectin, hemicellulose, and cellulose contents of roots; blocks nutrient uptake; and interferes with water use of plants [31,47,105,106]. Because of the high positive charges of the toxic molecular forms—Al^{n+} (Al^{3+}, Al_{13}, and others), Al damage in plant cells is mainly associated with binding affinity for enzymes such as calmodulin, cellulose microfibrils, ATP, GTP, and DNA [36,37,47,48,50,79].

23.5 ALUMINUM TOLERANCE IN TURFGRASSES

Plants have evolved multiple mechanisms of Al resistance. These mechanisms, either genetically controlled or induced include (1) altering toxic Al forms to nontoxic or less toxic forms; (2) excluding toxic Al from entering into roots; (3) efflux of Al outside the root or transporting it to less sensitive cellular compartments; and (4) tolerating Al by keeping high Al concentrations in different tissues (a very small group of plants has been identified as Al accumulators). In the past two decades, more and more evidence has been obtained for the involvement of organic acids, including citrate, malate, and oxalate, in facilitating the above mechanisms depending on species [50]. Either in plants or the rhizosphere, toxic Al species and organic acids can form chelating compounds that are not toxic to plants. The interesting recent evidences reported include the following: (1) Al-tolerant genotypes can react strongly to activate exudation of Al-chelating organic acids, while such an action is absent in Al-sensitive genotypes; (2) different plants tend to have their own dominant organic acid species to reduce Al toxicity; (3) to have enough organic acids to chelate Al, anion channels of plasma membrane must be further enhanced and the controlling or signaling mechanisms remain unclear; and (4) inside plants, the chelated compound with Al can be substitute among the three organic acids [50]. In addition, symbiotic relationships with microorganisms associated with Al resistance have been reported in different groups of plants [58,77]. The possible resistant mechanisms are summarized in Table 23.3.

Limited research has been published on acid soil–field response in turfgrasses. The available information on Al resistance among turfgrasses (Table 23.4) emphasizes that potential enhancement of Al resistance in turfgrasses is promising. The newly released tall fescue cultivar with Al resistance will have great market value as a transition zone turfgrass [20], where soil acidity and drought stress are dual constraints. More studies are needed to further understand the potential and value of Al resistance in other turfgrasses.

Low soil pH and Mn have a suppression effect on patch diseases (*Magnoporthe poae* Landshoot and Jackson; *Gaeumannomyces gramini; Ophiosphaerella herpotricha; Leptosphaeria spp.*) for both cool- and warm-season turfgrasses [39,109,111]. However, whether Al is toxic to pathogens or whether the turfgrasses studied have better adaptability with a slightly acid soil (5.2–5.6) is

TABLE 23.3
Possible Al-Tolerant Mechanisms of Plants

Mechanisms	Evidence	Reference
External Exclusion Mechanisms		
Exclusion: the Al-resistant root exudes organic acid into the rhizosphere to chelate Al in a soluble but nontoxic form	Organic acid excretion: mainly citrate, malate, and oxalate	[19,68,69,84]
Internal Tolerant Mechanisms		
Al accumulation in different tissues	Plants of the tea family (*Theaceae*) and rhododendrons (*Rhododendron spp.*). Buckwheat (*Fagopyrum esculentum* Moench. cv Jianxi) accumulates aluminum in leaves but not in seeds	[69,77,98]
Internal binding in apoplast and symplast	Cell walls, membranes, organic acid chelates, proteins, and other compounds	[46,47,106]
Abnormal of cell growth and development	Callose production Compartmentation in vacuoles	[80,114]
Plant hormone regulations	Signaling pathway in Al resistance through basipetal auxin transport and abscisic acid (ABA) involvement to open anion channel for organic acids to chelate Al	[51,70]
Environmental Detoxification Mechanisms		
Alternation of rhizosphere pH	Increased H^+ influx	[18]
Symbiotic partners: an ecological mechanism	Mycorrhizae, N_2-fixing bacteria, endophytes can alleviate Al toxicity, but the mechanisms are unclear	[8,18,63,75,102,117]

TABLE 23.4

Acid Soil and Al Tolerance among Turfgrass Species

Turfgrasses	Findings	References	Potentials for Al Tolerance Improvement
Cool-Season Turfgrasses			
Tall fescue (*Festuca arundinacea* Schreb.)	Acid soil–resistant cultivar: "Southeast"; diversity among cultivars	[20,24,29,60,83,86,87]	Medium to high
Hard fescue (*Festuca longifolia* Thuill)	More Al resistance	[63]	High
Chewings fescue (*Festuca rubra* L. subsp. *fallax* [Thuill.] Nyman.)	More Al resistance and some endophyte-infected cultivars with enhanced Al resistance	[24,63,87,117]	High
Creeping red fescue (*Festuca rubra* L. ssp. *rubra*)	Less Al resistance and some endophyte-infected cultivars with enhanced Al resistance	[24,63,87,117]	High
Sheep fescue (*Festuca ovina* L. spp. *Ovina*)	Least Al resistance among 58 cultivars compared (four species or subspecies compared)	[63]	Low
Creeping bentgrass (*Agrostis stolonifera* L.)	Diversity among 16 cultivars; comparison between creeping bentgrass and annual bluegrass in Al resistance	[52,53,61]	Medium
Colonial bentgrass (*Agrostis capillaries* L.)	Greater Al resistance	[24,61,74]	High
Kentucky bluegrass (*Poa pratensis* L.)	Diversity in Al resistance among cultivars and the least Al resistant among three *Poa* species	[24,30,62,64,86,87]	High
Annual bluegrass (*Poa annua* L.)	More acid soil resistance than Kentucky bluegrass	[52,53,64]	High
Supina bluegrass (*Poa supina* Schrad.)	Greatest Al resistance among three *Poa* species compared	[64]	Very high
Perennial ryegrass (*Lolium perenne* L.)	Diversity among cultivars	[63,72,90–92]	Medium
Annual ryegrass (*Lolium multiflorum* Lam.)	Greater Al resistance than perennial ryegrass	[89,90–92]	Medium
Intermediate ryegrass (*Lolium xhybridium* Hausskn.)	Diversity in Al resistance comparing perennial ryegrass and annual ryegrass	[24]	Medium
Warm-Season Turfgrasses			
Bermudagrass (*Cynodon dactylon* [L.] Pers.)	Difference in root length and root mass was found among cultivars	[3,24,62]	High
"TifEagle" hybrid bermudagrass (*Cynodon dactylon* [L.] Pers. X C. *transvaalensis* Burtt-Davy)	Relatively Al sensitive among 10 other warm-season turfgrasses	[3]	Low
Zoysiagrass (*Zoysia japonica* Steud.)	Cultivar diversity under acid soil complex conditions. Al tolerant	[3,21]	High
Bahiagrass (*Paspalum notatum* Flugge)	Neutral	[3]	Medium
Centipedegrass (*Eremochloa ophiuroides* [Munro] Hack)	Neutral	[3]	Medium
Buffalograss (*Buchloe dactyloide* [Nutt.] Engelm.)	Neutral	[3]	Medium

(Continued)

TABLE 23.4
(Continued)

Turfgrasses	Findings	References	Potentials for Al Tolerance Improvement
Common carpetgrass (*Axonopus affinis* Chase)	The most Al-tolerant species among 10 warm-season turfgrasses	[3]	Very high
Seashore Paspalum (*Paspalum vaginatum* Swartz.)	Some resistance to acid soils	[3,22]	Medium
St. Augustinegrass (*Stenotaphrum secundatum* [Walt.] Kuntze.)	Neutral	[3]	Medium

not clear. Winter-kill of bermudagrass associated with low soil pH in the upper transition zone has been observed by turfgrass scientists and superintendents in different locations in Kentucky (A. J. Powell, Personal Communication), but studies are needed to identify the specific effects of low soil pH on winter-kill in turfgrasses, such as decreased rhizome and stolon production, which is one of the symptoms associated with winter-kill.

Tremendous research effort has been contributed to the two key elements Ca and P for reducing Al toxicity in plants [47,48,105]. In cells, calcium mainly competes with Al^{n+} for binding sites, while phosphorus in phosphate form precipitates Al^{3+} to nontoxic forms [47]. Silicon and boron have been reported to reduce Al toxicity [12,34,56,57,77,108] in different plants.

23.6 TURFGRASS IMPROVEMENT FOR ACID SOIL AND Al RESISTANCE

23.6.1 DEVELOPMENT OF Al-RESISTANT TURFGRASSES

The possibility of developing Al-resistant turfgrasses exists and a more likelihood of developing turfgrasses with multiple resistances to unfavorable edaphic conditions including Mn tolerance and better P uptake efficiency. Research and identification of Al-resistant genes are poorly documented in turfgrasses. Field breeding techniques for enhancing acid soil stress resistance are well described by Duncan [20]. For Al resistance screening, the parameters may include root growth and biomass, root hair production, organic acid excretion, and rhizosphere pH adjustments. Cellular and ultrastructure screening methods include bioassays of Al concentrations in organelles; monitoring Al symplasmic and apolasmic movements; plant hormone involvement and regulation; membrane transport of Al; and cell wall Al-binding capacity. Due to the lack of Al isotopes, accurate Al concentration measurements of Al in plant tissues are difficult. Field screening is essential, especially when combined with other abiotic stresses and severe environmental conditions. Multiple breeding methodology and selection strategies using actual field sites are effective for acid soil resistance of crops including turfgrasses [20,22]. Due to the combined effects of acid soil toxicity and nutrient use efficiency [20], multiple trait selection in the field can be done and detailed breeding techniques for Al toxicity were summarized by Duncan [20].

The most needed turfgrass types for acid soil and Al tolerance include both cool- and warm-season turfgrasses. Bluegrass, the most popular cool-season turfgrass based on the total area used, and bermudagrass, the most popular warm-season turfgrass, are two most wanted and promising turfgrasses [59,60] for enhanced acid soil and Al tolerance due to their popularities as home-lawn turfgrasses in the world. Difficulties were found to screen tall fescue and perennial ryegrass for Al tolerance (Liu, unpublished data) and partial reasons might be the relative narrow genetic pools of the two species used for screening and the origins of both turfgrasses with neutral soil conditions.

23.6.2　Possibilities and Dilemma of Al-Tolerant Genes

Development of transgenic Al-resistant turfgrasses is possible. However, according to recent reports on different crops, both multiple genes and single genes of Al resistance have been identified [2,5,19,20,22]. Most of these genes identified are directly Al tolerant and they are secondarily related to organic acid chelating of toxic Al species [49,50,112]. Al induced genes can be isolated by different molecular strategies and genetic engineering techniques and these techniques can be used to develop Al-tolerant cultivars [16,27]. Over 50 Al-tolerant genes, most of them related to enhancement of organic acid production and chelation of toxic Al species, have been isolated in several plants or crops. Transgenic Al-tolerant plants demonstrated enhanced Al tolerance but the dilemma is lower rate of positive expression and many other stressful factors triggering gene expressions [112]. Recently, most attention has been turned to anion channel transporter genes related to citrate, oxalate, and malate but a very valid question still remains: why use a four-carbon organic acid molecule to "trap" one molecule of Al when the "carbon" costs are too expensive for plants, especially annual crops and grasses?

The production of transgenic turfgrasses with acid tolerance is feasible [7,16,27,49,50,112] with new breakthrough knowledge and technologies such as functional genomics and molecular techniques. A common feature of Al-tolerant plants is the exudation of organic acids in the rhizosphere to detoxify Al at the root–soil interaction or the detoxification internally [49,50,112]. Among these organic acids, citrate is the most commonly reported [16,47,106]. Studies of several such plants have successfully produced transgenic tobacco (*Nicotiana tabacum* L.) and *papaya* (*Carica papaya*) plants that overexpress the citrate synthase gene from *Pseudomonas aeruginosa* in their cytoplasm [15,49,50,112]. Aluminum-tolerant mutants of *Arabidopsis thaliana* L. provide important tools to study Al tolerance as well [27,49,50,54]. In a recent review by Duncan and Carrow [22], more than a dozen genes, proteins, and enzymes have been listed. These genetic materials can be used for genetic improvement of turfgrasses for acid soil and Al tolerance.

23.7　TURFGRASS MANAGEMENT FOR ACID SOIL CONDITIONS

23.7.1　Liming

Liming is commonly used to reduce soil acidity and Al toxicity for crops. For turfgrass management, liming materials are very common soil amendments normally applied after soil aerification or during the nongrowing season to reduce phytotoxicity to turfgrasses. Consistent lime application can cause nutrient imbalance, or deficiency of other cations such as K^+. Liming is often inefficient due to the slow movement and reactions in the soil [38,58,112].

23.7.2　Select Acid Soil and Al-Resistant Turfgrasses

Using acid soil and Al-tolerant turfgrass is a promising alternative because of the following reasons or advantages:

- All turfgrasses used are a group of grasses adapted by human beings as turfgrasses without limitations for yield and total biomass production.
- Significant genetic diversities exist among several important species such in genera of *Poa*, *Cynodon*, *Festuca*, *Agrostis*, and *Lolium*.
- The possibility exists that acid soil– and Al-tolerant cultivars of a certain species may also have enhanced resistance to other stresses such as drought, temperature extremes, and even salinity [3]. At least, the hypothesis that a superior turfgrass cultivar resistant to acid soil and Al has enhanced resistances to other edaphic stresses should be further tested.

23.7.3   Improve P Supply Plus Proper Ca and Si Supplies

Next to nitrogen, soils can easily become deficient in phosphorous although turfgrasses need amounts of P only a little over one-tenth of the nitrogen required by tissue concentrations. P availability in soil depends on its pH, and is significantly reduced when the soil pH is <5.0. During the seedling stage of turfgrass establishment, for rigorous root growth supplemental P will benefit the turf because there is poor absorption of and high demand for P. Soluble Al reacts with phosphate and additional P will reduce Al toxicity.

One of the possible Al toxicities is that Al^{3+} substitutes Ca^{2+} in plants, thereby interrupting Ca^{2+} functions as an essential nutrient. Additional Ca supply will reduce Al toxicity in addition to the replacement of Al at the cation exchange capacity (CEC) sites.

Silicon (Si), the most abundant element in soils contributing to about 27% soil mass, as a beneficial nutrient has been intensively investigated [25], and recently, the Si transporters in rice (*Oryza sativa* L.) across root plasma membrane have been identified [66]. Although the Si mechanisms that reduce Al toxicity are not clear, the positive impacts on turfgrasses, in particular in disease control have been reported [14]. Further research, including on the long-term impacts on thatch, is needed.

23.7.4   Symbiotic Relationships with Endophyte or Mycorrhizae

Symbiotic relationships with endophyte or mycorrhizae exist among turfgrass species, and such relationships are helpful to alleviate Al toxicity although the mechanisms are still not fully understood [63,117]. The symbiotic relationships with mycorrhizae are benefited primarily by better P acquisition and maybe by other nutrients that can alleviate Al toxicity. The significant amounts of alkaloids produced in host plants with endophyte infection may play a role in reducing Al toxicity either directly or indirectly. There may also be other environmental factors affecting the symbiotic relationships and thereby altering Al tolerance.

23.7.5   Reduce NH_4^+-Type Fertilizers

Ammonium-type fertilizers reduce soil pH and consistent use of NH_4^+-type fertilizers will cause acid soil problems. In addition to the use of nitrate-type fertilizers to alleviate acid soil conditions, foliar fertilization may be another alternative to turfgrasses under acid soil and Al stress. Excessive NH_4^+ uptake by plants causes physiological disorders termed ammonium toxicity [9], and more research is needed to investigate relationships of cation toxicities to plants.

23.8   SUMMARY AND PROSPECTS

Aluminum toxicity to crops and other higher plants is a global issue and Al toxicity and tolerance among plants are much better understood in some major agronomic cereal crops. To date, relatively limited research and data have been reported on both warm- and cool-season turfgrass species on Al tolerance. Although several secondary Al-tolerant mechanisms have been reported for a wide range of plants, the primary ones still remain unknown, such as the groups of Al accumulators. Despite the first acid soil–resistant turfgrass being released using multiple field breeding strategies, both traditional breeding for acid soil– and Al-tolerant turfgrasses and application of biochemical, genetic, and physiological approaches to develop turfgrasses with Al resistance have been used. Mechanisms governing acid soil resistance among turfgrasses must be better understood before major advances can be made in genetic improvement programs. However, screening superior varieties or cultivars among turfgrasses for acid soil and Al tolerance has always been needed, and there is a strong possibility of developing turfgrasses with multiple stress resistances, including Al toxicity, Mn toxicity, P stress, and drought.

REFERENCES

1. Andersson, M. 1988. Toxicity and tolerance of aluminum in vascular plants. *Water Air Soil Pollut.* 39: 449–462.
2. Aniol, A. and J. P. Gustafon. 1984. Chromosome location of genes controlling aluminum tolerance in wheat, rye, and triticale. *Can. J. Gen. Cytol.* 26: 701–705.
3. Baldwin, C.M., H. Liu, L. B. McCarty, W. B. Bauerle, and J. E. Toler. 2005. Aluminum tolerance of warm-season turfgrasses. *Inter. Turfgrass Soc. Res. J.* 10: 811–817.
4. Bennet, R. J., C. M. Breen, and M. V. Fey. 1985. The primary site of aluminum injury in the root of *Zea mays* L. *S. Afr. J. Plant Soil* 2(1): 8–17.
5. Bernzonsky, W. A. 1992. The genomic inheritance of aluminum tolerance in "Altas 66" wheat. *Genome* 35: 689–693.
6. Blancaflor, E. B., D. L. Jones, and S. Gilroy. 1998. Alterations in the cytoskeleton accompany aluminum-induced growth inhibition and morphological changes in primary roots of maize. *Plant Physiol.* 118: 159–172.
7. Bonos, S. A., B. B. Clarke, and W. A. Meyer. 2006. Breeding for disease resistance in the major cool-season turfgrasses. *Annu. Rev. Phytopath.* 44: 213–234.
8. Borrie, F. and R. Rubio. 1999. Effects of arbuscular mycorrhizae and liming on growth and mineral acquisition of aluminum-tolerant and aluminum-sensitive barley cultivars. *J. Plant Nutr.* 22(1): 121–137.
9. Britto, D. T. and H. J. Kronzuker. 2002. NH_4^+ toxicity in higher plants: a critical review. *J. Plant Physiol.* 159: 567–584.
10. Care, D. A. 1995. The effect of aluminum concentration on root hairs in white clover (*Trifolium repens* L.). *Plant Soil* 171: 159–162.
11. Casler, M. D. and R. R. Duncan. 2003. *Turfgrass Biology, Genetics, and Cytotaxonomy.* Sleeping Bear Press, Chelsea, MI.
12. Cocker, K. M., D. E. Evans, and M. J. Hodson. 1998. The amelioration of aluminum toxicity by silicon in wheat (*Triticum aestivum* L.) malate exudation as evidence for an in planta mechanism. *Planta* 204(3): 318–323.
13. Comin, J. J., J. Barloy, G. Bourrie, and F. Troland. 1999. Different effects of monomeric and polymeric aluminum on the root growth and on the biomass production of root and shoot of corn in solution culture. *Eur. J. Agron.* 11(2): 115–122.
14. Datnoff, L. E. and B. A. Rutherford. 2003. Accumulation of silicon by bermudagrass to enhance disease suppression of leaf spot and melting out. *USGA Turfgrass Environ. Res. Online* 2: 1–6.
15. De la Fuente, J. M., V. Ramrez-Rodriguez, J. L. Cabrera-Ponce, and L. Herrera-Estrella. 1997. Aluminum tolerance in transgenic plants by alteration of citrate synthesis. *Science* 276: 1566–1568.
16. De la Fuente-Martinez, J. M. and L. Herrera-Estrella. 1999. Advances in the understanding of aluminum toxicity and the development of aluminum tolerant transgenic plants. *Adv. Agron.* 66: 103–120.
17. De Lima, M. and L. Copeland. 1994. The effect of aluminum on respiration of wheat roots. *Physiol. Plantar.* 90: 51–58.
18. Degenhardt, J., P. B. Larsen, S. H. Howell, and L. V. Kochian. 1998. Aluminum resistance in the arabidopsis mutant *alr*-104 is caused by an aluminum-induced increase in rhizosphere pH. *Plant Physiol.* 117: 19–27.
19. Delhaize, E., S. Craig, C. D. Beaton, R. J. Bennet, V. D. Jagadish, and P. J. Randall. 1993. Aluminum tolerance in wheat (*Triticum aestivum* L.). I. Uptake and distribution of aluminum in root apices. *Plant Physiol.* 103: 685–693.
20. Duncan, R. R. 2000. Plant tolerance to acid soil constraints: genetic resources, breeding methodology, and plant improvement. In: *Plant Environment Interaction* (2nd Edition). R. E. Wilkinson (Ed.). Marcel Dekker, NY, pp. 1–38.
21. Duncan, R. R. and L. M. Shuman. 1993. Acid soil stress response of zoysiagrass. *Intern. Turfgrass Soc. Res. J.* 7: 805–811.
22. Duncan, R. R. and R. N. Carrow. 1999. Turfgrass molecular genetic improvement for abiotic/edaphic stress resistance. *Adv. Agron.* 67: 233–305.
23. Duncan, R. R. and R. N. Carrow. 2000. *Seashore Paspalum—The Environmental Turfgrass.* Ann Arbor Press, Chelsea, MI, 281 pp.
24. Edmeades, D. C., D. M. Wheeler, and R. A. Christie. 1991. The effects of aluminum and pH on the growth of a range of temperate grass species and cultivars. In: *Plant–Soil Interactions at Low Soil pH.* R. J. Wright, V. C. Baligar, and R. P. Murrmann (Eds.). Kluwer Academic Publishers, Netherlands, pp. 925–930.

25. Epstein, E. 1999. Silicon. *Annu. Rev. Plant Physiol. Plant Mol. Biol.*, 50: 641–664.
26. Ezaki, B., M. Sivaguru, Y. Ezaki, H. Matsomoto, and R. C. Gardner. 1999. Acquisition of aluminum tolerance in *Saccharomyces cerevisiae* by expression of the BcB or NtGDI1 gene derived from plants. *FEMS Microb. Lett.* 171: 81–87.
27. Ezaki, B., R. C. Gardner, Y. Ezaki, and H. Matsumoto. 2000. Expression of aluminum-induced genes in transgenic arabidopsis plants can ameliorate aluminum stress and/or oxidative stress. *Plant Physiol.* 122: 657–665.
28. Foy, C. D. 1988. Plant adaptation to acid, aluminum-toxic soils. *Commn. Soil Sci. Plant Anal.* 19: 959–987.
29. Foy, C. D. and J. J. Murray. 1998. Developing aluminum-tolerant strains of tall fescue for acid soils. *J. Plant Nutr.* 21(6): 1301–1325.
30. Foy, C. D. and J. J. Murray. 1998. Responses of Kentucky bluegrass cultivars to excess aluminum in nutrient solutions. *J. Plant Nutr.* 21(9): 1967–1983.
31. Foy, C. D., R. L. Chaney, and M. C. White. 1978. The physiology of metal toxicity in plants. *Annu. Rev. Plant Physiol.* 29: 511–566.
32. Foy, C. D. 1992. Soil chemical factors limiting plant root growth. In: *Advances in Soil Science: Limitation to Plant Root Growth.* Vol. 19. J. L. Hatfield and B. A. Stewart (Eds.). Springer-Verlag, New York, pp. 97–149.
33. Foy, C. D., B. Scott, and J. A. Fisher. 1988. Genetic differences in plant tolerance to manganese toxicity. In: *Manganese in Soils and Plants.* R. D. Graham, R. J. Hannam, and N. C. Uren (Eds.). Kluwer, the Netherlands, pp. 293–307.
34. Hammond, K. E., D. E. Vens, and M. J. Hodson. 1995. Aluminum/silicon interactions in barley (Hordeum vulgare L.) seedlings. *Plant Soil* 173: 89–95.
35. Hartwell, B. L. and R. F. Pember. 1918. The presence of aluminum as a reason for the difference in the effect of so-called acid soils on barley and rye. *Soil Sci.* 6: 259–279.
36. Haug, A. 1984. Molecular aspects of aluminum toxicity. *CRC Crit. Rev. Plant Sci.* 1: 345–373.
37. Haug, A., B. Shi, and V. Vitorelloa. 1994. Aluminum interactions with phosphoinositide-associated signal transduction. *Arch. Toxicol.* 68: 1–7.
38. Havlin, J. L., J. D. Beaton, S. L. Tisdale, and W. L. Nelson. 2005. *Soil Fertility and Fertilizers.* 7th Edition. Prentice Hall, Upper Saddle River, NJ.
39. Heckman, J. R., B. B. Clarke, and J. A. Murphy. 2003. Optimizing manganese fertilization for the suppression of take-all patch disease on creeping bentgrass. *Crop Sci.* 43: 1395–1398.
40. Hodson, M. J. and D. E. Evans. 1995. Aluminium/silicon interactions in higher plants. *J. Exp. Bot.* 46(2): 161–171.
41. Horst, W. J., C. J. Asher, I. Cakmak, P. Szulkiewicz, and A. H. Wissemeier. 1992. Short-term responses of soybean roots to aluminum. *J. Plant Physiol.* 140: 174–178.
42. Huang, J. W., J. E. Shaff, D. L. Grunes, and L. V. Kochian. 1992. Aluminum effects on calcium fluxes at the root apex of aluminum-tolerant and aluminum-sensitive wheat cultivars. *Plant Physiol.* 98: 230–237.
43. Kenworthy, K. E., C. M. Taliaferro, B. F. Carver, D. L. Martin, J. A. Anderson, and G. E. Bell. 2006. Genetic variation in *Cynodon transvaalensis* Burtt-Davy. *Crop Sci.* 46: 2376–2381.
44. Kinraide, T. B. 1990. Assessing the rhizotoxicity of the aluminate ion, $Al(OH)_4^-$. *Plant Physiol.* 94: 1620–1625.
45. Kinraide, T. B. 1991. Identify of the rhizotoxic aluminum species. *Plant Soil* 134: 167–178.
46. Kinraide, T. B. 1997. Reconsidering the rhizotoxicity of hydroxyl, sulphate, and fluoride complexes of aluminum. *J. Exp. Bot.* 48(310): 1115–1124.
47. Kochian, L. V. 1995. Cellular mechanisms of aluminum toxicity and resistance in plants. *Annu. Rev. Plant Physiol. Plant Mol. Biol.* 46: 237–260.
48. Kochian, L. V. and D. L. Jones. 1997. Aluminum and resistance in plants. In: *Research Issues in Aluminum Toxicity.* R. A. Yokel and M. S. Golub (Eds.). Taylor & Francis Publishers, Washington, DC, pp. 69–89.
49. Kochian, L. V., M. A. Piñeros, and O. A. Hoekenga. 2005. The physiology, genetics and molecular biology of plant aluminum resistance and toxicity. *Plant Soil* 274(1–2): 175–195.
50. Kochian, L. V., O. A. Hoekenga, and M. A. Piñeros. 2004. How do crop plants tolerate acid soils? Mechanisms of aluminum tolerance and phosphorus efficiency. *Annu. Rev. Plant Biol.* 55: 459–493.
51. Kollmeier, M., H. H. Felle, and W. J. Horst. 2000. Genotypical differences in aluminum resistance of maize are expressed in the distal part of the transition zone. Is reduced basipetal auxin flow involved in inhibition of root elongation by aluminum? *Plant Physiol.* 122: 945–956.

52. Kuo, S. 1993. Effect of lime and phosphate on the growth of creeping bentgrass and annual bluegrass in two acid soils. *Soil Sci.* 156: 95–100.

53. Kuo, S. 1993. Calcium and phosphorus influence creeping bentgrass and annual bluegrass growth in acid soils. *HortScience* 28: 713–716.

54. Larson, P. B., J. Degenhardt, C. Y. Tai, L. M. Senzler, S. H. Howell, and L. V. Kochian. 1998. Aluminum-resistant Arabidopsis mutants that exhibit altered patterns of aluminum accumulation and organic acid release from roots. *Plant Physiol.* 117: 9–18.

55. Le Van, H., S. Kuraishi, and N. Sakurai. 1994. Aluminum-induced rapid root inhibition an changes in cell-wall components of squash seedlings. *Plant Physiol.* 106: 971–976.

56. Lenoble, M. E., D. G. Blevins, and R. J. Miles. 1996. Prevention of aluminum toxicity with supplemental boron. II. Stimulation of root growth in acidic, high-aluminum subsoil. *Plant, Cell and Environment* 19(10): 1143–1148.

57. Lenoble, M. E., D. G. Blevins, R. E. Sharp, and B. G. Cumbie. 1996. Prevention of aluminum toxicity with supplemental boron. I. Maintenance of root elongation and cellular structure. *Plant Cell Environ.* 19(10): 1132–1142.

58. Liu, H. 2001. Soil acidity and aluminum toxicity response in turfgrass. *Int. Turfgrass Soc. Res. J.* 9: 180–188.

59. Liu, H. 2005. Aluminum toxicity of seeded bermudagrass cultivars. *HortScience* 40(1): 221–223.

60. Liu, H., J. R. Heckman, and J. A. Murphy. 1997. Greenhouse screening of turfgrasses for aluminum tolerance. *Int. Turfgrass Soc. Res. J.* 8: 719–728.

61. Liu, H., J. R. Heckman, and J. A. Murphy. 1997. Aluminum tolerance among genotypes of *Agrostis* species. *Int. Turfgrass Soc. Res. J.* 8: 729–734.

62. Liu, H., J. R. Heckman, and J. A. Murphy. 1995. Screening Kentucky bluegrass for aluminum tolerance. *J. Plant Nutr.* 18(9): 1797–1814.

63. Liu, H., J. R. Heckman, and J. A. Murphy. 1996. Screening fine fescues for aluminum tolerance. *J. Plant Nutr.* 19(5): 677–688.

64. Liu, H., Y. He, and J. W. Adelberg. 2002. Aluminum tolerance among three *Poa* species. *Agronomy Abstracts*, Annual meeting, November 10–14, Indianapolis, IN.

65. Liu. H., J. R. Heckman, and J. A. Murphy. 1997. Screening bermudagrasses for aluminum tolerance. In: *Agronomy Abstracts*, p. 135, Annual meeting, October 26–31, Anaheim, CA.

66. Ma, J. F., K. Tamai, N. Yamaji, M. Mitani, S. Konishi, M. Katsuhara, M. Ishiguro, Y. Murata, and M. Yano. 2006. Silicon transporter in rice. *Nature* 440: 688–691.

67. Ma, J. F., R. Shen, S. Nagao, and E. Tanimoto. 2004. Aluminum targets elongating cells by reducing cell wall extensibility in wheat roots. *Plant Cell Physiol.* 45: 583–589.

68. Ma, J. F., S Taketa, and Z. M. Yang. 2000. Aluminum tolerance genes on the short arm of chromosome 3R are linked to organic acid release in Triticle. *Plant Physiol.* 122: 687–694.

69. Ma, J. F., S. J. Zheng, S. Hiradate, and H. Matsumoto. 1997. Detoxifying aluminum with buckwheat. *Nature* 390: 569–570.

70. Ma, J. F., W. Zhang, and Z. Zhao, Z. 2001. Regulatory mechanisms of Al-induced secretion of organic acids anions-Involvement of ABA in the Al-induced secretion of oxalate in buckwheat. In: *Plant Nutrition-Food Security and Sustainability of Agro-Ecosystems through Basic and Applied Research*. W. J. Horst (Ed.). Kluwer Academic Publishers, Hannover, pp. 486–487.

71. Ma, Q. F., Z. Rengel, and L. Kuo. 2002. Aluminum toxicity in rye (*Secale cereale*): root growth and dynamics of cytoplasmic Ca^{2+} in intact root tips. *Annals Bot.* 89: 241–244.

72. Mackey, A. D., J. R. Caradus, and S. Wewala. 1991. Aluminum tolerance of forage species. In: *Plant–Soil Interactions at Low Soil pH*. R. J. Wright, V. C. Baligar and R. P. Murrmann (Eds). Kluwer Academic Publishers, the Netherlands, pp. 925–930.

73. Macklon, A. E. S. and A. Sim. 1992. Modifying effects of a non-toxic level of aluminum on phosphate fluxes and compartmentation in root cortex cells of intact ryegrass seedlings. *J. Exp. Bot.* 43(256): 1483–1490.

74. Macklon, A. E. S., D. G. Lumsdon, and A. Sim. 1994. Phosphate uptake and transport in Agrostis capillaries L.: effects of non-toxic levels of aluminum and the significance of P and Al speciation. *J. Exp. Bot.* 45(276): 887–894.

75. Malinowski, D. P. and D. P. Belesky. 2000. Tall fescue aluminum tolerance is affected by *Neotyphodium coenophialum* endophyte. *J. Plant Nutr.* 22(8): 1335–1349.

76. Marienfeld, S., H. Lehmann, and R. Stelzer. 1995. Ultrastructured investigations and EDX-analyses of Al-treated oat (*Avena sativa*) roots. *Plant Soil* 171: 167–173.

77. Marschner, H. 1995. *Mineral Nutrition of Higher Plants.* 2nd Edition. Academic Press, New York, 889 pp.
78. Marschner, H., M. Haussling, and E. George. 1991. Mechanisms of adaptation of plants to acid soils. *Plant Soil* 134: 1–20.
79. Martin, R. 1988. Bioinorganic chemistry of aluminum. In: *Metal Ions in Biological Systems.* Vol. 24. H. Sigel and A. Sigel (Eds.). Marcel Dekker, New York, pp. 1–57.
80. Massot, N., M. Llugang, C. Proschenrieder, and J. Barcelo. 1999. Callose production as indicator of aluminum toxicity in bean cultivars. *J. Plant Nutr.* 22(1): 1–10.
81. Meyer, W. A. and C. R. Funk. 1989. Progress and benefits to humanity from breeding cool-season grasses for turf. In: *Contributions from Breeding Forage and Turfgrasses.* D. A. Spleper, K. H. Asay, and J. F. Pederson (Eds.). CSSA Special Publication No. 15, Madison, WI, pp. 31–48.
82. Meyer, W. A. and F. C. Belanger. 1997. The role of convention breeding as biotechnical approaches to improve disease resistance in cool-season turfgrasses. *Int. Turfgrass Soc. Res. J.* 8: 777–790.
83. Miles, N. and A. D. Manson. 1995. Effects of soil acidity and phosphorus on the yield and chemical composition of tall fescue. *Commn. Soil Sci. Plant Anal.* 26: 843–860.
84. Miyasaka, S. C., J. G. Buta, R. K. Howell, and C. D. Foy. 1991. Mechanism of aluminum tolerance in snapbeans. *Plant Physiol.* 96: 737–743.
85. Mukherjee, S. K. and S. Asanuma. 1998. Possible role of cellular phosphate pool and subsequent accumulation of inorganic phosphate on the aluminum tolerance in *Pradyrhizobium japonicum. Soil Biol. Biochem.* 30(12): 1511–1516.
86. Murray, J. J. and C. D. Foy. 1980. Lime responses of Kentucky bluegrass and tall fescue cultivars on an acid aluminum-toxic soil. *Proceedings Third International Turfgrass Research Conference,* pp. 175–183.
87. Murray, J. J., and C. D. Foy. 1978. Differential tolerances of turfgrass cultivars to an acid soil high in exchangeable aluminum. *Agron. J.* 70: 769–774.
88. Ramgareeb, S., M. P. Watt, C. Marsh, and J. A. Cooke. 1999. Assessment of Al[13] availability in callus culture media for screening tolerant genotypes of *Cynodon dactylon. Plant Cell Tissue Organ Culture* 56(10): 65–68.
89. Rengel, Z. 1990. Competitive Al^{3+} inhibition of net Mg^{2+} uptake by intact *Lolium multiflorum* roots. II. Plant age effects. *Plant Physiol.* 93: 1261–1267.
90. Rengel, Z. and D. L. Robinson 1989. Aluminum and plant age effects on adsorption of actions in Donnan free space of ryegrass roots. *Plant Soil* 116: 223–227.
91. Rengel, Z. and D. L. Robinson 1989. Aluminum effects on growth and micronutrient uptake in annual ryegrass. *Agron. J.* 81: 208–215.
92. Rengel, Z. and D. L. Robinson. 1989. Competitive Al^{3+} inhibition of net Mg^{2+} uptake by intact *Lolium multiflorum* roots. I. Kinetics. *Plant Physiol.* 91: 1407–1413.
93. Rengel, Z. and W. H. Zhang. 2003. Role of dynamics of intracellular calcium in aluminum toxicity syndrome. *New Phytologist.* 159: 295–314.
94. Roy, A. K., A. Sharma, and G. Talukder. 1988. Some aspects of aluminum toxicity in plants. *Bot. Rev.* 54: 145–178.
95. Ruemmle, B. A., L. A. Brilman, and D. R. Huff. 1995. Fine fescue germplasm diversity and vulnerability. *Crop Sci.* 35: 313–316.
96. Shearman, R. C. 2006. Fifty years of splendor in the grass. *Crop Sci.* 46: 2218–2229.
97. Shen, H., A. Ligaba, M. Yamaguchi, H. Osawa, K. Shibata, X. Yan, and H. Matsumoto. 2004. Effect of K-252a and abscisic acid on the efflux of citrate from soybean roots. *J. Exp. Bot.* 55: 663–671.
98. Shen, R. F., R. F. Chen, and J. F. Ma. 2006. Buckwheat accumulates aluminum in leaves but not in seeds. *Plant Soil* 284: 265–271.
99. Sivaguru, M. and W. J. Horst. 1998. The distal part of the transition zone is the most aluminum-sensitive apical root zone of maize. *Plant Physiol.* 116: 155–163.
100. Sivaguru, M., F. Baluska, D. Volkmann, H. H. Felle, and W. J. Horst. 1999. Impacts of aluminum on the cytoskeleton of the maize root apex, short-term effects on the distal part of the transition zone. *Plant Physiol.* 119: 1073–1082.
101. Slaski, J. J. 1995. NAD$^+$ kinase activity in root tips of nearly isogenic lines of wheat (*Triticum aestivum* L.) that differ in their tolerance to aluminum. *J. Plant Physiol.* 145: 143–147.
102. Slattery, J. F. and D. R. Coventry. 1999. Persistence of introduced strains of *Rhizobium legumisosarum* by trifolii in acid soils of north-eastern Victoria. *Aust. J. Exp. Agric.* 37(7): 829–837.
103. Sticklen, M. B. and M. P. Kenna. 1998. *Turfgrass Biotechnology: Cell and Molecular Genetic Approaches to Turfgrass Improvement.* Ann Arbor Press, Chelsea, MI, 256 pp.

104. Taliaferro, C. M. 1995. Diversity and vulnerability of bermuda turfgrass species. *Crop Sci.* 35: 327–332.

105. Taylor, G. J. 1988. The physiology of aluminum tolerance in higher plants. *Commn. Soil Sci. Plant Anal.* 19: 1179–1194.

106. Taylor, G. J. 1991. Current views of the aluminum stress response: the physiological basis of tolerance. *Curr. Top. Plant Biochem. Physiol.* 10: 57–93.

107. Taylor, G. J., A. Basu, U. Basu, J. J. Slaski, G. Zhang, and A. Good. 1997. Al-induced, 51-kilodalton, membrane-bound proteins are associated with resistance to Al in a segregating population of wheat. *Plant Physiol.* 114: 363–372.

108. Taylor, G. J. and S. M. Macfie. 1994. Modeling the potential for boron amelioration of aluminum toxicity using the Weibull function. *Can. J. Bot.* 72: 1187–1196.

109. Thompson, D. C., B. B. Clarke, J. R. Heckman, and J. A. Murphy. 1993. Influence of nitrogen source and soil pH on summer patch development in Kentucky bluegrass. *Int. Turfgrass Soc. Res. J.* 7: 317–323.

110. Vazquez, M. D., C. Poschenrieder, I. Corrales, and J. Barcelo. 1999. Change in apolastic aluminum during the initial growth response to aluminum by roots of a tolerant maize variety. *Plant Physiol.* 119: 435–444.

111. Vinceli, P. and D. Williams. 1998. Managing spring dead spot of bermudagrass. *Golf Course Manage. GCSAA* May: 49–53.

112. Wang, J. P., H. Raman, G. P. Zhang, N. Mendham, and M. X. Zhou. 2006. Aluminum tolerance in barley (*Hordeum vulgare* L.): physiological mechanisms, genetics and screening methods. *J. Zhejiang Univ. Sci.* 7(10): 769–787.

113. Wang, Z. Y., A. Hopkins, and R. Mian. 2001. Forage and turf grass biotechnology. *Crit. Rev. Plant Sci.* 20(6): 573–619.

114. Wissemeier, A. H., G. Hahn, and H. Marschner. 1998. Callose in roots of Norway spruce (*Picea abies* (L.) Karst.) is a sensitive parameter for aluminum supply at a forest site (Hoglwald). *Plant Soil* 199: 53–57.

115. Yamamoto, Y., Y. Kobayashi, and H. Matsumoto. 2001. Lipid peroxidation is an early symptom triggered by aluminum, but not the primary cause of elongation inhibition in pea roots. *Plant Physiol.* 125: 199–208.

116. Yokoma, S. and K. Ojima. 1995. Physiological response of root tip of alfalfa to low pH and aluminum stress in water culture. *Plant Soil* 171: 163–165.

117. Zaurov, D. E., S. Bonos, J. A. Murphy, M. Richardson, and F. C. Belanger. 2001. Endophyte infection can contribute to aluminum tolerance in fine fescues. *Crop Sci.* 41: 1981–1984.

118. Zhang, G. C. and G. J. Taylor. 1990. Kinetics of aluminum uptake in *Triticum aestivum* L. Identity of the linear phase of aluminum uptake by excised roots of aluminum-tolerant and aluminum-sensitive cultivars. *Plant Physiol.* 94: 577–584.

24 Relative Salinity Tolerance of Turfgrass Species and Cultivars

Kenneth B. Marcum

CONTENTS

24.1 INTRODUCTION

Turfgrass industries worldwide have grown rapidly in recent decades. Current estimates place the land area planted to turfgrass in the United States at over 160,000 km^2, an area three times larger than that of any irrigated crop [1]. Annual expenditure for turfgrass maintenance in the United States was estimated to be $45 billion in 1993, double the amount in 1985 [2]. This growth trend is also occurring in other rapidly developing countries, such as Australia and China [3,4].

Growth of irrigated turfgrass landscape industries has accompanied rapid urban/suburban development, contributing to increased pressure on limited freshwater supplies. Local, state, and national governments have responded by placing restrictions on potable water use for landscape turf irrigation, and encouraging or requiring use of secondary, saline water sources, such as recycled water [5–8]. Rapidly developing coastal areas suffer the additional problem of saltwater intrusion due to freshwater aquifer overpumping; many of the affected (brackish) wells are utilized for turf landscape irrigation [9,10]. For these reasons, the need for salt-tolerant turfgrass cultivars has substantially increased in recent decades.

24.2 DEFINING RELATIVE SALINITY TOLERANCE AMONG TURFGRASSES

A broad range in salinity tolerance exists among both C_4 and C_3 turfgrass genera and species [11–13]. However, many environmental, edaphic, and plant factors also interact with root-exposed salinity level to influence absolute plant salinity tolerance [14–17]. Salinity tolerance often differs with the stage of plant development (e.g., seedling, juvenile, mature) [18]. Environmental factors such as temperature, light, and relative humidity can influence plant salinity response [14]. For example, plants are typically more sensitive to salinity under hot, dry conditions than under cool, humid ones, probably due to increased evapotranspirational demand, favoring increased salt uptake [19]. Edaphic factors also influence plant response to salinity [15,16]. Soil water–content changes (matric potential) have a direct effect on root-zone salinity (osmotic potential). Soil salinity varies with time and depth, increasing as the soil dries between irrigations, and also as depth increases, due to differential leaching in soil matrices [14,20]. Relative proportions of saline ions present in the root zone (e.g., ratio of sodium and calcium salts) also influence plant salinity response [14]. Sodicity problems can occur in saline situations, sodium being a primary ion in many saline waters and soils. Particularly in finer textured soils, high exchangeable sodium percentages may cause structural dispersion, resulting in root-zone anaerobic conditions, exacerbating salinity effects [16,20]. To minimize effects of variable edaphic conditions on plant responses to salinity, researchers have utilized porous, nonsoil rooting media coupled with solution culture (e.g., hydroponics, ebb-and-flow). To minimize climactic variable effects, plants are grown under controlled environmental conditions (ex. glasshouses, growth chambers).

Due to interacting factors discussed above, the "absolute" salinity tolerance level of a particular genotype or cultivar cannot be determined [14,15]. For example, salt tolerance of *Cynodon* spp. cultivar (cv.) Tifway, indicated as the salinity level resulting in 50% shoot dry-weight reduction was reported to be 33 [21], 32 [22], 18.6 [23], and 12 dS m^{-1} [24]. Use of different criteria to measure salinity tolerance further complicates comparisons. For example, shoot weight [18], relative (to nonsaline conditions) shoot weight reduction [25], root weight or length [26,27], shoot/leaf length [28,29], shoot visual injury [30], plant survival [31], and seed germination or seedling growth [32] have been used as measures of salinity tolerance in turfgrasses. Finally, units used in reporting salinity vary across studies. Units of measurement frequently used include salts on a weight basis (total dissolved solids [TDS mg L^{-1} = ppm], or [TDS meq L^{-1}]), or on an electrical conductivity basis (millimhos per centimeter [mmhos cm^{-1}] or decisiemens per meter [dS m^{-1}]). When possible, units have been converted in this chapter into dS m^{-1} to facilitate comparison of studies.

Though with these limitations, relative salinity tolerance of turfgrass species and cultivars can be compared among studies having common genotype(s). This chapter presents existing turfgrass salinity tolerance studies to-date for each turfgrass species. Comparisons are made of the turfgrass to other species, and where information is available, cultivar differences within species are noted. This is followed by a table, providing a summary comparison of relative salinity tolerance across all turfgrass species.

24.2.1 Relative Salinity Tolerance of C$_4$ Turfgrasses

24.2.1.1 *Axonopus* spp. Beauv. (Carpetgrasses)

Two *Axonopus* spp. have been used to a limited extent for turfgrass: *A. affinis* Chase (common carpetgrass) and *A. compressus* [Sw.] Beauv. (tropical carpetgrass) [33]. *Axonopus* spp. has been classified as having poor salt tolerance [34]. At germination, *A. affinis* accession (acc.) was less tolerant to salinity than *Cynodon dactylon* cv. Common, but more tolerant than *Paspalum notatum* cvs. Argentine and Pensacola [35].

24.2.1.2 *Bouteloua* spp. Lag. (Gramagrasses)

Bouteloua spp., native to the plains region of North America, are sometimes utilized as low-maintenance drought-tolerant turfgrasses, including *B. curtipendula* [Michx.] Torr. (sideoats grama), *B. eriopoda* [Torr.] Torr. (black grama), and *B. gracilis* [Willd. ex Kunth] Lag. ex Griffiths (blue grama) [12,18,36]. *Bouteloua* spp. have been ranked as salt sensitive [15], and limited studies of the individual spp. have supported this assessment [37–39].

24.2.1.3 *Buchloë dactyloides* [Nutt.] Englem. (Buffalograss)

B. dactyloides salinity tolerance has been classified as moderate (electrical conductivity of soil saturated paste extract [ECe] 6–10 dS m^{-1}) [11]. In a glasshouse study, *B. dactyloides* cv. Prairie was less salt tolerant than *Puccinellia distans* cv. Fults, *Festuca arundinaceae* cv. K-31, and *Poa pratensis* cv. Nugget, but more salt tolerant than *Bouteloua gracilis* acc. and *Poa pratensis* cv. Adelphi [37]. In another study, salinity tolerance declined in the order *Distichlis spicata* acc. > *Cynodon dactylon* cv. Arizona Common = *Zoysia japonica* cv. Meyer > *Bouteloua curtipendula* cv. Haskell = *Bouteloua eriopoda* acc. = *B. dactyloides* cv. Prairie [38].

Research has found little difference in salinity tolerance among *B. dactyloides* cultivars. No differences were found in seedling shoot growth reduction among cvs. Sharp Improved, Texoka, Topgun, and Plains, with salinity resulting in 50% shoot growth reported as ECw (electrical conductivity of water) 15 dS m^{-1} [40]. In another study, no differences in salinity tolerance at the seedling stage were found among nine *B. dactyloides* cvs. of three different ploidy levels [41].

Differences in salt tolerance at germination were reported among diploid and polyploid *B. dactyloides* acc., though differences were slight, 50% shoot dry-weight reductions occurring at 8–10 dS m^{-1} [42]. Germination percentages of two *B. dactyloides* acc. were higher than *Poa pratensis* cv. SD Common, with the upper limit for germination at 2.8% NaCl [43].

24.2.1.4 *Cynodon* spp. [L.] Rich. (Bermudagrasses)

Cynodon spp. generally used for turfgrass consist of two species. *Cynodon dactylon* var. *dactylon* [L.] Pers. (common bermudagrass), being cosmopolitan in distribution [44], is also the most commonly used C$_4$ turfgrass worldwide [45]. Interspecific hybrids of *C. dactylon* var. *dactylon* and *C. transvaalensis* Burt-Davy (African bermudagrass), often called "hybrid bermudagrasses," are sterile triploid, fine-textured grasses used for sports turf, golf courses, and other high-value areas. Though usually produced in turfgrass breeding programs, there also are some occurrences of natural crosses of *C. dactylon* var. *dactylon* and *C. transvaalensis*, commonly known as *C. x magennisii* Hurcombe [46]. These species are often referred to and grouped together as *Cynodon* spp. (bermudagrass) in the literature; however, where possible I will differentiate species of individual cvs. and accs. as "*dactylon*," "hb." (interspecific hybrid [hb.] cultivar), or "*x magennisii*" (interspecific hybrid variety).

Cynodon spp. are ranked as having excellent salinity tolerance, tolerating ECe 8–16 [13,16,47,34], >10 [11], 12–18 [20], or 16–18 dS m^{-1} [48]. Shoot growth of cv. Santa Ana hb. was reduced 50% relative to control plants when exposed to ECe 16 dS m^{-1} (as 50/50 mix of NaCl and CaCl$_2$) for 6 weeks [49]. Studies have shown a great deal of genetic variability in salinity tolerance within *Cynodon* spp. In India, an alkaline-soil *C. dactylon* acc. was found to have greater salinity tolerance than a normal-soil acc. [50]. Two *C. dactylon* accs. collected from salt-spray areas along the windward coast of Oahu, Hawaii, were more salt tolerant than other cvs. studied (Tifgreen hb., Sunturf hb., Tifdwarf hb., and FB-137) [51]. *C. dactylon* cv. Suwannee was reported to be more salt tolerant than *Paspalum vaginatum* acc. in a field experiment irrigated with ECw 14 dS m^{-1} [52].

There have been a number of studies comparing salinity tolerance of several *Cynodon* spp. cultivars. Dudeck and Peacock [21,25] compared *Cynodon* spp. cvs. Tifway hb. and Tifway II hb. with other C_4 turfgrasses. Tifway was more salt tolerant than Tifway II, 50% shoot growth reduction occurring at ECw 33 and 24 dS m^{-1}, respectively. Smith et al. [24] also compared Tifway with Tifway II in solution culture. Tifway was slightly more salt tolerant than Tifway II, with 50% shoot growth reductions occurring at ECw 12 and 11 dS m^{-1}, respectively. Peacock et al. [53] compared six *Cynodon* spp. cvs. in solution culture. Cultivars grouped into two relative salinity tolerance categories: tolerant—cvs. Navy Blue, GN-1 hb. and TifSport hb., and sensitive—cvs. Quickstand, Tifway hb., and Tifton 10 hb. In another study, the range of salinity tolerance among cultivars was relatively narrow, with 50% reductions in shoot growth occuring from ECw 17.4 to 22.5 dS m^{-1}. Salinity tolerance decreased in the order Tifdwarf hb., Tifgreen hb., FB-137, Tifway hb., Tiflawn hb., Everglades, Common, and Ormond [23]. In sand culture, tolerance decreased in the order Tifton 86 hb., Tifway II hb., and Tifton 10 hb., with a 50% shoot growth reduction ranging from ECw 31 to 24 dS m^{-1} [54]. More recently, 35 *Cynodon* spp. cvs. were compared, salinity tolerance (salinity causing 50% shoot growth reduction) ranging from ECw 26–40 dS m^{-1}, with cvs. FloraTex hb., Cheyenne, MS Supreme hb., and Blue-muda in the highest statistical group for salt tolerance [22].

Germination of *Cynodon dactylon* cv. Common bermudagrass was not affected by salinity up to ECw 10 dS m^{-1} as synthetic sea salt [35].

24.2.1.5 *Distichlis spicata* [L.] Greene ssp. *stricta* [Torr.] Thorne (Inland Saltgrass)

D. spicata ssp. *stricta* (*D. spicata*), native to deserts of North America, is considered a low-maintenance turfgrass in saline areas [12,18,55]. It thrives in extremely dry, saline environments, including saline salt flats having salinity levels above that of seawater [56,57]. The U.S.D.A. Salinity Laboratory ranked several turfgrasses in the salt-tolerant category (tolerating ECe 12–18 dS m^{-1}), in order of decreasing salt tolerance: *D. spicata* > *Puccinellia airoides* > *Cynodon* spp. [20]. Moderate salinities (200 mM NaCl) stimulated shoot and root growth in *Distichlis spicata* acc. WA-12, with 50% shoot growth reduction occurring at 600 mM NaCl (~45 dS m^{-1}) [58]. *D. spicata* acc. was more salt tolerant than *Paspalum vaginatum* or *Cynodon dactylon* acc. in field plot experiments [52]. In another experiment, *D. spicata* acc. A55 was more salt tolerant than *Paspalum vaginatum* cv. Sea Isle 1 and *Cynodon* spp. cv. Tifway hb. Shoot and root growth of A55 increased at all salinity levels up to 30,000 mg L^{-1} NaCl (~40 dS m^{-1}), with no loss in canopy color. In contrast, shoot and root growth as well as canopy color of Sea Isle 1 and Tifway declined with increasing salinity [59]. *D. spicata* acc. was more salt tolerant than *Cynodon dactylon* cv. Arizona Common, *Zoysia japonica* cv. Meyer, *Buchloë dactyloides* cv. Prairie, and *Bouteloua* spp. *D. spicata* acc. continued to grow roots and shoots in solutions containing 50 dS m^{-1}, equivalent to full-strength seawater [38].

Relative salinity tolerance of 21 *D. spicata* accs. were compared with *Cynodon* spp. cv. Midiron hb. All *D. spicata* accs. had at least twice the salinity tolerance of Midiron, though there were large differences among *D. spicata* accs. [60]. In another study, 12 *D. spicata* accs. were compared for salinity tolerance for 10 weeks. Again, substantial variability existed among accs., with two maintaining good turf quality at ECw 34 [61]. Among eight *D. spicata* accs. under turfgrass management, percent green-leaf canopy area ranged from 5% to 90% under high salinity stress (46 dS m^{-1}) depending on genotype [62].

24.2.1.6 *Eremochloa ophiuroides* [Munro] Hack. (Centipedegrass)

E. ophiuroides is invariably classified as having poor salinity tolerance, tolerating ECe <4 [13,16,34,47], or <3 dS m^{-1} [11]. *E. ophiuroides* cv. Common and *Paspalum notatum* cv. Argentine were least tolerant among C_4 turfgrasses studied, with 50% shoot growth reduction occurring at ECw 9 dS m^{-1}, more tolerant turfgrasses including *Stenotaphrum secundatum* cv. Floralawn, *Cynodon* sp. cvs. Tifway and Tifway 2 hbs., *Paspalum vaginatum* accs. FSP-1 and FSP-3, and *Zoysia* spp. cv. Emerald hb. [21,25]. In another study, *E. ophiuroides* cv. Common was least tolerant among six C_4 turfgrasses. Salinity tolerance declined in the order *Paspalum vaginatum* acc. Hawaii = *Zoysia matrella* cv. Manila > *Stenotaphrum secundatum* acc. Hawaii > *Cynodon* spp. cv. Tifway hb. > *Zoysia japonica* cv. Korean Common > *E. ophiuroides* cv. Common, with 50% shoot dry-weight reduction ranging from ECw 40 to 6 dS m^{-1} [63].

24.2.1.7 *Hilaria belangeri* [Steud.] Nash (Curly Mesquite)

H. belangeri, native to deserts of the U.S. southwest, has been considered a low-maintenance, unirrigated groundcover [64,65]. There are no data regarding salinity tolerance of this species.

24.2.1.8 *Paspalum notatum* Fluggé (Bahiagrass)

P. notatum has been classified as sensitive to salinity, tolerating ECe 3–6 dS m^{-1} [11]. However, in a pot study, shoot growth of *P. notatum* cv. Pensacola was reduced by ECw of only 0.4 dS m^{-1} and did not survive ECw higher than 0.8 dS m^{-1} (ECe 2.5 dS m^{-1}) over the 255-day trial [66]. *P. notatum* cv. Argentine and *Eremochloa ophiuroides* cv. Common were least salt tolerant compared with other C_4 turfgrasses, including *Zoysia* spp. cv. Emerald hb., *Paspalum vaginatum* accs. FSP-1 and FSP-3, *Cynodon* spp. cvs. Tifway 1 and Tifway 2 hbs., and *Stenotaphrum secundatum* cv. Floralawn, with 50% shoot yield of cv. Argentine occurring at ECiw 9.3 dS m^{-1} [25].

Germination of *P. notatum* cvs. Argentine and Pensacola were more adversely affected by salinity ranging from 0 to 5800 mg L^{-1} (0–11 dS m^{-1}) than were other C_4 turfgrasses, including *Cynodon dactylon* cv. Common, *Eremochloa ophiuroides* acc., and *Axonopus affinis* acc. [35].

24.2.1.9 *Paspalum vaginatum* Sw. (Seashore Paspalum)

Paspalum vaginatum has generally been ranked as the most salt-tolerant commonly-utilized C_4 turfgrass, tolerating ECe >16 dS m^{-1} [16,47]. *P. vaginatum* cv. Futurf was reported to exist in soils with ECe 40–45 dS m^{-1} [67], though the generally accepted range is that *P. vaginatum* can withstand up to ECe 22 dS m^{-1} [68]. Major [69] reported both *Puccinellia distans* cv. Fults and *P. vaginatum* cv. Futurf survived ECe greater than 50 dS m^{-1} in glasshouse pot culture. In glasshouse solution culture, *S. paspalum* acc. Hawaii was equal in salinity tolerance to *Stenotaphrum secundatum* acc. Hawaii and *Zoysia matrella* cv. Manila, with 50% shoot dry-weight reduction at ECw 40 dS m^{-1}, followed by (in decreasing tolerance) *Cynodon* spp. cv. Tifway hb., *Zoysia japonica* cv. Korean Common, and *Eremochloa ophiuroides* cv. Common [63]. In contrast, *P. vaginatum* acc. was reported to be less salt tolerant than *Cynodon dactylon* cv. Suwannee in field plots [52].

Differences in salinity tolerance among *P. vaginatum* genotypes have been noted. Dudeck and Peacock [25] compared two Florida accs. of *S. paspalum* (FSP-1 and FSP-3) with other C_4 turfgrasses in solution culture. *Zoysia* spp. cv. Emerald hb. was most salt tolerant, followed by FSP-3, then *Cynodon* spp. cv. Tifway hb. and FSP-1, then *Cynodon* spp. cv. Tifway II hb. and *Stenotaphrum secundatum* cv. Floralawn, and finally *Eremochloa ophiuroides* cv. Common and *Paspalum notatum* cv. Argentine. The authors subsequently compared four *P. vaginatum* cvs. [70]. FSP-1 was most tolerant, with 50% shoot growth reduction at ECw 28.6 dS m^{-1}, followed by Futurf and FSP-2, and finally Adalayd (ECw 18.4 dS m^{-1}). Lee et al. [71,72] evaluated a number of *S. paspalum* acc. for salinity tolerance in solution culture, using several shoot growth criteria. ECw resulting in 50% shoot growth ranged from 22 to 43 dS m^{-1}. Salinity tolerant acc. included SI 92, SI 93-1, SI 91, SI 93-2, and SI 89.

24.2.1.10 *Pennisetum clandestinum* Hochst. ex Chiov. (Kikuyugrass)

P. clandestinum has not been generally ranked among turfgrasses as to salinity tolerance. *P. clandestinum* acc. was moderately salt tolerant, with 50% shoot growth reduction occurring at 200 mM NaCl (~16 dS m^{-1}) [73]. In another study, shoot growth of *P. clandestinum* acc. was affected at 200 mM NaCl (~16 dS m^{-1}), but not at 150 mM NaCl (~13 dS m^{-1}) [74].

24.2.1.11   *Sporobolus virginicus* [L.] Kunth (Seashore Dropseed)

S. virginicus, a halophyte distributed in tropical and subtropical coastal zones worldwide [75], is currently considered a low-maintenance turfgrass in saline areas [76,77]. Moderate salinities stimulated shoot and root growth in both *S. virginicus* acc. and *Distichlis spicata* acc. WA-12, with 50% shoot growth reduction occurring at 600 mM NaCl (~45 dS m^{-1}) [58]. *S. virginicus* acc. Hawaii shoot growth was stimulated under moderate salinity up to 15 dS m^{-1}, then declined, though growth continued up to 45 dS m^{-1}. However, root growth continually increased, relative to control, to the highest salinity level (45 dS m^{-1}) [78]. In another study, *S. virginicus* acc. continued to grow without injury after 4 months' exposure to full-strength seawater (45 dS m^{-1}) [79]. Four *S. virginicus* acc. were compared for salinity tolerance, with initial shoot growth decline due to salinity ranging from 31 to 40 dS m^{-1}. Salinities less than 31 dS m^{-1} had no effect on shoot growth [80].

24.2.1.12   *Stenotaphrum secundatum* [Walt.] Kuntze (St. Augustinegrass)

S. secundatum has been ranked tolerant to salinity, tolerating ECe 8–16 [13,16,34,47] or >10 dS m^{-1} [11]. *S. secundatum* acc. Hawaii and cv. Floratine were reported equal in salinity tolerance to *Paspalum vaginatum* acc. Hawaii and *Zoysia matrella* cv. Manila, but more tolerant than *Cynodon* spp. cvs. FB-137, Tifgreen and Tifdwarf hbs., *Cynodon x magennisii* cv. Sunturf and *Zoysia japonica* cv. Korean Common, with 50% shoot growth reduction of *S. secundatum* acc. Hawaii occurring at ECw 40 dS m^{-1} [51]. *S. secundatum* cvs. Seville and Floratam were more salt tolerant than *Cynodon* spp. cvs. Tifway and Tifway 2 hbs., with 50% shoot dry-weight reductions occurring at ECw 30, 19, 12, and 11 dS m^{-1}, respectively [24]. In contrast, Dudeck and Peacock [25] reported *S. secundatum* cv. Floratam less salt tolerant than *Zoysia* spp. cv. Emerald hb., *Paspalum vaginatum* cv. FSP-1, and *Cynodon* spp. cv. Tifway hb., but more tolerant than *Eremochloa ophiuroides* and *Paspalum notatum*, with 50% shoot growth reduction of Floratam occurring at ECw 22 dS m^{-1}.

Differences in salinity tolerance among several *S. secundatum* cvs. have been reported. Shoot and root growth of Seville was less affected by salinity than Floratam in solution culture containing up to ECw 34 dS m^{-1}. [81]. In another study, Seville was more salt tolerant than Floratine, Floratam, and Floralawn, with a 50% shoot dry-weight reductions occurring within the range ECw 23–28 dS m^{-1} [82].

24.2.1.13   *Zoysia* spp. Willd. (Zoysiagrasses)

Zoysia spp. used as turfgrass consist of several interfertile species: *Zoysia japonica* Steud. (Japanese lawngrass), *Zoysia matrella* [L.] Merr. (Manilagrass), and *Zoysia pacifica* [Gaud.] Hotta & Kuroti (formerly *Z. tenuifolia*) (Mascarenegrass). Recent research has shown that *Zoysia* spp. readily hybridize; previously classified single sp. cultivars may in fact be interspecies hbs. [83]. Known interspecific hybrid cvs. are listed "hb."; e.g., cv. Emerald hb. (*Z. japonica* x *Z. pacifica*). Other *Zoysia* spp., including *Z. koreana* Mez, *Z. macrantha* Desv., *Z. macrostachya* Franch. and Sav., and *Z. sinica* Hance, are sometimes used, or are currently being considered for use as turfgrass, primarily in tropical/subtropical Asia and Pacific regions [76,84]. General salt-tolerance rankings comparing turfgrasses have traditionally not made a distinction in species, though recent studies have revealed large interspecific differences. *Zoysia* spp. in general have been ranked as salt tolerant, tolerating ECe 8–16 [13,16,34,47] or 6–10 dS m^{-1} [11]. *Z. japonica* cv. Meyer was reported equal in salt tolerance to *Cynodon dactylon* cv. Arizona Common, but more tolerant than *Buchloë dactyloides* cv. Prairie and *Bouteloua* spp. [38]. In another study, *Zoysia* spp. cv. Emerald hb. was more salt tolerant than other C$_4$ turfgrasses, with 50% shoot growth reduction occurring at ECw 37 dS m^{-1}. Salt tolerance decreased in the order cv. Emerald > *Seashore paspalum* cv. FSP-3 = *Cynodon* spp. cv. Tifway hb. > *Seashore paspalum* cv. FSP-1 > *Cynodon* spp. cv. Tifway II hb. = *Stenotaphrum secundatum* cv. Floralawn > *Eremochloa ophiuroides* cv. Common = *Paspalum notatum* cv. Argentine [25].

Major differences in salinity tolerance have been reported in *Zoysia* spp. on both inter- and intra-specific levels. *Z. matrella* cv. Manila was more salt tolerant than *Z. japonica* cv. Korean Common, with 50% shoot dry-weight reductions at ECw 40 and 12 dS m^{-1}, respectively [63]. In another study, *Z. koreana* acc. was most salt tolerant, followed by *Z. sinica* acc. and *Z. matrella* acc., and finally *Z. japonica* acc. [85]. In a study comparing 59 *Zoysia* ssp. genotypes, salinity tolerance decreased in the order *Z. pacifica (tenuifolia)* = *Z. matrella* > *Z. japonica* x *Z. pacifica (tenuifolia)* > *Z. sinica* > *Z. macrostachya* > *Z. japonica*. Among cultivars tested in the study, *Z. matrella* cv. Diamond was the most salt tolerant [86]. Twenty-nine Zoysia spp. accs./cvs. were compared for salinity tolerance, with fine-leaf textured (or Matrella) types, particularly Diamond cv., DALZ8501 acc., and their hybrids, being most salt tolerant [87].

24.2.2 Relative Salinity Tolerance of C$_3$ Turfgrasses

24.2.2.1 *Agropyron* spp. Gaertn. (Wheatgrasses)

A. cristatum [L.] Gaertn. (crested wheatgrass) and *A. desertorium* [Fisch. ex Link] J.A. Schultes (desert wheatgrass) have been used as low-maintenance, unirrigated turf in semiarid northern temperate climates [88]. *A. cristatum* has been generally rated as salt tolerant, tolerating ECe 7 [14], though there is currently no information regarding salinity tolerance of *A. desertorium*. Germination under salinity stress declined in the order *Lolium* spp. accs. = *Festuca* spp. accs > *A. cristatum* acc. > *Agrostis stolonifera* acc. > *Poa pratensis* acc. [89].

24.2.2.2 *Agrostis* spp. L. (Bentgrasses)

24.2.2.2.1 *A. stolonifera L. (Creeping Bentgrass)*
A. stolonifera has been ranked moderately salt tolerant, with a tolerance range to soil ECe from 8 to 16 dS m^{-1} [13,16,34,47], or from 6 to 10 dS m^{-1} [11]. *A. stolonifera* cv. Seaside was more salt tolerant than *Festuca arundinaceae* cv. Alta, *Agrostis capillaris* cv. Highland, and *Poa pratensis* acc. Cv. Seaside survived ECw 31 dS m^{-1}, with 50% reduced shoot growth occurring at ECw 18 dS m^{-1} [90]. Salinity tolerance of cv. Seaside, determined as visual leaf firing percentage, was greater than *Lolium perenne* cvs. Common and NK200, *Poa trivialis* acc. and *Poa pratensis* cvs. Fylking, Park, Pennstar, Nugget, Newport, and Merion [30].

There is substantial variability in salinity tolerance among *A. stolonifera* cvs. and accs. Substantial variation in salinity tolerance has been reported among natural populations (ecotypes) of *A. stolonifera*, with seaside selections being more salt tolerant than inland selections [27,91–93]. Differences in salinity tolerance among seven *A. stolonifera* cvs. were reported, with salinities resulting in 50% shoot growth ranging from ECw 9 to 26 dS m^{-1}. Salinity tolerance decreased in the order cvs. Seaside, Arlington, Pennlu, Old Orchard, Congressional, Cohansey, and Penncross [94]. More recently, large differences in salinity tolerance among 33 *A. stolonifera* cvs. were found when grasses were exposed to 8 dS m^{-1} for 10 weeks. Cvs. Mariner, Grand Prix, Seaside, and Seaside II were most salt tolerant [95].

24.2.2.2.2 *A. capillaris L. (Colonial Bentgrass), A. canina L. (Velvet Bentgrass), and A. gigantea Roth (Redtop)*
Very little salinity work has been done on these *Agrostis* spp. *A. capillaris* has been generally ranked as having poor salinity tolerance, tolerating ECe <4 [13,16,34] or <3 dS m^{-1} [11]. *A. canina* and *A. gigantea* have not been generally ranked for salinity tolerance. Under exposure to moderately high salinity, survival of *A. capillaris* cv. Astoria was lowest, followed by *Festuca rubra* cv. Ruby, then *Festuca arundinaceae* cvs. K-31 and Alta [96]. *A. canina* cv. Novobent and *A. capillaris* cv.

Bardot were equivalent in salinity tolerance, being less tolerant than *Festuca rubra* cvs. Dawson and Ensylva, *Poa pratensis* cv. Skofti, and *Lolium perenne* cvs. Loretta and Pelo [97]. Thirty-five *Agrostis* spp. cvs. were exposed to 8 dS m^{-1} for 10 weeks. *A. capillaris* cv. Ambrosia and *A. canina* cv. Avalon survived only 5 weeks at this salinity, whereas 29 of 34 *A. stolonifera* cvs. survived 10 weeks [95]. Salinity tolerance decreased in the order *Puccinellia lemmoni* acc. > *Lolium perenne* cvs. NK 200 and Common > *A. gigantea* acc. > *A. stolonifera* cv. Seaside > *Festuca arundinaceae* cv. Alta = *Festuca rubra* cv. Ruby [30].

Salinity levels required to reduce germination by 50% decreased in the order *A. gigantea* cv. Streaker (25 dS m^{-1}), *A. stolonifera* cv. Seaside (23 dS m^{-1}), *A. canina* cv. Kingston, *A. capillaris* cv. Exeter (23 dS m^{-1}), *A. capillaris* cv. Highland (22 dS m^{-1}), *A. stolonifera* cv. Penncross (21 dS m^{-1}), *A. stolonifera* cv. Pennlinks (20 dS m^{-1}), and *A. stolonifera* cv. Penneagle (18 dS m^{-1}) [9].

24.2.2.3 *Bromus inermis* Leyss. (Smooth Bromegrass)

B. inermis, native to North America, is sometimes used as a low-maintenance turf ground cover in temperate regions of arid western North America [33]. *B. inermis* has been generally ranked as moderately salt sensitive [14], though data are lacking regarding the specific tolerance range in this species.

24.2.2.4 *Festuca* spp. L. (Fescues)

24.2.2.4.1 *F. arundinaceae Schreb.;* Schedonorus phoenix *(Scop.) Holub (Tall Fescue)*
F. arundinaceae has been generally ranked as moderately salt tolerant, tolerating ECe 4–8 [13,16,34] or 6–10 dS m^{-1} [11,47,48]. Salinity tolerance decreased in the order *Puccinellia distans* acc., *Agrostis stolonifera* cv. Seaside, *F. arundinaceae* cv. Alta, *Poa pratensis* acc., and *Agrostis capillaris* cv. Highland. Cv. Alta shoot growth was reduced 50% at 160 meq L^{-1} (ECw 14 dS m^{-1}) [90]. In another study, salinity tolerance decreased in the order *F. arundinaceae* cv. K-31, *Poa pratensis* cv. Nugget, *Buchloë dactyloides* acc., *Bouteloua gracilis* acc., and *Poa pratensis* cv. Alelphi [34]. Greub and Drolsom [96] reported *F. arundinaceae* cvs. Alta and K-31 to be more salt tolerant than *Puccinellia aeroides* acc., as well as *F. rubra* ssp. *rubra* cv. Ruby, *Poa pratensis* cv. Merion, and *Agrostis capillaris* cv. Astoria. However, in a later study, Greub et al. [30] reported *Puccinellia aeroides* acc. to be more salt tolerant than *F. arundinaceae* cvs. Alta and K-31, whereas *Puccinellia lemmoni* acc. was equivalent to cvs. Alta and K-31. Grasses more sensitive to salinity included *Agrostis stolonifera* cv. Seaside, *F. rubra* ssp. *rubra* cv. Ruby, *Lolium perenne* cvs. Common and NK-200, and *Poa trivialis* acc. A salt screening of *F. arundinaceae* genotypes found cvs. Tar Heel II and Pure Gold most salt tolerant, measured as plant survival to 24,000 ppm (37 dS m^{-1}) [98].

Germination rates differed among 16 *F. arundinaceae* cvs. exposed to 15,000 ppm (ECw 25 dS m^{-1}) for 3 weeks. Highest germination rates occurred in cvs. Belt 16-1, Belt KPH-1, Belt TF-11, Belt TFp25, Houndog, Kenmont, Gallway, T-5, and K-31 [28].

24.2.2.4.2 *F. rubra L. (Creeping Red Fescues)*
Probably more salinity studies have been done on *F. rubra* than any other turfgrass species, particularly in Europe and Canada, due to their widespread use in cooler climates. There are two types of creeping red fescues: a strong creeping type with 56 chromosomes (*F. rubra* L. ssp. *rubra* Gaud.) and a slender creeping type with 42 chromosomes (*F. rubra* L. ssp. *trichophylla* Gaud. or ssp. *litoralis* [Meyer] Auguier) [33]. Most salinity research does not distinguish these two types, though I will make the distinction where possible.

F. rubra is generally ranked as having poor salinity tolerance, tolerating ECe <4 dS m^{-1} [13,16,34,47] or 3–6 dS m^{-1} [11]. However, Butler et al. [48] gave *F. rubra* a general ranking of medium salinity tolerance, tolerating ECe of 8–12 dS m^{-1}. A substantial range in salinity tolerance appears to exist within *F. rubra*. Shildrick [99] stated that *F. rubra* L. ssp. *trichophylla* cvs. are generally more salt tolerant than *F. rubra* ssp. *rubra* cvs. However, Humphreys [100] stated that differences among *F. rubra* are in fact more closely related to the point of origin than to the subspecies. Salt-tolerant naturally occurring coastal populations of *F. rubra* have been described [101,102]. A salt-marsh *F. rubra* acc. was reported more salinity tolerant than *F. rubra* ssp. *trichophylla* cvs. Dawson and Oasis [100]. *F. rubra* cv. Saltol, collected from a saline tidal marsh on the St. Lawrence River, was more tolerant to salt spray, as determined by visual appearance, than *F. rubra* cvs. Biljart, Carlawn, Highlight, and Ottawa 1, *Poa pratensis* cv. Baron, and *Lolium perenne* cv. Manhattan [103]. Accessions of both *F. rubra* ssp. *rubra* and *F. rubra* ssp. *trichophylla* were reported having greater salinity tolerance than *Lolium perenne* acc., *A. canina* acc., and *A. capillaris* acc. [104]. *F. rubra* ssp. *rubra* cv. Ruby was intermediate among C_3 grasses in salinity tolerance, relative salinity tolerance in decreasing order being *Festuca arundinaceae* cvs. Alta and K-31 > *Agrostis stolonifera* cv. Seaside > *Lolium perenne* cvs. Common and NK-200 = *F. rubra* cv. Ruby > *Poa trivialis* acc. [30]. In another study, *F. rubra* cvs. had poor salt tolerance, equivalent to *Agrostis capillaris* cvs., but less tolerant than *Poa pratensis* cvs. and *Lolium perenne* cvs., when grown in field plots having ECe averaging 11 dS m^{-1} [105].

Cultivar differences in salinity tolerance among *F. rubra* have been documented. Salinity tolerance of seven *F. rubra* cvs. was determined by measuring differences in seedling root length. Cvs. Oasis, Hawk, Polar, Merlin, and Dawson were more tolerant than S59 and Jupiter [106]. Among eighteen *F. rubra* cvs., *F. rubra* ssp. *trichophylla* cvs. Dawson and Golfrood were most salt tolerant, as determined by visual appearance in saline soils. Less tolerant were *F. rubra* ssp. *rubra* cvs. Ruby, Rainier, Steinacher, Illahee, Pennlawn, and Common [31].

Germination in 75% seawater (ECw 28 dS m^{-1}) of *F. rubra* spp. *trichophylla* cv. Dawson was superior to *Poa pratensis* cv. Merion and *Agrostis stolonifera* cv. Seaside, and equivalent to *Lolium perenne* cv. Pennfine and *Puccinellia lemmoni* cv. Common [107]. Germination was compared among 16 *F. rubra* cvs. following 7 weeks' exposure to 260 mM NaCl (25 dS m^{-1}). Germination decreased in the order Polar > Dawson > Koket = Novorubra > Erika > Famosa = Jamestown = Reptans [108].

24.2.2.4.3 F. rubra *L.* ssp. fallax *[Thuill.] Nyman (Chewings Fescue)*
F. rubra ssp. *fallax* has been generally ranked as sensitive to salinity, tolerating only ECe 3–6 dS m^{-1} [11]. Torello and Symington [109] ranked *F. rubra* ssp. *fallax* as less salt tolerant than either *F. rubra* ssp. *rubra* or *F. rubra* ssp. *trichophylla*. *F. rubra* ssp. *fallax* acc. had poor salt tolerance, equivalent to *Poa pratensis* cv. Merion and *Agrostis capillaris* cv. Astoria, but less tolerant than *Puccinellia aeroides* acc., *Festuca arundinaceae* cvs. Alta and K-31, and *F. rubra* cv. Ruby [96]. A maritime population of *F. rubra* ssp. *fallax* was more salt tolerant than an inland one, with 50% shoot growth reduction occurring at approximately 220 mM NaCl (ECw ~17 dS m^{-1}), and 70 mM (ECw ~7 dS m^{-1}), respectively [110]. *F. rubra* ssp. *fallax* cv. Highlight was found to be less salt tolerant than seven *F. rubra* ssp. *trichophylla* cvs. in a seedling root growth trial [106]. In a study comparing *Festuca* spp. cvs., salinity tolerance (determined as visual leaf firing) decreased in the order *F. rubra* ssp. *trichophylla* > *F. rubra* ssp. *rubra* > *F. rubra* ssp. *fallax* = *F. brevipila* [31].

24.2.2.4.4 F. brevipila *Tracey (Hard Fescue)*
F. brevipila has been generally ranked as sensitive to salinity, tolerating only ECe 3–6 dS m^{-1} [11]. In a study in which pots were subirrigated with 1.25% NaCl (ECw 32 dS m^{-1}) for 80 days, *F. brevipila* cvs. Scaldis, Centurion, and Durar had poorer salt tolerance than other *F. rubra* cvs., determined as visual leaf firing [31].

24.2.2.4.5 F. elatior *L.; Schedonorus pratensis (Huds.) Beauv. (Meadow Fescue)*
F. elatior has been generally ranked as having poor salt tolerance, tolerating ECe <4 dS m^{-1} [13,16,34,48]. *F. elatior* acc. was less salt tolerant than *F. arundinaceae* cvs. K-31, Falcon, Rebel, and Houndog when visually rated for salt injury over a 2-month period [37].

24.2.2.4.6 F. ovina *L. (Sheep Fescue)*
F. ovina has not yet been ranked in general terms for salinity tolerance. *F. ovina* cvs. Firmaula and Barok were equivalent in salinity tolerance to *F. brevipila* cvs., but less salt tolerant than *F. rubra* cvs. [31].

24.2.2.5 *Lolium* spp. L. (Ryegrasses)

24.2.2.5.1   L. perenne *L. (Perennial Ryegrass)*
L. perenne is generally ranked as moderately salt tolerant, tolerating ECe 4–8 [13,16,47] or 6–10 dS m^{-1} [11]. In a field trial with soil ECe's averaging approximately 11 dS m^{-1}, *Lolium perenne* cvs. Pelo and Manhattan maintained better visual quality than *F. rubra* cvs., *Poa pratensis* cvs., and *Agrostis capillaris* cvs. [105]. Salinity tolerance, determined as visual quality, decreased in the order *Agrostis stolonifera* cv. Seaside > *L. perenne* cvs. Common and NK 200 = *Poa pratensis* cvs. Park, Pennstar, and Nugget > *Poa trivialis* acc. [30].

Shoot dry-weight reduction, as a percentage of control plants, was not significantly different among *L. perenne* cvs. Vic. Cert., Tasdale, Barlata, and Linn. Grasses were exposed to 300 mM NaCl (ECw ~30 dS m^{-1}) for 2 weeks [32]. *L. perenne* cvs. and accs. were exposed to 17,000 ppm salinity (ECw ~26 dS m^{-1}) for 9 weeks, and genotypes visually rated for percent green-leaf area. Top entries included cvs. Brightstar SLT and Fiesta III, and accs. PST-216 and B-2 [111].

Germination of *L. perenne* cvs. exposed to 10,000 ppm (15 dS m^{-1}) declined in the order Pennant > Citation II = Palmer = Horizon = Derby > Fiesta [112].

24.2.2.5.2   L. multiflorum *Lam. (Italian, or Annual Ryegrass)*
and L. xhybridum *Hausskn. (Intermediate, or Hybrid Ryegrass)*
Neither *L. multiflorum* nor *L. xhybridum* has been generally ranked for salinity tolerance. *L. multiflorum* acc. was less salt tolerant than *L. perenne* cv. Vic. Cert., with 50% shoot growth reductions occurring at 100 and 150 mM NaCl, respectively (~ECw 10.5 and 15.5 dS m^{-1}). However, the effect on germination was reversed, with 50% germination occurring at 330 mM NaCl for annual ryegrass and at 250 mM NaCl for perennial ryegrass (~ECw 32 and 29 dS m^{-1}) [32].

24.2.2.6 *Poa* spp. L. (Bluegrasses)

24.2.2.6.1   Poa annua *L. (Annual Bluegrass)*
P. annua had been generally ranked salt-sensitive, tolerating <3 ECe [113]. However, there has been no documented research to quantify salinity tolerance of *P. annua*. Summer cultivation was found to improve permeability, reduce ECe, and improve visual quality of a *Poa annua* acc. golf green [114].

24.2.2.6.2   Poa compressa *L. (Canada Bluegrass)*
P. compressa is a low-maintenance, low-density turfgrass adapted to northern regions of cool humid climates [34]. Relative salinity tolerance of *P. compressa* is not currently known, nor has the species been generally ranked for salinity tolerance.

24.2.2.6.3   Poa pratensis *L. (Kentucky Bluegrass)*
A number of salinity studies have been done on *P. pratensis*, reflecting its wide use among C$_3$ turfgrasses. *P. pratensis* has been generally ranked as having poor salinity tolerance, tolerating ECe <4 [13,16,34,47,48] or <3 dS m^{-1} [11]. Lunt et al. [90] reported *P. pratensis* acc. to survive an ECw of only 8 dS m^{-1}, whereas *Agrostis stolonifera* cv. Seaside and *Festuca arundinaceae* cv. Alta survived an ECw of 19 and 13 dS m^{-1}, respectively. Butler et al. [12] reported that *P. pratensis* does not grow well in soils with an ECe >4 dS m^{-1}. *Agrostis stolonifera* cv. Seaside and *Lolium perenne* cvs. Common and NK200 had greater salinity tolerance (measured as relative shoot growth and leaf firing) than *Poa pratensis* cvs. Fylking, Park, Pennstar, Nugget, Newpart, and Merion when exposed to 4.5 Mg NaCl ha^{-1} per week over a 3-week period (actual concentrations of irrigation water were not given) [30].

The majority of studies show only a moderate range in salinity tolerance among *P. pratensis* cvs. Although there were no differences in the shoot dry matter yield, visual salt injury was significantly less for Nugget than other *Poa pratensis* cvs. (Fylking, Park, Pennstar, Newport, and Merion) [30]. Cvs. Nugget, Ram I, and Baron suffered less shoot visual injury than Adelphi when irrigated with ECw 15 dS m^{-1} over a 2-month period [37]. Salinity tolerance of 23 *P. pratensis* cvs. exposed to ECw 14 dS m^{-1} for 3 months, as indicated by visual quality, decreased in the order Nugget > KI-148 > Bristol > Parade, other cvs. being less tolerant than these four [31]. Authors of this study reported that there was not a very wide range of salinity tolerance among *P. pratensis* cvs. compared with other turfgrasses studied. Torello and Spokas [115] tested 37 *P. pratensis* cvs. by spraying NaCl at increasing weekly concentrations onto field plots. Following 9 weeks, cv. differences in visual turf quality were small. Majestic, Princeton, and Galaxy had the highest turf quality, whereas Haga, Plush, and Victa had the lowest. Salinity tolerance of five *P. pratensis* cv. seedlings grown in agar, measured as leaf and root length reduction, showed Adelphi and Ram 1 to be more salt tolerant than Nassau, Bensun, and Baron [109]. Cv. Fylking performed better than Merion in field plots having an average ECe of 11.4 dS m^{-1} [105]. Salinity tolerance of 44 *P. pratensis* cvs., indicated by leaf blade length, was determined at the seedling stage. Salt-tolerant cvs. included Arista, Nugget, Delta, Prato, Baron, Park, S-21, Pennstar, Fylking, Windsor, Victa, Birka, Banff, Cheri, and Oregon Common [29]. *P. pratensis* cvs. were visually screened for salt injury following exposure to approximately 15 dS m^{-1} for 8 weeks. Salt-tolerant cvs. included Northstar, Ascot, Moonlight, Wildwood, Sodnet, Dragon, and Blackstone [111]. Using absolute growth under salinity stress, salinity tolerance decreased in order Baron > Brilliant > Eagleton > Cabernet > Midnight; however relative growth (to control) revealed no differences [116].

In comparing cultivar salt tolerance ratings in the above studies, there are conflicting trends, perhaps due to a narrow range of salt tolerance within this species. Only Nugget was consistently in the top group in the four studies in which it appeared. In contrast, Fylking was in the top statistical group in two studies, but in two others, it had only intermediate salt tolerance. In two studies, Park was ranked fairly low in salt tolerance, but it was in the top group in the Horst and Taylor [29] study. Also, in some studies, Adelphi and Baron ranked as being salt tolerant, whereas they ranked as being salt sensitive in others. Merion consistently ranked as being salt sensitive in the three studies in which it was included.

Germination of *P. pratensis* cv. SD Common was completely inhibited by 0.6% NaCl (ECw 11 dS m^{-1}) [43]. Forty-four *P. pratensis* cvs. were tested for percentage of germination under a range of salinities from 7500 to 15,000 ppm (ECe 25 dS m^{-1}) [29]. There was a continuous decrease in the percentage of germination from 100% to 4%, with Delta, Park, Prato, Warrens 113, and Nugget having the highest percentage of germination.

24.2.2.6.4   Poa arachnifera *Torr. x P.* pratensis *L. (Texas x Kentucky Bluegrass hb.)*
Recently, interspecific hybrids of *P. pratensis* and *P. arachnifera* have been achieved, with cv. Reveille released to market [117]. *P. arachnifera*, native to south-central United States, provides the bluegrass hybrids with superior drought tolerance [118]. Limited research has not shown any consistent salinity tolerance differences between several hybrid crosses and their parents from either species [116,119].

24.2.2.6.5   Poa supina *Schrader (Supina Bluegrass)*
P. supina is a new turfgrass adapted to cool temperate and subarctic regions [33]. Relative salinity tolerance of *P. supina* is not currently known, nor has the species been generally ranked for salinity tolerance.

24.2.2.6.6   Poa trivialis *L. (Rough Bluegrass)*
P. trivialis is used as a perennial turf in cool temperate climates, as well as an overseeded winter turf in warmer climates [45]. *P. trivialis* has been generally ranked salt sensitive, tolerating < ECe 3 dS m^{-1} [113]. Salinity tolerance, rated as visual quality, of a *P. trivialis* acc. was lower than either a *Lolium perenne* acc. or an *Agrostis stolonifera* acc. [120]. In grasses exposed to salinity stress for 3 weeks (actual soil salinity level unknown), *P. trivialis* acc. was less salt tolerant (determined as visual quality) than *Agrostis stolonifera* cv. Seaside; was equal in salt tolerance to *Lolium perenne* cv. Common and three *P. pratensis* cultivars (Pennstar, Nugget, and Park); but was more salt tolerant than cultivars Fylking, Newport, and Merion [30].

Eight *P. trivialis* cvs. were compared for germination percent in up to 5 dS m^{-1} salinity. Cultivars Laser and Fuzzy tended to be most tolerant, and cultivar Winterplay the least, though this was influenced by seed lot [121].

24.2.2.7 *Puccinellia* spp. Parl. (Alkaligrasses)

Puccinellia spp. are found inhabiting saline and alkaline sites throughout the cooler portions of North America [44]. These low-growing, perennial bunchgrasses were first considered for use as turfgrass in Illinois [122] and Colorado [36] when *Puccinellia* spp. were found along roadsides where deicing salts had eliminated other grasses. Fults [123] listed three *Puccinellia* spp. as being the most valuable for turf: *Puccinellia distans* [Jacq.] Parl. (weeping alkaligrass), *P. airoides* [Nutt.] S. Wats. and Coult. (Nuttall's alkaligrass), and *P. lemmoni* [Vasey] Scribn. (Lemmon's alkaligrass). *Puccinellia* spp. are mainly suited for low-maintenance turf, and they have been used successfully along roadsides and on some residential and athletic grounds [12].

Puccinellia spp. are by far the most salt-tolerant cool-season grasses having turf-like growth characteristics. *P. distans* has been reported surviving in soils with ECe (sat. paste) over 46 dS m^{-1} [36]. U.S. Salinity Laboratory [20] listed *P. airoides* as having high salt tolerance (ECe 12–18 dS m^{-1}), being more salt tolerant than *Cynodon dactylon*, but less salt tolerant than *Distichlis spicata* var. *stricta*. Harivandi et al. [124] reported *P. distans* (acc.) more salt tolerant than *P. lemmoni* in terms of leaf yellowing and survival. Lunt et al. [90] reported *P. distans* acc. survived relatively well in sand culture when irrigated with ECw (irrigation water) 32 dS m^{-1} over a 4-month period, suffering less injury than *Agrostis stolonifera* cv. Seaside, *Festuca arundinacea* cv. Alta, *Agrostis capillaris* cv. Highland, and *Poa pratensis* acc. Hughes et al. [18] reported forage yields of *P. distans* acc. were reduced 33% over a 5-month period when irrigated with ECw 32 dS m^{-1} (NaCl) in a greenhouse pot trial. In contrast, *Lolium perenne* acc. was reduced 44% and *Poa pratensis* acc., 47%. Ahti et al. [31] reported *P. distans* cv. Fults as more salt tolerant than several *Festuca rubra* and *Poa pratensis* cultivars tested, continuing to exhibit healthy vigorous growth with essentially no leaf injury following 80 days of exposure to ECe 32 dS m^{-1}. Alshammary et al. [125] reported salinity tolerance to decline in the order *P. distans* cv. Fults, *Distichlis spicata* ssp. *stricta* acc., *Festuca arundinaceae* cv. Arid, and *Poa pratensis* cv. Challenger. Seedling root elongation of *P. distans* cv. Fults was inhibited to a lesser extent than in *Lolium perenne* cv. S23, *Agrostis stolonifera* cv. Prominent, or *Festuca rubra* cvs. Hawk and Dawson when exposed to 25 dS m^{-1} NaCl [126]. In a greenhouse pot trial irrigated with salinized (final levels not given) water for 5 weeks, *P. airoides* suffered no leaf chlorosis, whereas *P. distans* and *P. lemmoni* suffered slight leaf chlorosis. Other turfgrasses, including *Lolium perenne* cvs. Common and NK200; *Agrostis stolonifera* cv. Seaside, *Festuca arundinaceae* cv. Alta, *Festuca rubra* cv. Ruby; and *Poa pratensis* cvs. Fylking, Park, Pennstar, Nugget, Newpart, and Merion suffered higher degrees of leaf firing [30].

Germination percentages after a 15-day exposure to 75% seawater (ECw 28.5 dS m^{-1}) was greater for *P. distans* acc. and *P. lemmoni* acc. than for *Festuca rubra* cv. Dawson, *Lolium perenne* cv. Pennfine, *Agrostis stolonifera* cv. Seaside, and *Poa pratensis* cv. Merion [107].

24.3   RELATIVE SALINITY TOLERANCE RANKING OF TURFGRASSES

Precision in ranking relative salinity tolerance of turfgrasses is problematic due to factors discussed earlier. These include different methods of measuring relative salinity tolerance used in studies, including growth factors (e.g., relative shoot dry weight, root growth, leaf length), visual quality (e.g., percentage of leaf burning, shoot density), and germination percentage. Environmental factors, known to interact with salinity tolerance, vary widely among studies, including temperature, light, relative humidity, and soil/root media differences. Also, time of exposure to salinity stress as well as units used to quantify salinity levels differ among studies. As a result, there is a good deal of variability in results among studies, sometimes resulting in contradictory cultivar rankings within species, but at times also resulting in contradictory species rankings. Even more affected by these factors are the estimates of actual salinity levels (dS m^{-1}) tolerated by species or cultivars. Given such limitations, Table 24.1 is an attempt to summarize the current literature concerning turfgrass salt tolerance, ranking turfgrass species relative to one another. If information regarding cultivar differences is available, salt-tolerant cultivars are listed immediately below each turfgrass species.

TABLE 24.1
Relative Salinity Tolerance of Turfgrasses

C$_4$ (Warm-Season) Turfgrasses	Salinity Tolerance[a]	C$_3$ (Cool-Season) Turfgrasses
Distichlis spicata sp. *stricta*	35 dS m^{-1}	
Sporobolus virginicus		
Paspalum vaginatum	25 dS m^{-1}	*Puccinellia* spp.
'Sea Isle I'[b]		'Fults'
Zoysia matrella		'Salty'
'Diamond'		
Zoysia pacifica (*tenuifolia*)		
Stenotaphrum secundatum	20 dS m^{-1}	
'Seville'		
Cynodon spp.	18 dS m^{-1}	
'FloraTex'		
'MS Supreme'		
Zoysia japonica	14 dS m^{-1}	
'Meyer'		
Pennisetum clandestinum	12 dS m^{-1}	
	10 dS m^{-1}	*Agrostis stolonifera*
		'Mariner'
	8 dS m^{-1}	*Festuca arundinaceae*
		'Tar Heel II'
Buchloë dactyloides	7 dS m^{-1}	*Lolium perenne*
		'Paragon'
		'Fiesta III'
		Agropyron cristatum
	6 dS m^{-1}	*Festuca rubra* spp.
		'Dawson'
Bouteloua spp.	5 dS m^{-1}	*Bromus inermis*
		Lolium multiflorum
		Lolium xhybridum
	4 dS m^{-1}	*Poa pratensis*
		'Nugget'
		Poa arachnifera x *pratensis*
Axonopus spp.	3 dS m^{-1}	*Festuca rubra* ssp. *fallax*
Eremochloa ophiuroides		*Festuca brevipila*
Paspalum notatum		*Festuca elatior*
		Festuca ovina
	2 dS m^{-1}	*Agrostis capillaris*
		Agrostis canina
		Agrostis gigantea
		Poa trivialis
		Poa annua

[a] Relative salinity tolerance is an estimate (based on literature review) of the maximum salinity (ECe in dS m^{-1}) at which the grass can acceptably grow, or the salinity at which shoot growth is reduced by approximately 50%.

[b] Salt-tolerant cultivar within species. Indicates published information is available for comparisons to be made among cultivars within a species.

REFERENCES

1. C. Milesi, S.W. Running, C.D. Elvidge, J.B. Dietz, B.T. Tuttle, R.R. Nemani. Mapping and modeling the biogeochemical cycling of turf grasses in the United States. *Environ Manage* 36(3):426–438, 2005.
2. J.B. Beard, R.L. Green. The role of turfgrasses in environmental protection and their benefits to humans. *J Environ Qual* 23:452–460, 1994.
3. D.E. Aldous. Education and training opportunities for turf management in Australia. *Acta Horticulturae* 672:71–77, 2004.
4. K. Danneberger. China offers great growth for turfgrass industry. Turfgrass Trends, October 1, 2001, http://www.turfgrasstrends.com/turfgrasstrends/article/articleDetail.jsp?id=13081
5. California State Water Resources Control Board. Porter-Cologne Act provisions on reasonableness and reclamation promotion. California Water Code, Section 13552-13577. Available online with updates at http://www.swrcb.ca.gov/water_laws/index.html
6. Arizona Department of Water Resources. Modifications to the second management plan: 1990–2000. Phoenix, AZ. Available online with updates at http://azwater.gov/dwr/
7. Florida Department of Environmental Protection. Applicable rules for reuse projects. Chap. 62–610, Reuse of reclaimed water and land application. Available online with updates at http://www.dep.state.fl.us/water/reuse/index.html
8. Council of Australian Governments. National Water Initiative, 2004, http://www.coag.gov.au/meetings/250604/#nwi.
9. L.B. McCarty, A.E. Dudeck. Salinity effects on bentgrass germination. *HortScience* 28:15–17, 1993.
10. C.L. Murdoch. Water: the limiting factor for golf course development in Hawaii. *U.S.G.A. Green Section Record* 25:11–13, 1987.
11. M.A. Harivandi, J.D. Butler, L. Wu. Salinity and turfgrass culture. In: D.V. Waddington, R.N. Carrow, R.C. Shearman, eds., *Turfgrass*. Agronomy Monograph No. 32. Madison, WI: American Society of Agronomy, 1992:207–229.
12. J.D. Butler, I.L. Fults, G.D. Sanks. Review of grasses for saline and alkali areas. In: E.C. Roberts, ed., *Proc 2nd Int Turfgrass Res Conf*. Blacksburg, VA. ASA, Madison, WI., 1974:551–556.
13. G. Horst, J.B. Beard. Salinity in turf. *Grounds Maintenance*, April 1977:66, 69, 72, 73, 109.
14. E.V. Maas. Salt tolerance of plants. *Appl Agric Res* 1:12–26, 1986.
15. E.V. Maas, G.J. Hoffman. Crop salt tolerance-current assessment. *J Irrig Drain Div ASCE* 103:115,134, 1977.
16. A. Harivandi. Irrigation water quality and turfgrass management. *Calif Turfgrass Cult* 38(3,4):1–4, 1988.
17. F. Ghassemi, A.J. Jakeman, H.A. Nix. *Salinisation of Land and Water Resources: Human Causes, Extent, Management and Case Studies*. Sydney, Australia: University of New South Wales Press, 1995:20–22.
18. T.D. Hughes, J.D. Butler, G.D. Sanks. Salt tolerance and suitability of various grasses for saline roadsides. *J Environ Qual* 4:65–68, 1975.
19. G.J. Hoffman, S.L. Rawlins. Growth and water potential of root crops as influenced by salinity and relative humidity. *Agron J* 63:877–880, 1971.
20. U.S. Salinity Laboratory Staff. Diagnosis and improvement of saline and alkali soils. *U.S. Department of Agriculture Handbook* 60, 1969:67.
21. A.E. Dudeck, C.H. Peacock. Salinity effects on growth and nutrient uptake of selected warm-season turfgrasses. *Int Turfgrass Soc Res J* 7:680–686, 1993.
22. K.B. Marcum, M. Pessarakli. Salinity tolerance and salt gland excretion activity of *Cynodon* spp. (Bermudagrass) turf cultivars. *Crop Sci* 46(6):2571–2574, 2006.
23. A.E. Dudeck, S. Singh, C.E. Giordano, T.A. Nell, D.B. McConnell. Effects of sodium chloride on *Cynodon* turfgrasses. *Agron J* 75:927–930, 1983.
24. M.A.L. Smith, J.E. Meyer, S.L. Knight, G.S. Chen. Gauging turfgrass salinity responses in whole plant microculture and solution culture. *Crop Sci* 33:566–572, 1993.
25. A.E. Dudeck, C.H. Peacock. Salinity effects on warm-season turfgrasses. *Proc 33rd Ann Florida Turfgrass Conf* 33:22–24, 1985.
26. R. Chetelat, L. Wu. Contrasting response to salt stress of two salinity tolerant creeping bentgrass clones. *J Plant Nutr* 9:1185–1197, 1986.
27. C. Kik. Ecological genetics of salt resistance in the clonal perennial, *Agrostis stolonifera* L. *New Phytol* 113:453–458, 1989.

28. G.L. Horst, N.B. Beadle. Salinity affects germination and growth of tall fescue cultivars. *J Am Soc Hortic Sci* 109:419–422, 1984.
29. G.L. Horst, R.M. Taylor. Germination and initial growth of Kentucky bluegrass in soluble salts. *Agronomy J* 75:679–681, 1983.
30. L.J. Greub, P.N. Drolsom, D.A. Rohweder. Salt tolerance of grasses and legumes for roadside use. *Agronomy J* 77:76–80, 1985.
31. K. Ahti, A. Moustafa, H. Kaerwer. Tolerance of turfgrass cultivars to salt. In: J.B. Beard, ed., *Proc 3rd Int Turfgrass Res Conf*, Munich, Germany, 1980:165–171.
32. N.E. Marcar. Salt tolerance in the genus *Lolium* (ryegrass) during germination and growth. *Aust J Agr Res* 38:297–307, 1987.
33. A.J. Turgeon. *Turfgrass Management*, 7th ed. Englewood Cliffs, NJ: Prentice Hall, 2005.
34. J.B. Beard. *Turfgrass: Science and Culture*. Englewood Cliffs, NJ: Prentice Hall, 1973.
35. C.H. Peacock, A.E. Dudeck. Influence of salinity on warm season turfgrass germination. *Proc 6th Int Turfgrass Res Conf*, Tokyo, 1989:229–231.
36. J.D. Butler. Salt tolerant grasses for roadsides. *Highway Res Rec* 411:1–6, 1972.
37. E.J. Kinbacher, R.C. Shearman, T.P. Riordan, D.E. Vanderkolk. Salt tolerance of turfgrass species and cultivars. *Agron Abstr* 1981:88.
38. K.B. Marcum. Salinity tolerance mechanisms of grasses in the subfamily Chloridoideae. *Crop Sci* 39:1153–1160, 1999.
39. A. Mintenko, R. Smith. Native grasses vary in salinity tolerance. *Golf Course Manage* 69(4):55–59, 2001.
40. S.D. Reid, A.J. Koski, H.G. Hughes. Buffalograss seedling screening in vitro for NaCl tolerance. *HortScience* 28:536, 1993.
41. L. Wu, H. Lin. Salt concentration effects on buffalograss germplasm seed germination and seedling establishment. *Int Turfgrass Soc J* 7:823–828, 1993.
42. L. Wu, H. Lin. Salt tolerance and salt uptake on diploid and polyploid buffalograsses (*Buchloë dactyloides*). *J Plant Nutr* 17:1905–1928, 1994.
43. J.G. Reiten, C.W. Lee, Z.M. Cheng, R.C. Smith. Germination salt tolerance of Kentucky bluegrass (*Poa pratensis*) and buffalograss (*Buchloë dactyloides*) seeds. *HortScience* 27:676, 1992.
44. F.W. Gould, R.B. Shaw. *Grass Systematics*. College Station, TX: Texas A&M University Press, 1983.
45. R.L. Duble. *Turfgrasses: Their Management and Use in the Southern Zone*. College Station, TX: Texas A&M University Press, 1996.
46. N. Christians. *Fundamentals of Turfgrass Management*. Chelsea, MI: Ann Arbor Press, 1998:57–59.
47. A. Harivandi. Irrigation water quality: one key to success in golf course management. *Golf Course Manage*, January 1988:106–107, 189–193.
48. J.D. Butler, P.E. Rieke, D.D. Minner. Influence of water quality on turfgrass. In: V.A. Gibeault, S.T. Cockerham, eds., *Turfgrass Water Conservation*. University of California Publication 21405, 1985:71–84.
49. R.C. Ackerson, V.B. Younger. Responses of bermudagrass to salinity. *Agron J* 67:678–681, 1975.
50. P.S. Ramakrishnan, R Nagpal. Adaptation to excess salts in an alkaline soil population of *Cynodon dactylon* (L) Pers. *J Ecol* 61:369–381, 1973.
51. K.B. Marcum, C.L. Murdoch. Growth responses, ion relations, and osmotic adaptations of eleven C_4 turfgrasses to salinity. *Agronomy J* 82:892–896, 1990.
52. D. Pasternak, A. Nerd, Y. de Malach. Irrigation with brackish water under desert conditions. IX. The salt tolerance of six forage crops. *Agr Water Manage* 24:321–334, 1993.
53. C.H. Peacock, D.J. Lee, W.C. Reynolds, J.P. Gregg, R.J. Cooper, A.H. Bruneau. Effects of salinity on six bermudagrass turf cultivars. *Acta Horticulturae* 661:193–197, 2004.
54. L.E. Francois. Salinity effects on three turf bermudagrasses. *HortScience* 23:706–708, 1988.
55. D.M. Kopec, K.B. Marcum. Desert saltgrass: a potential new turfgrass species. *U.S.G.A. Green Section Record* 39(1):6–8, 2001.
56. Z.F. Zhao, J.W. Hayser, H.J. Bohnert. Gene expression in suspension culture cells of the halophyte *Distichlis spicata* during adaptation to high salt. *Plant Cell Physiol* 30:861–867, 1989.
57. F.W. Gould. *Grasses of the Southwestern United States*, 6th ed. University of Arizona Press, Tucson, 1993.
58. K.B. Marcum. Use of saline and non-potable water in the turfgrass industry: Constraints and developments. *Agr Water Manage* 80:132–146, 2006. Available online at www.elsevier.com/locate/agwat
59. M. Pessarakli, D.M. Kopec, A.J. Koski. Comparative responses of bermudagrass, seashore paspalum, and saltgrass to salinity stress. *Agron Abstr* 2004.

60. K.B. Marcum, M. Pessarkli, D.M. Kopec. Relative salinity tolerance of 21 turf-type desert saltgrasses compared to bermudagrass. *HortScience* 40(3):827–829, 2005.

61. M. Pessarkli, D.M. Kopec, J. Gilbert, A. Koski, Y.L. Qian, D. Christensen. Growth responses of twelve inland saltgrass clones to salt stress. *Agron Abstr* 2005.

62. K.B. Marcum, N.P. Yensen. Salinity tolerance of desert saltgrass (*Distichlis spicata* var. *stricta*) under turf-type management. International Salinity Forum, Riverside, California, April 2005:97–100.

63. K.B. Marcum, C.L. Murdoch. Salinity tolerance mechanisms of six C$_4$ turfgrasses. *J Am Soc Hort Sci* 119:779–784, 1994.

64. W.R. Kneebone. *Hilaria belangeri*, rediscovered potential low maintenance arid land turf. *Proc 5th Int Turfgrass Res Conf*, Avignon, France, 1985:285–288.

65. A.E. Ralowicz, C.F. Mancino. Afterripening in curly mesquite seeds. *J Range Manage* 45(1):85–87, 1992.

66. C.L. Dantzman. Salt tolerance of Pensacola bahiagrass (*Paspalum notatum* L Flugg.). *Agron Abstr* 68:145, 1976.

67. J.M. Henry, M.A. Gibeault, V.B. Youngner, S. Spaulding. *Paspalum vaginatum* 'Adalayd' and 'Futurf.' *Calif Turfgrass Cult* 29:9–12, 1979.

68. R.R. Duncan. The environmentally sound turfgrass of the future. *U.S.G.A. Green Section Record* 34: 9–11, 1996.

69. D.P. Major. The influence of salinity levels on two highly salt tolerant turfgrasses 'Fults' *Puccinellia distans* and *Paspalum vaginatum* 'Futurf.' MS thesis, California Polytechnic State University, San Luis Obispo, CA, 1978.

70. A.E. Dudeck, C.H. Peacock. Effects of salinity on seashore paspalum turfgrasses. *Agron J* 77:47–50, 1985.

71. G. Lee, R.N. Carrow, R.R. Duncan. Criteria for assessing salinity tolerance of the halophytic turfgrass seashore paspalum. *Crop Sci* 45(1):251–258, 2005.

72. G. Lee, R.R. Duncan, R.N. Carrow. Salinity tolerance of seashore paspalum ecotypes: shoot growth responses and criteria. *HortScience* 39(5):1138–1142, 2004.

73. M.R. Panuccio, M. Sidari, A. Muscolo. Effects of different salt concentrations and pH conditions on growth of *Pennisetum clandsetinum* HOCHST (kikuyu grass). *Fresenius Environ Bull* 11(6):295–299, 2002.

74. M. Radhakrishnan, Y. Waisel, M. Sternberg. Kikuyu grass: a valuable salt-tolerant fodder grass. *Commun Soil Sci Plant Anal* 37(9–10):1269–1279, 2006.

75. L.A. Donovan, J.L. Gallagher. Anaerobic substrate tolerance in *Sporobolus virginicus* (L.) KUNTH. *Am J Bot* 71(10):1424–1431, 1984.

76. D.S. Loch, B.K. Simon, R.E. Poulter. Taxonomy, distribution and ecology of *Zoysia macrantha* Desv., an Australian native species with turf breeding potential. *Int Turfgrass Soc Res J* 10(1):593–599, 2005.

77. M.W. Depew, P.H. Tillman. Commercial application of halophytic turfs for golf and landscape aevelopments utilizing hyper-saline irrigation. In: M.A. Khan, D.J. Weber, eds., *Tasks for Vegetation Science 40: Ecophysiology of High Salinity Tolerant Plants*, Springer, Netherlands, 2006:255–278.

78. K.B. Marcum, C.L. Murdoch. Salt tolerance of the coastal salt marsh grass, *Sporobolus virginicus* (L) Kunth. *New Phytol* 120:281–288, 1992.

79. K.C. Blits, J.L. Gallagher. Morphological and physiological responses to increases salinity on marsh and dune ecotypes of *Sporobolus virginicus* (L.) Kunth. *Oecologia* 87:330–335, 1991.

80. D.E. Aldous. Growth responses of four native seashore dropseed (*Sporobolus virginicus* (L.) Kunth) accessions to elevated salt concentrations. *Acta Horticulturae* 661:199–205, 2003.

81. M.J. Meyer, M.A.L. Smith, S.L. Knight. Salinity effects on St. Augustinegrass: a novel system to quantify stress response. *J Plant Nutr* 12:893–908, 1989.

82. A.E. Dudeck, C.H. Peacock, J.C. Wildmon. Physiological and growth responses of St. Augustinegrass cultivars to salinity. *HortScience* 28:46–48, 1993.

83. M.C. Engelke, S. Anderson. Zoysiagrasses. In: M.D. Casler, R.R. Duncan, eds., *Turfgrass Biology, Genetics, and Breeding.* Wiley, New York, 2003:271–276.

84. J.J. Murray, M.C. Engelke. Exploration for zoysiagrass in Eastern Asia. *U.S.G.A. Green Section Record*, May-June, 1983:8–12.

85. G.J. Lee, Y.K. Kweon, K.S. Kim. Comparative salt tolerance study in zoysiagrasses II. Interspecific comparison among eight zoysiagrasses (*Zoysia* spp.). *J Korean Soc Hort Sci* 35:178–185, 1994.

86. K.B. Marcum, S.J. Anderson, M.C. Engelke. Salt gland ion secretion: a Salinity tolerance mechanism among five zoysiagrass species. *Crop Sci* 38:806–810, 1998.

87. Y.L. Qian, M.C. Engelke, M.J.V. Foster. Salinity effects on zoysiagrass cultivars and experimental lines. *Crop Sci* 40(2):488–492, 2000.
88. A.J. Koski, Y. Qian, H.G. Hughes, D.K. Christensen, S. Reid, R.L. Cuany, S.J. Wilhelm. Alternative grasses for western U.S. Lawns. *Agron Abstr* 91:137, 1999.
89. H. Kang, C.W. Lee. Influence of NaCl on seed germination of selected cool-season turfgrass species. *HortScience* 34(3):490, 1999.
90. O.R. Lunt, V.B. Youngner, J.J. Oertli. Salinity tolerance of five turfgrass varieties. *Agron J* 53:247–249, 1961.
91. N. Hannon, A.D. Bradshaw. Evolution of salt tolerance in two coexisting species of grass. *Nature* 220:1342–1343, 1968.
92. L. Wu. The potential for evolution of salinity tolerance in *Agrostis stolonifera* L and *Agrostis tenuis* Sibth. *New Phytol* 89:471–486, 1981.
93. I. Ahmad, S.J. Wainwright. Tolerance to salt, partial anaerobiosis, and osmotic stress in *Agrostis stolonifera*. *New Phytol* 79:605–612, 1977.
94. V.B. Youngner, O.R. Lunt, F. Nudge. Salinity tolerance of seven varieties of creeping bentgrass, *Agrostis palustris* Huds. *Agron J* 59:335–336, 1967.
95. K.B. Marcum. Salinity tolerance of 35 bentgrass cultivars. *HortScience* 36(2):374–376, 2001.
96. L.J. Greub, P.N. Drolsom. Salt tolerance of selected grass species and cultivars as affected by soil type, soil phosphorus and level of salt application. *Agron Abstr* 69:111, 1977.
97. H.G. Brod, H.U. Preusse. The influence of deicing salts on soil and turf cover. In: J.B. Beard, ed., *Proc 3rd Int Turfgrass Res Conf*, Munich, Germany, 1980:461–468.
98. J.K. Wipff, C.A. Rose-Fricker. Salt tolerance comparisons among tall fescue cultivars. *Agron Abstr* 2003.
99. J.P. Shildrick. A provisional grouping of cultivars of *Festuca rubra* L. *J Sports Turf Res Inst* 52:9–13, 1976.
100. M.O. Humphreys. Response to salt spray in red fescue and perennial ryegrass. *Proc 4th Int Turfgrass Res Conf*, University of Guelph, Canada, 1980:47–54.
101. M.O. Humphreys. The genetic basis of tolerance to salt spray in populations of *Festuca rubra* L. *New Phytol* 91:287–296, 1982.
102. J. Rozema, M. Visser. The applicability of the rooting technique measuring salt resistance in populations of *Festuca rubra* and *Juncus* species. *Plant Soil* 62:479–485, 1981.
103. W.E. Cordukes. Saltol, a salt-tolerant red fescue. *Can J Plant Sci* 61:761–764, 1981.
104. G. Michelmann. Methoden zur selektion von rasengr asern auf salztoleranz im praktischen zuchtbetrieb. Rasen, Grunglachen, Begrunungen 9:45–48, 1978.
105. V.A. Gibeault, D. Hanson, D. Lancaster, E. Johnson. Final research report: Cool season variety study in high salt location. *Calif Turfgrass Cult* 27:11–12, 1977.
106. S. McHugh, M.S. Johnson, T. McNeilly, M.O. Humphreys, R. Rowling. Tolerance to salt in *Festuca rubra* L. *Proc 5th Int Turfgrass Res Conf*, Avignon, France, 1985:185–193.
107. M.A. Harivandi, J.D. Butler, P.M. Soltanpour. Salt influence on germination and seedling survival of six cool season turfgrass species. *Commun Soil Sci Plant Anal* 13:519–529, 1982.
108. H.A. Jönsson, C. Nilsson. Tolerans hos rodsvingel mot forhojd salthalt (Tolerance to a high salt concentration in *Festuca rubra*). *Weibulls Gras-tips* 20(12): 29–31, 1977.
109. W.A. Torello, A.G. Symington. Screening of turfgrass species and cultivars for NaCl tolerance. *Plant Soil* 82:155–161, 1984.
110. A.H. Khan, C. Marshall. Salt tolerance within populations of Chewings fescue (*Festuca rubra* L [ssp. *commutata* Gaud.]). *Commun Soil Sci Plant Anal* 12:1271–1281, 1981.
111. C. Rose-Fricker, J.K. Wipff. Breeding for salt tolerance in cool-season turf grasses. *Int Turfgrass Soc Res J* 9:206–212, 2001.
112. A.E. Dudeck, C.H. Peacock. Salinity effects on perennial ryegrass germination. *HortScience* 20: 268–269, 1985.
113. M.A. Harivandi, K.B. Marcum, Y. Qian. Recycled, gray, and saline water irrigation. In: *Water Quality and Quantity Issues for Turfgrasses in Urban Landscapes: Proceedings*, Special Pub. 26. Council for Agricultural Science and Technology, Ames, IA, 2007 (in press).
114. R.L. Green, L. Wu, G.J. Klein. Summer cultivation increases field infiltration rates of water and reduces soil electrical conductivity on annual bluegrass golf greens. *HortScience* 36(4):776–779, 2001.
115. W.A. Torello, L.A. Spokas. Evaluation of Kentucky bluegrass (*Poa pratensis* L) cultivars for salt tolerance. *Rasen, Grunflachen, Begrunungen* 14:71–73, 1983.

116. C.M. Grieve, S.A. Bonos, J.A. Poss. Salt tolerance assessment of Kentucky bluegrass cultivars selected for drought tolerance. *HortScience* 41(4):1057, 2006.
117. J.C. Read, J.A. Reinhart, P.E. Colbaugh, W.E. Knoop. Registration of 'Reveille' hybrid bluegrass. *Crop Sci* 39(2):590.
118. J.C. Read, S.J. Anderson. Texas bluegrass. In: M.D. Casler, R.R. Duncan, ed., *Turfgrass Biology, Genetics, and Breeding*. Hoboken, NJ: Wiley, 2003:61–66.
119. M.R. Suplick-Ploense, Y.L. Qian, J.C. Read. Relative NaCl tolerance of Kentucky bluegrass, Texas bluegrass, and their hybrids. *Crop Sci* 42:2025–2030, 2002.
120. M. Pessarakli, D. Kopec, J. Gilbert. Competitive growth responses of three cool-season turfgrasses to salinity & drought stresses. *Agron Abstr* 2005.
121. J. Camberato, M.S. Bruce, A.V. Turner. Cultivar differences in *Poa trivialis* germination with increased salinity. Clemson University Turfgrass Program Home Page http://virtual.clemson.edu/groups/turfornamental/tmi/irrigwatqualrr.html
122. G.D. Sanks. Adaptability of alkaligrass for roadside use. MS thesis, University of Illinois, Urbana, Champaign, 1971.
123. J. Fults. Grasses for saline and alkali areas. *Proc 18th Rocky Mountain Regional Turfgrass Conf*, Colorado Springs, CO, 1972:44–55.
124. M.A. Harivandi, J.D. Butler, P.N. Soltanpour. Effects of soluble salts on ion accumulation in *Puccinellia* spp. *J Plant Nutr* 6:255–266, 1983.
125. S. Alshammary, Y.L, Qian, S.J. Wallner. Salinity tolerance of four turfgrasses. *HortScience* 35(3):414, 2000.
126. M. Ashraf, T. McNeilly, A.D. Bradshaw. The potential for evolution of salt (NaCl) tolerance in seven grass species. *New Phytol* 103:299–309, 1986.

25 Physiological Adaptations of Turfgrasses to Salinity Stress

Kenneth B. Marcum

CONTENTS

25.1 INTRODUCTION

Poaceae, represented by over 7500 species, inhabit the earth in greater numbers and have a greater range of climatic adaptation than any other plant family [1,2]. Therefore, it is not surprising that grasses show an extreme range in salinity tolerance, from salt-sensitive to extremely salt-tolerant (halophytic). Though turf-type grasses represent only a small percentage of the total grass species, they encompass the full range of salinity tolerance found within the family. Examples range from salt-sensitive (*Poa annua*—C_3, *Eremochloa ophiuroides*—C_4), moderately salt-sensitive (*Lolium perenne*—C_3, *Buchloë dactyloides*—C_4), moderately salt-tolerant (*Agrostis stolonifera*—C_3, *Zoysia japonica*—C_4), salt-tolerant (*Cynodon* spp.—C_4), to highly salt-tolerant (*Puccinellia* spp.—C_3, *Paspalum vaginatum*—C_4) [3–6]. Turf-type species recognized as true halophytes (thriving in full-strength seawater) include *Distichlis spicata* ssp. *stricta* and *Sporobolus virginicus*, both C_4 grasses [7–10].

25.2 SHOOT AND ROOT GROWTH RESPONSES

Salt-sensitive plants (glycophytes) and moderately salt-tolerant plants (mesophytes) generally have linearly-declining shoot growth response to salinity. At low salinity, there may be a flat (non-declining) response prior to a threshold salinity level, beyond which shoot growth declines. In contrast, halophytic plants and often highly salt-tolerant plants as well, display stimulated shoot growth at moderate salinity levels, followed by decline [3,11–13]. These shoot growth characteristics are conserved in turfgrasses. Salt-sensitive to moderately-tolerant turfgrasses with shoot growth declining linearly with increasing salinity stress include, in order of increasing salt tolerance, *Poa pratensis* [14], *Paspalum notatum* [15], *Eremochloa ophiuroides* [16], *Agrostis stolonifera* [17], *Zoysia japonica* [18], and *Cynodon* spp. [19]. Highly salt-tolerant to halophytic turfgrasses with stimulated shoot growth under moderate salinities, followed by growth decline include, in order of increasing salt tolerance, *Stenotaphrum secundatum* [16,20], *Zoysia matrella* [16], *Paspalum vaginatum* [15,16,21], *Distichlis spicata* ssp. *stricta* [22], and *Sporobolus virginicus* [23,24].

Root growth stimulation (increased root mass or rooting depth) in salt tolerant turfgrasses is typically a more common, and often more accentuated response to moderate salinity stress than shoot growth stimulation [25,26]. The net result is an increase in root/shoot ratio, which may be a salinity tolerance mechanism to counter external root media osmotic stress by increasing root absorptive to shoot transpirational area [26–28]. Increased rooting under low to moderate salinity stress has been observed in *Cynodon* spp. [27,29], *Paspalum vaginatum* [18,30], *Puccinellia distans* [31], and *Zoysia matrella* [18]. However, rooting decline under moderate salinity stress has been observed in more salt-sensitive turfgrasses, including *Bouteloua curtipendula* [32], *Buchloë dactyloides* [33], *Festuca rubra* [34], *Paspalum notatum* [15], and *Poa pratensis* [35]. In contrast, progressive root growth stimulation under high salinity, relative to control, has been observed in halophytic turfgrasses *Distichlis spicata* ssp. *stricta* [32,36,37] and *Sporobolus virginicus* [8,23,24]. Root dry weights linearly increased with increasing salinity up to 450 mM NaCl (35 dS m^{-1}) in *Sporobolus virginicus*, resulting in a root/shoot ratio approximately twice that of control (nonsalinized treatment) [23]. In another study, total root mass of *Sporobolus virginicus* grown in seawater was twice that of control plants [8].

25.3 ION EXCLUSION

It has generally been accepted that the major causes of plant-growth inhibition under salinity stress are osmotic stress (osmotic inhibition of plant water absorption) and specific ion effects, including toxicities and imbalances [25,38,39]. In comparison with salt tolerant or halophytic dicotyledonous plants, monocots (including Poaceae) tend to exclude saline ions from shoots, thereby minimizing toxic effects [26,40,41]. Shoot saline ion exclusion has been associated with salinity tolerance across both C$_3$ [42,43] and C$_4$ [16,30,32] turfgrass species.

Shoot saline ion exclusion is also an important factor influencing intraspecies salinity tolerance, i.e., at the cultivar or accession level. For example, salt-sensitive inland accessions were found to have higher shoot Na$^+$ and Cl$^-$ levels than saline-site, salt-tolerant accessions in *Festuca rubra* [34,44], *Cynodon dactylon* [45], and *Agrostis* spp. [46]. Relative salinity tolerance of cultivars and accessions have successfully been predicted on the basis of shoot Na$^+$ concentrations in *Cynodon* spp. [19] and *Zoysia* spp. [47,48] (Figure 25.1).

25.4 OSMOTIC ADJUSTMENT AND ION REGULATION

Maintenance of cell turgor and plant growth requires sufficient increase in sap osmolarity to compensate for external osmotic stress, the process of osmoregulation, or osmotic adjustment [49,50]. Historically, salinity tolerance has been associated with osmotic adjustment and avoidance of "physiological drought" [6,25], but recent studies have cast doubt on this. Shoot saline ion exclusion, coupled with minimal yet adequate osmotic adjustment, is now considered central to salinity tolerance in most plant species [51–53]. Salt-sensitive plants in general have been found to accumulate saline ions to toxic levels, well above those required for osmotic adjustment, resulting in shoot sap osmolarities well in excess of the saline growing soil/media [54–56]. This trend is conserved in turfgrasses. Shoot saline ion exclusion, coupled with minimal yet adequate osmotic adjustment is strongly correlated to salinity tolerance among both C$_3$ [42,43] and C$_4$ [32,36,57,58] turfgrasses. In a study comparing eight C$_4$ grass spp., salinity tolerance was positively correlated with shoot Cl$^-$ ion exclusion ($r = 0.80$) and negatively correlated with leaf osmolarity (osmotic adjustment) ($r = -0.85$) [32]. Though all turfgrasses maintained complete osmotic adjustment relative to root-media salinity, osmotic adjustment of salt-tolerant grasses was minimal (i.e., shoot osmolarity was maintained close to saline media levels) compared with salt-sensitive ones (Figure 25.2).

Though salinity tolerance in turfgrasses is clearly associated with saline ion exclusion, Na$^+$ and Cl$^-$ are instrumental for shoot osmotic adjustment, comprising the majority of osmotically

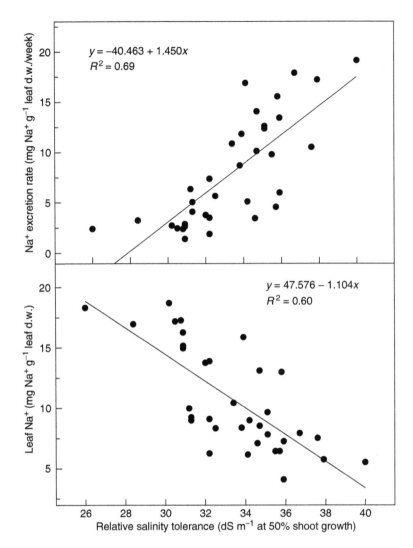

FIGURE 25.1 Leaf salt gland Na^+ excretion rate, and leaf sap Na^+ concentration versus relative salinity tolerance of 35 *Cynodon* spp. Cultivars. (From Marcum, K.B., Pessarakli, M., *Crop Sci.*, 46(6), 2571, 2006.)

active solutes [18,23,32,59]. Shoot Na^+ and Cl^- concentrations were highly correlated with osmotic adjustment among eight C_4 turfgrass spp. ($r = 0.9$) [32]. Therefore, though saline ion exclusion is critical for salinity tolerance in turfgrasses, saline ion *regulation*, or selectivity, rather than exclusion, may be a more apt description of the salinity tolerance mechanism operating in turfgrasses.

Maintenance of high shoot K^+/Na^+ ratios, necessary for proper cellular enzyme function [60], has been linked to salinity tolerance in turfgrasses [18,36,43]. Saline ion regulation, and selectivity for K^+ over Na^+, may occur via root or shoot mechanisms in turfgrasses. Selective ion uptake and transport processes occur at the apoplastic barrier of the root endodermis in turf and other grasses [41,61–63]. Selectivity occurs via low-affinity K^+ channels in plasma membrane root cells, enhanced by the Casparian band, in *Puccinellia* spp. [64]. Ion selectivity may also occur by selective sequestering of Na^+ and K^+/Na^+ exchange at the tonoplast via vacuolar compartmentation, coupled with compatible solute accumulation [62,65,66]. Finally, Chloridoid turfgrasses have specialized salt glands that selectively extrude saline ions from shoots [1,67,68]. These last two mechanisms will be further discussed below.

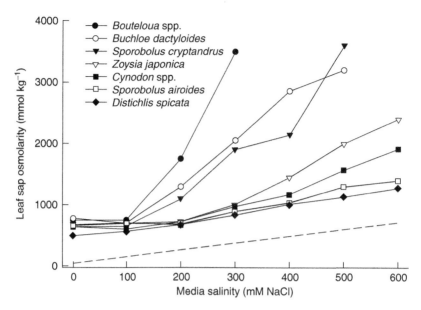

FIGURE 25.2   Leaf sap osmolarity of turfgrasses exposed to increasing salinity. Root-media salinity is indicated by dotted line. (From Marcum, K.B., *Crop Sci.*, 39, 1153, 1999.)

25.5   ION COMPARTMENTATION AND COMPATIBLE SOLUTES

In vitro studies have shown that enzymes of both glycophytes and halophytes have similar sensitivities to salt, being inhibited at concentrations above 100–200 mM (approximately 8–17 dS m^{-1}) [60,69]. Therefore, salt-tolerant plants growing under saline conditions must restrict the level of ions in the cytoplasm. Salt-tolerant turfgrasses utilize inorganic ions for a large part of their osmotic adjustment under saline growing conditions, as the ability to accumulate organic solutes on a whole-cell basis is metabolically expensive, and therefore limited [49,65]. Salt-tolerant plants that successfully accumulate saline ions for osmotic adjustment above concentrations of 100–200 mM do so by compartmentalizing them within the vacuole, which typically makes up 90–95% of a mature plant cell's volume [70]. Evidence exists for salinity inducing a K$^+$/Na$^+$ exchange across the tonoplast mediated by Na$^+$/H$^+$ antiport activity, resulting in saline ion compartmentation in vacuoles [62,66]. Under these conditions, osmotic potential of the cytoplasm is maintained by the accumulation of organic solutes that are compatible with enzyme activity, termed "compatible solutes" [71,72]. Under highly saline conditions, relatively few organic solutes, including glycinebetaine, proline, trigonelline, and certain polyols and cyclitols, can be accumulated in sufficient concentrations to osmotically adjust the cytoplasm without inhibiting enzymes [52]. Evidence exists for the cytoplasmic localization of these compounds [71,73,74]. Of these, glycinebetaine and proline typically accumulate in turf and other grasses [16,75,76].

Of possible compatible solutes, only glycinebetaine is typically associated with salinity tolerance in turfgrasses, though proline may play a minor role in *Puccinellia distans* [42]. Glycinebetaine, but not trigonelline or proline, has been correlated to salinity tolerance among C$_4$ turfgrasses, including *Bouteloua* spp., *Buchloë dactyloides*, *Cynodon* spp., *Distichlis spicata* ssp. *stricta*, *Eremochloa ophiuroides*, *Paspalum vaginatum*, *Sporobolus virginicus*, *Stenotaphrum secondatum*, and *Zoysia* spp. [16,23,32]. As glycinebetaine exists in the cytoplasm, which occupies approximately 10% of total cell volume [71,73,74], its contribution to cytoplasmic osmotic adjust can be estimated. Glycinebetaine made substantial contributions to cytoplasmic osmotic adjustment in salt-tolerant turfgrasses (*Cynodon* spp., *Distichlis spicata* ssp. *stricta*, *Paspalum vaginatum*, *Sporobolus virginicus*, *Stenotaphrum secondatum*, *Zoysia* spp.), but not in salt-sensitive ones (*Bouteloua* spp.,

TABLE 25.1

Estimated Contribution to Cytoplasmic Osmotic Adjustment of Glycinebetaine and Proline, in mosmol kg^{-1} (Osml), and as Percentage (%) of Total Cytoplasmic Osmolarity, of Plants Grown at 300 mM NaCl

Turfgrass	Glycinebetaine		Proline	
	Osml	%	Osml	%
Sporobolus virginicus	822	93	31	3
Distichlis spicata ssp. *Stricta*	625	73	17	2
Zoysia matrella	513	47	67	6
Cynodon spp.	378	39	27	3
Buchloë dactyloides	209	10	59	3
Bouteloua spp.	125	4	56	2

Note: Estimate assumes glycinebetaine and proline are located in the cytoplasm, comprising 10% of total cell volume, with an osmotic coefficient of 1.0 for each compound.

Source: Marcum, K.B., *J. Am. Soc. Hortic. Sci.*, 119, 779, 1994; Marcum, K.B., *New Phytol.*, 120, 281, 1992; and Marcum, K.B., *Crop Sci.*, 39, 1153, 1999.

Buchloë dactyloides, and *Eremochloa ophiuroides*) [16,32] (Table 25.1). In halophytic turfgrasses, glycinebetaine accumulates to very high levels. For example, in *Sporobolus virginicus* grown at high salinity (37 dS m^{-1}), glycinebetaine accumulated to 126 mM in shoots, potentially providing 93% of total cytoplasmic osmotic adjustment [23]. At similar salinities, *Distichlis spicata* ssp. *stricta* accumulated 62 mM glycinebetaine, potentially providing 73% of total cytoplasmic osmotic adjustment [32] (Table 25.1).

While shoot glycinebetaine concentrations have been positively correlated with salinity tolerance, proline concentrations have been negatively correlated, suggesting that glycinebetaine, but not proline, acts as a compatible solute in turfgrasses [16,23,32,43,77]. In addition, proline contributions were too small to contribute to cytoplasmic osmotic adjustment [16,23,32,43], this being especially evident in turfgrass halophytes *Sporobolus virginicus* and *Distichlis spicata* ssp. *stricta* [8,23,32,78].

Though both glycinebetaine and proline have traditionally been considered compatible solutes, recent evidence has favored the role of glycinebetaine. For example, (i) glycinebetaine is excluded from the hydration sphere of enzyme proteins and thus tends to stabilize their tertiary structure [79]; (ii) *Zea mays* mutants lacking a critical enzyme for glycinebetaine biosynthesis also lack salt tolerance [80]; (iii) exogenously applied glycinebetaine has enhanced the salinity tolerance of glycophytes such as *Oryza sativa* [81]; and (iv) transformation of *Poa pratensis*, which lacks glycinebetaine [43], with betaine aldehyde dehydrogenase gene (glycinebetaine synthesis gene) improves salinity tolerance [82]. In contrast, proline accumulation has been considered an indicator of plant injury, due to its universally rapid appearance following any type of stress [83,84].

25.6 GLANDULAR ION EXCRETION

Salt glands or bladders, which eliminate excess saline ions from shoots by active excretion, are present in a number of salt-adapted species [9,13,85]. Multicellular epidermal salt glands are present in several families of dicotyledons, e.g., Frankeniaceae, Plumbaginaceae, Aviceniaceae, and Tamaricaceae [9,86]. Within the Poaceae, bicellular leaf epidermal salt glands have been reported to occur in over 30 species within the tribes Chlorideae, Eragrosteae, Aeluropodeae, and Pappophoreae [67,68,87], members of the subfamily Chloridoideae [1,88]. Turfgrass genera having salt glands include *Bouteloua*, *Buchloë*, *Cynodon*, *Distichlis*, *Sporobolus*, and *Zoysia* [23,32,58].

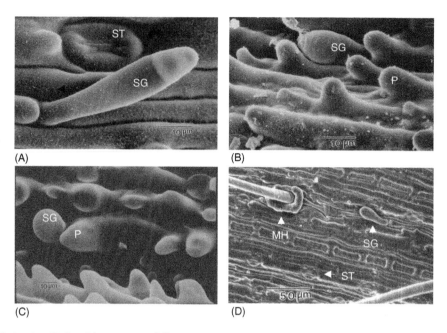

FIGURE 25.3 (**Color Figure 25.3 follows p. 262.**) Scanning electron micrographs of adaxial leaf surfaces. (A) Salt gland of *Bouteloua* spp. SG, salt gland; ST, stomate. (B) Salt gland of *Cynodon* spp. SG, salt gland; P, papillae. (C) Salt gland of *Distichlis spicata* var. *stricta*. Only cap cell is visible—basal cell is imbedded in epidermis. SG, salt gland; P, papillae. (D) Overview of *Buchloë dactyloides* leaf surface, showing location of salt gland relative to other structures. SG, salt gland; MH, macrohair; ST, stomate.

Bicellular salt glands of the Poaceae are distinct from the multicellular glands of dicots, being similar in appearance to leaf epidermal microhairs (trichomes). Poaceae glands are characterized by a cutinized basal cell, attached, or embedded, into the leaf epidermis, and a cap cell [67,86]. Though the basic, bicellular structure is the same in all Chloridoid spp., their appearance varies [85]. In some species, glands are sunken into the epidermis, with the basal cell totally embedded, e.g., *Distichlis spicata* ssp. *stricta*. In others, the basal cell is semi-embedded, e.g., *Cynodon* spp. Finally, the basal cell may extend out from the epidermis, with the gland lying recumbent to the leaf surface, e.g., *Bouteloua* spp., *Buchloë dactyloides*, *Sporobolus virginicus*, and *Zoysia* spp. (Figure 25.3). Salt glands have been found on both abaxial and adaxial leaf surfaces of excreting Chloridoid spp. [32,58,67]. Glands are longitudinally arranged in parallel rows atop intercostal regions of leaves, adjacent to rows of stomata (Figure 25.3).

Though microhairs resembling salt glands have been observed in all grass subfamilies except Pooideae [85,89], functioning salt glands have been found only within the subfamily Chloridoideae [67,68]. Intracellular ultrastructural modifications are hypothesized to be responsible for salt excretion in the Chloridoideae, a series of parallel, invaginated plasma membrane channels within the gland's basal cell [85,90,91]. These "partitioning" membranes are actually infoldings of the plasmalemma that originate adjacent to the wall separating the cap and basal cells, forming extracytoplasmic channels in the direction of ion flow. Ultracytochemical localization of ATPase activity within salt gland basal cells of *Sporobolus virginicus* supports the hypothesis of active ion loading at these sites [92]. Numerous mitochondria are intimately associated with the parallel membranes, involved in providing energy for channel ion-loading [90,91,93].

Evidence that salt gland ion excretion in the Poaceae is an active, metabolically driven process is varied, including effects of temperature [94], light [95], oxygen pressure [85], and metabolic

TABLE 25.2
Leaf Salt-Gland Cl⁻ Excretion Rates of Chloridoid Turfgrasses

Turfgrass	Cl⁻ Excretion Rate
Sporobolus virginicus	2104
Distichlis spicata ssp. *stricta*	1268
Zoysia matrella	423
Cynodon spp.	394
Zoysia japonica	225
Bouteloua spp.	56
Buchloë dactyloides	42

Note: Excretion rates in micromole ion per gram leaf dry weight per week. Ion excretion measured in plants exposed to 200 mM NaCl.

Source: Marcum, K.B., *New Phytol.*, 120, 281, 1992; Marcum, K.B., *Crop Sci.*, 39, 1153, 1999; and Marcum, K.B., *Ann. Bot.*, 66, 1, 1990.

inhibitors [96] on excretion rate. In addition, ion excretion in turfgrasses is highly selective for Na⁺ and Cl⁻ over K⁺, Ca²⁺, and Mg²⁺ [16,58,78,97]. Comparison of salt gland excretion rates among studies is difficult, due to the varying influence of environmental factors such as light and temperature, cumulative days of exposure to salt stress, and plant factors such as leaf age [61]. Also, units of measurement differ, one fundamental difference being whether excretion rates are based on leaf area or leaf weight. Finally, the excretion rate is not static, but is influenced by saline ion concentrations in the growing media. Increasing media salinity generally stimulates excretion up to an optimal level, depending on species, above which the excretion rate may decline [85]. Maximum excretion rate was reported to occur at <200 mM (17 dS m⁻¹) root-media NaCl in moderately salt-tolerant *Cynodon* spp. [67,97]. However, excretion rate was maximal at above 200 mM NaCl in *Distichlis spicata* ssp. *stricta* [85], and at 300 mM NaCl (23 dS m⁻¹) in *Sporobolus virginicus* [8].

Salinity tolerance among Chloridoid turfgrasses has been correlated with salt gland saline ion excretion rates, on both an inter- and intraspecific level. Salinity tolerance was directly related to saline ion excretion rates among eight turfgrass species belonging to six genera [32]. Salt gland ion excretion rates for Chloridoid turfgrass spp. are shown in Table 25.2. Note that halophytic *Sporobolus virginicus* has salt gland Cl⁻ excretion rates five times higher than moderately salt-tolerant *Cynodon* spp., and 50 times higher than salt-sensitive *Buchloë dactyloides* (Table 25.2). Salt gland excretion rates were highly correlated with salinity tolerance among 57 *Zoysia* spp. [47], and again among 35 *Cynodon* spp. genotypes [19] (Figure 25.1). Salt gland excretion rates and salinity tolerance, were highly correlated to leaf salt gland density in *Zoysia* spp. [47,58]. Gland excretion rates ranged from 130 μmol Na⁺/g leaf dry weight per week in salt-sensitive *Zoysia japonica* to 730 μmol Na⁺/g leaf dry weight per week in salt-tolerant *Zoysia matrella*, with gland densities ranging from 28 mm⁻² leaf surface in salt-sensitive *Zoysia japonica* to 100 mm⁻² in salt-tolerant *Zoysia macrostachya* Franch. & Sav. [47].

25.7   SUMMARY

Turfgrasses show extreme range in salinity tolerance, from salt-sensitive to extremely salt-tolerant (halophytic). Salinity tolerance of turfgrasses, indicated by salinity causing 50% shoot growth reduction, ranges from approximately 4 dS m⁻¹ (e.g., *Poa annua* and *Eremochloa ophiuroides*) to 40+ dS m⁻¹, essentially seawater (e.g., *Distichlis spicata* ssp. *stricta* and *Sporobolus virginicus*).

Though shoot growth decline with increasing salinity is typical, shoot growth may be stimulated by moderate salinity in highly salt tolerant, or halophytic grasses. However, root growth stimulation under moderate salinity is much more common in salt-tolerant turfgrasses, resulting in increased root/shoot ratios, and therefore increased water absorption/transpiration area, a potential adaptive mechanism to saline osmotic stress.

It has long been accepted that major causes of plant growth inhibition under salinity stress are osmotic stress (osmotic inhibition of plant water absorption), and specific ion effects, including toxicities and imbalances. Salinity tolerance in turfgrasses has been strongly correlated to shoot saline ion exclusion in a number of studies. However, studies have also shown that complete osmotic adjustment does occur under salt stress, even in salt-sensitive turfgrasses, and that salinity tolerance is *negatively* correlated to shoot osmolarity levels. Since the predominant osmotica utilized are typically saline ions, ion regulation, rather than ion exclusion, may be a more apt description of the mechanism of salt tolerance occurring in turfgrasses.

Saline ion regulation and selectivity for K^+ over Na^+ may occur via root or shoot mechanisms in turfgrasses. Ion regulation occurs at the apoplastic barrier of the root endodermis, by vacuolar ion compartmentation within root or shoot cells, or by saline ion excretion via leaf salt glands, which are present in Chloridoid turfgrasses. Excretion rates may be substantial, and are highly selective for Na^+ and Cl^-. Salinity tolerance of Chloridoid turfgrasses has been correlated to salt gland excretion rate and leaf salt gland density.

Enzymes of higher plants, salt-sensitive and tolerant alike, are inhibited by saline ion concentrations above 100–200 mM. Under salt stress, turfgrasses typically accumulate saline ions to well above these levels for shoot osmotic adjustment, necessitating Na^+ and Cl^- compartmentation in vacuoles, which comprise approximately 90% of mature cell volume. Remaining cytoplasmic osmotic adjustment is achieved by certain organic osmotica compatible with cell enzymes, termed "compatible solutes." Of potential compatible solutes, only glycinebetaine accumulates to levels sufficient for cytoplasmic osmotic adjustment in salt tolerant turfgrasses.

REFERENCES

1. FW Gould, RB Shaw. *Grass Systematics*, 2nd ed. College Station, TX: Texas A&M University Press, 1983, pp. 1–15.
2. AS Hitchcock. *Manual of the Grasses of the United States*, 2nd ed. New York: Dover Pub., Inc., 1971, pp. 1–14.
3. EV Maas. Salt tolerance of plants. *Appl Agr Res* 1(1):12–26, 1986.
4. KB Marcum. Salinity tolerance in turfgrasses. In: M Pessarakli, Ed. *Handbook of Plant and Crop Stress*. 2nd ed. New York: Marcel Dekker, 1999, pp. 891–906.
5. JD Butler, JL Fults, GD Sanks. Review of grasses for saline and alkali areas. *Int Turfgrass Res J* 2:551–556, 1974.
6. MA Harivandi, JD Butler, L Wu. Salinity and turfgrass culture. In: DV Waddington, RN Carrow, RC Shearman Eds. *Turfgrass*. Madison, WI: ASA, CSSA, and SSSA, 1992, pp. 207–229.
7. JA Aronson. Haloph: *A Data Base of Salt Tolerant Plants of the World*. Tucson, AZ: Office of Arid Land Studies, University of Arizona, 1989, pp. 68–70.
8. KC Blits, JL Gallagher. Morphological and physiological responses to increased salinity in marsh and dune ecotypes of *Sporoblus virginicus* (L.) Kunth. *Oecologia* 87:330–335.
9. Y Waisel. *Biology of Halophytes*. New York: Academic Press, 1972, pp. 141–165.
10. ZF Zhao, JW Hayser, HJ Bohnert. Gene expression in suspension culture cells of the halophyte *Distichlis spicata* during adaptation to high salt. *Plant Cell Physiol* 30:861–867, 1989.
11. EV Maas, GJ Hoffman. Crop salt tolerance–current assessment. *J Irrig Drainage Div* ASCE 103:115–132, 1977.
12. RN Carrow, RR Duncan. *Salt-Affected Turfgrass Sites—Assessment and Management*. Chelsea, Michigan: Ann Arbor Press, 1998, pp. 83–99.
13. TJ Flowers, PF Troke, AR Yeo. The mechanisms of salt tolerance in halophytes. *Ann Rev Plant Physiol* 28:89–121, 1977.

14. MR Suplick–Ploense, YL Qian, JC Read. Relative NaCl tolerance of Kentucky bluegrass, Texas bluegrass, and their hybrids. *Crop Sci* 42:2025–2030, 2002.
15. AE Dudeck, CH Peacock. Salinity effects on growth and nutrient uptake of selected warm-season turf. *Int Turfgrass Soc Res J* 7:680–686, 1993.
16. KB Marcum, CL Murdoch. Salinity tolerance mechanisms of six C_4 turfgrasses. *J Am Soc Hort Sci* 119:779–784, 1994.
17. KB Marcum. Salinity tolerance of 35 bentgrass cultivars. *HortScience* 36(2):374–376, 2001.
18. KB Marcum, CL Murdoch. Growth responses, ion relations, and osmotic adaptations of eleven C_4 turfgrasses to salinity. *Agron J* 82:892–896, 1990.
19. KB Marcum, M Pessarakli. Salinity tolerance and salt gland excretion activity of bermudagrass turf cultivars. *Crop Sci* 46(6):2571–2574, 2006.
20. AE Dudeck, CH Peacock, JC Wildmon. Physiological and growth responses of St. Augustinegrass cultivars to salinity. *HortScience* 28(1):46–48, 1993.
21. G Lee, RR Duncan, RN Carrow. Salinity tolerance of seashore paspalum ecotypes: shoot growth responses and criteria. *HortScience* 39(5):1138–1142, 2004.
22. M Pessarakli, DM Kopec, AJ Koski. Comparative responses of bermudagrass, seashore paspalum, and saltgrass to salinity stress. *Agron Abstr* 2004.
23. KB Marcum, CL Murdoch. Salt tolerance of the coastal salt marsh grass, *Sporobolus virginicus* (L.) Kunth. *New Phytol* 120:281–288, 1992.
24. HL Bell, JW O'Leary. Effects of salinity on growth and cation accumulation of Sporobolus virginicus (Poaceae). *Am J Bot* 90(10):1416–1424, 2003.
25. L Bernstein, HE Hayward. Physiology of salt tolerance. *Ann Rev Plant Physiol* 9:25–46, 1958.
26. J Gorham, RG Wyn Jones, E McDonnell. Some mechanisms of salt tolerance in crop plants. *Plant Soil* 89:15–40, 1985.
27. AE Dudeck, S Singh, CE Giordano, TA Nell, DB McConnell. Effects of sodium chloride on *Cynodon* turfgrasses. *Agron J* 75:927–930, 1983.
28. LA Donovan, JL Gallagher. Morphological responses of a marsh grass, *Sporobolus virginicus* (L.) Kunth., to saline and anaerobic stresses. *Wetlands* 5:1–13, 1985.
29. RC Ackerson, VB Youngner. Responses of bermudagrass to salinity. *Agron J* 67:678–681, 1975.
30. AE Dudeck, CH Peacock. Effects of salinity on seashore paspalum turfgrasses. *Agron J* 77:47–50, 1985.
31. SF Alshammary, YL Qian, SJ Wallner. Growth response of four turfgrass species to salinity. *Agr Water Manage* 66:97–111, 2004.
32. KB Marcum. Salinity tolerance mechanisms of grasses in the subfamily Chloridoideae. *Crop Sci* 39:1153–1160, 1999.
33. L Wu, H Lin. Salt concentration effects on buffalograss germplasm seed germination and seedling establishment. *Int Turfgrass Res J* 7:883–828, 1993.
34. AH Khan, C Marshall. Salt tolerance within populations of chewing fescue (*Festuca rubra* L.). *Commun Soil Sci Plant Anal* 12(12):1271–1281, 1981.
35. WA Torello, AG Symington. Screening of turfgrass species and cultivars for NaCl tolerance. *Plant Soil* 82(2):155–161, 1984.
36. KB Marcum, NP Yensen. Salinity tolerance of desert saltgrass (*Distichlis spicata* var. *stricta*) under turf–type management. *International Salinity Forum*, Riverside, California, April, 2005:97–100.
37. M Pessarakli, KB Marcum, DM Kopec. Growth responses and nitrogen–15 absorption of desert saltgrass under salt stress. *J Plant Nutr* 28(8):1441–1452, 2005.
38. H Greenway, A Gunn, DA Thomas. Plant response to saline substrates. VIII. Regulation of ion concentration in salt sensitive and halophytic species. *Aust J Biol Sci* 19:741–756, 1966.
39. JW O'Leary. Physiological basis for plant growth inhibition due to salinity. In: WG McGinnies, BJ Goldman, P Paylore, Eds. Food, *Fiber and the Arid Lands*. Tucson, AZ: University of Arizona Press, 1971, pp. 331–336.
40. R Albert, M Popp. Chemical composition of halopytes from the Neusiedler Lake region in Austria. *Oecologia* 27:157–170, 1977.
41. J Gorham, PJ Randall, E Delhaize, RA Richards, R Munns. Genetics and physiology of enhanced K/Na discrimination. *Genetic Aspects of Plant Mineral Nutrition: Developments in Plant and Soil Sciences* 50:151–158, 1993.
42. WA Torello, LA Rice. Effects of NaCl stress on proline and cation accumulation in salt sensitive and tolerant turfgrasses. *Plant Soil* 93:241–247, 1986.

43. YL Qian, SJ Wilhelm, KB Marcum. Comparative responses of two Kentucky bluegrass cultivars to salinity stress. *Crop Sci* 41:1895–1900, 2001.

44. NJ Hannon, HN Barber. The mechanism of salt tolerance in naturally selected populations of grasses. *Search* 3:259–260, 1972.

45. PS Ramakrishnan, R. Nagpal. Adaptation to excess salts in an alkaline soil population of *Cynodon dactylon* (L.) Pers. *J Ecol* 61:369–381, 1973.

46. L Wu. The potential for evolution of salinity tolerance in *Agrostis stolonifera* L. and *Agrostis tenuis* Sibth. *New Phytol* 89:471–486, 1981.

47. KB Marcum, SJ Anderson, MC Engelke. Salt gland ion secretion: A salinity tolerance mechanism among five zoysiagrass species. *Crop Sci* 38:806–810, 1998.

48. YL Qian, MC Engelke, MJV Foster. Salinity effects on zoysiagrass cultivars and experimental lines. *Crop Sci* 40:488–492, 2000.

49. J Levitt. *Responses of Plants to Environmental Stresses*, Vol. II. New York: Academic Press, 1980, pp. 35–50.

50. JA Hellebust. Osmoregulation. *Ann Rev Plant Physiol* 27:485–505, 1976.

51. RG Wyn Jones, J Gorham. Use of physiological trains in breeding for salinity tolerance. In: FWG Barer, Ed. *Drought Resistance in Cereals*. Wallingford, UK: CAB International, 1989, pp. 95–106.

52. J Gorham. Mechanisms of salt tolerance of halophytes. In: R Choukr-Allah, CV Malcolm, A Hamdy, Eds. *Halophytes and Biosaline Agriculture*. New York: Marcel Dekker, 1996, pp. 31–53.

53. Y Hu, H Steppuhn, KM Volkmar. 1998. Physiological responses of plants to salinity: A review. *Can J Plant Sci* 78:19–27, 1998.

54. DB Headley, N Bassuk, RG Mower. Sodium chloride resistance in selected cultivars of *Hedera helix*. *HortScience* 27:249–252, 1992.

55. MA Khan, IA Ungar, AM Showalter. Effects of salinity on growth, ion content, and osmotic relations in *Halopyrum mucronatum* (L.) Stapf. *J Plant Nutr* 22:191–204, 1999.

56. AR Yeo. Salinity resistance: Physiologies and prices. *Physiol Plant* 58:214–222, 1983.

57. CH Peacock, AE Dudeck. A comparative study of turfgrass physiological responses to salinity. *Int Turfgrass Res J* 5:821–829, 1985.

58. KB Marcum, CL Murdoch. Salt glands in the Zoysieae. *Ann Bot* 66:1–7, 1990.

59. RJ Daines, AR Gould. The cellular basis of salt tolerance studied with tissue cultures of the halophytic grass *Distichlis spicata*. *J Plant Physiol* 119:269–280, 1985.

60. RG Wyn Jones, CJ Brady, J Speirs. Ionic and osmotic relations in plant cells. In: DL Laidman, RG Wyn Jones, Eds. *Recent Advances in the Biochemistry of Cereals*. New York: Academic Press, 1979, pp. 63–103.

61. WD Jeschke, S Klagges, A Hilpert, AS Bhatti, G Sarwar. Partitioning and flows of ions and nutrients in salt–treated plants of *Leptochloa fusca* L. Kunth. I. Cations and chloride. *New Phytol* 130:23–35, 1995.

62. J Garbarino, FM Dupont. NaCl induces a Salinity induces K^+/Na^+ antiport in tonoplast vesicles from barley roots. *Plant Physiol* 86:231–236, 1988.

63. MK Leonard. Salinity tolerance of the turfgrass *Paspalum vaginatum* Swartz 'Adalayd': Mechanisms and growth responses. PhD dissertation, University of California, Riverside (Dissertation Abstract 84–05537), 1983.

64. YH Peng, YF Zhu, YQ Mao, SM Wang, WA Su, ZC Tang. Alkali grass resists salt stress through high $[K^+]$ and an endodermis barrier to Na^+. *J Exp Bot* 55(398):939–949, 2004.

65. D Kramer. Cytological aspects of salt tolerance in higher plants. In: RC Staples, GH Toenniessen, Eds. *Salinity Tolerance in Plants*. New York: Wiley, 1984, pp. 3–15.

66. WD Jeschke. $K^+–Na^+$ exchange at cellular membranes, intracellular compartmentation of cations, and salt tolerance. In: RC Staples, GH Toenniessen, Eds. *Salinity Tolerance in Plants*. New York: Wiley, 1984, pp. 37–66.

67. N Liphshchitz, Y. Waisel. Existence of salt glands in various genera of the Gramineae. *New Phytol* 73:507–513, 1974.

68. V Amarasinghe, L Watson. Variation in salt secretory activity of microhairs in grasses. *Aust J Plant Physiol* 16:219–229, 1989.

69. H Greenway, R Munns. Mechanisms of salt tolerance in nonhalophytes. *Ann Rev Plant Physiol* 31:149–190, 1980.

70. TJ Flowers. Physiology of halophytes. *Plant Soil* 89:41–56, 1985.

71. RG Wyn Jones. Phytochemical aspects of osmotic adaptation. In: BN Timmerman, Ed. *Recent Advances in Phytochemistry*, Vol. 3, Phytochemical Adaptations to Stress. New York: Plenum Press, 1984, pp. 55–78.

72. RG Wyn Jones, J Gorham. Osmoregulation. In: OL Lange, PS Nobel, CB Osmond, H Ziegler, Eds. Physiological Plant Ecology III. *Responses to the Chemical and Biological Environment.* Berlin: Springer-Verlag, 1983, pp. 35–58.

73. RA Leigh, N Ahmad, RG Wyn Jones. Assessment of glycinebetaine and proline compartmentation by analysis of isolated beet vacuoles. *Planta* 153:34–41, 1981.

74. D Aspinall, LG Paleg. Proline accumulation: Physiological aspects. In: LG Paleg, D. Aspinall, Eds., *Physiology and Biochemistry of Drought Resistance in Plants.* Sydney: Academic Press, 1981, pp. 205–241.

75. D Rhodes, AD Hanson. Quaternary ammonium and tertiary sulfonium compounds in higher plants. *Ann Rev Plant Physiol Plant Mol Biol* 44:357–384, 1993.

76. J Gorham, LL Hughes, RG Wyn Jones. Chemical composition of salt-marsh plants from Ynys-Mon (Anglesey): the concept of physiotypes. *Plant Cell Environ* 3:309–318, 1980.

77. CH Liu, JK Su, WH Huang. Studies of the physiological indicators of salt tolerance in grasses. *Acta Prataculturae Sinica* 2(1):45–54, 1993.

78. G Naidoo, Y Naidoo. Salt tolerance in *Sporobolus virginicus*: the importance of ion relations and salt secretion. *Flora-Jena* 193(4):337–344, 1998.

79. PH Yancy. Compatible and counteracting solutes. In: K Strange, Ed. *Cellular and Molecular Physiology of Cell Volume Regulation.* Boca Raton, FL: CRC Press, 1994.

80. HC Saneoka, C Nagasaka, DT Hahn, WJ Yang, GS Premachandra, RJ Joly, D Rhodes. Salt tolerance of glycinebetaine-deficient and -containing maize lines. *Plant Physiol* 107:631–638, 1995.

81. P Harinasut, K Tsutsui, T Takabe, M Nomura, S Kishitani, T Takabe. Glycinebetaine enhances rice salt tolerance. In: P Mathis, Ed. *Photosynthesis: From Light to Biosphere*, Vol IV. Dordrecht, The Netherlands: Kluwer Academic Press, 1995, pp. 733–736.

82. W Meyer, GY Zhang, S Lu, SY Chen, TA Chen, R Funk. Transformation of Kentucky bluegrass (*Poa pratensis* L.) with betaine aldehyde dehydrogenase gene for salt and drought tolerance. *Agron Abstr* p. 167, 2000.

83. TD Colmer, E Epstein, J Dvorak. Differential solute regulation in leaf blades of various ages in salt-sensitive wheat and a salt-tolerant wheat X *Lophopyrum elongatum* (Host) A. Löve amphiploid. *Plant Physiol* 108:1715–1724, 1995.

84. S Mumtaz, SSM Maqvi, A Shereen, MA Khan. Proline accumulation in wheat seedlings subjected to various stresses. *Acta Physiol Plant* 17:17–20, 1995.

85. N. Liphschitz, Y. Waisel. Adaptation of plants to saline environments: salt excretion and glandular structure. In: DN Sen, KS Rajpurohit, Eds. *Tasks for Vegetation Science*, Vol. 2: Contributions to the Ecology of Halophytes. The Hague: W. Junk Publishers, 1982, pp. 197–214.

86. A Fahn. Secretory tissues in vascular plants. *New Phytol* 108:229–257, 1988.

87. EL Taleisnik, AM Anton. Salt glands in *Pappophorum* (Poaceae). *Ann Bot* 62:383–388, 1988.

88. WD Clayton, SA Renvoize. *Genera Graminum, Grasses of the World.* London: HMSO Books, 1986, 389 pp.

89. V Amarasinghe, L Watson. Comparative ultrastructure of microhairs in grasses. *Bot J Linnean Soc* 98:303–319, 1988.

90. JW Oross, WW Thomson. The ultrastructure of *Cynodon* salt glands: the apoplast. *Eur J Cell Biol* 28:257–263, 1982.

91. JW Oross, WW Thomson. The ultrastructure of the salt glands of *Cynodon* and *Distichlis* (Poaceae). *Am J Bot* 69(6):939–949, 1982.

92. Y Naidoo, G Naidoo. Cytochemical localisation of adenosine triphosphatase activity in salt glands of *Sporobolus virginicus* (L.) Kunth. *S Afr J Bot* 65:370–373, 1999.

93. CA Levering, WW Thomson. The ultrastructure of the salt gland of *Spartina foliosa*. *Planta* 97:183–196, 1971.

94. G Pollak, Y Waisel. Ecophysiological aspects of salt excretion in *Aeluropus litoralis*. *Physiol Plant* 47:177–184, 1979.

95. G Pollak, Y Waisel. Salt secretion in *Aeluropus litoralis* (Willd.) Parl. *Ann Bot* 34:879–888, 1970.

96. J Wieneke, G Sarwar, M Roeb. Existence of salt glands on leaves of Kallar grass (*Leptochloa fusca* L. Kunth.). *J Plant Nutr* 10(7):805–820, 1987.

97. W Worku, GP Chapman. The salt secretion physiology of a Chloridoid grass, *Cynodon dactylon* (L.) Pers., and its implications. *Sinet* 21:1–16, 1998.

26 Salinity Issues Associated with Recycled Wastewater Irrigation of Turfgrass Landscapes

Yaling Qian and Ali Harivandi

CONTENTS

26.1 WATER REUSE IN URBAN LANDSCAPES

Rapid population growth in arid and semiarid regions of the United States continues to place increasing demand on finite and limited water supplies. Many cities and districts are struggling to balance water use among municipal, industrial, agricultural, and recreational users. Along with an increase in freshwater demand comes an increase in the volume of wastewater generated. Treated wastewater is the only water resource increasing as other sources dwindle. Reuse of treated wastewater for turf and landscape irrigation is often viewed as one way to maximize the existing urban water resources (U.S. EPA, 2004). In addition to the growing concerns of the future water supply, the more stringent wastewater discharge standards make the use of recycled wastewater increasingly attractive.

To date, the contribution of water reuse to water conservation varies by location. Water reuse satisfied 25% of the water demand in Israel, for example, where 66% of total treated sewage is reused (Avnimelech, 1993). Water reuse is expected to reach 10–13% of water demand in the next few years in Australia and California (Lazarova and Asano, 2005). Throughout the United States, large volumes of municipal recycled water is used to irrigate golf courses, community parks, cemeteries, athletic fields, schoolyards, roadsides, street medians, industrial and residential landscapes, and other urban landscape sites. For example, in the greater Denver area, approximately 30–40 million gallons of recycled wastewater is used for landscape irrigation every day during the growing season. Golf courses are by far the leading urban landscape users of recycled wastewater because intensively managed turf can use nutrients in the wastewater efficiently, golf courses require a high volume of irrigation water, and it is easier to implement recycled wastewater irrigation systems on golf courses than other systems (i.e., parks, school playground, athletic field, etc.). A 1978 survey reported that 26 golf courses across the country were using recycled wastewater. In contrast, recently the National Golf Foundation (NGF) reported that approximately 13% of golf courses (approximately 2000 golf courses) nationwide now use recycled water for irrigation, with 34% of golf courses in the southwest doing so (GCSAA, 2003).

26.2 RECYCLED WATER AND GRAY WATER

Most reuse water in landscape irrigation is treated municipal sewage water, also known as recycled water, effluent water, and reclaimed water. To a lesser extent, untreated household gray water is also used. It is anticipated that the use of these waters for irrigation would accelerate in the future.

26.2.1 RECYCLED WASTEWATER

"Recycled water" refers to any water that after residential and sometimes industrial use undergoes significant treatment at a sewage treatment plant to meet standards set by federal or state water laws and regulations. This water is usually suitable for various reuse purposes including irrigation. The most common treatment process includes primary treatment (such as settling and screening), secondary treatment (such as oxidation, activated sludge, filtration, and UV or chlorine disinfection), and tertiary treatment (such as clarification, coagulation/flocculation, sedimentation, filtration, adsorption of compounds by a bed of activated charcoal, UV or chlorine disinfection etc). UV disinfection is becoming one of the most popular and cost-effective disinfection alternatives. During treatments, suspended solids (SS) are removed, pathogens disinfected, and partial to substantial reduction in nutrients occurs, depending on treatment stage (Harivandi, 1994; Pettygrove and Asano, 1985). However, recycled wastewater may still contain different levels of dissolved solids, ions, nutrients (NO_3 and P_2O_4), and other elements. Currently, recycled water used for turf and landscape irrigation must be at least secondary effluent water.

To ensure human health and to protect the environment, comprehensive regulations for water reclamation and reuse have been established in many states with water shortages (such as Arizona, California, Florida, Colorado, and Texas) (State of Arizona, 1987; State of California, 1978; State of Florida, 1989; State of Colorado, 2000, and State of Texas, 1990). These state regulations typically use human-health-related parameters such as *Escherichia coli* count, turbidity, total suspended solids, and nitrogen and phosphorous content to evaluate water quality.

26.2.2 GRAY WATER

"Gray water" refers to water that has gone through one cycle of use in laundry, shower, or bath; it does not include water from toilets and dishwashers. Gray water used in landscape irrigation is typically without significant treatment. While recycled water is mostly used in large landscape facilities, such as golf courses, parks, etc., gray water irrigation offers a way in which people can reuse wastewater generated in their own homes. Several drought-stricken states, including California, Arizona, New Mexico, and Texas, have legalized the practice. However, the use of gray water is largely prohibited in other states due to the concerns over the potential existence of human disease-causing organisms.

Due to the limitations of gray water reuse in turfgrass systems, we will focus our discussion on recycled wastewater irrigation in the rest of this chapter.

TABLE 26.1
Benefits and Disadvantages of Recycled Wastewater Irrigation

Benefits	Disadvantages
Conserving water and increasing regional water-use efficiency	Negative salinity/sodicity effect on plants and soils associated with recycled water quality
Avoiding cost of new freshwater resource development and reducing the cost for wastewater disposal	Cost associated with the installation of recycled water infrastructure (pipes, cross-connection control)
Reducing fertilizer use	Leaching and runoff concerns of contaminants in recycled wastewater

26.3 ADVANTAGES AND CONCERNS

Both opportunities and problems exist in using recycled wastewater for landscape irrigation (Table 26.1). Water reuse for irrigation in urban landscapes is a powerful means of water conservation, water reclamation, and nutrient recycling. Due to the dense plant canopy and active root systems, turfgrass landscapes are increasingly viewed as environmentally desirable disposal sites for wastewater (Pepper and Mancino, 1993; Anderson et al., 1981). In fact, dense, well-managed turfgrass areas are among the best biofiltration systems available for removal of excess nutrients and further reclamation of treated wastewater (Hayes et al., 1990; Pepper and Mancino, 1993).

While the environmental and conservation benefits of wastewater reuse in landscape and turfgrass irrigation are clear, concerns associated with wastewater reuse include (1) the costs of irrigation pipeline installation and irrigation equipment maintenance (e.g., prevention of nozzle plugging); (2) health risk due to the possible presence of pathogens; (3) salt damage to landscape plants and salt accumulation on the soil surface and in the soil profile; and (4) potential contamination of groundwater caused by leaching of excess nutrients.

From an agronomic standpoint, the greatest water quality concerns associated with recycled water irrigation are salinity/sodicity and excessive nutrients. The higher Na content relative to Ca and Mg in such water can cause long-term reductions in soil hydraulic conductivity and infiltration rate in soil with high clay content. In turn, salt leaching is less effective when soil hydraulic conductivity and infiltration rate are reduced. Concurrently, an increased nutrient content increases algae populations in irrigation ponds, causing secondary problems, such as decreased clarity and aesthetic appearance, change in water pH, reduction in water aeration, and increased maintenance needs (Devitt et al., 2005).

26.4 WATER QUALITY CONSIDERATIONS

Current state regulations typically use human-health-related parameters such as *E. coli* count, turbidity, total suspended solids, and nitrogen and phosphorous content to evaluate water quality (State of Arizona, 1987; State of California, 1978; State of Florida, 1989; State of Colorado, 2000, and State of Texas, 1990). These parameters do not address the water chemistry that affects the suitability of treated wastewater for landscape irrigation. Landscape managers are often concerned about salinity, sodicity, and other chemical constituents. Generally, in the context of landscape irrigation, five aspects of water govern its quality: salinity, sodium absorption ratio (SAR), pH, nutrient levels, and specific ions.

26.4.1 SALINITY

During the municipal water use process, water salinity increases. Residential use of water typically adds 200–400 mg L^{-1} of dissolved salts from sources such as food processing, water softening, the use of soap and detergent, etc. These salts remain after wastewater treatment. Conventional sewage treatment is aimed at removing solids, decreasing organic matter, disinfecting pathogens, and reducing nutrient levels, not at reducing or removing inorganic salts.

Water salinity is most commonly measured by electrical conductivity (EC_{iw}), and is reported in units of decisiemens per meter (dS m^{-1}) or millimhos per centimeter (mmhos cm^{-1}). These units' values are equal; only their names differ. Water salinity can be reported as total dissolved solids (TDS) where EC_{iw} data are transformed to TDS in parts per million (ppm) or milligrams per liter (mg L^{-1}) by

$$\text{TDS (in ppm or mg L}^{-1}) = 640 \times EC_{iw} \text{ (in dS m}^{-1} \text{ or mmhos cm}^{-1})$$

Generally speaking, water with an EC_{iw} value above 0.75 dS m^{-1} (or TDS above 500 mg L^{-1}) increases salinity problems. Most waters of acceptable quality for turfgrass irrigation contain from 200 to 800 mg L^{-1} soluble salts, i.e., water with an EC_{iw} value from 0.3 to 1.2 dS m^{-1}. Water with an EC_{iw} above 3 dS m^{-1} should be avoided or diluted with less saline water before use.

However, the accumulation of excessive salts in the soil profile is not a function of water salinity alone. The rate at which salts accumulate to potentially toxic levels in a soil also depends on the amount of water applied annually, annual precipitation (rain plus snow), and a particular soil's physical/chemical characteristics. Good permeability and drainage allow a turfgrass manager to leach excessive salt from the root zone by periodic heavy irrigations. For example, water with an EC_{iw} of 1.5 dS m^{-1} may be successfully used on grass grown on sandy soil with good drainage, but prove injurious within a very short period of time if used to irrigate the same grass grown on a clay soil or soil with limited drainage due to salt buildup in the root zone. Sand-based golf greens amplify the soil structure that allows such relatively straightforward salinity management. Only careful management can prevent deleterious salt accumulation in a soil irrigated with high EC_{iw} water. Soil physical characteristics and drainage, both important factors in determining root zone salinity, must also be considered in determining the suitability of a given recycled irrigation water.

26.4.2 Sodium Adsorption Ratio

Although sodium can be directly toxic to plants, its most frequent damage is due to its effect on soil structure. The high sodium content common to recycled water can cause deflocculation or breakdown of soil clay particles, making such a soil less permeable to both water and air. The likely effect of a particular irrigation water on soil permeability is best gauged by the water's sodium adsorption ratio (SAR), a measure of the concentration of sodium in water relative to that of calcium and magnesium.

Complicating the SAR calculation for recycled water is the often elevated bicarbonate level in such water. In addition to increasing soil pH, bicarbonate combines with calcium and magnesium and precipitates, thereby lowering calcium and magnesium concentrations in the soil solution. For recycled wastewater high in bicarbonate, we therefore calculate an "adjusted SAR." Adjusted SAR is calculated by considering bicarbonate, carbonate, and the water's total salinity, in addition to the water's calcium, magnesium, and sodium content (Westcot and Ayers, 1985). Adjusted SAR decreases with increasing calcium, magnesium, and the water's total salinity and rises as sodium, bicarbonate, and carbonate concentrations increase.

We can use adjusted SAR to assess sodium permeability hazard. An adjusted SAR >6–9 gradually leads to soil structure deterioration, including soil sealing, crusting, and reduced water penetration after long-term use for irrigation. Such effects are more severe in fine-textured soil and heavy traffic areas than they are on sandy soil.

26.4.3 HYDROGEN ION ACTIVITY (pH)

pH is easily determined and provides useful information about water's chemical properties. Although seldom a problem in itself, a very high or low pH warns users that the water needs evaluation for other constituents. For example, increased pH may signal a high bicarbonate concentration in the water. The pH of most irrigation water ranges from 6.5 to 8.4. Waters with a pH outside the desirable range must be carefully evaluated for other chemical constituents. Consistently high pH can cause iron and manganese deficiencies in landscape plants, resulting in yellowing of leaves (chlorosis).

26.4.4 SPECIFIC IONS

In addition to contributing to the total soluble salt concentration of irrigation water, particular ions (sodium, chloride, and boron) may be directly toxic to plants, especially some trees and shrubs. Chloride (Cl^-), sodium (Na^+), and boron (B^+) can be absorbed by plant roots and translocated to leaves, where they accumulate in sensitive plants. This accumulation leads to leaf margin scorch in minor cases and in severe situations, total leaf kill and abscission. Irrigation waters with Na and Cl content above 100 and 355 mg L^{-1}, respectively, are considered toxic to sensitive plants when absorbed by roots.

With sprinkler irrigation, Na and Cl frequently accumulate by direct absorption through the leaves that are moistened by irrigation water. Trees and shrubs are more sensitive to sodium and chloride sprayed directly onto plant leaves. Sodium and Cl toxicity could occur on sensitive plants when Na and Cl concentrations exceed 70 and 100 mg L^{-1}, respectively, when water is sprayed directly on the foliage.

Boron is mainly absorbed from the root systems. Boron concentrations as low as 1–2 mg L^{-1} in irrigation water can be toxic to many sensitive ornamental plants.

26.4.5 NUTRIENTS

Recycled waters contain a range of microelements at levels sufficient to satisfy the need of most turfgrasses for these substances. They may also contain enough macronutrients, nitrogen (N), phosphorus (P), and potassium (K) to figure significantly in a fertilization program. The economic value of these nutrients can be substantial. When recycled wastewater is used for irrigation, regular testing will allow adjustment in N and P applications. For example, an N content in recycled wastewater of 10 mg L^{-1} adds 27 lb nitrogen per acre with each acre-foot of irrigation water. This amount of N should be deducted from the fertilization program.

26.4.6 VARIATIONS IN RECYCLED WATER CHEMICAL COMPOSITION

The chemical composition of recycled water varies significantly, even within the same geological area. The suitability of a given recycled water for irrigation must, therefore, be evaluated separately for each site. To demonstrate such variations, the result of one study from Colorado is described below.

This study was done to assess chemical properties of recycled wastewater used for urban landscape irrigation (Qian, 2004). Testing results of 37 recycled wastewater samples collected from six landscape sites indicated the average EC_{iw} of 0.84 dS m^{-1} and a range of 0.47 to 1.32 dS m^{-1} (Table 26.2). Adjusted SAR of the recycled wastewater ranged from 1.6 to 8.3. On the basis of the interactive effect of salinity and sodicity (Ayers and Westcot, 1985; Oster and Schroer, 1979), most of the water samples collected would likely have a slight to moderate effect on soil infiltration and permeability (Figure 26.1).

TABLE 26.2

Average Water-Quality Values of Recycled Wastewater from Advanced Wastewater Treatment Plants in Colorado

Parameter	Average	Minimum	Maximum
Total dissolved salts (mg L^{-1})	614	300	847
Conductivity (dS m^{-1})	0.84	0.47	1.32
SAR	3.1	1.1	5.8
Adjusted SAR	5.0	1.6	8.3
Sodium (mg L^{-1})	99	30	170
Chloride (mg L^{-1})	95	53	222
Bicarbonate (mg L^{-1})	112	14	269
Calcium (mg L^{-1})	61	38	101
Magnesium (mg L^{-1})	15	6.9	32
Sulfate (mg L^{-1})	160	71	280
Boron (mg L^{-1})	0.23	0.02	0.41
Iron (mg L^{-1})	0.35	0.04	1.52
Potassium (mg L^{-1})	12.7	3.5	93
pH	8.1	6.3	10.0
Total suspended solid (mg L^{-1})	11.7	1.0	50.0
NH$_4$-N (mg L^{-1})	0.76	0.23	2.05
NO$_3$-N (mg L^{-1})	3.62	0.2	9.5
Total P (mg L^{-1})	0.47	0.01	1.8

Note: Water samples were taken from irrigation ponds and irrigation sprinkler outlets on six reuse sites.

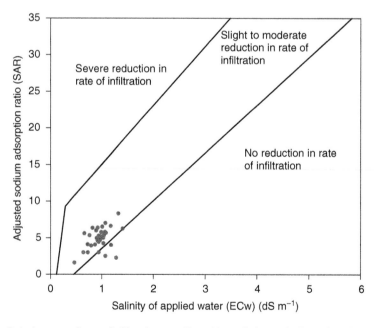

FIGURE 26.1 Relative rate of water infiltration as affected by salinity and adjusted sodium adsorption ratio of irrigation water. The dots are the data points of water samples collected from Colorado water reuse sites. (Adapted from Ayers, R.S. and D.W. Westcot, *Water Quality for Agriculture*, FAO Irrigation and Drainage Paper No. 29, 1985; Oster, J.D. and F.W. Schroer, *Soil Sci. Soc. Am. J.,* 43, 444, 1979.)

The average Na, Cl, B, and HCO_3 concentrations of the 37 recycled wasterwater samples were 99, 95, 0.23 , and 112 mg L^{-1}, respectively, with Na, Cl, B, and bicarbonate ranging from 30 to 170, 53 to 222, 0.02 to 0.41, and 14 to 269 mg L^{-1}, respectively.

26.5 IMPACT OF RECYCLED WASTEWATER IRRIGATION ON SOIL

26.5.1 Soil Infiltration Rate and Hydraulic Conductivity

Recycled wastewater irrigation has profound impacts on soil infiltration rate and hydraulic conductivity. A study conducted by Cook et al. (1994) investigated the effect of wastewater land application on infiltration rate of a highly permeable volcanic-ash soil. The study site was a 75-year-old stand of pine trees with dense understory vegetation. The authors found a 50% decrease in infiltration rate after nearly 3 years of wastewater irrigation.

Several mechanisms may contribute to reductions of soil infiltration rate and hydraulic conductivity. Interaction between dissolved salts and sodium in the water and soil may cause collapse of soil structure, decreased pore diameter, and consequently lower permeability. Growth of microorganisms in soil pores can reduce pore diameter, also resulting permeability reduction. This biological clogging is frequently associated with anaerobic conditions. Physical permeability reduction is often the result of suspended solid deposit narrowing of soil pores and by physical blockage of soil pores.

Changes in pore geometry of topsoil layers have a greater effect on soil infiltration rate than on hydraulic conductivity in subsoil. Rice (1974) studied the effect of secondary wastewater application on infiltration rate and hydraulic conductivity of soil columns. He found that the reduction in soil infiltration rate is a function of the SS concentration of irrigation water. Little evidence of biological clogging occurred at the soil surface. Therefore, maintaining low SS in the treated wastewater is most important in optimizing infiltration. Rice (1974) also found that after long periods of intermittent inundation, clogging developed below the surface because entrapped gases blocked soil pores, which was likely caused by microbial activity. The authors suggested that physical clogging caused by the deposition of SS on the soil surface was the predominant cause of infiltration reduction. However, biological clogging seemed to be a greater factor at deeper depths.

Mancino and Pepper (1992) studied the effect of secondary effluent on bacterial populations in sandy loam soil covered by bermudagrass. After 3.2 years of wastewater irrigation, there was no change in the aerobic bacterial population, indicating that microbes were neither promoted nor inhibited by use of wastewater.

26.5.2 Soil Chemical Properties

Recycled wastewater may contain appreciable concentrations of sodium (Na), bicarbonates (HCO_3), carbonates (CO_3), boron (B), and chloride (Cl), as well as nitrogen (N), phosphorous (P), calcium (Ca), magnesium (Mg), and potassium (K), which can have significant impacts on soil chemical properties by altering soil pH, salinity, sodicity, and fertility (Harivandi, 1991). Qian and Mecham (2005) compared golf-course fairways that used recycled wastewater irrigation with those with nonsaline surface water irrigation. Most of the surface water was stream and ditch water and from melting snow of the Rocky Mountains and exhibited good quality. The study found that soil pH was 0.3 units higher in sites irrigated with recycled wastewater than in sites irrigated with surface water. The soil pH increase likely resulted from the higher bicarbonate concentration in reuse water than surface water. In New Zealand, Schipper et al. (1996) found an increase in soil pH of 0.8 after applying tertiary-treated domestic wastewater to a forest site for 3 years at 4.9 cm per week. The author suggested that the rise in soil pH was likely related to a high rate of denitrification that produced hydroxyl ions. Mancino and Pepper (1992) found that recycled wastewater irrigation increased soil pH by 0.1–0.2 units when compared with potable water irrigation.

The rate at which salts accumulate to undesirable levels in soil depends on their concentration in irrigation water, amount of water applied annually, annual precipitation, the soil's physical and chemical properties, and drainage effectiveness (Harivandi, 1999). Careful management is required to prevent the accumulation of salts in the soil if water with high EC is used for irrigation. Hayes et al. (1990) found an increase in soil EC of 0.5 dS m^{-1} after 16 months of irrigating turfgrass planted on sandy loam soil with recycled water for irrigation (EC_{iw} in the range 0.65–0.91 dS m^{-1}). Qian and Mecham (2005) found that 5–33 years irrigation of turfgrass on a fine-textured soil with recycled wastewater (EC_{iw} in the range 0.47–1.32 dS m^{-1}) resulted in an increase of soil salinity by 3.0 dS m^{-1}, that was 187% higher than sites irrigated with surface water (EC_{iw} in the range 0.15–0.30 dS m^{-1}). Miyamoto et al. (2005) studied salt accumulation patterns on highly managed sports fields using recycled wastewater for irrigation and found no significant salt accumulation on sites with deep and well-drained sandy or gravelly soils. However, high levels of salts were accumulated in soils with thick thatch layers, in compacted clayey sites and in areas where undisturbed hardened caliche layers restricted drainage (Miyamoto et al., 2005; Miyamoto and Chacon, 2006).

The soil exchangeable sodium percentage (ESP) and sodium absorption ratio (SAR) for recycled wastewater-irrigated sites are typically higher due to the higher sodium to calcium and magnesium ratio in recycled wastewater. Qian and Mecham (2005) reported that sites irrigated with recycled wastewater for more than 5 years had 230–481% higher ESP and SAR than the surface-water-irrigated soil, although the ESP and SAR values on fairways are not high enough to be classified as a sodic soil. Mancino and Pepper (1992) observed that after 3.2 years, recycled wastewater irrigation increased soil sodium levels from 25 to 250 mg kg^{-1}, while potable water irrigation increased the soil sodium level from 25 to 100 mg kg^{-1}.

26.6 IMPACT OF RECYCLED WASTEWATER IRRIGATION ON PLANTS' PERFORMANCE

In most areas of the semiarid and arid United States, turfgrass and many other landscape plants need routine irrigation to maintain desirable quality and growth. During periods of water shortage, when use of potable water for landscape irrigation may be restricted, recycled wastewater provides a reliable irrigation supply.

Relative to conventional water sources (domestic, lake, or river water), an important caveat to the long-term use of recycled wastewater for turfgrass and landscape irrigation is the potential for salinity problems. Many turfgrass and landscape plants are tolerant of recycled water salinity. Some species of trees and shrubs, however, are susceptible to certain attributes of recycled water, especially after several years of exposure. In a recent survey among golf-course superintendents in the southwestern United States, 73% of respondents who irrigated with recycled wastewater reported that recycled water irrigation had a negative impact on shrubs and trees growing on their golf course (Devitt et al., 2004).

Many plants can tolerate recycled water salinity if this water is applied directly to soil. However, the majority of recycled wastewater is applied via sprinkler irrigation in large landscapes. In these conditions, foliar absorption of salts, especially salts with high solubility such as Cl and Na, may occur. Of course, roots can absorb high levels of dissolved salts in soil solution. These ions are carried through the sap stream to leaves (such as leaf margins and shoot tips) where they may accumulate to toxic levels. Salts that accumulate to a high level can result in characteristics of marginal (or tip) scorch.

Proper species selection at the time of landscape planning and during the subsequent replacement of damaged plants is important to mitigate the potential negative effects of recycled wastewater irrigation. Selection of plant species must consider the levels of salt load in recycled wastewater, soil types, plant species, and climate conditions, which vary from region to region. Those factors play important roles in plants' responses to recycled wastewater irrigation. In California, Wu et al. (2001) reported that species showing high tolerance to recycled wastewater included Japanese boxwood, oleander, juniper, dwarf olive, Mexican piñon, and California fan palm. Abelia, butterfly bush, Chinese hackberry, trumpet vine, marguerite, ginkgo, and Chinese pistache exhibited low

TABLE 26.3
Relative Tolerances of Turfgrass Species to Soil Salinity

Sensitive (<3 dS m⁻¹)	Moderately Sensitive (3–6 dS m⁻¹)	Moderately Tolerant (6–10 dS m⁻¹)	Tolerant (>10 dS m⁻¹)
Annual bluegrass	Annual ryegrass	Perennial ryegrass	Alkaligrass
Bahiagrass	Buffalograss	Coarse-leaf (Japanica type) zoysiagrass	Bermudagrass
Carpetgrass	Creeping bentgrass	Tall fescue	Fine leaf (Matrella type) zoysiagrasses
Centipedegrass	Slendor creeping, red and chewings fescues		Seashore paspalum
Colonial bentgrass			St. Augustinegrass
Hard fescues			
Kentucky bluegrass			
Rough bluegrass			

Source: Harivandi, M.A., K.B. Marcum, and Y.L. Qian in *Water Quality and Quantity Issues in Turfgrass in Urban Landscapes*, 2007.

tolerance to irrigation with recycled water. In Nevada, Jordan et al. (2001) reported that only 6 out of 20 tree species spray-irrigated with recycled wastewater had an acceptable visual rating. Foliar damage was linked to species-dependent sodium response (Jordan et al., 2001; Qian et al., 2005).

Salt tolerance of turfgrasses is often expressed in terms of the acceptable salt content of the soil root zone, measured as the EC of saturated soil paste. Table 26.3 is a general guide to the salt tolerance of individual turfgrasses (Harivandi et al., 2007). As the table indicates, soils with an EC below 3 dS m⁻¹ are considered satisfactory for growing most turfgrasses. Soils with an EC between 3 and 10 dS m⁻¹ successfully support only a few salt-tolerant turfgrass species.

Salt tolerance of a given turfgrass is also influenced by management practices. In a study to examine the effect of mowing regime on salt tolerance, Fu et al. (2005) reported that the salinity damage of creeping bentgrass becomes more severe under close mowing conditions. The quality of creeping bentgrass mowed at 6.4, 12.7, and 25.4 mm fell to unacceptable levels when soil EC reached 4.1, 12.5, or 13.9 dS m⁻¹, respectively. The reduction of creeping bentgrass salt tolerance under low mowing height was associated with carbohydrate depletion that reduced the plant's genetic abilities to generate osmoprotectants (such as reducing sugar), reduce Na^+ accumulation in shoots, and selectively uptake and transport K^+ (Qian and Fu, 2005).

26.7 MANAGEMENT PRACTICES

Many years of practice and field observation confirm that in most cases, recycled wastewater can be used successfully to irrigate turfgrasses. Nonetheless, challenges and concerns exist. The challenge of water reuse is to maintain long-term sustainability. Agronomically, two main concerns for turfgrass managers are (1) potential problems caused by excessive sodium and salinity and (2) excessive nutrients or nutrient imbalance.

A systemic discussion related to management of turfgrasses under recycled wastewater irrigation is beyond the scope of this chapter. Below is a list of various management practices to consider when managing turfgrass sites irrigated with recycled water:

- Regular monitoring of water and soil quality using a soil and water testing program. Water quality must also be evaluated thoroughly before use to develop appropriate management strategies.

- Adequate leaching and sufficient drainage to remove excess Na and salts from the root zone.
- Adding chemical amendments (e.g., gypsum or other soluble calcium products) to soil or injecting into irrigation water to reduce sodicity problems.
- Careful irrigation based on evapotranspiration and leaching requirements.
- Use of better-quality water to blend with recycled wastewater or in rotation with the same to reduce salinity impact.
- Dual plumbing to irrigate high-value sites, such as golf-course greens, with nonsaline water in cases of excessively high SAR or salinity in recycled water.
- Intensive cultivation programs (aeration and water injection) to provide for adequate soil permeability and infiltration.
- Vigorous traffic-control programs.
- Reduced nitrogen and phosphorus fertilization, accounting for the fertilizer present in recycled wastewater.
- Fertilizing based on soil chemical tests to alleviate nutrient imbalance.
- Replacing susceptible plants with adapted, salt-tolerant species and cultivars.
- Modifying root zone for better drainage and salt leaching. Turfgrasses grown on a well-drained sand-based root zone are less susceptible to soil sodicity problems than those grown on clay soils.

REFERENCES

Anderson, E.L., I.L. Pepper, and W.R. Kneebone. 1981. Reclamation of wastewater with a soil turf filter: I. Removal of nitrogen. *J. Water Poll. Control Fed.* 53:1402–1407.

Avnimelech Y. 1993. Irrigation with sewage effluent: The Israeli experience. *Environ. Sci. Technol.* 27:1278–1281.

Ayers, R.S. and D.W. Westcot. 1985. *Water Quality for Agriculture.* FAO Irrigation and Drainage Paper No. 29. Rome: Food and Agriculture Organization (FAO) of the United Nation, Rome.

Cook, F.J., F.M. Kelliher, and S.D. McMahon. 1994. Changes in infiltration and drainage during wastewater irrigation of highly permeable soil. *J. Environ. Qual.* 23:476–482.

Devitt, D.A., R.L. Morris, M. Baghzouz, M. Lockett, and L.K. Fenstermaker. 2005. Water quality changes in golf course irrigation ponds transitioning to reuse. *HortScience* 40:2151–2156.

Devitt, D.A., R.L. Morris, D. Kopec, and M. Henry. 2004. Golf course superintendents attitudes and perceptions toward using reuse water for irrigation in the southwestern United States. *Hort. Technol.* 14:577–583.

Fu, J.M., A.J. Koski, and Y.L. Qian. 2005. Responses of creeping bentgrass to salinity and mowing management: growth and turf quality. *HortScience* 40:463–467.

Golf Course Superintendents Association of America (GCSAA). 2003. *Water Woes: A New Solution for Golf Courses.* At http://www.gcsaa.org/news/releases/2003/june/effluent.asp

Harivandi, M.A. 1991. *Effluent Water for Turfgrass Irrigation.* University of California Cooperative Extension.

Harivandi, M.A. 1994. *Wastewater Reuse for Golf Course Irrigation.* Lewis Publishers, Ann Arbor.

Harivandi, M.A. 1999. Interpreting turfgrass irrigation water test results. *Calif. Turfgrass Cult.* 49:1–8.

Harivandi, M.A., K.B. Marcum, and Y.L. Qian. 2007. Recycled, gray, and saline water irrigation. In M.P. Kenna and J.B. Beard (eds.), *Water Quality and Quantity Issues in Turfgrass in Urban landscapes.* Council for Agricultural Science and Technology, Ames. IA (in press).

Hayes A.R., C.F. Mancino, W.Y. Forden, D.M. Kopec, and I.L. Pepper. 1990. Irrigation of turfgrass with secondary sewage effluents. II. Turf quality. *Agron. J.* 82:943–946.

Jordan, L.A., D.A. Devitt, R.L. Morris, and D.S. Neuman, 2001. Foliar damage to ornamental trees sprinkler-irrigated with reuse water. *Irrig. Sci.* 21:17–25.

Lazarova, V. and T. Asano. 2005. Challenges of sustainable irrigation with recycled water. In V. Lazarova and A. Bahri (eds.), *Water Reuse for Irrigation: Agriculture, Landscapes and Turfgrass.* CRC Press, Boca Raton, FL.

Mancino, C.F. and I.L. Pepper. 1992. Irrigation of turfgrass with secondary sewage effluent. *Agron. J.* 84:650–654.

Miyamoto, S. and A. Chacon. 2006. Soil salinity of urban turf areas irrigated with saline water: II. Soil factors. *Landscape Urban Plan.* 72:28–38.

Miyamoto S, A. Chacon, M. Hossain, and I. Martinez. 2005. Soil salinity of urban turf areas irrigated with saline water I. Spatial variability. *Landscape Urban Plan.* 71:233–241.

Oster, J.D. and F.W. Schroer. 1979. Infiltration as influenced by irrigation water quality. *Soil Sci. Soc. Am. J.* 43:444–447.

Pepper, I.L. and C.F. Mancino. 1993. Irrigation of turf with effluent water. In M. Pessarakli (ed.), *Handbook of Plant and Crop Stress*, Marcel Dekker, New York, pp. 623–641.

Pettygrove, G.S. and T. Asano. 1985. *Irrigation with Reclaimed Municipal Wastewater – A Guidance Manual.* Lewis Publishers, Chelsea, MI.

Qian, Y.L. 2004. Urban landscape irrigation with recycled wastewater: preliminary findings. *Colorado Water.* 21(1):7–11.

Qian Y.L. and J.M. Fu. 2005. Response of creeping bentgrass to salinity and mowing management: carbohydrate availability and ion accumulation. *HortScience* 40:2170–2174.

Qian Y.L., J.M. Fu, J. Klett, and S.E. Newman. 2005. Effects of long-term recycled wastewater irrigation on visual quality and ion content of ponderosa pine. *J. Environ. Hort.* 23:185–189.

Qian Y.L. and B. Mecham. 2005. Long-term effects of recycled wastewater irrigation on soil chemical properties on golf course fairways. *Agron. J.* 97:717–721.

Rice, R.C. 1974. Soil clogging during infiltration of secondary effluent. *J. Water Pollute Control Fed.* 46:708–716.

Schipper, L.A., J.C. Williamson, H.A. Kettles, and T.W. Speir. 1996. Impact of land-applied tertiary-treated effluent on soil biochemical properties. *J. Environ. Qual.* 25:1073–1077.

State of Arizona. 1987. Regulations for the reuse of wastewater. Arizona Administrative Code, Chapter 9, Article 7, Arizona Department of Environmental Quality, Phoenix, AZ.

State of California. 1978. Wastewater reclamation criteria. California Administrative Code, Title 22, Division 4, Calif. Department of Health Services. Sanitary Engineering Section, Berkeley, CA.

State of Colorado. 2000. Reclaimed domestic wastewater control regulation. Colorado Department of Public Health and Environment Water Quality Control Commission, Regulation No. 84. Denver, CO.

State of Florida. 1989. Reuse of reclaimed water and land application. Florida Administrative Code, Chapter 17–610. Florida Department of Environmental Regulation, Tallahassee, FL.

State of Texas. 1990. Use of reclaimed water. Texas Administrative Code, Chapter 310, Subchapter A, Texas Water Commission, Austin, TX.

US EPA. 2004. Manual: Guidelines for water reuse. EPA/625/R-04/108.

Westcot, D.W. and R.S. Ayers. 1985. Irrigation water quality criteria. In G.S. Pettygrove and T. Asano (eds.), *Irrigation with Reclaimed Municipal Wastewater – A Guidance Manual.* Lewis Publishers, Chelsea, MI, pp. 3:1–36.

Wu, L., X. Gao, and A. Harivandi. 2001. Salt tolerance and salt accumulation of landscape plants irrigated by sprinkler and drip irrigation systems. *J. Plant Nutr.* 24:1473–1490.

27 Turfgrass Drought Physiology and Irrigation Management

Stephen E. McCann and Bingru Huang

CONTENTS

27.1 INTRODUCTION

Water for use in turfgrass irrigation is becoming increasingly limited, which often leads to decline in turf quality, particularly in semiarid and arid regions. Therefore, water conservation becomes a prime concern of turfgrass managers and growers as communities place greater restriction on the use of potable water for irrigating turfgrass in home lawns, parks, athletic fields, and golf courses.[1–3] This situation is complicated by weather patterns that lead to extreme weather such as extended drought and heat. The combination of these factors puts great demand on water conservation strategies in turfgrass management, which decreases the overall quantity of irrigation water required to maintain healthy turf, as well as decreasing freshwater use by increasing the use of reclaimed or recycled water.

Developing efficient, water-saving management practices for any particular site, whether it is a golf course or an athletic field, requires an understanding of environmental factors that influence turfgrass growth and water use, water-use characteristics of plants, and how plants respond to water deficit. Extensive research efforts have been undertaken to investigate drought-resistance mechanisms for various turfgrass species. During the process of plant adaptation to drought stress, plants have developed various mechanisms, including drought avoidance, drought tolerance, and drought escape. Many water-conservation practices are based on physiological responses of plants to water stress, including the use of plant-growth regulators (PGRs) to improve stress tolerance and reallocate plant water use, applying antitranspirants to reduce evaporative losses, applying osmoregulants to improve osmotic adjustment, and adjusting fertilizer applications to improve drought tolerance. Plant water absorption can also be increased by avoiding establishment in heavy soils, limiting soil compaction, irrigating that avoids overwatering, as well as restricting low mowing heights that would result in shallow root systems. Other water conservation strategies may be developed based on soil and climatic conditions, including irrigation delivery systems that have reduced evaporative losses, application of surfactants to evenly distribute soil water and reduce surface water runoff, and irrigation at minimum frequency and quantity and at times when evapotranspirational demand is low.

As potable water supply for irrigation decreases, the use of reclaimed water to irrigate turf has been increasing over the past 30 years. The use of recycled water for irrigation increased from 8% to 21% over the period 1974–1994 in Florida, which has the greatest number and acreage golf courses in the United States.[4] Harivandi[5] suggests that recycled water has application in golf courses, parks, cemeteries, and other venues of nonfood urban horticulture. Irrigation with reclaimed water would greatly reduce potable water consumption in turfgrass management, although turf quality may be adversely affected. Reclaimed water may contain high salt content, which can result in soil salinity and induce osmotic stress in plants. Such problems may be intensified by abiotic factors leading to high evapotranspirational demand, such as high temperature, wind, and low humidity. Water-conservation strategies that are developed with a comprehensive understanding of interaction between plants, soils, and climates are therefore of great importance in turfgrass management, whether potable or reclaimed water is used for irrigation.

This chapter provides a review of recent literature on mechanisms of drought resistance in various turfgrass species, including drought tolerance, avoidance, and escape. Physiological factors and management practices that can reduce water use or improve turfgrass drought resistance, including use of PGRs, osmo-regulants, and antitranspirants will also be discussed. Efficient irrigation management practices that can be implemented for water conservation and improving plant drought resistance will conclude the chapter.

27.2 MECHANISMS OF DROUGHT RESISTANCE

Generally, all plants including turfgrasses that grow in water-limiting environments utilize various adaptive mechanisms. Plants that are capable of growing and surviving under drought-stress conditions are broadly considered drought resistant.[6] The literature suggested classification of drought resistance into three separate categories: drought avoidance, drought tolerance, and drought escape. These drought-resistance strategies are not mutually exclusive. A plant may exhibit more than one of these strategies to cope with drought stress. For an orderly discussion, the three drought-resistance strategies are discussed separately in this section.

27.2.1 DROUGHT TOLERANCE

Drought tolerance can be defined as a plant's ability to maintain physiological functions when very little or no water is available to the plant. One of the important factors controlling cell tolerance to desiccation or dehydration is the cell's capability to maintain adequate turgor pressure during drought stress. Generally, this is done by increasing the concentration of compatible solutes within the cell. Compatible solutes or osmoregulants include inorganic solutes such as potassium (K^+), calcium (Ca^{2+}), and sodium (Na^+) and organic solutes such as soluble sugars (e.g., sucrose), polyols (e.g., mannitol), organic acids (e.g., malate), ammonium compounds (e.g., glycine betaine), and nonprotein amino acids (e.g., proline).[8] Compatible solutes accumulate in cells subjected to water deficit, serving to lower the osmotic potential of the cell without causing toxic effects. The accumulation of compatible solutes triggered by dehydration differs from the accumulation of solutes simply due to the concentration effect resulting from water loss in that it is an active accumulation rather than a passive one, respectively.[7] This active accumulation of solutes is known as osmotic adjustment (OA), and has been shown to be positively correlated to drought tolerance. By increasing compatible solutes, a plant cell can lower its osmotic potential and, thus, water potential, which prevents water loss to intercellular spaces or can increase water movement into cells with low water potential due to OA.

There are various compounds and pathways that plants can utilize to adjust osmolality, each having differing energy requirements. Compatible solutes or osmoregulants include inorganic solutes such as potassium (K^+), calcium (Ca^{2+}), and sodium (Na^+) and organic solutes such as soluble sugars (e.g., sucrose), polyols (e.g., mannitol), organic acids (e.g., malate), ammonium compounds (e.g., glycine betaine), and nonprotein amino acids (e.g., proline).[8]

Many studies have demonstrated the importance of OA in drought tolerance in turfgrass species. For example, DaCosta and Huang[9] have shown the importance of OA in creeping bentgrass (*Agrostis stolonifera* L.) and velvet bentgrass (*A. canina* L.). Compared with creeping bentgrass, velvet bentgrass showed a 50–60% higher magnitude of OA under water deficit. The increase in OA was associated with better drought tolerance in velvet bentgrass, as determined by higher visual turf quality (TQ) and leaf relative water content (RWC) under drought-stress conditions, compared with bentgrass. In another study, Kentucky bluegrass (*Poa pretensis* L.) plants that were pre-exposed to moderate drought stress maintained higher RWC and TQ during subsequent stress exposure, compared with plants with no prior exposure to drought; the improved stress tolerance following preconditioning was at least partially attributed to increases in OA.[10] Qian and Fry[11] have shown a positive correlation between OA and recuperative ability from drought stress for buffalograss (*Buchloe dactyloides*), zoysiagrass (*Zoysia japonica*), bermudagrass (*Cynodon dactylon* [Nut.] Engelm.), and tall fescue (*Festuca arundinacea* [Schreb.]). However, studies with nonturfgrass species have shown that increased accumulation of osmotic solutes is not always correlated with increased stress tolerance,[12] and the degree of OA can vary with species and with genotypes of the same species.[13–15] Genetic variation in OA capacity has been reported in four turfgrass species, with the magnitude of OA ranking as buffalograss = zoysiagrass > bermudagrass > tall fescue.[11]

In addition to OA, other mechanisms such as cell-wall elasticity also play important roles in the maintenance of cell turgor pressure and desiccation tolerance. The general response of plants to drought stress is to increase cell-wall elasticity. Cells that are more elastic can maintain turgor pressure at low water potential. As a result of increased cell-wall elasticity, plants can maintain turgor for a longer period of time under drought stress. Research investigating the drought response of buffalograss ("Biloela") has shown that cell-wall elasticity, along with OA, functions to maintain turgor under water deficit.[16] White et al.[17] reported that zoysiagrass genotypes with high RWC at zero turgor and tissue elasticity recovered better from stress and required less supplemental irrigation. This result implies that the improvement in biophysical properties is also an important factor contributing to drought tolerance in turfgrass. Some plant species may rely solely on increasing cell-wall elasticity without OA involvement in turgor maintenance.[18] Research conducted by Barker et al.[19] of five forage grasses showed that C_3 grasses tend to have more elastic cell walls and are less reliant on OA for maintenance of turgor pressure.

Drought-tolerance mechanisms also involve changes in various metabolic processes, which help protect cells from further injury and help maintain metabolic functioning. A relatively drought-sensitive metabolic function is photosynthesis; drought stress inhibits carbon fixation. While light absorption continues under drought stress, reduced electron transport to carbon fixation leads to an accumulation of excessive energy that can be dissipated by reducing molecular oxygen, thus generating active oxygen species. Active oxygen species include singlet oxygen ($^1O_2$), superoxide ($O_2^{\cdot-}$), hydrogen peroxide (H_2O_2), and hydroxyl radical ($OH^{\cdot}$). These species can interact with lipids, nucleic acids, and proteins, and thus cause cellular damage. It has also been suggested that oxidative stress resulting from drought can negatively impact TQ and physiological functions.[20,21]

Many plant species have antioxidant defense systems to protect cells from oxidative stress induced by drought. One of the defense mechanisms against oxidative stress is to increase the activity of superoxide dismutase (SOD), catalase (CAT), hydrogen peroxidase (POD), and ascorbate peroxidase (APX). Through the increase of antioxidant enzyme activity, active oxygen species are scavenged, thereby reducing the amount of oxidative damage done to cells. Zhang and Schmidt[20] reported that Kentucky bluegrass plants that had higher antioxidant levels performed better under drought stress. Fu and Huang[21] found that tolerance of tall fescue and Kentucky bluegrass to soil surface drying was associated with the maintenance or enhancement of antioxidant enzyme activity. Exogenous application of PGRs to creeping bentgrass increased SOD activity and improved turf performance under drought stress.[22]

In response to drought stress, plants exhibit significant changes in protein composition, synthesis, and expression, including synthesis of stress-inducible proteins.[23,24] Synthesis of stress-induced proteins such as late embryogenesis abundant (LEA) proteins in response to drought stress has been associated with increased adaptive ability and tolerance to drought stress in various nonturfgrass species.[24–26] Limited information is available on the association of protein induction with drought tolerance in turfgrass.[27] A class of stress-induced LEA proteins, known as dehydrins, plays a positive role in drought tolerance. Dehydrins have been shown to accumulate under drought stress and in response to increased abscisic acid (ABA) concentrations in tall fescue.[27] The function of these proteins involved in drought tolerance remains unclear. However, some studies suggest that these proteins may act in the stabilization of membranes and as molecular chaperones, which prevent denaturation of other proteins.[28,29] Dehydrins may also act as OA regulators, thereby enhancing desiccation tolerance.[30] Transgenic plants overexpressing the genes that code for these proteins have increased tolerance to drought and high salinity.[31] The authors suggest that the known physical properties of dehydrins make them candidates for participation in stabilizing nuclear or cytoplasmic macromolecules, thus preserving structural integrity.

27.2.2 Drought Avoidance

Drought avoidance is the ability of a plant to maintain normal physiological function by postponing tissue dehydration. This mechanism may be achieved by increasing water uptake of the root system and reducing water loss from transpiring leaves.

Plants have developed multiple ways of procuring water, including developing extensive, deep root systems, increasing root branching and surface areas to exploit large soil volumes for water absorption. Research in turfgrass has shown that a shallow-rooted species, *Poa supina* Schrad, had lower drought tolerance than species with well-developed root systems, such as *A. stolonifera* and *Festuca rubra*.[32] Huang and Gao[33] reported that drought-resistant tall fescue cultivars had deeper root systems than sensitive cultivars. A study evaluating *Zoysia japonica* (cv. Meyer) showed that turfgrass plants exposed to drought developed more extensive rooting at deeper soil depths than well-watered turf.[34] A study done by Marcum et al.[35] also indicates the importance of rooting depth, weight, and branching at lower depths in zoysiagrass adaptation to drought stress. Tall fescue is an excellent example of a drought avoider by developing deep root system. It exhibited maximum root extension of 33–60% greater than buffalograss, zoysiagrass, or bermudagrass in a greenhouse study, and extracted over 50% more water at a 90-cm depth than zoysiagrass.[36] Various studies of tall fescue cultivars investigating the relationship between root development and drought avoidance have shown a strong correlation between the two, with greater total root length and density delaying water deficit and prolonging plant survival in drying soil.[37,38] Additional research has shown that cultivars of tall fescue with extensive root systems are capable of delivering more water to the plant, and as such avoid desiccation and wilting.[36,39,40] Highly plastic root systems were found to contribute to the persistence of tall fescue under prolonged periods of drought stress.[41–43] Huang et al.[42] reported that drought resistance of tall fescue is more closely related to root plasticity and root viability than total root mass or length.

Some grasses are able to access more water through deeper root systems, which can result in water reallocation through the root zone. Water deep in the profile can be transported at night by roots to dry, shallow areas where water leaks from roots into the drying surface soil. This phenomenon is referred to as hydraulic lift. Huang[44] showed that in buffalograss, water absorbed by roots deep in the profile at night helped to support root growth and function in nutrient uptake in the upper soil profile, resulting in delaying drought-stress injury. This reallocation of water in the soil profile can lead to efficient use of water under surface-soil drying conditions.

Differences in root distribution during drought stress may be due to carbon reallocation from shoots to roots for formation of a more extensive root system into deep soil. In a study investigating carbon metabolic responses under surface-soil drying, both Kentucky bluegrass and tall fescue exhibited decreased respiration rates in shoots and roots in the upper drying layer. In contrast, there was enhanced carbon allocation to those roots in the lower, wet soil layer.[45] In general, there may be more carbon allocated to roots, with a greater proportion of newly fixed carbon for roots in soil layers where water may be more available.[33]

Another important avoidance mechanism is the ability of plants to reduce water loss through transpiration. Most transpirational water loss is through stomata in leaf epidermal surfaces. Stomatal closure is one of the most sensitive responses to drought stress, which increases resistance for water diffusion out of leaves, and thus results in water conservation.

Stomatal closure can be affected by many factors. It has been found to be induced by increasing leaf concentration of ABA that is transported from roots exposed to drying soils.[46,47] Roots are important sites for the synthesis of ABA, which is transported to shoots and initiates a signal cascade in guard cells that alters the membrane transport of several ions, and as a result, guard cells lose their turgor and stomata close. This results in changes of stomatal conductance, transpiration rate, and photosynthesis.[48–51] The importance of ABA as a metabolic factor in the regulation of plant tolerance to stresses has received great attention in recent years in other species.[52,53] However, limited information is available on the association of ABA signaling and drought tolerance in turfgrass. Wang et al.[54] reported that Kentucky bluegrass cultivars tolerant of drought exhibited slower ABA accumulation rate than drought-sensitive cultivars during short-term drought stress, suggesting that low accumulation rate of ABA in leaves would be beneficial for the maintenance of photosynthesis during short-term drought, and allow dry matter to accumulate to support plant survival during prolonged drought. Drought-tolerant cultivars of Kentucky bluegrass were characterized by lower ABA accumulation and less severe decline in leaf water potential (ψ_{leaf}), photosynthesis (P_n), stomatal conductance (g_s), and TQ during drought stress. The stomates of drought-tolerant cultivars were more sensitive to changes in ABA level in leaves during drought.[55]

In addition to stomatal regulation of water loss, modification of shoot characteristics such as leaf shedding or folding, or the development of a thick cuticle reduces leaf transpiration. Research in both tall fescue and *Eragrostis curvula* (Schrad.) Nees. complex cv. Consol., a temperate-zone C_4 grass, has shown that transpiration can also be reduced by decreasing light intensity via rolling leaves.[56,57] Better tall fescue performance during drought stress was positively related to leaf thickness, epicuticular wax content, and tissue density but negatively related to stomatal density and leaf width.[58] Beard and Sifers[59] reported that better dehydration avoidance among *Cynodon* species compared with *Zoysia* species was attributed to lower water-use rate due to a faster rate of wax formation over the stomata during progressive drought stress.

Drought-avoidant plants control water loss to maintain growth under drought stress, which, therefore, may lead to increases in water-use efficiency (WUE). WUE is often used to assess the expense of water loss or consumption in terms of carbon gain. WUE can be defined as the ratio between the amount of CO_2 assimilated and the amount of water lost. A key issue limiting WUE is an inescapable trade-off between controlling water use and carbon assimilation. The opposing need of the plant to both assimilate CO_2 and avoid excessive water loss must ultimately be resolved in the control of leaf gas exchange by the stomata, water loss, and efficient expenditure of carbon, especially under drought conditions. Turfgrass species that performed better under drought stress maintained higher WUE than drought-sensitive species.[9]

Low-water-use turfgrass species and cultivars may postpone tissue desiccation or survive longer when water supply is limited. Turfgrass water use can be estimated by a measurement of the evapotranspiration (ET) rate, the sum of the amount of water transpired from leaves and evaporated from soil under the turf canopy within a given period of time. Water-use rate varies with plant species and cultivars within a species, depending largely on shoot growth characteristics in addition to stomatal behaviors. For example, Beard[60] summarized that tall fescue, Kentucky bluegrass, annual bluegrass (*Poa annua* L.), and creeping bentgrass had the highest ET rates; rough bluegrass (*Poa trivialis* L.) and perennial ryegrass (*Lolium perenne* L.) ranked intermediate; and chewings fescue (*Festuca rubra* L. subsp. *fallax* [Thuill.] Nyman), hard fescue (*Festuca brevipila* R. Tracey), and red fescue (*Festuca rubra* L. subsp. *rubra*) had the lowest ET rates. Among warm-season grasses, bermudagrass (*Cynodon* spp.), zoysiagrass (*Zoysia* spp.), and buffalograss have relatively low-water-use rates, while centipedegrass (*Eremochloa ophiuroides* [Munro] Hack.), seashore paspalum (*Paspalum vaginatum* Swartz.), and St. Augustinegrass (*Stenotaphrum secundatum* [walt.] Kuntze.) have relatively high water-use rates.[60] Within a species, cultivars also vary in water-use rate, and the difference may range from 20% to 60%.[61–67] It is important to point out that the comparative water-use rankings for different species and cultivars may change across different environments, climatic conditions, and cultural regimes. Thus, one strategy for reducing turfgrass irrigation requirements is to utilize grasses with reduced ET rates that perform well in a given environment. Available water supplies could be extended for longer periods of time to maintain favorable water status and also decrease the irrigation requirements with use of the appropriate species or cultivar.

27.2.3 DROUGHT ESCAPE

Drought escape, the third category of drought mechanisms discussed here, is when a plant completes its life cycle prior to drought exposure or becomes dormant during drought stress. An example of early reproduction to escape drought stress would be of an annual plant that germinates in the fall, over-winters, quickly matures through the spring and bears seed prior to summer stress. The plant then dies, and seeds germinate after the summer heat and drought, or germination possibly results from such stresses. Evolution of such a life cycle is simple and avoids the need to incorporate more complicated systems into its overall functioning.

Drought escape is used by some weedy annual grasses as a drought resistance approach. Annual bluegrass is a winter annual that germinates in autumn and grows vegetatively until late spring when seeds are produced and the mother plants die. Although efficient in nature, grasses that escape drought are not reliable selections for perennial stands. However, this mechanism has been exploited by crop breeders to develop lines that mature early or that are planted early, before drought occurs.[68,69]

When not supplied with water for an extended period, turfgrass may stop growing and leaves may become desiccated, but crowns, stolons, or rhizomes are still alive. This is often referred to as dormancy, which preserves the vital parts of the plant for regeneration of shoots and roots when water becomes available. By becoming dormant, turfgrasses can escape drought stress by reducing demand on water until soil water is replenished. The length of turfgrass survival in a dormant condition depends on many factors, including soil water availability, plant growth rate, and the health of the plants at the onset of dormancy. Turfgrasses generally can maintain dormancy for weeks with limited damage.

27.2.4 GENETIC VARIATION IN DROUGHT RESISTANCE

There are large genetic variations in drought resistance and water-use characteristics among turfgrass species and cultivars. A particular species or cultivar may possess multiple drought-resistance traits.[70] In a summarized literature ranking, Fry and Huang[70] reported that overall drought resistance for warm-season turfgrass species is ranked from excellent to fair in the following order: buffalograss, bermudagrass, zoysiagrass, seashore paspalum (*Paspalum vaginatum*), St. Augustine-grass (*Stenotaphrum secundatum* [Walt.] Kuntze), and centipedegrass (*Eremochloa ophiuroides* [Munro] Hack). For cool-season grass, overall drought resistance ranges from excellent to poor level in the order tall fescue, fine fescue, Kentucky bluegrass, perennial ryegrass, creeping bentgrass, and annual bluegrass.

TABLE 27.1
Drought-Resistant[a] Varieties of Commonly Used Turfgrass Species

Bentgrass (Fairway/Tee)[b]	**Tall Fescue[b]**
Seaside	Lion
SR 7200	Kitty Hawk S.S.T.
Glory	Marksman
Golfstar	Rembrandt
Chewings Fescue[b]	**Bermudagrass[b]**
Ambrose	Riviera
Ambassador	Shanghai
Treazure	Blackjack
Bridgeport	Blue-muda
Kentucky Bluegrass[b]	**Buffalograss[b]**
Unique	Cody
Apollo	Tatanka
Brilliant	Bison
Showcase	BAM-1000
Perennial Ryegrass[c]	**Zoysiagrass[b]**
Passport	Emerald
Affinity	Victoria
Calypso II	Zeon
Edge	El Toro

Note: The top four commercially available cultivars are listed for each species (not including experimental cultivars).

[a] Drought resistance was assessed as wilting, leaf firing, dormancy or recovery. A 1 to 9 visual rating scale is used with 1 being complete wilting, 100% leaf firing, complete dormancy or no plant recovery; and 9 being no wilting, no leaf firing, 100% green-no dormancy, or 100% recovery.

[b] Assessed via dormancy.

[c] Assessed via wilting.

Source: Based on NTEP data between 1996 and 2003.

Cultivars within a species also vary in drought resistance. The National Turfgrass Evaluation Program (NTEP) was developed to evaluate various turfgrasses, including both existing and new varieties, grown in various climates across the United States. Full reports are based on 4–5 years of field data, and summary reports are generated each year. Recent NTEP data comparing drought resistance are summarized in Table 27.1.

Among different water-conservation practices, use of grass species and cultivars that are capable of tolerating, avoiding or escaping drought stress, as well as utilizing different strategies is critical for water conservation and maintenance of high-quality turfgrass in water-limiting environments. Planting drought-resistant grasses can result in significant water saving and reduce the overhead costs of irrigation water.

27.3 GROWTH REGULATION AND DROUGHT-STRESS RESPONSES

Various chemical and hormonal compounds have been shown to impact plant drought resistance. These compounds can be divided into three broad classifications that correspond to plant drought-response mechanisms: (1) PGRs, (2) osmoregulants, and (3) antitranspirants. Growth regulators are synthetic compounds with hormone-like functions, which may affect plant growth, morphology, and metabolic activities, or can have a combined effect and can result in either drought avoidance or tolerance. Osmoregulants influence the osmotic potential of a cell by modifying the production of various solutes, which can change a cell's water potential. Cells with the ability to adjust osmotically can maintain positive turgor pressure by increasing concentration of various solutes in the vacuole or cytosol. Maintenance of turgor pressure has been shown to correlate to maintenance of physiological functions and overall plant health, as discussed in the above section. Antitranspirants are chemicals that function in inhibiting transpiration by causing stomatal closure or by forming a film covering the leaf epidermal surface. Use of antitranspirants helps reduce water loss through transpiration and may enhance drought avoidance.

Turfgrasses with rapid vertical shoot extension rate tend to have higher water-use rates than slower growing or dwarf-type grasses, because of increasing growth demand on water. Shearman[65] reported that shoot vertical extension rate was positively correlated with water-use rate for 20 Kentucky bluegrass cultivars with upright growth pattern. While many compounds can influence plant growth and morphology, possibly the most widely used PGR in turfgrass management are growth inhibitors, such as trinexapac-ethyl (TE). TE inhibits vertical shoot extension by inhibiting synthesis of gibberilic acid (GA), a hormone produced by plants. Numerous studies in turfgrass have shown that TE not only reduces vertical shoot growth rate, but also increases cell density in leaves, promotes darker green leaves, and in some cases increases tiller density.[71–73] This growth inhibition effect may result in reduction of water consumption, and thus may enhance plant drought resistance. Exploration of the specific mechanisms related to improved stress tolerance has suggested involvement of multiple mechanisms. Application of TE, which blocks GA formation, has improved drought-stress tolerance by reducing the rate of ET resulting in available water supply for an extended duration.[74,75] TE has also been shown to reduce sod heating during storage[76] and disease severity.[73,77,78]

Numerous studies have cited the importance of maintenance of cell turgor pressure in plants under drought stress.[9,79,80] Plants can maintain cell turgor pressure through OA, which is accomplished by increasing solute concentration, such as sugars and carbohydrates, in the cytosol. Increases in cell solute concentration decrease the cell's osmotic potential, which greatly reduces cell water loss. Many studies with agronomic crops have suggested that exogenous foliar application of glycine betaine (GB) or crop varieties that accumulate GB can maintain cell turgor pressure under water deficit.[81] GB application has been shown to enhance OA, net photosynthesis, photochemical efficiency of PS II (Fv/Fm), and yield in various crops.[82–84] There is limited information available on the use of GB in turfgrass management for improving drought tolerance. A study with creeping bentgrass by Saneoka et al.[85] has also shown that GB levels increased when the plant was exposed to drought stress, suggesting a possible link between GB and drought adaptation.

Strategies that can reduce water loss through transpiration can have significant impact on water conservation and promoting plants to avoid drought stress. Of the amount of water absorbed by plants, the largest portion (>90%) is transpired and only <3% of the absorbed water is used in photosynthesis and other metabolic activities. If transpirational water loss through stomata can be controlled to any extent, it is possible to maintain plant growth with the use of much less water. Antitranspirants, such as ABA, have been shown to reduce transpiration rates more than photosynthesis, causing significant increases in WUE and helping plants survive for an extended period of drought. Several controlled-environment studies with turfgrasses demonstrated reduced water use with foliar spray of ABA.[9,54,86,87] Stahnke[88] reported that ABA application reduced transpiration in creeping bentgrass by 59% by inducing stomatal closure.

Recent research conducted by Wang et al.[79] showed that exogenous ABA application to Kentucky bluegrass in drought conditions resulted in higher TQ; higher cell-membrane stability, as indicated by less electrolyte leakage; and higher photochemical efficiency (Fv/Fm) when compared with untreated turf. ABA has been widely used in field crops and vegetables for reducing water loss from plants, but little has been done on the practical use of ABA to reduce water requirements and enhance drought tolerance in turfgrasses under natural field conditions. This could be largely due to the unavailability of a synthetic form of ABA. Nevertheless, available data suggest use of antitranspirants, including ABA, in combination with routine cultural practices would be beneficial in water conservation for turfgrass in water-limiting environments.

27.4 WATER-SAVING IRRIGATION MANAGEMENT

As discussed in the previous sections, turfgrass water use is a function of plant-growth characteristics and environmental conditions. Therefore, a sound, efficient irrigation program should be developed based on plant needs and environmental conditions. Factors such as water quality, soil type, local weather, and turf type, all play a role in developing a water-conservation irrigation program.

The most important consideration in developing an irrigation program for any particular site is the frequency and amount of water that should be applied. Monitoring weather conditions, daily water-use rate, and soil water availability is critical for determining the appropriate quantity and timing of watering. In the case of using reclaimed water for irrigation, knowledge of dissolved salt content in irrigation water is also important. To design and manage an efficient irrigation program, it is imperative to determine the minimum water requirements of plants. The irrigation requirement may be estimated through visual evaluation of both the canopy and soil cores extracted with a soil probe. Leaf wilting or firing indicates water deficit, which is a good indicator of when to water. However, visual assessment can be highly subjective and may lead to excess or inadequate irrigation without knowing how much to water. Soil dry-down is also a good indication of drought stress and when and how much to irrigate to maintain turfgrass growth. The availability of soil water can be estimated using two parameters, field capacity and the permanent wilting point. Field capacity is the maximum water-holding capacity of a soil when drainage ceases. The permanent wilting point is the level of soil water content at which plants become permanently wilted. The amount of water available for plant growth in the soil is the difference between field capacity and permanent wilting point, which varies with soil types and plant species. Fine-textured soils, such as clay, have a greater water-holding capacity, and thus irrigation may be applied less frequently. In contrast, light, frequent irrigation may be required to replenish water reservoir in sandy soils due to low field capacity. Various methods are used to measure soil water content, including time-domain reflectometry, tensiometers, and gypsum blocks.[70] Daily ET measurement enables precise determination of how much water has been lost within a day, and the ET rate can be used to decide how much to irrigate in the next irrigation cycle. Canopy ET rate can be estimated using tools such as lysimeters, weather station data, and ET pans.[70] Weather stations gather data such as air temperature, humidity, barometric pressure, level of photosynthetically active radiation, and wind speed to calculate ET rates.

Light and frequent water application versus heavy and infrequent application has had much scientific light shed on the question over the past 20 years. It is now generally accepted that deep and infrequent irrigation promotes plant tolerance to drought stress and may result in water savings. Fry and Huang[70] reviewed extensive literature of previous research that has proved the benefits of deep, infrequent irrigation in various turfgrass species. Decreasing irrigation frequency resulted in a 6–18% reduction in water use for some warm-season grasses, and a 24–34% reduction in water use for some cool-season grasses.[89] However, optimum irrigation frequency varies with plant species, climatic conditions, and soil types. Research conducted by Jordan et al.[90] evaluating various cultivars of creeping bentgrass maintained under putting green conditions concluded that an irrigation frequency of every 4 days produced a larger and deeper root system, with higher TQ and shoot density when compared with treatments irrigated every 1–2 days. Zoysiagrass irrigated at the onset of leaf wilting resulted in a 40% reduction in shoot vertical growth rate compared with that watered daily, which decreases water demand.[91] Tall fescue with an irrigation frequency of twice a week produced higher-quality turf than that watered three or four times a week.[92]

Further proof of deep and infrequent irrigation has been offered in our most recent study with the evaluation of three species of creeping bentgrass maintained under golf-course fairway conditions and replacing 100% ET. Our study showed that turf generally performed best when irrigated twice a week in summer, and once a week in spring and fall, when compared with plots watered three times a week.[92] The experiment showed that ideal soil volumetric water content for turf growing in a native soil is approximately 25%. However, under light and frequent irrigation (watering three times per week and replacing 100% ET), soil water content averaged 30% or greater. By increasing the amount of water in the soil, there is a reduction in the amount of air. For example, a loam-type soil at field capacity generally consists of 50% soil, 25% water, and 25% air. If water content is increased to 30%, then air content is decreased to 20%. This can create an imbalance, where the plant has more than enough water for physiological functioning, but begins to experience limited O_2 supply to roots because of the decreased air content. Problems that have generally been associated with light and frequent water applications in turf are increased moss and algae, increased disease pressure, decreased O_2 availability in the rootzone, shallow rooting, and thinning turf canopy.

The quantity of irrigation applied should also be determined through the evaluation of water needs of a particular turfgrass species and soil water content. Various studies have demonstrated that it may not be necessary, or beneficial, to replace 100% of ET water loss, and many turfgrass species such as Kentucky bluegrass, perennial ryegrass, tall fescue, creeping bentgrass, velvet bentgrass, colonial bentgrass, zoysiagrass, and bermudagrass are able to maintain acceptable TQ when irrigated in an amount below that of the plant's maximum water demand (deficit irrigation). Deficit irrigation practice maintains soil water content below field capacity but above permanent wilting points, and thus limits water use by plants. Deficit irrigation can result in overall water savings and increases in WUE without loss of TQ in various turfgrass species. Fu et al.[93] found that tall fescue and bermudagrass maintained similar quality levels irrigated to replace 60% and 40% of actual ET compared with the same species under well-watered (replace 100% ET) conditions. Irrigating bermudagrass, buffalograss, and St. Augustinegrass at rates greater than 55% of actual water loss did not result in higher TQ.[94] In a 2-year field project conducted by DaCosta and Huang[9] evaluating three bentgrass species, data showed that turf receiving 80% of ET had similar or higher quality than turf replaced with 100% ET. An explanation for these results is that in replacing 100% of ET, the top few centimeters of the soil become saturated for a period long enough to prevent O_2 diffusion, increasing algae and moss content, and creating an environment favorable for disease development. Other research exploring irrigation quantity has been conducted in sod establishment and has shown that spring establishment requires less water (45 mm per week) than sod established in summer months (63 mm per week) during the first 7 weeks.[95] The research also showed that increasing irrigation volume did not improve TQ.

Reducing irrigation quantity and frequency can also be accomplished by incorporating PGRs and antitranspirants into management programs. As previously discussed, application of such compounds can moderate ET losses of plants. Measures that modify soil physical properties, such as use of surfactants, to improve water-holding capacity or infiltration could also be practiced for water conservation. Soil surfactants decrease surface tension in soil and reduce water runoff, and create matrix water flow, making water more uniformly and consistently distributed throughout the soil profile. In cases of practicing deep, infrequent irrigation that allows soil to dry down between irrigation events, surfactants may be required to minimize localized dry spots, particularly on sandy soils.

Effluent or reclaimed water is a potential source of irrigation water for turfgrass during periods of drought or restrictions on potable water. Use of reclaimed water can result in significant reduction of potable water consumption. In addition, reclaimed water may contain more nutrients than potable water and can enhance plant growth. Mujeriego et al.[96] reported that reclaimed water at high application rates can maintain TQ comparable to potable water with N application up to 25 kg ha^{-1} per month. Reduced N requirement has been shown in a study evaluating reclaimed water use in bermudagrass and seashore paspalum.[97] A study evaluating two common turfgrasses, *Cynodon dactylon* and *Lolium perenne*, showed that effluent-treated turf could be grown at a high-quality level, with reduced fertilizer needs, when compared with treatments only irrigated with tap water.[98] A wide range of ornamental plants (including *Cupressus sempervirens*, *Asparagus densiflorus*,

Pococarpus macrophyllus, Strelitzia reginae, Juniperus procumbens, Philodendron williamsii, Ilex vomitoria cv. Schellings, chrysanthemums, *Hibiscus rosa-sinensis* and its hybrids, and various hododendron hybrids) irrigated with reclaimed water showed no detrimental effect and were generally determined to respond best to irrigation rates of 1.0–1.5 in. per week.[99]

The practical application of this research can be evaluated at a site like Meadow Lakes Golf Course in Prineville, Oregon. The golf course covers hundreds of acres and has used treated wastewater in management of the facility to much success.[100] Mujeriego et al.[96] evaluated the agronomic and human health impact of using reclaimed water on a golf course in Spain (Mas Nou). They concluded that reclaimed water containing fecal coliforms and fecal streptococci concentrations below 100 c.f.u. per 100 ml had no detrimental effects to human health, while use of reclaimed water resulted in significant reductions in potable water use and also achieved management savings through reduced need for fertilizer inputs.

While use of reclaimed water can be a viable strategy for water conservation, care should be taken in areas irrigated frequently with large quantity. In addition to beneficial nutrients, all reclaimed water also contains dissolved mineral salts such as sodium, which may have adverse effects when accumulated in large amount in the soil. Water quality should be monitored closely during the course of reclaimed water irrigation. Soils should be irrigated periodically with potable water to leach out excess salt and prevent toxic effects. If soil salinity is expected to be a problem with reclaimed water irrigation, salt-tolerant grasses, such as seashore paspalum and weeping alkaligrass (*Puccinellia distans* [Jacq.] Parl.), may be used to maintain quality turfgrass in salt-affected areas.

27.5 CONCLUSIONS

The twenty-first century will present numerous challenges in turfgrass management, including limited water resources for irrigation. Increased human demand on water consumption, declines in potable water supply, and extreme weather resulting in drought stress are putting great demand on water conservation in all sectors of the turfgrass industry. Developing drought-resistant turfgrass species and efficient irrigation practices has become imperative to growing turfgrass in water-limiting environments.

As discussed throughout this chapter, turfgrass water use is a function of plant growth characteristics and environmental conditions. Therefore, an effective conservation program should be developed based on water-use characteristics of plants, soil, and climatic conditions. Turfgrasses have developed various survival mechanisms in response to drought stress, including tolerance, avoidance, and escape of drought stress. These mechanisms are not mutually exclusive, and in fact, one species may posses multiple mechanisms. Turfgrass species and cultivars vary in drought resistance. Using low-water-use or drought-resistant turfgrass species and cultivars is a primary means of reducing water needs, as well as improving water-use efficiencies. Growing warm-season turfgrass in arid and semiarid regions may provide a better TQ and more water savings than using cool-season turfgrasses. In addition to utilization of genetic variation, a myriad of other possibilities exist to maintain acceptable-quality turf with reduced water consumption. Monitoring actual water use by measuring ET rate under different environmental conditions and different times of the year is important for developing efficient, plant-based irrigation programs under changing environmental conditions. Knowledge of plant growth conditions and soil moisture status or available water content of different types of soils is necessary to decide when and how much to irrigate. Deficit irrigation can be effective in water savings without resulting in significant loss of turfgass quality. Deep, infrequent irrigation can be practiced to promote drought resistance of plants and reduce water use. Management programs that incorporate surfactants and PGRs, osmoregulants, and antitranspirants can be effective for water conservation in low-maintenance turf or water-limiting environments. In dry regions where potable water use is restricted, there is increasing use of reclaimed water as an alternative water source. Irrigation with reclaimed water has been proven to be a viable approach in dealing with potable water shortage.

Implementing such programs and designing programs based on the numerous scientific studies evaluating drought-resistance mechanisms and various water-conservation irrigation strategies, such as those referenced in this chapter, would greatly benefit the turfgrass industry and the sustainability of water resources.

REFERENCES

1. Lyon, B., Christie-Blick, N., and Gluzberg, Y., Water shortages, development, and drought in Rockland County, New York, *J. Am. Wat. Res. Assoc.*, 41, 1457, 2005.
2. Carbone, G.J. and Dow, K., Water resource management and drought forcasts in South Carolina, *J. Am. Wat. Res. Assoc.*, 41, 145, 2005.
3. Kenney, D.S., Klein, R.A., and Clark, M.P., Use and effectiveness of municipal water restrictions during drought in Colorado, *J. Am. Wat. Res. Assoc.*, 40, 77, 2004.
4. Haydu, J.J. et al., Economic and environmental adaptations in Florida's golf course industry: 1974–1994, *Int. Turf. Soc. Res. J.*, 8, 1109, 1997.
5. Harivandi, M.A., Irrigating turfgrass and landscape plants with municipal recycled water, *Acta Hort.*, 537, 697, 2000.
6. Turner, N.C., Crop water deficits: a decade of progress, *Adv. Agro.*, 39, 1, 1986.
7. Blum, A., *Plant Breeding for Stress Environments*, CRC Press, Boca Raton, 1988.
8. Nilsen, E.T. and Orcutt, D.M., *Physiology of Plants under Stress: Abiotic Factors*, Wiley, New York, 1996, chap. 5.
9. DaCosta, M. and Huang, B., Minimum water requirements for creeping, colonial, and velvet bentgrass under fairway conditions, *Crop Sci.*, in press.
10. Jiang, Y. and Huang, B., Osmotic adjustment and root growth associated with drought preconditioning-enhanced heat tolerance in Kentucky bluegrass, *Crop Sci.*, 41, 1168, 2001.
11. Qian, Y. and Fry, J.D., Water relations and drought tolerance of four turfgrasses, *J. Am. Soc. Hort. Sci.*, 122, 129, 1997.
12. Maggio, A. et al., Moderately increased constitutive proline does not alter osmotic stress tolerance, *Phys. Plant.*, 101, 240, 1997.
13. Zhang, J., Nguyen, H.T., and Blum, A., Genetic analysis of osmotic adjustment in crop plants, *J. Exp. Bot.*, 50, 291, 1999.
14. Morgan, J.M., Osmoregulation and water stress in higher plants, *Ann. Rev. Plant Phys.*, 35, 299, 1984.
15. Rhodes, D. and Samars, Y., Genetic control of osmoregulation in plants, in *Cellular and Molecular Physiology of Cell Volume Regulation*, Strange, K., Eds., CRC Press, Boca Raton, 347, 1994.
16. Wilson, J.R. and Ludlow, M.M., Time trends for change in osmotic adjustment and water relations of leaves of *Cenchrus ciliaris* during and after water stress, *Aust. J. Plant Phys.*, 10, 15, 1983.
17. White, R.H. et al., Zoysiagrass water relations, *Crop Sci.*, 41, 133, 2001.
18. Auge, R.M., Schekel, K.A., and Wample, R.L., Rose leaf elasticity changes in response to mycorrhizal colonization and drought acclimation, *Physiol. Plant.*, 70, 175, 1987.
19. Barker D.J., Sullivan C.Y., and Moser L.E., Water deficit effects on osmotic potential, cell wall elasticity, and proline in five forage grasses, *Agron. J.*, 85, 270, 1993.
20. Zhang, X. and Schmidt, R.E., Antioxidant response to hormone-containing product in Kentucky bluegrass subjected to drought, *Crop Sci.*, 39, 545, 1999.
21. Huang, B. and Fu, J., Growth and physiological responses of tall fescue to surface soil drying, *Int. Turfgrass Soc. Res. J.*, 9, 291, 2001.
22. Zhang, X. and Schmidt, R.E., Application of trinexapac-ethyl and propiconazole enhances super-oxide dismutase and photochemical activity in creeping bentgrass (*Agrostis stoloniferous* var. palustris), *J. Am. Soc. Hort. Sci.*, 125, 47, 2000.
23. Ouvraud, O. et al., Identification and expression of water stress- and abscisic acid-regulated genes in a drought-tolerant sunflower genotype, *Plant Molec. Bio.*, 31, 819, 1996.
24. Riccardi, F. et al., Protein changes in responses to progressive water deficit in maize, *Plant Physiol.*, 117, 1253, 1998.
25. Bewley, J.D. and Oliver, M.J., Desiccation tolerance in vegetative plant tissues and seeds: protein synthesis in relation to desiccation and a potential role for protection, in *Water and Life: Comparative Analysis of Water Relationships at the Organismic, Cellular, and Molecular Levels*, Somero, G.N., Osmond, C.B., and Bolis, C.L., Eds., Springer-Verlag, Berlin, 141, 1992.

26. Han, B. and Kermode, A.R., Dehydrin-like proteins in castor bean seeds and seedlings are differentially produced in response to ABA and water-deficit-related stresses, *J. Exp. Bot.*, 47, 933, 1996.

27. Jiang, Y. and Huang, B., Protein alterations in tall fescue in response to drought stress and abscisic acid, *Crop Sci.*, 42, 202, 2002.

28. Close, T.J., Dehydrins: Emergence of a biochemical role of a family of plant dehydration proteins, *Physiol. Plant.*, 97, 795, 1996.

29. Dure, L., Structural motifs in lea proteins, in *Plant Responses to Cellular Dehydration During Environmental Stress*, Close, T.J. and Bray, E.A., Eds., American Society of Plant Physiologists, 10, 91, 1993.

30. Nylander, M. et al., Stress-induced accumulation and tissue-specific localization of dehydrins in Arabidopsis thaliana, *Plant Molec. Bio.*, 45, 263, 2001.

31. Brini, F. et al., Cloning and characterization of a wheat vacuolar cation/protein antiporter and pyrophosphatase proton pump, *Plant Physiol. Biochem.*, 43, 347, 2005.

32. Leinauer, B. et al., Poa supina Schrad.: A new species for turf, *Int. Turfgrass Soc. Res. J.*, 8, 345, 1997.

33. Huang, B. and Gao, H., Root physiological characteristics associated with drought resistance in tall fescue cultivars, *Crop Sci.*, 40, 196, 2000.

34. Qian, Y., Irrigation frequency affects zoysiagrass rooting and plant water status, *HortScience*, 31, 234, 1996.

35. McAinsh, M.R., Brownlee, C., and Hetherington, A.M., Calcium ions as second messengers in guard cell signal transduction, *Physiol. Plant.*, 100, 16, 1997.

36. Qian, Y.L., Fry, J.D., and Upham, W.S., Rooting and drought avoidance of warm-season turfgrasses and tall fescue in Kansas, *Crop Sci.*, 37, 905, 1997.

37. White, R.H. et al., Competitive turgor maintenance in tall fescue, *Crop Sci.*, 32, 251, 1992.

38. Carrow, R.N., Drought avoidance characteristics of diverse tall fescue cultivars, *Crop Sci.*, 36, 371, 1996.

39. Sheffer, K.M., Dunn, J.H., and Minner, D.D., Summer drought response and rooting depth of three cool-season turfgrasses, *HortScience*, 22, 296, 1987.

40. Bonos, S. and Murphy, J.A., Growth responses and performance of Kentucky bluegrass under summer stress, *Crop Sci.*, 39, 770, 1999.

41. Duncan, R.R. and Carrow, R.N., Stress resistant turf-type tall fescue (*Festuca arundinacea* Schreb.): Developing multiple abiotic stress tolerance, *Int. Turfgrass Soc. Res. J.*, 8, 653, 1997.

42. Huang, B., Duncan, R.R., and Carrow, R.N., Root spatial distribution and activity for seven turfgrasses in response to localized drought stress, *Int. Turfgrass Soc. Res. J.*, 7, 681, 1997.

43. Huang, B., Duncan, R.R., and Carrow, R.N., Drought-resistance mechanisms of seven warm-season turfgrasses under surface soil drying: II. Root aspects, *Crop Sci.*, 37, 1863, 1997.

44. Huang, B., Water relations and root activities of *Buchloe dactyloides* and *Zoysia japonica* in response to localized soil drying, *Plant Soil*, 208, 179, 1999.

45. Huang, B. and Fu, J., Photosynthesis, respiration, and carbon allocation of two cool-season perennial grasses in response to surface soil drying, *Plant Soil*, 227, 17, 2000.

46. Davies, W.J., Tardieu, F., and Trejo, C.L., How do chemical signals work in plants that grow in drying soil? *Plant Physiol.*, 104, 309, 1994.

47. Davies, W.J., Wilkinson, S., and Loveys, B., Stomatal control by chemical signaling and the exploitation of this mechanism to increase water use efficiency in agriculture, *New Phytol.*, 153, 449, 2002.

48. Bohnert, H.J. and Jensen, R.G., Strategies for engineering water-stress tolerance in plants, *Trends Biotech.*, 14, 89, 1996.

49. Bray, E.A., Molecular responses to water deficit, *Plant Physiol.*, 103, 1035, 1993.

50. Ludewig, M., Dorffling, K., and Seifert, H., Abscisic acid and water transport in plants, *Planta*, 175, 325, 1988.

51. Zhang, J. and Davies, W.J., Increased synthesis of ABA in partially dehydrated root tips and ABA transport from roots to leaves, *J. Exp. Bot.*, 38, 2015, 1987.

52. Chen, J. et al., Involvement of endogenous plant hormones in the effect of mixed nitrogen source on growth and tillering of wheat, *J. Plant Nutr.*, 21, 87, 1998.

53. Quarrie, S.A., Understanding plant responses to stress and breeding for improved stress resistance-the generation gap, in *Plant Responses to Cellular Dehydration During Environmental Stress, Proceedings of the 16th Annual Riverside Symposium in Plant Physiology*, Close, T.J. and Bray, E.A., Eds., American Society of Plant Physiologists, 10, 1993.

54. Wang, Z., Huang, B., and Xu, Q., Effects of abscisic acid on drought responses of Kentucky bluegrass, *J. Am. Soc. Hort. Sci.*, 128, 36, 2003.

55. Wang, Z. and Huang, B., Genotypic variation in abscisic acid accumulation, water relations, and gas exchange for Kentucky bluegrass exposed to drought stress, *J. Am. Soc. Hort. Sci.*, 128, 349, 2003.

56. Renard, C. and Francois, J., Effects of increasing water stress on simulated swards of Festuca arundinacea Schreb. under wind tunnel conditions, *Ann. Bot.*, 55, 869, 1985.

57. Johnston, W.H., Koen, T.B., and Shoemark, V.F., Water use, competition, and a temperate-zone C4 grass (Eragrostis curvula (Schrad.) Nees. Complex) cv. Consol, *Aust. J. Agr. Res.*, 53, 715, 2002.

58. Fu, J. and Huang, B., Leaf characteristics associated with drought resistance in tall fescue cultivars, *Acta Hort.*, 661, 233, 2004.

59. Beard, J.B. and Sifers, S.I., Genetic diversity in dehydration avoidance and drought resistance within the Cynodon and Zoysia species, *Int. Turfgrass Soc. Res. J.*, 8, 603, 1997.

60. Beard, J.B., The water-use rate of turfgrasses, *TurfCraft Aust.*, 39, 79, 1994.

61. Bowman, D.C. and Macaulay, L., Comparative evapotranspiration rates of tall fescue cultivars, *HortScience*, 26, 122, 1991.

62. Ebdon, J.S. and Petrovic, A.M., Morphological and growth characteristics of low- and high-water use Kentucky bluegrass cultivars, *Crop Sci.*, 38, 143, 1998.

63. Kopec, D.M., Shearman, R.C., and Riordan, T.P., Evapotranspiration of tall fescue turf, *HortScience*, 23, 300, 1988.

64. Salaiz, T.A. et al., Creeping bentgrass cultivar water use and rooting responses, *Crop Sci.*, 31, 1331, 1991.

65. Shearman, R.C., Kentucky bluegrass cultivar evapotranspiration rates, *HortScience*, 21, 455, 1986.

66. Shearman, R.C., Perennial ryegrass cultivar evapotranspiration rates, *HortScience*, 24, 767, 1989.

67. Kjelgren, R., Rupp, L., and Kilgren, D., Water conservation in urban landscapes, *HortScience*, 35, 1037, 2000.

68. Rose, I.A., McWhirter, K.S., and Spurway, R.A., Identification of drought tolerance in early-maturing indeterminate soybeans (*Glycine max* (l.) Merr.), *Aust. J. Agr. Res.*, 43, 645, 1992.

69. Gimeno, V., Fernandez-Martinez, J.M., and Fereres, E., Winter planting as a means of drought escape in sunflower, *Field Crop Res.*, 22, 307, 1989.

70. Fry, J. and Huang, B., Drought, in *Applied Turfgrass Science and Culture*, Wiley, Hoboken, 2004, chap. 3.

71. Heckman, N.L., Horst, G.L., and Gaussoin, R.E., Influence of trinexapac-ethyl on specific leaf weight and chlorophyll content of *Poa pratensis*, *Int. Turfgrass Soc. Res. J.*, 9, 287, 2001.

72. Stier, J.C. and Rogers, J.N., Trinexapac-ethyl and iron effects on Supina and Kentucky bluegrass under low irradiance, *Crop Sci.*, 41, 457, 2001.

73. Pannacci, E., Covarelli, G., and Tei, F., Evaluation of trinexapac-ethyl for growth regulation of five cool-season turfgrass species, *Int. Soc. Hort. Sci.*, 661, 349, 2004.

74. Ervin, E.H. and Koski, A.J., Trinexapac-ethyl effects on Kentucky bluegrass evapotranspiration, *Crop Sci.*, 41, 247, 2001.

75. McCann, S. and Huang, B., Effects of Trinexapac-Ethyl on Creeping Bentgrass Tolerance to Combined Drought and Heat, unpublished data, 2006.

76. Heckman, N.L., Gaussoin, R.E., and Horst, G.L., Multiple trinexapac-ethyl applications reduce Kentucky bluegrass sod storage temperatures, *HortTechnology,* 11, 595, 2001.

77. Costa, G. et al., Incidence of scab (*Venturia Inaequalis*) in apple as affected by different plant growth retardants, *Acta Hort.*, 653, 133, 2004.

78. Maxson, K.L. and Jones, A.L., Management of fire blight with gibberellin inhibitors and SAR inducers, *Acta Hort.*, 590, 217, 2002.

79. Wang, Z., Huang, B., Bonos, S.S., and Meyer, W.A., Abscisic acid accumulation in relation to drought tolerance in Kentucky bluegrass, *HortScience*, 39, 1133, 2004.

80. Gaxiola, R.A. et al., Drought- and salt-tolerant plants result from overexpression of the AVP1 H$^+$-pump, *PNAS*, 98, 11, 444, 2001.

81. Agboma, P.C. et al., Effect of foliar application of glycinebetaine on yield components of drought-stressed tobacco plants, *Exp. Agr.*, 33, 345, 1997.

82. Makela, P. et al., Effect of foliar applications of glycinebetaine on stress tolerance, growth, and yield of spring cereals and summer turnip rape in Finland, *J. Agron. Crop Sci.*, 176, 223, 1996.

83. Diaz-Zorata, M., Fernandez-Canigia, M.V., and Grosso, G.A., Applications of foliar fertilizers containing glycinebetaine improve wheat yields, *J. Agron. Crop Sci.*, 186, 209, 2001.

84. Agboma, P.C. et al., Exogenous glycinebetaine enhances grain yield of maize, sorghum, and wheat grown under two supplementary watering regimes, *J. Agron. Crop Sci.*, 178, 29, 1997.

85. Saneoka, H. et al., Nitrogen nutrition and water stress effects on cell membrane stability and leaf water relations in *Agrostis palustris* Huds., *Environ. Exp. Bot.*, 52, 131, 2004.

86. Wang, Z. and Huang, B., Genotypic variation in abscisic acid accumulation, water relations, and gas exchange for Kentucky bluegrass exposed to drought stress, *J. Am. Soc. Hort. Sci.*, 128, 349, 2003.

87. Stahnke, G.K., Evaluation of antitranspirants on creeping bentgrass (Agrostis palustris Huds., cv. 'Penncross') and bermudegrass (Cynodon dactylon (L.) Pers. X Cynodon transvaalensis Burtt-Davy, cv. 'Tifway'), PhD thesis, Texas A and M University, College Station, TX, 1981.

88. DaCosta, M., Wang, Z., and Huang, B., Physiological adaptation of Kentucky bluegrass to localized soil drying, *Crop Sci.*, 44, 1307, 2004.

89. Biran, I., Bravdo, B., Bushkin-Harav, I., and Rawitz, E., Water consumption and growth rate of 11 turfgrasses as affected by mowing height, irrigation frequency, and soil moisture, *Agron. J.*, 73, 85, 1981.

90. Jordan, J. et al., Effect of irrigation frequency on turf quality, shoot density, and root length density of five bentgrass cultivars, *Crop Sci.*, 43, 282, 2003.

91. Qian, Y. and Fry, J., Irrigation frequency affects zoysiagrass rooting and plant water status, *HortScience*, 31, 234, 1996.

92. Richie, W. et al., Tall fescue performance influenced by irrigation scheduling, cultivar, and mowing height, *Crop Sci.*, 42, 2011, 2002.

93. Fu, J., Fry, J., and Huang, B., Water savings and performance of four turfgrasses under deficit irrigation, in Kansas State University Research and Extension, Report of Progress, 894, 1, 2002.

94. Qian, Y. and Engelke, M., Performance of five turfgrasses under linear gradient irrigation, *HortScience*, 34, 893, 1999.

95. Peacock, C.H., Irrigation requirements for turf establishment under supraoptimal temperature conditions, *Int. Turfgrass Soc. Res. J.*, 9, 900, 2001.

96. Mujeriego, R. et al., Agronomic and public health assessment of reclaimed water quality for landscape irrigation, *Water Sci. Tech.*, 33, 335, 1996.

97. Beltrao et al., Grass response to municipal waste water as compared to nitrogen and water application, in *Improved Crop Quality by Nutrient Management*, Anac, D. and Martin-Prevel, P., Eds., Kluwer Academic Publishers, The Netherlands, 1999, chap. 60.

98. Hayes, A.R., Mancino, C.F., and Pepper, I.L., Irrigation of turfgrass with secondary sewage effluent: II. Turf quality, *Agron. J.*, 82, 939, 1990.

99. Parnell, J.R., Irrigation of landscape ornamental using reclaimed water, *Proceedings of the Florida State Horticultural Society*, 101, 107, 1988.

100. Boss, S., Good natured golf, *Audubon*, 96, 20, 1994.

28 Shade Stress and Management

John Clinton Stier and David S. Gardner

CONTENTS

28.1 ORIGIN AND ADAPTATION OF TURFGRASSES TO SHADE

Grasses, members of the order Poales, are thought to have evolved during the upper Cretaceous (roughly 60 million years ago) in Gondwanaland [1]. Although the true evolution of grasses is unknown, their ancestors may have been from the order Commelinales [2]. These ancestral species perhaps were somewhat similar in appearance to the *Tradescantia* or *Zebrina* species that are common as houseplants today. The earliest grasses were probably herbaceous, rhizomatous, and broad-leaved inhabitants of tropical or subtropical forests or forest margins and were adapted to lower levels of light found in understory canopies [1].

The genera containing the species we use as turfgrasses evolved from these primitive forest-dwelling plants around 30–45 million years ago. During the course of this evolution, several changes were necessary to make grasses capable of withstanding savannah environments. Among these changes was a reduction in the morphology of the flowers, from showy petaled flowers pollinated by insects seen in the Commelinales to the reduced wind-pollinated flowers we see within the Poales. Routine defoliation and traffic from grazing animals conferred the traffic and mowing tolerances characteristic of turfgrasses [3,4].

The other major change was in the adaptation of the grasses to the open sunny environments of savannahs. Changes in the architecture of the leaf were necessary to both reduce the amount of transpiration, thus conserving water, and to deal with the much greater amount of photosynthetic photon flux density (PPFD) seen in a savannah environment. These changes are thought to have occurred over millions of years as the grasses became adapted to open sunny environments. Today, when we attempt to manage turfgrasses in shaded environments, we are, in a sense, placing those grasses back into the light environment from which they have been evolving away for millions of years.

Beard [4] estimated that 20–25% of turfgrass use is in shaded environments. Increasing areas of sports turf are being subjected to detrimental shading levels through the use of fully and semi-enclosed stadium structures [5,6]. Unlike a production agriculture system in which plants are space-planted to maximize yield, turf culture attempts to maximize density to produce a contiguous plant community, which in turn exacerbates competition for light in shaded environments [7].

Shade affects both the intensity and quality of light, which results in morphological and physiological changes to the turfgrass [8–10]. Many of these changes occur within 4–7 days of the onset of shade stress [11,12]. Few of the species we use as turfgrass are adapted to growth in shade [13]. Thus recommending a single turfgrass species, cultivar, or blend for a particular shaded location and the subsequent management of shaded turf is a challenge for turf professionals [14].

28.1.1 Shade Microclimate

The two most significant differences between a sunny and a shaded environment are the reduction in PPFD (light quantity) and the reduction in the red:far-red ratio (R:FR, or light quality) in shade. The R:FR ratio is the ratio of photon flux of light in the 10-nm band around 660 nm (red light) to light in the 10-nm band around 730 nm (far-red light) [15]. There are, however, other differences in the microclimate of a shade environment that contribute to turf decline. Compared with open, full-sun environments, shaded environments have moderated temperatures, reduced wind speeds, higher relative humidity, and an increase in duration of leaf wetness. Each of these factors is implicated in increasing the frequency and severity of disease as well as moss and algae invasion [16,17].

28.1.1.1 Moderated Temperatures

Daily temperature fluctuation and temperature extremes are moderated in shade, which may contribute to a more favorable climate for development of certain turfgrass pathogens [18]. Bell and Danneberger [9] found that canopy temperatures of turf in morning shade were on average 3°C cooler than air temperature. Turf in afternoon shade was on average 6°C cooler than air temperature. However, the authors found that canopy temperature was not a component of shade stress when comparing morning and afternoon shade. Koh et al. [19] found that both canopy and soil temperatures were lower in shade versus sun. Blad and Lemeur [20] observed a 1–3°C decrease in canopy temperatures of field crops grown under artificial shade versus full sun. Geisler et al. [21] found that maximum temperatures in a tall fescue foliage under 70% shade cloth were 3–5°C lower than in full sun foliage. The minimum temperature was 1.0–1.5°C less than in full sun. In addition to moderating the air temperature, soils in a shaded environment tend to remain cooler and more moist for a longer period of time and may not reach maximum levels until mid to late summer [22].

The combined effects of reduced light and temperature moderation have not been extensively studied. However, Stanford et al. [11] found that temperature and light level regulated the expression of dwarfism in "Tifdwarf" bermudagrass (*Cynodon dactylon* [L.] Pers. × *C. transvaalensis* Burtt-Davy) when grown for extended periods with daytime temperatures <28°C and PPFD <600 μmol m^2 s^{-1}. They further note that many of the changes observed occurred within 3–4 days of treatment initiation.

28.1.1.2 Reduced Wind Speeds

Wind movement decreases under tree canopies [23]. The reduction in wind speed, or airflow, in a shaded environment has several detrimental effects. Shading reduces evapotranspiration (ET) in a linear fashion [24]. During hot weather, reduced evapotranspirational cooling may increase turf temperatures, which could enhance respiration and photorespiration rates. Koh et al. [19] found turf canopy temperatures to be slightly greater when airflow was restricted, although soil temperatures were not affected. Decreased ET rates could presumably result in greater soil moisture in shade though data from turf situations are lacking. Koh et al. [19] found that no difference in soil moisture existed among creeping bentgrass putting greens under 80% shade compared to full sun, but irrigation may have interfered with soil drying rates. Restricted airflow was more detrimental to turf density of "L-93" and "SR-1020" creeping bentgrass (*Agrostis stolonifera* L.) than shade. Brown patch (*Rhizoctonia solani* Kühn) and dollar spot (*Sclerotinia homeocarpa* F.T. Bennett) diseases were enhanced more by restricted airflow than by shade [19].

28.1.1.3 Relative Humidity and Leaf Wetness

In shade, there is typically an increase in relative humidity [25]. Because of the reduction in light intensity and wind speed and the increase in relative humidity in shade, the amount of time the leaf remains wet is greater compared with grass in full sun [16]. In one study, leaf wetness duration was 1.5 h longer and the duration of time in which relative humidity exceeded 90% was longer in a shaded tall fescue (*Festuca arundinacea* Schreb.) canopy compared with a canopy in full sun [21]. These differences were of relatively low magnitude but are thought to contribute to microclimate changes within the shade environment, which contribute to an increase in disease activity.

28.1.2 The Light Environment in the Shade

More significant than microclimate factors such as increased humidity and leaf wetness are the changes in the both the quantity, or PPFD, and quality of light available for photosynthesis in a shaded environment. Growth and development of turfgrasses is greatly influenced by both the quality and the quantity of light available for photosynthesis [10]. Cool-season turfgrasses may survive in as little as 5% of full sunlight [16]. However, depending on species and management, 25–35% of sunlight is required for normal turf growth [26]. In environments with reduced PPFD, the quality of light may become the most important factor [10,12].

During photosynthesis, light energy from the sun is converted into chemical energy by photoautotrophic plants. Light energy can be expressed both as particle energy and as wave energy. The light spectrum is divided into wavelengths. As the wavelength of light increases, the total energy decreases. Wavelengths of light between 400 and 700 nm are referred to as photosynthetically active radiation (PAR) and are that portion of the solar spectrum visible to the human eye [27]. Wavelengths below 300 nm lie within the ultraviolet portion of the spectrum, while wavelengths between 700 and 3000 nm are referred to as the infrared portion of the spectrum [8,32]. Approximately

1–5% of the total solar energy reaching earth is utilized for photosynthesis [28]. PPFD, when used in association with plant growth, is a measure of the number of photons occurring between wavelengths 400 and 700 nm and the current SI standard is micromoles per square meter per unit time (μmol m^{-2} time^{-1}). Historically, though, various units have been used for monitoring turf response to light. Many of the older devices measured wavebands greater than PAR, sometimes undefined, providing a variety of units including photometric units of footcandles (ftc), kilolux (klux), and radiometric units (energy) such as watts per square meter or Langley [29]. Though equations designed to estimate PAR from non-SI measurements have been published [29], we have used the originally published units to ensure accuracy except for μE, which is equivalent to the modern μmol. A good estimate of PAR when only solar radiation (SR) is known is to multiply SR by 0.45 [30].

Chlorophyll *a* selectively absorbs light within the visible spectrum, peaking at 410, 430 (blue light), and 660 nm (red light). Absorption by chlorophyll *b* peaks at 430, 455, and 640 nm [31]. While the energy associated with blue light is greater than that of red light, this extra energy is lost to radiationless transfer during photosynthesis [32].

Light in the range 500–600 nm is referred to as the green wavelengths of light and are not especially absorbed by chlorophyll for photosynthesis but instead are reflected, giving plants their green color. Far-red light is the region of the spectrum between 700 and 800 nm. While not absorbed by chlorophyll for photosynthesis, it has a strong influence on photomorphogenesis, which is a term that describes how light influences plant architecture [10,33].

28.1.2.1 Light Quantity

A variety of factors affect the amount of sunlight available for photosynthesis including time of day, cloud cover and thickness, and latitude [8]. As the latitude increases, the seasonal fluctuation in light levels also increases [34]. The PPFD on a clear day at solar noon in the middle of the United States during the summer solstice is approximately 1900 μmol m^{-2} s^{-1} [35] while the diurnal variation resembles a bell curve (Figure 28.1). Light available for photosynthesis in the shade of a deciduous tree can be as little as 1–5% of this amount [35,36]. As latitude increases from the equator, the total daily PPFD decreases as time moves away from the summer solstice (Figure 28.2).

The rate of CO_2 uptake by plants generally increases proportionate to the amount of light. The intensity of light at which the rate of CO_2 uptake equals the rate of CO_2 release by respiration is referred to as the light compensation point. Therefore, the light compensation point of a plant greatly influences its ability to adapt to shade. Leaves of plants that are not adapted to shaded environments

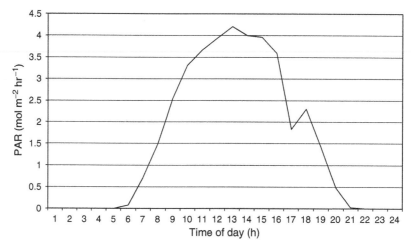

FIGURE 28.1 Diurnal variation of PAR, May 28, 2006, Madison, WI. The dip at 1700 h was due to cloud cover.

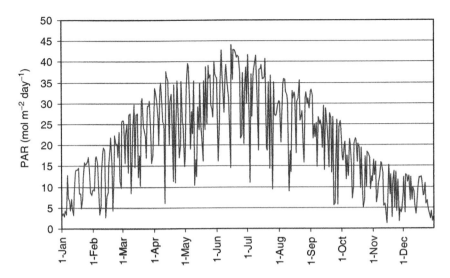

FIGURE 28.2 Annual variation of PAR, Madison, WI, 2005.

have light compensation points typically between 10 and 20 μmol PAR m^{-2} s^{-1} while shade-adapted plant leaves have light compensation points as low as 1 μmol PAR m^{-2} s^{-1} [32].

Several factors affect the amount of light required for growth and development, including species of plant, temperature, and CO_2 level [37]. Cool-season turfgrasses generally reach light saturation at 50% of full sunlight, while most warm-season grasses require full sunlight to reach light saturation [38].

A few studies have attempted to determine the minimum amount of light required to sustain turf, though the value clearly depends on a number of factors including turf type, cutting height, and traffic. Most studies report radiant energy as instantaneous measurements (e.g., μmol m^{-2} s^{-1}) or percent of full sunlight. Normal solar radiation varies throughout the day, however, with much of the irradiance within several hours of solar noon possibly excessive for photosynthesis. Information on the actual daily PPFD requirement would be useful for maintaining turf in consistently shaded conditions so that managers can accurately assess the effects of tree removal, roof openings, or supplemental irradiance. Light sensors capable of measuring PPFD up to full sunlight are now readily available, inexpensive, and simple enough for most educated turf managers to use.

28.1.2.1.1 Cool-Season Turfgrasses

Goss et al. [39] concluded that 60% shade had negligible effects on creeping bentgrass maintained at putting green height, while 80% shade was too severe for acceptable putting green turf. Rogers et al. [40] found approximately 30% of the radiant energy of full sun, achieved by high-pressure sodium lamps and sunlight filtered through Sheerfill™ fiberglass fabric (Chemical Fabrics, Buffalo, NY), necessary to maintain Kentucky bluegrass (*Poa pratensis* L.) turf using normal maintenance procedures. Cockerham et al. [41] estimated that Kentucky bluegrass could be maintained at <11.1 mol PAR m^{-2} d^{-1} but perennial ryegrass required more than 20 mol PAR m^{-2} d^{-1}. Stier et al. [5] found that approximately 9 mol PAR m^{-2} d^{-1} was just insufficient to maintain Kentucky bluegrass during a 4-month period without using gibberellic acid (GA) inhibitors. Wherley et al. [10] managed tall fescue in 92% shade caused by either deciduous trees or artificial shade structures. It was found that grass growing in perpetual 92% artificial shade maintained acceptable to good quality, color, and density while turf growing under 92% tree shade displayed many of the adverse physical changes associated with the decline of shaded turf. In the second year of their study, the tall fescue grown under 92% tree shade did not maintain adequate quality, but the tall fescue grown under artificial shade did maintain quality, color, and density. Gardner and Wherley [15] also grew acceptable tall fescue in 92% tree shade. However, Sheep fescue (*Festuca ovina* L.) and rough bluegrass (*Poa trivialis* L.) performance was marginal. Ultimately, the amount of light needed for turf growth depends on the amount of light available as a function of time, since high amounts of PAR for a short time in excess of that utilized for photosynthesis and development are not as important as having a sufficiency over a longer time. Bell and Danneberger [9] noted that "Penncross" creeping bentgrass grown in full sun for 40% of the day was able to maintain sufficient quality. They also observed that turfgrass receiving up to 6 h of shade per day did not differ from turf grown in full sun. Perpetual shade caused the turf quality to decline but 80–100% shade for up to 6 h per day did not have an effect on quality.

28.1.2.1.2 Warm-Season Turfgrasses

McBee and Holt [26] reported that a minimum of 25–35% incident light was required for optimum turf quality of "No-mow" bermudagrass. In actuality, both the duration and the time of day of shading impacts the quality of warm-season turfgrasses. Bunell et al. [42] found that "Tifeagle" bermudagrass could tolerate as little as 8 h of full sunlight and still maintain acceptable quality. However, bermudagrass that received 4 h of sunlight only maintained acceptable quality if treated with the GA inhibitor trinexapac-ethyl at 3-week intervals when mowed at 4.7 mm. Bunnel et al. [43] reported quality of Tifeagle bermudagrass significantly declined between approximately 42 and 38 mol PPFD m^{-2} d^{-1} but acceptable turf quality was maintained above 28 mol PPFD m^{-2} d^{-1} depending on the timing of diurnal shading. Afternoon shade was more detrimental to bermuda-grass growth than was morning shade. This may have been due to decrease in light intensity in the afternoon shade (26 mol m^{-2} d^{-1}) as compared to the morning shade (28 mol m^{-2} d^{-1}).

"Sea Isle 1" seashore paspalum (*Paspalum vaginatum* Swartz.) maintained similar quality to grass in full sunlight when grown under 5 h of either morning or afternoon 90% shade. However, when subjected to traffic stress, the shaded turf declined considerably [44]. Jiang et al. [45] found that many cultivars of seashore paspalum performed better in shade compared with bermudagrass. Miller et al. [46] found that "Floradwarf" and Tifdwarf bermudagrass cover under 4 h of 63% shade declined to 93%, but under 6 h of 63% shade both cultivars declined to 71% shade cover. Intensity also affected performance of the cultivars. Under 12 h of 30% shade, bermudagrass cover was 91%. However, under 12 h of 63% shade, bermudagrass cover declined to around 55%. Miller et al. [46] concluded that Floradwarf and Tifdwarf bermudagrasses need at least 45.6 mol PAR m^{-2} d^{-1}. Ultimately, mea-surements and models will need to account for diurnal and three-dimensional variations across and within the turf canopy to determine the amount of PPFD required for various turf situations.

28.1.2.2 Light Quality

The quality of light, as it relates to plant growth and development, refers to the ratio of red light (R; 600–700 nm) to far-red light (FR; 700–800 nm). Comparisons of R:FR are typically made at 660:730 nm. The ratio of blue (400–500 nm) to red light may also be important. While the spectral distribution of sunlight varies with geographic location and time of day, it is not affected by changes in climate or cloud cover [47]. The mean daily R:FR ratio is about 1.15 regardless of time of year [47]. At dusk, the R:FR ratio declines from 1.15 to 0.7 [48].

The turfgrass plant's perception of the R:FR ratio is an important aspect of shade acclimation [47]. There is a difference in growth and development as well as plant metabolism between turf grown in natural sun light, tree shade, and building shade [10,49]. Under dense deciduous shade, blue irradiance is reduced more than red irradiance [12]. Bermudagrass grown under red light had reduced quality compared with when grown under blue light.

Changes in spectral quality occur between morning and afternoon; however, the total amount of photosynthetically active red + blue light is the same in the morning sun and afternoon sun [49]. The amount of red light likewise did not change significantly between morning and afternoon. The ratio of blue light (400–500 nm) to total photosynthetic photon flux plus far red light (400–800 nm) was lower in the morning than in the afternoon, while the ratio of far-red light (700–800 nm) to total PAR was higher in the morning and lower in the afternoon.

Several studies have been conducted to determine the effect of foliage and structural shade on the R:FR ratio of light reaching the turfgrass canopy [9,49–51]. Buildings and many artificial covers do not selectively absorb light [10,49,52]. Wherley et al. [10] reported an R:FR ratio of 1.02 in artificial shade compared with 0.43 in deciduous shade. Tree leaves of both coniferous and deciduous species alter the R:FR ratio of light by selectively absorbing red and blue light while reflecting or transmitting proportionately more green and far-red light [10,40]. R:FR ratios in deciduous foliage shade range between 0.36 and 0.97 [53]. Sweet gum (*Liquidamber styraciflua* L.) transmitted fivefold less FR than Norway maple (*Acer platanoides* L.) when shading effects of brown patch disease (*Rhizoctonia* spp.) on tall fescue were studied [54]. However, there is some debate over whether deciduous or coniferous shade results in lower R:FR ratios. Beard [4] stated that R:FR is more severely reduced under deciduous shade, while Bell et al. [49] reported an R:FR ratio of 0.91 in deciduous tree shade and 0.80 in coniferous tree shade.

28.1.2.3   Morning versus Afternoon Shade

The effects of the duration and timing of shade on cool-season turfgrass are somewhat unclear. Golf-course superintendents have routinely reported that turf receiving shade only in the afternoon grows better than turf shaded in the morning. The decrease in morning sunlight further results prolonged periods of leaf wetness, which increases disease incidence. Additionally, higher afternoon temperatures when the turf is in full sun could increase respiration and photorespiration, leading to insufficient carbohydrates for desirable growth and development. Bell and Danneberger [9] tested morning and afternoon shade effects on Penncross creeping bentgrass using vertically oriented shade cloths to provide 6 h of shade (40% of the day) in either the morning or afternoon. They found that there were no significant differences in turfgrass growth whether subjected to morning or afternoon shade. Turf canopy temperatures differed only by 2–3°C between plots shaded in the morning and those shaded in the afternoon. Unlike full-day shading, temporal shading did not affect any of the turf parameters tested (color, density, root mass, total nonstructural carbohydrates (TNCs), or pigments). The lack of differences may have been due to sufficient PPFD as the part-time shading still provided >30% of full sun during the shading period, which may be insufficient to cause shade stress [4,39]. Unlike most naturally shaded conditions air movement was not restricted, which has been shown to significantly affect turf growth [19]. Diseases were controlled by fungicides and there other common stresses (e.g., traffic, heat, tree-root competition) were absent. Traffic will negatively impact turf quality under situations of moderately reduced PPFD, which is otherwise insufficient to reduce turf quality [44].

Other plant research indicates that light incites a photosynthetic circadian clock at the beginning of a diurnal period. The D1 protein of photosystem II in plant nuclei binds P when stimulated to light, with the process peaking approximately 4 h after exposure to light [55]. The process may cause adjustments in plant metabolism that repress negative responses to excess solar radiation near solar noon [56]. Additional research may be warranted on the effects of morning and afternoon shade, utilizing shading of >70% daily PPFD or the imposition of other stresses (e.g., temperature or traffic). C_4 grasses may actually respond differently to shade as Jiang et al. [44] reported that seashore paspalum actually suffered less from simulated sports traffic when exposed to morning shade compared with afternoon shade. The authors speculated that the potential changes in blue:far-red wavelengths and improved moisture availability in the afternoon due to morning shade's reduction of evapotranspiration may have positively influenced seashore paspalum growth and development.

28.1.2.4   Photomorphogenesis

Photomorphogenesis refers to the nonphotosynthetic influence of light on the growth and development of plants and is governed by the R:FR ratio [57]. Many plant responses, including germination, plant form and orientation, chloroplast development, cell elongation, carotenoid biosynthesis, tillering, and reproduction are influenced by the plant's perception of the relative amounts of blue, red, and far-red light [58,59]. Altered R:FR, either due to trees or other grass plants in the stand, is an important regulator of tillering [9,60].

Most photomorphogenic responses are controlled by the phytochrome system, which provides plants with a mechanism for adapting to changes in the light environment [60]. Phytochrome is a photoreceptor that consists of a protein and an attached chromophore. Two isomeric forms of phytochrome exist and are intraconvertible by red and far-red light [57]. Phytochrome is most sensitive to red light at 660 nm and to far-red light at 735 nm. Its conversion initiates a chain of metabolic and physiological changes within the plant that manifest as response by the plant to the light environment [61,62]. Plants that are not adapted to shaded environments commonly exhibit shade-avoidance symptoms, such as etiolation, when grown under the low R:FR ratios typical of shaded environments. The most dramatic shade-avoidance responses are usually observed in plants adapted to open environments, such as most of the species used for turfgrass [63].

Reductions in both the PPFD and the quality of light in a shade environment will impact growth and development of turfgrass. In tall fescue, Wherley et al. [10] found that the reductions in PPFD caused many of the photomorphogenic responses observed in shade. However, these changes were further influenced by light quality.

28.2 CHANGES IN TURF GROWTH IN SHADE VERSUS SUN

The negative effects of reduced irradiance have long been scientifically documented. Harrison [64] published the first modern comprehensive report on the impact of low irradiance on Kentucky bluegrass, showing that supplemental light in low irradiance was more important for growth than supplemental nitrogen. A few reports were published in the 1950s and 1960s, with increasing amounts in the 1970s. Shade research waned in the 1980s. Propelled by the utility of GA inhibitors for turf management, interest in covered sports stadiums, and increasing recognition of the importance of turf for urban landscapes, new research on shaded turf physiology and management has been published by several programs on both cool- and warm-season turfgrasses since the mid-1990s. Due to funding sources and amounts, turfgrass shade research has been largely limited to management techniques with most true physiology aspects extrapolated from other plant systems.

28.2.1 PHYSIOLOGICAL AND BIOCHEMICAL RESPONSES TO SHADE

Long-term plant survival depends on production of more photosynthate than is used for respiration, i.e., maintaining a positive photosynthesis:respiration ratio. Respiration utilizes photosynthate (nonstructural carbohydrates) throughout the year, day and night, even when a turf is "dormant" due to summer or winter stresses. Respiration rates are directly related to temperature and may exceed photosynthesis rates at excessive temperatures [32]. Photosynthesis occurs only during daytime and under conditions of sufficient temperature, which are usually slightly greater than the optimal temperatures for growth and development. The optimal temperature range for cool-season turfgrasses, which have a C_3 metabolism, is presumably 20–25°C [65] though photosynthesis can in some species function to near freezing. Warm-season grasses (C_4) have optimal photosynthesis between 30°C and 35°C. C_3 grasses have lower light compensation and saturation points than C_4 plants, which allows C_3 grasses on a whole to survive better in less light than most C_4 turfgrasses. Compensation and saturation points for individual leaves are different than those reported for turf swards due to interleaf competition for light and CO_2. Alexander and McCloud [66] reported that the light saturation point for individual bermudagrass leaves ranged from 2500 to 3000 footcandles (ftc), which is approximately 25–30% full sunlight. Conversely, the light saturation point for turf maintained at 20 cm height was 5000 ftc while turf maintained at 2.5 or 5 cm was not saturated at 7000 ftc, approximately 70% of full sunlight. Turf maintained at 20 cm actually had less interleaf shading as lower leaves died and leaf angles were more horizontal compared with lower heights of cut. Species and varieties also have different compensation and saturation points. Light saturation points for photosynthesis have been reported to be as low as 0.25% of full sunlight in C_3 grasses

TABLE 28.1
Light Compensation and Saturation Points for Turfgrasses

Grass	Situation	Growing Condition	Saturation Point	Compensation Point	Reference
Tall fescue	Individual leaves	30% full sun	1000–1200 μmol m^{-2} s^{-1}	—	[79]
		Full sun	1500 μmol m^{-2} s^{-1}	—	
KBG Merion	Individual leaves	43 klux	38 klux	4.2 klux	[86]
Red fescue Pennlawn			35 klux	4.0 klux	
KBG Merion		2.7 klux	18 klux	1.3 klux	
Red fescue Pennlawn			19 klux	1.5 klux	
KBG Merion	Swards	43 klux	>64 klux	6.0 klux	
Red fescue Pennlawn			>64 klux	6.2 klux	
KBG Merion		2.7 klux	43 klux	1.2 klux	
Red fescue Pennlawn			43 klux	1.35 klux	
Bermudagrass	Leaves	8 h 1540 μmol m^{-2} s^{-1} PPFD + 4 h 63% shade	—	265–282 μmol m^{-2} s^{-1} PPFD	[46]
Diamond zoysiagrass	Swards	86% shade	1231 μmol m^{-2} s^{-1} PPFD	270 μmol m^{-2} s^{-1} PPFD	[112]

Note: While the unit klux does not directly translate into the SI unit of μmol m^{-2} s^{-1}, full summer sunlight is approximately 2000 μmol PAR m^{-2} s^{-1} and 100 klux.

while C$_4$ grasses saturate at full sunlight [67]. The actual quantity of light required for achieving the compensation point and saturation depends on several factors. In general, leaves developed under low irradiance will have lesser compensation and saturation points than leaves developed under full solar irradiation (Table 28.1). As temperatures increase, C$_3$ plants may lose as much as 50% of their potential photosynthate production to a competing process termed photorespiration as ribulose-1,5-bisphosphate (RuBP) carboxylase/oxygenase binds oxygen instead of CO$_2$. C$_4$ plants avoid photorespiration by utilizing a different enzyme, phosphoenol pyruvate carboxylase, combined with a special intraleaf architecture known as the Kranz anatomy to more efficiently capture CO$_2$ at high temperatures [32]. Adverse temperatures during summer and autumn may exacerbate the ability of shaded turf to produce sufficient TNCs to use for respiration during the winter, though the necessary research to prove the point has not been conducted.

Part of the photosynthetic reduction in shade is linked to decreased amounts of enzymes used in photosynthesis. RuBP carboxylase declines as chloroplast stromal area and membrane content are decreased due to flattening of the plastids [68]. Leaves of annual ryegrass (*Lolium multiflorum* L.) grown at 75 μmol PAR m^{-2} s^{-1} fixed no more than 200 μg CO$_2$ m^{-2} s^{-1}, while leaves grown at 500 μmol PAR m^{-2} s^{-1} fixed approximately 500 μg CO$_2$ m^{-2} s^{-1} [68]. On a leaf-area basis, leaves developing under low irradiance had fewer cells and chloroplasts. Chlorophyll is subsequently less in shade than full sun at least on a leaf-area basis. Conversely, when assayed on a weight or volume basis, chlorophyll often increases, at least under moderate shade [10,69], due to differences in leaf morphology [70].

Loss of photosynthetic activity decreases the reserve of total TNCs more than structural carbohydrates. In a comparison of tall fescue growth rate and assimilate partitioning, TNC deposition was decreased 43% at 60 μmol PAR compared with 300 μmol, while dry-matter development decreased only 13% at low PPFD [71]. In bermudagrass, shading caused lignin content, P, Ca, Mg, and proteins to increase as a function of dry weight while cellulose remained constant and TNCs decreased [72].

Shading also stimulates the production of GA, which enhances cell enlargement without a concomitant increase in cell-wall thickness [73]. Affected plants have a weakened, etiolated growth habit (Figures 28.3 and 28.4). The development of GA inhibitors has profound implications for turf management, as discussed in Section 28.4.6.

FIGURE 28.3 **(Color Figure 28.3 follows p. 262.)** Cultivar Plantation tall fescue grown under different light treatments 5 weeks following seeding in Columbus, Ohio. Light intensity was the average of measurements at 15-min intervals. The R:FR ratio was measured using a spectroradiometer. (From Wherley, B.G., Turfgrass photomorphogenesis as influenced by changes in spectral composition and intensity of shadelight, Ph.D. thesis, The Ohio State University, Columbus, 2003. With permission.)

FIGURE 28.4 **(Color Figure 28.4 follows p. 262.)** Cultivar Plantation tall fescue grown under different light treatments 14 months following seeding in Columbus, Ohio. Light intensity was the average of measurements at 15-min intervals. The R:FR ratio was measured using a spectroradiometer. (From Wherley, B.G., Turfgrass photomorphogenesis as influenced by changes in spectral composition and intensity of shadelight, Ph.D. thesis, The Ohio State University, Columbus, 2003. With permission.)

28.2.2 Physical and Anatomical Responses to Shade

28.2.2.1 Cell/Leaf Development

The altered biochemical and physiological status of shaded turfs leads to differences in anatomy and the physical makeup compared with turf grown under optimal light conditions (Table 28.2). Leaves at the top of the plant usually receive more irradiance than leaves lower in the canopy and so respond differently based on the amount of irradiance received [74,75]. In a dense turf, much of the light impinging on the turf surface may be filtered or reflected before reaching lower leaves. Excessive shading reduces the number of leaves and tillers supported by single plants in addition to reducing plant density [10,19,40,45,76–78].

Stomatal density of C_3 turfgrasses is often decreased as PPFD declines though length remains unchanged [13,79]. Low PPFD also causes a dispersal of chloroplast grana [10] and reduction or loss of the palisade layer, presumably to maximize light absorption [70]. Varieties with more shade tolerance are less affected than shade-intolerant varieties. For example, Wherley et al. [10] showed the number of grana in a shade-tolerant cultivar of tall fescue ("Plantation") was not affected by shading while grana increased in a less shade-tolerant cultivar ("Equinox") as shading increased. Likewise, the ultrastructure of shade-tolerant "Pennlawn" red fescue did not change due to shading unlike shade-intolerant "Merion" Kentucky bluegrass [13,76]. Shade-intolerant grasses such as "Merion" Kentucky bluegrass suffered a decrease in vascular and support tissues under reduced light [13].

Leaf cuticle, epidermal layer, and overall leaf thickness of turfgrasses are reduced at low PPFD [10,13,71]. Intercellular space is typically increased in leaves subjected to suboptimal PPFD [10,79]. Specific leaf dry weight decreases at low irradiance as water content is increased [68], resulting in succulent leaves that are more prone to traffic damage, fungal pathogen ingress, and have less tolerance to environmental stresses such as drought or cold. The decrease in specific leaf dry weight and increased succulence is due to excessive elongation of leaf and internodes in both C_3 and C_4 grasses. Winstead and Ward [69] reported 75% shading increased leaf and internode elongation of bermudagrass and St. Augustinegrass (*Stenotaphrum secundatum* [Walt.] Kunz.) even though St. Augustinegrass was deemed shade tolerant compared with bermudagrass. In tall fescue, low irradiance stimulated 33% greater leaf elongation compared with high irradiance. Subsequent leaf area increases were codependent on changes in leaf width, which sometimes decreased at low PPFD, mitigating changes in leaf area [71,79]. Leaf and shoot elongation caused by shading is due to excessive production of GA [73] and possibly auxins. Production of these plant hormones is mediated by specific photoreceptors. Phototropins and cryptochromes respond to blue light and phytochromes

TABLE 28.2
Physical and Physiological Changes in Turf Subjected to Shade

Plant Basis		
Physical	**Physiological/Biochemical**	**Turf Basis**
Reduced tillering	Reduced pigments	Reduced rigidity
Thinner cuticle	Increased gibberellic acid	Decreased density
Thinner epidermis	Decreased evapotranspiration	Decreased traffic tolerance
Thinner leaves	Reduced nonstructural carbohydrates	Increased propensity for powdery
Increased leaf extension rate	Decreased light compensation/saturation	mildew, leaf spot, *Microdochium*
	points	patch, and brown patch diseases
Increased intercellular spaces		
Narrower leaves		
Reduced root mass		
Fewer stomata		

respond to red and far red light [80,81]. The natural dwarf, compact growth habit of "Nuglade" Kentucky bluegrass is due to inherently lesser GA_1 levels compared with other Kentucky bluegrass cultivars [73]. The Scotts Co. developed transgenic Kentucky bluegrass in the early 2000s with repressed GA biosynthesis and subsequent low growth though the product was not commercialized by 2006 (E. Nelson, personal communication).

28.2.2.2 Tiller and Root Production

The leaf area index (LAI) per plant is usually reduced as fewer tillers and leaves are produced and turf density declines as plants die from attrition. Individual plants or groups of plants may survive in sun fleck areas otherwise shaded by tree canopies [82]. Tiller production is usually reduced at low PPFD [5,10,71,79] though shade-tolerant grasses such as Pennlawn red fescue may maintain tiller density as light intensity decreases [76]. Low PPFD reduced TNC supplies in bermudagrass and zoysiagrass roots and rhizomes [42] and reduced stolon and rhizome numbers of St. Augustinegrass [83]. In addition to decreased TNC for growth and development, low R:FR ratios under tree shade may also be responsible for reduced tillering and leaf width [10]. Study conditions may affect results as fine fescue tiller production was decreased at moderate levels of supplemental light, supplied 24 h d^{-1}, in a greenhouse compared with higher intensity supplemental light while tiller production in ambient light was similar to both supplemental light conditions [77].

Root:shoot ratios for both C_3 and C_4 grasses decline when PPFD is growth-limiting [10,76,79,84] as shoots are stronger sinks for photosynthate. Loss of roots reduces the turf's ability to extract moisture and nutrients from the soil, requiring different management techniques for maintaining turf.

Ultimately, low PPFD and altered ratios of light wavelengths combine to produce weak, etiolated turf plants that are more subject to damage from traffic [44,85] and diseases [18] than turf provided optimal PPFD. Turf recovery rates in low PPFD are reduced compared with optimal PPFD conditions due to insufficient TNC, the lack of which restricts root development followed by tillering.

28.2.3 Characteristics of Shade-Tolerant Turfgrasses

Various characteristics may assist a given species or variety to tolerate shade. Ultimately, these characteristics will have a genetic basis that is expressed on either a physiological or anatomical level. Characteristics found or suspected to be beneficial for shade tolerance include greater photosynthetic rates [46], decreased dark respiration [86], improved resistance to specific diseases such as powdery mildew or leaf spot [16,18], decreased GA production [73], better carbohydrate allocation to roots, earlier spring green-up [22], and wide leaf angles [76]. Shade-tolerant Plantation had superior tillering compared with less shade-tolerant Equinox tall fescue when grown at 7% to 8% full sun [10]. Wilkinson and Beard [13] determined that superior cuticle, vascular and support tissue condition, and chloroplast ultrastructure helped Pennlawn red fescue tolerate shade better than Kentucky bluegrass. Leaf angles of Kentucky bluegrass decreased as irradiation diminished while Pennlawn leaf angles remained horizontal, presumably enhancing light gathering and minimizing loss of new tissue by mowing [76]. Further study indicated Pennlawn maintained a greater photosynthesis:respiration ratio by substantially reducing dark respiration rates at low irradiance compared with Kentucky bluegrass [86].

Jiang et al. [44] examined eight varieties of seashore paspalum and two varieties of bermudagrass maintained at 1.7 cm height and exposed to full, 30%, or 10% sunlight for up to 40 days. Apparent rates of canopy photosynthesis ranged from 28.7 to 35.8 μmol CO_2 m^{-2} s^{-1} with varietal differences greater than species differences. In one study, the authors were unable to determine why St. Augustinegrass had better shade tolerance than Tiflawn bermudagrass (*Cynodon dactylon* [L.] Pers.) as both species had similar net photosynthesis and dark respiration rates at 75% shade [69].

28.3 FACTORS AFFECTING THE LOSS OF TURFGRASS IN SHADE

Few species of turfgrass are adapted to growth in shade [14]. Selection of shade-tolerant species and cultivars and adjusting cultural practices are the two primary methods of managing turfgrass under shade. The environmental factors that contribute to the decline of turf in shade include competition from trees for water and nutrients, increased disease pressure due to microclimate differences, and reduction in quality and quantity of light available for photosynthesis [4].

28.3.1 TREE-ROOT COMPETITION

Tree-root competition has been implicated in the loss of turf in shade, even if water and nutrients are maintained at optimum levels for turfgrass growth [87,88]. Whitcomb [87] found that Kentucky bluegrass was more sensitive to tree-root competition compared with perennial ryegrass (*Lolium perenne* L.), rough bluegrass, and red fescue (*Festuca rubra* L.). In Whitcomb's study [87], red fescue was competitive with tree roots and grew equally well with or without tree-root competition. Gardner and Taylor [14] reported the performance of several species and cultivars of turfgrass 8 years after establishment under a dense tree canopy. In this study, there was no attempt to prune the roots of the trees. They found that certain cultivars of tall fescue were able to maintain and even improve in quality, despite the presumed competition from the trees for water and nutrients. Similar to previous findings [87], the authors found that the poorest performing species was Kentucky bluegrass [14].

The type of tree or shrub that is near the turfgrass can also affect turfgrass performance. Whitcomb and Roberts [88] found that Kentucky bluegrass is more severely restricted when grown in proximity to trees with shallow roots and trees that produce allelopathic compounds such as silver maple (*Acer saccharinum* L.) or honeylocust (*Gleditsia triacanthos* L.).

28.3.2 DISEASE ACTIVITY

The shade environment allows increased microbial activity and therefore increased frequency and severity of certain turfgrass diseases, most notably rust (*Puccinia* spp.), powdery mildew (causal agent *Erysiphe graminis* DC), leaf spot (causal agents *Bipolaris* and *Drechslera* spp.), brown patch, and *Microdochium* patch (causal agent *Microdochium nivale* [Fr.] Samuels & I.C. Hallett) [16,18,21,54,89]. Shading of the internal turf canopy can result in higher humidity, longer periods of leaf wetness, and moderated temperatures, which in turn lead to increased disease activity. Giesler et al. [21] found that as the density of a tall fescue canopy increased, moisture within the canopy and severity of brown patch also increased.

While the microclimate in shade makes conditions more conducive for disease activity, it is the many morphological and physiological modifications in the plant, such as the increased succulence of foliage and decreased carbohydrate reserves, that also act to make the plant less tolerant of disease. The epidermis is a single layer of cells on the outside of the plant leaf that protects plants from desiccation losses, herbivores, and plant pathogens. Tall fescue leaves exposed to full sun had thicker epidermal cell layers compared with leaves grown either in neutral or deciduous tree shade [90]. The epidermal thickness of leaves in neutral shade was significantly greater than those grown under tree shade [10]. In fact, leaves of tall fescue grown under tree shade had several breaks in the integrity of the epidermis, and it was speculated that this could result in much easier penetration by fungal hyphae into the leaves of grass under tree shade [90]. Grasses that are more tolerant of shade have structural features that increase resistance to infection, compared with less shade-tolerant grasses. Red fescue, for example, was found to have a thicker cuticle than Kentucky bluegrass, which can be severely limited by diseases in shaded environments [4,13]. Varietal differences to brown patch disease in the shade were presumably related to morphological and physiological effects of shading on tall fescue [54].

28.3.3 Reduced Light Quality and Quantity

Most turfgrass species are negatively affected when grown in fewer than 4–5 h of direct sunlight per day [26,39]. However, there are considerable differences in growth between cool-season turf species under moderate or dense shade [14,78]. Differences in the morphology and physiology of the panicoid and choridoid grasses generally result in decreased ability to adapt to shaded conditions compared with the pooid grasses. Warm-season grasses have higher light compensation points compared with cool-season grasses [32]. Additionally, the lower respiratory rates of cool-season grasses may further enable their survival in reduced light environments.

28.3.3.1 Cool-Season Turfgrasses

In most cool-season turfgrasses, considerable reductions in growth and turf quality are observed at less than 30% full sunlight. The lower limit of light necessary for maintaining a turfgrass sward varies by species and is greater than that needed for individual leaves or space plants. Beard [4] indicated that a minimum of 2–5% full sunlight is necessary to maintain turf but the time of year and amount of cloud cover confounds using a simple percentage of sunlight, as Kentucky bluegrass died at 2.5% sunlight in the northern hemisphere between December and July 1992–1993 [40]. Ultimately, information on the daily light integral of PAR is needed to define light requirements for different turf species as percent full sun is a nebulous descriptor. Cockerham et al. [41] determined that the threshold for Kentucky bluegrass was between 2.2 and 11.1 mol PAR m^{-2} d^{-1} while perennial ryegrass required between 11.1 and 20 mol PAR m^{-2} d^{-1}. Stier et al. [5] found that a minimum of 5.5–9 mol PAR m^{-2} d^{-1} was needed to maintain a Kentucky bluegrass sward mowed at 3 cm height.

Fescues are more shade tolerant than other commonly used cool-season turfgrasses [14]. Certain cultivars were even able to increase in visual quality over an 8-year period in moderate to dense shade (40–300 μmol m^{-2} s^{-1}). Other studies found that certain tall fescue cultivars could be maintained at acceptable quality in 92% tree shade [10,15]. Tegg and Lane [6] found that tall fescue produced high-quality turf under 56% and 65% shade.

In contrast, the Kentucky bluegrass and perennial ryegrass cultivars perform poorly in shade. Cultivars of the two species tested by Gardner and Taylor [14] had mean quality ratings of 1.9–2.9 (1, dead; 9, highest) after 8 years. Tegg and Lane [6] found that Kentucky bluegrass and perennial reyegrass had the greatest increase in vertical elongation when placed in shade.

Supina bluegrass (*Poa supina* Schrad.) is also capable of maintaining high-quality turf in moderate to dense shade [6,22,91]. Steinke and Stier [22] suggest that the earlier spring green-up observed with supina bluegrass, as compared with Kentucky bluegrass or creeping bentgrass, may result in comparatively greater annual photosynthate production, due to the increased time for photosynthesis early in the season before the tree leaves have fully expanded. This could lead to the increased shade tolerance observed in supina bluegrass. Further, supina bluegrass exhibits significant growth between autumn and spring, which could improve the species ability to recover from injury during late fall and early spring sporting events.

Differences in shade tolerance among species can affect population dynamics in a mixed stand. Stiegler et al. [92] conducted a study on creeping bentgrass that was shaded until 1 h after solar noon. In this study, annual bluegrass (*Poa annua* L.) growth and development was favored over that of creeping bentgrass. Quality of creeping bentgrass, when maintained under 60% artificial shade, was not significantly affected. Mixing a strongly shade-tolerant grass with those of lesser tolerance can improve the density of the turf sward, though the overall appearance will be primarily affected by the most dominant species, which can vary over time [16]. The relative shade tolerance of supina bluegrass allows it to outcompete other turfgrasses when planted as a mixture in moist, shaded conditions (J. Stier, personal observation).

28.3.3.2 Warm-Season Turfgrass

St. Augustinegrass is the warm-season turfgrass counterpart to the fescues with respect to shade tolerance. However, most warm-season turfgrasses, such as bermudagrass, require high irradiance to maintain a favorable carbon balance. Dudeck and Peacock [8] report that photosynthetic carbon production increases in bermudagrass as photosynthetic irradiance increases from 1782 to 2125 μmol m^{-2} s^{-1}. Thus, shade is implicated as a major factor limiting the growth of warm-season grasses such as bermudagrass and zoysiagrass (*Zoysia* spp.) [26,42,93,94]. Reduction in the growth of warm-season grasses occurs when the level of light decreases below 25 mol PAR per day.

There is considerable variation in shade tolerance among bermudagrass and zoysiagrass cultivars. "Celebration" bermudagrass maintained acceptable quality under 58% shade cloth, while Tifway and Tifsport maintained quality under a maximum of 41% shade cloth [42]. Bunnell et al. [43] determined that a minimum of 32.6 mol m^{-2} d^{-1} is required to maintain Tifeagle bermudagrass of acceptable quality. Miller et al. [46] found that Floradwarf bermudagrass has a slight advantage over Tifdwarf bermudagrass in reduced light because Floradwarf had 8–66% greater net photosynthetic rates in reduced light. Floradwarf also exhibits a more prostrate growth habit under moderate levels of shade. Within the bermudagrass cultivars, those with more prostrate growth habits are considered more shade tolerant [26].

Significant losses of Meyer zoysiagrass (*Zoysia japonica* Steudel) were observed on turf under 77–89% shade [95]. Qian and Engleke [96] found that "Diamond" zoysiagrass (*Z. matrella* [L.] Merrill) maintained acceptable quality when subjected to 73% shade cloth for 3 months but not under 86% shade. However, Riffell et al. [97] found that Diamond zoysiagrass performed better in shade versus other zoysiagrass cultivars.

28.3.4 Shading and Turf Establishment

Studies have shown that both turf and tree shade can impact seed germination and establishment [49,98]. Miltner et al. [99] studied establishment of *Poa annua* var. *reptans*, which is generally considered a relatively shade-tolerant turf species. Their work showed that seedling emergence occurred more quickly in the sun compared with shade. However, there were no differences in rate of cover due to shade. Similarly, Murphy et al. [100] found no differences between an open sun environment and one bordered by trees when establishing creeping bentgrass. Shading by established turfgrass plants can also affect establishment vigor during an overseeding operation. Zuk et al. [98] found that increasing light penetration in established perennial reyegrass, either by scalping or treating the ryegrass with glyphosate, resulted in increased germination of overseeded zoysiagrass.

Establishment vigor alone may not be the best indicator of future stand success in shade. Gardner and Taylor [14] report that best germination and cover under shade were observed in perennial ryegrass, fine fescue (*Festuca* spp.), and rough bluegrass. However, after 6 years the percent cover of perennial ryegrass had declined from an average of 75% to an average of 9%. "Laser" rough bluegrass declined from 58% to 13% ground cover.

28.3.5 Attrition Due to Reduction of Light Quality and Light Quantity

In a shaded environment, plants will exhibit either shade-tolerance responses or shade-avoidance responses. Shade avoidance is characterized by increased internode elongation and higher photosynthetic and respiration rates. Shade-tolerant species generally have reduced growth rates and lower respiration and stem elongation rates [48,90]. Most turfgrasses that adapt to shaded environments have shade-avoidance responses. However, there are differences in the shade-tolerance responses within the turfgrasses that exhibit shade tolerance [13].

Turfgrass stand density decreases over time in shaded environments because of the competition from tree roots, increased disease pressure, and reduced light quality and quantity. Plants are more likely to die and less likely to be replaced than those in optimal sun conditions. Several factors contribute to the overall decline including low TNC reserves. This is thought to reduce the turf's ability to withstand or recover from stresses or injuries. Shade rarely kills turfgrasses quickly; rather, turf density thins over time as individual tillers or plants die and are not replaced due to insufficient TNC. Occasionally, severe climatic conditions or other stresses may kill a large percentage of the plants over several weeks.

Turfgrass response to tree shade is a combination of the effects of reduced R:FR and reduced PPF [10,90]. Shade-tolerant species, when grown in neutral shade that has similar R:FR as full sunlight, will develop similarly to a shade-tolerant plant that is grown in deciduous shade [48]. However, shade-intolerant species may not develop similarly when neutral and non-neutral shade conditions are compared. Many of the responses of turf to shading are controlled by phytochrome reacting to altered R:FR light, indicating that caution must be exercised if using artificial shade structures to simulate shade when investigating potential management strategies for tree-shaded turfgrass [10,15,96]. Consequently, increasing the quantity of light under the canopy by selective pruning may not be as effective as previously thought. For example, Wherley et al. [10] found that tall fescue could maintain acceptable density in 92% neutral shade but quality declined considerably in 92% tree shade primarily due to phytochrome-mediated plant responses. The authors speculated that selective pruning in this case would increase the quantity of light, but would not have ameliorated the quality of the light and thus the morphological changes that lead to turf decline in shade would still have been present. Light quality can also be decreased as red and blue light are selectively absorbed by neighboring turfgrass foliage, which results in changes in the growth habit of the grass, compared with growing individual plants in full sunlight [90]. Self-shading by turfgrasses can cause longer leaves and a decrease in tillering [101]. Information on the amount of PPFD needed to overcome the impact of reduced light quality is needed.

Several turfgrass that may have acceptable quality when grown under shade are often placed in management situations that exacerbate the problems posed by the low light environment. For example, the fine fescues and rough bluegrass are typically considered tolerant of shade. However, they can not withstand the combination of a shaded environment, traffic, and low heights of cut, such as would be present if used on a golf-course tee [102]. Supina bluegrass has good shade tolerance, even under traffic and when mowed at tee height [22,89].

There are several methods of ameliorating a shade environment to favor turfgrass survival, including addition of light and increasing air movement. In the next section, several management practices that have been investigated for improving the quality of shaded turfgrass are reviewed.

28.4 MANAGING TURF IN THE SHADE

A number of factors influence the ability to grow and manage a turf in shade. As discussed previously, many biotic and abiotic variables combine to reduce turf's ability to survive in shade. Depending on the turf performance expectations and severity of conditions, some of or all the factors need to be addressed to maintain turf in the shade. Management of shaded turf is inherently different than turf managed under full sun.

28.4.1 TURFGRASS SELECTION

Turfgrass selection is often a critical element in success for a grass to survive shade as some grasses are more shade adapted than others. Table 28.3 gives the relative shade tolerance of common turfgrass species as determined from the literature and personal observations of the authors. Unfortunately, no single study has compared the relative shade tolerance of all turfgrasses due to the wide variation of climatic and maintenance requirements. In addition, individual cultivars can differ greatly from the mean shade tolerance of a species [14,45]. Diamond zoysiagrass has excellent shade tolerance compared with most zoysiagrass cultivars [97]. St. Augustinegrass has excellent shade tolerance with "Floratam" listed as having poor shade tolerance in some situations [103,104] but not in others [83].

TABLE 28.3
Relative Shade Tolerance of Cool- and Warm-Season Turfgrass Species

Species	Poor			Fair			Good			Excellent		
Cool-Season												
Annual bluegrass										X	X	X
Supina bluegrass										X	X	X
Tall fescue							X	X	X	X	X	X
Fine fescues							X	X	X	X	X	X
Velvet bentgrass							X	X	X	X		
Creeping bentgrass							X	X	X			
Rough bluegrass							X	X	X			
Perennial ryegrass			X	X	X	X						
Kentucky bluegrass	X	X	X	X	X	X						
Warm-Season												
St. Augustinegrass										X	X	X
Zoysiagrass							X	X	X			
Seashore paspalum							X	X	X			
Bahiagrass				X	X	X						
Centipedegrass				X	X	X						
Carpetgrass				X	X	X						
Bermudagrass	X	X	X	X	X							
Buffalograss	X	X	X									

Note: Individual cultivars can differ significantly from the species.

Gaussoin et al. [105] concluded that sufficient variation existed among bermudagrass clones to develop shade-tolerant varieties. Disease resistance to powdery mildew, leafspot, and other diseases may be important for allowing certain cultivars of Kentucky bluegrass to survive shaded conditions [18,93]. Climatic conditions can also affect the relative shade tolerance of species [14]. For example, Gardner and Taylor [14] found tall fescue to be superior under dense shade (72% cover) compared with other cool-season turfgrasses, with rough bluegrass and perennial ryegrass maintaing ≤10% cover after three years while Kentucky bluegrass maintained 50% cover. They concluded that temperature and moisture extremes combined to favor the tall fescue. In some cases, species such as rough bluegrass survive better when used in a mix than when planted as a monostand [16]. Management also affects relative shade tolerance of a species. Bentgrasses are generally deemed to have good shade tolerance, yet the suboptimal mowing heights employed for golf-course greens severely impair bentgrass quality in the shade. While good management may help a relatively shade-tolerant turfgrass, it will often not prevent a grass with poor shade tolerance from dying.

28.4.2 MOWING

Most publications regarding shade management include generic statements that increasing mowing height will improve shade survival of turfgrasses as leaf area should be increased, leading to greater carbon uptake [8,65,106,107]. Photosynthetic apparatus and activity decline from tip to base of mature leaves regardless of irradiance during development [68]. On a leaf-area basis in perennial ryegrass, chlorophyll content and RuBP carboxylase increased two- to threefold from the leaf base to tip. Mowing practices that strip the turf of its youngest leaf tissue may cause even more damage to shaded turf than turf growing in full sun because turfgrasses affected by shading have less TNC reserves to initiate new growth. Data from mowing-height studies in full sun may be suitable to extrapolate to shaded turf but the necessary studies to allow extrapolation have not been conducted. Relatively few refereed publications provide data linking the effect of mowing height to shade tolerance. Penncross creeping bentgrass maintained on a sand-based root zone and mowed at 4 mm failed to provide acceptable turf quality at 80% shade [39], while a separate study showed that Penncross managed at 13 mm height on a silt loam soil maintained acceptable turf quality under 80% shade [22]. Reducing the mowing height from 25 to 16 mm significantly ($P < 0.05$) decreased turf quality of three bermudagrasses in a South Carolina study at 71% shade though not at 41% shade or no shade [42,43]. Increased leaf area may enhance photosynthetic rates but also increase respiratory demand for CO_2 and may increase free moisture on turf surfaces by inhibiting evaporation, thereby promoting disease. Higher mowing heights may cause sufficient interleaf shading that exacerbates the impact of reduced irradiance. In bermudagrass, 5-cm mowing height resulted in less carbon dioxide uptake on a sward basis compared to 2.5 or 20 cm height [66]. Additional studies are needed to determine optimal mowing heights under reduced PPFD for various species, varieties, and conditions including traffic stress.

28.4.3 Fungicides

Fungicides are usually required to maintain high-quality turfgrass in the shade. Leafspot and powdery mildew diseases are capable of overcoming genetic resistance in shaded conditions [18]. *Microdochium* patch (sometimes termed pink snow mold) is considered a winter/spring disease of turf in northern climates but the turf disease diagnostic lab at the University of Wisconsin has diagnosed *Microdochium* patch disease in turf samples in every month of a year except August, with summer samples coming from shaded areas (J. Stier, unpublished data). Left untreated, *M. nivale* can cause additional stress to shaded turf by ramifying throughout the turf verdure, hastening senescence of lower leaves and causing a general thinning of the turf (J. Stier, unpublished data). Temperature can be less of a consideration than moisture for pathogen development, as sufficient wind movement dried turf and prevented the need for fungicides when *Microdochium* patch was the primary disease [89].

28.4.4 Traffic Effects

The etiolated nature of shaded turf and reduced carbohydrate availability for recovery from damage decreases traffic tolerance. While information on traffic effects of shaded turf is mostly empirical or anecdotal, a few studies have investigated the effect of regulated traffic on shaded turf. Simulated soccer traffic reduced Kentucky bluegrass turf quality from an average of 3.4 quality to 2.2 on a 1–9 scale (9—ideal turf) at 1.0–1.7 mol PAR m^{-2} d^{-1} over 2–3 months [5]. At an average 5.6–9.0 mol PAR m^{-2} d^{-1}, the same amount of traffic had less of an effect, reducing turf quality from 5 to 4.65. In another study, simulated soccer traffic caused Kentucky bluegrass to almost completely die (quality rating <2.0) 6–9 weeks sooner than untrafficked turf when placed in a simulated covered stadium transmitting 1.9–3.2 mol PAR m^{-2} d^{-1} in two different years [89]. Jiang et al. [44] showed that mechanical traffic significantly damaged seashore paspalum subjected to either 90% morning or afternoon shade, while no damage occurred to unshaded turf. Cockerham et al. [85] simulated four football games weekly for 7 months on tall fescue, perennial ryegrass with and without creeping red fescue, and "El Toro" zoysiagrass (*Zoysia japonica* Steudel) under full sun, 33%, 54%, and 73% shade. The zoysiagrass provided the best turf quality, approximately 9.0 on a 1–9 scale, 2 months after traffic began, and was the only grass not affected by shade at that time. Zoysiagrass quality declined with increasing shade by 4 months after traffic began. Over time, the quality of all other grasses declined due primarily to traffic at 0%, 33%, and 54% shading, though all grasses deteriorated most extensively under 73% shade plus traffic.

28.4.5 Fertilization

Empirically, nitrogen should normally be applied at approximately half the rate used for nonshaded turf. Burton et al. [72] showed that bermudagrass biomass decreased 20–25% for each of three levels of under shade at nitrogen rates of 1792 kg ha^{-1} compared with 224 kg ha^{-1}. Schmidt and Blaser [108] showed that Tifgreen bermudagrass foliage biomass increased 41% less at 16 klux as N rates increased, compared with 32 klux. They concluded that nitrate reduction and N use were inhibited by the low light intensity. Increasing nitrogen rates above sufficiency levels decreased TNC of bermudagrasses subjected to shade [72,94]. Tripling the monthly nitrogen rate from 24 to 96 kg ha^{-1} reduced supina bluegrass LAI by one third when maintained at 5 mol PAR d^{-1} [109]. Doubling the N rate applied to creeping bentgrass maintained at putting green height in 80% shade decreased turf density though TNCs were not affected [39].

Turf species react differently to N forms in the shade. Steinke and Stier [22] showed that creeping bentgrass under 80% shade had better turf quality when fertilized with liquid-applied urea than granular urea, while Kentucky bluegrass quality was better when granular urea was used. Supina bluegrass reaction to the form of urea depended on the season, with better turf quality resulting from granular applications during spring and early summer and liquid applications providing better turf quality in late summer through autumn. Surprisingly little work has been reported on the effect of other macroelements (P and K) on turf in shade, though K applications on sand-based root zones with marginal ambient K levels improved turf quality at approximately 5.6 mol PAR m^{-2} day^{-1} (J. Stier, unpublished data).

Foliar treatments of iron and silicon have been tested to improve turf color and response to low irradiance. Iron is a component of chlorophyll and additions might presumably enhance photosynthesis by increasing chlorophyll production. While iron applications have been shown to increase chlorophyll content in Kentucky bluegrass while reducing chlorophyll a:b ratios under normal irradiance [110], sequential iron applications did not routinely increase chlorophyll content or alter chlorophyll a:b; iron also failed to effect turf density or growth in bluegrasses [89]. Silicon, which is deposited as crystals in epidermal cells, has been used to treat St. Augustinegrass subjected to shade but without apparent benefit [111].

28.4.6 PLANT-GROWTH REGULATORS

Plant-growth regulators that restrict GA biosynthesis have been particularly effective for enhancing turf quality in low irradiance conditions [5,22,39,89,112]. As previously stated, excess GA production is a natural response of turfgrasses to shaded conditions but the resulting etiolated growth habit weakens the turf. Research on the use of GA inhibitors for use on shaded turf was accelerated by the need to develop a management system for an indoor grass field used in the 1994 World Cup [113,114]. GA inhibitors labeled for turf include paclobutrazol, flurprimidol, and trinexapac-ethyl (TE). One benefit of TE is its foliar absorption; paclobutrazol and flurprimidol need to be watered into the turf for uptake, which can potentially incite diseases resulting from wet leaf surfaces. The root absorption of paclobutrazol and flurprimidol, however, allow them to be used in conjunction with fertilizer for shaded areas. The Scotts Company (Marysville, Ohio) previously marketed a fertilizer specifically for shaded lawns, which contained paclobutrazol (J. Stier, personal observation).

Both flurprimidol and TE increased the number of tillers and leaves per plant and plant density in reduced irradiance [5,89,112]. TE increased root mass of Diamond zoysiagrass at 80% shade, but neither flurprimidol nor TE affected root mass of Kentucky bluegrass, supina bluegrass, or creeping bentgrass at 60% or more shade [5,22,39]. Chlorophyll content and color of shaded turfgrass was increased in both warm- and cool-season turfgrasses when routinely treated with TE [5,22,39,89]. Photosynthetic rates were increased by TE applications to zoysiagrass [112] but did not affect carbon exchange rates of supina bluegrass maintained at 3–5 mol PAR d^{-1} [109] or photochemical efficiency of bluegrasses or creeping bentgrass at 80% shade [22]. The TNC levels can sometimes be increased through TE application [39,112] although the effect may depend on environmental conditions [22]. Sufficient TNC levels are needed for respiration during dormancy periods and practices that enhance TNC under shaded conditions should reduce turf attrition. Routine applications of TE improved cold tolerance of supina bluegrass, maintained at 80% shade during the growing season, when conditions allowed dehardening during the winter [115]. TE may have partially enhanced cold tolerance by decreasing the rate of respiration as evidenced by a slower loss of nonstructural carbohydrates compared with untreated turf during winter conditions when photosynthesis was inactive or negligible.

TE-treated bluegrass had better wear tolerance than untreated turf [89]. The enhanced wear tolerance was possibly due to thicker cell walls [116] as specific weight of bluegrasses was increased, perhaps by reducing cell enlargement without a concomitant decrease in cell wall deposition. Divot recovery under 80% shade was not inhibited by TE applications to creeping bentgrass, supina bluegrass, or Kentucky bluegrass because horizontal growth from stolons or rhizomes was not inhibited [22]. TE also improved the mowing quality of turfgrasses under reduced irradiance by increasing their rigidity [89].

Not all grasses respond similarly to GA inhibitors. Supina bluegrass quality at reduced light was enhanced by TE more than Kentucky bluegrass [89]. In the absence of simulated soccer traffic, supina bluegrass provided acceptable turf quality at approximately 3–5 mol PAR m^{-2} d^{-1} when treated with TE while Kentucky bluegrass turf died [89].

28.4.7 ENVIRONMENTAL MODIFICATION

Tree and shrub canopies restrict airflow and block sunlight, increasing relative air humidity and leaf surface wetness periods on shaded turf. Certain tree species such as oaks (*Quercus* spp.) and maples (*Acer* spp.) have denser canopies than trees such as lindens (*Tilia* spp.) and birches (*Betula* spp.) and consequently are more likely to cause shading problems. Turf management guides routinely suggest that tree canopies be pruned to a height of 2.4–3.0 m [8,106,117], which may allow sunlight penetration for part of the day and increase air movement. Selective removal of individual branches or trees may provide sufficient sunlight penetration to enhance turf growth but knowing which trees or branches to remove is often not clear. Computer-based technology may be used to model the effect of individual tree removal [118]. Pruning branches of specific trees may help sunlight penetration for a time but regrowth will necessitate subsequent pruning. Pruning of tree roots that interfere with nutrient and water uptake or mowing is sometimes recommended [8], though others recommend against this practice as it may severely impact tree health [119].

Shaded areas typically have reduced airflow due to blockage of wind movement from trees or buildings. The effects of restricted airflow can be ameliorated by removing trees, shrubs, or objects that block airflow. Some intensively managed turf can benefit from the addition of fans to circulate air, such as those at Trappers Turn golf course (Wisconsin Dells, Wisconsin).

28.5 SPORTS TURF IN COVERED OR HIGHLY SHADED STADIUMS

The desire to eliminate weather as a factor in sporting events and to increase human comfort has led to the development of many enclosed facilities worldwide. A few enclosed or partially enclosed facilities have, with mixed success, maintained turfgrass fields. In 1993, the Pontiac Silverdome (Pontiac, Michigan) became the first covered stadium to host a major sporting event, the U.S. Cup final between England and Germany, on natural grass. One year later, the field was used for the 1994 World Cup. The success of the turf depended on the use of GA inhibitors, fans to circulate air, and a portable field that allowed the turf to be removed from the stadium after 4 weeks for recovery in full sun [120]. Other enclosed fields such as Bank One Ballpark (Phoenix, Arizona) and Ajax Stadium in Holland have utilized turfgrass on a permanent basis, though turf replacement is a regular occurrence. Stadium covers range from Teflon®*-coated fiberglass to polycarbonate. Polycarbonate transmits nearly complete solar radiation and allows reradiation of infrared wavelengths, which reduces the greenhouse effect. Unfortunately, polycarbonate is expensive and requires a solid support

---

* Registered trademark of E.I. du Pont de Nemours & Company, Inc., Wilmington, Delaware.

framework that reduces the influx of PAR. Fiberglass coverings such as that used in Pontiac were supported by air pressure, eliminating the need for a solid framework, but the best coverings allow no more than 20% solar transmission. Several new stadiums have moveable rooftops such as Miller Park (Milwaukee, Wisconsin) or partially enclosed roofs such as Texas Stadium (Dallas, Texas). However, the infrastructure of such buildings prevent full sunlight from reaching the entire turf, causing some or much of the turf to regularly suffer shade stress. At Miller Park baseball field, the lack of solar radiation in right field delays thawing of soil in the spring: combined with the proximity to the field entrance, turf in right field requires extra management and more frequent replacement than other field areas (R. Volkening, personal communication). An increasing number of unenclosed stadiums have fields experiencing shade problems due to seating structures. Supplemental lighting from high-intensity discharge lamps may help sustain shaded athletic turf. The number of lamps required to provide sufficient irradiance may be minimized by keeping the lamps on 24 h d^{-1}, though turf color may be reduced at higher intensities, possibly due to photodegradation of chlorophyll [40].

28.6 CONCLUSION

Turfgrasses are sun-adapted plants and perform poorly under intense shade, with noticeable effects at <30% sunlight for cool-season grasses and <50% sunlight for warm-season grasses. Lack of sufficient light quantity and quality is the single greatest factor affecting shaded turf followed by disease activity and tree-root competition for water and nutrients. Turf management in the shade requires different approaches than in full sun due to changes in grass physiology, which result in weaker plants less able to resist and recover from damage or disease. To maintain turf in the shade, shade-tolerant grasses should be selected. Tree branches should be pruned to at least 2.5–3.0 m aboveground to enhance sunlight transmission and airflow. Mowing heights should be raised to the upper end of the optimal height range. Irrigation should be applied only as needed, with rates sufficient to thoroughly moisten the rooting depth so as to increase the time needed between irrigation events. Nitrogen fertilization should be applied at the same schedule as nonshaded turf but at half the rate; different species may require either foliar or granular applications. Fungicides are needed to maintain high-quality turf on shaded golf courses and athletic fields. Traffic should be reduced whenever possible. Application of GA-inhibiting growth regulators will increase the quality and survivability of most turf species in all but the most extreme shaded conditions. In some cases, the amount of shade will be too great to sustain acceptable turf cover, and other ground covers such as shade-adapted broadleaves or mulches should be used.

REFERENCES

1. Clark, L.G., Zhang W., and Wendel, J.F., A phylogeny of the grass family (Poaceae) based on ndhf sequence data. *System. Bot.*, 20:436–460, 1995.
2. Clayton, W.D., Evolution and distribution of grasses, *Ann. Missouri Bot. Gard.*, 68:5–14, 1981.
3. Casler, M.D. and Duncan R.R., Origins of the turfgrasses, in *Turfgrass Biology, Genetics, and Breeding*, Casler, M.D. and Duncan, R.R., Eds., Wiley, Hoboken, NJ, 2003, 5–23.
4. Beard, J.B., *Turfgrass: Science and Culture*, Prentice Hall, Englewood Cliffs, NJ, 1973.
5. Stier, J.C. et al., Flurprimidol effects on Kentucky bluegrass under reduced irradiance, *Crop Sci.*, 39:1423–1430, 1999.
6. Tegg, R.S. and Lane, P.A., A comparison of the performance and growth of a range of turfgrass species under shade, *Aust. J. Exp. Agric.*, 44:353–358, 2004.
7. Madison, J.H., *Practical Turfgrass Management.*, Van Nostrand Reinhold Co., New York, 1971.
8. Dudeck, A.E. and Peacock, C.H., Shade and turfgrass culture, in *Turfgrass*, Agronomy Monograph 32, Waddington, D.V., Carrow, R.N., and Shearman, R.C., Eds., ASA-CSSA-SSSA, Madison, WI, 1992, 269.
9. Bell, G.E. and Danneberger, T.K., Temporal shade on creeping bentgrass turf, *Crop Sci.*, 39:1142–1146, 1999.

10. Wherley, B.G., Gardner, D.S., and Metzger, J.D., Tall fescue photomorphogenesis as influenced by changes in the spectral composition and light intensity, *Crop Sci.*, 45:562–568, 2005.

11. Stanford, R.L. et al., Temperature, nitrogen and light effects on hybrid bermudagrass growth and development, *Crop Sci.*, 45:2491–2496, 2005.

12. McBee, G.G., Association of certain variations in light quality with the performance of selected turfgrasses, *Crop Sci.*, 9:14–17, 1969.

13. Wilkinson, J.F. and Beard, J.B., Anatomical responses of 'Merion' Kentucky bluegrass and 'Pennlawn' red fescue at reduced light intensities, *Crop Sci.*, 15:189–194, 1975.

14. Gardner, D.S. and Taylor, J.A., Change over time in quality and cover of various turfgrass species and cultivars maintained in shade, *HortTechnology*, 12:465–469, 2002.

15. Gardner, D.S. and Wherley, B.G., Growth response of three turfgrass species to nitrogen and trinexapac-ethyl in shade, *HortScience*, 40:1911–1915, 2005.

16. Beard, J.B., Factors in adaptation of turfgrasses to shade, *Agron. J.*, 57:457–459, 1965.

17. Danneberger, T.K., *Turfgrass Ecology and Management*, Franzak and Foster, div. of G.I.E. Publishers, Cleveland, OH, 1993.

18. Vargas, J.M. and Beard, J.B., Shade environment-disease relationships of Kentucky bluegrass cultivars, in *Proc. 4th Int. Turfgrass Res. Conf.*, Sheard, R.W., Ed., International Turfgrass Society and Ontario Agricultural College, University of Guelph, Guelph, ON, Canada, 1981, 391–395.

19. Koh, K.J. et al., Shade and airflow restriction effects on creeping bentgrass golf greens, *Crop Sci.*, 43:2182–2188, 2003.

20. Blad, B.L. and Lemeur, R., Miscellaneous techniques for alleviating heat and moisture stress, in *Modifications of the Aerial Environment of Crops*, ASAE Monograph 2, Bartfield, B.J., and Gerber, J.F., Eds., ASAE, St. Joseph, MI, 1979, 409–425.

21. Giesler, L.J., Yuen, G.Y., and Horst, G.L., Canopy microenvironments and applied bacteria population dynamics in shaded tall fescue, *Crop Sci.*, 40:1325–1332, 2000.

22. Steinke, K. and Stier, J.C., Nitrogen selection and growth regulator applications for improving shaded turf performance, *Crop Sci.*, 43:1399–1406, 2003.

23. Fons, W.L., Influence of forest cover on wind velocity, *J. Forest.*, 38:481–486, 1940.

24. Feldhake, C.M., Danielson, R.E., and Butler, J.D., Turfgrass evapotranspiration. I. Factors influencing rate in urban environments, *Agron. J.*, 75:824–830, 1983.

25. Denmead, O.T., Evaporative sources and apparent diffusivities in a forest canopy, *J. Appl. Meteorol.*, 3:383–389, 1964.

26. McBee, G.G. and Holt, E.C., Shade tolerance studies on bermudagrass and other turfgrasses, *Agron J.*, 58:523–525, 1966.

27. Monteith, J.L., The reflection of short-wave radiation by vegetation, *Quart. J. Royal Meteorol. Soc.*, 85:386–392, 1959.

28. Cooper, J.D., Potential production and energy conversion in temperate and tropical grasses, *Herb. Abstr.*, 40:1–15, 1970.

29. Thimijan, R.W. and Heins, R.D., Photometric, radiometric, and quantum light units of measure: a review of procedures for interconversion, *HortScience*, 18:818–822, 1983.

30. Meek, D.W. et al., A generalized relationship between photosynthetically active radiation and solar radiation, *Agron. J.*, 76:939–945, 1984.

31. French, C.S., Light pigments and photosynthesis, in *A Symposium on Light and Life*, McElroy, W.D. and Blas, B., Eds., John Hopkins Press, Baltimore, MD, 1961, 441–470.

32. Taiz, L. and Zeiger, E., *Plant Physiology*, Sinauer Associates, Inc., Sunderland, MA, 1998.

33. Casal, J.J. and Sanchez, R.A., Impaired stem-growth responses of blue-light radiance in light-grown transgenic tobacco seedlings overexpressing *Avena* phytochrome A, *Physiol. Plant.*, 91:268–272, 1994.

34. Frankland, B., Perception of light quantity, in *Photomorphogenesis in Plants*, Kendrick, R.E. and Kronenberg, G.H.M., Eds., Martinus Nijhoff, Dordrecht, 1986, 219–236.

35. Shirley, L.H., Light as an ecological factor and its measurement, *Bot. Rev.*, 11:463–524, 1945.

36. Beard, J.B., Turfgrass shade adaptation, in *Proc. of the 1st Int. Turfgrass Res. Conf.*, Roberts, E.C., Ed., Alf. Smith and Co., Bradford, UK, 1969, 273–282.

37. Salisbury, F.B. and Ross, C., *Plant Physiology*, Wadsworth Publishing Co., Inc., Belmont, CA, 1969.

38. McCarty, L.B., *Best Golf Course Management Practices*, Prentice Hall, Upper Saddle River, NJ, 2001.

39. Goss, R.M. et al., Trinexapac-ethyl and nitrogen effects on creeping bentgrass grown under reduced light conditions, *Crop Sci.*, 42:472–479, 2002.

40. Rogers, J.N. et al., The sports turf management research program at Michigan State University, in *Safety in American Football, ASTM STP 1305,* Hoerner, E.F., Ed., American Society for Testing and Materials, West Conshohocken, PA, 1996, 132–144.
41. Cockerham, S.T. et al., Turfgrass growth response under restricted light: growth chamber studies, *Calif. Turfgrass Cult.,* 52:13–17, 2002.
42. Bunnell, B.T., McCarty, L.B., and Bridges, W.C., Jr., Evaluation of three bermudagrass cultivars and Meyer Japanese zoysiagrass grown in shade, *Int. Soc. Res. J.,* 10:826–833, 2005.
43. Bunnell, B.T. et al., Quantifying a daily light integral requirement of a 'Tifeagle' bermudagrass golf green, *Crop Sci.,* 45:569–574, 2005.
44. Jiang, Y., Carrow, R.N., and Duncan, R.R., Effects of morning and afternoon shade in combination with traffic stress on seashore paspalum, *HortScience,* 38:1218–1222, 2003.
45. Jiang, Y., Duncan, R.R., and Carrow, R.N., Assessment of low light tolerance of seashore paspalum and bermudagrass, *Crop Sci.,* 44:587–594, 2004.
46. Miller, G.L., Edenfield, J.T., and Nagata, R.T., Growth parameters of Floradwarf and Tifdwarf bermudagrass exposed to various light regimes, *Int. Soc. Res. J.,* 10:879–884, 2005.
47. Holmes, M.G. and Smith, H., The function of phytochrome in the natural environment. I. Characterisation of daylight for studies in photomorphogenesis and photoperiodism, *Photochem. Photobiol.,* 25:533–538, 1977.
48. Smith, H., Light quality, photoperception, and plant strategy. *Ann. Rev. Plant Physiol.,* 33:481–518, 1982.
49. Bell, G.E., Danneberger, T.K., and McMahon, M.J., Spectral irradiance available for turfgrass growth in sun and shade, *Crop Sci.,* 40:189–195, 2000.
50. Federer, C.A. and Tanner, C.B., Spectral distribution of light in the forest, *Ecology,* 47:555–560, 1966.
51. Vezina, P.E. and Boulter, D.W.K., The spectral distribution of near ultraviolet and visible radiation beneath forest canopies, *Can. J. Bot.,* 44:1267–1284, 1966.
52. Lee, D.W., Duplicating foliage shade for research on plant development, *HortScience,* 20:116–118, 1985.
53. Goodfellow, S. and Barkham, J.P., Spectral transmission curves for a beech (*Fagus sylvatica* L.) canopy, *Acta Bot. Neerl.,* 23:225–230, 1974.
54. Zarlengo, P.J., Rothrock, C.S., and King, J.W., Influence of shading on the response of tall fescue cultivars to *Rhizoctonia solani* AG-1 IA, *Plant Dis.,* 78:126–129, 1994.
55. Booij-James, I.S. et al., Phosphorylation of the D1 Photosystem II reaction center protein is controlled by an endogenous Circadian rhythm, *Plant Physiol.,* 130:2069–2075, 2002.
56. Elstein, D., Plant's biological clocks help them prepare for the day, *Agric. Res.,* 51:11, 2003.
57. Hart, J.W., *Light and Plant Growth,* Unwin Hyman, London, 1988.
58. Harding, R.W. and Shropshire, W., Jr., Photocontrol of carotenoid biosynthesis, *Ann. Rev. Plant Physiol.,* 31:217–238, 1980.
59. Casal, J.J., Sanchez, R.A., and Deregibus, V.A., Tillering responses of *Lolium multiflorum* plants to changes of red/far red ratio typical of sparse canopies, *J. Exp. Bot.,* 38:1432–1439, 1987.
60. Casal, J.J., Sanchez, R.A., and Gibson, D., The significance of changes in red/far-red ratio, associated with either neighbor plants or twilight, for tillering in *Lolium multiflorum* Lam., *New Phytol.,* 116:565–572, 1990.
61. Butler, W.L. et al., Detection, assay, and preliminary purification of the pigment controlling photoresponsive development of plants, *Proc. Natl. Acad. Sci.,* 45:1703–1708, 1959.
62. Grant, R.H., Partitioning biologically active radiation in plant canopies, *Int. J. Biometeorol.,* 40:26–40, 1997.
63. Morgan, D.C. and Smith, H., Control of development in *Chenopodium album* L. by shadelight: The effect of light quantity (total fluence rate) and light quality (red:far-red ratio), *New Phytol.,* 88:239–248, 1981.
64. Harrison, C., Responses of Kentucky bluegrass to variations in temperature, light, cutting, and fertilization, *Plant Physiol.,* 9:83–106, 1934.
65. Fry, J. and Huang, B., *Applied Turfgrass Science and Ohysiology,* Wiley, Hoboken, NJ, 2004.
66. Alexander, C.W. and McCloud, D.E., CO^2 uptake (net photosynthesis) as influenced by light intensity of isolated bermudagrass leaves contrasted to that of swards under various clipping regimes, *Crop Sci.,* 2:132–135, 1962.
67. Jones, C.A., C_4 *Grasses and Cereals,* Wiley Interscience, New York, 1985.
68. Prioul, J-L., Brangeon, J., and Reyss, A., Interaction between external and internal conditions in the development of photosynthetic features in a grass leaf. I. Regional responses along a leaf during and after low-light or high-light acclimation, *Plant Physiol.,* 66:762–769, 1980.

69. Winstead, C.W. and Ward, C.Y., Persistence of southern turfgrasses in a shade environment, in *Proc. 2nd Int. Turfgrass Res. Conf.*, Roberts, E.C., Ed., ASA and CSSA, Madison, WI, 1974, 221–230.

70. Boardman, N.K., Comparative photosynthesis of sun and shade plants, *Annu. Rev. Plant Physiol.*, 28:355–377, 1977.

71. Schnyder, H. and Nelson, C.J., Growth rate and assimilate partitioning in the elongation zone of tall fescue leaf blades at high and low irradiance, *Plant Physiol.*, 90:1201–1206, 1989.

72. Burton, G.W., Jackson, J.E., and Knox, F.E., The influence of light reduction upon the production, persistence, and chemical composition of coastal bermudagrass, *Cynodon dactylon*, *Agron. J.*, 51:537–542, 1959.

73. Tan, Z.G. and Qian, Y.L., Light intensity affects gibberellic acid content in Kentucky bluegrass, *HortScience*, 38:113–116, 2003.

74. Nobel, P.S., Photosynthetic rates of sun versus shade leaves of *Hyptis emoryi* Torr., *Plant Physiol.*, 58:218–223, 1976.

75. Graham, P.L., Steiner, J.L., and Wiese, A.F., Light absorption and competition in mixed sorghum-pigweed communities, *Agron. J.*, 80:415–418, 1988.

76. Wilkinson, J.F. and Beard, J.B., Morphological responses of *Poa pratensis* and *Festuca rubra* to reduced light intensity, in *Proc. 2nd Int. Turfgrass Res. Conf.*, Roberts, E.C., Ed., ASA and CSSA, Madison, WI, 1974, 231–240.

77. Smalley, R.R., Tillering response of five fine-leaf fescue cultivars to variations of light intensity, in *Proc. 4th Int. Turfgrass Res. Conf.*, Sheard, R.W., Ed., International Turfgrass Society and Ontario Agricultural College, University of Guelph, Guelph, ON., 1981, 487–492.

78. Van Huylenbroeck, J.M. and Van Bockstaele, E., Effects of shading on photosynthetic capacity and growth of turfgrass species, *Int. Turfgrass Soc. Res. J.*, 9:353–359, 2001.

79. Allard, G., Nelson, C.J., and Pallardy, S.G., Shade effects on growth of tall fescue: I. Leaf anatomy and dry matter partitioning, *Crop Sci.*, 31:163–167, 1991.

80. Folta, K.M. et al., Genomic and physiological studies of early cryptochrome 1 action demonstrate roles for auxin and gibberellin in the control of hypocotyls growth by blue light, *Plant J.*, 36:203–214, 2003.

81. Kawai, H. et al., Responses of ferns to red light are mediated by an unconventional photoreceptor, *Nature*, 421:287–290, 2003.

82. Evans, G.C., An area survey method of investigating the distribution of light intensity in woodlands, with particular reference to sunflecks, *J. Ecol.*, 44:391–427, 1956.

83. Peacock, C.H. and Dudeck, A.E., The effects of shade on morphological and physiological parameters of St. Augustinegrass cultivars, in *Proc. 4th Int. Turfgrass Res. Conf.*, Sheard, R.W., Ed., International Turfgrass Society and Ontario Agricultural College, University of Guelph, Guelph, ON, 1981, 493–501.

84. Eriksen, F.I. and Whitney, A.S., Effects of light intensity on growth of some tropical forage species. I. Interaction of light intensity and nitrogen fertilization on six forage grasses, *Agron. J.*, 73:427–433, 1981.

85. Cockerham, S.T., Gibeault, V.A., and Borgonovo, M., Traffic effects on turfgrasses under restricted light, *Calif. Turfgrass Cult.*, 44:1–3, 1994.

86. Wilkinson, J.F., Beard, J.B., and Krans, J.V., Photosynthetic-respiratory responses of Merion' Kentucky bluegrass and 'Pennlawn' red fescue at reduced light intensities, *Crop Sci.*, 15:165–168, 1975.

87. Whitcomb, C.E., Influence of tree root competition on growth of four cool-season turfgrasses, *Agro. J.*, 64:355–359, 1972.

88. Whitcomb, C.E. and Roberts, E.C., Competition between established tee roots and newly seeded Kentucky bluegrass, *Agron. J.*, 65:126–129, 1973.

89. Stier, J.C. and Rogers, J.N. III., Trinexapac-ethyl and iron effects on supina and Kentucky bluegrass under reduced irradiance, *Crop Sci.*, 39:1423–1430, 2001.

90. Wherley, B.G., Turfgrass photomorphogenesis as influenced by changes in spectral composition and intensity of shadelight, Ph.D. thesis, The Ohio State University, Columbus, 2003.

91. Leinauer, B. et al., *Poa Supina* Schrad: A new species for turf, *Int. Turfgrass Soc. Res. J.*, 8:345–351, 1997.

92. Stiegler, J.C., Bell, G.E., and Martin, D.L., Foliar applications of magnesium and iron encourage annual bluegrass in shaded creeping bentgrass putting greens, *Crop Manage.* (on-line), doi:10.1094/CM-2003-0821-01-RS, 2003.

93. Beard, J.B., Shade stresses and adaptation mechanisms of turfgrasses, *Int. Turfgrass Soc. Res. J.*, 8:1186–1195, 1997.
94. Bunnell, B.T., McCarty, L.B., and Bridges, W.C. Jr., 'Tifeagle' bermudagrass response to growth factors and mowing height when grown at various hours of sunlight, *Crop Sci.*, 45:575–581, 2005.
95. Ervin, E.H. et al., Trinexapac-ethyl restricts shoot growth and prolongs stand density of 'Meyer' zoysiagrass fairway under shade, *HortScience*, 37:502–505, 2002.
96. Qian, Y.L. and Engelke, M.C., Diamond zoysiagrass as affected by light intensity, *J. Turfgrass Manage.*, 3:1–15, 1999.
97. Riffell, S.K., Engelke, M.C., and Morton, S.J., Performance of three warm-season turfgrasses cultured in shade:zoysiagrass, in *Texas Turfgrass Res.*, Texas Agric. Exp. Stn. PR-turf 95-1, 1995, 60-65.
98. Zuk, A.J., Brenner, D.J., and Fry, J.D., Establishment of seeded zoysiagrass in a perennial ryegrass sward:effects of soil-surface irradiance and temperature, *Int. Turfgrass Soc. Res. J.*, 10:302–309, 2005.
99. Miltner, E.D. et al., Establishment of *Poa annua* var. *reptans* from seed under golf course conditions in the Pacific Northwest, *Crop Sci.*, 44:2154–2159, 2004.
100. Murphy, J.A. et al., Creeping bentgrass establishment on root zones varying in sand sizes, *Int. Turfgrass Soc. Res. J.*, 9:573–579, 2001.
101. Bahmani, I. et al., Differences in tillering of long- and short-leaved perennial ryegrass genetic lines under full light and shade treatments, *Crop Sci.*, 40:1095–1102, 2000.
102. Beard, J.B., *Turf Management for Golf Courses*, 2nd ed., Ann Arbor Press, Chelsea, MI, 2002.
103. Barrios, E.P. et al., Quality and yield response of four warm-season lawngrasses to shade conditions, *Agron. J.*, 78:270–273, 1986.
104. Landry, G., *Lawns in Georgia*, University of Georgia Cooperative Extension Bulletin, 773, 2000.
105. Gaussoin, R.E., Baltensperger, A.A., and Coffey, B.N., Response of 32 bermudagrass clones to reduced light intensity, *HortScience*, 23:178–179, 1988.
106. Harivandi, M.A. and Gibeault, V.A., Turfgrass management in shade. *Calif. Turfgrass Cult.*, 47:1–3, 1997.
107. Bell, G. and Danneberger, K., Managing creeping bentgrass in shade, *Golf Course Manage.*, 10:56–60, 1999.
108. Schmidt, R.E. and Blaser, R.E., Effect of temperature, light, and nitrogen on growth and metabolism of 'Tifgreen' bermudagrass (*Cynodon* spp.), *Crop Sci.*, 9:5–9, 1969.
109. Stier, J.C., The effects of plant growth regulators on Kentucky bluegrass (*Poa pratensis* L.) and supina bluegrass (*P. supina* Schrad.) in reduced light conditions, Ph.D. Dissertation, Michigan State University, East Lansing, 1997.
110. Lee, C.W. et al., Induced micronutrient toxicity in 'Touchdown' Kentucky bluegrass, *Crop Sci.*, 36:705–712, 1996.
111. Trenholm, L.E., Datnoff, L.E., and Nagata, R.T., Influence of silicon on drought and shade tolerance of St. Augustinegrass, *HortTechnology*, 14:487–490, 2004.
112. Qian, Y.L. and Engelke, M.C., Influence of trinexapac-ethyl on Diamond zoysiagrass in a shade environment, *Crop Sci.*, 39:202–208, 1999.
113. Rogers, J.N., III and Stier, J.C., Effect of plant growth regulators on traffic tolerance of indoor sports turf, in *Agron. Abstr.*, ASA, Madison, WI, 1993, 163.
114. Stier, J.C. et al., An indoor sports turf research facility for World Cup 1994, in *Agron. Abstr.*, ASA, Madison, WI, 1993, 164.
115. Steinke, K. and Stier, J.C., Influence of trinexapac-ethyl on cold tolerance and non-structural carbohydrates of shaded supina bluegrass, *Acta Hort (ISIS)*, 661:207–215, 2004.
116. Shearman, R.C. and Beard, J.B., Turfgrass wear tolerance mechanisms. II. Effects of cell wall constituents on turfgrass wear tolerance, *Agron. J.*, 67:211–215, 1975.
117. Stier, J.C., *Growing Grass in Shade*, University of Wisconsin Extension Bulletin A3700, 1999.
118. Anonymous, Eliminating the guesswork of shade, *TurfNet Monthly*, 5:1–4, 1998.
119. Harivandi, A. and Gibeault, V.A., *Managing Lawns in Shade*, University of California Cooperative Extension Bulletin 7214, 2002.
120. Rogers, J.N., III., In from the cold, *The World and I*, 9:192–199, 1994.

29 Cold-Stress Physiology and Management of Turfgrasses

John Clinton Stier and Shui-zhang Fei

CONTENTS

29.1 CLIMATIC ADAPTATION OF TURFGRASSES

Temperature is one of the primary factors that limit geographic adaptation of a plant species [1]. Turfgrass species are classified as either cool- or warm-season grasses based on their growth and development response to varied temperatures. The optimum growth temperature for cool-season grasses is 18–24°C (~65–75°F) while the optimum growth temperature for warm-season grasses is 27–35°C (~80–95°F) [2]. Cool-season turfgrasses including bluegrasses (*Poa* spp.), bentgrasses (*Agrostis* spp.), fescues (*Festuca* spp.), and ryegrasses (*Lolium* spp.) have a C_3 photosynthetic pathway and are well adapted to cool temperate regions. Alternatively, warm-season turfgrasses such as bermudagrass (*Cynodon* spp.), zoysiagrass (*Zoysia* spp.), St. Augustinegrass (*Stenotaphrum secundatum* [Walt.] Kuntze), centipedegrass (*Eremochloa ophiuroides* [Munro.] Hack.), buffalograss (*Buchloë dactyloides* [Nutt.] Engelm.), and bahiagrass (*Paspalum notatum* Flugge.) have a C_4 photosynthetic pathway and grow primarily in tropical and subtropical climates. Some of the cool-season grasses such as tall fescue (*F. arundinacea* Schreb.), Kentucky bluegrass (*Poa pratensis* L.), perennial ryegrass (*Lolium perenne* L.), and creeping bentgrass (*Agrostis stolonifera* L.) as well as the warm-season grasses zoysiagrass, bermudagrass, and buffalograss are frequently grown in the transitional zone between the cool and warm regions.

Although cool-season turfgrasses are well adapted to the cool temperate region where severe winter conditions often occur, freezing tolerance may vary considerably among different species. Some cool-season turfgrass species such as perennial ryegrass and annual bluegrass (*Poa annua* L.) may suffer serious winter injuries during harsh winters due to their poor low-temperature resistance [3,4]. In addition, warm-season turfgrasses grown in the transitional regions may also experience winter injuries or even winter kill [5,6]. In this chapter, we will discuss the types of low-temperature injuries, mechanisms for turfgrass to cope with low temperature injuries, genetic variability of freezing tolerance among turfgrasses species, environmental factors, and management practices that influence freezing tolerance and winter survival of turfgrass species.

29.2 COLD ACCLIMATION AND DEACCLIMATION

To effectively cope with low-temperature stresses, cool-season turfgrass species have evolved mechanisms to perceive the climatic changes signaling the approach of winter. These signals trigger a series of physiological and biochemical events that minimize freeze injury and enhance winter survival of turfgrasses. Exposure to low, nonfreezing temperatures induces cold acclimation (hardening), which is maximized by concomitant exposure to light [7]. Cold acclimation greatly enhances freezing tolerance, making it possible for plants to withstand low or even subfreezing temperatures that would otherwise kill nonacclimated plants. For example, crowns of nonacclimated Chewings fescue (*Festuca rubra* var. *commutata* [Thuill.] Nyman) "Wintergreen" can tolerate low temperature to −12.4°C, but tolerance increases to −27°C after cold acclimation at 5°C for 6 weeks [8].

Cool-season turfgrasses have a great capacity for cold acclimation and as a result their freezing tolerance is greatly enhanced through cold acclimation. Tropical warm-season grasses are not well characterized for cold acclimation and may have limited capacity for cold acclimation. For example, St. Augustinegrass "Floratam" exhibits little variability in freezing tolerance between autumn and spring, indicating that the grass is unable to cold acclimate [9]. However, some warm-season grasses, particularly those that are native to the temperate regions, can cold acclimate and exhibit good freezing tolerance. In a study of freezing tolerance in saltgrass (*Distichlis spicata* var. stricta [L] Greene), Shahba et al. [10] collected field-grown materials from autumn through the following spring and measured freezing tolerance at each collection date. The authors observed that freezing tolerance of all the accessions examined in this study increased in fall, reaching a maximal freezing tolerance in December and January, a clear indication of its ability to cold acclimate. Similar observations were reported for other warm-season grasses including buffalograss [6], bermudagrass [11,12], and seashore paspalum (*Paspalum vaginatum* Sw.) [13].

Acclimating temperatures vary with species. To achieve cold acclimation under controlled environment, most researchers use constant temperatures ranging from 2–5°C, coupled with a short photoperiod, typically 8-h day light and a low light intensity (~150 μmol photosynthetically active radiation [PAR] m^{-2}s^{-1}). In bermudagrass and perennial ryegrass, a day or night temperature of 2°C coupled with another higher temperature provides superior cold hardening compared with a constant temperature [12,14,15]. The duration of cold acclimation also varies considerably depending on whether plants are acclimated naturally in the field or under experimental conditions. Many studies with field-grown and field-acclimated plants reported that peak freezing tolerance occurs in early to mid-winter (December and January), indicating that several months from the onset of acclimation are needed to reach the fully acclimated status [6,10,16]. The effective duration of cold acclimation under experimental conditions varies with species from a few days [17] to a few weeks for most studies [3,8,12,14]. We tested freezing tolerance of a turf-type perennial ryegrass over a period of up to 6 weeks at 4–5°C with photoperiod of 8 h and a photosynthetic photo flux density (PPFD) of ~120 μmol photons m^{-2}s^{-1} and observed that freezing tolerance is greatly enhanced during the first 2 weeks of cold acclimation treatment, but improved little thereafter (S. Fei, unpublished data).

Cold acclimation in cereal crops occurs in two distinct phases, with the first phase occurring at a relatively high temperature followed by a second phase occurring at a lower temperature [18,19]. A study with Chewings fescue indicated that its cold acclimation also occurs in at least two phases with exposure to 5°C conducive to the initial phase followed by 0°C for the second phase [8]. Similar results were obtained in annual bluegrass cultivars with contrasting freezing tolerance [3]. Freezing tolerance was significantly increased when annual bluegrasses were exposed to 2°C for 2 weeks compared with nonacclimated controls; however, freezing tolerance was further increased when the acclimated plants were exposed to a nonlethal subfreezing temperature of −2°C. These results suggest that the two-phase cold acclimation mechanism may exist in many perennial turfgrass species. The need to experience exposure to a freezing temperature may explain why field-grown materials do not achieve maximal hardiness until early winter, as surface temperatures may not experience sufficient freezing temperatures until that time. Exposure to temperatures conducive to growth can negate cold acclimation, though acclimation may be regained if the event occurs in the autumn. Cold acclimation is not likely to be regained if high-temperature exposure occurs in mid to late winter.

Maximal freezing tolerance may not be realized in many species until new development of shoots occur under low temperatures. In the model plant species *Arabidopsis*, many studies have shown that freezing tolerance can be significantly increased from approximately −3°C to −8°C after only a few days of exposure to a nonfreezing low temperature [17,20]. However, these studies were only concerned with the leaves that were developed under warm temperatures and later shifted to cold-acclimating temperatures. Subsequent research indicated that leaves that were developed under cold-acclimating conditions had additional freezing tolerance of 2–3°C [21,22]. These studies showed that full realization of freezing tolerance needed for winter survival is not achieved until foliage developed under cold-acclimating temperatures replaces foliage developed during temperatures optimal for growth [23].

Cold hardiness cannot be achieved without light [24]. Many of the low-temperature-induced acclimation changes are facilitated by light. For example, *Wcs* genes are produced in response to low temperatures but are maximized in the presence of light [25]. Specific weights of plant tissues developed under low-temperature conditions are typically greater than those developed under higher temperatures, a result that is mediated by excitation pressure from photosystem II (PSII) [7]. Cold-hardened rye (*Secale cereale* L.) plants had superior freezing tolerance when exposed to 250 μmol PAR m^{-2} s^{-1} compared with 50 μmol PAR m^{-2} s^{-1} [7].

29.2.1 Physiological and Biochemical Changes

Cold acclimation is a complex process involving changes at molecular, physiological, and biochemical levels. Numerous studies have shown that membrane systems are the primary targets of freezing injury [26–28]. As discussed later, an extended period of exposure to low freezing temperature will cause water within a cell to move out to intercellular spaces where the water potential has lowered due to ice formation. This freeze-induced dehydration is the primary cause for injury to the cell membrane systems [1]. Cell membranes are the sites where many important metabolic activities including photosynthesis and respiration occur; consequently, cold and freezing injury will impair these important physiological processes. Thus, a key function of cold acclimation is to stabilize membrane systems against low-temperature injuries. This stabilization may involve multiple mechanisms including changes in lipid composition [1,29,30] and accumulation of proteins or sugars such as sucrose that have been shown to protect membranes *in vitro* [31,32]. The following section briefly reviews the changes in photosynthesis, carbohydrate accumulation, cell membrane lipid composition, and protein synthesis during cold acclimation.

29.2.1.1 Photosynthesis

Photosynthesis is severely reduced in herbaceous plants upon exposure to low temperatures [33]. This may be due to either photoinhibition, reduction of activities for carbon assimilation enzymes, or carbohydrate accumulation, which initiates feedback inhibition. Photoinhibition occurs when light harvested by PSII exceeds the capacity for electron transport [34]. Photoinhibition damages the photosynthetic system, thereby decreasing photosynthetic activity. Photoinhibition occurs primarily in PSII with the D1 core protein as the primary target [35] although it may also occur in photosystem I (PSI) [36]. Under high light intensity the D1 protein is inactivated, inhibiting the photochemical reactions of the PSII [37]. Plant sensitivity to photoinhibition is increased at low temperatures [38]. Excessive photoinhibition causes bleaching of the chlorophyll, a condition termed *photooxidation* [39], which results in susceptible turfs turning brown during the winter. Cold-hardy plants are more resistant to cold-induced inactivation of PSII [40]. They are able to reduce potential photoinhibition by dissipating excess energy harvested by PSII, thereby maintaining an adequate photosynthesis rate [41]. Leaves of certain turfgrass varieties, particularly among creeping bentgrass, appear red or purple in the autumn. The effect is likely due to anthocyanin accumulation to protect against photoinhibition of PSII [42], though accumulation of sugars has also been proposed as the cause [43].

Besides photoinhibition, enzymes of the Calvin cycle are inactivated when plants are exposed to low temperatures and the total activity of Rubisco is inhibited by light at low temperatures [44]. In addition, the short-term accumulation of soluble carbohydrates resulting from an exposure to growth-inhibiting low temperatures causes feedback regulation of photosynthesis [45,46] and represses the expression of photosynthetic proteins such as the small subunit of the Rubisco, rbcS. In winter annuals and biennials, in parallel to cold acclimation, new growth and development are initiated as part of their natural life cycle. Leaves developed at low temperatures are able to fully recover photosynthesis capacity in *Arabidopsis* [21], wheat (*Triticum aestivum* L.) [47], and rye [48]. This acclimation of photosynthetic carbon metabolism to low temperature in newly developed leaves is attributed to increased activities of Calvin cycle enzymes and the selective stimulation of sucrose synthesis, which is important for essential metabolism and possible cryoprotection [23].

29.2.1.2 Nonstructural Carbohydrate (NSC)

The dynamics of carbohydrates during cold acclimation and overwintering is one of the most studied aspects of cold tolerance in turfgrasses. Nonstructural carbohydrates (NSC) are an important energy source and can be stored in vegetative parts of turfgrasses. They are utilized for maintenance respiration as well as new growth and development when photosynthesis cannot supply sufficient NSC. Fructans are the primary nonstructural carbohydrates present in cool-season grasses while starch is the primary nonstructural carbohydrate present in warm-season grasses [49]. Soluble carbohydrates can regulate cell osmolarity because their presence reduces cell water potential, mitigating water loss during extracellular ice formation or drought [2]. The accumulation of soluble carbohydrates can also depress the freezing point and delay or avoid cell freezing [2]. For these reasons, carbohydrates have been suggested to act as cryoprotectants. The accumulation of sucrose and other simple sugars has been shown to protect membranes from freeze-induced injuries *in vitro*, perhaps playing a role in stabilizing cell membrane systems [31,32].

Numerous studies with turfgrass species have shown that carbohydrates are increased during cold acclimation under either experimental conditions or in the field [50]. Positive correlations between increased carbohydrates and enhanced freezing tolerance have been reported. Rogers et al. [51] measured the NSC and freezing tolerance in zoysiagrass (*Zoysia japonica* [Steud] Meyer) during fall, winter, and the following spring and found that NSC increased from September to December in both stolons and rhizomes but decreased sharply by March when freezing tolerance also declined. Increased sucrose, fructose, glucose, and raffinose were observed in two buffalograss cultivars with contrasting freezing tolerance during cold acclimation when freezing tolerance also increased [52]. Similarly, Shahba et al. [10] reported that fructose, glucose, raffinose, and stachyose exhibited clear seasonal changes in field-grown saltgrass, reaching the highest concentration during midwinter commensurate with freezing tolerance. Fry et al. [53] reported sucrose concentrations correlated to winter hardiness in centipedegrass. In a cool-season turfgrass, *P. annua*, Dionne et al. [3] measured the changes of fructan and mono- and disaccharides during cold acclimation in crowns of three ecotypes under both environmentally controlled and simulated winter conditions in an unheated greenhouse. They showed a positive correlation between the level of sucrose and fructan accumulation and freezing tolerance, though freezing tolerance differences among ecotypes were not associated with fructan and sucrose levels.

Carbohydrate accumulation does not always correlate with freezing tolerance. Fry et al. [9] reported that the total NSC or soluble sugars did not seem to affect the freezing tolerance in St. Augustinegrass "Floratam." Later, Maier et al. [54] reported that starch and sucrose, two primary storage carbohydrates in St. Augustinegrass, were not correlated with freezing tolerance in the three cultivars of St. Augustinegrass examined including "Floratam." This may be due to the fact that St. Augustinegrass does not cold acclimate as its freezing tolerance does not change from fall to the spring.

29.2.1.3 Lipid Composition

Membrane systems are the primary site of freezing injury. An important function of cold acclimation is to stabilize the cell membrane systems. Plant cell membrane systems consist of a lipid bilayer interspersed with proteins including $H^+ATPases$, electron carriers, and channel-forming proteins that regulate the permeability of the membrane systems and other molecules [55]. Cold-tolerant plant species have a high proportion of unsaturated fatty acids in their membranes, which help maintain fluidity during low temperatures for proper protein functioning. Plants with genetically altered unsaturated fatty acids improved the rates of photosynthesis and growth at low temperature and freezing tolerance [56,57]. The unsaturated fatty acids are primarily 16- and 18-carbon acyl groups with up to three double bonds, e.g., 16:3 [58]. During cold acclimation, desaturase activity increases, which increases the unsaturation of membranous fatty acids [59,60]. Cyril et al. [61] examined the changes of lipid composition in three cultivars of seashore paspalum with contrasting cold tolerance during cold treatment and found that the two saturated fatty acids, palmitic acid and stearic acid, did not change. However, linolenic acid (18:3), a triunsaturated fatty acid, increased significantly during the treatment and the magnitude of change was greater in the cold-tolerant seashore paspalum cultivar SeaIsle1 than less cold-tolerant cultivars or accessions. Similarly, Samala et al. [62] reported that the proportion of linolenic acid in the crown tissue of a relatively cold-hardy bermudagrass cultivar "Midiron" increased significantly during a 3-week cold acclimation period while the saturated or the double-bond unsaturated fatty acid content remained steady or decreased.

29.2.1.4 Protein and Amino Acids

Many cold-regulated (COR) genes are expressed during cold acclimation [63,64]. Examples include *kin* genes, some of which encode proteins that contribute to drought and salinity stress in addition to their role in freezing tolerance [65]. Other genes may encode regulatory proteins that apparently contribute to freezing tolerance by regulating expression of freezing-tolerant genes or by regulating the activities of proteins involved in freezing tolerance [1,4]. Chitinases are often synthesized during cold acclimation and may also contribute to freezing tolerance [66]. Some low-temperature-induced proteins share homology with the RAB/LEA/DHN (responsive to ABA, late embryogenesis abundant, and dehydrin, respectively) protein family. Members of this protein family are very hydrophilic and highly stable at boiling temperature. They are synthesized in seeds during seed maturation, and are induced in vegetative tissues by water deficit or exposure to ABA. These proteins are thought to protect cells from dehydration caused by water deficit or freezing, by stabilizing other proteins and membranes [55]. Sugar-binding proteins known as lectins may serve as cryoprotectants by binding to functional groups on membranes. Binding of lectins alters membrane permeability to inhibit excessive solute efflux/influx during freeze–thaw processes [67].

Antifreeze proteins (AFPs) are a unique group of proteins that bind to ice crystals and restrict their growth and recrystallization, enabling plants to survive freezing conditions [68–71]. AFPs are produced by some freezing-tolerant plants such as winter rye [69], perennial ryegrass [72], and meadow fescue (*Festuca pratensis* L.) [73]. Kuiper et al. [70] reported that the antifreeze protein isolated from perennial ryegrass has superior ability to inhibit ice recrystallization. Similarly, Griffith et al. [69] concluded that AFPs are free in solution in cold-acclimated winter rye and therefore readily transported. AFPs have no effect on the supercooling temperature of the leaves in the absence of an ice nucleator. In the presence of an ice nucleator, however, AFPs lowered the leaf freezing temperature by 0.3–1.2°C. The authors concluded that winter rye AFPs have no specific cryoprotective activity; rather, they interact directly with ice *in planta* and reduce freezing injury by slowing the growth and recrystallization of ice.

Synthesis of proteins with no known function in response to cold temperatures has also been reported. In bermudagrass, cold acclimation induced greater synthesis of proteins having intermediate molecular weight (32–37 kDa) or low molecular weight (20–26 kDa) in crowns of the more cold-tolerant "Midiron" compared with the cold-sensitive "Tifgreen" [74]. In annual bluegrass, production of proteins ranging in size from 13 to 31 kDa coincided with maximum freezing tolerance and the accumulation of these proteins was higher in a cold-tolerant ecotype than in a cold-sensitive ecotype [3]. Cold acclimation also induced major amino acid changes in crowns of annual bluegrass. Proline, glutamine, and glutamic acid all increased after cold acclimation, particularly following a two-phase acclimation period with the initial one at 2°C for 2 weeks followed by 2 weeks at −2°C. Amino acid levels were not, however, related to the differential freezing tolerance among the three annual bluegrass ecotypes examined [3].

Low temperature generally increases the stability of proteins; however, there is evidence that protein denaturation still occurs, particularly for some "cold-labile" proteins [75]. Some aspects of translocation and assembly may also be negatively influenced by low temperature. It is proposed that chilling injury may arise, in part, from an impairment of normal protein biogenesis; therefore induction of genes that encode molecular chaperones would prevent proteins from being degraded [76].

The role of many other cold-induced proteins has not yet been elucidated [77]. Other biochemical reactions, as yet not fully understood, are also likely to affect cold acclimation and freezing tolerance. For example, Ca^{2+} appears to play an important role in cold acclimation as it may instigate a series of reactions, possibly along with protein phosphorylation [78].

29.2.2 MORPHOLOGICAL CHANGES IN RESPONSE TO LOW TEMPERATURES

As winter approaches, day length becomes shorter and temperatures decrease below the optimum range for growth and development for both cool- and warm-season grasses. Low temperatures significantly reduce or halt chlorophyll synthesis while existing chlorophyll is degraded by high light intensity [24,79]. Consequently, turfgrass leaves gradually lose color and senesce. If temperatures remain above 0°C, shoot and root growth may continue for cool-season grasses, although at a much reduced rate [24,80]. Continued growth and production/protection of chlorophyll allow cool-season grasses to maintain green color during the winter as long as the temperature remains above freezing or plants are protected from light by snow cover. Although the growth rate of turfgrasses is greatly reduced during the winter, the crowns of most cold-hardy cool-season grasses remain active and are capable of resuming growth when favorable conditions are met [24,81]. The new growth may assume different morphology. *Arabidopsis* leaves developed under low temperatures (5°C) are smaller than warm-temperature counterparts [23]. In winter rye, it was reported that new development under low temperatures resulted in more compact plants and the crowns become larger [7,42]. Low temperature is often thought to be responsible for the development of prostrate or rosette growth morphology [82,83]. The change to prostrate growth habit was highly correlated with increased freezing tolerance of perennial ryegrass [84]. Gray et al. [7], however, provided evidence that low temperature alone was insufficient for growth differences accompanying low temperatures and hypothesized a role for light, probably as a function of PSII excitation pressure. More importantly, these newly developed leaves have greater freezing tolerance as indicated by their lower ion leakage than those observed in preexisting leaves that have been shifted to and fully cold acclimated at 5°C.

Warm-season grasses do not significantly alter their morphology with the onset of long-term low temperatures; instead, the plants partially escape potentially harmful conditions by entering a dormant state. Dormancy significantly slows down the metabolic activities and gives the grasses a distinct brown yellow color as chlorophyll in senescing leaves is degraded and not replaced [24,79]. Physiological changes inside the plant provide additional cold tolerance as previously discussed.

29.2.3 SEASONAL VARIATION OF COLD TOLERANCE

Understanding the seasonal variation of cold hardiness for both warm- and cool-season grasses is important for properly managing turfgrass to avoid winter kill. This is particularly true for temperate to arctic regions where severe winter conditions are often encountered and extreme temperature fluctuations frequently occur in early spring. Beard [85] reported that turfgrasses reached the highest freezing tolerance in December and declined slightly in January followed by a sharp decline in April. In a study of seasonal change of freezing tolerance in cool-season turfgrass species in St. Paul, MN, White and Smithberg [16] described the monthly changes of freezing tolerance during 1972–1976 for creeping bentgrass, Kentucky bluegrass, creeping red fescue (*F. rubra* L.), and perennial ryegrass (Figures 29.1 and 29.2). The monthly data reported in the figures are averages of different cultivars within a species over the period 1972–1976. The authors reported that cold acclimation of the cool-season grasses began in July and August and increased steadily to a peak in January. Loss of hardiness starts in midwinter and is initially slow. However, this slow decline of freezing tolerance is followed by rapid loss of hardiness in the months of March and April when new growth is being initiated [16,85–88]. Consequently, perennial grasses are particularly susceptible to low-temperature injury in late winter and early spring. Temperature fluctuations above 0°C may accelerate deacclimation [15]; once acclimation is lost, it cannot be fully recovered in the same season [88]. Deacclimation occurs relatively fast compared with the acclimation process [89] and is controlled more by soil than by air temperatures [88].

Deacclimation is accompanied by changes in NSC [3,90] and an increase in crown moisture, which predisposes plants to low-temperature kill if followed by rapid decreases to subfreezing temperatures [85,88]. Dionne et al. [3] reported that freezing tolerance of annual bluegrass increased from October until midwinter (February) and subsequently decreased in spring (late March), coinciding with a rise and decline of sucrose. Warm-season grasses such as buffalograss and saltgrass,

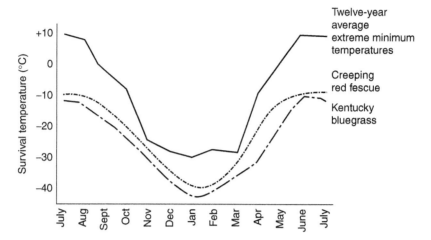

FIGURE 29.1   Seasonal variation of freezing tolerance of creeping red fescue and Kentucky bluegrass. Each data point is an average of different cultivars used in the experiment during 1972–1976. Creeping red fescue included cultivars of "Ruby," "Wintergreen," "Arctared," and "Jamestown" Chewings fescue. Kentucky bluegrass included cultivars of "Nugget," "Pennstar," "Baron," and "Kenblue." (After White, D.B., and Smithberg, M.H., in *Proc. 3rd Int. Turfgrass Res. Conf.,* ASA-CSSA-SSSA and Int. Turfgrass Soc., Madison, WI, 1980, 149–154. With permission.)

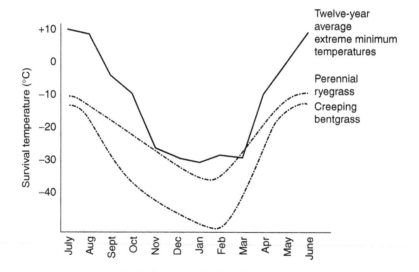

FIGURE 29.2   Seasonal variation of freezing tolerance of perennial ryegrass and creeping bentgrass. Each data point is an average of different cultivars used in the experiment during 1972–1976. Perennial ryegrass included cultivars of "NK-100" and "NK-200." The cultivar of "Penncross" is the only cultivar used for creeping bentgrass in this experiment. (After White, D.B., and Smithberg, M.H., in *Proc. 3rd Int. Turfgrass Res. Conf.*, ASA-CSSA-SSSA and Int. Turfgrass Soc., Madison, WI, 1980, 149–154. With permission.)

adapted to areas subject to freezing, also reach peak freezing tolerance in December and January and start to deacclimate in March [6,10].

29.2.4   DORMANCY

Dormancy is defined as temporary suspension of visible growth of any plant structure containing a meristem [91]. Plants use dormancy as an escape mechanism to avoid potentially fatal but temporary conditions (e.g., cold, drought, heat). Most of the cool-season turfgrasses do not have a well-characterized winter dormancy. Leaves often retain functioning chlorophyll and may photosynthesize during winter thaws [24]. In temperate and some subtropical areas, warm-season grasses enter winter dormancy when they are exposed to an extended period of low temperature, e.g., soil temperatures below 10°C [92]. Chlorophylls in these dormant grasses are degraded; consequently fully dormant grasses have a distinct brown appearance [24,79]. Dormancy confers maximum freezing tolerance to plants, enabling them to survive the winter [93]. Early dormancy is correlated with increased freezing tolerance in buffalograss [6]. For example, the hexaploid buffalograss "Texoka" enters dormancy earlier, acclimates faster, and possesses a greater capacity to cold acclimate than tetraploid or diploid cultivars, which can retain green color longer and enter dormancy at a later time, if at all, potentially becoming exposed to lethally low temperatures. Dormant grasses initiate new growth from meristems of stolons, rhizomes and crown when soil temperatures rise above 10°C in the spring [24].

29.3   LOW-TEMPERATURE INJURY

Grasses may be injured by either chilling or ice formation within the plant [50]. Both warm- (C_4) and cool-season (C_3) grasses are susceptible to either type of injury. Cool-season turfgrasses are more likely to be injured by ice formation within the plant as few warm-season turfgrasses can survive in climates where the mean minimal temperature is below freezing. Grasses survive low-temperature injury either by tolerating or escaping the condition in time or space. "Nugget" and "Merion" Kentucky bluegrass escape cold injury by entering winter dormancy earlier than other cultivars, which allows them to have superior winter hardiness [24]. Kentucky bluegrasses have delayed spring greenup compared with perennial ryegrass and annual bluegrass, which may help them avoid crown hydration-related freezing injury in late winter/early spring. Certain varieties of perennial ryegrass are less susceptible to cold damage as their crowns are lower in the ground [94]. Warm-season grasses such as common zoysiagrass (*Zoysia japonica* Steud.) or bermudagrass (*Cynodon dactylon* [L.] Pers.) may survive in freezing climates over steam tunnels in urban areas or in otherwise protected sites.

29.3.1   CHILLING INJURY

Chilling injury occurs to susceptible leaves at temperatures above freezing to 12°C and has been mostly explored in annual crops and tropical and subtropical plants. Injury occurring within 24 h of exposure is considered acute (direct) while chronic (indirect) injury occurs only after several days of low temperature [27]. For example, exposure of chill-sensitive centipedegrass to 5°C resulted in electrolyte leakage after 10 but not 5 days [95]. Warm-season grasses are usually more susceptible to chilling injury than cool-season grasses though both types may be affected. Differences exist among both turfgrass species and cultivars within species [96]. Physiological damage caused by chilling is sometimes reversible but may require several days under suitable growing conditions [97,98]. In some cases, prolonged exposure to chilling temperatures may kill sensitive species or cultivars [96,99]. Whole plant symptoms include wilting, leaf pitting, growth cessation, water-soaked leaf lesions, and reduced chlorophyll, particularly in leaves most exposed to light [96,99,100]. The necrosis observed among warm-season turfgrasses following temperature reductions below 10°C to 15°C is symptomatic of chilling injury in warm-season grasses.

Chilling injury disrupts cell membrane activity, leading to electrolyte loss from the cytoplasm, release of vacuolar substances, and loss of protein activity [39,100]. Chloroplast and mitochondrial membranes may also be affected [39,101]. Evidence suggests chilling may affect only discrete portions of membranes, causing a phase shift from a fluid to a gel-like condition that affects metabolism [27,102]. Photosynthesis rates declined 85–90% when "Derby" perennial ryegrass was exposed to a single 24-h period of chilling temperatures [97]. Data indicated reduction of photosynthesis was primarily due to internal factors involving CO_2 assimilation with relatively minor effects due to stomatal changes. In *Z. japonica*, exposure to 10/7°C (day/night) for 2 weeks dramatically reduced RUBPCase and C_4 enzyme activity, particularly phosphoenolpyruvate carboxylkinase and phosphoenolpyruvate carboxlyase, commensurate with a 75% reduction in photosynthesis. Comparatively, the chill-insensitive C_4 species *Spartina anglica* was significantly less affected [103]. Chill-damaged tissues may experience a buildup of toxic by-products resulting from disrupted metabolic processes may occur [99].

Chlorophyll loss occurs first in the youngest leaf tissues or those most exposed to light, resulting in bleaching or purple-colored foliage [96]. The increased chilling injury observed as light exposure is increased [96,104] indicates that part of the response may be due to photoinhibition, which increases free radical activity (O_2^-, H_2O_2, and $^1O_2$) causing accelerated chlorophyll degradation and cell membrane damage [105–107]. Some cold-adapted C_4 plants may utilize high levels of carotenoids such as zeaxanthin to protect against photoinhibition [98]. In centipedegrass, activity of the antioxidant superoxide dismutase was decreased when exposed to chilling temperatures of 5°C although catalase (CAT) and peroxidase (POD) activities were increased [95]. Hormones exert some control over chilling injury. Abscisic acid has been associated with increased chill tolerance [108,109] apparently by stimulating antioxidant activity [95,110,111]. Exogenous applications of abscisic acid to centipedegrass decreased the amount of electrolyte leakage from chilling injury, presumably at least partially due to enhance CAT and POD activity [95]. Foliar applications of gibberellic acid$_3$ (GA_3) reduced chilling injury in bermudagrass but hastened injury in St. Augustinegrass possibly due to phytotoxicity [96].

The prevailing theory for the limitation of C_4 plants to perform well in cold climates is a reduction of enzymes associated with C_4 photosynthesis, particularly pyruvate Pi-dikinase which helps regenerate phosphoenolpyruvate (PEP) [98]. This may be true for species adapted to climates where chilling temperatures do not occur during the growing season. For example, photosynthetic inhibition of *Z. japonica* following chilling exposure was likely due to reductions of both PEPCase and PEP carboxylase activity [103]. In other situations, different enzymes may be affected. Controlled experiments with the high altitude C_4 grass *Muhlenbergia montana* (Nutt.) A.S. Hitch. indicated that exposure to chilling temperature during the dark period decreased photosynthetic capacity during the following light period despite sufficient temperatures for photosynthesis (20°C) [98]. Reduced activity of C_4-associated photosynthetic enzymes due to night chilling was likely responsible for reduced photosynthetic activity. Sufficient cold acclimation overcame these inhibitions. However, Rubisco activity was reduced upon exposure to chilling temperatures regardless of acclimation. The authors concluded that the low amount of Rubisco in C_4 plants combined with decreased Rubisco activity at chilling temperatures limits C_4 photosynthesis in cooler climates. Some C_4 species adapted to experiencing chilling temperatures during the growing season appear to have overcome chill-sensitive enzymatic limitations. Photosynthesis and biomass production in some *Miscanthus* spp. from SE Asia montane areas remains stable despite low-temperature conditions [112], while carbon assimilation in *Spartina anglica* is stimulated following chilling exposure [103].

29.3.2 FREEZING INJURY

Freezing damage can occur to both warm- and cool-season grasses. Tolerance to freezing is a complex, quantitative genetic trait encoded by multiple genes and loci [113,114]. Soil temperature near the surface is more important than air temperature because turf crowns at or below the soil surface are the ultimate meristems for new shoots and roots. Low-cut annual bluegrass on golf courses will sometimes appear to survive winter only to die several weeks afterward as other grasses resume growth; crown examination reveals lack of new root growth presumably due to freezing damage of the roots and lower crown tissues [115]. Lack of functional roots and crown vascular elements causes affected plants to desiccate as temperatures warm and transpirational demands exceed moisture absorption.

Not all the tissues of the same plant cold acclimate at the same rate and not all tissues achieve the same degree of freezing tolerance. Differential freezing tolerance has been observed for root, leaf, and crown tissues in perennial ryegrass, Kentucky bluegrass, and red and Chewings fescue, with roots being the least tolerant and crowns and leaves having comparable degrees of freezing tolerance [8]. Pearce [116] showed that tall fescue roots were less cold hardy than shoots when tissues were placed in a freezing bath of methylated spirits; leaves had the greatest freezing tolerance of the three tissue types.

Infrared thermography has been used to show that freezing begins in small roots of perennial ryegrass and supina bluegrass (*Poa supina* Schrad.), then ramifies quickly throughout the root system and into the lower crown [117]. Freezing of the crown is relatively slow, perhaps due to small, disjointed xylem elements. Once the upper crown is frozen, ice rapidly spreads upward into leaf blades. Ice-nucleating bacteria (*Pseudomonas syringae*) placed on the surface of roots, crowns, and leaves failed to incite freezing events, indicating that freezing begins within cool-season turfgrasses [117] rather than spreading into the plant from outside through foliage or buds like many dicot species [118,119]. The speed of ice progression [117] and extensive discoloration of xylem tissues reported by Beard and Olien [115] suggest that water in the xylem is responsible for rapid progression of intercellular ice in plants subjected to freezing.

29.3.2.1 Intercellular and Intracellular Ice Formation

Most freezing damage to cells occurs indirectly as intercellular water freezes. Above $-40°C$, ice initiation requires a nucleator such as bacteria, carbohydrates, or phospholipids [39,118,120]. Water inside plant cells has a lower water potential than intercellular water due to the presence of dissolved solutes [121]. The water potential of the intercellular water decreases as ice forms, rapidly drawing water out of adjoining plant cells until an equilibrium is achieved [39]. Cell death is essentially caused by dehydration as too much moisture is lost from the cytoplasm, causing loss of membrane integrity, protein denaturation, and solute precipitation [26,39]. Not all the water in a plant need necessarily freeze for damage to occur: Gusta et al. [122] reported approximately 5–10% of the water in cereals remained unfrozen even at $-40°C$. Upon thawing, affected leaves and shoots lose turgidity and appear water-soaked, followed by necrosis.

Tissues with lower total water or water potential generally freeze at lower temperatures than those with greater water or water potential [54,122]. Gusta et al. [122] reported that wheat varieties with inherently lesser amounts of water were more freeze tolerant than varieties with greater amounts of water. Freezing-intolerant varieties were killed when a relatively low percentage of their water froze compared with freezing-tolerant varieties. Adaptations to lesser water potentials may be one of the reasons why turfgrasses exposed to moderate drought stress tend to be more freeze-resistant than well-watered turfgrasses [86,123].

Other factors can influence the extent of freezing damage. Injury is more severe as the length of time the tissue is frozen extends [14,53,124]. Freezing tolerance increases with plant age to a certain point then declines [125,126].

Intracellular freezing or puncturing of cell membranes from intercellular ice formation is rare but possible in sensitive species [39], most likely occurring at freezing rates of 10 to $100°C$ min^{-1} [121]. In nature, subzero air temperature changes rarely if ever occur at greater than $2°C$ hr^{-1} [127].

29.3.2.2 Species Tolerance to Freezing Damage

Freezing tolerance of various turfgrass species is the result of the interaction between genotype and the environment. The genetic makeup of a turfgrass species is ultimately responsible for its potential to cold acclimate and develop freezing tolerance. Thus, the freezing tolerance difference between species is usually much larger than the differences exhibited between cultivars within a species. However, large differences may exist within species when cultivars have different ploidy levels. For example, buffalograss has at least three ploidy levels: diploid, tetraploid, and hexaploid [128]. Qian et al. [6] reported that freezing tolerance of the hexaploid cultivar is much higher than the diploid cultivars.

TABLE 29.1

Relative Low-Temperature Tolerance for Both Cool- and Warm-Season Grasses

	Ranking of Low-Temperature Tolerance	Species
Cool-season	Excellent	Rough bluegrass (*Poa trivialis* L.)
		Creeping bentgrass (*Agrostis stolonifera* L.)
	Good	Kentucky bluegrass (*Poa pratensis* L.)
		Colonial bentgrass (*Agrostis capillaris* L.)
		Red top (*Agrostis alba* L.)
	Medium	Annual bluegrass (*Poa annua* L.)
		Red fescue (*Fetsuca rubra* L.)
		Tall fescue (*Fetuca arundinaceae* Schreb.)
		Meadow fescue (*Festuca elatior* L.)
	Poor	Perennial ryegrass (*Lolium perenne* L.)
	Very poor	Italian ryegrass (*Lolium multiflorum* Lam.)
Warm-season	Excellent	Buffalograss (*Buchloe dactyloides* [Nutt.] Engelm.)
		Blue grama (*Bouteloua gracilis* [H.B.K.] Lag. ex Steud.)
	Good	Zoysiagrass (*Zoysia japonica*)
	Medium	Bermudagrass (*Cynodon* spp.)
	Poor	Centipedegrass (*Eremochloa ophiuroides* [Munro.] Hack.)
		Seashore paspalum (*Paspalum vaginatum* Sw.)
	Very poor	St. Augustinegrass (*Stenotaphrum secundatum* [Walt.] Kuntze)
		Bahiagrass (*Paspalum notatum* Flugge.)

Source: After Beard, J.B., *Turfgrass: Science and Culture*, Prentice Hall, Englewood Cliffs, NJ, 1973; Fry, J.D. and Huang, B., *Applied Turfgrass Science and Physiology*, Wiley, Hoboken, NJ, 2004.

Freezing tolerance can be evaluated by assessing survival of whole plants following exposure to subfreezing temperatures in either field [3,53,88] or laboratory conditions [53,129]. Many researchers have found it expedient to gauge relative cold tolerance by exposing crown or other meristematic tissue pieces to subfreezing temperatures in controlled laboratory conditions [10,130]. Survival can be assessed by measuring regrowth from meristems [6,10,130] or more quickly estimated by measuring electrolyte leakage [8,54,131,132]. While the electrolyte leakage test may be useful for determining relative cold tolerance among species or prefreezing treatments, it does not always provide reliable data [8]. Gusta et al. [130] used regrowth from crowns to determine relatively precise lethal temperatures for 50% of the test populations (LT_{50} values) for several grass species. Varieties within perennial ryegrass and Kentucky bluegrass exhibited a range of LT_{50} values, with ryegrass ranging from −5°C to −15°C and Kentucky bluegrass from −21°C to −30°C. Three creeping bentgrass varieties all had LT_{50} values of −35°C [130]. Table 29.1 lists the relative ranking of low-temperature tolerance of both warm- and cool-season grasses. The ranking is primarily based on data of the killing temperatures after grasses are cold acclimated.

29.3.2.3 Shade Stress Effect on Cold Tolerance

Excessive shading may reduce plant cold tolerance by reducing plant vigor [133–135]. Shading of turf is commonplace as up to 25% of turf in the United States may be subject to shading [24]. The exact mechanism(s) of shade effects on cold tolerance have not been elucidated but a combination of factors may be important. Shading can restrict nonstructural carbohydrates in tissues if photosynthesis is limited by light quality or quantity [134,136]. Insufficient carbohydrates may decrease cold tolerance by providing too little osmoticum for cell water adjustment [10]. Shading may also cause carbohydrate supply to be insufficient for maintenance respiration during nonphotosynthetic periods of winter. Turf attrition occurs as weakened shoots or plants are killed and not replaced.

29.4 OTHER CAUSES OF WINTERKILL

Winterkill is a generic term for turf death that occurs during winter. Both warm- and cool-season grasses may be affected. In addition to low-temperature injury, there are four other primary causes of winterkill, which are preventable to varying degrees.

29.4.1 FROST HEAVING

Frost heaving occurs on newly established turf in areas subject to freezing soil [137,138]. Soil upheaval occurs as water freezes and expands, pushing the soil upward toward the surface. While it can occur on established turf, young turfstands are most susceptible to dying because the lack of root mass allows the seedlings to be pushed out of the soil. The exposed roots cause the plants to die from either low-temperature kill or desiccation. On established turf, frost heaving is primarily a nuisance issue as the turf is readily pushed back into the soil void after thawing occurs. Frost heaving is most likely to occur in soils with high water content. Soils with high amounts of clay and silt are most susceptible because they can hold more water than sandy soils. Turf death from frost heaving can be prevented by planting early enough before winter to ensure a sufficiently dense root mass and by providing sufficient surface drainage to quickly allow the soil to reach field capacity or below should a late autumn or winter rainfall occur [24].

29.4.2 WINTER DESICCATION

Winter desiccation affects turf similarly to desiccation or drought stress at any time of the year. As excessive water is lost from plant cells, organelles and membranes are disrupted and some or all the critical functions of plant cell metabolism are irrevocably lost [26,39]. Winter desiccation occurs frequently but in most cases is limited to leaves that are replaced in the spring [24,139]. The death of turf leaves may promote crown survival as evapotranspiration nearly ceases due to the absence of functional stomata and the dead leaves act as a cover to inhibit moisture loss from below the turf canopy. Crown desiccation results in dead turf and is most likely to occur on elevated sites with exposed green, actively transpiring leaves during windy conditions when air temperatures are above freezing, but frozen soil prevents moisture absorption by roots [24,140]. Bunch-type grasses are more susceptible than species capable of recovering from rhizomes or stolons [24]. Desiccation is more likely to occur on sandy soils than mineral soils that have a greater moisture holding capacity and consequently are slower to freeze. Snow cover throughout the winter prevents desiccation and moderates temperature to reduce low temperature injury [141–143].

29.4.3 LOW-TEMPERATURE FUNGI

Low-temperature (psycrophilic) fungi can damage or kill both warm- and cool-season grasses during the winter. Spring dead spot (SDS) disease is a root rot of bermudagrass that occurs primarily if not entirely during the winter dormancy period and is the most important disease affecting *Cynodon* spp. in the United States and Australia [144]. Affected turf fails to develop new leaf tissue in spring, resulting in patches of dead turf ranging from several cm to one meter diameter [145]. *Gauemannnomyces graminis* (Sacc.) Arx & Olivier var. *graminis* was originally reported to cause spring dead spot [146] followed by *Leptosphaeria korrae* J.C. Walker & A.M. Smith [147,148]. *L. korrae* has since been renamed as *Ophiosphaerella korrae* while *O. herpotricha* and *O. narmari* have also been identified as causal agents of spring dead spot [149–152]. Infection by SDS pathogens reduces bermudagrass cold tolerance [153–155]. Cold acclimation further decreases the freezing tolerance of infected cultivars compared with nonacclimated plants [156]. Low-temperature tolerance of specific cultivars can be negated if infected by *Ophiosphaerella* spp. [155].

Cool-season grasses are susceptible to a group of fungi causing several diseases collectively known as snow molds. Snow mold diseases can be severe and kill large areas of turf. Although ranges of the snow mold fungi often overlap, the pathogens do exhibit some preferences for certain climates or conditions [157–160]. Snow scald, caused by *Myriosclerotinia borealis* (Bubák & Vleugel) L. M. Kohn = *Sclerotinia borealis* Bub. and Vleug., and speckled snow mold (causal agent *Typhula ishikariensis* Imai) are most likely to occur when snow cover exceeds 200 days [161]. At least three varieties of *T. ishikariensis* exist: var. *ishikariensis*, var *idahoensis*, and var. *canadensis* [158,162]. Gray snow mold (causal agent *Typhula incarnata* Lasch ex Fr.) occurs when snow cover exceeds 70 days and often in milder climates than speckled snow mold or snow scald [163]. For example, only *T. incarnata* occurs in the eastern United States although it often overlaps the range of *T. ishikariensis*, which has been reported in the upper Midwest of the United States, Canada, Scandanavia, and northern Japan [157,158,162,164,165]. The causal agent of *Microdochium* patch, sometimes known as pink snow mold, has had several names during the twentieth century, including *Gerlachia* patch and *Fusarium* patch though it is currently is known as *Microdochium nivale* (Fr.) Samuels & Hallet [144]. *M. nivale* has the widest distribution of the snow mold fungi as it does not require snow cover and occurs over a much wider temperature range, capable of infecting turf at temperatures up to approximately 15°C [165]. *Microdochium* patch disease may extend long into the growing season, particularly where temperatures are moderated by shade [144].

Several other pyscrophilic fungi have also been implicated as low-temperature grass pathogens but are rarely reported [144,166]. All are primarily foliar diseases though in some situations the turf may be thoroughly killed [144,165]. Most snow mold pathogens can grow to −5°C due to antifreeze proteins and solutes that depress cell freezing temperature [167].

Immature grasses have less snow mold resistance than mature grasses [168]. Cold acclimation reduces the severity of snow mold diseases on grasses [169–171]. The exact relationship of snow mold resistance and cold tolerance is not well understood. Snow mold resistance is not lost to the extent of cold acclimation when plants are subjected to temporary thaws during the winter [27,88,169,172]. Host resistance may be due in part to decreases in cell water potential and changes in plant carbohydrates during cold acclimation [166,171]. Greater accumulation of carbohydrates during cold acclimation and slower carbohydrate metabolism during winter, particularly of fructans, is related to snow mold resistance [166]. The mechanism of resistance is likely related to activation of host defense genes mediated by hexokinase [166]. Genes for pathogenesis-related proteins (PR proteins), including chitinase, β-1,3-glucanase, peroxidase, and pathogenesis-related (PR) protein 1a were upregulated more in cold-hardened plants when infected by *M. nivale* than in nonhardened plants [66,170]. Chitinases and glucanases can degrade cell walls of fungal pathogens while peroxidases affect a number of plant defense reactions including lignification and phytoalexin production [173,174].

29.4.4 Ice Cover and Encasement

Ice damage is relatively infrequent compared with other forms of winterkill but can be devastating when it does occur. Ice cover occurs when surface water freezes. The underlying soil may or may not be frozen if surface water from rainfall or snowmelt is quickly frozen by a sudden decrease in air temperature while unfrozen soil continues to drain [175]. Ice encasement is a more severe form of ice cover. Liquid precipitation infiltrating into poorly draining unfrozen soil followed rapidly by freezing may cause a continuous ice profile from below to above ground.

Acclimation to ice cover is enhanced by low-temperature acclimation [175]. However, low-temperature tolerance does not necessarily correspond to ice cover tolerance [176]. On January 12, 2005, a 2.5-cm rainfall over partially frozen soil, followed by rapid and complete freezing of the soil, caused ice encasement and complete ice cover of the 2001 National Turfgrass Evaluation Program tall fescue test at Madison, WI. All 159 varieties of tall fescue died when encased in ice for approximately 60 days [177] even though air and soil temperatures were similar to the previous winter (Figure 29.3). Both ice cover and encasement decrease cold hardiness at a greater rate than when grasses are covered with snow [178]. In a perennial type of annual bluegrass (*Poa annua* f. *reptans* [Hausskn.] T. Koyama) maintained at putting green height, the LD_{50} decreased fourfold between 45 and 90 days under ice cover but remained unchanged when under snow; creeping bentgrass cold tolerance was less severely reduced [178].

Various mechanisms for turf kill under ice cover may exist. The rate of plant death increases as the freezing temperature decreases during development of ice encasement [175]. In alfalfa, oxygen radicals from primary freezing lesions have been proposed as a mechanism leading to larger-scale damage as antioxidants such as superoxide dismutase are lost over time when plants are

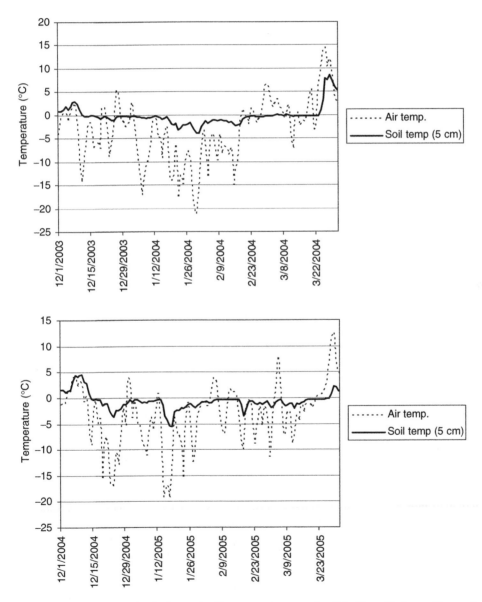

FIGURE 29.3 Mean air and soil (5 cm depth) temperatures at the O.J. Noer Turfgrass Research and Educational Facility, Madison, WI, for two winters, the second of which resulted in the death of all tall fescue varieties in the 2001 National Turfgrass Evaluation Program test due to ice cover, which lasted from January 13 to approximately March 10, 2005.

frozen [179]. Oxygen radicals may come from frozen microsomal membranes [180], reduction of oxygen by electrons as freezing interrupts electron transport mechanisms, or other sources [179]. Toxic levels of compounds such as malate, citrate, and succinate may be released as plant cells collapse from freezing, confounding the effect of ice cover on grass death [181].

However, low-temperature exposure is not likely a sole or even direct cause of death as temperatures in some cases are similar under ice and snow cover even when turf is damaged more under ice than snow [175,143]. Plant death under ice is also not likely due to lack of plant carbohydrates for respiration [182,183]. Instead, death may occur from one or more indirect causes due to lack of air exchange between the turf/soil profile and the aboveground atmosphere [175,178,184].

Both ice cover and encasement limit gas exchange between the turf/soil profile and the aboveground atmosphere [175,185,186]. Ice encasement decreases survival more than simple ice cover because solid ice more effectively prevents gas exchange, which increases the likelihood or extent of hypoxic or anoxic conditions [178,186,187]. Soil–atmosphere gas exchange may still occur when an ice sheet develops over a frozen but unsaturated soil or into the air space between the ice and soil created when ice forms from standing water over unfrozen and still-draining soils [175,188]. Irregular, granulated ice formation may allow sufficient aeration for localized plant survival [175]. Penetration of plant tissues through an ice layer further reduce ice cover's deleterious effects, presumably by enhancing gas exchange [189]. Likewise, ice over snow-covered turf is less damaging than ice directly over turf because snow porosity allows gas exchange to occur [187,190].

Anoxia (lack of oxygen), buildup of toxic metabolites in plants, and toxic gases have been proposed as primary causes of grass damage under ice [181,184,187,191–193]. Under field conditions, anoxia is likely preceded by a hypoxic period (limited oxygen), which can increase tolerance to anoxia [194–196]. Anoxia induces anaerobic metabolism (fermentation) in plants, which produces less energy than aerobic respiration [193].

In addition to the potentially limiting amounts of ATP and other energy compounds, the resulting fermentation cycles produce lactic acid, ethanol, alanine, and lesser by-products such as aminobutyrate, succinate, tartarate, and acetate [192,197,121]. Annual cereal grasses are less tolerant to ice than perennial grasses and produce greater amounts, in sufficient quantities to destroy cells and kill plants, of ethanol, CO_2, and lactic acid during ice encasement [198–200]. Ice-tolerant perennial grasses such as timothy (*Phleum pretense* L.) and berings hairgrass (*Deschampsia berengensis*) do not accumulate high levels of ethanol during ice encasement but instead accumulate organic acids, acetate, and butyrate [200,201]. Perennial types of annual bluegrass (*Poa annua* var. *reptans*) accumulated alanine, serine, and tryptophan while fructan levels declined under anoxic conditions although it was unknown if these changes were directly toxic [202]. Perennial grass ice tolerance may be due to slower fermentation rates and less subsequent accumulation of toxic compounds compared with annual cereals [193,200]. Other research suggests that buildup of toxic plant metabolites is irrelevant and O_2 supply is most important for survival during ice encasement [192]. Grasses have greater freezing tolerance when light is present during ice encasement, perhaps due to oxygen generation during the freezing process [199].

Toxic levels of CO_2 may be just as, or more, important for turf death under ice, than anoxia alone though anoxic conditions eventually cause high levels of CO_2 as conditions become anaerobic [184,203]. Grasses continue to respire throughout the winter though rates decrease with temperature and are minimal at $-5°C$ [204]. Microbial respiration can occur at soil temperatures as low as $-6°C$, and temperatures of $0°C$ allow respiratory rates close to those observed at $6°C$ [185,205,206]. Soil moisture is directly proportional to respiration, even at subzero temperatures [185,205,207]. Respiration rates are also directly related to soil organic matter content [203]. High soil CO_2 concentrations limit O_2/CO_2 flux between plants and soil. The resulting CO_2 from anaerobic respiration within plants may cause significant damage to plants under ice and kills cells by causing cytoplasmic acidosis [175,184,198]. High-density clear ice may inhibit gas exchange more than low-density "white" or "milk ice" [24]. However, clear ice may transmit more PAR to stimulate photosynthesis which can occur at temperatures as low as $-1°C$ [199]. Subsequent light/dark periods may produce alternating day/night cycles of hypoxia/anoxia [175,199].

Cyanide production by Basidiomycete fungi under ice has been held responsible for turf injury in western Canada [191]. A small proportion of soil bacteria may also cause plant-inhibiting levels of cyanide [208,209]. Ice encasement caused toxic rates of butyrate to develop in hay-fields of orchardgrass (*Dactylis glomerata* L.) and colonial bentgrass (*Agrostis capillaris* L.), with butyrate concentrations and plant death increasing over time [181]. It was unknown whether the butyrate originated from plant or microbial sources though butyrate and other potentially toxic by-products from microbial anaerobic respiration have been reported [210]. Toxic levels of volatile fatty acids resulting from fermentation under ice may also be important [181,211]. When ice cover causes anaerobiosis, oxidation of CH_4 by soil microbes may decrease and CH_4 production may increase although methane has not been reported in association with turf damage [185].

Species differ in their survival under ice though most data focus only on creeping bentgrass and annual bluegrass, as extensive turf damage is usually restricted to closely mowed putting greens [141,143,178,212]. In each study, creeping bentgrass has been found to survive ice significantly better than annual bluegrass [178,187]. Creeping bentgrass was also more tolerant of ice cover than an old variety of colonial bentgrass, "Astoria," while the vegetative bentgrasses "Cohansey" and "Toronto" were superior to seeded bentgrasses [213]. Creeping bentgrass typically is able to survive under ice for >120 days [178,213] while annual bluegrass survival declines after 45–60 days depending on the manner of ice cover [178,187]. Kentucky bluegrass ice tolerance is similar to creeping bentgrass, though Kentucky bluegrass may be killed if traffic occurs over slushy conditions followed by refreezing [190].

29.5 MANAGING TURF TO AVOID OR MITIGATE LOW-TEMPERATURE INJURY

29.5.1 PREVENTING WINTER DESICCATION

Turf death due to winter desiccation may be reduced by decreasing evaporative losses during the winter. Moisture can be lost from the turf as dry winter air sweeps over the turf, particularly when the air temperature is above freezing and frozen soil conditions prevent water movement into turf roots. Desiccation potential has long been reduced by application of topdressing (approximately 0.6 cm depth) on putting greens or mulches at the end of the growing season or [24,140]. Synthetic turf covers are gaining in popularity as an alternative to avoid damage and playability problems associated with a thick layer of topdressing. Synthetic turf covers, including types that allow some air passage between the turf and the atmosphere, can function as well as snow cover to prevent desiccation injury [24,139,143].

Antitranspirants are chemicals applied to plant foliage to decrease transpirational water loss [214–216]. Some antitranspirants function by forming a film over the leaf surface while others induce stomatal closure [24]. Though antitranspirants are sometimes used to reduce winter desiccation of fine turf [217], studies have not shown them to be effective [218,219].

In the absence of snow or other covers, high-value turf can be irrigated during winter thaws [220]. However, for reasons not totally understood, irrigation during winter thaws may reduce turf survival and recovery in the spring [143].

29.5.2 ENHANCING FREEZING TOLERANCE AND WINTER SURVIVAL

Turf survival may be aided by planting the most cold-tolerant species suitable for the location (Table 29.1). Choices will be guided by site conditions and requirements; for example, while creeping bentgrass has better cold tolerance than perennial ryegrass, it is not suited for athletic fields [142]. Turf should be planted early enough in the growing season so that plants have reached maturity before growth ceases in autumn/winter and sufficient root development has occurred to prevent frost heaving [142]. For certain cool-season grasses such as tall fescue, a minimum of 60 days between germination and winter dormancy are required [126]. Warm-season grasses should be seeded in the spring or summer [142]. Although less successful, turf may be dormant-seeded: seeding is conducted after the growing season so that germination does not occur until the following spring [142].

Turf management during the autumn must not conflict with the natural cold acclimation metabolism and growth of the turf. The primary cultural practices of mowing, fertilization, and irrigation all must be managed with a view to cold tolerance and winter survival. A number of secondary cultural practices may be combined with each other and the primary cultural practices to enhance winter hardiness.

29.5.2.1 Mowing

Mowing should continue until vertical foliage growth has ceased in order to prevent excessively long turf prior to winter, which might encourage some snow mold diseases [16,144]. Mowing heights should not be significantly reduced during autumn: this situation removes significant amounts of nonstructural carbohydrates as the majority is stored in the shoots and leaves [221–223]. In situations where the turf is maintained below the optimal mowing height during the growing season, increasing the mowing height to the upper limit of the optimal range prior to winter may be beneficial by increasing leaf area for photosynthate production. The extra foliage may also further shield turf crowns from freezing injury [24]. Supraoptimal mowing heights should be avoided. In creeping bentgrass, a supraoptimal height of 5 cm reduced cold tolerance, possibly due to aboveground elevation of crowns and stolons [16]. The same authors reported that the cold tolerance of Kentucky bluegrass and red fescue was similar whether turf was maintained at 1.5 or 5 cm height over a 3-year period. However, data in Beard [24] showed Kentucky bluegrass survival increased with mowing height across a range from 1.5 to 5 cm following exposure to both 0°C and −5°C.

29.5.2.2 Fertilization

29.5.2.2.1 Nitrogen Fertility for Cool-Season Turf
Root growth of Kentucky bluegrass continues until soil temperatures approach 0°C allowing appropriately timed autumn fertilizations to enhance root but not foliage growth [81,87,224]. Late fall fertilization also enhances turf color and growth in the spring and sometimes autumn [16,87,225–228]. However, N applied just prior to or during the cold acclimation period reduces cold hardiness of many grasses compared with unfertilized situations [86,87,229,230]. Excessively high N rates during the year and autumn also reduce cold tolerance [86,227,229]. Reduction of cold tolerance resulting from nitrogen applications to cool-season grasses in autumn may be related to the subsequent production of long-chain carbohydrates instead of short-chain polymers [231]. Additionally, crown moisture levels during deacclimating periods in late winter/early spring correspond directly to annual and autumn N rates, with higher crown moisture associated with reduced cold hardiness [86,172].

Nitrogen application to creeping bentgrass after cold-hardening can enhance cold hardiness [16]. Maximum cold hardiness in perennial ryegrass was obtained at low to moderate annual N rates of 49 to 147 kg ha^{-1}, with the final application of the year occurring in late fall after shoot growth ceased and no N applied in October during the first stage of cold acclimation [172]. These N rates were similar to those currently recommended for cool-season turf [92,142,232]. Consequently, N applications should be timed so that foliage growth is not stimulated in late autumn while excessive annual N rates should be avoided [16,172,231].

29.5.2.2.2 Nitrogen Fertility for Warm-Season Turf
High rates of N fertilization (245 kg ha^{-1}) applied throughout autumn were harmful to zoysiagrass growing in the transition zone [227]. Other warm-season turfgrasses appear to be less fastidious

regarding timing of N fertility. Cold tolerance of centipedegrass [*Eremochloa ophiuroides* (Munro) Hack] was slightly enhanced at an autumn-applied N rate of 196 kg ha^{-1} [233]. Cold tolerance of bermudagrass appears to be relatively unaffected by autumn N applications while spring greenup is improved [225,228,234,235].

29.5.2.2.3 Potassium Fertility

Potassium fertilization has often been reported to enhance cold tolerance [172,227,233,236,237] but not all studies have shown an effect. Cook and Duff [238] failed to see a relationship between K and freezing tolerance of tall fescue while Miller and Dickens [239] reported that increased K fertilization and tissue concentrations did not affect cold tolerance of "Tifdwarf" and "Tifway" bermudagrasses. Dunn et al. [227] reported autumn applications of K actually increased winter injury of zoysiagrass. Potassium's ability to enhance cold tolerance in cool- and warm-season grasses may depend on the assessment method and N:K as ratios appear to influence both crown moisture and cold tolerance [172,233]. Some reports indicate that a 2:1 fertilizer ratio of N:K is desirable for maximum winter hardiness of cool-season grasses [236,237]. Gilbert and Davis [234] reported that a 2:3 ratio of N:K applied in late summer provided the best winter hardiness of bermudagrass with a ratio of 4-1-6 N-P-K being ideal. The best ratio may depend on the rates of N and K being used [172]. Annual K additions of ≥196 kg ha^{-1} increased centipedgrass winter survival compared with lower rates when N was supplied at less than a 1:1 ratio of N:K [233]. Cold tolerance of perennial ryegrass was greatest at an N:K ratio of 1:1.7 when annual N rates ranged from 49 to 147 kg ha^{-1} and K rates ranged from 245 to 441 kg ha^{-1} [172].

29.5.2.3 Irrigation

Irrigation should be reduced or halted during the autumn to facilitate hardening and improve cold stress tolerance by avoiding high water content in the crowns [85,88]. Irrigation should be supplied only to avoid obvious drought stress as a reduction in plant cell hydration is part of the cold acclimation process and mild drought stress increases cold hardiness [1,86]. Sites predisposed to holding excess water should be renovated to avoid potentially lethal ice cover or encasement during the winter, particularly where sensitive species such as tall fescue and annual bluegrass exist. Cold hardiness of bermudagrass may be increased by applying moderate rates of salt or effluent water prior to cold acclimation [240]. Proline content, a useful osmolyte for enhancing cold tolerance, increased in "Princess" bermudagrass when irrigated with water having either 5 or 20 dS m^{-1} compared with either 0 or 40 dS m^{-1} [240]. Regrowth of cold acclimated plants following exposure to $-7°C$ was enhanced in plants irrigated with water having either 5 or 20 dS m^{-1}.

29.5.2.4 Plant-Growth Regulators

Plant-growth regulators that suppress gibberellic acid (GA) biosynthesis can enhance NSC supplies [241,242] that are necessary for maximum cold tolerance [3,10,31,32,51,52]. Application of GA inhibitors throughout the growing season improves cold tolerance of cool-season grasses, particularly when mild winter temperatures cause deacclimation [243,244]. The mechanism of cold tolerance may be due at least in part to decreased respiration, which minimizes NSC loss during winter [244,245]. Late-autumn application of GA inhibitors to cool-season grasses, which prevents full metabolism of the regulators prior to winter dormancy, causes delayed spring greenup (*J. Stier, unpublished data*). In bermudagrass, summer applications of the GA inhibitor trinexapac-ethyl (TE) delayed winter dormancy while late-season applications hastened winter dormancy, though neither set of treatments affected freezing tolerance [246]. Richardson [235] reported TE extended the growing season of bermudagrass by delaying winter dormancy and promoting early spring greenup without affecting freezing tolerance of rhizomes.

Exogenous application of synthetic or natural sources of plant hormones such as cytokinins and auxins can inhibit senescence and reduce abiotic stress damage to turf, particularly drought

[247,248]. However, the synthetic cytokinin benzyladenine did not affect carbon exchange rates, total NSC, or color when treated bermudagrass was subjected to chilling temperatures [100]. Likewise, Munshaw et al. [228] reported no effects of seaweed extracts containing cytokinins and other hormones on cold tolerance, proline, fatty acids, or spring greenup of four bermudagrass cultivars representing seeded and vegetative types.

29.5.2.5 Synthetic Covers

Sufficient snow cover protects turf from low-temperature kill by moderating temperature [24,141]. Rainfall that penetrates and compacts the snow may reduce the efficacy of snow to moderate temperatures [175,212]. In the absence of snow cover, synthetic covers may reduce cold injury, moderate temperature, and reduce the length of winter dormancy [139,212,249,250]. Under most conditions, spun-bonded or loosely woven polyester covers provide better results than solid plastic covers though both can moderate temperature extremes and hastening postdormancy recovery [139,212,250,251]. Spun-bonded and woven polyester covers have the added benefits of allowing gas exchange and better penetration of PAR than solid polypropylene covers, thus enhancing photosynthetic activity during suitable temperatures and reducing buildup of noxious gases [139,175]. Covers must be used cautiously to avoid conditions that incite disease or stimulate turf growth while decreasing hardiness at critical times. Shashikumar and Nus [250] reported that spun-bonded covers increased moisture content of bermudagrass crowns during the winter and also decreased the LD_{50} by 1–2°C for the first 8 weeks of winter cover, though the higher moisture did promote root growth. Roberts [139] found bentgrass leaf moisture in springtime was increased 10–20% and up to 24% greater root length occurred compared with uncovered turf. Soil temperatures were 10°C greater and spring greenup occurred 5–12 days earlier when turf was covered with a spun-bonded polyester blanket.

Use of clear, impermeable polypropylene covers in early spring only avoided a lengthy period of gas exchange limitations and hastened the postdormancy recovery of bermudagrass [249]. In some instances, impermeable covers result in anoxic conditions though annual bluegrass may survive for at least 40–50 days [203,252]. Impermeable covers may also increase disease incidence when used over porous organic mulches or covers [212].

Ultimately, any given cover will provide varying results depending on the environmental conditions to which the turf is exposed immediately before, during, and after use of the cover [141]. In most cases, synthetic covers should be applied as late in the winter as possible, preferably after the ground is frozen and snow mold fungicides applied, to prevent inducing turf growth and fungal pathogens by creating a "greenhouse effect." Covers should be removed in late winter before soil thaws to avoid artificially raising temperatures, which leads to turf and fungal pathogen growth. If covers are left on too long, growth-inducing temperatures precipitate lush turf growth commensurate with loss of cold hardiness, predisposing turf to low temperature injury.

29.5.2.6 Thatch Management

Turf should be managed so as to avoid excessive thatch [135]. The large amount of roots, crowns, stolons, and rhizomes in thatch are less protected from temperature extremes than when growing in the soil. Freezing, which begins in the roots, quickly migrates to the crown, killing meristematic tissue [117]. Protection of roots from freezing temperatures may therefore aid in plant survival.

29.5.2.7 Tree Leaf Removal

Tree leaves should be removed from turf as soon as possible after they fall to the ground in the autumn. This is a critical time for turf to photosynthesize and store surplus carbohydrates for winter maintenance respiration and cold tolerance. Optimizing the ability for photosynthesis in the autumn may be especially important for turf subjected to shade during the growing season in order to enhance carbohydrate production.

29.5.2.8 Winter Traffic

Winter traffic that comes into direct contact with dormant or frozen tissues should be avoided. Schmidt et al. [249] showed that traffic on dormant bermudagrass reduced growth the following growing season. In climates subject to snow, snowmobile and other traffic is not of concern as long as a minimum of 3 cm of snow covers the turf [253]. However, traffic over wet, slushy ground followed by freezing is more likely to kill turf [190]. Perennial ryegrass, tall fescue, red fescue, and Kentucky bluegrass are all more sensitive to traffic injury over wet, slushy ground followed by freezing compared with rough bluegrass (*Poa trivialis* L.), creeping bentgrass, and colonial bentgrass [85,190].

29.5.2.9 Ice Removal

Turf loss from ice cover is best prevented by ensuring good drainage, especially during thaws [24]. Surface drainage may be sufficient if internal drainage is insufficient. Poorly drained areas should be reconstructed during the growing season to minimize the potential for standing water to freeze.

When ice forms, it should be removed as soon as possible if the turf is sensitive to ice. Most or all warm-season grasses, annual bluegrass, tall fescue, and perennial ryegrass are noticeably more susceptible to ice-related damage than bentgrass, rough and Kentucky bluegrasses, and fine fescues. Though some reports indicate annual bluegrass survives without injury for at least 45 days under ice cover [178,187], severe injury to putting green turf may occur within the first 15 days of ice cover [143]. Avoid mechanical damage to turf when removing ice. Aerators have been used to make holes for gas exchange but depending on the thickness of ice and the force of the aerator is it often difficult to penetrate the ice. Superintendents often apply a covering of black particles such as soot, black sand, or Milorganite® (Milwaukee Metropolitan Sewage District, Milwaukee, Wisconsin) to melt holes in ice due to solar heat absorption by the particles [24,135]. Synthetic salt-based, granular fertilizers can melt holes in ice but salt accumulation may kill turf; so these are best avoided. Various alcohol and other formulations have been tried which effectively melt ice but most still severely injure turf (J. Stier, unpublished data).

29.5.2.10 Disease Management

All cool-season turfgrasses are susceptible to snow mold diseases although a range of susceptibility exists among species [165]. *M. nivale* is more severe on annual bluegrass and Kentucky bluegrass than on creeping bentgrass [165]. Fine fescues are more resistant to *Typhula* and *Microdochium* snow molds than colonial bentgrass (*A. capillaris* L.), which is more resistant than creeping bentgrass [254,255]. Annual bluegrass is especially susceptible to *Typhula* and *Microdochium* snow molds [165]. No cultivars have sufficient resistance to preclude the use of fungicides in high-value turf [165]. However, individual creeping bentgrass plants collected from old golf courses in Wisconsin not treated with fungicides exhibit good to excellent resistance, indicating that the potential for development of disease-resistant varieties does exist [256].

Snow mold diseases on fine turf (golf-course greens/tees/fairways and high end sports turf) are primarily managed with fungicides. Preventive treatments are applied in autumn and curative applications in late winter/early spring as snow melts [144]. While many fungicides are labeled for several types of snow mold, efficacy varies even between *Typhula* spp. and varieties [144,257]. In general, combinations of iprodione, pentachloronitrobenzene (PCNB), and chlorothalonil applied just prior to snowfall, but before leaf growth completely stops, provide the best overall control of

Typhula and *Microdochium* snow molds [144,165,258]. Demethylation inhibiting fungicides (DMIs, including triadimefon, propiconazole, and fenarimol) are not effective against *T. ishikariensis* [165]. *M. nivale* resistance to benzimidazole and dicarboximide fungicides has been reported [165].

Spring dead spot disease is controlled through a combination of approaches. The disease is typically associated with close-cut turf and increasing the mowing height may reduce severity in vegetatively propagated cultivars [144,259]. Among seeded cultivars, patch size may be decreased by mowing at 1.3 cm height rather than 3.8 cm [260]. Ammonia-based N fertilizers that acidify the soil reduce the severity of SDS, while both ammonia- and urea-based N sources enhanced recovery [261]. Penetrant-type fungicides applied in autumn and TE reduce the amount of affected turf [144,262]. Tisserat and Fry [263] reported that aeration plus vertical mowing reduced SDS severity as did removing affected sod, which was replaced by new growth from remaining rhizomes.

While dormant nitrogen application for turf growth and winter hardiness is desirable, both excessive N applications during the hardening period and applications after the first frost are undesirable from a standpoint of controlling snow molds [144,165,258]. Potassium applications may suppress *Microdochium* snow mold [258]. Turf should be managed to avoid excessive thatch as these pathogens are capable of surviving as saprophytes on senescent turf tissues [258]. Avoid excessively long grass prior to winter as leaves lying over each other increase humidity under the canopy and favor pink snow mold [264]. Traffic that compacts snow over turf and large snowdrifts should be prevented [144].

29.5.2.11 Artificial Soil Heating

Underground heating systems are sometimes used to maintain green color and turf growth on putting greens and sports fields during suboptimal temperatures [24,92,265]. Systems used include electric cables, forced-heated air, and hot water/glycol in subterranean pipes [24,266–270]. Cables and pipes need to be as close to the surface as possible but deep enough to avoid damage from turf cultivation practices such as core cultivators, usually being placed at a 15- to 25-cm depth [92,266,271]. Soil temperatures for cool-season grasses must be maintained above 2°C and above 16°C for warm-season grasses to avoid low-temperature dormancy [24]. Soil heating may also encourage seed germination and sod rooting [24,272].

Electrical systems have been based on wires spread 15 to 23 cm apart using 107 W m^{-2} [266]. Hot water/glycol systems that pump heated solution in a closed system through a 1.6-cm-diameter polypropylene pipe, typically spaced 15 to 23 cm apart, have become the most common form of soil heating in recent years [92,271]. The system at Lambeau Field in Green Bay, Wisconsin, uses five boilers to heat the solution totaling 7.5 million Btu when operating at maximum capacity [271]. At full capacity, the system heated a water–glycol solution to 16°C; when piped belowground at 25 cm depth, a surface temperature under an Evergreen® (Covermaster, Rexdale, Ontario, Canada) cover of 10°C was maintained even when air temperatures declined to −6°C for several days [272].

Heating requirements can be expensive; the cost to heat Lambeau Field during December 1997 was approximately $12,000 (T. Edlebeck, personal communication). Tarps or mulch covers may be used to reduce heat loss [24]. Made of woven polyethylene, Evergreen covers allow limited air exchange between turf and the atmosphere and are the most commonly used turf covers. Straw mulch, Tyvek® (E.I. du Pont de Nemours & Company, Wilmington, Delaware), and straw plus Tyvek overlaying an Evergreen turf cover caused temperatures at the turf surface and 2.5 cm subsurface to be as much as 5°C greater compared with Evergreen cover alone on a sand-based sports field heated with a hot water/glycol pipe system [272]. Stier et al. [272] found that Evergreen cover alone was more desirable than addition of other insulation as bermudagrass and perennial ryegrass survival was reduced when straw or Tyvek was placed over the Evergreen cover (Table 29.2).

TABLE 29.2

Winter Survival of Common Bermudagrass (*Cynodon dactylon* [L.] Pers.; Cd) and Perennial Ryegrass (*Lolium perenne* L.; Lp) under Varying Covers in a Subsurface-Heated Sand-Based Sports Field, Green Bay, WI, April 23, 1998

	Temperature of Water–Glycol Solution 25 cm below Surface							
	Unheated		4.4°C		10°C		15.5°C	
	Number of Plants 81 cm^{-2}							
Type of cover	Cd	Lp	Cd	Lp	Cd	Lp	Cd	Lp
Straw (S) + E	7.0	3.0	13.5	0.0	14.5	0.5	18.0	0.5
Tyvek (T) + E	10.5	20.5	10.0	11.5	13.0	7.5	12.5	7.5
Evergreen only (E)	26.5	7.5	18.0	6.0	14.0	11.0	28.5	18.5
S + T + E	10.0	0.0	4.5	4.0	12.5	1.0	23.0	1.5
LSD (0.05)	10.9	ns[a]	4.9	ns	ns	2.9	ns	5.8

Note: Covers included Evergreen™ (E) woven polyethylene tarp, Tyvek® (T), and 5-cm depth wheat straw (*Triticum aestivum* L.).

[a] Not significant at $P \leq 0.05$.

29.6 CONCLUSION

All grasses are potentially susceptible to cold stress injury. Most grasses utilize a combination of physical, anatomical, physiological, and biochemical characteristics to avoid or tolerate low temperatures. An increase in nonstructural carbohydrates, unsaturated membrane lipids, and specialized proteins during cold acclimation are common traits for low-temperature tolerance among warm- and cool-season grasses. A decrease in free water, particularly in the meristematic crowns, is universal during the cold hardening process and varieties with greater cold tolerance generally have lesser water potential than more susceptible grasses. Cold acclimation is a two-phase process, with the first phase occurring over several days to weeks at above-freezing temperatures and the final phase occurring at or just below freezing temperatures.

A number of management practices affect cold tolerance. Correct N and K fertilization rates and ratios are important for a number of species. Nitrogen applications and excessive moisture during cold acclimation decrease cold tolerance. Plant-growth regulators that inhibit gibberellic acid production may enhance cold tolerance, particularly among cool-season grasses. Ice cover and encasement can be deadly for some turfgrasses with potential causes ranging from anoxia to accumulation of toxic substances or gases. Ice removal is recommended for sensitive species though avoiding conditions conducive to ice formation is ultimately the best management strategy.

REFERENCES

1. Thomashow, M.F., Plant cold acclimation: Freezing tolerance genes and regulatory mechanisms, *Annu. Rev. Plant Physiol. Plant Mol. Biol.*, 50, 571–599, 1999.
2. Fry, J.D., and Huang, B., *Applied Turfgrass Science and Physiology*, Wiley, Hoboken, NJ, 2004.
3. Dionne, J. et al., Freezing tolerance and carbohydrate changes during cold acclimation of green–type annual bluegrass (*Poa annua* L.) ecotypes, *Crop Sci.*, 41, 443–451, 2001.
4. Xiong Y., and Fei, S., Functional and phylogenetic analysis of a DREB/CBF–like gene in perennial ryegrass (*Lolium perenne* L.), *Planta*, DOI: 10.1007/s00425-006-0273-5, 2006.
5. Ibitayo O.O., and Butler, J.D., Cold hardiness of bermudagrass and *Paspalum vaginatum* Sw., *HortScience*, 16, 683–684, 1981.
6. Qian, Y.L. et al., Freezing tolerance of six cultivars of buffalograss, *Crop Sci.*, 41, 1174–1178, 2001.

7. Gray, R.G. et al., Cold acclimation and freezing tolerance—A complex interaction of light and temperature, *Plant Physiol.*, 114, 467–474, 1997.

8. Rajashekar, C., Tao, D., and Li, P., Freezing resistance and cold acclimation in turfgrasses, *HortScience*, 18, 91–93, 1983.

9. Fry, J.D., Lang, N.S., and Clifton, R.G.P., Freezing resistance and carbohydrates composition of 'Floratam' St. Augustinegarss, *HortScience*, 26,1537–1539, 1991.

10. Shahba, M.A. et al., Relationships of soluble carbohydrates and freeze tolerance in Saltgrass, *Crop Sci.*, 43, 2148–2153, 2003.

11. Anderson, J.A., Kenna, M.P., and Taliaferro, C.M., Cold hardiness of "Midiron" and 'Tifgreen' bermudagrass, *HortScience*, 23, 748–750, 1988.

12. Anderson, J.A., Taliaferro, C.M., and Martin, D., Evaluating freeze tolerance of bermudagrass in a controlled environment, *HortScience*, 28, 955, 1993.

13. Cardona, C.A., Duncan, R.R., and Lindstrom, O., Low temperature tolerance assessment in paspalum, *Crop Sci.*, 37, 1283–1289, 1997.

14. Anderson, J.A., Taliaferro, C.M., and Martin, D., Longer exposure durations increase freeze damage to turf bermudagrasses, *Crop Sci.*, 43, 973–977, 2003.

15. Eagles, C.F., and Williams, J., Hardening and dehardening of *Lolium perenne* in response to fluctuating temperatures, *Ann. Bot.*, 70, 333–338, 1992.

16. White, D.B., and Smithberg, M.H., Cold acclimation and deacclimation in cool-season grasses, in *Proc. 3rd Int. Turfgrass Res. Conf.*, Beard, J.B., Ed., ASA-CSSA-SSSA and Int. Turfgrass Soc., Madison, WI, 1980, 149–154.

17. Uemura, M., Joseph, R.A., and Steponkus, P.L., Cold acclimation of *Arabidopsis thaliana*: Effect on plasma membrane lipid composition and freeze-induced lesions, *Plant Physiol.*, 109, 15–30, 1995.

18. Olien, C.R., An adaptive response of rye to freezing, *Crop Sci.*, 24, 51–54, 1984.

19. Livingston, D.P. III., The second phase of cold hardening: Freezing tolerance and fructan isomer changes in winter cereal crowns, *Crop Sci.*, 36, 1568–1573, 1996.

20. Gilmour, S.J., Hajela, R.K., and Thomashow, M.F., Cold acclimation in *Arabidopsis thaliana*, *Plant Physiol.*, 87, 745–750, 1988.

21. Strand, A. et al., Development of *Arabidopsis thaliana* leaves at low temperatures releases the suppression of photosynthesis and photosynthetic gene expression despite the accumulation of soluble carbohydrates, *Plant J.*, 12, 605–14, 1997.

22. Hurry, V. et al., The role of inorganic phosphate in the development of freezing tolerance and the acclimatization of photosynthesis to low temperature is revealed by the photomutants of *Arabidopsis thaliana*, *Plant J.*, 24, 383–96, 2000.

23. Hurry, V. et al., Photosynthesis at low temperatures: A case study with *Arabidopsis*, in *Plant Cold Hardiness: Molecular Biology, Biochemistry, and Physiology*, Li, P., and Palva, T., Eds., Kluwer Academic/Plenum Publishers, New York, 2002, 161–180.

24. Beard, J.B., *Turfgrass: Science and Culture*, Prentice Hall, Englewood Cliffs, NJ, 1973.

25. Chauvin, L.P., Houde, M., and Sarhan, F., A leaf-specific gene stimulated by light during wheat acclimation to low temperature, *Plant Mol. Biol.*, 23, 255–265, 1993.

26. Pearce, R.S., and McDonald, I., Ultrastructure damage due to freezing followed by thawing in shoot meristem and leaf mesophyll cells of tall fescue (*Festuca arundinacea* Schreb.), *Planta*, 134, 159–168, 1977.

27. Levitt, J., *Responses of Plants to Environmental Stresses*, 2nd ed., Vol. I & II, Academic Press, Inc. New York, 1980.

28. Steponkus, P.L., Role of plasma membrane in freezing injury and cold acclimation, *Ann. Rev. Plant Physiol.*, 35, 543–584, 1984.

29. Steponkus, P.L., Uemura, M., and Webb, M.S., A contrast of cryostability of the plasma membrane of winter rye and spring oat-two species that widely different in their freezing tolerance and plasma membrane lipid composition, in *Advances in Low-Temperature Biology*, Steponkus, P.L., Ed., JAI Press, London, 1993, chap. 2.

30. Uemura, M., and Steponkus, P.L., Effect of cold acclimation on membrane lipid composition and freeze-induced membrane destabilizationm, in *Plant Cold Hardiness: Molecular Biology, Biochemistry and Physiology*, Li, P.H., and Chen, T.H.H., Eds., Plenum Press, New York, 1997, chap. 15.

31. Anchordoguy, T.J. et al., Modes of interaction of cryoprotectants with membrane phospholipids during freezing, *Cryobiology*, 24, 324–31, 1987.

32. Strauss, G., and Hauser, H., Stabilization of lipid bilayer vesicles by sucrose during freezing, *Proc. Natl. Acad. Sci. USA*, 83, 2422–2426, 1986.

33. Hurry, V., and Huner, N., Low growth temperature effects a differential inhibition of photosynthesis in spring and winter wheat, *Plant Physiol.*, 96, 491–497, 1991.

34. Powles, S.B., Photoinhibition of photosynthesis induced by visible light, *Ann. Rev. Plant Physiol.*, 35, 15–44, 1984.

35. Russell, A.W. et al., Photosystem II regulation and dynamics of the chloroplast D1 protein in *Arabidopsis* leaves during photosynthesis and photoinhibition, *Plant Physiol.*, 107, 943–952, 1995.

36. Sonoike, K. et al., Destruction of photosystem I iron-sulfur centers in leaves of *Cucumis sativus* L. by weak illumination at chilling temperatures, *FEBS Lett.*, 362, 235–238, 1995.

37. Murata, N., and Nishiyama, Y., Molecular mechanisms of the low-temperature tolerance of the photosynthetic machinery, in *Stress Responses of Photosynthetic Organisms*, Satoh, K., and Murata, N., Eds., Elsevier Science B.V., 1998, 93–112.

38. Aro, E.M., Virgin, I., and Andersson, B., Photoinhibition of Photosystem II. Inactivation, protein damage and turnover. *Biochim. Biophys. Acta.*, 1143, 113–34, 1993.

39. Jones, H., *Plants and Microclimate*, 2nd ed., 1992. Cambridge University Press, Cambridge, Great Britain. 1992.

40. Rapacz, M. et al., Changes in cold tolerance and the mechanisms of acclimation of photosystem II to cold hardening generated by anther culture of *Festuca pratensis × Lolium multiflorum* cultivars, *New Phytol.*, 162, 105–114, 2004.

41. Huner, N.P.A., Öquist, G., and Sarhan, F., Energy balance and acclimation to light and cold, in *Trends in Plant Science*, Vol. III, 1998, 224–230.

42. Huner, N.P.A. et al., Photosynthesis, photoinhibition and low temperature acclimation in cold tolerant plants, *Photosynth. Res.*, 37, 19–39, 1993.

43. Dernoeden, P.H., Why putting greens appear red or purple in the winter and spring, *Turfax*, 7, 4, 1999.

44. Sassenrath, G.F., Ort, D.R., and Portis, A.R. Jr., Impaired reductive activation of stromal bisphosphatases in tomato leaves following low-temperature exposure at high light, *Arch. Biochem. Biophys.*, 282, 302–308, 1990.

45. Foyer, C. et al., The mechanisms contributing to photosynthetic control of electron transport by carbon assimilation in leaves, *Photosynth. Res.*, 25, 83–100, 1990.

46. Goldschmidt, E.E., and Huber, S.C., Regulation of photosynthesis by end-product accumulation in leaves of plants storing starch, sucrose, and hexose sugars, *Plant Physiol.*, 99, 1443–1448, 1992.

47. Hurry, V., and Huner, N., Effect of cold hardening on sensitivity of winter and spring wheat leaves to short-term photoinhibition and recovery of photosynthesis, *Plant Physiol.*, 100, 1283–1290, 1992.

48. Hurry V.M., Gardeström, P., and Öquist, G., Reduced sensitivity to photoinhibition following frosthardening of winter rye is due to increased phosphate availability, *Planta*, 190, 484–490, 1993.

49. Smith, D., Carbohydrate reserves of grasses, in *The Biology and Utilization of Grasses*, Youngner, V.B., and McKell, C.M., Eds., Academic Press, New York, 1972, 318–333.

50. DiPaola, J.M., and Beard, J.B., Physiological effects of temperature stress, in Waddington, D.V., Carrow, R.N., and Shearman, R.C., Eds., *Turfgrass*, Agron. Monogr. 32, ASA-CSSA-SSSA, Madison, WI, 1992, 231–267.

51. Rogers, R.A., Dunn, J.H., and Nelson, C.J., Cold hardening and carbohydrates composition of 'Meyer'zoysia, *Agron. J.*, 67, 836–838, 1975.

52. Ball, S., Qian, Y.L., and Stushnoff, C., Soluble carbohydrates in two buffalograss cultivars with contrasting freezing tolerance, *J. Am. Soc. Hort. Sci.*, 127, 45–49, 2002.

53. Fry, J.D. et al., Freezing tolerance and carbohydrate content of low-temperature-acclimated and nonacclimated centipedegrass, *Crop Sci.*, 33, 1051–1055, 1993.

54. Maier, F.P., Lang, N.S., and Fry, J.D., Freezing tolerance of three St. Augustinegrass cultivars as affected by stonlon carbohydrates and water content, *J. Am. Soc. Hort. Sci.*, 119, 473–476, 1994.

55. Taiz L., and Zeiger, E., *Plant Physiology*, Sinauer Associates, Sunderland, MA, 1998.

56. Ariizumi, T. et al., An increase in unsaturation of fatty acids in phosphatidylglycerol from leaves improves the rates of photosynthesis and growth at low temperatures in transgenic rice seedlings, *Plant Cell Physiol.*, 43, 751–758, 2002.

57. Murata, N. et al., Genetically engineered alteration in the chilling sensitivity of plants, *Nature*, 356, 710–713, 1992.

58. Tokuhisa, J. et al., Investigating the role of lipid metabolism in chilling and freezing tolerance, in *Plant Cold Hardiness*, Li, P.H. and Chen, T.H.H., Eds., Plenum Press, New York, 1997, chap. 14.

59. Palta, J.P., Whitaker, B.D., and Weiss, L.S., Plasma membrane lipids associated with genetic variability in freezing tolerance and cold acclimation of *Solanum* species, *Plant Physiol.*, 103, 793–803, 1993.

60. Vega, S.E. et al., Evidence for the up–regulation of stearoyl–ACP (Δ9) desaturase gene expression during cold acclimation, *Am. J. Potato Res.*, 81, 125–135, 2004.

61. Cyril, J. et al., Changes in membrane polar lipid fatty acids of seashore paspalum in response to low temperature exposure, *Crop Sci.*, 42, 2031–2037, 2002.

62. Samala, S., Yan, J., and Baird, W.V., Changes in polar lipid fatty acid composition during cold acclimation in "Midiron" and "U3" bermudagrass, *Crop Sci.*, 38, 188–195, 1998.

63. Guy, C.L., Niemi, K.J., and Brambl, R., Altered gene expression during cold acclimation of spinach, *Proc. Natl. Acad. Sci. USA*, 82, 3673–3667, 1985.

64. Jaglo-Ottosen, K.R. et al., *Arabidopsis* CBF1 overexpression induces COR genes and enhances freezing tolerance, *Science*, 280, 104–106, 1998.

65. Kurkela, S., and Borg-Franck, M., Structure and expression of kin2, one of two cold– and ABA–induced genes of *Arabidopsis thaliana*, *Plant Mol. Biol.*, 19, 689–692, 1992.

66. Gatschet, M.J. et al., A cold-regulated protein from bermuda grass crowns is a chitinase, *Crop Sci.*, 36, 712–718, 1996.

67. Hincha, D.K., Sieg, F., and Schmitt, J.M., Protection of thylakoid membranes from freeze-thaw damage by proteins, in *Plant Cold Hardiness*, Li, P.H., and Chen, T.H.H., Eds., Plenum Press, New York, 1997, chap. 13.

68. Yu, X-M., and Griffith, M., Antifreeze proteins in winter rye leaves form oligomeric complexes, *Plant Physiol.*, 119, 1361–1369, 1999.

69. Griffith, M. et al., Antifreeze proteins modify the freezing process in planta, *Plant Physiol.*, 138, 330–340, 2005.

70. Kuiper, M.J., Davies, P.L., and Walker, V.K., A theoretical model of a plant antifreeze protein from *Lolium perenne*, *Biophys. J.*, 81, 3560–3565, 2001.

71. Urrutia, M.E., Duman, J.G., and Knight, C.A., Plant thermal hysteresis proteins, *Biochim. Biophys. Acta.*, 1121, 199–206, 1992.

72. Sidebottom, C. et al., Heat-stable antifreeze protein from grass, *Nature,* 406, 256, 2000.

73. Humphreys, M.W. et al., Molecular breeding and functional genomics for tolerance to abiotic stress, in *Molecular Breeding of Forage and Turf*, Hopkins, A., Wang, Z., Mian, R., Sledge, M., and Barker, R.E., Eds., Kluwer Academic Publisher, Dordrecht, The Netherlands, 2004, 61–80.

74. Gatschet, M.J. et al., Cold acclimation and alterations in protein synthesis in bermudagrass crowns, *J. Am. Soc. Hort Sci.*, 119, 477–480, 1994.

75. Guy, C.L., Haskell, D., and Li, Q.B., Association of proteins with the stress 70 molecular chaperons at low temperature: evidence for the existence of cold labile proteins in spinach, *Cryobiology*, 36, 301–14, 1998.

76. Guy, C.L., and Li, Q.B., The organization and evolution of the spinach stress 70 molecular chaperone gene family, *Plant Cell*, 10, 539–56, 1998.

77. Boothe, J.G. et al., Purification, characterization, and structural analysis of a plant low-temperature-induced protein, *Plant Physiol.*, 113, 367–376, 1997.

78. Monroy, A.F., Sarhan, F., and Dhindsa, R.S., Cold-induced changes in freezing tolerance, protein phosphorylation, and gene expression, *Plant Physiol.*, 102, 1227–1235.

79. Hendry G.A.F., Houghton J.D., and Brown, S.B., The degradation of chlorophyll – a biological enigma, *New Phytol.*, 107, 255–302, 1987.

80. Brown, M.E., Some effects of temperature on the growth and chemical composition of certain pasture grasses. *Missouri Agr. Exp. Stn. Res. Bull.*, 360, 1–56, 1939.

81. Hanson, A.A., and Juska, F.V., Winter root activity in Kentucky bluegrass (*Poa pratensis* L.), *Agron. J.*, 53, 372–374, 1961.

82. Vasil'yev, I.M., *Wintering of Plants*. American Institute of Biological Sciences, Washington, DC, 1961.

83. Roberts, D.W.A., The effect of light on development of the rosette growth habit of winter wheat, *Can. J. Bot.*, 62, 818–822, 1984.

84. Waldron, B.L. et al., Genetic variation and predicted gain from selection for winterhardiness and turf quality in a perennial ryegrass topcross population, *Crop Sci.*, 38, 817–822, 1998.

85. Beard, J.B., Direct low temperature injury of nineteen turfgrasses, *Q. Bull. Michigan Agr. Exp. Stn.*, 48, 377–383, 1966.

86. Welterlen, M.S., and Watschke, T.L., Influence of drought stress and fall nitrogen fertilization on cold deacclimation and tissue components of perennial ryegrass turf, in *Proc. 5th Int. Turfgrass Res. Conf.*, F. Lemaire, Ed., Inst. Natl. de la Recherche Agron., Paris, 1985, 831–840.

87. Wilkinson, J.F., and Duff, D.T., Effects of fall fertilization on cold resistance, color, and growth of Kentucky bluegrass, *Agron. J.*, 64, 345–348, 1972.

88. Tompkins, D.K., Ross, J.B., and Munoz, D.L., Dehardening of annual bluegrass and creeping bentgrass during late winter and early spring, *Agron. J.*, 92, 5–9, 2000.

89. Gray, A.P., and Eagles, C.F., Quantitative analysis of cold hardening and dehardening in *Lolium*, *Ann. Bot.*, 67, 339–345, 1991.

90. Steinke, K., and Stier, J.C., Influence of trinexapac–ethyl on cold tolerance and nonstructural carbohydrates of shaded supina bluegrass, *Acta Hort.*, 661, 207–215, 2004.

91. Lang, G.A. et al., Endo-, para-, and eco-dormancy: physiological terminology and classfication for dormancy research, *HortScience*, 22, 371–77, 1987.

92. Christians, N.E., *Fundamentals of turfgrass management*, 2nd ed., Wiley, Hoboken, NJ, 2004.

93. Sakai, A., and Larcher, W., *Frost Survival of Plants: Responses and Adaptation to Freezing Stress*, Springer-Verlag, NewYork, 1987.

94. Wood, G.M., and Cohen, R.P., Predicting cold tolerance in perennial ryegrass from subcrown internode length, *Agron. J.*, 76, 516–517, 1984.

95. Lu, S. et al., Effect of abscisic acid on chilling injury of centipedegrass, *Int. Turfgrass Soc. Res. J.*, 10, 862–866, 2005.

96. Karnok, K.J., and Beard, J.B., Morphological responses of *Cynodon* and *Stenotaphrum* to chilling temperatures as affected by gibberellic acid, *HortScience*, 18, 95–97, 1983.

97. Moon, J.W. Jr. et al., Limitations of photosynthesis in *Lolium perenne* after chilling, *J. Amer. Soc. Hort. Sci.*, 115, 478–481, 1990.

98. Pittermann, J., and Sage, R.F., The response of the high altitude C_4 grass *Muhlenbergia montana* (Nutt.) A.S. Hitch. to long- and short-term chilling, *J. Exper. Bot.*, 52, 829–838, 2001.

99. Nobel, P.S., *Physicochemical and Environmental Plant Physiology*, Academic Press, San Diego, CA, 1991.

100. White, R.H., and Schmidt, R.E., Bermudagrass response to chilling temperatures as influenced by iron and benzyladenine, *Crop Sci.*, 29, 768–773, 1989.

101. Rogers, R.A., Dunn, J.H., and Nelson, C.J., Photosynthesis and cold hardening in zoysia and Bermudagrass, *Crop Sci.*, 17, 727–732, 1977.

102. Lyons, J.M., Graham, D., and Raison, J.K., *Low Temperatures Stress in Crop Plants. The Role of the Membrane*, Academic Press, New York, 1979.

103. Matsuba, K. et al., Photosynthetic responses to temperature of phosphoenolpyruvate carboxylkinase type C_4 species differing in cold sensitivity, *Plant Cell Environ.*, 20, 268–274, 1997.

104. Youngner, V.B., Growth of U-3 bermudagrass under various day and night temperatures and light intensities, *Agron. J.*, 51, 557–559, 1959.

105. Fadzillah, N.M. et al., Chilling, oxidative stress and antioxidant responses in shoot cultures of rice, *Planta*, 199, 552–556, 1996.

106. Kang, H.M., and Saltveit, M.E., Reduced chilling tolerance in elongating cucumber seedling radicles is related to their reduced antioxidant enzyme and DPPH-radical scavenging activity, *Physiol. Plant.*, 115, 244–250, 2002.

107. O'Kane, D. et al., Chilling, oxidative stress and antioxidant responses in *Arabidopsis thaliana* callus, *Planta*, 198, 371–377, 1996.

108. Bravo, L.A. et al., The role of ABA in freezing tolerance and cold acclimation, *Physiol. Plant.*, 103, 12–23, 1998.

109. Machackova, I., Hanisova, A., and Krekule, J., Levels of ethylene, ACC, MACC, ABA and praline as indicators of cold hardening and frost resistance in winter wheat, *Physiol. Plant.*, 76, 603–607, 1989.

110. Lu, S., Guo, Z., and Peng, X., Effects of AGA and S-3307 on drought resistance and antioxidative enzyme activity of turfgrass, *J. Hort. Sci. Biotech.*, 78, 663–666, 2003.

111. Zhou, B., Guo, Z., and Lin, L., Effects of abscisic acid on antioxidant systems of *Stylosanthes guianensis* (Aublet) Sw. under chilling stress, *Crop Sci.*, 45, 599–605, 2005.

112. Long, S.P., Environmental responses, in C_4 *Plant Biology*, Sage, R.F., and Monson, R.K., Eds., Academic Press, San Diego, CA, 1999, 215–249.

113. Fowler, D.B., Lumin, A.E., and Gusta, L.V., Breeding for winterhardiness in wheat, in *New Frontiers in Winter Wheat Production*, Fowler, D.B., Gusta, L.V., Slinkard, A.E., and Hobin, B.A., Eds., University of Saskatchewan, Saskatchewan, Canada, 1983, 136–184.

114. Grant, M.N., Winter wheat breeding objectives for Western Canada, in *New Frontiers in Winter Wheat Production*, Fowler, D.B., Gusta, L.V., Slinkard, A.E., and Hobin, B.A., Eds., University of Saskatchewan, Saskatchewan, Canada, 1983, 89–101.

115. Beard, J.B., and Olien, C.R., Low temperature injury in the lower portion of *Poa annua* crowns, *Crop Sci.*, 3, 362–363, 1963.

116. Pearce, R.S., Relative hardiness to freezing of laminae, roots and tillers of tall fescue, *New Phytol.*, 84, 449–463, 1980.

117. Stier, J.C. et al., Visualization of freezing progression in turfgrasses using infrared video thermography, *Crop Sci.*, 43, 415–420, 2003.

118. Lindow, S.E., Arny, D.C., and Upper, C.D., Bacterial ice nucleation: a factor in frost injury to plants, *Plant Physiol.*, 70, 1084–1089, 1982.

119. Workmaster, B.A., Palta, J.P., and Wisniewski, M., Ice nucleation and propagation in cranberry uprights and fruit using infrared video thermography, *J. Am. Soc. Hortic. Sci.*, 124, 619–625, 1999.

120. Brush, R.A., Griffith, M., and Mlynarz, A., Characterization and quantification of intrinsic ice nucleators in winter rye (*Secale cereale*) leaves, *Plant Physiol.*, 104, 725–735, 1994.

121. Hopkins, W.G., and Hüner, N.P.A., *Introduction to Plant Physiology*, 3rd ed., Wiley, Hoboken, NJ., 2004, 459–491.

122. Gusta, L.V., Burke, M.J., and Kapoor, A.C., Determination of unfrozen water in winter cereals at subfreezing temperatures, *Plant Physiol.*, 56, 707–709, 1975.

123. Humphreys, M. et al., Dissecting drought- and cold-tolerance traits in the *Lolium-Festuca* complex by introgression mapping, *New Phytol.*, 137, 55–60, 1997.

124. Gusta, L.V., and Fowler, D.B., Factors affecting the cold survival of winter cereals, *Can. J. Plant Sci.*, 57, 213–219, 1977.

125. Andrews, J.E., Horricks, J.S., and Roberts, D.W.A., Interrelationships between plant age, root-rot infection, and cold hardiness in winter wheat, *Can. J. Bot.*, 38, 601–611, 1960.

126. Hollman, A.B., and Stier, J.C. Tall fescue maturity and cold tolerance, in *Agronomy Abstracts*. Am. Soc. Agron. (ASA), Madison, WI, 2003.

127. Steffen, K.L., Arora, R., and Palta, J.P., Relative sensitivity of photosynthesis and respiration to freeze-thaw stress in herbaceous species, *Plant Physiol.*, 89, 1372–1379, 1989.

128. Johnson, P.G., Riordan, T.P., and Arumuganathan, K., Ploidy level determinations in buffalograss clones and populations, *Crop Sci.*, 38, 478–482, 1998.

129. Anderson, J., Taliaferro, C., and Martin, D., Freeze tolerance of bermudagrasses: vegetatively propagated cultivars intended for fairway and putting green use, and seed–propagated cultivars, *Crop Sci.*, 42, 975–977, 2002.

130. Gusta, L.V. et al., Freezing resistance of perennial turfgrasses, *HortScience*, 15, 494–496, 1980.

131. Sukumaran, N.P., and Weiser, C.J., An excised leaflet test for evaluating potato frost tolerance, *HortScience*, 7, 467–468, 1972.

132. Ebdon, J.S., Gagne, R.A., and Manly, R.C., Comparative cold tolerance in diverse turf quality genotypes of perennial ryegrass, *HortScience*, 37, 826–830, 2002.

133. Kjelgren, R., Montague, D.T., and Rupp, L.A., Establishment in treeshelter. II. Effect of shelter color on gas exchange and hardiness, *HortScience*, 32, 1284–1287, 1997.

134. Qian, Y.L., and Engelke, M.C., 'Diamond' zoysiagrass as affected by light intensity, *J. Turfgrass Manage.*, 3, 1–13, 1999.

135. Happ, K., Winter damage: Control the variables that can minimize the potential for winter turf loss, *USGA Green Sec. Rec.*, 42, 1–6, 2004.

136. Schnyder, H., and Nelson, C.J., Growth rate and assimilate partitioning in the elongation zone of tall fescue leaf blades at high and low irradiance, *Plant Physiol.*, 90, 1201–1206, 1989.

137. Kinbacher, E.J., and Laude, H.M., Frost heaving of seedlings in the laboratory, *Agron. J.*, 47, 415–418, 1955.

138. Kinbacher, E.J., Resistance of seedlings to frost heaving injury, *Agron. J.*, 48, 166–170, 1956.

139. Roberts, J.M., Influence of protective covers on reducing winter desiccation of turf, *Agron. J.*, 78, 145–147, 1986.

140. Koski, T., Winter prep, *Golf Course Manage.*, 64, 8, 1996.

141. Dionne, J. et al., Golf green soil and crown–level temperatures under winter protective covers, *Agron. J.*, 91, 227–233, 1999.

142. Turgeon, A.J., *Turfgrass Management*, 7th ed., Prentice Hall, Upper Saddle River, NJ, 2005.

143. Minner, D. et al., Field assessment of winter injury on creeping bentgrass and annual bluegrass putting greens. Golf Course Super. Assoc. Am. and Iowa Golf Course Super. Assoc. Coop. Res. Proj. final rpt., www.iowaturfgrass.org/research/turfresearch.htm, 2006 (verified 1 July 2006).

144. Smiley, R.W., Dernoeden, P.H., and Clarke, B.B., Snow molds, in *Compendium of Turfgrass Diseases*, 3rd ed., American Phytopathological Society, St. Paul, MN, 2005, 46–53.
145. Martin, D., Resistant cultivars are bermuda's best battle vs. spring dead spot. *TurfGrass Trends*, 2005, http://www.turfgrasstrends.com/turfgrasstrends/article/articleDetail.jsp?id=145153 (verified 1 July 2006).
146. McCarty, L.B., and Lucas, L.T., *Gaumannomyces graminis* associated with spring dead spot of bermudagrass in the southeastern United States, *Plant Dis.*, 73, 659–661, 1989.
147. Endo, R.M., Ohr, H.D., and Krausman, E.M., *Leptosphaeria korrae*, a cause of the spring dead spot disease of bermudagrass in California, *Plant Dis.*, 69, 235–237, 1985.
148. Crahay, J.N., Dernoeden, P.H., and O'Neill, N.R., Growth and pathogenicity of *Leptosphaeria korrae* in bermudagrass, *Plant Dis.*, 72, 945–949, 1988.
149. Tisserat, N., Pair, J.C., and Nus, J.A., *Ophiosphaerella herpotricha*, a cause of spring dead spot of bermudagrass in Kansas, *Plant Dis.*, 73, 933–937, 1989.
150. Wetzel, H.C. III, Hulbert, S.H., and Tisserat, N.A., Molecular evidence for the presence of *Ophiosphaerella narmari* n. comb., a cause of spring dead spot of bermudagrass, in North America, *Mycol. Res.*, 103, 981–989, 1999.
151. Wetzel, H.C. III, Skinner, D.Z., and Tisserat, N.A., Geographic distribution and genetic diversity of three *Ophiosphaerella* species that cause spring dead spot of bermudagrass, *Plant Dis.*, 83, 1160–1166, 1999.
152. Iriarte, F.B. et al., Genetic diversity and aggressiveness of *Ophiosphaerella korrae*, a cause of spring dead spot of bermudagrass, *Plant Dis.*, 88, 1341–1346, 2004.
153. McCarty, L.B., DiPaola, J.M., and Lucas, L.T., Regrowth of bermudagrass infected with spring dead spot following low temperature exposure, *Crop Sci.*, 31, 182–184, 1991.
154. Nus, J.L., and Shashikumar, K., Fungi associated with spring dead spot reduces freezing resistance in bermudagrass, *HortScience*, 28, 306–307, 1993.
155. Iriarte, F.B. et al., Effect of cold acclimation and freezing on spring dead spot severity in bermudagrass, *HortScience*, 40, 421–423, 2005.
156. Iriarte, F.B. et al., Aggresiveness of spring dead spot pathogens to bermudagrass cultivars exposed to low temperatures, *Phytopathology*, 92, S138, 2002.
157. Millett, S., Distribution, biological and molecular characterization, and aggressiveness of *Typhula* snow molds of Wisconsin golf courses, PhD dissertation, University of Wisconsin, Madison, WI, 1999.
158. Scheef, E.A., Burke-Scroll, K., Gregos, G., and Gung, J., Geographical distribution of *Typhula* snow molds in Wisconsin using species-specific PCR markers and GIS, *Phytopathology*, 92, S72–73, 2002.
159. Jung, G., Application of DNA marker technology to turfgrass pathology research, *The Grass Roots*, 29, 38–41, 2000.
160. Jung, G., Burke-Scroll, K., and Gregos, J., Distribution of *Typhula* species and their sensitivity to fungicides *in vitro* and under field conditions in Wisconsin, *2000 Wisc. Turfgrass Res. Rpts.*, 18, 28–31, www.plantpath.wisc.edu/tdl, 2000 (verified 6 July 2006).
161. Nissinen, O., Analysis of climatic factors affecting snow mould injury in first-year timothy (*Phleum pratense* L.) with special reference to *Sclerotinia borealis*, *Acta Univ Oulu A*, 289, 1–115, 1996.
162. Årsvoll, K. and Smith, J.D., *Typhula ishikariensis* and its varieties, var. *idahoensis* comb. nov. and var. *canadensis* var. nov., *Can. J. Bot.*, 56, 348–364, 1978.
163. Bruehl, G.W., Developing wheat resistant to snow mold in Washington State, *Plant Dis.*, 66, 1090–1095, 1982.
164. Matsumoto, N., Sato, T., and Araki, T., Biotype differentiation in the *Typhula ishikariensis* complex and allopatry in Hokkaido, *Ann. Phytopath. Soc. Jpn.*, 48, 275–280, 1982.
165. Vargas, J.M., Jr., *Management of Turfgrass Diseases,* 3rd ed., Wiley, Hoboken, NJ, 2005, chap. 4.
166. Gaudet, D.A., Laroche, A., and Yoshida, M., Low temperature-wheat-fungal interactions: A carbohydrate connection, *Physiol. Plant.*, 106, 437–444, 1999.
167. Snider, C.S. et al., Role of ice nucleation and antifreeze activities in pathogenesis and growth of snow molds, *Phytopathology*, 90, 354–361, 2000.
168. Årsvoll, K., Effects of hardening, plant age, and development in *Phleum pratense* and *Festuca pratensis* on resistance to snow mould fungi, *Meld Nor Landbrogsk*, 56, 1–14, 1977.
169. Tronsmo, A.M., Effects of dehardening on resistance to freezing and infection by *Typhula ishikariensis* in *Phleum pretense*, *Acta Agr. Scandinavica*, 35, 113–116, 1985.
170. Ergon, Å, Klemsdal, S.S., and Tronsmo, A.M., Interactions between cold hardening and *Microdochium nivale* infection on expression of pathogenesis-related genes in winter wheat, *Physiol. Mol. Plant Path.*, 53, 301–310, 1998.

171. Gaudet, D.A., Progress towards understanding interactions between cold hardiness and snow mold resistance and development of resistant cultivars, *Can. J. Plant Path.*, 16, 241–246, 1994.

172. Webster, D.E., and Ebdon, J.S., Effects of nitrogen and potassium fertilization on perennial ryegrass cold tolerance during deacclimation in late winter and early spring, *HortScience*, 40, 842–849, 2005.

173. Mauch, F., Mauch-Mani, B., and Boller, T., Antifungal hydrolases in pea tissue II. Inhibitions of fungal growth by combinations of chitinase and β-1,3-glucanase, *Plant Physiol.*, 88, 936–942, 1988.

174. Moerschbacher, B.M., Plant peroxidases: involvement in response to pathogens, in *Plant Peroxidases 1980–1990*, Penel, C., Gaspar, T., Grebbin, T., Eds., University of Geneva, Switzerland, 1992, 91–99.

175. Andrews, C.J., How do plants survive ice? *Ann. Bot.*, 78, 529–536, 1996.

176. Gudleifsson, B.E., Andrews, C.J., and Bjornsson, H., Cold hardiness and ice tolerance of pasture grasses grown and tested in controlled environments, *Can. J. Plant Sci.*, 66, 601–608, 1986.

177. National Turfgrass Evaluation Program (NTEP), Tall fescue progress report, 2005, www.ntep.org. (verified 1 July 2006).

178. Tompkins, D.K., Ross, J.B., and Munoz, D.L., Effects of ice cover on annual bluegrass and creeping bentgrass putting greens, *Crop Sci.*, 44, 2175–2179, 2004.

179. McKersie, B.D., and Bowley, S.R., Active oxygen and freezing tolerance in transgenic plants, in *Plant Cold Hardiness: Molecular Biology, Biochemistry, and Physiology*, Li, P.H. and Chen, T.H.H., Eds., Plenum Press, New York, 1997, 203–214.

180. Kendall, E.J., and McKersie, B.D., Free radical and freezing injury to cell membranes of winter wheat, *Physiol. Plant.*, 76, 86–94, 1989.

181. Brandsæter, L.O. et al., Identification of phytotoxic substances in soils following winter injury of grasses as estimated by bioassay, *Can. J. Plant Sci.*, 85, 115–123, 2005.

182. McKersie, B.D. et al., Changes in carbohydrate levels during ice encasement and flooding of winter cereals, *Can. J. Bot.*, 60, 1822–1826, 1982.

183. Gao, J-Y., Andrews, C.J., and Pomeroy, M.K., Interactions among flooding, freezing, and ice encasement in winter wheat, *Plant Physiol.*, 72, 303–307, 1983.

184. Dionne, J. et al., Annual bluegrass and creeping bentgrass responses to contrasting O_2 and CO_2 levels at two near-freezing temperatures, in *Agron. Abstr.*, ASA, Madison, WI, 2002.

185. Clein, J.S., and Schimel, J.P., Microbial activity of tundra and taiga soils at −subzero temperatures, *Soil Biol. Biochem.*, 27, 1231–1234, 1995.

186. Rakatina, Z.G., The permeability of ice for O_2 and CO_2 in connection with the study of the reasons for winter cereal mortality under the ice crust, *Soviet Plant Physiol.*, 12, 795–803, 1965.

187. Beard, J.B., The effects of ice, snow, and water covers on Kentucky bluegrass, annual bluegrass, and creeping bentgrass, *Crop Sci.*, 4, 638–640, 1964.

188. Cary, J.W., and Mayland, H.F., Salt and water movement in unsaturated frozen soil, *Proc. Soil Sci. Soc. Am.*, 36, 549–555, 1972.

189. Beard, J.B., Effects of ice covers in the field on two perennial grasses, *Crop Sci.*, 5, 139–140, 1965.

190. Andrews, C.J., and Pomeroy, M.K., Survival and cold hardiness of winter wheats during partial and total ice immersion, *Crop Sci.*, 15, 561–566, 1975.

191. Lebeau, J.B., Pathology of winter injured grasses and legumes in western Canada, *Crop Sci.*, 6, 23–25, 1966.

192. Gudleifsson, B.E., Survival and metabolite accumulation by seedlings and mature plants of timothy grass during ice encasement, *Ann. Bot.*, 79 (Suppl. A), 93–96, 1997.

193. Bertrand, A. et al., Molecular and biochemical responses of perennial forage crops to oxygen deprivation at low temperature, *Plant Cell Environ.*, 24, 1085–1093, 2001.

194. Saglio, P.H., Drew, M.C., and Pradet, A., Metabolic acclimation to anoxia by low (2–4 kPa partial pressure) oxygen pretreatment (hypoxia) in root tips of *Zea mays*, *Plant Physiol.*, 86, 61–66, 1988.

195. Waters, I. et al., Effects of anoxia on wheat seedlings. 2. Influence of O_2 supply prior to anoxia on tolerance to anoxia, alcoholic fermentation, and sugar levels, *J. Exp. Bot.*, 42, 1437–1447, 1991.

196. Zhang, Q., and Greenway, H., Anoxia tolerance and anaerobic catabolism of aged beetroot storage tissues, *J. Exp. Bot.*, 45, 567–575, 1994.

197. Dolferus, R. et al., Strategies of gene action in *Arabidopsis* during hypoxia. *Ann. Bot.*, 79 (Suppl.), 21–31, 1997.

198. Andrews, C.J., and Pomeroy, M.K., Toxicity of anaerobic metabolites accumulating in winter wheat seedlings during ice encasement, *Plant Physiol.*, 64, 120–125, 1979.

199. Andrews, C.J., The increase in survival of winter cereal seedlings due to light exposure during ice encasement, *Can. J. Bot.*, 66, 409–413, 1988.
200. Andrews, C.J., A comparison of glycolytic activity in winter wheat and two forage grasses in relation to their tolerance to ice encasement, *Ann. Bot.*, 79 (Suppl.), 87–91, 1997.
201. Gudleifsson, B.E., Metabolite accumulation during ice encasement of timothy grass (*Phleum pratense* L.), *Proc. Royal Soc. Edinburgh* 102B, 373–380,1994.
202. Dionne, J. et al., Physiological effects of anoxia on annual bluegrass during cold acclimation, *Agron. Abstr.*, 153, ASA, Madison, WI, 2000.
203. Rochette, P. et al., Anoxia-related damage to annual bluegrass golf greens under winter protective covers, in *Agron. Abstr.*, 151, ASA, Madison, WI, 2000.
204. Dormaar, J.F., and Sauerbeck, D.R., Seasonal effects on photoassimilated carbon-14 in the root system of blue grama and associated soil matter, *Soil Biol. Biochem.*, 15, 475–479, 1983.
205. Coxson, D.S., and Parkinson, D., Winter respiratory activity in aspen woodland forest floor litter and soils, *Soil Biol. Biochem.*, 19, 49–59, 1987.
206. Sommerfield, R.A., Mosier, A.R., and Musselman, R.C., CO_2, CH_4, and N_2O flux through a Wyoming snowpack and implications for global budgets, *Nature*, 361, 140–142, 1993.
207. Coxson, D.S., and Parkinson, D., The pattern of winter respiratory response to temperature, moisture, and freeze-thaw exposure in *Bouteloua gracilis* dominated grassland soils of southwestern Alberta, *Can. J. Bot.*, 65, 1716–1725, 1987.
208. Bakker, A.W., and Schippers, B., Microbial cyanide production in the rhizosphere in relation to potato yield reduction and *Pseudomonas* spp.–mediated plant growth-stimulation, *Soil Biol. Biochem.*, 19, 451–457, 1987.
209. Alstrom, S., and Burns, R.G., Cyanide production by rhizobacteria as a possible mechanism of plant growth inhibition, *Biol. Fert. Soils*, 7, 232–238, 1989.
210. Lynch, J.M., Phytotoxicity of acetic acid produced in the anaerobic decomposition of wheat straw, *J. Appl. Bacteriol.*, 42, 81–87, 1977.
211. Lynch, J.M., Products of soil-microorganisms in relation to plant growth, *CRC Crit. Rev. Microbiol.*, 5, 67–107, 1976.
212. Dionne, J., Laganière, M., and Desjardins, Y., Evaluation of the efficiency of winter protections against winter damage on *Poa annua* golf greens, *HortScience*, 30, 897, 1995.
213. Beard, J.B., Bentgrass (*Agrostis* spp.) varietal tolerance to ice cover injury, *Agron. J.*, 57, 513, 1965.
214. Beard, J.B., and Stahnke, G.K., Evaluation of antitranspirants on creeping bentgrass (*Agrostis palustris* cv. 'Penncross') and bermudagrass (*Cynodon dactylon* (L.) Pers. X *Cynodon transvaalensis* Burtt-Davy, cv. 'Tifway'), *Agron. Abstr.*, 128, 1981.
215. Mancino, C.F. et al., Influence of antitranspirants on turfgrass growth and physiology, Agron. abstracts, 137, *Am. Soc. Agron.*, Madison, WI, 1987.
216. Gu, S. et al., Effects of short-term water stress, hydrophilic polymer amendment, and antitranspirant on stomatal status, transpiration, water loss, and growth in 'Better Boy' tomato plants, *J. Am. Soc. Hort. Sci.*, 121, 831–837, 1996.
217. Aylward, L., Dealing with desiccation: superintendent applies antitranspirant to protect turf from winter woes, *Golfdom*, 56, 66, 2000.
218. Beard, J.B., Covers for the protection of turfgrasses against winter desiccation and direct low temperature injury, *Agron. Abstr.*, 61, 52, 1969.
219. Minner, D.D., and Valverde, F.J., Anti-desiccant winter protection of creeping bentgrass putting greens, *2002 Iowa Turfgrass Res. Rpt.*, 2002, 96–97, http://turfgrass.hort.iastate.edu/pubs/turfrpt/2002/30-Anti-Desiccant.html (verified 1 July 2006).
220. Latham, J.M., Winter water wagons minimize Dakota desiccation, *USGA Green Sctn. Rec.*, 29, 17, 1991.
221. Hull, R.J., A carbon-14 technique for measuring photosynthate distribution in field grown turf, *Agron. J.*, 68, 99–102, 1976.
222. Hull, R.J., Kentucky bluegrass photosynthate partitioning following scheduled mowing, *J. Am. Soc. Hort. Sci.*, 112, 829–834, 1987.
223. Hull, R.J., and Smith, L.M., Photosynthate translocation and metabolism in Kentucky bluegrass turf as a function of fertility, in *Proc. 2nd Int. Turfgrass Res. Conf.*, Roberts, E.C., Ed., Am. Soc. Agron., Madison, WI, 1974, 186–195.
224. Stuckey, I.H., Seasonal growth of grass roots, *Am. J. Bot.*, 28, 486–491, 1941.

225. Reeves, S.A. et al., Effect of N, P, and K tissue levels and late fall fertilization on the cold hardiness of Tifgreen bermudagrass, *Agron. J.*, 62, 659–662, 1970.
226. Wehner, D.J., Haley, J.E., and Martin, D.L., Late fall fertilization of Kentucky bluegrass, *Agron. J.*, 80, 466–471, 1988.
227. Dunn, J.H. et al., Fall fertilization of zoysiagrass, *Int. Turfgrass Soc. Res. J.*, 7, 565–571, 1993.
228. Munshaw, G.C. et al., Influence of late-season iron, nitrogen, and seaweed extract on fall color retention and cold tolerance of four bermudagrass cultivars, *Crop Sci.*, 46, 273–283, 2006.
229. Carroll, J.C., and Welton, F.A., Effect of heavy and late applications of nitrogenous fertilizer on the cold resistance of Kentucky bluegrass, *Plant Physiol.*, 14, 297–308, 1939.
230. Jung, G.A., and Kocher, R.E., Influence of applied nitrogen and clipping treatments on winter survival of cool-season grasses, *Agron. J.*, 66, 62–65, 1974.
231. Duff, D.T., Influence of fall nitrogenous fertilization on the carbohydrates of Kentucky bluegrass, in *Proc. 2nd Int. Turfgrass Res. Conf.*, Roberts, E.C., Ed., Am. Soc. Agron., Madison, WI, 1974, 112–119.
232. Kussow, W.R., Combs, S.M., and Sausen, A.J., *Lawn fertilization*, University of Wisconsin Extension Bulletin A2303, Madison, WI, 4 pp., 1997 (also available at http://s142412519.onlinehome.us/uw/pdfs/A2303.PDF; verified 22 July 2006).
233. Palmertree, H.D., Ward, C.Y., and Pluenneke, R.H., Influence of mineral nutrition on the cold tolerance and soluble protein fraction of centipedegrass, in *Proc. 2nd Int. Turfgrass Res. Conf.*, Roberts, E.C., Ed., American Soc. of Agron. and Crop Science Soc. Am. with Intl. Turfgrass Soc., Madison, WI, 1974, 500–507.
234. Gilbert, W.B., and Davis, D.L., Influence of fertility ratios on winter hardiness of bermudagrass, *Agron. J.*, 63, 591–593, 1971.
235. Richardson, M.D., Turf quality and freezing tolerance of 'Tifway' bermudagrass as affected by late-season nitrogen and trinexapac-ethyl, *Crop Sci.*, 42, 1621–1626, 2002.
236. Beard, J.B., and Rieke, P.E., The influence of nitrogen, potassium and cutting height on the low temperature survival of grasses, *Agron. Abstr.*, Am. Soc. Agron., Madison, WI, 34, 1966.
237. Hurto, K.A., and Troll, J., The influence of nitrogen:potassium ratios on nutrient content and low temperature hardiness of perennial ryegrass, *Agron. Abstr.*, Am. Soc. Agron., Madison, WI, 114, 1980.
238. Cook, T.W., and Duff, D.T., Effects of K fertilization on the freezing tolerance and carbohydrate content of *Festuca arundinacea* Schreb. maintained as turf, *Agron. J.*, 68, 116–119, 1976.
239. Miller, G.L., and Dickens, R., Potassium fertilization related to cold resistance in Bermudagrass, *Crop Sci.*, 36, 1290–1295, 1996.
240. Munshaw, G.C., Zhang, X., and Ervin, E.H., Effect of salinity on bermudagrass cold hardiness, *HortScience*, 39, 420–423, 2004.
241. Qian, Y.L., and Engelke, M.C., Influence of trinexapac–ethyl on Diamond zoysiagrass in a shade environment, *Crop Sci.*, 39, 202–208, 1999.
242. Goss, R.M. et al., Trinexapac-ethyl and nitrogen effects on creeping bentgrass grown under reduced light conditions, *Crop Sci.*, 42, 472–479, 2002.
243. Rossi, F.S., and Buelow, E.J., Exploring the use of plant growth regulators to reduce winter injury on annual bluegrass (*Poa annua* L.), *USGA Green Sect. Rec.*, 35, 12–15, 1997.
244. Steinke, K., and Stier, J.C., Nitrogen selection and growth regulator applications for improving shaded turf performance, *Crop Sci.*, 43, 1399–1406, 2003.
245. Heckman, N.L., Trinexapac–ethyl influences plant metabolism and development, MS thesis, Department of Agronomy and Horticulture, University of Nebraska-Lincoln, 2000.
246. Fagerness, M.J. et al., Temperature and trinexapac-ethyl effects on bermudagrass growth, dormancy, and freezing tolerance, *Crop Sci.*, 42, 853–858, 2002.
247. Goatley, J.M., and Schmidt, R.E., Anti-senescence activity of chemicals applied to Kentucky bluegrass, *J. Am. Soc. Hort. Sci.*, 115, 654–656, 1990.
248. Zhang, X., and Ervin, E.H., Cytokinin-containing seaweed and humic acid extracts associated with creeping bentgrass leaf cytokinins and drought resistance, *Crop Sci.*, 44, 1737–1745, 2004.
249. Schmidt, R.E., Henry, M.L., and Shoulders, J.F., Clear plastic cover influence of postdormancy growth of *Cynodon dactylon* (L.) Pers. grown under different managements, in *Proc. Sixth Int. Turfgrass Res. Conf.*, 157–159, 1989.
250. Shashikumar, K., and Nus, J.L., Cultivar and winter cover effects on bermudagrass cold acclimation and crown moisture content, *Crop Sci.*, 33, 813–817, 1993.

251. Schmidt, R.E., Evaluation of spun-bonded covers for winter protection and enhancement of Tifgreen bermudagrass post-dormancy growth, in *Proc. 5th Int. Turfgrass Res. Conf.*, F. LeMaire, Ed., INRA, Versailles, France, 1985, 167–170.
252. Rochette, P. et al., Atmospheric composition under impermeable winter golf green protections, *Crop Sci.*, 46, 1644–1655, 2006.
253. Eaton, W.J., and Beard, J.B., Snowmobile traffic relationships on turfgrasses. *HortScience*, 21, 531–532, 1986.
254. Gregos, J.S., Casler, M.D., and Stier, J.C., Snow mold tolerance of bentgrass and fine fescue fairway collections, *Agron. Abstr.*, Am. Soc. Agron., Madison, 158, WI, 2000.
255. Casler, M., Gregos, J., and Stier, J., Seeking snow mold–tolerant turfgrasses, *Golf Course Manage.*, 69, 49–52, 2001.
256. Wang, Z., Casler, M.D., Stier, J.C., Gregos, J.S., and Millett, S.M., Genotypic variation for snow mold reaction among creeping bentgrass clones, *Crop Sci.*, 45, 399–406, 2005.
257. Jung, G., Gregos, J.S., and Abler, S.W., Results of fungicide evaluation for 2003–04 snow mold control in Wisconsin and Minnesota, *Grass Roots*, 33, 33, 2004.
258. Abler, S. and Jung, G., Microdochium patch, *Grass Roots*, 32, 21, 2003.
259. Vincelli, P., and Williams, D., Managing spring dead spot of bermudagrass, *Golf Course Manage.*, 66, 49–53, 1998.
260. Martin, D.L. et al., Spring dead spot resistance and quality of seeded bermudagrasses under different mowing heights, *Crop Sci.*, 41, 451–456, 2001.
261. Dernoeden, P.H., Crahay, J.N., and Davis, D.B., Spring dead spot and bermudagrass quality as influenced by nitrogen source and potassium, *Crop Sci.*, 31, 1674–1680, 1991.
262. Iriarte, F., Fry, J., and Tisserat, N., Evaluating best management practices for spring dead spot suppression in bermudagrass, *HortScience*, 35, 390–391, 2000.
263. Tisserat, N., and Fry, J., Cultural practices to reduce spring dead spot (*Ophiosphaerella herpotricha*) severity in *Cynodon dactylon*, *Int. Turfgrass Soc. Res. J.*, 8, 931–936, 1997.
264. McBeath, J.H., Snow mold: winter turfgrass nemesis, *Golf Course Manage.*, 71, 121–124, 2003.
265. Miwa, T., Kihara, H., and Tonogi, H., The methodological study of under–soil heating system (USHS) for warm-season grass, *HortScience*, 34, 553, 1999.
266. Escritt, J.R., Electrical soil warming as an anti–frost measure for sports turf – a further report, *J. Sports Turf Res. Inst.*, 10, 29–41, 1959.
267. Lebeau, J.B., Soil warming and winter survival of turfgrass, *J. Sports Turf Res. Inst.*, 43, 5–11, 1967.
268. Ede, A.N., Soil heating system using warm air, *J. Sports Turf Res. Inst.*, 46, 76–91, 1970.
269. Westwood, J., The heat is on!: Keeping winter pitches playable. *Int. Turfgrass Bull.*, 210, 5–7, 2000.
270. Vastyan, J., Pipe dreams: Gillete Stadium's turf conditioning system. *SportsTurf*, 18, 18–20, 2002.
271. Kiefer, M., Natural gas comes to Lambeau Field. American Gas, http://www.aga.org/Content/ContentGroups/American_Gas_Magazine1/October_1997/Natural_Gas_Comes_to_Lambeau_F_4917.htm, 1997 (verified 21 August 2006).
272. Stier, J.C. et al., Winter covers and soil heating to sustain bermudagrass at Lambeau Field. *Wisc. Turf Res.*, 16, 79–82, 1999.

30 Cold-Stress Response of Cool-Season Turfgrass: Antioxidant Mechanism

Prasanta C. Bhowmik, Kalidas Shetty, and Dipayan Sarkar

CONTENTS

30.1   INTRODUCTION

Growth and development of plants are governed by different environmental conditions. Temperature (high and low) is one of the most impacting and important environmental factors that limit the productivity and geographical distribution of plants in large areas of the world (Larcher 1981). Optimal temperature range determines mainly the biological activity of a plant cell. Physiological or metabolic reaction rates either decline or stop above and below the optimum temperature range in most plant species.

Plants vary greatly in their responses to cold temperature (Levitt 1980). Negative heat balance of a body or larger area of the earth's surface causes low temperature on earth. The energy balance of earth is the difference between energy input as short-wave irradiation and energy losses due to the emission of long-wave radiation, and this balance largely depends on

1. The angle of incidence of irradiation (depends on the geographical latitude, season and slope of the terrain)
2. The ratio of day length to night length
3. Topographical variations in net radiation

Important physical factors related to the effect of low temperature in plants are *cold, frost, ground frost,* and *snow.* Cold is a low-kinetic-energy (thermodynamic) state of the molecules, and biologically it causes retardation of chemical processes, displacement of equilibrium reactions in the direction of energy liberation, and alterations of biologically important structures of the cell (lipid components of biomembrane) (Sakai and Larcher 1987). Frost is a condition when temperatures fall below 0°C. The detrimental effect of frost on plants varies widely, and different plant species have different threshold values below which cellular functions are jeopardized. Ground frost occurs when soil temperature falls below 0°C and it directly affects the underground organs of a plant (root, rhizome, tuber, etc.) by hindering uptake of water and mineral nutrients from the soil (Sigafoos 1952). Snow is a low-temperature event and it has both helpful and harmful effects on plants during winter (Sakai and Larcher 1987).

Turfgrasses are herbaceous perennial plants from the family *Gramineae* and are used widely in different parts of the world, in golf courses, sports fields, lawns, parks, and other recreational areas (Turgeon 1996). Due to its huge commercial importance, different turfgrass species are adapted in different regions of the world. Out of 5000 grass species, less than 30 species had been adapted as turfgrass due to some special characteristics (Beard 1973). In North America, the majority of turfgrass species are non-native but have become naturalized with continuous use over a long period of time. The geographical distribution pattern of turfgrass largely depends on the temperature and precipitation gradient of a region. Depending on its tolerance to temperature, turfgrass species are divided into two major groups: (i) cool-season and (ii) warm-season turfgrass (Figure 30.1).

Turfgrasses growing under cool humid, cool subhumid, and cool semiarid climates are known as cool-season turfgrasses. The optimum temperature ranges for shoot and root growth of cool-season turfgrass are 64–75°F (16–24°C) and 50–65°F (10–18°C), respectively (Fry and Huang 2004). Most of the cool-season turfgrass species have originated in Eurasia as forest-margin species (Emmons 1995). There are around 20 cool-season turfgrass species utilized throughout the world.

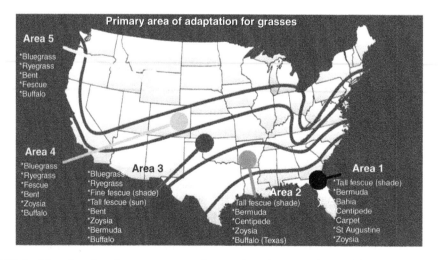

FIGURE 30.1 Map of cool- and warm-season turfgrass adaptation zones in the United States. (From http://www.baileyseed.com.)

Generally, cool-season turfgrass species have C_3 photosynthesis pathway (Fry and Huang 2004). In the northern zones of the United States, optimum growth of shoot and root of cool-season grasses is observed during spring and fall. The optimum air and soil temperature during this period favor this growth characteristic, and maximum growth of both root and shoot takes place during spring.

30.2 EFFECTS OF COLD STRESS ON TURFGRASS

Cool-season turfgrasses have originated and adapted predominately in cooler regions of the world (Beard 1973). In general, cool-season turfgrass species have higher ability to tolerate low temperatures and maintain good-quality turf below freezing, where most plant species are unable to sustain their biological functionality (Beard 1973; Turgeon 1996). Although most cool-season turfgrass species have cold-hardiness characteristics, a few are susceptible to cold injury just below freezing. Among cool-season grasses, perennial ryegrass, tall fescue, and annual bluegrass are most susceptible to low-temperature injury while creeping bentgrass is the most tolerant species (Table 30.1). The cold tolerance varies not only with species but widely with cultivars of different cool-season turfgrasses. Different parts of the turfgrass plant also have wide tolerance ranges of low temperature, which plays a significant role in turfgrass performance (Beard 1973).

The temperature balance of turf largely depends on heat gain and heat loss from the aboveground canopy. Solar radiation is the major contributor to the heat gain, where thermal radiation from the atmosphere and earth's surface also contributes to the heat load of the turfgrass system. The heat loss of turfgrass is mainly by evapotranspiration, while reradiation, conduction, and convection are other paths to heat loss (Beard 1973). During winter overall heat gain is very low, and thus increasing heat loss from any of these mechanisms causes rapid decrease in temperature. Along with these, temperature loss varies widely with latitude, altitude, topography, season, and time of the day.

Winter injury of cool-season turfgrass is generally observed in the temperate zone of the United States and in Canada, which experience intense cold during the prolonged winter. Episodic drops of temperature during spring or summer in temperate climates are more dangerous than periodic cold, because the plants are in an active growth stage (Alberdi and Corcuera 1991).

TABLE 30.1
Cold Hardiness of Cool-Season Turfgrasses

Turfgrass	Cold Hardiness Ranking	Range of Lethal Temperatures, °F (°C)
Rough bluegrass	Excellent	–
Creeping bentgrass		−31 (−35)
Timothy		–
Kentucky bluegrass	Good	−22 to −5.8 (−30 to −21)
Canada Bluegrass		–
Colonial bentgrass		–
Redtop		–
Annual bluegrass	Medium	–
Red fescue		−11.2 (−24)
Tall fescue		–
Hard fescue		−5.8 (−21)
Perennial ryegrass	Poor	5 to 23 (−15 to −5)

Sources: Beard, J.B., *Turfgrass: Science and Culture*, Prentice Hall, Englewood Cliffs, NJ, 1973; Gusta, L.V., J.D. Butler, C. Rajashekar, and M.J. Burke, *HortScience* 15, 494, 1980; Fry, J. and B. Huang, *Applied Turfgrass Science and Physiology*, Wiley, Hoboken, NJ, 2004.

Thus, winter-kill or winter injury is more pronounced during late winter, rather than mid and early winter. There are different kinds of winter stress on turfgrass, and it is very important to identify the specific cause for winter stress and its proper management. Major winter stresses for cool-season turfgrass in the northern latitudes are direct low-temperature kill (freezing stress), winter desiccation, and low-temperature fungi (snow molds) (Alberdi and Corcuera 1991; Fry and Huang 2004). Suffocation due to ice encasement and winter injury due to traffic are indirect factors affecting cool-season turfgrass during winter (Beard 1973).

30.2.1 Low-Temperature Kill

Direct low-temperature kill is a typical problem for both cool- and warm-season turfgrasses during late winter to early spring. It occurs most commonly during periods of alternating freezing and thawing. In late winter, when the deacclimation process starts, and plants come out from the dormant to the semidormant condition, sudden fall of temperature below 20°F (−7°C) causes severe damage to cells (Beard 1973). Low-temperature kill increases with (i) more rapid freezing or thawing events, (ii) greater number of freeze/thaw cycles, and (iii) the time the tissue is frozen. Crown hydration of cool-season turfgrass due to standing water also accelerates low-temperature kill by increasing the potential of extracellular ice formation (Beard 1973). Crown hydration actually predisposes the plant to freezing injury because of a loss of cold hardiness. Crowns are in general more winter-hardy than leaves and roots, and upper crown region and apex is more hardy compared with the basal crown region. Freezing injury involves ice formation either within cells (intracellular) or between cell walls (extracellular) (Beard 1973; Fry and Huang 2004). Intracellular freezing is usually a nonequilibrium process, whereas extracellular ice formation is an equilibrium freezing process. In intracellular freezing, mechanical disruption of protoplasmic structure and eventual death of the tissue occurs, due to the explosive growth of ice crystal within cells with high hydration level. Formation of extracellular ice in the equilibrium freezing process lowers the water potential in the extracellular region and thus draws water from inside cell, leading to the desiccation of the protoplasm (Beard 1973). If the desiccation process takes place for a long time, then contraction of the cell and the protoplasm occurs, which results in mechanical damage to the protoplasm. During thawing, cell walls expand more rapidly than the protoplasts creating tension, which also creates severe damage to cells and tissues (Beard 1973; Sakai and Larcher 1987).

30.2.2 Winter Desiccation

Winter desiccation is another primary cause of winter injury to the turfgrass. This type of injury may be the result of soil or atmospheric drought. Winter desiccation is most severe (i) with insufficient snow cover, (ii) on elevated sites, (iii) in areas exposed to excessive wind movement, favoring rapid evapotranspiration, (iv) when air temperatures are above 32°F, (v) where surface runoff of precipitation is high, causing insufficient recharge or restricted rooting, and (vi) in the case of unavailability of water due to frozen soil (Beard 1973). Two most important factors controlling winter desiccation are rapid evapotranspiration and unavailability of soil water. Winter desiccation is more prominent in regions of low rainfall or with insufficient distribution of rain during winter (Beard 1973). Soil with less available water restricts root growth and decreases root membrane permeability and thus induces moisture stress inside the plant. Mild winter desiccation is very common in turfgrass and it is not harmful, unless it causes serious damage to the crown. Turfgrass species or cultivars that maintain or initiate physiological activities are more susceptible to winter desiccation injury compared with dormant turfgrass species. Turfgrass species with noncreeping, bunch-type growth are also more prone to winter desiccation damage, as such a growth characteristic favors maximum transpiration (Beard 1973).

30.2.3 Low-Temperature Fungi

Other than those abiotic factors biotic factors, like low-temperature disease, are an important winter stress to cool-season turfgrass. Pink snow mold (caused by *Microdochium nivale* Fr. Samuels and

Hallett) and gray snow mold (caused by *Typhula* spp. Lasch ex. Fr.) are major fungal diseases that cause severe damage to cool-season turfgrass during late winter to spring (Hsiang and Cook 2001). Pink snow mold is also known as *Fusarium patch*, and it is often associated with snow cover. In contrast, gray snow mold is most pronounced under snow cover. The infestation of that fungal pathogen is more severe in temperature ranges between 32°F to 40°F, along with wet weather (Beard 1973). Application of fungicides in early winter is the only solution to such biotic stress.

30.2.4 ICE ENCASEMENT

Snow covers act as an insulator and protect turf from winter injury during late winter. In this way, snow directly protects plants from low temperature; also, when temperature rises, the snow cover helps plants prolong cold hardiness by retarding the dehardening process (Tompkins et al. 2002). Snow particularly helps against low night temperatures during late and early spring. Although snow or ice cover is beneficial to turfgrass to some extent, prolonged ice cover causes serious damage to cool-season turfgrass. Cool-season turfgrasses vary widely in terms of tolerance to ice cover (or encasement). Creeping bentgrass can generally survive up to 120 days under ice cover without damage, whereas annual bluegrass shows significant damage symptoms when ice covers are present for more than 75 days (Danneberger 2006). Kentucky bluegrass is intermediate in ice-cover tolerance to creeping bentgrass and annual bluegrass. Ice-cover injury is mainly due to the ice acting as a physical barrier to the exchange of gases between turfgrass tissue and the atmosphere. The turfgrass plants suffocate with lack of oxygen (anoxia) that is required for the respiration process (Tompkins et al. 2002). The accumulation of toxic gasses (carbon dioxide and cyanide) due to ice encasement also damages turfgrass severely (Beard 1964). The carbohydrate depletion under ice cover is another reason for the turfgrass injury (Rossi 2003). Ice formation can result from (i) water accumulation in low areas, (ii) uniform ice buildup from repeated sleet storms, and (iii) freezing and thawing of snow cover. Clear ice (high density) may impair gaseous exchange more than milk or white ice (low density); however, physical appearance may not be good indicator of permeability. Sometimes, ice encasement injury is confused with low-temperature kill, as both are associated with ice cover.

30.2.5 TRAFFIC INJURY

Traffic injury of cool-season turfgrass during winter is also very common. Two different kinds of traffic injuries occur in cool-season turfgrass. The most common is mechanical injury due to foot and vehicular traffic on turf when leaf and stem tissues are frozen (brittle), resulting in disruption of the protoplasm and cell death (Beard 1973). Footprinting and traffic patterns on turf result from this kind of injury. This type of injury is temporary and new growth from viable crown can replace damaged tissues. Limiting or redirecting traffic from frozen turf after thaw, and irrigating frozen tissues early morning to accelerate thawing before traffic movement can prevent this type of damage during winter. The most serious type of damage occurs due to traffic on turf that is covered with wet slush. Rapid drops in air temperature to 20°F (−6.7°C) following traffic over slush-covered turf are the primary reason behind this kind of damage (Beard 1973). This type of injury may cause permanent damage to crown tissues and meristems of rhizomes and stolons and negate the recuperative potential of turfgrass. Withholding traffic under this situation is the only way to avoid such serious damage of turfgrass.

30.3 COLD HARDENING OR ACCLIMATION

Cold hardening or acclimation is a complex developmental process through a nonheritable modification of structures and functions as a response to cold, which minimizes cold injuries and improves the fitness of an individual plant (Alberdi and Corcuera 1991). During the cold acclimation period, there is an orchestration of many biochemical and physiological events that are required to achieve maximum cold hardiness (Steponkus 1990). In regions with a seasonal climate, perennial plants undergo a sequence of processes that are mutually interdependent, each stage preparing the way for the next. Due to the complexity of external factors involved in acclimation process, the same species may exhibit unequal degrees of cold hardiness at different localities (Guy 2003). The first stage of cold acclimation is initiated after the cessation of the plant growth during fall. The changes in the photoperiod (short day) and falling temperatures (10–20°C) induce the acclimation process in most plants (Sakai and Larcher 1987). Cool-season turfgrasses require temperatures between 32°F and 45°F for 2–3 weeks to initiate the cold hardening process. During this period, organic substances are stored and the most important synthetic processes are accumulation of reserve starch and neutral lipids. These are essential substrates and the energy source for the metabolic changes occurring during the second stage of acclimation. The second stage of cold acclimation is induced by subzero temperatures. During this stage of hardening, proteins and membrane lipids are synthesized or undergo structural changes, ultimately leading to maximal hardiness (Sakai and Larcher 1987).

30.3.1 GENE REGULATION

Earlier theories of cold acclimation were developed on the basis of the nonheritable modification, but mounting evidence points to the involvement of a genetic component associated with the ability to cold acclimation and the induction of freezing tolerance. The reason behind this new approach is the inheritance pattern of hardiness in progeny of parents with different tolerance levels, but it also includes (i) accumulation of soluble proteins, (ii) changing in protein electrophoretic patterns, (iii) alteration in isoenzyme composition, (iv) appearance of new proteins, (v) accumulation of rRNA and soluble RNAs, (vi) changes in RNA base composition, (vii) altered mRNA content, and (viii) induction of freezing tolerance by ABA (Guy 1990). Cold tolerance or freezing tolerance of the plant is a multigenic trait, generating a multifaceted complexity in the cellular level under subzero temperature (Smallwood and Bowles 2002).

30.3.2 ROLE OF CRYOPROTECTANTS AND PROTEINS

Although controversies exist regarding the relationship between accumulation of cryoprotective substances and enhanced cold hardiness of plants, several studies showed the role of those substances as protectants of enzymes and membranes susceptible to chilling. Carbohydrates, amino acids, glycinebetaine, and polyamines (compatible solutes) as well as proteins and lipids are accumulated with low temperature induction in plants (Alberdi and Corcuera 1991; Kishitani et al. 1994; Guy 1990). Compatible solutes act as cryoprotectants by maintaining the native conformation of protein at low-water activities by decreasing the protein–solvent interaction (Alberdi and Corcuera 1991). The role of protein metabolism on cold acclimation is well documented by several researchers (Graham and Patterson 1982; Guy 1990; Levitt 1980). Accumulations of soluble proteins were observed in cold-acclimated wheat (Rochat and Therrien 1975). The protein synthesis pattern of the plant also changes significantly with low-temperature exposure compared with protein synthesis at warm temperature. A study of plasma membrane proteins of plants in response to low temperature has shown synthesis of new polypeptides with cold acclimation treatments (Yoshida and Uemura 1984). In cold-acclimated tissues, changes were also observed in the glycoprotein and lipid composition of the plasma membrane. With low temperature, proteins also directly act as cryoprotective substances, and heat-stable, soluble proteins are capable of protecting the membrane against freezing-stress damage. Information regarding accumulation of protein and cold hardiness of cool-season turfgrass is very limited. Dionne et al. (2001a) studied the changes of protein in three different annual bluegrass (*Poa annua* L.) ecotypes during cold acclimation. They found the cold responsiveness of specific soluble polypeptides and thermostable proteins in annual bluegrass. The peak accumulation of those proteins also showed association with maximum freezing tolerance. The accumulation of polypeptide increased from fall until midwinter while it decreased during the spring. These proteins may act as antifreeze agents and can prevent the growth of ice crystal within cells.

30.3.3 ROLE OF AMINO ACIDS

Amino acids also play a significant role in cold hardiness of plants. Accumulation of some specific amino acids during cold acclimation was documented in several plants (Sakai and Larcher 1987; Sagisaka and Araki 1983). The studies conducted on annual bluegrass showed higher accumulation of amino acids with cold acclimation (Dionne et al. 2001a). At subfreezing temperatures, higher accumulation of proline, glutamine, and glutamic acid in annual bluegrass crown were observed. Although they were unable to find the relationship between proline accumulation and freezing tolerance of annual bluegrass ecotypes, several studies showed accumulation of proline during cold acclimation. The role of proline in osmotic stressed plant is well documented in various studies (Dorffling et al. 1997). It is not clear whether proline accumulation is a direct result of low temperature or if it is induced by the low-water activities within the cell during cold.

The specific role of other amino acids on freezing tolerance is also not clearly portrayed and it varies widely with different plant species. The activity of several enzymes also increased in manyfold with cold acclimation. Isozymic variants of ATPases, esterases, acid phosphatases, and peroxidases were observed in alfalfa during winter (Krasnuk et al. 1976). Higher activity of glutathione reductase, NADP reductase, and RUBPcase (ribulose biphosphate carboxylase/oxygenase) was observed in different plants with cold acclimation (Alberdi and Corcuera 1991). These enzymes are generally stable at low temperature and help the plants to maintain their biological function and energy synthesis during cold stress.

30.3.4 ROLE OF CARBOHYDRATES

Another important factor determining freezing tolerance is carbohydrate reserve in plant organs. In terms of cold hardiness of cool-season turfgrass, carbohydrate plays a dominant role. The major carbohydrates stored in cool-season turfgrass are starch, fructans, and sucrose. They are mainly stored in the crown and stolons, and provide energy for respiration during winter. The relation of fructans and cold tolerance was observed in some plant groups under low temperature (Pollock et al. 1989; Pontis 1989). Dionne et al. (2001b) found fructans to be the most abundant carbohydrate in cold-acclimated annual bluegrass ecotypes. Higher amounts of the soluble carbohydrate sucrose was also observed after exposure to freezing temperature, and its association with freezing tolerance of annual bluegrass was also observed. Sucrose acts as a cryoprotectant and lowers the water potential within the cell by increasing solute concentration and thus prevents cell dehydration under cold stress. Another study with perennial ryegrass (*Lolium perenne* L.) showed lower levels of sugars in cultivars of inferior freezing tolerance compared with hardy cultivars (Bredemeijer and Esselink 1995). However, many studies reported no relationship between nonstructural carbohydrate and low-temperature tolerance in turfgrass. Agronomic practices like higher nitrogen fertilization or low mowing height, which reduce the carbohydrate reserve in the crown and roots during the period of acclimation, also adversely affect the cold resistance in cool-season turfgrass.

Other than carbohydrate, potassium also acts as an antifreeze agent by lowering the freezing point within the cell and thus restricts ice formation at subfreezing temperature (Fry and Huang 2004). Proper potassium fertilization during fall can increase the cold tolerance in cool-season turfgrass. For turfgrass, any management that increases the water content in crown and leaf, such as high soluble-nitrogen application, shade, low-lying areas, poor subsurface drainage, and higher freeze-thaw cycle predispose turfgrass to more winter injury. Potassium deficiency is inversely related with cold hardiness. A high level of thatch accumulation also reduces cold tolerance, as it exposes the crown to low temperature during winter. Moderate amounts of thatch give better protection to cool-season turfgrass during winter.

30.3.5 ROLE OF CELL MEMBRANE

For any plant, the cell membrane is the main site that is susceptible to cold injury. All metabolic and structural functions that increase the membrane fluidity give better cold hardiness to plants. The most accepted hypothesis for general cold injury was proposed by Lyons, Raison, and co-workers, popularly known as the "membrane hypothesis" (Lyons 1973). The hypothesis primarily states that reversible change in the physical state of a membrane is the initial response of cold injury (Lynch 1990). The alteration of membrane properties causes changes in membrane-associated enzyme activities and membrane permeability, and consequently decreases in ATP levels, ion-solute leakage, loss of compartmentation, and metabolic imbalances. Later, the theory was further developed by Murata and colleagues, who proposed the correlation between chilling sensitivity and molecular species composition of phosphatidylglycerol (Murata 1983). The respiratory and photosynthetic activity of the plant is affected by the disruption of the mitochondria and chloroplast and such changes of physiological and biochemical activities with abrupt thermotropic changes are related to changes in the molecular ordering of the membrane lipid. Many studies observed that plants undergo phase transition from a liquid crystalline state to a solid gel phase at a critical temperature at which chilling injury occurs (Lyons 1973). At low temperature, the biomembrane undergoes a phase transition, and the membrane becomes leaky to small electrolytes, thereby disrupting the ion gradient across the membrane, which is essential for the maintenance of cellular homeostasis (Paliyath and Droillard 1992). A variety of membranous lipolytic enzymes like phospholipase D, phosphatidate phosphatase, lipolytic acyl hydrolase, and lipoxygenase play a crucial role in the degradation of membrane lipids and thus dictate the chilling injury in both chilling-sensitive and chilling-tolerant plants (Pinhero et al. 1998). Research shows that phospholipase D was released from the vacuole in wheat crown and roots during its exposure to frost and causes severe lipid degradation and membrane deterioration (Willemot 1983). The saturation and unsaturation of fatty acids of the membrane also determine the freezing sensitivity of plants but several controversies exist with this theory. Cyril et al. (2002) found higher amounts of unsaturated fatty acids in the membrane of freeze-tolerant seashore paspalum (*Paspalum vaginatum* Swarz.) plants compared with freezing-sensitive plants. The information regarding the membrane composition and molecular changes of the membrane with cold and frost in cool-season turfgrass is very limited. Various studies showed higher electrolyte leakage with low temperature in chilling-sensitive turfgrass cultivars, and thus proved the role of membrane lipid in turfgrass during cold stress. Measurement of electrolyte leakage is the most reliable way to understand the chilling sensitivity of turfgrass. A study on low-temperature tolerance of perennial ryegrass showed less leakage loss from the high cold-stress-tolerant cultivars compared with cultivars of low tolerance (Ebdon et al. 2002). Similar results were also observed during assessment of cold injury of two warm-season turfgrasses, seashore paspalum and bermudagrass (*Cynodon* spp.) (Cardona et al. 1997; Anderson et al. 1988).

30.3.6 ROLE OF PLANT HORMONES

Cold acclimation also causes changes in hormonal levels in the plant. In general, during cold acclimation the abscisic acid (ABA) level increases and that of gibberellin decreases (Smallwood and Bowles 2002). Biosynthesis of plant hormone auxin and sensitivity to gibberilins is affected by cold (Gray et al. 1998; Hisamatsu and Koshioka 2000). Studies with the cell culture of some plants suggests that ABA may play a significant role in the induction of freezing tolerance and control the expression of genes responsible for increased hardiness. The concentrations of ABA are transiently elevated in response to cold temperature, and exogenous application of ABA at nonacclimating temperature can enhance freezing tolerance as well as inducing many of the genes that respond to low temperature (Chen et al. 1983; Lang et al. 1994; Heino et al. 1990). Association of higher concentrations of abscisic acid and freezing tolerance was observed in smooth bromegrass (*Bromus inermis* Leyss) cell suspension culture (Lee and Chen 1993). The induction of freezing tolerance by abscisic acid is governed either by the expression of novel polypeptides or by causing changes in protein synthesis. Abscisic acid also helps the plant with cold stress by increasing dehydration tolerance (Fry and Huang 2004).

30.4 SURVIVAL MECHANISM OF COLD TOLERANCE

The frost-survival capacity of the plant primarily depends on the various frost-tolerance mechanisms involved that help the plant to mitigate, prevent, or escape excessive low-temperature stress. Two major strategies of plant survival at subzero temperatures are freeze avoidance and freeze tolerance. In the case of freeze avoidance, plants that survive in a vegetative state during winter are able to prevent the crystallization of ice within their tissues, while in the freeze-tolerance mechanism plants allow ice to crystallize in the apoplast (Smallwood and Bowles 2002). Freezing avoidance involves supercooling that represents a transient, unstable state, which may provide protection against brief radiation frost to an extent of 3–8 K below the freezing point of tissue (Hatakeyama 1960; Kaku 1975). Floral bud meristems and xylem parenchyma of many temperate plants supercool to low temperatures, in some cases below the homogeneous nucleation point. Plants with persistant supercooling are able to remain in the supercooled state for a long time (Sakai and Larcher 1987). In the case of extraorgan supercooling, buds and seeds attain a high degree of freezing tolerance by translocation of water from supercooled tissues or organs to nucleation centers in adjacent tissues or outside the organs (Sakai 1979). There is a gradient in the rate of dehydration or water migration from the cell, tissue, or organ; this is rapid in extracellular freezing, slow in extraorgan freezing, and very little in supercooling (Sakai and Larcher 1987). The gradient also exists in the effect of slow freezing in enhancing the cold hardiness in plants.

Turfgrass also survives freezing temperatures and maintains its function, which includes growth, development, and productivity through two survival mechanisms: avoidance and tolerance. In the avoidance mechanism, low-temperature exposure is either prevented or reduced. Although supercooling is an important avoidance mechanism in many woody plants, its role in turfgrass is not clear (Levitt 1980; DiPaola and Beard 1992). The significant depression of freezing point was not observed in turfgrass cells during cold stress (Thomas and James 1993; Rossi 1997). Extensive research is required to understand the avoidance mechanism of turfgrass. The freezing-stress tolerance of turfgrass depends on the tolerance to extracellular ice formation and resulting dehydration (Levitt 1980). The overall tolerance level largely depends on the acclimation and deacclimation process of turfgrass (Fry et al. 1993; Rossi 1997). The cold hardiness is a result of multiple factors and requires either a few of or all the mechanisms mentioned earlier. The regrowth or recuperative potential of turfgrass also determines the survival of specific turfgrass species. Tiller production or regrowth from lateral tiller buds dictates the whole plant survival to freeze-stress temperature. Superior cultivars of cool-season turfgrass can produce more tillers after cold stress compared with inferior cultivars. Thus, the turfgrass crown plays the most important role in this survival mechanism by regenerating root and shoots after extensive cold stress (DiPaola and Beard 1992). Smallwood and Bowles (2002) described biochemical alterations of plant during cold acclimation as follows:

- The accumulation of compatible solutes or cryoprotectant substances
- Alterations in membrane lipid composition
- Alteration of plant hormonal levels
- Changing composition and strength of cell wall
- Altered pattern of gene expression including those encoding soluble proteins
- Alteration of protein synthesis and function including nucleic acid and enzymes
- Increased antioxidant activity

30.5 ROLE OF ANTIOXIDANTS

Environmental conditions that induce or favor photooxidative stress are common events in the growth and developmental processes of the plant (Polle 1997; Thompson et al, 1987; Paliyath and Droillard 1992). Within various biochemical pathways, particularly for respiratory energy production, O_2 is an important electron acceptor. However, this mechanism is a mixed blessing as while it permits the biosphere's versatility and productivity, it is also life threatening due to oxidative stress. As a result of its oxidizing capacity, O_2 acts primarily as an electron acceptor, which leads to the formation of a variety of reactive oxygen intermediates. These reactive oxygen species with unpaired electrons, such as superoxide anion ($O_2^{\bullet-}$) and $^{\bullet}OH$, or species like hydrogen peroxide (H_2O_2), and hypochlorous acid (HOCl) with the ability to abstract electrons from other molecules, can react immediately with cellular macromolecules and cause damage (Asada 1993; Fuchs et al. 1997). The active oxygen species can attack cell membranes by a cascade of free-radical chain reactions, resulting in extensive damage of cell membrane and other cellular structure. These intermediates can react with the polyunsaturated fatty acids and cholesterol present in cell membrane and generate oxidized lipids, which finally leads to apoptosis or cell death (Halliwell and Gutteridge 1989; Christ et al. 1993). These reactive oxygen species may also cause damage to DNA or can induce rapid depletion of cellular NAD/NADH pools, leading to the depletion of ATP reserves and cell death (Christ et al. 1993).

As a part of their aerobic existence, plants had to develop an effective defensive system to deal with these reactive intermediates. Due to this reason, self-protection, the cell initiates various antioxidant mechanisms. Balancing reactive oxygen intermediates and antioxidants is the key function of the cell to counter oxidative challenges (Fuchs et al. 1997). Antioxidative systems are present in all subcellular compartments of the plants, including the apoplastic region (Figure 30.2). The antioxidant defense system is dynamic and is composed of hydrophilic and lipophilic metabolites and antioxidant enzymes (Polle 1997). These include (1) enzymatic scavengers such as SOD (super oxide dismutase), which hasten the dismutation of $O_2^{\bullet-}$ to H_2O_2, and catalase and glutathione peroxidase (GPX), which convert H_2O_2 to water, (2) hydrophilic radical scavengers such as ascorbate, urate, and glutathione (GSH), (3) lipophilic radical scavengers such as tocopherols (e.g., α-tocopherol, γ-tocopherol), flavonoids (e.g., quercetin, epigallocatechin gallate), carotenoides (β-carotene, lycopene) and ubiquinone, (4) enzymes involved in the regeneration of oxidized forms of the small

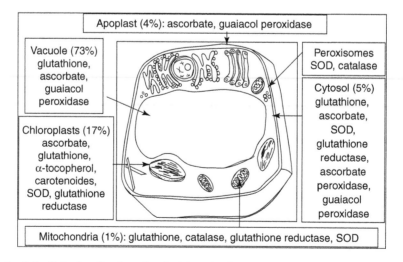

FIGURE 30.2 Subcellular localization of antioxidants in plant cells. (Modified from Polle, A., in *Oxidative Stress and the Molecular Biology of Antioxidant Defenses*, Cold Spring Harbor Laboratory Press, Plainview, NY, 1997, 623–666.)

molecular antioxidants (GSH reductase, dehydroascorbate reductase) or responsible for the maintenance of protein thioles (thioredoxin reductase), (5) the cellular machinery that maintains a reducing
environment (e.g., glucose-6-phosphate dehydrogenase, which regenerates NADPH), and (6) secondary metabolites like phenolics, which scavenge free radicals and also trigger the activity of antioxidant enzymes (Beckman and Ames 1997; Jorgensen et al. 1999; Shetty 1997; Shetty and Wahlqvist
2004). In different cellular compartments, different antioxidants play a crucial role. Carotenoids are
the most competent scavengers in chloroplast, while ascorbate is a major water-soluble antioxidant in
plant leaves. Multiple roles of antioxidants and their subcellular localization determine the fitness of
a plant to several oxidative stresses (Polle 1997).

The potential role of antioxidant enzymes in protecting plants from cold injury is well established (Christie et al. 1994). During high illumination, if low temperature limits the utilization
of reductant produced in photosynthesis, then it enhances photooxidative stress. With decreasing
temperature, solubility of the gas increases, which leads to a higher concentration of oxygen and
thus enhances the risk of oxidative stress at low temperature (Polle 1997). Plant injury caused by
freeze stress along with illumination may be mediated by reactive oxygen species such as superoxide radicals, singlet oxygen, hydrogen peroxide, and hydroxyl radicals (Wise and Naylor 1987).
A higher concentration of $O_2^{\bullet-}$ was reported from lethally frozen crown of winter wheat (*Triticum
aestivum* L.) compared with unstressed or sublethally stressed wheat plant (Kendall and McKersie
1989). Other than dehydration and mechanical injury, activated oxygen species participate in freezing stress, causing lipid peroxidation and a collapse of antioxidative systems in unhardened tissues
(Sagisaka 1985; Kuroda et al. 1992; Walker and McKersie 1993). Antioxidant systems can maintain
a cellular homeostasis that governs the level of stress tolerance in the plant at low temperature. In
field conditions, the cold acclimation process generally invokes important changes in plant physiology, including increases in antioxidative protection (Pinhero et al. 1997).

In chilling-sensitive plants, significant reduction of catalase activity was observed with low-
temperature stress. Chilling-induced reduction in catalase activity along with increase in H_2O_2
concentration was observed in cucumber (*Cucumis sativus* L.) (Feierabend et al. 1992). Catalase
and guaiacol peroxidase activity increased with H_2O_2-induced acclimation in maize (*Zea mays* L.)
(Prasad et al. 1994). In maize and spinach, high SOD activity in response to high light in combination with low temperature was observed (Schöner and Krause 1990; Jahnke et al. 1991). Higher
SOD activity corresponded with higher protection from cold injury in alfalfa (*Medicago sativa* L.)
(McKersie et al. 1993). Numerous studies reported higher GSH content in leaves of perennial plants,
during winter (Smith et al. 1990; Polle and Rennenberg 1994). Direct role of GSH in cold tolerance is
not manifested clearly. The involvement of GSH in cold tolerance is probably due to its role as a substrate of glutathione reductase (Esterbauer and Grill 1978). Higher concentration of ascorbate and
ascorbate peroxidase was also observed in evergreens in winter (Anderson et al. 1992). The importance of a balanced adjustment of electron transport and antioxidative systems for avoidance and
compensation of oxidative stress was documented in maize (Massacci et al. 1995). Down-regulation
of electron transport and adjustment of SOD and ascorbate peroxidase activity is crucial in oxidative stress-tolerance mechanism (Massacci et al. 1995). Antioxidant enzymes not only scavenge
free radicals but also induce important signals to protect plants from low-temperature stress
(Polle 1997).

The direct role of antioxidants in low-temperature tolerance of cool-season turfgrass was not
investigated. The effect of other oxidative stresses, such as high temperature, low light, ultraviolet-beta
radiation on turfgrass, and role of the antioxidative system in stress tolerance were investigated (Wang
et al. 2003; Larkindale and Huang 2004; Xu and Huang 2004; Ervin et al. 2004; Jiang et al. 2005).
Transient increase in ascorbate peroxidase (APX) and peroxidase (POX), and the long-term increase
in SOD were observed in creeping bentgrass under heat stress. They also observed reduced catalase
activity with heat stress in creeping bentgrass (*Agrostis stolonifera* L.) (Larkindale and Huang 2004).
Zhang and Schmidt (1999, 2000) reported higher antioxidant content in drought-induced Kentucky
bluegrass (*Poa pratensis* L.), and creeping bentgrass plants. They found that seaweed extract, either

alone or in combination with humic acid and iron, enhanced the photochemical and SOD activity in creeping bentgrass (Zhang et al. 2002). Wang et al. (2003) reported that SOD and catalase activity decreased with higher root-zone temperature in heat-tolerant "Penn-A-4" and heat-susceptible "Putter" cultivars of creeping bentgrass. The enhancement of the antioxidant protection system through the application of phytohormone to mitigate UV-B stress was observed in Kentucky bluegrass (Ervin et al. 2004). Xu and Huang (2004) observed changes in the antioxidant system in creeping bentgrass under field conditions during summer. They found that activities of SOD, catalase, hydrogen peroxidase, and ascorbate peroxidase increased from May to July, when the temperature increased from optimum to moderate, and then declined to the lowest level in August, when temperature reached the highest level of tolerance. As low temperature in combination with high light-intensity-induced oxidative stress in the plant, one can assume the role of antioxidant system in the stress-tolerance mechanism of cool-season turfgrass during winter. Antioxidant enzymes either directly enhance low-temperature tolerance or induce defensive systems through signaling mechanism in cool-season turfgrass. The role and exact mechanism of antioxidants in relation to cold tolerance needs to be investigated thoroughly in cool-season turfgrass for better understanding.

30.6   PHENOLIC ANTIOXIDANTS

Phenolic compounds are secondary metabolites, distributed widely in plants (Javanmardi et al. 2003). The role of phenolics as antioxidants has been well documented. The focus on phenolic antioxidants in fruits and vegetables is mainly due to their role in protecting human health (Shetty 1997, 1999, 2004). The role of phenolic antioxidants in plant abiotic stress has received less attention until now. Studies on the free-radical scavenging properties of flavonoids have opened the characterization of other phenolic compounds as antioxidants. The structural chemistry of polyphenols suggests its free-radical scavenging properties (Rice-Evans et al. 1997). The antioxidant activity of phenolics is actually governed by their redox properties, which play a crucial role in adsorbing and neutralizing free radicals, quenching single and triplet oxygen, or decomposing peroxides (Rice-Evans et al. 1997; Shetty 2004). The activity of the antioxidant is determined by its reduction potential, ability to stabilize and delocalize the unpaired electron, reactivity with other antioxidants, and transition-metal chelating potential.

Polyphenols have all the above characteristics and thus play a significant role in the antioxidative defense systems in plants. The most widely distributed phenolic components in plant tissues are the hydroxycinamic acids, p-coumaric, caffeic, and ferulic acids, which are synthesized via the shikimate pathway (Rice-Evans et al. 1997). In response to biotic and abiotic stress, biosynthesis of phenolic antioxidants takes place through the stimulation of plant secondary metabolite pathways. Many phenylpropanoid compounds such as flavonoids, isoflavonoids, anthocyanins, and polyphenols are induced in response to wounding, nutritional stress, cold stress, and high visible light. UV radiation also induces phenolic antioxidants in plants (Hahlbrock and Scheel 1989; Graham 1991; Christie et al. 1994; Beggs et al. 1987). A few studies have examined increases in phenolic acids relative to low-temperature stress (Prasad 1996; Rice-Evans et al. 1997). Pennycooke et al. (2005) found that cold acclimation induced accumulation of total phenolics, which had positive correlation to the antioxidant capacity in petunia. The recovery from the chilling stress indicates some kind of antioxidant protection, including anthocyanin and phenylpropanoid biosynthesis (Christie et al. 1994).

30.7   METABOLIC PATHWAYS AND REGULATION UNDER COLD STRESS

The metabolic pathway shifts associated with cold acclimation of dormancy-developing and nondormant tissues have been reported (Sakai and Larcher 1987). Sagisaka (1972, 1974) reported a shift of glucose-6-phosphate metabolism from glycolysis to the pentose phosphate cycle in poplar (*Populus gelrica*) twigs in early autumn. The pentose phosphate cycle provides various substrates and NADPH for important biological reactions, including energy synthesis in plant cells.

With exposure to cold, increased activity of glucose-6-phosphate dehydrogenase and important changes in the catalytic properties of pyruvate kinase was observed in leaves of winter rape (*Brassica napus* L.) (Sobczyk and Kacperska-Palacz 1980). High activity of glucose-6-phosphate dehydrogenase was also observed in *Chlorella ellipsoidea* (Gerneck) cells, when it was exposed to 3°C for 2 days (Sadakane et al. 1980; Hatano and Kabata 1982). Other than the important shift in glucose catabolism, it was observed that in cold-acclimated cells reducing reactions are favored. In wintering perennial plants, glutathione reductase appears to be a ubiquitous enzyme (Sagisaka 1982). During winter, ascorbate not only just detoxifies the peroxides but also serves for regulating the NADPH level. A significant increase of ATP and of adenylate energy charge has also been reported in cold-acclimated winter rape leaves. It seems that high availability of ATP and NADPH is necessary for the hardening process as it allows for the synthesis of RNA, protein phospholipid, and other substances in low temperature (Kacperska-Palacz 1978).

Christie et al. (1994) studied the impact of low-temperature stress on general phenylpropanoid and anthocyanin pathways in maize seedling. They found rapid deposition of anthocyanin in maize seedling, when it was exposed to low temperature. Several genes are required for the expression of anthocyanin and those genes can be considered to be *cor* (cold regulation) genes. They also observed accumulation of phenylpropanoid transcripts during cold stress, although each transcript showed a distinct pattern with respect to the magnitude of change and the kinetics of accumulation. Significance of PAL (phenylalanine ammonia-lyase) and CHS (chalcone synthase) for catalyzing the first steps in the synthesis of a variety of protective molecules was observed with profound increases of their transcripts in maize (Christie et al. 1994). Induction of PAL was observed in cold-tolerant cultivars of *Brassica napus* (L.) compared with cold-sensitive cultivars (Parra et al. 1990). An influx of calcium from the apoplast into the cytosol is an early response of the plant to low temperature (Knight et al. 1991). Elevation of calcium has been found to be necessary to promote the expression of low-temperature responsive genes in alfalfa and *Arabidopsis* (Knight et al. 1996; Monroy and Dhindsa 1995). Calcium influx under low temperature is actually regulated by reorganization of the cytoskeleton triggered by rigidification of the membrane (Orvar et al. 2000).

30.8 PROLINE-LINKED PENTOSE PHOSPHATE PATHWAY AND ANTIOXIDANT MECHANISM UNDER STRESS

Stimulation of protective secondary metabolite pathways, such as biosynthesis of phenolic antioxidants, is a natural response of the plant to biotic and abiotic stresses (Rice-Evans et al. 1997; Shetty 1997, 1999, 2004). Induction of the phenylpropanoid pathway and production of proline in plants during abiotic stress is well documented (Christie et al. 1994; Dorffling et al. 1997). Studies in wheat indicated a higher accumulation of proline during acclimation and its association in frost tolerance improvement (Dorffling et al. 1997). The role of proline in protection against freeze-induced lesion includes specific effects on the plasma membrane during osmotically induced contraction and reduces loss of osmotic responsiveness to proline compared with sorbitol, as well as reduced solute loading into thylakoid (Steponkus 1984; Popova et al. 1998). Proline is synthesized from glutamate by a series of reduction reactions. In this synthesis process, proline and pyrroline-5-carboxylate (P5C) regulate redox and hydride ion–mediated stimulation of pentose phosphate pathway (Hagedorn and Phang 1983; Phang 1985). During respiration, oxidation reactions produce hydride ions, which help reduction of P5C to proline into cytosol. Through the reaction of proline dehydrogenase, proline can enter mitochondria. Within mitochondria, instead of NADH, proline acts as a reducing equivalent and can support oxidative phosphorylation. The reduction of P5C in cytosol provides $NADP^+$, which is a cofactor for glucose-6-phosphate dehydrogenase (G6PDH). G6PDH plays crucial role, by catalyzing the rate-limiting step of the pentose phosphate pathway. Phang (1985) first proposed this model and stated its role in the stimulation of purine metabolism via ribose-5-phosphate in the animal cell.

On the basis of this hypothesis, Shetty (1997) proposed a model that proline-linked pentose phosphate pathway could stimulate both the shikimate and phenylpropanoid pathways, and therefore, the modulation of this pathway could lead to the stimulation of phenolic phytochemicals (Figure 30.3). Using this model, proline, proline precursors, and proline analogs were effectively utilized to stimulate total phenolic content in plants (Kwok and Shetty 1998; Yang and Shetty 1998). It was also proposed that demand for NADPH for proline synthesis during microbial interaction and proline analog treatment may increase cellular NADP$^+$/NADPH ratio, which should

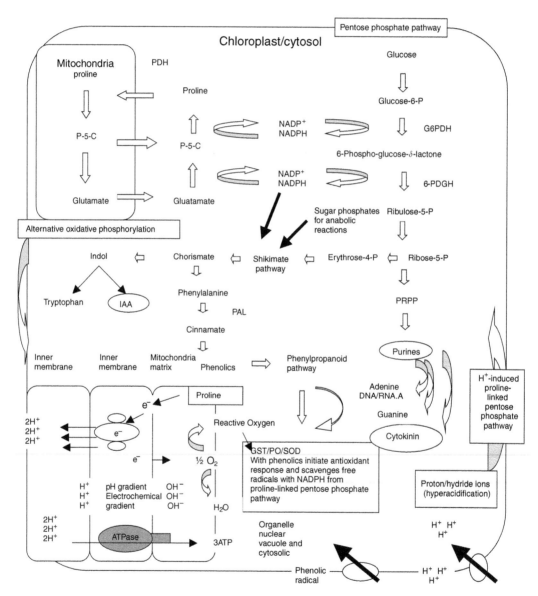

FIGURE 30.3 Model for the role of proline-linked pentose phosphate pathway in regulating phenolic biosynthesis in plants, which also accommodates the action mechanism of external phenolic phytochemicals. Abbreviations: Glucose-6-phosphate dehydrogenase, G6PDH; 6-phosphogluconate dehydrogenase, 6-PDGH; Glutathione-s-transferase, GST; indole acetic acid, IAA; Phenylalanine ammonia lyase, PAL; peroxidase, PO; proline dehydrogenase, PDH; pyrroline-5-carboxylate, P-5-C; superoxide dismutase, SOD. (From Shetty, K. *Process Biochem.*, 39:789–803, 2004.)

activate G6PDH. The proline analog, azetidine-2-carboxylate (A2C), is an inhibitor of proline dehydrogenase and therefore tolerance to the analog could stimulate proline synthesis, which drives the demand for NADPH (Elthon and Stewart 1984). Another analog, hydroxyproline is a competitive inhibitor of proline for incorporation of proteins. Either analog at low level should deregulate proline synthesis from feedback inhibition (allow proline synthesis). Therefore, deregulation of the pentose phosphate pathway may drive metabolic flux toward erythrose-4-phosphate for biosyntheis of shikimate and phenylpropanoid metabolites. At the same time, proline serves as a reducing equivalent, instead of NADH for oxidative phosphorylation (ATP synthesis) in mitochondria (Shetty and McCue 2003; Shetty 2004).

This model has been proposed for the mode of action of phenolic metabolites based on the correlation between stress-stimulated phenolic biosynthesis and stimulation of the antioxidant enzyme response pathways in plants. Within the plant system model, acid, exogenous phenolic, proline analogs, and precursor combinations and microbial elicitors were used to stimulate phenolic biosynthesis and key enzyme responses (Shetty 2004). Proline/G6PDH correlations during phenolic response were also associated with phenolic content, potential polymerization of phenolics by GPX, and antioxidant activity based on free-radical scavenging activity of phenolics and SOD (McCue et al. 2000; McCue and Shetty 2002; Bowler et al. 1994). In this above-mentioned model, elicitors through various independent or common pathways directly or indirectly mediate a proton/hydride ion (acidification)–linked redox cycle in the cytosol, which is then coupled to the stimulation of proline-linked pentose phosphate pathway. This stimulation mechanism results in proton recycling through NADPH for proline biosynthesis and other biosynthetic pathways, including phenylpropanoid and antioxidant response pathways involving SOD and peroxidases. Since NADH production would be reduced as carbon flux is diverted to proline via α-ketogluterate and glutamic acid, proline can be used as an alternative reducing equivalent instead of NADH to drive mitochondrial oxidative phosphorylation for energy (ATP) synthesis (Shetty 2004). The synthesis of phenolic metabolites and stimulation of the antioxidant response pathway can minimize the oxidation-induced damage within tissues where it occurs. Phenolic antioxidants can behave as antioxidants by trapping free radicals in direct interactions or scavenge them through a series of coupled, antioxidant enzyme defense system reactions. A coupled enzymatic defense system could involve low-molecular-weight antioxidants, such as ascorbate, glutathione (GSH), α-tocopherol, carotenoids, and phenylpropanoids, in conjunction with several enzymes such as SOD, catalase, peroxidases, glutathione reductase, and ascorbate peroxidase (Rao et al. 1996; Bowler et al. 1994; Pinhero et al. 1997). It is clear from current theories that the antioxidant response pathway in plants is dependent upon key NADPH-requiring enzymes, similar to proline biosynthesis. This defensive mechanism is believed to work through antioxidant response pathways involving peroxidases, as well as biosynthesis of polymeric phenolics that leads to protective lignification or smaller polymers that act as antioxidants (Figure 30.4) (Shetty 2004). We can assume that the proline-linked pentose phosphate pathway may play a crucial role under low-temperature stress in plants and as a result of this, phenolic phytochemicals can effectively counter oxidative stress within the cell during winter.

30.9 MODERN BIOTECHNOLOGICAL TOOLS TO IMPROVE COLD-STRESS TOLERANCE IN PLANTS

Significant developments in the field of molecular biology and biotechnology in past 3 decades have opened the prospects of new cultivars with superior cold-tolerance characteristics. The understanding of complex molecular and biochemical mechanisms of plants in response to cold stress is essential for biological advances. Plants generally undergo many physiological and biochemical changes during exposure to low temperature as discussed previously. Modified levels and activities of enzymes from various metabolic pathways, accumulation of carbohydrates (fructans or sucrose),

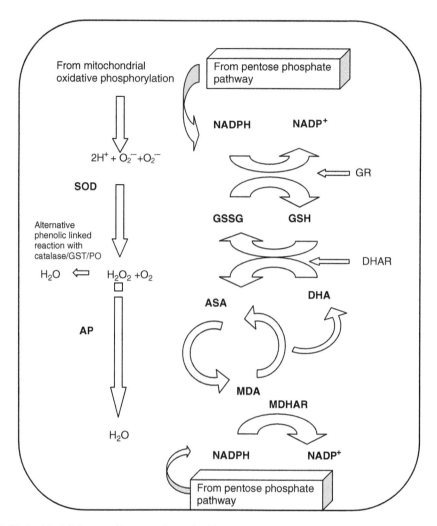

FIGURE 30.4 Model for specific steps in antioxidant response pathway in plants. Abbreviations: SOD, superoxide dismutase; AP, ascorbate peroxidase; GR, glutathione reductase; GSSG, reduced glutathione; GSH, glutathione; DHAR, dehydroascorbate reductase; ASA, reduced ascorbate; DHA, dehydroascorbate; MDA, monodehydroascorbate; MDHA, monodehydroascorbate reductase. (From Shetty, K. *Process Biochem.*, 39:789–803, 2004.)

amino acids (proline), or polyamines, glycinebetaine, and altered lipid composition of cell membrane are the major areas targeted for cold tolerance improvements (Kishitani et al. 1994; Hughes and Dunn 1996; Smallwood and Bowles 2002).

Low-temperature stress leads to differential gene expression and action. Many cold-inducible genes have been isolated from different plants and the numbers are growing constantly. From the information of the isolated genes, different inferences can be made as follows: (i) several characterized genes seem to occur in a wide spectrum of plant species and contain conserved structural elements suggesting conservation for functional reasons; (ii) a group of low-temperature-induced genes can be isolated, which are homologous to *LEA* genes and preferentially expressed during embryo maturation and mainly encode hydrophilic proteins (Dure et al. 1989; Bartels et al. 1993); (iii) some isolated genes are exclusive to cold-tolerance mechanisms, where some are involved in osmotic response, which is common to cold, water, and salt stress. On the basis of the findings of several studies, Bartels and Nelson (1994) proposed some strategies of gene technology to elucidate the mechanism of cold acclimation in plants:

1. Overexpression of specific gene products in plants targeted to particular cell compartments. Depending upon the nature of the expressed polypeptides, either the contribution of protein molecules can be assessed or if the protein encodes a functional enzyme, then the effect of specific metabolites should be studied.
2. If function and regulation of major metabolic pathways related to cold acclimation can be identified, then key enzymes can be modified by overexpression or by suppression (antisense approach) for cold tolerance improvement.
3. Easy purification of large amounts of biologically active or specifically modified molecules through specific system like *E. coli*, yeast, or others can be tested for functions *in vitro*.

Murata et al. (1982) first demonstrated a strong correlation in different plant species between the degree of chilling sensitivity and the proportion of disaturated phosphatidylglycerol (PG) molecules of chloroplast membrane. Chloroplast enzyme glycerol-3-phosphate acetyltransferase (GPAT) determined the degree of unsaturation in fatty acids (Murata 1983). *Arabidopsis* plants transformed with the *E. coli* GPAT contained higher amounts of saturated fatty acids in the chloroplast membrane lipids than the wild type (Wolter et al. 1992). Overexpressing GPAT in tobacco plants (*Nicotiana tabacum* L.) from *Arabidopsis* showed increase cold tolerance (Murata et al. 1992). From the different investigations, it is clear that reducing the proportion of disaturated PG, and perhaps increasing overall lipid unsaturation, can measurably improve cold tolerance of the tobacco plant, by facilitating turnover of D1 protein and therefore allowing faster recovery from photoinhibition (Kanervo et al. 1995; Thomashow and Browse 1999). But disaturated PG is not the sole reason for chilling sensitivity in plants. Overexpression of chloroplastic Cu/Zn superoxide dismutase (SOD) in tobacco plants showed higher photosynthetic activity at 10°C and greater recovery at 25°C after photoinhibition at 3°C for 4 h (Gupta et al. 1993). Transgenic tobacco line, transformed with *E. coli* pyrophosphatase showed higher cold tolerance as induction of pyrophosphatase leads to the induction of other genes and these gene products contribute to cold tolerance. Many cold-responsive genes that have been isolated from the cold-acclimated plant encode proteins with known activities, but the majority does not (Thomashow and Browse 1999).

Maximum numbers of cold-responsive genes were isolated from *Arabidopsis*. Xin and Brows (1998) have identified an important gene (*esk1*) that has a major role in cold tolerance. High proline concentration and higher freezing tolerance characteristics were observed in *esk1* mutant compared with wild-type plants. All studies so far were conducted with a few specific plants (e.g., *Arabidopsis*, tobacco), and therefore to understand overall response, research should be undertaken with a wide range of plant species. This challenge is more daunting in cross-pollinating species such as turfgrass, where model used for herbs (oregano, thyme) can be applied.

On the basis of the proposed alternative models for the role of proline in reducing acidification during stress, proline-linked phenolic synthesis in clonal shoot culture was achieved in different herb species (Shetty 2004). The major drawback for consistent phytochemicals production in cross-pollinated plants is the genetic heterogenecity, which leads to inconsistency in phenolic and other phytochemical synthesis. To overcome such a problem, plant tissue-culture technique has been developed to isolate a clonal pool of plants originating from a single heterozygous seed (Shetty et al. 1995). A single, elite clonal line with superior phenolic profile can be screened and selected based on tolerance to *Pseudomonas* spp. (Eguchi et al. 1996; Shetty et al. 1996). Large-scale clonal propagation (micropropagation) and evaluation of functionality is the last step for the development of elite clones with high antioxidant property. Elite, high phenolic clonal line of thyme (*Thymus valguris* L.), oregano (*Origanum vulgare* L.), and rosemary (*Rosmarinus officinalis* L.) were developed by using this approach (Shetty 1997; Shetty and Labbe 1998). Although all plants, and specifically phenolic synthesis, are targeted against major oxidation-linked diseases, the same method can be useful to counter oxidative damage caused by environmental stress in plants (Shetty 2004). The antioxidative defense system is an efficient mechanism of plants against low-temperature stress; so production of phenolic antioxidants and stimulation of the enzyme-based antioxidant defense system by using this approach may be an effective tool for development of cold-tolerant cultivars.

30.10 FUTURE DIRECTIONS

There is paucity of information regarding the biochemical and molecular response of cool-season turfgrass to cold stress. In the northern United States and Canada, the aesthetic value of the cool-season turfgrass is very much dependent on their cold-stress-tolerance mechanism during winter. Intensive research is essential to evaluate the biochemical and genetic factors related to cold stress in turfgrass. Research should focus on species with low cold tolerance, like perennial ryegrass, annual bluegrass, and tall fescue, and their improvement should be initiated. The understanding of the cold-tolerance mechanism of superior species like creeping bentgrass, rough bluegrass (*Poa trivialis* L.), and Kentucky bluegrass can also be helpful to evaluate key characteristics that regulate better fitness of such species under cold stress. Several strategies can be undertaken to improve our basic knowledge of cold tolerance of turfgrass, and to provide better solutions for further improvement.

1. First, detailed understanding of the exact mechanism of the damage or injuries at a cellular level by different cold-related factors (low temperature kill, desiccation, ice encasement, snow mold) are essential.
2. Biochemical and molecular shift during the acclimation period should be studied in detail (carbohydrate metabolism, alternative anabolic pathways, lipid metabolism).
3. Protein synthesis and function of the protein in turfgrass during cold stress.
4. Identification of cold-related genes and their expression would provide a better solution for developing new cultivars.
5. Understanding the role of the membrane and evaluation of membrane protein, membrane lipid, and unsaturation of fatty acids related with the membrane may provide a clear picture for cold stress regulation.
6. Accumulation of different cryoprotectants in turfgrass during the cold stress period also needs to be investigated.
7. The regulation of different pathways (metabolic and anabolic) and the role of key enzymes under cold stress.
8. Evaluation of the overall cold tolerance level of turfgrass, and the role of antioxidants and their signal induction during cold stress.

Such extensive knowledge of the cold-tolerance mechanism of turfgrass can provide the basis and rationale for advancement of turfgrass research. With the help of modern biotechnological tools, it is easier to develop superior cold-tolerant cultivars, which will improve the performance of cool-season turfgrass during winter and spring. The field evaluation of different superior cultivars and the selection of proper agronomic management practices for different cool-season turfgrasses during the period of acclimation and deacclimation is also very important. Several new chemicals for plant protection against stress (plant growth regulators, antioxidants chemicals, antidesiccants) have been developed and used in other crops and plants. Evaluations and performance of such chemicals in the turfgrass system can also provide a cost-effective management option during the cold-stress period. Finally, there is a need to transfer advanced technologies from the research station to commercial areas to maintain high-quality turfgrass with greater economic benefit.

30.11 CONCLUSIONS

Each winter, in the northern United States, substantial turfgrass losses occur due to damage from severe and prolonged cold stress. Cool-season turfgrasses generally have higher cold tolerance characteristics; still huge energy inputs are imperative for the preparation and recovery of turfgrass during winter (Rossi 1997). Freezing injury of turfgrass can have an economic and environmental impact as well reduce the aesthetic value of commercial turfgrass (DiPaola and Beard 1992). The performance and function of cool-season turfgrass during spring depend on the severity and site of the freezing injury, where the crown plays a crucial role for turfgrass recovery. Although maximum damage occurs during late winter to early spring, in the time of deacclimation, the biological process during acclimation (late fall to early winter) plays a significant role in the cold-tolerance mechanism. Different agronomic practices during fall can affect the cold hardening process, thus making proper management decisions very important for turfgrass experts.

An understanding of the mechanism and nature of damage due to cold stress can provide the right scientific insight to achieve better results. Molecular and biochemical studies carried out in other plants suggest several possible ways to improve cold tolerance characteristics in cool-season turfgrass. The use of modern biotechnological tools to create new, cold-hardy cultivars is promising. Like most living organisms of the earth, turfgrasses also live in an oxygen-rich environment, and are thus susceptible to oxidative damage during stress periods. Emerging studies in redox biology indicate that the antioxidative defense system of plants and animals is one of the most important research concepts for countering biotic and abiotic stress in the oxygen-dependent environment. Although the role of phenolic antioxidants against abiotic stress of turfgrass is not known, we can assume its significance in turfgrass system from studies conducted in other plants. As cold injuries of turfgrass occur by the interaction of many environmental, mechanical, and biological factors, an integrated approach involving the system biology is necessary rather than isolating and dealing with an individual problem. Knowledge of the antioxidative defense system of turfgrass related to cold stress will not only provide solutions to turfgrass scientists, but also give broad insights to other plant and grass physiologists. Interdisciplinary and dynamic research that integrates the molecular, biochemical, and genetic responses of turfgrass under cold stress will provide insights for enhancing the overall performance and commercial value of turfgrass. High-quality turfgrass is important not only for turfgrass managers, golf-course superintendents, sport personalities, or turfgrass industries, but also has certain benefits for society at large with respect to its role in recreation and ecosystem management.

REFERENCES

Alberdi, M. and L.J. Corcuera. 1991. Cold acclimation in plants. *Phytochemistry* 30:3177–3184.

Anderson, J., B. Chevone, and J. Hess. 1992. Seasonal variation in antioxidant system of eastern white pine needles. *Plant Physiol.* 98:501–508.

Anderson, J.A., M.P. Kenna, and C.M. Taliaferro. 1988. Cold hardiness of 'Midron' and 'Tifgreen' bermudagrass. *HortScience* 23:748–750.

Asada, K. 1993. Divergence of peroxide-scavenging peroxidases in organisms. In Yagi, K. (ed.) *Active Oxygens, Lipid Peroxides and Antioxidants.* CRC Press, Boca Raton, FL, pp. 289–298.

Bartels, D. and D. Nelson. 1994. Approaches to improve stress tolerance using molecular genetics. *Plant Cell Environ* 17:659–667.

Bartels, D., R. Alexander, K. Schneider, R. Elster, R. Velasco, J. Alamillo, G. Bianchi, D. Nelson, and F. Salamini. 1993. Desiccation-related gene products analyzed in a resurrection plant and in barley embryos. In Close, T.J. and E.A. Bray (eds.) *Plant Responses to Cellular Dehydration During Environmental Stress. Current Topics in Plant Physiology: An American Society of Plant Physiologist Series,* Vol. 10. Rockville, Maryland, pp. 119–127.

Beard, J.B. 1964. Effects of ice, snow and water covers on Kentucky bluegrass, annual bluegrass and creeping bentgrass. *Crop Sci* 4:638–640.

Beard, J.B. 1973. *Turfgrass: Science and Culture.* Prentice Hall, Englewood Cliffs, NJ.

Beckman, K.B. and B.N. Ames. 1997. Oxidants, antioxidants, and aging. In Scandalios, J.G. (ed.) *Oxidative Stress and the Molecular Biology of Antioxidant Defenses*. Cold Spring Harbor Laboratory Press. Plainview, NY., pp. 201–246.

Beggs, C.J., K. Kuhn, R. Bocker, and E. Wellmann. 1987. Phytochrome-induced flavonoid biosynthesis in mustard *Sinapsis alba* L. cotyledons: enzymatic control and differential regulation of anthocyanin and quercitin formation. *Planta* 172:121–126.

Bowler, C., W. Van Camp, M. Van Montagu, and D. Inze. 1984. Superoxide dismutase in plants. *CRC Crit Rev Physiol* 13:199–218.

Bredemeijer, G.M.M. and G. Esselink. 1995. Sugar metabolism in cold hardened *Lolium perenne* varieties. *Plant VarSeeds* 8:187–195.

Cardona, C.A., R.R. Duncan, and O. Lindstrom. 1997. Low temperature tolerance assessment in paspalum. *Crop Sci* 37:1283–1291.

Chen, H., M. Brenner, and P. Li. 1983. Involvement of abscisic acid in potato cold–acclimation. *Plant Physiol* 71:362–365.

Christ, M., B. Luu, J.E. Mejia, I. Moosbrugger, and P. Bischoff. 1993. Apoptosis induced by oxysterols in murine lyphoma cells and in normal thymocytes. *Immunology* 78:455–460.

Christie, P.J., M.R. Alfenito, and V. Walbot. 1994. Impact of low-temperature stress on general phenyl-propanoid and anthocyanin pathways: Enhancement of transcript abundance and anthocyanin pigmentation in maize seedlings. *Planta* 194:541–549.

Cyril, J., G.L. Powell, R.R. Duncan, and W.V. Baird. 2002. Changes in membrane polar lipid fatty acids of seashore paspalum in response to low temperature exposure. *Crop Sci* 42:2031–2037.

Danneberger, T.K. 2006. Ice cover: Its role in winter injury. The British and International Golf Greenkeepers Association, Online Magazine. Website: http://www.bigga.org.uk/magazine/back-issues/04-2005/2388/ice-cover-its-role-in-winter-injury.html

Dionne, J., Y. Castonguay, P. Nadeau, and Y. Desjardins. 2001a. Amino acids and protein changes during cold acclimation of green-type annual bluegrass (*Poa annua* L.) ecotypes. *Crop Sci* 41:1862–1869.

Dionne, J., Y. Castonguay, P. Nadeau, and Y. Desjardins. 2001b. Freezing tolerance and carbohydrate changes during cold acclimation of green-type annual bluegrass (*Poa annua* L.) ecotypes. *Crop Sci* 41:443–451.

DiPaola, J.M. and J.B. Beard. 1992. Physiological effects of temperature stress. In Waddington, D.V., R.N. Carrow, and R.C. Shearman (eds.) *Turfgrass-Agronomy Monograph* no. 32. ASA, CSSA, and SSSA, Madison, WI.

Dorffling, K., H. Dorffling, G. Lesselich, E. Luke, C. Zimmerman, G. Melz, and H.U. Jurgens. 1997. Heritable improvement of frost tolerance in winter wheat by in vitro selection of hydroxyproline-resistant proline overproducing mutants. *Plant Mol Biol* 23:221–225.

Dure, L. III, M. Crouch, J. Harada, T.H. Ho, R. Quatrano, T. Thomas, and Z.R. Sung. 1989. Common amino acid sequence domains among the LEA proteins of higher plants. *Plant Mol Biol* 12:475–486.

Ebdon, J.S., R.A. Gagne, and R.C. Manley. 2002. Comparative cold tolerance in diverse turf quality genotypes of perennial ryegrass. *HortScience* 37:826–830.

Eguchi, Y., O.F. Curtis, and K. Shetty. 1996. Interaction of hyperhydricity-preventing *Pseudomonas* spp. with oregano (*Origanum vulgare*) and selection of high rosmarinic acid-producing clones. *Food Biotechnol.* 10:191–202.

Elthon, T.E. and C.R. Stewart. 1984. Effects of proline analog L-thiazolidine-4-carboxylic acid on proline metabolism. *Plant Physiol* 74:213–218.

Emmons, R. 1995. *Turfgrass Science and Management*, 2nd edition. Delmar Publishers. Albany, NY.

Ervin, E.H., X. Zhang, and J.H. Fike. 2004. Ultraviolet-B radiation damage on Kentucky bluegrass II: Hormone supplement effects. *HortScience* 39:1471–1474.

Esterbauer, H. and D. Grill. 1978. Seasonal variation of glutathione and glutathione reductase in needles of *Picea abies*. *Plant Physiol* 61:119–121.

Feierabend, J., C. Schaan, and B. Hertwig. 1992. Photoinactivation of catalase occurs under both high- and low-temperature stress conditions and accompanies photoinhibition of photosystem II. *Plant Physiol* 100:1554–1561.

Fry, J. and B. Huang. 2004. *Applied Turfgrass Science and Physiology*. Wiley, Hoboken, NJ.

Fry, J.D., N.S. Lang, R.G.P. Clifton, and F.P. Maier. 1993. Freezing tolerance and carbohydrate content of low temperature acclimated and non-acclimated centipedegrass. *Crop Sci* 33:1051–1055.

Fuchs, D., G.B. Bitterlich, I. Wede, and H. Wachter. 1997. Reactive oxygen and apoptosis. In Scandalios, J.G. (ed.) *Oxidative Stress and the Molecular Biology of Antioxidant Defenses*. Cold Spring Harbor Laboratory Press, Plainview, NY, pp. 139–167.

Graham, D. and B.D. Patterson. 1982. Responses of plants to low, non-freezing temperature- proteins, metabolism, and acclimation. *Ann Rev Plant Physiol Plant Mol Biol* 33:347–372.

Graham, T.L. 1991. Flavonoid and isoflavonoid distribution in developing soybean seedling tissue and in seed and root exudates. *Plant Physiol* 95:594–603.

Gray, W., A. Ostin, G. Sandberg, C. Romano, and M. Estelle. 1998. High temperature promotes auxin mediated hypocotyls elongation in *Arabidopsis. Proc Natl Acad Sci USA* 95:7197–7202.

Gupta, A.S., J.L. Heinen, A.S. Holaday, J.J. Burke, and R.D. Allen. 1993. Increased resistance to oxidative stress in transgenic plants that overexpress chloroplastic Cu/Zn superoxide dismutase. *Proc Natl Acad Sci USA* 90:1629–1633.

Gusta, L.V., J.D. Butler, C. Rajashekar, and M.J. Burke. 1980. Freezing resistance of perennial turfgrasses. *HortScience* 15:494–496.

Guy, C.L. 1990. Molecular mechanism of cold acclimation. In Katterman, F. (ed.) *Environmental Injury to Plants.* Academic Press, San Diego, CA, pp. 35–61.

Guy, C.L. 2003. Freezing tolerance of plants: current understanding and selected emerging concepts. *Can J Bot* 81:1216–1223.

Hagedorn, C.H. and J.M. Phang. 1983. Transfer of reducing equivalents into mitochondria by the interconversions of proline and pyrroline-5-carboxylate. *Arch Biochem Biophys* 225:95–101.

Hahlbrock, K. and D. Scheel. 1989. Physiology and molecular biology of phenylpropanoid metabolism. *Plant Mol Biol* 40:347–369.

Halliwell, B. and J.M.C. Gutteridge. 1989. *Free Radicals in Biology and Medicine*, 2nd edition. Clarendon Press, Oxford.

Hatakeyama, I. 1960. The ralation between growth and cold hardiness of leaves of *Camellia sinensis. Biol J Nara Women's Univ* 10:65–69.

Hatano, S. and K. Kabata. 1982. Transition of lipid metabolism in relation to frost hardiness in Chlorella elipsoidea. In Li, P.H. and A. Sakai (eds.) *Plant Cold Hardiness and Freezing Stress*, Vol II. Academic Press, London, pp. 145–156.

Heino, P., G. Sandman, V. Lang, K. Nordin, and E.T. Palva. 1990. Abscisic acid deficiency prevents development of freezing tolerance in *Arabidopsis thaliana. Theor Appl Genet* 79:801–806.

Hisamatsu, T. and M. Koshioka. 2000. Cold treatments enhance resposiveness to gibberellin in stock (*Matthiola incana* L.). *J Hort Sci Biotechnol* 75:672–678.

Hsiang, T. and S. Cook. 2001. Effect of *Typhula phacorrhiza* on winter injury in field trials across Canada. *Int Turf Soc Res J* 9:669–673.

Hughes, M.A. and M.A. Dunn. 1996. The molecular biology of plant acclimation to low temperature. *J Exp Bot* 47:291–305.

Jahnke, L.S., M.R. Hull, and S.P. Long. 1991. Chilling stress and oxygen metabolizing enzymes in *Zea mays* and *Zea diploperennis. Plant Cell Environ* 14:97–104.

Javanmardi, J., C. Stushnoff, E. Locke, and J.M. Vivanco. 2003. Antioxidant activity and total phenolic content of Iranian *Ocimum* accessions. *Food Chem* 83:547–550.

Jiang, Y., R.N. Carrow, and R.R. Duncan. 2005. Physiological acclimation of seashore paspalum and bermudagrass to low light. *Scientia Horticulturae* 105:101–115.

Jorgensen, L.V., H.L. Madsen, M.K. Thomsen, L.O. Dragsted, and L.H. Skibsted. 1999. Regulation of phenolic antioxidants from phenoxyl radicals: an ESR and electrochemical study of antioxidant hierarchy. *Free Rad Res* 30:207–220.

Kacperska-Palacz, A. 1978. Mechanism of cold acclimation in herbaceous plants. In Li, P.H. and A. Sakai (eds.) *Plant Cold Hardiness and Freezing Stress*, Vol I. Academic Press, London, pp. 139–152.

Kaku, S. 1975. Analysis of freezing temperature distribution in plants. *Cryobiology* 12:154–159.

Kanervo, E., E-M. Aro, and N. Murata. 1995. Low unsaturation level of thylakoid membrane lipids limits turnover of D1 protein of photosystem II at high irradiance. *FEBS Lett* 364:239–242.

Kendall, B.J. and B.D. McKersie. 1989. Free radical and freezing injury to cell membranes of winter wheat. *Plant Physiol* 76:86–94.

Kishitani, S., K. Wtanabe, S. Yasuda, K. Arakawa, and T. Takabe. 1994. Accumulation of glycinebetaine during cold acclimation and freezing tolerance in leaves of winter and spring barley plants. *Plant Cell Environ* 17:89–95.

Knight, M.R., A.K. Campbell, S.M. Smith, and A.J. Trewavas. 1991. Transgenic plant *Aequorin* reports the effects of touch and cold shock and elicitors on cytoplasmic calcium. *Nature* 352:524–526.

Knight, H., A.J. Trewavas, and M.R. Knight. 1996. Cold calcium signaling in *Arabidopsis* involves two cellular pools and a change in calcium signature after acclimation. *Plant Cell* 8:489–503.

Krasnuk, M., F.H. Witham, and G.A. Jung. 1976. Electrophoretic studies of several hydrolytic enzymes in relation to cold tolerance of alfalfa. *Cryobiology* 13:225–242.

Kuroda, H., S. Sagisaka, and K. Chiba. 1992. Collapse of peroxide-scavenging systems in apple flower-buds associated with freezing injury. *Plant Cell Physiol* 33:743–750.

Kwok, D. and K. Shetty. 1998. Effects of proline and proline analogs on total phenolic and rosmarinic acid levels in shoot clone of thyme (*Thymus vulgaris* L.). *J Food Biochem* 22:37–51.

Lang, V., E. Mantyla, B. Welin, B. Sundberg, and E. Palva. 1994. Alterations in water status, endogenous abscisic acid content and expression of rab18 gene during the development of freezing tolerance in *Arabidopsis thaliana*. *Plant Physiol* 104:1341–1349.

Larcher, W. 1981. Effects of low temperature stress and frost injury on plant productivity. In Johnson, C.D. (ed.) *Physiological Processes Limiting Plant Productivity*. Butterworths, London, pp. 253–269.

Larkindale, J. and B. Huang. 2004. Thermotolerance and antioxidant systems in *Agrostis stolonifera*: Involvement of salicylic acid, abscisic acid, calcium, hydrogen peroxide, and ethylene. *J Plant Physiol* 161:405–413.

Lee, S.P. and T.H. Chen. 1993. Molecular cloning of abscisic acid-responsive mRNAs expressed during the induction of freezing tolerance in bromegrass (*Bromus inermis* Leyss) suspension culture. *Plant Physiol* 101:1089–1096.

Levitt, J. 1980. *Responses of Plant to Environmental Stresses*. Vol I. *Chilling, Freezing and High Temperature Stresses*, 2nd edition. Academic Press, New York.

Lynch, D.V. 1990. Chilling injury in plants: The relevance of membrane lipids. In Katterman, F. (ed.) *Environmental Injury to Plants*. Academic Press, San Diego, CA, pp. 17–34.

Lyons, J.M. 1973. Chilling injury in plants. *Ann Rev Plant Physiol* 24:445–466.

Massacci, A., M.A. Iannelli, F. Pietrini, and F. Loreto. 1995. The effect of growth at low temperature on photosynthetic characteristics and mechanisms of photoprotection of maize leaves. *J Exp Bot* 46:119–127.

McCue, P. and K. Shetty. 2002. Clonal herbal extracts as elicitors of phenolic synthesis in dark germinated mungbean for improving nutritional value with implications for food safety. *J Food Biochem* 26:209–232.

McCue, P., Z. Zheng, J.L. Pinkham, and K. Shetty. 2000. A model for enhanced pea seedlings vigor following low pH and salicylic acid treatments. *Process Biochem* 35:603–613.

McKersie, B., Y. Chen, M. de Beus, S. Bowley, C. Bowler, D. Inzé, K. D'Halluin, and J. Botterman. 1993. Superoxide dismutase enhances tolerance of freezing stress in transgenic alfalfa (*Medicago sativa* L.). *Plant Physiol* 103:1155–1163.

Monroy, A.F. and R.S. Dhindsa. 1995. Low temperature signal transduction-induction of genes of alfalfa by calcium at 25°C. *Plant Cell* 7:321–331.

Murata, N. 1983. Molecular species composition of phosphatidylglycerols from chilling-sensitive and chilling-resistant plants. *Plant Cell Physiol* 24:81–86.

Murata, N., O. Ishizaki-Nishizawa, S. Higashi, H. Hayashi, Y. Tasaka, and J. Nishida. 1992. Genetically engineered alteration in the chilling sensitivity of plants. *Nature* 356:313–326.

Murata, N., N. Sato, N. Takahashi, and Y. Hamazaki. 1982. Compositions and positional distributions of fatty acids in phospholipids from leaves of chilling-sensitive and chilling-resistant plants. *Plant Cell Physiol* 23:1071–1079.

Orvar, B.L., V. Sangwan, F. Omann, and R.S. Dhindsa. 2000. Early steps in cold sensing by plant cells: the role of actin cytoskeleton and membrane fluidity. *Plant J* 23:785–794.

Paliyath, G. and M.J. Droillard. 1992. The mechanism of membrane deterioration and disassembly during senescence. *Plant Physiol Biochem* 30:789–812.

Parra, C., J. Saez, H. Perez, M. Alberdi, M. Delseny, E. Hubert, and L. Mezabasso. 1990. Cold resistance in rapeseed (*Brassica napus*) seedlings-Searching a biochemical marker of cold-tolerance. *Arch Biol Med Exp* 23:187–194.

Pennycooke, J.C., S. Cox, and C. Stushnoff. 2005. Relationship of cold acclimation, total phenolic content and antioxidant capacity with chilling tolerance in petunia (*Petunia* X *hybrida*). *Environ Exp Bot* 53:225–232.

Phang, J.M. 1985. The regulatory functions of proline and pyrroline-5-carboxylic acid. *Curr Top Cell Regul* 25:91–132.

Pinhero, R.G., G. Paliyath, R.Y. Yada, and D.P. Murr. 1998. Modulation of phospholipase D and lypoxygenase activities during chilling. Relation to chilling tolerance of maize seedlings. *Plant Physiol Biochem* 36:213–224.

Pinhero, R.G., M.V. Rao, G. Paliyath, D.P. Murr, and R.A. Fletcher. 1997. Changes in activities of antioxidant enzymes and their relationship to genetic and paclobutrazol-induced chilling tolerance of maize seedlings. *Plant Physiol* 114:695–704.

Polle, A. 1997. Defense against photooxidative damage in plants. In Scandalios, J.G. (ed.) *Oxidative Stress and the Molecular Biology of Antioxidant Defenses*. Cold Spring Harbor Laboratory Press, Plainview, NY, pp. 623–666.

Polle, A. and H. Rennenberg. 1994. Photooxidative stress in trees. In Foyer, C.H. and P.M. Mullineaux (eds.) *Causes of Oxidative Stress and Amelioration of Defense Systems in Plants*. CRC Press, Boca Raton, FL, pp. 199–218.

Pollock, C.J., A.J. Cairns, and B.E. Collis. 1989. Direct effects of low temperature upon components of fructan metabolism in leaves of *Lolium temulentum* L. *J Plant Physiol* 134:203–208.

Pontis, H.G. 1989. Fructans and cold stress. *J Plant Physiol* 134:148–150.

Popova, A.V., J.M. Schmitt, and D.K. Hincha. 1998. Interactions of proline, serine and leucine with isolated spinach thylakoids: solute loading during freezing is not related to membrane fluidity. *Cryobiology* 37:92–99.

Prasad, T.K. 1996. Mechanism of chilling induced oxidative stress injury and tolerance in developing maize seedlings: changes in antioxidant system, oxidation of proteins and lipids and protease activities. *Plant J* 10:1017–1026.

Prasad, T.K., M.D. Anderson, B.A. Martin, and C.R. Stewart. 1994. Evidence for chilling-induced oxidative stress in maize seedlings and a regulatory role for hydrogen peroxide. *Plant Cell* 6:65–74.

Rao, M.V., G. Paliyath, and D.P. Ormrod. 1996. Ultraviolet-B and ozone-induced biochemical changes in antioxidant enzymes of *Arabidopsis thaliana*. *Plant Physiol* 110:125–136.

Rice-Evans, C.A., N.J. Miller, and G. Paganga. 1997. Antioxidant property of phenolic compounds. *Trends Plant Sci* 2:152–159.

Rochat, E. and H.P. Therrien. 1975. Proteins of hardy wheat (Kharkov) and less hardy wheat (Selkirk) during cold hardening process. 1. Soluble proteins. *Can J Bot* 53(21): 2411–2416.

Rossi, F.S. 1997. Physiology of turfgrass freezing stress injury. *TurfGrass Trends* 6:1–8.

Rossi, F.S. 2003. New light on freeze stress. *CUTT* 14(3):1,4,5.

Sadakane, H., K. Kabata, K. Ishibashi, T. Watanabe, and S. Hatano. 1980. Studies on frost hardiness in *Chlorella ellipsoidea*. V. The role of glucose and related compounds. *Environ Exp Bot* 20:297–305.

Sagisaka, S. 1972. Decrease of glucose-6-phosphate and 6-phosphogluconate dehydrogenase activities in the xylem of *Populus gelrica* on budding. *Plant Physiol* 50:750–755.

Sagisaka, S. 1974. Transition of metabolisms in living poplar bark from growing to wintering stage and vice versa. *Plant Physiol* 54:544–549.

Sagisaka, S. 1982. Comparative studies on the metabolic function of differentiated xylem and living bark of wintering perennials. *Plant Cell Physiol* 23:1337–1346.

Sagisaka, S. 1985. Injuries of cold acclimatized poplar twigs resulting from enzyme inactivation and substrate depression during frozen storage at ambient temperatures for a long period. *Plant Cell Physiol* 26:1135–1145.

Sagisaka, S. and T. Araki. 1983. Amino acids pools in perennial plants at the wintering stage and at the beginning of growth. *Plant Cell Physiol* 24:479–494.

Sakai, A. 1979. Deep supercooling of winter flower buds of *Cornus florida* L. *HortScience* 14:69–70.

Sakai, A. and W. Larcher. 1987. *Frost Survival of Plants: Responses and Adaptation to Freezing Stress*. Springer-Verlag, Berlin.

Schöner, S. and H. Krause. 1990. Protective systems against active oxygen species in spinach: Response to cold acclimation in excess light. *Planta* 180:383–389.

Shetty, K. 1997. Biotechnology to harness the benefits of dietary phenolics; focus on Lamiaceae. *Asia Pac J Clin Nutr* 6:162–171.

Shetty, K. 1999. Phytochemicals: biotechnology of phenolic phytochemicals for food preservatives and functional food applications. In Francis, F.J. (ed.) *Wiley Encyclopedia of Food Science and Technology*, 2nd edition. Wiley, New York, pp. 1901–1909.

Shetty, K. 2004. Role of proline-linked pentose phosphate pathway in biosynthesis of plant phenolics for functional food and environmental applications: a review. *Process Biochem* 39:789–803.

Shetty, K. and R.G. Labbe. 1998. Forborne pathogens, health and role of dietary phytochemicals. *Asia Pac J Clin Nutr* 7:270–276.

Shetty, K. and P. McCue. 2003. Phenolic antioxidant biosynthesis in plants for functional food application: Integration of system biology and biotechnological approaches. *Food Biotechnol* 17:67–97.

Shetty, K. and M. Wahlqvist. 2004. A model for the role of proline-linked pentose phosphate pathway in phenolic phytochemical biosynthesis and mechanism of action for human health and environmental application. *Asia Pac J Clin Nutr* 13:1–24.

Shetty, K., O.F. Curtis, and R.E. Levin. 1996. Specific interaction of mucoid stains of *Pseudomonas* spp. with oregano (*Origanum vulgare*) clones and the relationship to prevention of hyperhydricity in tissue culture. *J Plant Physiol* 149:605–611.

Shetty, K., O.F. Curtis, R.E. Levin, R. Witkowsky, and W. Ang. 1995. Prevention of virtification associated with in vitro shoot culture of oregano (*Origanum vulgare*) by *Pseudomonas* spp. *J Plant Physiol* 147:447–451.

Sigafoos, R.S. 1952. Frost action as a primary physical factor in tundra plant communities. *Ecology* 33:480–487.

Smallwood, M. and D.J. Bowles. 2002. Plants in cold climate. *Phil Trans Roy Soc Lond* 357:831–847.

Smith, I.K., A. Polle, and H. Rennenberg. 1990. Glutathione. In Alscher, R.G., and J.R. Cumming (eds.) *Stress Responses in Plants: Adaptation and Acclimation Mechanisms*. Wiley-Liss, New York, pp. 201–217.

Sobczyk, E.A. and A. Kacperska-Palacz. 1980. Changes in some enzyme activities during cold acclimation of winter rape plants. *Acta Physiol Plant* 2:123–131.

Steponkus, P.L. 1984. Role of plasma membrane in freezing injury and cold-acclimation. *A Rev Plant Physiol Plant Mol Biol* 35:543–584.

Steponkus, P.L. 1990. Cold acclimation and freezing injury from a perspective of plasma membrane. In Katterman, F. (ed.) *Environmental Injury to Plants*. Academic Press, San Diego, CA, pp. 1–16.

Thomas, H. and A.R. James. 1993. Freezing tolerance and solute changes in contrasting genotypes of *Lolium perenne* L. acclimated to cold and drought. *Ann Bot* 72:249–254.

Thomashow, M.F. and J. Browse. 1999. Plant cold tolerance. In Shinozaki, K., and K. Yamaguchi-Shinozaki (eds.) *Molecular Responses to Cold, Drought, Heat and Salt Stress in Higher Plants*. R.G. Landes Company, Austin, Texas, pp. 61–80.

Thompson, J.E., R.L. Legge, and R.F. Barber. 1987. The role of free radicals in senescence and wounding. *New Phytol* 105:317–344.

Tompkins, D.K., J.B. Ross, and D.L. Moroz. 2002. Dehardening of annual bluegrass and creeping bentgrass during late winter and early spring. *Agron J* 92:925–929.

Turgeon, A.J. 1996. *Turfgrass Management*, 4th edition. Prentice Hall, Upper Saddle River, NJ.

Walker, M. and B. McKersie. 1993. Role of the ascorbate-glutathione antioxidant system in chilling resistance to tomato. *J Plant Physiol* 141:234–239.

Wang, Z., J. Pote, and B. Huang. 2003. Responses of cytokinins, antioxidant enzymes, and lipid peroxidation in shoots of creeping bentgrass to high root-zone temperature. *J Am Soc Hort Sci* 128:648–655.

Willemot, C. 1983. Lipid degradation of polar lipids in frost damaged winter wheat crown and root tissue. *Phytocehmistry* 22:861–863.

Wise, R. and A. Naylor. 1987. Chilling enhanced photooxidation. Evidence for the role of singlet oxygen and superoxide in the breakdown of pigments and endogenous antioxidants. *Plant Physiol* 83:278–282.

Wolter, F.P., R. Schmidt, and E. Heinz. 1992. Chilling sensitivity of *Arabidopsis thaliana* with genetically engineered membrane lipids. *EMBO J* 11:4685–4692.

Xin, Z. and J. Browse. 1998. eskimo1 mutants of *Arabidopsis* are constitutively freezing tolerant. *Proc Natl Acad Sci USA* 95:7799–7804.

Xu, Q. and B. Huang. 2004. Antioxidant metabolism associated with summer leaf senescence and turf quality decline for creeping bentgrass. *Crop Sci* 44:553–558.

Yang, R. and K. Shetty. 1998. Stimulation of rosmarinic acid in shoot cultures of oregano (*Oreganum vulgare*) clonal line in response to proline, proline analog and proline precursors. *J Agr Food Chem* 46:2888–2893.

Yoshida, S. and M. Uemura. 1984. Protein and lipid compositions of isolated plasma-membranes from orchardgrass (*Dactylis glomerata*) and changes during cold acclimation. *Plant Physiol* 75:31–37.

Zhang, X. and R.E. Schmidt. 1999. Antioxidant response to hormone containing product in Kentucky bluegrass subjected to drought. *Crop Sci* 39:545–551.

Zhang, X. and R.E. Schmidt. 2000. Hormone containing product impact on antioxidant status of tall fescue and creeping bentgrass subjected to drought. *Crop Sci* 40:1344–1349.

Zhang, X., R.E. Schmidt, E.H. Ervin, and S. Doak. 2002. Creeping bentgrass physiological responses to natural plant growth regulators and iron under two regimes. *HortScience* 37: 898–902.

31 Responses of Turfgrass to Low-Oxygen Stress

Yiwei Jiang

CONTENTS

31.1 INTRODUCTION

Oxygen is crucial for plants since their roots require oxygen for respiration to provide energy for the growth of the whole plant. Low-oxygen stress occurs in turfgrass as a result of high water input such as overirrigation or excessive rainfall. In regions where freshwater is readily available, golf courses may be given an excessive amount of water during summer time to prevent drought stress. Excess water in the root environment or compacted soil blocks the transfer of oxygen and other grasses between the soil and the atmosphere; as a result, root injury occurs due to oxygen deficiency, and the growth and development of the plant is inhibited. Even temporal or short-term oxygen deficiency can be detrimental to the grass, particularly when temperatures are high. Turfgrass can also suffer from low-oxygen stress without standing water, including poor soil quality with slow drainage, soil compaction, and ice cover. All these factors influence grass growth and quality.

Low oxygen changes soil properties and influences plant growth and survival. Responses of plants to various low-oxygen conditions are diverse, including anatomical, morphological, physiological, biochemical, and molecular adaptations. Some of these changes allow plants to become tolerant to oxygen deficiency, depending on the plant species and cultivar, duration and depth of stress, and other environmental factors such as temperature and light. Responses and adaptations of turfgrass species and cultivars to low-oxygen stress have been documented [1–5]; however, the mechanisms of low-oxygen tolerance are poorly understood. A better understanding of turfgrass responses and mechanisms of tolerance would help in the selection of tolerant grass materials through breeding programs. Once the tolerant and intolerant cultivars have been identified, they can be used to explore the mechanisms influencing low-oxygen tolerance at the whole-plant, cellular, and molecular levels. Furthermore, knowledge of mechanisms of stress tolerance would benefit management practices in achieving better turf quality.

31.2   EFFECTS OF LOW OXYGEN ON SOILS

Low oxygen (e.g., waterlogging) influences the physical, chemical, and biological processes in soil that affect the growth of the plant. Major chemical changes in the soil include decrease in O_2 availability, accumulation of CO_2, reduction of iron (Fe) and manganese (Mn), anaerobic decomposition of organic matter, and formation of toxic compounds such as sulfides, CO_2, and soluble Fe and Mn [6]. Oxygen deficiency is one of the primary root stresses in waterlogged or flooded soils [7]. The changes in soil properties can inhibit root growth and adversely influence the whole plant. Soil redox potential (Eh) measures the current state of oxidation–reduction in a soil and can be used as an indicator of soil oxygenation status [8]. Well-drained soils have Eh of +300 mV or greater, whereas flooded soils have Eh of −300 mV or lower [9]. Soil Eh drops from 350 to 170 mV in creeping bentgrass (*Agrostis stolonifera* L.) subjected to waterlogged soil for 21 days [5]. The reduced soil Eh could further influence grass performance, as demonstrated by a decrease in grass quality observed simultaneously [5]. In wheat (*Triticum aestivum* L.), grown in waterlogged soil, the soil Eh decreases from 600 to 200 mV over the first 14 days and then to 40 mV after 28 days of treatment [10]. Kludze and DeLaune [11] reported that root air space, radial O_2 loss from plant roots to atmosphere, biomass production of cattail (*Typha domingensis* L.), and sawgrass (*Cladium jamaicense* L.) were controlled by soil Eh.

The oxygen diffusion rate (ODR) is another indicator of potential O_2 supply by the soil. Latey [12] reported that ODR above 0.2 $\mu g\ cm^{-2}\ min^{-1}$ was critical for the growth of Kentucky bluegrass (*Poa pratensis* L.) and common bermudagrass (*Cynodon dactylon* L.). Similar results of ODR were observed in creeping bentgrass under waterlogged soil, whereas well-aerated soil had ODR of 1.5 $\mu g\ cm^{-2}\ min^{-1}$ [3]. The study by Wadding and Baker [13] demonstrated that creeping bentgrass grew well at ODR below 0.5, while root growth of Kentucky bluegrass ceased at ODR of 0.5–0.9. A value of 0.15 $g\ min^{-1}$ appears to be the lower limit for root growth in turfgrass [14]. Although variations in soil Eh or ODR have been observed in different plant species under hypoxia, soil Eh and ODR are good indicators of soil oxygenation status and determine the responses of plants to low-oxygen conditions.

31.3   RESPONSES OF TURFGRASS TO LOW OXYGEN

Turfgrass is often exposed to various low-oxygen conditions. The responses of turfgrass species or cultivars to low-oxygen stress are diverse and involve morphological and anatomical changes [5], growth responses [1,15], and physiological characteristics such as nutrients [4,16]; stomatal resistance [1]; photosynthesis [3]; and chlorophyll, protein, and antioxidants [5,17].

31.3.1   Low-Oxygen Stress Injury

The adverse effects of low-oxygen stress on the growth and development of roots and shoots have been well documented in other plant species [18]. Oxygen deficiency (e.g., anoxia) causes cell death that is closely associated with the acidification of the cytoplasm in maize (*Zea mays* L.) root tips [19]. Plants that show earlier cytoplasmic acidosis die earlier under hypoxia conditions [20]. The responses of cytoplasmic pH depend on the proton-releasing metabolization of the nucleoside triphosphate pool (short term) and lactate synthesis, succinate, malate, amino acid metabolism, and ethanolic fermentation (long term) [21]. Low oxygen limits aerobic respiration and results in low ATP production. A lack of ATP to drive root metabolism leads to the injury of root cells in waterlogging intolerant wheat [22]. Anaerobic respiration of roots also enhances some products (e.g., ethanol and acetaldehyde) that are potentially harmful in a large concentration [23]. Other toxic substances (e.g., organic acids, NO_2^-, Mn^{2+}, Fe^{2+}, and H_2S) that accumulate in anaerobic soil or roots also damage plants [23]. Long-term waterlogging (after 7 days) can cause leaf chlorosis in Kentucky bluegrass cultivars, particularly in older leaves [17]. In some plants, leaf wilting was observed under waterlogging due to a decline in leaf water potential [24].

In plant cells, O_2 participates in more than 200 different reactions [25]. Several reactive oxygen species (ROS) such as superoxide ($O_2^{\cdot-}$), hydrogen peroxide (H_2O_2), and hydroxyl radicals (·OH) are continuously produced as by-products of various metabolic processes in different cellular compartments [26]. Oxidative damage occurs in plant cells as a result of the toxicity of ROS to proteins, DNA, and lipids [27]. Accumulation of ROS has been detected in plant tissue under low-oxygen conditions, including hypoxia and flooding [28–30].

Lipid peroxidation is one of the most investigated oxidative damages, where ROS acts on membrane structure and function. Flooding causes a significant increase in lipid peroxidation, membrane permeability, and production of $O_2^{\cdot-}$ and H_2O_2 in corn leaves [28]. Accumulation of H_2O_2 has also been observed under hypoxia conditions in the roots and leaves of barley (*Hoedrum vulgare* L.) [31] and in the roots of wheat (*Triticum sativum* L.) [29]. At the intracellular level, mitochondria are major sites of ROS generation in nonphotosynthetic plant cells [32]. Szal et al. [33] reported that hypoxia did not increase generation of $O_2^{\cdot-}$ in root mitochondria of barley, while lipid peroxidation of membranes increased in the whole barley root tissue. Waterlogging increases H_2O_2 content in roots of creeping bentgrass [5]. At day 21 of treatment, H_2O_2 content increased 25% for Penncross and 33% for G-6 when the water level was 1 cm below the soil surface. Root mitochondria of creeping bentgrass swelled and lost ultrastructures under waterlogging conditions, compared with the well-drained control [5]. The adverse effects of waterlogging on mitochondria may contribute to the decline in root growth, metabolic activity, and the overall quality observed in their study.

Re-aeration following a period of flooding or waterlogging can be detrimental to plants in addition to the damage caused by the oxygen deficiency itself. The increased production of ROS after re-aeration induces both cellular damage and protective responses. Re-aeration following hypoxia results in an increased concentration of $O_2^{\cdot-}$ and H_2O_2 in plant tissues [28,34,35]. As a result of the higher level of ROS, lipid peroxidation accelerates upon reoxygenation [28,36].

31.3.2 Depths of Waterlogging

In the field, the water level may not always appear at or above the soil surface after flooding or overirrigation. The depths of water in the soil can provide a useful indication as to whether the soil is aerobic or anaerobic and how these conditions affect plant growth and root metabolic activity. Malik et al. [10] reported that the growth rate of wheat (*Triticum aestivum* L.) was less affected by waterlogging at 20 cm depth compared with waterlogging at 10 cm depth or at the soil surface. In their study, the photosynthesis was reduced by 70–80% for the two most severe waterlogging treatments, but was little affected when the water was at 20 cm below the soil surface. They also found that the growth rate recovered when waterlogging occurred at 20 cm below but not at 10 cm or at the soil surface. Jiang and Wang [5] found that turf quality and chlorophyll content decreased with the increasing water level from 15 to 1 cm below the soil surface in creeping bentgrass, and the total soluble protein content of roots decreased with the increasing water level in the sensitive cultivar Penncross. Although soil Eh declined when the water level was 15 cm below the soil surface, it did not cause a reduction in protein content in the tolerant cultivar Penn G-6. The results indicate that partial waterlogging has similar effects to full waterlogging on growth and physiological activities in turfgrass, particularly to the intolerant cultivars.

31.3.3 ANATOMICAL, MORPHOLOGICAL, AND GROWTH RESPONSES

Low oxygen promotes the formation of aerenchyma in many plant species, including turfgrass [5]. Aerenchyma provides a low-resistance internal diffusion pathway for the exchange of gases between the plant parts above the water and the submerged tissues [37,38], thus facilitating plant survival from oxygen deficiency. Jiang and Wang [5] found that the formation of aerenchyma was enhanced under waterlogging conditions in creeping bentgrass. An increased internal ethylene concentration induced the formation of aerenchyma in the roots of several species [39]. Waterlogging increases root porosity in wheat [40] and creeping bentgrass [3]. Some species develop roots with a complete or partial "barrier" to radial oxygen loss (ROL) in their epidermis, exodermis, or subepidermal layers to prevent excessive oxygen loss [41,42]. Low oxygen increases the development of adventitious roots [43,10]. Setter et al. [44] reported that there was a positive correlation between percentage of aerenchyma in adventitious roots and the yield of 17 wheat cultivars subjected to intermittent waterlogging, suggesting roles of aerenchyma and adventitious roots on waterlogging adaptation and tolerance.

Poor aeration reduces oxygen availability for the growth of turfgrass. Root growth of Kentucky bluegrass was greatly reduced when the ODR was less than 5×10^{-8} to 9×10^{-8} g cm^{-2} min^{-1} [13]. Agnew and Carrow [1] reported that low oxygen caused by short-term soil compaction decreased the root weight of Kentucky bluegrass at 15–20-cm soil depths. They also found that low soil O_2 and compaction caused higher stomatal diffusive resistance, which was associated with lower photosynthesis and greater high-temperature stress. A reduction in root growth was also observed in creeping bentgrass under low-oxygen conditions [3]. Exposure of dark-grown seedlings of barnyard grass (*Echinochloa oryzoides*) to oxygen deficiency for 3 days (0 or 5 kPa) reduced the growth of mesocotyl and increased shoot extension, changes which were not consequences of increased ethylene production [45]. Soil waterlogging induced variations in the allocation pattern in two waterlogging-tolerant grasses, and the reduction in root:shoot ratio caused by waterlogging did not have a cost in terms of capacity for nutrient uptake [46]. Low oxygen increased adaxial leaf angle in tall fescue (*Festuca arundinacea* L.) and aerenchyma formed in the cortical tissue of the root-elongation zone when plants were grown under water-saturated conditions [47].

31.3.4 PHYSIOLOGICAL RESPONSES

31.3.4.1 Nutrients

Hypoxia alters nutrient uptake and transport through the roots; hence, plants may suffer from nutrient deficiency [48]. Waterlogging reduces the concentration of nitrogen (N), phosphorus (P), potassium (K), magnesium (Mg), and zinc (Zn) in leaves and stems of wheat, and increases the concentration of these elements in the root system to a greater extent in the sensitive cultivar [49]. Supplies of important nutrients increase waterlogging tolerance by reducing the rate of decline in photosynthesis, chlorophyll content, and root and shoot growth [43]. Flooding reduces the concentration of N, P, K, calcium (Ca), Mg in shoots and the concentration of Ca and Mg in the roots of maize, but K concentrations increase in roots [50]. Nutrient uptake was higher in shoots of flooded plants compared with nonflooded ones after a postdrainage recovery period with the exception of P uptake. Flooding causes a complete depletion of NO_3-N, but NH_4-N significantly increases in the soil [51]. The contents of P and K increase in the flooded soil, whereas Ca and Mg contents remain unchanged in their study. In tree species, Pezeshki et al. [52] reported that flooding did not cause significant changes in element uptake except for Fe decreases in the roots of a tolerant species, but uptake of aluminum (Al), boron (B), K, Mg, P, and Zn was significantly lower in the shoot and roots of an intolerant species.

Waterlogging increases tissue iron (Fe) and Manganese (Mn) content, particularly in root tissue in creeping bentgrass and red fescue (*Festuca rubra* L.) [19]. These results suggest that an increased accumulation of Fe and Mn level within or on roots can reduce or limit toxic levels of Fe in the leaf tissue. The contents of Mn and Fe also increased in corn subjected to flooding [51].

Bush et al. [4] found similar results in carpetgrass (*Axonopus affinis* Chase) and centipedegrass (*Ermochloa ophuiroides* Munro. [Kunz.]) in response to soil waterlogging. Exclusion of leaf Fe and Mn uptake below toxic levels could be an adaptive mechanism for the survival of the grass plant under waterlogging conditions. Collectively, changes in the growth of shoots and roots are associated with different patterns of nutrient uptake and accumulation under low-oxygen conditions.

31.3.4.2 Carbohydrates

Carbohydrate metabolisms are vital in the adaptation of plants to low-oxygen stress. Root-water-soluble carbohydrate (RWSC) decreases and shoot-soluble carbohydrate content increases with decreasing soil Eh in creeping bentgrass [5] and Kentucky bluegrass [17]. An approximate 45% reduction in RWSC content was observed under waterlogging compared with the drained control in creeping bentgrass [5]. RWSC was not affected by different depths of waterlogging and a similar pattern was also observed in root dry weight. These results suggest that reduction in RWSC under waterlogging is one of the major factors contributing to decreased root growth. Fructan is a major carbohydrate reserve in cool-season turfgrass. Wang and Jiang [17] reported that fructan concentration was significantly increased in shoots and decreased in roots of Kentucky bluegrass exposed to waterlogging conditions. In the roots of Kentucky bluegrass, the content of total soluble sugar and total nonstructure carbohydrates was reduced respectively by 26% and 37% for the intolerant cultivar and 10% and 29% for the tolerant cultivar [17]. The results indicate that root carbohydrate status is important to waterlogging tolerance in turfgrass.

In other plant species, waterlogging or flooding increases sucrose and glucose concentration in shoots [53,54] or in roots [55–57], and increases starch concentration in shoots [57]. Seedling survival is strongly associated with nonstructure carbohydrates maintained after submergence in rice (*Oryza sativa* L.) [58]. Huang and Johnson [55] found that root respiration is greater in waterlogging-tolerant wheat than the sensitive one with 8 days of hypoxia, and the concentration of glucose, sucrose, and fructose decreased in leaves but increased in roots [55]. Albrecht and Wiedenroth [59] found that the total carbohydrate content of roots increased fourfold during 6 days of hypoxia, with a 17-fold increase in fructans in wheat. In another study, fructans accumulated in both flooding-tolerant and intolerant plant species, but starch fraction marginally increased or remained constant [60]. The results indicate that accumulation of fructan shows advantages in comparison to starch synthesis in terms of location, storage, energy-using processes of these carbohydrates, and frunctan metabolisms, and could play a role in the tolerance of oxygen deficiency.

Studies in sucrose metabolisms in other plants indicate that the essential sucrose cleavage by either invertase or sucrose synthase is affected by low-oxygen stress [61]. The activity of sucrose synthase typically increased and invertase decreased under extended periods of low oxygen (up to 7 days) [62,63]. The activities of sucrose-phosphate synthase, sucrose synthase, and acid invertase increased during anaerobic growth of pondweed (*Potamogeton distinctus* A. Benn.), but the induction of sucrose synthase was the most significant [64].

Zeng et al. [65] reported that the invertase gene expression and invertase–sucrose synthase ratio decreased in response to low oxygen in maize root tips. A rapid down-regulation of invertase could have potential advantages for the conservation of sucrose and possibly ATP, as well as for sugar signaling [65]. The results suggest that patterns of carbohydrate allocation and translocation appear to be critical for low-oxygen tolerance.

31.3.4.3 Anaerobic Metabolism

Metabolic modification is one of the adaptive strategies of plants in response to oxygen shortage; however, the mechanisms of metabolic adaptation are not fully understood, particularly in the turfgrass species. In some plant species, the role of the induction of certain enzyme activities during low oxygen has been investigated, including enzymes in glysolysis, fermentation, and antioxidation. The enzymes of alcohol dehydrogenase (ADH), lactate dehydrogenase (LDH), and pyruvate decarboxylase (PDC) are involved in plant anaerobic metabolisms that produce energy for the survival of the plant in anaerobic environments. Activities of enzymes involved in glycolysis and fermentation exhibit different patterns in response to low-oxygen stress. In Kentucky bluegrass, root activities of ADH, LDH and PDC increased under 3 and 5d of waterlogging [134]. Burdick and Mendelssohn [66] reported that ADH activity of the grass *Spartina petens* dramatically increased within 3 days of flooding but exhibited a significant decline as root internal aeration increased. The activities of fermentative enzymes, such as ADH and LDH, significantly increased in the roots of *Lepidium latifoluim*, while PDC and cytochrome *c* oxidase (CCO) activities remained stable during anaerobic treatment when Eh was about 130–200 mV in the growing solution [67]. However, Vantoai et al. [68] found that flooding increased the activities of ADH, LDH, and PDC in maize seeds, and no correlations in enzyme activity and flooding tolerance were observed across genotype, seed quality, and flooding temperature. Strong overexpressions of LDH and ADH have no effect on improving survival in low-oxygen stress in arabisopsis, but overexpression of PDC enhances survival [69]. Fukao et al. [70] found that aldehyde dehydrogenase (ALDH) was strongly induced in tolerant families of *Echinochloa crusgalli*, and the results suggest that ALDH played a role in anaerobic tolerance during seed germination. Collectively, anaerobic metabolism could play an important role in plant tolerance to low-oxygen stress, depending on growth stage, species and cultivars, as well as on depth and duration of stress.

31.3.4.4　Antioxidant Metabolism

The antioxidant system is vital for metabolic adaptations of plants to low-oxygen conditions. Plants have evolved well-developed defense systems to protect their cells against oxidative injury by removing, decomposing, or scavenging ROS, including both enzymatic and nonenzymatic defense systems. The formation and removal of ROS is in balance, but not under stress conditions. In enzymatic systems, superoxide dismutase (SOD) is in the first line of defense against ROS by catalyzing the dismutation of $O_2^{\cdot-}$ to H_2O_2 and oxygen (O_2) [71]. Superoxide is produced during electron transport; therefore, it may occur in any of the cells, including mitochondria, chloroplasts, peroxisomes, and the cytosol [72]. SOD plays a central role in the enzymatic defense system and is a major enzymatic component of the cellular defense system [71,73]. Three types of SODs have been identified: manganese SOD (Mn-SOD), iron SOD (Fe-SOD) and copper-zinc SOD (Cu/Zn-SOD), which are localized in plants in mitochondria and peroxisome, chloroplasts, cytosol, and chloroplasts, respectively. In the antioxidant system, the produced H_2O_2 is decomposed by peroxidase (POD) and catalase (CAT). The ascorbate–glutathione cycle is an important and efficient enzymatic defense for decomposing H_2O_2 and maintaining the balance of antioxidants. This cycle involves several enzymes, including ascorbate peroxidase (AP), monodehydroascorbate reductase (MR), dehydroascorbate reductase (DR), and glutathione reductase (GR) [26,74]. AP is the first enzyme in this pathway and its major function is catalyzing the H_2O_2 to H_2O. GR is the last step in the pathway, which plays a crucial role in protection against oxidative stress by maintaining a reduced glutathione level [75].

The expression and activity of SOD, AP, and GR under low-oxygen conditions have been investigated in a number of plant species, but the results are inconsistent. When plant roots are subjected to hypoxia or waterlogging conditions, SOD activity increases in barley [31] and creeping bentgrass [17], decreases in maize [28], or remains unaffected in wheat [29] and barley roots [33]. Biemelt et al. [29] reported that hypoxia did not cause a significant change in the activity and patterns of individual SOD isoforms in wheat roots. Their results also indicated that no significant alterations in transcripts and protein levels of SOD isoforms were observed under hypoxia compared with the aerated control. At the mRNA level, Biemelt et al. [29] found that Mn-SOD decreased after 2 and 4 days of hypoxia, but Cu/Zn-SOD increased after 2 days of hypoxia and increased after 4 days of hypoxia. Wang and Jiang [17] reported that the activity of SOD in roots of creeping bentgrass increased by 32% and 26% for Penncross and 83% and 44% for PennG-6 when waterlogging occurred at 15 and 1 cm below soil level, respectively. Protein gel revealed that five SOD isoforms were detected in the roots, including two manganese SODs (Mn-SODs) and three copper-zinc SODs (Cu/Zn-SODs). Levels of Mn-SOD and Cu/Zn-SOD increased under waterlogging conditions in the sensitive Penncross, especially for Mn-SOD. High levels of Mn-SOD under waterlogging conditions could result from a combination of changes in transcription, translation, and increased stability of transcript and protein, which deserves further investigation. High levels of SOD expression could be associated with protection of root tissue from further oxidative damage induced by low-oxygen conditions. Hypoxia increases AP activity but reduces the activities of GR, DR, and MR in wheat roots [34]. Yan et al. [28] reported that short-term flooding enhanced SOD, AP, and GR activity in maize leaves, while extended periods of flooding decreased the activities of these antioxidant enzymes with increasing lipid peroxidation. The activity of AP in creeping bentgrass roots significantly decreased by 46% and 40% when water levels were 1 and 15 cm below the soil surface for the intolerant Penncross, respectively; the AP activity remained unchanged in the tolerant G-6 [17]. One isoform of AP has been identified with strong intensity, and waterlogging increases expression of this band in both cultivars. The activity of POD in roots increases with increasing waterlogging level for both cultivars, and to a greater extent when the water level occurs at 1 cm below the soil surface for both cultivars. Waterlogging slightly increases the activity of GR in roots in both cultivars, and two major isoforms of GR are identified and their intensities increase under waterlogging [17]. Collectively, these results demonstrated different patterns of antioxidant enzymes in response to hypoxia caused by waterlogging or flooding conditions. The antioxidative enzymes of SOD and AP seem to play an important role in low-oxygen tolerance.

31.3.4.5 Protein

Waterlogging reduces root-soluble protein content in creeping bentgrass. The protein content decreased 16%, 18%, and 22% when waterlogging occurred at 15, 5, and 1 cm below the soil, respectively, at 21 days [5]. Similar results of protein reduction appeared in beech (*Fagus sylvatica* L.) [76] and rice [77] in response to waterlogging or submergence. However, the protein content remained unchanged in flood-tolerant oak (*Quercus robur* L.) and poplar (*Populus tremula* × *P. alba*) [76] or increased in whitetop [66] under flooding conditions. Franklin et al. [78] found that hypoxia down-regulated one protein and up-regulated three proteins in canola (*Brassica napus* L.) leaves. A synthesis of 20 anaerobic proteins has been found during low-oxygen treatment in maize roots [79], and many of these proteins are involved in sugar metabolisms such as glycolysis and fermentation pathways [79–81]. A recent study by Huang et al. [82] indicated that anoxic rice coleoptiles synthesized a pyruvate orthophosphate dikinase (PPDK), in addition to the "classical anaerobic proteins" involved in glycolysis and fermentation. This newly synthesized protein in combination with other proteins (e.g., pyruvate kinase) could enhance PPi production and thus increase ATP production to provide energy for anoxic rice coleoptile.

Haemoglobins (Hbs) are proteins widely distributed in higher plants. Class 1 nonsymbiotic Hbs have dramatically different oxygen-binding properties and are induced in plant cells by hypoxic stress [83]. The rate of binding and releasing oxygen varies with the type of Hbs, which determines the cellular functions [83]. Although the functions of Hbs have not been well elucidated, they are closely associated with plant adaptation to hypoxia. They have a high affinity for oxygen, permitting utilization of oxygen at extremely low-oxygen concentration [84], thus affecting plant metabolisms and growth under low-oxygen conditions. Overexpression class 1 Hbs has been shown to improve the growth and energy status in transgenic maize cells and alfalfa (*Medicago sativa* L.) root cultures [83,85] and to enhance the survival of hypoxia stress in *Arabidopsis thaliana* [86]. Hbs may be a potential source for respiratory metabolisms at a low-oxygen level [84], in addition to the well-known glycolytic fermentation pathway. One main role of Hbs is to modulate the nitric oxide (NO) produced in the hypoxic stresses [85], and modulation of NO can lead to significant changes in hormonal responses in plants [87]. Induction of Hbs under hypoxia and regulation of NO through Hbs and the way they are linked to ATP production and hormone changes deserve further investigation.

31.3.4.6 Hormones

Hypoxia alters hormone balance in other plant species, and research on the role of hormones in low-oxygen tolerance in turfgrass is lacking. Ethylene is one of the hormones that is closely associated with flooding stress. Accumulation of ethylene has been found in plant species in response to hypoxia or flooding stress [88–90], and the production of ethylene has advantages in the low-oxygen tolerance of plants. Ethylene stimulates the development of adventitious roots and influences the rate of root extension [91,90]. Ethylene also promotes a greater formation of aerenchyma in hypoxic maize roots [39] and in waterlogging-tolerant wheat than the intolerant kind by increasing cellulase activity [88]. The formation of aerenchyma in roots is an important adaptive trait to low-oxygen stress [92–94].

Ethylene biosynthesis is mediated by two key enzymes in higher plants: 1-aminocyclopropane-1-carboxylate (ACC) synthase and ACC oxidase, both of which are limiting factors for ethylene biosynthesis. The diurnal changes in ACC synthase activity results in diurnal production of ethylene [95,96]. The enhanced ethylene levels following hypoxia involve the production of the ethylene precursor ACC. The ACC synthesis increased in roots, then was transported to shoots where it was rapidly oxidized to ethylene [97]. Hypoxia roots have a higher activity of ACC synthase compared with that of well-aerated roots [98], and this increased ACC production is associated with flooding-induced activity of ACC synthase in roots [99]. Analysis of RNA revealed that the ACC synthase gene was rapidly induced in the roots of the tomato (*Lycopersicon esculentum* L.). Manac'h-Little et al. [100] indicated that suppression of haemoglobin gene expression in maize cell suspensions resulted in an increased level of ethylene synthesis. It has been reported that haemoglobin and nitric oxide (NO) in plants involves sensing and signaling during flooding [101]. These results suggest a role of haemoglobin/NO cycle on flooding tolerance and hormone signaling.

Low-oxygen stress affects other important hormones and influences the survival of plants. Hypoxia or flooding decreases cytokinin concentration in shoots and roots [102,103]. Hypoxia reduces fluxes of cytokinin (CK) out of the roots of the poplar (*Populus trichocarpa* × *Populus detoides*), and application of CK has no effects on alleviating symptoms of the decreased growth and closed stomata [102]. This is because CK is synthesized at the roots, and a decline in CK synthesis under flooding conditions interferes more rapidly with root metabolic activity and cell death than with other tissues [104].

Waterlogging largely decreases levels of indole-3-acetic acid (IAA) in wheat [105,106], but endogenous concentration of IAA remains unchanged in the roots of *Rumex palustris* during flooding stress [90]. An accumulation of abscisic acid (ABA) is found in flooded plants [103,107–109], and exogenous ABA application increases anoxia tolerance in maize and Arabidopsis [109,110]. Bruxelles et al. [111] indicated that the level of ABA also remained unchanged during hypoxia stress. Xie et al. [106] suggested that decreased levels of IAA, zeatin riboside (ZR), and gibberellins (GA) and increased ABA in source and sink organs could be associated with changes in the yield and content of grain starch and protein under waterlogging conditions. During recovery following flooding stress, Olivella et al. [103] found that the IAA and CK contents did not change in shoot and roots but root ABA content rapidly decreased during recovery, which was related to root death. All these changes in hormones were accompanied with reduced leaf water potential and increased root hydraulic resistance in their study.

New genes to control hormone production have been introduced into plants to enhance low-oxygen tolerance. Transgenic lines of arabidopsis thaliana with autoregulated CK production are more tolerant to flooding than wild-type plants [112]. The results indicate that endogenously increasing CK can regulate senescence caused by flooding and enhance flooding tolerance. Huynh et al. [133] reported that transgenic arabidopsis exhibited the delayed senescence after introducing a gene to increase CK biosynthesis. Accumulation of CK was found simultaneously with chlorophyll retention and increased biomass and carbohydrate content in transgenic plants relative to wild-type plants. Transgenic plants showed a greater recoverability than wild-type plants. They also demonstrated that ABA content accumulated rapidly in both transgenic and wild-type plants under waterlogging. Thus, transgenic plants provide good materials for investigating the mechanisms of low-oxygen tolerance associated with hormone changes.

31.4 LOW-OXYGEN AND ENVIRONMENTAL INTERACTION

31.4.1 TEMPERATURE

The effects of low-oxygen stress on plants may be confounded by their interaction with temperatures. High temperature significantly reduces root growth and photosynthesis [113] and increases oxygen demand in turfgrass [114,115]. The combination of high temperature with poor soil aeration causes more reduction in photosynthesis, chlorophyll content, and root viability than either factor alone [3]. The survival rate of cool-season turfgrass decreases when the soil is flooded or the turfgrass is submerged during periods of high temperatures [2,116].

Low oxygen under cold temperatures dramatically affects plant growth and physiological responses. Franklin et al. [78] found that canola with roots in a cold and hypoxic solution accumulated starch in roots, and had greater reductions in fresh weight and leaf area than either a cold or hypoxic solution alone. The changes in 17 proteins were only observed in cold hypoxic treatment in their study. Plants flooded at low temperatures accumulate more ethanol in ice and have a higher activity of ADH than those not previously flooded, and even a low concentration of ethanol can cause a reduction in freezing tolerance [117]. But Andrews [118] found that especially high-tolerance forage grasses accumulated low levels of ethanol, CO_2, and lactate and had a lower activity of PDC in ice than wheat, demonstrating a different strategy that contributes to their high level of tolerance. New protein synthesis is displayed when ice encasement is preceded by flooding [119]. The synthesized proteins include new forms of glycolytic and fermentation enzyme protein, a function that is associated with the survival of the acclimated plants. Dionne et al. [120] found that freezing tolerance was decreased by up to 7°C for annual bluegrass (*Poa annua* L.) exposed to anoxia conditions. Also, a decrease in high-molecular-weight fructan concentration and an accumulation of specific amino acids were observed under anoxia conditions. Their results suggest that hypoxia/anoxia interferes with winter survival potential in turfgrass by affecting metabolic activities.

31.4.2 SALINITY

The interaction between low oxygen and salinity affects ion relations and the growth and survival of higher plants. Waterlogging under saline conditions increases concentrations of Na^+ or Cl^- in the leaves [121]. Barret-Lennard [122] summarized the effects of waterlogging plus salinity on concentrations of Na^+ and Cl^- in shoot tissue in 24 annual crop plants. The results showed that plants grown under waterlogging and salinity had 0.9 to 55.5% higher concentration of Na^+ and 22–191% higher concentrations of Cl^- in shoots or leaves than plants grown at salinity but under drained conditions. The increased concentrations of Na^+ and/or Cl^- in shoots under waterlogging/saline are partly due to decreases in shoot growth [122]. Waterlogging and salinity increase leaf senescence to a greater extent to older leaves [123]. Formation of aerenchyma for gas transport and development of salt glands for salt removal in some species is possible physiological adaptations to interaction of waterlogging and salinity.

31.4.3 LIGHT

Low oxygen affects the response of plants to shade. Flooding injury was greater when flooding occurred under high irradiance than under low irradiance in alfalfa (*Medicago sativa* L.) [124]. Similar results have been found in wetland species: soil flooding decreased specific leaf area only when plants were grown under full sunlight [125]. The limitation of using additional energy under high irradiance may cause this greater injury to flooded plants. Although severe carbon deprivation can occur under low light, it may not cause injury under flooding conditions [126,127]. In tree species, short-term flooding increases leaf mass per unit area, reduces foliar N concentration and net photosynthesis in both full and partial (27%) sunlight, but a greater reduction in photosynthesis is shown in leaves receiving full sunlight [132]. These results indicate that the effects of flooding on photosynthetic activity are determined partly by a light environment. A review by Pierik [128] proposed that shade avoidance provided an excellent signaling system that determined plant responses to submergence.

31.5 MANAGEMENT PRACTICES

Selection of low-oxygen-tolerant cultivars is important for turfgrass management. Research indicates that large variations are exhibited in different species and cultivars. A wide range in flood damage among perennial ryegrass cultivars is observed, particularly when the temperature is high [2]. In their study, All Star and Premier showed relatively tolerant cultivars to flooding. Jiang and Wang [5] found that G-6 and L-93 had better quality than A-4, Penncross, and Pennlinks creeping bentgrass under waterlogging conditions. Kentucky bluegrass cultivars differing with growth habits also demonstrated variations in waterlogging tolerance [17]. The results showed that Moonlight was the most tolerant cultivar, followed by Serene, Limousine, Champagne, Midnight II, Awesome, and Julia, particularly when grasses were exposed to short-term waterlogging. Kenblue and Eagleton were the least tolerant cultivars.

Deep line cultivation improves the physical condition of soil. Ten weeks after the cultivation treatment, the oxygen concentration in the soil significantly increased in the sandy loam plots compared to in the untreated plots in the soccer field [129]. Oxygen values are higher (8.3%) 1 week after cultivation than in the untreated plots in the sandy root zone. Although the benefit of deep-line cultivation treatment is temporal in their study, it raises oxygen concentration in trafficked turf areas and would improve turf quality. Thatch management is also important for improving drainage efficiency and oxidation status. Thompson et al. [130] reported that low soil Eh and O_2 concentrations and accumulation of C_2H_4 were found in the thatch of Kentucky bluegrass on poorly drained soils in the field and in the greenhouse. However, application of lime and calcium arsenate amplified the extent of poor oxidation, while calcium nitrate improved oxidation status.

Golf courses adjacent to rivers may suffer from flooding as a result of heavy and frequent rainfall. When flooding occurs, good drainage systems can help get the water away soon after the river water recedes. Improving drainage and subdrainage systems would enhance grass survival from flooding damage. Irrigation management is important for maintaining healthy grass. Overirrigation can cause low-oxygen problems on golf courses or other turfgrass sites where construction is taking place with poor underlying soil profiles. The damage becomes more severe when high temperatures simultaneously occur with low oxygen in these sites. Hand-cutting greens, collars, and tees may be worthwhile under wet conditions to avoid the potential damage from more machines on the golf courses [131]. Improving drainage and surface firmness through sound management could be a medium or long-term project, and modification of current programs to address climate changes should be carefully considered.

31.6 SUMMARY

Low-oxygen stress adversely affects turfgrass. Mechanisms of stress tolerance are diverse and are associated with anatomical, morphological, physiological, and molecular changes; however, information about low-oxygen tolerance in turfgrass is lacking. Identification of stress-tolerant species and cultivars is essential for turfgrass management. Once physiological mechanisms controlling low-oxygen tolerance have been identified, they can be used as selection criteria for breeding programs to improve the stress tolerance of turfgrass. They can also be used for further molecular research, including stress gene expression and regulation. A better understanding of the low-oxygen tolerance of turfgrass will aid managers in developing effective strategies to maximize grass performance under wet and hot conditions.

REFERENCES

1. Agnew, M.L. and Carrow, R.N., Soil compaction and moisture stress preconditioning in Kentucky bluegrass. I. Soil aeration, water use, and root responses, *Agron. J.*, 77, 872, 1985.
2. Razmjoo, K., Kaneko, S., and Imada, T., Varietal differences of some cool-season turfgrass species in relation to heat and flood stress, *Int. Turf. Res. J.*, 7, 636, 1993.
3. Huang, B.R., Liu, X.Z., and Fry, J.D., Effects of high temperature and poor soil aeration on growth and viability of creeping bentgrass, *Crop Sci.*, 38, 1618, 1998.
4. Bush, E.D. et al., Carpetgrass and centipedegrass tissue iron and manganese accumulation in responses to soil waterlogging, *J. Plant Nutr.*, 22, 435, 1999.
5. Jiang, Y.W. and Wang, K.H., Growth, physiological and anatomical responses of creeping bentgrass cultivars to different depths of waterlogging, *Crop Sci.*, 46, 2420, 2006.
6. Janiesch, P., Ecophysiological adaptation of higher plants in natural communities to waterlogging. In *Ecological Responses to Environmental Stresses*, Rozema, J. and Verkleij, J.A.C., Eds., Kluwer, The Netherlands, 1991, p. 50.
7. Kozlowski, T.T., Extent, causes, and impact of flooding. In *Flooding and Plant Growth*, Kozlowski, T.T., Ed., Academic Press, New York, 1984, 1.
8. Setter, T.L. and Waters, I., Review of prospects for germplasm improvement for waterlogging tolerance in wheat, barley and oats, *Plant Soil*, 53, 1, 2003.
9. Pezeshki, S.R. and Chmabers, J.L., Stomatal and photosynthetic responses of sweet gum (*Liquidambar styraciflua*) to flooding, *Can. J. For. Res.*, 15, 371, 1985.
10. Malik, Al.I. et al., Changes in physiological and morphological traits of roots and shoots of wheat in response to different depths of waterlogging, *Aust. J. Plant Physiol.*, 28, 1121, 2001.
11. Kludze, H.K. and DeLaune, R.D., Soil redox potential effects on oxygen exchange and growth of cattail and sawgrass, *Soil Sci. Soc. Am. J.*, 60, 616, 1996.
12. Latey, J., Aeration, compaction, and drainage, *Calf. Turfgrass Cult.*, 14, 9, 1961.
13. Waddington, D.V. and Baker, J.H., Influence of soil aeration on the growth and chemical composition of three grass species, *Agron. J.*, 57, 253, 1965.
14. Latey, J. et al., Physical soil amendments, soil compaction, irrigation, and wetting agents in turfgrass management III. Effects oxygen diffusion rate and root growth, *Agron. J.*, 58, 531, 1966.
15. Fry, J.K., Submission tolerance of warm-season turfgrasses, *HortScience*, 26, 927, 1991.
16. Jones, R., Comparative studies of plant growth and distribution in relation to waterlogging, *J. Ecol.*, 60, 131, 1972.
17. Wang, K.H. and Jiang, Y.W., Waterlogging tolerance of Kentucky bluegrass cultivars, *HortScience*, 42, 386, 2007.
18. Drew, M.C., Soil aeration and plant root metabolism, *Soil Sci.*, 154, 259, 1992.
19. Robert, J.K.M., Andrade, F.H., and Anderson, I.C., Further evidence that cytoplasmic acidosis is a determinant of flooding intolerance in plants, *Plant Physiol.*, 77, 492, 1985.
20. Robert, J.K.M. et al., Mechanisms of cytoplasmic pH regulation in hypoxia maize root tips and its role in survival under hypoxia, *Natl. Acad. Sci. USA*, 81, 6029, 1984.
21. Gout, E. et al., Origin of the cytoplasmic pH changes during anaerobic stress in higher plant cells. Carbon-13 and phosphorous-31 nuclear magnetic resonance studies, *Plant Physiol.*, 125, 912, 2001.
22. Waters, I. et al., Effects of anoxia on wheat seedlings. I. Interaction between anoxia and other environmental factors, *J. Exp. Bot.*, 42, 1427, 1991.
23. Drew, M.C., Plant injury and adaptation to oxygen deficiency in the root environment: A review, *Plant Soil*, 75, 179, 1983.
24. Kramer, P.J. and Jackson, W.T., Causes of injury to flooded tobacco plants, *Plant Physiol.*, 29, 241, 1954.
25. Hendry, G.A.F., Oxygen and environmental stress in plants: an evolutionary context, *Proc. Royal Soc. Edinburgh*, 102B, 155, 1994.
26. Foyer, C.H. and Harbinson, J., Oxygen metabolism and the regulation of photosynthetic electron transport. In *Causes of Photooxidative Stress and Amelioration of Defense Systems in Plants*, Foyer, C.H. and Mullineaux, P.M., Eds., CRC Press, London, UK, 1994, p. 1.
27. Smirnoff, N., The role of active oxygen in the response of plants to water deficit and desiccation, *New Phytol.*, 125, 27, 1993.

28. Yan, B. et al., Flooding-induced membrane damage, lipid oxidation and activated oxygen generation in corn leaves, *Plant Soil*, 179, 261, 1996.

29. Biemelt, S. et al., Expression and activity of isoenzymes of superoxide dismutase in wheat roots in response to hypoxia and anoxia, *Plant Cell Environ.*, 23, 135, 2000.

30. Garnczarska, M. and Bednarski, W., Effect of a short-term hypoxia treatment followed by re-aeration on free radicals level and antioxidant enzymes in lupine roots, *Plant Physiol. Biochem.*, 42, 233, 2004.

31. Kalashnikov, J.E., Balakhnina, T.I., and Zakrzhevsky, D.A., Effect of soil hypoxia on activation of oxygen and the system of protection from oxidative destruction in roots and leaves of *Hordeum vulgare*, Russ, *J. Plant Physiol.*, 41, 583, 1994.

32. Möller, I.M., Plant mitochondria and oxidative stress: electron transport, NADPH turnover, and metabolism of reactive oxygen species, *Annu. Rev. Plant Physiol. Plant Mol. Biol.*, 52, 561, 2001.

33. Szal, B., Drozd, M., and Rychter, A.M., Factors affecting determination of superoxide anion generated by mitochondria from barley roots after anaerobiosis, *J. Plant Physiol.*, 161, 1339, 2004.

34. Biemelt, S., Keetman, U., and Albrecht, G., Re-aeration following Hypoxia or Anoxia leads to activation of the antioxidative defense system in roots of wheat seedlings, *Plant Physiol.*, 116, 651, 1998.

35. Garnczarska, M., Responses of the ascorbate-glutathione cycle to re-aeration following hypoxia in lupine roots, *Plant Physiol. Biochem.*, 43, 583, 2005.

36. Albrecht, G., Mustroph, A., and Fox, T.C., Sugar and fructan accumulation during metabolic adjustment between respiration and fermentation under low oxygen conditions in wheat roots, *Physiol. Plant.*, 120, 90, 2004.

37. Armstrong, W., Aeration in higher plants, *Adv. Bot. Res.*, 7, 225, 1979.

38. Lann, P. et al., Internal oxygen transport in *Rumex* species and its significance for respiration under hypoxic conditions, *Plant Soil*, 122, 39, 1990.

39. Drew, M.C., He, C.J., and Morgan, P.W., Programmed cell death and aerenchyma formation in roots, *Trends Plant Sci.*, 5, 123, 2000.

40. Thomson, C.J., Aerenchyma formation and associated oxygen movement in seminal and nodal roots of wheat, *Plant Cell Environ.*, 13, 395, 1990.

41. Jackson, M.B. and Armstrong, W., Formation of aerenchyma and the process of plant ventilation in relation to soil flooding and submergence, *Plant Biol.*, 1, 274, 1999.

42. Visser, E.J.W. et al., Flooding and plant growth, *Ann. Bot.*, 91, 107, 2003.

43. Huang, B.R. et al., Root and shoot growth of wheat genotypes in response to hypoxia and subsequent resumption of aeration, *Crop Sci.*, 34, 1538, 1994.

44. Setter, T.L. et al., Genetic diversity of barley and wheat for waterlogging tolerance in western Australia. In *9th Barley Technical Symposium*, Melbourne, Australia, 1999, 2.17.1.

45. Pearce, D.M.E., Hall, K.C., and Jackson, M.B., The effects of oxygen, carbon dioxide and ethylene on ethylene biosynthesis in relation to shoot extension in seedlings of rice (*Oryza stavia*) and barnyard grass (*Echinochloa oryzoides*), *Ann. Bot.*, 69, 441, 1992.

46. Rubio, G. and Lavado, R.S., Acquistion and allocation of resources in two waterlogging-tolerant grasses, *New Phytol.*, 143, 539, 1999.

47. Li, D.Y. et al., Morphological changes of tall fescue I response to saturated soil conditions, Annual Turfgrass Report of Iowa Sate University, 2000.

48. Trought, M.C.T. and Drew, M.C., Effects of waterlogging on young wheat (*Triticum aestivum* L.) plants and on soil solutes at different temperatures, *Plant Soil*, 69, 311, 1980.

49. Huang, B.R. et al., Nutrient accumulation and distribution of wheat genotypes in response to waterlogging and nutrient supply, *Plant Soil*, 173, 47, 1995.

50. Lizaso, J.I., Melendez, L.M., and Ramirez, R., Early flooding of two cultivars if tropical maize. II. Nutritional responses, *J. Plant Nutr.*, 24, 997, 2001.

51. Ashrf, M. and Rehman, H., Mineral nutrient status of corn in relation to nitrate and long-term waterlogging, *J. Plant Nutr.*, 22, 1253, 1999.

52. Pezeshki, S.R., DeLaune, R.D., and Anderson, P.H., Effect of flooding on elemental uptake and biomass allocation in seedlings of three bottomland tree species, *J. Plant Nutr.*, 22, 1481, 1999.

53. Barta, A.L., Response of field grown alfalfa to root waterlogging and shoot removal. I. Plant injury and carbohydrate and mineral content of roots, *Agron. J.*, 80, 889, 1988.

54. Bertrand, A. et al., Oxygen deficiency affects carbohydrate reserves in overwintering forage crops, *J. Exp. Bot.*, 54, 1721, 2003.

55. Huang, B.R. and Johnson, J.W., Root respiration and carbohydrate status of two wheat genotypes in response to hypoxia, *Ann. Bot.*, 75, 427, 1995.

56. Su, P.H., Wu, T.H., and Lin, C-H., Root sugar level in luffa and bitter melon is not referential to their flooding tolerance, *Bot. Bull. Acad. Sin.*, 39, 175, 1998.
57. Castonguay, Y., Nadeau, P., and Simard, R.R., Effects of flooding on carbohydrate and ABA levels in roots and shoots of alfalfa, *Plant, Cell Environ.*, 16, 695, 1993.
58. Das, K.K., Sarker, R.K., and Ismail, A.M., Elongation ability and non-structure carbohydrate levels in relation to submergence tolerance in rice, *Plant Sci.*, 168, 131, 2005.
59. Albrecht, G. and Wiedenroth, E-M., Protection against activated oxygen following re-aeration of hypoxically pretreated wheat roots. The response of the glutathione system, *J. Exp. Bot.*, 45, 449, 1994.
60. Albrecht, G., Biemelt, S., and Baumgartner, S., Accumulation of fructans following oxygen deficiency stress in related plant species with different flooding tolerance, *New Phytol.*, 136, 137, 1997.
61. Zeng, Y. et al., Rapid repression of maize invertase by low oxygen. Invertase/sucrose synthase balance, sugar signaling potential, and seedling survival, *Plant Physiol.*, 121, 599, 1999.
62. Guglielminetti, L. et al., Effects of anocia on sucrose degrading enzymes in cereal seeds, *J. Plant Physiol.*, 150, 251, 1997.
63. Perata, P., Guglielminetti, L., and Alpi, A., Mobilization of endosperm reserves in cereal seeds under anoxia, *Ann. Bot.*, 79, 49, 1997.
64. Harada, T. and Ishizawa, K., Starch degradation and sucrose metabolism during anaerobic growth of pondweed (*Potamogeton distinctus* A. Benn.) turions, *Plant Soil*, 253, 125, 2003.
65. Zeng, Y. et al., Differential regulation of sugar-sensitive sucrose-synthase by hypoxia and anoxia indicate complementary transcriptional and posttranscriptional responses, *Plant Physiol.*, 116, 1573, 1998.
66. Burdick, D.M. and Mendelssohn, I.A., Relationship between anatomical and metabolic responses to soil waterlogging in the coastal grass *Spartina patens*, *J. Exp. Bot.*, 41, 223, 1990.
67. Chen, H. and Qualls, R.G., Anaerobic metabolism in the roots of seedlings of the invasive exotic *Lepidium latifolium*, *Environ. Exp. Bot.*, 50, 29, 2003.
68. Vantoai, T.T., Fausey, N.R., and McDonald, M.B., Anaerobic metabolism enzymes as markers of flooding stress in maize seeds, *Plant Soil*, 102, 33, 1987.
69. Dolferus, R. et al., Enhancing the anaerobic response, *Ann. Bot.*, 91, 111, 2003.
70. Fukao, T. et al., Genetic and biochemical analysis of anaerobically-induced enzymes during seed germination of *Echinochloa crus-galli* varieties tolerant and intolerant of anoxia, *J. Exp. Bot.*, 54, 1421, 2003.
71. Scandalios, J.G., Oxygen stress and superoxide dismutases, *Plant Physiol.*, 101, 7, 1993.
72. Elstner, E.F., Mechanisms of oxygen activation in different compartments of plant cells. In *Active Oxygen/Oxidative Stress and Plant Metabolism*, Pell, E.J. and Steffen, K.L. Eds., *Am. Soc. Plant Physiol.*, Rockville, MD, 13, 1991.
73. Bowler, C., Van Montagu, M., and Inze, D., Superoxide dismutase and stress tolerance, *Ann. Rev. Plant Physiol. Plant Mol. Biol.*, 43, 83, 1992.
74. Asada, K., Ascorbate peroxidase—a hydrogen peroxide scavenging enzyme in plants, *Physiol. Plant*, 85, 235, 1992.
75. Asada, K., The water–water cycle in chloroplasts: scavenging of active oxygen and dissipation of excess photons, *Annu. Rev. Plant Physiol. Plant Mol. Biol.*, 50, 601, 1999.
76. Kreuzwieser, J., Fürniss, S., and Rennenberg, H., Impact of waterlogging on the N-metabolism of flood tolerant and non-tolerant tree species, *Plant Cell Environ.*, 25, 1039, 2002.
77. Mohanty, B. and Ong, B.L., Contrasting effects of submergence in light and dark on pyruvate decarboxylase activity in roots of rice lines differing in submergence tolerance, *Ann. Bot.*, 91, 291, 2003.
78. Franklin, J.A. et al., Root temperature and aeration affects on the protein profile of canola leaves, *Crop Sci.*, 45, 1379, 2005.
79. Sachs, M.M., Subbaiah, C.C., and Saab, I.N., Aanerobic gene expression and flooding tolerance in maize, *J. Exp. Bot.*, 47, 1, 1996.
80. Kelly, P.M., Maize pyruvate decarboxylase mRNA is induced anaerobically, *Plant Mol. Biol.*, 13, 213, 1989.
81. Dolferus, R. et al., Molecular basis of the anaerobic responses in plants, *IUBMB Life*, 51, 79, 2000.
82. Huang, S.B. et al., Protein synthesis by rice coleoptiles during prolonged anoxia: implications for glycolysis, growth, and energy utilization, *Ann. Bot.*, 96, 703, 2005.
83. Dordas, C. et al., Expression of a stress-induced hemoglobin affects NO levels produced by alfalfa under hypoxia stress, *Plant J.*, 35, 763, 2003.
84. Igamberdiev, A.U. et al., The haemoglobin/nitric oxide cycle: involvement in flooding stress and effects on hormone signaling, *Ann. Bot.*, 96, 557, 2005.

85. Igamberdiev, A.U. et al., NADH-dependent metabolism of nitric oxide in alfalfa root cultures expressing barley hemoglobin, *Planta*, 219, 95, 2004.

86. Hunt, P.W. et al., Increased level of haemoglobin 1 enhances survival of hypoxic stress and promotes early growth in *Arabidopsis thaliana*, *PNAS*, 99, 17197, 2002.

87. Wendehenne, D., Durner, J., and Klessig, D.F., Nitric oxide: a new player in plant signaling and defense responses, *Curr. Opin. Plant Biol.*, 7, 449, 2004.

88. Huang, B.R. et al., Root characteristics and physiological activities of wheat in response to hypoxia and ethylene, *Crop Sci.*, 37, 812, 1997.

89. Schravendijk, W-V. and Van Andel, O.M., The role of ethylene during flooding of *Phaseolus vulgaris*, *Physiol. Plant.*, 66, 257, 1986.

90. Visser, E.J.W. et al., An ethylene-mediated increase in sensitivity to auxin induces adventitious root formation in flooded *Rumex palustris* Sm, *Plant Physiol.*, 1112, 1687, 1996.

91. Blom, C.W.P.M. et al., Physiological ecology of riverside species: adaptive responses of plants to submergence, *Ann. Bot.*, 74, 253, 1994.

92. Lann, P. et al., Root morphology and aerenchyma formation as indicators of the flooding-tolerance of *Rumex* species, *J. Ecol.*, 77, 693, 1989.

93. Jackson, M.B. and Drew, M.C., Effects of flooding on growth and metabolism of herbaceous plants. In *Flooding and Plant Growth*, Kozlowski, T.T., Ed., Academic Press, New York, US, 1984, p. 47.

94. Justin, S.H.F.W. and Armstrong, W., The anatomical characteristics of roots and plant response to soil flooding, *New Phytol.*, 106, 465, 1987.

95. Machckova, I. et al., Diurnal fluctuations in ethylene formation in *Chenopodium rubrun*, *Plant Physiol.*, 113, 981, 1997.

96. Shiu, O.Y. et al., The promoter of *LE-ACS7*, an early flooding-induced 1-aminocyclopropane-1-carboxylate synthase gene of the tomato, is tagged by a *Sol3* transposon, *PNAS*, 95, 10334, 1998.

97. Jackson, M.B., Hormones from roots as signal for the shoots of stressed plants, *Trends Plant Sci.*, 2, 22, 1997.

98. Drew, M.C., Oxygen deficiency and root metabolism: injury and acclimation under hypoxia and anoxia, *Annu. Rec. Plant Physiol. Plant Mol. Biol.*, 48, 233, 1997.

99. Olson, D.C., Oetiker, J.H., and Yang, S.F., Analysis of *LE-ACS3*, a 1-aminocyclopropane-1-carboxylic-acid synthase gene expressed during flooding in the roots of tomato plants, *J. Bio. Chem.*, 270, 14056, 1995.

100. Manac'h-Little N., Igamberdiev, N., and Hill, R.D., Hemoglobin expression affects ethylene production in maize cell cultures, *Plant Physiol. Biochem.*, 43, 485, 2005.

101. Dat, J.F. et al., Sensing and signaling during plant flooding, *Plant Physiol. Biochem.*, 42, 273, 2004.

102. Neuman, D.S., Rood, S.B., and Smith, B.A., Does cytokinin transport from root-to-shoot in the xylem sap regulate leaf responses to root hypoxia? *J. Exp. Bot.*, 41, 1325, 1990.

103. Olivella, C. et al., Hormonal and physiological responses of *Gerbera jamesonii* to flooding stress, *HortScience*, 35, 222, 2000.

104. VanToai, T.T. et al., Developmental regulation of anoxic stress tolerance in maize, *Plant Cell Environ.*, 18, 937, 1995.

105. Nan, R., Carman, G.J., and Salisbury, F.B., Water stress, CO_2, and photoperiod influence hormone levels in wheat, *J. Plant Physiol.*, 159, 307, 2002.

106. Xie, Z. et al., Relationships of endogenous plant hormones to accumulation of grain protein and starch in winter wheat under different post-anthesis soil water statuses, *Plant Growth Regul.*, 41, 2003.

107. Zhang, J.H. and Davis, W.J., ABA in root and leaves of flooded pea plants, *J. Exp. Bot.*, 38, 649, 1987.

108. Zhang, J.H. and Zhang, X.P., Can early wilting of old leaves account for much of the ABA accumulation in flooded pea plants? *J. Exp. Bot.*, 45, 1335, 1994.

109. Hwang, S.-Y. and Vantoai, T.T., Abscisic acid induces anaerobiosis tolerance in corn, *Plant Physiol.*, 97, 593, 1991.

110. Ellis, M.H., Dennis, E.S., and Peacock, W.J., Arabidopsis roots and shoots have different mechanism for hypoxia stress tolerance, *Plant Physiol.*, 119, 57, 1999.

111. Bruxelles, G.L. et al., Abscisic acid induces the alcohol dehydrogenase gene in Arabidopsis, *Plant Physiol.*, 111, 381, 1996.

112. Zhang, J. et al., Development of flooding tolerant *Arabidopsis thaliana* by autoregulated cytokinin production, *Mol. Breeding*, 6, 135, 2000.

113. Jiang, Y.W. and Huang, B.R., Effects of drought and heat stress alone and in combination on Kentucky bluegrass, *Crop Sci.*, 40, 1358, 2000.

114. Ralston, D.S and Daniel, W.H., Effects of temperatures and water table depth on the growth of creeping bentgrass roots, *Agron. J.*, 64, 709, 1972.

115. Kurtz, K.W. and Kneebone, W.R., Influence of aeration and genotype upon root growth of creeping bentgrass at supra-optimal temperatures, *Int. Turf. Res. J.*, 3, 145, 1980.

116. Beard, J.B. and Martin, D.P., Influence of water temperature on submersion tolerance of four grasses, *Agron. J.*, 62, 257, 1970.

117. Andrews, C.J. and Pomeroy, M.K., The influence of flooding pretreatment on metabolite changes in winter cereal seedlings during ice encasement, *Can. J. Bot.*, 61, 142, 1983.

118. Andrews, C.J., A comparison of glycolytics activity in winter wheat and two forage grasses in relation to their tolerance to ice encasement, *Ann. Bot.*, 79, 1997.

119. Andrew, C.J., How do plants survive ice? *Ann. Bot.*, 78, 529, 1996.

120. Dionne, J. et al., Physiological effects of anoxia on annual bluegrass during cold acclimation, Annual Meeting of Crop Science Society of America, Abstract, 2000.

121. John, C.D., The structure of rice roots grown in aerobic and anaerobic environments, *Plant Soil*, 47, 269, 1977.

122. Barrett-Lennard, E.G., The interaction between waterlogging and salinity in higher plants—causes, consequences and implications, *Plant Soil*, 253, 35, 2003.

123. Marcar, N.E., Waterlogging, modified growth, water use, and ion concentrations of seedlings of salt-treated *Eucalyptus camaldulensis*, *E. tereticornis*, *E. robusta* and *E. globules*, *Aust. J. Plant Physiol.*, 20, 1, 1993.

124. Barta, A.L. and Sulc, R.M., Interaction between waterlogging injury and irradiance level in alfalfa, *Crop Sci.*, 42, 1529, 2002.

125. Lenssen, J.P.M., Menting, F.B.J., and Van der Putter, W.H., Plant responses to simultaneous stress of waterlogging and shade: amplified or hierarchical effects? *New Phytol.*, 157, 281, 2003.

126. Barta, A.L., Supply and partitioning of assimilates to roots of *Medicago sativa* L., and *Lotus Corniculatus* L., under anoxia, *Plant Cell Environ.*, 10, 151, 1987.

127. Castonguay, Y., Nadeau, P., and Simard, R.R., Effects of waterlogging on carbohydrate and ABA levels in roots and shoots of alfalfa, *Plant Cell Environ.*, 16, 695, 1993.

128. Pierik, R. et al., New perspectives in flooding research: the use of shade avoidance and *Arabidopsis thaliana*, *Ann. Bot.*, 96, 533, 2005.

129. Morhard, J.M. and Kleisinger, S., Short-term effects of deep line cultivation on soil oxygen, penetration resistance and turf quality of two soccer fields, *Acta Horticulturae*, 661, 343, 2004

130. Thompson, D.C., Smiley, R.W., and Fowler, M. C., Oxidation status and gas composition of wet turfgrass thatch and soil, *Agron. J.*, 75, 603, 1983.

131. Isaac, S., Hell is a wet place, The British and International Golf Greenkeepers Association, Greenkeeper International, 2002.

132. Gardiner, E.S. and Krauss, K.W., Photosynthetic light responses of flooded cherrybark oat (*Quercus pagoda*) seedling grown in two light regimes, *Tree Physiol.*, 21, 1103, 2001.

133. Huynh, L.N. et al., Regulation of flooding tolerance of *SAG12:ipt* Arabidopsis plants by cytokinin, *J. Exp. Bot.*, 56, 1397, 2005.

134. Wang, K.H. and Jiang, Y.W., unpublished data, 2007.

32 Identification of Turfgrass Stress Utilizing Spectral Reflectance

Yiwei Jiang

CONTENTS

32.1 INTRODUCTION

Leaf or canopy spectral reflectance has been widely used for the estimation of a plant's physiological status and environmental stress tolerance. Stress affects the spectral reflectance of a plant, and changes in spectral reflectance are closely associated with stress injury and alterations of physiological functions. Thus, reflectance signals from plant canopy can be potentially used for stress monitoring and management. Turfgrass stress management, such as evaluation of drought and nutrient status, requires frequent assessment of water and nutrient variability across a turfgrass site (e.g., golf course). This requires large and extensive sampling of soil and plant tissues. Also, a rapid, nondestructive, accurate method of monitoring stress occurrence is desirable and important for the management of turfgrass. Spectral-related technology provides such a tool for estimating the physiological status of a plant and identifying stressed plants.

Spectral reflectance has been found to be highly correlated with leaf pigment content [1–3], green biomass [4], and water status [5,6] in turfgrass. Thus, it could be used to help screen large germplasms for breeding programs and to improve turf quality by enhancing management practices. To date, the application of spectral reflectance to indicate performance of different turfgrass species and cultivars has been investigated under biotic and abiotic stress conditions, including drought [7–10], wear and compaction [11–14], nutrient deficiency [15,16], disease [17–19], and insects [20]. Reflectance can also be used to analyze turfgrass soil properties [21]. On the basis of the relationship between turf canopy variables and spectral reflectance, a linear or more complicated model can be developed to predict the severity of stress injury and other physiological parameters at the whole canopy level. In addition to visual observation of turf quality, research on spectral reflectance in turfgrass provides valuable information for field mapping of stress, correcting a problem in a timely manner, and managing turf plots under unfavorable conditions.

32.2 CHARACTERISTICS OF SPECTRAL REFLECTANCE

Spectral reflectance is the percentage of reflected light from the plant leaf or canopy relative to the incident light. When light strikes a leaf, only part of the incident energy is reflected from the leaf, while the remainder is either absorbed or transmitted. Leaf reflectance is controlled by surface properties, leaf internal structure, and concentration and distribution of biochemical components [22], and can be collected using various spectral radiometers. When plants start experiencing stress, the reflectance of individual leaves changes more dramatically in the visible spectral regions than in the infrared regions because of the sensitivity of chlorophyll to physiological disturbance [23]. A healthy plant leaf exhibits low reflectance in the visible spectral region (400–700 nm) because of strong absorption by photosynthetic pigments (e.g., chlorophyll). It also exhibits a relatively high reflectance in the near-infrared region (NIR; 700–1300 nm) because of internal leaf scarring and no strong features, and a relatively low reflectance in the middle infrared or infrared region (>1300 nm) due to a strong absorption of water and other compounds [23]. Figure 32.1 illustrates turfgrass canopy reflectance in the 400–1100-nm region. Canopy spectral reflectance is a combination of vegetation and soil reflectance, and factors such as illumination and canopy structure determine the value of the reflectance. Turfgrass typically provides a good and uniform vegetation structure; thus, it can be a good model for a study of the relationship between canopy reflectance and plant physiological status.

In general, reflectance in the visible wavelength (e.g., 675, 680, and 700 nm) is highly correlated with leaf chlorophyll content [1,24,25], and reflectance in the NIR (e.g., 810, 900, 970, and 1150 nm) is associated with leaf water content [5,6,26]. Additionally, various indices based on reflectance in the multiple wavelengths have been used to indicate leaf or canopy status under normal and stress conditions. The most widely used index is the normalized difference vegetation index (NDVI), calculated from the reflectance in the NIR and the red (R) region (NIR − R)/(NIR + R). The NDVI is highly correlated with leaf chlorophyll content [27,28], green biomass [4], turf quality [8,11,12], and turf color and percent of live cover [29]. The leaf area index (NIR/R) is also correlated with turf quality and canopy temperature [12] and turf density [11]. The water index (900 nm/970 nm) has been used as an indicator of plant water content [5]. Since biotic and abiotic stresses often interfere with leaf pigment and water content, these correlations between reflectance or reflectance indices and leaf or canopy variables provide a useful tool to identify stress injury and assess stress tolerance.

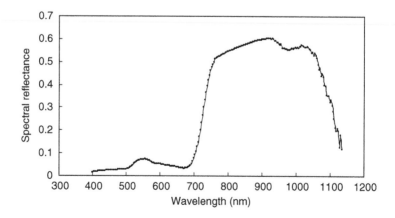

FIGURE 32.1   Turfgrass canopy reflectance in the 400–1100 nm region.

32.3 STRESS RESPONSE AND SPECTRAL REFLECTANCE

32.3.1 DROUGHT STRESS

Turfgrass is frequently subjected to drought stress, and decline in turf quality and physiological activities occur due to water deficit. Leaf wilting and leaf firing (chlorosis, tan-brown color) are common symptoms of drought stress injury in turfgrass. Visual observation and rating of these symptoms are often used for evaluation of drought resistance, but early detection and quantitative assessment of stress symptoms or injury by using spectral reflectance are necessary for drought stress management. Jiang and Carrow [9] tested the changes in canopy reflectance in the range 400–1100 nm and their relationships to turf quality and leaf firing under drought stress in five turfgrass species. They reported that canopy reflectance from 667 to 693 nm had the highest correlations with turf quality or leaf firing in bermudagrass (*Cynodon dactylon* L. × C. *transvaalensis*), while near-infrared reflectance at 750, 775, or 870 nm exhibited the peak correlations with both ratings in seashore paspalum (*Paspalum vaginatum* Swartz). Reflectance at 687–693 nm was correlated with quality and leaf firing in zoysiagrass (*Zoysia japonica*) and St. Augustinegrass (*Stenotaphrum secundatum*), and the highest correlation was observed at 671 nm in tall fescue (*Festuca arundinacea*). These results indicated that reflectance at photosynthetic regions from 664 to 687 nm is relatively important in determining turf quality and leaf firing in selected bermudagrass, tall fescue, zoysiagrass, and St. Augustinegrass. Another study by Suplick-Ploense and Qian [30] found that near-infrared reflectance at 736–874 nm was correlated with decreasing leaf water content in Kentucky bluegrass (*Poa pratensis* L.) but not in perennial ryegrass (*Lolium perenne* L.). Huang et al. [8] reported that Kentucky-31 tall fescue had better drought tolerance than MIC-18, whereas NDVI (R935−R661)/(R935+R661) was higher in Kentucky-31 than in MIC-18.

Research on spectral reflectance under drought stress has been reported in many plant species [26,31–34]. Yield loss in corn (*Zea Mays* L.) due to water stress is highly correlated with red reflectance ($r = 0.58$) and NDVI ($r = -0.61$) [31]. Fan and Claupein [32] found that reflectance significantly changed when leaf water content was below 75%. In rice, canopy reflectance at 486, 553, 660, 1376, 1435, 1676, 1810, and 2245 nm had significant linear relationships with the water stress level. However, the best correlation was found at 2245 nm ($r = 0.80$), and the relationship between NDVI and soil water potential was significant [35]. Reflectance at 1450 nm was correlated with water content in corn, spinach (*Spinacia oleracea* L.), and snap beans (*Phaseolus vulgaris* L.) [34]. Cure et al. [36] found that canopy temperature was correlated with reflectance at 620 and 850 nm in soybean (*Glycine max* L.) in response to drought stress. Collectively, the best region of reflectance indicating drought stress symptoms may shift with individual wavelength, depending on different plant species and the severity and duration of stress. Reflectance can be used to assess and quantify the water status of plants and to further develop appropriate models for crop management and monitoring drought stress in turfgrass.

Linear or multiple linear regression models based on spectral reflectance can be used to predict changes in turf quality and canopy temperature under drought stress, but the most important wavelength in a model varied with the turfgrass species or cultivar [10]. The best multiple linear regression models for five turfgrass species contained three wavelengths either in chlorophyll (660–700 nm) or water absorption regions (810–1480 nm), which exhibited a significant coefficient of determination (R^2) values ranging from 0.40 to 0.74 [10]. Reflectance near water-absorption regions (900–1200 nm) was important for evaluating turf quality in tall fescue, a species exhibiting a large reduction in turf quality under drought stress compared with the other four warm-season grasses used in the experiment [10]. Furthermore, this region also accounted for a large variability in models for cultivars in different species with a higher reduction in turf quality, such as TifEagle bermudagrass, Sea Isle 1 seashore paspalum, and Meyer zoysiagrass. These results suggested that reflectance near 900–1200 nm was vital in determining a turf-quality model for those grasses with less drought resistance. Other studies in different plant species demonstrated that reflectance near

940, 970, and 1150 nm was highly correlated with leaf or canopy water content [3,5], which could be used for an estimate of plant water status under drought stress. Fenstermaker-Shaulis et al. [7] reported that reflectance in the 600–650 and 800–890-nm ranges had a high correlation with tissue moisture content (31–68% range) in tall fescue under drought stress. The model to predict NDVI from turf color using visual rating for creeping bentgrass (*Agrostis palustris* Huds.) was not as accurate as that for tall fescue [29]. The results indicated that different models should be developed and used in the evaluation of particular turfgrass species under drought stress. A more detailed study is needed to determine a suitable mode for better predicting turf quality and leaf or canopy water content simultaneously in different grass species in response to soil drying.

Deficit irrigation is one of the strategies for water conservation and drought stress management in turfgrass. Lee at al. [37] reported that five warm-season turfgrasses irrigated at 100% and 80% of evapotranspiration (ET) showed no differences in NDVI and IR/R; however, grasses irrigated at 60% and 40% of ET exhibited differences in these reflectance indices. In their study, the infrared spectral region was more sensitive to drought stress than the visible region, which was associated with changes in the water content of leaves. All wavelengths tested were highly correlated with turf quality, soil moisture, leaf rolling, leaf firing, and chlorophyll index. These results suggested that data from spectral reflectance over turf canopy could be used to help determine irrigation scheduling and provide an understanding of the water status of the plant.

32.3.2 Traffic Stress

Turfgrass traffic stress generally consists of wear stress alone or a combination of wear and soil compaction, resulting in direct injury to the shoot and increasing the degree of soil compaction, respectively [38]. Both traffic stresses cause a reduction in turf quality and decreases in shoot density and uniformity. Analysis of spectral reflectance may aid in evaluating stress injury and recoverability for the plant. Trenholm et al. [13] found that reflectance at 661 and 813 nm, NDVI (R935 − R661)/(R935 + R661), IR/R (R935/R661), and stress index (R706/R760) and (R706/R813) were highly correlated with turf quality, shoot density, and shoot tissue injury for seashore paspalum and bermudagrass under wear stress. Jiang et al. [12] demonstrated that all NDVI, stress index, leaf-area index, and water-band index were highly correlated with canopy temperature and turf quality in seashore paspalum exposed to wear or wear plus soil compaction. The two indices R936/R661 and R693/R759 had the strongest correlations with canopy temperature for wear ($r = -0.63$) and wear plus soil compaction ($r = 0.66$), while a ratio of R693/R759 had the highest correlations with turf quality for wear ($r = -0.89$) and wear plus soil compaction ($r = -0.82$), respectively. Although soil compaction may interfere with reflectance collected from turf canopy, the stress index (R693/R759) was a good indicator of turf quality and canopy temperature under both traffic stress conditions in seashore paspalum. Guertal and Shaw [14] evaluated the impact of levels of soil traffic on spectral reflectance, soil resistance, and bulk density and their relationships in compacted bermudagrass turf over multiple years. They concluded that the correlations between reflectance data (from 507 to 935 nm) or NDVI and IR/R, soil resistance, or bulk density were not consistent over the sampling date, although the treatments showed differences in soil resistance and bulk density. The best correlation was observed in the visible region; however, no unique wavelength that provided the strongest correlation between reflectance and soil resistance was identified. Many factors could contribute to the variability of reflectance, such as sunlight, turf color, leaf age, and undetected disease. More wavelengths may be needed to better determine the correlations among reflectance, soil resistance, bulk density, and turf quality in different grass species under traffic stress.

32.3.3 Nutrient Stress

Evaluation and characterization of nutrient availability and variability are important components in turfgrass site-specific management. Nutrient deficiency normally affects turf color, and therefore, changes in nutrient status can be detected using spectral reflectance. Kruse et al. [15] found that the spectral index (NSI, R695/R760) accurately identified nitrogen-stressed creeping bentgrass. There was a strong correlation between nitrogen concentration in plant tissue and the NSI index ($r = -0.8$ to -0.9). Leaf N content in creeping bentgrass and annual bluegrass (*Poa annua* var. *reptans* Hausskn) is correlated with canopy reflectance, but one of the best correlations shows up in the spectrum at 670 nm, corresponding to chlorophyll A transmission [39]. Keskin et al. [40] found that reflectance at the green band (520–580 nm) and NIR region (770–1050 nm) increased as N concentration increased in creeping bentgrass, and regression models using either single wavelengths 550, 680, 770, and 810 nm or multiple wavelengths predicted well for N content. But in their study, no general trends were observed in bermudagrass and perennial ryegrass blends. Sembiring et al. [14] reported that reflectance at 435 nm was independent of nitrogen and phosphorus treatment in bermudagrass, and N uptake, P uptake, and N concentration could be predicted using R695/R405 (435 nm as a covariate). These results suggested that spectral reflectance indices have the potential to be a valuable tool for predicting N and P nutrient status in turfgrasses. However, further work is needed to document the relationship between other nutrients and reflectance, or the test nutrient response of turfgrass in different environments.

In crop plants, deficiency of an essential element may significantly reduce growth rate and yield, and spectral reflectance has been used for the evaluation of nutrient status in different plant species. Near-infrared reflectance is positively correlated with plant biomass and N uptake in cotton (*Gossypium hirsutum* L.) [41]. Al-abbas et al. [42] found that corn plants grown under N-, Mg-, S-, K-, P-, and Ca-deficiency conditions had higher leaf reflectance in the 400–700-nm regions than the normal plants [42]. Milton et al. [43,44] reported that reflectance at 500–650 nm increased in Co-, Ni-, and Zn-deficient *Hosta Ventricosa* plants and in As-, P-, and Se-deficient soybeans. A decreased concentration of Mn in soybean leaves also increased reflectance [45]. Masoni et al. [46] reported that Fe, S, Mg, and Mn deficiency reduced leaf chlorophyll content, increased leaf reflectance, decreased leaf absorbance, and shortened the red edge position in barley (*Hordeum vulgare* L.), wheat, corn, and sunflower (*Helianthus annuus* L.). Read et al. [47] reported that N stress increased reflectance at 695 nm and decreased reflectance at 410 nm, and leaf N was correlated with R659 or R755 and canopy N status was correlated with R410 or R700. They concluded that R415 was the most stable spectral feature under N stress in corn, compared with the changes in the region of 690–730 nm.

A number of reflectance indices have been proposed to assess N stress. Corn yield loss due to N is correlated with standardized green normalized difference vegetation index ([(NIR − green)/ (NIR + green)]/[(NIR$_r$ − green$_r$)/(NIR$_r$ + green$_r$)]) (NIR$_r$ and green$_r$ are values from well-fertilized and watered control); normalized difference water index (NIR−MIR)/(NIR+MIR) (MIR is mid-infrared); nitrogen reflectance index (NIR/green)/(NIR$_r$/green$_r$); chlorophyll green index (R810/R568); and chlorophyll red edge index (R810/R710) [41]. A model based on yield loss due to N is more accurate at predicting N requirements than that based on yield or yield loss due to water [31]. Although soil N test using near-infrared spectrophotometer could not accurately predict relative grain yield or N-supplying capacity in response to N fertilizer, NIR offered a convenient, rapid, and inexpensive alternative compared with other tests [48].

32.3.4 DISEASES

The multispectral approach has been used in plant pathology and crop protection in many plant species [49]. Early and accurate identification of disease occurrence and severity are important to disease management in turfgrass. The relationship between reflectance and disease assessment has been reported in turfgrass species affected by various diseases, including dollar spot (*Sclerotinia homoeocarpa*) [18], rhizoctonia blight (*Rhizoctonia solani*) [16], gray leaf spot (*Pyricularia grisea*) [50], brown patch [18,19], and pythium blight (*Pythium aphanidermatum*) [19]. Green et al. [50] found that reflectance at 810 nm had a strong relationship with the visual severity of rhizoctonia blight and gray leaf spot in tall fescue. The percentage of reflectance at 760 and 810 nm decreased as the severity of rhizoctonia blight increased [17]. They further suggested that the reflectance was particularly useful in assessing differences in disease severity at discrete points in time rather than in an entire epidemic. Rinehart et al. [18] demonstrated that 90–92% of the data taken by reflectance were accurate in indicating the severity of dollar spot and brown patch in creeping bentgrass and annual bluegrass. Also, the first derivative of reflectance indicated that 700, 1400, and 1930 nm showed the greatest correlations between disease rating and reflectance. These results suggested that reflectance provides a useful method and high degree of precision in disease rating, although factors other than disease severity, such as crop architecture, soil moisture, and reflectance, can influence canopy reflectance [50].

The utilization of reflectance in the visible and near-infrared regions in disease detection or yield loss caused by diseases has been tested in other plant species [51–55]. The increased reflectance at the near-infrared region and decreased reflectance in the visible region are related to fungal disease severity in wheat [54]. NIR could accurately predict the incidence of kernel of maize infected by fungi and by *Fusarium verticillioides* ($r^2 = 0.80$) and monitor mold contamination in postharvest maize [51]. The reflectance method shows a higher correlation with yield and demonstrates the ability to measure downy mildew–induced defoliation in quinoa (*Chenopodium quinoa* L.) [53]. Guan and Nutter [52] reported that canopy reflectance at 810 nm can be used to quantitatively estimate yield in the alfalfa leaf-spot pathosystem. A similar region at 800 nm has been found to be a rapid and precise measurement to evaluate fungicide efficacy to control late leaf spot in peanuts (*Arachis hypogea* L.) [55]. Collectively, remote sensing is a sensitive tool for the discrimination and quantification of disease severity in crop plants.

32.3.5 OTHER APPLICATIONS

Spectral reflectance has also been applied to other turfgrass management practices, including analysis of soil properties, insect detection, wetness assessment, and salinity responses [20,21,56,57]. Couillard et al. [21] reported that reflectance in the NIR region showed the highest accuracy in predicting soil moisture content, organic matter, density, particle-size distribution, pH, and potassium from the turf soil profiles. Li et al. [41] found that the red and NIR reflectance and NDVI were cross-correlated with soil water, sand, and clay across a distance of 60–80m. Wetness is a key factor contributing to the development of foliar disease on turfgrass. Madeira et al. [56] found that reflectance at 1165 nm exhibited the strongest responses to increasing wetness for dew or rainfall in creeping bentgrass and bluegrass. For detection of insect damage in a greenhouse, using wavelength at 650–700 nm, NDVI, and stress indices clearly identify white grub–induced damage, but using NIR (800–950 nm) cannot separate treatment differences [20]. A field study is needed to determine the insect infestation and damage that can be evaluated using spectra analysis.

Visible and near-infrared reflectance has been used in the assessment of salinity effects on seashore paspalum [58] and barley [57]. Lee et al. [58] found that the tolerant seashore paspalum SI 93-2 and HI 101 had higher NDVI and IR/R and lower stress index (R706/760, R706/R813) than the least tolerant Adalayd. They concluded that IR/R and R706/R760 were useful tools for the evaluation of salinity tolerance, accounting for 85% and 51% of shoot and root growth variations, respectively. In barley, Peñuelas et al. [57] reported that near-infrared reflectance decreased and visible reflectance increased in response to increasing salinity levels, and NDVI and water index were good indicators of salinity effects.

Spectral reflectance has also been used in estimating different leaf biochemical concentrations other than chlorophyll content. Near-infrared reflectance is correlated with total nonstructural carbohydrate (TNC) in Tifdwarf and Tifway bermudagrass with an r^2 value of 0.86 [59]. Narra et al. [60] found that the predicted value of glucose, fructose, fructan, and TNC using NIR were highly correlated ($r^2 > 0.9$) with the laboratory value in creeping bentgrass. Curran et al. [61] reported that the first derivative of reflectance at certain regions was causally related to concentrations of foliar chemicals: 1124 nm (lignin), 2352 and 2172 nm (N), 1800 nm (cellulose), 1008 and 2352 nm (protein), 906 and 1496 nm (amino acids), 1484 and 1800 nm (sugar), 978 and 1208 nm (starch). The accuracy of estimation decreased from total chlorophyll, nitrogen, sugar, chlorophyll A, cellulose, chlorophyll B, lignin, water, phosphorus, protein, and amino acids to starch. These results suggested that NIR can determine biochemical compounds with good accuracy.

32.4 SPECTRAL DATA PROCESSING

Canopy spectral reflectance data are often collected in the field using ambient illumination. The accuracy of the data can be influenced by several factors, such as sun angle, cloud cover, or canopy structure, which may contribute to inconsistency or variations observed in some studies [14,17]. To enhance the accuracy of the experiment and eliminate background noise, reflectance data can be appropriately transformed and manipulated prior to use in correlation analysis. The first derivative of reflectance is a commonly used form of data transformation that can reduce the variations in illumination and canopy architecture [62–64]. The first derivative of reflectance is highly correlated with fresh and dry biomass, canopy water content, and photosynthetic pigment [25].

Reflectance data are normally smoothed before they are used for analysis of the first derivative. The process of data smoothing is removing random noise from spectra and reducing data fluctuation. The degree of smoothing should be considered since smoothing may cause a shift in the peak correlation from one wavelength position to another, depending on the sensitivity of the canopy variable collected in each individual study. Rollin and Milton [6] found that the peak correlation between the first derivative of reflectance at 1150 nm and water content in hay meadows was sensitive to the degree of smoothing. They suggested that the position of peak correlation should be interpreted with caution due to wavelength shifting. Jiang et al. [12] reported that correlations were not sensitive to the degrees of smoothing of reflectance from 400 to 1100 nm, and smoothing by 5 points and 9 points had no difference in peak correlation between canopy temperature and turf quality, and reflectance in wear and wear plus soil compaction in seashore paspalum, respectively.

The analysis of the first derivative of reflectance significantly increased the correlation between reflectance and canopy variables in turfgrass under both wear and wear plus soil compaction, particularly under wear and soil compaction [12]. When grasses are exposed to wear and soil compaction, the first derivative of reflectance showed the higher correlation with turf quality ($r = -0.81$) than the original reflectance ($r = -0.70$) and higher correlation with canopy temperature ($r = 0.70$) than the original reflectance ($r = 0.50$). Since wear and soil compaction may involve more soil background, these results demonstrated that the first derivative of reflectance could be used to reduce error and increase correlations in turfgrass.

In addition, when using original reflectance data, single or multiple wavelengths cannot be identified to detect the best correlations between reflectance with both canopy temperature and turf quality under both wear and wear plus soil compaction stress. However, the first derivative of reflectance at 667 nm is the optimum position to use in determining all these correlations simultaneously. Thus, the first derivative of reflectance can provide an alternative for data processing of reflectance to indicate canopy status in turfgrasses when subjected to stress conditions such as traffic.

32.5 SUMMARY

Spectral reflectance can provide a fast, easy, and accurate tool for identification of biotic and abiotic stress, assessment of plant physiological status, prediction of nutrient availability, and determination of water use and irrigation scheduling in turfgrasses. Table 32.1 summarizes the correlations between reflectance and canopy variables in different turfgrass species under stress conditions. Environmental factors such as sun angle, cloud cover, and canopy structure affect reflectance by increasing background noise when sampling; therefore, data on reflectance can be transformed and manipulated (e.g., the first derivative of reflectance) to increase the correlations between reflectance and

TABLE 32.1

Stress Detection Using Spectral Reflectance in Different Turfgrass Species

Stress	Species	Wavelength (nm) or Indices	References
Drought	Tall fescue	671	[9]
	Kentucky bluegrass	736–874	[29]
	Bermudagrass	667–693	[9]
	Zoysiagrass	687–693	[9]
	St. Augustinegrass	687–693	[9]
	Seashore paspalum	750, 775, 870	[9]
Traffic	Seashore paspalum	706/760, 935/661	[13]
		$(935 - 661)/(935 + 661)$	[13]
		693/759	[12]
Nutrients	Creeping bentgrass	695/760	[15]
		550, 680, 770, 810	[40]
		670	[39]
	Annual bluegrass	670	[39]
	Bermudagrass	695/405	[14]
Salinity	Seashore paspalum	706/760, 706/813	[58]
Diseases	Creeping bentgrass	760, 810	[17]
	Tall fescue	810	[50]
Insects	Kentucky bluegrass	650–700	[20]
Wetness	Creeping bentgrass	1165	[56]
	Kentucky bluegrass	1165	[56]

canopy variables. In addition, spectrometry using internal illumination has become commercially available, which may provide a better use of reflectance over turf canopy. Once the best correlation between canopy reflectance and canopy variables has been identified for a particular turf species, an adequate model can be developed to better predict turfgrass status under stress conditions. Sensors near these regions could be developed and installed into a machine for turfgrass stress monitoring and field practices. Future studies are needed to identify stress-specific wavelengths or indices for particular species or cultivars under various stress conditions.

REFERENCES

1. Carter, G.A. and Spiering, B.A., Optical properties of intact leaves for estimating chlorophyll concentration., *J. Environ. Qual.*, 31, 1424, 2002.
2. Bell, G.E. et al., Optical sensing of turfgrass chlorophyll content and tissue nitrogen, *HortScience*, 39, 1130, 2004.
3. Stiegler, J.C. et al., Spectral detection of pigment concentrations in creeping bentgrass golf greens, *Int. Turfgrass Soc. Res. J.*, 10, 818, 2005.
4. Gamon, J.A. et al., Relationship between NDVI, canopy structure and photosynthesis in three California vegetation types, *Ecol. Appl.*, 5, 28, 1995.
5. Penuelas, J. et al., Estimation of plant water concentration by the reflectance water index WI (R900/R970). *Int. J. Remote Sens.*, 18, 2863, 1997.
6. Rollin, E.M. and Milton, E.J., Processing of high spectral resolution reflectance data for the retrieval of canopy water content information, *Remote Sens. Environ.*, 65, 86, 1998.
7. Fenstermaker-Shaulis, L.K., Leskys, A., and Devitt, D.A., Utilization of remotely sensed data to map and evaluate turfgrass stress associated with drought, *J. Turfgrass Manage.*, 2, 65, 1997.

8. Huang, B.R., Fry, J.D., and Wang, B., Water relations and canopy characteristics of tall fescue cultivars during and after drought stress, *HortScience*, 33, 837, 1998.

9. Jiang, Y.W. and Carrow, R.N., Assessment of narrow-band canopy spectral reflectance and turfgrass performance under drought stress, *HortScience*, 40, 242, 2005.

10. Jiang, Y.W., Carrow, R.N., and Duncan, R.R., Broad-band spectral reflectance models for monitoring drought stress in different turfgrass species, Annual Meeting of Crop Science Society of America, Abstract, 2004.

11. Jiang, Y.W., Carrow, R.N., and Duncan, R.R., Effects of morning and afternoon shade in combination with traffic stress on seashore paspalum, *HortScience*, 36, 1218, 2003.

12. Jiang, Y.W., Carrow, R.N., and Duncan, R.R., Correlation analysis procedures for canopy spectral reflectance data of seashore paspalum under traffic stress, *J. Amer. Soc. Hort. Sci.*, 128, 343, 2003.

13. Trenholm, L.E., Carrow, R.N., and Duncan, R.R., Relationship of multispectral radiometry data to qualitative data in turfgrass research, *Crop Sci.*, 39, 763, 1999.

14. Guertal, E.A. and Shaw, J.N., Multispectral radiometer signatures for stress evaluation in compacted bermudagrass turf, *Hortscience*, 39, 403, 2004.

15. Sembiring, H. et al., Detection of nitrogen and phosphorus nutrient status in bermudagrass using spectral radiance, *J. Plant Nutr.*, 21, 1189, 1998.

16. Kruse, J.K., Christians, N.E., and Chaplin, M.H., Nitrogen stress level in creeping bentgrass can be detected via remote sensing, Annual Meeting of Crop Science Society of America, Abstract, 2004.

17. Raikes, C. and Burpee, L.L., Use of multispectral radiometry for assessment of rhizoctonia blight in creeping bentgrass, *Phytopathology*, 88, 446, 1998.

18. Rinehart, G.J. et al., Remote sensing of brown patch and dollar spot on creeping bentgrass and annual bluegrass using visible and near-infrared, spectroscopy, *Int. Turfrgass Res. J.*, 9, 705, 2001.

19. Anderson, Z. et al., Methods of direct sensing to measure turf diseases, Annual Meeting of Crop Science Society of America, Abstract, 2004.

20. Hamilton, R.M. and Gibb, T.J., Detection of white grub damage in turfgrass using remote sensing, Annual report, Purdue University Turfgrass Science Program, 2000.

21. Couillard, A. et al., Near infrared reflectance spectroscopy for analysis of turf soil profiles, *Crop Sci.*, 37, 1554, 1997.

22. Penuelas, J. and I. Filella, Visible and near-infrared reflectance techniques for diagnosing plant physiological status, *Trends Plant Sci.*, 3, 151, 1998.

23. Knipling, E.B., Physical and physiological basis for the reflectance of visible and near-infrared radiation from vegetation, *Remote Sen. Environ.*, 1, 155, 1970.

24. Gitelson, A.A., Merzlyak, M.N., and Lichtenthaler, H.K., Detection of red edge position and chlorophyll content by reflectance measurements near 700 nm, *J. Plant Physiol.*, 148, 501, 1996.

25. Blackburn, G.A., Quantifying chlorophylls and carotenoids at leaf and canopy scales: An evaluation of some hyperspectral approaches, *Remote Sens. Environ.*, 66, 273, 1998.

26. Bowman, W.D., The relationship between leaf water status, gas exchange, and spectral reflectance in cotton leaves, *Remote Sens. Environ.*, 30, 249, 1989.

27. Carter, G.A., Ratios of leaf reflectance in narrow wavelength as indicators of plant stress, *Int. J. Remote Sens.*, 15, 697, 1994.

28. Gamon, J.A. and Surfus, J.S., Assessing leaf pigment content and activity with a reflectometer, *New Phytol.*, 143, 105, 1999.

29. Bell, G.E. et al., Vehicle-mounted optical sensing: an objective means for evaluating turf quality, *Crop Sci.*, 42, 197, 2002.

30. Suplick-Ploense, M.R. and Qian, Y.L., Spectral reflectance responses of three turfgrasses to leaf dehydration, Annual Meeting of Crop Science Society of America, Abstract, 2002.

31. Clay, D.E. et al., Characterizing water and nitrogen stress in corn using remote sensing, *Agron. J.*, 98, 579, 2006.

32. Graeff, S., Fan, Z.X., and Claupein, W., Use of reflectance measurements to clearly identify water stress in wheat (*Triticum aestivum* L.), *Proc. 4th Intl. Crop Sci. Congress*, 2004.

33. Jones, C.L. et al., Estimating water stress in plants using hyperspectral sensing, ASAE/CSAE Annual International Meeting, 2004.

34. Kimura, R. et al., Relationships among the leaf area index, moisture availability, and spectral reflectance in an upland rice field, *Agr. Water Manage.*, 69, 83, 2004.

35. Yang, C-M. and Su, M-R., Analysis of spectral characteristics of rice canopy under water deficiency, *Proc. GIS Develop.*, ACRS, 2000.

36. Cure, W.W., Flagler, R.B., and Heagle, A.S., Correlation between canopy reflectance and leaf temperature in irrigated and droughted soybeans, *Remote Sens. Environ.*, 29, 273, 1989.

37. Lee, J.H., Unruh, J.B., and Trenholm, L.E., Multi-spectral reflectance of warm-season turfgrasses as influenced by deficit irrigation, Annual Meeting of Crop Science Society of America, Abstract, 2004.

38. Carrow, R.N. et al., Turfgrass traffic (soil compaction plus wear) simulator: response of Paspalum vaginatum and cynodon spp., *Int. Turfgrass Soc. Res. J.*, 9, 253, 2001.

39. Rinehart, G.J., Remote sensing of leaf tissue nitrogen content and disease severity in creeping bentgrass and annual bluegrass using infrared spectroscopy, MS Thesis, Michigan State University, 2000.

40. Keskin, M. et al., Assessing nitrogen content of golf course turfgrass clippings using spectral reflectance, Annual American Society of Agricultural Engineers, Abstract, 2001.

41. Li, H. et al., Multispectral reflectance of cotton related to plant growth, soil water and texture, and site elevation, *Agron. J.*, 93, 1327, 2001.

42. Al-Abbas, A.H. et al., Spectra of normal and nutrient-deficient maize leaves, *Agron. J.*, 66, 16, 1974.

43. Milton, N.W., Arsenic- and selenium-induced changes in spectral reflectance and morphology of soybean plants, *Remote Sens. Environ.*, 30, 263, 1989.

44. Milton, N.W., Eiswerth, B.A., and Ager, C.M., Effect of phosphorus deficiency on spectral reflectance and morphology of soybean plants. *Remote Sens. Environ.*, 26, 121, 1991.

45. Adams, M.L. et al., Fluorescence and reflectance characteristics of manganese deficit soybean leaves: effect of leaf age and choice of leaflet, *Plant Soil*, 155/156, 235, 1993.

46. Masoni, A., Ercoli, L., and Mariotti, M., Spectral properties of leaves deficient in iron, sulfur, magnesium, and manganese, *Agron. J.*, 88, 937, 1996.

47. Read, J.J. et al., Narrow-waveband reflectance ratios fro remote estimation of nitrogen status of cotton, *J. environ. Qual.*, 31, 1442, 2002.

48. Fox, R.H. et al., Comparison of near-infrared spectroscopy and other soil nitrogen availability quick tests for corn, *Agron. J.*, 85, 1049, 1993.

49. Nilsson, Hans-E., Remote sensing and image analysis in plant pathology, *Can. J. Plant Pathol.* 17, 154, 1995.

50. Green II, D.E., Burpee, L.L., and Stevenson, K.L., Canopy reflectance as a measure of disease in tall fescue, *Crop Sci.*, 38, 1603, 1998.

51. Berardo, N. et al., Rapid detection of kernel rots and mycotoxins in maize by near-infrared reflectance spectroscopy, *J. Agr. Food Chem.*, 53, 8128, 2005.

52. Guan, J. and Nutter, F.W., Jr., Relationships between defoliation, leaf area index, canopy reflectance, and forage yield in the alfalfa-leaf spot pathosystem, *Comput. Electro. Agr.*, 37, 97, 2002.

53. Danielsen, S. and Munk, L., Evaluation of disease assessment methods in quinoa for their ability to predict yield loss caused by downy mildew, *Crop Prot.*, 23, 219, 2004.

54. Muhammed, H.H. and Larsolle, A., Feature vector based analysis of hyperspectral crop reflectance data for discrimination and quantification of fungal disease severity in wheat, *Biosyst. Eng.*, 86, 125, 2003.

55. Nutter, F.W., Jr., Littrell, R.H., and Brenneman, T.B., Utilization of a multispectral radiometer to evaluate fungicide efficacy to control late leaf spot in peanut, *Phytopathology*, 80, 102, 1990.

56. Madeira, A.C., Gillespie, T.J., and Duke, C.L., Effects of wetness on turfgrass canopy reflectance, *Agr. Forest Meteorol.*, 107, 117, 2001.

57. Peñuelas, J. et al., Visible and near-infrared reflectance assessment of salinity effects on barley, *Crop Sci.*, 37, 198, 1997.

58. Lee, G.J., Carrow, R.N., and Duncan, R.R., Photosynthetic responses to salinity stress of halophytic seashore paspalum ecotypes, *Plant Sci.*, 166, 1417, 2004.

59. Miller, G.L. and Dickens, R., Bermudagrass carbohydrate levels as influence by potassium fertilization and cultivar, *Crop Sci.*, 36, 1283, 1996.

60. Narra, S., Fermanian, T.W., and Swiader, J.M., Analysis of mono-and polysaccharides in creeping bentgrass turf using near infrared reflectance spectroscopy, *Crop Sci.*, 45, 266, 2005.

61. Curran, P.J., Dungan, J.L., and Peterson, D.L., Estimating the foliar biochemical concentration of leaves with reflectance spectrometry testing the Kokaly and Clark methodologies, *Remote Sens. Environ.*, 76, 349, 2001.

62. Danson, F.M. et al., High spectral resolution data for monitoring leaf water content, *Int. J. Remote Sens.*, 13, 3045, 1992.

63. Malthus, T.J. and Madeira, A.C., High resolution spectroradiometry: spectral reflectance of field beans leaves infected by *Botrytis fabae*, *Remote Sens. Environ.*, 45, 107, 1993.

64. Tsai, F. and Philpot, W., Derivative analysis of hyperspectral data, *Remote Sens. Environ.*, 66, 41, 1998.

33 Enhancing Turfgrass Nitrogen Use under Stresses

Haibo Liu, Christian M. Baldwin, Hong Luo,
and Mohammad Pessarakli

CONTENTS

33.1 INTRODUCTION

Nonfood and nonfiber grasses including ornamental grasses and turfgrasses are important to human beings. Those grasses came from their origins by natural spreading or artificial collections and require relatively less nutrient input in comparison with other agricultural crops with focuses on yields. Turfgrass has been well accepted by human beings since the 1300s or an earlier time [25,27,28]. Home lawns show the closest relationship between today's human life and grasses—hundred millions of home lawns surrounding their homes on this earth, which comprise the largest turf area in the world. Currently, it is estimated that there are more than 30,000 golf courses, several hundred millions of home lawns, and several millions of sports fields in the world, among which the United States alone has about 17,000, 60 million, and 775,000, respectively [25,27,434]. In addition, the world has multiple billions of acres of natural grasslands and forage grass areas [518]. The entire turfgrass industry in the United States is conservatively estimated as being a $60–$90 billion per year industry with a total area of more than 50 million acres, in which home lawns comprise about 65%, and the rest 35% include utility turf areas, golf courses, sports fields, and turf production (turfgrass seed and sod) areas [25,27,30,325]. Most of these managed turfgrass areas experience an unfavorable stress such as drought, low fertility, poor soil conditions, temperature extremes, salinity, shade, low soil pH, traffic compaction, weed, disease, insect and mite pests, and others at least one time per growing season. Some may have to overcome multiple stresses at a time and some may have to face the permanent stresses year-round [115,119,159]. Intensive studies related to turfgrasses, particularly in the past 50 years, have been focused on the mentioned stresses for enhanced turfgrass management, turfgrass improvement, and environmental stewardship [247,379,434]. Findings [25,27,28,100,115,119,134,159,325,492] have been applied in turfgrass management, while main future challenges are still to maximize turf use to human beings with minimized resource costs and input including nitrogen (N) fertilizers.

Major turfgrass species (Table 33.1) are in the grass family, which includes major agricultural crops such as rice (*Oryza sativa* L.), wheat (*Triticum aestivum* L.), corn (*Zea mays* L.), sorghum (*Sorghum bicolor* [L.] Moench), barley (*Hordeum vulgare* L.), rye (*Secale cereale* L.), and other forage crops. Unlike legumes, most gasses cannot fix N through N fixation, and N is one of the most limiting factors for any types of grass production for yields, forage, or recreational and functional ground covers as turfgrasses [25,27,28,325,492]. By comparison with other crops, turfgrasses relatively require much less N but it is still needed in the largest quantity than any other essential mineral nutrients [95,223,224]. Turfgrass quality can be enhanced by proper N application with adequate chlorophyll content, stress and pest resistance, and shoot and root growth under different levels of maintenance to meet the turf use. Reviews on turfgrass N use, physiology, and impacts to the environment have been well documented in the last two decades [95,115,159,223,224,325,379, 494,513]. However, reviews and research in areas such as turfgrass N use and performance under stressful conditions are rather limited. Therefore, this chapter focuses on enhancing turfgrass N use under major stressful conditions.

33.2   A BRIEF OVERALL REVIEW OF PLANT NITROGEN

Nitrogen, the most needed mineral nutrient for plants, is also the most commonly deficient nutrient for crops. Each year the total global N input for crops is more than 100 billion kg [175]. The trend of total N input has increased following the population growth since the first synthesized N fertilizers were used in the early 1900s [193]. Today, the N use and cycle are not only managed just for crop needs but also with management regulations and implementations to the environment and other living things [511]. The plant characteristics of N absorption; assimilation to amino acids; N metabolisms; the secondary N metabolisms; N fixation; and synergetic, symbiotic, and antagonistic relationships associated with other livings are well studied and understood, although these processes are very complicated [322]. A few recent reviews and books may serve as examples to demonstrate the major findings [5,34,136,148,149,181,193,194,266,267,322,339,347,351, 454,501,502,530]:

- Plant tissues (excluding seeds, flowers, and fruits) normally contain about 3 to 6% N of their total dry matter to function properly without suffering N deficiency. These N requirements must come from soils, water, and air [194,322,339,530].

TABLE 33.1
Main Turfgrass Species

Cool-Season Turfgrasses
Alkaligrass (*Puccinellia* spp.)
Annual bluegrass (*Poa annua* L.)
Annual ryegrass (*Lolium multiflorum* Lam.)
Canada bluegrass (*Poa compressa* L.)
Chewings fescue (*Festuca rubra* L. ssp. fallax [Thuill.] Nyman.)
Colonial bentgrass (*Agrostis capillaries* L.)
Creeping bentgrass (*Agrostis stolonifera* L.)
Creeping red fescue (*Festuca rubra* L. ssp. rubra)
Hard fescue (*Festuca trachyphylla* [Hackel] Krajina)
Intermediate ryegrass (*Lolium xhybridium* Hausskn.)
Kentucky bluegrass (*Poa pratensis* L.)
Meadow fescue (*Festuca elatior* L.)
Perennial ryegrass (*Lolium perenne* L.)
Retop (*Agrostis gigantea* Roth.)
Rough bluegrass (*Poa trivialis* L.)
Sheep fescue (*Festuca ovina* L. spp. Ovina)
Supina bluegrass (*Poa supina* Schrad.)
Tall fescue (*Festuca arundinacea* Schreb.)
Texas bluegrass (*Poa arachnifera* Torr.)
Velvet bentgrass (*Agrostis canina* L.)

Warm-Season Turfgrasses
African bermudagrass (*Cynodon transvaalensis* Burtt-Davy)
Bahiagrass (*Paspalum notatum* Flugge)
Buffalograss (*Buchloe dactyloides* [Nutt.] Engelm.)
Centipedegrass (*Eremochloa ophiuroides* [Munro] Hack)
Chinese lawngrass (*Zoysia sinica* Hance.)
Common carpetgrass (*Axonopus affinis* Chase)
Commopn bermudagrass (*Cynodon dactylon* [L.] Pers.)
Hybrid bermudagrass (*Cynodon dactylon* [L.] Pers. × *C. transvaalensis* Burtt-Davy)
Hybrid zoysiagrasses (*Zoysia japonica* × *Z. matrella*, *Z. japonica* × *Z. pacinica*, *Z. matrella* ×
    Z. pacifica, *Z. japonica* × *Z. tenuifolia*)
Inland saltgrass (*Distichlis spicata* var. stricta [L.] Greene)
Japanese lawngrass (*Zoysia japonica* Steud.)
Kikuyugrass (*Pennisetum clandestine* Hochst ex Chiov.)
Largespike lawngrass (*Zoysia macrostachya* French.)
Magennis bermudagrass (*Cynodon* × *magennissi* Hurcombe)
Manilagrass (*Zoysia matrella* [L] Merr.)
Mascarenegrass (*Zoysia pacifica* [Gaud.] Hotta and Kuroti)
Seashore paspalum (*Paspalum vaginatum* Swartz.)
St. Augustinegrass (*Stenotaphrum secundatum* [Walt.] Kuntze.)
Thinleaf lawngrass (*Zoysia tenuifolia* Willd. ex Trin.)
Tropical carpetgrass (*Axonopus compressus* [Swarty] Beauv.)

- Nitrogen cycle is one of the most important element cycle in the world and extremely important to livings in which plants play the most important role regardless of human beings [454,511].
- Human beings have been more and more involved in the N cycle regardless of the healthy or unhealthy impacts to the global ecosystems [454,511].
- Nitrogen can be absorbed through roots, leaves, stems, and other organs with roots serving as the main N absorption organs [322,339].
- Nitrogen can be absorbed as nitrate (NO_3^-), ammonium (NH_4^+), urea ($NH_2\text{-}CO\text{-}NH_2$), amino acids, amides, and other N compounds with NO_3^- and NH_4^+ serving as the most common N forms being absorbed [86,148,172,176,194,206,274,322,339].
- The overall N use efficiency (recovery by plants/N input $\times$ 100%) is between 30 and 60% with an average lower than 50% because of multiple factors [175].
- Plants use N differently in rates, forms, and time during their growth and development. Normally, younger plants and tissues require more N than older plants and tissues [322,339].
- During a life span of any individual plant, a plant at least experiences one time of N stress either under deficiency or excessiveness [322,339].
- Plant N uptake physiology, kinetics, genetics, and the regulations are complicated and related to all major plant metabolic pathways [148,149,322,339]. However, genetic control approaches (conventional or molecular biological techniques) of enhancing N use efficiency (NUE) have been undertaken more and more intensively due to the increased pressure of food, fiber, and recreational requirements [454,511].

33.3   NITROGEN AND TURFGRASS QUALITY

Similar to other agricultural crops, turfgrasses normally contain 3–6% N of the total dry biomass. This depends on species and cultivars [52,101,224,239,286,288], turfgrass age, and the season plus environmental influences. In general, cool-season turfgrasses (C_3 plants) contain a little higher N than warm-season turfgrasses (C_4 plants); younger turfgrasses or tissues contain higher N than older turfgrasses or tissues; and the N concentrations in tissues in the spring season for both cool- and warm-season turfgrasses are higher than other seasons [95]. Nitrogen can be absorbed both by roots and above ground parts including leaves and shoots [95,157,322,424]. The primary absorbed N forms are nitrate (NO_3^-), ammonium (NH_4^+) for roots. For foliar absorption, urea is absorbed more than NO_3^- and NH_4^+ (Figure 33.1) with improved understanding cuticular penetrations [410,428]. Table 33.2 demonstrate the nitrogen input, output, and cycling within a typical turf–atmosphere–soil system indicating multiple factors involved in such a system in addition to routine management and use of the turf [42,52,53,68,95,99,102,115,159,223,224,272,379].

Nitrogen is important for turf quality including turf color, appearance, shoot growth and density, root growth, recuperative capability, wear tolerance, thatch accumulation, resistance to stressful conditions and pests [25,27,223,224,325]. Since N is a part of chlorophyll molecule, proteins, nucleic acids, plant hormones, and secondary metabolites, N stress will cause weakened metabolic processes including chlorophyll formation, enzymatic formations and reactions, hormone formation and regulations, and carbohydrate reserves, which result in overall poor turf quality and performance [25,27,95,115,159]. Deficiency of N profoundly influences the morphology and physiology of turfgrasses [46,95,463]. Under N deficiency, a turfgrass exhibits a greater root:shoot ratio with yellowish leaves and shoots inducing chloroplast disintegration and loss of chlorophyll [95]. The older leaves show chlorosis compared to young leaves because of high mobility of N in the plants [322,325,339]. In contrast, excessive N supply will weaken the turfgrass by producing imbalanced shoots and roots, more succulent tissues, and thicker leaves to face stressful conditions leaving the turf more vulnerable to pests in addition to increased potentials of N losses to the environment [25,27,325]. Therefore, proper N supply is an essential practice to keep the turfgrass

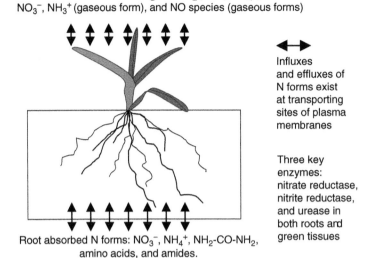

Foliar absorbed N forms: $NH_2\text{-}CO\text{-}NH_2$ (urea), NH_4^+, NO_3^-, NH_3^+ (gaseous form), and NO species (gaseous forms)

Influxes and effluxes of N forms exist at transporting sites of plasma membranes

Three key enzymes: nitrate reductase, nitrite reductase, and urease in both roots and green tissues

Root absorbed N forms: NO_3^-, NH_4^+, $NH_2\text{-}CO\text{-}NH_2$, amino acids, and amides.

FIGURE 33.1 The forms of N absorbed by turfgrass roots and green tissues.

TABLE 33.2
Nitrogen Input, Output, and Cycling in a Turf–Soil–Atmosphere System

N Input	N Output	N Cycling
Fertilizers	Plant uptake	Immobilization
Clipping returning	Clipping removal	Mineralization
Thatch decomposition	Thatch removal	Nitrification
Dead roots	Denitrification	
Lightening	Volatilization	
Other chemicals containing N	Leaching	
	Ammonium fixation	

under the optimum condition to meet all characteristics for the turf use, particularly under stressful conditions. Tables 33.3 and 33.4 provide the normal N range for the most common types of turfgrass uses. Differences exist due to species and cultivars, locations, climatic conditions, soil conditions, management levels, uses, and environmental factors and conditions.

In addition to color and appearance, turfgrass quality can be specifically evaluated and measured by shoot growth and density, root growth, rhizome and stolon growth, and carbohydrate reserves, among which carbohydrate reserves are the key factor to overcome stresses [95,395,414]. Nitrogen plays a significant role shift in carbohydrate reserves for both cool- and warm-season turfgrasses. The carbohydrate reserves for C_3 cool-season turfgrasses are mainly fructosans while starch is the main carbohydrate reserves for C_4 warm-season turfgrasses [25,27,72,95,325]. Before reaching an optimum N status, turfgrass stress tolerance increases following an increased N input and carbohydrate reserves [95]. However, excessive N supplies will shift turfgrasses to more shoot growth with a cost of carbohydrate reserves, and therefore, turfgrasses lose stress tolerances [159]. Thatch accumulation stops at certain level when N application continues [95,159].

TABLE 33.3

The Normal Annual N Requirement Range for Different Types of Cool-Season Turfgrass Uses

Cool-Season Turfgrasses Type of Turf	Turfgrass Species	Soil Type	Total Annual N Input Range (N kg/ha/year)	Turf Use and Management Remarks	References
Putting greens with mowing heights of 2.5–4.8 mm	Creeping bentgrass	Sandy soils	120–250	Clippings removal	[27,95,99,100,147,225,242, 269,325,348,349,380,387, 388,421,422,425,455, 505,534]
		Native soils	80–200	Clippings removal	
		Sandy or native soils	50–150 (winter months)	Winter overseeding and clipping removal	
	Colonial bentgrass	Sandy soils	120–200	Clippings removal	
		Native soils	80–200	Clippings removal	
		Sandy or native soils	50–120 (winter months)	Winter overseeding uncommonly used and clipping removal	
	Annual bluegrass	Sandy soils	200–300	Clippings removal	
		Native soils	80–200	Clippings removal	
		Sandy or native soils	50–120 (winter months)	Winter overseeding uncommonly used and clipping removal	
	Fine fescue	Sandy or native soils	120–200	Winter overseeding and clippings removal	
	Perennial ryegrass	Sandy or native soils	50–150 (winter months)	Winter overseeding and clippings returned	
Fairways and tees with mowing heights of 6.0–20.0 mm	Creeping bentgrass	Sandy soils	120–200	Clippings returned	
		Native soils	80–200	Clippings returned	
		Sandy or native soils	50–100 (winter months)	Winter overseeding and clipping returned	
	Colonial bentgrass	Sandy soils	120–200	Clippings returned	
		Native soils	80–200	Clippings returned	
		Sandy or native soils	50–120 (winter months)	Winter overseeding uncommonly used and clipping removal	

	Species	Soil	N rate	Clippings
	Annual bluegrass	Sandy soils	150–250	Clippings returned
		Native soils	80–180	Clippings returned
		Sandy or native soils	50–120 (winter months)	Winter overseeding uncommonly used and clipping returned
	Fine fescue	Sandy soils	100–180	Clippings returned
	Fine fescue	Native soils	100–150	Clippings returned
	Fine fescue	Sandy or native soils	100–120	Winter overseeding and clippings returned
	Perennial ryegrass	Sandy soils	150–200	Clippings returned
	Perennial ryegrass	Native soils	150–180	Clippings returned
	Perennial ryegrass	Sandy or native soils	100–120	Winter overseeding and clippings returned
	Kentucky bluegrass	Sandy soils	180–220	Clippings returned
	Kentucky bluegrass	Native soils	150–200	Clippings returned
	Kentucky bluegrass	Sandy or native soils	100–120	Winter overseeding and clippings returned
Sports fields including baseball fields, soccer fields, and football fields with mowing height range 12–36 mm	Perennial ryegrass	Sandy soils	150–200	Clippings returned
	Perennial ryegrass	Native soils	150–180	Clippings returned
	Perennial ryegrass	Sandy or native soils	100–120	Winter overseeding and clippings returned
	Kentucky bluegrass	Sandy soils	180–220	Clippings returned
	Kentucky bluegrass	Native soils	150–200	Clippings returned
	Kentucky bluegrass	Sandy or native soils	100–120	Winter overseeding and clippings returned
Roughs and home lawns with mowing heights between 36 and 100 mm	Perennial ryegrass	Sandy or native soils	150–200	Clippings returned
	Kentucky bluegrass	Sandy or native soils	180–220	Clippings returned
	Tall fescue	Sandy or native soils	150–200	Clippings returned
	Fine fescues	Sandy or native soils	100–150	Clippings returned

TABLE 33.4

The Normal Annual N Requirement Range for Different Types of Warm-Season Turfgrass Uses

Warm-Season Turfgrasses Type of Turf	Turfgrass Species	Soil Type	Total Annual N Input Range (N kg/ha/year)	Turf Use and Management Remarks	References
Putting greens with mowing height of 2.5–4.8 mm	Hybrid bermudagrasses	Sandy soils	250–1000	Clippings removal	[27,95,100,102,125,153, 177,184,208,225,254, 268,325,348,380,381, 385,386,415,416,421, 438,439,450,486, 526,534]
		Native soils	250–450	Clippings removal	
	Seashore paspalum	Sandy soils	200–450	Clippings removal	
		Native soils	200–400	Clippings removal	
	Zoysiagrass	Sandy soils	200–350	Clippings removal	
		Native soils	200–300	Clippings removal	
Fairways and tees with mowing heights of 6.0–20.0 mm	Hybrid bermudagrasses	Sandy soils	200–350	Clippings returned	
		Clayey soils	180–300	Clippings returned	
	Common bermudagrasses	Sandy soils	180–300	Clippings returned	
		Clayey soils	150–280	Clippings returned	
	Zoysiagrass	Sandy soils	200–300	Clippings returned	
		Clayey soils	180–280	Clippings returned	
	Seashore paspalum	Sandy soils	200–300	Clippings returned	
		Clayey soils	180–250	Clippings returned	
	Buffalograss	Sandy soils	180–220	Clippings returned	
		Clayey soils	150–200	Clippings returned	
Sports fields including baseball fields, soccer fields, and football fields with mowing height range 12–36 mm	Hybrid bermudagrasses	Sandy soils	200–300	Clippings returned	
		Clayey soils	180–280	Clippings returned	
	Common bermudagrasses	Sandy soils	180–250	Clippings returned	
		Clayey soils	150–200	Clippings returned	
	Zoysiagrass	Sandy soils	180–220	Clippings returned	
		Clayey soils	150–200	Clippings returned	
	Buffalograss	Sandy soils	150–200	Clippings returned	
		Clayey soils	120–200	Clippings returned	
Roughs and home lawns with mowing heights between 36 and 100 mm	Hybrid bermudagrass	Native soils	200–300	Clippings returned	
	Common bermudagrass	Native soils	100–250	Clippings returned	
	Zoysiagrass	Native soils	100–250	Clippings returned	
	Centipedegrass	Native soils	50–120	Clippings returned	
	St. Augustinegrass	Native soils	150–200	Clippings returned	
	Seashore paspalum	Native soils	150–200	Clippings returned	
	Bahiagrass	Native soils	80–150	Clippings returned	
	Carpetgrass	Native soils	100–150	Clippings returned	
	Buffalograss	Native soils	80–120	Clippings returned	

33.4 TURFGRASS NITROGEN USE EFFICIENCY (TNUE)

Unlike normal crops, N use efficiency of a given nutrient can be easily determined by the yield (total biomass) produced per unit of nutrient applied [18,193,322]. Turfgrass nutrient use efficiency can be mainly determined by performance and appearance of the turf [95]. The total biomass can be collected and calculated based on the input of a nutrient. However, the high yield of total biomass produced by a turf does not necessarily mean a high quality of turf because turf quality is also determined by its overall intended use [27,325]. For example, a better playability of golf putting greens may not mean a higher N input, but a slight N stress is recommended to reduce the potential of diseases and increase of ball speed. In addition, the plant growth regulator (PGR) applications help to reduce mowing frequency for a turf with reduced clipping yields [138,139,141,142,144,168,180,330–333,394,459,514,544]. Therefore, the general term "N use efficiency" normally used for yield crops [18] has its limitations for turfgrass management and science.

TNUE can be defined as the amount of N used to reach the acceptable turf quality for a given turf area at a given time period. A golf course putting green obviously requires more N for an acceptable playability than a fairway if the same turfgrass is used since two different mowing heights and mowing frequencies are used for the different levels of maintenance [27,325]. For a putting green, mowing frequency is a daily basis with clipping removal requiring more N input during a growing season. A creeping bentgrass putting green in a northern state of the country has a better N use efficiency than a bermudagrass putting green in a southern state during the summer because of more growth and more N requirement of the bermudagrass and significant removal of clippings. Therefore, based on the example provided, it is difficult to evaluate TNUE of a turfgrass just based on clippings removal.

The TNUE has been studied for many years. Approaches include N uptake efficiency, N recovery in different tissues (typically in clippings) [223,224], N losses from the system, and total N budget for a given turf [3,36,46,51,61,68,128,137,171,173,178,179,185,202,209,210,233,237, 243,244,275,276,286,288,307,309,342,346,350,372,379,443–445,460,462,477,480,525].

Turfgrasses differ at both cultivar and species levels in N uptake and N recovery in tissues, which provide potentials for future improvement for turfgrass N use [79–81,101,223,224,226–240, 463].

The improvement for TNUE challenges both the biological improvement and the environmental protection. The approaches of reduced total N input, minimized N losses, controlled released N fertilizers, and returned clippings [223,224,379,494] to the turf have been commonly used to enhance turfgrass N use efficiency in addition to genetic and breeding improvements in turfgrass N use [66,101,287].

33.5 TURFGRASS NITROGEN USE UNDER STRESSES

33.5.1 BIOTICAL STRESSES

33.5.1.1 Weeds

Weeds can take advantages of a turf with a better resistance to stresses including N and other nutrients [121,145,326,493,522]. The combination of proper nutrient balance associated with other cultural practices ensures long-term turf vigor, even weeds free turf [77,115,159,328,522]. It is very common that weed invasion happens when a turf is under stress or at the early stage of establishment [115,145,493,522]. Documented research is limited on competitions between turfgrass species and cultivars and weed species; and among multiple turfgrass and weed species on N forms, supplies, and sources. However, under N deficiency, legumes such as clovers (*Trifolium* spp.), black medic (*Medicago lupulina* L.), common lespedeza (*Kummerowia striata* [Thunb.]), along with grassy weeds such as bromesedge (*Andropogon virginicus* L.), quackgrass (*Agropyron repens* L.), bahiagrass (*Paspalum notatum* L.), centepedegrass (*Eremochloa ophiuroides* [Munro] Hack.) become troubling weeds for the turf (Table 33.5). These weeds do not need or require very little N input and can survive well under N stress [212].

TABLE 33.5

Weed Occurrence Potential under N Deficiency or Excessiveness for Turf

Under Severe Nitrogen-Deficiency Stress
Black medic (*Medicago lupulina* L.)
White clover (*Trifolium repens* L.)
Lespedeza (*Kummerowia striata* [Thunb.])
Other legumes

Under Moderate to Slight Nitrogen-Deficiency Stress
Carpet weed (*Mollugo verticillata* L.)
Crabgrasses (*Digitaria* spp.)
Common Chickweed (*Stellaria media* [L.] Cyrillo.)
Centipedegrass (*Eremochloa ophiuroides* [Munro] Hack.)
Bahiagrass (*Paspalum notatum* L.)
Quackgrass (*Agropyron repens* L.)
Bromesedge (*Andropogon virginicus* L.)
Yellow woodsorrel (*Oxalis stricta* L.)

Under Optimum to Excessive Nitrogen Supply
Bentgrasses (*Agrostis* spp.)
Bermudagrasses (*Cynodon* spp.)
Annual bluegrass (*Poa annua* L.)

Under N deficiency, turfgrasses lose their vigor to compete with weeds and become worse under stresses of drought, nutrient deficiency, other pests, and traffic impacts [115,121,159]. With weed existence, the actual available N to the turfgrass will be reduced depending on the degree of weed coverage. Most likely the worse the weed coverage, the worse the N stress to the turf in addition to the differences between the turfgrass and weed N uptake capabilities [358]. Some broadleaf weeds may have further advantages with deeper tap root systems for N acquisition than grasses having relatively shorter fibrous root systems. Although there is a lack of research, it is not difficult to predict that some weeds may have advantages to absorb N from soils at very low concentrations with a high-affinity transport system (HATS) functions when N ions are present in low concentrations (<0.5 mM) with very low Km values [322,474], where some turfgrass species may already stop to absorb N. Furthermore, recent studies by Hossain et al. [212] and Sistani et al. [446] showed much aggressive N and phosphorus (P) uptake by grassy weeds of torpedograss (*Panicum repens* L.) and southern crabgrass (*Digitaria ciliaris* [Retz.] Koel.). The capability to accumulate N in the green tissues also further indicates a high N uptake efficiency by crabgrass species [471].

The optimum to excessive N supplies encourage several weed species occurrence including annual bluegrass (*Poa annua* L.), bentgrasses (*Agrostis* spp.), bermudagrasses (*Cynodon* spp.), ryegrasses (*Lolium* spp.), and crabgrasses (*Digitaria* spp.) plus several annual broadleaf weeds (Table 33.5). Encroachments between turfgrass species are also encouraged by favorable nutrient supply for the more aggressive species in addition to other conditions and their own characteristics. For example, the more commonly observed encroachment is the surrounding bermudagrass invasion to creeping bentgrass putting greens and however, the opposite invasion was observed partially due to the aggressive N fertilization program to the creeping bentgrass green versus the N stressed bermudagrass surrounded. Excessive N supply to turf promotes shoot, stolon, and rhizome growth with relative smaller root systems, less carbohydrate reserves, more succulent tissues, and more vulnerable to some diseases and insects, which contribute to a weakened vigor for the turfgrass to compete with weeds.

Turfgrasses also can lose their competitiveness with weeds under excessive N supplies. At high external N ion concentrations (>1.0 mM), the Vmax plateau of ion uptake kinetics [322] that is exceeded as a low-affinity transport system (LATS) becomes functional. Therefore, the uptake rates increase in a linear function with ambient nutrient concentrations (between 5 and 100 mM NO_3^-) and exhibit no evidence of saturation kinetics [224]. LATS for NO_3^- influx maybe NH_4^+ influx as well are inducible among plants [146,224]. Some weeds have superior LATS systems in absorbing N ions than turfgrasses when excessive N is applied. Mowing practice adds complexity for N effects on turfgrass and weed competitions [182,199,255,257,285,305,324,343,358,388,491]. Although both turfgrasses and weeds are normally encouraged by N input, proper mowing practices enhance turf quality and suppress weeds more significantly, particularly, for broadleaf weeds [115] depending on turfgrass species and weed species.

Turf thatch accumulation is encouraged by optimum to heavy N applications [27,95,325]. Thatch normally provides a physical barrier from weed invasion particularly seed germination [349,439]. Under proper N supply, thatch functions positively to suppress weed invasion. A turf becomes more vulnerable to weed invasion after aerifiactions [27,325]. Weeds germinate much easier on bared soils than on a turf. Some turfgrass species may naturally have allelopathic effects to weeds by producing toxins or having aggressive growing habits. However, there is a lack of evidence on N roles for allelophathic effects of turfgrasses and the relationship between N supply and enhancement of allelophathy.

Nitrogen sources and forms also have ecological impacts on turf and weed competition [62,120,522]. In general, annual summer grassy weeds will be encouraged with inorganic and quick release N fertilizers. Perennial grassy weeds with the closest similarity to turfgrasses are more difficult to control by cultural practices. Nitrogen forms themselves (NO_3^-, NH_4^+, urea) may not be as significant as the chemical properties of N carriers for the consequences to the turfgrass and weed competition. Ammonium type N fertilizers have been encouraged to use with a potential to reduce several patch diseases for turfgrasses. An extended period of application of the same ammonium type fertilizer may lead to a lower soil pH with reduced available P; under such a soil condition, annual bluegrass as a weed is discouraged since it needs more P and a neutral soil pH than the desired creeping bentgrass [120,264].

Overall, N itself cannot serve as a remedy mean, neither other nutrients, to control weeds effectively for turf but it plays an extremely important role associated other cultural practices to keep a strong turf vigor for healthy conditions to minimize weed invasion. Corn gluten meal has been used for both grassy and broadleaf weed control by suppressing weed seedling germination [335,339]. Corn gluten meal on average contains 60% protein and 10% N by weight and also functions as N fertilizers [332,335]. The amino acids as dipeptide forms in corn gluten meal mainly inhibit new weed seedling root germination and growth with negligible phototoxic effects on the existing turf. Excessive ammonium supply can be toxic to plants by interrupting cell membrane pH gradients and reduce membrane ion transportation [65]. The sensibility to NH_4^+ toxicity between turfgrasses and weeds deserves more attention [64].

The future N-related research enlightens areas such as turf–weed ecology with a better understanding of N uptake efficiency and N nutrition physiology between turfgrasses and weeds; impacts of N application rates, N forms and sources, and time of application to the turf and weed competition; and helpful N managements for turfgrasses when they are under weed stress.

33.5.1.2 Diseases

Nitrogen form, source, rate, and time of application play important roles in turfgrass diseases [82,117,146,245,263,503]. The relationships between N application practice and disease severity and disease occurrence among turfgrasses are rather complicated by the existence of host vertical and horizontal resistances to diseases [115,145,448,503], epidemiology of different diseases, coexistence of multiple diseases, and different host recovery mechanisms [40,145,448,503]. In general, two groups of turf diseases are affected by N application rate: disease severity increased and disease severity reduced, plus a group of diseases not affected by N application rate or lack of information (Table 33.6).

TABLE 33.6
Effects of Nitrogen Application on Turfgrass Diseases

Disease Severity Decreased by Nitrogen Fertilization (Most Common Foliar Diseases)
Anthracnose (*Colletotrichum graminicola* [Ces.] G. W. Wils.)
Dollar spot (*Sclerotinia homoeocarpa* F. T. Bennett)
Fairy ring (*Basidiomycetes*)
Leaf spot (*Drechslera* spp.)
Melting-out (*Drechslera* and *Marielliottia*)
Necrotic patch (*Ophiosphaerella korrae* [J. Walker and A.M Sm.])
Pink patch (*Limonomyces roseipellis* Stalpers and Loerakker)
Red thread (*Laetisaria fuciformis* [McAlpine] Burdsall)
Rusts (*Puccinia* spp.)
Take-all patch (*Gaeumannomyces graminis* [Sacc.] Arx and D. Oliver var. Avenae)

Disease Severity Unaffected by Nitrogen Fertilization or Lack of Information
A group of fungous pathogenic diseases
Bacterial wilts
Nematodes (orders of *Triplonchida*, *Dorylaimida*, and *Tylenchida*)
Slime molds (*Myxomycete*)
Viral diseases

Disease Severity Increased by Nitrogen Fertilization
Brown patch (*Rhizoctonia solani* Kühn)
Gray leaf spot (*Pyricularia grisea* [Cooke] Sacc.)
Microdochium patch (*Monographella nivalis* [Schaffnit])
Pythium blight (*Pythium aphanidermatum* [Edson Fitzp.])
Spring dead spot (*Ophiosphaerella* spp.)
Stripe smuts (*Ustilago* spp.)
Summer patch (*Magnaporthe poae* Landschoot and Jackson)
Typhula blight (*Typhula incarnata* Fr.)

Although limited information is available on mechanisms of reduced severity of diseases by N application, turfgrasses have been reported [82,117,146,245,263,503] either with improved resistance or speedy recovery of foliar diseases including dollar spot (*Sclerotinia homoeocarpa* F. T. Bennett), leaf spot (*Drechslera* spp.), red thread (*Laetisaria fuciformis* [McAlpine] Burdsall), rusts (*Puccinia* spp.), and foliar anthracnose (*Colletotrichum graminicola* [Ces.] G. W. Wils.). Recent studies indicate foliar N application can even more efficiently reduce or prevent these foliar diseases [283].

The worst group of turfgrass diseases are soil-borne crown diseases including pythium blight (*Pythium aphanidermatum* [Edson Fitzp.]), brown patch (*Rhizoctonia solani* Kühn), gray snow molds (*Typhula incarnata* Fr.), gray leaf spot (*Pyricularia grisea* [Cooke] Sacc.), dead spot (*Ophiosphaerella agrostis*), and spring dead spot (*Ophiosphaerella* spp.), which are easily induced or worsened by heavy N rates, particularly when inorganic quick-release N fertilizers are used [41,146,189,207,245,262,352,484,487,488,497,508,521,527]. Excessive application of N promotes thinner cell walls, more succulent tissues, and less carbohydrate reserves, which encourage pathogen's penetration to the host plants [119,448,503].

The influence of the two major N forms of NO_3^- and NH_4^+ to crop diseases have been intensively studied and the two N forms normally have opposite effects to particular diseases [221]. The influences of N form on disease severity include impacts to the host, pathogen, and the soil environment [221]. The effect of specific N form on disease severity depends on many factors [221]. Susceptible cultivars or species are easily affected by N forms, while resistant turfgrasses are not affected by diseases regardless of N forms [221].

Practices of N application to turf are rather more important than N nutrient itself for encouragement or discouragement of turf diseases. Improper application time, rate, and N form and sources will promote disease occurrence [27,119,325,448,503]. For example, N fertilizers effect soil pH and some turfgrass diseases are soil pH sensitive [103,204,473]. Alkaline soil conditions promote take-all patch (*Gaeumannomyces graminis* [Sacc.] Arx & D. Oliver var. avenae) and slight acidic soils suppress most of the patch diseases [103,198,203,447]. There is a lack of evidence of N forms on the aggressiveness of turfgrass pathogens.

Limited information is available for the interaction between N application and nonfungal pathogenic turf diseases such as nematodes [417], viral diseases, and bacterial wilts for turfgrasses. However, recent findings of the other symbiotic relationship between host and endophyte may provide another possible dollar spot control method [104].

Although short of quantitative data, golf course putting greens have the highest disease potential whether using creeping bentgrass, bermudagrass, or zoysiagrass because of the lowest mowing height and intensive traffic in comparison with any other types of turf [27,119,325]. Nitrogen plays important roles in physical damage recovery to turf. Typically, golf course tees and sports fields receive more damages than golf putting greens but tees and sports fields can recover quicker than greens due to the higher mowing height and much less pressure for diseases [27,325]. The better reserves in roots, crowns, and other parts of higher mowing turf benefit a quicker recovery from damages with a relatively less N requirement comparing with a lower and more frequently mowed turf [27,325]. Owing to the controversial responses to N for turf diseases, the zone for optimum N application is very restricted for most turfgrasses. In the zone, there is still a group of diseases independent to N status and supplies or short of information (Figure 33.2).

Other mineral nutrient management has been applied to reduce crop and turfgrass diseases or as a remedy to cure crop or turfgrass diseases including copper, manganese, silicon, and potassium by either enhancing turf resistance to pathogens or by suppressing pathogen infections [116]. However, N itself may never be an efficient tool to cure turfgrass diseases directly but its complicated roles to different diseases deserves much more attention for management and future turfgrass improvement [38,40,321]. For example, a selected genetic line of creeping bentgrass with natural resistance

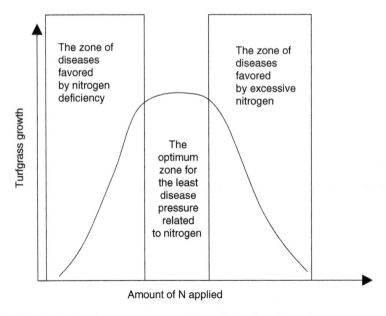

FIGURE 33.2 The relationship between amount of N applied and turfgrass-disease occurrence and *y*-axis represents turfgrass growth.

to dollar spot may also have greater capability to absorb N efficiently by roots and leaves. Nitrogen effects to nonpathogenic microbial activities exist and these effects may have an influence to the turf quality [133,191,310,427].

33.5.1.3 Insects and Mites

Insects and mites require high N dietary for their need for amino acids and proteins. Normally, high N contents in plants benefit insect and mite pests [14,33,192,323,340,432,475]. High N fertility levels result in rapid growth and succulent tissues, which may also increase the chance of damages from insect and mite pests [60,111,145,383,429,512,522]. Excessive N fertilization associated with improper mowing, watering, and fungicide application encourage accumulation of thatch layer in some turfgrasses [27,325]. A thick layer of thatch is a good habitat for chinch bugs (*Blissus* spp.), billbugs (*Sphenophorus* spp.), sod webworms (*Crambus* spp.), two-lined spittlebug (*Prosapia bicincta* [Say]) and several other insects and mites [31,115,383,512]. Heavy N fertilization of turfgrasses stimulates rapid leaf and shoot growth [95] with the costs of root growth and the relatively reduced root growth [491] and development. Under such a situation, the turfgrass will have to overcome more severe stress when the environmental conditions are not favorable for roots such as the summer time for cool-season turfgrasses and cool springtime for warm-season turfgrasses [27,329]. In addition, the turfgrasses with poor root systems will suffer more severe damages from subsurface feeders [383,512]. However, a slightly higher N supply may encourage foliar growth and speedy recovery from foliar feeders' damages.

Endophyte infections with some turfgrasses including *Fescue, Lolium* species provide one of the most important natural resistance to surface feeders—maybe subsurface feeders [165,371]—and N deficiency to the host turfgrasses can discourage the symbiotic relationship since most of the alkaloids produced between the endophyte and hosts are N-dependent secondary N metabolites [14,201,261,419].

There is a lack of information on N form influences in turf insect and mite pests including earthworms as pests. However, recent reports indicate the black turfgrass ataenius (*Ataenius spretulus* [Haldeman]) density was encouraged by using N fertilizers as sewage sludge [384] but had no effects when manure-rich N fertilizers were used [382]. Turfgrass species with resistance to turf insects and mites vary at both the species and cultivar levels [13,55–59,165,200,211,384,403,406, 409,442,531,532] and the resistances include speedy recovery, which is highly related to the efficiency of nutrient acquisition and production of the N related metabolites [14,201,261,419].

Multiple-factor ecosystems including host N status, pest, and natural enemy are complicated and interesting since plant N status can affect insect consumption rates and population dynamics of herbivorous insects and mites [14,33,115,192,323,340,432,475]. The changes of such ecosystems depend heavily on carbon deposit in addition to the plant N statuses [323,340]. For a turfgrass ecosystem, multiple factors are involved including consistent disturbing of mowing and traffic on the insect and mite pests.

Development of insect-resistant turfgrass species or cultivars is promising since diversified resistances to insect and mite pests exist among turfgrass species [13,55–59,200,403]. The resistant turfgrass species or cultivars need sufficient N supply to produce enough toxins, which are most likely the case for the insect and mite resistance, and unhealthy turfgrasses either due to N deficiency or excessiveness will significantly lose the resistance.

33.5.2 ENVIRONMENTAL STRESSES

33.5.2.1 Water

Water and N are the most important controlling factors for turfgrass growth, development, and performance [25,27,95,159,325]. Soil-moisture conditions directly affect N availability, movement, and absorption by turfgrasses [1,92,93,159,215,227,228,344,397–399,519]. Adequate water supplies in soils maximize root N absorption and generate healthy turf. When turfgrasses are under water stresses of drought or waterlogging conditions, N uptake and use are significantly affected [115,159]. The unhealthy relationships for turfgrasses between water and N can be summarized as the followings due to improper N management or restrictions:

- Drought with excessive N input
- Drought under N deficiency
- Waterlogging with excessive N input
- Waterlogging under N deficiency

When a turf is under water stress, reduced N and other fertilizer input can alleviate drought stress, and excessive N input will worsen the situation due to increased osmotic stresses in soils [95]. Turfgrasses under drought stresses still need fertility input but in a reduced rate. The degree of reduction of N rate will depend on the sereneness of drought, turfgrass used, and other environmental conditions. Drought periods can last for hours to months even in yearly or multiple-year cycles. The N-fertility program needs to be adjusted to meet the optimum N need without further worsening drought stresses. Light and frequent fertilizations are normally recommended for turfgrasses to avoid N deficiency during drought stress times. If the drought stress is severe, the N application is totally avoided [25,27,325,492].

In comparison to longevity of drought, waterlogging situations for turfgrasses may be shorter particularly for sand-based fine turf areas [230]. However, even temporary waterlogging will change the N supply in soils, and denitrification and leaching are encouraged [350]. During the waterlogging period, any N applications to soils seem helpless and the critical time is postwaterlogging with proper N supply to replace the N lost during the waterlogging stress. Extended waterlogging periods more likely result in nutrient deficiency, particularly N deficiency in addition to the root damages. Owing to the weak root systems immediately following a waterlogging period time, light and foliar N application is encouraged with caution of hot temperatures [289].

33.5.2.2 Salinity

Salinity is an important growth-limiting factor for most turfgrasses, which are nonhalophytic plants. Excessive salts in soils inhibit turfgrass growth by osmotic stress, nutritional imbalance, and specific ion toxicity in addition to the structural damages to soils [301,302,356]. Soil salinity can be progressively exacerbated by practices such as irrigation and fertilization, especially in arid regions [94]. The proper use of N fertilization in saline soils is important to sustain N supply to turfgrasses. However, over fertilization with N may contribute to soil salinization and increase the negative effects of soil salinity on turf performance [94].

The two major salinity sources for turfgrasses are natural salinity conditions of soils and irrigation water with high salt contents applied to turfgrasses [94]. The relationship between turfgrass salinity stress and N fertilization deserves much more attention because of the total areas of turf under salinity stress and the complicated soil–chemical interactions among Na^+, Cl^-, NO_3^-, NH_4^+, other N forms, cations, and anions in soils [94]. Turfgrasses demonstrated diversified salinity tolerance [4,6,17,44,45,47,83,84,127,152,183,190,270,311–320,366,375–378,389,393,400,401,464,479], and salinity-tolerant plants seem less affected in N uptake and metabolism than salinity-sensitive plants [94].

Turfgrass responses to salinity may change with mowing height, turf use, seasons, and N form and sources. To date, there is lack of proper recommendation of N fertilizers to either salinity sensitive or tolerant turfgrass species. Studies of turfgrass growth responses to N and soil salinity for different turfgrass species and turf use are important to reveal whether the amount of N applied alleviates or aggravates the detrimental effects of salinity. In addition, examining turfgrass growth during different season associated with other stresses may provide profound information to practitioners to enhance salt tolerance over time such as winter overseeding and association with other pest stresses [83,84]. Under salinity stresses, both NO_3^- and NH_4^+ uptake were affected in tall fescues [44,45]. Ammonium compared to NO_3^- nutrition is known to increase the salinity sensitivity in maize, wheat, and peas [301,356]. Under both N and salinity stress, shoot growth is much reduced than roots [94].

33.5.2.3 Heat

High temperature stress is a major environmental factor limiting turfgrass growth and performance during the summer months in tropical, subtropical, and other areas [25,29,119,159]. Cool-season turfgrasses (C_3) including *Agrostis, Poa, Festuca, Lolium,* and others often experience a summer decline under heat or radiation stresses [74,118,163,195–197,216–220,229,248,264,284,311,373,413,465,507,519, 537–540,545] and the heat stress can last from hours to weeks. High temperatures not only inhibit photosynthesis, limit carbohydrate accumulation, damage cell membranes, cause protein folding, and lead to cell death [214], but also increase N losses from the turf-soil systems such as NH_4^+ volatilization, denitrification N losses, and NO_3^- leaching potential [51,128,209,210,224,477,480]. Therefore, the overall NUE for most of turfgrasses is reduced under heat stress due to reduced N uptake, less available N, and weakened root systems [95,217]. Under high temperature stresses, N uptake and use for cool-season turfgrasses are significantly reduced and fertilization burn potentials also increased [523]. However, all cool-season turfgrasses have the ability to acquire heat tolerance in certain degrees by exposure to a gradual increase in temperature, which happens naturally, also called heat acclimation [119,214]. Extremes to avoid any N input during heat stress tend to further worsen the stress leading to further carbon depletion [164] during the hot summer months. Foliar and light N applications are highly recommended for highly maintained turf including golf-course putting greens [164] (Tables 33.3 and 33.4).

Heat-shock proteins (HSP), hormone regulations, and antioxidant enzymes [228,248,507] are all directly or indirectly associated with proper N supply and the N supply in enhancing these heat-resistance mechanisms need to be further investigated for turfgrasses.

A recent study reported that corn (*Zea mays* L.) was treated under different N levels and then exposed to acute heat stress in result of high-N plants with greater amounts of mitochondrial HSP and the chloroplastic HSP per unit protein that their low-N counterparts. The patterns of HSP production were related to photosynthetic system II efficiency [423]. The results indicate that N availability influences HSP production in higher plants suggesting that HSP production might be resource-limited, and that among other benefits, chloroplast HSPs may limit damage to PSII function during heat stress [423]. However, a similar result needs to be tested for C3 grasses. There is lack of information on N forms related to heat stress for turfgrasses. A recent study reported that tomato plants fed with NH_4^+ as the N source show higher tolerance to heat stress (35°C) than plants fed with NO_3^- [413].

33.5.2.4 Cold

It has been well documented that increased cold acclimation could improve freeze tolerance of turfgrasses and other crops [8–12,19,78,90,105,112,122,123,130,158,160,161,170,222,246,258,304, 336,337,341,353,354,377,387,404,408,418,426,430–433,472,543,546], but the rate and level of cold acclimation primarily affected by temperature are also influenced by factors such as light intensity, day length, cultural practices, and other abiotic stresses such as drought and salinity [115,159]. Turfgrasses possess various adaptive mechanisms for surviving freezing temperatures, such as increases in certain sugars or amino acids, synthesis of novel proteins, and the degree of unsaturation of membrane lipid fatty acids [546]. Cold stresses for turfgrasses include winter month N management for both cool- and warm-season turfgrasses [25,27,159,325,492]. When a soil is colder than optimum temperatures for root activities, N uptake is significantly reduced and the demand for N is to the minimum and plant growth ceases. However, turfgrasses are perennial crops and winter green color is ideal. Several reports stated that late-season N improved fall and spring color of bermudagrass and had little effect on total nonstructural carbohydrate (TNC) levels without any negative effects on postdormancy recovery in the spring [353,354,408,426]. However, aggressive late-fall-season N applications to turfgrasses may increase disease potentials including spring dead spot for bermudagrass and gray snow molds for creeping bentgrasses and winter kill potential for warm-season turfgrasses [108,109,448].

By observations for two winters, foliar and lower rate N enhanced turf quality of overseeded rough bluegrass on bermudagrasses than 100% granular N fertilizers [Liu, unpublished data]. Warm-season turfgrasses possess various N-rich defensive metabolites and enzymes to cope with low temperature stress including some proteins. "Midiron" bermudagrass during cold acclimation was correlated with its superior freezing tolerance compared to "Tifway" [170] in protein contents [413,479], and cultivars with greater stolon proline content exhibited greater freezing tolerance than those with less proline during the winter [170,543]. Zhang et al. [543] compared "Riviera" (cold-tolerant) and "Princess-77" (cold-sensitive) bermudagrasses selected and either subjected to cold acclimation at 8/4°C (day/night) with a light intensity of 200 µmol m^{-2} s^{-1} over a 10 h photoperiod for 21 days or maintained at 25/23°C (day/night) with natural sunlight in a glasshouse. Cold acclimation induced accumulation of sugars, proline, and TNC in both cultivars but only protein accumulation was found in Riviera not in Princess-77. Superoxide dismutase (SOD) [187,246] increased during the first 7 days and then declined, while catalase (CAT) and ascorbate peroxidase (APX) activity decreased in response to cold acclimation in both cultivars. Significant correlations of LT50 with sugars, proline, protein, CAT, and APX were obtained in Riviera, but only with proline and the antioxidant enzymes in Princess-77. These results suggest screening cold-tolerant cultivars among bermudargasses with rapid accumulation of C- and N-rich compounds during cold acclimation is possible. However, the direct N nutrient impacts were not included in these reports and future research is needed. Numerous reports have stated that unsaturated membrane lipids play a significant role in both chilling and freezing tolerance with little effects or no effects with saturated fatty acids [112,214]. There are no direct evidences of N-nutrient levels in relation to saturation and unsaturation of membrane lipids; even transgenic plants have further approved the importance of unstaurated membrane lipids [216,222].

33.5.2.5 Shade

Reducing N has been recommended for both warm- and cool-season turfgrasses under shade stress because of a low ratio of photosynthesis:respiration, which reduces turf growth and divot recovery [115,157]. Selecting shade-tolerant species and adjusting management techniques are the two approaches commonly recommended for maintaining turf in reduced-light conditions [23,24,26,32, 70–74,76,126,139,174,226,362,390–392,405,457,468,470,504]. Raising mowing height to alleviate shade stress is commonly recommended [25,27,115,159,325,492].

One management technique to improve turf in shaded conditions is the application of plant-growth retardants (PGRs), particularly trinexapac-ethyl, which inhibits gibberellic acid biosynthesis [139,391,459]. In addition to reducing clipping yields, multiple applications of PGRs increased turf density, color, and quality of both cool- and warm-season turfgrasses [330–333,401,459,514,544]. In addition, increased tillering and density of turf mowed at golf-green height under shade were found for both warm- and cool-season turfgrasses, but turf quality was still unsatisfactory due to low irradiance for most warm-season turfgrasses, particularly hybrid bermudagrasses, when the light reduction is beyond 60% [70–73].

Reduced fertility, particularly N, has long been a recommended practice for maintaining turf in shade [115,126,159]. Goss et al. [180] compared liquid applications of N to confirm that lower N rates (150–185 kg ha^{-1} annually) resulted in better quality turf than higher N rates (212–235 kg ha^{-1}). Granular forms of N are absorbed through the roots and transported to the shoots of the plant: a process that could be energy inefficient by forcing roots to use their carbohydrates for energy to assimilate and transport N to the shoots [223,224,322]. In a shaded environment, turfgrass root development and energy budgets are stressed due to low photosynthetic-active radiation. Since the majority of N is utilized in the shoots, increased foliar absorption from liquid applications of N may increase N-use efficiency and allow more photosynthate to be allocated to the roots of the plant, enabling the turfgrass to attain more nutrients and water [159]. Although research reports indicate that 100% liquid N and a combination of liquid N and granular N helped for improved turf quality [481–483], comparisons between granular and liquid N applications, forms of N on shaded turf have been rarely reported.

33.5.2.6 Acid Soils

Plants and turfgrasses differ in their acidsoil and aluminum (Al) tolerance [this book; 16,129,150, 151,249–253,260,277,278,280–282,322,355]. The available information on Al resistance among turfgrasses [16,129,150,151,260,277–279,282,322,355] emphasizes that potential enhancement of Al resistant in turfgrasses is promising. The newly released tall fescue cultivar with Al resistance will have great market value as a transition-zone turfgrass [129], where soil acidity and drought stress are dual constraints. More studies are needed to further understand the potential and value Al resistance in other turfgrasses. There is lack of N-form impacts on turfgrass acid soil and Al tolerance although ammonium type fertilizers will worsen the stress by lowering soil pH [95,194,322]. With poor root growth under acid soil and Al stress, foliar N application may be beneficial for the turfgrass to obtain needed N to survive.

Low soil pH and Mn have a suppression effect on patch diseases (*Magnoporthe poae Landshoot and Jackson; Gaeumannomyces gramini; Ophiosphaerella herpotricha; Leptosphaeria* spp.) for both cool- and warm-season turfgrasses [198]. However, whether Al plays a toxic role to pathogens or the turfgrasses studied with a better adaptability [281] with a slightly acid soil (pH 5.2–5.6) was not clear. Winterkill of bermudagrass associated with low soil pH in the upper transition zone have been observed by turfgrass scientists and superintendents in different locations, but studies are needed to identify the specific low soil pH effects on winterkill in turfgrasses such as decreased roots, rhizome and stolon production, which is one of the symptoms found associated with winterkill problem [159].

Tremendous research effort has been contributed to the two key elements: Ca and P for reducing Al toxicity in plants [249–253]. In cells, calcium mainly competes with Al^{n+} for the binding sites and phosphorus in phosphate form precipitates Al^{3+} to nontoxic forms [249–253]. Silicon and boron have been reported to reduce Al toxicity [249–253] in different plants. (For further information, refer to Chapter 23 of this book.)

33.5.2.7 Imbalanced Nutrients

Nitrogen is available mainly in two major forms as NO_3^- and NH_4^+ and both can interact with other soil cations and anions [15,194,322]. Optimum K, Mg, and Fe supplies enhance N uptake but excessive cations and anions will slow N absorption and even cause nutrient imbalance [194,322]. More research is needed for N/K, N/Ca, N/S, N/Mg, N/other micronutrients for turfgrasses, and imbalanced nutrient soils can be caused by low soil pH, high-leaching potential, excessive lime application, and improper management of soils. Under imbalanced nutrient stress, foliar-applied N may be applied to correct N deficiency [289,481–483]. To date the solo N form turfgrass has not been identified and excessive NH_4^+ as the solo N source can be toxic to plants [224].

33.5.2.8 Traffic Stress

Traffic causes wear injury to turfgrass aboveground parts and soil compaction with a weakened root system [25,325]. Wear injury can be physical abrasion and tearing of aboveground turfgrass parts [25,91,138]. Increased N application increases aboveground biomass with greater shoot density and faster growth [25] and promotes recovery from traffic damages [87–89,138]. Higher N rates during establishment increase tiller numbers and shoot density [25,89]. Severe traffic damages turf plants [435–440] and a moderate N fertilization substantially aid in recovery [88,89,138,485]. Canaway [88,89] reported that established perennial ryegrass plots, receiving high annual N rates (>290 kg ha^{-1}), deteriorated faster under artificial wear than those receiving moderate levels (100–225 kg ha^{-1}) due to more succulent tissues with high levels of N supply. However, there were no differences in wear tolerance or recovery of Kentucky bluegrass that received annual N rates of 96 or 192 kg ha^{-1} [91]. Compacted soils will lose space for air and water and the root systems in such compacted soils will lose vigor for efficient nutrient uptake. In addition, compacted soils will contribute more for nutrient runoff losses due to the reduced soil–water percolation rates [275,276,492]. The alleviation of traffic stress is the primary remedy for the situation and N can function as an essential recovery element.

33.5.2.9 Cultivation

All types of turf require cultivation practices to improve soil compaction, reduce thatch accumulation [25,27,325,492]. The frequency, intensity, and type of cultivation heavily depend on the turf use, turfgrass species and cultivar, and the environmental conditions [27,325]. Golf-course putting greens, tees, and fairways require at least once a year of hollow-tine core cultivation (in most cases multiple cultivations per year) plus other types of cultivation to improve the soil and root-zone conditions and reduce thatch accumulation [27,225,325,362,381,421,534]. Sports fields and lawns are also recommended at least one cultivation per year [25,100,385,386,420,492]. Nitrogen and other nutrients play a significant role to help turfgrass recover from the temporary disruption caused by cultivation [27,95,325]. Turfgrass cultivation includes hollow tines, solid tines, slicers, groomers, hydrojet (high-pressure water beams vertically penetrating turf soil profile), and vertical cutting. These processes temporarily stress the turf and proper nutrient supplies help the turf recovery. The soil cultivation with opening holes provides better contacts between fertilizers and roots, and controlled release of N fertilizers are normally recommended. However, if cultivation is conducted under stresses, more caution is needed with proper N supply. Light and frequent N applications in many such cases help the turf growth without uneven growth [25,325]. Proper N application can minimize the time requirement to resume the playability for golf courses and sports fields. Nitrogen deficiency slows the recovery from cultivation.

33.5.2.10 Mowing

Proper mowing does not generate severe stress to a turf but it consistently removes parts of the green tissues for photosynthesis and loses N and other nutrients in clippings [25,27,325,492], and clipping returning can save N fertilization as much as 50% [199,255,257,285,343,388]. Tables 33.3 and 33.4 show more N requirements for putting greens although they are mowed at lowest heights because of daily mowing with clipping removal [27,325]. The immediate following mowing after a either granular or foliar N fertilization will remove parts of N just applied, and sometimes it can be significant [305]. Mowing under heat stress weakens the turf by increasing ET and water stress, and under such stresses N applications should be avoided. The relationship between mowing cuts and leaf conditions in responding N fertilizers, particularly foliar N requires more attention [289,481–483].

33.6 ENHANCING TURFGRASS NITROGEN USE UNDER STRESSES

33.6.1 Enhance Nitrogen Absorption

Nitrogen can be absorbed by roots and above ground tissues (Figure 33.1). The best turfgrass management practice to enhance N absorption during a stressful period is to raise mowing height to encourage shoot and root growth in addition to the efforts to release or improve the stressful condition. Nitrogen dynamics are complicated in a turf environment and sound N management programs (Table 33.1) require the management of all aspects of N input, out, and cycling to reach the ultimate goal of high-quality turf with minimum N input and impacts to the environment. Nitrogen-absorption efficiency can be improved by:

- Providing optimum water supply to the soil
- Reducing salinity, nutrient imbalance, shade, temperature extremes, and other stresses
- Using frequent and light rate N fertilization
- Improving root growth
- Choosing proper application time and methods
- Selecting high N–use efficient turfgrass: (1) high efficiency in N uptake at lower N concentrations (HATs for NO_3^- and NH_4^+) [224]; (2) high efficiencies in nitrate reductase and nitrite reductase activities [234]; and (3) high efficiency of urease activity to convert foliar absorbed urea to ammonium before N assimilation [289]

The general N management strategies under different stressful conditions are listed in Table 33.7. The key is to incorporate other factors for stress release, with N as an enhancing factor to regain the turf vigor.

33.6.2 BALANCE NH_4^+ AND NO_3^- AND UREA TYPES OF FERTILIZERS

Turfgrasses can obtain three N forms under different conditions depending on the availability and environmental conditions [223,224]. Ammonium-type fertilizers should be reduced or avoided when a soil pH is already low and under salinity stress [301,356], however, ammonium fertilizers should be encouraged when patch disease pressure is high [103,198,203,447]. When a foliar fertilization is chosen, urea should be the main N form applied [48–49]. There is lack of information on urease activity [322,533], it is also reported inducible in several other plants and the enzyme activity is crucial for foliar-absorbed urea becoming available forms through urea hydrolysis to ammonium as the ready form for N assimilation in leaves [322,533]. Root-absorbed urea is very limited and further research is needed.

33.6.3 USE CONTROLLED-RELEASE FERTILIZERS

The main advantages of using controlled-release N fertilizers include the controlled rate of immediate available N to plants and saving labor from frequent applications [95,194]; increasing longevity of N availability in soils and long-term controlled turfgrass response to N [95,492]; reducing N loss potentials from the turfgrass-soil systems [106,135,156,224,256,303,306,308,456]. The controlled N forms include slowly soluble types such as urea formaldehyde (UF), isobutylidine diurea (IBDU); slow-release such as sulfur-coated urea (SCU), polymer-coated urea (PCU), and sulfur–polymer coated urea (PSCU), and natural organic types (Milorganite) [95,322,492]. When turfgrasses are under stresses such as under temperature extremes, controlled-release N fertilizers benefit the turfgrasses more than quick-release fertilizers [115,159]. During the slow growing season, controlled N fertilizers serve the purpose to provide N needs and save labors from frequent applications. Other controlled methods applied to turfgrasses use types of inhibitors to control the reactions related to N such as urease inhibitors and nitrification inhibitors [18].

33.6.4 USE AND BALANCE FOLIAR AND GRANULAR N FERTILIZERS

Turfgrasses have unique advantages to absorb foliar fertilizers because of the dense and short canopy, relatively thin cuticle layers, and the intensive fibrous root systems than any other crops [25,27,100,325,492]. Higher-canopy crops including woody plants tend to have unstable leaf movement (during foliar application), thick cuticle layers, and unfavorable leaf orientation for efficient foliar N absorptions. The foliar absorbed N forms following such an order in absorption efficiency Urea $> NH_4^+ > NO_3^-$ [48–50]. Spraying techniques also affect the foliar-absorption efficiency by the size of nuzzles, pressure, and volume of the N-fertilizer carriers.

When radiolabeled isotopes first became available in the 1950s for research, scientists already started to track the chemical movement in parts of plant based on the labeled element's radioactivity. The most commonly labeled plant nutrient has been ^{15}N, which has a much longer half-life and minimum hazardous impact in comparison with ^{13}N, with a half-life of only 7 min [35,54,175]. In addition to N, K, Fe, Mg, and Ca have been applied as liquid fertilizers to turfgrasses [95]. The following list provides general characteristics of foliar fertilization [157,228,322,410,424,458]:

- Younger leaves have better foliar absorption.
- Lower (underneath) leaf surfaces (with more stomata) absorb more nutrients than upper side of the leaf.
- Neutral ion absorption seems more efficient than cation (positively charged ions) and anion (negatively charged ions) absorption.
- All 16 plant nutrients including some beneficial elements have been reported to be absorbed by leaves.
- Cuticle penetration is possible and is genetically regulated.

Studies and reviews on crops and both cool- and warm-season turfgrasses indicate a foliar-absorption rate normally between 30 and 60% of the nitrogen applied [48–50,131,132,157,164, 241,289,322,327,410,412,424,428,453,458,481–483,523–525,533,535,536,541]. For P and K, absorption efficiencies are even lower, between 20 and 30% [322,536]. So another question arises: Where does the rest of the liquid fertilizer go? The rest may be left in the soil and the turf–soil system, lost by removal of clippings, or held in the thatch layer similar to a granular fertilizer. However, the unabsorbed liquid fertilizer in the turf–soil system is still available to turfgrasses simply because it does not require a process to be dissolved in the soil solution before being absorbed by roots.

33.6.5 PLANT GROWTH REGULATORS AND BIOSTIMULANTS

PGR and biostimulants are beneficial to release stresses for turfgrasses [159,325,492]. Proper use of N fertilization and combination with PGRs will further enhance turf quality, color, and resistance to stresses [330–333,394,459,514,544]. The benefits of PGR and biostimulants for turfgrasses are summarized as follows:

- Enhanced shade tolerance for both cool- and warm-season turfgrasses [330–333,401,459, 514,544]
- Enhanced cold tolerance for warm-season turfgrasses [354]
- Enhanced turf quality, rooting, salinity resistance [17,330–333]
- Enhanced heat tolerance [330]

Nitrogen plays a significant role in plant hormone metabolisms [159,466], and N deficiency affects plant hormone production and influences turfgrass performance. Turfgrass PGRs work best to enhance turf quality under a moderate N application. Severe N deficiencies can lose the beneficial functions of PGRs and result in phyto-toxic effects [330–333].

33.6.6 TURFGRASS IMPROVEMENT FOR NITROGEN USE

Turfgrass improvement for N-use efficiency is promising due to the genetic diversities among turfgrass species and cultivars (Table 33.1) and the technology availability [21,22,38,39,63,96,213,

TABLE 33.7

General Recommendations and N Uses under Stresses for Turfgrasses

Stresses	General Management Recommendations to Reduce Stresses	Nitrogen Recommendations	Updated Evidence for Genetic Improvements (Turfgrasses)	References
		Biotic Stresses		
Weeds	Sound cultural control plus necessary chemical control and biological control is promising	Increased N input will alleviate weed pressure associated with other weed controls since few weeds can compete with most of host turfgrass species	Does not exist yet	[77,145,326,338,493,498]
Diseases	Sound cultural controls include resistant species or cultivars plus necessary chemical and biological control is difficult but with possibilities. Mineral nutrients and other beneficial elements function as remedies	Reduced N input is recommended, although some foliar diseases can be alleviated by adding more N, but the consequences may be more costs of fungicides and control with other more serious diseases. The key is to keep optimum N supply with a slight N stress keeping a caution of soil pH and N fertilizer effects on soil pH	Exist and promising	[39,41,75,82,85,97,103,104,107–110,116, 117,146,169,188,189,198,203,204,207, 245,262,263,265,271,287,345,352, 359–361,447,469,473,476,484,487,488, 495–497,499,500,508–510,521,527–529]
Insects and mites	Sound cultural control including resistant species and cultivars plus promising biological control and IPM approaches	Reduced N may not be beneficial for insect control but may have a negative impact on the insect or mite pest population particularly for large scale of sod production and seed production. However, reduced N input will cause higher sugar content in plants, which may attract more pests	Exist and promising	[13,14,31,33,55–59,165,166,192, 200,211,323,340,363,384,403,406, 409,432,441,442,475,531,532]
		Environmental Stresses		
Drought	Drought-resistant species and cultivars and cultural management with potential for further root-zone improvement	Under drought stress, the N input needs to be reduced and the light and frequent N application is recommended	Exist and promising	[1,20,92,93,114,162,216,227,228,390, 395,396,519]
Waterlogging	Soil condition improvement and resist species and cultivars	Waterlogging damages roots first and the root N absorption is retarded, therefore, foliar N application is recommended during the stress and poststress.	Exist	[113,230,338,396,449]

Stress	Approach	Recommendation	Status	References
Salinity	Soil and water salinity condition improvement and resistant species and cultivars	Cautions should be paid to N forms. Ammonium N forms in most cases will worsen the salinity stresses. Soil urea hydrolysis and nitrification are significantly reduced in addition to the reduced root N absorption capabilities under salinity stresses	Exist and very promising	4,6,17,44,45,47,83,84,127,152, 183,190,270,311–320,366, 375–378,389,393,400,401,464,479
Heat	Cultural management, water management, and resistant species and cultivars	Overall N input should be reduced and light foliar N applications are recommended. However, volatilization of urea and ammonium-type N fertilizers can be significant under hot and windy weather conditions	Exist and promising	70,120,124,159,163,195,196,215, 217–220,229,264,288,312,373, 413,465,518,537–540,545
Cold	Cultural management, protection, and resistant species and cultivars	Controlled release N forms should be used and light frequent foliar N applications are beneficial under chilling conditions for both cool- and warm-season turfgrasses	Exist and promising	8–12,19,115,411,443,444,446
Shade	Cultural management with plant growth regulators and resistant species and cultivars	Total N should be reduced and lack of information on N forms and quick and controlled release of N forms. The combination of N and PGR, particularly, trinexapac-ethyl to alleviate shade stress	Exist	23,24,26,32,70,71,73,76,126,139, 174,226,390–392,405,457, 468,470,504
Acid soils	Cultural management, enhancement of symbiotic relationships, and resistant species and cultivars	Ammonium-type fertilizers should be avoided and increased K and Ca will be beneficial.	Exist	16,154,155,283,284,290–293,365
Nutrient imbalances	Cultural management and specific nutrient management	Both nitrate and ammonium should be applied to improve extremes for N absorption and foliar N applications are also recommended	Unknown	95,136,148,181,193,194,266, 322,339,347,351,454,502
Turfgrass Use and Management Stresses				
Traffic stresses	Cultural management plus Si and K applications and resistant species and cultivars	Total N should be reduced to avoid more succulent tissues and other nutrient element applications should be enhanced, including K and Si	Exist	67,87,88,89,91,98,138, 362,435,436–440,485
Cultivations	Cultural management plus biostimulants	Immediate N application is beneficial and light quick release N tends to help recovery. PGRs normally negatively impact recovery from cultivations	Unknown	25,27,100,337,492
Mowing	Cultural management plus with plant growth regulators	N should be applied regularly to replace losses of clippings and under extremely low mowing heights, light foliar N, and other nutrient elements including Fe, Mg, and S will beneficial. Try returning clippings	Unknown	27,30,103,137,337,504,505

TABLE 33.8

Selected Genetic Evidence of Other Crops in the Grass Family for NO_3^- and NH_4^+ Uptake Control and Efficiency Improvement

Crops or Plants	Evidence	References
Rice (*Oryza sativa* L.)	NH_4^+ tranporter gene expressions of *OsAMT1.1*, *OsAMT1.2*, *OsAMT1.3*, *OsAMT2.1*, *OsAMT2.2*, *OsAMT2.3*, *OsAMT3.1*, *OsAMT3.2*, *OsAMT3.3*, and *OsAMT4*	[66,292,451,452,461]
Corn (*Zea mays* L.)	QTL mapping and *ZmNRT2.1* gene expression	[66,140,167,402,423]
Wheat (Triticum aestivum L.)	Genes of *TaNRT2.1*, *TaNRT2.2*, and *TaNRT2.3* were identified; mapping QTLs for nitrogen uptake	[7,66]
Barley (*Hordeum vulgare* L.)	The first *NRT2* genes were cloned in 1996; root exudation is responding to localized nitrate supply; signal that can regulate nitrate uptake and assimilation	[66,143,374,478,489,506]
Sorghum (*Sorghum bicolor* [L.] Moench)	Varietal differences in root growth related to nitrogen uptake	[66,357]
Arabidopsis thaliana: two gene families encode proteins that are involved in either the low (NRT1) or high (NRT2)-affinity NO_3^- transporting. Five related genes were identified for NH_4^+ transporters: *AtAMT1.1*, *AtAMT1.2*, *AtAMT1.3*, *AtAMT1.4*, and *AtAMT1.5*	*AtNRT1.1* (*CHL1*) was first cloned in 1993; nitrate reductase-null mutant was identified; nitrate transporter gene expressions for high NO_3^- affinity including *AtNRT2.1*, *AtNRT2.2*, *AtNRT2.3*, *AtNRT2.4*, *AtNRT2.5*, *AtNRT2.6*, and *AtNRT2.7* and most of these genes are expressed in roots except *AtNRT2.7* with greater expression in shoots. Identification amino acid transporter gene; identification of protein–protein interaction in nitrate uptake; regulation of the high-affinity NO_3^- uptake system by *NRT1.1*-mediated NO_3^- demand signaling; inducible high-affinity nitrate transporter gene cluster	[66,172,186,205,259,273, 292,347,364,365–367–370, 407,490,515,516,542]

294–300,467,517,518]. Table 33.8 lists some examples of genetic controls of other crops for N-use efficiency although it is still in an early stage for genetic improvements [66]. Nitrogen dynamics in a crop-soil-environment system are multiple factor–driven and even with the most N-efficiency varieties, the influencing factors can significantly change the overall outcome and consequences. For example, most of the N ranges provided in Tables 33.3 and 33.4 are in two- to threefold ranges and it is impossible to further narrow down because: (1) the maintenance levels can be different and (2) the soil and location conditions can be different. The genetic variability among genetic lines for any given turfgrass species may hardly demonstrate two- to threefolds of differences in N-use efficiency in most cases [66]. With such challenges, the most significant improvement for turf N-use efficiency may lead to the improvement of the high-affinity systems of turfgrasses by focusing on the N-absorption efficiency when N concentrations in the soils are lower than 1 mM as HATs [224]. The specific morphological and molecular improvements of turfgrasses for more N-use efficiency may include (1) deeper roots, improved root architecture, and more actual root absorption surface areas [37,39,46,230,334,463] and (2) enhanced urease activity inside leaves in hydrolyzing urea more efficiently, which is one of the three N forms dominant as foliar N applied to turfgrasses, particularly C4 warm-season turfgrasses [481–483].

33.7 SUMMARY AND PROSPECTS

Nitrogen, the most limiting nutrient factor, has been investigated thoroughly for different turfgrasses. The balanced use of N and other nutrients (not in this chapter) can assure profound fertility programs for turf use to successfully overcome stresses. Nitrogen itself may not be an effective remedy to control turfgrass stresses but it can significantly affect stresses either positively or negatively. The future stress-resistant turfgrasses need to have multiple-stress resistances being a better N user with improved HATs. The future turfgrass improvement on multiple-stress-resistance enhancement seems more likely than just a narrowed and vertical resistance, which may be the biggest challenge for environmental stress–resistant transgenic plants. Furthermore, research is needed to explore the relationship and complexity of N associated with stresses for different turf uses with precise guidelines for practitioners to further enhance N-use efficiency and minimize N losses. The turfgrass improvement for N-use efficiency is promising and needed; and it will heavily rely on the progress of further understanding of N-use physiology, genetics, and ecology of turfgrasses. Both traditional breeding and modern technology of genomic and genetic engineering are important for turfgrass improvement. Pests to turfgrasses associated with other abiotic stresses including water stress, salinity stress, poor soil conditions, and even heat stress in most of the turf areas will continue becoming more severe than ever and traditional remedies to release these stresses rather more limited due to shortages of resources and more restricted regulations.

REFERENCES

1. Abraham, E. M., B. Huang, S. A. Bonos, and W. A. Meyer. 2004. Evaluation of drought resistance for Texas bluegrass, Kentucky bluegrass, and their hybrids. *Crop Sci.* 44:1746–1753.
2. Addisscott, T. M. 2005. *Nitrate, Agriculture and the Environment.* CABI Publishing. Cambridge, MA.
3. Allen, S. E., G. L. Terman, and H. G. Kennedy. 1978. Nutrient uptake by grass and leaching losses from soluble and sulfur-coated urea, and KCl. *Agron. J.* 70:264–268.
4. Alshammary, S., Y. L. Qian, and S. J. Wallner. 2004. Growth response of four turfgrasses to salinity. *Agric. Water Manage.* 66:97–111.
5. Amancio, S., and I. Stulen. 2004. *Nitrogen Acquisition and Assimilation in Higher Plants.* Kluwer Academic Publishers. Boston, MA.
6. Amarasinghe, V., and L. Watson. 1989. Variation in Asalt secretory activity of microhairs in grasses. *Aust. J. Plant Physiol.* 16:219–229.
7. An, D., J. Su, Q. Liu, Y. Zhu, Y. Tong, J. Li, R. Jing, B. Li, and Z. Li. 2006. Mapping QTLs for nitrogen uptake in relation to the early growth of wheat (*Triticum aestivum* L.). *Plant and Soil* 284:73–84.
8. Anderson, J., C. Taliaferro, M. Anderson, D. Martin, and A. Guenzi. 2005. Freeze tolerance and low temperature-induced genes in bermudagrass plants. *USGA Turfgrass Environ. Res. Online* 4(1):1–7.
9. Anderson, J. A., C. M. Taliaferro, and D. L. Martin. 1993. Evaluating freeze tolerance of bermudagrass in a controlled environment. *HortScience* 28:955–959.
10. Anderson, J. A., C. M. Taliaferro, and D. L. Martin. 2002. Freezing tolerance of bermudagrass: Vegetatively propagated cultivars intended for fairway and putting green use, and seed-propagated cultivars. *Crop Sci.* 42:975–977.
11. Anderson, J. A., M. P. Kenna, and C. M. Taliaferro. 1988. Cold hardiness of 'Midiron' and 'Tifgreen' bermudagrass. *HortScience* 23:748–750.
12. Anderson, M. P., C.M. Taliaferro, and J. A. Anderson. 1997. The cold facts on bermudagrass. *Golf Course Manage.* 65:59–63.
13. Anderson, W. G., T. M. Heng-moss, and F. P. Baxendale. 2006. Evaluation of cool- and warm-season grasses for resistance to multiple chinch bug (*Hemiptera: Blissidae*) species. *J. Econ. Entomol.* 99(1):203–211.
14. Ashihara, H., and A. Crozier. 1999. Biosynthesis and metabolism of caffeine and related pruine alkaloids in plants. *Adv. Bot. Res.* 30:116–205.
15. Aulakh, M. S., and S. S. Malhi. 2005. Interactions of nitrogen with other nutrients and water: effect on crop yield and quality, nutrient use efficiency, carbon sequestration, and environmental pollution. *Adv. Agron.* 86:341–409.

16. Baldwin, C. M., H. Liu, L. B. McCarty, W. B. Bauerle, and J.E. Toler. 2005. Aluminum tolerance of warm-season turfgrasses. *Int. Turfgrass Soc. Res. J.* 10:811–817.

17. Baldwin, C. M., H. Liu, L. B. McCarty, W. L. Bauerle, and J. E. Toler. 2006. Effects of trinexapac-ethyl on the salinity tolerance of two ultradwarf bermudagrass cultivars. *HortScience* 41(3):808–814.

18. Baligar, V. C., and N. K. Fageria. 2005. Enhancing nitrogen efficiency in crop plants. *Adv. Agron.* 88:97–185.

19. Ball, S., Y. Qian, and C. Stushnoff. 2002. Soluble carbohydrates in two buffalograss cultivars with contrasting freezing tolerance. *J. Am. Soc. Hortic. Sci.* 127:45–49.

20. Barrett, J., B. Vinchesi, R. Dobson, P. Roche, and D. Zoldoske. 2003. *Golf Course Irrigation: Environmental Design and Management Practices.* John Wiley & Sons. Hoboken, NJ.

21. Basu, C., H. Luo, A. P. Kausch, and J. M. Chandlee. 2003. Promoter analysis in transient assays using a GUS reporter gene construct in creeping bentgrass (*Agrostis palustris* L.). *J. Plant Physiol.* 160:1233–1239.

22. Basu, C., H. Luo, A. P. Kausch, and J. M. Chandlee. 2003. Transient reporter gene (GUS) expression in creeping bentgrass (*Agrostis palustris*) is affected by *in vivo* nucleolytic activity. *Biotechnol. Lett.* 25:939–944.

23. Beard, J. B. 1965. Factors in the adaptation of turfgrasses to shade. *Agron J.* 57:457–459.

24. Beard, J. B. 1969. Turfgrass shade adaptation. pp. 273–282. In E. C. Roberts (ed.), *Proceedings of the 1st International Turfgrass Research Conference.* Alf Smith and Co. Bradford, UK.

25. Beard, J. B. 1973. *Turfgrass: Science and Culture.* Prentice-Hall, Inc. Englewood Cliffs, NJ.

26. Beard, J. B. 1997. Shade stresses and adaptation mechanisms of turfgrasses. *Int. Turfgrass Soc. Res. J.* 8:1186–1195.

27. Beard, J. B. 2001. *Turf Management for Golf Courses.* 2nd Ed. John Wiley & Sons. Chelsea, MI.

28. Beard, J. B., and H. J. Beard. 2005. *Beard's Turfgrass Encyclopedia for Golf Courses—Grounds— Lawns—Sports Fields.* Michigan State University Press. East Lansing, MI.

29. Beard, J. B., and W. H. Daniel. 1966. Relationship of creeping bentgrass (*Agrostis palustris* Huds.) root growth to environmental factors in the field. *Agron. J.* 58:337–339.

30. Beard, J. B., and R. L. Green. 1994. The role of turfgrasses in environmental protection and their benefits to humans. *J. Environ. Qual.* 23(3):452–460.

31. Beck, E. W. 1963. Observation on the biology and cultural-insecticidal control of *Prosapia bicincta,* a spittlebug, on coastal bermudarass. *J. Econ. Entomol.* 56:747–752.

32. Bell, G., and T. K. Danneberger. 1999. Temporal shade on creeping bentgrass turf. *Crop Sci.* 39:1142–1146.

33. Belovsky, G. E., and J.B. Slade. 2000. Insect berbivory accelerates nutrient cycling and increases plant production. *Proc Natl. Acad. Sci.* 97(26):14412–14417.

34. Below, F. E. 2001. Nitrogen metabolism and crop productivity. pp. 385–406, In M. Pessarakli (ed.), *Handbook of Plant and Crop Physiology.* 2nd Ed. Marcel Dekker, Inc. New York.

35. Below, F. E., S. J. Crafts-Brandner, J. E. Harper, and R. H. Hageman. 1985. Uptake, distribution, and remobilization of 15N-labeled urea applied to maize canopies. *Agron. J.* 77:412–415.

36. Bigelow, C. A., D. C. Bowman, and D. K. Cassel. 2001. Nitrogen leaching in sand-based rootzones amended with inorganic soil amendments and sphagnum peat. *HortScience* 126:151–156.

37. Bloom, A. J., P. A. Meyerhoff, A. R. Taylor, and T. L. Rost. 2003. Root development and absorption of ammonium and nitrate from the rhizosphere. *J. Plant Growth Regul.* 21:416–431.

38. Bonos, S. A., B. B. Clarke, and W. A. Meyer. 2006. Breeding for disease resistance in the major cool-season turfgrasses. *Annu. Rev. Phytopathol.* 44:213–234.

39. Bonos, S. A., and B. Huang. 2006. Breeding and genomic approaches to improving abiotic stress tolerance in plants. In B. Huang (ed.), *Plant-Environment Interactions.* 3rd Ed. CRC Taylor & Francis. New York.

40. Bonos, S. A., C. Kubik, B. B. Clarke, and W. A. Meyer. 2004. Breeding perennial ryegrass for resistance to gray leaf spot. *Crop Sci.* 44:575–580.

41. Bonos, S. A., D. Rush, K. Hignigh, S. Langlois, and W. A. Meyer. 2005. The effect of selection on gray leaf spot resistance in perennial ryegrass. *Int. Turfgrass Soc. Res. J.* 10:501–507.

42. Bowman, D. C. 2003. Daily vs. periodic nitrogen addition affects growth and tissue nitrogen in perennial ryegrass turf. *Crop Sci.* 43:631–638.

43. Bowman, D. C., C. T. Cherney, and T. W. Rufty, Jr. 2002. Fate and transport of nitrogen applied to six warm-season turfgrasses. *Crop Sci.* 42:833–841.

44. Bowman, D. C., G. R. Cramer, and D. A. Devitt. 2006. Effects of nitrogen status on salinity tolerance of tall fescue. *J. Plant Nutr.* 29:1491–1497.

45. Bowman, D. C., G. R. Cramer, and D. A. Devitt. 2006. Effects of salinity and nitrogen status on nitrogen uptake by tall fescue. *J. Plant Nutr.* 29:1481–1490.

46. Bowman, D. C., D. A. Devitt, M. C. Engelke, and T. W. Ruffy, Jr. 1998. Root architecture affects nitrate leaching from bentgrass turf. *Crop Sci.* 38:1633–1639.

47. Bowman, D. C., D. A. Devitt, and W. W. Miller. 1995. The effect of salinity on nitrate leaching from turfgrass. *USGA Green Sect. Rec.* 33(1):45–49.

48. Bowman, D. C., and J. L. Paul. 1989. The foliar absorption of urea-N by Kentucky bluegrass turf. *J. Plant Nutr.* 13(5):659–673.

49. Bowman, D. C., and J. L. Paul. 1989. The foliar absorption of urea-N by tall fescue and creeping bentgrass turf. *J. of Plant Nutr.* 13(9):1095–1113.

50. Bowman, D. C., and J. L. Paul. 1992. Foliar absorption of urea, ammonium, and nitrate by perennial ryegrass turf. *J. Am. Soc. Hort. Sci.* 117(1):75–79.

51. Bowman, D. C., J. L. Paul, W. B. Davis, and S. H. Nelson. 1987. Reducing ammonia volatilization from Kentucky bluegrass turf by irrigation. *HortScience* 22:84–87.

52. Bowman, D. C., J. L. Paul, and W. B. Davis. 1989. Nitrate and ammonium uptake by nitrogen-deficient perennial ryegrass and Kentucky bluegrass turf. *J. Am. Soc. Hort. Sci.* 114:421–426.

53. Bowman, D. C., J. L. Paul, W. B. Davis, and S. H. Nelson. 1989. Rapid depletion of nitrogen applied to Kentucky bluegrass turf. *J. Am. Soc. Hortic. Sci.* 114:229–233.

54. Boynton, D. 1954. Nutrition by foliar application. *Annu. Rev. of Plant Physiol.* 5:31–54.

55. Braman, K. 2004. Resistant turf: Front line defense for insect pests. *USGA Turfgrass Environ. Res. Online* 3(2):1–8.

56. Braman, S. K., R. R. Duncan, and M. C. Engelke. 2000. Evaluation of turfgrasses for resistance to fall armyworms (*Lepidoptera: Noctuidae*). *HortScience* 35:1268–1270.

57. Braman, S. K., R. R. Duncan, W. W. Hanna, and W. G. Hudson. 2000. Evaluation of turfgrasses for resistance to mole crickets (*Orthoptera: Gryllotalpidae*). *HortScience* 35:665–668.

58. Braman, S. K., A. F. Pendley, R. N. Carrow, and M. C. Engelke. 1994. Potential resistance in zoysia-grasses to tawny mole crickets (*Orthoptera: Gryllotalpidae*). *Fla. Entomol.* 77:302–305.

59. Braman, S. K., and P. L. Raymer. 2006. Impact of Japanese beetle (*Coleoptera: Scarabaeidae*) feeding on *Seashore paspalum. J. Econ. Entomol.* 99(5):1699–1704.

60. Brandenburg, R. L., and M.G. Villani. 1995. *Handbook of Turfgrass Insect Pests.* The Entomological Society of America. Hyattsville, MD.

61. Brauen, S. E., and G. K. Stahnke. 1995. Leaching of nitrate from sand putting greens. *USGA Green Sect. Rec.* 33:29–32.

62. Brede, D. 2000. *Turfgrass Maintenance Reduction Handbook: Sports, Lawns, and Golf.* John Wiley & Sons. Chelsea, MI.

63. Brilman. L. A. 2005. Turfgrass breeding in the United States: public and private, cool and warm season. *Int. Turfgrass Soc. Res. J.* 10:508–515.

64. Britto, D. T., A. D. M. Glass, H. J. Kronzucker, and M. Y. Siddiqi. 2001. Cytosolic concentrations and transmembrane fluxes of NH_4^+/NH_3. An evaluation of recent proposals. *Plant Physiol.* 125:523–526.

65. Britto, D. T., and H. J. Kronzuker. 2002. NH_4^+ toxicity in higher plants: a critical review. *J. Plant Physiol.* 159:567–584.

66. Broadley, M. R., and P. J. White. 2005. *Plant Nutritional Genomics.* Blackwell Publishing CRC Press. Boca Raton, FL.

67. Brosnan, J. T., J. S. Ebdon, and W. M. Dest. 2005. Characteristics in diverse wear tolerant genotypes of Kentucky bluegrass. *Crop Sci.* 45:1917–1926.

68. Brown, K. W., R. L. Duble, and J. C. Thomas. 1977. Influence of management and season on fate of N applied to golf greens. *Agron. J.* 69:667–671.

69. Brown, K. W., J. C. Thomas, and R. L. Duble. 1982. Nitrogen source effect on nitrate and ammonium leaching and runoff losses from greens. *Agron. J.* 74:947–950.

70. Bunnell, B. T., L. B. McCarty, and W. C. Bridges, Jr. 2005. Evaluation of three bermudagrass cultivars and Meyer Japanese zoysiagrass grown in shade. *Int. Turfgrass Soc. Res. J.* 10:826–833.

71. Bunnell, B. T., L. B. McCarty, and W. C. Bridges, Jr. 2005. 'TifEagle' bermudagrass responses to growth factors and mowing height when grown at various hours of sunlight. *Crop Sci.* 45:575–581.

72. Bunnell, B. T., L. B. McCarty, R. B. Dodd, H. S. Hill, and J. J. Camberato. 2002. Creeping bentgrass growth response to elevated soil carbon dioxide. *HortScience* 37(2):367–370.

73. Bunnell, B. T., L. B. McCarty, J. E. Faust, W. C. Bridges, Jr., and N. C. Rajapakse. 2005. Quantifying a daily light integral requirement of a 'TifEagle' bermudagrass golf green. *Crop Sci.* 45:569–574.

74. Bunnell, B. T., L. B. McCarty, and H. S. Hill. 2004. Soil gas, temperature, matric potential, and creeping bentgrass growth response to subsurface air movement on a sand-based golf green. *HortScience* 39(2):415–419.

75. Burpee, L. L. 1994. *A Guide to Integrated Control of Turfgrass Diseases: Vol. 1—Cool Season Turfgrasses.* GCSAA Reference Materials. Lawrence, KS.

76. Burton, G. W., J. E. Jackson, and F. E. Knox. 1959. The influence of light reduction upon the production, persistence and chemical composition of coastal bermudagrass, *Cynodon dactylon. Agron. J.* 51:537–542.

77. Busey, P. 2003. Cultural management of weeds in turfgrass: a review. *Crop Sci.* 43:1899–1911.

78. Bush, E., P. Wilson, D. Phepard, and J. McCrimmon. 2000. Freezing tolerance and nonstructural carbohydrate composition of carpetgrass (*Axonopus affinis* Chase). *HortScience* 35:187–189.

79. Bushoven, J. T., and R. J. Hull. 2001. Nitrogen use efficiency is linked to nitrate reductase activity and biomass partitioning between roots and shoots of perennial ryegrass and creeping bentgrass. *Int. Turfgrass Soc. Res. J.* 9:245–252.

80. Bushoven, J. T., and R. J. Hull. 2005. The role of nitrate in modulating growth and partitioning of nitrate assimilation between roots and leaves of perennial ryegrass (*Lolium perenne* L.). *Int. Turfgrass Soc. Res. J.* 10:834–840.

81. Bushoven, J. T., Z. Jiang, H. J. Ford, C. D. Sawyer, R. J. Hull, and J. A. Amador. 2000. Stabilization of soil nitrate by reseeding with perennial ryegrass following sudden turf death. *J. Environ. Qual.* 29:1657–1661.

82. Cahill, J. V., J. J. Murray, and P. H. Dernoeden. 1983. Interrelationship between fertility and red thread fungal disease of turfgrasses. *Plant Dis.* 67:1080–1083.

83. Camberato, J. J., and S. B. Martin. 2004. Salinity slows germination of rough bluegrass. *HortScience* 39(2):394–397.

84. Camberato, J. J., P. D. Peterson, and S. B. Martin. 2005. Salinity alters rapid blight disease occurrence. *USGA Turfgrass Environ. Res. Online* 4(16):1–7.

85. Campbell, C. L., and L. C. Madden. 1990. *Introduction to Plant Disease Epidemiology.* John Wiley & Sons, Inc. New York.

86. Campbell, W. H. 1999. Nitrate reductase structure, function and regulation. *Ann. Rev. Plant Physiol. Plant Mol. Biol.* 50:277–303.

87. Canaway, P. M. 1981. Wear tolerance of turfgrass species. *J. Sports Turf Res. Inst.* 57:65–83.

88. Canaway, P. M. 1984. The response of Lolium perenne turf grown on sand and soil to fertilizer nitrogen. I. Ground cover response as affected by football-type wear. *J. Sports Turf Res. Inst.* 60:8–18.

89. Canaway, P. M. 1984. The response of Lolium perenne turf grown on sand and soil to fertilizer nitrogen. II. Above ground biomass, tiller numbers and root biomass. *J. Sports Turf Res. Inst.* 60:19–26.

90. Cardona, C. A., R. R. Duncan, and O. Lindstrom. 1997. Low temperature tolerance assessment in paspalum. *Crop Sci.* 37:1283–1291.

91. Carroll, M. J., and M. A. Petrovic. 1991. Wear tolerance of Kentucky bluegrass and creeping bentgrass following nitrogen and potassium application. *HortScience* 26:851–853.

92. Carrow, R. N. 1996. Drought resistance aspects of turfgrasses in the southeast: Root-shoot responses. *Crop Sci.* 36:687–694.

93. Carrow, R. N. 2005. Seashore paspalum ecotype responses to drought and root limiting stresses. *USGA Turfgrass Environ. Res. Online* 4(13):1–9.

94. Carrow, R. N., and R. R. Duncan.1998. *Salt-Affected Turfgrass Sites: Assessment and Management.* John Wiley & Sons. Chelsea, MI.

95. Carrow, R. N., D. V. Waddington, and P. E. Rieke. 2001. *Turfgrass Soil fertility and Chemical Problems—Assessment and Management.* Ann Arbor Press. Chelsea, MI.

96. Casler, M. D., and R. R. Duncan. 2003. *Turfgrass Biology, Genetics, and Cytotaxonomy.* Sleeping Bear Press. Chelsea, MI.

97. Casler, M., G. Jung, S. Bughrara, A. Hamblin, C. Williamson, and T. Voigt. 2006. Development of creeping bentgrass with resistance to snow mold and dollar spot. *USGA Turfgrass Environ. Res. Online* 5(18):1–10.

98. Cereti, C. E., F. Rossini, and F. Nassetti. 2005. Wear tolerance characteristics of 110 turfgrass varieties. *Int. Turfgrass Soc. Res. J.* 10:508–515.

99. Chapman, J. E. 1994. The effects of nitrogen fertilizer rate, application timing, and root-zone modification on nitrate concentrations and organic matter accumulation in sand-based putting greens. M.S. thesis. Washington State University. Pullman, WA.

100. Christians, N., 2003. *Fundamentals of Turfgrass Management—2nd ed.* John Wiley & Sons. Hoboken, NJ.
101. Cisar, J. L., R. J. Hull, and D. T. Duff. 1989. Ion uptake kinetics of cool season turfgrasses. pp. 233–235. In H. Takatoh (ed.), *Proc. Int. Turfgrass Res. Conf., 6th, Tokyo,* Japan. 31 July–5 August 1989. *Int. Turfgrass Soc.Japanese Soc. of Turfgrass Sci.,* Tokyo.
102. Cisar, J., G. Snyder, and D. Park. 2005. The effect of nitrogen rates on ultradwarf bermudagrass quality. *USGA Turfgrass Environ. Res. Online* 4(17):1–6.
103. Clarke, B. B., and A. B. Gould. 1993. *Turfgrass Patch Diseases Caused by Ectotrophic Root-Infecting Fungi.* APS Press. St. Paul, MN.
104. Clarke, B. B., J. F. White, R. H. Hurley, M. S. Torres, S. Sun, and D. R. Huff. 2006. Endophyte-mediated suppression of dollar spot disease in fine fescues. *Plant Dis.* 90:994–998.
105. Cloutier, Y. 1983. Changes in the electrophoretic patterns of the soluble proteins of winter wheat and rye following cold acclimation and desiccation stress. *Plant Physiol.* 69:256–258.
106. Cohen, S., A. Svrjcek, T. Durborow, and N. L. Barnes. 1999. Water quality impacts by golf courses. *J. Environ. Qual.* 28:798–809.
107. Cook, R. N., R. E. Engel, and S. Bachelder. 1964. A study of the effect of nitrogen carriers on turfgrass disease. *Plant Dis. Rep.* 48:254–255.
108. Couch, H. B. 1995. *Diseases of turfgrasses.* 3rd Ed. Krieger Publishing Company. Malabar, FL.
109. Couch, H. B. 2000. *The Turfgrass Disease Handbook.* Krieger Publishing Company Melbourne, FL.
110. Craft, C., and E. Nelson. 1996. Microbial properties of composts that suppress damping-off and root rot of creeping bentgrass caused by *Pythium graminicola. Appl. Environ. Microbiol.* 62:1550–1557.
111. Crafts-Brander, S. J. 2002. Plant nitrogen status rapidly alters amino acid metabolism and excretion in Bemisia tabaci. *J. Insect Physiol.* 48(1):33–41.
112. Cyril, J., G. L. Powell, R. R. Duncan, and W. V. Baird. 2002. Changes in polar lipid fatty acids of seashore paspalum in response to low temperature exposure. *Crop Sci.* 42:2031–2037.
113. DaCosta, M., and B. Huang. 2006. Deficit irrigation effects on water use characteristics of bentgrass species. *Crop Sci.* 46:1779–1786.
114. DaCosta, M., and B. Huang. 2006. Minimum water requirements for creeping, colonial, and velvet bentgrasses under fairway conditions. *Crop Sci.* 46:81–89.
115. Danneberger, T. K., 1993. *Turfgrass Ecology and Management.* G.I.E. Media Inc. Cleveland, OH.
116. Datnoff, L. E., and B. A. Rutherford. 2003. Accumulation of silicon by bermudagrass to enhance disease suppression of leaf spot and melting out. *USGA Turfgrass Environ. Res. Online* 2:1–6.
117. Davis, J. G., and P. H. Dernoeden. 2002. Dollar spot severity, tissue nitrogen, and soil microbial activity in bentgrass as influenced by nitrogen source. *Crop Sci.* 42:480–488.
118. Demirevska-kepova, K., R. Holzer, L. Simova-stoilova, and U. Feller. 2005. Heat stress effects on ribulose-1,5-bisphosphate carboxylase/oxygenase, rubisco binding protein and rubisco activase in wheat leaves. *Biologia Plantarum* 49(4):521–525.
119. Dernoeden, P. H. 2000. *Creeping Bentgrass Management: Summer Stresses, Weeds and Selected Maladies.* Ann Arbor Press. Chelsea, MI.
120. Dernoeden, P. H., M. J. Carroll, and J. M. Krouse. 1993. Weed management and tall fescue quality as influenced by mowing, nitrogen and herbicides. *Crop Sci.* 33:1055–1061.
121. Dest, W. M., and K. Guillard. 1987. Nitrogen and phosphorus nutritional influence on bentgrass-annual bluegrass community composition. *J. Am. Soc. Hortic. Sci.* 112:769–773.
122. Dionne, J., Y. Castonguay, P. Nadeau, and Y. Desjardins. 2001. Amino acid and protein changes during cold acclimation of green-type annual bluegrass (*Poa annua* L.) ecotypes. *Crop Sci.* 41:1862–1870.
123. Dionne, J., Y. Castonguay, P. Nadeau, and Y. Desjardins. 2001. Freezing tolerance and carbohydrate changes during cold acclimation of green-type annual bluegrass (*Poa annua* L.) ecotypes. *Crop Sci.* 41:443–451.
124. DiPaola, J. M., and J. B. Beard. 1992. Physiological effects of temperature stress. In D. V. Waddington et al. (eds.), *Turfgrass.* Vol. 32. ASA. Madison, WI.
125. Duble, R. 1996. *Turfgrasses: Their Management and Use in the Southern Zone.* Texas A&M University Press. College Station, TX.
126. Dudeck, A. E., and C. H. Peacock. 1992. Shade and turfgrass culture. pp. 269–284. In D. V. Waddingtion, R. N. Carrow and R. C. Shearman (eds.), *Turfgrass.* American Society of Agronomy Monograph 32. Madison, WI.
127. Dudeck, A. E., S. Singh, C. E. Giordano, T. A. Nell, and D. B. McConnell. 1983. Effects of Sodium chloride on *Cynodon* turfgrasses. *Agron. J.* 75:927–930.

128. Duff, D. T., H. Liu, R. J. Hull, and C. D. Sawyer. 1997. Nitrate leaching from long established Kentucky bluegrass turf. *Int. Turfgrass Soc. Res. J.* 8:175–186.

129. Duncan, R. R. 2000. Plant tolerance to acid soil constraints: genetic resources, breeding methodology, and plant improvement. pp. 1–38. In R. E. Wilkinson (ed.), *Plant Environment Interaction (2nd Edition)*. Marcel Dekker Inc. NY.

130. Dunn, J. H., and C. J. Nelson. 1974. Chemical changes occurring in three bermudagrass turf cultivars in relation to cold hardiness. *Agron. J.* 66:28–31.

131. Eliot, R. C. 1960. Relationship between foliar and root development of turfgrass species and strains. *Illinois Turfgrass Conference Proceedings* 32–35.

132. Eliot, R. C. 1972. The use of soluble fertilizers for lawn care. *Proc. 42nd Ann. Michigan Turfgrass.* 1:61–71.

133. Elliott, M. L., E. A. Guertal, and H. D. Skipper. 2004. Rhizosphere bacterial population flux in golf course putting greens in the southeastern United States. *HortScience* 39:1754–1758.

134. Emmons, R. D. 2000. *Turfgrass Science and Management—3rd Edition*. Delmar Learning. New York.

135. Engelsjord, M. E., B. E. Branham, and B. P. Horgan. 2004. The fate of nitrogen-15 ammonium sulfate applied to Kentucky bluegrass and perennial ryegrass turfs. *Crop Sci.* 44:1341–1347.

136. Epstein, E. 2005. *Nutrition of Plants: Principles and Perspectives*. Sinauer Associates, Inc. Publishers. Sunderland, MA.

137. Erickson, J. E., J. L. Cisar, J. C. Volin, and G. H. Snyder. 2001. Comparing nitrogen runoff and leaching between newly established St. Augustinegrass turf and an alternative residential landscape. *Crop Sci.* 41:1889–1895.

138. Ervin, E. H., and A. J. Koski. 2001. Kentucky bluegrass growth responses to trinexapac-ethyl, traffic, and nitrogen. *Crop Sci.* 41:1871–1877.

139. Ervin, E. H., X. Zhang, S. D. Askew, and J. M. Goatly, Jr. 2004. Trinexapac-ethyl, propiconazole, iron, and biostimulant effects on shaded creeping bentgrass. *HortTechnology* 14(4):500–506.

140. Espen, L., F. F. Nocito, and M. Cocucci. 2005. Effect of NO_3^- transport and reduction on intercellular pH: an in vivo NMR study in maize roots. *J. Exp. Bot.* 55:2053–2061.

141. Fagerness, M. J., D. C. Bowman, F. H. Yelverton, and T. W. Rufty. 2004. Nitrogen use in Tifway bermudagrass, as influenced by trinexapac-ethyl. *Crop Sci.* 44:595–599.

142. Fagerness, M. J., F. H. Yelverton, D. P. Livingston, III, and T. W. Rufty, Jr. 2002. Temperature and trinexapac-ethyl effects on bermudagrass growth, dormancy, and freezing tolerance. *Crop Sci.* 42:853–858.

143. Fan, X., R. Gordon-Weeks, Q. Shen, and A. J. Miller. 2006. Glutamine transport and feedback regulation of nitrate reductase activity in barley roots leads to changes in cytosolic nitrate pools. *J. Exp. Bot.* 57:1333–1340.

144. Feng, Y., D. M. Stoeckel, E. van Santen, and R. H. Walker. 2002. Effects of subsurface aeration and trinexapac-ethyl application on soil microbial communities in a creeping bentgrass putting green. *Biol. Fertil. Soils* 36:456–460.

145. Fermanian, T. W., M. C. Shurtleff, R. Randell, H. T. Wilkinson, and P. L. Nixon. 2002. *Controlling Turfgrass Pests*. 3rd Ed. Prentice-Hall. Upper Saddle River, NJ.

146. Fidanza, M. A., and P. H. Dernoeden. 1996. Brown patch severity in perennial ryegrass as influenced by irrigation, fungicide, and fertilizers. *Crop Sci.* 36:1631–1638.

147. Fitzpatrick, R. J. M., and K. Guillard. 2004. Kentucky bluegrass response to potassium and nitrogen fertilization. *Crop Sci.* 44:1721–1728.

148. Follett, R. F., and J. L. Hatfield. 2001. *Nitrogen and Environment: Sources, Problems, and Management*. Elsevier, New York.

149. Forde, B. G., and D. T. Clarkson. 1999. Nitrate and ammonium nutrition of plants: Physiological and molecular perspectives. *Adv. Bot. Res.* 30:1–90.

150. Foy, C. D., and J. J. Murray. 1998. Developing aluminum-tolerant strains of tall fescue for acid soils. *J. Plant Nutr.* 21(6):1301–1325.

151. Foy, C. D., and J. J. Murray. 1998. Responses of Kentucky bluegrass cultivars to excess aluminum in nutrient solutions. *J. Plant Nutr.* 21(9):1967–1983.

152. Francois, L. E. 1988. Salinity effects on three turf bermudagrasses. *HortScience* 23(4):706–708.

153. Frank, K.W. 2000. Nitrogen allocation of three turfgrass species and turf-type buffalograss management. Ph.D. dissertation. University of Nebraska, Lincoln, NE.

154. Frank, K. W., R. E. Gaussoin, T. P. Riordan, R. C. Shearman, J. D. Fry, E. D. Miltner, and P. G. Johnson. 2004. Nitrogen rate and mowing height effects on turf-type buffalograss. *Crop Sci.* 44:1615–1621.

155. Frank, K. W., K. O'Reilly, J. Crum, and R. Calhoun. 2006. Nitrogen fate in a mature Kentucky bluegrass turf. *USGA Turfgrass Environ. Res. Online* 5(2):1–6.

156. Frank, K.W., K. M. O'Reilly, J. R. Crum, and R. N. Calhoun. 2006. The fate of nitrogen applied to a mature Kentucky bluegrass turf. *Crop Sci.* 46:209–215.

157. Franke, W. 1967. Mechanisms of foliar penetration of solutions. *Ann. Rev. Plant Physiol.* 18:281–300.

158. Fry, J. D. 1990. Cold temperature tolerance of bermudagrass. *Golf Course Manage.* 58:26–32.

159. Fry, J. D., and B. Huang. 2004. *Applied Turfgrass Science and Physiology.* John Wiley & Sons. Hoboken, NJ.

160. Fry, J. D., N. S. Lang, and R. G. P. Clifton. 1991. Freezing resistance and carbohydrate composition of 'Floratam' St. Augustinegrass. *HortScience* 26:1537–1539.

161. Fry, J. D., N. S. Lang, R. G. P. Clifton, and F. P. Maier. 1993. Freezing tolerance and carbohydrate content of low temperature-acclimated and non-acclimated centipedegrass. *Crop Sci.* 33:1051–1055.

162. Fry, J. D., S. Wiest, Y. Qian, and W. Upham. 1997. Evaluation of empirical models for estimating turfgrass water use. *Int. Turfgrass Soc. Res. J.* 8:1268–1273.

163. Fu, J., and B. Huang. 2003. Growth and physiological response of creeping bentgrass to elevated night temperature. *HortScience* 38(2):299–301.

164. Fu, J., and J. Huang. 2003. Effects of foliar application of nutrients on heat tolerance of creeping bentgrass. *J. Plant Nutr.* 26(1):81–96.

165. Funk, C. R., P. M. Halisky, S. Ahmad, and R. H. Hurley. 1985. How endophytes modify turfgrass performance and response to insect pests in turfgrass breeding and evaluation trials. pp. 127–145. In F. Lemaire (ed.), *Proc. Int. Turf. Res. Conf.*, 5th Ed. Avignon, 1–5 July. Versailles, France.

166. Funk C. R., P. M. Halisky, M. C. Johnson, M. R. Siegel, and A.V. Stewart. 1983. An endophytic fungus and resistance to sod webworms: association in *Lolium perenne* L. *Biotechnology* 1:189–91.

167. Gallais, A., and B. Hirel. 2004. An approach to the genetics of nitrogen use efficiency in maize. *J. Exp. Bot.* 55:295–306.

168. Gardner, D. S., and B. G. Wherley. 2005. Growth response of three turfgrass species to nitrogen and trinexapac-ethyl in shade. *HortScience* 40(6):1911–1915.

169. Garling, D. D., J. W. Rimelspach, J. R. Street, and M. J. Boehm. 1999. The impact of foliar nitrogen concentration on dollar spot severity and turfgrass quality. *Ann. Meet. Abstr. [ASA/CSSA/SSSA]* 91:135.

170. Gatschet, M. J., C. M. Taliaferro, D. R. Porter, M. P. Anderson, J. A. Anderson, and K. W. Jackson. 1996. A cold-regulated protein from bermudagrass crowns is a chitinase. *Crop Sci.* 36:712–718.

171. Gaudreau, J. E., D. M. Vietor, R. H. White, T. L. Provin, and C. L. Munster. 2002. Response of turf and quality of water runoff to manure and fertilizer. *J. Environ. Qual.* 31:1316–1322.

172. Gazzarrini, S., L. Lejay, A Gojon, O. Ninnemann, W. B. Frommer, and N. von Wiren. 1999. Three functional transporters for constitutive, diurnally regulated, and starvation-induced uptake of ammonium into *Arabidopsis* roots. *Plant Cell.* 11:937–947.

173. Geron, C. A., K. Danneberger, S. J. Traina, T. J. Logan, and J. R. Street. 1993. The effects of establishment methods and fertilization practices on nitrate leaching from turfgrass. *J. Environ. Qual.* 22:119–125.

174. Giesler, L. J., G. Y. Yuen, and G. L. Horst. 2000. Canopy microenvironments and applied bacteria population dynamics in shaded tall fescue. *Crop Sci.* 40:1325–1332.

175. Glass, A. D. M. 2003. Nitrogen use efficiency of crop plants: physiological constraints upon nitrogen absorption. *Crit. Rev. Plant Sci.* 22:453–470.

176. Glass, A. D. M., D. T. Britto, B. N. Laiser, J. R. Kinghorn, H. J. Kronzucker, A. Kumar, M. Okamoto, S. Rawat, M. Y. Siddiqi, S. E. Unkles, and J. J. Vidmar. 2002. The regulation of nitrate and ammonium transport systems in plants. *J. Exp. Bot.* 53:855–864.

177. Goatley, J. M., Jr., V. Maddox, D. J. Lang, and K. K. Crouse. 1994. 'Tifgreen' bermudagrass response to late-season application of nitrogen and potassium. *Agron. J.* 86:7–10.

178. Gold, A. J., W. R. DeRagon, W. M. Sullivan, and J. L Lemunyon. 1990. Nitrate-nitrogen losses to groundwater from rural and suburban land uses. *J. Soil Water Conser.* 45:305–310.

179. Gold, A. J., and P. M. Groffman. 1993. Leaching of agrichemicals from suburban areas. pp. 182–190. In K. D. Racke and A. R. Leslie (eds.), *Pesticides in Urban Environments: Fate and Significance.* ACS symposium series 522. ACS. Washington, DC.

180. Goss, R. M., J. H. Baird, S. L. Kelm, and R. N. Calhoun. 2002. Trinexapac-ethyl and nitrogen effects on creeping bentgrass grown under reduced light conditions. *Crop Sci.* 42:472–479.

181. Goyal, S. S., R. Tischner, and A. S. Basra. 2005. *Enhancing the Efficiency of Nitrogen Utilization in Plants.* Food Products Press. New York.

182. Gray, E., and N. M. Call. 1993. Fertilization and mowing persistence of Indian mockstrawberry (*Duchesnea indica*) and common blue violet (*Viola papilionacea*) in a tall fescue (*Festuca arundinacea*) lawn. *Weed Sci.* 41:548–550.

183. Grieve, C. M., J. A. Poss, S. R. Grattan, D. L. Suarez, S. E. Benes, and P. H. Robinson. 2004. Evaluation of salt-tolerant forages for sequential water reuse systems II. Plant ion relations. *Agric. Water Manage.* 70:121–135.

184. Guertal, E. A., and D. L. Evans. 2006. Nitrogen rate and mowing height effects on TifEagle bermuda-grass establishment. *Crop Sci.* 46:1772–1778.

185. Guillard, K., and K. L. Kopp. 2004. Nitrogen fertilizer form and associated nitrate leaching from cool-season lawn turf. *J. Environ. Qual.* 33:1822–1827.

186. Guo, F. Q., J. Young, and N. M. Crawford. 2003. The nitrate transporter *AtNRT1.1 (CHL1)* functions in stomatal opening and contributes to drought susceptibility in Arabidopsis. *Plant Cell* 15:107–117.

187. Gupta, A. S., J. L. Heinen, A. S. Holaday, J. J. Burke, and R. D. Allen. 1993. Increased resistance to oxidative stress in transgenic plants that overexpress chloroplastic Cu/Zn superoxide dismutase. *Proc. Natl. Acad. Sci. USA* 90:1629–1633.

188. Gussack, E., and F. S. Rossi. 2001. *Turfgrass Problems: Picture Clues and Management Options.* Cornell University Natural Resource, Agriculture, and Engineering Service (NRAES). Ithaca, NY.

189. Han, Y., S. A. Bonos, B. B. Clarke, and W. A. Meyer. 2003. Inoculation techniques for selection of gray leaf spot resistance in perennial ryegrass. *USGA Turfgrass Environ. Res. Online* 2(9):1–7.

190. Harivandi, M. A., J. D. Butler, and L. Wu. 1992. Salinity and turfgrass culture. pp. 207–229. In D. V. Waddington et al. (eds.), *Turfgrass.* ASA, CSSA, and SSSA. Madison, WI.

191. Harman, G. E. 1991. The development and benefits of rhizosphere competent fungi for biological control for plant pathogens. *J. Plant Nutr.* 15:835–843.

192. Hatcher, P. E., N. D. Paul, P. G Ayres, and J. B. Whittaker. 1997. Nitrogen fertilization affects interactions between the components of an insect-fungus-plant triparitite system. *Functional Ecol.* 11:537–544.

193. Hauck, R. D. 1984. *Nitrogen in Crop Production.* American Society of Agronomy, Crop Science Society of America, Soil Science Society of America. Madison, WI.

194. Havlin, J. L., J. D. Beaton, S. L. Tisdale, and W. L. Nelson. 2005. *Soil Fertility and Fertilizers.* 7th Ed. Prentice Hall. Upper Saddle River, NJ.

195. He, Y., Y. Liu, W. Cao, M. Huai, B. Xu, and B. Huang. 2005. Effects of salicylic acid on heat tolerance associated with antioxidant metabolism in Kentucky bluegrass. *Crop Sci.* 45:988–995.

196. He, Y., X. Liu, and B. Huang. 2005. Protein changes in response to heat stress in acclimated and nonac-climated creeping bentgrass. *J. Am. Soc. Hort. Sci.* 130(4):521–526.

197. Heckathorn, S. A., J. Gretchen, J. Poeller, J. S. Coleman, and R. L. Hallberg. 1996. Nitrogen availability alters patterns of accumulation of heat stress-induced proteins in plants. *Oecologia* 105(3):413–418.

198. Heckman, J. R., B. B. Clarke, and J. A. Murphy. 2003. Optimizing manganese fertilization for the suppression of take-all patch disease on creeping bentgrass. *Crop Sci.* 43:1395–1398.

199. Heckman, J. R., H. Liu, W. Hill, M. DeMilia, and W. L. Anastasia. 2000. Kentucky bluegrass responses to mowing and nitrogen fertility management. *J. Sustainable Agri.* 15:25–33.

200. Heng-Moss, T. M., F. P. Baxendale, T. E. Eickhoff, and R. C. Shearman. 2006. Physiological and biochemical responses of resistant and susceptible buffalograsses to chinch bug feeding. *USGA Turfgrass Environ. Res. Online* 5(20):1–11.

201. Herrmann, K. M. 1995. The shikimate pathway as an entry point to aromatic secondary metabolism. *Plant Physiol.* 107:7–12.

202. Hesketh, E. S., R. J. Hull, and A. J. Gold. 1995. Estimating non-gaseous nitrogen losses from established turf. *J. Turfgrass Mange.* 1:17–30.

203. Hill, W. J., J. R. Heckman, B. B. Clarke, and J. A. Murphy. 1999. Take-all patch suppression in creeping bentgrass with manganese and copper. *HortScience* 34:891–892.

204. Hill, W. J., J. R. Heckman, B. B. Clarke, and J. A. Murphy. 2001. Influence of liming and nitrogen on severity of summer patch of Kentucky bluegrass. *Int. Turfgrass Soc. Res. J.* 9:388–393.

205. Hirner, A., F. Ladwig, H. Stransky, S. Okumoto, M. Keinath, A. Harms, W. B. Frommer, and W. Koch. 2006. *Arabidopsis LHT1* is a high-affinity transporter for cellular amino acid uptake in both root epidermis and leaf mesophyll. *Plant Cell* 18:1931–1946.

206. Hodges, M. 2002. Enzyme redundancy and the importance of 2-oxoglutarate in plant ammonium assimilation. *J. Exp. Bot.* 53:905–916.

207. Hoffman, N. E., and A. M. Hamblin. 2000. Progress in identifying resistance to gray leaf spot in perennial ryegrass. pp. 161–162. In *Agronomy Abstracts.* ASA. Madison, WI.

208. Hollingsworth, B. S., E. A. Guertal, and R. H. Walker. 2005. Cultural management and nitrogen source effects on ultradwarf bermudagrass cultivars. *Crop Sci.* 45:486–493.

209. Horgan, B. P., B. E . Branham, and R. L. Mulvaney. 2002. Direct measurement of denitrification using ^{15}N-labeled fertilizer applied to turfgrass. *Crop Sci.* 42:1602–1610.

210. Horgan, B. P., B. E. Branham, and R. L. Mulvaney. 2002. Mass balance of ^{15}N applied to Kentucky bluegrass including direct measurement of denitrification. *Crop Sci.* 42:1595–1601.

211. Horn, G. C. 1962. Chinch bugs and fertilizers, is there a relationships? *Fla. Turfgrass Assoc. Bull.* 9(4):3,5.

212. Hossain, M. A., Y. Ishimine, H. Akamine, and H. Kuramochi. 2004. Effect of nitrogen fertilizer application on growth, biomass production and N-uptake of torpedograss (*Panicum repens* L.). *Weed Biol. Manage.* 4:86–94.

213. Hu, Q., K. Nelson, and H. Luo. 2006. FLP-mediated site-specific recombination for genome modification in turfgrass. *Biotechnol. Lett.* 28:1793–1804.

214. Huang, B. 2006. Cellular membranes in stress sensing and regulation of plant adaptation to abiotic stresses. In B. Huang (ed.), *Plant-Environment Interactions.* 3rd Ed. CRC Taylor & Francis. New York.

215. Huang, B., and J. D. Fry. 1998. Root anatomical, physiological, and morphological responses to drought stress for tall fescue cultivars. *Crop Sci.* 38:1017–1022.

216. Huang, B., and H. Gao. 2000. Growth and carbohydrate metabolism of creeping bentgrass cultivars in response to increasing temperatures. *Crop Sci.* 40:1115–1120.

217. Huang, B., and X. Liu. 2003. Summer root decline: production and mortality for four cultivars of creeping bentgrass. *Crop Sci.* 43:258–265.

218. Huang, B., X. Liu, and J. D. Fry. 1998. Shoot physiological responses of two bentgrass cultivars to high temperature and poor soil aeration. *Crop Sci.* 38:1219–1224.

219. Huang, B., J. Pote, and Q. Xu. 2004. Soil temperatures controlling creeping bentgrass growth. *USGA Turfgrass Environ. Res. Online* 3(18):1–5.

220. Huang, B., and Z. Wang. 2005. Cultivar variation and physiological factors associated with heat tolerance for Kentucky bluegrass. *Int. Turfgrass Soc. Res. J.* 10:559–564.

221. Huber, D. M., and R. D. Watson. 1974. Nitrogen form and plant diseases. *Ann. Rev. Phytopathol.* 12:139–165.

222. Hughes, M. A., and M. A. Dunn. 1996. The molecular biology of plant acclimation to low temperature. *J. Exp. Bot.* 47:291–305.

223. Hull, R. J. 1996. N usage by turfgrass. *Turfgrass Trends* 5(11):6–15.

224. Hull, R. J., and H. Liu. 2005. Turfgrass nitrogen: physiology and environmental impacts. *Int. Turfgrass Soc. Res. J.* 10:962–975.

225. Hurdzan, M. J. 2004. Golf Greens: History, Design, and Construction. John Wiley & Sons. Hoboken, NJ.

226. Jiang, Y., R. R. Duncan, and R. N. Carrow. 2004. Assessment of low light tolerance of seashore paspalum and bermudagrass. *Crop Sci.* 44:587–594.

227. Jiang, Y., and B. Huang. 2000. Effects of drought or heat stress alone and in combination on Kentucky bluegrass. *Crop Sci.* 40:1358–1362.

228. Jiang, Y., and B. Huang. 2001. Drought and heat stress injury of two cool-season turfgrasses in relation to antioxidant metabolism and lipid peroxidation. *Crop Sci.* 41:436–442.

229. Jiang, Y., and B. Huang. 2001. Physiological responses to heat stress alone or in combination with drought: a comparison between tall fescue and perennial ryegrass. *HortScience* 36:682–686.

230. Jiang, Y., and K. Wang. 2006. Growth, physiological, and anatomical responses of creeping bentgrass cultivars to different depths of waterlogging. *Crop Sci.* 46:2420–2426.

231. Jiang, Z. 2005. Biologically Based Fertilizers for Turf Ecosystems. pp. 329–334. In *Third International Nitrogen Conference Proceedings Contributed Papers.* Science Press USA, Inc.

232. Jiang, Z. 2005. Nutrient concentrations in turfgrass clippings and groundwater as affected by fertilizers. *Int. Turfgrass Soc. Res. J.* 10:944–951.

233. Jiang, Z., J. T. Bushoven, H. J. Ford, C. D. Sawyer, J. A. Amador, and R. J. Hull. 2000. Mobility of soil nitrogen and microbial responses following the sudden death of established turf. *J. Environ. Qual.* 29:1625–1631.

234. Jiang, Z., and R. J. Hull. 1998. Interrelationships of nitrate uptake, nitrate reductase, and nitrogen use efficiency in selected Kentucky bluegrass cultivars. *Crop Sci.* 38:1623–1632.

235. Jiang, Z., and R. J. Hull. 1999. Partitioning of nitrate assimilation between shoots and roots of Kentucky bluegrass. *Crop Sci.* 39:746–754.

236. Jiang, Z., and R. J. Hull. 2000. Diurnal patterns of nitrate assimilation in Kentucky bluegrass. *J. Plant Nutr.* 23:443–456.

237. Jiang, Z., R. J. Hull, and W. M. Sullivan. 2002. Nitrate uptake and reduction in C3 and C4 grasses. *J. Plant Nutr.* 25:1303–1314.

238. Jiang, Z., and W. M. Sullivan. 2004. Nitrate uptake of seedling and mature Kentucky bluegrass plants. *Crop Sci.* 44:567–574.

239. Jiang, Z., W. M. Sullivan, and R. J. Hull. 2000. Nitrate uptake and nitrogen use efficiency by Kentucky bluegrass cultivars. *HortScience* 35:1350–1354.

240. Jiang, Z., W. M. Sullivan, and R. J. Hull. 2001. Nitrate uptake and metabolism in Kentucky bluegrass as affected by nitrate levels. *Int. Turfgrass Soc. Res. J.* 9:303–310.

241. Johnson, S. J., and N. E. Christians. 1984. Fertilizer burn comparison of concentrated liquid fertilizers applied to Kentucky bluegrass turf. *J. Am. Soc. Hort. Sci.* 109(6):890–893.

242. Johnson, P. G., R. T. Koenig, and K. L. Kopp. 2003. Nitrogen, phosphorus, and potassium responses and requirements in calcareous sand greens. *Agron. J.* 95:697–702.

243. Johnston, W. J., and C. T. Golob. 2002. Nitrogen leaching through a sand-based golf green. *USGA Turfgrass Environ. Res. Online* 1(19):1–7.

244. Johnston, W. J., C. T. Golob, C. M. Kleene, W. L. Pan, and E. D. Miltner. 2001. Nitrogen leaching through a floating sand-based golf green under golf course play and management. *Int. Turfgrass Soc. Res. J.* 9:19–24.

245. Kaminski, J. E., and P. H. Dernoedon. 2005. Nitrogen source impact on dead spot (*Ophiosphaerella agrostis*) recovery in creeping bentgrass. *Int. Turfgrass Soc. Res. J.* 10:214–223.

246. Karpinski, S., G. Wingsle, B. Karpinska, and J. Hallgren. 2002. Low-temperature stress and antioxidant defense mechanisms in higher plants. pp. 69–104. In D. Inze and M. Van Montagu (eds.), *Oxidative Stress in Plants*. Taylor & Francis. London.

247. Kenna, M. P., and J. T. Snow. 2002. Environmental research: past and future. *USGA Turfgrass Environ. Res. Online* 1(3):1–27.

248. Key, J. L., C. Y. Lin, and Y. M. Chen. 1981. Heat shock proteins in higher plants. *Proc. Natl. Acad. Sci. U.S.A.* 78:3526–3530.

249. Kochian, L. V. 1995. Cellular mechanisms of aluminum toxicity and resistance in plants. *Annu. Rev. Plant Physiol. Plant Mol. Biol.* 46:237–260.

250. Kochian, L. V. 2000. Molecular physiology of mineral nutrient acquisition, transport and utilization. In B. Buchanan, W. Gruissem and R. Jones (eds.), *Biochemistry and Molecular Biology of Plants*. American Society of Plant Physiologists. Rockville, MD.

251. Kochian, L. V., O. A. Hoekenga, and M. A. Piñeros. 2004. How do crop plants tolerate acid soils? Mechanisms of aluminum tolerance and phosphorus efficiency. *Annu. Rev. Plant Physiol. Plant Mol. Biol.* 55:459–493.

252. Kochian, L. V., and D. L. Jones. 1997. Aluminum and resistance in plants. pp. 69–89. In R. A. Yokel and M. S. Golub (eds.), *Research Issues in Aluminum Toxicity*. Taylor & Francis Publishers. Washington DC.

253. Kochian, L. V., M. A. Piñeros, and O. A. Hoekenga. 2005. The physiology, genetics and molecular biology of plant aluminum resistance and toxicity. *Plant and Soil* 274(1–2):175–195.

254. Kopec, D. M., J. H. Walworth, J. J. Gilbert, G. M. Sower, and M. Pessarakli. 2005. Ball roll distance of 'Sea Isle 2000' paspalum in response to mowing height and nitrogen fertility. *USGA Turfgrass Environ. Res. Online* 4(7):1–6.

255. Kopp, K. L., and K. Guillard. 2002. Clipping management and nitrogen fertilization of turfgrass: growth, nitrogen utilization, and quality. *Crop Sci.* 42:1225–1231.

256. Kopp, K. L., and K. Guillard. 2002. Relationship of turfgrass growth and quality to soil nitrate desorbed from anion exchange membranes. *Crop Sci.* 42:1232–1240.

257. Kopp, K. L., and K. Guillard. 2005. Clipping contributions to nitrate leaching from creeping bentgrass under varying irrigation and N rates. *Intl. Turfgrass Soc. Res. J.* 10:80–85.

258. Koster, K. L., and A. C. Leopold. 1988. Sugars and desiccation tolerance. *Plant Physiol.* 88:829–832.

259. Krouk, G., P. Tillard, and A. Gojon. 2006. Regulation of the high-affinity NO_3^- uptake system by *NRT1.1*-mediated NO_3^- demand signaling in *Arabidopsis*. *Plant Physiol.* 142:1075–1086.

260. Kuo, S., S. E. Brauen, and E. J. Jellum. 1992. Phosphorus availability in some acid soils influences bentgrass and annual bluegrass growth. *HortScience* 27:370.

261. Kutchan, T. M. 1995. Alkaloid biosynthesis—the basis for metabolic engineering of medicinal plants. *Plant Cell* 7:1059–1070.

262. Landschoot, P. J., and B. F. Hoyland. 1992. Gray leaf spot of perennial ryegrass turf in Pennsylvania. *Plant Dis.* 76:1280–1282.

263. Landschoot, P. J., and A. S. McNitt. 1997. Effect of nitrogen fertilizers on suppression of dollar spot disease of *Agrostis stolonifera* L. *Int. Turfgrass Soc. Res. J.* 8:905–911.

264. Larkindale, J., and B. Huang. 2004. Changes of lipid composition and saturation level in leave and roots for heat-stressed and heat-acclimated creeping bentgrass (*Agrostis stolonifera*). *Environ. Expt. Bot.* 51(1):57–67.

265. Lawton, M. B., and L. L. Burpee. 1990. Seed treatments for Typhula blight and pink snow mold of winter wheat relationships among disease intensity, crop recovery, and yield. *Can. J. Plant Pathol.* 12:63–74.

266. Lea, U. S., M. T. Leydecker, I. Quilleré, C. Meyer, and C. Lillo. 2006. Posttranslational regulation of nitrate reductase strongly affects the levels of free amino acids and nitrate, whereas transcriptional regulation has only minor influence. *Plant Physiol.* 140:1085–1094.

267. Lea, P. J., and J. F. Morot-Gaudry. 2001. *Plant Nitrogen.* Springer. New York.

268. Lee, D. J., D. C. Bowman, D. K. Cassel, C. H. Peacock, and T. W. Rufty, Jr. 2003. Soil inorganic nitrogen under fertilized bermudagrass turf. *Crop Sci.* 43:247–257.

269. Lehman, V. G., and M. C. Engelke. 1991. Heritability estimates of creeping bentgrass root systems grown in flexible tubes. *Crop Sci.* 31:1680–1684.

270. Leonard, M. K. 1983. Salinity tolerance of the turfgrass Paspalum vaginatum Swartz 'Adalayd': mechanisms of growth responses. Ph.D. dissertation. University of California, Riverside.

271. Leslie, A. R. 1994. *Handbook of Integrated Pest Management for Turf and Ornamentals.* Lewis Publishers. Ann Arbor, MI.

272. Lestienne, F., B. Thornton, and F. Gastal. 2006. Impact of defoliation intensity and frequency on N uptake and mobilization in *Lolium perenne. J. Exp. Bot.* 57:997–1006.

273. Li, W., Y. Wang, M. Okamoto, N. M. Crawford, M. Y. Siddiqi, and A. D. M. Glass. 2006. Dissection of the *AtNRT2.1:AtNRT2.2* inducible high-affinity nitrate transporter gene cluster. *Plant Physiol.* First published on December 8, 2006; 10.1104/pp.106.091223.

274. Lillo, C., C. Meyer, U. S. Lea, F. Provan, and S. Oltedal. 2004. Mechanism and importance of posttranslational regulation of nitrate reductase. *J. Exp. Bot.* 55:1275–1282.

275. Linde, D. T., and T. L. Watschke. 1997. Nutrients and sediment in runoff from creeping bentgrass and perennial ryegrass turfs. *J. Environ. Qual.* 26:1248–1254.

276. Linde, D. T., T. L. Watschke, A. R. Jarrett, and J. A. Borger. 1995. Surface runoff assessment from creeping bentgrass and perennial ryegrass turf. *Agron. J.* 87:176–182.

277. Liu, H. 2001. Soil acidity and aluminum toxicity response in turfgrass. *Int. Turfgrass Soc. Res. J.* 9:180–188.

278. Liu, H. 2005. Aluminum toxicity of seeded bermudagrass cultivars. *HortScience* 40(1):221–223.

279. Liu, H., J. R. Heckman, and J. A. Murphy. 1995. Screening Kentucky bluegrass for aluminum tolerance. *J. Plant Nutr.* 18:1797–1814.

280. Liu, H., J. R. Heckman, and J. A. Murphy. 1996. Screening fine fescues for aluminum tolerance. *J. Plant Nutr.* 19:677–688.

281. Liu, H., J. R. Heckman, and J. A. Murphy. 1997. Aluminum tolerance among genotypes of Agrostis species. *Int. Turfgrass Soc. Res. J.* 8:729–734.

282. Liu, H., J. R. Heckman, and J. A. Murphy. 1997. Greenhouse screening of turfgrasses for aluminum tolerance. *Int. Turfgrass Soc. Res. J.* 8:719–728.

283. Liu, L. X., T. Hsiang, K. Carey, and J. L. Eggens. 1995. Microbial populations and suppression of dollar spot disease in creeping bentgrass with inorganic and organic amendments. *Plant Dis.* 79:144–147.

284. Liu, X., and B. Huang. 2000. Heat stress injury in relation to membrane lipid peroxidation in creeping bentgrass. *Crop Sci.* 40:503–510.

285. Liu, X., and B. Huang. 2003. Mowing height effects on summer turf growth and physiological activities for two creeping bentgrass cultivars. *HortScience* 38(3):444–448.

286. Liu, H., and R. J. Hull. 2006. Comparing cultivars of three cool-season turfgrasses for nitrogen recovery in clippings. *HortScience* 41(3):827–831.

287. Liu, H., R. J. Hull, and D. T. Duff. 1993. Comparing cultivars of three cool-season turfgrasses for nitrate uptake kinetics and nitrogen recovery in the field. *Int. Turfgrass Soc. Res. J.* 7:546–552.

288. Liu, H., R. J. Hull, and D. T. Duff. 1997. Comparing cultivars of three cool-season turfgrass for soil water nitrate concentration and nitrate losses through percolation. *Crop Sci.* 37:526–534.

289. Liu, H., L. B. McCarty, F. W. Totten, and C. M. Baldwin. 2006. Fertilizing golf greens, foliar versus granular—research results and perspectives. *Carolinas Green*, January/February.

290. Long, D. H., F. N. Lee, and D. O. TeBeest. 2000. Effect of nitrogen fertilization on disease progress of rice blast on susceptible and resistant cultivars. *Plant Dis.* 84:403–409.

291. Longo, C., C. Lickwar, Q. Hu, K. Nelson, D. Viola, J. Hague, J. M. Chandlee, H. Luo, and A. P. Kausch. 2006. *Agrobacterium tumefaciens*—mediated transformation of turf grasses In K. Wang (ed.), *Methods in Molecular Biology—Agrobacterium Protocols (2nd edition)*, Vol. 2. The Humana Press Inc. Totowa, NJ.

292. Loque, D., and N. von Wiren. 2004. Regulatory levels for transport of ammonium in plant roots. *J. Exp. Bot.* 55:1293–1305.

293. Lovvorn, R. L. 1945. The effect of defoliation, soil fertility, temperature, and length of day on the growth of some perennial grasses. *J. Am. Soc. Agron.* 37:570–582.

294. Luo, H., Q. Hu, K. Nelson, C. Longo, and A. P. Kausch. 2004. Controlling transgene escape in genetically modified grasses. In A. Hopkins, Z. Y. Wang, R. Mian, M. Sledge and R. Barker (eds.), *Molecular Breeding of Forage and Turf*, Kluwer Academic Publishers. Dordrecht.

295. Luo, H., Q. Hu, K. Nelson, C. Longo, A. P. Kausch, J. M. Chandlee, J. K. Wipff, and C. R. Fricker. 2004. *Agrobacterium* tumefaciens-mediated creeping bentgrass (*Agrostis stolonifera* L.) transformation using phosphinothricin selection results in a high frequency of single-copy transgene integration. *Plant Cell Rep.* 22:645–652.

296. Luo, H., and A. P. Kausch. 2002. Application of FLP/ FRT site-specific DNA recombination system in plants. In J. K. Setlow (ed.), *Genetic Engineering, Principles and Methods*, Vol. 24. Kluwer Academic/Pleum Publishers. New York.

297. Luo, H., A. P. Kausch, Q. Hu, K. Nelson, J. K. Wipff, C. C. R. Fricker, T. P. Owen, M. A. Moreno, J. Y. Lee, and T. K. Hodges. 2005. Controlling transgene escape in GM creeping bentgrass. *Mol. Breed.* 16:185–188.

298. Luo, H., J. Y. Lee, Q. Hu, K. Nelson-Vasilchik, T. K. Eitas, C. Lickwar, A. P. Kausch, and J. M. Chandlee, T. K. Hodges. 2006. RTS, a rice anther-specific gene is required for male fertility and its promoter sequence directs tissue-specific gene expression in different plant species. *Plant Mol. Biol.* 62:397–408.

299. Luo, H., L. A. Lyznik, D. Gidoni, and T. K. Hodges. 2000. FLP-mediated recombination for use in hybrid plant production. *Plant J.* 23:423–430.

300. Luo, H., P. Morsomme, and M. Boutry. 1999. The two major types of plant plasma membrane H+-ATPases show different enzymatic properties and confer differential pH sensitivity of yeast growth. *Plant Physiol.* 119:627–634.

301. Maas, E. V. 1986. Salt tolerance of plants. *Appl. Agric. Res.* 1:12–26.

302. Maas, E. V., and G. J. Hoffman. 1977. Crop salt tolerance-current assessment. *ASCE J. Irr. Drain. Div.* 103:115–134.

303. Maggiotto, S. R., J. A. Webb, C. Wagner-Riddle, and G. W. Thurtell. 2000. Nitrous and nitrogen oxide emissions from turfgrass receiving different forms of nitrogen fertilizer. *J. Environ. Qual.* 29:621–630.

304. Maier, F. P., N. S. Lang, and J. D. Fry. 1994. Freezing tolerance of three St. Augustinegrass cultivars as affected by stolon carbohydrate and water content. *J. Am. Soc. Hortic. Sci.* 119:473–476.

305. Mancino, C. F., D. M. Petrunak, and D. Wilkinson. 2001. Loss of putting greens-grade fertilizer granules due to mowing. *HortScience* 36(6):1123–1126.

306. Mancino, C. F., W. A. Torello, and D. J. Wehner. 1988. Denitrification losses from Kentucky bluegrass sod. *Agron. J.* 80:148–153.

307. Mancino, C. F., and J. Troll. 1990. Nitrate and ammonium leaching losses from N fertilizers applied to 'Penncross' creeping bentgrass. *HortScience* 25:194–196.

308. Mangiafico, S. S., and K. Guillard. 2005. Turfgrass reflectance measurements, chlorophyll, and soil nitrate desorbed from anion exchange membranes. *Crop Sci.* 45:259–265.

309. Mangiafico, S. S., and K. Guillard. 2006. Fall fertilization timing effects on nitrate leaching and turfgrass color and growth. *J. Environ. Qual.* 35:163–171.

310. Maqbool, S. B., and M. B. Sticklen. 2003. Genetic engineering turfgrasses for pest resistance. *USGA Turfgrass Environ. Res. Online* 2(21):1–13.

311. Marcum, K. B. 1998. Cell membrane thermostability and whole-plant heat tolerance of Kentucky bluegrass. *Crop Sci.* 38:1214–1218.

312. Marcum, K. B. 1999. Salinity tolerance mechanisms of grasses in the subfamily Chloridoideae. *Crop Sci.* 39:1153–1160.

313. Marcum, K. B. 2006. Use of saline and non-potable water in the turfgrass industry: Constraints and developments. *Agric. Water Manage.* 80:132–146.

314. Marcum, K. B., S. J. Anderson, and M. C. Engelke. 1998. Salt gland ion secretion: a salinity tolerance mechanism among five zoysiagrass species. *Crop Sci.* 38:806–810.

315. Marcum, K. B., and C. L. Murdoch. 1990. Growth responses, ion relations, and osmotic adaptations of eleven C4 turfgrasses to salinity. *Agron. J.* 82:892–896.
316. Marcum, K. B., and C . L. Murdoch. 1990. Salt glands in the Zoysieae. *Ann. Bot.* (Lond.) 66:1–7.
317. Marcum, K. B., and C. L. Murdoch. 1994. Salinity tolerance mechanisms of six C4 turfgrasses. *J. Am. Soc. Hortic. Sci.* 119:779–784.
318. Marcum, K. B., and M. Pessarakli. 2006. Salinity tolerance and salt gland excretion efficiency of bermudagrass turf cultivars. *Crop Sci.* 46:2571–2574.
319. Marcum, K. B., M. Pessarakli, and D. M. Kopec. 2005. Relative salinity tolerance of 21 turf-type desert saltgrasses compared to bermudagrass. *HortScience* 40:827–829.
320. Marcum, K. B., G. Wess, D. T. Ray, and M. C. Engelke. 2003. Zoysiagrass, salt glands, and salt tolerance. *USGA Turfgrass Environ. Res. Online* 2:1–6.
321. Markland, F. E., E. C. Roberts, and I. R. Fredrick. 1969. Influence of nitrogen fertilizers on Washington creeping bentgrass, *Agrostis palustris* Huds. II. Incidence of dollar spot, *Sclerotinia homoeocarpa*, infection. *Agron. J.* 61:701–705.
322. Marschner, H. 1995. *Mineral Nutrition of Higher Plants.* 2nd Ed. Academic Press. London.
323. Mattson, W. J., Jr. 1980. Herbivory in relation to plant nitrogen content. *Ann. Rev. Ecol. Syst.* 11:119–161.
324. Mattsson, M., and J. K. Schjoerring. 2002. Dynamic and steady-state responses of inorganic nitrogen pools and NH_3 exchange in leaves of *Lolium perenne* and *Bromus erectus* to changes in root nitrogen supply. *Plant Physiol.* 128:742–750.
325. McCarty, L. B. 2005. *Best Golf Course Management Practices.* 2nd Ed. Prentice-Hall Inc. Upper Saddle River, NJ.
326. McCarty, L. B., J. W. Everest, D. W. Hall, T. R. Murphy, and F. Yelverton. 2001. *Color Atlas of Turfgrass Weeds.* Ann Arbor, Chelsea, MI.
327. McCarty, L. B., G. H. Snyder, and J. B. Sartain. 1994. Foliar feeding: How effective and safe is it? *Florida Turf Digest.* 11(3):16–19.
328. McCarty, L. B., and B. J. Tucker. 2005. Prospects for managing turf weeds without prospective chemicals. *Int. Turfgrass Soc. Res. J.* 10:34–41.
329. McCaslin, B. D., and C. E. Watson. 1977. Fertilization of common bermudagrass *Cynodon dactylon* L. with foliar application of iron. *New Mexico Agricultural Experiment Station, State College* No. 334. p. 4.
330. McCullough, P. E., H. Liu, and L. B. McCarty. 2005. Response of creeping bentgrass to nitrogen and ethephon. *HortScience* 40(3):836–838.
331. McCullough, P. E., H. Liu, and L. B. McCarty. 2006. Bermudagrass putting green performance influenced by nitrogen and trinexapac-ethyl. *HortScience* 41(3):802–804.
332. McCullough, P. E., H. Liu, L. B. McCarty, T. Whitwell, and J. E. Toler. 2006. Growth and nutrient partitioning of 'TifEagle' bermudagrass as influenced by nitrogen and trinexapac-ethyl. *HortScience* 41(2):453–458.
333. McCullough, P. E., H. Liu, L. B. McCarty, T. Whitwell, and J. E. Toler. 2006. Bermudagrass putting green growth, quality, and nutrient partitioning influenced by nitrogen and trinexapac-ethyl. *Crop Sci.* 46:1515–1525.
334. McCully, M. E. 1995. How do real roots work? Some new views of root structure. *Plant Physiol.* 109:1–6.
335. McDade, M. C., and N. E. Christians. 2001. Corn gluten hydrolysate for crabgrass (*Digitaria* spp.) control in turf. *Int. Turfgrass Soc. Res. J.* 9:1026–1029.
336. McKenzie, J. S., R. Paquin, and S. H. Duke. 1988. Cold and heat tolerance. pp. 259–302. In A. A. Hanson (ed.), *Alfalfa and alfalfa improvement.* Agron. Monogr. 29. ASA. Madison, WI.
337. McKersie, B. D., and S. R. Bowley. 1997. Active oxygen and freezing tolerance in transgenic plants. pp. 203–214. In P. H. Li and T. H. H. Chen (eds.), *Plant Cold Hardiness.* Plenum Press. New York.
338 Melvin, B. P., and J. M. Vargas, Jr. 1994. Irrigation frequency and fertilizer type influence necrotic ring spot of Kentucky bluegrass. *HortScience* 29:1028–1030.
339. Mengel, K., E. A. Kirby, H. Kosegarten, and T. Appel. 2001. *Principles of Plant Nutrition.* Kluwer Academic Publishers. Boston.
340. Meyers, J. H., and B. J. Post. 1981. Plant nitrogen and fluctuations of insect populations: a test with the cinnabar moth-tansy ragwort system. *Oecologia* 48(2):1432–1439.
341. Miller, G. L., and R. Dickens. 1996. Potassium fertilization related to cold resistance in bermudagrass. *Crop Sci.* 36:1290–1295.

342. Miltner, E. D., B. E. Branham, E. A. Paul, and P. E. Rieke. 1996. Leaching and mass balance of 15N-labeled urea applied to Kentucky bluegrass turf. *Crop Sci.* 36:1427–1433.

343. Miltner, E. D., G. K. Stahnke, and G. J. Rinehart. 2005. Mowing height, nitrogen rate, and organic and synthetic fertilizer effects on perennial ryegrass quality and pest occurrence. *Int. Turfgrass Soc. Res. J.* 10:982–988.

344. Monteith, J. L. 1963. *Dew, Facts and Fallacies: Water Relations of Plants.* Blackwell. London.

345. Monteith, J., Jr., and A. S. Dahl. 1932. Turf diseases and their control. *Bull. U.S. Golf Assoc. Green Sect.* 12:86–186.

346. Moran, K. K., R. L. Mulvaney, and S. A. Khan. 2002. A technique to facilitate diffusions for nitrogen-isotope analysis by direct combustion. *Soil Sci. Soc. Am. J.* 66:1008–1011.

347. Morot-Gaudry, J. F. 2001. *Nitrogen Assimilation by Plants—Physiological, Biochemical and Molecular Aspects.* Science Publishers, Inc. Enfield, NH.

348. Morris, K. 2003. Bentgrasses and bermudagrasses for today's putting greens. *USGA Turfgrass Environ. Res. Online* 2(1):1–7.

349. Morris, K. 2004. Grasses for overseeding bermudagrass fairways. *USGA Turfgrass Environ. Res. Online* 3(1):1–8.

350. Morton, T. G., A. J. Gold, and W. M. Sullivan. 1988. Influence of overwatering and fertilization on nitrogen losses from home lawns. *J. Environ. Qual.* 17:124–130.

351. Mosier, A. R., J. K. Syers, and J. R. Freney. 2004. *Agriculture and the Nitrogen Cycle—Assessing the Impact of Fertilizer Use on Food Production and the Environment.* Island Press. Washington.

352. Moss, M. A., and L. E. Trevathan. 1987. Environmental conditions conducive to infection of ryegrass by Pyricularia grisea. *Phytopathology* 77:863–866.

353. Munshaw, G. C. 2004. Nutritional and PGR effects on lipid unsaturation, osmoregulant content, and relation to bermudagrass cold hardiness. Ph.D. dissertation. Virginia Polytechnic Inst. and State Univ., Blacksburg, VA.

354. Munshaw, G. C., E. H. Ervin, C. Shang, S. D. Askew, X. Zhang, and R. W. Lemus. 2006. Influence of late-season iron, nitrogen, and seaweed extract on fall color retention and cold tolerance of four bermudagrass cultivars. *Crop Sci.* 46:273–283.

355. Murray, J. J., and C. D. Foy. 1978. Differential tolerances of turfgrass cultivars to an acid soil high in exchangeable aluminum. *Agron. J.* 70:769–774.

356. Naidoo, Y., and G. Naidoo. 1998. Salt tolerance in Sporobolus virginicus: the importance of ion relations and salt secretion. *Flora* 193:337–344.

357. Nakamura, T., J. J. Adu-Gyamfi, A. Yamamoto, S. Ishikawa, H. Nakano, and O. Ito. 2002. Varietal differences in root growth as related to nitrogen uptake by sorghum plants in low-nitrogen environment. *Plant and Soil* 245:17–24.

358. Nam-il, P., M. Ogasawara, K. Yoneyama, and Y. Takeuchi. 2001. Responses of annual bluegrass (*Poa annua* L.) and creeping bentgrass (*Agrostis palustris* Huds.) seedlings to nitrogen, phosphorus and potassium. *Weed Biol. Manage.* 1(4):222–225.

359. Nelson, E. B., and C. M. Craft. 1989. Suppression of dollar spot with topdressings amended with composts and organic fertilizers. *Biol. Cult. Tests Control Plant Dis.* 6:93.

360. Nelson, E. B., and C. M. Craft. 1991. Suppression of Pythium root rot with top dressings amended with composts and organic fertilizers. *Biol. Cult. Tests Control Plant Dis.* 7:104.

361. Nelson, E. B., and C. M. Craft. 1992. Suppression of dollar spot on creeping bentgrass and annual bluegrass turf with composted-amended topdressings. *Plant Dis.* 76:954–957.

362. Nelson, K. E., A. J. Turgeon, and J. R. Street. 1980. Thatch influence on mobility and transformation of nitrogen carriers applied to turf. *Agron. J.* 72:487–492.

363. Niemczyk, H. D. 2001. *Destructive Turf Insects—Second Edition.* G.I.E. Media Inc. Cleveland, OH.

364. Ninnemann, O., J. C. Jauniaux, and W. B. Frommer. 1994. Identification of a high affinity NH_4^+ transporter from plants. *EMBO J.* 13:3464–3471.

365. Okamoto, M., A. Kumar, W. Li, Y. Wang, M. Y. Siddiqi, N. M. Crawford, and A. D. M. Glass. 2006. High-affinity nitrate transport in roots of *Arabidopsis* depends on expression of the *NAR2*-like gene *AtNRT3.1*. *Plant Physiol.* 140:1036–1046.

366. Oross, J. W., and W. W. Thomson. 1982. The ultrastructure of the salt glands of *Cynodon* and *Distichlis*. *Am. J. Bot.* 69:939–949.

367. Orsel, M., K. Boivin, H. Roussel, C. Thibault, A. Krapp, F. Daniel-Vedele, and C. Meyer. 2004. Functional genomics of plant nitrogen metabolism. pp 431–450. In D. Leister (ed.), *Plant Functional Genomics.* Haworth Press. Binghamton, NY.

368. Orsel, M., F. Chopin, O. Leleu, S. J. Smith, A. Krapp, F. Daniel-Vedele, and A. J. Miller. 2006. Characterization of a two-component high-affinity nitrate uptake system in *Arabidopsis*. physiology and protein-protein interaction. *Plant Physiol.* 142:1304–1317.

369. Orsel, M., S. Filleur, V. Fraisier, and F. Daniel-Vedele. 2002. Nitrate transport in plants: which gene and which control? *J. Exp. Bot.* 53:825–833.

370. Orsel, M., A. Krapp, and F. Daniel-Vedele. 2002. Analysis of the NRT2 nitrate transporter family in *Arabidopsis*. structure and gene expression. *Plant Physiol.* 129:886–896.

371. Panaccione, D. G., R. D. Johnson, J. Wang, C. A. Young, B. Scott, and C. L. Schardl. 2001. Elimination of ergovaline from a grass—*Neotyphodium* endophyte symbiosis by genetic modification of the endophyte. *Proc. Natl. Acad. Sci. U.S.A* 98:12820–12825.

372. Paré, K., M. H. Chantigny, K. Carey, W. J. Johnston, and J. Dionne. 2006. Nitrogen uptake and leaching under annual bluegrass ecotypes and bentgrass species: a lysimeter experiment. *Crop Sci.* 46:847–853.

373. Park, S. Y., R. Shivaji, J. V. Krans, and D. S. Luthe. 1996. Heat-shock response in heat-tolerant and nontolerant variants of Agrostis palustris Huds. *Plant Physiol.* 111:515–524.

374. Paterson, E., A. Sim, D. Standing, M. Dorward, and A. J. S. McDonald. 2006. Root exudation from *Hordeum vulgare* in response to localized nitrate supply. *J. Exp. Bot.* 57:2413–2420.

375. Peacock, C. H., and A. E. Dudeck. 1985. A comparative study of turfgrass physiological responses to salinity. pp. 821–830. In F. Lemaire (ed.), *Proceedings of Fifth International Turfgrass Research Conference*. Avignon, France. 1–5 July 1985. Institut National de la Recherche Agronomique, Paris.

376. Peacock, C. H., D. J. Lee, W. C. Reynolds, J. P. Gregg, R. J. Cooper, and A. H. Bruneau. 2004. Effects of salinity on six bermudagrass turf cultivars. *Acta Hortic.* 661:193–197.

377. Perras, M., and F. Sarhan. 1989. Synthesis of freezing tolerance proteins in leaves, crown, and roots during cold acclimation of wheat. *Plant Physiol.* 89:577–585.

378. Pessarakli, M., and D. M. Kopec. 2005. Responses of twelve inland saltgrass accessions to salt stress. *USGA Turfgrass Environ. Res. Online* 4(20):1–5.

379. Petrovic, A. M. 1990. The fate of nitrogenous fertilizers applied to turfgrass. *J. Environ. Qual.* 19:1–14.

380. Picchioni, G. A., and H. M. Quiroga-Garza. 1999. Growth and nitrogen partitioning, recovery, and losses in bermudagrass receiving soluble sources of labeled 15nitrogen. *J. Am. Soc. Hortic. Sci.* 124:719–725.

381. Pira, E. 1997. *A Guide to Golf Course Irrigation System Design and Drainage*. Ann Arbor Press, Inc. Chelsea, MI.

382. Podeszfinski, C., Y. Dalpe, and C. Charest. 2002. In situ turfgrass establishment: I. Responses to arbuscular mycorrhizae and fertilization. *J. Sust. Agric.* 20:57–74.

383. Potter, D. A. 1998. *Destructive Turfgrass Insects: Biology, Diagnosis, and Control*. Ann Arbor Press, Inc. Chelsea, MI.

384. Potter, D. A., D. W. Held, and M. E. Roger. 2005. Natural organic fertilizers as a risk factor for ataenius spretulus (*Coleoptera: scarabaeidae*) infestation on golf course. *Int. Turfgrass Soc. Res. J.* 10:753–760.

385. Puhalla, J., J. Krans, and M. Goatley. 2002. *Sports Fields: A Manual for Construction and Maintenance*. John Wiley & Sons. Hoboken, NJ.

386. Puhalla, J., J. Krans, and M. Goatley. 2003. *Baseball and Softball Fields: Design, Construction, Renovation, and Maintenance*. John Wiley & Sons. Hoboken, NJ.

387. Qian, Y. L., S. Ball, Z. Tan, A. J. Koski, and S. J. Wilhelm. 2001. Freezing tolerance of six cultivars of buffalograss. *Crop Sci.* 41:1174–1178.

388. Qian, Y., W. Bandaranayake, W. J. Parton, B. Mecham, M. A. Harivandi and A. R. Mosier. 2003. Long-term effects of clipping and nitrogen management in turfgrass on soil organic carbon and nitrogen dynamics: The CENTURY model simulation. *J. Environ. Qual.* 32:1694–1700.

389. Qian, Y. L., J. A. Cosenza, S. J. Wilhelm, and D. Christensen. 2006. Techniques for enhancing saltgrass seed germination and establishment. *Crop Sci.* 46:2613–2616.

390. Qian, Y. L., and M. C. Engelke. 1999. Diamond zoysiagrass as affected by light intensity. *J. Turfgrass Manage.* 3:1–13.

391. Qian, Y. L., and M. C. Engelke. 1999. Influence of trinexapac-ethyl on Diamond zoysiagrass in a shade environment. *Crop Sci.* 39:202–208.

392. Qian, Y. L., and M. C. Engelke. 1999. Performance of five turfgrasses under linear irrigation. *HortScience* 34:893–896.

393. Qian, Y. L., M. C. Engelke, and M. J. V. Foster. 2000. Salinity effects on zoysiagrass cultivars and experimental lines. *Crop Sci.* 40:488–492.

394. Qian, Y. L., M. C. Engelke, M. J. V. Foster, and S. Reynolds. 1998. Trinexapac-ethyl restricts shoot growth and improves quality of 'Diamond' zoysiagrass under shade. *HortScience* 33:1019–1022.
395. Qian Y. L., R. F. Follett, S. Wilhelm, A. J. Koski, and M. A. Shahba. 2004. Carbon isotope discrimination of three Kentucky bluegrass cultivars with contrasting salinity tolerance. *Agron. J.* 96:571–575.
396. Qian, Y. L., and J. D. Fry. 1996. Zoysiagrass rooting and plant water status as affected by irrigation frequency. *HortScience* 31:234–237.
397. Qian, Y. L., and J. D. Fry. 1997. Water relations and drought tolerance of four turfgrasses. *J. Am. Soc. Hort. Sci.* 122:129–133.
398. Qian, Y. L., J. Fry, and W. Upham. 1997. Rooting and drought avoidance of warm-season turfgrasses and tall fescue in Kansas. *Crop Sci.* 37:699–704.
399. Qian, Y. L., J. Fry, S. Wiest, and W. Upham. 1996. Estimation of turfgrass evapotranspiration using atmometers and an empirical model. *Crop Sci.* 36:669–674.
400. Qian, Y. L., and M. Suplick. 2001. Interactive effect of salinity and temperature on Kentucky bluegrass and tall fescue germination. *Int. Turfgrass Soc. Res. J.* 9:334–339.
401. Qian, Y. L., S. J. Wilhelm, and K. B. Marcum. 2001. Comparative responses of two Kentucky bluegrass cultivars to salinity stress. *Crop Sci.* 41:1895–1900.
402. Quaggiotti, S., B. Ruperti, P. Borsa, T. Destro, and M. Malagoli. 2003. Expression of putative high-affinity NO_3^- transporter and of an H^+-ATPase in relation to whole plant nitrate transport physiology in two maize genotypes differently responsive to low nitrogen availability. *J. Exp. Bot.* 54:1023–1031.
403. Quisenberry, S. S. 1990. Plant resistance to insects and mites in forage and turf grasses. *Fla. Entomol.* 73:411–421.
404. Reeves, S. A., G. C. McBee, and M. E. Bloodworth. 1970. Effect of N, P, and K tissue levels and late fall fertilization on the cold hardiness of Tifgreen bermudagrass (*Cynodon dactylon x C. transvaalensis*). *Agron. J.* 62:659–662.
405. Reid, M. E. 1933. Effects of shade of the growth of velvet bent and metropolitan creeping bent. *USGA Green Sect.* 13:131–135.
406. Reinert, J. A., J. C. Read, M. E. McCoy, J. J. Heitholt, S. P. Metz, and R. J. Bauernfeind. 2006. Susceptibility of bluegrasses to bluegrass billbug. *USGA Turfgrass Environ. Res. Online* 5(5):1–9.
407. Remans, T., P. Nacry, M. Pervent, T. Girin, P. Tillard, M. Lepetit, and A. Gojon. 2006. A central role for the nitrate transporter *NRT2.1* in the integrated morphological and physiological responses of the root system to nitrogen limitation in *Arabidopsis*. *Plant Physiol.* 140:909–921.
408. Richardson, M. D. 2002. Turf quality and freezing tolerance of 'Tifway' bermudagrass as affected by late-season nitrogen and trinexapac-ethyl. *Crop Sci.* 42:1621–1626.
409. Richmond, D. S., P. S. Grewal, and J. Cardina. 2004. Influence of Japanese beetle *Popillia japonica* larvae and fungal endophytes on competition between turfgrasses and dandelion. *Crop Sci.* 44:600–606.
410. Riederer, M., and C. Mullar. 2006. *Biology of Plant Cuticle*. Blackwell Publishing. Oxford, UK.
411. Rieke, P. E., and G. T. Lyman. 2002. Turf tips for the homeowner: fertilizing home lawns to preserve water quality. *MSUE Bulletin E05TURF*. Mich. St. Univ. Coop. Ext. Serv., East Lansing, MI.
412. Rieke, P. E., S. L. McBurney, and R. A. Bay. 1982. Turfgrass soils report—effect of foliar applications of urea nitrogen and ferrous sulfate on quality of Adelphi Kentucky bluegrass. *Proc. 53rd Ann. Mich. Turfgrass.* 12:24–25.
413. Rivero, R. M., J. M. Ruiz, and L. M. Romero. 2004. Importance of N source on heat stress tolerance due to the accumulation of proline and quaternary ammonium compounds in tomato plants. *Plant Biol (Stuttg).* 6:702–707.
414. Rodriguez, I. R., L. B. McCarty, J. E. Toler, and R. B. Dodd. 2005. Soil CO_2 concentration effects on creeping bentgrass grown under various soil moisture and temperature conditions. *HortScience* 40(3):839–841.
415. Rodriguez, I. R., and G. L. Miller. 2000. Using near infrared reflectance spectroscopy to schedule nitrogen applications on dwarf-type bermudagrasses. *Agron. J.* 92:423–427.
416. Rodriguez, I. R., G. L. Miller, and L. B. McCarty. 2002. Bermudagrass establishment on high sand-content soils using various N–P–K ratios. *HortScience* 37:208–209.
417. Rodriguez-Kabana, R. 1986. Organic and inorganic nitrogen amendments to soil as nematode suppressants. *J. Nematol.* 18:129–135.
418. Rogers, R. A., J. H. Dunn, and C. J. Nelson. 1975. Cold hardening and carbohydrate composition of 'Meyer' zoysia. *Agron. J.* 67:836–838.

419. Rosenthal, G. A. 1982. *Plant Nonprotein Amino and Amino Acids. Biological, Biochemical and Toxicological Properties.* Academic Press. New York.
420. Sachs, P. D. 2004. *Managing Healthy Sports Fields: A Guide to Using Organic Materials for Low-Maintenance and Chemical-Free Playing Fields.* John Wiley & Sons. Hoboken, NJ.
421. Sachs, P. D., and R. T. Luff, 2002. *Ecological Golf Course Management.* Ann Arbor, MI.
422. Salvatore, S. M., and K. Guillard. 2006. Anion exchange membrane soil nitrate predicts turfgrass color and yield. *Crop Sci.* 46:569–577.
423. Santi, S., G. Locci, R. Monte, R. Pinton, and Z. Varanini. 2003. Induction of nitrate uptake in maize roots: expression of a putative high-affinity nitrate transporter and plasma membrane H^+-ATPase isoforms. *J. Exp. Bot.* 54:1851–1864.
424. Sargent, J.A. 1965. The penetration of growth regulators into leaves. *Annu. Rev. Plant Physiol.* 16:1–12.
425. Schlossberg, M. J., and K. J. Karnok. 2001. Root and shoot performance of three creeping bentgrass cultivars as affected by nitrogen fertility. *J. Plant Nutr.* 24(3):535–548.
426. Schmidt, R. E., and R. E. Blaser. 1969. Effect of temperature, light, and nitrogen on growth and metabolism of Tifgreen bermudagrass. *Crop Sci.* 9:5–9.
427. Schnrer, J., and T. Rosswall. 1982. Fluorescein diacetate hydrolysis as a measure of total microbial activity in soil and litter. *Appl. Environ. Microbiol.* 43:1256–1261.
428. Schönherr, J. 2006. Characterization of aqueous pores in plant cuticles and permeation of ionic solutes. *J. Exp. Bot.* 57:2471–2491.
429. Schumann, G. L., P. J. Vittum, M. L. Elliott, and P. P. Cobb. 1998. *IPM Handbook for Golf Courses.* Ann Arbor. Chelsea, MI.
430. Shahba, M. A., Y. L. Qian, H. G. Hughes, A. J. Koski, and D. Christensen. 2003. Relationship of carbohydrates and cold hardiness in six saltgrass accessions. *Crop Sci.* 43:2148–2153.
431. Shahba, M. A., Y. L. Qian, H. G. Hughes, D. Christensen, and A. J. Koski. 2003. Cold hardiness of saltgrass accessions. *Crop Sci.* 43:2142–2147.
432. Sharma, H. C., K. K. Sharma, and J. H. Couch. 2004. Genetic transformation of crops for insect resistance: potential and limitations. *Crit. Rev. Plant Sci.* 23(1):47–72.
433. Shashikumar, K., and J. L. Nus. 1993. Cultivar and winter cover effects on bermudagrass cold acclimation and crown moisture content. *Crop Sci.* 33:813–817.
434. Shearman, R. C. 2006. Fifty years of splendor in the grass. *Crop Sci.* 46:2218–2229.
435. Shearman, R. C., and J. B. Beard. 1975. Turfgrass wear tolerance mechanisms: I. Wear tolerance of seven turfgrass species and quantitative methods for determining turfgrass wear injury. *Agron. J.* 67(2):208–211.
436. Shearman, R. C., and J. B. Beard. 1975. Turfgrass wear tolerance mechanisms: II. Effects of cell wall constituents on turfgrass wear tolerance. *Agron. J.* 67(2):211–215.
437. Shearman, R. C., and J. B. Beard. 1975. Turfgrass wear tolerance mechanisms: III. Physiological, morphological, and anatomical characteristics associated with turfgrass wear tolerance. *Agron. J.* 67(2):215–218.
438. Shearman, R. C., T. P. Riordan, B. G. Abeyo, T. M. Heng-Moss, D. J. Lee, and R. E. Gaussoin. 2006. Buffalograss: tough native turfgrass. *USGA Turfgrass Environ. Res. Online* 5(21):1–13.
439. Shearman, R. C., S. Severmutlu, T. P. Riordan, and U. Bilgili. 2006. Overseeding fine fescue in buffalograss turfs. *USGA Turfgrass Environ. Res. Online* 5(9):1–8.
440. Shearman, R. C., and J. E. Watkins. 1985. Kentucky bluegrass lateral growth and stem rust response to soil compaction stress. *HortScience* 20:388–390.
441. Shetlar, D., P. Heller, and P. Irish. 1990. *Turfgrass Insect and Mite Manual—Third Edition.* Pennsylvania Turfgrass Council. Bellefonte, PA.
442. Shortman, S. L., S. K. Braman, R. R. Duncan, W. W. Hanna, and M. C. Engelke. 2002. Evaluation of turfgrass species and cultivars for potential resistance to twolined spittlebug, *Prosapia bicincta* (Say) (*Homoptera:Cercopidae*). *J. Econ. Entomol.* 95:478–486.
443. Shuman, L. M. 2001. Phosphate and nitrate movement through simulated golf greens. *Water Air Soil Pollut.* 129:305–318.
444. Shuman, L. M. 2002. Phosphorus and nitrate nitrogen in runoff following fertilizer application to turfgrass. *J. Environ. Qual.* 31:1710–1715.
445. Sims, G. K., T. R. Ellsworth, and R. L. Mulvaney. 1995. Microscale determination of inorganic nitrogen in water and soil extracts. *Commun. Soil Sci. Plant Anal.* 26:303–316.
446. Sistani, K. R., G. A. Pederson, G. E. Brink, and D. E. Rowe. 2003. Nutrient uptake by ryegrass cultivars and crabgrass from a highly phosphorus-enriched soil. *J. Plant Nutr.* 26(12):2521–2535.

447. Smiley, R. W., and R. J. Cook. 1973. Relationship between take-all of wheat and rhizosphere pH in soils fertilized with ammonium vs. nitrate-nitrogen. *Phytopathology* 68:882–890.

448. Smiley, R. W., P. H. Dernoeden, and B. B. Clarke. 2005. *Compendium of Turfgrass Diseases.* 3rd Ed. APS Press. Pilot Knob, MN.

449. Snyder, G. H., B. J. Augustin, and J. M. Davidson. 1984. Moisture sensor-controlled irrigation for reducing N leaching in bermudagrass turf. *Agron. J.* 76:964–969.

450. Snyder, G. H., and J. L. Cisar. 2000. Nitrogen/potassium fertilization ratios for bermudagrass turf. *Crop Sci.* 40:1719–1723.

451. Sonoda Y., A. Ikeda, S. Saiki, N. von Wirén, T. Yamaya, and J. Yamaguchi. 2003. Distinct expression and function of three ammonium transporter genes (*OsAMT1;1–1;3*) in rice. *Plant Cell Physiol.* 44:726–734.

452. Sonoda, Y, A. Ikeda, S. Saiki, T. Yamaya, and J. Yamaguchi. 2003. Feedback regulation of the ammonium transporter gene family AMT1 by glutamine in rice. *Plant Cell Physiol.* 44:1396–1402.

453. Spangenberg, B. G., T. W. Fermanian, and D. J. Wehner. 1986. Evaluation of liquid-applied nitrogen fertilizers on Kentucky bluegrass turf. *Agron. J.* 78:1002–1006.

454. Sprent, J. I. 1987. *The Ecology of the Nitrogen Cycle.* Cambridge University Press. New York.

455. Stanford, R. L., R. H. White, J. P. Krausz, J. C. Thomas, P. Colbaugh, and S. D. Abernathy. 2005. Temperature, nitrogen and light effects on hybrid bermudagrass growth and development. *Crop Sci.* 45:2491–2496.

456. Starr, J. L., and H. C. DeRoo. 1981. The fate of nitrogen fertilizer applied to turfgrass. *Crop Sci.* 21:531–536.

457. Steinke K., and J. C. Stier. 2003. Nitrogen selection and growth regulator applications for improving shaded turf performance. *Crop Sci.* 43:1399–1406.

458. Stewart, I. 1963. Chelation in the Absorption and Translocation of Mineral Elements. *Annu. Rev. Plant Physiol.* 14:295–310.

459. Stier, J. C., and J. N. Rogers, III. 2001. Trinexapac-ethyl and iron effects on supina and Kentucky bluegrasses under low irradiance. *Crop Sci.* 41:457–465.

460. Storer, D. A. 1984. A simple high volume ashing procedure for determining soil organic matter. *Comm. Soil Sci. Plant Anal.* 15:759–772.

461. Suenaga, A., K. Moriya, Y. Sonoda, A. Ikeda, N. von Wirén, T. Hayakawa, J. Yamaguchi, and T. Yamaya. 2003. Constitutive expression of a novel-type ammonium transporter *OsAMT2* in rice plants. *Plant Cell Physiol.* 44:206–211.

462. Sullivan, W. M., and Z. Jiang. 2003. Soil nitrate bioavailability monitoring in production and use environments. *Acta Horticulturae* 661:413–419.

463. Sullivan, W. M., Z. Jiang, and R. J. Hull. 2000. Root morphology and its relationship with nitrate uptake in Kentucky bluegrass. *Crop Sci.* 40:765–772.

464. Suplick-Ploense, M. R., Y. L. Qian, and J. C. Read. 2002. Relative NaCl tolerance of Kentucky bluegrass, Texas bluegrass, and their hybrids. *Crop Sci.* 42:2025–2030.

465. Tahir, I. S. A., and N. Nakata. 2005. Remobilization of nitrogen and carbohydrate from stems of bread wheat in response of heat during grain filling. *J. Agron. Crop Sci.* 191:106–115.

466. Taiz, L., and E. Zeiger. 2005. *Plant Physiology.* 4th Ed. The Benjamin/Cummings Publishing Company, Inc. Redwood City, CA.

467. Taliaferro, C. M., D. L. Martin, J. A. Anderson, M. P. Anderson, and A. C. Guenzi. 2004. Broadening the horizons of turf bermudagrass. *USGA Turfgrass Environ. Res. Online* 3(20):1–9.

468. Tan, Z. G., and Y. L. Qian. 2003. Effects of light intensity on gibberellic acid content in Kentucky bluegrass. *HortScience* 38:113–116.

469. Tani, T., and J. B. Beard. 2002. *Color Atlas of Turfgrass Diseases.* John Wiley & Sons. New York.

470. Tegg, R. S., and P. A. Lane. 2004. Shade performance of a range of turfgrass species improved by trinexapac-ethyl. *Aust J. Exp. Ag.* 44:939–945.

471. Teutsch, C. D., and W. M. Tilson. 2005. Nitrate accumulation in crabgrass as impacted by nitrogen fertilization rate and source *Online Forage and Grazinglands:* 30 August 2005.

472. Thomashow, M. F. 1999. Plant cold acclimation: freezing tolerance genes and regulatory mechanisms. *Annu. Rev. Plant Physiol. Plant Mol. Biol.* 50:571–599.

473. Thompson, D. C., B. B. Clarke, and J. R. Heckman. 1995. Nitrogen form and rate of nitrogen and chloride application for the control of summer patch in Kentucky bluegrass. *Plant Dis.* 79:51–56.

474. Thornton, B. 2004. Inhibition of nitrate influx by glutamine in *Lolium perenne* depends upon the contribution of the HATS to the total influx. *J. Exp. Bot.* 55:761–769.

475. Throop, H. L., E. A. Holland, W. J. Parton, D. S. Ojima, and C. A. Keough. 2004. Effects of nitrogen deposition and insect herbivory on patterns of ecosystem-level carbon and nitrogen dynamics: results from the CENRURY model. *Global Change Biol.* 10:1092–1105.

476. Tisserat, N. A., F. B. Iriarte, H. C. Wetzel III, J. D. Fry, and D. L. Martin. 2003. Identification, distribution, and aggressiveness of spring dead spot pathogens of bermudagrass. *USGA Turfgrass Environ. Res. Online* 2(20):1–8.

477. Titko III, S., J. R. Street, and T. J. Logan. 1987. Volatilization of ammonia from granular and dissolved urea applied to turfgrass. *Agron. J.* 79:535–540.

478. Tong. Y., J. J. Zhou, Z. Li, and A. J. Miller. 2005. A two-component high-affinity nitrate uptake system in barley. *Plant J.* 41:442–450.

479. Torello, W. A., and L. A. Rice. 1986. Effects of NaCl stress on proline and cation accumulation in salt sensitive and tolerant turfgrasses. *Plant Soil* 93:241–247.

480. Torello, W. A., D. J. Wehner, and A. J. Turgeon. 1983. Ammonia volatilization from fertilized turfgrass stands. *Agron. J.* 75:454–456.

481. Totten, F. W. 2006. Long-term evaluation of liquid vs. granular nitrogen fertilization in creeping bentgrass. Ph.D. dissertation. Clemson University, Clemson, SC.

482. Totten, F. W., H. Liu, L. B. McCarty, and C. M. Baldwin. 2005. Comparison of granular and liquid fertility on creeping bentgrass. *Agronomy Abstracts.*

483. Totten, F. W., H. Liu, L. B. McCarty, and C. M. Baldwin. 2005. Comparison of granular and liquid fertility on "TifEagle" bermudagrass. *Agronomy Abstracts.*

484. Tredway, L. P., M. D. Soika, and B. B. Clarke. 2005. Red thread development in perennial ryegrass in response to nitrogen phosphorus, and potassium fertilizer applications. *Int. Turfgrass Soc. Res. J.* 10:715–722.

485. Trenholm, L. E., R. N. Carrow, and R. R. Duncan. 2000. Mechanisms of wear tolerance in seashore paspalum and bermudagrass. *Crop Sci.* 40:1350–1357.

486. Trenholm, L. E., A. E. Dudeck, J. B. Sartain, and J. L. Cisar. 1998. Bermudagrass growth, total nonstructural carbohydrate concentration, and quality as influenced by nitrogen and potassium. *Crop Sci.* 38:168–174.

487. Trevathan, L. E. 1982. Response of ryegrass introductions to artificial inoculation with Pyricularia grisea under greenhouse conditions. *Plant Dis.* 66:696–697.

488. Trevathan, L. E., M. A. Moss, and D. Blasingame. 1994. Ryegrass blast. *Plant Dis.* 78:113–117.

489. Trueman, L. J., A. Richardson, and B. G. Forde. 1996. Molecular cloning of higher plant homologues of the high-affinity nitrate transporters of *Chlamydomonas reinhardtii* and *Aspergillus nidulans. Gene.* 175:223–231.

490. Tsay, Y. E., J. I. Schroeder, K. A. Feldmann, and N. M. Carwford. 1993. The herbicide sensitivity gene *CHL1* of *Arabidopsis* encodes a nitrate-inducible nitrate transporter. *Cell.* 72:705–713.

491. Tucker, B. J., L. B. McCarty, H. Liu, C. E. Wells, and J. R. Rieck. 2006. Mowing height, nitrogen rate, and biostimulant influence root development of field-grown 'TifEagle' bermudagrass. *HortScience* 41(3):805–807.

492. Turgeon, A. J. 2005. *Turfgrass Management—7th Edition.* Prentice-Hall. Upper Saddle River, NJ.

493. Turgron, A. J. 1994. *Turf Weeds and Their Control.* American Society of Agronomy, Inc. Crop Science Society of America, Inc. Madison WI.

494. Turner, T. R., and N. W. Hummel, Jr. 1992. Nutritional requirements and fertilization. In D. V. Waddington, R. N. Carrow and R. C. Shearman, (eds.), *Turfgrass.* Agronomy Monograph 32. Madison, WI.

495. Uddin, W., L. L. Burpee, and K. L. Stevenson. 1997. Influence of temperature and leaf wetness duration on development of gray leaf spot (blast) of tall fescue. pp. 137–138. In *Agronomy Abstracts.* ASA. Madison, WI.

496. Uddin, W., M. D. Soika, F .E. Moorman, and G. Viji. 1999. A serious outbreak of blast disease (gray leaf spot) of perennial ryegrass in golf course fairways in Pennsylvania. *Plant Dis.* 83:783.

497. Uddin, W., G. Viji, and P. Vincelli. 2003. Gray leaf spot of perennial ryegrass turf: an emerging problem for the turfgrass industry. *USGA Turfgrass Environ. Res. Online* 2(11):1–18.

498. Uva, R. H., J. C. Neal, and J. M. Ditomaso. 1997. *Weeds of the Northeast.* Cornell University Press. Ithaca, NY.

499. Vaiciunas, S., and B. B. Clarke. 1998. Impact of cultural practices and genotype on the development of gray leaf spot in cool-season turfgrasses. p. 140. In *Agronomy Abstracts.* ASA. Madison, WI.

500. Vaiciunas, S., and B. B. Clarke. 2000. Impact of management practices on the development of gray leaf spot in cool-season turfgrasses. *Proceedings of the Rutgers Turfgrass Symposium.* New Brunswick, NJ.

501. Valiela, I., G. Collins, J. Kremer, K. Lajtha, M. Geist, B. Seely, J. Brawley, and C. H. Sham. 1997. Nitrogen loading from coastal watersheds to receiving estuaries: new method and application. *Ecol. Appl.* 7:358–380.

502. Van Der Mear, H. G., J. C. Ryden, and G. C. Ennik. 1986. *Nitrogen Fluxes in Intensive Grassland Systems.* Martinus Nijhoff Publishers. Boston, MA.

503. Vargas, J. M. 2005. *Management of Turfgrass Diseases—Third Edition.* John Wiley & Sons. Hoboken, NJ.

504. Vargas, J. M., and J. B. Beard. 1981. Shade environment-disease relationships of Kentucky bluegrass cultivars. *Int. Turf Soc. Res. J.* 4:391–395.

505. Vargas, J. M., and A. J. Turgeon. 2003. *Poa Annua: Physiology, Culture, and Control of Annual Bluegrass.* John Wiley & Sons. Hoboken, NJ.

506. Vidmar, J. J., D. Zhuo, M. Y. Siddiqi, J. K. Schjoerring, B. Touraine, and A. D. M. Glass. 2000. Regulation of high-affinity nitrate transporter genes and high-affinity nitrate influx by nitrogen pools in roots of barley. *Plant Physiol.* 123:307–318.

507. Vierling, E. 1991. The role of heat shock proteins in plants. *Annu. Rev. Plant Physiol. Plant Mol. Biol.* 42:579–620.

508. Vincelli, P. 1999. Gray leaf spot. An emerging disease of perennial ryegrass. *Turfgrass Trends* 7:1–8.

509. Vinceli, P., and A. J. Powell. 1994. Impact of mowing height and nitrogen fertility on brown patch in tall fescue. *Biol. Cult. Tests Control Plant Dis.* 10:42.

510. Vinceli, P., and D. Williams. 1998. Managing spring dead spot of bermudagrass. *Golf Course Manage.* 66(5):49–53.

511. Vitousek, P. M., J. A. Aber, R. W. Howarth, G. E. Likens, P. A. Matson, D. W. Schindler, W. H. Schlesinger, and G. D. Tilman. 1997. Human alterations of the global nitrogen cycle: causes and consequences. *Ecol. Appl.* 7:737–750.

512. Vittum, P., M. G. Villani, and H. Tashiro. 1999. *Turfgrass Insects of the United States and Canada.* Cornell University Press. Ithaca, NY.

513. Walker, W. J., and B. Branham. 1992. Environmental impacts of turfgrass fertilization. pp. 105–219. In J. C. Balogh and W. J. Walker (eds.), *Golf Course Management and Construction-Environmental Issues.* USGA, Lewis Publishers. Chelsea, MI.

514. Waltz, F. C., Jr. 1997. The effects of trinexapac-ethyl on total nonstructural carbohydrates in field-grown hybrid bermudagrass. MS Thesis. Clemson University. Clemson, SC.

515. Wang, R., D. Liu, and N. M. Crawford. 1998 The *Arabidopsis CHL1* protein plays a major role in high-affinity nitrate uptake. *Proc. Natl. Acad. Sci. U.S.A.* 95:15134–15139.

516. Wang, R., R. Tischner, R. A. Gutierrez, M. Hoffman, X. Xing, M. Chen, G. Coruzzi, and N. M. Crawford. 2004. Genomic analysis of the nitrate response using a nitrate reductase-null mutant of Arabidopsis. *Plant Physiol.* 136:2512–2522.

517. Wang, Y., A. P. Kausch, H. Luo, J. M. Chandlee, B. A. Ruemmele, M. Browning, N. Jackson, and M. R. Goldsmith. 2003. Co-transfer and expression of chitinase, glucanase, and bar genes in creeping bentgrass for conferring fungal disease resistance. *Plant Sci.* 165:497–506.

518. Wang, Z. Y., A. Hopkins, and R. Mian. 2001. Forage and turf grass biotechnology. *Crit. Rev. Plant Sci.* 20(6):573–619.

519. Wang, Z., Q. Xu, and B. Huang. 2004. Endogenous cytokinin levels and growth responses to extended photoperiods for creeping bentgrass under heat Stress. *Crop Sci.* 44:209–213.

520. Wang, Z. and B. Huang. 2005. Physiological recovery of Kentucky bluegrass from simultaneous drought and heat stress. *Crop Sci.* 44:1729–1736.

521. Watkins, J. E., R. E. Gaussoin, W. K. Frank, and L. A. Wit. 2005. Brown patch severity and perennial ryegrass quality as influenced by nitrogen rate and source and cultivar. *Int. Turfgrass Soc. Res. J.* 10:723–728.

522. Watschke, T. L., P. H. Dernoeden, and D. J. Shetlar. 1994. *Managing Turfgrass Pests.* Lewis Publishers. Ann Arbor, MI.

523. Wehner, D. J., and J. E. Haley. 1988. Foliar burn from turfgrass fertilization. *Agronomy Abstracts* p. 157.

524. Wesly, R. W., R. C. Shearman, and E. J. Kinbacker. 1985. Foliar N-uptake by eight turfgrass grown in controlled environment. *J. Amer. Soc. Hort. Sci.* 110(5):612–614.

525. Wesly, R. W., R. C. Shearman, E. J. Kinbacher, and S. R. Lowry. 1987. Ammonia volatilization from foliar-applied urea on field-grown Kentucky bluegrass. *HortScience* 22(6):1278–1280.

526. White, R. H. 2004. Environment and culture affect bermudagrass growth and decline. *USGA Greens Sect. Rec.* 42(6):21–24.
527. Williams, D. W., P. B. Burrus, and P. Vincelli. 2001. Severity of gray leaf spot in perennial ryegrass as influenced by mowing height and nitrogen level. *Crop Sci.* 41:1207–1211.
528. Williams, D. W., A. J. Powell, Jr., C. T. Dougherty, and P. Vincelli. 1998. Separation and quantization of the sources of dew on creeping bentgrass. *Crop Sci.* 38:1613–1617.
529. Williams, D. W., A. J. Powell, Jr., P. Vincelli, and C. T. Dougherty. 1996. Dollar spot on bentgrass influenced by displacement of leaf surface moisture, nitrogen, and clipping removal. *Crop Sci.* 36:1304–1309.
530. Williams, L. E., and A. J. Miller. 2001. Transporters responding for the uptake and partitioning of nitrogenous solutes. *Annu. Rev. Plant Biol.* 52:659–688.
531. Williamson, R. C., and S. C. Hong. 2005. Identification of mechanism(s) of resistance in Kentucky bluegrass for control of black cutworms. *USGA Turfgrass Environ. Res. Online* 4(22):1–6.
532. Williamson, R. C., A. T. Walston, and D. Soldot. 2005. Influence of organic-based fertilizers and root zone mixes on the incidence of black turfgrass ataenius (Coleoptera: Scarabaeidae) infestations on golf courses. *Int. Turfgrass Soc. Res. J.* 10:803–810.
533. Witte, C. P., S. A. Tiller, M. A. Taylor, and H. V. Davies. 2002. Leaf urea metabolism in potato. Urease activity profile and patterns of recovery and distribution of ^{15}N after foliar urea application in wild-type and urease-antisense transgenics. *Plant Physiol.* 128:1129–1136.
534. Witteveen, G., and M. Bavier. 1998. *Practical Golf Course Maintenance: The Magic of Greenkeeping.* 2nd Ed. John Wiley & Sons. Chelsea, MI.
535. Wittwer, S. H., and F. G. Teubner. 1959. Foliar absorption of mineral nutrients. *Ann. Rev. Plant Physiol.* 10:13–30.
536. Xie, S., and Q. Zhang. 2004. Kinetics of uptake and export of foliar-applied radio-labeled phosphorus by leaf and fruit rind of Satsuma mandarin during fruit development. *J. Plant Nutr.* 27(2):223–237.
537. Xu, Q., and B. Huang. 2000. Growth and physiological responses of creeping bentgrass to changes in air and soil temperatures. *Crop Sci.* 40:1363–1368.
538. Xu, Q., and B. Huang. 2001. Lowering soil temperature improves creeping bentgrass growth under heat stress. *Crop Sci.* 41:1878–1883.
539. Xu, Q., and B. Huang. 2003. Seasonal changes in carbohydrate accumulation for two creeping bentgrass cultivars. *Crop Sci.* 43:266–271.
540. Xu, Q., and B. Huang. 2004. Antioxidant metabolism associated with summer leaf senescence and turf quality decline for creeping bentgrass. *Crop Sci.* 44:553–560.
541. Yust, A. K., D. J. Wehner, and T. W. Fermanian. 1984. Foliar application of N and Fe to Kentucky bluegrass. *Agron. J.* 76:934–938.
542. Zhang, H., and B. G. Forde. 2000. Regulation of *Arabidopsis* root development by nitrate availability. *J. Exp. Bot.* 51:51–59.
543. Zhang, X., E. H. Ervin, and A. J. LaBranche. 2006. Metabolic defense responses of seeded bermudagrass during acclimation to freezing stress. *Crop Sci.* 46:2598–2605.
544. Zhang, X., E. H. Ervin, and R. E. Schmidt. 2003. Plant growth regulators can enhance the recovery of Kentucky bluegrass sod from heat injury. *Crop Sci.* 43:952–956.
545. Zhang, X., E. H. Ervin, and R. E. Schmidt. 2005. The role of leaf pigment and antioxidant levels in UV-B resistance of dark- and light-green Kentucky bluegrass cultivars. *J. Am. Soc. Hortic. Sci.* 130:836–841.
546. Zhang, X., G. C. Munshaw, and E. H. Ervin. 2004. Influence of late-season jasmonic acid and sodium chloride treatments on seeded bermudagrass cold hardiness. *Agronomy Abstracts.* ASA, CSSA, SSSA. Madison, WI.

Section IX

Turfgrass Management and Physiological Issues in the Future

34 Saltgrass (*Distichlis spicata*), a Potential Future Turfgrass Species with Minimum Maintenance/Management Practice Requirements

Mohammad Pessarakli

CONTENTS

34.1 INTRODUCTION

Desert saltgrass (*Distichlis spicata* [L.] Greene var. *stricta* [Gray] Beetle) [4], a potential feed plant and turf species, grows in very poor- to fair-condition soils, in both salt-affected soils and soils under drought conditions. It grows in arid and semiarid regions. The plant is abundantly found in areas of the western parts of the United States as well as on the sea-shores of the United Arab Emirates and several other Middle Eastern countries, Africa, and South and Central American countries. Although better adapted to arid and semiarid regions, the grass is found in a wide range of climatic and geographical regions. The plant can be manipulated to increase its yield and productivity or control its growth to produce turf-type grasses. This plant has multipurpose usages. It can be substituted for animal feeds like alfalfa or can be used in soil conservation for covering

605

roadsides and soil surfaces in areas with high risk of soil erosion. Recently, the United States Golf Association (USGA) has shown a great deal of interest in financing research work on this plant for use as turfgrass. Most of these research works have been conducted at the University of Arizona and Colorado State University. Consequently, the USGA funds for the investigations on this grass species have been allocated to these institutions.

Positive and promising results have already been obtained from these studies [7–11,13,19, 20,23–31].

To our knowledge, there is no work reported in the literature regarding the optimum nutrient and water requirements of saltgrass. The degrees of salinity and drought tolerances of this grass have not been fully established either. The above reports [7–11,13,19, 20,23–31] and those of Enberg and Wu [3], Miller et al., [14] Miyamoto et al., [15] Rossi et al., [36] Sigua and Hudnall, [39] and Sowa and Towill [40] are concerned only with the growth of this species.

These issues are best addressed in the research reports of this author and the other investigators conducting research on this grass species at the University of Arizona and Colorado State University.

The objectives of this chapter are to present new findings on saltgrass and compare this grass with other common warm-season turfgrasses (i.e., bermudagrass [*Cynodon dactylon* L.] and seashore paspalum [*Paspalum vaginatum*]). These comparisons will be made on the growth responses in terms of shoot and root length, fresh and dry weights, and nitrogen content and concentration of the shoot tissues of desert saltgrass grown under control and drought or salinity conditions.

Pessarakli and Kopec [23], Pessarakli et al., [24–26,28–31] and Pessarakli and Marcum [27] studied growth responses of several accessions of saltgrass and compared them with bermudagrass (*Cynodon dactylon* L.). These investigators found that various accessions of saltgrass were superior to bermudagrass in respect to both salinity and drought tolerance. Saltgrass survived 4 months of drought stress and over twice the salinity level of seawater in these studies. Amazingly, all the accessions of saltgrass completely recovered after salinity and drought stresses were alleviated. These investigations proved that this grass can survive under minimum water availability or very harsh stressful environmental conditions. Therefore, it requires minimum water or nutrients for its growth and development compared with any other grasses [19,20,23–25,28–30].

In another study, Pessarakli et al. [26] compared establishment of saltgrass, bermudagrass, and seashore paspalum (*Paspalum vaginatum*) under salinity stress conditions and found that for any plant parts used for the establishment, saltgrass performed better than both bermudagrass and paspalum under stress conditions. However, under normal (nonstressed) and lower levels of salinity stress both bermudagrass and paspalum performed better than saltgrass, particularly in color and quality.

In an ongoing salinity and drought stress investigation at the University of Arizona, Pessarakli et al. [24,25] found a wide range in the degrees of salinity and drought tolerances of saltgrass.

Kopec et al. [7,8] found significant differences in traffic tolerance and mowing height of various saltgrass accessions. These investigators found that rolling (traffic stress), on average, generally increased the quality of saltgrass. However, rolling did decrease percent plot green cover in early summer, but not so afterward.

Entries A138 and A65 where the only two clones to produce season-long high-quality turfs whether rolled regularly or not, and mowed at 2.2 cm height. Pessarakli et al. [24] in another study examined interactive effects of salinity and mowing of saltgrass under various mowing heights. They found that various saltgrass clones performed differently. The grasses generally responded very well under salinity stress at 5 and 2.5 cm mowing height. However, they did not show satisfactory growth at 0.5 cm mowing height.

The results of several research investigations [9–11,13,19,20,23–31] conducted under extremely high levels of salinity and extended drought conditions by the author and the other investigators on this grass species at the University of Arizona showed that although at longer exposure times to salinity stress the growth of saltgrass in terms of shoot and root lengths and dry weights ceased, the grass still maintained some green color and survived the stress condition. This phenomenon suggests that saltgrass growing even under poor soil conditions (salt-affected soils) can be a beneficial turfgrass, and still show a favorable growth.

Pessarakli et al. [24,25,28–30] in numerous salinity stress and drought studies identified some superior accessions of saltgrass among several tested grasses. For example, the results of several of their investigations suggest that A138 and A137 accessions show clear superiority over others under salinity and drought-stress conditions. These accessions may be recommended for further tests and use as turfgrass under low-maintenance conditions. In most of the investigations, when bermudagrass was compared with saltgrass regarding stress tolerance, bermudagrass reached full dormancy before most of the saltgrass accessions did.

Pessarakli et al. [24,25] compared salinity stress tolerance of 12 new clones of saltgrass and concluded the following:

- Clones differed greatly in their maintenance of green color retention (quality) as electrical conductivity (EC) levels (salinity) increased.
- Shoot (clipping) dry-matter (DM) weights decreased linearly with an increase in EC level.
- Two clones that produced acceptable quality at the EC level of 34 were clones 239 and 240.
- No clones could maintain adequate quality turf at EC 48 after 7 weeks of exposure at this EC level. It should be noted that the EC 48 is equal to or greater than the EC of most seawaters.
- Salinity tolerance level between clones was significant.
- The grasses were separated in several groups with different degrees of salt tolerance.

These investigators evaluated several parameters (shoot and root lengths and DMs, grass quality; Tables 34.1 through 34.3, Figure 34.1) for identifying the salinity tolerance of saltgrass and

TABLE 34.1
Saltgrass Shoot (Clippings) DM under Salinity Stress

	Shoot DM (g) at EC			
Grass ID	6	20	34	48
A37	1.10cde	0.57bcde	0.27cde	0.15c
A49	1.26bcd	0.77ab	0.32bcde	0.13c
A50	1.65ab	0.60bcd	0.21de	0.17bc
A60	1.03cde	0.38e	0.17e	0.13c
72	1.38bc	0.82a	0.38abc	0.19bc
A86	1.66ab	0.86a	0.26cde	0.14c
A107	0.95de	0.52cde	0.30bcde	0.20bc
A126	0.83e	0.41de	0.18e	0.15c
A128	1.37bc	0.73abc	0.52a	0.30a
A138	1.09cde	0.46de	0.36abcd	0.25ab
239	1.67ab	0.88a	0.44ab	0.15c
240	1.94a	0.91a	0.49a	0.24ab

Note: The values are the means of three replications of each treatment. The values in each column followed by the same letters are not statistically significant at the 0.05 probability level.

Source: Pessarakli, M. et al., *Growth Responses of Twelve Inland Saltgrass Clones to Salt Stress.* ASA-CSSA-SSSA International Annual Meetings, Nov. 6–10, Salt Lake City, UT, 2005.

TABLE 34.2
Saltgrass Root DM (Cumulative Values) under Salinity Stress

Grass ID	Root DM (g) at EC			
	6	20	34	48
A37	0.74cde	0.99def	1.10cdef	0.78cd
A49	1.61b	1.11cdef	1.56bcd	1.03bcd
A50	1.83b	1.65a	1.94abc	0.74cd
A60	1.46bc	1.71a	1.31bcde	0.84bcd
72	0.77cde	0.93def	0.72def	0.50d
A86	1.06bcde	1.18bcde	0.76def	0.81bcd
A107	0.68de	0.84ef	0.53ef	0.68cd
A126	0.50e	0.68f	0.26f	0.48d
A128	3.46a	1.50abc	2.05ab	1.18bc
A138	1.17bcde	0.88def	0.43ef	2.28a
239	1.31bcd	1.30abcd	2.82a	1.21bc
240	3.36a	1.63ab	1.25bcde	1.42b

Note: The values are the means of three replications of each treatment. The values in each column followed by the same letters are not statistically significant at the 0.05 probability level.

Source: Pessarakli, M. et al., *Growth Responses of Twelve Inland Saltgrass Clones to Salt Stress.* ASA-CSSA-SSSA International Annual Meetings, Nov. 6–10, Salt Lake City, UT, 2005.

TABLE 34.3
Saltgrass General Quality under Salinity Stress

Grass ID	General Quality at EC			
	6	20	34	48
A37	8.0cde	5.1f	3.3g	2.6e
A49	7.7def	6.4d	4.3ef	2.8e
A50	8.6abc	7.2bc	5.0cd	4.0bc
A60	8.2bcd	5.5ef	3.9fg	3.5cd
72	9.0a	7.4bc	5.9b	4.8a
A86	8.5abc	6.7cd	5.7b	3.9bc
A107	7.5def	5.9def	5.4bc	4.4ab
A126	6.7g	5.3f	4.6de	3.9bc
A128	7.1fg	6.2de	5.0cd	3.0de
A138	8.6abc	7.9b	5.4bc	4.2ab
239	8.9ab	9.3a	6.6a	4.2ab
240	9.2a	9.7a	7.1a	2.8e

Note: The values are the means of three replications of each treatment. The values in each column followed by the same letters are not statistically significant at the 0.05 probability level.

Source: Pessarakli, M. et al., *Growth Responses of Twelve Inland Saltgrass Clones to Salt Stress.* ASA-CSSA-SSSA International Annual Meetings, Nov. 6–10, Salt Lake City, UT, 2005.

reported that in terms of salinity tolerance (quality), green foliage retention was empirically the best assessment of clonal response to increased salinity stress level. Therefore, for large-scale screening of saltgrass germplasm, the maintenance of green tissue at a specific EC level would seem to be adequate as a simple selection method for salinity tolerance.

Tubs in each row, left to right:
EC = 6, 20, 34, and 48 dS/m, respectively

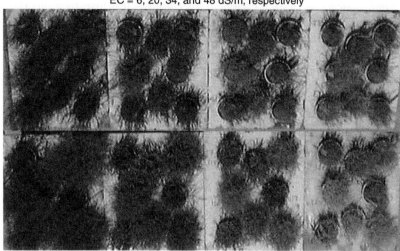

FIGURE 34.1 **(Color Figure 34.1 follows p. 262.)** Saltgrass growth under various salinity levels of the culture medium. (From Pessarakli, M. et al., *Growth Responses of Twelve Inland Saltgrass Clones to Salt Stress*. ASA-CSSA-SSSA International Annual Meetings, Nov. 6–10, Salt Lake City, UT, 2005.)

Kopec et al., [10] in an attempt to find water requirement of saltgrass and comparison with that of bermudagrass and paspalum under greenhouse conditions when soil moisture was not limiting, found the following:

1. Seashore paspalum had a higher evapotranspiration (ET) rate than A119 saltgrass in terms of millimeters per day, which was reflected in the greatest ET water loss (consumptive water use) throughout the test.
2. A48 saltgrass and Tifway bermudagrass had similar daily ET rates and similar total water use.
3. Total water use between two select saltgrass clones was not significantly different. Saltgrass A48 and A119 had a total consumptive water use of 84.2 and 76.5 mm, respectively, over the 19-day test period.

Pessarakli et al. [31] conducted an experiment using ^{15}N to assess nitrogen uptake of saltgrass under salinity stress conditions. This study is reported briefly here.

34.2 MATERIALS AND METHODS

34.2.1 PLANT MATERIALS

Desert saltgrass, accession WA 12, a prostrate physiotype collected in 1999 from a salt playa in Wilcox, Arizona, was used in a greenhouse experiment to evaluate its DM yield and nitrogen (regular and ^{15}N) absorption rates under control and salt (NaCl) stress conditions, using a hydroponics technique.

34.2.2 PLANT ESTABLISHMENT

The plants were vegetatively grown in cups, 9 cm in diameter, and cut to 7 cm height. Silica sand was used as the plant anchor medium. The cups were fitted in plywood lid holes and the lids were placed on $42 \times 34 \times 12$-cm Carb-X polyethelene tubs containing half-strength Hoagland nutrient solution [5]. Four replications of each treatment were used in this investigation. The experiment was conducted in two different seasons to take the possible seasonal variations on the studied parameters into account. In each experiment, the plants were allowed to grow in this nutrient solution for 42 days. During this period, the plant shoots were harvested weekly for them to reach full maturity and develop uniform and equal size plants. The harvested plant materials were discarded. The culture solutions were changed biweekly to ensure adequate amounts of plant essential-nutrient elements for normal growth and development. At the last harvest, day 42, the roots were also cut to 2.5 cm length to have plants with uniform roots and shoots for the stress phase of the experiment.

34.2.3 SALT TREATMENT

The salt treatments were initiated by adding 50-mM 2.925 g of NaCl per liter of the culture solution per day. Four treatments were used, including control (no salt addition), 100, 200, and 400 mM salinity levels. The culture solution levels in the tubs were 10 L in volume, and solution conductivity was monitored/adjusted to maintain prescribed treatment levels. After the final salinity levels were reached, the shoots were harvested and the harvested plant materials discarded prior to beginning the nitrogen-15 treatment.

34.2.4 NITROGEN-15 TREATMENT

The nitrogen-15 treatment was started by adding 5 mg ^{15}N, following procedures outlined by Pessarakli and Tucker [32–35] and Pessarakli [16–18], as 22.931 mg ammonium sulfate $[(^{15}NH_4)_2SO_4]$, 53% ^{15}N (atom percent ^{15}N) per liter of the culture solution per day. After the ^{15}N addition, plant shoots were harvested weekly for the evaluation of the DM production and determination of the ^{15}N absorption. At each weekly harvest, both shoot and root lengths were measured and recorded. The harvested plant materials were oven-dried at 60°C and dry weights measured and recorded. The recorded data were considered to be the weekly plant DM production. At the termination of the experiment, the last harvest, plant roots were also harvested, oven-dried at 60°C, and dry weights determined and recorded. Plant shoots were analyzed for total N and ^{15}N contents.

The data were subjected to analysis of variance, using the SAS statistical package [38]. The means were separated, using the Duncan multiple range test.

34.3 RESULTS AND DISCUSSION

34.3.1 ROOT LENGTH

The root length decreased under sodium chloride stress conditions. This decrease was more pronounced as the exposure time to salinity increased. However, for the first five harvests, the reduction in root length was not statistically significant (Table 34.4). After 6 weeks' exposure to NaCl stress, the reduction in root length due to salt stress was statistically significant.

34.3.2 SHOOT LENGTH

The shoot length was more severely affected under high levels (400 mM) of sodium chloride stress compared with root length (Table 34.4). The effect of stress on shoot length at this level of salinity was shown from the first harvest. Only at the last harvest, the effect of the medium (200 mM) salinity level on the shoot length was statistically significant compared with the control and the low (100 mM) level of NaCl.

TABLE 34.4
Root and Shoot Length, Shoot Fresh and Dry Weights, and Root Dry Weight
(for Only the Last Harvest) of 7-Weekly Harvests of Desert Saltgrass under Four
Salinity Levels

Harvest	Salinity (mM)	Length (cm)		Shoot Weight (g)		Root Weight (g)
		Root	Shoot	Fresh	Dry	Dry
1	Control (0)	21.3a	11.4a	0.487a	0.201a	
	100	19.3a	10.6a	0.457a	0.197a	
	200	20.0a	8.1ab	0.281ab	0.124ab	
	400	20.2a	5.9b	0.137b	0.068b	
2	Control (0)	28.0a	10.2a	0.484a	0.211a	
	100	26.2a	10.1a	0.444a	0.193a	
	200	24.5a	9.0a	0.529a	0.222a	
	400	26.2a	5.8b	0.187b	0.086b	
3	Control (0)	32.0a	12.8a	0.563ab	0.252ab	
	100	32.8a	12.1a	0.536ab	0.240ab	
	200	28.9a	11.1a	0.742a	0.304a	
	400	31.7a	6.7b	0.350b	0.168b	
4	Control (0)	37.6a	12.7a	0.801ab	0.309ab	
	100	34.7a	12.6a	0.740ab	0.281a	
	200	34.1a	12.1a	0.961a	0.349a	
	400	35.0a	7.8b	0.501b	0.202b	
5	Control (0)	42.5a	9.5a	0.386b	0.199b	
	100	37.2a	10.0a	0.447b	0.203b	
	200	37.0a	9.7a	0.651a	0.282a	
	400	36.7a	6.5b	0.420b	0.193b	
6	Control (0)	48.4a	7.8a	0.296b	0.129b	
	100	42.4ab	7.8a	0.396b	0.162b	
	200	39.7ab	7.6a	0.525a	0.206a	
	400	37.1b	5.8b	0.384b	0.157b	
7	Control (0)	50.8a	7.3a	0.301a	0.164a	0.441a
	100	43.9ab	7.2a	0.310a	0.166a	0.454a
	200	40.6b	6.5b	0.404a	0.205a	0.549a
	400	38.2b	5.6c	0.376a	0.183a	0.539a

Note: The values are means of six replications. All the values followed by the same letter in each column are not statistically different at the 0.05 probability level.

Source: Pessarakli, M. et al., *J. Plant Nutr.*, 28(8), 1441–1452, 2005.

34.3.3 SHOOT FRESH WEIGHT

For the first two harvests, the high (400 mM) level of NaCl stress significantly reduced the shoot fresh weight compared with the other levels of salinity (Table 34.4). However, starting from harvest 3 to the last harvest, no significant differences were detected on the shoot fresh weights between the high (400 mM) salinity level compared with the low (100 mM) NaCl stress and the control. As the stress period progressed, the medium (200 mM) sodium chloride stress enhanced the shoot fresh weight. This beneficial effect of NaCl on the shoot fresh weight was statistically significant at harvests 5 and 6 (Table 34.4). This phenomenon is probably due to the fact that saltgrass is a true halophyte, and true halophytes use Na as an essential nutrient element. The Pessarakli and Tucker [32,33] findings of the enhancement of cotton (*Gossypium hirsutum*), a salt-tolerant plant, growth and protein synthesis by this plant under low level of NaCl stress supports this result.

Growth stimulation under moderate salinity was also reported in the salt-tolerant grass, *Sporobolus virginicus* [12]. At the last harvest, harvest 7, despite the higher numerical values of the shoot fresh weight for the medium (200 mM) stressed plants, there was no statistically significant difference between any of the treatments. This is probably because as the plants grew older, they physiologically developed more tolerance by adjusting themselves to stress. This observation was recorded by Pessarakli and Tucker [32,33] for cotton and by Al-Rawahey et al. [1] as well as by Pessarakli [21] for tomatoes.

34.3.4 SHOOT DRY WEIGHT

The shoot dry weights of the plants followed essentially the same pattern as the shoot fresh weights (Table 34.4). For both fresh and dry weights of the shoots, except for the last harvest, the gap between the means of the medium (200 mM) stressed plants and the other treatments was wider as the exposure time to stress progressed.

34.3.5 SHOOT SUCCULENCE

Shoot succulence (fresh wt./dry wt.) increased from control to 200 mM NaCl salinity, then declined. These values (average of all the harvests) were 1.94, 2.20, 2.34, and 2.18 for the control, 100, 200, and 400 mM NaCl-treated plants, respectively.

34.3.6 ROOT DRY WEIGHT

Although the root dry weight was enhanced at the medium (200 mM) and the high (400 mM) sodium chloride levels, there was no statistically significant difference detected between the means of the different treatments. Sagi et al. [37] also found that the adverse effects of salinity stress was more pronounced on shoot than the root growth.

34.3.7 CUMULATIVE AND AVERAGE VALUES

An attempt was made to examine the cumulative and the average values of the studied parameters. This observation indicated that for both cumulative and the average values, all the parameters (shoot length and shoot fresh and dry weights) were similarly affected by sodium chloride stress. Only the high level of stress (400 mM) had a statistically significant adverse effect on these parameters (Table 34.5). The shoot and the root lengths and weights under salinity stress were calculated

TABLE 34.5
Cumulative and Average Values of Shoot Length, Fresh, and Dry Weights of Desert Saltgrass at Four Salinity Levels for All Harvests

	Shoot					
			Weight (g)			
	Length (cm)		Fresh		Dry	
Salinity (mM)	Cum.	Avg.	Cum.	Avg.	Cum.	Avg.
Control (0)	71.8a	10.3a	3.318a	0.474a	1.465a	0.209a
100	70.4a	10.1a	3.330a	0.476a	1.442a	0.206a
200	64.2a	9.2a	4.093a	0.585a	1.692a	0.242a
400	44.0b	6.3b	2.355b	0.336b	1.057b	0.151b

Note: The values are means of six replications. All the values followed by the same letter in each column are not statistically different at the 0.05 probability level.

Source: Pessarakli, M. et al., *J. Plant Nutr.*, 28(8), 1441–1452, 2005.

TABLE 34.6
Root and Shoot Length, Shoot Fresh and Dry Weights, and Root Dry Weight Values of Desert Saltgrass at Four Salinity Levels Based on the Percent of Control

	Length		Weight		
			Shoot		Root
	Root	Shoot	Fresh	Dry	Dry
Salinity (mM)			% Control		
Control (0)	100.0	100.0	100.0	100.0	100.0
100	86.3	98.1	100.4	98.6	103.0
200	79.8	89.4	123.4	115.8	124.5
400	75.2	61.3	70.9	72.3	122.2

Note: The values are means of six replications.
Source: Pessarakli, M. et al., *J. Plant Nutr.*, 28(8), 1441–1452, 2005.

TABLE 34.7
Total Nitrogen (Regular-N) Concentration of Desert Saltgrass Shoot Tissues for Four Harvests (Harvests 2, 3, 4, and 5) at Four Salinity Levels

	Total N concentration (g kg^{-1})			
	Harvests			
Salinity (mM)	2	3	4	5
Control (0)	32.7b	33.1b	36.3ab	33.4b
100	33.9b	33.7b	34.8b	35.6ab
200	37.1a	36.4a	39.5a	38.0a
400	37.5a	38.7a	38.3a	39.2a

Note: The values are means of four replications. All the values followed by the same letter in each column are not statistically different at the 0.05 probability level.
Source: Pessarakli, M. et al., *J. Plant Nutr.*, 28(8), 1441–1452, 2005.

based on the percent of the control and are presented in Table 34.6. Both root and shoot lengths decreased as salinity levels increased. However, the shoot dry weights substantially increased under the medium (200 mM) sodium chloride stress compared with any other treatment. All NaCl levels enhanced the root dry weight (Table 34.6).

34.3.8 TOTAL NITROGEN AND ^{15}N CONCENTRATION AND CONTENT OF THE SHOOTS

Total N concentration of the shoots for four harvests (harvests 2, 3, 4, and 5) are presented in Table 34.7. At each harvest, total N concentration of shoots generally increased by increasing the level of NaCl. This is supported by the Pessarakli and Tucker [32,33], Khalil et al. [6], Pessarakli and Fardad [22], and Pessarakli [18] findings that various plants had higher total N concentration in shoots and roots under salinity stress conditions compared with the controls. The magnitudes of the atom% ^{15}N in plant shoots treated with NaCl were higher than the controls (Table 34.8). However, the differences were not statistically significant.

TABLE 34.8
Atom Percent ^{15}N of Desert Saltgrass Shoot Tissues for Four Harvests (Harvests 2, 3, 4, and 5) at Four Salinity Levels

| Salinity (mM) | Atom Percent ^{15}N (%) | | | |
| | Harvests | | | |
	2	3	4	5
Control (0)	2.62ab	2.84ab	2.73ab	2.78ab
100	3.12a	3.36a	3.47a	3.45a
200	3.60a	3.86a	3.89a	3.94a
400	3.18a	3.76a	3.86a	3.96a

Note: The values are means of four replications. All the values followed by the same letter in each
column are not statistically different at the 0.05 probability level.
Source: Pessarakli, M. et al., *J. Plant Nutr.*, 28(8), 1441–1452, 2005.

TABLE 34.9
Total Nitrogen (Regular-N) Content of Desert Saltgrass Shoot Tissues for Four Harvests (Harvests 2, 3, 4, and 5) at Four Salinity Levels

| Salinity (mM) | Total N Content (mg) | | | |
| | Harvests | | | |
	2	3	4	5
Control (0)	6.54b	6.63b	7.27b	6.68b
100	6.77b	6.75b	6.95b	7.12b
200	10.38a	10.20a	11.06a	10.63a
400	7.13b	7.36b	7.28b	7.45b

Note: The values are means of four replications. All the values followed by the same letter in each
column are not statistically different at the 0.05 probability level.
Soruce: Pessarakli, M. et al., *J. Plant Nutr.*, 28(8), 1441–1452, 2005.

TABLE 34.10
Nitrogen-15 (^{15}N) Content of Desert Saltgrass Shoot Tissues for Four Harvests (Harvests 2, 3, 4, and 5) at Four Salinity Levels

| Salinity (mM) | ^{15}N Content (mg) | | | |
| | Harvests | | | |
	2	3	4	5
Control (0)	0.171b	0.188b	0.198b	0.186b
100	0.212b	0.227b	0.241b	0.246b
200	0.374a	0.393a	0.431a	0.419a
400	0.227b	0.277b	0.281b	0.295b

Note: The values are means of four replications. All the values followed by the same letter in each
column are not statistically different at the 0.05 probability level.
Source: Pessarakli, M. et al., *J. Plant Nutr.*, 28(8), 1441–1452, 2005.

Total N contents of the shoots were significantly higher for the medium (200 mM) NaCl-treated plants compared with the control or any other levels of salinity (Table 34.9). Several investigators have reported that salt-tolerant plants accumulate more nitrogen under low or moderate levels of salinity compared with the controls [1,2,6,18,21,22,32–34]. Since saltgrass is a true halophyte, it is expected to behave better or as the salt-tolerant plants and thereby accumulate nitrogen under salt-stress conditions. Nitrogen-15 content of plants followed the same pattern as the total N. Plants under the medium (200 mM) NaCl salinity contained significantly higher amounts of ^{15}N compared with the control or any other salinity treatments (Table 34.10).

34.4 CONCLUSIONS

Shoot and root lengths decreased with increased salinity levels of the culture medium. However, both shoot fresh and dry weights significantly increased at 200 mM NaCl salinity compared with the control. Root dry weights at both 200 and 400 mM NaCl salinity were significantly higher than the control. Shoot succulence (fresh wt./dry wt.) increased from control to 200 mM NaCl salinity, then declined. Concentrations of both total N and ^{15}N in the shoots were higher in the NaCl-treated plants compared with the controls. Both total N and ^{15}N content of the shoots were significantly higher in the 200 mM NaCl-treated plants compared with the control or any other levels of salinity.

The results of this investigation suggest that saltgrass growing even under poor soil conditions (salt-affected soils, drought) can be benefited from fertilizer (i.e., nitrogen) application/management, and still yield a favorable economical return.

REFERENCES

1. Al-Rawahy, S.A., J.L. Stroehlein, and M. Pessarakli. 1992. Dry matter yield and nitrogen-15, Na$^+$, Cl$^-$, and K$^+$ content of tomatoes under sodium chloride stress. *J. Plant Nutr.*, 15(3):341–358.
2. Dubey, R.S. and M. Pessarakli. 2001. Physiological Mechanisms of Nitrogen Absorption and Assimilation in Plants under Stressful Conditions. In: *Handbook of Plant and Crop Physiology*, 2nd Edition, Revised and Expanded (M. Pessarakli, Ed.), pp. 637–655, Marcel Dekker, Inc., New York.
3. Enberg, A. and L. Wu. 1995. Selenium assimilation and differential response to elevated sulfate and chloride salt concentrations in two saltgrass ecotypes. *Ecotoxicol. Environ. Safe.*, 32(2):71–178.
4. Gould, F.W. 1993. *Grasses of the Southwestern United States*. 6th Edition. University of Arizona Press, Tucson.
5. Hoagland, D.F. and D.I. Arnon. 1950. *The Water Culture Method for Growing Plants without Soil*. Calif. Agr. Exp. Sta. Circ. 347 (Rev.).
6. Khalil, M.A., A. Fathi, and M.M. Elgabaly. 1967. A Salinity Fertility Interaction Study on Corn and Cotton. *Soil Sci. Soc. Am. Proc.*, 31:683–686.
7. Kopec, D.M., A. Adams, C. Bourn, J.J. Gilbert, K. Marcum, and M. Pessarakli. 2001. *Field Performance of Selected Mowed Distichlis Clones*, USGA Res. Report #3. Turfgrass Landscape & Urban IPM Res. Summary 2001, Coop. Ext., Agric. Exp. Sta., Univ. of Ariz., Tucson, U.S. Dept. Agr., AZ1246 Series P-126, pp. 295–304.
8. Kopec, D.M., A. Adams, C. Bourn, J.J. Gilbert, K. Marcum, and M. Pessarakli. 2001. *Field Performance of Selected Mowed Distichlis Clones*, USGA Res. Report #4. Turfgrass Landscape & Urban IPM Res. Summary 2001, Coop. Ext., Agric. Exp. Sta., Univ. Ariz., Tucson, U.S. Dept. Agr., AZ1246 Series P-126, pp. 305–312.
9. Kopec, D.M., K.B. Marcum, and M. Pessarakli. 2000. Collection and Evaluation of Diverse Geographical Accessions of *Distichlis* for Turf-Type Growth Habit, Salinity and Drought Tolerance. Report #2, Univ. of Ariz., Coop. Ext. Service, 11p.
10. Kopec, D.M., A. Suarez, M. Pessarakli, and J.J. Gilbert. 2005. *ET Rates of Distichlis (Inland Saltgrass) Clones A119, A48, Sea Isle 1 Sea Shore Paspalum and Tifway Bermudagrass*. Turfgrass Landscape and Urban IPM Research Summary 2005, Coop. Ext., Agric. Exp. Sta., The Univ. of Ariz., Tucson, U.S. Dept. of Agric., Series P-141, pp. 162–166.

11. Marcum, K.B., D.M. Kopec, and M. Pessarakli. 2001. *Salinity Tolerance of 17 Turf-type Saltgrass (Distichlis spicata) Accessions.* Int. Turfgrass Res. Conf., July 15–21, 2001, Toronto, Ontario, Canada.

12. Marcum, K.B. and C.L. Murdoch. 1992. Salt tolerance of the coastal salt marsh grass, *Sporobolus virginicus* (L.) kunth. *New Phytol.,* 120:281–288.

13. Marcum, K.B., M. Pessarakli, and D.M. Kopec. 2005. Relative salinity tolerance of 21 turf-type desert saltgrasses compared to bermudagrass. *HortScience,* 40(3):827–829.

14. Miller, Deborah L., Fred E. Smeins, and James W. Webb. 1998. Response of a Texas *Distichlis spicata* coastal marsh following lesser snow goose herbivory. *Aquatic Bot.,* 61(4):301–307.

15. Miyamoto, S., E.P. Glenn, and M.W. Olsen. 1996. Growth, water use and salt uptake of four halophytes irrigated with highly saline water. *J. Arid Environ.,* 32(2):141–159.

16. Pessarakli, M. 1991. Dry matter yield, nitrogen-15 absorption, and water uptake by green beans under sodium chloride stress. *Crop Sci. Soc. Am. J.,* 31(6):1633–1640.

17. Pessarakli, M. 1999. Response of Green Beans (*Phaseolus vulgaris* L.) to Salt Stress. In: *Handbook of Plant and Crop Stress,* 2nd Edition, Revised and Expanded (M. Pessarakli, Ed.), pp. 827–842, Marcel Dekker, Inc., New York.

18. Pessarakli, M. 2001. Physiological Responses of Cotton (*Gossypium hirsutum* L.) to Salt Stress. In: *Handbook of Plant and Crop Physiology,* 2nd Edition, Revised and Expanded, (M. Pessarakli, Ed.), pp. 681–696, Marcel Dekker, Inc., New York.

19. Pessarakli, M. 2005. Supergrass: Drought-tolerant turf might be adaptable for golf course use. Golfweek's SuperNews Magazine, November 16, 2005, p. 21 and cover page. http://www.supernewsmag.com/news/golfweek/supernews/20051116/p21.asp?st=p21_s1.htm

20. Pessarakli, M. 2005. Gardener's delight: Low-maintenance grass. Tucson Citizen, Arizona, Newspaper Article, September 15, 2005, Tucson, AZ, Gardener's delight: Low-maintenance grass http://www.tucsoncitizen.com/

21. Pessarakli, M. 2006. Growth Responses and Nitrogen-15 Uptake of Tomatoes under Salt Stress. In: *Vegetables: Growing Environment and Mineral Nutrition* (R. Dris, Ed.), pp. 80–89, WFL Publisher, Helsinki, Finland.

22. Pessarakli, M. and H. Fardad. 1995. Nitrogen (Total and ^{15}N) uptake by barley and wheat under two irrigation regimes. *J. Plant Nutr.,* 18(12):2655–2667.

23. Pessarakli, M. and D.M. Kopec. 2003. *Response of Saltgrass to Environmental Stress.* USGA, New or Native Grasses Annual Meeting, June 17–19, 2003, Omaha, NE.

24. Pessarakli, M., D.M. Kopec, J.J. Gilbert, and A.J. Koski. 2006. *Interactive Effects of Salinity and Mowing Heights on the Growth of Various Inland Saltgrass Clones.* ASA-CSSA-SSSA International Annual Meetings, Nov. 12–16, 2006, Indianapolis, IN.

25. Pessarakli, M., D.M. Kopec, J.J. Gilbert, A.J. Koski, Y.L. Qian, and Dana Christensen. 2005. *Growth Responses of Twelve Inland Saltgrass Clones to Salt Stress.* ASA-CSSA-SSSA International Annual Meetings, Nov. 6–10, 2005, Salt Lake City, UT.

26. Pessarakli, M., D.M. Kopec, and A.J. Koski. 2003. *Establishment of Warm-Season Grasses under Salinity Stress.* ASA-CSSA-SSSA Annual Meetings, Nov. 2–6, 2003, Denver, CO.

27. Pessarakli, M. and K.B. Marcum. 2000. *Growth Responses and Nitrogen-15 Absorption of Distichlis under Sodium Chloride Stress.* ASA-CSSA-SSSA Annual Meetings, Nov. 5–9, 2000, Minneapolis, Minnesota.

28. Pessarakli, M., K.B. Marcum, and D.M. Kopec. 2001. *Drought Tolerance of Twenty One Saltgrass (Distichlis) Accessions Compared to Bermudagrass.* Turfgrass Landscape and Urban IPM Res. Summary 2001, Coop. Ext., Agric. Exp. Sta., Univ. of Ariz., Tucson, U.S. Dept. of Agr., AZ1246 Series P-126, pp. 65–69.

29. Pessarakli, M., K.B. Marcum, and D.M. Kopec. 2001. *Growth Responses of Desert Saltgrass under Salt Stress.* Turfgrass Landscape and Urban IPM Res. Summary 2001, Coop. Ext., Agric. Exp. Sta., Univ. of Ariz., Tucson, U.S. Dept. of Agric., AZ1246 Series P-126, pp. 70–73.

30. Pessarakli, M., K.B. Marcum, and D.M. Kopec. 2001. *Drought Tolerance of Turf-Type Inland Saltgrasses and Bermudagrass.* ASA-CSSA-SSSA Annual Meetings, Oct. 27–Nov 2, 2001, Charlotte, North Carolina.

31. Pessarakli, M., K.B. Marcum, and D.M. Kopec. 2005. Growth responses and nitrogen-15 absorption of desert saltgrass (*Distichlis spicata*) to Salinity Stress. *J. Plant Nutr.,* 28(8):1441–1452.

32. Pessarakli, M. and T.C. Tucker. 1985. Uptake of nitrogen-15 by cotton under salt stress. *Soil Sci. Soc. Amer. J.,* 49:149–152.

33. Pessarakli, M. and T.C. Tucker. 1985. Ammonium (^{15}N) metabolism in cotton under salt stress. *J. of Plant Nutr.*, 8:1025–1045.
34. Pessarakli, M. and T.C. Tucker. 1988. Dry matter yield and nitrogen-15 uptake by tomatoes under sodium chloride stress. *Soil Sci. Soc. Amer. J.*, 52:698–700.
35 Pessarakli, M. and T.C. Tucker. 1988. Nitrogen-15 Uptake by eggplant under sodium chloride stress. *Soil Sci. Soc. Amer. J.*, 52:1673–1676.
36. Rossi, A.M., B.V. Brodbeck, and D.R. Strong. 1996. Response of xylem-feeding leafhopper to host plant species and plant quality. *J. Chem. Ecol.*, 22(4):653–671.
37. Sagi, M., N.A. Savidov, N.P. L'vov, and S.H. Lips. 1997. Nitrate reductase and molybdenum cofactor in annual ryegrass as affected by salinity and nitrogen source. *Physiol. Plant.*, 99:546–553.
38. SAS Institute, Inc. 1991. SAS/STAT User's guide. SAS Inst., Inc., Cary, NC.
39. Sigua, G.C. and W.H. Hudnall. 1991. Gypsum and water management interactions for revegetation and productivity improvement of brackish marsh in Louisiana. *Commun. Soil Sci. Plant Anal.*, 22(15/16):1721–1739.
40. Sowa, S. and L.E. Towill. 1991. Effects of nitrous oxide on mitochondrial and cell respiration and growth in *Distichlis spicata* suspension cultures. *Plant Cell Tiss. Org. Cult.* (Netherlands), 27(2):197–201.

35 Native Grasses as Drought-Tolerant Turfgrasses of the Future

Paul G. Johnson

CONTENTS

35.1   INTRODUCTION

What is a turfgrass? A common definition is a grass that forms a dense canopy of leaves that can be mowed frequently and creates a root system that forms a sod [1–3]. In horticultural circles, turfgrass seems to have a variety of definitions but usually refers to a highly managed, uniform, short-mowed area of grasses. Likewise, in horticultural and botanical circles, turfgrasses are seen with a variety of viewpoints ranging from very important and useful to unnecessary and wasteful. With respect to water, turfgrass is often synonymous with the wasting of water and therefore is not considered sustainable [4]. But in reality, turfgrass does not require inordinate amounts of water. Instead, it often requires less water than some agricultural crops, even in arid regions. Irrigation needs are often related to how an area is managed rather than the grasses themselves, but there are many opportunities to reduce irrigation (and other management inputs) on turfgrasses by using native or well-adapted species.

Water conservation is one of the primary goals for the development of low-maintenance turfgrasses. In many areas of the United States and the world where drought is an occasional or regular occurrence, ordinances or government mandates are imposed to limit water use on landscapes [5]. Such ordinances have been implemented in many areas of the western United States; however, demand for recreational space in urban areas continues to increase. To meet both the need to conserve water and increase recreation space, turfgrass managers need to maintain acceptable turf areas, but use less irrigation [6]. Using native and adapted grasses is one important strategy. In using these grasses, however, very close attention is required to their adaptation to various types of turf areas.

Unique expectations apply to each turfgrass area and depending on these expectations, varying levels of inputs and maintenance are required [7]. For most low-maintenance turf, these expectations include stabilizing the soil, a relatively uniform appearance, and a perennial stand that competes well with weeds [8]. In many cases, turf quality is also a major priority. Acceptable quality can mean that the turf meets the expectations for how an area is used, and also provides acceptable color, fine leaf texture, high plant shoot density, and overall aesthetic appeal.

35.1.1 Where Native Species Might Be Adapted

Introduced species are widely grown throughout North America because of their competitiveness under turfgrass conditions, primarily traffic and mowing and their high turfgrass quality. This is especially the case for intentionally bred varieties of bluegrasses, ryegrasses, and fescues, among others, with increased traffic tolerance and vigor [9]. Most grasses native to North America are not as highly tolerant to these unique stresses. There are exceptions, but in general, most native grasses are less competitive under highly managed conditions and are best suited for lower traffic areas and higher mowing heights than those practiced on most turfs.

Expectations for a turf area are an important part of the equation to determine which turfgrass species to use. If a landscape manager must have a highly traffic-tolerant turf, dark green color, and fine texture, then not many natives will meet those needs. In those situations, introduced species may be wonderfully adapted. If we change management practices slightly to suit the native grasses, or permit somewhat different quality characteristics, many more native and adapted grasses could be successfully used.

There are many turfgrass landscapes where very high expectations are not needed, not wanted, or not realistic. In these situations, native grasses may fit the needs very well. Native species often have somewhat less uniform appearance, coarser leaf texture, or a light green, blue, or gray-green color. While the appearance of natives may be different than commonly used turfgrasses, these differences can also be opportunities for unique landscape designs and greater diversity of plant materials.

Native grasses are well adapted to less managed locations, specifically in the semiarid and arid regions of North America. While many introduced turfgrasses grow well with management in those regions, specifically irrigation, without management many introduced species either cannot survive or will slowly die out. Grasses native to the area can tolerate the stressful conditions in a number of ways to provide a long-lasting turf cover that enhances aesthetics, prevents soil erosion, allows occasional use of the area, and in some cases, brings the serenity and beauty of native grassland communities closer to home.

35.1.2 Turfgrasses Are Well Adapted to Many Urban Areas

As population centers grow and populations become more urban, the demand for plants to put in those landscapes keeps growing, in addition to demands for recreation areas of all types. Urban areas present unique growing conditions that landscape plants have to deal with, especially the turfgrasses. Temperatures are often higher due to the urban heat island effect, soil conditions are extremely variable ranging from undisturbed, deep, well-drained soils to compacted subsoil clay, to very thin layers of what might be called soil [3,10], with the latter two being most common. In addition, challenges unique to turfgrasses include the wear and soil compaction effects of traffic, specifically walking or machines driving on the landscapes [1–3]. Hardscape (pavers, concrete, gravel) is often used in especially difficult-to-manage areas, but there are many cases where plants are needed and essential.

Turfgrasses are rather unique in their ability to tolerate frequent traffic that would damage or kill most other plants, thanks to the subapical meristems and lateral growth [1–3]. Turfgrasses can survive, even thrive, under these conditions. As a result, turfgrasses are one of the most widely used plants in urban landscapes with approximately 163,000 km^2 nationwide, making irrigated turfgrass the largest irrigated crop in North America [11]. Because of the unique conditions of urban turfgrass environments, the use of native grasses in *all* turfgrass areas is not realistic. While many of the native species have evolved for thousands of years under local climatic conditions, the urban areas where managers intend to use these plants are not natural [2]. High population density results in significant traffic, altered climate, and unique site expectations [9].

Most of the species typically used as turfgrasses in North America and other temperate regions of the world originate in Europe and Asia. Many of these same grasses with very wide adaptation have spread over much of the continent and become naturalized, either due to their own competitiveness or through human-assisted movement [14]. Examples of such species are perennial ryegrass (*Lolium perenne* L.) and Kentucky bluegrass (*Poa pratensis* L.) [9].

If the traditionally used introduced grasses are so well adapted to much of North America and the rest of the temperate regions of the world, why is there a desire, or need, to work with North American native grasses and adapted grasses as turfgrasses? Most often, the answer is water. Introduced grasses are being grown in a very wide range of climates and in regions such as the western United States, with climates much different than those in the grasses' native range. For example, climates in the Intermountain West regions (the region between the Rocky Mountains and the Sierra Nevada) often have similar temperature ranges to those in the eastern United States, but precipitation patterns are dramatically different. Total precipitation is lower (150–500 mm yr^{-1} in Utah compared with >1000 mm/yr in the northeast), much of that precipitation falls during the winter, and summer growing seasons are very dry. Most introduced grasses are not well adapted to this type of climate. Other stresses include high pH and saline soils. High pH is typical in arid regions because of reduced leaching of soils and reduced weathering of parent materials [12]. Saline soils occur for several reasons including soil parent materials, but their frequency is increasing because of poor irrigation practices, and low quality irrigation water [13].

Environmental conditions in much of the western United States and Canada require irrigation to keep most turfgrasses green throughout the summer. With ample quantities of good-quality irrigation water, introduced turfgrasses are relatively easy to grow. However, due to rapid population growth, water supplies are being stretched to meet the demand [15,16].

Up to 75% of potable water in the West is currently applied to outdoor landscapes [17]—much of that is applied to turfgrasses. Therefore, water districts, municipalities, states, and conservation groups look to reduce water use in the region to postpone additional water storage infrastructure development. As mentioned previously, more efficient irrigation is important, but water-efficient grasses are also important for water conservation. Salt-tolerant grasses are also needed. As irrigation water becomes tighter, lower-quality water is often used for irrigation, including brackish water sources, reclaimed water, and lower-quality stream flows. Many native grasses can meet both these important needs.

The need to identify and develop new turfgrass species to conserve water and other resources has not gone unnoticed. The United States Golf Association (USGA) has funded a number of breeding programs of native and introduced but adapted grasses since the mid 1980s [18]. Most notable are the breeding programs for buffalograss (*Buchloë dactyloides* [Nutt.] Engelm.), seashore paspalum (*Paspalum vaginatum* L.), zoysiagrass (*Zoysia japonica* Steud.), blue grama (*Bouteloua gracilis* [Kunth] Lag. *Ex* Griffiths), inland saltgrass (*Distichlis, spicata* L. Greene) and wheatgrasses. Several species (some of these mentioned plus many others) were also evaluated in a regional research program in the United States Midwest with low-input management as the primary goal [19]. Many of these turfgrass species continue to experience breeding improvements, but numerous other species offer additional potential for water savings in the landscape.

35.2 INTEREST IN USING NATIVE SPECIES IN LANDSCAPES

There has been an increasing amount of interest among homeowners and landscape managers in the use of native plants in urban landscapes in recent years. This includes grasses, forbs, shrubs, and trees. The reason for using these plants rather than the more common introduced species is multifaceted. The primary reason homeowners and landscape designers may use native plants is to reduce water and nutrient inputs, as mentioned previously. But many urban landscapers also use native plants to establish a unique sense of place—an environment that fits or blends with the natural landscapes or those originally there [20]. Immigrants typically brought plants and seeds with them, even though these plants were not perfectly adapted. They grew the plants that they were familiar with [14]. In many cases, they probably chose plants for their own survival, as is the case with the forage grasses. Some plants were planted simply because they were reminders of "home." Today, urban dwellers are many years removed from an agricultural way of life, but might wish to be closer to the natural environment. Rather than changing the landscapes to fit their needs, they wish to make their piece of the city more like the native environment [20].

However, while one goal is to bring "nature" into the city, most consumers wish to bring only the good parts, or a tamed down version of nature, to the urban landscapes. They look to the mountains and plains vistas and say, "We want that here," but in reality, most want shorter grasses, complete ground cover, and something green. They also don't want the fauna along with the native grassland flora. What is most often desired is a naturalized landscape with grasses that are low to moderately low, uniform with complete cover, and the ability to mow the area to reduce dead leaf material and weeds, and increase functionality.

Many grasses, both native and introduced, can meet goals for more naturalized landscapes. Using natives is not necessarily a "better" option, but is a viable option that landscape managers can use. Because urban areas are not natural environments, it is not realistic to use only native plants in urban landscapes. However, the use of native grasses can be a good option for a site where the user desires to reduce inputs, reduce pest problems, and is content with somewhat lower turf quality than more typically used grasses provide [19]. In this way, a native turfgrass area can have benefits beyond those of usual of turf, which include cooling the environment, controlling dust, providing recreational space, and controlling soil erosion [21].

Sometimes native grasses are used simply to increase diversity in appearance and to have diversity for the sake of diversity. Native plants can be used to provide unique colors, textures, and even scents. In some cases, natives are selected because using them makes the manager feel better—as though they are doing something good for the environment.

35.2.1 What Is Native?

Before we enter a discussion and description of native grasses that may be appropriate for turf situations, it is important to address the issue or definition of native. The important questions are the following: (1) What is native? (2) How important is using a native species?

One definition of native is a plant indigenous to North America—New World versus Old World. This is the broadest definition. A good example is the use of buffalograss as a turfgrass. Buffalograss is native to the Great Plains region of North America, but it can be used quite effectively in other areas of North America. Because buffalograss is not native to the Intermountain West, where summers are nearly rain-free, some irrigation is needed for it to survive or successfully compete against other plants. However, buffalograss offers very high turf quality during the summer season and requires significantly less irrigation than more commonly used species [22]. Buffalograss can also be successfully grown east of the Missouri River in North America where rainfall is higher, but extra weed control may be needed for buffalograss to compete with plants better adapted to the higher rainfall environment [19].

A somewhat narrower definition is a species native to a geographical region, i.e., the Great Plains, the West, or New England. This narrower definition often results in better adaptability and competitiveness, but does not take into account soil type, elevation, and other factors that vary within a geographical region and those due to urban development.

Being native to an individual state or province is often a defining characteristic, but since states or provinces, etc., are not defined by environments, being native to a state or province does not assure growing success. For example, a species that is native to southern Utah might not be at all adapted to the area around Salt Lake City because of very large environmental differences between the two locations, including temperature, rainfall, soils, and elevation.

35.3 SPECIES NATIVE TO NORTH AMERICA

The amount of information available on native grasses, their adaptation to turf situations, and their management is highly variable. A great deal of information is available for some species, including numerous tests under turf conditions. Other species are mentioned in the literature as adapted to turf situations, but have not been rigorously tested. Other species appear to have turf adaptation based on their description in taxonomic literature, but have not been evaluated for adaptation to turf situations. In this review, we focus primarily on species for the western United States, particularly the arid and semiarid regions. In this region, rainfall is often very limited, and soils often have high pH and high salinity. In addition, commonly used introduced species of turfgrasses are not ideally suited and require more management for their survival.

35.3.1 BUFFALOGRASS (*BUCHLOË DACTYLOIDES* [NUTT.] ENGELM.)

Buffalograss is one of the most commonly used and most evaluated North American native species in turfgrass applications. This popularity comes from its excellent adaptation to turfgrass use, including good recuperative ability, relatively quick establishment, and tolerance to a variety of mowing heights. It also has been the subject of more breeding work than most other native grasses, with many varieties developed specifically for turfgrass use. Detailed reviews of the species, genetics, and its use have been published by Riordan and Browning [23] and Shearman et al. [22].

Buffalograss is a warm-season grass species native to the Great Plains region of North America, ranging from Canada to the north and Mexico to the south [24,25]. It is the dominant or codominant grass species in the arid short-grass prairies of North America where rainfall is less than 500 mm annually and grazing is present [24,26]. Buffalograss appears to grow best on medium and fine-textured soils [25,27], but it will grow on sandy sites as well. Buffalograss is not adapted to highly saline soils [28].

Buffalograss is a member of the Chlorideae tribe and the only species in the *Buchloë* genus. Other genera within the tribe include *Bouteloua, Chloris,* and *Trichloris* [22]. Some question exists whether the genus for buffalograss should be *Bouteloua* [29], but for this publication, *Buchloë* will be used.

Buffalograss tends to have a dull gray-green color, but some varieties have moderate to dark green color. Buffalograss has a vigorous stoloniferous habit, which is desirable in turfgrass as it creates a uniform and complete turf in a short time [30]. Buffalograss is morphologically unique because of its dioecious flowering where female flowers, or burs, are low within the canopy and male flowers are expressed above the canopy. Monecious forms also exist, and very rarely hermaphroditic plants [25].

Buffalograss can be established using sod or plugs (usually female-only cultivars), stolons (less frequently), or from seed. Female plants have been considered more desirable [22,23] in formal turf applications and as a result, have been sold as vegetative cultivars. Stands of a female clone tend to create a denser turf than a mixed stand and do not have male flowers extending above the turf canopy [30]. Seeded stands of buffalograss generally have an equal number of male and female plants. The appearance of the male flowers above the canopy is desirable to users for a more informal or "prairie" look [22].

Weed control is often difficult during establishment from seed, but some pre-emergent herbicides can be used to control annual weeds at the time of seeding, including imazapic and simazine. Goss [31] reported that a number of other herbicides can be applied when buffalograss seedlings are at the 1–2 leaf stage, including imazapic, metsulfuron, pendimethalin, and prodiamine. Once established, pre-emergent herbicides as well as many broadleaf herbicides can be used. Once fully established, buffalograss competes well with weeds and results in a well-adapted low-input turf [22].

Mowing heights for buffalograss can be as low as 2 cm [32] or can be left unmowed, which will provide a turf about 15–20 cm tall. Most commonly, buffalograss is mowed at about 7 cm to provide good lawn quality [22].

Buffalograss is frequently reported to require very little nitrogen fertilization. While this might be the case in rangelands, buffalograss responds quickly in terms of increased growth and darker green color to applications of moderate amounts of N fertilizer. Extra growth allows for improved traffic tolerance, improved color, and recuperative ability. Levels of 98 kg N ha^{-1} are sufficient, since higher rates of N did not significantly improve turf quality [33].

In the Great Plains region, buffalograss does not require supplemental irrigation once established. Its C4 metabolism, very deep rooting, and summer dormancy under extremely dry conditions enables buffalograss to survive in dry and hot climates [30]. However, for best turf quality, some supplemental irrigation in the hottest part of the year, especially during periods of drought, is beneficial [22,30]. But in regions with little summer precipitation, some supplemental irrigation is required. In Utah, irrigation to replace 40% of ET$_0$ is adequate to maintain active growth, recovery from wear, and desirable color during the summer months [34].

While the C4 metabolism of buffalograss improves its tolerance to heat and drought during summer months, it limits or prevents active growth during cooler periods of the growing season. This is the primary detriment to wider use of buffalograss in the northern half of the United States. Mixing buffalograss with cool-season species has been attempted and may be a method to increase the overall growing season for a resulting mixture turf. Fine leaf fescues (*Festuca* spp.) seeded into buffalograss improve the fall and spring color of the resulting turf, but the fine leaf fescues tended to out-compete the buffalograss after 2 years [35–37]. Other cool-season grasses that may be less competitive include mutton grass (*Poa fendleriana)* and western wheatgrass (*Pascopyrum smithii*) [37].

A number of breeding programs have developed turf-adapted varieties, primarily for turfgrass applications with specific efforts in improving turf quality, color, density, mowing tolerance, seed production, length of growing season, and sod strength [22,30]. Seeded varieties, with area of origin, include "Texoka" (southern Great Plains), "Cody" (northern Great Plains), "Bowie" (northern Great Plains), "Bison" (southern Great Plains), "Sharps Improved" (southern and northern Great Plains), "Topgun" (southern Great Plains), and "Plains" (southern Great Plains). Vegetatively propagated varieties include "Legacy" (northern Great Plains), "609" (southern Great Plains), "Prairie" (southern Great Plains), "Prestige" (northern Great Plains), "Density" (southern Great Plains), and UC-Verde (southern Great Plains).

35.3.2 Grama Grasses (*Bouteloua* spp.)

The gramas may be the most significant genus in the natural vegetation of the grasslands of central and southern North America [38]. The gramas are members of the Chlorideae tribe and are native to the western Great Plains of the United States and Canada and the deserts of Mexico and the southwestern United States. The native range of sideoats grama (*Bouteloua curtipendula* [Michx.] Torr.) is from north central Mexico, Arizona, New Mexico, and Texas, and north to Ontario and Manitoba. Black grama (*Bouteloua eriopoda* [Torr.] Torr.) is native to drier regions, specifically northern Mexico and Arizona, as well as Texas north to Wyoming and Colorado. The most important grama in turfgrass situations is blue grama (*Bouteloua gracilis* [Kunth] Lag. *Ex* Griffiths). It exists in the prairies from central Mexico to the southwestern United States to central Canada and often in association with buffalograss [38]. Black grama and side oats grama are rarely planted alone in turfgrass settings, but can be used with blue grama to add diversity and visual interest [20].

35.3.2.1 Blue Grama (*Bouteloua gracilis*)

Blue grama is a bunchgrass but can produce a uniform turf through numerous tillers [39]. Leaves tend to be bluish green. In tests conducted in southern Canada, blue grama was the best-performing species with high turfgrass-quality ratings throughout the test [40]. In the same study, blue grama was tolerant of a wide range of mowing heights from 18 to 62 mm. Preliminary trials in Utah also indicate very competitive turf of good quality [37]. Blue grama has very good seedling vigor, so when seeded together with buffalograss, blue grama will dominate the stand [41]. Like most other low-maintenance turfgrasses, little fertilization is required, but up to 98 kg N ha^{-1} per year will improve color and growth [37]. Little other information exists on management of blue grama specifically as a turfgrass.

Most of the available varieties of blue grama are ecotypes that have been selected only for seed yield or plant vigor [40]. Few or none are available that have been selected under turfgrass conditions. However, several of the varieties (and the state of plant material origin) available do perform well. Those include "Alma" (New Mexico), "Hatchita" (New Mexico), and "Bad River" (South Dakota). Blue grama is also sold as sod or plugs that can be shipped from commercial sources [41].

35.3.2.2 Side Oats Grama (*Bouteloua curtipendula*)

Side oats grama is primarily a bunchgrass that results in a clumpy stand, but sometimes may produce stolons or rhizomes, creating an open sod [38]. Side oats grama rarely forms monostands, even in natural sites, and as a result, side oats grama is not used by itself as a turf species. Side oats grama is best used in conjunction with blue grama and buffalograss to add diversity of species and appearance in a naturalized turf.

As a solid stand, turfgrass quality of side oats grama is relatively low, mostly because of poor mowing tolerance, coarse texture and low percent ground cover. It does, however, have desirable color and is very drought tolerant [40]. A number of released varieties (with the state or country of origin) are available, including "Butte" (Nebraska), "El Reno" (Oklahoma), "Haskell" (Texas), "Niner" (New Mexico), "Premier" (Mexico), "Killdeer" (North Dakota), and "Pierre" (South Dakota).

35.3.2.3 Black Grama (*Bouteloua eriopoda*)

Black grama also has a bunchgrass habit, but does have wiry stolons [38]. While black grama has been and continues to be an important forage species in the southwest United States, and can tolerate at least moderate grazing [38], it has not been evaluated for turfgrass characteristics, especially mowing tolerance and stand establishment. However, because of the reports of moderate grazing tolerance, black grama may be a potential turf in the driest areas of the American Southwest. Released varieties (and state of origin) include "Nogal" (New Mexico) and "Sonora" (Arizona).

35.3.3 BLUEGRASSES (*POA* SPP.)

The bluegrasses are among the most diverse genera of the grasses, consisting of as many as 500 species. They encompass a range of life cycles from annual to perennial, and growth habits from caespitose to rhizomatous. Although Eurasia is considered the center of origin for many of the bluegrasses, they are distributed worldwide in temperate, arctic, subtropical, and alpine areas of the tropics [42]. Because of this wide adaptation, it is one of the most widely dispersed of all grass genera [43]. In the western United States and other parts of the world, native bluegrasses are an important understory species of the sagebrush steppe vegetation [44].

Poa is very complex taxonomically because of considerable overlap of morphological characteristics [45,46] and genetic makeup [47]. This continuous variation has been explained by hybridization among bluegrass species and sharing of different combinations of ancestral genomes. *Poa* in general is characterized by wide variation in ploidy levels, and variability within species [48–50].

Most bluegrasses are not considered native to North America, but several are native to the semiarid West and may have some adaptability to turfgrass uses. These include Texas bluegrass (*Poa arachnifera*), muttongrass (*Poa fendleriana*), and *Poa secunda*. Kentucky bluegrass (*Poa pratensis*) is not usually considered to be native, but there is some taxonomic debate on this subject [47]. In this chapter, Kentucky bluegrass will be discussed with the adapted but non-native species. This is because most varieties of Kentucky bluegrass being grown in North America have European origin [51].

35.3.3.1 Texas Bluegrass (*Poa arachnifera*)

Texas bluegrass, typically considered a range grass, native to the southern Great Plains region [52] and not possessing many high quality turfgrass traits, is now very much on the forefront of turfgrass management and genetics. This is mostly due to the development and success of interspecific hybrids of Texas bluegrass and Kentucky bluegrass. Texas bluegrass is rhizomatous, variable in height from 20 to 92 cm, has highly variable leaf width, and is dioecious [52]. It is highly drought tolerant in its native range, primarily through summer dormancy, but does maintain green color when some supplemental irrigation is supplied [52].

Hybrids of Texas bluegrass with Kentucky bluegrass have combined the turfgrass quality of Kentucky bluegrass with the heat tolerance of Texas bluegrass [52]. The first marketed hybrid was "Reveille" [53], but several other similar hybrids have been developed in various public and private breeding programs as well. These varieties have performed well in trials around the country, and drought resistance appears to be improved, compared with many Kentucky bluegrass varieties [54]. One Texas bluegrass release "Tejas 1" is available and may have adaptation to low-maintenance turfgrass uses [55].

35.3.3.2 *Poa secunda*

Poa secunda is an extraordinarily diverse species [45,46] that is very important for forage, soil stabilization, and in some cases, lawns in the western United States [56]. It is often the dominant grass species in dry to mesic regions [57], especially in sagebrush grassland vegetation [56,58]. *Poa secunda* is a bunch-type species with narrow leaves. Like muttongrass, *Poa secunda* is one of the first grasses to begin active growth in early spring, matures early, and enters summer dormancy. *Poa secunda* is very drought tolerant and tolerates relatively heavy grazing [56,58]. Members of *Poa secunda* that have been classified as separate species include big bluegrass (*Poa ampla* Merr.), canby bluegrass (*Poa canbyi* Scribn.), and sandberg bluegrass (*Poa sandbergii* Vasey) [46]. Apomixis in many members of *Poa secunda* increases the uniformity of stands.

Although *Poa secunda* has not been evaluated as a low-maintenance turf for western regions because of its short stature (in some of its members or subspecies). However, its ability to stay green with some additional moisture, and to tolerate grazing [27], appear to make *Poa secunda* a good candidate for a native and drought tolerant turf. Varieties currently available include "Sherman" and "Canbar."

35.3.3.3 Muttongrass (*Poa fendleriana*)

Muttongrass, or mutton bluegrass, is native to the western United States. It is relatively well known in rangelands, but is less common and less economically important compared with *Poa secunda*.

Muttongrass is native in mesic to arid climates, and is a bunch-type species with short rhizomes and basal leaves [27]. It is incompletely dioecious [56], and can be apomictic [59,60]. Because of frequent hybridization, it is an exceptionally polymorphic species [57]. Like *Poa secunda*, muttongrass starts growing early in the year, its seed matures early, then enters summer dormancy. It survives well under moderately heavy grazing and produces relatively small amounts of seed [57].

Muttongrass has recently been studied for turfgrass characteristics in Utah, primarily in mixes with other grass species [37]. By itself, muttongrass does not create a desirable turf because of the bunch type growth habit and nonuniform stand. However, in combination with other species (blue grama, streambank wheatgrass, western wheatgrass), muttongrass provides a naturalized appearance, is very tolerant of the dry conditions, maintains green color throughout the growing season, and can tolerate mowing to 4 in. [37]. No improved varieties are currently available.

35.3.4 WHEATGRASSES

Wheatgrasses are one of the most important groups of cool-season grasses for forage, soil stabilization, and aesthetics in temperate regions of the world [61]. Up to 150 species of wheatgrasses have been named, with most native to Eurasia, but up to 30 are native to North America. In North America, wheatgrasses are most common in the northern Great Plains and on semiarid and arid rangelands of the Intermountain West and Great Basin [61]. Traditionally, wheatgrasses have been included in the genus *Agropyron* [25]. More recently, several of the wheatgrasses have been moved into other genera, including *Elymus*, *Pseudoroegneria*, and *Thinopyrum* [61].

While most wheatgrasses are best adapted to rangeland uses, a few species are adaptable to turfgrass situations, the most limiting factors being mowing tolerance and extended and sometimes unavoidable summer dormancy [62]. Those species with the most promise for turf include crested wheatgrass (*Agropyron cristatum* [L.] Gaertn.), thickspike wheatgrass (*Elymus lanceolatus* [Scribn. & J.G. Sm.] Gould), western wheatgrass (*Pascopyrum smithii* [Rydb.] A. Löve), and bluebunch wheatgrass (*Pseudoroegneri spicata* [Pursh] A. Löve). Crested wheatgrass is an introduced species and will be discussed in the introduced species section of this chapter. The others are considered native to North America.

Most of the wheatgrass breeding conducted to date has been by the USDA-ARS with forage and reclamation uses as the primary breeding goals. However, varieties of crested wheatgrass for roadsides have also been identified and studied further for turfgrass potential [62–65]. Varieties of wheatgrasses are best used as a low-maintenance turf where irrigation is not available, and in areas of less than 18–20 in. of rainfall. They thrive in areas with dry, hot summers by going dormant, and surviving dormancy better than most introduced species [63]. If irrigated, more aggressive introduced species and weeds tend to outcompete the wheatgrasses (unpublished data).

35.3.4.1 Western Wheatgrass (*Pascopyrum smithii*, *Agropyron smithii*, *Elymus smithii*)

Western wheatgrass is a well-known wheatgrass species with an extensive rhizome system, deep rooting, and blue-green glaucous leaf blades. It is very similar to streambank wheatgrass or thickspike wheatgrass, but has a coarser texture, more aggressive rhizome system, and is not considered drought tolerant. Western wheatgrass is native to North America and found throughout the western two-thirds of the United States and Canada, from Ontario west to Alberta and south to New Mexico and Texas. Western wheatgrasses are better suited to finer-textured soils while thickspike or streambank wheatgrasses are best suited to coarser-textured soils [61,66]. Western wheatgrass is also more tolerant to saline soils than the other native wheatgrasses [61].

Although western wheatgrass has a number of attributes needed in low-maintenance semiarid landscapes, its primary detriment is lack of close-mowing tolerance. When mowed relatively tall (>12–15 cm), western wheatgrass offers a quality turf. However, under more typical turf mowing heights of 10 cm, turf density is relatively low because of coarse leaf texture and greater culm elongation [37]. Although western wheatgrass is not considered to be a "weedy species," it can spread through rhizomes into adjacent plant communities [66]. Varieties (and location of origin) that have been released include "Rosana" (Montana), "Rodan" (North Dakota), "Arriba" (Colorado), "Flintlock" (Kansas and Nebraska), and "Barton" (Kansas).

35.3.4.2 Thickspike Wheatgrass (*Elymus lanceolatus*, *Agropyron riparium*)

Thickspike wheatgrass and streambank wheatgrass are often listed separately in the literature [25], but to be most accurate taxonomically, both should be listed under thickspike wheatgrass [67]. This species is frequently used for dryland forage, erosion control, and reclamation in the Intermountain West and Great Plains [61,68]. Thickspike wheatgrass is rhizomatous, has a green to blue-green color (with variation among cultivars), and can produce a quality low-maintenance turf [69]. It is somewhat finer in texture compared with western wheatgrass, and maintains a denser leaf canopy as well [37].

As a low-maintenance turf, thickspike wheatgrass is acceptable, but best used in areas where irrigation is not available. Thickspike wheatgrass can also be mixed with western wheatgrass to improve the shoot density [37], but may also mix with other native species, such as Snake River wheatgrass (*Elymus wawawaiensis* J. Carlson & Barkworth), bluebunch wheatgrass, western wheatgrass, and needlegrass (*Achnatherum* Beauv.) [68]. Thickspike wheatgrass does not compete well with many introduced species, but has been successfully used in a mixture with western wheatgrass and sheep fescue for sod production [70]. While thickspike wheatgrass can be mowed lower than western wheatgrass due to less culm elongation and denser leaf cover, leaves on some cultivars, especially "Sodar" (frequently called streambank wheatgrass), tend to shred rather extensively when mowed, creating a whitish cast over the turf. Other releases that are available include "Bannock" (Washington, Oregon, Idaho) and "Critana" (Montana).

35.3.4.3 Bluebunch Wheatgrass (*Pseudoroegneria spicata, Agropyron spicatum, Elymus spicatus*)

Bluebunch wheatgrass is native to the northern Great Plains and Intermountain regions of the western United States [71,61] and an important component of the Palouse prairie region in the state of Washington [61]. Bluebunch wheatgrass is an important native forage species, but has not been used in turf settings, likely because of its bunch-type habit and more open canopy. However, it can be used as a companion with other native wheatgrasses listed here [71] to increase diversity in color and general appearance. Releases of bluebunch wheatgrass that are available include "Anatone" (Washington), "Goldar" (Washington), and "P-7" (Idaho, Nevada, Oregon, Utah, Washington, British Columbia).

A closely related, but distinct, species is Snake River wheatgrass (*Elymus wawawaiensis* J. Carlson & Barkworth). "Secar," a cultivar of this species originating near Lewiston, Idaho, is considered one of the most drought-tolerant perennial grasses available [72].

35.3.5 Seashore Paspalum (*Paspalum vaginatum* Sw.)

Seashore paspalum is becoming one of the most important and widely grown warm-season turfgrasses. This is primarily due to its high salinity tolerance [73,13], drought resistance [74], wear tolerance [74,75], and low-light tolerance [74,76]. Seashore paspalum is considered a native species [77], but may have been introduced to North America from South America or Africa [74,78]. The native environment for seashore paspalum is coastal regions in tropical and subtropical areas of the southeast United States on sandy beaches, estuaries, and costal rivers. Its range extends from Mexico to Argentina and the West Indies [78].

Seashore paspalum is a relatively aggressive species with lateral growth by both stolons and rhizomes, which produces a strong sod [74,78]. Seashore paspalum has been used as a utility turf for erosion control on sites, including sand dunes, athletic fields, and now golf courses throughout the southern United States, South Africa, Australia, Israel, Argentina, Brazil, and islands in the Pacific Ocean. It has also been used as forage where salinity prevents the growth of other species [74].

Salinity tolerance is the most important characteristic of this species, making it ideal where reclaimed water must be used or if the soil is naturally high in salts. The species is tolerant enough to withstand irrigation with seawater [74]. Lack of cold hardiness, however, prevents the species from being grown north of the southernmost tier of states of the United States. Cultivars of seashore paspalum have been available since the 1950s but, fine-turf varieties became available for use in golf courses and athletic fields beginning in the late 1990s. These new varieties include "Sea Isle 2000," "Sea Isle 1," "Seaway," "Seadwarf," "Seagreen," and "Seafine" [74].

35.3.6 Inland Saltgrass (*Distichlis spicata* L. Greene)

Although inland saltgrass is little known as a cultivated species, it has many characteristics that are highly desirable in turfgrass areas today, most importantly, salt tolerance. This warm-season perennial is native to the western United States, Canada, and Mexico, especially in salt marshes; however, it also is common in upland areas and is highly drought tolerant [79]. Inland saltgrass is also highly tolerant of compacted soil conditions [27] as observed along many roadsides, as well as grazing lands [18]. Inland saltgrass is a dioecious species with extensive and aggressive rhizomes, bright green color, and moderate to fine leaf texture. It can produce a very dense turf, depending on the selections. If the root system is not restricted, it can be extremely drought tolerant. Because of these characteristics, inland saltgrass has been the subject of an extensive breeding program in Colorado and Arizona [18].

35.3.7 FINE LEAF FESCUES (*FESTUCA* SPP.)

Fine fescues are a broad group of grasses with very fine leaf blades (<1 mm wide) that are well suited for low-maintenance turfgrass sites. As a general rule, they can tolerate dry, infertile soils and moderate shade, and are often used where irrigation will be minimal [80]. This group includes species such as creeping red fescue (*Festuca rubra* L. ssp. *rubra* Gaudin), chewings fescue (*Festuca rubra* L. ssp. *commutata* [Thuill.] Nyman), and hard fescue (*Festuca trachyphylla* [Hackel] Krajina), all three of which are introduced to North America and are native to Europe. Idaho fescue (*Festuca idahoensis* Elmer) and sheep fescue (*Festuca ovina* L. ssp. *hirtula* [Hackel *Ex* Travis] M. Wilkinson) both are indigenous to North America [1,80]. This group does not include tall fescue (*Festuca arundinacea* Schreb.), which has much wider leaf blades and a coarser texture.

As a general rule, most of the fine-leaved fescues have had relatively poor heat tolerance, especially among the drought-tolerant species [1,80]. However, heat and drought tolerance in these grasses has already been improved through the addition of beneficial endophytes [81]. More research on this and breeding work with endophytes will likely further improve stress tolerance of this group of grasses.

35.3.7.1 Sheep Fescue (*Festuca ovina*)

Sheep fescue (sometimes called "sheeps fescue" or "sheep's fescue") has not been widely used for turfgrass purposes, usually overlooked in favor of the introduced species that typically have higher quality. However, sheep fescue can also provide a reasonably high-quality turf [82], is more drought tolerant than the introduced species, and can provide a more naturalized look. The resulting turf is sometimes "swirly" in appearance, or the leaves "curve" rather than laying in one direction. Because of its bunch-type habit, older sheep fescue turfs can be somewhat clumpy [1], but management and some supplementary irrigation can help prevent this. Heat tolerance is reported to be relatively poor [1], so sheep fescue is best suited to the cooler parts of the semiarid regions. Compared with introduced fine fescues, there are relatively few varieties of sheep fescue, but they include "Quatro," "Covar," and "Bighorn." Some selections or cultivars of sheep fescue have a distinct blue color [80], which can offer some unique colors to grass landscapes.

35.3.7.2 Idaho Fescue (*Festuca idahoensis*)

Like sheep fescue, Idaho fescue is also a member of the *Festuca ovina* complex. It is native to North America with a native range of the northern Great Plains and northern Intermountain West. Idaho fescue lacks rhizomes, has a blue or yellow-green color, and like other fine leaf fescues, very fine leaf blades [82]. It is very tolerant to cold, shade, and drought; however, Idaho fescue is not as tolerant to these stresses as sheep fescue [83] and may be slower to establish [84]. Like the other species in the *Festuca ovina* complex, Idaho fescue has been used in lawns, grazing lands [27], and for wildflower gardens [80], but use in these situations has been limited to date.

A number of other native fine leaf fescues are described and may be appropriate for some native turf areas. These include Arizona fescue (*Festuca arizonica* Vasey), Colorado fescue (*Festuca brachyphylla* J.A. Shultes ex J.A. & J.H. Shultes spp. *coloradensis* Frederiksen), and

Rocky Mountain fescue (*Festuca saximontana* Rydb.). For most of these species, however, it may be difficult to obtain the quantities of seed needed for establishment of turf areas.

35.3.8 Hairgrass (*Deschampsia* spp.)

The hairgrasses have only recently been considered for use as turfgrasses, but have potential as low-maintenance grasses. Hairgrasses are considered native wetland grasses, but are found in many temperate regions and have adapted to marshes, streambanks, mountain meadows, and have been described as useful for restoration and reclamation plantings [88].

Deschampsia is a rather small genus, with about 24 species [85], but the most commonly used hairgrass in turfgrass areas is *Deschampsia caespitosa* or tufted hairgrass. It is a common species in the Rocky Mountain region, especially in alpine tundra plant communities [85], but is also present throughout Europe and temperate regions of Asia [86]. In turfgrass evaluations conducted in the Great Plains region of Canada, tufted hairgrass provided excellent texture, density and mowing tolerance, but was significantly damaged by leaf rust infestations [40]. Similar results were observed in Utah, but summer heat and drought reduced overall quality compared with other grasses [34]. Tufted hairgrass is reported to have good wear tolerance and is very shade tolerant, possibly making it well adapted to many landscape and athletic field situations [85], including the variety "Barchampsia." For a detailed review of the genetics and breeding history of the hairgrasses, the reader is referred to Brilman and Watkins [85].

35.3.9 Junegrass (*Koeleria cristata* or *K. macrantha*)

Koeleria cristata (*K. macrantha*), or junegrass, is one of the most widely distributed native grasses in North America, ranging throughout the United States and into Mexico, except the extreme southeast [87]. In Europe, *Koeleria* grows in dry to dry-mesic areas, primarily in meadows, grasslands, and rocky slopes. In the Western United States, it grows on semiarid to dry sites in meadows, slopes, and often in conjunction with sagebrush and pinyon juniper [88].

In turfgrass trials of an introduced cultivar from Europe ("Barkoel"), *Koleria* showed acceptable quality with a minimum of maintenance across a range of mowing heights [40]. In those trials, the color and density were very good, but the variety did not green-up as early as North American *Koeleria*. The leaves also were prone to shredding [40]. *Koleria* turf thrives on low amounts of fertilizer and grows slowly, both of which are highly desirable traits for naturalized turfs, but drought tolerance is only moderate [87]—not as drought tolerant as most bluegrasses or fescues and similar to perennial ryegrass [34]. *Koeleria* may be relatively difficult to establish and is not highly tolerant of traffic [84]. Another limitation of wider use in turfgrass areas is seed cost, which is relatively high due to low seed yields. *Koeleria* has also been available as sod under the trade name "Turtleturf."

35.3.10 Idaho Bentgrass (*Agrostis idahoensis*)

Numerous *Agrostis* species are used in turfgrass areas throughout the world, especially creeping bentgrass (*Agrostis palustris* Huds.) and colonial bentgrass (*Agrostis capillaris* L.). Most of these are introduced species from Europe, but Idaho bentgrass is native to North America and has been considered for use as low-maintenance golf turf. Idaho bentgrass is a bunchgrass with narrow and basal leaves, and is somewhat broader and darker than colonial bentgrass [25]. It is a wetland species, but can be used as a low-maintenance turf species and on golf course fairways [40,89]. One variety, "Golfstar," is currently being sold [90].

35.3.11 Other Native Species

There are many grass species that are not typically considered turfgrasses, but either have characteristics that might lend themselves to use as a turf in special situations, or that have not been evaluated for turfgrass uses but offer some potential. These include a number of range grasses that could survive on almost no maintenance at all. The following is by no means an exhaustive list of additional potential turf species, but instead a list of species that have either been used by seed companies as a component of low-maintenance lawn mixes, mentioned in scientific literature, or other sources. For more comprehensive lists, refer to Brede [84], numerous seed companies, and the USDA, National Resources Conservation Service (NRCS) PLANTS database at http://plants.usda.gov/. Information on most of these species is limited, relying on rangeland testing and observations, popular literature, and anecdotal information. To determine adaptation, testing in a local area is highly recommended. If used, most of these will not develop a satisfactory turf in monoculture, but are best used in a mix of several species.

35.3.11.1   Galleta (*Hilaria jamesii*)

Galleta is a warm-season native rhizomatous grass that is a desirable forage species [27]. It is also useful for roadsides, campgrounds, and picnic areas, since it is quite traffic tolerant [91]. It has been included in some seed mixes available from seed companies, too.

35.3.11.2   Alkali Sacaton (*Sporobolus airoides*)

Alkali sacaton is a warm-season species adapted to very tough sites, including those with high salinity, high pH, heat, drought, and flooding [84]. It is also relatively mowing tolerant.

35.3.11.3   Little Bluestem (*Schizachyrium scoparium*)

Little bluestem is a warm-season species, very common in the short to mid-grass prairie regions of the Great Plains and now grown in many places in North America. This species is sometimes included in native turf meadow mixes to provide species diversity and ornamental value because of its rusty red color in fall and fluffy inflorescences. However, little bluestem is not tolerant to typical mowing heights and will gradually disappear from a species mixture (unpublished data).

35.3.11.4   Alpine Bluegrass (*Poa alpina* L.)

Alpine bluegrass is a native Poa that is frequently found in mountain meadows, alpine summits, and artic regions of the northern hemisphere [25]. While alpine bluegrass is not highly drought tolerant, since it enters summer dormancy [40], it is very slow growing, significantly reducing the need for mowing [84].

35.3.11.5   Fowl Bluegrass (*Poa palustris* L.)

Another bluegrass that has not been evaluated widely as a turfgrass is fowl bluegrass. This bunch-grass is native to North America and has exhibited reasonably good turfgrass quality, especially during summer stresses in Canada [40]. It is also adaptable to wetland sites; however, fowl bluegrass does not tolerate high-pH soils or saline conditions [84]. Adaptation to the arid or semiarid climates of the American West is unknown.

35.3.11.6   Sand Dropseed (*Sporobolus cryptandrus* [Torr.] Gray)

Sand dropseed is native to the North American Great Plains and intermountain region. It is a bunch-grass and spreads through seeds and tillers [27]. It is tolerant of heavy grazing [27] and mowing [84].

35.3.11.7   Indian Ricegrass (*Oryzopsis hymenoides* [R. & S.] Ricker)

This species is also native to the North American Great Plains and intermountain regions. Indian ricegrass is a bunchgrass [27] and is one of the more widely adapted native grasses [91]. Indian rice-grass is well adapted to generally shallow, high pH, and moderately saline soils [91].

35.4 SPECIES ADAPTED BUT NOT NATIVE TO NORTH AMERICA

35.4.1 CRESTED WHEATGRASS (*AGROPYRON CRISTATUM*)

Crested wheatgrass is well adapted to, and very commonly grown in, arid sections of the American West. It is an introduced species, native to European Russia and southwestern Siberia and first introduced to North America in 1892 [61]. Like the other wheatgrasses, it is most often used for forage production, and helped to reclaim large areas of rangelands and croplands following the extreme drought of the 1930s, commonly called the Dust Bowl [61]. Crested wheatgrass's ability to stabilize and reclaim areas is due to its drought tolerance, extensive root system, and seedling vigor where rainfall is at or above 9 in. per year. Crested wheatgrass is normally a bunch-type species, but some cultivars ("Roadcrest" and "Ephraim") also produce rhizomes when provided with slightly more than the minimum amount of water (at least 14 in. per year) [92].

Newer collections from the Middle East have finer leaves and a shorter growth habit compared with the more traditional varieties [61], and further breeding efforts with these collections have evaluated a number of turfgrass traits and identified improved populations [64,93]. These species' characteristics can meet the demands of turfgrass sites, but like other wheatgrasses, summer dormancy is common and canopy cover is not comparable to traditional turfgrasses with typical mowing heights of 3 in. [94]. Higher mowing heights, along with nitrogen fertilization, might enhance the longevity and quality of crested wheatgrass under turfgrass conditions [94].

Because crested wheatgrass is so widely planted and exists over such large areas in the American West, there is concern over its being invasive and its effects on grassland ecosystems. Some literature present various indications of concern, including the strong competitiveness of crested wheatgrass that creates a monoculture where planted, increases in insect herbivores, more exposed soil, and reduced organic matter which have been measured in crested wheatgrass stands [95]. However, crested wheatgrass is not considered to be an invasive species in USDA publications, and planting with other species is recommended to avoid monocultures [92].

35.4.2 CANADA BLUEGRASS (*POA COMPRESSA* L.)

Canada bluegrass is native to Eurasia, but has been introduced to North America and is widely distributed throughout mesic to wet sites [57,96]. Canada bluegrass has a rhizomatous growth habit and is apomictic. It forms a turf that is rather open and of low density compared with Kentucky bluegrass [1]. *Poa compressa* begins growth early in the growing year and has a lush appearance. However, under mowing and summer stresses, plants lose most of their leaves [40], developing a stemmy appearance and a stiff feel [1,37]. When Canada bluegrass is left unmowed, some green color is retained throughout the growing season [37].

In Utah, it can be quite competitive in mixtures with other grasses and plants [97], and quickly dominated a stand when mixed with western wheatgrass and crested wheatgrass [37]. Because of this competitiveness, Canada bluegrass is sometimes considered to be invasive [98]. In Connecticut it is labeled a noxious weed, preventing its sale in that state. However, in the West, it is frequently sold as part of a low-maintenance turf mixture, designed for higher-elevation plantings [99].

35.4.3 KENTUCKY BLUEGRASS (*POA PRATENSIS* L.)

Kentucky bluegrass has been called the most widely distributed plant in the temperate northern hemisphere [100]. It has been seeded in lawns, pastures, revegetation projects, and has spread on its own accord. As mentioned previously, Kentucky bluegrass is most often described as an introduced species to North America with the center of origin in Eurasia [51]. The first mention of Kentucky bluegrass in North America was in 1685 by William Penn, who planted seed from England [14]. While most of the varieties grown as turfgrass in North America and those that have become naturalized originated from European sources, some Kentucky bluegrass is quite probably native or migrated to North America prior to European settlement [101]. Morphological and genetic characteristics of native Kentucky bluegrass overlap with introduced species, making a solid delineation difficult [42,47].

The taxonomy of bluegrasses, as mentioned previously, is very difficult, and Kentucky bluegrass has the most complex taxonomy of them all. Kentucky bluegrass has the most shared genomes with other bluegrasses due to interspecific hybridization [50], which enables a wide variety of interspecific hybrids in nature [48]. This genomic complexity might be one reason for the species' wide adaptation. In addition to being adapted to a wide range of conditions, it is also grown because of its high quality as a turfgrass and forage species. It has a highly desired color, recovers from traffic and wear through rhizomatous growth, has a soft feel and a relatively fine texture. Kentucky bluegrass does require more water than some species (i.e., tall fescue, buffalograss, fine fescues) to remain green and actively growing in semiarid regions of North America, typically requiring approximately 70–80% [102,103] of reference evapotranspiration [104].

This relatively high water use often overshadows the fact that Kentucky bluegrass is actually quite drought tolerant and can survive extended periods, through dormancy, without rain or irrigation, then resume growth when temperature and moisture conditions allow [1,51]. In preliminary tests conducted in Utah, Kentucky bluegrass survived at least 120 days without irrigation before stand loss began to occur upon regrowth. This drought survival was longer than perennial ryegrass, creeping red fescue, and *Koleria cristata*. Tall fescue survived longer, but did so because of deep rooting and a very deep soil, allowing rooting to extend more than 100 cm deep [34]. These drought-tolerance characteristics can make Kentucky bluegrass a turf requiring little irrigation, as long as dormancy and a few weeds can be tolerated. In this way, Kentucky bluegrass could be used in naturalized settings, possibly as a companion with other grasses. However, Kentucky bluegrass may be too competitive, especially when irrigated, and may dominate a planting. Breeding efforts underway in a number of public and private breeding programs are working to further improve the drought tolerance of Kentucky bluegrass through hybridization with native bluegrasses, followed by selection for turfgrass quality traits. One example of the results of such a program are the hybrids with Texas bluegrass [52].

Wide adaptability and stress tolerance, while a benefit for pasture and turfgrass use in many situations, are a detriment to some plant communities. Kentucky bluegrass has been shown to outcompete native species in some national parks and national forests [105], and has been listed as an invasive species or potentially invasive species in some areas [106].

35.5 RESTORATION GENE POOL CONCEPT AND TURFGRASSES

As described in earlier sections of this chapter, the expectations and demands on turfgrass communities are rather unique, which often limits which species can or cannot be used, both native and non-native. However, the desire to use the most appropriate species and to best correspond with native sites is important. In these situations, the decision process is improved by the restoration gene pool concept [107]. This concept has initially been applied to the restoration of rangelands and damaged natural plant communities, but can be applied to decisions in turf areas as well. In this concept, a hierarchy is proposed on species and populations to be used in choosing plant materials. The following is a summary of the concept as described by Jones [107] with adaptation to turfgrass situations, with a site in the Intermountain West as an example.

35.5.1 PRIMARY GENE POOL

The goal is restoration of a population of grasses that previously existed on a site using seed from a plant population with the same or very similar genetic identity as that of population(s) originally in that area. However, an appropriate source of seed for primary gene pool restoration is usually difficult to obtain for several reasons. The landscape manager may not know which species, let alone what populations, were originally present in the area. Another limitation of the original populations is their possible lack of adaptation to the current site conditions due to changes in soil, water, and how the area will be used. The ability to use primary gene pool restoration populations in turfgrass sites is likely to be very difficult. An example of primary gene pool restoration might be to use seed of thickspike wheatgrass and other wheatgrasses on a site in the Intermountain West from a nearby native source to establish a turf.

35.5.2 SECONDARY GENE POOL

In using a secondary gene pool for restoration, the goal would be to use the same species of grasses as those originally on the site, but from different and disjunct populations that exhibit enhanced adaptation. These populations could originate in other areas of the region or be cultivated varieties of the same species. These varieties, however, still need to meet the expectations desired for the area. An example for the situation described previously would be using cultivated varieties of thickspike wheatgrass on the site.

35.5.3 TERTIARY GENE POOL

If varieties of the indigenous species are not available or acceptable for the site, the landscape manager could consider using other closely related species, but these might not be native to the local area. An example might be to use a Snake River wheatgrass x thickspike wheatgrass hybrid with tiller density superior to thickspike wheatgrass.

35.5.4 QUATERNARY GENE POOL

Finally, if the other potential gene pools are not available or applicable to the conditions or needs of the site, the quaternary gene pool might include either native or non-native species that are highly adapted to the site, considering both growing environment and expectations. An example here might be the use of crested wheatgrass, sheep fescue, buffalograss, blue grama, western wheatgrass, even Kentucky bluegrass.

Most uses of "native grasses" in turfgrass situations would be in the quaternary classification. Disturbed growing environment and limitations imposed by how the turf area will be used and what is expected of the area may preclude using species native to an adjacent area or even close relatives.

In this review, the listing of species is separated into two very broad categories, but the two categories that are most typically used are those native to North America, and those that are not native but are well adapted to naturalized situations. As described by the gene pool concept, a native species is sometimes desired in a particular location, but the environment is so disturbed or is used in a manner that does not allow truly native species to compete. In these cases, other species may be available that can provide many of the benefits of native species, including reduced management and a naturalized appearance.

35.6 MANAGEMENT OF NATIVE GRASSES AS A TURF

Information has been published on how to best manage some native and adapted species as turfgrasses. This is the case for buffalograss and seashore paspalum, and somewhat for crested wheatgrass. However, this is not the case for most of the species. Some management information may be available from seed companies in the form of fact sheets, and some may be available from local native plant groups. This lack of management information is probably the largest hurdle to the effective use of these grasses. Published research is slowly becoming available (e.g., [37,65,94]), but for many species, only general recommendations are available until personal expertise can be obtained by the user.

35.6.1 ESTABLISHMENT

One of the most difficult stages of managing native grasses as a turf is establishment. Many of these species have moderate to low seedling vigor and therefore, do not compete well with weeds. While these grasses may be adapted to the regional climatic conditions, urban soils may be quite different and likely have a large seed bank of introduced weeds, such as crabgrass, foxtails, etc. Some pre-emergent herbicides have been effectively used at seeding for some species [31], but many of these herbicides have not been tested on a wide variety of native and introduced species. In some cases, mowing can be effective to suppress some weeds, but weed competition may still be great. Work is ongoing at many research locations to evaluate the use of cover crops, transplanting as sods, and new herbicides.

35.6.2 WEED CONTROL

Once native species are established, weed control becomes easier as many typically used pre-emergent and post-emergent herbicides can be effective. Once full canopy coverage has been obtained with the grasses, weeds may be effectively out-competed, especially if managed to favor the native species.

35.6.3 MOWING

In most native turf situations, mowing will be done at least occasionally to control weeds, reduce dead plant material, discourage rodents and other animals, or to increase the functionality of the turf area (allow for recreation). As a general rule, mowing heights need to be relatively high (>10 cm). Some species may tolerate occasional mowing, but not with the same frequency as traditionally used turfgrasses.

35.6.4 FERTILITY

Native turfgrasses are frequently used to reduce fertilizer inputs, but some fertilization is usually needed to maintain a desirable color, growth, and to compete with introduced weeds. However, excessive fertilization will often decrease the competitiveness of many native or low-maintenance species through increased thatch development and increased competitiveness of weeds. Generally, 98 kg N ha^{-1} is the maximum required [33,37]. Optimal levels of P and K are not reported, but adequate soil test levels would be expected to be similar to the needs of other turf species and in the ranges of 10–20 mg kg^{-1} P and 175–250 mg kg^{-1} K [3].

35.6.5 IRRIGATION

Water has the most impact on overall growth of many native turf species. Most of the species mentioned can either survive without supplemental irrigation or require less than most other turfgrasses. Warm-season grasses require some water during the summer growing season, so if some rainfall is not experienced during that season, supplemental irrigation will be needed. Buffalograss in the intermountain region is an example of this, as previously discussed. Cool-season grasses may require some water to survive or provide optimal quality throughout the year, but many may be able to survive extended dry periods by entering summer dormancy. This is the case for native bluegrasses, Kentucky bluegrass, and the wheatgrasses, to name a few. In some cases where summer dormancy is unavoidable, as with some crested wheatgrasses, supplemental irrigation will reduce competitiveness.

35.7 CONCLUSION

Native and adapted grasses have many attributes to offer landscape managers. While they are not appropriate in the most highly maintained and highly used areas, native and adapted grasses could be used on many other, lesser-used turfgrass sites. The qualities they offer range from practical ones, such as reduced irrigation, mowing, and fertilizer inputs, to creative ones, such as achieving new appearances in landscapes that may replicate the natural surroundings or native areas.

There are a number of species that could be used, but information on many of them is lacking, both in where they can be used and how they can be managed. Experience in growing native species in a particular location is the best recommendation for many native species. Experiment, try new things, be creative.

REFERENCES

1. Beard, J.B., *Turfgrass: Science and Culture*. Prentice Hall, Englewood Cliffs, NJ, 1973.
2. Turgeon, A.J., *Turfgrass Management*, 6th ed. Prentice Hall. Upper Saddle River, NJ, 2002.
3. Christians, N. *Fundamentals of Turfgrass Management*, 2nd Ed. Wiley, Hoboken, NJ, 2004.
4. Jenkins, V.S., *The Lawn: A History of an American Obsession*. Smithsonian Institute Press. Washington, DC, 1994.
5. Pleban, S., Living with water-conservation laws. *Grounds Maint.* 28(5), 44, 1993.
6. Garrott, D.J., Jr. and Mancino, C.F., Consumptive water use of three intensively managed bermudagrass growing under arid conditions. *Crop Sci.* 34, 215, 1994.
7. Emmons, R., *Turfgrass Science and Management*. 3rd ed. Delmar Publishers, Albany, NY, 2000.
8. Dernoeden, P.D., Carroll, M.J., and Crouse, J.M., Mowing of three fescue species for low-management turf sites. *Crop Sci.* 34, 1645, 1994.
9. Casler, M.D. and Duncan, R.R., Origins of turfgrasses, in *Turfgrass Biology, Genetics, and Breeding*. Casler, M.D. and Duncan, R.R. Eds., Wiley, Hoboken, NJ, 2003, chap. 1.
10. Fenton, T.E. and Collins, M.E., The soil resource and its inventory, in *Managing Soils in an Urban Environment*. Brown, R.B., Huddleston, J.H., and Anderson, J.L. Eds., ASA-CSSA-SSSA. Madison, WI, 2000.
11. Milesi, C., Running, S.W., Elvidge, C.D., Tuttle, B.T., and Nemani, R.R., Mapping and modeling the biogeochemical cycling of turf grasses in the United States. *Environ. Manage.* 36, 426, 2005.
12. Plaster, E.J., *Soil Science and Management*, 3rd ed., Delmar Publishers, Albany, NY, 1997.
13. Carrow, R.N. and Duncan, R.R., *Salt-Affected Turfgrass Sites: Assessment and Management,* Ann Arbor Press, Chelsea, MI, 1998.
14. Forest Service-USDA, *Range Plant Handbook*, United States Government Printing Office, Washington, DC, 1937.
15. Vickers, A., *Handbook of Water Use and Conservation*, Waterplow Press. Amherst, MA, 2001.
16. US Department of Interior, Water 2025: Preventing crises and conflicts in the West. Online at http://www.doi.gov/water2025/, 2005. Verified July 15, 2006.
17. Kjelgren, R., Rupp, L., and Kilgren, D., Water conservation in urban landscapes, *HortScience* 35, 1037, 2000.
18. USGA, USGA turfgrass and environmental research online, Online at http://www.usga.org/turf/green_research/green_research.html, 2006. Verified July 16, 2006.
19. Diesburg, K.L., Christians, N.E., Moore, R., Branham, B., Danneberger, T.K., Reicher, Z.J., Voigt, T., Minner, D.D., and Newman, R., Species for low-input sustainable turf in the U.S. upper Midwest, *Agron. J.* 89, 690, 1997.
20. Mee, W., Barnes, J., Kjelgren, R., Sutton, R., Cerny, T., and Johnson, C., *Water Wise: Native Plants for Intermountain Landscapes*, Utah State University Press, Logan, UT, 2003.
21. Beard, J.B. and Green, R.L., The role of turfgrasses in environmental protection and their benefits to humans, *J. Environ. Qual.* 23, 452, 1994.
22. Shearman, R.C., Riordan, T.P., and Johnson, P.G., Buffalograss, in *Warm-Season Grasses*. ASA/CSSA/SSSA Monograph No. 45. Madison, WI, 2004.
23. Riordan, T.P. and Browning, S.J., Buffalograss, in *Turfgrass Biology, Genetics, and Breeding*, Casler, M.D. and Duncan, R.R. Eds, Wiley, Hoboken, NJ, 2003.
24. Beetle, A.A., 1950, Buffalograss: native of the shortgrass plains, Bull. 293, University of Wyoming Agr. Exp. Stn., Laramie, WY, 1950.
25. Hitchcock, A.S., *Manual of the Grasses of the United States,* Vol. 1, Dover, New York, 1951.
26. Wenger, L.E., Buffalograss. Bull. 321. Kans. Agr. Exp. Stn., Manhattan, KS, 1943.
27. Stubbendieck, J., Hatch, S.L., and Butterfield, C.H., *North American Range Plants*, 5th ed., University of Nebraska Press, Lincoln, Nebraska, 1997.
28. Wu, L. and Lin, H., 1994, Salt tolerance and salt uptake in diploid and polyploid buffalograsses (*Buchloë dactyloides*), *J. Plant Nutr.* 17, 1905, 1994.

29. Columbus, J.T., An expanded circumscription of *Bouteloua* (Graminae: Chloridoideae): new combinations and names, *Aliso.* 18, 61, 1999.

30. Riordan, T.P., DeShazer, S.A., Johnson-Cicalese, J.M., and Shearman, R.C., An overview of breeding and development of buffalograss, *Int. Turf. Soc. Res. J.* 7, 816, 1993.

31. Goss, R.M., McCalla, J.H., Gaussoin, R.E., and Richardson, M.D., Herbicide tolerance of buffalograss, *Applied Turfgrass Sci.* (Online), doi:10.1094/ATS-2006-0621-01-RS, 2006.

32. Johnson, P.G., Riordan, T.P., and Johnson-Cicalese, J., Low mowing tolerance in buffalograss, *Crop Sci.* 40, 1339, 2000.

33. Frank, K.W., Gaussoin, R.E., Riordan, T.P., Shearman, R.C., Fry, J.D., Miltner, E.D., and Johnson, P.G., Nitrogen rate and mowing height effects on turf-type buffalograss, *Crop Sci.* 44, 1615, 2004.

34. Johnson, P.G., Maximum length of summer dormancy in turfgrass species with or without minimal irrigation. Center for Water Efficient Landscaping Research report, Online at http://www.hort.usu.edu/pdf/paul/DormancyWriteup.pdf, 2004. Verified June 14, 2006,

35. Johnson, P.G., Mixtures of buffalograss and fine fescue or streambank wheatgrass as a low-maintenance turf, *HortScience* 38, 1214, 2003.

36. Severmutlu, S., Riordan, T.P., Shearman, R.C., Gaussoin, R.E., and Moser, L.E., Overseeding buffalograss turf with fine-leaved fescues, *Crop Sci.* 45, 704, 2005.

37. Bunderson, L., Johnson, P.G., Kopp, K.L., and Dougher, T., Evaluation of native and adapted grass species and their management for turfgrass applications in the Intermountain West, in Proc. ASA-CSSA-SSSA Annual Meetings, Indianapolis, 2006.

38. Smith, S.E., Haferkamp, M.R., and Voigt, P.W., 2004. Gramas. in *Warm-Season Grasses*, ASA/CSSA/SSSA Monograph No. 45. Madison, WI, 2004.

39. Stubbendieck, J. and Burzlaff, D.F., Nature of phytomer growth of blue grama, *J. Range Manage.* 24, 154, 1971.

40. Mintenko, A.S., Smith, S.R., and Cattani, D.J., Turfgrass evaluation of native grasses for the northern Great Plains region, *Crop Sci.* 42, 2018, 2002.

41. Koski, T., 2005. Blue grama for low maintenance lawns. Colorado State University Cooperative Extension. Online at http://csuturf.colostate.edu/pdffiles/Blue%20Grama%20Lawns.pdf, 2005.

42. Soreng, R.J., *Poa* L. in New Mexico, with a key to Middle and Southern Rocky Mountain species (*Poaceae*), *Gt. Basin Nat.* 45, 395, 1985.

43. Hartley, W., Studies on the origin, evolution, and distribution of the Gramineae 4: the genus *Poa, Aust. J. Bot.* 9, 152, 1961.

44. Jones, T.A. and Larson, S.R., Status and use of important native grasses adapted to sagebrush communities, USDA Forest Service Proceedings, RMRS-P-38, pp. 49–55, 2005.

45. Kellogg, E.A., A biosystematic study of the *Poa secunda* complex, *J. Arnold Arboretum* 66, 201, 1985.

46. Kellogg, E.A., Variation and names in the *Poa secunda* complex, *J. Range Manage.* 38, 516, 1985.

47. Soreng, R.J., Chloroplast-DNA phylogenetics and biogeography in a reticulating group: study in *Poa* (*Poaceae*), *Am. J. Bot.* 77, 1383, 1990.

48. Clausen, J., Introgression facilitated by apomixis in polyploid *Poas, Euphytica* 10, 87, 1961.

49. Soreng, R.J., An infrageneric classification for *Poa* in North America, and other notes on sections, species, and subspecies of *Poa, Puccinellia*, and *Dissanthelium* (Poaceae), *Novon* 8, 187, 1998.

50. Patterson, J.T., Larson, S.R., and Johnson, P.G., Genome relationships in polyploid *Poa pratensis* and other *Poa* species inferred from phylogenetic analysis of nuclear and chloroplast DNA sequences, *Genome* 48, 76, 2005.

51. Huff, D.R., Kentucky bluegrass, in *Turfgrass Biology, Genetics, and Breeding*, Casler, M.D. and Duncan, R.R. Eds., Wiley, Hoboken, NJ, 2003.

52. Read, J.C. and Anderson, S.J., Texas bluegrass (*Poa arachnifera* Torr.), in *Turfgrass Biology, Genetics, and Breeding*, Casler, M.D. and Duncan, R.R. Eds., Wiley, Hoboken, NJ, 2003.

53. Read, J.C., Reinert, J.A., Colbaugh, P.F., and Knoop, W.E., Registration of 'Reveille' hybrid bluegrass, *Crop Sci.* 39, 590, 1999.

54. Abraham, E.M., Huang, B., Bonos, S.A., and Meyer, W.A., Evaluation of drought resistance for Texas bluegrass, Kentucky bluegrass, and their hybrids, *Crop Sci.* 44, 1746, 2004.

55. Read, J.C., Reinert, J.A., Evers, G.W., Ocumpaugh, W.R., Sanderson, M.A., and Hopkins, A.A., Registration of 'Tegas 1' Texas bluegrass, *Crop Sci.* 45, 2124, 2005.

56. Larson, S.R., Waldron, B.L., Monsen, S.B., St. John, L., Palazzo, A.J., McCracken, C.L., and Harrison, R.D., AFLP variation in agamospermous and dioecious bluegrasses of western North America, *Crop Sci.* 41, 1300, 2001.

57. Welsh, S.L., Atwood, N.D., Higgins, L.C., and Goodrich, S., *A Utah Flora*, Brigham Young University Press, Provo, UT, 1987.
58. NRCS, Sandberg bluegrass *Poa secunda* J. Presl., United States Department of Agriculture, National Resource Conservation Service Plant Fact Sheet, Online at http://plants.usda.gov/factsheet/pdf/fs_pose.pdf, 2002. Verified August 21, 2006.
59. Soreng, R.J. and Van Devender, T.R., Late quaternary fossils of *Poa fendleriana* (muttongrass): Holocene expansions of apomicts, *Southwest. Naturalist* 34, 35, 1989.
60. Soreng, R.J., Apomixis and amphimixis comparative biogeography: a study in *Poa* (Poaceae), in *Grasses: Systematics and Evolution.*, Jacobs, S.W.L. and Everett, J. Eds., CSIRO, Melbourne, 2000.
61. Asay, K.H. and Jensen, K.B., Wheatgrasses, in *Cool-Season Forage Grasses,* Agronomy Monograph no. 34, Moser, L.E., Buxton, D.R., and Casler, M.D. Eds., American Society of Agronomy, Madison, WI, 1996.
62. Johnson, P.G., An overview of North American native grasses adapted to meet the demand for low-maintenance turf, *Diversity* 16(1,2), 40, 2000.
63. Waldron, B.L., Asay, K.H., Jensen, K.B., and Johnson, P.G., RoadCrest crested wheatgrass: A new alternative low-maintenance grass for semiarid regions, *Golf Course Manage.* 69(6), 71, 2001.
64. Hanks, J.D., Waldron, B.L., Johnson, P.G., Jensen, K.B., and Asay, K.H., Breeding CWG-R crested wheatgrass for reduced maintenance turf, *Crop Sci.* 45, 524, 2005.
65. Hanks, J.D., Johnson, P.G., and Waldron, B.L., Recommended seeding rates for reduced-maintenance, turf-type wheatgrasses, *Appl. Turfgrass Sci.* doi:10.1094/ATS-2006-0808-01-RS, 2006.
66. NRCS, Western wheatgrass *Pascopyrum smithii* (Rydb.) A. Love., United States Department of Agriculture, National Resource Conservation Service Plant Guide, Online at http://plants.nrcs.usda.gov/plantguide/pdf/pg_pasm.pdf, 2003. Verified July 16, 2006.
67. Dewey, D.R., 1969, Synthetic hybrids of *Agropyron albicans* X *A. dasystachyum, Sitanion hystrix*, and *Elymus canadensis, Am. J. Bot.* 62, 524, 1969.
68. NRCS, Streambank wheatgrass *Elymus lanceolatus* (Scribn. & J.G. Sm.) Gould., United States Department of Agriculture, National Resource Conservation Service Plant Guide, Online at http://plants.nrcs.usda.gov/plantguide/pdf/pg_ella3.pdf, 2000. Verified July 16, 2006.
69. McKernan, D. and Ross, J.B., The evaluation of various grasses grown under low maintenance conditions – Edmonton/Fort Saskatchewan trial, Prairie Turfgrass Research Centre Research Report, Online at http://ptrc.oldscollege.ca/1997_ar/9507.html 1997. Verified July 16, 2006.
70. Bell, W., personal communication, 2006.
71. NRCS, Bluebunch wheatgrass *Pseudoroegneria spicata* (Pursh.) A. Love., United States Department of Agriculture, National Resource Conservation Service Plant Guide, Online at http://plants.nrcs.usda.gov/plantguide/pdf/pg_pssp6.pdf, 2003. Verified July 16, 2006.
72. NRCS, Snake river wheatgrass *Elymus wawawaiensis* J. Carlson & Barkworth., United States Department of Agriculture, National Resource Conservation Service Plant Fact Sheet, Online at http://plants.nrcs.usda.gov/plantguide/pdf/pg_pssp.pdf 2002. Verified July 16, 2006.
73. Dudeck, A.E. and Peacock, C.H., Effects of salinity on seashore paspalum turfgrasses, *Agron. J.* 77, 47, 1985.
74. Duncan, R.R., Seashore paspalum, in *Turfgrass Biology, Genetics, and Breeding*, Casler, M.D. and Duncan, R.R. Eds., Wiley, Hoboken, NJ, 2003.
75. Trenholm, L.E., Carrow, R.N., and Duncan, R.R., Wear tolerance, growth, and quality of seashore paspalum in response to nitrogen and potassium, *HortScience* 36, 780, 2001.
76. Jiang, Y., Duncan, R.R., and Carrow, R.N., Assessment of low light tolerance of seashore paspalum and bermudagrass, *Crop Sci.* 44, 587, 2004.
77. NRCS, Seashore paspalum *Paspalum vaginatum* Sw., United States Department of Agriculture, National Resource Conservation Service Plant Guide, Online at http://plants.nrcs.usda.gov/plantguide/pdf/pg_pava.pdf, 2001. Verified July 16, 2006.
78. Evers, G.W. and Burson, B.L., Dallisgrass and other *Paspalum* species, in *Warm-Season Grasses*, ASA/CSSA/SSSA Monograph No. 45. Madison, WI, 2004.
79. Hansen, D.J., Dayanandan, P., Kaufman, P.B., and Brotherson, J.D., Ecological adaptations of salt marsh grass *Distichlis spicata* (Gramineae), and environmental factors affecting its growth and distribution, *Am. J. Bot.* 63, 635, 1976.
80. Ruemmele, B.A., Wipff, J.K., Brilman, L., and Hignight, K.W., Fine-leaved *Festuca* species, in *Turfgrass Biology, Genetics, and Breeding,* Casler, M.D. and Duncan, R.R. Eds., Wiley, Hoboken, NJ, 2003.

81. Brilman, L., Breeding improves fine fescue varieties, *Turfgrass Trends*, Sept. 1, 2002, Online at http://www.turfgrasstrends.com/turfgrasstrends/article/articleDetail.jsp?id=32496, 2002. Verified August 18, 2006.

82. NTEP, National Turfgrass Evaluation Program fine fescue test, http://www.ntep.org, 2006. Verified July 16, 2006.

83. NRCS, Idaho fescue *Festuca idahoensis* Elmer., United States Department of Agriculture, National Resource Conservation Service Plant Guide, Online at http://plants.nrcs.usda.gov/plantguide/pdf/pg_feid.pdf, 2003. Verified July 16, 2006.

84. Brede, D., *Turfgrass Maintenance Reduction Handbook: Sports, Lawns, and Golf*, Sleeping Bear Press, Chelsea, MI, 2000.

85. Brilman, L. and Watkins, E., 2003, Hairgrasses in *Turfgrass Biology, Genetics, and Breeding*, Casler, M.D. and Duncan, R.R. Eds., Wiley, Hoboken, NJ, 2003.

86. Davy, A.J., 1980, Biological flora of the British Isles-*Deschampsia caespitose*, *J. Ecol.* 68, 1075, 1980.

87. Coupland, R.T., 1950, Ecology of mixed prairie in Canada, *Ecol. Monogr.* 20, 271, 1950.

88. Arnow, L.A., 1994, *Koeleria macrantha* and *K. pyramidata* (Poaceae): Nomenclatural problems and biological distinctions, *Syst. Bot.* 19, 6, 1994.

89. Brede, A.D. and Sellman, M.J., Three minor *Agrostis* species: Redtop, highland bentgrass, and Idaho bentgrass, in *Turfgrass Biology, Genetics, and Breeding*, Casler, M.D. and Duncan, R.R. Eds., Wiley, Hoboken, NJ, 2003.

90. Brede, A.D., 'GolfStar: A turf, ornamental, and reclamation cultivar of North American Native Idaho Bentgrass, *HortScience* 39, 188, 2004.

91. USU Extension, Range plants of Utah Website, Online at http://extension.usu.edu/rangeplants//, 2006. Verified August 18, 2006.

92. NRCS, Crested wheatgrass *Agropyron cristatum* (L.) Gaertn., United States Department of Agriculture, National Resource Conservation Service Plant Guide, Online at http://plants.nrcs.usda.gov/plantguide/pdf/pg_agcr.pdf, 2003. Verified July 16, 2006.

93. Asay, K.H., Jensen, K.B., Horton, W.H., Johnson, D.A., Chatterton, N.J., and Young, S.A., Registration of 'Roadcrest' crested wheatgrass, *Crop Sci.* 39, 1535, 1999.

94. Robins, J.G., Waldron, B.L., Cook, D.W., Jensen, K.B., and Asay, K.H., Evaluation of crested wheatgrass managed as turfgrass, *Appl. Turfgrass Sci.* doi:10.1094/ATS-2006-0523-01-RS, 2006.

95. Lesica, P. and DeLuca, T.H., Long-term harmful effects of crested wheatgrass on Great Plains grassland ecosystems, *J. Soil Water Conserv.* 51, 408, 1996.

96. NRCS, Canada bluegrass *Poa compressa* L., United States Department of Agriculture, National Resource Conservation Service Plants Profile, Online at http://plants.usda.gov 2006. Verified July 16, 2006.

97. Dewey, D.W., Johnson, P.G., and Kjelgren, R.K., Species composition changes in a rooftop grass and wildflower meadow: Implications for designing successful mixtures, *Native Plants J.* 5, 57, 2004.

98. Hoffman, R. and Kearns, K., Wisconsin Manual of Control Recommendations for Ecologically Invasive Plants, Wisconsin Dept. Natural Resources, Madison, WI, 1997.

99. Boyce, O., personal communication, 2005.

100. Smoliak, S., Ditterline, R.L., Scheetz, S.D., Holzworth, L.K., Sims, J.R., Wiesner, L.E., Baldridge, D.E., and Tibke, G.L., 1990, Plant species, in *Montana Interagency Plant Materials Handbook (EB-69)*, Baldridge, D.E. and Lohmiller, R.G. Eds., Montana State University Extension, Bozeman, MT, 1990.

101. Welsh, S.L., *Anderson's Flora of Alaska and Adjacent Parts of Canada*, Brigham Young University Press, Provo, UT, 1974.

102. Ervin, E.H. and Koski, A.J., 1998, Drought avoidance aspects and crop coefficients of Kentucky bluegrass and tall fescue turfs in the semiarid West, *Crop Sci.* 38, 788, 1998.

103. Meyer, J.L. and Gibeault, V.A., Turfgrass performance when underirrigated, *Appl. Agr. Res.* 2, 117, 1987.

104. Allen, R.G., Pereira, L.S., Raes, D., and Smith, M., *Crop evapotranspiration: Guidelines for computing crop water requirements*. United Nations Food and Agr. Organ., Irr. Drain. Paper 56. Rome. 1998.

105. Tyser, R.W. and Worley, C.A., Alien flora in grasslands adjacent to road and trail corridors in Glacier National Park, Montana (U.S.A.), *Conserv. Biol.* 6, 253, 1992.

106. NRCS, Kentucky bluegrass, *Poa pratensis* L., United States Department of Agriculture, National Resource Conservation Service Plant Guide, Online at http://plants.nrcs.usda.gov/plantguide/pdf/pg_popr.pdf, (verified 16 July06), 2004.

107. Jones, T.A., The restoration gene pool concept: Beyond the native vs. non-native debate, *Restor. Ecol.* 11, 281, 2003.

36 The History, Role, and Potential of Optical Sensing for Practical Turf Management

Gregory E. Bell and Xi Xiong

CONTENTS

36.1 THE HISTORY OF OPTICAL SENSING FOR TURF MANAGEMENT

36.1.1 WHAT IS OPTICAL SENSING

Roughly defined, optical sensing is the use of reflectance detectors to evaluate parameters that are normally assessed by visual means. Turfgrass managers generally use a visual quality evaluation to estimate turfgrass health status, including fertilizer or irrigation need, damage, or environmental stress. Evaluation of turfgrass quality and health is a routine but important job for turfgrass managers, researchers, and breeders. Turf quality includes both aesthetic and functional characteristics [1]. Qualities such as density, color, and texture are especially important to turfgrass managers and clientele. An experienced turfgrass manager can accurately schedule fertilizer applications and irrigation based on visual quality evaluations.

Visual quality ratings are the standard evaluation method for turfgrass researchers. Visual rating is used to measure overall turfgrass quality, density, genetic color, turfgrass texture, and percent living ground cover. Except for percent living ground cover, the precedented quality evaluation system is based on a 1–9 rating using a whole-number scale [2]. Number 1 is the poorest or lowest and 9 is the best or highest. Other visually rated attributes include spring green-up, winter color, pest problems, drought stress, winter injury, and traffic tolerance. The practice of visual rating is not complicated, but it does require a properly trained and experienced evaluator. Evaluation timing is also important, because direct solar radiation coming from a nearly horizontal direction may disturb the evaluator. The time between mid-morning and early afternoon is usually the best choice for rating [3]. Another limitation of visual rating is its subjective nature. Ratings may be inconsistent from person to person and from day to day for the same person. Bell et al. [3] demonstrated that the ratings of individual evaluators differed significantly for turf color, texture, and percent live cover among three researchers with 12, 7, and 3 years of experience and that individual evaluators did not rate consistently from day to day. Furthermore, some minor or subtle differences in turf response to different treatments or early stages of environmental stress are not easily discerned by human eyes or simply may not be discriminated using a whole-number scale [4,5]. In contrast, reflectance detectors provide an unbiased, highly consistent method for turfgrass quality evaluation that requires minimal training and experience.

Reflectance detectors measure the irradiance reflected from a plant canopy or any target surface. Irradiance is not the same thing as radiation or radiance and the terms can be confusing. Radiance, solar radiance, for instance, includes a directional component and is usually used to refer to direct solar radiation. Irradiance is the term used to describe the total amount of radiation striking a surface from all directions. Solar irradiance is a combination of direct radiation, diffuse radiation, and reflected radiation striking, in this case, a plant canopy. Solar irradiance may be reflected from the plant canopy, transmitted through the canopy to the soil, or absorbed by the plants. Little radiant energy can penetrate the normally dense turf canopy; so transmittance is usually low compared with reflectance or absorbance [5,6]. In the visible spectrum (400–700 nm), referred to as photosynthetically active radiation (PAR) or light, the amount of available irradiance reflected from a plant canopy is relatively low. Most of the PAR is strongly absorbed by plant pigments, especially chlorophylls. However, near-infrared radiation in a band from approximately 700–1300 nm is highly reflected because of internal leaf scattering and low absorption [6,7]. Leaf absorption increases in the infrared wavelengths beyond 1300 nm primarily due to absorption by water. Within the PAR, chlorophyll absorbs light with peaks at 430 nm (blue light) and 680 nm (red light) wavelengths. Green light (500–600 nm) is not as highly absorbed as red and blue, and a comparatively large portion is reflected from a plant canopy [6,8–10]. This accounts for the green color perceived by the human eye.

Leaf physical characteristics such as cell structure, tissue water content, and pigment concentration may affect plant canopy reflectance, transmittance, and absorption [11]. Knipling [6] indicated that a spectrum change in leaf reflectance within the visible wavelengths can be caused by the sensitivity of chlorophyll to metabolic disturbances under stress. This research further indicated that leaf chlorophyll content has a negative relationship with green light reflection (500–600 nm) and a positive relationship with near-infrared reflection [10,12]. Consequently, as leaf chlorophyll content increases, more of the green light striking a leaf is absorbed and more of the near-infrared is reflected. In the near- and middle-infrared wavelengths (1300–3000 nm), stress-induced leaf reflectance variations have been attributed to altered leaf mesophyll structure and water content [6,13,14]. The reflectance of red wavelengths (600–700 nm) increased while near-infrared reflectance decreased from nitrogen-deprived canopies [15]. These studies suggest that reflectance detectors have potential for monitoring the growth and development of plants.

It is important to point out that reflectance, the term most commonly used in the preceding paragraph, is not the same as simple reflection or reflected irradiance. Reflectance is the fraction of available irradiance reflected from a target of interest, in this case, a plant canopy. To determine reflectance, the amount of irradiance striking the plant canopy must be known. Reflectance is determined by dividing the amount of irradiance reflected from the plant canopy by the amount of irradiance striking the plant canopy. This procedure is sometimes overlooked in casual research but is critical to obtaining accurate observations.

A simple means of obtaining reflectance is to place a detector in the vicinity of the target plant canopy facing the source of radiance, then reversing the detector so that it faces the plant canopy.

The irradiance reflected is then divided by the irradiance available to determine reflectance. A more accurate method is to use a white plate or other media specifically designed to measure reflectance. The white plate is placed in the vicinity of the plant canopy and the amount of irradiance reflected from the white plate is measured. Next, the same detector is used to measure the irradiance reflected from the plant canopy and reflectance is determined.

Because the use of reflectance detectors is based on similar physical principles as those used for detection by the human eye and because these detectors measure similar parameters as those that are visually observed, the use of these detectors for plant evaluation and other purposes is normally referred to as optical sensing. Another term commonly used to describe this procedure is *remote sensing*, simply meaning that the observing device need not be in the same location as the target to make observations normally assessed visually. Remote sensing is generally used to describe satellite imagery but also fits with most reflectance and photographic observations. Precision sensing refers to the use of reflectance sensing for the determination of plant status in small sections of a large area. For instance, precision sensing could be used to measure plant status in 1-ft^2 increments of a 1-acre field, resulting in 43,560 different measurements. Precision sensing can be used to prescribe variable rate fertilizer treatments in very small increments over a very large area. Other terms commonly used are spectral analysis, multispectral analysis, and hyperspectral analysis. Multispectral analysis refers to the use of detectors to assess reflectance at multiple (normally five to eight) energy levels called wavelengths. The term *spectral analysis* refers to the use of detectors to assess all or nearly all wavelengths within an energy band; such as the detection of reflectance from 400 to 1100 nm in 2-nm increments, and hyperspectral analysis refers to wavelength detection in a very broad band such as 400–3000 nm.

36.1.2 The Research and Development of Optical Sensing for Use in Crops

In the 1980s, optical sensing was introduced into agriculture as a complementary monitoring tool to estimate crop health status, growth, environmental stress, and crop yield [16]. Many scientists have contributed in the effort to find the relationship between the spectral and agronomic characteristics of plant canopies and ultimately attempt to determine useful vegetation indices based on spectral reflection. In the early 1980s, Walburg et al. [15] used an Exotech 20C spectroradiometer to detect reflected radiance from 400 to 2400 nm on corn (*Zea mays* L.) grown under four nitrogen treatment levels. All the wavelengths measured responded to nitrogen treatment effects. Research on soybean (*Glycine max* [L.] Merr) indicated that reflectance near the 550-nm wavelength was best for distinguishing between the three nitrogen treatments [17]. This result was supported by Blackmer et al. [10] on corn. Filella et al. [18] measured several reflected wavebands on wheat (*Triticum aestivum* L.) receiving five different fertilization treatments and found that reflectance at 550 and 680 nm were significantly correlated with canopy chlorophyll content. Osborne et al. [19] also suggested that plant nitrogen concentration was best predicted using reflectance in the red and green regions of the spectrum. However, the blue region was found best for predicting early season P stress on corn. Riedell et al. [20] measured leaf reflectance from 350 to 1075 nm on wheat and found that reflectance in the 625- to 635-nm and the 680- to 695-nm bands were good indicators of chlorophyll loss and leaf senescence caused by Russian wheat aphids (*Diuraphis noxia* Mordvilko). Daughtry et al. [21] demonstrated that the fraction of absorbed PAR could be estimated with remotely sensed multispectral (seven wavebands from 450 to 2350 nm) data and that phytomass production could be estimated as a function of accumulated absorbed PAR in corn and soybean. The researchers also proposed that since grain yield of corn crops is closely linked to the accumulation of dry phytomass, it may be possible to estimate grain yields from total dry phytomass. The combination of these two concepts could provide a basis for estimating crop yield by monitoring canopy multispectral reflectance. Since chlorophyll content is highly correlated with leaf nitrogen concentration [18,22,23], finding the best way to estimate leaf nitrogen concentration became a major thrust for agronomic optical sensing.

To help achieve rapid field estimations of leaf nitrogen concentration, new detectors called chlorophyll meters were developed and became commercially available. One of the earliest of these was the Minolta SPAD 502 chlorophyll meter (Spectrum Technologies, Plainfield, Illinois). The SPAD 502 measures the amount of chlorophyll in a leaf, that is, the greenness of the leaf, by transmitting light from light-emitting diodes through the leaf at wavelengths of 650 and 940 nm. The meter calculates an index normalized for variables such as leaf thickness and cuticle reflectance properties, which are not directly related to pigment concentration [24]. Chlorophyll meters and canopy reflectance detectors provide two nondestructive methods for estimating leaf nitrogen concentration and for determining fertilizer requirements. Some researchers recommended the chlorophyll meter as a good choice for nitrogen management [25,26], but others disagreed. Blackmer et al. [10] measured light reflectance (400–700 nm) on corn and demonstrated that reflectance near 550 nm was better able to distinguish nitrogen treatment differences than the chlorophyll meter. However, the authors believed that a better relationship with yield could be expected if the chlorophyll meter readings were taken from 30 leaves as recommended, a time-intensive process, instead of collecting 30 readings from only 10 leaves, the method used in the study. Again, in similar research conducted on corn, canopy reflectance recorded from eight wavelengths between 450 and 800 nm was more strongly correlated with field greenness at almost all growth stages compared with the SPAD 502 chlorophyll meter [27]. It was suggested by the authors that light reflectance measurements may provide better in-season indications of N deficiency. Other researchers were not only concerned about the large number of random chlorophyll meter observations needed to make statistical comparisons meaningful, but were also concerned about the physical contact with the leaf by the meter that made damage inevitable [24]. Therefore, canopy reflectance was deemed a more reliable method for estimating plant nitrogen status.

Although many researchers successfully demonstrated that several different canopy reflectance wavebands had strong relationships with plant status, dissension still existed concerning methods of reflectance measurement and derived plant indices. Some researchers believed that a single wavelength was inaccurate in some cases because of its sensitivity to biomass, variable irradiance, and background effects [28]. Filella et al. [18] confirmed the limitations of single wavebands for chlorophyll estimation when sensitivity was reduced due to wavelength saturation at medium chlorophyll concentrations. Walburg et al. [15] demonstrated that near-infrared reflectance divided by red reflectance was a better measure of nitrogen status in corn than any single waveband from 400 to 2400 nm. The researchers also found that this near infrared divided by red ratio enhanced the difference in treatment effects in canopy reflectance by reducing reflectance variability caused by extraneous factors, such as soil moisture. Wanjura et al. [29] indicated that vegetation indices, defined as the combination of observations from two or more spectral wavelengths according to mathematical formulas to derive single spectrally based numbers, usually had greater sensitivity to plant vegetation reflectance than did the reflectance of a single wavelength. Many different indices have been used to estimate crop quality or yield, including R750/R550 (maximum reflectance near 750 nm divided by maximum reflectance near 550 nm); R750/R650 (maximum reflectance near 750 nm divided by minimum reflectance near 650 nm); simple ratios like SR = R900/R680; the photochemical reflectance index, (R531 − R570)/(R531 + R570); the relative nitrogen vegetation index (near-infrared divided by green reflectance); the normalized total pigment to chlorophyll a ratio index, (R380 − R430)/(R680 + R430); the water-band index (R950/R900); the normalized difference vegetation index, ([Rnear infrared − Rred]/[Rnear infrared + Rred]) and other indices such as the yellowness index that estimates the degree of leaf chlorosis from the concavity–convexity of the reflectance spectrum at a wavelength near the midpoint between the maximum at 550 nm and the minimum at 670 nm [30]. Most of these indices involve the green (500–600 nm), red (600–700 nm), and near-infrared (700–900 nm) wave bands [31]. Among them, normalized difference vegetation index (NDVI) is the most commonly used [32].

According to Perry et al. [33], the first proposal for use of the NDVI was in the 1970s. Compared with soil, plant reflectance is very high in near-infrared wavelengths, while most visible radiation is absorbed [7]. Consequently, combining visible and near-infrared reflectance would estimate the fraction of incident radiation absorbed by the plant canopy as a function of leaf area index. Since chlorophyll absorption is maximum at about 680 nm and because red reflectance increases and near-infrared (NIR) reflectance decreases when plant greenness declines, the comparison of red and NIR wavebands is very effective. The NDVI is defined as (RNIR − Rred)/(RNIR + Rred) where R refers to reflectance. The specific near-infrared and red wavelength range used for the calculation varies with individual studies and applications but often corresponds to the bands used in resource-monitoring satellites [24]. Normalized difference vegetation indices can theoretically vary from −1.0 to 1.0 but typically range from 0.1 on nearly bare soil up to 0.9 on dense turf. Higher NDVI values are associated with greater density and greenness of a plant canopy. Soil values are close to zero. When plants are under water stress, disease, or other environmental stress, their leaves reflect significantly less radiation in the near-infrared bands and more in the red bands. As a result, the NDVI value is lower than normal when plants are stressed.

A number of experiments suggested that NDVI was a good indicator of plant status. Adamsen et al. [24] compared NDVI and Rgreen/Rred on wheat and found that both the indices were good chlorophyll concentration indicators, but NDVI was more sensitive at low values than Rgreen/Rred. Adams et al. [30] compared NDVI and R750/R650 on soybean with micronutrient deficiency treatments and found that NDVI was more sensitive than R750/R650 at lower chlorophyll concentrations. They concluded that NDVI was a more valuable measure for detecting deficiencies, because it would plateau when the plant was adequately supplied with nutrients. Other research demonstrated that NDVI was a good indicator of barley (*Hordeum vulgare* L.) biomass and yield affected by salinity treatments [34]. On corn, Ma et al. [27] demonstrated that NDVI was strongly correlated with field greenness, therefore NDVI could be used to estimate corn yield at harvest. On cotton, NDVI indicated the presence of nitrogen stress even in those cases where the stress did not result in a significant yield reduction and also showed NDVI decline approximately coincident with the onset of measurable water stress [31]. In summary, NDVI has been related to many important crops and crop stresses and has become the most widely used reflectance index for agronomic assessment.

36.1.3 The Research and Development of Optical Sensing for Use in Turf

On turfgrass, Schuerger et al. [35] used NDVI, (R760 − R695)/(R760 + R695), to detect zinc stress in bahiagrass (*Paspalum notatum* Flugge) and demonstrated that it was effective for predicting the concentrations of chlorophyll in canopies grown at various levels of Zn. Green et al. [36] and Raikes et al. [37] related NDVI to disease measurement. Bell et al. [38] indicated that NDVI, (R780 − R671)/(R780 + R671), was effective for measuring herbicide damage on bermudagrass turf. Trenholm et al. [5] measured NDVI (R935 − R661)/(R935 + R661) on seven seashore paspalum (*Paspalum vaginatum* Swartz) ecotypes and three hybrid Bermudagrasses (*Cynodon dactylon* L. × *C. transvaalensis* Burtt-Davy) and regressed against visual quality scores. The results demonstrated that NDVI was highly correlated with visual turf quality, shoot density, and shoot tissue injury rating, and the relationship between NDVI and visual quality was approximately linear. This relationship was supported by Fitz-Rodríguez et al. [16]. Bell et al. [3] measured NDVI on a National Turfgrass Evaluation Program (NTEP) tall fescue (*Festuca arundinacea* Schreb) trial and a creeping bentgrass (*Agrostis stolonifera* L.) trial. They found that NDVI was closely correlated with visual evaluations for turf color, moderately correlated with percent live cover, and independent of texture, and developed an empirical prediction equation for determining turf quality from NDVI.

Optical sensing has proved reliable for discrimination of nitrogen fertilization rates in creeping bentgrass [4] and African bermudagrass [39]. Most recently, turfgrass reflectance sensing research has focused on relating NDVI, GNDVI (green normalized difference vegetation index), and other indices with leaf tissue concentrations of chlorophyll, nitrogen, and phosphorus. Bell et al. [40] demonstrated that NDVI and GNDVI, (R780 − R550)/(R780 + R550), were closely related with chlorophyll and nitrogen concentrations in common bermudagrass (*Cynodon dactylon* L.) leaf tissue. The close relationships between optical sensing indices with fertilization rate and optical sensing with tissue nitrogen implied that optical sensing could be used for prescribing nitrogen fertilization in turf. Kruse et al. [41] used a combination of four wavelengths to predict phosphorus content in P-deficient creeping bentgrass suggesting that reflectance sensing may also be useful for prescribing phosphorus fertilizer applications.

36.1.4 HISTORICAL SUMMARY

The use of turf reflectance to estimate turf quality and turf response to irrigation and fertility treatments was first reported in 1966 [42]. Interestingly, the purpose of this work was to attempt to find a simple, objective means for determining irrigation and fertilizer need. In this writing, these same objectives are the basis for reflectance research at several universities. This original research was followed closely by another report published in 1968 by researchers attempting to find an objective means for determining turfgrass color [43]. Both of these projects found that reflectance was an effective means for determining some of the optical characteristics of turf that were normally evaluated visually. Although crop production scientists continued to be active in reflectance research through the 1970s, 1980s, and early 1990s, the development of reflectance techniques for use in turf was nearly ignored. The instruments used in early optical sensing efforts were not sensitive enough to differentiate among turf species and cultivars [44]. However, the relationships between NIR and red reflectance with percent turf canopy cover [45] and turf color [46,47] were found to be potentially meaningful during the period from 1970 to 1995. The use of optical sensing for quantifying turf cover during establishment and herbicide damage was also investigated during that period [48]. Nutter et al. [49] used reflectance to assess dollar spot symptoms on creeping bentgrass turf and published the work in 1993. The late 1990s seemed to signal a return to reflectance consideration as an alternative to visual evaluation of turfgrass (Table 36.1). The successes of crop production scientists in this area may have had a role in the resurgence of reflectance research in turf. Regardless, this area of study has maintained a strong international following during the period from 1998 to 2006 and companies have begun to manufacture reflectance sensors specifically for use in both crop production and turf.

36.2   THE ROLE OF OPTICAL SENSING FOR TURF MANAGEMENT

Optical sensing of turfgrass has three main purposes: turf quality evaluation and mapping, precision fertilization, and precision pest control. A single, high-quality sensor can accomplish all these tasks individually and, in some cases, simultaneously. It is commonly believed that reflectance sensors that measure NDVI and other vegetation indices are primarily affected by the amount of plant chlorophyll contained in a given area of turf. This is probably true. However, little is known about how many other contributing factors may also affect turf reflectance. Many conditions can affect the visual quality of a turfgrass stand. As visual quality increases or decreases, reflectance is also affected. Consequently, optical sensing is not a diagnostic tool. Reflectance sensors can measure turfgrass performance in quantitative terms but these detectors cannot discriminate among the factors that affect turfgrass performance. Therefore, a competent turfgrass specialist is required to set the parameters for use of the instrument in fertilization or pest control or to determine the cause of a decline in turfgrass quality indicated by a sensor map.

At the time of this writing, there are very few sensors available for commercial purchase. In fact, only one or two companies are producing sensors for practical use. The GreenSeeker® sensor (NTech Industries, Ukiah, California) detects reflectance in an area approximately 0.375 in. long in the direction the unit is facing and 24 in. wide perpendicular to it. This area of detection can be referred to as a frame. When this particular sensor is mounted on a vehicle and activated, it detects up to 900 frames per second. So, as the sensor is propelled across the turf, it can be used to record the reflectance properties of the entire area or the reflectance properties of samples within the area. Sensors mounted to a vehicle are set to acquire frames when the vehicle travels a specific distance. At 900 frames per second, the sensor can be set to scan 100% of the area covered at speeds up to 19 mph. However, electronic processing slows signal transfer, and according to casual estimates, the fastest commercially available system can sample 85% of the area traveled at 15 mph.

Sensors such as the GreenSeeker can also be used with handheld devices for determining the reflectance properties of turf at a walking speed (Figure 36.1). Since handheld sensors must be configured to sample frames based on time rather than distance, the number of samples recorded within a given area of turf is determined by the speed at which the sensor records samples and by the speed at which the sensor is moving. A sensor that records frames that are 0.375 in. in the direction of travel by 2 ft wide every 110 ms can record four subsamples in a 2-ft^2 area of turf when moving at 3 mph or six to seven subsamples when moving at 2 mph. A rate of 110 ms per frame is a slow sampling rate that does not take full advantage of the sensor capabilities. Slow sampling rates have proved sufficient for scientific studies [60,61], but faster sampling rates that allow the user to sample the entire area of interest would probably be more accurate. Today's sensors can sample a given area of turf much faster than a person can walk the area. However, at least two factors significantly affect the speed at which a sensor can record and display results. The sensor has to transfer data over a serial interface and may not be acquiring data during the time that it is transferring characters. The speed of transfer depends on the baud rate and significantly affects the speed at which a sensor mechanism can record data. In addition, today's video displays cannot keep pace with the sensor. Consequently, if the user wishes to see the reflectance results as the unit passes across the turf, the sampling rate must be considerably slower than the sensor's peak capabilities. If the user does not require reports of reflectance results in the field, the unit can be programmed for a high sampling rate and the reflectance record can be reviewed at a later time. A sensor with a smaller viewing area would take more samples over a given area of turf than one with a large viewing area and a sensor that records frames rapidly would record more samples than a slow one. Consequently, the fastest sensor, with the fastest baud rate, and the smallest viewing area (greatest resolution) is likely to be the most accurate provided it is manufactured from high-quality components.

36.2.1 TURFGRASS-QUALITY EVALUATION AND MAPPING

Reflectance sensors are quite sensitive to changes in turfgrass optical quality and can be used to create reflectance maps that indicate areas of poor to excellent visual quality [4]. The sensor provides a quantitative measure of greenness in a given area and records that measure to a data logger or similar media. The target area could be a square yard, a square foot, or a substantially smaller area based on the focus range of the detector. For mapping purposes, a geographic positioning system (GPS), is attached to the datalogger to record the position of the detector each time a reflectance frame is acquired. The greater the resolution of the sensor, the finer the mapping detail. When mapping, however, the resolution of the sensor usually exceeds the resolution of the GPS used to record position. Currently, most commercially available GPS are only accurate to about one square yard. More accurate systems are extremely expensive. Consequently, the use of a sensor system that provides high resolution is not necessary for mapping purposes.

Reflectance mapping can be used to measure the optical qualities of large turfgrass areas (Figure 36.2). By viewing a reflectance map, a turfgrass manager can pinpoint locations that demonstrate poor visual quality and take steps to diagnose and improve those areas. A reflectance map, for instance, that provides average reflectance in 3-ft^2 increments over an acre of turf provides 4840 squares or pixels for observation. If the area is divided into increments of 2 ft^2 per measurement, that would result in 10,890 pixels in the same map, improving resolution by a factor of 2.25. The higher resolution map would also require that 2.25 times more data points be recorded but probably not increase the practical value of the map.

TABLE 36.1
Recent Optical Sensing Studies Conducted on Turf from 1998 to 2006

Turf Species	Stress or Characteristics Studied	Optical Sensing Indices or Wavebands	Main Finding or Emphasis	Reference
Festuca arundinacea (tall fescue)	Rhizoctonia blight, and gray leaf spot	Bands between 430 and 840 nm, infrared/red, blue, green or yellow, NDVI	Canopy reflectance within the 810-nm band can be used for modeling disease epidemics, but abiotic factors can influence the measurement	[36]
Agrostis stolonifera (creeping bentgrass)	Rhizoctonia blight	Reflectance between 460 and 810 nm at 50-nm intervals	Near-infrared reflectance had significant decline as disease severity increased	[37]
Paspalum vaginatum (seashore paspalum) and *Cynodon dactylon* × *C. transvaalensis* (hybrid bermudagrasses)	Turf quality, wear stress	Reflectance at 507, 559, 661, 706, 760, 813, and 935 nm and NDVI, IR/R, R_{706}/R_{760}, R_{706}/R_{813}	Vegetation indices were highly correlated with turf quality, and able to distinguish wear stressed turf	[4]
Paspalum vaginatum (seashore paspalum) and *Cynodon dactylon* × *C. transvaalensis* (hybrid bermudagrasses)	Wear stress	Reflectance at 507, 559, 661, 706, 760, 813, and 935 nm and NDVI, IR/R, R_{706}/R_{760}, R_{706}/R_{813}	Reflectance was able to distinguish wear stressed turf and differentiate stress response among cultivars	[50]
Cynodon spp. (bermudagrasses)	Herbicide tolerance	NDVI	NDVI is effective in measuring herbicide damage on bermudagrass turf	[38]
Agrostis stolonifera (creeping bentgrass) and *Cynodon dactylon* × *C. transvaalensis* (hybrid bermudagrasses)	Nitrogen fertilization, turf quality	Reflectance at 507, 559, 661, 706, 760, 813, and 935 nm and NDVI, IR/R, R_{706}/R_{760}, R_{706}/R_{813}	C_4 bermudagrass had greater reflectance than C_3 bentgrass in visible range	[51]
Paspalum vaginatum (seashore paspalum)	Nitrogen and potassium fertilization	Reflectance at 507, 559, 661, 706, 760, 813, and 935 nm and NDVI, IR/R, R_{706}/R_{760}, R_{706}/R_{813}	Increased nitrogen fertilization reduced visible range reflectance but K fertilization had minimal effect	[52]
Festuca arundinacea (tall fescue) and *Agrostis stolonifera* (creeping bentgrass)	Turf quality	NDVI	NDVI was closely correlated with turf color, moderately correlated with percent live cover, and independent of texture	[3]
Agrostis stolonifera (creeping bentgrass)	Nitrogen fertilization	NDVI	NDVI is reliable in discriminating N fertilization rates. NDVI mapping is used in large turf areas with clearly indicated poor nutrition and sparse areas	[4]
Cynodon dactylon × *C. transvaalensis* (hybrid bermudagrasses)	Turf quality	NDVI	Vegetation indices were correlated well with visual quality rating	[16]
Paspalum vaginatum (seashore paspalum)	Traffic stress	Narrow bands between 400 and 1100 nm, R_{936}/R_{661}, R_{693}/R_{759}	Canopy reflectance at 667 nm and reflectance ratio R_{693}/R_{759} were good indicators of canopy temperature stress	[53]

Species	Stress/Application	Index/Method	Description	Reference
Paspalum notatum (Bahiagrass)	Zinc stress	NDVI, $(R760 - R695)/(R760 + R695)$	Optical sensing is effective in predicting the concentrations of chlorophyll at various Zn levels	[35]
Cynodon dactylon (common bermudagrass)	Leaf chlorophyll and nitrogen concentrations	NDVI, GNDVI	NDVI and GNDVI were closely related with chlorophyll and nitrogen concentrations in leaf tissue. Suggested that optical sensing could be used for prescribing N fertilization	[40]
Cynodon dactylon × *C. transvaalensis* Burtt-Davy (hybrid bermudagrass)	Soil compaction	Bands between 310 and 1050 nm	Use of reflectance showed some potential for detection of compaction but was affected by seasonal variation	[54]
Paspalum vaginatum (seashore paspalum) and *Cynodon dactylon* × *C. transvaalensis* (hybrid bermudagrasses)	Shade stress	NDVI	NDVI decreased under low light stress	[55]
Cynodon dactylon × *C. transvaalensis* (hybrid bermudagrasses), *Paspalum vaginatum* (seashore paspalum), *Zoysia japonica* (zoysiagrass), *Stenotaphrum secundatum* (St. Augustinegrass), and *Festuca arundinacea* (tall fescue)	Drought stress, turf quality	Narrow bands between 400 and 1100 nm	The correlations between canopy reflectance and turf quality varied with turf species and cultivars	[56]
Agrostis stolonifera (creeping bentgrass)	Phosphorus deficiency	Four wavelength combination	Suggested that optical sensing could be used for prescribing phosphorus fertilization	[41]
Cynodon dactylon × *C. transvaalensis* (hybrid bermudagrass)	Water deficit, localized dry spots	Red and infrared reflectance	Optical sensing was effective for discriminating drought stress and localized dry spots	[57]
Agrostis stolonifera (creeping bentgrass)	Pigment concentration	NDVI	Chlorophyll and carotenoid concentrations were significantly correlated with NDVI. However, another factor or combination of factors accounted for a stronger influence on NDVI than plant pigments	[58]
C. transvaalensis (African bermudagrass)	Nitrogen fertilization	Multiple wavebands and indices	Optical sensing was able to discriminate the nitrogen fertilization rate	[39]
Agrostis stolonifera (creeping bentgrass)	Water deficit	Individual bands in near-infrared wavelength	Individual wavebands were sensitive to early season drought stress	[59]
Buchloe dactyloides (buffalograss), *Zoysia* spp. (zoysiagrasses), *Cynodon* spp. (bermudagrasses)	Turf quality	NDVI	Optical sensing provided an accurate quantitative measure of overall turf quality and provided similar results to visual rating in less time	[60]
Cynodon dactylon (common bermudagrass)	Nitrogen fertilization and irrigation, turf quality	NDVI, GNDVI, R/NIR, G/NIR	Bermudagrass growth seasonal pattern were defined and described, and minimum and optimal turf quality were adjusted based on NDVI responses	[61]

FIGURE 36.1   A commercially available GreenSeeker® handheld sensor (NTech Industries, Ukiah, California). The handheld unit records reflectance measurements to an IPAQ in a format that allows direct transfer to the spreadsheet software commonly used on a personal computer. Reflectance data are collected at a normal walking speed and at a sampling rate determined by the user. The first time the trigger is depressed, the unit begins recording samples under a heading called Plot 1. When the trigger is released and depressed again, the samples are recorded to Plot 2. The plot numbers increase by one each time the trigger is released and depressed again.

Reflectance sensors are also sensitive to greenness produced by combinations of turfgrass cover and color and can be used to provide a quantitative measure of genetic quality [3]. Turfgrass cover, color, texture, and uniformity are the primary components that influence the overall visual quality of a turfgrass stand. Reflectance is influenced by a combination of cover and color and can also provide a quantitative measure of uniformity. The variation in reflectance among individual frames collected over a given area provides an accurate measure of uniformity. Reflectance is not affected by turfgrass texture but is affected by other optical qualities. Reflectance is affected when turfgrass is under stress from traffic [5], drought, disease [36,49], and other factors. Most factors that affect visual quality also affect reflectance. If a phosphorus-deficient turf is fertilized with phosphorus, visual quality and reflectance is affected [41]. If nitrogen fertilization causes a change in visual quality of a turfgrass stand, the reflectance characteristics also change [40]. The relationship between reflectance and other turfgrass visual traits offers a possibility for quantitative characterization of visual performance that has not yet been widely accepted by scientists and practitioners.

36.2.2   PRECISION FERTILIZATION OF TURFGRASS

Optical sensing could play an important role in turfgrass fertilization. For instance, a reflectance sensor attached to a variable rate sprayer has the ability to sense the nitrogen status of a small area of turf based on a visual-like assessment. The reflectance value can be used as a prescription of fertilizer need that signals a variable rate sprayer to apply fertilizer at predetermined rates. This procedure has definite advantages compared with broadcast fertilization. A reflectance-sensing unit that averages samples over a 2-ft^2 area divides an acre of turf into 10,890 small plots, each fertilized independently according to need. This type of fertilization practice could result in improved visual and functional uniformity compared with a typical broadcast system. It is also likely to reduce fertilizer use and reduce the amount of excess fertilizer available for off-site transport in surface runoff or groundwater.

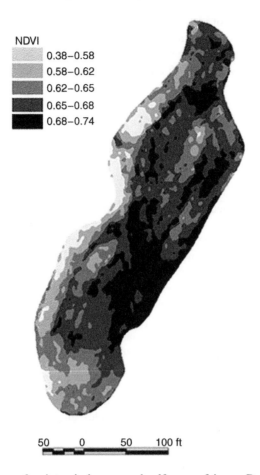

FIGURE 36.2 A sensor map of an intensively managed golf-course fairway. Darker areas indicate greener turf. NDVI values greater than 0.58 indicate that this "Meyer" zoysiagrass turf is reasonably healthy. Areas of NDVI less than 0.58 should be investigated.

Optical sensing systems are currently being used on a limited basis for prescription nitrogen fertilization of commercially produced cotton (*Gossypium* spp.), wheat (*Triticum aestivum* L.), and corn (*Zea mays* L.). Typical techniques employ a nitrogen-rich strip as an indicator of fertilizer response and determination of fertilizer need calculated from a predicted yield response curve developed over several years of research [62]. Developing similar systems for turf is possible but may be more difficult. Because of the aesthetic value of turfgrass, a nitrogen-rich strip for a baseline seasonal comparison is not practical. Turfgrass is a perennial crop and may be fertilized frequently during the growing season. For that reason, it requires a season-long response curve. Some progress has already been made in that area. Xiong et al. [61] described a seasonal pattern in bermudagrass reflectance that followed the same general trend during two growing seasons. This seasonal response pattern must be investigated further and several other parameters tested and included before a procedure for bermudagrass fertilization can be developed. Park et al. [57], for instance, found some evidence that bermudagrass reflectance can vary throughout the course of a single day. The more we learn about reflectance from turf, the more complicated the modeling procedures seem to become. Once a procedure is developed for one turfgrass species, similar procedures will have to be developed for 14 or more additional species and some characterization of mixed species stands may also have to be investigated.

Although the development of techniques for precision fertilization of turfgrass seems complicated, the problems are not insurmountable. In fact, precision sprayers for turfgrass could be widely available within the next few years. Recent advances in system management have produced equipment that can sense and spray at up to 15 mph. It is now up to the turfgrass specialists to develop systems that can accurately prescribe fertilizer need.

36.2.3 PRECISION PEST CONTROL

The role of optical sensing for precision pest control may be applicable to diseases and insects but it is probably best suited for weed control. An optical sensor can easily detect turfgrass stress caused by disease or insect damage. However, it is not known if spot spraying areas of turf where symptoms appear would be an effective means of controlling disease and insect predators. The predators may be equally active in surrounding areas of turf that do not show symptoms at the time of application.

Reflectance sensors are sensitive to different species of plants. Different species [51], even cultivars [60], can be identified by reflectance characteristics. The differences among cultivar reflectance, however, are slight and a sensing system can be adjusted to ignore cultivar differences. Attempting to adjust a system to ignore species differences is difficult and may be counterproductive for purposes of nitrogen fertilization or stress detection. The easiest difference for a reflectance detector to identify is the difference between areas containing green plant material and those containing no green plants.

The WeedSeeker® optical sensor (NTech Industries, Ukiah, California) is the most widely used reflectance sensor for vegetation management and is one of only a few optical sensors available for commercial use internationally. The primary purpose of a WeedSeeker system is nonselective vegetation management. The sensor is designed to activate a spray nozzle or multiple nozzles when green vegetation is identified. WeedSeeker systems are currently in use for weed control in both horticultural and agronomic crops as well as on roadsides, airport runways, railroad right of ways, and other areas. The system has proved effective for vegetation control using reduced amounts of herbicide compared with typical broadcast applications. One major advantage to this system and other optical sensing systems is that nontarget areas do not receive product applications. This system and systems designed for fertilizer application and other uses shut down the sprayer when concrete, bare soil, or other nonvegetative surfaces are detected.

The WeedSeeker sensor requires regular on-the-go baseline calibration. This characteristic makes it poorly suited for agronomic or horticultural fertilization. The system was tested at Oklahoma State University for turf use during a period from 1998 to 2001 and was not considered acceptable for purposes other than sensor mapping or same-day turf-quality measurements [3,4,38]. Hence the GreenSeeker sensor was developed and has, so far, proved acceptable for most agronomic and horticultural uses [63]. Neither of these sensors, however, is acceptable for selective weed control in turf. The problem is not one of sensitivity but resolution. The reflectance sensors currently being used in various forms of agriculture and agricultural research do not scan an acceptable target area to be effective for selective weed control.

Of the sensors being used commercially or for research, the WeedSeeker probably has the greatest resolution. It scans vegetation reflectance in a frame of approximately 12×0.375 in., an area amounting to 4.5 in.^{2} This frame area is quite adequate for nonselective vegetation management against a baseline background of bare soil, gravel, or concrete but it is not adequate for selective weed control against a background of turfgrass plants. Researchers at Oklahoma State University tested the WeedSeeker for its ability to detect small (approximately 2-in. diameter) cool-season weeds in dormant bermudagrass turf over the winter of 1999–2000 (unpublished data). The sensor effectively discriminated weeds when two or more weeds were captured in a frame simultaneously but seldom identified the presence of a single weed in a frame. The difficulty of detecting weed species in green turf is greater than the difficulty of detecting green weeds in dormant turf. Consequently, the performance of the sensor was not deemed acceptable for nonselective weed

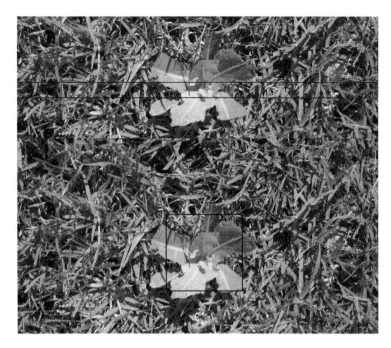

FIGURE 36.3 One of the reflectance frames in these photographs is long and narrow, resulting in a reflectance measurement that is only slightly affected by the target weed and is predominately affected by the turfgrass to the right and left of the weed. This same weed would occupy the majority of the area included in a square reflectance frame and have a greater effect on the reflectance from the area measured. Although both the narrow frame and the square frame cover approximately the same area, the square frame would detect the weed but the narrow frame would not.

control. That does not indicate a problem with contemporary sensor technology. The WeedSeeker has proved effective for nonselective weed control in a number of different situations over a period of several years. The problem occurs because of the shape of the frame and, more importantly, economic practicality.

A scanned frame of 4.5 in.² is probably sufficient for weed detection and control in green turf. A frame shape, however, that is long perpendicular to the direction of travel and short in the direction of travel does not suit the purpose (Figure 36.3). A reflectance detector measures the photon flux reflected from a scanned frame and compares it with the photon flux striking the frame to determine reflectance at a specific wavelength or wavelength band. The most sophisticated sensors contain integrated light sources and filtering systems so that the detectors can only sense the radiance provided by the integrated sources. Sensors that measure ambient radiation and compare the ambient radiation with reflected radiation also work reasonably well under most sky conditions but tend to be less accurate when solar radiation is very low such as early morning, late afternoon, and in shade, especially dappled shade. A reflectance detector determines the reflectance contained in a single frame in a particular wavelength band and converts the radiant energy to an electrical charge that is measured by a processor. Consequently, if a weed 2 in. in diameter is contained in a scanned frame of turfgrass that is 12 in. wide by 0.375 in. long, the weed will account for less than 17% of the reflectance from that frame. Since the reflectance signature of the turf is very close to that of the weed, and because the reflectance of the turf and weeds changes slightly over any given area due to multiple factors, it is very difficult to determine when a weed is present in a narrow frame. However, if this same weed were scanned in a frame of turf that measured 2 × 2 in., the weed reflectance would account for approximately 79% of the total reflectance and be detectable. Although, the 2 × 2 in. frame has a scanned area nearly the same size as the 12 × 0.375 in. frame, the weed detection possibilities increase dramatically using the square frame. This square frame concept would be effective but is highly impractical.

The most efficient weed control strategy is a sprayer system controlled by optical sensors that are programmed to fire individual nozzles when a weed is detected. Fertilizer sprayers work in the same manner and a single system could function for both applications. The scanning area perpendicular to the direction of travel of an individual sensor needs to cover approximately the same width as the nozzle pattern so that only a small area of turf surrounding the weed is sprayed. Such a procedure potentially provides considerable savings in herbicide. The financial savings in herbicide and fertilizer must be great enough to justify the expense of the sensor system. If multiple sensors were required to scan the area covered by a single nozzle, the price of the system would quickly become cost prohibitive. For instance, if a boom of sensors each scanning a 12-in. width were used to individually control an equal number of overlapping flat fan nozzles, the sensor requirement would be one sensor per boom-foot. If each sensor covered a 2-in.^{2} area as previously discussed, the sensor requirement would be six sensors per boom-foot. That would make a boom sprayer for weed control up to six times more expensive than a boom sprayer for precision fertilization or precision pest control. Clearly, a relatively inexpensive sensor that scans multiple focal points rapidly and independently is needed before an economical selective weed-control system is developed. Theoretically, that technology is available but, to date, has not been actively pursued [64].

36.2.4 IRRIGATION MANAGEMENT

Many believe that irrigation management is another feature that could be controlled using reflectance detectors. Since reflectance appears to be affected by environmental conditions that affect visual quality, turfgrass wilt should, logically, be detectable. Irrigation patterns producing stressed and nonstressed turf are detectable and sensor maps of large turfgrass areas appear to indicate drought-stressed areas. It is likely that reflectance sensors can be used to indicate a need for irrigation but it may or may not be possible for the sensor to prescribe the irrigation rate necessary to service a turfgrass area where water is needed.

36.3 THE POTENTIAL OF OPTICAL SENSING FOR TURF MANAGEMENT

Scientists are just beginning to unlock the potential of reflectance sensor technology for use in agronomic and horticultural crops including turf. The technology is not extremely expensive and could function to fill a variety of needs. For instance, a handheld optical sensor currently costs about $3500 and an existing sprayer can be outfitted with an optical detection system for approximately $1000 per boom-foot. If the sprayer is outfitted to sense and spray on a boom width basis rather than by individual nozzles, the cost is substantially less. Presently, there are no spray units commercially available for use in turf but at least one university is in the process of testing sense and spray systems for turf. Unlike many complex electronic instruments, optical sensors are outwardly relatively simple and, consequently, quite stable. A well-made unit requires very little care and does not require factory calibration unless it is damaged or over 5 years old. Whether, handheld or vehicle-mounted, the systems are easy to use and require little training. The complexities of programming and operation are built into the hardware and software of the systems requiring little user input. As these units are updated, their operation becomes increasingly user-friendly. It is likely that optical sensing systems will be used more frequently and for more operations in turfgrass management as more people become aware of their existence and more options become available.

36.3.1 SENSOR MAPPING FOR TURFGRASS MANAGEMENT

Sensor mapping is a potential function that serves managers of large turfgrass areas such as golf courses, parks, or grounds, especially well. A golf-course superintendent, for instance, generally manages 150 acres or more of highly maintained turfgrass. The superintendent is expected to maintain these acres at management levels that are aesthetically pleasing and highly functional. The high demands of the turf require daily inspections for areas of decline and early warning of potential problems. Sensor mapping fits this role exceptionally well and provides a means for unbiased assessment of all turfgrass areas maintained by the superintendent.

Logically, the best means of sensor mapping over large turfgrass areas would be to mount sensors on the mowers used for normal maintenance. Each time the turf is mowed, a sensor map would be produced. Consequently, no additional labor would be required to produce these maps and some labor would be saved because a daily visual assessment would not be required. The cost of mounting sensor systems onto mowers or other equipment for mapping purposes would have to be reasonable. Other than the labor saved for visual assessment, there is no return on investment for sensors that are dedicated to mapping functions only. A turfgrass manager would have to determine if the sensor mapping was worth the expense of adding the sensor systems to the mowers or other maintenance equipment.

Sensor mapping provides a means for early detection of turfgrass stress. Generally, an optical sensor will detect declines in visual quality before visual assessment identifies the decline [57]. The human eye tends to blend the landscape but the sensor targets specific small areas and reports data without bias or blending. Sensor mapping can also identify irrigation patterns where coverage is compromised and, in general, detect areas that are more prone to stress than others. One of the advantages of sensor mapping on a regular basis is the ability to determine areas that are declining or improving in visual quality with time. Because reflectance is a quantitative rather than a qualitative measure, computer software can compare data by location over time, automatically providing an alarm function that alerts the user to study recent maps and visit the affected sites for diagnosis.

Sensor data for mapping can be handled in a number of ways. Sensor systems provide data more rapidly than video monitors can display it, so there is always a delay in the map making process. However, it is possible to accumulate data in the field and send it to a data logger or computer by wireless communication, resulting in a map-making process that occurs almost simultaneously with sensor collection. The advantage of wireless communication is the nearly simultaneous collection and, more importantly, the automatic saving of data at a central location. The wireless process avoids potential losses of data due to operator error or media failure that can occur using a manual transfer system. However, manual transfer systems, although slower than wireless and more labor intensive, are basically reliable and less complicated to design. Onboard devices such as simple data loggers, data cards, flash cards, or IPAQs with sufficient memory for data storage can be used to manually transfer data from the sensor system to a computer for storage and processing. The sensor system can be designed to record data in a format that can be downloaded directly into common computer software programs in a matter of seconds or minutes. Regardless of the design specifics of individual systems, sensor mapping can be a powerful tool for turfgrass management. The mounting of optical sensors to mowers, golf carts, sprayers, spreaders, or other equipment that regularly traverses a turfgrass area can supplement or replace the visual assessment systems that are normally employed to define areas of decline or areas in need of general improvement.

36.3.2 PRECISION FERTILIZATION FOR TURFGRASS MANAGEMENT

The visual affects of nitrogen fertilization status of turfgrass can be measured by optical sensors [3,40]. Other nutrient deficiencies can also be measured [41]. Nitrogen, however, is the nutrient used most widely and in greatest proportions for turfgrass fertilization. Consequently, sensor systems that are used to sense nitrogen status and spread or spray accordingly are likely to be of greatest benefit to the turfgrass industry. A boom of variable-rate spray nozzles, or potentially spreaders, individually controlled by a boom of optical sensors with an integrated GPS locator and software system can fertilize very small areas (2 ft^2 or less) of a large turfgrass stand at the fertilizer rate prescribed by the sensor and software, record the location of the area sprayed, record the amount of fertilizer sprayed in that area, and provide a map of fertilizer use and a map of turfgrass optical status in a single application. Given high sensor resolution, this same sprayer could be used for weed control, providing only the fertilizer or herbicide needed and theoretically a meaningful savings in product applied over a given area compared with broadcast techniques.

Fertilization by prescription in small sections of turf may provide an increase in turf uniformity and functional value (supporting data not available) but it is the fertilizer and pesticide savings compared with broadcast applications that is most desirable. The potential for product savings is currently unsubstantiated in turf but crop production results and unpublished turfgrass research lead the authors to believe that product savings are a very likely outcome of precision turfgrass management using optical sensing devices.

Unlike sensor mapping, precision fertilization provides a quantitative return on investment. Fertilizer production is an energy-rich process and the cost of fertilizer will continue to increase with increasing energy costs. As fertilizer becomes more expensive, it will take turfgrass managers less and less time to recover the cost of a sensor-based system with the savings provided by reduced fertilizer use. If, for example, a turfgrass manager could outfit a broadcast sprayer with a sensor-based precision system for $1000 per boom-foot, an 18-foot boom would cost $18,000. An educated guess based on a few simple experiments conducted by the authors, certainly not conclusive evidence, suggests a fertilizer savings using the precision system in the neighborhood of 10–30%. Assuming that the cost of relatively inexpensive urea fertilizer (46-0-0) is $.30 per pound and a golf-course superintendent needs to fertilize 100 acres at 150 pounds nitrogen per acre per season, the fertilizer cost using broadcast techniques would be $32,609 per season. Using a precision system, the superintendent would reduce fertilizer use by at least 10%, a savings of $3260 per season or by 30%, a savings of $9780 per season. Under those terms, the precision fertilizer unit would pay for itself in 2–6 years. If a lawn-care professional were to employ precision fertilization, the return on investment would probably be greater. The lawn-care professional purchases a mixed formulation fertilizer (25-5-15) with a soluble slow release component that also costs $.30 per pound. The lawn-care professional uses a 5-ft boom sprayer to fertilize 100 acres of home lawns at a rate of 200 pounds of nitrogen per acre per season. The cost of the fertilizer is $80,000 and the cost of the precision equipment is $5000. If the lawn-care professional were to realize a 10% reduction in fertilizer use, the savings would be $8000 per season or three times that if the unit reduced fertilizer use by 30%. In either case, the lawn-care professional would pay for the precision equipment in less than a single season. A sod grower could probably realize a return that would pay for the equipment very quickly but an athletic field manager would not realize a return for several years.

In addition to economic savings, precision fertilization also provides a favorable environmental component. The so-called dead zones in the Mississippi Delta and Chesapeake Bay, as well as many other water features throughout the world, are caused by a process called eutrophication. Eutrophication occurs when excess nitrogen or phosphorus concentrations encourage rapid algal growth that results in a loss of oxygen from the water. The filtering effects of turfgrass areas that help prevent nitrogen leaching and nitrogen and phosphorus runoff are well documented. However, off-site movement of nutrients from turf can and does occur. Nutrient losses from turf are minimal compared with nutrient losses from other surfaces, but given that turfgrasses cover millions of acres in the United States and billions of acres throughout the world, even small nutrient losses can be damaging. Broadcast fertilization is highly effective but sense and spray practices reduce the likelihood that excess fertilizer will be applied to a given area and increase the likelihood that the fertilizer will be taken up by plants before leaching or runoff can occur. As described, the potential use of precision fertilization in turf appears to be both economically and environmentally practical. Although sensor-based precision sprayers are not commercially available at present, that may change very soon (Figure 36.4).

FIGURE 36.4 A prototype prescription sprayer for turfgrass designed by researchers at Oklahoma State University. Variable-rate spray nozzles apply up to seven different fertilizer rates based on a prescription determined by reflectance sensors at speeds of up to 15 mph. The sensors divide the area covered into plots that measure 2 ft^2 and apply the appropriate fertilizer amount to each plot according to need. This sprayer can fertilize, record data for mapping based on geographic position, and record the amount of fertilizer sprayed to each 2-ft^2 plot, in a single pass across the turf.

36.3.3 PRECISION PEST CONTROL FOR TURFGRASS MANAGEMENT

Sensor-based nonselective vegetation control is already a reality and precision weed, insect, and disease control may be possible. Turfgrass declines in visual quality caused by disease or insect damage is sensor detectable [49]. In fact, any stress that causes a visual quality decline can be sensor detected. However, the potential for precision pest control of diseases and insects has not been investigated. Large patch diseases such as spring dead spot (*Ophiosphaerella herpotrica* [Fr.] Walker; and other pathogens), take-all patch (*Gaeumannomyces graminis* [Sacc.] von Arx and Oliver var *avenae*), and summer patch (*Magnaporthe poae* Landschoot and Jackson) may be prime candidates for sensor-based precision pest control. These diseases tend to occur in relatively isolated patches with less than half the turfgrass in a given large area normally showing visual disease symptoms. Consequently, if a fungicide spray is required, only the areas with visual symptoms would be sprayed providing a very large savings in fungicide compared with a broadcast application. Spraying areas of visual symptoms, however, may not provide control. Patch diseases, for instance, start as a small affected area that grows progressively larger if environmental conditions or management do not check its progress. The infection area may be much larger than the area showing symptoms. Consequently, spraying only the area with symptoms may not be effective. Insect infestations may provide similar problems. These are areas that need to be investigated before precision pest control of diseases and insects can be utilized.

Sense and spray precision weed control is an area of potential sensor use that needs to be thoroughly investigated and developed for commercial application. The herbicide savings on highly managed turf areas where weed cover is only 10–20% could be enormous. Current sensor technology is capable of nonselective precision weed control but the number of sensors required to control a single spray nozzle and perform selective weed control adequately is cost-prohibitive. As new sensor technology is developed, turfgrass researchers and industry should be positioned to take advantage of new opportunities. Development of precision weed-control systems at reasonable costs could provide huge economic savings and severely reduce the environmental impact of herbicide use.

36.4 SUMMARY

The first report of researchers investigating reflectance sensing for use in turf occurred over 40 years ago [42]. However, few turfgrass researchers conducted subsequent studies until the late 1990s. The first of the Trenholm et al. papers in 1999 seemed to mark a resurgence of interest in turfgrass reflectance. Since that time, thanks to substantial advancements made by crop production researchers in wheat, soybeans, corn, cotton, and other crops and to a few interested turfgrass researchers, reflectance sensing of turf has rapidly gained popularity. The development of new, technically advanced sensors and sensor-based mechanical systems over the past few years suggests that optical sensing has a future in practical turfgrass management. High integrity, sensor-based mapping, fertilization, and weed-control systems are commercially available for crop production and nonselective vegetation management. The turfgrass industry appears to be poised to take advantage of the economic and environmental advantages of the newest technology.

REFERENCES

1. Turgeon, A.J. *Turfgrass Management*. Prentice-Hall, Englewood Cliffs, NJ. 2002.
2. Morris, K.N. and Shearman, R.C. NTEP turfgrass evaluation guidelines. National Turfgrass Evaluation Program, Beltsville, MD. 2005. Available at http://www.ntep.org/pdf/ratings.pdf (verified 23 Mar. 2006).
3. Bell, G.E., Martin, D.L., Wiese, S.G., Dobson, D.D., Smith, M.W., Stone, M.L. and Solie, J.B. Vehicle-mounted optical sensing: An objective means for evaluating turf quality. *Crop Sci.* 42, 197, 2002.
4. Bell, G.E., Martin, D.L., Stone, M.L., Solie, J.B. and Johnson, G.V., Turf area mapping using vehicle-mounted optical sensors. *Crop Sci.* 42, 648, 2002.
5. Trenholm, L.E., Carrow, R.N. and Duncan, R.R. Relationship of multispectral radiometry data to qualitative data in turfgrass research. *Crop Sci.* 39, 763, 1999.
6. Knipling, E.B. Physical and physiological basis for the reflectance of visible and near-infrared radiation from vegetation. *Remote Sens. Environ.* 1, 155, 1970.
7. Asrar, G., Fuchs, M., Kanemaru, E.T. and Hatfield, J.L. Estimating absorbed photosynthetic radiation and leaf area index from spectral reflectance in wheat. *Agron. J.* 76, 300, 1984.
8. Bell, G.E., Danneberger, T.K. and McMahon, M.J. Spectral irradiance available for turfgrass growth in sun and shade. *Crop. Sci.* 40, 189, 2000b.
9. Buchanan, B.B., Gruissem, W. and Jones, R.L. *Biochemistry and Molecular Biology of Plants*. American Society of Plant Physiologists, Rockville, Maryland. 2000.
10. Blackmer, T.M., Schepers, J.S. and Varvel, G.E. Light reflectance compared with other nitrogen stress measurements in corn leaves. *Agron. J.* 86, 934, 1994.
11. Maas, S.J. and Dunlap, J.R. Reflectance, transmittance, and absorptance of light by normal, etiolated, and albino corn leaves. *Agron. J.* 81, 105, 1989.
12. Adcodk, T.E., Nutter, F.W., Jr. and Banks, P.A. Measuring herbicide injury to soybean (*Glycine max*) using a radiometer. *Weed Sci.* 38, 625, 1990.
13. Carlson, R.E., Yarger, D.N. and Shaw, R.H. Factors affecting the spectral properties of leaves with special emphasis on leaf water status. *Agron. J.* 63, 486, 1971.
14. Gausman, H.W., Allen, W.A., Myers, V.I. and Cardenas, R. Reflectance and internal structure of cotton leaves, *Gossypium hirsutum* (L.) *Agron. J.* 61, 374, 1969.
15. Walburg, G., Bauer, M.E., Daughtry, C.S.T. and Housley, T.L. Effects of nitrogen nutrition on the growth, yield, and reflectance characteristics of corn canopies. *Agron. J.* 74, 677, 1981.
16. Fitz-Rodríguez, E. and Choi, C.Y. Monitoring turfgrass quality using multispectral radiometry. *Trans. ASAE* 45, 865, 2002.
17. Chappelle, E.W., Kim, M.S. and McMurtrey, F.E. Ratio analysis of reflectance spectra (RARS), An algorithm for the remote estimation of the concentrations of chlorophyll a, chlorophyll b, and carotenoids in soybean leaves. *Remote Sens. Environ.* 39, 239, 1992.
18. Filella, I., Serrano, L., Serra, J. and Peñuelas, J. Evaluating wheat nitrogen status with canopy reflectance indices and discriminant analysis. *Crop Sci.* 35, 1400, 1995.
19. Osborne, S.L., Schepers, J.S., Francis, D.D. and Schlemmer, M.R. Detection of phosphorus and nitrogen deficiencies in corn using spectral radiance measurements. *Agron. J.* 94, 1215, 2002.

20. Riedell, W.E. and Blackmer, T.M. Leaf reflectance spectra of cereal aphid-damaged wheat. *Crop Sci.* 39, 1835, 1999.

21. Daughtry, C.S.T., Callo, K.P., Goward, S.N., Prince, S.D. and Kustas, W.P. Spectral estimates of absorbed radiation and phytomass production in corn and soybean canopies. *Remote Sens. Environ.* 39, 141, 1992.

22. Serrano, L., Filella, I. and Peñuelas, J. Remote sensing of biomass and yield of winter wheat under different nitrogen supplies. *Crop. Sci.* 40, 723, 2000.

23. Daughtry, C.S.T., Walthall, C.L., Kim, M.S., Brown de Colstoun, E. and McMurtrey, J.E. Estimating corn leaf chlorophyll concentration from leaf and canopy reflectance. *Remote Sens. Environ.* 74, 229, 2000.

24. Adamsen, F.J., Pinter, P.J., Jr., Barnes, E.M., LaMorte, R.L., Wall, G.W., Leavitt, S.W. and Kimball, B.A. 1999. Measuring wheat senescence with a digital camera. *Crop Sci.* 39, 719, 1999.

25. Hussain, F., Bronson, K.F., Yadvinder-Singh, Bijay-Singh and Peng, S. Use of chlorophyll meter sufficiency indices for nitrogen management of irrigated rice in Asia. *Agron. J.* 92, 875, 2000.

26. Turner, F.T. and Jund, M.F. Chlorophyll meter to predict nitrogen topdress requirement for semidwarf rice. *Agron. J.* 83, 926, 1991.

27. Ma, B.L., Morrison, M.J. and Dwyer, L.M. Canopy light reflectance and field greenness to assess nitrogen fertilization and yield of maize. *Agron. J.* 88, 915, 1996.

28. Munden, R., Curran, P.J. and Catt, F.A. The relationship between the red edge and chlorophyll concentration in the Broadbalk winter wheat experiment at Rothamsted. *Int. J. Remote Sens.* 15, 705, 1994.

29. Wanjura, D.F. and Hatfield, J.L. Sensitivity of spectral vegetative indices to crop biomass. *Trans. ASAE* 30, 810, 1987.

30. Adams, M.L., Norvell, W.A., Philpot, W.D. and Peverly, J.H. Spectral detection of micronutrient deficiency in 'Bragg' soybean. *Agron. J.* 92, 261, 2000.

31. Plant, R.E., Munk, D.S., Roberts, B.R., Vargas, R.L., Rains, D.W., Travis, R.L. and Hutmacher, R.B. Relationship between remotely sensed reflectance data and cotton growth and yield. *Trans. ASAE* 43, 535, 1999.

32. Tucker, C.J. Red and photographic infrared linear combinations for monitoring vegetation. *Remote Sens. Environ.* 8, 127, 1979.

33. Perry, C.R., Jr. and Lautenschlager, L.F. Functional equivalence of spectral vegetation indices. *Remote Sens. Environ.* 14, 169, 1984.

34. Peñuelas, J.R., Isla, Filella, I. and Araus, J.L. Visible and near-infrared reflectance assessment of salinity effects on barley. *Crop. Sci.* 37, 198, 1997.

35. Schuerger, A.C., Capeele, G.A., Di Benedetto, J.A., Mao, C., Thai, C.N., Evans, M.D., Richards, F.T., Blank, T.A. and Stryjewski, E.C. Comparison of two hyperspectral imaging and two laser-induced fluorescence instruments for the detection of zinc stress and chlorophyll concentration in bahiagrass (*Paspalum notatum* Flugge.). *Remote Sens. Environ.* 84, 572, 2003.

36. Green, D.E., Burpee, L.L. and Stevenson, K.L. Canopy reflectance as a measure of disease in tall fescue. *Crop Sci.* 38, 1603, 1998.

37. Raikes, C. and Burpee, L.L. Use of multispectral radiometry for assessment of rhizoctonia blight in creeping bentgrass. *Phytopathology* 88, 446, 1998.

38. Bell, G.E., Martin, D.L., Kuzmic, R.M., Stone, M.L. and Solie, J.B. Herbicide tolerance of two cold-resistant bermudagrass (*Cynodon* spp.) cultivars determined by visual assessment and vehicle-mounted optical sensing. *Weed Technol.* 14, 635, 2000.

39. Volterrani, M., Grossi, N., Foschi, L. and Miele, S. Effects of nitrogen nutrition on bermudagrass spectral reflectance. *Int. Turfgrass Res. J.* 10, 1005, 2005.

40. Bell, G.E., Howell, B.M., Johnson, G.V., Solie, J.B., Raun, W.R. and Stone, M.L. A comparison of measurements obtained using optical sensing with turf growth, chlorophyll content, and tissue nitrogen. *HortScience* 39, 1130, 2004.

41. Kruse, J.K., Christians, N.E. and Chaplin, M.H. Remote sensing of phosphorus deficiencies in *Agrostis stolonifera*. *Int. Turfgrass Soc. Res. J.* 10, 923, 2005.

42. Mantell, A. and Stanhill, G. Comparison of methods for evaluating the response of lawngrass to irrigation and nitrogen treatments. *Agon. J.* 58, 465, 1966.

43. Birth, G.S. and McVey, G.R. Measuring the color of growing turf with a reflectance spectrophotometer. *Agron. J.* 60, 640, 1968.

44. Sheehy, J.E. Some optical properties of leaves of eight temperate grasses. *Ann. Bot.* 39, 377, 1975.

45. Biran, I. and Bushkin-Harav, I. Green coverage and color evaluation of turfgrass by means of light reflection. *HortScience* 16, 76, 1981.
46. Gooding, M.J. and Gamble, L.J. Colour evaluation of *Poa pratensis* Cultivars. *J. Sports Turf Res. Inst.* 66, 134, 1990.
47. Lodge, T.A., Baker, S.W., Canaway, P.M. and Lawson, D.M. The construction, irrigation and fertilizer nutrition of golf greens. I. Botanical and reflectance assessments after establishment and during the first year of differential irrigation and nutrition treatments. *J. Sports Turf Res. Inst.* 67, 32, 1991.
48. Haggar, R.J. and Isaac, S.P. The use of a reflectance ratio meter to monitor grass establishment and herbicide damage. *Grass Forage Sci.* 40, 331, 1985.
49. Nutter, F.W., Jr., Gleason, M.L., Henco, J.H. and Christians, N.C. Assessing the accuracy, intra-rater repeatability, and inter-rater reliability of disease assessment systems. *Phytopathology* 83, 806, 1993.
50. Trenholm, L.E., Duncan, R.R. and Carrow, R.N. Wear tolerance, shoot performance, and spectral reflectance of seashore paspalum and Bermudagrass. *Crop Sci.* 39, 1147, 1999.
51. Trenholm, L.E., Schlossberg, M.J., Lee, G., Geer, S.A. and Parks, W. An evaluation of multispectral responses on selected turfgrass species. *Int. J. Remote Sens.* 21, 709, 2000.
52. Trenholm, L.E., Carrow, R.N. and Duncan, R.R. Wear tolerance, growth, and quality of seashore paspalum in response to nitrogen and potassium. *HortScience* 36, 780, 2001.
53. Jiang, Y., Carrow, R.N. and Duncan, R.R. Correlation analysis procedures for canopy spectral reflectance data of seashore paspalum under traffic stress. *J. Am. Soc. Hort. Sci.* 128, 343, 2003.
54. Guertal, E.A. and Shaw, J.N. Multispectral radiometer signaturwes for stress evaluation in compacted bermudagrass turf. *HortScience* 39, 403, 2004.
55. Jiang, Y., Duncan, R.R. and Carrow, R.N. Assessment of low light tolerance of seashore paspalum and bermudagrass. *Crop Sci.* 44, 587, 2004.
56. Jiang, Y. and Carrow, R.N. Assessment of narrow-band canopy spectral reflectance and turfgrass performance under drought stress. *HortScience* 40, 242, 2005.
57. Park, D.M., Cisar, J.L., McDermitt, D.K., Williams, K.E., Haydu, J.J. and Miller, W.P. Using red and infrared reflectance and visual observation to monitor turf quality and water stress in surfactant-treated bermudagrass under reduced irrigation. *Int. Turfgrass Soc. Res. J.* 10, 115, 2005.
58. Stiegler, J.C., Bell, G.E., Maness, N.O. and Smith, M.W. Spectral detection of pigment concentrations in creeping bentgrass putting greens. *Int. Turfgrass Soc. Res. J.* 10, 818, 2005.
59. Hutto, K.C., King, R.L., Byrd, J.D., Jr. and Shaw, D.R. Implementation of hyperspectral radiometry in irrigation management of creeping bentgrass putting greens. *Crop Sci.* 46, 1564, 2006.
60. Koh, K.J., Bell, G.E., Martin, D.L. and Han, H.R. Comparison of turfgrass visual quality ratings with ratings determined using the GreenSeeker® handheld optical sensor. *Crop Sci.* (submitted June 30, 2006).
61. Xiong, X., Bell, G.E., Solie, J.B., Smith, M.W. and Martin, B. Bermudagrass seasonal responses to n fertilization and irrigation detected using optical sensing. *Crop Sci.* (submitted June 26, 2006).
62. Raun, W.R., Solie, J.B., Stone, M.L., Martin, K.L., Freeman, K.W., Mullen, R.W., Zhang, H., Schepers, J.S. and Johnson, G.V. Optical Sensor-Based Algorithm for Crop Nitrogen Fertilization. *Commun. Soil Sci. Plant Anal.* 36, 2759, 2005.
63. Stone, M.L., Needham, D., Solie, J.B., Raun, W.R. and Johnson, G.V. Optical spectral reflectance sensor and controller. U.S. Patent No. 6, 596, 996. 2003.
64. Jayasekara, H.K.S.R. Design and validation for laser based scanning reflectometers. Unpublished PhD Dissertation, Oklahoma State University, Stillwater, OK, 2006.

Index

A

Printed and bound by CPI Group (UK) Ltd, Croydon, CR0 4YY

23/10/2024

01778257-0008